Economic Analysis
for Highways

Economic Analysis for Highways

ROBLEY WINFREY

Formerly Professor of Civil Engineering
Iowa State University

Presently Highway Engineering Consultant

INTERNATIONAL TEXTBOOK COMPANY
Scranton, Pennsylvania

International Textbooks in

CIVIL ENGINEERING

Consulting Editor

RUSSELL C. BRINKER

New Mexico State University

Standard Book Number 7002 2244 8

COPYRIGHT ©, 1969, BY INTERNATIONAL TEXTBOOK COMPANY

All rights reserved. No part of the material protected by this copyright notice may be reproduced or utilized in any form or by any means, electronic or mechanical, including photocopying, recording, or by any informational storage and retrieval system, without written permission from the copyright owner. Printed in the United States of America by The Haddon Craftsmen, Inc., Scranton, Pennsylvania. Library of Congress Catalog Card Number: 69-16620.

Preface

The universal aspects of highways, their many facets, the $200 billion accumulation of highway needs, the limits on construction funds, and the widespread concern of practically every citizen give highway transportation a natural and compelling position in transportation. This position demands close and expert attention to the economy and to the general social and economic consequences of highways. This book is designed to serve these purposes through making available to the practicing engineers, economists, and analysts a source of theory, procedures, and applied data. It may be thought of as serving highway transportation in a manner comparable to the way similar books on engineering economy serve industrial engineering.

Within the past fifteen years there has been much published on highway economy and highway economics. Many a case has come to my attention in which the analyst did not follow the correct concept, theory, and practice, at least as I understand them. Most of these cases were prepared by individuals having had only a brief contact with the subject, and who, therefore, lacked the understanding necessary to achieve a correct solution. Through this book I hope to contribute toward a better understanding and application of economic analysis as a decision-making tool. One phase of my position may be described by stating first that there is a marked difference between my concept of "economy" and of "economics," and second, economic analysis is not wholly scientific, and therefore, engineers, public officials, and others should always keep in mind that the total process is inexact and enmeshed in professional judgment.

To many persons the subject of highway engineering economy is a simple one to be mastered in an hour's reading or by listening to a 50-minute lecture. But the truth is that engineering economy is a vital and extensive subject in industrial engineering–both academically and professionally. Highway engineering economy analysis should be just as vital to highway and general transportation, but in fact it has not been.

At the end of most chapters there is a generous list of references to the literature. These lists are intended for graduate students and others who may wish to penetrate more deeply into the subject of the chapter, or for the professor who may care to assign outside reading.

Although this book stresses economic analysis on a project basis, the same principles, methods, concepts, and cost and benefit data apply equally well to analyses of highway systems. The main differences lie in the selection of the input data. For instance, motor vehicle running cost, travel time, and traffic accidents must be used on a general rather than on a specific basis when applied to an area served by a highway system.

This book has been materially influenced by some forty-five years of contact with the subject and its many first and second cousins through teaching, research, and consulting. This book was materially aided by three major experiences. In the fall of 1961 I was honored by Stanford University with an appointment to teach as a guest professor in their graduate program in Engineering Economic Planning, partially financed by a grant from the Ford Foundation. This term of teaching at Stanford and its professional contacts did much to crystalize my thinking on many concepts and procedures. At the University of Cape Town in February 1964 I was privileged to give a group of professional engineers, professors, and economists a two-week lecture course in highway economics, under the sponsorship of the South African Road Federation. Preparing the lecture notes and the stimulating response from this class afforded me new points of view. In the graduate school of the Catholic University of America, Washington, D.C., I have taught "Transportation Economics and Finance," starting in 1961. These three academic activities afforded opportunities for trial of ideas and methods in the classroom and often with mature individuals in active engineering practice who were my students.

Although the preparation of this book has been wholly a personal professional venture unsupported through my employment with the Bureau of Public Roads, U.S. Department of Transportation, nevertheless this employment aided me in many ways. The normal flow of work across my desk and many official duties brought me in touch with helpful ideas, research reports, management problems, technical problems, analyses, and teaching experiences. The Bureau of Public Roads library staff gave me much personal help for which I extend full acknowledgment and thanks.

Among those individuals to whom I extend my personal thanks for material contributions are Professors Eugene L. Grant and Clarkson H. Oglesby, both of Stanford University. These two professors, leaders, and book authors in engineering economy and highway engineering, respectively, were responsible for my guest professorship at Stanford in 1961 and ever since have served as my two most helpful consultants. They have my deep professional and personal thanks. Nathan D. Lieder, of the Bureau of Public Roads, gave me much needed help with the mathematical derivations in Chapter 6, specifically those on continuous compounding. Chapter 25, Highway Finance and Taxation, bears the strong imprint of Thomas R. Todd, of Wilbur Smith and Associates, who expanded my rough ideas into final shape.

Thanks are extended to the following for reviewing the chapters indicated: Chapters 2, 3, 4, and 7, Eugene L. Grant; Chapter 15, Charles W. Prisk and Charles Billingsley; Chapter 24, Clinton H. Burnes and Niel H. Wilson; and

Preface

Chapter 25, Fred W. Haxton. The problems at the ends of chapters were proofed by Charles W. Dale and Howard Duke Niebur. Mr. Dale also assisted with final calculations and summary of the motor vehicle running cost tables in Appendix A.

Having been teaching since 1922 (thirty years at Iowa State University) I have learned to place high value on my student contacts. I extend to my former students my personal appreciation for their challenging questions, patience, and stimulating motivation. Especially, I mention my academic students and professional training groups of the past seventeen years.

Many authors of text and reference books acknowledge the assistance and support of their wives last in the sequence of acknowledgments. I follow that same custom in extending to Verne C. Winfrey my husbandly appreciation for her moral support, editorial contribution, typing, and secretarial efforts. In placing Mrs. Winfrey last it is with the thought that the bottom item is the support for all that is above, as is true with the foundations of engineering works.

ROBLEY WINFREY

May, 1969
Arlington, Virginia

Contents

Chapter 1   **Highways in the Motor Vehicle Age . . . 1**

Early development of land transportation. Highways and the motor vehicle age. The public and its highways. Highway design versus economy. Current setting of highway economy.

Chapter 2   **Concepts and Principles of Engineering Economy . . . 8**

Economics and economy. Nature of engineering economy. Early recognition of engineering economy. The basic premise of engineering economy. Project evaluation and project formulation. Principles of analysis. Problems. References.

Chapter 3   **Highways and Engineering Economy . . . 34**

Service character of highways. Classification of the consequences of highway improvement. Applications of the analysis for economy. Problems. References.

Chapter 4   **Identification and Measurement of Highway Benefits . . . 49**

Understanding highway benefits. Identification of benefits. Road-user consumer surplus. Road-user benefits and traffic components. Problems. References.

Chapter 5   **Interest and Vestcharge . . . 67**

Concept of interest. Concept of vestcharge. Rate of vestcharge. Problems. References.

Chapter 6   **Compound Interest Equations . . . 82**

The six standard compounding equations. Compounding a uniformly increasing series. Compounding an exponential series. Nominal and effective rates of interest. Continuous compounding. Problems.

Chapter 7   Methods of Economic Analysis . . . 123

Toward an understanding of economic analysis. Basic characteristics of the methods of analysis. Definitions of the six methods. Basic equations for the six methods of analysis. Illustrative applications of the six basic equations. Comparison of the methods of analysis when applied to a group of mutually exclusive alternatives. Economic evaluation of the null alternative. Summary discussions of the six methods and the numerical solutions. Characteristics and limitations of the methods of economic analysis. Ranking of independent projects. Factors for special attention. Sensitivity of factors. Problems. References.

Chapter 8   Depreciation Concepts . . . 176

Meaning of the word depreciation. Depreciation cost accounting methods. Cost accounting and depreciation. Depreciation, taxes, and government accounting. Problems. References.

Chapter 9   Service Life of Physical Property . . . 200

Some definitions and concepts. Factors which lead to retirement of property. Methods of determining service lives. Some general service lives of highway components. Problems. References.

Chapter 10   Highway Transportation Costs . . . 228

Annual highway and transportation costs. Annual economic cost components and equations. Factors of annual capital cost. Annual recurring expenses. The price base and inflation. Sunk cost. Interest during construction. General construction and maintenance costs per highway mile. Problems. References.

Chapter 11   Road-User Consequences . . . 262

Highways and the road user. The factor of travel time. Passenger-car travel time. Commercial vehicle travel time. Highway cost to reduce passenger-car travel time. Personal-preference satisfactions. Problems. References.

Chapter 12   Power Performance of Motor Vehicles . . . 284

Power system of the engine. Use of power developed by the engine. References.

Contents

Chapter 13   Vehicle Costs and the Highway . . . 298

Variable factors in motor vehicle costs. Motor vehicle operating cost. Highway and vehicle factors compared. Problems. References.

Chapter 14   Vehicle Running Costs for Economy Studies . . . 331

Running cost tables–an explanation. Exclusion of fuel tax in economy studies. Explanation of each running cost item. Problems. References.

Chapter 15   Traffic Accidents . . . 360

Reporting and classifying accidents. Accidents and economy. Cost of traffic accidents. Death and permanent disability. Unit accident costs for economy studies. Accident rates for economy studies. Cornell aeronautical laboratory study. Problems. References.

Chapter 16   Traffic Characteristics . . . 428

Composition of traffic by vehicle classes. Factors affecting traffic speeds. Observed speeds of traffic. Speed changes of vehicles. Problems. References.

Chapter 17   The Traffic Volume Estimate . . . 474

Components of total traffic. Discussion of the seven traffic sources. The traffic forecast. Summary of traffic volume estimate. Problems. References.

Chapter 18   Consequences of Right-of-Way Takings . . . 484

Real estate taxes. Economic cost of land taken for right-of-way. Problems. References.

Chapter 19   Nonuser Consequences . . . 493

Classification of consequences of highway improvement. Nonuser factors. Problems. References.

Chapter 20   The Urban Bypass . . . 510

Types of bypasses. Reasons for the urban bypass. Consequences of the urban bypass. Typical case histories and results. Summary of findings of the economic consequences of the urban bypass. The basic factors to consider. Problems. References.

Chapter 21   Economic Analyses in Developing Countries . . . 525

The common situation. Basic types of highway proposals. Objectives of the road improvements. Factors pertinent to the economic analysis. Probable consequences. Method of analysis. The final decision. Problems. References.

Contents

Chapter 22   Formal Report on the Economic Analysis . . . 542

The formal report. Preparing the report. Specific content of the report. Problems. References.

Chapter 23   The Management Decision . . . 552

Management defined. Factors to be considered by management. Characteristics of the decision process. Resumé of a typical situation for decision. Management decision case no. 1. The total process leading to the decision. Scalar devices and indexes. Management decision case no. 2. Problems. References.

Chapter 24   Highway Needs Studies . . . 584

Scope and objectives of needs studies. Content of a highways needs study. The final report. Organizing for a needs study. Conducting a needs study. Continuing needs studies. References.

Chapter 25   Highway Finance and Taxation . . . 602

The financial program. Tax support of highways. Highway cost allocation to road users and nonusers. Allocation of highway user share of highway cost to classes of vehicles. Discussion on allocation of highway cost. References.

Chapter 26   Construction Programming and Scheduling . . . 638

Definitions. Advantages of a construction program. The functions of programming and scheduling. Organization and procedures for programming. Formulating the construction program. Construction scheduling and control. References.

Chapter 27   Illustrative Problems and Solutions . . . 658

Illustrative problem 27-1 (economy of size of culvert). References. Illustrative problem 27-2 (economy of intersection traffic control methods). References. Illustrative problem 27-3 (stage construction, 2-lanes to 4-lanes). Illustrative problem 27-4 (project formulation for high level bridge vs. bascule bridge). Illustrative problem 27-5 (analysis for economical maximum vertical grade).

Appendix A   Tables of the Running Cost of Motor Vehicles . . . 679

Appendix B   Standard Compound Interest Factors . . . 728

Appendix C   Arithmetic Gradient Factors . . . 773

Appendix D   Exponential Growth Factors . . . 784

Appendix E   Continuous Compounding Factors . . . 875

Index . . . 905

Chapter **1**

Highways in the Motor Vehicle Age

The history of transportation is the history of the world. Social and economic changes thrive on improvements in the transportation of people and goods. Exploitation of the natural resources and subsequent development of man-made wealth have come about through transportation, most particularly land transportation. Even before the coming of the motor vehicle, a country's development could be measured by its network of roads and railroads.

EARLY DEVELOPMENT OF LAND TRANSPORTATION

In the United States, following its first settlement in 1607 by Europeans at Jamestown, Virginia, the movement to expand north and south, then westward to new land, was controlled by transportation. Typical of all countries, the early development in America was dependent upon the overland trails for pony and horse and wagon. To facilitate these movements turnpikes (toll roads) were constructed with both public and private capital. So the development continued until 1830 when the steam locomotive and the railroad began to take over except for the local land service roads.

From 1830 to 1914 railroad building predominated. During this period railroads were built to serve the entire country. Throughout the rich farming land in the midwest no town was more than 10 miles from a railroad. The wagon-road construction program during this period was centered on local land access roads and roads leading to the railroad stations. Urban settlements and goods distribution centers were located strictly along the railroad lines.

Few persons took the motor car seriously between its first appearance in the United States in 1893 and up to World War I (1914-1918). But following that war a determined effort began to construct highways[1] for the motor vehicle. This new effort called for surfacing dirt roads with gravel and stone or hardtop pavement. So it has been ever since and will continue because the road and street mileage in the United States is far from being fully improved for the high-speed motor vehicle.

[1] Throughout this book the word "highways" is used to include all highways, roads, streets, parkways, tunnels, and bridges–rural and urban, without distinction.

The status in 1965 is given in Table 1-1, where it is seen that 24.8 percent of rural and urban mileage is nonsurfaced, and 35.8 percent is surfaced with loose material such as gravel, stone, slag, or selected soil.

TABLE 1-1
TOTAL ROAD AND STREET MILEAGE IN THE UNITED STATES 1965

Roadway Surface Type	Rural Systems		Municipal Systems		All Systems	
	Miles	Percent	Miles	Percent	Miles	Percent
Nonsurfaced miles	881,101	27.7	32,508	6.4	913,609	24.8
Gravel, slag, stone, soil, etc. surfaced	1,240,566	39.0	80,891	16.0	1,321,457	35.8
Low and medium type bituminous	586,971	18.4	171,326	33.8	758,297	20.6
High type pavement ...	474,582	14.9	221,721	43.8	696,303	18.9
Total	3,183,220	100.0	506,446	100.0	3,689,666	100.0

Source: U.S. Bureau of Public Roads, *Highway Statistics* (1965), Table M-2p. 141.

HIGHWAYS AND THE MOTOR VEHICLE AGE

The rapid increase in ownership and use of motor vehicles provide a two-pronged highway problem: the desirability of (1) improving 24.8 percent of the road and street mileage not yet surfaced, and (2) at the same time improving most of the other mileage to higher levels of service, mainly to increase speed, vehicular capacity, and to reduce motor vehicle travel costs. This motor vehicle growth is illustrated by Table 1-2 where population, motor vehicles, and road mileage are related. The population per registered motor vehicle is still decreasing and the population per highway mile is increasing.

The steady increase in the number of vehicles registered combined with an increase in the average number of miles driven per year per vehicle has provided a continuous higway problem to all local, state, and national lawmaking bodies and to all highway, road, and street departments. The lawmakers are continually faced with such policy questions as the financing of highways, tax allocation, system jurisdiction, and regulation of motor vehicles as a machine and their use as a mobile vehicle. The highway departments have the annual task of getting a construction program organized which will satisfy the public (but the public is never satisfied), allocating the limited funds to gain the maximum economy of transportation, and providing the proper highway service for the general welfare of the community. Thus comes the need for serious and continuous study of highway engineering economy, transportation economics, social economics, high-

TABLE 1-2
UNITED STATES POPULATION, MOTOR VEHICLE REGISTRATION
AND HIGHWAY MILEAGE AND CONSTRUCTION COST

Year	Population (Thousands)	Total Public and Private Motor Vehicle Registration		Miles of Highways, Roads, and Streets (Thousands)	Free and Toll Highway Construction Expenditure from Public Funds Only, (Million Dollars)	Population Per Registered Vehicle		Population Per Highway Mile	Construction Expenditures (Dollars)	
		Passenger Cars (Thousands)	Total Vehicles (Thousands)			Cars	All Vehicles		Per Person	Per Highway Mile
1920	106,466	8,132	9,239	3,160*R	832*	13.09	11.52	33.7	7.81	263
1925	115,832	17,481	20,069	3,246 R	1,036	6.63	5.77	35.7	8.94	319
1930	123,077	23,035	26,750	3,259 R	1,521	5.34	4.60	37.8	12.36	467
1935	127,250	22,568	26,546	3,310 R	1,143	5.64	4.79	38.4	8.98	345
1940	132,594	27,466	32,453	3,287	1,984	4.83	4.09	40.3	14.96	604
1945	140,468	25,793	31,035	3,319	368	5.45	4.53	42.3	2.54	111
1950	152,271	40,334	49,162	3,313	2,503	3.78	3.10	46.0	16.44	756
1955	165,931	52,145	62,689	3,418	5,230	3.18	2.64	48.5	31.52	1,530
1960	180,684	61,682	73,869	3,546	6,451	2.93	2.45	50.9	35.70	1,819
1965	194,032	75,251	90,361	3,690	7,000	2.58	2.15	54.7	36.08	1,974

* For 1921. R is rural only.
Source: U.S. Bureau of Public Roads, *Highway Statistics, 1920-65*, not corrected for inflation or value of dollar.

way finance, and general administrative management. These problems are complex and getting more complex each year. Our transportation requirements call for the execution of the highest possible skill in management and foresight. The economy of highway transportation and the economic results of highway construction and highway use call for a broader and deeper understanding of land transportation than now prevails.

THE PUBLIC AND ITS HIGHWAYS

The financing of road improvements in the United States has been almost entirely an undertaking of the separate political subdivisions of government. That is, the building and maintenance of highways has been a public task.

The highways are still being built and maintained by public agencies, but under the prevailing systems of taxation the users of the roads furnish most of the money for financing them. The users, therefore, are directly interested in the design and construction of a highway transportation system over which their vehicles can be operated in the most economical manner. The development of the principles and methods by which this result can be accomplished is one of the main objectives of this book.

Highway transportation is a necessary link and adjunct to other forms of transportation. For both passenger and commodity hauling the motor vehicle is generally used in the initial and final movement to and from other transportation modes. Thus passengers travel to and from railway, air, and water terminals by automobile and bus; both raw materials and manufactured goods are hauled either to or from freight stations, pipelines or docks, by motor trucks. Trails and roads (highways) were the earliest mode of land transportation and were an essential link in the early development of water transportation.

Without doubt, ground transportation over highways, roads, and streets will continue to be the key mode of transportation, regardless of the development of air, pipeline, rail, or as yet some undeveloped mode, because individuals are mainly dependent upon ground transportation for local movements related to business, social life, and recreation.

Despite its great importance and magnitude, highway transportation is not likely ever to become the single transportation system of the country. The railroads, pipelines, waterways, and airways have important places in the economic life of the United States. It is the task of government and the authorities connected with each mode of transportation to coordinate all systems so that the public may have the most efficient and economical transportation in total.

HIGHWAY DESIGN VERSUS ECONOMY

In the earlier days (1920-42) of highway construction for the motor vehicle the emphasis was placed on design for the lowest construction cost. The cost of operating vehicles on the highways as well as discomfort and inconvenience to the users, was often neglected. Although research on economy of highway design and development of the analysis for economy was started in the early 1920's, such efforts were almost entirely devoted to the roadway surface–improving dirt, gravel, and stone surfaces to some higher type surface.

When driving today on highways designed as late as 1942, one readily observes that little or no consideration was given to the economy of vertical and horizontal alignment as these factors affect motor vehicle running cost and travel time. In rolling country and mountain country the standard design guides called for balancing cuts and fills, almost regardless of passing sight distances, speed changes for horizontal curves, and factors pertaining to prevention of accidents. Recent research and computation of motor vehicle running costs (Chapter 14) show that the cost of changing the speed of motor vehicles is high in comparison to travel at uniform speed. Continuous uniform speed is the most economical manner in which to operate a motor vehicle.

The economy in motor vehicle operation produced by the access-controlled interstate highway system and toll highways is mostly produced by providing for uniform driving speeds as compared to the almost continuous changing of speed on other highway systems for horizontal curves, overtaking and passing, variable legal speed limits (mostly for urban settlements), wide differentials in plus and minus gradients, traffic densities, and–in urban areas–traffic control devices.

In the interests of achieving the most desirable overall transportation economy (highway cost plus motor vehicle cost), attention in highway design is now being given the motor vehicle operating factors. Such procedure will bring to light the choice of design alternatives which will identify that design of greatest overall transportation economy for a given situation.

The economics of highway improvement has long been a subject studied by highway officials and legislative interests. The growth of highway transportation, particularly since 1920, has affected the social and economic structure of the nation in practically every activity. Everyone in the United States is affected in some way everyday by highway transportation. The social and economic forces activated by highway transportation are of increasing importance to highway system improvement. Successful management of a highway system necessitates specific attention to the forces. This book gives only minor attention to these social and broad economic factors, but the detailed attention in this book to the economy of highway management is a part of the same transportation economic complex. As the operation, maintenance, and reconstruction of highways assume greater importance in relation to the original construction of highways, the economics of highway transportation likewise becomes of greater importance.

Motor vehicles are owned and operated by the public as individuals or as corporations; the publicly owned highway plant is operated by an agency of the public. But these facts offer no basis for the managements of the highways not to consider the economies to be achieved in the overall costs of highway transportation. The railway transportation plants are owned by the same organization that owns and operates the rolling equipment. Logically, the same management gives the proper attention to the overall cost of rail transportation, considering the railroad track and the rolling equipment.

With the highways, it is somewhat difficult for the highway administrators to see tangible results of their efforts to achieve the utmost in economy in the overall highway transportation. The reason is that there is no common financial statement of profit and loss made to cover the roadway operated by the public highway agency and the private cost of operating the motor vehicles. Nevertheless, highway officials have long applied the principles of engineering economy to the development of highway facilities in much the same manner that the railways apply the same principles and procedures.

Since about 1955 highway engineers have been developing increasing attention to analyses for highway transportation economy, but have not yet adopted the standard procedure of checking all proposals for highway and traffic improvements for their economy of highway cost and of motor vehicle running cost.

CURRENT SETTING OF HIGHWAY ECONOMY

The foregoing summary of the growth of highway transportation, the use of the motor vehicle, and highway economy leads us to the present setting. The people of the country want highways that will give even better service than the high level they enjoy today. Yet these people are reluctant to provide the money with which to gain the total improved service they want. Money for highway construction is therefore a scarce resource in terms of the overall need. In the earlier days of road building for the motor vehicle in the United States, say beginning in 1920, improvement of most any logically chosen road paid acceptable dividends. But now, with the heavy demand to further improve much of the 3.7 million miles of highways, roads, and streets, highway authorities need economic guides to aid them in selecting the projects to authorize for construction just as the engineer needs similar guides to achieve economy of design.

Highway engineering economy analysis is one guide that is available but not fully utilized. Industry is much further advanced in its understanding and utilization of the tool of engineering economy analysis than are highway departments. It is as desirable to construct public works with full consideration to engineering economy as it is to construct private works with such consideration.

To cope with the complexities of the demands for improved highway facilities, most highway departments have instituted planning divisions. These divisions perform two main functions: (1) data gathering and record keeping, and (2) developing short-range and long-range plans for highway system construction. Construction planning involves highway system plans, route plans, and project plans. These plans utilize the traffic, financial, motor vehicle use, land use,

Highways in the Motor Vehicle Age

population trends, business trends, and other social and economic data collected and recorded yearly as a part of the planning process.

In urban areas transportation planning has become a part of the overall general planning process. All modes of travel, land-use zoning, and civic development are coordinated. Urban planning is a complex, extensive, and slow process, but essential to perform on a continuous basis if economy, satisfactory service, and desirable civic goals are to be accomplished.

To achieve these goals requires a highly competent technical staff, and a bold, dedicated legislative group. Public works managers and their staffs come into direct contact with such disciplines as public administration, political science, public finance, most phases of civil engineering, sociology, economics, economy, and others. This book is designed to lay the foundation for work in certain of these areas, particularly the economy of highway transportation and related subjects.

Chapter **2**

Concepts and Principles of Engineering Economy

To understand the application of engineering economy to the economic analysis of proposed highway facilities, it is first necessary to understand engineering economy as it has been developed and applied over the years. Engineering economy is a special phase of engineering devised to help guide the engineer toward the most economical designs for specific tools, equipment, work, construction, and processes. Over the years during which the concepts and applications of engineering economy have developed there has evolved a set of principles that have wide application in both private and public works. Each analyst for the economy of proposed private and public works should be familiar with these principles.

ECONOMICS AND ECONOMY

Top management of a highway department has two responsibilities to the public which are closely related, often confused, and yet distinctly different. First, in planning and in design he must give attention to the general social and general economic consequences of highway improvement which consequences the highway administrator cannot fully control, but he can be guided by them in his decision process. Second, he must give attention to the economy of highway design which he can control. These two responsibilities are often in conflict, but in the final decision process their respective values are to be properly weighted. The above statements point out a difference in the meaning of the words *economics* and *economy*. In Webster's[1] Third New International Dictionary the definition of economics is:

> **Economics,** *n.* 2a: a social science that studies the production, distribution, and consumption of commodities; b: ... considerations of cost and return...[1]

This is a broad all-inclusive definition that includes all forms of transpor-

---

[1] By permission. From Webster's Third New International Dictionary. Copyright 1966 by G. & C. Merriam Company, Springfield, Mass., publishers of the Merriam-Webster Dictionaries (1968) p. 720.

tations manufacturing, exploitation of natural resources, business, and land use. The improvement of highways usually has a direct consequence upon the economics of the community and of the nation. These economic consequences affect business volume, business location, employment, land use, recreation, and in fact most every aspect of living in this present motor vehicle age. Both the location of a highway route and its vehicular capacity affect the economic development resulting from the highway. Such development comes about because of the decision of many persons to use the highway and from contacts by others who use the highway. These developments are not within the control of highway officials but highway decisions greatly affect the social and economic factors of a community.

Economy of highway transportation, on the other hand, is a factor largely within the control of highway officials because it is largely controlled by the factors of highway design.

Economy is defined as follows:

Economy, *n.* 2a: thrifty or economical use or administration of material resources: frugality in expenditures sometimes verging on parsimony...[2]

Highway officials in their choice of highway location, construction materials, geometrics of design, and design traffic volume, directly affect the economy of highway travel. First, the cost of highway construction and second, the cost of operating motor vehicles over the highway are both affected by the highway design. The objective is to so design the highway that in the long run the sum of the highway cost plus the motor vehicle running cost is a minimum. This practice is the husbanding of resources; it is economy practiced through thrift in the use of resources. But in the final decision consideration is given to other than transportation factors when they are present.

NATURE OF ENGINEERING ECONOMY

Engineering economy may be defined as that phase of engineering which has to do with the analysis of proposed engineering works, equipment, and processes to determine the relative worth of the net economic gains[3] to be expected from the proposals in relation to the net economic costs required to produce the gains.

Stated in terms of the process used to determine the economy, the analysis for the economy of a proposed engineering work is the process of comparing by use of the principles of compound interest, the cash flows–outwards flows, and inward flows or their equivalent values–over time to arrive at a measure of the profita-

[2] Webster, op. cit., p. 720.

[3] The term "net economic gains" is used to imply that there may be some adverse consequences which reduce the dollar value of the favorable consequences. All consequences are to be considered whether plus or minus. There is not yet a settled terminology in the literature. For instance, the adverse consequences may be indicated by such other terms as unfavorable consequences, undesirable consequences, detriments, disbenefits, negative benefits, and malefics.

10 Concepts and Principles of Engineering Economy

bility of the proposal. In business, management seeks a fair net return on the moneys invested; in public works, the public seeks values in services that exceed their costs.

In some types of public ventures such as public buildings, parks, libraries, and art galleries, there are no specific cash monetary returns that can be used as a measure of the economy, but the decision to build is made on grounds of personal and community satisfactions to be gained from such facilities. But for highways, utility systems, and water-resource projects there usually can be found, to some extent at least, a sound cash-flow basis for analysis for their economy. Thus the consequences of some proposals for expenditure of public money are nonmarket[4] in character, i.e., the factors that measure their success cannot be directly priced in dollar terms. In other projects, such as highways, the products are both market (road-user consequences) and nonmarket (general social and economic consequences).

Because the analysis for economy is most often applied to proposals in which engineering is involved to some extent at least, the process is often called analysis for "engineering economy." It then follows that the analysis can include only those factors that can be dollar-priced so the final answer may be one that applies only to part of the consequences. Top management, then, in its final judgment gives such weight to the economy analysis and to all other factors as is deemed proper. Thus it is readily seen that the analysis for economy should not be the sole factor controlling the decision—it is rather a useful guide or tool that should be available to the decision-making authority.

EARLY RECOGNITION OF ENGINEERING ECONOMY

Economy as a phase of engineering has long been recognized and widely practiced. The history of the subject generally or as applied to transportation need not be developed in this book, but a brief mention of two notable references will help the reader to realize that transportation engineering economy was practiced long before the motor vehicle made its appearance.

EARLY ECONOMY IN WAGON ROADS

Gillespie's book of 1847 [2-20][5] recognizes many of the factors and principles of engineering economy as applied to road making for horse-drawn vehicles that are equally applicable to the motor vehicle highway. His book discusses the

---

[4] The terms "market" and "nonmarket" are introduced to distinguish the factors that can be priced on the public market and those that cannot. In the nonmarket group are some factors that are actually bought on the market, but to date no one has devised a method of determining the price paid. Personal preferences are such factors. Nonmarket factors are also called extra market factors, irreducibles, and secondary consequences.

[5] Numbers in brackets refer to references at the end of each chapter.

Concepts and Principles of Engineering Economy 11

economy of line, grade, horizontal curvature, roadway surface, construction methods, and maintenance work. It is significant that he recognizes that interest on the investment (pp. 18, 26, 28) is a necessary factor. In one illustration (p. 68) he uses a rate of 6 percent per year. A few quotations from Gillespie are appropriate:

> But roads belong to that unappreciated class of blessings, of which the value and importance are not fully felt, because of the very greatness of their advantages, which are so manifold and indispensable, as to have rendered their extent almost universal and their origin forgotten. Perhaps we will better appreciate them, if we endeavor to imagine what would be our condition if none had ever been constructed. (p. 15)

> ... the expenses of transportation will be reduced to one-third of their former amount, so that two-thirds will be completely saved, and two out of three of all the horses formerly employed can then be dispensed with. If such an improvement can be made for a sum of money, the interest of which will be less than the total amount of the annual saving of labor, it will be true economy to make it, however great the original outlay; for the decision of all such questions depends on consideration of comparative profits. (p. 18)

> While, therefore, it would be an inexcusable waste of money to construct a costly road to connect two small towns which had little intercourse, it would be equally wasteful, and is a much more frequent short-sightedness of economy, to leave unimproved and almost in a state of nature, the communications between a great city and the interior regions from which its daily sustenance is drawn, and into which its own manufactures are conveyed. (p. 20)

> Among the most remarkable consequences of the improvements of roads, is the rapidly increasing proportion in which their benefits extend and radiate in every direction, as impartially and benignantly as the similarly diverging rays of the sun. (p. 22)

> For these reasons, even if a railroad came to every man's door, he could more economically use a good common road; but since on the contrary, the expense of the construction of railroads must always restrict them to important lines of communication, (where, indeed, their value can scarcely be estimated too highly) in every other situation, the greatest good of the greatest number, and the most universal benefits with the fewest accompanying evils, will be most effectually secured, by improving (in accordance with the principles to be presently set forth) the people's highways–the common roads of the country. (p. 24)

> Rapidity, safety, and economy of carriage are the objects of roads. They should therefore be so located and constructed as to enable burdens, of goods and of passengers, to be transported from one place to another, in the least possible time, with the least possible labor, and, consequently, with the least possible expense. (p. 25)

> A minimum of expense is, of course, highly desirable; but the road which is truly cheapest is not the one which has cost the least money, but the one which makes the most profitable returns in proportion to the amount which has been expended upon it. (p. 65)

Of historical significance is the fact that Gillespie calculated his road-user benefits in horse-days saved by improvements such as harder road surfaces, distance reduction, and grade reduction. His horses were priced at 75 cents per day for a 10-hour day. He considered the rate of their doing work (speed of travel, about 3 mph) and literally priced "horsepower." For many reasons this book by Gillespie is both profitable and interesting reading. The sixth edition was published in 1853.

ENGINEERING ECONOMY IN RAILROADS

The subject of engineering economy in railroad engineering goes back at least to 1877 and 1888 when Wellington [2-11, p. 1] wrote:

> It would be well if engineering were less generally thought of, and even defined, as the art of constructing. In a certain important sense it (engineering) is rather the art of not constructing; or, to define it rudely but not inaptly, it is the art of doing that well with one dollar, which any bungler can do with two after a fashion.

In his "little" first edition (about 1877) Wellington states in the preface:

> The various problems of location, in fact, have been discussed or neglected by technical writers with an airy lightness which would convince an unskillful reader that they were either too simple, or too unimportant, or too well understood, for any careful analysis. And yet there is no field of professional labor in which a limited amount of modest incompetency, at $150 per month, can set so many picks and shovels and locomotives at work to no purpose whatever.
>
> As a natural consequence of this general negligence, all our railways are uneconomically located, most of them in respect to the minor details of alignment, and in many cases these errors are shockingly evident.... In the care taken in this respect we are not advancing beyond, but rather falling below, the standard set up forty years ago, when the art of designing railways started out in this country with such brilliant promise.

Judging from the number of miles of railroads which have been relocated since these early days of railroad construction this statement by Wellington in 1877 was well said. Perhaps something similar can be said about some of the early location and design of highways.

Surely the highway engineer should take the time and suffer the small expense to make certain of good economy of design. The know-how is available, but as now applied perhaps some highway officials have not followed the advice of Dr. Thomas Arnold [2-1, p. 304]:

> For it is clear that in whatever it is our duty to act those matters also it is our duty to study.

ENGINEERING ECONOMY IN HIGHWAYS

The study of the economy of highway improvement dates back to the early

Concepts and Principles of Engineering Economy

1920's. Research on motor vehicle fuel consumption and rolling resistance began about the same time. The early literature dealt frequently with the advantage of gravel and pavement surfaces over dirt surfaces. The early literature contains a continuous series of technical papers and a few books on the economy and economics of highways [2-12, 2-13, 2-14, 2-15, 2-17, and 2-18].

A statement by E. W. James [2-13, p. 120] was made before 1927 in a series of papers in the engineering journal *Ingenieria Internacional* which presents the economy approach to highway engineering. At this early date and because James was writing for a foreign journal, the following quotation is significant:

> Every piece of construction should be planned with an eye to the future and to the possibility, indeed to the probability, that a betterment of type will be required. The advantage of permanent line and grade should be developed always, even on the first and cheapest work. Local material supplies should be studied and designs made to utilize such materials.
>
> Roads should be built only to the extent and of such types as will pay for themselves. Sound economic, financial, and technical principles should prevail over the exigences of transitory political conditions.
>
> Finally, it must be clear from a reading of the fundamentals incident to the laying out of a national highway system, discussed in the first chapter, that any highway expenditure to be justified must be earned by the road in the form of cheaper transportation. This means that there must be enough traffic, and the type of improvement shall be such that the actual saving in cost of transportation shall at least equal the cost of the improvement. This alone justifies the cost of highway construction.
>
> During its life a highway must pay for itself; otherwise it will be a luxury, whereas our entire discussion of the creation, design, construction, maintenance, and financing of a national highway system has been from the point of view that highways are fundamental requirements in a healthy, progressing, prosperous, and ambitious nation.

The literature on the engineering economy of highways since 1920 and more recently on the general economics of highways affords a rich source of reading on the subject, and has made possible the development of the subject to a high level of concepts, principles, and procedures upon which this book is based.

RESPONSIBILITY OF THE ENGINEER

The practice of engineering is replete with instances of making decisions on the basis of economy. Involved are decisions on choices of materials, processes, sizes and shapes, functions, capacities, locations, and–most of all–whether to do something or to do nothing. Within engineering there has been developed well-accepted and well-applied concepts and theories of engineering economy. Industry practices engineering economy extensively; governments less so. But within the public works area of government functions there is reason to practice engineering economy extensively, as is done in private enterprise. Highways are

public works. There is great need to examine the highway function of governments on the basis of the concepts and principles of engineering economy.

Engineering economy analyses are performed to determine the economic efficiency of proposals, so far as the input and output can be measured in common terms—the monetary dollar. Although the engineer has the responsibility of making the analysis for economy, and designating that choice of alternatives which will perform the desired service with greatest economy consistent with safety, service, and function, top management or top government officials may elect to choose another of the alternatives because of other consequences or factors involved that could not be reduced to money values. These other values may pertain to general public welfare, social values, public policy, or aesthetics.

The widespread net effects of highway improvement on the overall economics of the nation or of a community are most difficult to determine because of transfers and offsetting consequences. Economics is generally concerned with those consequences involving wealth and its many manifestations about which the engineer has little control. On the other hand, he can measure the consequences of his highway design on the road user and on his highway costs, both factors of economy. Therefore, the engineer's analysis for economy should be restricted to economy; the economics can be a separate consideration by the economist. Top management then may give proper consideration to both economy and economics.

The highway engineer can control vertical and horizontal alignment of his route, the features of geometric cross section, and traffic operations. He has no control over the use of the land adjacent to the highway, who will use the highway and when, nor the effect of the highway upon the creation or consumption of commodities off of the highway. But these factors do affect the use and the economy of the highways, so therefore these factors are within the area of concern of the highway engineer.

Because the consequences of highway improvement to whomsoever they may accrue are of concern to highway officials, though not controllable, highway management should be governed accordingly. On the other hand, in order that the tool of engineering economy can best serve the ultimate decision maker, the analysis of economy should be just that, and that alone. The final choice can then be made by those responsible for the decision, with just and proper weight given to all factors, each factor separated from the others so it can be seen, measured, and weighed with enlightened and clear vision.

THE BASIC PREMISE OF ENGINEERING ECONOMY

Conservation of resources is the main end objective of the analysis for economy. What is saved today is available for tomorrow. In public works there is a strong obligation on the part of public officials to conserve resources and tax

Concepts and Principles of Engineering Economy 15

dollars because these officials act as the agents of the public. People are inclined to want to save and to spend wisely, but they are especially concerned when their money (tax money) is being spent by their public officials.

INSTINCTIVE DESIRE TO SAVE

Man saves through instinct. But why does he save? When commodities and material resources are saved it is with the thought that they will have future usefulness. "Future" in this sense means in the next moment of time, next day, next year, or next generation. This future usefulness is based upon the assumption that there will be a future desirable use of the commodity which has been saved. This future use may be: (1) doing more of the same, such as driving farther on the motor fuel saved, (2) utilizing the commodity saved for an entirely different purpose, such as industrial use of the tire rubber saved because of more economical use of motor vehicles, or (3) just plain saving to conserve resources.

CONSERVATION OF COMMODITIES

Individuals desire to save both commodities and money. This is part of our natural, selfish instinct. When an individual saves a commodity by not consuming it at the time, he can make future use of the commodity; when he saves money by not buying so much of a commodity (motor vehicle fuel, for instance) he creates a two-part saving–the commodity which he does not consume and his money which he does not spend. He then devotes his money saved to other satisfactions and leaves the commodity saved to be used by others as they may desire. But when he did not buy the gasoline, some petroleum company did not sell it. Thus, in this sense the saving to the road user became a decrease in sales to the petroleum industry. But not consuming the gasoline is desirable when the objective of highway travel can be achieved at a consumption of less gasoline. It is universally recognized as a part of our social and economic system that to conserve natural resources is desirable, regardless of the fact that in conserving such resources we are at the moment reducing economic activity, including the demand for labor.

Even with surplus commodities it is against public and individual policy to deliberately waste them. Generally what is one person's gain is another person's loss, expense–or at least not a gain. In this respect it is public policy to save commodities through economical management, design, and operation, despite the fact that to save commodities is to deny business income and profit to others. Likewise, business and industry save their own resources and let their suppliers take the consequences.

Part of this instinctive saving is on the basis that the economy in the long run will balance out between the total of all economic activity as it continues its shifting. Consider what the motor vehicle and the airplane have done to the railways, the effect of television on the motion picture industry, and how power farming and highways have transformed rural living.

The goal of the analysis for economy to save commodities will always remain. Its premise is that whatever is saved today will find beneficial application elsewhere in the future. Certainly no one wants to adopt a public policy of consuming all possible commodities by building pavements 10-ft thick, and bridges for 100,000-lb axle weights. In the long run the consequences of saving are superior to the consequences of wastefulness.

CONSERVATION OF LABOR

The same premise applied to the saving of commodities applies to the saving of labor. Labor, not utilized because less highway is built and because smooth, dustless highways require less mechanic labor to maintain and repair vehicles, becomes a saving to the highway user but at the sacrifice of less wage to particular types of labor. Here again, however, the premise is that there is a demand for labor and what is saved by a certain highway improvement will be utilized to advantage by increased use of highways, the manufacture of more vehicles, or in similar worthwhile activities.

Even in periods of economic depression, our sense of economy demands that we use labor and commodities economically. Even work authorized for the express purpose of putting unemployed persons to work calls for the exercise of the principle of economy; the objective should be always to spend wisely and to get the greatest return possible for the dollar expended. Government subsidy for relief in depression periods is no excuse to be wasteful of labor; rather, do more work with the money and labor available.

An improvement in procedure or process which will effect a decrease in the amount of labor or the amount of professional service required is considered a benefit, even though such action may result in less wage to labor or less compensation to professional persons. Both the medical and legal professions earn sizeable compensation by reason of motor vehicle accidents. To eliminate all accidents would reduce the professional fees accordingly. Nevertheless, such reduction is to be sought. And such savings are to be included in economy studies. It is assumed that all labor and professional services saved can be applied to other desirable activities.

LONG-RANGE RESULT OF CONSERVATION OF RESOURCES

The country has reached its present state of high level of living, of income, and of technology by following the principle of economy. Our progress has been achieved year to year as we have brought on the new for the old, as we saved commodities and labor, and as we have reduced price to the purchaser and sales to the seller. Evidently, economy has paid off in spite of the fact in the saving of commodities and of labor by economical design and operation we have still grown economically, and at a rate faster than the population has grown.

Concepts and Principles of Engineering Economy 17

Our social-economic system has great powers of adjustment. Economic support of workhorses and their feed oats has gone, television has come, the computer age is here, coal consumption is down, steel is manufactured automatically, and so on. All these changes have been in the interest of economy. The losses that have accompanied the gains have been absorbed by transfers of economic support to other functions and commodities, by increased consumption, and by the natural shifting of personal preferences. Therefore in highway economy management should look for the most economical solution from the viewpoint of the community as a whole considering all consequences and not solely from the viewpoint of a particular small group of people who temporarily may suffer some loss.

PUBLIC VERSUS PRIVATE VIEWPOINT

One of the principles of economy in public works at all levels of government is that all consequences are to be considered to whomsoever they may accrue. When all consequences are considered and when the viewpoint is that of the public as a whole, there is little chance that overall injustice will result by following the laws of economy. But all public decision makers should weigh carefully local injustices, which usually prevail with every major public undertaking. Private viewpoints, however, are private viewpoints for self-preservation and for selfish objectives. These viewpoints are normal to humans and are to be respected and considered, but not allowed to outweigh the community interest and welfare.

In private industry there prevails a much narrower viewpoint. Although a well-managed business does not ignore the public, the main interest of the management is centered upon company gain. Certainly, one business is not prone to consider the welfare of its competitors. Businessmen when considering matters of engineering economy are prone to limit their viewpoint to their company's interests and to consider the consequences only as they affect their business. But within business the same human instinct prevails to save and to spend a dollar only when there are good prospects of getting the dollar back plus a reasonable return thereon.

PROJECT EVALUATION AND PROJECT FORMULATION

Engineering economy may be separated into two basic objectives or applications: (1) analyses to determine the economic feasibility *(economic evaluation)* of a proposal to construct an engineering structure or machine, and (2) analyses to determine the *economic formulation* of the features of design and use of the engineering structure or machine. Preliminary economic evaluation is followed by economic formulation. The initial economic evaluation is often made with estimates less exact than those used in project formulation and by using only the more controlling factors. In fact, formulation often involves many suboptimizations of individual features of the project as a whole, that are not specifically identified in the analysis for economic evaluation. A use of the results of the analysis for economic evaluation is to assist in ranking independent projects for priority of construction as discussed in Chapter 26.

ECONOMIC EVALUATION AND PROJECT FORMULATION COMPARED

Practically all of the literature on the subject of economic analysis of proposed investment improvements deals with the subject on the basis that economic evaluation (analysis to determine the degree of economic feasibility) is the only application of the several methods of analysis. Yet testing the alternative in design (project formulation) is the objective more often than is economic evaluation. Although the principles of analysis hold for both objectives, the required data, quantification of factors, and choice of method of analysis often differ.

Many of the details of designs when alternatives are considered offer a choice only on the basis of total cost, there being no income or benefits applicable except the difference in costs. This situation generally prevails with choices of construction materials (steel, concrete, asphalt), geometrics of design (circular culverts versus box culverts), basic design (earth fill versus a longer bridge), and methods and sequences of construction.

Wherein vehicular traffic is affected by the choices of design, an "income" factor prevails in the form of differences in the running cost of the motor vehicles so that analysis methods that require a benefit or income can be used in project formulation. A difference, however, between such application and the application in the analysis to determine economic feasibility is that in the design process the traffic volume to use is that which is forecast for that particular alternative. The source of that traffic and its specific pricing for project formulation are immaterial so long as the same pricing is used for all the alternatives of design in the mutually exclusive group of alternatives. Even the specific volume of traffic is not too significant in project formulation as long as the traffic volume is comparable between the alternatives of design.

RELATIVE REQUIRED EXACTNESS OF ESTIMATES

In project economic evaluation usually there is no need to work with final designs and cost estimates, but preliminary designs and rough estimates can be used, except possibly in those cases near the margin of economy and when decision to build the project will be based largely upon its economic feasibility. In the highway field there are many projects, large and small, that are programmed for construction on the basis of their being (1) a subproject of a large project or system, (2) needed despite their economy, or (3) such that their economy is self-evident. Replacing damages to a highway system caused by natural catastrophe, rebuilding

Concepts and Principles of Engineering Economy 19

an unsafe bridge, or widening a trafficway to eliminate traffic congestion are examples of projects the need for which is usually self-evident, and for which the economic analysis would consist only of finding the most economical design.

Project formulation analysis is necessary in all designs of highway improvements regardless of whether the project itself has been proved to be economically sound based solely on the priceable consequences. As long as competent authority has authorized the designs to be prepared, they should be prepared using to the fullest extent feasible the principles of engineering economy.

ILLUSTRATIVE EXAMPLE OF PROCEDURE

An analysis could be made to determine whether building a highway bridge across a river mainly to save travel distance is justified on the basis of economy. The general situation could be analyzed by assuming (1) a general location, (2) the cost of the bridge at so many dollars a square foot of roadway, and (3) general average motor vehicle running costs. Should this analysis determine that a new bridge would be, without question, a desirable project, the engineer could then move on to its design. Should this first analysis point to some uncertainty as to the economy of the proposed bridge, a second or third analysis should be made with stepped-up accuracies and certainties in the design features.

After determining that the proposed new river crossing is desirable from the economy viewpoint, the formulation of design might call for the following several analyses of economy, among others:

Type of design–truss, girder, suspension
Type of material–steel, concrete, aluminum
Length and spacing of spans–waterway capacity
Approach–earth fill or trestle

In keeping with the principle that separable decisions are to be made separately, each of these factors would be analyzed as individual alternatives.

PRINCIPLES OF ANALYSIS

The analysis for the economy of a proposed highway improvement is based upon a set of principles and concepts that have evolved over the years, primarily from similar analyses in industry. An understanding of these principles and concepts is a requisite to a proper analysis. Assembling the numerical factors, determining their magnitude, and performing the step-by-step procedure involved in a proper analysis for the economy of proposed works are dependent upon an accurate understanding of all the aspects of engineering economy.

These concepts, principles, and standards are set forth in the following statements. These statements go somewhat beyond the strict scope of the analysis for economy and cover as well the final decision process. The final decision process includes consideration of social and economic factors that are not included in the analysis for economy, mainly for the reason that these factors cannot be reduced to a market-priced basis.

The guidelines, principles, and concepts in the analysis for the economy of a proposed highway improvement may be set forth under the following headings:

1. *Complete objectivity–not subjectivity*–is required by the analyst in estimating, forecasting, selecting the factors and their magnitude, and in pricing the factors.

2. The analysis for the economy of a proposed work is *not the decision process.*

3. *"Hunch" decisions have no place in an economy study*–not in project evaluation and not in project formulation and not by the analyst and not by the final decision maker.

4. Consideration should be given to *all possible alternatives:* the analysis is a comparison of alternatives to seek out the best one.

5. The basic concept is one expressed by the words *"with and without,"* or, *"to do or not to do."*

6. The analysis for economy is *wholly money based;* therefore, the factors that are priceable on the market are separated from those nonmarket factors that cannot be market-priced.

7. The analysis, being based on expected consequences, is a *study of the future;* estimates and forecasts should be as good and reliable as those responsible for them can make them.

8. All *past events and investments are irrelevant.*

9. The input (costs) and the output (adverse and beneficial consequences) must each be *considered over the exact same period of time.*

10. The analysis *period into the future must be restricted* to that period of time over which the forecasts of key factors can be regarded as reliable.

11. All factors in the analysis are *discounted to the same time date* by the appropriate interest rate (vestcharge rate, or time-discount factor).

12. It is the *differences in alternatives* that are significant and relevant.

13. Common factors *of equal magnitude* in the alternatives being compared may be omitted from the analysis.

14. The analysis should be based upon all *net costs and net consequences.*

15. Since the results of the analysis for economy are factors that bear upon the decision to finance or not, and since economy itself is not a part of financing, the analysis is *independent of the method of financing the construction* and by whom the construction is financed.

16. The final decision, and to the extent possible the analysis for economy, should *give just weight to the uncertainties.*

17. Since within the organization (highway department) responsible for the decision, *separate levels of management may make separate decisions,* the total analysis and reports should give proper recognition to these levels of decision making.

Concepts and Principles of Engineering Economy 21

18. The viewpoint should be clearly established. *From whose viewpoint* needs to be answered before the final decision can be made.

19. The *criteria for decision making* should be established.

20. All consequences to *whomsoever they may accrue* and all the factors in the analysis itself should be given consideration in the final decision.

21. The final decision should *give proper weight to the nonmarket factors* (secondary consequences, irreducibles) not included in the analysis for economy.

EXPLANATION OF THE PRINCIPLES

These twenty-one items encompass the major philosophy of the analysis for the economy of proposed engineering improvements, as well as the ultimate decision on which the economy is a factor. Each item is next explained in some detail.

1. Complete Objectivity Is Required

The analysis for economy is strictly an objective analysis. It is a hunt for an answer of unknown magnitude and unknown direction. Although professional judgments must be exercised in making the analysis, the necessity of making these judgments is not a warrant for injecting bias into the analysis. The judgments involve such items as identifying the factors to be included and determining their magnitudes, pricing the outlay of cost and income of benefits, estimating future consequences, and identifying all alternatives. This principle of objectivity calls for intellectual honesty and ethical direction.

The whole of the analysis for economy is for the purpose of bringing to light the relative position of outlays of cost and incomes of benefit as a basis of aiding management in making final decisions on the questions: Why construct any such improvement at all? What design is the better one? and, What is the better time for construction or purchase? Management cannot use the analysis for economy to the best of advantage unless it is the result of an unbiased objective inquiry. It is unfortunate that the analyst's judgments can control the final answer of economy to a considerable extent, but this condition should be an incentive to the analyst to strive for estimates, selections, and judgments that can stand the tests of logic, reason, reality, and objectivity.

2. The Analysis for Economy Is Not the Management Decision

The analysis is an unbiased, objective analysis of the comparisons of cash outflows and inflows of all factors that can be reduced to a price, or a market, basis. Further, the analyst when performing in that capacity is not the decision maker. When he does wear both hats he must be careful to wear them only one at a time. In project formulation, often the alternative details of design may be tested for economy by the designer rather than by a higher-level official. The tool–analysis for economy–can serve the decision maker properly only when it is divorced from the other factors of nonmarket character, public attitudes, and political influences. These factors are weighed against the results of the economy study as the end action, and then by the decision maker–not by the analyst.

3. "Hunch" Decisions Have No Place in an Economy Study

Important decisions are made during the process of making the analysis for economy and by management in deciding whether to construct, or purchase, or to do nothing. Unfortunately, some managements make important decisions on the investment of capital by "hunch," or intuition. To quote Samuel Butler (1835-1902): "Life is the art of drawing sufficient conclusions from insufficient premises." In any case, the final decision of management is strictly a professional decision of the mind, not the numerical answer of a measuring, weighing, or calculating procedure. But the decision of the mind has a greater probability of being a good decision when the results of measuring, weighing, and calculating are available as guides.

The analysis for economy as a guide to decision can be compared to the physician's diagnosis, including chemical and microscopic analyses, and the lawyer's search of the law and of court decisions. In the medical, legal, and engineering professions the objective is to bring into light all possible evidence, facts, probabilities, and professional judgments that can aid in making the decision on a future course of action. This procedure certainly is preferable to "hunch" decisions and purely offhand intuitive judgments.

4. Study All Possible Alternatives

The purpose of the analysis of the factors of economy, sociology, and economics of a proposal for a highway improvement is to reach the best possible solution to achieve the objective of safe, fast, convenient, and economical transportation. Therefore if management is to be assured that the best overall solution has been found, every possible alternative of attaining the objective of the proposed transportation facility being studied must be considered.

Although the words "every possible alternative" are used, they do not mean that a detailed analysis would be required of each possible solution. Certain possible solutions can be eliminated after briefly considering them when sound engineering judgment shows that they may be physically undesirable because of location, construction difficulties, or level of vehicular service. Other alternatives may be set aside as being far beyond the available funds. Still other proposals may not meet the aesthetic requirements, or would result in the abandonment of civic facilities (parks, monuments, waterfronts) which the community would not give up. Even so, these answers cannot be reached until each separate alternative has been examined in the detail required.

Those alternatives that meet the physical, financial, and aesthetic requirements can then pass on to the more detailed phase of analysis. Here again certain

Concepts and Principles of Engineering Economy 23

proposals may be set aside after preliminary or approximate analysis warrants a decision as to their acceptability.

In studies of highway route location it sometimes happens that attention is given solely to locating the route under consideration without considering the possibility of relocating, extending, or altering the adjacent feeder, or secondary routes. In crossing navigable waterways consideration frequently must be given to low-level movable bridges, high-level fixed spans, underwater tubes, and abandonment of navigation on the waterway.

The sound answer can be reached only when the analyses for economy and for economics are made for every practicable proposal for reaching the objective.

5. The "With or Without" Concept Is Basic

One of the alternatives to consider in any proposal is getting along "without" the improvement–that is, continuing the existing situation. This alternative is the "do nothing" choice. The other choice is the "with" situation, i.e., the analysis of the consequences with the proposed improvement. There is need always to consider the economy of the several proposals to do something, but also there is a need to compare the proposals with the existing condition. The existing condition is sometimes referred to as the base condition, or the null alternative. Its continuation or abandonment needs also to be examined.

Comparing the economy of a proposal with the existing situation results in determining the degree of *economic justification* of making the change; but when the several alternatives of "doing something" are compared with each other the objective is usually *project formulation*. Here the goal is to measure the economy of the engineering differences in the several proposals; that is, with or without certain features of design. The "with or without" concept is an important one in economy studies because it briefly expresses the basic consideration to be given every proposal for a change in existing facilities or in proposals encountered in the several stages of design.

6. Separate Market and Nonmarket Factors

In the analysis for economy only factors of economy are included; the factors of general economic and social consequences are omitted from the calculations. This separation is made primarily on the assumption that the items of economy can be dollar-priced and therefore are market items, while the general economic factors often cannot be dollar-priced and therefore are nonmarket items. The final decision of management can be reached on a sound basis only when the analysis for economy is kept on a sound pricing basis free from conjecture and unwarranted assumptions. For instance, to include a dollar value for the personal preference of comfort and convenience to passenger car travelers would be including an unwarranted and unsound assumption. Such factor is not one of economy in the sense of the saving of commodities, reduction of costs, or reduction of travel time, and not one that has been priced on any sound basis of market value.

However, highway facilities should be designed for comfort and convenience of the highway users, and in the final decision process these factors should be considered.

7. The Analysis Is a Study of the Future

The analysis for economy is wholly an analysis of the future. The cash flows for costs and for consequences are estimates, projections, and forecasts of what the analyst expects in the future, *with* and *without* the proposed improvement. The futuristic character of the analysis does not warrant loose estimates or guesses. Each factor is to be forecast using sound logical reasoning, calculations, and assumptions. The use of a range of values is frequently desirable as a means of indicating the sensitivity of a given factor upon the final result. The traffic volume forecast factor, for instance, may be used on both an optimistic and pessimistic basis.

8. All Past Events and Investments Are Irrelevant

Closely associated with the fact that the analysis for economy is based on the future is the fact that all past actions, past cash flows, and past costs are irrelevant. What is past is past and common to each proposed alternative; therefore all past actions are embobied in the existing situation. To this purpose all past cash flows are sunk–that is, beyond recall. The decision to be made now is what to do to create future consequences favorable to economy.

This concept is not to imply that the past is not useful in predicting the future, for it most often is. Further, all past work may have some bearing on the cost of certain proposed alternatives. In one alternative an existing bridge may be considered for widening and strengthening as compared to constructing a completely new bridge. But in such comparison the past money cost of constructing the existing bridge is not considered.

9. Use Identical Time Periods for All Factors Included

Analysis of the consequences (benefits and costs) with and without the improvement must be made over the same time period and discounted to the same time date for comparison. The cash flows of the future are the critical factors that are being compared in the analysis of the relative economy of proposed improvements. They are compared by reducing the cash flows to the same time basis and over the same future spread in time, say the next 5 or next 20 years. When comparing the economy of a proposed urban bypass, for instance, with the existing route through the urban area it is essential to forecast the probable traffic use of the existing urban route and other consequences for the same period of future time as is used in the analysis of the proposed bypass. Only when both proposals–the existing facility and the proposed new one, or two proposed ones–are analyzed over the exact identical time period can comparable results be obtained.

Concepts and Principles of Engineering Economy

10. The Analysis Period Should Not Extend Beyond the Period of Reliable Forecasts

Because the analysis for economy is wholly an analysis of the future it follows that the analysis period should extend short of that future date beyond which forecasts are unreliable or without an acceptable foundation. Of particular importance with respect to this principle is the traffic forecast. Certainly, 20 years or so is a maximum future period for reliable forecasts of traffic volume, composition, and performance. It is not good practice to base the analysis on a time period of, say 40 years, when a 20-year traffic forecast period is used as a basis of design.

The analysis of economy is not a calculation of the economic cost of owning and operating a machine, a process, or a highway facility. It is an economic analysis of alternatives, strictly for the purpose of making a decision now relative to a future course of action. Forecasts, then, should be held to that period of time over which they can be regarded as sound. The analysis for economy, in order to be equally sound, should not extend beyond the forecast period.

11. Discount All Factors in the Analysis to the Same Time Date

The outward and the inward cash-flow factors in the comparison of the alternatives usually occur at different time dates and in different magnitudes or changing magnitudes over the analysis period. In order to compare them on a sound basis these factors are reduced to equivalent or comparable values at a common date. This so-called discounting procedure is accomplished by using the appropriate interest rate (vestcharge rate, or discount factor) in accordance with the established principles of compound interest and present-worth concepts.

12. The Differences in Alternatives Are Controlling

It cannot be repeated too often that it is the *difference in alternatives* that should control the decision between them. Differences are the relevant factors in all decisions between alternatives. If all factors pertaining to two proposals are identical and the predicted consequences the same, then there is no difference in the two proposals and either one may be chosen. But if either the inputs or outputs differ or if both differ between the two proposals, then it is these differences that must be examined critically.

In many situations for highway improvements there exists several proposals, one of which is to do nothing–i.e., keep the existing situation. One alternative, however, is to abandon the existing situation and not replace it. It could also be contracted as opposed to expanded. In comparing the relative economy of the several proposals, a common but incorrect procedure is to compare each with the existing situation and stop there. But such stopping is in error. Each of the proposals in turn must be compared with one another. In so doing the differences in cost and in net benefits are compared. This step is sometimes called the *incremental solution*. It is a necessary step if correct answers are to be reached.

In a grade-reduction study, the alternatives are many, nine in fact, when the existing grade is 8 percent and only grades to the even integer in percent of grade are to be considered. But there are ten alternatives if abandonment of the grade is considered. Thus the question is: Should the existing 8 percent grade be retained or should the grade be reduced to 7, 6, 5, 4, 3, 2, 1, or 0 percent? A comparison of each percent grade with the 8 percent existing grade could show an economic justification of reducing the grade all the way to level. But when the increments of costs and increments of benefits are compared grade by grade, it may be discovered that any grade reduction from 3 percent to less than 3 percent produces more costs than it does benefits. In such case, the adverse relation of benefit to cost below a 3 percent grade is overbalanced by the favorable relation in reduction from 8 down to 3 percent. It is only through this incremental analysis that the most economical grade can be found. The same is true for any other analysis; the separate grades of 7, 6, 5, 4, 3, 2, 1, and 0 may be thought of as alternatives in design such as eight different locations of a proposed highway route.

13. Common Factors of Equal Magnitude May Be Omitted

Another variation of the principle that only differences in alternatives are relevant is found in certain factors of costs and consequences. For instance, the general annual expenses of administrating and policing a highway route may be the same for all alternatives considered including the existing alternative–at least so closely the same that the increment of differences cannot be reliably determined. Therefore this item of annual highway expense generally may be omitted in the analysis for economy. Similarly, when considering the nonmarket items of general social and community benefits of highway improvements, they may be found to be so nearly the same for each proposal that it is unnecessary to evaluate these nonmarket items in detail.

14. Use the Net Basis for All Costs and All Consequences

Although sometimes difficult to arrive at in both the study of construction cost and of monetary consequences, data wanted for the analysis are the net of each, and without any double counting, transfers, or omissions of either positive or negative amounts, but including both market-priced factors and opportunity costs and gains.

In rights-of-way taking, the net cost would be the total price paid for the land, buildings, and damages, less any income from sales of land, buildings, or crops, and any rents received prior to start of construction.

Expressways and interstate routes frequently attract motor vehicles to them, but at the cost of adverse distance of travel. This adverse travel is a negative consequence and should be accounted for in the analysis. Likewise, all consequences to all traffic in any way affected by the improvements under consideration should be fully evaluated and brought into the analysis.

Concepts and Principles of Engineering Economy 27

Often the consequences to business, markets, and real estate are affected adversely by highway improvements. Studies of the economic impacts of highway improvements generally itemize the economic gains within the land areas closely associated with the highway improvement. But seldom are the adverse consequences brought into vision, mainly because they are apt to be farther away geographically and scattered. When, however, these general economic consequences whether opportunity costs or gains or market-priced costs or benefits are to be used as a factor in the final decision, care is warranted in getting the total net consequences–the beneficial less the adverse.

15. The Analysis for Economy Is Independent of Financing

The decision on economy is not a decision of financing. Financing is a separate function to be analyzed separately. Financing involves factors and decisions differing from those included in the analysis for economy. The analysis for the economic evaluation of a proposal produces a result used by management in reaching a decision whether to construct the project, and (when project formulation is involved) to what design to construct the project. The financing of the project is also a factor considered by management in reaching the decision to construct or not to construct, but such an analysis is made separately from the analysis for economy. The economic evaluation of a proposal and the economic formulation of the features of design are both independent of how or by whom the project proposed may be financed if and when constructed. Financing is a subject of available funds and their allocation, not a subject of economic feasibility or of engineering design.

It is true, however, that the amount of money available may alter the design and therefore the estimated economy, but in such case there is introduced another alternative. It is doubtful that any increase in road-user taxes as a means of financing the improvement would alter the use of the facility sufficiently to switch it from an economical to an uneconomical improvement.

16. Uncertainties Need To Be Acknowledged

In spite of care in estimating and the use of low and high values for certain factors, uncertainties will still prevail, more so in certain studies than in others. The final decision therefore will give weight to these uncertainties and probabilities. Future land use is always an uncertainty; so is flood; so is technology. Today many highway improvements are not adequate only because of past changes in land use that affected the volume and character of traffic attracted to the highway. One advantage of a high vestcharge rate is that it discounts the future uncertainties much more than does a low rate.

17. Separate Decisions Are Made at Separate Levels of Management

Within the objectives of the analysis for economy there are separate levels of decision making to which economic evaluation and project formulation may apply. Within a highway department, the analyst and the design engineers consider technical elements as subalternatives and endeavor to reach separate decisions on separable factors. The choice between bituminous pavement and portland cement concrete pavement is an example. In so doing their decisions are to be in conformity with the policies and viewpoints of top management who will make the major and final decisions. The design for a tunnel under a mountain ridge, its approach grade, the tunnel grade, the tunnel elevation, and the number of lanes to build in the first stage should each be analyzed for economy as separate factors. Restraints on both overdesigning and underdesigning are in order. Further, the final decision of top management is made upon the results of the analysis for economy weighed against the pertinent factors not included in the analysis.

The magnitude and direction of such factors as value of time and consequences to truck traffic as opposed to passenger-car traffic need to be set out for individual consideration. Decision making at the separate levels on questions of economy becomes difficult because of the involvement of designing engineers, technical specialists, consultants, top management, and–somewhere within this order–the analysts who are familiar with the concepts, principles, and procedures of analysis for economy. As analysis is accomplished toward economic evaluation and project formulation, decision at the separate levels can be kept in harmony and on the correct technical and policy basis by coordination and full exchange of information, through conference, record keeping, and reporting.

18. From Whose Viewpoint Important to Final Decision

The value of commodities, services, experiences, and satisfactions varies from individual to individual according to each individual's preference. Furthermore, the individual will change his values as the time, place, and circumstance change. A person not owning an automobile will not value good highways at the same level as will an owner of an automobile. A given section of a highway traveled at least twice daily by a person is valued by that person at a higher level than by a person who travels that section only once a year. An offer of a steak dinner has little value to one who has just finished a hearty meal, but is worth the proverbial king's ransom if he has just come in after a long day's hike in the woods.

In highway improvement the viewpoint affects the values that individuals place on the proposals for improvements. Truck owners and passenger car owners have explicitly different viewpoints; owners of land adjacent to the highway maintain a viewpoint different from those landowners 10 blocks or 2 miles away. Similarly, there is a change in the viewpoint of public officials and citizens as the level of government moves upward from the town, county, and state to the federal government. Thus federal officials maintain (or should) the viewpoint of the national objective and national consequences rather than that of evaluating the consequences in Arizona as compared to those in California. The

Concepts and Principles of Engineering Economy

viewpoint of the public in adjacent counties is often opposite; each wants the highway route regardless of the unfavorable consequences upon the adjacent county.

The viewpoint is particularly relevant when considering such nonuser consequences as land values, business development, whose land is to be taken for rights-of-way, and resource development. In the final selection of a proposal to construct a highway facility, weight must be given to all factors and to all viewpoints as may be judged to be just and right.

When considering the factors of economy easily reducible to money and included in the analysis, there is not much choice of viewpoint. Commercial travel versus passenger car use, local road users versus the through traveler, and the peak-hour traffic versus the off-peak traffic do offer some basis of differing viewpoints, however. In grade reduction, for instance, the heavy vehicle combinations would be favored much more than would the passenger car. Scenic aspects would favor the vacationists more than the local people and the passenger car more so than the truck.

19. Establish Criteria for Decision Making

The analysis for the economy of proposed highway improvements affords the decision maker another factor, or tool, to use in combination with all other pertinent factors in reaching his decision. But in his decisioning process he must be guided by proper and sound criteria. He must have a policy, a viewpoint, and objective, and some guidelines.

One important criterion is how much net benefit is to be demanded. A measure of how much benefit is possible is given in the prospective rate of return from the improvement. Certainly for a given improvement project there is a rate of return below which it would not be in the interests of the public to go. Similarly, in the benefit/cost ratio solution, does a ratio of 1.0 justify the improvement, or should a ratio of 1.25 be adopted as the minimum ratio of net benefits to net costs to justify a choice of alternatives? Within the benefit/cost ratio method the rate of vestcharge (interest) used in the solution could be regarded as the minimum attractive rate of return. When the vestcharge rate is soundly high enough and when other factors are amply conservative, a benefit/cost ratio of slightly more than 1.0 could be used as a criterion. Full consideration of the factor of vestcharge rate should be given ahead of making the analysis in all procedures, regardless of which of the six methods may be used. Even in the rate-of-return solution, the minimum attractive rate of return is a decision-making criterion that must be set.

In addition, a time period for analysis becomes an important criterion. Shall the analysis be made on the basis of a forecast for 10, 20, or 30 years? A short period has the advantage of holding the effects of uncertainties to a minimum. But the time horizon used in every case should depend upon the circumstances of that case.

20. Consider All Consequences to Whomsoever They May Accrue

The analysis for economy is an attempt to weigh the costs of producing the desired consequences against the probable worth of those consequences. All consequences to whomsoever they may accrue are therefore subject to examination. Any attempt to be selective of the expected consequences could inject a bias or subjective bent and warp the result. At the outset of the analysis all consequences should be recognized; they need to be evaluated by the market prices when practical. Once the consequences are identified, regardless of whom may be affected by them–beneficially or adversely–they are evaluated. They enter into either the analysis directly or the decision process according to their proper place. The viewpoint adopted for the decision may rule out or rule in certain consequences.

Often in highway transportation studies the consequences to other modes of transportation are overlooked, such as to railroad transportation. Such omission is questionable from the viewpoint of public policy. Likewise the consequences to commercial interests and to public functions resulting from different geographical locations may be overlooked sometimes. It is one thing to overlook an item of consequence but quite another to look at it, then rule it out as irrelevant, inconsequential, or immaterial.

21. Final Decision Gives Weight to Nonmarket Factors

The analysis for economy is made on the basis of all factors which are truly ones of economy and which can be reduced to reliable money values. The consequences of highway improvement, however, are far-reaching into the secondary, nonmarket, or nonuser areas of social, economic, political, community, and personal aspects. Therefore in the decision process of management the just and right weight is given to these consequences in accordance with their worth as determined by judgment and public policy. Here again the viewpoint is an important factor; what is an economic gain to one geographical area may be a loss to another. A highway improvement may be of special benefit to an industrial area, but a detriment to a residential area.

PROBLEMS

2-1. Discuss the concepts of "engineering economy" and "welfare economics" from the viewpoint of the engineer's responsibilities.

2-2. Write your own version of the origin and evolution of economy as developed over the years as a distinctive phase of the practice of engineering. Why and how did the subject come about? Compare engineering economy with household economy.

2-3. In the normal operation of your family affairs, list actions and decisions that are based on economy. In what ways does your family contribute to the economic forces and conditions of your community?

2-4. Give your understandings of the following words as they might be applied to

Concepts and Principles of Engineering Economy 31

economy analyses: (1) estimate; (2) forecast; (3) opinion; (4) judgment; and (5) guess.

2-5. For your city, town, or urban community, list the specific public works improvements, civic developments, public services, and similar governmental activities that could be studied from the viewpoint of engineering economy, cost effectiveness, or benefit-cost analyses.

2-6. In your family activities, discuss those activities and decisions that are related (1) to project evaluation and (2) to project formulation.

2-7. In public works development in your community or state, list specific types of projects (past, present, or future) in which the factor of project evaluation would not be a major consideration, and those projects in which project formulation would be a major factor.

2-8. List all the alternatives you can think of (really every one) that could be employed in reaching each of the following objectives:

1. A means for getting people and vehicles to a point on the opposite shore of a navigable stream where no means (except swimming) now exist.

2. Improving the traffic flow at a four-way street or highway intersection.

3. Providing a new highway service to the opposite side of a mountain ridge.

4. Increasing the traffic flow at peak hour on an urban arterial which now caters to bus traffic and local parking.

5. Reducing or eliminating highway traffic delays and accidents at a railroad grade crossing.

2-9. For the following types of proposed improvements in highways and their operation, make a list of the important items–calculated, observed, and assumed–that you would need to consider in determining their relative economy:

1. Urban bypass route.

2. River bridge versus a longer no-bridge route.

3. Wood versus steel versus concrete for a bridge.

4. Traffic control devices for an intersection.

5. Constructing two lanes now and two lanes more in the future versus constructing all four lanes now.

6. Whether to tunnel a mountain ridge or use open cut.

7. Detouring traffic around reconstruction versus carrying traffic through the reconstruction site.

8. Overtime or double shifts versus longer construction time with 40-hour week of one shift.

9. Power-applied chemical weed and brush killers versus power and hand mowing and hand clearing.

10. Spacing distance apart of and number of maintenance shops (supply and service centers) for highway maintenance operations.

2-10. For the following general proposals for improvements, ventures, or projects, list both the main types of alternatives that may be appropriate, if any, and the data and factors to consider in the analyses necessary to the final decision of whether to go ahead, and if so, how.

1. A toll highway east and west across the state of Wisconsin, financed by bonds.

2. A state-owned or a privately-owned bond-financed toll bridge across the Mississippi River at Keokuk, Iowa.

3. Using a company-owned central concrete mixing plant in lieu of buying ready-

mixed concrete from a transit-mix company. The plant will be used for manufacturing precast, prestressed concrete shapes for bridges and buildings.

4. Letting to contract the patching and other repair work on pavements rather than performing the work with department crews and equipment.

5. When to trade in automobiles used by the staff for new ones.

REFERENCES

2-1. Thomas Arnold. *The Miscellaneous Works of Thomas Arnold, D.D.* 2d American ed., Appleton, 1846.

2-2. Norman N. Barish. *Economic Analysis (For Engineering and Managerial Decision Making)*. McGraw-Hill, New York, 1962.

2-3. E. Paul DeIarmo. *Engineering Economy.* 3d ed., Macmillan, New York, 1960.

2-4. Otto Eckstein. *Water Resource Development (The Economics of Project Evaluation).* Harvard U. P., Cambridge, Mass. 1958.

2-5. Eugene L. Grant. *Concepts and Applications of Engineering Economy.* Highway Research Board, National Academy of Sciences, Washington, D.C., Special Report 56, 1960, pp. 8-18.

2-6. Eugene L. Grant and W. Grant Ireson. *Principles of Engineering Economy.* 4th ed., Ronald, New York, 1960.

2-7. John V. Krutilia and Otto Eckstein. *Multiple-Purpose River Development.* Johns Hopkins Press, Baltimore, 1958.

2-8. Tillo E. Kuhn. *Public Enterprise Economics and Transport Problems.* U. of California Press, Berkeley, 1962.

2-9. Edwin Scott Roscoe. *Project Economy.* Richard D. Irwin, Homewood, Ill., 1960.

2-10. U.S. Government Interagency Committee on Water Resources. Subcommittee on Evaluation Standards. *Proposed Practices for Economic Analysis of River Basin Projects (The Green Book)* Government Printing Office, Washington, D.C., May 1958.

2-11. Arthur M. Wellington. *Economic Theory of the Location of Railways.* 2d ed., Wiley, New York, 1887.

2-12. Robley Winfrey. *Concepts and Applications of Engineering Economy in the Highway Field.* Highway Research Board, National Academy of Sciences, Washington, D.C., Special Report 56, *Economic Analysis in Highway Programming, Location, and Design* September 17-18, 1959, pp. 19-33.

2-13. E. W. James. *Highway Construction, Administration, and Finance.* Highway Education Board, Washington, D.C. Reprinted from a series of articles c. 1927 in Spanish from *Ingenieria Internacional.* McGraw-Hill, New York.

2-14. Sigvald Johannesson. *Highway Economics.* McGraw-Hill, New York, 1931.

2-15. C. B. McCullough. *Economics of Highway Bridge Types.* Gillette, Chicago, 1929.

2-16. C. B. McCullough and John Beakey. *The Economics of Highway Planning.* Oregon State Highway Commission, Salem, September 1937.

2-17. James J. Tobin and A. R. Losh. *Highway Cost Keeping.* U.S. Department of Agriculture, Washington, D.C., Bulletin 60, September 1918.

2-18. Harry Tucker and Marc Leager. *Highway Economics.* International Textbook Co., Scranton, Pa., 1942.

2-19. John C. L. Fish. *Engineering Economics.* 2d ed., McGraw-Hill, New York, 1923.

2-20. William M. Gillespie. *A Manual of the Principles and Practice of Road-Making, Comprising the Location, Construction, and Improvement of Roads and Rail-Roads.* 2d ed., A. S. Barnes & Co., New York, 1848.

Chapter **3**

Highways and Engineering Economy

Engineering economy is an integral feature of engineering design and of the responsibility of the engineer in advising management of the economy of engineering-based proposals. The principles of engineering economy, developed largely in private industry, apply equally well to public works.

Highways and highway transportation are unlike private business and most other forms of public business and functions of government, yet many of the same principles of business management apply. An understanding of the special character of highways is an essential foundation to understanding the concepts and procedures of economic analysis of proposed improvements of highways. The function of highways is to produce that transportation service wanted by the public at the lowest cost consistent with the quality of service desired. Unlike industry, highway departments have no product to sell. The departments are merely the agents of the public performing under the direction of the public for the benefit of the public. As such, the beneficial and adverse consequences of highways are not simple costs and sales income as found in industry.

SERVICE CHARACTER OF HIGHWAYS

Highways are also unlike private industry when considered in economic analysis of proposed capital investments. Highways must serve the dynamic vehicles whose operating and ownership costs are about seven times the cost of the highway facilities. Many features of the design of the highway affect the running cost of the vehicles–an important consideration in project formulation. Highways do not produce a direct cash income. The benefits from improvements of highways have to be measured as cost reductions or increases in personal preferences rather than as sales income.

CHARACTER OF HIGHWAY SERVICE

In private business, investments are made for the purpose of producing a service or commodity for sale to the general public, which has no special interest in the selling organization, and the selling organization has no interest in the

Highways and Engineering Economy

buying individual other than to get him to buy as much of its product as possible. When viewed in light of these differences, highways cannot be examined for economy of their improvement with the same procedures and concepts as are usually applied to private industry.

With highways, the designers, managers, and operators are highway departments that are working not for themselves but as the official agents of the public. The public in turn, is both the owner and user of the public highways. True, there is a small percentage of the public which is not a part of the highway-user group, but this percentage is directly dependent upon the highway services and does indirectly contribute to the support of highways. The road users are the main financial support of the highways. So, in effect, the whole highway transportation business is a closed business that is owned, operated, and used by the same set of people. Therefore any reductions in cost of highway transportation accrue to these owner-users, and any increases in costs likewise must be absorbed by these same owner-users. The highway department does not seek to induce or to increase highway travel (sales) but strives to supply the service wanted by the public.

The highway department has a direct obligation to design highway facilities to render the desired level of service wanted by the owners and at the lowest total cost for transportation that is consistent with this level of service. These owners have no second party (customers) on which to unload any increases in cost, and no stockholders, other than themselves, to receive the benefit of cost reductions. In other words, there just cannot be any sales involved because the seller and purchaser are one and the same person.

NONROAD-USER CONSEQUENCES

But in achieving the minimum possible annual cost of transportation, proper control must be exercised in all investments so that the new facilities do not run counter to the nonhighway interests of the people as a whole, including the highway users. Thus the highway designer, in striving to achieve the minimum transportation cost, may have to stop short of minimum cost in order to protect the nonhighway factors desired by the people. Often a compromise is in order which provides an acceptable balance of these seemingly conflicting desires of the public.

As discussed in Chapters 4 and 19, the consequences of highways that accrue to the nonhighway user and the public are real, identifiable, and generally desirable. But difficulty is encountered in dollar pricing these consequences so that they may be merged with the dollar quantities of highway cost and road-user consequences. Further, some of these nonroad-user consequences are most difficult to separate from road-user consequences as well as to determine their net value when considering transfers, overlapping, offsets, substitutions, and double counting within the total economic system. For these reasons in the economic analysis these general and nonroad-user consequences are not included in the calculations of the index of economy–annual cost, present worth, benefit/cost ratio, or rate of return.

CLASSIFICATION OF THE CONSEQUENCES OF HIGHWAY IMPROVEMENT

Highway improvements may be viewed through four groups of consequences and the relation of these four groups to the financing required and to the community economics. The four groups are: (1) commodity savings, (2) travel time reductions, (3) personal preferences, and (4) community consequences.

COMMODITY SAVINGS

The saving of commodities is an objective of good management despite the fact that such saving may lower the sales volume of suppliers. In highway design, one objective is to reduce the comsumption of commodities such as construction materials, motor fuel, oil, tires, repair parts, medical supplies, and hospital services. As viewed by the road user these reductions in consumption would reduce his motor vehicle running costs and total road-user costs. Perhaps these commodity savings may result in a total saving of say 1.0 cent a vehicle-mile by using the new highway. If the highway cost to the road user for the new highway is increased 0.7 cent a vehicle-mile he benefits directly at the rate of 0.3 cent a vehicle-mile. In theory, the road user could be taxed an additional 1.0 cent a vehicle mile, and in total he would break even. On the basis of the 0.7 cent additional tax, he pays for the new highway and still is money ahead. Here, then, there is a fully realized economy to the road user. The new highway is paid for, the highway user has money left in his pocket to spend otherwise, and no adjustments in the other factors of his economic life are necessary. Commodity savings, then, are real, tangible, direct money savings.

In highway economy studies, commodity savings should be given the highest weight, because the dollar value of these savings is fully realized in actual dollars mile by mile of highway use. When a highway improvement can be economically justified by savings in commodities purchased by the highway users, there can be no doubt about the wisdom of constructing the improvement from the viewpoint of highway use. The improvement definitely will pay for itself without requiring the road users to alter their pattern of economic support of society by transferring some of their support of other activities to highways in order to pay for the new facility.

TRAVEL TIME REDUCTIONS

To reduce travel time is one thing; to convert the time not used in travel into dollars is quite another thing. Time has value, because, through time we accom-

Highways and Engineering Economy

plish that which we desire to accomplish. But time is not money; it is only the medium through which wages and salary may be earned. When a highway improvement is justified by the value of the highway travel time reduction, the highway user does not for certain gain the dollars needed to pay for the highway. In fact he merely is afforded the time in which to earn the money needed to pay the highway cost or to pursue other activities of his choice. When he is successful in converting the time into money, he will have that much money to apply on the highway cost. But when he is unsuccessful in making sufficient conversion of the reduced travel time into money to pay the full highway cost, he must find another source of his highway tax dollar.

The new highway facility itself probably does not increase the personal income of the highway user nor reduce his cost of living (neglecting for the moment any decrease in motor vehicle running cost not balanced by increased highway taxes). The source of the highway tax dollar must then be obtained by curtailment of some other economic activity. The road user can travel fewer miles, spend less on recreation, eat cheaper foods, postpone buying new household appliances, or reduce other of his normal expenditures. But the highway tax must be paid.

Considering only the factor of travel time, the new highway facility is economical only when the travel time is reduced and as a result the traveler can earn a greater income. When the reduced travel time leads to activities that do not produce more income, there is no true net overall social economy, but only a transfer of economic support from one activity to another. Note the difference between the economic consequences of saving commodities and of reducing travel time. Saved commodities are real dollar savings; reducing travel time is creating only an opportunity to earn dollars through a different usage of time or to pursue other activities of choice. Therefore, less weight should be given to travel time reduction than to commodity saving in the final decision on highway construction. For this reason, it is good procedure to analyze proposed highway projects for their economy of travel time separately from the economy of commodities. But it is well established that the highway traveler wants to reduce travel time and is willing to pay the costs to achieve that goal. This discussion simply strives to point out that there is a difference between a true saving and a transfer of one's spending pattern from other activity to highway travel.

PERSONAL PREFERENCES

When commodities are saved, the road user travels at less cash cost; when travel time is reduced the road user has the opportunity to increase his income through productive use of the time not used in travel; but when the new highway provides the road user only the benefit of comfort, convenience, uniform speed, and other purely personal nonmonetary satisfactions, his benefits are other than monetary. He must, therefore, find his highway tax dollar through curtailment of his economic support given to other aspects of his daily economic life. He pays his increased road-user taxes with money obtained by reducing his spending for food, shelter, clothing, recreation, travel, or savings.

These personal-preference factors of the highway traveler are not factors of economy that may be used to justify highway construction, but are only indicators that show the preference of the highway traveler in spending money. Except for one possible source, they contribute no real economic saving to society. A motor vehicle trip of such mental and physical comfort that the driver and passengers reach their destination in such relaxed and rested state that they are more productive than they would be otherwise (by using the old highway route) may be said to have some net economy. To identify this economy quantitatively is most difficult and to price it in dollars is equally difficult. In economy studies to evaluate the degree of economic justification of constructing highway facilities, there seems to be no place for personal preference items. Certainly, the road user who must pay the highway costs can find no source for his highway tax dollar that is directly traceable to the personal preference consequences of the new facility. He therefore must make other economic sacrifices in accordance with his personal choice.

Nothing in the foregoing discussion about personal preferences of the road user is intended to imply that the road builder should avoid building comfort, convenience, uniform speed, and aesthetics into highways. If the road user desires these preferences and is willing to pay for them, certainly they should be provided. The result is that the road user desires a higher level of service at higher cost and is willing to sacrifice other utilizations of his dollar in favor of these higher highway costs. Certain road users drive high-priced cars in preference to low-priced cars; some women wear mink coats rather than the less expensive wool coats; and some families live in $100,000 homes rather than in $25,000 homes. These are privileges and means of satisfying personal preferences; they are not actions to achieve economy—that is, husbanding the dollar—but they are actions that affect both family and community economics.

The basic premise of the analysis for economy is that the cost of acquiring and operating a highway facility should, over time, return to the public its initial cost and annual expense of operation plus compensation for the sacrifices (vestcharge) of not being permitted to use the money to gain other satisfactions. When a proposed highway facility will provide these returns it is economically justified. The road-user benefits to achieve these two returns must be real, priced on the market, and positively identifiable.

As mentioned in the next section on community consequences, there may be benefits and desires of a nonmarket character that justify highway construction. When they exist, these benefits should be identified as such and not confused with economy of transportation. Personal preferences of the road user should likewise be identified as nonmarket consequences.

Highways and Engineering Economy

COMMUNITY CONSEQUENCES

Highways, roads, and streets provide the means of achieving many economies and satisfactions for the communities served. These beneficial consequences are not reliably reducible to the monetary dollar so are in the nonmarket group of consequences excluded from the analysis for economy for both project evaluation and project formulation. The community consequences may pertain to the following and the economic and social effects associated with them:

Aesthetics	Postal services
Business and trade	Protective services of fire, police, and health
Community pride	Recreation
Education	Social interchange
Land values and land uses	Utility services
National defense	

To the extent that this group of items and the associated other social and economic factors involve the use of motor vehicles, such highway use would be included in the road-user consequences. But beyond the use of the highways, roads, and streets, such community services as fire, police, and health protection are aided beyond their vehicular use. Services can be more frequent, more quickly given, and they may reach to more distant areas. Good rural roads made possible the consolidated school system; in urban areas, school buildings may be fewer and farther apart because bus transportation of pupils is available. Highway transportation affords more use of zoos, parks, and recreational areas. All these benefits are gained by the community as a result of highways, roads, and streets. But to evaluate these benefits in dollar terms is beyond our know-how. Some items of this group are economic, others are personal preference, but all are nonmarket items. They must be appraised by personal judgment based upon what the people of the community want and are willing to pay for.

The business and trade items are in a slightly different category, yet are community in character. Highway improvement affects land values because of resulting changes in land use. Business volume is also directly affected. These items are more fully discussed in Chapter 19. Here, though, it is pointed out that land values and business volume may result in benefits to the community or to select individuals in the community. Such benefits, however, are nonroad-user consequences and usually of a nonmarket character–that is, they are not priceable on the market. To include both the road-user benefits and the benefits to land and business leads to double counting. Reduction in travel costs is transferred to these nonuser consequences.

The most difficult aspect of these business consequences, however, is in determining the net consequences and in adopting the proper point of view. Usually a highway facility will attract business to adjacent land or even to the general community. This attraction may have a net local benefit above the cost of supporting the business and its people with protective services and education. But if the economic consequences are traced to their faraway points of beginning and ending, adverse consequences may be uncovered to offset most all of the local beneficial consequences. Good highways do not cause people to spend more money; the main effect is just to change the place of spending and to some extent what the spending is for. The traveler eats and sleeps the same regardless of the highway location; manufacturing produces the same total product, more or less, regardless of the state of highway improvement. Thus when a local community gains a larger business volume because of a highway improvement, that gain may be at the loss of another community or that gain may have been a natural gain brought on by overall increase in population or by changing economics wholly independent of highways.

The point of view now becomes an important factor. To a local city, new business may exceed the local public costs of supporting that business by a positive measurable sum. To this city such gain may be a net benefit, but a nearby city may have actually lost the business that the other city gained. The overall economic net gain is zero. Here the two cities are in a competitive position, similar to that of two business competitors. Usually in these situations the two cities or communities battle it out in open forum and trust that the public officials responsible for the final decision will be fair, just, and act in the best interest of the largest public. But here again the consequences are not of the direct market and monetary character which permits them to be included in the analysis for justification through economy. These community factors are held separate and given just and right weight as the circumstances dictate.

One of the troublesome aspects of the whole area of community consequences lies in the difficulty of determining all the consequences–beneficial and adverse–to whomsoever they may accrue. The tendency of analysts and management is to look only close at hand and to consider what they see there. Good procedure, however, calls for a faraway look as well and getting at the real net of all consequences to consider in the final decision process.

APPLICATIONS OF THE ANALYSIS FOR ECONOMY

Within the highway field, the analyses for the economy of proposed improvements may be applied to the following objectives:

1. To measure the economic justification of proposed highway facilities–project evaluation.
2. To measure the economic choice between mutually exclusive features of design–project formulation.
3. To aid in scheduling a given year's program of construction–project priority selection.

Highways and Engineering Economy

4. To measure the relative cost or tax responsibility of the beneficiaries–cost allocation.

5. To measure relative justification of highways with other public works or public functions including other modes of transportation–public financial policy.

DISCUSSION OF THE OBJECTIVES

Project evaluation and project formulation serve the same purposes in highway and other public works programs as they do in private industry. These two objectives of the analyses for engineering economy are the most frequently applied. In fact, they should be applied to most all proposals for investments in the improvements of highways, including construction and maintenance equipment. By using the results of the analyses for economy, the final decision maker will find himself on a much firmer base than he will be without these results.

The scheduling or programming of a year's construction offers a good application of the analysis for economy (see Chapter 26). In the interest of maximizing the favorable consequences of highway improvements, those projects that return the larger net benefits, i.e., have the highest prospective rate of return, should be scheduled for construction first. The rate-of-return solution, to the extent that it can be applied to all possible independent proposals for use of construction funds, offers a good basis of ranking proposed projects as to priority. Management, however, will adopt the final construction program only after considering other factors such as geographical distribution of work; distribution of types of work between earthwork, paving, and structures; continuity of route improvement; availability of rights-of-way; and availability of materials and contractors.

Within the total analysis for the economy of a proposed facility, costs and net benefits are often assembled in such a manner that they can be associated with chosen classes of highway users and nonusers. When this can be done, the data and solution may be used in studies of tax rates and tax allocation among those who should pay for the highways.

Lastly, highways are only one function of government. The highway construction program and, therefore, the tax program through which the funds for highway construction are raised, should not be undertaken unmindful of the other needs of the community. The taxpayer's dollar once committed to highways is not available for general government, health, education, parks, or water resources. Rates of return from highways can be compared with rates of return from other public functions including other transportation systems. When rates of return cannot be computed, at least a comparison of project costs with project benefits is helpful in determining public policy.

The many functions of a government compete for the tax-income dollar as a scarce commodity. Careful study and comparison of proposed total budgets of governments is highly desirable to assure that the total tax income of the community is utilized to achieve the greatest satisfactions and economy. The situation is similar to that of the family budget. A family must weigh its costs for alternative uses of its income with the probable satisfactions achieved. For instance, the family must choose between a new car, some new furniture, a long vacation trip, advance payments on its house mortgage, or investment in securities. In its own way it may not make an economy analysis, but it does of certainty reach a decision after weighing in some fashion the probable satisfactions from each possible choice of expenditure.

UNIQUE CHARACTER OF HIGHWAYS

Because the direct costs and benefits associated with highways can be measured with acceptable reliability, highways, as a function of government, lend themselves to the analysis for economy in a way similar to that applied by private business and regulated public utilities. There is the same reason for exercising wise husbanding of public monies as there is in private industries. In the end, the citizens are burdened with all costs and receive all benefits, so their money should be invested on a basis of sound principles of economy.

Highways, as compared to most government functions, are unusual. They are often thought of as a public utility, publicly owned and operated, but without regulation other than by direct legislative action. In the United States highways are about the only function of government largely supported by tax imposed upon the direct beneficiaries of the function. The receipts from motor vehicle license fees and fuel taxes, with small exception, are dedicated to highway use. In contrast, but little of the general tax returns are appropriated to highway construction. The city streets receive considerable support from general property tax, and sometimes from special assessments, but rural roads and highways do not.

This earmarking of the road-user tax payments gives rise to considering the fuel tax as a direct service fee, proportional to use, and the license fee as a readiness-to-serve charge. Thus the road user can trace his tax payments directly to his benefits. He expects a favorable relationship of benefits to costs. The highway departments, therefore, are in a favorable position to use engineering economy as one of their tools of management.

In the whole of the highway system under a given political administration it is common to find proposals for improvements requiring several times the amount of construction money usually available for a given year. Analysis of economy is therefore one good guide for the commitment of the limited money available.

TYPES OF IMPROVEMENTS TO WHICH ECONOMIC ANALYSIS SHOULD BE APPLIED

Throughout the highway engineering profession the general thought often prevails that the cost-benefit or cost-effective type of analysis is applicable only to highway location proposals, or at least the practice is often to apply the analysis

Highways and Engineering Economy

to location proposals only. The truth is that the analysis for economy of investment of public resources in highways is applicable to every proposal for improvement, regardless of its character.

Another misconception is that the analysis of economy is used only in the sense of justifying on an economic basis (project evaluation) the basic improvement proposed. The application of the analysis to the details of design (project formulation) is an important application of the analysis, that usually follows the decision to construct the improvement. In project formulation, the analysis for the engineering economy is applied to subfeatures of the overall proposal, such as the optimum maximum or minimum vertical gradient, the optimum elevation of a tunnel under a mountain ridge, or planned stage construction over a period of years.

It is easy to say that all proposed investments in highway improvements should be analyzed from the standpoint of engineering economy, but quite another matter to get the required data to make a sufficient analysis. This fact does not negate the necessity and usefulness of economy analyses; it only points up one of the weaknesses inherent in our statistical data-gathering techniques.

The following listing of types of highway improvements which could be analyzed for their economy does not include all possible applications, but these 53 items do imply that the opportunities for such studies are extensive.

1. Route location
    A. Arterial and freeway system spacing and location
    B. Bridge versus a longer no-bridge route
    C. Highway system development such as an arterial or freeway system for an urban area
    D. Rural route location
    E. Special conditions[1]
    F. Special routes for selected traffic[2]
    G. Toll facility
    H. Urban bypass location
2. Geometrics and other details of design
    A. Grade crossing elimination
    B. Elevated versus depressed freeway
    C. Horizontal curvature
    D. Length and location of frontage roads
    E. Median and shoulder width
    F. Navigational clearance at waterway crossings
    G. Number of lanes

---

[1] For example: (a) Reconstruction of the existing route on essentially the same right-of-way. (b) Existing route abandoned as a public highway. (c) Existing route continued as a public highway, but may be transferred to another highway system.

[2] For example, routes which are to be used only by trucks, or parkways which allow only passenger cars to travel on them.

- H. Open cut or tunnel
- I. Size of culverts and bridge waterways
- J. Size and spacing of drainage inlets
- K. Spacing of interchanges and grade separations
- L. Travel distance
- M. Uphill truck lanes
- N. Vertical gradients
3. Materials and types of design
    - A. Roadway or pavement surface material
    - B. Type of bridge structure[3]
    - C. Wood versus steel versus concrete versus aluminum
4. Traffic controls
    - A. Computerized traffic signal controls
    - B. Legal speed limit determination
    - C. One-way traffic
    - D. Railroad-crossing traffic protection
    - E. Reversible lanes
    - F. Type of intersection control
5. Modernization of highways and traffic-relief improvements
    - A. Add lanes or develop new route
    - B. Change to higher type roadway surface
    - C. Install median barrier
    - D. Lengthen and flatten horizontal curvature
    - E. Lengthen sight distances
    - F. Relief route to reduce traffic congestion on a parallel route
    - G Spot improvements to reduce accidents
    - H. Widen roadway and resurface
6. Stage construction and future provisions
    - A. Buy R-O-W (rights-of-way) now for future use
    - B. Construct 4 lanes now versus 2 lanes now and 2 lanes in the future
    - C. Grade and drain now for future use
    - D. Provide now for double-decking a bridge in the future
7. Management policies
    - A. Desirable maximum dimensions and weights of motor vehicles
    - B. Handling traffic via detour route versus routing traffic through construction
    - C. Length of haul from central plant to work site
    - D. Location and size of maintenance shops and service centers
    - E. Long or large construction projects versus short or small projects
    - F. Overtime and night constructions versus 40-hour week and one crew
    - G. Short contract time versus long contract time
8. Construction and maintenance equipment and operations
    - A. Highway maintenance procedures and processes
    - B. Replacement of automotive equipment machinery, and other equipment
    - C. Substitution of electrical and mechanical devices for manual labor

---

[3] For example: truss, through girder, draw span, fixed span, tunnel, high-level structure, etc.

FACTORS OF ANALYSIS FOR HIGHWAY ENGINEERING ECONOMY

The ultimate decision of management is first to approve or to disapprove construction, and second, to approve a specific design of the public works. The first decision is highly dependent upon both the analysis for the engineering economy and the probable nonmarket or nonroad-user consequences that cannot be dollar priced. The second decision will likewise depend upon both market (the dollar-priceable factors) and nonmarket factors, depending upon the character of the proposed improvement. A proposed new highway, because of alternatives in centerline location, will often involve many nonmarket factors affecting the community, so the first decision will be weighted accordingly. But the subfeatures of design about which the public is not much concerned will be decided mainly upon the results of the analysis for engineering design. Whether to elevate or to sink an urban freeway is an example where often the factors of economy and of community choice are in conflict. The responsible official, or

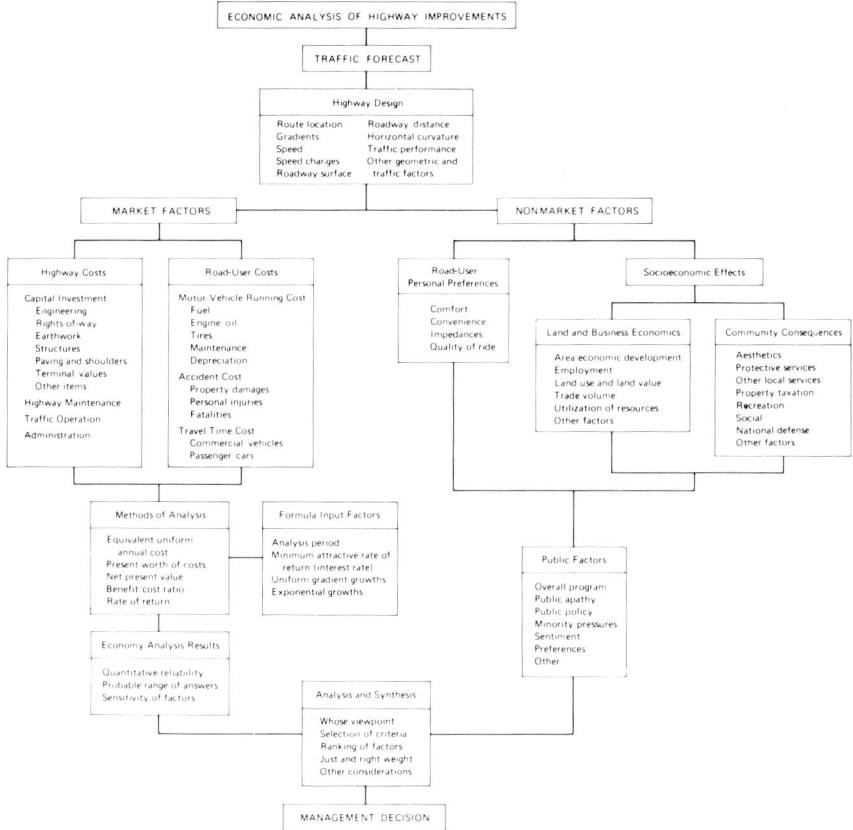

FIG. 3-1. The factors of analysis for economy and other consequences of highway improvement.

board, in such instances, must make the decision giving just weight to the many factors involved.

The diagram in Fig. 3-1 illustrates how the market and nonmarket factors enter the analysis and come together at the desk of top management. It is to be noted that the market factors are those that pertain to the physical highway and to the road users and their motor vehicles. The market factors are thus those included in the analysis for engineering economy; the nonmarket factors are evaluated only on a quantitative basis in light of what is understood about the public's desires and interests.

SEQUENCE OF STEPS IN THE ANALYSIS FOR ECONOMY AND OTHER CONSEQUENCES OF HIGHWAY IMPROVEMENT

Figure 3-1 gives the main factors and methods that enter the process of determining the engineering economy of proposed highway capital improvements. These are the factors and methods to be dealt with in the following chapters.

The total process of analysis follows essentially the sequence of steps as follows:

1. Conceive proposed improvement.
2. Pinpoint objectives and requirements of the improvement.
3. Identify all possible alternatives including stage construction.
4. Select alternatives worthy of detailed analysis.
5. Gather facts by observations, surveys, and search of files.
6. Examine the existing facility or function to determine whether it should be abolished, retracted, or continued.
7. Select (1) Analysis period, (2) Traffic growth rate, (3) Unit values of travel time, (4) Vestcharge rate (MARR), and (5) Terminal values.
8. Forecast and classify traffic for each alternative, based on design period and traffic growth rate.
9. Design each alternative for preliminary cost estimate.
10. Estimate construction cost and annual maintenance and operating cost.
11. Compute zero time motor vehicle running, accident, and time costs.
12. Select method of analysis.
13. Compute highway costs for analysis period and method.
14. Compute road-user costs for analysis period and method.
15. Compute answer–equivalent uniform annual cost, present worth of cost, net present value, benefit/cost ratio, or rate of return.
16. Recalculate answer for other values of sensitive factors.
17. Estimate consequences related to personal preferences, land and business economics, and the community.
18. Write report.
19. Review report (by competent authority).
20. Make final decision (a management responsibility).

Highways and Engineering Economy

The sequence itself is not rigid, but is variable to fit the situation. The steps, however, do illustrate the overall process, the factors involved, and one practical sequence of steps. When calculating the economy of one engineering material compared to another or other features of design that do not affect the running cost of motor vehicles, those steps pertaining to the traffic use of the facility would be omitted.

The factor diagram and the sequence of steps taken together afford a good preview of the whole process of analysis for the economy and for the economic consequences of proposed highway capital improvements. The following chapters enlarge upon this brief preview.

PROBLEMS

3-1. What are the major specific objectives of a highway department and by what criteria can the management of a highway department measure its success toward those objectives?

3-2. With respect to economy studies of proposed works, compare the situations and objectives of private enterprise to those of highway departments.

3-3. For your community, compare your street and highway service to the other major services (education, courts, parks, health, recreation, libraries, etc.) supported from tax incomes. How does each service affect the individual, contribute to economic activity, social development, etc. Could the other services be placed on a user tax basis similar to that practiced for streets and highways?

3-4. Discuss some family actions and decisions on the basis of their bearing on commodity-savings economy and personal preference.

3-5. Reflect upon your own driving by automobile and discuss the basis of trip and route choices with respect to (1) total travel time, (2) environmental factors, (3) motor vehicle running costs, and (4) your personal preferences.

3-6. Considering your routine work and family trips by automobile, to what use would you devote your time immediately before or after the trips if the travel time could be shortened by 2 minutes, 5 minutes, and 10 minutes respectively.

3-7. As a citizen and a road user, is your concern strongest for (1) reducing motor vehicle running costs, (2) reducing travel time, or (3) for increasing the general community benefits through highway improvements? Discuss your position and attitude.

REFERENCES

3-1. Howard W. Bevis. *The Application of Benefit-Cost Ratios to an Expressway System.* Highway Research Board, National Academy of Sciences, Washington, D.C., *Proc.,* Vol. 35, 1956, pp. 63-75.

3-2. District of Columbia Department of Highways and Traffic. *Detour Impact Study.* Washington, D.C., July 1967. (An economy study of handling traffic during construction of about 2,500 feet of an urban street.)

3-3. Eugene L. Grant and W. Grant Ireson. *Principles of Engineering Economy.* 4th ed., Ronald, New York, 1960.

3-4. George Haikalis. *Economic Analysis of Roadway Improvements.* Chicago Area Transportation Study (36,500-VI), December 31, 1962.

3-5. Highway Research Board. National Academy of Sciences, Washington, D.C., Special Report 56. *Economic Analysis in Highway Programming, Location, and Design.* Workshop Conference Proceedings, Sept. 17-18, 1959.

3-6. Highway Research Board. National Academy of Sciences, Washington, D.C., Highway Research Record No. 180, *Transportation System Analysis and Evaluation of Alternate Plans,* 1967:

    1. Marvin L. Manheim. "Principles of Transport Systems Analysis", pp. 1-10.

    2. Morris Hill. "A Method for the Evaluation of Transportation Plans," pp. 21-34.

    3. William Jessiman, Daniel Brand, Alfred Tumminia, and C. Roger Brussee. "A Rational Decision-Making Technique for Transportation Planning," pp. 71-80.

    4. Stephen H. Putnam. "Modeling and Evaluating the Indirect Impacts of Alternative Northeast Corridor Transportation Systems," pp. 81-93.

    5. Howard Duke Niebur. "Preliminary Engineering Economy Analysis of Puget Sound Regional Transportation Systems," pp. 94-119.

3-7. C. H. Oglesby and Eugene L. Grant. *Economic Analysis–The Fundamental Approach to Decisions in Highway Planning and Design.* Highway Research Board, National Academy of Sciences, Washington, D.C., *Proc.,* Vol. 37, 1958, pp. 45-57.

3-8. James A. Thompson, and Bender I. Fansler. *Economic Study of Various Mounting Heights for Highway Lighting.* Highway Research Board, National Academy of Sciences, Washington, D.C., Highway Research Record No. 179, *Night Visibility,* 1967.

3-9. U.S. Government Inter-Agency Committee on Water Resources. *Proposed Practices for Economic Analysis of River Basin Projects.* Prepared by the subcommittee on Evaluation Standards, Revised May 1958. U.S. Government Printing Office, Washington, D.C., 1958.

3-10. Richard M. Zettel. *Highway Benefit and Cost Analysis As An Aid to Investment Decision.* Institute of Transportation and Traffic Engineering, Reprint No. 49, University of California, Berkeley, 1956.

3-11. Richard M. Zettel. *Highway Benefits and the Cost Allocation Problem.* American Association of State Highway Officials, Washington, D.C., *Proc.,* 1957, pp. 25-38.

Chapter **4**

Identification and Measurement of Highway Benefits

In private industry there usually is a sales income to use as an index to the value of a monetary investment. In highways there is no sales income as such. The measurement of the value of highways and their improvement must be approached by an analysis of the consequences of the use of highways, including the consequences to both road users and nonusers. Any reductions in the highway costs, the costs to the users of the highway and the costs and gains to others, are pertinent. The word "benefit" as a descriptive word applying to the beneficial consequences is widely used in highway literature, but with varying meanings. It is appropriate to establish the concepts and usages of the term "highway benefits" and their identifications before getting into the economic analysis of proposed highway improvements.

UNDERSTANDING HIGHWAY BENEFITS

The highway literature is replete with uses of the term "highway benefits," under conditions where it is more appropriate to use the words "consequences," or "net benefits." The phrase, highway benefits, is often loosely used. An understanding of the word *benefits* is necessary to an understanding of highway cost-benefit analysis. The general consequences resulting from improving highways are discussed here in Chapter 4 so that the reader will be in a better position to understand the discussions on economic analysis in Chapter 7. The subject of road user and nonuser consequences, however, is presented in more detail in later chapters.

MEANING OF THE WORD "BENEFIT"

The word "benefit" has many meanings, which depend upon the specific context. With reference to the consequence of highway improvements the resulting benefits are generally considered to include those consequences that are desired and those that reduce transportation cost, other social or economic cost, or which produce satisfactions not otherwise enjoyed. Included, also, in benefits are those gains, profits, and advantages related to business, land use, land values, welfare, and community goals. Basically, benefits from highway improvements are the gains, profits, and advantages realized therefrom by people and society in general.

Again, quoting from Webster's Third New International Dictionary "gain" and "profit" are found to mean:

> **Gain,** *n.* 1: an increase in or addition to what is of profit, advantage, or benefit: resources or advantage acquired or increased... (p. 928)
>
> **Profit,** *n.* 1: an advantage, benefit, accession of good, gain or valuable return esp. in financial matters... 2: the excess of returns over expenditure in a transaction or series of transactions: as a): the excess of the price received over the price paid for goods sold - opposed to loss; b): the excess of the price received over the cost of purchasing and handling, or of producing and marketing goods. 3a (1): net income (as in a business usu. for a given period of time); (2): a benefit or advantage accruing from the management, use, or sale of property... (p. 1811)[1]

It can be concluded that there is a great similarity in the meanings of the words *benefit, profit, gain,* and *savings,* and that they all can be applied to the concept of the enhancement to society or any of its members produced by the investment in any public work, including highway facilities.

It is this profit or gain or enhancement that is meant by the benefit resulting from highway improvements, regardless of whether this beneficial consequence is expressed in dollars or just in words. Benefit is that increment of satisfaction, wealth, or gain that accrues above the sacrifices (costs) that were necessary to produce and that were incurred as a means of producing that benefit.

Benefit is also stated in terms of net benefit, though the word is generally assumed to mean net of costs and net of adverse consequences. Net benefit may be taken to mean that care has been taken to deduct those adverse consequences and costs which might be closely related to the benefit or possibly confused with the benefit.

Economists' Meaning of "Benefit"

It is emphasized that the foregoing definition of benefit is not that generally used by economists when discussing price, demand, cost, and consumer surplus. The concept of the economist is that the benefit is the total cost one pays for a good, a service, or a satisfaction. In this sense, benefit can be more nearly regarded as a value than as a benefit. The value that one places upon a purchase of a good, favor, or service is its cost plus at least a fraction more. Certainly in

---

[1] By permission. From Webster's Third New International Dictionary, Copyright 1966, by G. & C. Merriam Company, Springfield, Mass., publishers of the Merriam-Webster Dictionaries.

Identification and Measurement of Highway Benefits 51

all transactions value received must be greater than the cost suffered. The economist's use of the word benefit need not be confusing once its special definition is recognized, but such definition must be clearly distinguished from the sense of "profit or gain" as used in engineering economy analyses.

Savings Versus Benefits

The words *savings* and *benefits* infer a before and after type of comparison. A monetary saving is defined as the reduction in cost of a product or service effected by a change to a new process, procedure, or device. For instance, when a new highway constructed from A to B reduces the motor vehicle running cost to 4.9 cents per vehicle-mile as compared to 5.3 cents on the highway before the new construction, there is a true saving of 0.4 cent a vehicle-mile. This is a direct before and after comparison. The concept of saving is on the basis that you accomplish the same objective as before, but at less cost.

A benefit may be realized without having gained a true saving by reason of not having a "before base" from which to calculate the monetary saving. On your first trip to a city you drive on the new urban expressway. You gained the benefit of the expressway, but no actual saving. The saving is hypothetical only, because to have achieved it in reality you would have had to drive the former route, which no longer exists. A saving can be computed for you, however, on the basis that had you driven the former route you would have experienced a saving on your current trip over the former trip. These hypothetical savings do not accrue in a manner that could be used to pay for the facility. Generated traffic is the best example of traffic receiving a benefit but no saving.

IDENTIFICATION OF BENEFITS

In the industrial world there is generally no difficulty in identifying benefits, profits, or returns to use as the measure of the overall profitability of a proposal to invest capital. In the public works sector, incomes, in the sense of sales as found in industry, usually do not exist. The substitute for sales is in the form of general benefits or in the form of reductions in operating expense. In this concept the question sometimes arises as to whether a certain factor is a positive cost, a negative cost, a positive benefit, or a negative benefit. With straight thinking where this question arises, there can result no real uncertainty.

Benefits must be reductions in costs, reductions in travel time, increases in net incomes, increases in creation of comforts and conveniences, and other favorable net consequences or net incomes. An outlay of cash for investment or operating expense cannot be a negative benefit. It is a positive cost.

Some of this uncertainty comes from the practice of placing the annual highway maintenance expense in the numerator in the benefit/cost ratio method of economic analysis (which is correct, see page 147) as a subtraction from road-user benefits and then calling this expense a negative benefit. Such terminology is confusing and accomplishes nothing. Benefits and costs are not classified by their position in an equation or formula, but by their true nature as being favorable or adverse, or outgo or income.

ROAD-USER BENEFITS

In highways there are the road-user motor vehicle running cost reductions (including travel time and accidents) to use as a measure of benefit, but there is no sales income as such. There are also social and economic benefits and costs to the nonusers and the community at large. Tests of the economic feasibility of a highway proposal are most often dependent upon having available information by which a cost reduction can be computed and then using this cost reduction as the equivalent of an income. This procedure is possible where there is an existing situation the operating cost of which can be used as the base transportation cost from which any reduction in cost resulting from a proposed improvement can be measured.

Road-User Taxes Versus Benefits

Wohl and Martin [4-12] used the road-user taxes generated by the project as the equivalent of a sales income. Thus they were able to compute separately an answer of economy (net present value, benefit/cost ratio, and rate of return) for each alternative without considering the cost difference between a pair of alternatives. But this technique is not recommended; often it is not possible and often it will give misleading answers. In order to use the user tax generated as a sales income and the main source of road-user benefits (neglecting nonuser benefits) some estimates must be made of the pro rata share of license fees, driver licenses, special fees on trucks, none of which is related directly to the use of a specific facility. The fuel tax generated can be reliably estimated, however. But a valid objection to using the fuel tax as a measure of the benefits of a highway project or system is that the elements of highway design affect the fuel consumption. Often one of the objectives of the improvement is to reduce motor vehicle running cost, including a reduction in motor vehicle fuel consumption. A proposed improvement, highly successful in reducing the fuel consumption of motor vehicles, would be penalized by the corresponding reduction in the fuel tax generated. Such a device would be quite hazardous to use in determining the economy of those elements of highway design which materially affect the fuel consumption of the traffic.

Another objection to considering road-user taxes as an income is that such taxes are a cost to the road user in the same sense as is his motor vehicle running expense. Since the road user is the owner of the highways and since his road-user tax is used to pay highway costs, such tax is hardly an income to him.

Many of the project formulation applications of engineering economy depend

Identification and Measurement of Highway Benefits 53

wholly on the comparison of transportation cost differences between the alternatives, including the total motor vehicle running costs when they apply. Road-user tax payments have no place at all as the factor by which to measure the economy of highway design alternatives.

The Sales Concept

Public works are considered somewhat differently than private competitive business. Private business is concerned with sales volume, operating expense, and profits. When anticipated profits indicate a high rate of return, the management of the business is inclined to make the necessary investment to produce this net profit. The management is not concerned whether the sales come from existing customers (unless the sales volume of the management's other products would decrease), are diverted from competitors, are transferred from sales of other products (not their own), arise out of normal increase in population, or are sales generated only because of the attractiveness of the product.

As a public function, the construction of highways should not be accomplished without recognition of the fact that the traffic to use an improved facility may be transferred from other modes of transportation or generated.[2] Transferred traffic is a loss of business to the modes from which the traffic is transferred. Highways are a public function that should not be undertaken with disregard to private business. Further, highways are not supplied as a service to be "sold" to the public but are made available because of the public demand.

The conceptual difference is great between producing an article for sale on the market and producing a service for one's self or supplying through public funds a service or function which the public has asked for. Many persons wrongly accept the concept that highway departments are producing highways for the purpose of selling highway travel. The highway departments supply the facilities for highway transportation in the same concept that a public school board supplies a public educational system for the good of its community, or that libraries are supplied for public use. Highway departments do not operate under a policy which says that their objective is to increase highway travel or to increase road-user taxes. Their policy, as agents of the public, is simply to provide the services the public wants with the money the people supply.

To regard highway service as something to be sold on the market in the sense of commercial sales for a profit leads to confusion and wrong conclusions in economic analyses because the sales income simply does not exist, and there is no intent to maximize profit. However, the traffic is used as an index to quantification of benefits because traffic represents a direct measure of the public demand for the service; to a lesser extent, the traffic is a measure of the nonroad-user demand for the highway facility.

[2] See Chapter 17 for definitions of source components of the traffic stream.

Project Evaluation

In project evaluation the question is raised as to whether transferred and generated traffic should be included in the traffic volume used in computation of road-user benefits. The answer is found in the viewpoint adopted. If the sales viewpoint is to be used, then all traffic (sales) using the proposed facility is considered regardless of source and regardless of other economic consequences. This approach is the same as is followed in business. As a matter of fact, it is the one most often followed by highway managements, knowingly or unknowingly.

The other viewpoint would be to eliminate transferred and generated traffic from the forecast (but not for design purposes) and justify the new improvement wholly upon true savings, except for the benefit to normal growth traffic and developed traffic. Public policy is involved. How far should public policy be extended to gain what the public wants, or seems to want, at the expense of private property? Perhaps as long as legislative authority and highway department management follow the will of the people, they would be correct in basing economic evaluation on the whole traffic stream regardless of differences in true savings to different components of the traffic. The situation is not dissimilar to the normal advance of technology in private business. One product after another is forced off the market by the pressures of public acceptance of new products. Street railways and the interurban electric railways are obsolescent or defunct as transportation modes. The public has shown its preference for the motor vehicle. Likewise, the farm tractor and truck have replaced the workhorses.

The pricing of the road-user costs for the existing situation is an important factor in economic evaluation. The total net benefits to be had from a proposed new facility are directly dependent upon the prices and resulting calculated cost experienced by the existing road users. And the calculated net benefit is one of the two base factors upon which the proposed facility is determined to be economically feasible or not feasible.

The source of traffic may be considered in computing the motor vehicle running cost. Existing, diverted, transferred, and community traffic may each have a different base running cost from which to measure gain or loss on the new facility. The assumption usually made, but not always the best assumption, is that all traffic forecasted to use the new facility has the same base (before) unit running cost, including accident costs. Less uncertainty is introduced by separating the traffic into source groups and assigning to each group its base unit running cost and accident cost appropriate to their sources.

Project Formulation

In the analysis for project formulation, the source of the total traffic forecasted to use the new facility is immaterial except to the extent that the traffic source affects the alternatives differently with respect to the traffic volume on the new facility and the traffic flow in the community, because both of these factors

Identification and Measurement of Highway Benefits 55

will affect the relative overall road user economy. The facility must be designed for whatever traffic will use it. This design should be that producing the greatest economy for that forecasted traffic, considering all consequences within the area affected.

When there are two or more alternatives in a design factor of the proposed facility, the unit pricing for such items as vertical grades, horizontal curves, and speed changes for each alternative would be wholly comparable and related solely to proposed designs as contrasted to project evaluation which often compares a proposed facility to an existing situation. In Chapter 7 there is a discussion of how to handle unequal traffic volumes in the comparison of mutually exclusive alternatives. As a general rule each alternative is analyzed on the basis of the traffic forecasted to use that alternative design during the analysis period. The pricing problem is relatively easy; unit prices are used that apply to each factor of design to be found in each alternative.

SOCIAL, GENERAL ECONOMIC, AND COMMUNITY BENEFITS

Generally the benefits of highway improvements accruing to the social activities, general land, business and trade, and the local community are not included with the road-user benefits in the economic analysis. The reason for this is mainly that these nonroad-user benefits are not easily or reliably reduced to dollar values comparable with the road-user dollars. Also to be considered is the probability that many of these nonuser benefits are already reflected in the road-user cost reductions or that they are offsets or transfers when a larger area is considered. But here again are important factors involved in highway transportation economics that are not found in industry and, though found in some other public works, do not usually loom so important to the public officials and civic leaders.

So far as the methods of economic analysis are concerned, these nonuser benefits offer no particular problem. The problem comes largely at the management decision level with respect to route location alternatives and such design alternatives as elevated or depressed gradeline and tunnel or bridge structure which are factors involving both engineering and aesthetics. Since the majority of nonuser consequences of highways are most difficult to price or to isolate from the consequences to the road user, the decision maker usually must weigh the desirability of such consequences against the cost of the highway by means other than arithmetical calculation of the type involved in economic analysis. See Chapter 23 on management decision.

HIGHWAY COST FACTORS AND BENEFITS

In designing, constructing, maintaining, and operating highways, the cost and expense factors are quite similar to those of private industry. For proposed works, the investment cost and the annual maintenance and operating expenses can be estimated within acceptable limits of reliability. So can the motor vehicle running costs. However, unlike industry, the highway department has no control over the operation of the motor vehicles, except as the prevailing law may regulate the type of vehicles and their use of the highways. Allowances, therefore, are in order in all forecasts and analyses for economy of highway projects on the basis that the designing agency and operating agency cannot control the volume of "customers" or just how the highway will be used. Moreover, not every highway improvement project will be involved with traffic or at least the traffic will not be controlling the factors of economy, which are mainly factors of design.

In the economic analysis of proposed highway improvements, specifically for project evaluation, the usual situation is that the highway cost involved and the motor vehicle running cost involved can be acceptably priced. But the external factors, both adverse and beneficial—the nonuser consequences—usually cannot be reliably priced, even when identifiable.

ROAD-USER CONSUMER SURPLUS

Price, cost, and value are not synonymous terms. Neither is benefit the equivalent of value. The concept of consumer surplus as set forth by the economist is a useful device for reaching a better understanding and evaluation of the benefits the road-user gains from highway improvements.

CONCEPT OF CONSUMER SURPLUS

When one pays the asking price for a commodity, service, or favor, his total cost is usually in excess of the price paid to the vendor or supplier. Price and cost are not equivalent because the purchaser or receiver usually has some expense other than price paid to the vendor attached to his activity and effort to acquire title or possession of that which he purchases. These expenses may include transportation (riding, walking, or driving), ordering, phone calls, overheads, and a host of other possibilities of large and small items of expense. There is even wear and tear on pockets and purses in carrying money in person and the expense of operating a checking account. Few persons, however, stop to think of these expenses, although business managers do.

Furthermore, the cost of the article, service, or favor procured or received is not its value to the purchaser. At the moment of dealing, the purchaser values that which he is receiving in excess of his total cost. If he did not he would not take the time, trouble, or expense to acquire the article or service. Likewise, the seller, or the person performing the service, values the article or service at less than his price. Thus both seller and purchaser get the best of the transaction— from their individual viewpoints. The economists have a name—consumer surplus—for this amount of value in excess of the cost of the article purchased.

Identification and Measurement of Highway Benefits

PRICE-VOLUME DEMAND CURVES

Consumer surplus may be demonstrated from the demand-price relationship in Fig. 4-1. The concept is that when price is lowered the volume of sales will increase and vice versa. A commodity which so reacts in the free market is said to be elastic with respect to price. If the sales volume (or consumption) does not react with changes in price, the commodity is said to be inelastic with respect to its price.

In Fig. 4-1 when the unit price is lowered from P_0 to P_1 unit sales increase from V_0 to V_1. The demand curve D_0 represents the importance of the product to current and potential customers with respect to the price they must pay. This demand curve extended to the right and downward, would intersect the sales volume axis at that sales volume which would represent the maximum distribution of the product when it was free for the asking. But vertically there is no known end point of the demand curve. It is possible that one or more customers would be found for the product at any high price. At least the demand curve tends to be asymptotic to the price axis. The area above the line $P_0 A$ and between the vertical price axis and the demand curve is called the *consumer surplus*. It represents the value received in excess of the price paid by the V_0 customers. The excess value materializes because all customers except the one at V_0 paid less than their marginal price.

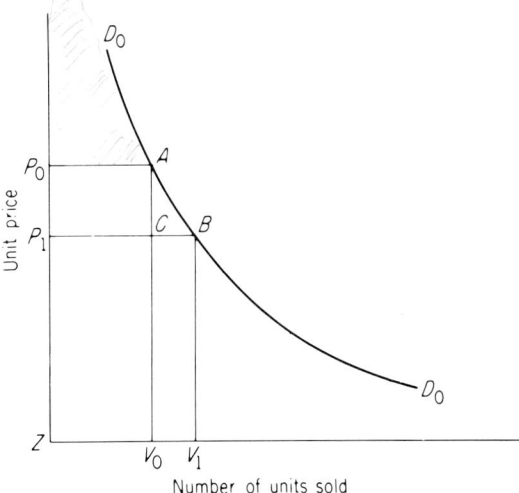

FIG. 4-1. A demand curve for a single commodity unchanged in quality at the time of a decrease in unit price, from P_0 to P_1.

It is important to bear in mind that a demand curve as normally drawn applies to one specific product or service at a specific time and place. When the product is altered in quality, shape, and function the demand curve also changes. For instance, the demand curve for a four-lane fully access-controlled highway is quite different from the demand curve for a conventional two-lane bidirectional highway with many points of access. The combined changes in motor vehicle design and in highway design continually result in changes in the demand curve for automobiles. The demand curve for air-conditioned cars in Maine is vastly different from that in Texas.

In Fig. 4-1 the price P_0 is the marginal price for the V_0 sales volume. In theory, a price raise would reduce the number of sales and a price reduction would bring in more customers. There are customers at the P_0 price who are just barely willing to pay that price. The product of P_0 times V_0 is the total sales income at price P_0. Similarly, P_1 times V_1 is the total dollar sales volume at price P_1.

When the price is lowered to P_1, the V_0 customers make a net total saving of $V_0 (P_0 - P_1)$. But the $V_1 - V_0$ individual customers each received a varied amount of benefit (gain, or saving). The marginal customer at V_0 will benefit by the price reduction of $P_0 - P_1$, while the marginal customer at V_1, because the price P_1 is just barely low enough to attract him, is almost indifferent and his benefit is infinitesimal. For practical purposes the area ABC may be considered as a triangle so the total net benefit to the $V_1 - V_0$ customers is $1/2 (P_0 - P_1)(V_1 - V_0)$. Therefore, as a result of the price reduction, the total net benefit received by the total number of customers is $V_0 (P_0 - P_1) + 1/2 (P_0 - P_1)(V_1 - V_0)$.

DEMAND CURVES APPLIED TO HIGHWAYS

The economic relationship of price and demand can be applied to highway transportation [4-6, 4-10, 4-12] with particular respect to economic analyses of proposed improvements. Generally it is assumed that the proposed improved facility will reduce the total cost of vehicle transportation, including all elements, such as running costs, travel time, and traffic accidents. It is common experience that improved highways result in higher traffic volumes than existed just before the improvement. When the before and after costs per trip are known, two points on a demand curve have been determined. See points A and B on demand curve D_0 of Fig. 4-1.

Shift of Demand Curve with Highway Improvement

But with highways, any improvement in a segment of a route or system changes the quality of the service available and a new price-demand curve results, such as illustrated by curve D_1 in Fig. 4-2. In fact, because of the changes in quality of service, the new facility could incur higher travel cost and yet attract a larger number of trips than the former facility attracted.

D_0 is the demand curve before the improvement, and D_1 the demand curve after the improvement in the highway facility. Should the facility have offered the satisfactions before being improved that it did afterwards, and without any

Identification and Measurement of Highway Benefits

decrease in user costs, it would have attracted V_2 traffic volume. However, with these increased personal satisfactions plus a reduction in road-user cost per vehicle-trip of $P_0 - P_1$, the new facility attracted a traffic volume of V_3. Note that with only the decrease in cost per vehicle-trip and without any change in the other qualities of traffic service to alter the demand curve, the new traffic volume would have been only V_1. But with the increase in personal satisfactions no longer does the triangle ABC represent the net gain (benefit) to the new traffic. The net total cost reduction for the V_0 original trips is $V_0 (P_0 - P_1)$. The $(V_3 - V_0)$ new trips attracted to the improved facility made a reduction in total cost of their trips of only $1/2 (P_0 - P_1)(V_3 - V_0)$, on the basis that P_0 is the highest price per trip that anyone of the new users would have paid, and P_1 the lowest price actually paid by any of the new users. These total gains (cost reductions) do not include any values attached to consumer surplus.

This analysis is on the basis that the highway and traffic conditions fix the cost of the trip (price paid) and that the road user has the option of using the facility or not. In Fig. 4-2 the marginal nonusers at price P_0 and demand curve D_0 were looking for either a price reduction, a better highway, more favorable traffic conditions, or a combination of these inducements as a condition to their using the specific facility. When these conditions were supplied, marginal and not-so-

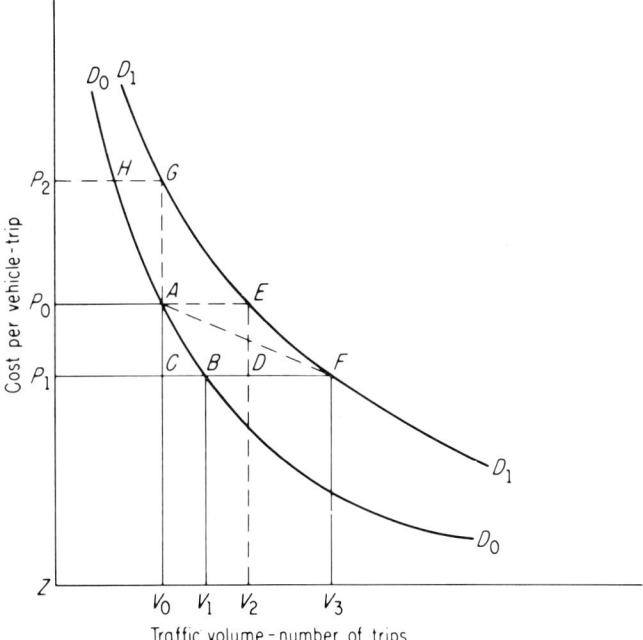

FIG. 4-2. Diagram to illustrate the price-volume relationship (demand curves D_0 and D_1) of traffic before and after improvement of a highway facility.

marginal potential users reevaluated their willingness to pay and their demands for service so that points E and F were established to form a new demand curve.

The above discussion neglects any change in the supply of transportation affected by the highway improvement. Any increase in the capacity of the highway would attract additional traffic not so much because of the increased hourly capacity, but because of a higher level of service-faster speed and freedom of movement. The price was lowered to P_1 not to attract more customers to buy an increased supply; rather, the lowered price was the object of the improvement, and the intent was not to attract more customers, as is the object of lowering sales price in business. Some of the new trips may be transfers from other modes or highway facilities of unknown price.

An examination of Fig. 4-2 will disclose that the original V_0 traffic added materially to its consumer surplus. Because of the higher level of travel service the original V_0 traffic would now be willing to pay a price of P_2. It is likely that the new V_0 trips do not encompass exactly the same persons comprising the original V_0 trips under the D_0 demand curve. The new demand curve D_1 results in a consumer-surplus addition for the V_0 traffic equal to the area between the two demand curves and above and to the left of the line AG. This additional consumer surplus is the value to the V_0 traffic above its cost, but it is not considered to be a benefit in the form of a gain or saving.

Road-User Benefit and Consumer Surplus

Many writers, such as Wohl [4-12, p. 8], consider the benefit to be the total price paid; other writers, as Kuhn [4-6, p. 37], include the consumer surplus as a part of the benefits. In engineering economy the general practice of most writers is to reserve the word "benefit" for the difference when costs are subtracted from income, or the estimated reduction in costs effected by the proposed new facility or machine. Another way of looking at benefit is that the total net benefit or gain is the amount added to consumer surplus through price reduction, not including the effects of a new demand curve. On this basis the area P_1P_0AB in Fig. 4-2 for the D_0 demand curve is generally the benefit or gain considered in highway engineering economy studies. Often the triangle ABC is mishandled by considering it to be a rectangle with sides AC and CB, thus overstating the benefits.

In the case of highways, however, an improvement to a facility usually results in an improvement in the quality of service as well as in a reduction of travel cost per trip. The improvement in quality of the ride or trip results in the travelers changing their individual appraisal of trip cost and trip value to effect a new demand curve as shown in Fig. 4-2. With this change in the price-demand relationship, the amount of increase (or decrease) in consumer surplus cannot be directly arrived at. Some assumptions are required. From Fig. 4-2 it can be seen that the total area added to the original consumer surplus is the irregular area between the two demand curves above AE plus the area P_1P_0EF.

Since the total consumer surplus corresponding to demand curve D_0 and the

Identification and Measurement of Highway Benefits 61

amount of consumer surplus added by demand curve D_1 cannot be specifically determined, and since they have only academic interest, the practical approach in highway studies is to consider that the area $P_1 P_0 A F$ is the net gain generated by the new facility as compared to the existing facility. This assumption is in line with the objectives of the analysis for engineering economy and is conservative and practical. The quantity sought is really the total net gain in terms of cost reduction to the highway users (and others) and not values as measured by what the users would be willing to pay. It is true, however, that an improved old highway facility might result in higher direct costs to the road user than he paid on the facility before improvement, yet be so attractive in personal preferences or nonhighway satisfactions that it would attract more trips than before. Thus this particular improved facility would have a negative road-user cost reduction but this increase in cost would be overbalanced by nonpriceable satisfactions. In this case, an important consideration is where did the increase in number of trips come from? It is possible that they were transfers from other facilities where the direct road-user costs were even higher than on the improved facility to which they transferred.

The consumer surplus associated with existing highway facilities can be regarded as a sunk surplus and ignored in the same way and for the same reasons that sunk investment costs are ignored.

ELASTICITY OF DEMAND OF HIGHWAY TRAVEL

A conventional demand curve such as Fig. 4-1 implies that demand is influenced solely by price. Such a curve is appropriate when there are no changes in the product, service, or convenience of purchasing the product, a situation that does not often exist with improvements to highways, but does exist with many market commodities. An improved highway facility changes the quality of the travel service as compared to the older facility or existing facility and, therefore, changes the demand curve. Even highway beautification will change the demand curve for passenger car traffic although there is no change in road-user costs. The added scenic value (no junkyards, no billboards, but good views of artistic landscape) may attract some highway travel because of this added attraction.

When improvements are made to existing highway systems by altering specific facilities in that system, often the effective supply of transportation is increased along with increasing the quality of the transportation. But the supply of highway transportation is difficult to quantify. Is it to be measured in miles of highway, lane-miles, 24-hour capacity, hourly capacity, peak-hour flow, travel time, travel speed, traffic safety, comfort, or convenience? Traffic control devices often reduce the capacity of an intersection in vehicles per hour. Increasing a two-lane highway to four lanes increases the total capacity in vehicles per hour. For certain types of highway improvements, therefore, both the quality of the ride, or service, and the supply of service measured in vehicle speed or total vehicles per hour, may be increased or decreased. The combination of the two factors is represented in any change in the traffic volume using a facility after its improvement as compared to before its improvement. But the most important factor for increased use is the greater satisfaction to be had rather than a capacity to handle more vehicles per hour.

The demand curve for highway travel is continually changing. It changes hourly, daily, seasonally, and yearly; it changes as the traffic volume changes and with location and quality of highway travel facilities. Such factors as environment, traffic strains, quality of ride, and other satisfactions obtainable have their influence on the acceptable price a traveler will pay for a given trip. Figure 4-2 is a representation of a set of before and after demand curves for a highway facility which has undergone improvement.

Inelastic Character of Highway Travel Demand

It is worth noting that the demand for highway travel is highly inelastic within the price levels normally experienced. Travel, having become a necessity of life, does not vary appreciably in volume with changes in price. Trips by motor vehicle are unlike the purchase of luxury goods or goods and services purchased as a result of conscious decision, in which price plays a powerful role. Motor vehicle fuel is a commodity the demand for which is highly inelastic. Studies of gasoline sales before and after an increase in gas tax rates indicate no measurable change in total gallons purchased. Further, as shown by Cook [4-2], gasoline is purchased commonly by car owners without deliberate reference to total price per gallon or the road-user tax per gallon.

This inelasticity of the demand for highway travel with respect to the cost of a trip is an important factor bearing upon trip forecasting and the use of a specific highway facility. A new route, a new bridge, and a materially improved old route will be used heavily, because of the ever-present demand for highway facilities. Many newly opened street and highway improvements are heavily used immediately and often by increased traffic volumes. This traffic volume is not all generated traffic but it comes from other facilities (diverted traffic) and perhaps other modes (transfers from bus and rapid transit). Because of offering higher values in quality of service (comfort, convenience, lower travel time) with or without a reduction in motor vehicle running cost, the new facility experiences immediate heavy use. Thus it appears that travel demand is largely inelastic with respect to price considering the total of all consequences of the new improvement. But perhaps for a specific facility the behavior of traffic is in line with a commodity or service which is truly elastic, such as rosebushes and theater shows. On the other hand, if the whole of the travel routes within a traffic corridor are considered as a total highway system, the number of daily trips will not have changed appreciably because of the improvement of one sector of a route within the system.

Reversal of Price-Volume Relationship

This switch of traffic from one route to another within a system gives rise to

Identification and Measurement of Highway Benefits 63

cases of the reverse relationship between price and amount of goods sold. When an arterial route is paralleled by a newly opened freeway, the general result is that traffic is diverted from the arterial to the freeway. When this happens, the traffic remaining on the arterial experiences a reduction in price through faster travel and lower running cost while at the same time there is a reduction in traffic volume, i.e., a reduction in amount of goods sold. It might be said that the arterial takes on the characteristics of an "inferior good," like hamburger, for example. A fall in the price of one good, with no change in the price of other goods, has the same effect as a rise in real income. With a rise in income, consumers substitute better cuts of meat such as sirloin steak for hamburger–and such as a new freeway for an arterial route.

This discussion of price, demand, and traffic volume, emphasizes the complexity of the analysis for economy of highway improvements and the importance of considering all consequences to whomsoever they may accrue.

ROAD-USER BENEFITS AND TRAFFIC COMPONENTS

In project evaluation the estimate of dollar net benefits is a most important factor in the analysis for economy. Because the seven components[3] of traffic may experience benefits of different degrees, it is desirable to examine each traffic component separately and to estimate its net benefits carefully. Price-volume, or demand, curves may be used as a conceptual framework for the process of pricing the benefits to traffic.

BENEFITS TO EXISTING TRAFFIC

When there is an existing traffic using a highway facility and that facility is improved by reconstruction or betterments, such traffic will probably use the facility after its improvement. This traffic, therefore, gains full benefit from any travel cost reduction afforded by the improvement. Since the costs to the existing traffic before and after improvement of the facility can be determined, the net benefit resulting is readily calculated. In this case the "before" cost base is established from which to measure true savings or net benefits.

BENEFITS TO GENERATED TRAFFIC

Estimating the benefits to generated traffic and putting a dollar value on these benefits call for some decision on what concept prevails and what the objective is

By definition, generated traffic did not exist in any form or mode until the improved highway facility came into use. Therefore this traffic has no base from which to compute its savings or net benefit. True, generated traffic is benefited by use of the new facility and it receives a value thereby. The existing

---
[3] See Chapter 17 for identification of these components of a traffic volume.

traffic makes a direct monetary savings which could be captured in part for payment of the added highway cost, so there is true and overall economy effected. The generated traffic, on the other hand, experiences no direct savings in costs, so its share of the highway cost and its other costs of the trip, must be taken from less spending in other activities. From the broad economic picture there is no economy achieved, merely a transfer of expenditures from other than highway travel to highway travel.

From a practical viewpoint and considering that highways serve a public demand, there seems to be no great injustice to follow from adopting the sales concept with respect to generated traffic and considering it in the same class as existing traffic. When so handled in economic evaluation analysis, the index of economic feasibility may be somewhat overstated (depending upon one's viewpoint). When the public demonstrates its preference for the facility in lieu of whatever other activity it would have supported with the money spent for the trip on the new facility, no serious error in judgment can result in treating generated traffic in the same manner as existing traffic.

BENEFITS TO DIVERTED AND TRANSFERRED TRAFFIC

Diverted and transferred traffic have a before price base, but their demand curves are not the same as that for existing traffic on the facility under consideration. Knowing the price base for the diverted and transferred traffic, their net benefits could be calculated. Their total net benefit would be the difference (plus or minus) in the before price base on their other facilities and the price on the improved facility times the number of vehicles that made the change to the new facility.

BENEFITS TO DEVELOPED AND NORMAL GROWTH TRAFFIC

The developed and normal growth traffic have no prior price base and have not in the past refused to use any mode of travel for the reasons that there was no purpose in travel for the developed traffic and the normal growth traffic did not exist. Thus these two components of the forecasted traffic volume are in a class by themselves. Generated traffic is unlike these two components, for the generated traffic had the opportunity to travel, but refused to do so because the price was not right or the conditions were not attractive.

There exists neither a marginal price nor a price base for the developed traffic and normal growth traffic. These two classes of traffic receive the same gross benefit from the new facility as does the existing traffic. A considerable percentage of the forecasted increase in traffic, especially that after the first year of use of the improved facility, will be accounted for by developed and normal growth traffic. This increase will continue to the year when the traffic capacity of the facility is reached or to the end of the analysis period, whichever comes first.

In the analysis for economic evaluation it seems reasonable to price the net benefits to developed and normal growth traffic on the same basis as applied to existing traffic. This procedure may be liberal, but it may be justified on the basis that one of the main objectives of many facilities is to serve the land development in the area and to provide for future population growth and motor vehicle use.

PRICING BENEFITS TO COMMUNITY TRAFFIC

Community traffic is all traffic in the area that is affected by the improved facility but does not use it. See for example Table 15-14 for the North Carolina example where the travel on an arterial was reduced from 104 to 65 million vehicle-miles by the opening of a new nearby expressway, and the number of accidents was reduced from 330 to 260. Because of the reduced average daily traffic on the arterial, this residual traffic moved at higher running speed and had fewer speed changes, which reduced its cost in cents per vehicle-mile. The net benefit to consider in cases of this kind is the direct net reduction in road-user costs effected by the highway improvement and applied to all traffic so affected.

Admittedly, estimating the change in road-user costs for community traffic as affected by any specific highway improvement is a difficult task. The total cost change, however, may be considerable in amount and importance in project evaluation. Until research comes forth with some specific cases as workable guides, the analysis will perhaps be confined to traffic for only those nearby routes or facilities for which the traffic changes can be forecast. It is to be kept in mind, however, that much of urban highway improvement has as a main objective the lessening of traffic congestion in the whole corridor affected. Thus the benefits from a specific facility to the traffic residual on the other routes can be material.

PROBLEMS

4-1. Select a highway improvement or traffic control improvement with which you are familiar. Describe on a qualitative basis the benefits and adverse consequences, for both the road user and nonroad user, resulting from the improvement.

4-2. Discuss the role of road-user net benefits (1) in project evaluation and (2) in project formulation. What is the relative importance of the quantification of benefits in these two types of analyses?

4-3. What is the role of the nonuser consequences in (1) project evaluation and (2) project formulation?

4-4. Compare road-user benefits to commercial sales as related to economic evaluation analyses of proposed investments. Compare them also with respect to the concept of consumer surplus and price-demand curves.

4-5. In the economic analysis of highway improvements for economic evaluation, how do you think generated traffic should be handled? In the end objective, what injustice if any may accrue because of (1) giving generated traffic equal consideration to existing traffic, and (2) giving generated traffic less weight?

REFERENCES

4-1. William J. Baumol. *Economic Theory and Operation Analysis.* Prentice-Hall, Englewood Cliffs, N.J., 1961.

4-2. Kenneth E. Cook and Patrick A. Rush. *Consumer Awareness of Motor Fuel Tax Rates and Prices.* Highway Research Board, National Academy of Sciences, Washington, D.C., Highway Research Record No. 138, 1966.

4-3. Otto Eckstein. *Water Resource Development (The Economics of Project Evaluation).* Harvard U. P., Cambridge, 1958.

4-4. Eugene L. Grant. *Concepts and Applications of Engineering Economy.* Highway Research Board, National Academy of Sciences, Washington, D.C., Special Report 56, 1960, pp. 8-18.

4-5. John V. Krutilia and Otto Eckstein. *Multiple-Purpose River Development.* Johns Hopkins Press, Baltimore, Md., 1958.

4-6. Tillo E. Kuhn. *Public Enterprise Economics and Transport Problems.* U. of California Press, Berkeley, 1962.

4-7. C. B. McCullough and John Beakey. *The Economics of Highway Planning.* Oregon State Highway Commission, Salem, September 1937.

4-8. Herbert Mohring and Mitchell Harwitz. *Highway Benefits–An Analytical Framework.* Northwestern U. P., Evanston, Ill., 1962.

4-8A. Paul A. Samuelson. *Economics–An Introductory Analysis.* 7th ed., McGraw-Hill, New York, 1967.

4-9. U.S. Government Inter-Agency Committee on Water Resources. *Proposed Practices for Economic Analysis of River Basin Projects.* Prepared by the Subcommittee on Evaluation Standards, Revised May 1958. U.S. Government Printing Office, Washington, D.C., 1958.

4-10. David W. Winch. *The Economics of Highway Planning.* University of Toronto Press, Toronto, 1963.

4-11. Robley Winfrey. *Concepts and Applications of Engineering Economy in the Highway Field.* Highway Research Board, National Academy Of Sciences, Washington, D.C., Special Report 56, *Economic Analysis in Highway Programming, Location, and Design,* 1960, pp. 19-34.

4-12. Martin Wohl and Brian Martin. *Evaluation of Mutually Exclusive Design Projects.* Highway Research Board, National Academy of Sciences, Washington, D.C., Special Report No. 92, 1967.

4-13. Richard M. Zettel. *Highway Benefit and Cost Analysis as an Aid to Investment Decision.* Institute of Transportation and Traffic Engineering, Reprint No. 49, University of California, Berkeley, 1956.

4-14. Richard M. Zettel. *Highway Benefits and the Cost Allocation Problem.* American Association of State Highway Officials, Washington, D.C., *Proc.,* 1957, pp. 25-38.

4-15. See chapter 7 for additional references.

2 cap. 4 trozos día 8-4

Chapter **5**

Interest and Vestcharge

The economic cost of owning and operating a machine, a business, or a home and the economy of one machine, one process, one product, or one construction plan, as compared to similar alternatives, can be estimated correctly only by including a factor for the economic cost of money. This statement is accepted by most economists and all professionals who have thoroughly studied engineering economy. Yet many public officials, among other groups, do not accept this principle unless the cost of money is a reality in the form of a cash disbursement for interest on borrowed money. It is desirable, therefore, to discuss in depth the concept, theory, and principles of interest as a cost of money, and as a device to transfer values from one date in time to equivalent values at other dates.

CONCEPT OF INTEREST

An understanding of the concept, theory, and meaning of interest, vestcharge, discount factor, rate of return, profit, and dividends is essential to an understanding of engineering economy. This understanding is fundamental to making the analysis for economy as well as to the decision process in which the solution for economy may be a key factor.

DEFINITIONS AND CONCEPTS

With the exception of "vestcharge," a coined word, the following definitions are well accepted in standard references and are those meanings applicable to the terms as used herein.

Interest — money paid by the borrower for the use of money loaned to him. *Rate of interest* is the rate of the interest based on the amount of the loan per unit of time. The time period is most often one year.

Return — monetary net return (exclusive of depreciation return) received from operating an enterprise. It is the *net profit* received. *Rate of return* is the quotient of the net return for one year divided by the net invested capital at the beginning of that year.

Dividends — stock-type ventures pay dividends to the stockholders from the net profits, or net earnings, of the company. Generally only a part of the year's net profit is paid out in dividends. That profit not paid to the stockholders is retained in the company as "surplus," or undistributed earnings.

Vestcharge – charge for the use of money invested in physical assets; it is in lieu of cash net return on invested capital wherein operational money returns are received from business ventures, but the concept of vestcharge assumes that there is no direct money return. *Rate of vestcharge* is used in the same sense as rate of interest or rate of return.

The terms, *discount factor, cost of money,* and *minimum attractive rate of return* (MARR) are used in a similar sense. Each term is an expression which implies that invested capital involves by the investor thereof, some consideration of return or interest as compensation for investing. The term *discount factor* implies that the element of time is involved. Cost of money implies that money invested in the business or enterprise must be paid for. The term *cost of money* is commonly used in public utility rate making when considering what is the fair rate of return to be earned. The *minimum attractive rate of return* is an expression to designate that specific interest rate, cost of money, or vestcharge rate below which rate the investor would not invest his money.

In the foregoing definitions, it is important to recognize that *interest* is defined only in its narrow sense as applied to borrowed money. But the word *interest* has long been used in the general sense of the cost of money, discount rate, charge for investment, and return on investment. Books on engineering economy, and other books as well, use *interest* in its several meanings.

The word "vestcharge" is introduced so that the word *interest* may be used only in its narrow sense of rent paid (and received) on borrowed money. Supporting the use of the word vestcharge is the fact that those persons who think of interest as meaning only interest on borrowed money in the form of notes, bonds, and mortgages, may, upon encountering the word vestcharge, immediately recognize that the general economic value of money is the concept intended.

The Merriam-Webster New *International Dictionary* (1959) defines interest rather restrictedly, then follows with some of its broader meanings:

> **Interest** ... 3a: The price paid for borrowing money generally expressed as a percentage of the amount borrowed paid in one year ... (p. 1178)[1]

In the 1959 Second Edition of the New International Dictionary Webster includes the idea that interest on the use of borrowed money is a forbearance in lieu of paying the debt. Also, interest as considered by most economists includes two segments: first the pure interest on the loan, and second, a separate fraction of the total to represent payment for risk and uncertainties. Interest is also a factor that measures the relative value of present goods compared to future goods.

The term *imputed interest,* found in the highway economics literature, is used by some authors to apply to the use of an interest rate in the calculations of high-

---

[1] By permission. From Webster's Third New International Dictionary, Copyright 1966, by G & C. Merriam Company, Springfield, Mass., publishers of the Merriam-Webster Dictionaries (1968).

Interest and Vestcharge   69

way economic cost, economy studies, or other applications wherein there is no interest actually paid or no money borrowed. Imputed interest is interest that is imposed in concept rather than existing in reality. Often these authors take the position that the charge for imputed interest should not be included in the consideration at hand. Again, usage of the term of imputed interest illustrates how difficult it is for some persons to see the concept or accept the concept of an interest charge, or interest cost, when no interest is actually paid or collected.

From these definitions and explanations from Webster and others, the concepts of the word *interest* may be summarized as follows:
 1. Interest is rent earned and rent paid on borrowed money.
    A. Interest is compensation to the creditor for his having forgone other satisfactions obtainable with the money he loaned; the return forgone concept.
    B. Interest is the price paid by the borrower in order that he could gain immediate satisfaction, not obtainable with his current resources.
 2. Interest is a concept of return on productive capital–investments in physical assets.
    A. Return is compensation to the investor for his having denied himself other probable satisfactions to be achieved with the same money; the return forgone concept.
    B. Return is a penalty for delayed consumption of goods; goods consumed upon purchase, bear no return charge, but goods which become long-term assets do bear the return charge.
 3. Interest is a mathematical concept by which values at one point in time may be converted to equivalent values at another point in time.
    A. Present goods and services have a value different than the value of future goods and services.
    B. The interest rate may be viewed as a time-discount factor.

Saliers sums up the whole concept of interest and return in his definition [5 10]: "Interest is expense or income resulting from the use of wealth over a period of time."

These concepts bring out that, in addition to rent on borrowed money, the word *interest* is used to convey the idea that money (capital) when invested in plant, goods, or enterprise, should return the investor some compensation, and that when so invested the investor is forgoing his personal use of the money. A deeper insight into these concepts is in order to bring out their implications and applications, and the fact that there is value attached to the mere possession of money as money. To be deprived of this possession results in a sacrifice–an economic cost.

JUSTIFICATION OF INTEREST ON BORROWED MONEY

Value (and price) is a result of measuring present sacrifice in favor of future satisfactions anticipated as a result of the sacrifice. Therefore interest on a loan is a price (present sacrifice) we pay for a future satisfaction. When we borrow money we are exhibiting our preference for current satisfactions (benefits).

Likewise, the lender of the money charges the price (interest) because of his willingness to sacrifice that current satisfaction he could gain by using the money himself rather than lending it. In this case his value of the interest to be gained is in excess of his value of the possible and probable other future satisfactions had he not loaned the money.

The charge of interest on borrowed money or goods is a just charge, recognized in history at least as early as 1800 B.C. Payment of interest was often in kind–grain for the loan of grain.[2] Considering both parties to the act of lending–borrowing of money–the following concepts explain the factor of interest:

1. The market will pay a profit on the purchase of money for the same reason the market will pay a profit on the purchase of consumable goods and services. The satisfactions gained in both transactions are worth the cost.

2. The borrower gains immediate satisfactions which he could not gain without the borrowed money. This fact is related to the time value of money.

RETURN ON INVESTMENT

Return is distinguished from interest by the fact that the amount of return and the date of return are unknown in advance. The interest on notes, bonds, and mortgages is paid at a rate and on dates fixed by terms of the contract between lender and borrower. On general investments in business ventures, the net return (profit) is unknown beforehand. The return is compensation to the investor for tying up his money in a venture which is expected to produce a monetary return thereon. This return is over and above what he puts into the venture; the return is after any depreciation return (return of capital).

This monetary profit, or net profit, is the attraction that leads to the investment. The probability that the investor will get back his investment and a return thereon is one factor on which depends the decision to invest or not to invest. A second factor is the probable rate of return with respect to other opportunities for investments that will probably pay a monetary return–or other immediate satisfactions to be gained by consumptive spending. Thus the higher the risk of not getting a net return from the investment and the higher the probable rates of return from other uses of the money–actual or prospective–the higher the rate of return that one demands from a prospective investment.

From these two factors come the phrase "minimum attractive rate of return." This MARR varies with time and opportunity; it can and likely does vary with each consideration of a possible use of money.

---

[2] In ancient times it was considered immoral or sinful to charge interest. The subject of charging interest was debated in high levels of religious organizations.

Interest and Vestcharge   **71**

MATHEMATICAL CONCEPT OF INTEREST

Money has a time value. This statement is proved by the fact that people pay interest in order to get the use of money now; lenders charge interest because they must postpone to the future other satisfactions to be gained with the money loaned. The mathematical concepts of interest–an exponential function–may be used for the transformation of any quantities from one time base to another. The six standard equations (see Chapter 6) are general mathematical relationships, not restricted to consideration of interest and return on money. The factors are equally applicable to compounding traffic growth, population increase, or biological change.

In analyses for economy, application of this mathematical concept is necessary in order to bring cash flows of money to common points of time. Also, the concept is necessary to convert an annual recurring series (annual maintenance expense, for example) to a single sum in order that it may be added to a single sum such as a onetime cash disbursement for construction investment. The reverse conversion is also desirable.

CONCEPT OF VESTCHARGE

Interest may be considered as monetary rent on borrowed money; *return* may be considered as expected monetary compensation for investing money in an enterprise. Needed is a third term to represent the economic cost and sacrifice of investing money in property–particularly public works–for which no return in monetary form can be expected. The return in this third case is in satisfactions, general benefits, or economic savings which are not measured by direct money income.

The word *vestcharge* is coined for this expressed concept. It means an economic charge against the investment–an investment charge. It represents the sacrifice on the part of those who supplied the money to be invested, insofar as these investors are denied any possible receipt of monetary interest or of monetary return on these particular monies. The term "vestcharge" has particular significance to government activities, but applies as well to personal investments, such as real estate, appliances, motor vehicles, and works of art.

The word *vestcharge* is in no way related to depreciation, return of capital, or the books of accounts–balance sheet and profit and loss statement. It is intended to be applied only as an economic cost factor in those considerations when some recognition of the economic charge against investments is desirable and there is no expectation of receiving cash interest, cash return on investment, or cash dividends. In economic analyses the word is used exactly as are the terms *discount factor*, *minimum attractive rate of return*, and *minimum acceptable discount rate*.

CONCEPTS MAINTAINED BY GOVERNMENT OFFICIALS

In government activities it is difficult to get public officials to understand the concept of the economic cost of money. These officials are familiar with interest paid on borrowed money. But an economic charge on the investment in a public works not supported by outstanding bonds is a concept they do not understand. The main difficulty lies in the fact that these officials are accustomed to dealing only with cash income and cash disbursement accounting and cash budgets. Cost accounting, investment accounting, depreciation accounting, and balance sheets are not part of their business. Every cash disbursement is a current cost of operating their government activity, and once the dollar has been audited as honestly and legally disbursed, no further accounting is required. Thus interest is of concern only when tax revenue must be collected to pay for the interest on the public debt as a cash disbursement. Furthermore, these government officials recognize the time value of money only when borrowing is concerned. Let it be said that there are professional engineers, businessmen, and just plain citizens who also share this conceptual block.

Vestcharge is a penalty paid for not consuming goods currently. Fixed assets–long-lifed property–bear the vestcharge, but consumed goods–operating necessities of fuel, raw materials, custodial supplies, maintenance, and repair–do not bear the charge. It might be assumed that the object is to consume goods rather than to preserve them. In terms of double-entry bookkeeping, disbursements for property for future use, charged to the fixed property investment accounts, become invested capital that bears a vestcharge burden. The disbursements to pay for daily supplies consumed in the course of operations are written off to operating expense at the time of disbursement (except for some items which pass through an inventory account) and thus they are not invested. The accounting for them is complete. Expenditures to maintain existing highway facilities in an acceptable condition for use are current expenses, but the cash disbursements to create the property (highways) which is maintained in future years is an investment cost subject to vestcharge. The distinguishing difference between investment cost and operating expense is the time period of consumption of the property.

WHY USE VESTCHARGE

It is appropriate to use the factor of vestcharge in economy studies of highway improvements to adjust for the differences in the time flow of money and account for the investment in highways in terms of the cost of money. The following four statements support this use of the concept of vestcharge:

1. Money invested in durable, long-lasting goods should pay the penalty for failure to be consumed immediately.

2. The investor–the public–has alternative uses of the highway tax money for dividend or profit-returning investments or for immediate satisfactions

Interest and Vestcharge 73

which he forgoes when his money is taken for highway construction. This factor is sometimes called "the opportunity cost" of investment.

3. The risks of all alternative uses of money in highways or in nonhighway uses are variable and there is need of a factor in the analysis to provide some relative measure of these risks.

4. The time dates of proposed disbursements (cash flows) will vary with the disbursements, so that a mathematical discount factor is required to bring all monetary disbursements to the same time basis. Construction investment and annual maintenance disbursements cannot be added directly; they must be brought to the same time basis. This time basis is either an equivalent uniform annual series or a single sum at a specific time date.

Any analysis that uses a zero rate of vestcharge (none at all) is misleading because such procedure does not weigh the relative desirability of unequal cash flows over time and says that a dollar today has the same value now as a dollar to be received, say, 10 years from today would have now. Even the government officials recognize the truth and acceptance of this time value of money.

Of course, vestcharge as an economic cost is not a factor in the preparation of a budget request to the legislature or city council. Interest is a budget item when it must be paid on outstanding debt. But in the analyses for economy or in estimating the annual economic cost of owning and operating a public works, there is no connection at all with budgets and appropriations. Confusion of these distinct items is one reason why certain public officials do not accept the concept of vestcharge as an economic factor of cost. They erroneously reject the factor in favor of considering interest only when it is a cash-disbursement item.

RATE OF VESTCHARGE

Upon accepting the justice of vestcharge as a factor in the analysis for economy of proposed highway improvements, the rate of vestcharge must be determined at the beginning of the analysis except when the rate-of-return solution is used. The rate is usually expressed in percent per year–that is, a percentage of the invested amount at the beginning of the year. The mathematical concept of compound interest is applied because the highways are used continuously over a period of years. The rate is a judgment value, but in exercising the judgment and understanding of the principles, theory, and factors involved will provide a reasonable basis of exercising judgment.

There is no formula by which to calculate the vestcharge rate and no reference table from which the rate may be selected. There are, however, some general guides that the analyst may use as aids in exercising his judgment.

GOVERNMENT CONTROL OF INTEREST RATES

Interest rates on borrowings, whether in the form of cash loans, bonds, or mortgages, are based on the market price as reached by negotiation between lender and borrower, in much the same way that the market price of commodities are set. In many states the maximum interest rate that may be charged by a loaning agency is fixed by law or regulation. This setting of legal interest rates dates back many centuries [5-8].

Over the centuries during which interest payments on loans in the form of either money or goods have been a standard business practice, the rate of interest has moved up and down from country to country, century to century, and year to year. The rate has been controlled by law or decree of the rulers, and left uncontrolled at other times and other places.

About 1800 B.C. in the time of Hammurabi, of Babylonia, the rate of interest was decreed to be not more than 33 1/3 percent per annum on loans of grain, repayable in kind, and 20 percent per annum for loans of silver by weight [5-8].

In Rome, about 450 B.C., the legal maximum interest rate was 8 1/3 percent per annum. The interest rates on loan-shark loans, starting as early as 287 B.C., have varied from 25 percent per week to 25 percent per day. Even as of 1963, unlicensed loan sharks were charging interest rates which, theoretically, would compute to 1300 percent per annum. Rates on bank loans, household finance types of loans, and department store credit are controlled by law in many places.

GENERAL FACTORS

The rate of vestcharge appropriate to each analysis may depend upon the following factors, among others:

1. The price that citizens are currently paying on the money they borrow for personal or business purposes.
2. The probable earning rate of return in private investments or the concurrent alternative uses of the money.
3. The relative probable returns available in other public works improvement projects.
4. The probable rate of interest to be paid on current borrowings by the government concerned. The cumulative rates on all existing loans and bonds are not a factor, because, being wholly in the past, they are irrelevant.
5. The risks and uncertainties involved in the particular proposed improvement being studied.

It is fundamental that the vestcharge rate to use in economy studies is not a universal all-time permanent rate. The rate varies with the going price of money, the local situation with respect to economic conditions, opportunities for public works improvements, and the risk and uncertainty involved within the proposal being analyzed. It is generally, but incorrectly, accepted that there is just one vestcharge rate to use and that this rate is to be used by everyone for all purposes. Shopping around any community will disclose many going effective interest rates

Interest and Vestcharge 75

on borrowed money. There also exists a difference in interest rates geographically.

The private opportunities for profitable investment of personal money is relevant because these opportunities represent alternative uses by the public of the money to be provided for highways. These rates determine what the citizens would forgo.

The best approach to selecting a vestcharge rate for a given application is to give the task the same attention as given other factors involved–costs, benefits, and analysis period.

The rate of vestcharge should be selected for each proposed improvement in accordance with the risks and uncertainties as foreseen for that particular proposal. Route location studies may justify a lower vestcharge rate than a grade-separation project or a traffic signal installation. A grade reduction on right-of-way of permanent location could justify a low rate as compared to a river crossing where water shipping is a factor related to heighth of navigational clearance.

The vestcharge rate should be determined from the viewpoint of those individuals who will furnish the capital, either through pay-as-you-go taxes or in the form of servicing a debt. Certainly it is these persons who forgo other uses of their money and who make the personal sacrifices.

Bierman and Smidt [5-2, p. 331] and other writers have suggested that a risk-free discount rate be used in the analysis, and that risk and uncertainty be taken into account separately. In such practice the vestcharge rate would be that rate comparable to the rate on a default-free investment of the type available to the suppliers of the invested money (the road users). The risk and uncertainty would be measured in the magnitude of the estimates of cash flow. Presumably high factors of contingencies and high unit prices would be used in estimating all costs, and conservative modest estimates would be made for benefits. This method has the advantage of some degree of certainty of the minimum vestcharge rate, but this advantage is canceled by the disadvantages of estimating the uncertainties and their dollar values of the cash flows. This method is not consistent with the general public acceptance of paying higher interest rates for greater risks and of demanding greater rates of return on business ventures of greater risk and uncertainty. A vestcharge rate that includes both the bare interest rate and an allowance for the risks and uncertainties is recommended.

MISCONCEPTIONS ABOUT THE VESTCHARGE RATE

There is a tendency on the part of some unexperienced analysts to want to vary the vestcharge rate, or the minimum attractive rate of return, in accordance with unrelated factors. For instance, the author has received statements in which the minimum attractive rate of return was indicated to depend somewhat upon such factors as the need of the project, its probable degree of acceptance by the public, overall public attitude, obvious desirability of the project, the probable magnitude of the probable benefits, project size, project total dollar cost, and, finally, and most startling, whether or not (in the mind of the analyst) the project should be economically justified. Certainly not a single one of the above factors is pertinent to the selection of the vestcharge rate unless it directly affects the risk and uncertainty.

There is a pronounced tendency for those government officials and other persons who accept the principle of using a vestcharge rate for economic cost and analysis of economy to select a rate that is either the average interest rate the agency is paying on all outstanding debts or the rate at which bonds could be sold currently. Either of these rates, as such, is incorrect. They reflect the cost of financing or the cost of borrowing, and not the economic cost to the people who will furnish the taxes to service the debt. The bond interest rates do not reflect the worth of money or the worth of a proposed highway improvement to the individual citizens.

A citizen who is paying 6 percent per annum on his home mortgage certainly would not consider it appropriate for a government to tax him for payment of a highway facility or a water resource project which was economically justified by using interest rates of 3 percent per annum because that was the average rate on the particular government's bonded debt. It would be wonderful for the citizen if he could borrow money for personal use at a rate of 3 percent. He then could pay off his 6 percent mortgage, his 9 percent automobile loan, and his 18 percent revolving charge account at the department store. But when he is forced to maintain these debts with high interest rates in order to pay his taxes it is logical for him to want his tax money invested in public works to earn returns on his investment more nearly equivalent to his cost of money.

Risk and Uncertainty

The risks involved are factors bearing upon the rate of vestcharge. In agreement with the normal practice of charging higher interest rates on loans of greater risk, expecting greater returns for greater risks in business, and charging greater insurance premiums for greater risks, the vestcharge rates should be higher for greater risks and uncertainties. But what are the risks and uncertainties? They may be measured by:

1. Will the highway improvement remain useful for the time period considered? Is it likely to become obsolete because of advancement of technology, increase in use, or decrease in use? A railroad grade separation may not be necessary in 10 years if railroad service is abandoned or materially reduced.

2. How accurate and how reliable are the estimates of future cash costs, traffic volumes, and the motor vehicle running costs used in the analysis? What unaccounted for changes may take place in traffic volumes, motor vehicle performance in traffic, and motor vehicle design which may affect running costs?

3. The probability and uncertainty of the consequences happening as

estimated are dependent upon the natural hazards of flood, fire, and other disaster. Also, how stable are the business use of the highway and purpose of trips? Is the traffic dependent upon land use, recreation, or industry?

4. How reliable are the estimates of construction cost and maintenance cost? What are the hazards of construction?

The higher rates of vestcharge are advisable to reduce the effects of the uncertainties and to reduce the weight of those monetary consequences far into the future. The faraway future has a great uncertainty. A high rate of vestcharge gives the greater weight to the present and immediate future as compared to the far future.

Highway improvement projects as a general class require the use of vestcharge rates higher than would be used in most water resource projects. The reason for these higher rates lies in the factors of risks and uncertainties. Highway design and highway use are yet undergoing continuous and significant yearly change. Therefore, the risk of early inadequacy or obsolescence is relatively high. An examination of the highway systems now in use will disclose the fact that much of the mileage lacks the features of design now considered essential to satisfactory service. Overcrowding, especially at peak hours of use, has lowered the speeds and increased the impedances[3] to uniform travel speed to the extent that benefits provided by the last construction no longer exist. The studies of the need for highway improvement conducted by the several states, show great needs for reconstruction of many routes and sections of routes built or last reconstructed only 10 to 15 years ago.

The difference in water resource projects and highway projects lies mainly in that a dam, once built, continues to perform at the design rate; the dam does not become obsolete or inadequate from a continuously changing volume and character of use, as has been (and will be) true of highway facilities.

SPECIFIC RATES

At some risk of leading the reader, at least part way, to a specific vestcharge rate for his analysis, a few specific rates will be mentioned. However, the emphasis of this section on the choice of rate of vestcharge is on the fact that the rate for every set of alternatives being analyzed is to be chosen explicitly for that analysis. Even though a highway department once determines that x percent is its choice of vestcharge rate for economic studies, such rate should be regarded as a guideline and to hold only when there are not factors of specific weight to warrant a rate lower or higher than the guideline rate.

Hirshleifer, DeHaven, and Milliman [5-7, p. 161] recommend a discount rate of 10 percent per year for public projects. The basis of this 10 percent rate, among other factors, is the overoptimism that usually prevails in the public invest-

---

Impedance is the electrical term related to the resistance to the flow of an alternating current. The word is used here in a similar sense, but related to the flow of motor vehicle traffic.

ment decision process. Also, these authors take the position that the rate on private investments and on public investments should be the same. If the rates are not the same at the margin, the money available for investment would be attracted unequally between the private and public sectors. The use of low vestcharge rates, such as bond interest rates, for economic evaluation of public works, will justify large public works programs and induce larger tax imposts, where such tax money could be invested at higher rates of returns if attracted to private works.

Perhaps the analyses for the economy of highway improvements would call for vestcharge rates varying between 6 and 15 percent per year. Grant, in 1959, suggested 7 percent [5-5]. A rate somewhat higher than the fair rate of return allowed regulated public utilities could be justified for many types of highway facilities.

In the rate-of-return solution except when the differential solution is used, the selection of the vestcharge rate may be delayed to the time of the final decision. At this point, however, a decision must be made as to what is the minimum attractive rate of return that is acceptable, which minimum attractive rate

TABLE 5-1

METHOD OF DETERMINING MINIMUM ATTRACTIVE RATE OF RETURN BASED UPON THE FUNDS AVAILABLE FOR CONSTRUCTION

Project Number	Estimated Cost (Thousands)	Accumulated Cost (Thousands)	Probable Rate of Return (Percent)
1	$ 540	$ 540	22
2	1,610	2,150	22
3	854	3,104	21
4	360	3,464	20
5	1,200	4,664	20
6	4,640	9,304	19
—	—	—	—
36	612	21,916	13
37	1,410	23,326	13
38	3,610	26,936	12
39	200	27,136	12
40	805	27,941 *	11 *
41	512	28,453	11
42	3,900	32,353	11
43	2,600	34,953	10
—	—	—	—
92	726	65,762	7

* The funds available for construction for the year are $28,000,000. Therefore, on the basis of only the probable rate of return, no project No. 41 to 92 would be undertaken, and the minimum attractive rate of return for this budget year would be 11 percent per annum.

Interest and Vestcharge 79

of return becomes the bench mark against which the adequacy of the solved for rate of return is measured. A selection of the minimum attractive rate of return depends upon the factors just discussed. There is a device, however, which can be used as a guide to selection of the vestcharge rate, or the minimum attractive rate of return.

Highway departments usually have the need for more money for construction than is available. This scarcity of funds means that the construction program for a given year will not include all the projects that are desirable and beneficial. To the extent that the separate proposals for construction can be analyzed for their probable rate of return, an array of projects can be listed in the form of Table 5-1

In Table 5-1, assuming that only $28,000,000 is available for construction projects for the year, the minimum attractive rate of return for the year would be 11 percent. All proposed projects numbered higher than No. 40 would earn 11 percent or less; to construct them would require more than the $28,000,000 available. Not considering other appropriate factors such as continuity of route development, geographical distribution of construction work, contractor availability and consequences not included in the analysis for rate of return, the construction program for the year woud be held to projects 1 through 40. The calculated rate of return (vestcharge rate) of 11 percent would be the minimum attractive rate of return for the year.

This discussion on rate of vestcharge is fittingly closed by the following poem[4] written several years ago by Professor Kenneth Boulding of the University of Michigan, during a conference he was attending:

> Around the mysteries of finance
>   We must perform a ritual dance
> Because the long-term interest rate
>   Determines any project's fate.
> At four percent the case is clear;
>   At five some sneaking doubts appear;
> At six it draws its final breath,
>   While seven percent is certain death.

PROBLEMS

5-1. List the items of family finance and economics upon which compound interest and the principles of engineering economy could be applied in the decision process.

5-2. From Prob. 5-1, or for a separate listing, select four of the more important items of family finance and business. List the main factors that should be considered in the decision process.

5-3. As a consumer and as a borrower through a home mortgage, automobile financing, a department store revolving or charge account, or other source, what has been (1)

---

[4] Courtesy of Prof. Eugene L. Grant. The interest rates were upgraded by the author to reflect current conditions.

your concept of interest, (2) your attitude toward the practice, and (3) to what extent have you given attention to the rate of interest charged?

5-4. Name any differences you think appropriate in the application of the concept of interest to cost and economic analyses pertaining to (1) personal family business, (2) private commercial business, (3) corporation business, (4) general governmental functions, (5) government water resource projects, and (6) highway and street projects.

5-5. List the specific factors of (1) risk and (2) uncertainty that may exist in highway economy studies and name a type of improvement or highway element to which the risk or uncertainty would pertain.

5-6. From your library or other source, select two or three reports, papers, or analyses in which a vestcharge rate has been used in an economic analysis or a cost analysis. Discuss the appropriateness of the rate used. What factors did the author or analyst probably consider when selecting his specific rate?

5-7. In a certain economic analysis of an inland waterway used by ocean vessels and involving overhead highway structures, the analyst used three different vestcharge rates: (a) 3 percent for federal government funds, (b) 4 percent for state highway funds, and (c) 6 percent for funds from private industry (ocean shipping companies). Comment upon the scheme used and the specific rates.

5-8. For your own current economic and financial situation, what minimum attractive rate of return would you use? To the extent you care to do so (there is no demand that you expose your financial activities) support your choice by discussion of the factors involved.

5-9. Considering your personal position, that of your friends and acquaintances, and your community as a whole, what vestcharge rate would you recommend for use in the economic analysis of a local proposal for an urban freeway? Support your recommendation by appropriate discussion.

REFERENCES

5-1. E. B. Berman. *The Normative Interest Rate.* The Rand Corp., Santa Monica, Calif., 1959, p. 1796.

5-2. Harold J. Bierman and Seymour Smidt. *The Capital Budgeting Decision.* 2d ed., Macmillan, New York, 1966.

5-3. Joseph W. Conrad. *An Introduction to the Theory of Interest.* U. of California Press, Berkeley, 1959.

5-4. Irving Fisher. *The Theory of Interest.* Macmillan, New York, 1930.

5-5. Eugene L. Grant. *Interest and the Rate of Return on Investments.* Highway Research Board, National Academy of Sciences, Washington, D.C., Special Report 56, *Economic Analysis in Highway Programming, Location, and Design,* Sept. 17-18, 1959, pp. 82-90.

5-6. Eugene L. Grant and W. Grant Ireson. *Principles of Engineering Economy.* 4th ed., Ronald, New York, 1960.

5-7. Jack Hirshleifer, James C. DeHaven, and Jerome W. Milliman. *Water Supply Economics, Technology, and Policy.* U. of Chicago Press, Chicago, 1960. (*Note:* Strongly recommended as preferred reading on economic analysis, discount rates, and public works investments.)

5-8. Sidney Homer. *A History of Interest Rates.* Rutgers U. P., New Brunswick, N.J., 1963.

5-9. Frederick Charles Kent. *Mathematical Principles of Finance* (with tables). 2d ed., McGraw-Hill, New York, 1927.

5-10. E. A. Saliers (ed.). *Accountants' Handbook*. Ronald, New York, 1923, p. 112.

5-11. Jack Ochs. *Discount Rates for Public Investment Decisions.* Washington University, Institute For Urban and Regional Studies, St. Louis, Mo., October 1966.

5-12. "Planning-Programming-Budgeting System: A Symposium." *Public Administration Review,* Vol. XXVI, No. 4 (December 1966), quarterly journal of the American Society for Public Administration, 1329 Eighteenth Street, N.W., Washington, D.C. A reprint includes a collection of six papers on methods of analysis, interest rates, management, and budgeting.

Chapter **6**

Compound Interest Equations *

Chapter 5 sets forth the main theory of interest, return, and vestcharge. Next, it is in order to study the mathematics related to the computation of interest as expressed in the six standard compound interest equations. A familiarity with these equations and their use is essential to an understanding of the analyses for economy of alternative proposals to be given in the chapters to follow. Although commonly referred to as compound interest equations, they are, in reality, mathematical expressions that apply to any exponential growth function. Gradient and exponential growth compoundings and continuous compounding are also discussed.

THE SIX STANDARD COMPOUNDING EQUATIONS

The six standard compound interest equations are summation equations. Their use makes it possible to calculate the desired end sum without making the series of individual calculations time period by time period. Of the six equations, two—one the reciprocal of the other—apply to single sums. The other four equations apply to uniform periodic sums or series. Of these four series equations, two are reciprocals of the other two. Of the first pair of reciprocal series equations, one is used in calculating the compound amount of the series, and the other is used to calculate the present worth of the series. Of the last two equations, again, one the reciprocal of the other, one is used to calculate the uniform year-end sinking fund deposit, and the other for calculating the uniform annual capital cost—capital recovery with interest.

BASIC CONCEPT AND NOTATIONS

A sum at compound interest grows according to an exponential curve. Practically all of the analyses for engineering economy use the compound interest theory, rather than simple interest. Simple interest is interest on only the principal sum over any number of time periods. Thus, simple interest on

---

* Grateful acknowledgment is extended to Nathan Lieder for his material assistance with the mathematical derivations in this chapter.

Compound Interest Equations

$100 for three years at 6 percent per year would be $18; compound interest on the $100 under the same conditions would amount to $19.10, which includes interest on interest.

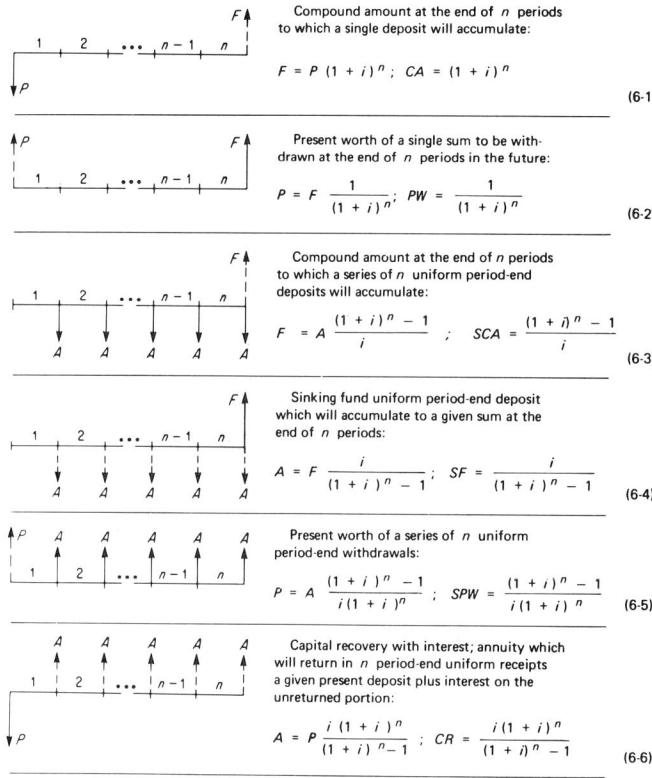

FIG. 6-1. Cash-flow diagrams and the six standard compound interest equations based on the period-end step convention.

The six basic compound interest equations in Fig. 6-1 are presented according to the following notations:

P = a present sum of money; a single payment; the present worth of an annuity or of a single sum

F = a sum at a future date, n interest payment periods from the present, which sum is equivalent to P with compound interest at i rate over n periods

A = end-of-period annuity; amortization payment

i = interest rate per period of time

n = number of interest periods; number of period-end payments; n is most often expressed in years

The six equations are derived by use of the period-end step convention. They are based on the flow of money (or other quantity) in lump sums at specific moments in time at the end of a period, or in some equations, at the beginning of the period. See Fig. 6-1 for the cash-flow diagrams and the six equations.

DERIVATION OF THE SIX COMPOUND INTEREST EQUATIONS

From the fundamental definitions of interest and the rate of interest, the earned interest at the end of one period is equal to the principal sum at the beginning of the period times the rate of interest per period, which is P times i. The total sum F_1 at the end of the period is:

$$F_1 = P_0 + P_0 i = P_0(1 + i)$$

For the end of the second year the sum F_2 would be:

$$F_2 = P_0(1 + i) + P_0(1 + i)i$$
$$= P_0(1 + i)(1 + i) = P_0(1 + i)^2$$

For the end of the third year the sum F_3 would be:

$$F_3 = P_0(1 + i)^2 + P_0(1 + i)^2 i$$
$$= P_0(1 + i)^2(1 + i) = P_0(1 + i)^3$$

It is obvious now that F_n for any specific number of periods n will be the initial sum P_0 times $(1 + i)^n$, or to generalize,

$$F_n \text{ or simply } F = P(1 + i)^n \qquad (6\text{-}1)$$

Solving Eq. 6-1 for P gives

$$P = \frac{F}{(1 + i)^n} \qquad (6\text{-}2)$$

which is the present worth of the sum F to be paid at the end of n periods.

Equation 6-3 for the accumulation of a uniform periodic period-end series for n periods is the equation of a geometric series in which the common ratio is $(1 + i)$. Equation 6-3 may be derived as follows, remembering that the uniform periodic sum A is a period-end payment.

The last payment is made at the end of the nth period (Fig. 6-1) and, therefore, it compounds no interest, not even simple interest. Prior equal payments are made at the end of each prior period. The first payment is made at the end of the first period or at $n = 1$. This first payment will accumulate compound interest for $n - 1$ periods. These uniform period-end single sum payments, A, grow according to Eq. 6-1, the equation for the compound accumulation of single sums. In reverse order, last to first, these separate accumulating sums form a geometric

Compound Interest Equations

series ranging from $A(1+i)^0$ to $A(1+i)^{n-1}$. For $n = 5$, the series of compounded single sums is as follows:

$$\begin{aligned}\text{Fifth and last payment} &= A(1+i)^0 \\ \text{Fourth payment} &= A(1+i)^1 \\ \text{Third payment} &= A(1+i)^2 \\ \text{Second payment} &= A(1+i)^3 \\ \text{First payment} &= A(1+i)^4\end{aligned}$$

The exponent 4 is equivalent to $n - 1$. The sum of this series may be written

$$F = A + A(1+i)^1 + A(1+i)^2 + A(1+i)^3 + A(1+i)^{n-1} \tag{6-A}$$

Multiplying the above Eq. 6-A by $(1 + i)$ gives

$$F(1+i) = A(1+i)^1 + A(1+i)^2 + A(1+i)^3 + A(1+i)^4 + A(1+i)^n \tag{6-B}$$

Subtracting Eq. 6-A from Eq. 6-B gives

$$F(1+i) - F = -A + A(1+i)^n$$

$$F = A \frac{(1+i)^n - 1}{i} \tag{6-3}$$

Equation 6-4 for the sinking fund year-end deposit (annuity) A is found by solving Eq. 6-3 for A:

$$A = F \frac{i}{(1+i)^n - 1} \tag{6-4}$$

Equation 6-3 gives the compound amount of n period-end uniform payments of amount A and at the end of n periods. This accumulated amount F, a single sum, may be converted to its present worth by multiplying by the present-worth single sum factor, Eq. 6-2. Thus

$$P = A \left(\frac{(1+i)^n - 1}{i} \right) \left(\frac{1}{(1+i)^n} \right) = A \frac{(1+i)^n - 1}{i(1+i)^n} \tag{6-5}$$

Lastly, Eq. 6-5 may be converted to the annuity form by solving it for A or

$$A = P \frac{i(1+i)^n}{(1+i)^n - 1} \tag{6-6}$$

These derivations of these six basic equations emphasize that of the six, three are reciprocals of the other three, and of the six, original derivations were necessary for only two, Eqs. 6-1 and 6-3, the other four being derived by using prior derivations. This interrelationship of the six equations is the key to the solving of many problems involving compound interest and of checking the solutions.

It is customary to print tables of the solutions of these equations based upon unity, that is, the coefficients *P, F,* and *A* are set equal to unity and the term involving $(1 + i)$ is then solved. These factors have been published over the years in many books, for a range of n from 1 to 100 periods or more and for rates of interest per period varying from fractions of 1 percent upward to 50 or more percent. See Appendix B for the tables accompanying this book.

MNEMONIC SYMBOLS

In equations involving compound interest factors, it is convenient to use mnemonic symbols to represent the several interest factors. These symbols provide the means of rapid checking of the correctness of computations and equations in which the factors are used without being concerned with the equations themselves. The symbols used are those (plus others) recommended by the Engineering Economy Division of the American Society for Engineering Education [7-3] plus three others for exponential growth factors. The complete list used in this book is:

1. *CA* = compound amount factor of single sum, Eq. 6-1.
2. *PW* = present-worth factor of single sum, Eq. 6-2.
3. *SCA* = compound amount factor of uniform series, Eq. 6-3.
4. *SPW* = present-worth factor of uniform series, Eq. 6-4.
5. *SF* = sinking-fund factor (uniform year-end deposit), Eq. 6-5.
6. *CR* = Capital recovery factor, Eq. 6-6.
7. *GCA* = gradient compound amount factor, Eq. 6-7.
8. *GPW* = gradient present-worth factor, Eq. 6-8.
9. *GUS* = gradient factor for conversion of a gradient series to an equivalent uniform series, Eq. 6-9.
10. E_tCA = exponential compound-amount factor; the subscript t is the rate of compound increase of the base quantity, Eq. 6-10.
11. E_tPW = exponential present-worth factor, Eq. 6-11.
12. E_tUS = exponential factor for conversión of an exponential series to an equivalent uniform series, Eq. 6-12.
13. The prefix *C* to any of the above symbols is used to denote continuous compounding, for example, *CCA*.

Equations involving these symbols may be written as follows:

$$F = P(1 + i)^n = P(CA)$$

For a *P* of 100, an interest rate of 7 percent, and 8 time periods, the following form would apply:

$$F = P(CA\text{-}i\%\text{-}n)$$
$$F = 100(CA\text{-}7\%\text{-}8)$$

or

$$F = 100(1.7182) = 171.82$$

or

$$F = 100(CA\text{-}7\%\text{-}8 = 1.7182) = 171.82$$

Compound Interest Equations

ILLUSTRATIVE APPLICATIONS

The applications and interrelationships of the standard six compound interest equations are explained by a few illustrations. For applications to highway economy studies, the period n is usually one or more years. Periods less than one year, will not be used, but fractional years are commonly used in the banking and loan business. The numerical values for the factors are taken from standard tables as given in Appendix B. Although the dollar sign, $, is used in these illustrations, it is to be kept in mind that the equations and theory supporting them are applicable to any form of quantitative units–motor vehicles, people, or guinea pigs.

Equation 6-1 gives the accumulation of a single amount at compound interest for n periods, in this illustration, n years. Thus $100 kept at compound interest at 6 percent per year[1] will amount to $150.36 in 7 years:

$$F = 100\,(CA\text{-}6\%\text{-}7 = 1.50363) = \$150.36$$

Equation 6-2 is used to find the present worth of a single sum due at a specific future date. Thus $100 due in 7 years from the present date is worth at the present time only $66.51:

$$P = 100\,(PW\text{-}6\%\text{-}7 = 0.66506) = \$66.51$$

This result means that if $66.51 is kept at 6 percent per annum compound interest for 7 years the accumulation will amount to $100.

Equation 6-3 is used to find the accumulation of a uniform year-end payment for a series of years. The formula would be used to find the compound amount of an equal annual saving for a given number of years. Thus an annual saving of $100 for 7 years at 6 percent per year amounts to $839.38:

$$F = 100\,(SCA\text{-}6\%\text{-}7 = 8.39384) = \$839.38$$

This result means that the series of seven $100 year-end amounts at 6 percent per year interest earned a total of $139.38 interest during the 7 years they were being deposited.

Equation 6-4 gives the annual uniform deposit in a year-end sinking fund that is needed to accumulate to a given sum F in n years at i rate of compound interest. Thus an annual deposit of $11.91 will accumulate to $100 at 6 percent compound interest in 7 years:

$$A = 100\,(SF\text{-}6\%\text{-}7 = 0.11914) = \$11.91$$

This equation would be used to determine the sinking-fund deposit required to build up a fund for the redemption of bonds issued for construction of a highway improvement. Equation 6-4 is the reciprocal of Eq. 6-3 which gives the accumulation of the deposit.

---

[1] It is customary to express the interest rate at x percent per annum. Unless specifically stated to the contrary all references in this book to rates of interest (or vestcharge) are rates per annum.

Equation 6-5 gives the present worth of the accumulation of a uniform annual year-end deposit for a term of years. It gives the present worth of the sum which is found by Eq. 6-3. The present worth at 6 percent interest of $100 receivable each year-end for 7 years is $558.24:

$$P = 100(SPW\text{-}6\%\text{-}7 = 5.58238) = \$558.24$$

Equation 6-5 is identical with Eq. 6-3 except for the factor $(1 + i)^n$ in the denominator. This factor being in the denominator converts the expression to the present worth.

Equation 6-6 gives the capital recovery with interest. It is an annuity equation used to calculate the uniform annual year-end payment that would exhaust a fund in a given number of years at the specified rate of interest. Thus a $100 initial fund that earns interest at 6 percent per year on its balance would be exactly paid out in 7 years should the payment be $17.91 each year end:

$$A = 100\ (CR\text{-}6\%\text{-}7 = 0.17914) = \$17.91$$

Equation 6-6 has an important application in economy and annual cost studies because this single factor may be used to express the combined uniform yearly charge (or cost) for depreciation and vestcharge on the initial (or other) investment.

The foregoing example may be broken down into these two components to illustrate the working of the capital recovery factor. The equation is the combination of the sinking-fund yearly increment (deposit plus interest earning on the fund for the year) and interest on the unamortized portion of the original sum P. Sinking fund Eq. 6-4 may be reduced to the capital recovery Eq. 6-6 by adding the term i:

$$\frac{i}{(1 + i)^n - 1} + i = \frac{i + i\left[(1 + i)^n - 1\right]}{(1 + i)^n - 1} = \frac{i(1 + i)^n}{(1 + i)^n - 1}$$

Table 6-1 illustrates how the depreciation charge and the interest charge are combined to produce a uniform annual sum. Note that the annual depreciation amount increases each year (column 5) and the annual vestcharge decreases each year, each at such a rate as to produce a constant annual sum (column 8).

Equation 6-6, capital recovery with interest, is the one used in economy analyses to reduce the capital investment (construction cost) to an equivalent uniform annual sum—the annual capital cost.

COMPOUNDING A UNIFORMLY INCREASING SERIES

As illustrated in Fig. 6-1, the four series compound interest Eqs. 6-3, 6-4, 6-5, and 6-6 are based upon uniform, or equal amount sums, each sum considered to flow instantly at each period end over n periods. Thus the equations do not give

Compound Interest Equations

TABLE 6-1
Interrelation of Sinking Fund and Capital Recovery Equations
($P = \$100.00$, $i = 6$ percent, $n = 7$, $(SF\text{-}6\%\text{-}7) = 0.11913$, $(CR\text{-}6\%\text{-}7) = 0.179135$)

Year-End, n	Sinking Fund Deposit at End of Year	Sinking Fund Balance at Beginning of Year	Interest Earning at End of Year on Sinking Fund Balance at Beginning of Year	Total Sinking Fund Increment at End of Year-Amount of Capital Recovered at End of Year	Balance of Unrecovered Capital at Beginning of Year	Interest Received at End of Year on Unrecovered Capital at Beginning of Year	Combined Capital Recovery and Interest Received at End of Year
	Eq. 6-4	$\Sigma[(2)+(4)]$	$0.06 \times (3)$	$(2)+(4)$	$100-(3)$	$0.06 \times (6)$	$(5)+(7)$
(1)	(2)	(3)	(4)	(5)	(6)	(7)	(8)
First	$11.914	$0.000	$0.000	$11.914	$100.000	$6.000	$17.914
Second	11.914	11.914	0.715	12.629	88.086	5.285	17.914
Third	11.914	24.543	1.473	13.387	75.457	4.527	17.914
Fourth	11.914	37.930	2.276	14.190	62.070	3.724	17.914
Fifth	11.914	52.120	3.127	15.041	47.880	2.873	17.914
Sixth	11.914	67.161	4.030	15.944	32.839	1.970	17.914
Seventh	11.914	83.105	4.986	16.900	16.895	1.014	17.914
Eighth	—	100.005	—	—	−0.005	—	—

consideration to any increase or decrease in the periodic period-end lump-sum cash flows.

In the practical world of business, however, the cash flows may tend to increase or decrease per year over a period of years. With reference to highways, three factors which may exhibit increasing or decreasing tendencies over time are annual highway (or machine) maintenance costs, average yearly traffic volume, and the running cost per mile of vehicles as traffic increases. Although in actuality these changes may be irregular from year to year, they may be approximated by a gradient increase of an equal amount each year or by an exponential increase of an increasing amount each year. The four series compound interest equations are derived in this section for uniformly increasing annual cash flows, and in the next section for exponential increases.

CONCEPT OF THE GRADIENT INCREASE

The more practical procedure for solution of compound interest applications involving an increasing series as opposed to a uniform series is to assume the series to follow a uniform gradient, or straight line. The slope of the line–that is, the increase per period (year)–is then uniform for each period. The period-end step convention is also assumed, so that the problem resolves itself into finding the compound amount, or, say, present worth, of a series of sums, equally spaced over time and increasing each period at a uniform amount per period. This periodic increase may be called the gradient factor, or slope, designated as G. Figure 6-2 illustrates the cash flow.

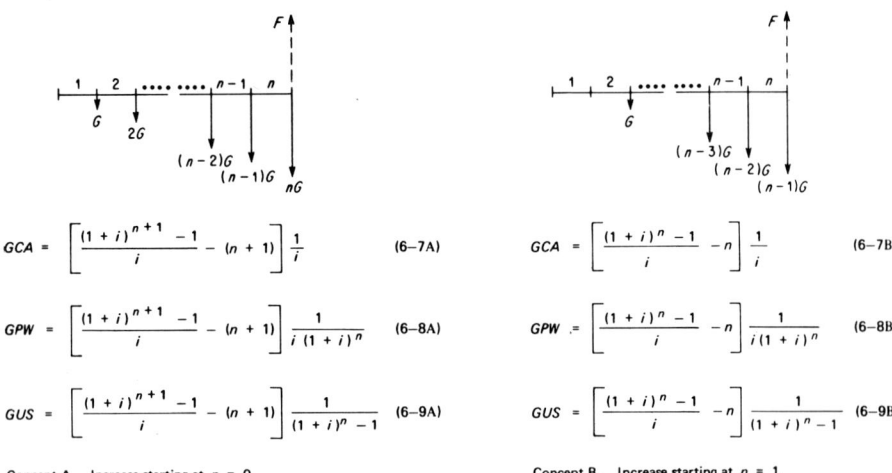

FIG. 6-2. Cash-flow diagrams and equations for the gradient growth factors. Note that the equations are for a G of unity and a zero quantity of A at time zero. Any uniform annual A is handled by the appropriate compound interest equation in Fig. 6-1 separately from the G factor.

Compound Interest Equations

There are two cash-flow diagrams in Fig. 6-2, and two sets of equations. Diagram A and its equations apply to the concept of starting the increase at time $n = 0$, so that the first gradient sum G comes at the end of the first period, or at $n = 1$. Diagram B and accompanying equations follow the concept of the increase starting at time $n = 1$ so that the first gradient sum is acknowledged at $n = 2$. In this book Concept A is used; Grant and Ireson [5-6] follow Concept B. Either concept is acceptable, but in a particular case one concept may be more applicable than the other. For instance, the traffic ADT may be estimated as of $n = 0$, in which case the increase would start at $n = 0$, so Concept A would apply. Because of the similarity of the equations for Concepts A and B, it is easy to convert from one concept to the other.

DERIVATION OF THE GRADIENT EQUATIONS

Derivation of the equations for the Concept A of starting the gradient increase at $n = 0$, beginning of the first period, can be achieved by proceeding to find the compound amount in a manner similar to that used previously for the compound amount of a uniform series. The period-end amount (say deposit) at the end of the first year is G, at the end of the second year it is 2 G, and so on to the nth year-end at which the deposit will be nG. This series of increasing amount may be summed on the basis that at the end of each year a new sum G starts a uniform series of deposits to carry through to the end of the nth year. The G that comes into being at the end of the first year will compound for n periods in accordance with Eq. 6-3, which is $F = P \dfrac{(1+i)^n - 1}{i}$ and the last G which builds up from time $n - 1$ to n will compound no interest. The sum of these several compounded amounts may be expressed as follows:

$$F_G = G\frac{(1+i)^n - 1}{i} + G\frac{(1+i)^{n-1} - 1}{i} + G\frac{(1+i)^{n-2} - 1}{i}$$
$$+ \ldots + G\frac{(1+i)^2 - 1}{i} + G\frac{(1+i)^1 - 1}{i}$$

$$F_G = \frac{G}{i}\Big[[(1+i)^n - 1] + [(1+i)^{n-1} - 1] + [(1+i)^{n-2} - 1]$$
$$+ \ldots + [(1+i)^2 - 1] + [(1+i)^1 - 1]\Big]$$

$$F_G = \frac{G}{i}\big[(1+i)^n + (1+i)^{n-1} + (1+i)^{n-2}$$
$$+ \ldots + (1+i)^2 + (1+i)^1\big] - \frac{G(n)}{i} \quad \text{(6-C)}$$

Multiplying Eq. 6-C by the common ratio $(1 + i)$ of the above geometric series:

$$F_G(1 + i) = \frac{G}{i}[(1 + i)^{n+1} + (1 + i)^n + (1 + i)^{n-1}$$
$$+ \ldots + (1 + i)^3 + (1 + i)^2] - \frac{Gn(1 + i)}{i} \quad (6\text{-D})$$

Subtracting Eq. 6-C from Eq. 6-D:

$$F_G(1 + i) - F_G = \frac{G}{i}[(1 + i)^{n+1} - (1 + i)] + \frac{Gn}{i} - \frac{G(1 + i)}{i}$$

$$F_G[(1 + i) - 1] = \frac{G}{i}[(1 + i)^{n+1} - 1 - i] + \frac{Gn}{i}[1 - 1 - i]$$

$$F_G = \frac{G}{i}\left\{\left[\frac{(1 + i)^{n+1} - 1}{i}\right] - 1\right\} - \frac{Gn}{i}$$

$$F_G = \frac{G}{i}\left[\left(\frac{(1 + i)^{n+1} - 1}{i}\right) - 1\right] - \frac{Gn}{i}$$

$$F_G = \left[\left(\frac{(1 + i)^{n+1} - 1}{i}\right) - (n + 1)\right]\frac{G}{i}$$

Letting $G =$ unity

$$GCA = \left[\left(\frac{(1 + i)^{n+1} - 1}{i}\right) - (n + 1)\right]\frac{1}{i} \quad (6\text{-7A})$$

The present worth of a gradient series is easily obtained from Eq. 6-7A by dividing by $(1 + i)^n$:

$$GPW = \left[\left(\frac{(1 + i)^{n+1} - 1}{i}\right) - (n + 1)\right]\frac{1}{i(1 + i)^n} \quad (6\text{-8A})$$

From the present-worth gradient factor (GPW), Eq. 6-8A, the GUS may be calculated by multiplying by the capital recovery factor, $\frac{i(1 + i)^n}{(1 + i)^n - 1}$. The result-

Compound Interest Equations

ing gradient factor is thus that uniform period-end amount which is equivalent to the series of period-end increasing gradient amounts, thus,

$$GUS = \left[\left(\frac{(1+i)^{n+1}-1}{i}\right) - (n+1)\right] \frac{1}{(1+i)^n - 1} \qquad (6\text{-}9A)$$

The gradient Eqs. 6-8A and 6-9A, present worth of an increasing series and the equivalent uniform factor, have been solved for a series of interest rates and for a range of n. The two tables are in Appendix C.

Caution is in order to remember that the gradient equations apply only to the gradient increase; they exclude any amount at $n = 0$, which continues to n as a uniform series.

COMPOUNDING AN EXPONENTIAL SERIES

Highway traffic and other growth phenomena, such as population, are often forecast on a yearly compound increasing basis. Instead of a uniform gradient increase, the increase is forecast, say at 4 percent per year, compounded. When growth is so forecasted, it is convenient to have available suitable tables of the compound interest series factors for compound (exponential) increases. The tables in Appendix D are for a series of compound rates of increases and a series of interest rates. The factors for present worth *(EPW)* and for the equivalent uniform amount *(EUS)* only are given in Appendix D.

Equations 6-10, 6-11, and 6-12, the three equations relating to Fig. 6-3, are based upon the exponential increase starting at zero time, $n = 0$, the beginning of the first period. The step, period-end convention is used, however, rather than continuous compounding.

The notation used is:

t = rate per period of exponential increase expressed in decimal form
i = interest rate per period (year) in decimal form
P = initial amount at zero time and the base on which the rate of increase t is calculated
A_x = amount deposited at the end of each period and is equal to $P(1+i)^x$
n = number of periods (years)

The equations for compounding of exponential increasing series may be derived in a manner similar to that followed for the uniform series and for the gradient series and adhering to the period-end step convention.

Figure 6-3 illustrates the cash-flow pattern used in deriving the equation for the compound amount factor of an exponential growing series. Equations for $E_t CA$, $E_t PW$, and $E_t US$ are given in Fig. 6-3.

The derivation is made on the basis that the amount P at time $n = 0$ is the base to which the exponential rate t is applied. But the single amount P at

Compound Interest Equations

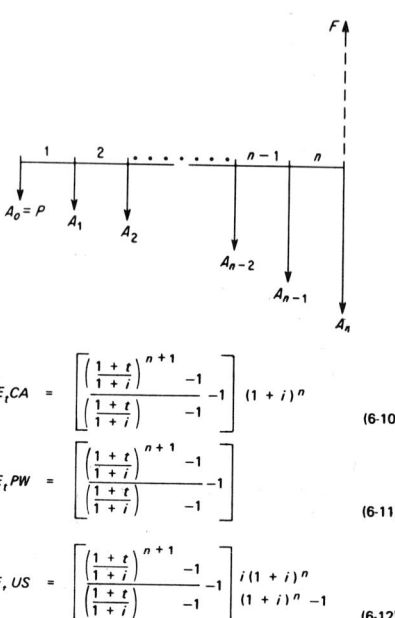

$$E_t CA = \left[\frac{\left(\frac{1+t}{1+i}\right)^{n+1} - 1}{\left(\frac{1+t}{1+i}\right) - 1} - 1 \right] (1+i)^n \tag{6-10}$$

$$E_t PW = \left[\frac{\left(\frac{1+t}{1+i}\right)^{n+1} - 1}{\left(\frac{1+t}{1+i}\right) - 1} - 1 \right] \tag{6-11}$$

$$E, US = \left[\frac{\left(\frac{1+t}{1+i}\right)^{n+1} - 1}{\left(\frac{1+t}{1+i}\right) - 1} - 1 \right] \frac{i(1+i)^n}{(1+i)^n - 1} \tag{6-12}$$

When $t = i$ the quantity within the brackets reduces to n.

FIG. 6-3. Cash-flow diagrams and equations for compounding an exponentially increasing series based upon an initial quantity P, the base of the increasing series. These equations are for a P of unity.

$n = 0$ is not included in the summation F, in conformity with the period-end step convention.

The increasing period-end series amount at the end of each period that develops from the growth factor t will be:

$A_1 = P(1 + t)^1$ = amount deposited at the end of the first period to accumulate at compound interest

$A_2 = P(1 + t)^2$ = amount deposited at the end of the second period to accumulate at compound interest

. . . .

$A_{n-1} = P(1 + t)^{n-1}$ = amount deposited at the end of the $(n-1)$th period to accumulate at compound interest

$A_n = P(1 + t)^n$ = amount deposited at the end of the nth period to accumulate at compound interest

Each of the above compound period-end amounts will compound interest at the rate i per period for their respective number of periods, so their accumulated amounts at the end of the nth period may be written:

$P(1 + t)^1 (1 + i)^{n-1}$ = compound amount at end of nth period of the first period-end deposit

Compound Interest Equations

$P(1 + t)^2 (1 + i)^{n-2}$ = compound amount at end of nth period of the second period-end deposit

. . . .

$P(1 + t)^{n-1} (1 + i)^1$ = compound amount at end of nth period of the $(n-1)$th period-end deposit

$P(1 + t)^n (1 + i)^0$ = compound amount at end of nth period of the nth period-end deposit

The equation for the sum of these n compound amounts at the end of the nth period may be written:

$$F_E = P(1 + t)^1 (1 + i)^{n-1} + P(1 + t)^2 (1 + i)^{n-2}$$
$$+ \ldots + P(1 + t)^{n+1}(1+i)^1 + P(1+t)^n(1+i)^0$$

$$F_E = P(1 + i)^n \left[\left(\frac{1+t}{1+i}\right)^1 + \left(\frac{1+t}{1+i}\right)^2 \right.$$
$$\left. + \ldots + \left(\frac{1+t}{1+i}\right)^{n+1} + \left(\frac{1+t}{1+i}\right)^n \right] \quad \text{(6-E)}$$

Multiplying Eq. 6-E by the common ratio $\frac{1+t}{1+i}$,

$$F_E \left(\frac{1+t}{1+i}\right) = P(1+i)^n \left[\left(\frac{1+t}{1+i}\right)^2 + \left(\frac{1+t}{1+i}\right)^3 \right.$$
$$\left. + \ldots + \left(\frac{1+t}{1+i}\right)^n + \left(\frac{1+t}{1+i}\right)^{n+1} \right] \quad \text{(6-F)}$$

Subtracting Eq. 6-E from Eq. 6-F:

$$F_E \left(\frac{1+t}{1+i}\right) - F_E = P(1+i)^n \left[-\left(\frac{1+t}{1+i}\right) + \left(\frac{1+t}{1+i}\right)^{n+1} \right]$$

$$F_E \left(\frac{1+t}{1+i} - 1\right) = P(1+i)^n \left[\left(\frac{1+t}{1+i}\right)^{n+1} - \left(\frac{1+t}{1+i}\right) \right]$$

$$F_E = P(1+i)^n \left[\frac{\left(\frac{1+t}{1+i}\right)^{n+1} - \left(\frac{1+t}{1+i}\right)}{\left(\frac{1+t}{1+i}\right) - 1} \right] \quad \text{(6-G)}$$

Equation 6-G may be rearranged to the following:

$$F_E = P(1+i)^n \left[\frac{\left(\frac{1+t}{1+i}\right)^{n+1} - 1}{\left(\frac{1+t}{1+i}\right) - 1} - 1 \right] \quad \text{(6-H)}$$

Setting $P = $ unity, the final equation for the compound amount of an exponential increasing series may be written:

$$E_tCA = (1 + i)^n \left[\frac{\left(\frac{1+t}{1+i}\right)^{n+1} - 1}{\left(\frac{1+t}{1+i}\right) - 1} - 1 \right] \qquad (6\text{-}10)$$

Dividing Eq. 6-10 by $(1 + i)^n$ will give the present worth of the accumulation:

$$E_tPW = \left[\frac{\left(\frac{1+t}{1+i}\right)^{n+1} - 1}{\left(\frac{1+t}{1+i}\right) - 1} - 1 \right] \qquad (6\text{-}11)$$

The equivalent uniform series amount may be obtained by multiplying the present-worth Eq. 6-11 by the capital recovery factor:

$$E_tUS = \left[\frac{\left(\frac{1+t}{1+i}\right)^{n+1} - 1}{\left(\frac{1+t}{1+i}\right) - 1} - 1 \right] \left[\frac{i(1+i)^n}{(1+i)^n - 1} \right] \qquad (6\text{-}12)$$

When $t = i$ the expression within the brackets common to Eqs. 6-10, 6-11, and 6-12 reduces to n as can readily be seen from the summation Eq. 6-E.

NOMINAL AND EFFECTIVE RATES OF INTEREST

Interest rates are nominally quoted on bank and other loans on the basis of interest per one year, even though the interest may be paid monthly, quarterly, or twice yearly. The interest rate per annum is the nominal rate; the effective rate is that rate corresponding to compounding the interest for the conversion periods of less than a year. The effective rate can be calculated for any conversion period.

DERIVATION OF THE EQUATION FOR FINDING THE EFFECTIVE RATE

The nominal and effective rates may be computed for certain conditions directly from standard compound interest tables. For instance, a nominal rate of 8 percent per annum amounts to 4 percent per period compounded semiannually, and 2 percent per period compounded quarterly. The effective rates would be found as follows for a principal amount of $100 at a nominal rate of 8 percent per year:

$$\begin{aligned}
\text{Annually, } F &= 100(1 + i)^n = 100(1.08)^1 = \$108.00 \\
\text{Semiannually, } F &= 100(1 + i)^n = 100(1.04)^2 = \$108.16 \\
\text{Quarterly, } F &= 100(1 + i)^n = 100(1.02)^4 = \$108.24
\end{aligned}$$

Compound Interest Equations

The effective rates per annum are thus 8.16 percent and 8.24 percent, respectively, for the semiannual and quarterly conversion compared to the nominal annual rate of 8.00 percent. In the above three cases, each scheme is based on an interest rate of 8 percent per year, or $8.00 per year, but the interest is paid (or collected) in one yearly payment, two semiannual payments, or four quarterly payments.

The rates of 8.00, 8.16, and 8.24 percent are equivalent because the effective rate of 8.16 percent for twice yearly payment of interest and the effective rate of 8.24 percent for quarterly payment of interest both result from the nominal rate of 8 percent for yearly payment of interest. These three rates merely reflect the effects of paying interest on three different schedules. Interest rates to be compared with each other must be based on the same base time period with interest payments on the same dates. When the interest is paid more frequently than indicated by the time period attached to the quoted interest rate, the effective rate will be higher than the nominal (quoted) rate. It is correct that such effective rate is higher than the quoted rate because when the interest is paid, say quarterly, the borrower is without the use of that much money for the remainder of the year. Being deprived of the use of that interest money reduces his working capital for the remainder of the year without reducing his total interest payments. Therefore in effect he is paying a higher rate of interest per year than the quoted, or nominal rate.

Based on the foregoing explanations it is in order to derive the equation for converting a nominal rate, say for one year and the year-end convention, to its effective rate for periods less than a year. The following notation applies:

i = interest rate per base conversion period, that is, the quoted interest rate which agrees with the interest payment dates
r = nominal rate per annum
j = effective rate per annum
m = times per year, or base period, the nominal rate is converted

It then follows that $i = r/m$ and $F = P(1 + i)^m$. Using the effective rate j, $F = P(1 + j)$ for one year. Therefore.

$$P(1 + j) = P(1 + i)^m$$
$$j = (1 + i)^m - 1 \qquad (6\text{-}13)$$

and since $i = \dfrac{r}{m}$,

$$j = \left(1 + \frac{r}{m}\right)^m - 1 \qquad (6\text{-}14)$$

That is, in terms of one year and a nominal interest rate per year of r, compounded m times a year, the rate i per conversion period is r/m and the number of periods is m. Generalized, we can write:

$$F = P(1 + i)^n \quad \text{or} \quad F = P\left(1 + \frac{r}{m}\right)^{mn} \qquad (6\text{-}15)$$

To illustrate Eq. 6-15, find the effective rate of interest for $100 for 1 year at a nominal rate of 12 percent per year, interest payable monthly:

$$F = 100\left(1 + \frac{0.12}{12}\right)^{(12)(1)} = 100(1.01)^{12} = 112.6825$$

$$j = \frac{112.6825 - 100}{100} = 0.126825 \text{ or } 12.68 \text{ percent}$$

EFFECTIVE RATES WHEN THE PRINCIPAL IS REDUCED PERIODICALLY

Not only is the interest on loans now paid on a monthly, quarterly, or semi-annual basis, but many financial loan or credit organizations require that the principal be repaid on a monthly basis. Usually one uniform monthly payment includes both interest and principal. A further complication is that the interest may be paid in advance, that is, the interest is deducted from the amount of the loan actually extended to the borrower, though the note signed is for the full amount. A third complication embraces charges other than for interest. These charges involve such things as service charges, recording of documents (such as the mortgage on an automobile), credit ratings, and life insurance premiums against death of the borrower.

Factors on Which the Effective Rate Depends

All of these practices vary from one loan agency to another, and from one geographical location to another. The result is that it is difficult to understand exactly what a loan will cost, or exactly what effective rate of interest per annum is being paid until all of the monetary facts are at hand and some calculations made. It is only by this process that a prospective borrower can make an intelligent decision of which loan offer serves his purpose the best and which one may be obtained at the least cost in money and at the lowest effective interest rate.

There are just three factors to consider in order that one may know his financial cost and which proposal offers the greatest economy in the use of his money. The first factor is the amount of money that is actually borrowed in net cash or the amount of credit that is advanced in the case of the charge account. The second factor is the total cost in dollars (for all items, including interest) that the borrower must pay for the loan of cash or for the availability of credit. The third factor is the time schedule over which the money is loaned or the credit advanced is repaid. The rate of interest quoted or used as a nominal rate in calculating the total interest to be paid is immaterial in calculating the true effective rate. The rate of interest, which is judged to be the true effective rate of interest, comes as a result of calculations involving the amount borrowed, the cost of borrowing the cash, and the time period over which the loan is extended.

Calculating the true effective rate of interest is somewhat complicated when the principal is repaid on a periodic basis more often than once a year such as the

Compound Interest Equations 99

now-popular monthly payment plan. Thus a principal of $100 is borrowed at 6 percent per annum, a note is signed for $106 and the monthly payment is $106 divided by 12, or $8.83.

On this loan the true effective rate of interest per annum is much greater than 6 percent for the reason that the borrower does not have the use of the full $100 for a full year. During the first month he does have the use of the full $100, but this principal reduces by $8.33 at the end of each month. For the twelfth month, the borrower has only the use of $8.33 of his borrowed $100. His average borrowed amount on a monthly basis is ($100 + 8.33) ÷ 2, or $54.16 for which he pays the interest at the rate of $0.50 a month for twelve months. The effective rate of interest is the 6 percent corrected upward to account for the average loan of only $54.16 and the fact that interest is paid monthly, not just once and at the year end.

Calculating Effective Rates

The true, or effective, rate per annum may be calculated as follows for loans repaid on a monthly (or other periodic) basis:

$$P = A(SPW \text{-} i \text{-} n)$$
$$\$100.00 = (106.00/12)(SPW \text{-} i\% \text{-} 12)$$
$$(SPW \text{-} i \text{-} 12) = 100.00 \div 8.833 = 11.3208$$

Solving by trial from interest tables:

$$(SPW \text{-} 1\% \text{-} 12) = 11.2551$$
$$(SPW \text{-} 5/6\% \text{-} 12) = 11.3745$$

By straight-line interpolation i is found to be 0.9082 percent or the rate for one month, which corresponds to a nominal rate of $(0.9082)(12) = 10.898$ percent per year, but with monthly payments of interest. By using Eq. 6-14 the effective rate j may be found:

$$j = \left(1 + \frac{0.10898}{12}\right)^{12} - 1$$
$$= (1.009082)^{12} - 1 = 1.11459 - 1 = 11.459 \text{ percent}$$

This solution for the effective rate is in two steps. First, the nominal rate per year was found to correct for the fact that the principal is repaid monthly, and, second, this nominal rate was adjusted to the equivalent rate based on one interest payment at the end of the year, rather than monthly.

An Equation for Finding the Approximate Effective Rate

An approximation of the effective rate of interest may be computed by the following equation:

$$r = \frac{2mQ}{P(n+1)} \qquad (6\text{-}16)$$

where r = approximate true effective rate of interest per annum

m = number of uniform periodic payments per year
Q = total cost of the loan–total of interest, service charges, recording fees, life insurance premiums, etc.
P = net cash received or net credit extended
n = number of uniform periodic payments over the life of the loan

Note that this equation does not include the amount of the monthly payment nor the "interest rate" used in computing the charges for rent on the money borrowed. The charges for any items other than for interest are included because they are as much a cost of the loan as is the interest itself. Further, the loan agency may quote a low "interest rate" and make up for its lowness by adding in other charges. An example of the use of the above formula follows:

1. Number of payments per year is 12.
2. The total cost of the loan is $6.00
3. The net cash received (borrowed) is $100.00
4. The number of uniform monthly payments is 12.

$$r = \frac{2(12)(6.00)}{(100.00)(12+1)} = \frac{144}{1300} = 0.1108, \text{ or } 11.08 \text{ percent}$$

This approximate rate of 11.08 percent compares with the true effective rate of 11.459 percent.

CONTINUOUS COMPOUNDING

The six standard compound interest equations are based upon three significant premises (1) step compounding, (2) the period-end convention, and (3) uniform (equal) periodic amounts within the series. Thus in compounding a single sum the amount of interest earned during the first period is added at the end of the period to the initial amount (principal) existing at the beginning of the first period to get the compound amount at the end of the first period. This end-of-first period sum is then compounded in a like manner to compute the compound amount at the end of the second period, and so on. The compound amount increases in lump sums instantly at the end of each successive period. This process is step compounding as contrasted to continuous compounding. Continuous compounding is compounding each moment of time so that the period-end steps are eliminated.

DERIVATION OF EQUATIONS FOR CONTINUOUS COMPOUNDING OF SINGLE SUMS

Equation 6-14 offers the means of determining the formula for continuous compounding–that is, compounding continuously and exponentially every moment throughout time. The six basic compound interest equations are based upon step compounding at the end of periods, and a rate of interest for the period. This same convention was used for the gradient series and the exponential series.

Compound Interest Equations

A result may be calculated by use of Eq. 6-14 which approximates continuous compounding. Starting with a true rate per annum of 8.00 percent, compounded annually, and then successively converting it to shorter conversion periods, j, the effective rate, is calculated from Eq. 6-14 with the results given in Table 6-2.

Note that in Table 6-2, j, the effective rate, increases as m increases, but at a decreasing rate, even though m increases rapidly. Thus if the compounding were at still shorter intervals say hourly, j would not increase much beyond that resulting from the daily frequency, or 365 compoundings a year. Continuous compounding at every moment in time would be the ultimate, and the corresponding effective rate j would be maximum. This maximum j can be calculated from Eq. 6-14,

$$j = \left(1 + \frac{r}{m}\right)^m - 1$$

when m becomes infinite.

From standard mathematics $e = 2.71828 +$ and e is defined as the limit approached by the term, $\left(1 + \frac{1}{k}\right)^k$, as k increases without limit. The symbol e is the base of the Napierian, or natural, logarithm system. The logarithm of e to base 10 (common logarithm) is 0.434 2944 819. The term $\left(1 + \frac{1}{k}\right)^k$ is in the same form as the term $\left(1 + \frac{r}{m}\right)^m$ in Eq. 6-14, so the effective interest rate j for continuous compounding may be found by relating $\left(1 + \frac{r}{m}\right)^m$ to $\left(1 + \frac{1}{k}\right)^k$.

From Eq. 6-1, a sum P will compound to F in n periods according to the equation, $F = P(1 + i)^n$. Therefore, substituting $\frac{r}{m}$ for i, the following expression for F may be written:

$$F = P\left(1 + \frac{r}{m}\right)^{mn} \tag{6-I}$$

for the compound sum for n periods when compounded m times a period. By letting $m/r = k$, $\frac{r}{m} = \frac{1}{k}$ and $m = rk$. Substituting in Eq. 6-I,

$$F_c = P\left(1 + \frac{1}{k}\right)^{rkn} = P\left[\left(1 + \frac{1}{k}\right)^k\right]^{rn} \tag{6-J}$$

But since $\left(1 + \frac{1}{k}\right)^k$ is equal to e as k increases without limit, the final expression for F_c under continuous compounding is

$$F_c = Pe^{rn} \tag{6-16}$$

Compound Interest Equations

TABLE 6-2

COMPARISON OF NOMINAL AND EFFECTIVE INTEREST RATES

Interest Rate, Percent	m = Times per Year Interest Is Compounded							
	1	2	4	8	12	52	365	∞
Nominal r	8	8	8	8	8	8	8	
Base period i	8	4	2	1	0.66667	0.15385	0.021918	→0.0
Effective j	8.0000	8.1600	8.2432	8.2857	8.30000	8.3220	8.3278	8.3287

When P = unity, we may write,

$$CCA = e^{rn} \tag{6-16A}$$

By letting P = unity, n = 1 year, and r = 8 percent, the value F_c is computed (by logarithms) to be 1.083287, from which j is to found to be 8.3287 percent. See Table 6-2 for comparison with the j value for other compounding times per year. For practical purposes, monthly compounding will approximate continuous compounding.

The present worth equation for a single sum is found from Eq. 6-16:

$$P_c = F\left(\frac{1}{e^{rn}}\right) \tag{6-17}$$

For F = unity,

$$CPW = \frac{1}{e^{rn}} \tag{6-17A}$$

DERIVATION OF EQUATIONS FOR CONTINUOUS COMPOUNDING OF UNIFORM SERIES OF SUMS

The equation $F_c = Pe^{rn}$ is the compounded accumulation of a single sum P over n periods (years) under continuous compounding. The concept for continuous compounding of a series includes a second feature. Not only is a given amount (single sum) continuously compounded, but under the series concept, there is a continuous stream of cash flow during the time period. Under the year-end step convention, the series cash flows of A amounts were lump sums at each period end. The concept of continuous compounding of a series is that infinitesimal amounts flow an infinite number of times during the time period, and each of these infinitesimal amounts compounds itself continuously.

For deriving the equation for the compound amount accumulation under continuous compounding let it be assumed that during one period (year) the total deposit will sum to R without interest and that there are m deposits of equal magnitude spread uniformly throughout the period at m infinitesimal intervals of time. Each deposit, therefore, is equal to R/m and over the n period of years mn deposits would be made. Figure 6-4 illustrates the cash-flow patterns.

Compound Interest Equations

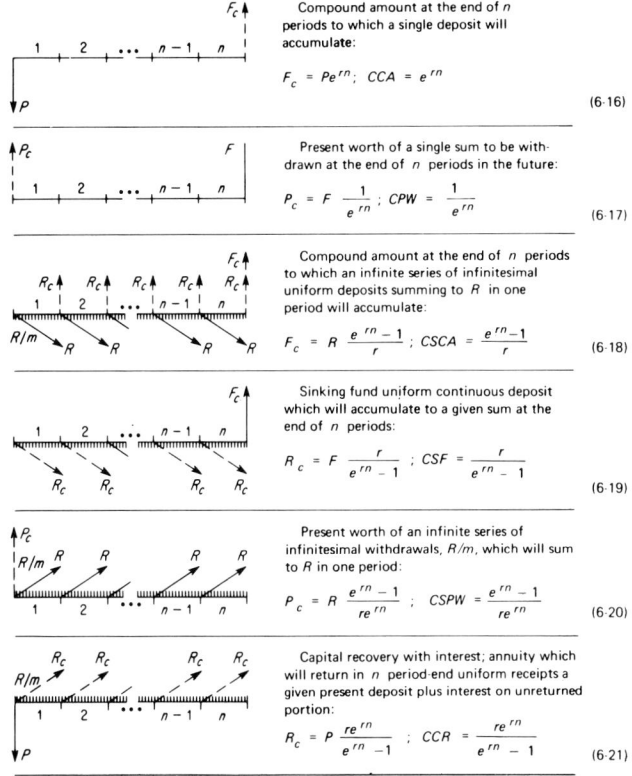

The cash-flow diagrams represent the position of the person owning the cash who deposits in or withdraws from an interest-bearing fund. Downward is outgo, or deposit, and upward is income, or withdrawal. Solid arrows represent known cash flows and dashed arrows represent unknown cash-flows, or solutions of the equations. Compounding is continuous and the uniform flows are infinitesimal in amount and infinite in number. The arrows at the beginning of the interest periods represent the R or R_c summations of the infinitesimal flows during the single period.

Compound amount at the end of n periods to which a single deposit will accumulate:

$$F_c = Pe^{rn}; \quad CCA = e^{rn} \quad (6\text{-}16)$$

Present worth of a single sum to be withdrawn at the end of n periods in the future:

$$P_c = F \frac{1}{e^{rn}}; \quad CPW = \frac{1}{e^{rn}} \quad (6\text{-}17)$$

Compound amount at the end of n periods to which an infinite series of infinitesimal uniform deposits summing to R in one period will accumulate:

$$F_c = R \frac{e^{rn}-1}{r}; \quad CSCA = \frac{e^{rn}-1}{r} \quad (6\text{-}18)$$

Sinking fund uniform continuous deposit which will accumulate to a given sum at the end of n periods:

$$R_c = F \frac{r}{e^{rn}-1}; \quad CSF = \frac{r}{e^{rn}-1} \quad (6\text{-}19)$$

Present worth of an infinite series of infinitesimal withdrawals, R/m, which will sum to R in one period:

$$P_c = R \frac{e^{rn}-1}{re^{rn}}; \quad CSPW = \frac{e^{rn}-1}{re^{rn}} \quad (6\text{-}20)$$

Capital recovery with interest; annuity which will return in n period-end uniform receipts a given present deposit plus interest on unreturned portion:

$$R_c = P \frac{re^{rn}}{e^{rn}-1}; \quad CCR = \frac{re^{rn}}{e^{rn}-1} \quad (6\text{-}21)$$

FIG. 6-4. Cash-flow diagrams and the six equations for continuous compounding over a series of n years, or periods.

R may be thought of as the total maintenance expense or the total traffic flow during a year and R/m as a uniform flow of the maintenance expense or traffic during the year which will accumulate to R at the end of the year without any compounding.

R_c is the accumulation of the infinite number of the R/m deposits during the period plus continuously compounded interest thereon. F_c is the similar accumulation for the full n periods, or its equivalent, the summation with continuously compounded interest of the n number of R_c yearly accumulations. The derivation will be made in terms of R/m for the full n periods consisting of mn total number of R/m deposits during the n periods. In each period (year) there are m number of intervals and m number of R/m deposits.

In reverse time order the R/m deposits with their compound interest may be expressed as follows, using the form of Eq. 6-15:

Last, or mnth, deposit	R/m (compounds no interest)
Next to last, or $(mn-1)$th deposit	$R/m\left(1+\dfrac{r}{m}\right)^1$
Second from last, or $(mn-2)$nd deposit	$R/m\left(1+\dfrac{r}{m}\right)^2$
. . . .	. . . .
Third deposit	$R/m\left(1+\dfrac{r}{m}\right)^{mn-3}$
Second deposit	$R/m\left(1+\dfrac{r}{m}\right)^{mn-2}$
First deposit	$R/m\left(1+\dfrac{r}{m}\right)^{mn-1}$

The above expressions for the first, second, and third deposits are based on the first deposit being made at the end of the first R/m interval rather than at time zero. This concept is probably correct in terms of highway operations, but in deposits to an interest-bearing fund the first deposit would probably be made at time zero. Fortunately, either concept leads to the same final equation. The derivation is somewhat simpler, though, with the interval-end concept as used.

The individual interval deposits with their compound interest may be summed as follows (the middle two of the six terms in the foregoing example are omitted):

$$F_c = \frac{R}{m} + \frac{R}{m}\left(1+\frac{r}{m}\right)^1$$
$$+ \ldots + \frac{R}{m}\left(1+\frac{r}{m}\right)^{mn-2} + \frac{R}{m}\left(1+\frac{r}{m}\right)^{mn-1} \quad (6\text{-}K)$$

Multiplying by $\left(1+\dfrac{r}{m}\right)$,

$$F_c\left(1+\frac{r}{m}\right) = \frac{R}{m}\left(1+\frac{r}{m}\right)^1 + \frac{R}{m}\left(1+\frac{r}{m}\right)^2$$
$$+ \ldots + \frac{R}{m}\left(1+\frac{r}{m}\right)^{mn-1} + \frac{R}{m}\left(1+\frac{r}{m}\right)^{mn} \quad (6\text{-}L)$$

Subtracting 6-K from 6-L,

$$F_c\left[\left(1+\frac{r}{m}\right)-1\right] = -\frac{R}{m} + \frac{R}{m}\left(1+\frac{r}{m}\right)^{mn}$$

$$F_c = \frac{\dfrac{R}{m}\left[\left(1+\dfrac{r}{m}\right)^{mn}-1\right]}{r/m} = \frac{R\left[\left(1+\dfrac{r}{m}\right)^{mn}-1\right]}{r}$$

$$F_c = R\left[\frac{\left(1+\dfrac{r}{m}\right)^{mn}}{r} - \frac{1}{r}\right]$$

Compound Interest Equations

$$\lim_{m \to \infty} F_c = R \left(\frac{e^{rn} - 1}{r} \right) \qquad (6\text{-}18)$$

For an R of unity:

$$CSCA = \frac{e^{rn} - 1}{r} \qquad (6\text{-}18\text{A})$$

The sinking-fund equation CSF may be written as the reciprocal of the $CSCA$ equation.

$$CSF = \frac{r}{e^{rn} - 1} \qquad (6\text{-}19)$$

The present-worth equation $CSPW$ may be obtained by multiplying the $CSCA$ factor by $1/e^{rn}$, the single sum present-worth factor, Eq. 6-17A:

$$CSPW = \frac{e^{rn} - 1}{re^{rn}} \qquad (6\text{-}20)$$

The capital-recovery factor CCR is the reciprocal of the series present-worth factor:

$$CCR = \frac{re^{rn}}{e^{rn} - 1} \qquad (6\text{-}21)$$

Equation 6-18 is the summation of the infinite number of the infinitesimal R/m deposits from time zero to the end of the nth period. For any one period, such as the 0-1 time period, the summation would be

$$F_{CR} = R \left(\frac{e^r - 1}{r} \right) \qquad (6\text{-}22)$$

which is Eq. 6-18 when $n = 1$.

It is to be remembered that the foregoing equations for $CSCA$, CSF, $CSPW$, and CCR, for continuous compounding of a continuous flow (deposits) of infinitesimal amounts are based upon the concept that this flow of principal infinitesimal amounts would, at the end of any one period, sum to unity without compounded interest. Thus, the equations are applicable in the same concept as are the period-end step equations, except for the features of continuous compounding and continuous infinitesimal flow.

Tables E-1 and E-2 of Appendix E give the factors for the six equations using continuous compounding. Table E-1 factors are the solutions for nominal integral interest rates that correspond to effective rates slightly greater than the integral rates used in calculating the factors. In Table E-2, the rates used in the calculations of the six factors are those nonintegral rates corresponding to their respective integral rates.

FURTHER EXPLANATION OF CONTINUOUS COMPOUNDING

Table 6-3 gives a comparison of the six compound interest factors for step compounding and for continuous compounding for an 8 percent interest rate and $n = 1$ and $n = 10$. Table 6-4 gives the effective continuous compounding rates corresponding to a series of nominal rates from 1 to 100 percent. These effective rates were calculated from the equation $F_c = Pe^{rn}$, when P and n are unity, and $j = F_c - 1$. It is shown in Table 6-4 that as the nominal rate of interest increases, the effective rate under continuous compounding increases at an increasing rate.

TABLE 6-3

COMPARISON OF FACTORS UNDER STEP COMPOUNDING AND CONTINUOUS COMPOUNDING

Factor	n	Step Compounding $i = 8$ Percent	Continuous Compounding $r = 8$ Percent
CA and CCA	$n = 1$	1.080 000	1.083 287
	$n = 10$	2.158 925	2.225 541
PW and CPW	$n = 1$	0.925 926	0.923 116
	$n = 10$	0.463 193	0.449 329
SCA and CSCA	$n = 1$	1.000 000	1.041 100
	$n = 10$	14.486 562	15.320 914
SF and CSF	$n = 1$	1.000 000	0.960 522
	$n = 10$	0.069 029	0.065 2703
SPW and CSPW	$n = 1$	0.925 926	0.961 045
	$n = 10$	6.710 081	6.883 386
CR and CCR	$n = 1$	1.080 000	1.040 534
	$n = 10$	0.149 029	0.145 277

The left-hand section of Table 6-4 gives the effective rate of interest (column 2) under continuous compounding that results from using integral nominal rates (column 1). In column 3, the nominal rates correspond to integral effective rates in column 4. Thus, when using the nominal rates of column 3 in continuous compounding the effective rates would be integral as shown in column 4. For single sums, the calculations of compound interest by continuous compounding using the nominal rates of column 3 would produce results identical with those obtained by using the effective rates in column 4 and the period-end step convention. Thus for $n = 10$, $r = 7.696014$ percent and $j = 8$ percent.

$$e^{rn} = (1 + j)^n; \quad e^{0.769\,6014} = (1.08)^{10}$$

$$\log_{10} e^{rn} = 0.769\,6014 (\log e) = 0.769\,6014 (0.4342945)$$

$$\log_{10} e^{rn} = 0.334\,2336\,552$$

$$e^{rn} = 2.1589$$

Compound Interest Equations

TABLE 6-4
Nominal and Effective Interest Rates under Continuous Compounding

Nominal Rate = r, Percent	Effective Rate = j, Percent	Nominal Rate = r, Percent	Effective Rate = j, Percent	$R_c = \dfrac{e^r - 1}{r}$ for r of Column 3
1	2	3	4	5
1	1.00501 67084	0.99503 30853	1	1.0049 9170
2	2.02013 40027	1.98026 27296	2	1.0099 6700
3	3.04545 33954	2.95588 02242	3	1.0149 2610
4	4.08107 74192	3.92207 13153	4	1.0198 6927
5	5.12710 96376	4.87901 64169	5	1.0247 9672
6	6.18365 46545	5.82689 08124	6	1.0297 0867
7	7.25081 81254	6.76586 48474	7	1.0346 0535
8	8.32870 67675	7.69610 41136	8	1.0394 8698
9	9.41742 83705	8.61776 96241	9	1.0443 5375
10	10.51709 18076	9.53101 79804	10	1.0492 0587
12	12.74968 51579	11.33286 85307	12	1.0588 6696
15	16.18342 42728	13.97619 42375	15	1.0732 5354
17	18.53048 51320	15.70037 48810	17	1.0827 7669
20	22.14027 58160	18.23215 56794	20	1.0969 6299
25	28.40254 16688	22.31435 51314	25	1.1203 5503
30	34.98588 07576	26.23642 64467	30	1.1434 4841
35	41.90675 48593	30.01045 92450	35	1.1662 6006
40	49.18246 97641	33.64722 36621	40	1.1888 0536
45	56.83121 85490	37.15635 56432	45	1.2110 9832
50	64.87212 70700	40.54651 08108	50	1.2331 5173
55	73.32530 17867	43.82549 30931	55	1.2549 7732
60	82.21188 00391	47.00036 29246	60	1.2765 8589
65	91.55408 29014	50.07752 87912	65	1.2979 8737
70	101.37527 07470	53.06282 51062	70	1.3191 9098
75	111.70000 16613	55.96157 87935	75	1.3402 0522
80	122.55409 28492	58.77866 64902	80	1.3610 3802
90	145.96031 11157	64.18538 86172	90	1.4021 8829
100	171.82818 28459	69.31471 80560	100	1.4426 9506

From standard interest tables $(1 + j)^{10} = (1.08)^{10} = 2.1589$ (checks the above 2.1589).

But when a continuous flow is continuously compounded, the comparison with step compounding is different. Again using $n = 10$, $r = 7.696\,014$·percent, and $j = 8$ percent,

$$\frac{e^{rn}-1}{r} \neq \frac{(1+j)^n-1}{j}$$

$$\frac{e^{0.769\ 6014}-1}{0.0769\ 6014} \neq \frac{(1.08)^{10}-1}{0.08}$$

$$\frac{2.1589-1}{0.0769\ 6014} \neq 14.4866 \text{ (from tables)}$$

$$15.0584 \neq 14.4866$$

This lack of equality is because of the fact that for step compounding factors the A uniform amount is unity and the R_c under continuous compounding is $\frac{e^r-1}{r}$, or, for the example at hand, from Table 6-4, $R_c = 1.039\ 487$. The R_c is more than unity by the accumulation of the continuously compounded interest. Thus $(14.4866)(1.039487) = 15.0586$, which agrees (within calculation accuracy) with the 15.0584 obtained from the CCA factor.

Under continuous compounding, this $\frac{e^r-1}{r}$ factor, it will be recalled, is the sum that the infinite number of infinitesimal deposits flowing uniformly throughout the first period accumulate to at the end of the period. This factor provides the main difference between year-end step compounding and continuous compounding of a series.

DERIVATION OF EQUATIONS FOR CONTINUOUS COMPOUNDING OF A GRADIENT SERIES

The equations in Fig. 6-4 apply continuous compounding to single sums and to uniform series. The step period-end convention is applied to a gradient-increasing series as illustrated in Fig. 6-2. In a similar way continuous compounding will be applied to the gradient-increasing function.

Figure 6-5 shows the cash-flow diagram for the compound amount of a gradient over n periods compounded continuously. The concept is that during the first period (year) the deposits will be made in uniformly increasing infinitesimal amounts such that at the end of this first period a total sum G will have been deposited. Each of these infinite number of deposits will continuously compound itself. The continuous stream of deposits will continue to increase in amount each period at the rate of G for each period. The total of deposits will be $G, 2G, 3G, ..., nG$, respectively for the first, second, third, ..., and nth periods. Note that the infinitesimal flows increase as a true gradient over the n periods. These flows each period are m in number for m intervals for each of the n periods.

An examination of illustrative cash flows with low finite values of m, such as 2 to 6, and a specific value of G, such as 1.0, will disclose that an initial deposit at time $m = 0 = n = 0$ is required to which the gradient increase is added. Both the initial deposit and the gradient deposit for each m interval depend upon m.

Compound Interest Equations

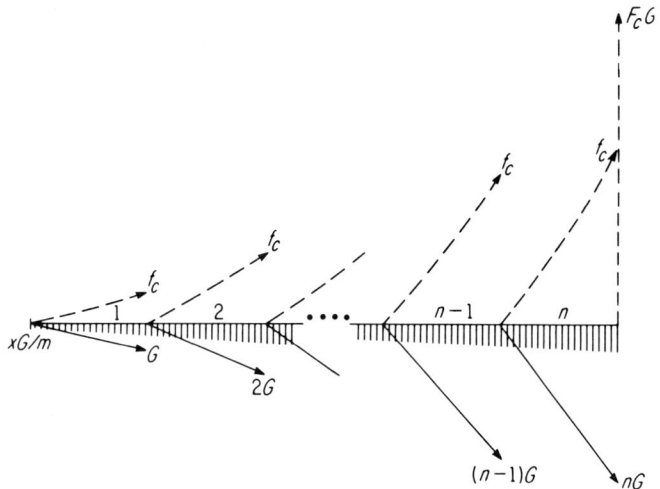

FIG. 6-5. Cash-flow diagrams for the compound amount of a uniformly (gradient) increasing series under continuous compounding.

This dependency may be derived as follows when the initial deposit is p and the gradient deposit increases each interval by the amount d.

For the first period from time $n = 0$ to $n = 1$ the summation of the m deposits, omitting the initial (or existing) deposit at time $m = 0$, will be

$$G_1 = (p + d) + (p + 2d) + \ldots + (p + id) + \ldots + (p + md) = G \quad \text{(6-M)}$$

The terms in this summation equation form a straight-line gradient, so their sum is one-half of the number of individual deposits times the sum of the first and last deposits, or

$$G_1 = \frac{m}{2}\left[(p + d) + (p + md)\right] = \frac{m}{2}\left[2p + d(m + 1)\right] = G$$

For the second and third periods, $n = 1$ to $n = 2$ and $n = 2$ to $n = 3$, the deposits will sum to

$$G_2 = \frac{m}{2}\left[2p + d(3m + 1)\right] = 2G$$

$$G_3 = \frac{m}{2}\left[2p + d(5m + 1)\right] = 3G$$

And for the ith period and ith $+ k$ period the summations will be:

$$G_i = \frac{m}{2}\left[2p + d\left[(2i - 1)m + 1\right]\right] = iG \quad \text{(6-N)}$$

$$G_{i+k} = \frac{m}{2}\left[2p + d\left[(2i + 2k - 1)m + 1\right]\right] = (i + k)G \quad \text{(6-O)}$$

Subtracting Eq. 6-N from Eq. 6-O gives

$$\frac{m}{2}[(2dk)m] = (i+k)G - iG = kG \tag{6-P}$$

$$m^2 d = G \text{ or } d = \frac{G}{m^2} \tag{6-PA}$$

Substituting G/m^2 for d in Eq. 6-N gives

$$\frac{m}{2}\left[2p + \frac{G}{m^2}(2i-1)m + \frac{G}{m^2}\right] = iG$$

$$2pm^2 + 2iGm - Gm + G = 2iGm$$

$$p = \frac{G}{2}\left(\frac{m-1}{m^2}\right) = \frac{G(m-1)}{2m^2} \tag{6-Q}$$

Thus the initial deposit p at time $m = 0 = n = 0$ is $\frac{G(m-1)}{2m^2}$ and the gradient deposit is $\frac{G}{m^2}$.

Since the rate of compounding is r per period, the rate is r/m for each interval within each period. The compound amount factor for the first interval is, therefore, $\left(1 + \frac{r}{m}\right)^m$ and $\left(1 + \frac{r}{m}\right)^{mn}$ for n periods. The next step is to set up the summation equation for the continuous compounding of the continuous deposits which were summed in Eq. 6-M.

$$F_{CG} = (p+d)\left(1 + \frac{r}{m}\right)^{mn-1} + (p+2d)\left(1 + \frac{r}{m}\right)^{mn-2}$$

$$+ \ldots + (p+id)\left(1 + \frac{r}{m}\right)^{mn-i} + \ldots + [p+(mn-1)d]\left(1 + \frac{r}{m}\right)^{mn-(mn-1)}$$

$$+ (p+mnd)\left(1 + \frac{r}{m}\right)^{mn-mn} \tag{6-R}$$

The compound amount for the deposit made at the end of the last interval before the ith period is

$$[p+(i-1)md]\left(1 + \frac{r}{m}\right)^{mn-(i-1)m}$$

and the general term for the jth interval in the ith period is

$$[p+[(i-1)m+j]d]\left(1 + \frac{r}{m}\right)^{mn-(i-1)m-j} \tag{6-S}$$

There are two independent simultaneous summations to account for–the i series for n periods and the j series for m intervals in each period, which may be written,

Compound Interest Equations

$$F_{CG} = \sum_{i=1}^{n} \sum_{j=1}^{m} \left[p + [(i-1)m + j]d \right] \left(1 + \frac{r}{m}\right)^{mn-(i-1)m-j}$$

$$= \sum_{i=1}^{n} \sum_{j=1}^{m} \left[(p - md) + (imd + jd) \right] \left(1 + \frac{r}{m}\right)^{mn-m} \left(1 + \frac{r}{m}\right)^{-im-j}$$

$$= (p - md)\left(1 + \frac{r}{m}\right)^{mn-m} \sum_{i=1}^{n} \sum_{j=1}^{m} \left(1 + \frac{r}{m}\right)^{-im-j} \quad \text{(A)(6-T)}$$

$$+ md \left(1 + \frac{r}{m}\right)^{mn-m} \sum_{i=1}^{n} i \left(1 + \frac{r}{m}\right)^{-im} \sum_{j=1}^{m} \left(1 + \frac{r}{m}\right)^{-j} \quad \text{(B)(6-T)}$$

$$+ d \left(1 + \frac{r}{m}\right)^{mn-m} \sum_{i=1}^{n} \left(1 + \frac{r}{m}\right)^{-im} \sum_{j=1}^{m} j \left(1 + \frac{r}{m}\right)^{-j} \quad \text{(C)(6-T)}$$

Let

$$\left(1 + \frac{r}{m}\right)^{-1} = X; \quad X^m = Y; \quad X = \frac{m}{m+r}; \quad (1-X) = \frac{r}{m+r}$$

and

$$\frac{X}{1-X} = \frac{m}{r}$$

From prior derivations (6-Q and 6-P), $p = \dfrac{G(m-1)}{2m^2}$ and $d = G/m^2$. Then Eq. 6-T may be rewritten as

$$F_{CG} = \left[\frac{G(m-1)}{2m^2} - \frac{G}{m}\right] \left[Y^{-(n+1)}\right] \sum_{i=1}^{n} Y^i \sum_{j=1}^{m} X^j \quad \text{(A)(6-U)}$$

$$+ \frac{G}{m}\left[Y^{-(n+1)}\right] \sum_{i=1}^{n} \sum_{j=1}^{m} i Y^i X^j \quad \text{(B)(6-U)}$$

$$+ \frac{G}{m^2}\left[Y^{-(n+1)}\right] \sum_{i=1}^{n} \sum_{j=1}^{m} j Y^i X^j \quad \text{(C)(6-U)}.$$

These three terms will be reduced separately beginning with A.

The A term is composed of two geometric series of the form $a + ab + ab^2 \ldots ab^{n-1}$, whose sum is $\dfrac{a(1-b^n)}{1-b}$, so the A term of Eq. 6-U may be written,

$$A = -\frac{G(m+1)}{2m^2}\left(Y^{-(n+1)}\right)\left(\frac{Y(1-Y^n)}{1-Y}\right)\left(\frac{X(1-X^m)}{1-X}\right)$$

$$A = -\frac{G(m+1)}{2m^2}\left(Y^{-(n+1)}\right)\frac{m}{r}Y(1-Y^n)$$

$$A = -\left(\frac{G(m+1)}{2m^2}\right)\frac{m}{r}\left(Y^{-n}-1\right) = -\frac{G(m+1)}{2rm}\left(Y^{-n}-1\right) \quad (6\text{-V})$$

In the B term the summation of the i series is of the form $z^1 + 2z^2 + 3z^3 \ldots nz^n$ so it will sum to[2]

$$\frac{z - (n+1)z^{n+1} + nz^{n+2}}{(1-z)^2}$$

Therefore the term B may be written

$$B = \frac{G}{m}\left(Y^{-(n+1)}\right)\left[\frac{Y - (n+1)Y^{n+1} + nY^{n+2}}{(1-Y)^2}\right]\left[\frac{X(1-X^m)}{1-X}\right]$$

$$B = \frac{G}{m}\left(Y^{-(n+1)}\right)\left[\frac{Y - Y^{n+1} - nY^{n+1} + nY^{n+2}}{(1-Y)^2}\right]\frac{m}{r}(1-Y)$$

$$B = \frac{G}{r}\left[\frac{Y^{-n} - 1 - n + nY}{1 - Y}\right] \quad (6\text{-W})$$

Similarly for the C term,

$$C = \frac{G}{m^2}\left(Y^{-(n+1)}\right)\left[\frac{Y(1-Y^n)}{1-Y}\right]\left[\frac{X - (m+1)X^{m+1} + mX^{m+2}}{(1-X)^2}\right]$$

$$C = \frac{G}{m^2}\left(\frac{Y^{-n}-1}{1-Y}\right)\left[\frac{X - (m+1)XY + mX^2Y}{(1-X)^2}\right]$$

$$C = \frac{G}{m^2}\left(\frac{Y^{-n}-1}{1-Y}\right)\left[\frac{X(1-Y)}{(1-X)^2} - \frac{mXY(1-X)}{(1-X)^2}\right]$$

$$C = \frac{G}{m^2}\left(\frac{Y^{-n}-1}{1-Y}\right)\left[\left(\frac{X}{1-X}\right)\left(\frac{1-Y}{1-X}\right) - \frac{mXY}{1-X}\right]$$

$$C = \frac{G}{m^2}\left(\frac{Y^{-n}-1}{1-Y}\right)\left[\frac{m}{r}\left(\frac{m+r}{r}\right)(1-Y) - \frac{m^2Y}{r}\right]$$

$$C = \frac{G}{m^2}\left(\frac{Y^{-n}-1}{1-Y}\right)\left[\left(\frac{m(m+r)}{r^2}\right)(1-Y) - \frac{m^2Y}{r}\right]$$

$$C = \frac{G}{r^2}\left(Y^{-n}-1\right) + \frac{G}{mr}\left(Y^{-n}-1\right) - \frac{G}{r}\left(Y^{-n}-1\right)\left(\frac{Y}{1-Y}\right) \quad (6\text{-X})$$

---

[2] Proof that the term within the brackets correctly sums the series $\sum_{i=1}^{mn} iX^i$ as i varies from 1 to mn may be had by solving a specific example. Letting $mn = 6$, the bracketed term in Eq. 6-P may be developed as follows:

$$\frac{X - (6+1)X^7 + 6X^8}{1 - 2X + X^2} = X + 2X^2 + 3X^3 + 4X^4 + 5X^5 + 6X^6$$

A similar substitution of $mn = 6$ in the bracketed term of Eq. 6-U yields an identical result:

$$X + 2X^2 \ldots 5X^5 + 6X^6$$

Compound Interest Equations

Putting terms A, B, and C together,

$$F_{CG} = -\frac{G(m+1)}{2rm}\left(Y^{-n}-1\right) + \frac{G}{r}\left(\frac{Y^{-n}-1-n+nY}{1-Y}\right) + \frac{G}{r^2}\left(Y^{-n}-1\right)$$

$$+ \frac{G}{mr}\left(Y^{-n}-1\right) - \frac{G}{r}\left(Y^{-n}-1\right)\left(\frac{Y}{1-Y}\right)$$

$$F_{CG} = -\frac{G}{2r}\left(Y^{-n}-1\right) - \frac{G}{2rm}\left(Y^{-n}-1\right) + \frac{G}{r}\left(\frac{Y^{-n}-1}{1-Y}\right) - \frac{G}{r}\left(\frac{1-Y}{1-Y}\right)$$

$$+ \frac{G}{r^2}\left(Y^{-n}-1\right) + \frac{G}{mr}\left(Y^{-n}-1\right) - \frac{G}{r}\left(Y^{-n}-1\right)\left(\frac{Y}{1-Y}\right)$$

$$F_{CG} = \frac{G}{r}\left(Y^{-n}-1\right)\left(-\frac{1}{2}-\frac{1}{2m}+\frac{1}{1-Y}+\frac{1}{r}+\frac{1}{m}-\frac{Y}{1-Y}\right) - \frac{nG}{r}$$

$$= \frac{G}{r}\left[\left(1-\frac{r}{m}\right)^{mn}-1\right]\left(-\frac{1}{2}+\frac{1}{r}+1-\frac{1}{2m}+\frac{1}{m}\right) - \frac{nG}{r}$$

$$\lim_{m\to\infty} F_{CG} = \frac{G}{r}\left(e^{rn}-1\right)\left(\frac{1}{2}+\frac{1}{r}\right) - \frac{nG}{r}$$

$$F_{CG} = \frac{G}{r}\left(e^{rn}-1\right)\left(\frac{1}{2}+\frac{1}{r}\right) - \frac{nG}{r} \tag{6-23}$$

$$CGCA = \left(\frac{e^{rn}-1}{r}\right)\left(\frac{1}{2}+\frac{1}{r}\right) - \frac{n}{r} \tag{6-23A}$$

The compound amount Eq. 6-23 may be converted to the present worth equation by multiplying by the single-sum compound amount factor $1/e^{rn}$, Eq. 6-17:

$$P_{CG} = \frac{G}{r}\left(\frac{e^{rn}-1}{e^{rn}}\right)\left(\frac{1}{2}+\frac{1}{r}\right) - \frac{nG}{re^{rn}} \tag{6-24}$$

$$CGSPW = \left(\frac{e^{rn}-1}{re^{rn}}\right)\left(\frac{1}{2}+\frac{1}{r}\right) - \frac{n}{re^{rn}} \tag{6-24A}$$

The capital recovery equation A_{CG} is obtained by multiplying the compound amount factor by the sinking fund factor, Eq. 6-19,

$$A_{CG} = G\left(\frac{1}{2}+\frac{1}{r}\right) - \left(\frac{nG}{e^{rn}-1}\right) \tag{6-25}$$

$$CGUS = \left(\frac{1}{2}+\frac{1}{r}\right) - \left(\frac{n}{e^{rn}-1}\right) \tag{6-25A}$$

In the process of deriving Eq. 6-23 for the compound amount of a gradient under continuous compounding the equation was derived for a uniform equal-size flow of the infinite number of infinitesimal deposits during each period. It is

$$F_{CGS} = G\left[\frac{e^{r(n+1)}-(n+1)e^r+n}{r(e^r-1)}\right] \tag{6-26}$$

CONTINUOUS COMPOUNDING OF AN EXPONENTIAL GROWTH SERIES

Continuous compounding can be applied to exponential growth in much the same manner as to a uniform series and to a gradient growth. The following derivations of the equations for the compound amount, present worth, and capital recovery (equivalent uniform amount factor), follow the basic procedure of the other derivations in Chapter 6. Figure 6-6 illustrates the cash flows and basic concepts.

At time zero ($n = 0$), the flow of deposits would be at the rate of P per period. The quantity P is the base for the t rate of growth per period. This rate of flow would compound itself such that at the end of the first period the rate of flow would be $P(1 + t)$ per period, and so on to the end of the nth period when the flow rate would be $P(1 + t)^n$ per period. During each period the rate of flow continuously increases such that by the end of the period, the accumulation of the infinite number of the infinitesimal series of deposits, without compound interest, would sum to $P(1 + t)$, $P(1 + t)^2$, ..., $P(1 + t)^n$, respectively, period by period, 0 to n.

The number of these infinitesimal deposits per period will be designated by m and the number of compounding intervals within a period will also be m. Therefore, the first infinitesimal deposit will be P/m and the compounding growth rate will be t/m. The nominal continuous compounding rate is designated as r per period, or r/m per infinitesimal interval. The first deposit is considered to be made at the end of the first interval on the basis that the P rate of deposit prevails at time zero and it will take the first interval of time to generate the first

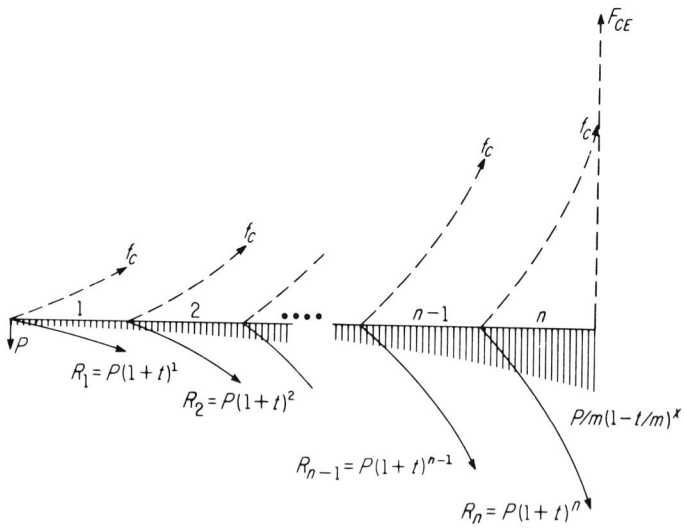

FIG. 6-6. Cash-flow diagram for the compound amount of an exponentially growing series under continuous compounding.

Compound Interest Equations   115

P/m deposit. Over the n periods the total number of deposits will be mn and there will be mn intervals of time elapsed.

The equation for summing the series of growing infinitesimal deposits to the end of n periods may be written:

$$F_{CE} = \frac{P}{m}\left(1 + \frac{t}{m}\right)^1 \left(1 + \frac{r}{m}\right)^{mn-1} + \frac{P}{m}\left(1 + \frac{t}{m}\right)^2 \left(1 + \frac{r}{m}\right)^{mn-2}$$

$$+ \ldots + \frac{P}{m}\left(1 + \frac{t}{m}\right)^{mn-1} \left(1 + \frac{r}{m}\right)^1 + \frac{P}{m}\left(1 + \frac{t}{m}\right)^{mn} \left(1 + \frac{r}{m}\right)^0$$

$$F_{CE} = \frac{P}{m}\left(1 + \frac{r}{m}\right)^{mn} \left[\left(\frac{1+\frac{t}{m}}{1+\frac{r}{m}}\right)^1 + \left(\frac{1+\frac{t}{m}}{1+\frac{r}{m}}\right)^2 \right.$$

$$\left. + \ldots + \left(\frac{1+\frac{t}{m}}{1+\frac{r}{m}}\right)^{mn-1} + \left(\frac{1+\frac{t}{m}}{1+\frac{r}{m}}\right)^{mn}\right] \quad (6\text{-}Y)$$

Multiplying by the common ratio $\dfrac{1+\frac{t}{m}}{1+\frac{r}{m}}$,

$$F_{CE}\left(\frac{1+\frac{t}{m}}{1+\frac{r}{m}}\right) = \frac{P}{m}\left(1 + \frac{r}{m}\right)^{mn} \left[\left(\frac{1+\frac{t}{m}}{1+\frac{r}{m}}\right)^2 + \left(\frac{1+\frac{t}{m}}{1+\frac{r}{m}}\right)^3 \right.$$

$$\left. + \ldots + \left(\frac{1+\frac{t}{m}}{1+\frac{r}{m}}\right)^{mn} + \left(\frac{1+\frac{t}{m}}{1+\frac{r}{m}}\right)^{mn+1}\right] \quad (6\text{-}Z)$$

Subtracting Eq. 6-Y from 6-Z,

$$F_{CE}\left(\frac{1+\frac{t}{m}}{1+\frac{r}{m}}\right) - F_{CE} = \frac{P}{m}\left(1 + \frac{r}{m}\right)^{mn} \left[\left(\frac{1+\frac{t}{m}}{1+\frac{r}{m}}\right) + \left(\frac{1+\frac{t}{m}}{1+\frac{r}{m}}\right)^{mn+1}\right]$$

$$F_{CE}\left(\frac{1+\frac{t}{m}}{1+\frac{r}{m}} - 1\right) = \frac{P}{m}\left(1 + \frac{r}{m}\right)^{mn} \left[\left(\frac{1+\frac{t}{m}}{1+\frac{r}{m}}\right)^{mn+1} - \left(\frac{1+\frac{t}{m}}{1+\frac{r}{m}}\right)\right]$$

$$F_{CE} = -\dfrac{\dfrac{P}{m}\left(1+\dfrac{r}{m}\right)^{mn}\left(\dfrac{1+\dfrac{t}{m}}{1+\dfrac{r}{m}}\right)\left[\left(\dfrac{1+\dfrac{t}{m}}{1+\dfrac{r}{m}}\right)^{mn}-1\right]}{\dfrac{1+\dfrac{t}{m}}{1+\dfrac{r}{m}}-1}$$

Multiplying and dividing by $\dfrac{1+\dfrac{r}{m}}{1+\dfrac{t}{m}}$,

$$F_{CE} = \dfrac{\dfrac{P}{m}\left(1+\dfrac{r}{m}\right)^{mn}\left[\left(\dfrac{1+\dfrac{t}{m}}{1+\dfrac{r}{m}}\right)^{mn}-1\right]}{\left(\dfrac{1+\dfrac{t}{m}}{1+\dfrac{r}{m}}-1\right)\left(\dfrac{1+\dfrac{r}{m}}{1+\dfrac{t}{m}}\right)} = \dfrac{\dfrac{P}{m}\left[\left(1+\dfrac{t}{m}\right)^{mn}-\left(1+\dfrac{r}{m}\right)^{mn}\right]}{1-\left(\dfrac{1+\dfrac{r}{m}}{1+\dfrac{t}{m}}\right)}$$

$$F_{CE} = \dfrac{P\left[\left(1+\dfrac{t}{m}\right)^{mn}-\left(1+\dfrac{r}{m}\right)^{mn}\right]}{m\left(\dfrac{t-r}{m+t}\right)}$$

$$F_{CE} = \dfrac{P\left[\left(1+\dfrac{t}{m}\right)^{mn}-\left(1+\dfrac{r}{m}\right)^{mn}\right]}{\dfrac{t-r}{1+\dfrac{t}{m}}}$$

The limit of F_{CE} as $m \to \infty$ is equal to the difference of the limits of the two terms in the numerator divided by the limit of the term in the denominator, provided there is no division by zero. Therefore

$$\lim_{m \to \infty} = F_{CE} = P\left(\dfrac{e^{tn}-e^{rn}}{t-r}\right) = P\left(\dfrac{e^{rn}-e^{tn}}{r-t}\right) \tag{6-27}$$

When $t = r$,

$$F_{CE} = Pne^{rn} \tag{6-27A}$$

$$CECA = \dfrac{e^{rn}-e^{tn}}{r-t} \tag{6-27B}$$

Compound Interest Equations

The present worth of an exponentially compounding series under continuous compounding would be Eq. 6-27 multiplied by the present-worth factor for a single sum which is Eq. 6-17:

$$P_{CE} = P \left(\frac{e^{rn} - e^{tn}}{e^{rn}(r-t)} \right) \tag{6-28}$$

$$CEPW = \frac{e^{rn} - e^{tn}}{e^{rn}(r-t)} \tag{6-28A}$$

The equivalent uniform series factor may be obtained by multiplying the present-worth factor, Eq. 6-28, by the capital recovery factor, Eq. 6-21:

$$A_{CE} = P \left(\frac{e^{rn} - e^{tn}}{e^{rn}(r-t)} \right) \left(\frac{re^{rn}}{e^{rn} - 1} \right) \tag{6-29}$$

$$CEUS = \left(\frac{e^{rn} - e^{tn}}{e^{rn}(r-t)} \right) \left(\frac{re^{rn}}{e^{rn} - 1} \right) \tag{6-29A}$$

The equations for gradient and exponential growth under continuous compounding as derived in the preceding pages may be used by making individual calculations for each application. Tables covering a suitable range of interest rates, growth rates, and analysis periods are too extensive to warrant their inclusion in this book. The currently increasing use of continuous compounding may indicate its widespread use in the future, in which case publication of a set of tables may be warranted.

PROBLEMS

When working problems involving cash flows which are to be combined or interrelated in the calculations to find the answer, the cash-flow diagram should be drawn in every case. The use of the cash-flow diagram will help materially in visualizing the problem as the first step in setting up the equation or series of calculations leading to the solution. Cash-flow diagrams are illustrated in Fig. 6-1. Use the scheme of drawing outward flows downward (negative) and inward flows upward (positive) from the horizontal time scale.

6-1. Given only a table of factors of $(1 + i)^n$ and $(1 + i)^{-n}$ how would you use the table to find factors for each of the four uniform series terms? Illustrate for an interest rate of 10 percent and 6 interest periods.

6-2. How would you convert the series present-worth factors *(SPW)* in Appendix B to an application in which the first deposit was made at time zero and the last deposit at the beginning of the last period? Work a specific example for an interest rate of 10 percent.

6-3. Derive each of the four series factors when the first payment (or cash-flow) is made at $n = 0$ rather than for $n = 1$, and the last payment is at $n - 1$. Prove the correctness of your derivations.

6-4. From Appendix B for a series of vestcharge rates starting at 2 percent and going upward, show the corresponding analysis periods all of which will give a capital recovery factor of approximately 0.15. Comment with respect to the selection of vestcharge rates relative to selection of analyses periods.

6-5. Interpolate between the *SPW* factors for 4 and 5 percent and compare the interpolated factor at $4\frac{1}{2}$ percent with the true value. Repeat for interest rates of 10 and 15 percent and interpolate for the *SPW* factor for 12 percent. Use 5 and 20 interest periods in your interpolations. Evaluate this use of straight-line interpolation in the analysis for economy.

6-6. Compare the advantages and disadvantages and types of applications of the compound amount factors *CA* and *SCA* with the present-worth factors *PW* and *SPW*.

6-7. Compute the present value of a $1,000 bond which is to pay 6 percent per year interest annually for 10 years. Prove your answer.

6-8. You are offered $28,000 for your patent on which you expect to receive royalties averaging $5,000 a year for the 10 years of remaining life of the patent. You are quite willing to sell since other than for the net income you have no special interest in the patent. You have excellent prospects of a new venture upon which you expect to make 12 percent return per year on the $28,000. Should you sell your patent? (*Ans.:* No.)

6-9. To provide a college education fund of $1,500 on each of your son's 18th, 19th, 20th, and 21st birthdays, you wish to deposit a uniform sum on each of his 5th through 18th birthdays. If the deposits will compound at 4 percent annually, what uniform amount must you deposit annually? (*Ans.:* $309.57.)

6-10. Compute the net rate of return on an investment in a vacant lot for which the owner paid $6,000 and sold it 6 years later for $8,600. Year-end taxes were paid as follows; first year, $170; second, $180; third, $200; fourth, $200; fifth, $240; and sixth year, $220. (*Ans.:* 3.31 percent.)

6-11. Rework Prob. 6-10 on the basis that the taxes increased uniformly $20 per year starting from $170 paid at the end of the first year. (*Ans.:* 3.05 percent.)

6-12. Compare the following plans for the expenditure of money by computing the present worth, the compound amount, and the equivalent uniform annual cost. Check your answers by converting each answer to the other of the two forms. Use 6 percent vestcharge. (For Prob. W-1: *Ans.:* Plan A: $PW = \$1,000$; $CA = \$1,791$; $EUAC = \$136$. *Ans.:* Plan B: $PW = \$1,023$; $CA = \$1,833$; $EUAC = \$139$.)

	Plan A	Plan B
Problem W-1		
Spend now	$1,000	$800
Spend 10 years hence	–	400
Problem W-2		
Spend now	1,000	800
Spend each year-end for 10 years	100	95
Spend 10 years hence (lump sum)	–	400
Problem W-3		
Spend now	1,000	800
Spend each year-end for 10 years	100	95
Spend at end 10th year hence	–	400
Spend each year-end, 11th through 18th year	120	110

6-13. Compare the single-sum compound amount factor (CA)–numerical value–on a monthly basis to the compound amount factor for 10 percent per annum for 12 years.

Compound Interest Equations 119

6-14. On January 1 each year you deposit $100 in a bank savings account to earn interest at 4 percent per annum compounded quarterly. What will be your account balance one year after your 5th deposit? (*Note:* Read the problem carefully, then draw the cash-flow diagram.)

6-15. What rate of interest per annum compounded semiannually will compound $1,000 to $1,104 in 5 years?

6-16. Calculate the number of months required for a principal sum of $1,000 to double at compound interest at the following nominal rates per year and interest periods:

4 percent	annually	semiannually	quarterly	monthly
6 percent	"	"	"	"
8 percent	"	"	"	"
10 percent	"	"	"	"

6-17. At a nominal interest rate of 6 percent per annum, what will monthly deposits of $50 accumulate to in 10 years, when the interest is credited quarterly? (*Note:* Solve on the basis that deposits of $150 are made quarterly.) Rework this problem on the basis that interest is credited monthly.

6-18. What period in months will it take for you to accumulate $5,000 from monthly deposits of $50 in a building and loan association which pays a nominal interest of 4 percent per year, interest credited monthly? Repeat the calculation for an interest rate of 6 percent.

6-19. What sum must you deposit monthly in a bank savings account paying a nominal interest of 4 percent per year compounded monthly to have an account balance in 5 years of $2,000?

6-20. If your minimum attractive rate of return is 5 percent, what price would you pay for a bond of $1,000, due in 10 years, and bearing semiannual interest coupons at 4 percent per year? Prove your answer by a check calculation.

6-21. At a certain date, a $100 West Virginia Turnpike Commission bond could be purchased on the open market for $65.50. This series bears semiannual interest coupons at $3\sqrt[3]{}_4$ percent per annum. Assume maturity in 20 years. (1) What nominal rate of return is earned at this price? (2) If your minimum attractive rate of return is 6 percent what price would you pay for this bond?

6-22. Your minimum attractive rate of return is 6 percent. There is on the market a $1,000 bond paying 4 percent interest per annum payable quarterly for 10 years which is priced at $900. Would you buy it? Why?

6-23. You are in the 26 percent income tax bracket. What rate of return would you net after income tax on a $1,000 bond, due in 8 years and bearing 6 percent semiannual interest coupons, which you can purchase at the market price of $888? (*Note:* Include the capital gains tax.)

Solution to Prob. 6-23:

After taxes the net interest return = $22.20. After taxes the net return of principal at the end of 8 years would be $985.44, which is the $1,000 principal less 26 percent tax on one-half of the capital gain: ($1,000 − $888)(0.5)(0.26) = $14.56.

Solve for nominal rate of return:

$0 = -888 + 22.20\,(SPW\text{-}i\%\text{-}16) + 985.44\,(PW\text{-}r\%\text{-}16)$

Try 3 percent: $0 = -888 + 22.20(12.561\ 102) + 985.44(0.623\ 167)$

$0 = -888.00 + 278.86 + 614.09$

$0 \neq -4.95$

Try 3½ percent: $0 = -888.00 + 22.20(12.094\ 117) + 985.44(0.576\ 706)$
$0 = -888.00 + 268.49 + 568.31$
$0 \neq +51.20$

By interpolation, $r = 3.044$ percent semiannually, or 6.088 percent per year. The effective rate is

$$i = \left(1 + \frac{r}{m}\right)^m - 1 = \left(1 + \frac{6.088}{2}\right)^2 - 1$$

$i = 1.0618 - 1 = 6.18$ percent per year

6-24. Derive Eq. 6-16 for computing the approximate effective interest rate on a monthly payment loan.

6-25. Compute the effective interest rate per annum on $1,200 borrowed at rate of 5 percent per year, interest and principal payable in equal amounts at end of month for 12 months. Assume that the $60 interest is added to the principal in computing the equal monthly payment.

6-26. Compute the effective interest rate per annum on $1,000 borrowed on a discount basis at 6 percent per year with principal sum of $1,000 payable in uniform amounts at end of month for 12 months. The discount basis means that the interest is deducted from the principal initially, so that the cash received on the note is (in this case) only $940.

6-27. Compute the effective interest rate per annum on $1,000 borrowed at an interest rate of 1 percent a month on the principal balance at the beginning of the month, and the principal is payable in 12 equal amounts at the end of the month. (*Ans.*: 12.68 percent.)

6.28. The federal government offers Series E bonds and Freedom Share bonds on the following basis:

	E Bond	FS Bond
Principal (face) value	$100	$100
Cost	$75	$81
Maturity period	7 years	4½ years
Redemption value	$100	$100

The Freedom Share bonds cannot be bought without buying an equal or greater amount of E bonds. On the basis that $100 of each is bought and each held to maturity, what effective interest rate is earned on each type of bond? What is the joint effective interest rate on the two bonds combined?

6-29. An employee credit union charges ¾ percent per month on the monthly beginning balance on automobile loans. A local bank advertises automobile loans at 4 percent per year. Compute the effective interest rates of these two loans on a $2,000 cash-received amount, repayable in 24 monthly payments. Consider that the bank will add $160 interest cost and $20 service and insurance cost to the $2,000 before calculating the uniform monthly payment.

6-30. The Central National Bank advertised that it pays 5 percent per annum interest on 90-day renewable certificates of deposit compounded quarterly to yield 5.094 percent. Prove the correctness of this rate.

6-31. For Prob. 6-16 calculate the effective interest rates for the quarterly and monthly interest periods.

6-32. Compute the effective interest rate per annum on $1,000 borrowed at a rate of 12 percent per year, one-twelfth of yearly interest payable at end of each month and

Compound Interest Equations

principal payable at end of 12th month. Repeat for (1) quarterly interest payments, (2) semiannual interest payments, and (3) annual interest payments.

6-33. Calculate the effective rate of interest on the following car loan. Remainder of purchase price borrowed is $2,000, to be paid in 36 uniform monthly payments. Loan costs, including term life insurance, are $30. Total interest charges at 5 percent per year are $300. Both the loan cost and interest are added to the principal sum to be repaid.

6-34. The Union Bank and Trust Company advertised a loan plan whereby $992.30 cash could be had by signing a 15-month note for a principal sum of $1,080 and paying thereon $72 a month for 15 payments. Or on an 18-month basis the $1,080 would yield a cash proceeds of $974.81 and require payments of $60 a month for 18 months. A footnote states that the cost of the loan includes life insurance to pay off the balance of the note in case of death of the borrower. What are the nominal and effective rates of interest charged on these loans?

6-35. The Second National Bank advertises, "$3\frac{1}{2}$ percent interest on new car financing—$3.50 per hundred per year on the amount borrowed." For a loan of $2,000 for 30 months the note signed is for a principal of $2,175.00 for which $72.50 is the monthly payment. How was the $2,175.00 arrived at, and what are the nominal and effective interest rates? Compare with Prob. 6-34 and discuss.

6-36. The First National Bank is soliciting automobile loans as follows: "New cars financed at 4 percent discount per annum. If you borrow $2,000 to purchase a car, your repayment is $63.13 per month for 36 months." Show how the $63.13 was arrived at, and calculate the nominal and effective interest rates.

6-37. From a daily newspaper story about two local court suits involving usery, the following data are taken:

Case 1. Mr. May borrowed $1,500 on a second trust on his home. He paid 60 monthly payments of $49 each, for a total of $2,940.

Case 2. Mr. Winter claims his defendant was charged 32 percent interest on a home loan trust of $2,200 on which he was forced to pay 60 monthly payments of $66 each, for a total of $3,960.

Compute the nominal and effective rates for these two cases and comment upon your answers.

6-38. What uniform monthly payment of principal and interest is required on a 6 percent, 25-year mortgage of $20,000?

6-39. Write out the two equations in terms of P, i, n, and x, for calculating (1) the amount of reduction in principal of a mortgage, or other debt, and (2) for the total amount of interest paid after x monthly uniform payments. Prove your equation by calculations for $P = \$1,000$, $i = 6\%$, $n = 20$ years, and $x = 3$ years. (*Note*: Mortgage financing is usually based upon uniform monthly payments, combining increasing principal payments and decreasing interest payments.)

6-40. You are reducing your 6 percent, 15-year, $6,000 home mortgage through uniform monthly payments of $50.63 for principal and interest. After 20 payments what is your balance of the principal assuming that monthly interest is charged on the principal balance? How much interest have you paid?

6-41. You have a 5 percent, 20-year, first mortgage on your home of $15,000 and a 6 percent, 5-year second mortgage of $3,000. Your 18 monthly payments of $99.00 and $56.61 respectively, have reduced the principal amounts to $14,319.30 and $2,177.17. You can replace both mortgages with one new 5 percent, 20-year, $16,500 first mortgage of monthly payments of $108.90. The cost of refinancing is $300. You have cash enough to refinance without borrowing the $300 refinancing expense less the $3.53 difference in principal balances. You plan on living in this home at least 10 years more. Should you refinance? Why?

Chapter **7**

Methods of Economic Analysis

Six arithmetical procedures will be discussed that may be used in the economic analysis of proposed investments for both economic evaluation and project formulation. The methods have grown up over the years as different individuals have studied economic analysis and applied it to public works and to private ventures. Because every method cannot be applied to every different type of proposal, an understanding of the characteristics and limitations of each of the six methods is essential for the analyst. When properly applied in accordance with their limitations, each method will give a reliable result for economic evaluation and for project formulation. Among the writers on the subject of economic analysis, engineering economy, and capital budgeting, there is not full agreement as to the relative worth of the several methods of analysis, how to handle certain factors, and the limitations of the methods. But understanding is growing, and the fields of application are widening, particularly in governments.

TOWARD AN UNDERSTANDING OF ECONOMIC ANALYSIS

As a foundation for a better understanding of the six methods of economic analysis, a general discussion of economic analysis is offered to bring into view some of the concepts, factors, principles, and objectives. Although the literature on the subject deals widely with industrial applications, there is a fair amount relating to water-resource developments. Other forms of public works, particularly transportation, are now being examined by the decision tool of economic analysis.

STATUS OF THE LITERATURE

The literature in economics, engineering, and public works since 1950 has contained many papers and books upon the subject of economic analysis of proposals for investment in capital goods, plant, works, and machines. Because this particular tool of management must involve selecting certain factors by subjective processes, and since there is no set arithmetical procedure, it is likely that discussion of the subject will continue for some years. Economic analysis leading to management decisions on the investment of new capital is not unlike the subject of depreciation. Since about 1830 the subject of depreciation has been in the literature and no doubt will continue to be. The handling of depreciation as related to physical properties in whatever concept and application are chosen is not an exact science. Likewise, economic analysis is an art and not a science, though some writers and appliers of the results of analyses endeavor to make out that economic analysis is an exact science. Forthcoming years should bring still further understanding and agreement upon the concepts, procedures, and applications of economic analysis as a management tool.

ENGINEERS VERSUS ECONOMISTS

The literature has brought the economists and the engineers into debate on the relative merits of the several methods of analyses and procedures. Most likely what is known as engineering economy was developed by engineers for use in their design and decision processes, largely in private industry. Later, the subject was applied to the public sector, particularly in the water resource and transportation areas. Now, the federal government of the United States is applying the general principles to its whole budgeting system and operating program. The planning, programming, and budgeting systems (PPBS) in the federal government are designed to provide for economic analysis, cost effectiveness analysis, or system analysis, to support the request for appropriations from Congress.

As the economist has come into the field he has brought some new concepts and procedures based upon classical economics, including the allocation of resources in the public sector, welfare economics, and theory of value. The interdisciplinary discussions have been helpful in gaining a better understanding of the subject, but agreement on many factors has not yet been achieved. These factors of questionable concept and procedure are discussed throughout this chapter.

THE DISCOUNT RATE

In the literature of the economist there is much discussion about the cost of capital (see 7-6 and 7-26 and their references) which discussion leads to the choice of discount rate to use in the analysis for economy. There are now the following terms to be found in the literature, each having more or less the same intended use: interest rate, rate of return, minimum attractive rate of return, vestcharge rate (introduced by the author of this book), discount rate, and cost of money. Yet there is a difference in meaning in each of these terms. In the literature of the economist the cost of money is usually defined as what the corporation pays for investment capital and is a combination rate based upon

Methods of Economic Analysis

the relative amounts of debt capital (bond issues) and equity capital (preferred and common stocks) in the financial structure. Such cost of capital is worth determining as a guide to setting the discount rate to use in the economic analysis, but the cost of capital itself is not always the best of choices. Each proposal for investment should be examined thoroughly and the discount rate chosen for it in accordance with its risks and uncertainties and nature of the proposal as compared to similar proposals and the normal functions of the corporation.

However the cost of capital is arrived at, there always remains some question as to whether the correct rate, the most desirable rate, or an acceptable rate has been determined. Some writers indicate that a given corporation has a definite cost of capital that can be directly calculated. Others are much more realistic and uncertain. Solomon [7-26-2, p 241] states.

> Quite apart from the practical difficulties of measuring capital costs for a specific company, there is today no explicit and generally acceptable account of what is the correct conceptual approach to its measurement. This gap represents the weakest link in the theory of capital budgeting, and, until it is filled, capital-budgeting theory will remain, at best, only a partial guide to decision making in this important area of business activity.

There is a conceptual difference between the term "cost of capital" as used by economists, and the term *minimum attractive rate of return* as found in the literature of engineering economy [7-15]. Cost of capital infers that a company undergoes a specific financial cost in procuring its invested capital and working capital and that this cost can be determined. The term *minimum attractive rate of return* implies a desired rate of return reached by judgment, often for a particular application. This rate need not be the same as the cost of capital.

Because of the many uncertainties related to calculating the cost of capital and in view of the objective of economic analysis—which is developing an aid to a management decision as to investment—there is really nothing gained by attempts to precisely determine a cost of capital. In any case, whatever is determined today will be different tomorrow. The preferred practice in all analyses, whether using a cost of capital, a minimum attractive rate of return, a discount rate, or a vestcharge rate, is to make the particular economic analysis with two or more rates. Management then has a basis of judging the sensitivity of the results with respect to the factor of rate of probable return, and then is in a position to make a more enlightened decision. For cases of unequal or irregular cash flows, a change in rate can materially affect the calculated differences between alternatives.

BASIC CHARACTERISTICS OF THE METHODS OF ANALYSIS

The six methods of economic analysis to be described have the common objective of comparing the future streams of costs and benefits in such a way that for a specific future period of time the analysis will disclose the probable net return on the proposed investment, or the most economical design required to produce the returns. The return here indicated is in the form of dollar net profits, dollar values of benefits, dollar values of satisfactions, or dollar worth of other sought-for consequences. Each method applies the principles and concepts of compound interest in a way to take into the calculations the differences in the worth of money over time. Each method also uses as input data the future negative and positive cash flows of money which are required to produce the returns and those that are a consequence of the investment in the property concerned.

In the development of these methods, their evolution can be traced largely to industry, except that the benefit/cost ratio method is largely attributed to public works. The methods grew out of the efforts of management to get some measure of the economic wisdom of investing capital in additional facilities, products, or processes, and the design engineer's desire of a method of measuring the economy of his design in both construction and use. Some of the methods of economic analysis developed in the early years are not discussed herein in detail because they either have little merit or are not suitable for use in the highway area of public works. The engineer and the economist each has had a hand in developing the concepts and procedures.

DEFINITIONS OF THE SIX METHODS

The following definitions of the six methods of economic analysis are offered at this point in order that the reader may get an early indication of their distinctions. Following the section on equations and example applications, these six methods are discussed in more detail.

EQUIVALENT UNIFORM ANNUAL COST METHOD

The equivalent uniform annual cost method combines all investment costs and all annual expenses into one single annual sum that is equivalent to all disbursements during the analysis period if spread uniformly over the period. When more than one alternative is being examined the one with the lowest equivalent uniform annual cost is the more economical. The present worth of this equivalent uniform annual cost will give the same answer as obtained by the present worth of costs method.

PRESENT WORTH OF COSTS METHOD

The present worth of costs method combines all investment costs and all

Methods of Economic Analysis

annual expenses into a single present-worth sum, which represents the sum necessary at time zero to finance the total disbursements over the analysis period. This present sum when multiplied by the capital recovery factor will give the equivalent uniform annual cost obtained by the equivalent uniform annual cost method. Of the alternatives compared the one with the lowest present worth is the more economical.

EQUIVALENT UNIFORM ANNUAL NET RETURN METHOD

The equivalent uniform annual net return method is the equivalent uniform annual cost method plus the inclusion of an income factor or benefit factor. The answer indicates the amount by which the equivalent uniform annual income exceeds (or is less than) the equivalent uniform annual cost. The alternative having the greatest equivalent uniform annual net return is the one of greatest economy.

NET PRESENT VALUE METHOD

The net present value method gives the algebraic difference in the present worths of both outward cash flows and inward flows of incomes or benefits. It is the same in principle as the present worth of costs method, but includes the factor of annual income. The alternative having the greater net present value is the one with greatest economy.

BENEFIT/COST RATIO METHOD

The benefit/cost ratio method expresses the ratio of equivalent uniform annual benefit (or its present worth) to the equivalent uniform annual cost (or its present worth). Any alternative that has a benefit/cost ratio above 1.0 is economically feasible and the alternative that has the highest incremental benefit/cost ratio is indicated as the preference. Usually in highway work the benefit/cost ratio method is applied in pairs of alternatives in order to develop a differential benefit.

RATE OF RETURN METHOD[1]

The rate of return method determines that vestcharge rate, or discount rate, which will equalize the negative costs and the positive returns or benefits. As in the benefit/cost ratio method, the rate of return method in highway proposals usually compares two alternatives in order to develop a differential benefit. The higher the rate of return the greater the economy.

[1] Although in this book the term "rate of return" is used, this method of analysis has other names, such as internal rate of return, discounted cash flow, yield rate, premium rate, marginal rate of return, and others.

BASIC EQUATIONS FOR THE SIX METHODS OF ANALYSIS

The equations to be given for solving for the numerical solution of the six methods of analysis for economy are written on the concept of the cash-flow diagrams wherein the outward flows of investments and expenses (disbursements) are negative and the inward flows of income or benefits are positive.

NOTATION SCHEME

The factors and terms in the several equations correspond to the following notations in addition to those given in Chapter 6 for compound interest terms:

$EUAC$	Equivalent uniform annual cost
$PWOC$	Present worth of costs and expenses
$EUANR$	Equivalent uniform annual net return
NPV	Net present value
B/C	Benefit/cost ratio
ROR	Rate of return
$MARR$	Minimum attractive rate of return
I	Original, or initial, investment, or equivalent investment at time zero including discounted investments subsequent to time zero
T	Terminal value at end of analysis period
K	Total uniform annual expense of administration A, traffic services and highway operations J, and highway maintenance M
K_G	Equivalent uniform annual K when K grows as a uniform gradient
K_E	Equivalent uniform annual K when K grows exponentially
U	Uniform annual road-user costs, exclusive of road-user taxes, but inclusive of travel time value, and accident costs when so designated
U_D	Equivalent uniform annual road-user benefits, being the difference in road-user costs between a pair of alternatives
U_G	Equivalent uniform annual road-user costs under a gradient growth of traffic volume
U_E	Equivalent uniform annual road-user costs under an exponential growth of traffic volume
U_T	Equivalent uniform annual road-user tax payment, or tax revenue from road users.
R	Uniform annual gross income from sales revenue, receipts or their equivalent, or gross benefits. R is inclusive of return of investment (depreciation) and return on investment (net profit).
R_G and R_E	Equivalent uniform annual R when either a gradient or exponential increase is present
R_D	Difference in the equivalent uniform annual recepts of a pair of alternatives

Methods of Economic Analysis

B and *P* Subscripts indicate, respectively, the base alternative (often the existing situation or defender) and the proposed alternative (the challenger).

EQUATIONS FOR THE SIX METHODS OF SOLUTION FOR ECONOMY

The six methods of solution for the economy of alternative proposals may be written in more than one form and by using different compound interest factors. Further, the particular cash flows involved will call for different compositions of the basic equation. For instance, more than one investment may be involved, annual expenses and road-user costs may be stated in terms of gradient or exponential increases (or decreases), and there may or may not be a terminal value.

The following equations do not include any capital investments put in place subsequent to the initial, or first, investment at $n = 0$. Subsequent capital investments, such as for stage construction, may be easily handled in the solutions by treating them in the same manner as the initial investment, just taking care to see that the money sums involved are transferred to the same point in time. This transfer is accomplished by applying the appropriate vestcharge factor.

The following twelve equations illustrate the general forms of the equations to use; where the gradient and exponential factors are shown it is for the purpose of illustrating one possible form of the equation.

$$EUAC = -I(CR\text{-}i\text{-}n) + T(SF\text{-}i\text{-}n) - K - U \tag{7-1}$$

$$EUAC = -I(CR\text{-}i\text{-}n) + T(SF\text{-}i\text{-}n) - K - G_K(GUS\text{-}i\text{-}n) - U_E \tag{7-1A}$$

$$PWOC = -I + T(PW\text{-}i\text{-}n) - K(SPW\text{-}i\text{-}n) - U(SPW\text{-}i\text{-}n) \tag{7-2}$$

$$PWOC = -I + T(PW\text{-}i\text{-}n) - K(EPW\text{-}i\text{-}n) - U(EPW\text{-}i\text{-}n) \tag{7-2A}$$

$$EUANR = -I(CR\text{-}i\text{-}n) + T(SF\text{-}i\text{-}n) - K + R \tag{7-3}$$

$$EUANR = -I(CR\text{-}i\text{-}n) + T(SF\text{-}i\text{-}n) - K - G_K(GUS\text{-}i\text{-}n) + R_G \tag{7-3A}$$

$$NPV = -I + T(PW\text{-}i\text{-}n) - K(SPW\text{-}i\text{-}n) + R(SPW\text{-}i\text{-}n) \tag{7-4}$$

$$NPV = -I + T(PW\text{-}i\text{-}n) - K(SPW\text{-}i\text{-}n) - G_K(GPW\text{-}i\text{-}n) + R(EPW\text{-}i\text{-}n) \tag{7-4A}$$

$$B/C = \frac{-(U_{GP} - U_{GB}) - (K_{GP} - K_{GB})}{-(I_P - I_B)(CR\text{-}i\text{-}n) + (T_P - T_B)(SF\text{-}i\text{-}n)} \tag{7-5}$$

$$B/C = \frac{-(U_{GP} - U_{GB})(SPW\text{-}i\text{-}n) - (K_{GP} - K_{GB})(SPW\text{-}i\text{-}n)}{-(I_P - I_B) + (T_P - T_B)(PW\text{-}i\text{-}n)} \tag{7-5A}$$

ROR (Solve for *i* by trial):

$$0 = -(I_P - I_B)(CR\text{-}i\text{-}n) + (T_P - T_B)(SF\text{-}i\text{-}n) - (U_P - U_B) - (K_P - K_B) \tag{7-6}$$

$$0 = -(I_P - I_B) + (T_P - T_B)(PW\text{-}i\text{-}n) - (U_P - U_B)(SPW\text{-}i\text{-}n) - (K_P - K_B)(SPW\text{-}i\text{-}n) \tag{7-6A}$$

Of the above equations, the *EUAC* and *PWOC* will give negative answers because all the cash flows are outward, that is, disbursements. When the *EUANR* and *NPV* are negative, disbursements exceed income and when positive the income exceeds the disbursements–all expressed on a present-worth basis. The *B/C* ratio will ordinarily be negative because the numerator will usually be positive and the denominator will be negative; when the numerator is positive affix a positive sign to the *B/C* ratio; when the numerator is negative it means that the net benefits are negative and the alternative would not be further considered.

Stating the *B/C* ratio equation as shown is preferable to other forms because it results in keeping the same signs an in the cash-flow diagram and the sequence of the base alternative *B* and of the proposed alternative *P* are the same in all terms. Note that the *B/C* ratio and the *ROR* equations are expressed in terms of the differences between a pair of alternatives; the other four equations are expressed in terms of a single alternative.

ILLUSTRATIVE APPLICATIONS OF THE SIX BASIC EQUATIONS

The six methods of solution for the economy of proposed improvements and the six basic equations may be illustrated by applying them to an example set of alternatives. In Table 7-1 are the basic cash-flow data for four proposed

TABLE 7-1
CASH-FLOW DATA FOR PROBLEMS TO ILLUSTRATE THE SIX METHODS OF ANALYSIS
(Equal Level of Service Is Assumed for Each of the Two Pairs of Mutually Exclusive Alternatives)

	Symbol and Cash-Flow Item	Base Alternative A_1	Proposed Alternative A_2	Base Alternative B_3	Proposed Alternative B_4
I	Initial investment or present worth of initial and any subsequent investments	140,000	160,000	120,000	200,000
T	Terminal value	40,000	50,000	10,000	18,000
A	Uniform annual expense of administration	–	–	–	–
J	Uniform annual expense of traffic operations	–	–	–	–
M	Uniform annual expense of highway (plant) maintenance	–	–	–	–
K	Total of A, J, and M–uniform annual expense	7,000	8,000	500,000*	485,500*
U	Uniform annual road-user costs exclusive of road-user taxes	74,000	70,000	–	–
U_T	Uniform annual road-user taxes generated	15,000	14,500	–	–
R	Uniform annual gross income from sales, or their equivalent, inclusive of depreciation	–	–	529,200	550,000
i	Rate of vestcharge per annum or minimum attractive rate of return	8%	8%	8%†	8%†
n	Analysis period in years	10	10	10	10

*For the industrial plant the K factor includes overheads, production expense, sales expense, plant maintenance and operations.

† Minimum attractive rate of return disregarding income taxes.

Methods of Economic Analysis

alternatives, two mutually exclusive pairs, one pair for a highway improvement and one pair for an industrial plant proposal. As appropriate, each of these two pairs of alternatives is solved by each of the six methods. Calculations for the two highway alternatives are made first, followed by solutions for the two industrial plant alternatives.

EQUIVALENT UNIFORM ANNUAL COST

The equivalent uniform annual cost of alternatives A_1 and A_2 as calculated by Eq. 7-1 is:

$EUAC_{A1} = -140,000(CR\text{-}8\%\text{-}10) + 40,000(SF\text{-}8\%\text{-}10) - 7,000 - 74,000$
$EUAC_{A1} = -140,000(0.149029) + 40,000(0.069029) - 81,000$
$EUAC_{A1} = -20,864 + 2,761 - 81,000$
$EUAC_{A1} = -99,103$
$EUAC_{A2} = -160,000(CR\text{-}8\%\text{-}10) - 50,000(SF\text{-}8\%\text{-}10) - 8,000 - 70,000$
$EUAC_{A2} = -160,000(0.149029) - 50,000(0.069029) - 78,000$
$EUAC_{A2} = -23,845 + 3,451 - 78,000$
$EUAC_{A2} = -98,394$

By difference, $-98,394 - (-99,103) = 709$, the proposed alternative A_2 is superior to the base alternative A_1 by an equivalent uniform cost of 709.

PRESENT WORTH OF COSTS

In a similar procedure, but using Eq. 7-2, the present worth of the costs of alternatives A_1 and A_2 are found:

$PWOC_{A1} = -140,000 + 40,000 \, (PW\text{-}8\%\text{-}10) - 7,000 \, (SPW\text{-}8\%\text{-}10)$
$\qquad\qquad\qquad\qquad\qquad\qquad\qquad\qquad - 74,000 \, (SPW\text{-}8\%\text{-}10)$
$PWOC_{A1} = -140,000 + 40,000 \, (0.463193) - 7,000 \, (6.710081)$
$\qquad\qquad\qquad\qquad\qquad\qquad\qquad\qquad - 74,000 \, (6.710081)$
$PWOC_{A1} = -140,000 + 18,528 - 46,971 - 496,546$
$PWOC_{A1} = -664,989$
$PWOC_{A2} = -160,000 + 50,000 \, (PW\text{-}8\%\text{-}10) - 8,000 \, (SPW\text{-}8\%\text{-}10)$
$\qquad\qquad\qquad\qquad\qquad\qquad\qquad\qquad - 70,000 \, (SPW)$
$PWOC_{A2} = -160,000 - 50,000 \, (0.463193) - 8,000 \, (6.710081)$
$\qquad\qquad\qquad\qquad\qquad\qquad\qquad\qquad - 70,000 \, (6.710081)$
$PWOC_{A2} = -160,000 + 23,160 - 53,681 - 469,706$
$PWOC_{A2} = -660,227$

The present worth of the costs for alternative A_2 is less by 4,762 [$-660,227 - (-664,989)$] than the present worth of the costs for A_1. A_2 is therefore the economical choice because less equivalent capital at age zero would be required to finance it. A check of the agreement between the $EUAC$ and the $PWOC$ may be had by converting one to the other:

$$709 \ (SPW\text{-}8\%\text{-}10 = 6.710081) = 4,757$$
$$4,762 \ (CR\text{-}8\%\text{-}10 = 0.149029) = 710$$

which check within the error resulting from rounding.

EQUIVALENT UNIFORM ANNUAL NET RETURN

Because the highway alternatives A_1 and A_2 do not have the equivalent of a sales income they cannot be examined separately by the equivalent uniform annual net return and the net present value methods. However, they can be compared by using their differences and letting the reduction in road-user costs equal the equivalent of a cash income.

$$EUANR_{A2} = -(I_P - I_B)(CR\text{-}8\%\text{-}10) + (T_P - T_B)(SF\text{-}8\%\text{-}10)$$
$$- (U_A - U_B) - (K_P - K_B)$$
$$EUANR_{A2} = -(160{,}000 - 140{,}000)(0.149029) + (50{,}000 - 40{,}000)(0.069029)$$
$$- (70{,}000 - 74{,}000) - (8{,}000 - 7{,}000)$$
$$EUANR_{A2} = -2{,}981 + 690 + 4{,}000 - 1{,}000 = 709$$

This equivalent uniform annual net return of 709 agrees with the differences in the equivalent uniform annual cost of the two alternatives as previously calculated.

NET PRESENT VALUE

In a like manner the net present value method may be applied to A_1 and A_2 with the following result:

$$NPV_{A2} = -(I_P - I_B) + (T_P - T_B)(PW\text{-}8\%\text{-}10) - [(U_A - U_B) + (K_P - K_B)]$$
$$(SPW\text{-}8\%\text{-}10)$$
$$NPV_{A2} = -20{,}000 + 10{,}000 (0.463193) + (4{,}000 - 1{,}000)(6.710081)$$
$$NPV_{A2} = -20{,}000 + 4{,}632 + 20{,}130$$
$$NPV_{A2} = 4{,}762$$

As compared to A_1, alternative A_2 has a net present worth of 4,762 above the present worths of the added disbursements required. Again, this present worth is equal to the difference between the present worth of costs (4,762) previously calculated for A_1 and A_2 separately.

BENEFIT/COST RATIO

The benefit/cost ratio method is applied to alternatives A_1 (the base) and A_2 (the proposed) by use of Eq. 7-5:

$$B/C_{A2} = \frac{-(70{,}000 - 74{,}000) - (8{,}000 - 7{,}000)}{-(160{,}000 - 140{,}000)(CR\text{-}8\%\text{-}10) + (50{,}000 - 40{,}000)(SF\text{-}8\%\text{-}10)}$$

Methods of Economic Analysis

$$B/C_{A2} = \frac{4,000 - 1,000}{-20,000\,(0.149029) + 10,000\,(0.069029)}$$

$$B/C_{A2} = \frac{+3,000}{-2,981 + 690} = \frac{+3,000}{-2,291} = 1.31$$

The benefit/cost ratio is considered to be positive because the numerator is positive, indicating that the benefits exceed the annual expenses. Should the present-worth procedure be chosen for the benefit/cost ratio calculation it will produce the following result:

$$B/C_{A2} = \frac{+20,130}{-15,368} = 1.31$$

RATE OF RETURN

The rate-of-return solution will be calculated by using the present-worth form illustrated by Eq. 7-6A. First try a vestcharge rate of 10 percent per year:

$$0 = -(160,000 - 140,000) + (50,000 - 40,000)(PW\text{-}10\%\text{-}10)$$
$$-(70,000 - 74,000)(SPW\text{-}10\%\text{-}10) - (8,000 - 7,000)(SPW\text{-}10\%\text{-}10)$$
$$0 = -20,000 + 10,000\,(0.385543) + 4,000\,(6.144567) - 1,000\,(6.144567)$$
$$0 = -20,000 + 3,855 + 24,578 - 6,145 = 2,288$$

Second trial is with a rate of 12 percent:

$$0 = -20,000 + 10,000\,(0.321973) + 4,000\,(5.65023) - 1,000\,(5.65023)$$
$$0 = -20,000 + 3,220 + 22,601 - 5,650 = 171$$

By extrapolation:

$$ROR_{A2} = 10.0 + 2.0\,\frac{2,288 - 171}{2,288} = 10.0 + 2.0\,(1.02)$$

$$ROR_{A2} = 12.1 \text{ percent}$$

The solutions indicate that the incremental investment of 20,000 in alternative A_2 over A_1 will produce a ratio of benefits to costs of 1.31 or a rate of return on the 20,000 of 12.1 percent.

Not one of the foregoing solutions tests the base alternative A_1 to see whether it is economically desirable. The solutions, in each case, compare alternative A_2 to alternative A_1 and determine the relative worth of A_2 over A_1. See page 137 for a discussion of the economic merit of the base or null alternative.

APPLICATION TO THE INDUSTRIAL PLANT

The equivalent uniform annual net return and the net present value methods are next applied individually to alternatives B_3 and B_4 for which there has been assumed to be an annual sales income.

$$EUANR_{B3} = -120{,}000\ (CR\text{-}8\%\text{-}10) + 10{,}000(SF\text{-}8\%\text{-}10) - 500{,}000$$
$$+ 529{,}200$$
$$EUANR_{B3} = -120{,}000\ (0.149029) + 10{,}000\ (0.069029) + 29{,}200$$
$$EUANR_{B3} = -17{,}883 + 690 + 29{,}200$$
$$EUANR_{B3} = 12{,}007$$

The solution for B_4 is:

$$EUANR_{B4} = -200{,}000\ (CR\text{-}8\%\text{-}10) + 18{,}000(SF\text{-}8\%\text{-}10) - 485{,}500$$
$$+ 550{,}000$$
$$EUANR_{B4} = -200{,}000\ (0.149029) + 18{,}000\ (0.069029) + 64{,}500$$
$$EUANR_{B4} = -29{,}806 + 1{,}243 + 64{,}500 = 35{,}937$$

The net present-value method seeks to find the net present worth of the negative and positive cost flows at the chosen discount rate, in this example 8 percent.

$$NPV_{B3} = -120{,}000 + 10{,}000\ (PW\text{-}8\%\text{-}10) + (529.200 - 500{,}000)\ (SPW\text{-}8\%\text{-}10)$$
$$NPV_{B3} = -120{,}000 + 10{,}000\ (0.463193) + 29{,}200\ (6.710081)$$
$$NVP_{B3} = -120{,}000 + 4{,}632 + 195{,}934 = 80{,}566$$

The solution for B_4 is:

$$NVP_{B4} = -200{,}000 + 18{,}000\ (PW\text{-}8\%\text{-}10) + (550{,}000 - 485{,}500)\ (SPW\text{-}8\%\text{-}10)$$
$$NVP_{B4} = -200{,}000 + 18{,}000\ (0.463193) + 64{,}500\ (6.710081)$$
$$NVP_{B4} = -200{,}000 + 8{,}337 + 432{,}800 = 241{,}137$$

A check on these four solutions may be made by conversion from net annual returns to net present value or vice versa:

$$12{,}007\ (SPW\text{-}8\%\text{-}10 = 6.710081) = 80{,}568$$
$$35{,}937\ (SPW\text{-}8\%\text{-}10 = 6.710081) = 241{,}140$$

Both the equivalent uniform net return and the net present-value methods identify alternative B_4 as the superior alternative.

Note that in the above solutions by the equivalent uniform annual net return and net present value methods each alternative, B_3 and B_4, is individually proved acceptable. This solution is possible here because each alternative has its own individual income or benefit, which situation does not exist with alternatives A_1 and A_2.

Alternatives B_3 and B_4 could be analyzed individually, also, by the benefit/cost ratio and the rate-of-return methods.

COMPARISON OF THE METHODS OF ANALYSIS WHEN APPLIED TO A GROUP OF MUTUALLY EXCLUSIVE ALTERNATIVES

The six methods when applied to a multigroup of mutually exclusive alternatives having the same level of service will each indicate that identical alternative

Methods of Economic Analysis 135

as the best choice when the proper procedures of application and calculation are followed. Thus in this respect there is no best method of analysis so far as the final ranking of alternatives is concerned. The basis of choice between methods must be based upon (1) the form of available data, (2) whether benefit or income amounts are available, and (3) the preferences of the analyst and the decision maker.

ILLUSTRATIVE SOLUTIONS

To illustrate the fact that the methods have equal ability and reliability in indicating the most economical alternative, the methods were applied to the mutually exclusive alternatives in Table 7-2, which table also gives the results. These analyses emphasize that it is the difference in alternatives that is the important factor rather than the gross quantities. The method of differences will also demonstrate that its proper use will select the alternative that will maximize benefits.

DISCUSSION OF THE SOLUTIONS

From Table 7-2 it will be observed that each of the four methods which compare the alternatives individually by computing the annual cost, annual return, present worth, or present value designates alternative E as the most economical. However, the benefit/cost ratio and rate-of-return methods select alternative B when each alternative is compared to alternative A. But it will be noticed that in this incorrect procedure, the comparison is entirely by comparing each alternative with alternative A and not each alternative with each other alternative. In the individual comparisons by the first four methods, each alternative is compared with each other by observing the differences in the final answers, thus adhering to the principle that it is the difference between alternatives that is the factor which distinguishes their relative merit. Consequently, the benefit/cost ratio and rate-of-return methods must be applied in accordance with this principle of differences.

It will be noted that Table 7-2 arrays the seven alternatives in order of increasing investment at zero time. By starting with alternative A as the base and taking sequential differences between pairs of adjacent alternatives the analysis can be made to indicate the relative economy of the differential increase of investments and operating expenses as compared to the differential increase in benefits.

This differential analysis is the procedure to determine that each additional investment produces an additional gain greater than its cost. Thus, as long as the increment of investment produces a benefit/cost ratio of 1.0 or greater, or a rate of return equal to or greater than the minimum attractive rate of return (8.0 percent in this case), the additional investment is warranted and will be accepted. The increment of investment is rejected, however, when these criteria are not met.

TABLE 7-2

COMPARISON OF METHODS OF ECONOMIC ANALYSIS WHEN APPLIED TO A GROUP OF MUTUALLY EXCLUSIVE ALTERNATIVES

Alternative	Initial Construction Cost, I	Annual Uniform Highway Expense, K	Annual Uniform Road-User Costs, U	Annual Road-User Benefits Measured from Alternative A	Calculated Results	
					Equivalent Uniform Annual Transportation Cost	Present Worth of Costs
A	none	60	500	—	560.0	5,498
B	800	70	280	220	431.5	4,236
C	1,000	55	250	250	406.8	3,995
D	1,300	52	225	275	409.4	4,020
E	1,350	48	220	280	405.5*	3,981*
F	1,500	46	210	290	408.8	4,013
G	1,650	46	195	305	409.1	4,016

Alternative	Calculated Results			
	Equivalent Uniform Annual Return Compared to Alternative A	Net Present Value Compared to Alternative A	Benefit/Cost Ratio Compared to Alternative A	Rate of Return Compared to Alternative A
A	—	—	—	—
B	128.5	1262	2.6*	26.0*
C	153.2	1503	2.5	25.2
D	150.6	1478	2.1	21.3
E	154.5*	1517*	2.1	21.1
F	151.2	1485	2.0	19.7
G	150.9	1482	1.9	18.6

Results of the Incremental Solutions

Paired Comparison	B/C Ratio	ROR	Conclusion †
B challenges A	2.6	26.0	B better than A
C challenges B	2.2	22.1	C better than B
D challenges C	0.9	6.9	C better than D
E challenges C	1.77	8.5	E better than C
F challenges E	0.8	5.0	E better than F
G challenges E	0.9	6.4	E better than G
			E best of all

* Indicated best choice. Note that the B/C and ROR methods when compared only to A indicate that B is best, but in the incremental solution the correct choice of E is indicated.

† The challenger is dropped whenever the B/C ratio is less than 1.0 or the ROR is less than 8.0 percent, the minimum attractive rate of return.

Note: These mutually exclusive alternatives apply to a highway improvement for motor vehicle traffic. Therefore, there is no sales income as such; the reduction in road-user costs is considered as the equivalent of cash income. The vestcharge rate, or minimum attractive rate of return, is 8 percent. Terminal value is zero; analysis period is 20 years.

Methods of Economic Analysis

This is the case with alternatives D, F, and G, as shown by their incremental benefit/cost ratios of 0.9, 0.8, and 0.9 and their incremental rates of return of 6.9 percent, 5.0 percent, and 6.4 percent.

Thus when the proper procedure is used each of the six methods of analysis will designate the same alternative as being the one having the greatest economy.

ECONOMIC EVALUATION OF THE NULL ALTERNATIVE

In all procedures when a pair of alternatives is compared by an analysis of their differences, this analysis in itself does not determine that either one of the pair of alternatives is economically desirable, but only that the difference between the pair is economical or not economical.

The most frequent case in economic analysis is that of examining a proposal for improvement (or changing) of an existing property, process, or function. Often the procedure is to test the proposal by comparing it with the costs and benefits of the existing or null situation without first determining that the existing facility is economically desirable. If the existing facility is not economically desirable in its function or its service there is generally nothing to be gained by improving it or replacing it with another facility. It is worthwhile to discuss how to determine whether existing highway facilities or operations are economically desirable.

Of the six methods of economic analysis, the methods of equivalent uniform annual cost and present worth of costs cannot result in an index of the profitability of the single alternative to which the calculations apply. These two methods result in just a quantification of cost without comparison with any base or standard of income or benefit. In the other four methods the economic desirability, or profitability, can be determined separately by individual alternatives only when the data available provide an income or dollar-priced benefit which results solely from that alternative. When this specific income or dollar benefit does not exist, other devices must be used to determine the economic desirability of a single alternative, such as the null alternative.

METHODS OF MEASURING THE ECONOMIC DESIRABILITY OF THE NULL ALTERNATIVE

Proof of the degree of economic desirability of an existing highway facility may be approached in the following ways.

1. Determine that public demand, welfare, or choice is to continue the function made possible by the existing facility. This determination may be based on willingness to pay or just public policy.

2. Compare the costs to society without the facility and with the costs with the facility as is.

3. Determine that the use of the facility and that the public's willingness to pay the costs associated with its use are evidence of its economic desirability.

As examples, consider the following specific proposals involving an existing facility or function (the null alternative) and a proposed improvement:

A. Replacing a river bridge with one more safe and more suitable to the traffic.
B. Improving the traffic flow through a street intersection.
C. Designing a highway culvert for cross flow of natural drainage.
D. Resurfacing a section of a secondary rural road.
E. Establishing a new system of mowing rights-of-way using a new and different type of equipment.

The economic value of the bridge in example A could be determined by comparing its annual cost with the costs to society without the bridge, which cost would include travel across the river by any alternative route or mode available. In effect this scheme is merely setting up other alternatives to compare against the existing bridge. In this type of crossing if canoes or ferries proved to be less costly than the bridge, presumably the bridge would not be replaced, but a crossing facility would be maintained. But the real question this approach does not answer is—are the trips across the river necessary in the first place. This case really resolves itself into determining whether the function is desirable, rather than whether the bridge is desirable.

The intersection example B is somewhat similar to the bridge example. If the intersection carries 1,000 vehicles a day, it must be useful to that extent at least. If the intersection were to be abandoned these 1,000 vehicles would have to travel other routes to reach their destinations. The travel costs by these other routes could be compared with the travel costs using the intersection. But here again there is the implied assumption that the trips through the intersection are economically desirable.

Example C involves project formulation. The basic question is: Is a culvert necessary? By judgment of the adverse consequences to the highway and to traffic without some device to keep the drainage water off the highway, the designer concluded that in some manner or other the water must be taken care of. Presumably the designer has considered all other schemes or will do so. Again, the approach is the "with and without" analysis.

A rural secondary road (example D) may carry light traffic. Does this traffic justify keeping the road and maintaining it? Instead of resurfacing the road why not abandon it? Perhaps its only service is to serve the adjacent land. Without this access the value of the land would likely materially decrease. The people living on the road want access to their property and their livelihood depends upon this access. This example may be a case where the combined social and economic values of the road must be accepted as being worth the cost of the road. The social and economic values of the road as land access and for such services as mail delivery and school bus routes are difficult to price, so compar-

Methods of Economic Analysis

isons of the value or costs of these services with the road costs are not practical. Perhaps the willingness to pay the costs of the road is proof enough to warrant its retention.

According to the principle that an existing facility or function should be proved desirable before improving it, the rights-of-way mowing proposal in example E calls for first proving that the mowing of rights-of-way is a desirable (and economically valuable) function. Here again pricing the value of the service is difficult. In this case perhaps there are factors other than those of cost comparisons to prove that rights-of-way should be mowed. Some states have laws requiring that roadsides be mowed to prevent weed seeds from spreading to farm land. From the viewpoint of aesthetics and custom, the public demands that rights-of-way be maintained in a state of reasonable attractiveness. Mowing of roadsides is practically equivalent to mowing one's front yard in a residential area. Mowing rights-of-way, then, must be concluded to be desirable because people demand that it be done.

EXAMPLE WITH SPECIFIC COST

The foregoing five examples are typical of the null alternatives that are found in highway economic analyses. Another way of examining the economic desirability may be illustrated. Assume the following data apply to a specific highway section being considered for improvement:

Initial cost–sunk (and the roadway is 100 years old)
Annual highway maintenance–$400
Annual road-user cost of travel excluding road-user taxes–$2,000
Annual road-user tax generated–$300
Value of road to adjacent property and the community–not determined

Because the road users are paying $2,300 a year to use this section of road and it costs only $400 a year to maintain it, one could conclude that they desire to keep it. But actually, as road users, they are being subsidized $100 a year from their property tax, other taxes, or by other sources. If the road is viewed solely on the basis of road-user tax generated it is not economically desirable because of the deficit of $100. But because of the value of the road as an access way to the property, the community without question would vote to keep the road in use.

Should the annual road maintenance costs exceed the road-user total personal costs, say $900 and $700, respectively, it could be concluded that the road was not economically sound. But here again, the real test is whether the service is worth its cost of $900 plus $700, or $1,600. In the minds of those who use the road the answer is yes, because they pay that cost. This is a good example that the test of highway transportation (and of other modes) is not the transportation itself, but it is the objective accomplished by using the facility of transportation. Roads are not for roads, but for accomplishing social and economic objectives.

THE ALTERNATIVE OF ABANDONMENT

The procedure to prove that an existing facility or function is economically or socially desirable to continue before testing it for probable improvement is a sound procedure. But with highways, such proof cannot be made by comparing costs with incomes, so resort must be to the "with and without" comparison of costs only, or by examining the value of the end product of road use in terms of the total transportation cost.

Proposals for improving highway facilities in which there may be some question of whether the facility or function could be abandoned or materially reduced are so infrequent that it is easy to forget this step in analysis. It is important in this step to distinguish between the existing facility and the function performed by the facility. Most often the examination will be directed toward the function or end objective which requires the use of the facility or a substitute for it. The analyst should be aware that abandonment or curtailment is one alternative. This consideration is in conformity with the principle that all alternatives are to be considered.

SUMMARY DISCUSSIONS OF THE SIX METHODS AND THE NUMERICAL SOLUTIONS

The foregoing series of applications of the six methods of economic analysis to the example proposals and the subsequent explanations bring out both the concept of each method and the limits of application. Summary of the characteristics will point up more clearly the main differences in the methods, their similarities, and their restrictions in application.

THE COST METHODS APPLY ONLY IN PROJECT FORMULATION

The equivalent uniform annual cost and the present worth of costs methods are directly equivalent; the solution by either method is convertible to the solution by the other method by dividing or multiplying by the capital recovery factor. The methods are applicable to each alternative separately; it is unnecessary to have a pair of alternatives or proposals.

Each of these two methods is conceptually designed to compare alternatives where there is no inward cash flow from sales or benefits, i.e., they compare capital investments plus annual operating expense. Therefore they are directly applicable to project formulation analyses and not to economic evaluation analyses.

APPLICATION TO INDIVIDUAL ALTERNATIVES

The equivalent uniform annual net return, the net present value, the benefit/cost ratio, and the rate of return methods are applicable to single alternatives or

Methods of Economic Analysis

proposals only when each alternative or proposal has its own individual income or dollar benefits, otherwise these methods are applied to pairs of alternatives or proposals. When the application is by pairs (the most usual case) the difference in cost is used as being equivalent of an income or dollar benefit.

FORM OF NUMERICAL SOLUTION

When the data for a pair of alternatives give the motor vehicle running cost or total road-user cost for each alternative, a solution is possible by each of the six methods; and each method will indicate the same alternative as the better one of the pair. The answers, however, are widely apart numerically because each method gives a solution in widely different concepts and units. For instance, in comparing proposals A_1 and A_2 the following answers were obtained (dollar signs are applied to facilitate the comparison).

	A_1	A_2	A_2 compared to A_1
EUAC	$ −99,103	$ −98,394	–
PWOC	−664,989	−660,227	–
EUANR	–	–	–
NPV	–	–	$ 709
B/C ratio	–	–	$ 4,762
ROR	–	–	1.31
			12.1%

In each solution alternative A_2 is superior to A_1. But not one of the solutions gives a direct index to economic feasibility because there is no measure of the economic feasibility of A_1, to which A_2 is compared. But as to the numerical answers of $4,762, 1.31, and 12.1 percent the decision maker must appreciate just how these answers were obtained before attempting to evaluate them. Probably to most decision makers the 12.1 percent return is the most meaningful because he is likely to be familiar with rates of return on investments.

IMPORTANCE OF DIFFERENCES

The most important result illustrated by the foregoing solutions is that every mutually exclusive alternative must be compared on a differential basis with each other alternative to make certain that the added increments of investment costs produce more benefits (or sales) than the incremental investment cost. For instance, the solution in Table 7-2 for benefit/cost ratio and the rate of return methods when compared solely with alternative A as the base indicate incorrectly that alternative B is preferred. The net present value and equivalent uniform annual net return methods make the correct selection of alternative E. The benefit/cost ratio and the rate of return incremental solutions also select alternative E.

CHARACTERISTICS AND LIMITATIONS OF THE METHODS OF ECONOMIC ANALYSIS

The foregoing equations and the solutions of illustrative alternatives offer a basis of further explanation and comparisons of the methods of analysis for economy. Each method has its own concept and logic, advantages, disadvantages, and procedures. The more important characteristics of each method of analysis are discussed in this section and other more general factors are discussed in later sections of this chapter.

EQUIVALENT UNIFORM ANNUAL COST METHOD

The method of analysis which is known as the equivalent uniform annual cost method requires as input data the investment costs, initial and those investments at any subsequent date within the analysis period, together with the year by year annual expenses for overheads, operations, maintenance, and production. The item of cash inflows, or benefits, is not considered. Terminal values of the investments are included, but, although positive in sign, are not considered as an income or benefit, but as a reduction in capital cost. The method has popularity because of its simplicity and because most persons readily understand the meaning of average annual costs.

The equivalent uniform annual cost method is applied mainly to those project formulation proposals for which there is no ready or usable way to establish a dollar measure of benefits; the general situation is that the benefits are equal as between the mutually exclusive alternatives. However, when this method is applied to a highway proposal which does involve motor vehicle running costs there is a measure of benefit between two or more alternatives in the form of any differential in the vehicle running costs or other form of consequences. This method is widely used in industry as a means of selecting the preferable machine, process, or design feature.

The Factor of Economic Evaluation

Another distinctive characteristic of the equivalent uniform annual cost method is that it cannot be used to measure the economic desirability of proposals. The equivalent uniform annual cost calculated for a proposed improvement or an existing facility gives no indication of whether the improvement is economically warranted except as the decision maker may intuitively decide that the calculated annual cost is below the marginal annual cost above which the facility would not be acceptable. The significance of the difference in equivalent uniform annual costs of a pair of alternatives is simply to identify the difference. The difference does not mean that the incremental investment and annual expenses contributing to the annual cost difference are economically warranted.

Methods of Economic Analysis 143

The equivalent uniform annual cost method is consistent and reliable. Since it does not utilize the factors of benefit or income, it is not subject to some of the cautions of reverse cash flows as is the rate of return method and the question in the benefit/cost ratio method of whether to place the annual expense factor of operation and maintenance in the numerator as a deduction from benefits or in the denominator as an addition to annual cost.

The Level-of-Service Factor

The one really important factor that must be carefully considered when using the equivalent uniform annual cost method is that in mutually exclusive alternatives, and other similar comparisons, it is essential that they be compared only under conditions of equality of service performed. The best illustration of this requirement is given by considering three locations for a highway route, of which only one route will be constructed. Often in this situation the location of the route affects the volume of traffic attracted, and, therefore, the annual total vehicle operating costs will be directly affected–often by large annual dollar amounts. Thus, if the *ADT*'s are 8,000, 10,000, and 13,000 for the three alternatives and the vehicle running costs average 5 cents per vehicle-mile (the running cost may vary somewhat between the three locations, however) the vehicle cost portion of the total transportation cost for one mile of the route would be $400, $500, and $650 a day respectively. Generally, under such situations the alternative with the lowest volume of traffic will have the lowest equivalent uniform annual cost, and only for the reason that it carries less traffic and not because it has greater economy in handling the traffic.

PRESENT-WORTH-OF-COSTS METHOD

The present-worth-of-costs method is directly convertible to the equivalent uniform annual cost method. The two methods are made numerically equal by multiplying either answer by the appropriate compound interest factor. In fact, the logical procedure in combining irregular flows of cost in the equivalent uniform annual cost method is to first calculate their present worth, then multiply by the capital recovery factor. Unless the annual cost is wanted for a particular reason, it is not necessary to calculate the equivalent uniform annual cost in those analyses wherein the present worth solution must be used in order to compute the annual cost. The present worth of the cash flows of cost is every bit as useful and reliable as a measure of the comparison of alternatives as is the equivalent uniform annual cost. Generally though, the present-worth-of-costs method is not as well understood as is the equivalent uniform annual cost method, especially by public officials.

The discussion and explanation already given for the equivalent uniform annual cost method applies equally to the present-worth-of-costs method.

EQUIVALENT UNIFORM ANNUAL NET RETURN METHOD

The method of equivalent uniform annual net return has the same general characteristics as does the equivalent uniform cost method. The significant difference is that the factor of income or benefit has been added so that the final answer is in terms of annual net return rather than annual gross cost. This additional term provides the measure of economy, or productivity, of the alternatives being compared. Since the minimum attractive rate of return is used in the calculations, the final equivalent uniform annual net return is a direct measure of economic desirability. The higher the equivalent uniform annual net return the greater the economic desirability so fas as mutually exclusive alternatives are concerned.

The inclusion of the benefit or income factor limits application of the method of equivalent uniform annual net return to those proposals that produce an annual income or annual benefit that can be dollar-priced.

The method is applicable to alternatives, each separately, and the comparison is made by the magnitude of the answers. It is equally applicable to economic evaluation and to project formulation. As long as the equivalent uniform annual net return for a proposal is positive, the proposal will be earning a rate of return more than the minimum attractive rate of return used in making the analysis.

The equivalent uniform annual net return method is reliable and consistent. Since it measures consequential benefits against costs, the level of service rendered is not a critical factor. Presumably, higher costs would produce higher incomes or benefits, and, therefore, the changes in cash flows are in harmony as between alternatives.

The discussion on the net present value method will point out that these two methods are equivalent; they merely express the answer in terms of annual net return in one method and net present worth in the other method.

The equivalent uniform annual net return method can be applied to mutually exclusive pairs of alternatives on the basis of differences. In such application, the reduction in annual expenses and costs is assumed to be equivalent to an annual cash income. However, in this application the method cannot be used to determine the economic profitability of either alternative. The answer merely determines the profitability of the incremental changes in cost of the proposed alternative as compared to the other alternative, or the base condition.

NET PRESENT VALUE METHOD

The method of net present value and the equivalent uniform annual net return method may be made numerically identical by multiplying by the appropriate compound interest factor.

The net present value method is the favorite of the economists [7-7, 7-23, 7-26-1, 7-50]. It is a highly useful method in business applications wherein there

Methods of Economic Analysis   145

results a stream of income dollars from an investment in production property or processes. The numerical answer obtained by the present net value method is simply the present worth of the net positive cash flows to be anticipated. The net present value is calculated at the assumed cost of capital rate or the minimum attractive rate of return.

General Applicability

The net present value method is applicable to single proposals wherein the other alternative is to do nothing. In such case it is assumed that the null alternative is actually to leave the money where it is. Presumably the money is currently earning at the minimum attractive rate of return. When by the net present value method the present value is positive, the proposal would be earning at some rate above the minimum attractive rate of return. When mutually exclusive alternatives are being considered, the better choice is that alternative producing the greatest net present value.

When there are two or more reversals of the sign of the net accumulated cash flow or there are highly irregular fluctuations in either the positive or negative cash flows, or both, it is possible to reach misleading conclusions with the net present value method, the present-worth method, and the rate-of-return method. See page 162 for a discussion of irregular cash flows.

With the net present value method of analysis it is unnecessary to make the incremental analysis when each alternative has its individual income or benefits. When the challenger alternative shows a net present value greater than that of the defender, it follows that the difference in investment will likewise have a positive net present value when computed at the identical discount rate.

The net present value method is applicable to pairs of alternatives in the same way that the equivalent uniform annual net return method must be applied. In this application, the answer shows the profitability of the incremental investment, which is a bit of helpful information to the decision maker, but whether either of two alternatives as a whole is profitable is not determined.

General Limitations

The limitation of the net present value method is that it cannot be applied to single alternatives except when the benefits or incomes can be estimated for each alternative. Where this estimate of income is not possible, the net present value method can be used in testing a pair of alternatives when there is a reduction in yearly operating expenses involved. In this situation, the reduction in annual expense can be considered to be an annual benefit, or income, to be compared to the incremental change in investment to achieve this reduction in annual expense. In this application the net present value method still produces the measure of desirability in the form of the present worth of the net returns, but it applies only to the difference in the pair of alternatives.

Although the net present value method gives a direct index to the overall profitability of the proposal, it does not express this profitability as a rate; it is expressed as a lump sum for the period of the analysis. On this score the method is not as enlightening as is the rate-of-return method. Managements are more prone to want to know the rate of return than they are the present-worth lump sum of net returns over time.

The level of service in the net present value method is not a factor in getting comparable answers between mutually exclusive alternatives, but it is a factor in interpreting the results. The method may be used for either project evaluation or project formulation (wherein there is an income stream which is priceable), but in project formulation care is to be exercised in comparing those alternatives which do not render the same service so that any increased net present value may be traced to those features of design that actually cause a greater flow of income or benefits as distinguished from those features of design that are only economy in the use of materials, construction methods, or future operation and maintenance.

Application to Highways

As a whole, the net present value method has no particular advantage in economy studies of highways, and in many instances it is not applicable. It can be used for estimating the economy of alternatives which reduce the motor vehicle operating costs, but in such applications it is necessary to consider the reduction in road-user costs between two alternatives as the equivalent of a sales income.

Basically, the decision to use the net present value method in preference to the benefit/cost ratio method or the rate-of-return method is simply in what qualitative form (unit) is the final answer more meaningful to the decision maker and whether the favorable consequences are priceable for each alternative.

BENEFIT/COST RATIO METHOD

The benefit/cost ratio method is one that is found mainly in the public works field, though it is as applicable to industrial applications as is any of the other methods. Likewise, the other methods are equally applicable to public works. Its use was suggested in the U.S. Flood Control Act of June 22, 1936, by the wording "...if the benefits to whomsoever they may accrue are in excess of the estimated costs, ..." The American Association of State Highway Officials [7-2] describes only the benefit/cost ratio method in its "Information Report by the Committee on Planning and Design Policies on Road-User Benefit Analyses for Highway Improvements." Because this publication has been the only one in the highway field, the benefit/cost ratio method has been used almost exclusively since 1952 in the application of economic analysis in the highway field.

Methods of Economic Analysis

Characteristics of the Method

The data required for use in the benefit/cost ratio method are identical with those required for other methods. The benefit/cost ratio method is distinctive only in that the form of the answer is an abstract number that represents the ratio of net benefits (usually the equivalent uniform annual net benefits) to the net costs (usually the equivalent uniform annual costs). In this form, the answer is a useful index to the profitability of the proposal as compared to its base.

In application, the benefit/cost ratio method must have a measure of the benefits of the particular project or alternative being analyzed. This benefit can be obtained for a single alternative only when that alternative provides a whole measure of its benefit in the form of its own generation of income or benefits. When one is to measure benefits, however, as distinguished from sales income, it is usually found that the benefits are measured from a prior or existing situation as compared to a proposed alternative. Thus in most applications of the benefit/cost ratio method as with the rate of return method, the numerical analyses requires a pair of alternatives in order to produce a measure of the benefits. In most of these cases the benefits are measured by reductions in annual expenses or by increase in the value of services as between the two alternatives.

When properly handled, including the use of the differential analysis, the benefit/cost ratio method is applicable to a group of more than two alternatives and to both project economic evaluation and to project formulation. Smith [7-43] reports differently on this point, but as stated in Fleischer's discussion, the proper procedure will produce results consistent with the rate of return method. Although most frequently applied as the ratio of equivalent uniform annual benefits to the equivalent uniform annual cost, the corresponding ratio of the present worths or of the compound amounts is equally reliable.

Because the answer is in the form of a ratio, or rate, the level of service as between the mutually exclusive alternatives under study does not affect the comparison in the end, as is true of the equivalent uniform annual cost and present-worth-of-cost methods.

Relation to Public Works

Taylor [7-45, pp. 386-389] offers the concept that the benefit/cost ratio method is especially adapted to public works on the basis that the method places the emphasis on the benefits received instead of on profitability, and also because the benefits may accrue to a group different from the investors. Taylor further expresses the view that one disadvantage of the rate-of-return method when applied to public works is that it gives the implication "that the rate of return is the return to the investors, which as we know is wrong because the direct beneficiaries, those who benefit from flood control for example, and the benefactors, the tax payers, may be distinctly separate." When it is considered that the public supplies all taxes and gains all benefits therefrom, who furnishes what in the way of support for public works is immaterial in calculating the economy of proposed public works.

Any separation of benefactors and beneficiaries, if it need be done, is a responsibility of the legislative body in connection with financing the project and allocating the cost responsibility. Such factors of finance are not a part of the process of determining the economic wisdom of constructing the facility. The concept of the people earning a return on their investments in public works is just as valid as it is when applied to their private investments. The fact that any "benefit" accruing to the people from a public work must first be converted to dollars before it can enter the process of economic analysis, destroys the concept that benefits are different from returns. In fact, perhaps the public and the legislative groups would understand public works much better if they did regard them in the sense of a business investment. It is largely by unintended wording of federal law that the benefit/cost ratio method has become popular in public works analyses and not because someone thought its concept more appropriate than the rate of return method or one of the other methods that might be applicable.

Position of Annual Expense Factor

Some writers [7-45, p. 386; 7-43, pp. 77-90; 7-50, p. 69] have criticized the benefit/cost ratio method because they claim that there is uncertainty of whether the annual highway maintenance and operating expense is a deduction from income, or benefits, in the numerator or is an addition to annual highway costs in the denominator. This factor in the procedure of solution for the benefit/cost ratio really offers no decision problem once the procedure is reduced to logic. Logic will indicate that the annual maintenance and operating expense is a deduction from annual income or benefits.

Benefits versus Expenses. The phrase "negative benefits" has been often used to refer to the annual maintenance expense. Such concept is illogical. In the use of the concept of negative benefits, it would be just as illogical to call the original investment "negative benefits," though such concept has probably never been suggested. Cost–capital or annual–cannot be called income or benefits. They are outward cash flows and must be so treated. Income and benefits are inward cash flows and must be treated as such.

Numerator versus Denominator. The concept of negative benefits comes from the procedure of putting the annual highway maintenance cost in the numerator of the benefit/cost ratio expression as a subtraction from the annual road-user benefits. Thus:

$$B/C \text{ ratio} = \frac{\text{annual road-user benefits minus annual highway maintenance expense}}{\text{annual highway capital cost}}$$

Methods of Economic Analysis

$$B/C \text{ ratio} = \frac{\text{annual road-user benefits}}{\text{annual highway capital costs plus annual highway maintenance expense}}$$

These two basic ratios have both been used. The second one with the annual highway maintenance expense in the denominator has been popular only because the publication by the Association of State Highway Officials [7-2] on road-user benefit analysis used this form. The logic of this form is to place the road-user costs (benefits) in the numerator and all the highway costs in the denominator. But in terms of economics, economy, and cost accounting it is much more logical to put the repetitive annual cash flows in the numerator and the capital investments in the denominator.

When the benefit/cost ratio is 1.0 both forms of the expression for the ratio yield the same result. But when by either form the ratio is not 1.0 the two forms will produce different ratios. For instance:

$$B/C \text{ ratio} = \frac{U_D - K}{I} = \frac{15 - 5}{-10} = 1.0$$

$$B/C \text{ ratio} = \frac{U_D}{I + K} = \frac{15}{-10 - 5} = 1.0$$

$$B/C \text{ ratio} = \frac{U_D - K}{I} = \frac{15 - 10}{-2} = 2.5$$

$$B/C \text{ ratio} = \frac{U_D}{I + K} = \frac{15}{-2 - 10} = 1.25$$

Logical Considerations. In the benefit/cost ratio method the position in the ratio fraction of the annual highway maintenance expense or certain factors of benefits may be analyzed by considering the following points of the analysis:

1. Conceptual logic–what is being sought
2. Normal handling of expense and income in profit and loss accounting
3. Position of the annual expense factor in the other methods of analysis

Objectives of Analyses. All six of the methods of analysis for economy have as their basic objective the search for the efficient use of money. The equivalent uniform annual cost and present-worth-of-costs methods each compare alternatives on the total outward flow of cash, but the net present value, benefit/cost ratio, and rate-of-return methods compare alternatives on the rate of inflow of returns or benefits as compared to outward flow of cash. The real base of the outflow of cash is the investment, or initial cost. The real objective of these three methods, therefore, is to see whether the original investment, plus any subsequent investments, will return their investments with sufficient vestcharge to make the investment economically desirable. The net present value, the benefit/cost ratio, and the rate of return of any pair of alternatives in terms of the objective, then, are best expressed by using the investment cost as the base upon which the net benefit is to be measured. In terms of the benefit/cost ratio this means that the ratio should be expressed as

$$\frac{\text{gross benefit} - \text{annual expense}}{\text{investment}}$$

Putting the annual maintenance expense in the denominator is illogical and conceptually unsound in terms of the objective.

Cost Accounting Procedure. In the ordinary cost accounting procedure and profit and loss statements the concept most often used is that operating expense is a deduction from income before gross profits are stated. It is true, however, where depreciation expense accounting is practiced, depreciation expense would be included in total operating expense. However, the basic concept is still income minus expense gives profits.

Consistency with Other Methods of Analysis. The normal form of equations for calculating the net present value and the rate of return utilizes the annual maintenance expense as a deduction from income rather than as an addition to capital costs. Since all six methods of analysis will indicate the same alternative as the first choice, it is logical to handle the terms in all of the equations of solution in a consistent manner.

Using a simplified form of the rate-of-return solution the following equations may be written:

$$0 = -I(CR\text{-}i\text{-}n) + U_D - K$$

$$(CR\text{-}i\text{-}n) = \frac{U_D - K}{I}$$

and it is logical to use the same form for the benefit/cost ratio.

Consistency with the Objectives of Analysis. Based upon the objective that the analysis is to indicate the profitability of the invested capital, the annual expenses must be placed in the numerator. For instance, large annual expenses in proportion to the invested capital lead to misleading ratios when the annual expenses are placed in the denominator.

$$B/C = \frac{U_D - K}{I(CR\text{-}i\text{-}n)} = \frac{100 - 20}{-1} = 80$$

$$B/C = \frac{U_D}{I(CR\text{-}i\text{-}n) + 20} = \frac{100}{-1 - 20} = 4.8$$

The ratio of 4.8 really has no meaning because essentially it means that the gross profits were 4.8 times the annual operating expense, and the return on invested capital is not calculated. But the ratio of 80 does reflect the size of the net return on the invested capital.

The foregoing explanations and reasoning point to the logical conclusion that

Methods of Economic Analysis 151

in the benefit/cost ratio method of analysis the annual expenses of highway maintenance and operation should be placed in the numerator as a subtraction from the gross benefits.

Position of the Terminal Value Factor

There has been no general suggestion by the writers on the subject of the benefit/cost ratio to place the terminal value in the numerator along with the road-user benefits. But such terminal income is just as logically a benefit as is any reduction in road-user costs. Further, the terminal value is more logically a positive benefit than is annual highway expense a negative benefit.

Logical analysis will lead to the conclusion that terminal value is a deduction from capital cost, and not to treat it as such leads to overstating capital costs upon which the degree of profitability is measured. If terminal value were 100 percent of capital investment and were placed in the numerator as an addition to benefits, the resulting benefits from the project would be grossly overstated; to place this 100 percent terminal value in the denominator reduces the capital cost to just the charge for use of the capital which is correct, for in reality there is no capital consumed.

RATE OF RETURN METHOD

In each of the five methods just discussed, it is necessary at the start of their application to select a vestcharge rate or discount rate by which to reduce the cash flows to an equivalent basis with respect to the time periods involved. In the annual rate method selection of the vestcharge rate or minimum attractive rate of return can be delayed, sometimes clear to the decision stage. The rate of return method has as its objective the finding of that discount rate that will exactly equate the negative to the positive cash flows so that their algebraic discounted sum is zero. In applications where there are two or more reversals in the sign of the summation of cash flow, there may be two or more solutions, that is, the basic equation setting the cash flows equal to zero may be satisfied with two or more discount rates. There can be situations where no solution is possible. This characteristic of the method is the only real objection offered to its use. The subject of irregular and reversing cash flows is discussed on page 159.

The rate-of-return method is applicable to the benefit and cost streams to the same extent as is the benefit/cost ratio method and the net present value method. It is necessary that the alternatives have a dollar-priced benefit upon which to compute the rate of return. This benefit can be actual cash incomes, as in business ventures or toll facilities, or it may be a cost reduction between a pair of alternatives used as the equivalent of income. The method may be applied to a single proposal when an income is identifiable, but generally in the public works area the application is to the differences in a pair of alternatives. It is equally applicable to a group of more than two mutually exclusive alternatives when applied to a series of pairs as is required in the benefit/cost ratio method.

The rate of return method is equally applicable to project evaluation and to project formulation as long as there are dollar-priceable benefits associated with each pair of alternatives. The method is especially suited to project evaluation because the end product is the percentage rate of return which gives a direct index as to the degree of profitability of the proposal. To most decision makers this rate of return is more meaningful than is a net present value or a benefit/cost ratio.

The level of service afforded by each alternative in a group of mutually exclusive alternatives is in no way a restrictive factor in the use of the rate of return method of analysis. In fact, one of the advantages of the method is that it is capable of reducing alternatives of different levels of service to a common base (rate of return) for comparison. The benefit/cost ratio makes a similar reduction, but the abstractness of the final ratio is more difficult to interpret than is a rate of return.

OTHER METHODS OF ANALYSIS

Over the years during which economic analysis has developed as a managerial tool there have been many methods developed for examining proposals for investment of capital in order to bring to light their economic worth. Many methods are variations of others, and some have but little merit, particularly in view of later improvements in the methods of economic analysis. The six methods described in this chapter cover the most important ones and offer reliable means of investigating the economic merit of a wide variety of proposals. Just for the sake of information three other methods are mentioned to round out the list.

Capitalized Cost

The capitalized cost method of comparison of alternatives is simply the present worth of costs method when the analysis period is assumed to be infinite, or the operation being examined is assumed to be perpetual. With high rates of vestcharge, the capitalized cost method will give results comparable to the present worth method when the analysis period is 50 or more years. The capitalized cost method is not now often used, but it was a popular method in the early days of railroad construction. When present-worth solutions are desired, a specific time period short of infinity is preferred. The capitalized cost method is equally suited to project evaluation and project formulation, but a poor choice of method in either application.

Methods of Economic Analysis

Payback Period

The payback period, or payout period, has been used as an index of the merit of investment proposals on the basis that the alternative having the shorter payback period is the better choice. Payback is defined as that period of operation required to accumulate net profits, or savings in costs, which equal the investment without including a discount factor. This method has two main objections. First, it does not recognize the time value of money, including the relative rates of payback within the payback period between the alternatives, and second, no recognition is given to the value of future possible operations after the end of the payback period. The method has little or no application to public works. In fact, about its only element of merit is found in situations where management desires early–within a year or so–return of its capital and the analysis is made more for project formulation than for economic evaluation.

Break-Even Analysis

Break-even analysis is a process of comparing alternatives by controlling certain factors while one factor is allowed to vary until a point of equality is reached between two situations. For instance, the relative merit of a gravel surfacing for roads as compared to a low-cost bituminous pavement can be measured by calculating the daily traffic volume at which the two equivalent uniform annual transportation costs are equal. Another example would be to determine the traffic volume required for a specific highway which would equate the road-user tax earnings and the equivalent uniform annual highway cost.

The break-even analysis has many applications in highway management in developing criteria for decision making and in locating situations where attention to highway operating costs is in order. The process of locating the break-even point requires no special procedures. It is essential, of course, to see that all factors are included and priced appropriately.

RANKING OF INDEPENDENT PROJECTS

The process of getting together a program of highway construction for a specific year or period of years should take into consideration the relative economic merit of the independent projects which will make up the total program. (This is discussed further in Chapter 26.) The analysis for economy is one possible ranking factor.

APPLICATION WHEN THE CAPITAL BUDGET IS NOT RESTRICTED

When the capital budget is not restricted, the independent projects may be programmed in order of their economic desirability as indicated by the net present value, the benefit/cost ratio, or the rate-of-return methods. Under this situation the total program will maximize the benefits for the specific total capital investment. Within each project the most economical alternative would be programmed.

APPLICATION UNDER A CAPITAL BUDGET RESTRAINT

The total construction program for a year may be restricted to a specific maximum capital budget. Under such circumstances it becomes necessary to compare alternatives within a given project to alternatives within other independent projects. The iterative procedure required for these multicomparisons can get heavily involved when there are numerous projects and several alternatives within each project. See Grant and Oglesby [7-17], Haney [7-19], and Curry [7-9] for conceptual discussions and applicable procedures.

When the conditions are just right, the application of the iterative procedure will select an alternative within a project less favorable than the most economical alternative because the differential investment is less attractive than the differential investment in a competing project, and the budget restraint will not permit both proposals to be adopted. A single large project that produces an acceptable return may have to give way to two or more less costly projects (or alternatives) with less favorable economy, because to accept the large project would require more capital funds than are available. The objective here is to maximize the total benefits within a total fixed capital budget as opposed to maximizing the benefits within a single project.

FACTORS FOR SPECIAL ATTENTION

There are a few situations that require special attention in the analysis for economy. They need to be recognized to avoid pitfalls and understood to achieve correct handling. Unequal levels of service among mutually exclusive alternatives have been mentioned in connection with the discussion of the six methods. At times the service lives or analysis periods among a group of alternatives will be unequal. Although the normal flow of cash for investment proposals usually begins with the negative flow of the initial investment followed by annual flows of expense and of income, there are times when this sequence is reversed at the beginning or within the analysis period. Annual cash flows are usually assumed as uniform, as gradients, or as exponentials, but more often than not in real life the annual flows fluctuate year to year. Some authorities state that the rate of earning on the reinvestment of the returns is a factor to consider. These special factors are discussed in terms of the different methods of analysis.

Methods of Economic Analysis

UNEQUAL LEVELS OF SERVICE

The books and journal papers on the subject of economic analyses rarely mention that mutually exclusive alternatives may render unequal service, and that because they do not accomplish identical goals their costs and benefits can differ widely in quality and quantity. For instance, there may be four mutually exclusive route locations being considered for an urban freeway. Because of their locations they each will attract different traffic volumes and from different sources. The ADT's may be 8,000, 13,500, 16,000, and 21,000. Thus the equivalent uniform annual transportation cost would not be an acceptable index of superiority because the lowest transportation cost route would most likely be that alternative with the 8,000 traffic volume. However, when the total area traffic affected is considered the effects of these ADT's on the proposed freeway may be lessened materially. The same deficiency is found in the present-worth-of-costs method and for the same reason.

The equivalent uniform annual net return and the net present value methods are applicable to mutually exclusive alternatives with unequal services because the ranking factors are in effect sales, less costs, or benefits, less costs. Therefore, any costs caused by added services are properly related as between alternatives.

The benefit/cost ratio method and the rate-of-return method will identify the most economical alternative with certainty and maximize the benefits. These two methods give solutions which are rates of benefits based upon investment inputs.

In project formulation the equality of service is usually not a factor to control because each alternative must be designed for that traffic, load, or service it will be expected to carry. In economic evaluation the incremental procedure will assure that the incremental costs and, therefore, the incremental services, are economical when compared by pairs of alternatives.

In the end, the final choice of alternative with respect to both economic evaluation and project formulation may be made by management on the basis of objectives. Is the route to serve local traffic or through traffic as its main purpose is an important question for management to answer.

When mutually exclusive alternatives have different levels of service–unequal ADT's–their economic comparison must be on some rate basis such as the benefit/cost ratio or rate of return methods. The net present value method is also acceptable when a reliable measure of the returns is available and when any increase in service provided is properly measured in both project costs and project benefits.

The critical procedural question in the analysis for economy of mutually exclusive alternatives when the ADT varies as between the alternatives is that of computing comparable cost reductions for the ADT. Usually, especially in urban freeways, the traffic on the new facility will be much greater than that on the facility replaced. Further, this traffic will be composed of vehicles diverted from other highways, transferred from other modes, generated travel, normal growth, and developed traffic. Thus, computing the net saving of these components of the ADT presents difficulties because of the absence of or the uncertainty of their running cost before their use of the new facility.

A proper procedure for analysis of a set of mutually exclusive proposals having different traffic volumes, or levels of service, is to compute the total transportation cost in the area or corridor affected separately for each alternative on a before and after basis. This procedure is in keeping with the principle of considering all consequences to whomsoever they may accrue. In effect, this procedure treats the mutually exclusive proposals as independent proposals.

ALTERNATIVES WITH UNEQUAL ANALYSIS PERIODS

Not only must the cost factors and the benefit, or income, factors within a given proposal be considered over the same time period, but the alternative proposals being compared likewise must be compared over equal time periods. Sometimes they are equal by assumption and under other situations the calculations must be adjusted to equalize any difference in time periods.

Effect of Repetition of Cash-Flow Cycles on Equivalent Uniform Annual Cost

Once a cycle of cash flow is established, its equivalent uniform annual cost, benefit/cost ratio, and rate of return will continue the same for any number of repetitions of the whole cash-flow cycle. The present worth and the net present value, of course, will continue to increase with each added cash-flow cycle. Using alternatives A_1 and A_2 (Table 7-1) as examples the following calculations result.

The equivalent uniform annual cost of alternative A_1 as previously developed is

$$EUAC_{A1} = -140{,}000\ (CR\text{-}8\%\text{-}10) + 40{,}000\ (SF\text{-}8\%\text{-}10) - 7{,}000 - 74{,}000 = -99{,}103$$

For two consecutive cycles the solution would be written:

$$EUAC_{A1\text{-}2} = [-140{,}000 - 140{,}000(PW\text{-}8\%\text{-}10)]\ (CR\text{-}8\%\text{-}20)$$
$$+ 40{,}000\ (SF\text{-}8\%\text{-}10) - 7{,}000 - 74{,}000$$
$$EUAC_{A1\text{-}2} = [-140{,}000 - 140{,}000\ (0.463193)]\ (0.101852) + 40{,}000\ (0.06903)$$
$$- 7{,}000 - 74{,}000$$
$$EUAC_{A1\text{-}2} = -20{,}864 + 2{,}761 - 7{,}000 - 74{,}000 = -99{,}103$$

Any number of consecutive cycles of the A_1 cash-flow proposal will produce the equivalent uniform annual cost of $-99{,}103$. This conclusion can be reached by simple logic from the fact that for each cycle, considered separately, the cash flows are the same, and, therefore, the equivalent uniform annual cost over any number of cycles must also be the same.

Methods of Economic Analysis

The correct conclusion is that for any calculation of equivalent uniform annual cost, the calculated answer for any number of consecutive identical cash-flow cycles is identical to that obtained from the first cycle alone. This same conclusion applies also to the equivalent uniform annual net return method.

Effect of Repetition of Cash-Flow Cycles on Present Worth of Costs

When the present-worth-of-costs method is applied to a series of cycles, the present worth of the cash flows will steadily increase with each additional cycle, but it will increase at a decreasing rate. The capital recovery factor decreases at a decreasing rate with an increasing number of interest periods n, with i as the limit. It follows, therefore, that the present worth of the cash flows must also increase with n if the equivalent annual cost is to remain constant with increasing number of cash-flow cycles.

More Than One Cash-Flow Cycle versus NPV, B/C, and ROR

From pages 132 and 133 the following answers were obtained by the methods of net present value, benefit/cost ratio, and rate of return when applied to the first cycle of cash flows of proposals A_1 and A_2 taken by differences:

$NPV = 4,762$
B/C ratio $= 1.31$
$ROR = 12.1$ percent

The solutions will next be examined for two or more cash-flow cycles. The simple way to prove the effect of repeated cycles of cash flows when applied to the methods of net present value, benefit/cost ratio, and rate of return is by logic based upon the calculations and conclusions with respect to the equivalent uniform annual cost method and present worth method.

Whatever the net present value is for the first cycle it will be exactly the same for every future cycle when the cycle cash flows are discounted to the beginning of their respective cycles. The net present value of a consecutive series of cycles is the sum of the present worths of the net present worth of each cycle from the beginning of its cycle back to zero date. Thus,

NPV for first cycle $= 4,762$
NPV for two cycles $= 4,762 + 4,762 \, (PW\text{-}8\%\text{-}10 = 0.463193)$
NPV for two cycles $= 4,762 + 2,206 = 6,968$
NPV for three cycles $= 6,968 + 4,762 \, (PW\text{-}8\%\text{-}20 = 0.214548)$
NPV for three cycles $= 6,968 + 1,022 = 7,990$

As the number of cycles is increased the net present value will also increase in lump sums by the amount of the net present value of each additional cycle discounted back to time zero. Therefore a comparison of alternatives by the net present value method must be for identical analysis periods.

The benefit/cost ratio for any number of consecutive cycles will be identical to the ratio calculated for the first cycle. This statement is true because if including additional future cash-flow cycles does not change the equivalent uniform annual cost or the equivalent uniform annual net returns there cannot be any change in the ratio of the benefits to the costs.

When each cycle of cash flows in a consecutive series of cycles is considered separately the calculated benefit/cost ratio and rate of return will be identical for each cycle. Therefore, benefit/cost ratio and the rate of return for any number of cycles combined will be identical with the solutions for the first cycle. For example, consider the simple cash flow per 5-year cycle of an initial investment of 100 and a net annual return of 25.04 for each of 5-year ends:

$$0 = -100 + 25.04 \, (SPW\text{-}i\text{-}5)$$

When i is 8 percent,

$$0 = -100 + 25.04 \, (3.992710) \text{ or } 0 = -100 + 100$$

For two cycles:

$$0 = -100 + 25.04 \, (SPW\text{-}i\text{-}5) - 100 \, (PW\text{-}i\text{-}5) + 25.04 \, (SPW\text{-}i\text{-}5)(PW\text{-}i\text{-}5)$$

When i is 8 percent,

$$0 = -100 + 25.04 \, (3.992710) - 100 \, (0.680583) + 25.04 \, (3.992710)(0.680583)$$
$$0 = -100 + 100 - 68 + 68 = 0$$

Conclusions with Respect to Unequal Analyses Periods

When an exact repetition of the cycle of cash flows is logical for a second or greater number of cycles, it may be correctly assumed that the equivalent uniform annual cost, the benefit/cost ratio, and the rate-of-return solutions for the first cycle will continue indefinitely for all future cycles.

The present worth and net present value solutions, however, will increase with each added cycle.

These characteristics of the methods may be interpreted to mean that a pair of mutually exclusive alternatives having different analysis periods (service lives) may be directly compared by the equivalent uniform annual cost, benefit/cost ratio, and the rate of return solutions for their first cash-flow cycle as long as (1) it is logical to assume the possibility of additional identical cycles of cash flow, and (2) when the two time periods have a common multiple. Of course, if perpetuity were to be assumed the common multiple would be infinity.

By using the common multiple assumption, the analysis of the pair of alternatives by implication covers an identical time period. For example, if one alternative had a service life of 8 years, the other one would have to be 4, 8, 16, 24, 32... years. But if 17 and 23 years, respectively, were the service lives to be used, and it would be illogical to assume an analysis period of 391 years, an adjustment in the procedure of solution is required. Perhaps also a repetition of the two cash-flow cycles could not be assumed.

Methods of Economic Analysis

Adjusting the Analysis for Unequal Time Periods

The adjustment in procedure to equalize the analysis periods between the alternatives is to use an analysis period equal to the shorter lived alternative and to allow a terminal value for the remainder of the life of the longer lived alternative. This terminal value would be equal to the value of the unexpired service, based upon a pro rata share of the original investment and subsequent investments or upon a reduction of the equivalent uniform annual cost by the value of the number of years of service remaining spread over the analysis period–that is, a reduction in the equivalent uniform annual cost from that annual cost as calculated for the full life period of the alternative.

This type of problem arises in comparing the economy of bituminous pavement with portland cement concrete pavement. The service life of the original pavement surfaces may be 17 and 20 years, respectively, and the resurfacings may have different lives and come at different ages. Overall cycles, including resurfacings, might be 32 and 35 years in which case the analysis period could be 32 years and the alternative having the 35-year cycle would receive a terminal value of 3 years of unexpired service. [2]

REVERSALS IN DIRECTION OF CASH FLOWS

Most of the applications of economic analysis of proposed capital investments involve proposals which follow the normal cash-flow sequence of an initial investment in constructed or purchased property at time zero followed by annual streams of cash expenses and cash incomes or benefits. When these flows of cash are uniform or uniformly changing and the main negative flow of the initial investment comes at time zero, there are no uncertainties attached to the solution and the alternatives will be properly ranked without question. However, when the flows are irregular–that is, fluctuating up and down as between the alternatives being compared–different vestcharge rates may result in different ranking of the alternatives. Further, when there is more than one reversal from negative to

---

[2] The following calculation illustrates one way of adjusting for this terminal value (last 3 years of service). Assume an initial investment of $40,000, a $15,500 resurfacing at age 20, $500 a year equivalent uniform maintenance, and a total service period of 35 years, with zero probability of the cycle repeating itself. The alternative being compared has an overall service period of 32 years.

$$EUAC = -40,000 \ (CR\text{-}8\%\text{-}35) - 15,000 \ (PW\text{-}8\%\text{-}20) \ (CR\text{-}8\%\text{-}35) - 500$$
$$= -40,000 \ (0.08580) - 15,000 \ (0.2145) \ (0.08580) - 500$$
$$= -3,432 - 276 - 500 = -4,208$$

Adjustment of the $4,208 from 35 years to 32 years:

$$SPW = 4,208 \ (SPW\text{-}8\%\text{-}3 = 2.577) = 10,844$$

Equivalent uniform annual adjustment to 32 years from 35:

$$10,844 \ (SF\text{-}8\%\text{-}32 = 0.00745) = \$81$$

Adjusted equivalent annual cost for 32 years = $4,208 - 81 = \$4,127$

positive or positive to negative in the stream of cash flows, the rate-of-return method will result in two or more rates [7-7, 7-15, 7-26-1]. In the other methods, two or more rankings of the alternatives are also possible in this situation should different discount rates be used in the calculations. In other words, all the methods which involve negative and positive flows of cash offer the possibility of different rankings of mutually exclusive alternatives with different choices of the vestcharge rate used in the calculations. A sample calculation will illustrate these possibilities.

Multiple Rates of Return with Reversals of Cash Flows

Table 7-3 gives the net present values (trial present worths in the rate-of-return solution) for a series of discount rates for the assumed example problem which has the cash-flow diagram given in the table. Note that the positive flows (incomes) at the end of the first and second period precede the first negative cumulative cash flow at the end of the third period. The ordinary form of the net present-value equation (Eq. 7-4) was solved with the results given in Table 7-3. Note that

TABLE 7-3
ILLUSTRATION OF HOW THE NET PRESENT VALUE VARIES WITH THE DISCOUNT RATE UNDER CONDITIONS OF REVERSAL OF DIRECTION OF CASH FLOWS
(Calculated Net Present Value for a Series of Discount Rates)

	Cash Flows			Discount Rate, Percent	NPV	Discount Rate, Percent	NPV
n	Income	Outgo	Summation				
1	200		200	0	100.00	40	−1.24
2	200		400	2	55.03	40.6	0.00
3	200	1,000	−400	4	21.90	45	8.77
4	200		−200	5.9	0.00	50	17.75
5	200		0	6	−2.06	60	32.51
6	200	900	−700	10	−30.41	70	43.47
7	200		−500	15	−42.86	80	51.37
8	200		−300	20	−41.61	90	56.94
9	200		−100	25	−33.83	100	60.73
10	200		100	30	−23.33	120	64.75
				35	−12.12		

at discount rates of 5.9 and 40.6 percent the net present value is zero. Thus, this case has two rates of return that satisfy the equation by reducing the net cash present worths of the cash flows to zero.

Two or more or no solutions are possible, depending upon the number of reversals of sign in the sequence of the cash flows. Table 7-3 shows a reversal of the accumulated cash flow at the end of the third and tenth periods.

Methods of Economic Analysis 161

Detection and Solution of Multirate Situations

In the industrial world there are a few examples where the benefits or profits from an operation may precede the investment cost, but in normal highway application of economic analysis it is rarely that such a type of problem will be encountered. In the petroleum industry a royalty type of lease may call for substantial investments in oil wells to increase the flow of oil some number of years after income from the well has been received. Also, in some types of plant operations or maintenance a savings in cost may be realized by failure to apply full maintenance or by heavy overloading. After a time duration of reduced expense (and more profit) a substantial cost is incurred to restore the property to acceptable operating condition. In both these types of situations a period of profitable income, or cost reduction, precedes the main investment outgo. These are the kind of situations where two or more rates of return could be found.

Cases of possible two rates of return can be detected by examining the cash flows for reversals in the sign (either direction) in the algebraic summation of the cash flows.

As suggested by Grant and Ireson [7-15, p. 510] and Taylor [7-45, p. 130] a single solution is possible and logical by considering that the net incomes preceding the outward flow for investment are considered as being invested at the minimum attractive rate of return, or other possible earnable rate, up to the date of the major investment which reverses the direction of the accumulated flow. These advance earnings at their compounded value may then be used, in effect, to reduce the amount of net investment at the later date. This procedure is logical and practical in view of the fact that the purpose of the economic analysis is to find the rate of return or the net present value of an investment.

Rate of Return and Net Present Value Methods Compared

Some writers [7-7, 7-23, 7-26-1, 7-50] recommend against the rate-of-return method of analysis solely on the ground that in this special situation two or more rates of return may result. Their preference is for the net present value method. But the two methods do not result in the same type of answer. When the decision maker wants a rate of return it should be supplied to him. Since the situation of two or more rates of return is so infrequent, there is no need to outlaw the rate-of-return method, a highly useful and understandable method of analysis.

The net present value method is also subject to the same type of question. The minimum attractive rate of return (cost of capital) gives the true net present worth of the proposal only at that rate. But the answer does not disclose what could be the evaluation at other rates. For example, in the problem in Table 7-3 the proposal might be accepted at a rate of 5.5 percent, but rejected at a rate of 6.0 percent, and accepted at any discount rate above 40.6 percent. Selection or determination of the minimum attractive rate of return or the cost of capital does not result in a rigid and only choice of rate. It should be considered as the approximate level at which the decision maker begins to look at other factors for guidance as to the acceptability of the proposal.

FLUCTUATING CASH FLOWS AND RANKING OF ALTERNATIVES

The same characteristics of compound interest mathematics that produce the two rates of return just discussed also may affect the comparison of mutually exclusive alternatives when analyzed by the rate of return method as compared to the other methods. The other methods use a chosen vestcharge rate in the analysis and presumably the same rate would be used in all methods when applied to the same set of alternatives. In the rate-of-return method, however, the rate that is found by the solution is that rate that reduces the positive and negative cash flows to zero on a discounted basis. The example in Table 7-4 illustrates how irregular flows of cash may affect the choice of the best alternative.

Alternative *A* has the greater net present value when computed at a minimum attractive rate of return of 10 percent, but it has the lower rate of return, 17 percent as compared to 20 percent for alternative *B*. This example again demonstrates the power of the discount rate in combination with the time period. The farther-into-the-future sum has much less present value at high discount rates as compared to its value at low discount rates.

REINVESTMENT RATE

The economists, when discussing the methods of economic analysis, often mention the reinvestment rate–that is, the stipulated or implicitly assumed rate of return that would be earned on the original investment, annual expenses, and profits when such negative cash flows are returned through sales income and reinvested [7-6, 7-8, 7-24]. Some of the authors [Wohl and Martin, 7-50, p. 43; Hunt, 7-24; and Solomon, 7-44, p. 74] state that the rate-of-return method makes the implicit assumption that the reinvestment of income is at the rate of return earned on the original investment. Others take the position that the reinvestment rate, however considered, is not a factor in project analysis and selection [7-25, 7-33]. This subject is introduced most often when comparing the merits of the net present value and rate of return methods. The papers on this subject do not generally explain the basis of the assumption that the rate-of-return method implicitly makes the assumption as to reinvestment rate.

This book on highway economics is not the place to enter the argument on reinvestment rate, but it is appropriate to acknowledge the problem and to make a brief comment.

It is most difficult to convince the layman that his rate of return on a given investment is dependent upon how he reinvests his return from that investment; neither does it seem logical when comparing possible investment alternatives that

Methods of Economic Analysis

TABLE 7-4

ILLUSTRATION OF THE EFFECT OF IRREGULAR CASH FLOWS ON COMPARISON OF ALTERNATIVES BY DIFFERENT METHODS OF ANALYSIS

n	Net Cash Flows		Net Present Value at 10 Percent MARR		Net Present Value	
	Alternative A	Alternative B	Alternative A	Alternative B	Alternative A at 17 Percent	Alternative B at 20 Percent
0	−210	−137	−210.00	−137.00	−210.00	−137.00
1	22	20	20.00	18.18	18.80	16.67
2	35	30	28.93	24.79	25.57	20.83
3	60	40	45.08	30.05	37.46	23.15
4	75	60	51.23	40.98	40.02	28.94
5	90	60	55.88	37.26	41.05	24.11
6	120	70	67.74	39.51	46.78	23.44
Total	192	143	58.86	53.77	−0.32	−0.14

Note: Alternative A has the greater net present value (58.86) at 10 percent discount rate, but the lower rate of return (17 Percent).

the choice of investment could depend upon how the return from each alternative would be reinvested.

In the highway field much of the consequential benefit results from cost reductions, travel time reductions, or certain noncash returns. These benefits accrue to a host of individuals including investors and noninvestors and to communities at large. The application here of economic analysis is different than in private commercial and industrial business. The final decision of management with respect to highway improvements is often based upon factors other than those included in the calculated net present value, benefit/cost ratio, or rate of return or in project formulation as distinguished from economic evaluation. As long as the decision maker has acceptable and reliable information usable in his decision process, it is not likely that the theory of reinvestment rate, regardless of how concluded, would alter his decision. Until the theorists can solve the subject of reinvestment in a practical manner, there will be no resulting great injustice to anyone by forgetting the subject as applied to public highways. The reader who may wish to pursue the subject of reinvestment is directed to the following references, which also contain helpful information on the general subject of economic analysis: 7-1, 7-6, 7-7, 7-8, 7-10, 7-13, 7-18, 7-22, 7-23, 7-24, 7-25, 7-26-1, 7-33, 7-34, 7-40, 7-44, 7-46, 7-48, and 7-50.

SENSITIVITY OF FACTORS

The analysis for economy produces an arithmetical answer, the magnitude of which depends upon engineering judgment in selecting factors and estimating the future. Unlike many aspects of engineering, such as structural design wherein the engineering answer will be approximately the same regardless of who is the designer, the solution for engineering economy may vary up to say 300 percent depending upon the factors chosen by the analyst. To gain some understanding of how certain factors affect the solution, a good practice is to solve for economy by using low, medium, and high values of the critical factors, and in different combinations. The sensitivity of each factor in controlling the result should be understood by each analyst and each decision maker.

When analyzing engineering proposals for their economic evaluation, certain factors of analysis may be really critical, especially for marginal proposals. When analyzing for project formulation the factors may not be so critical because of a common effect on all alternatives being compared. See Grant and Oglesby for a good discussion on sensitivity [7-17, pp. 34-38].

SENSITIVITY TO TERMINAL VALUE

The effect of the recovery of terminal value at the end of n years is to reduce the capital costs. The investment I is, in effect, reduced by the present worth of the terminal value. See Eqs. 7-1 to 7-6. Therefore, the effect of terminal value is dependent upon three factors; (1) the amount of terminal value, T, expressed as a percentage of the investment, I, (2) the number of years hence when the terminal value is to be recovered, n, and (3) the rate of vestcharge, i.

The percentage of terminal value would have an effect directly proportional to the percentage of I which is recovered because of its direct reduction of I, upon which the capital cost depends. The present worth of a future sum becomes less and less as the due date is pushed farther into the future and as the rate of vestcharge increases. Table 7-5 illustrates the effects the vestcharge rate and the analysis period have upon the effective I after it is adjusted for terminal value. Terminal recoveries of less than 30 percent for vestcharge rates of 7 percent and more when the analysis period is 30 years or more, would reduce I to an effective I by less than 5 percent. Even for a recovery rate of 50 percent, $n = 20$ years, and $i = 7$ percent, the effective I is 87 percent of the initial investment.

Terminal value is not often a strong critical factor in economy studies for highways. For most all analyses for economic evaluation, terminal value may be assumed to be zero, especially when the analysis period is taken as an arbitrary period much less than probable service life. For formulation of engineering design, terminal value may also be assumed to be zero, except for those pairs of alternatives in which the terminal value is a controlling factor, such as when comparing one construction material with another, one machine with another, or when determining the most favorable trade-in period of mechanical equipment such as motor vehicles.

TABLE 7-5
Effective *I* in Percent of Actual *I*
(Based upon the Reduction of *I* by the Present Worth of the Terminal Value)

Terminal Value, Percent of Initial Investment, *I*	n = 10			n = 20			n = 30			n = 50		
	i=4%	i=7%	i=10%	i=4%	i=7%	i=10%	i=4%	i=7%	i=10%	i=4%	i=7%	i=10%
10	93.24	94.92	96.14	95.44	97.42	98.51	96.92	98.69	99.43	98.59	99.66	99.91
20	86.49	89.83	92.29	90.87	94.83	97.02	93.83	97.37	98.85	97.19	99.32	99.83
30	79.73	84.75	88.43	86.31	92.25	95.54	90.75	96.06	98.28	95.78	98.98	99.74
40	72.98	79.67	84.58	81.74	89.66	94.05	87.67	94.75	97.71	94.37	98.64	99.66
50	66.22	74.58	80.72	77.18	87.08	92.57	84.58	93.43	97.13	92.96	98.30	99.57
100	32.44	49.17	61.45	54.36	74.16	85.14	69.17	86.86	94.27	85.93	96.61	99.15

SENSITIVITY TO LENGTH OF ANALYSIS PERIOD

The analysis period or service life is a critical factor in the analysis for economy when the vestcharge rate is relatively low and the analysis period is relatively short. The capital recovery factor *CR* approaches *i* as a limit as *n* increases, and the present worth series factor *SPW* approaches the reciprocal of *i* as a limit as *n* increases.

By examining these factors in a set of compound interest tables it is readily seen that when the vestcharge rate is 7 percent and greater there is not a greatly significant change in the *CR* or *SPW* factors when *n* is greater than 30 years. Thus, one advantage of high vestcharge rates is to nullify the influence of cash flows in the distant future when uncertainty is the greatest. In combination with *i*, the *n* factor is critical at its lower ranges, even to the extent of shifting a proposed project from the not justified to the justified category.

Table 7-6 is a comparison of benefit/cost ratios for a series of *n* factors for vestcharge rates of 4, 7, and 10 percent. Note that when the annual benefit is 10 percent of the initial investment, and *i* is 7 percent, the benefit/cost ratio shifts from 0.70 for $n = 10$ to 1.06 for $n = 20$. But the shift is only from 1.24 to 1.38 when *n* moves from 30 to 50 years. For $i = 10$ percent, the benefit/cost ratio rises from 0.61 to 0.99 as *n* moves from 10 years to 50 years. Note also, that at $i = 4$ percent there is a greater change in the benefit/cost ratio with increasing *n* than is true for the 7 percent and 10 percent vestcharge rates.

For reasons of conservatism the analysis period should be held low, even though it is more critical at low than at high values. In project formulation the *n* factor may be especially critical when analyzing for the economy of one material or one machine as compared to others. For overall highway location projects which may call for an analysis period of 20 to 30 years, *n* is not critical in project selection. The compensatory factor is to use the highest rate of vestcharge that is reasonable.

TABLE 7-6
COMPARISON OF BENEFIT/COST RATIOS* FOR THREE VESTCHARGE RATES, FOUR ANALYSIS PERIODS, AND ZERO TERMINAL VALUE
(Based upon Ratios of Net Road-User Annual Benefits to Initial Investment)

Ratio of Net Annual Benefits to Initial Investment, I	$n = 10$			$n = 20$			$n = 30$			$n = 50$		
	$i=4\%$	$i=7\%$	$i=10\%$	$i=4\%$	$i=7\%$	$i=10\%$	$i=4\%$	$i=7\%$	$i=10\%$	$i=4\%$	$i=7\%$	$i=10\%$
0.10	0.81	0.70	0.61	1.36	1.06	0.85	1.73	1.24	0.94	2.15	1.38	0.99
0.20	1.62	1.40	1.23	2.72	2.12	1.70	3.46	2.48	1.89	4.30	2.76	1.98
0.30	2.43	2.11	1.84	4.08	3.18	2.55	5.19	3.72	2.83	6.44	4.14	2.97
0.40	3.24	2.81	2.46	5.44	4.24	3.41	6.92	4.96	3.77	8.59	5.52	3.97

* These benefit/cost ratios are the solution to Eq. 7-5, when $T = 0$, and the ratio of net annual benefits to initial investment is as given in column 1.

SENSITIVITY TO THE VESTCHARGE RATE

The vestcharge rate is the most critical factor in the analysis for economy. It is a factor chosen by judgment; it is a factor to which the final result of the analysis is highly sensitive; it is a factor that causes the results to vary geometrically rather than linearly; therefore, the analyst should be fully aware of how the vestcharge rate affects his final answer. When analyzing by the annual cost, present worth, or benefit/cost ratio methods it is well to make at least two solutions, each with different vestcharge rates.

As is indicated in Table 7-6, the vestcharge rate may materially affect the benefit/cost ratio by shifting it from an unattractive ratio to an attractive one or vice versa. Even more critical, a change in the vestcharge rate may shift the selection from one alternative to another. Table 7-7 compares four alternative highway proposals under conditions of $n = 20$ years, $T = 0$, and 4, 7, 10, and 12 percent vestcharge rates. The comparison is on the basis of equivalent uniform annual transportation cost. In this solution, a different alternative is chosen for each of the four vestcharge rates. Note, also, that the order of rank of preference is directly reversed when the vestcharge rate is changed from 4 percent to 12 percent.

In project evaluation and in project priority selection, the vestcharge rate is highly critical. The ratio of initial investment cost to annual disbursements (administration, traffic services, and maintenance) is a controlling factor in causing a shift in choice from one alternative to another as the vestcharge rate is changed. The higher the vestcharge rate, the higher is the annual charge for capital recovery. In certain types of construction, higher investment costs are chosen in order to reduce annual maintenance costs; also, in highway design, higher investments are chosen in order to reduce the motor vehicle running costs. Thus as the initial investment is increased, the ratio of net benefit to

Methods of Economic Analysis

TABLE 7-7
EFFECT OF VESTCHARGE RATE ON SELECTION OF MUTUALLY EXCLUSIVE ALTERNATIVE WITH MOST FAVORABLE ECONOMY

Alternative Project Number	Initial Investment I, Dollars	Annual Uniform Maintenance and Operation Costs, K	Total of All Road-User Uniform Annual Costs, U, Dollars	Equivalent Uniform Annual Transportation Costs, Dollars, when $n = 20$ and $T = 0$			
				$i = 4\%$	$i = 7\%$	$i = 10\%$	$i = 12\%$
1	$3,500	$60	$1,790	2108(4)*	2180(4)*	2261(3)*	2319(1)*
2	5,000	50	1,610	2028(3)	2132(3)	2247(1)	2329(2)
3	6,500	40	1,450	1968(2)	2104(1)	2253(2)	2360(3)
4	8,000	25	1,350	1964(1)	2130(2)	2315(4)	2446(4)

* Numbers in parentheses are rank of project, lowest to highest annual transportation cost.

capital cost changes, and the rate of change is affected by the vestcharge rate. The consequence, then, is that in certain analyses the alternative or project proposal with the most favorable economy may shift between alternatives with a shift in the rate of vestcharge used in the analysis.

Although in project evaluation the vestcharge rate is particularly critical, in project formulation the vestcharge rate is not so critical except when the alternatives have a marked difference in initial investment cost or in the ratio of annual maintenance expense to investment cost.

Solving for the rate of return is perhaps the best solution because it avoids selecting a vestcharge rate. However, in the end it is necessary to adopt a minimum attractive rate of return to use as a guide in final selection from among the alternatives, as well as to determine the value of increments of investment in multialternatives in mutually exclusive groups.

PROBLEMS

7-1. Calculate the equivalent uniform annual cost at 7 percent per annum of the following outward cash flows: A first investment at $20,000 at $n = 0$; $5,000 secondary investment at $n = 10$, and an annual maintenance of $100 each year through $n = 15$. Terminal values are $2,000 for the first investment and $400 for the second investment. (*Ans.:* $2,479.)

7-2. A maintenance motor truck has been severely damaged in an accident. Should it be *(A)* repaired, or *(B)* traded on a new truck? The cost of repairs would be $1,200 after which the truck could be used for 3 years more and traded then at a terminal value of $300. A new truck now would cost $6,000, used for 6 years, and traded for $400. The annual operating costs of the two trucks are approximately equal. A vestcharge rate of 8 percent is appropriate.

7-3. A new member of a city council complained to the city manager that he had been spending too much on maintenance of paving and not enough for repaving. One case cited involved the expenditure for *(A)* patching and repairing of $4,600 a year for the past 8 years on one street. The councilman claimed that the street could have been *(B)* repaved 8 years ago for $30,000. The city could have borrowed on a general bond issue for 8 years at $4^1/_2$ percent. Did the councilman have a good case?

7-4. How much capital investment is justified at vestcharge rates of 6 and 10 percent for an auxiliary 4-speed transmission to reduce the annual operating expense by $300 and to extend the life of the main engine from an anticipated 4 years to 6 years? The main engine will cost new $1,500.

7-5. How many years will it take for the following flood prevention works to pay for itself? Cost of constructed works, $30,000; average past annual flood damage, $5,000; annual maintenance, $500; and a vestcharge rate of 8 percent compounded annually.

7-6. A highway reconstruction project has a first cost of $120,000 and an estimated terminal value after 25 years of $20,000. Estimated annual road-user net benefits are $15,840, and estimated average annual highway maintenance costs are $3,000. Assuming that annual road-user net benefits and annual highway maintenance costs will be a uniform amount over the years, compute the prospective rate of return using an analysis period of 25 years.

7-7. Using a discount rate of 8 percent per annum find the equivalent uniform annual maintenance expense corresponding to the following annual expenses, year by year: first, $500; second, $600; third, $650; fourth, $740; fifth, $880; and sixth, $950. (*Ans.:* $700.)

7-8. What is the equivalent uniform gradient for the maintenance expenses given in Prob. 7-7?

7-9. A highway department has been *(A)* maintaining a wooden bridge across a canyon river. Flash floods wash away the wooden superstructure once every two years. It costs $4,000 to replace the bridge after each washout. A permanent higher level steel bridge *(B)* good for 60 years, would cost $120,000 to construct and $700 a year maintenance. At a vestcharge rate of 10 percent which bridge is the least costly in the long run? Neglect any cost to the traffic.

7-10. Refer to Table 7-1. Compute the rates of return for Alternatives B_3, B_4, and for B_4 compared to B_3. Which alternative would you prefer to accept?

7-11. A contractor who has a minimum attractive rate of 12 percent after taxes is considering the following four schemes for paving a 16-mile length of highway with 42,000 tons of bituminous concrete: *(A)* central plant at location *N*, *(B)* central plant at location *S*, *(C)* contract to buy premixed bituminous from another producer and haul to job in own trucks, and *(D)* contract to buy from another producer who will deliver the mix to the job. The essential facts and estimates are:

	A	B	C	D
Average haul distance, miles	12	15	18	18
Monthly rent of plant site	$400	$100	–	–
Net mixing plant cost for this job	$6,000	$4,500	–	–
Operating and cost of materials, per ton	$0.70	$0.80	$1.20	1.85
Hauling cost dollars per truck-mile	0.42	0.40	0.38	
Months required to complete job	3	4	3	3

Which alternative should the contractor adopt, assuming management and other non-direct cost factors equal? (*Ans.:* Alternative *A*, because monthly cost of $22,320 is lowest.)

7-12. A motel in a mountain resort area needs to increase its domestic water suply. Three methods are being considered: *(A)* add a pump to the existing gravity line; *(B)* lay

Methods of Economic Analysis

a parallel gravity line and operate both lines; and (C) install a new well and pump at the motel site and abandon the existing line. The cash costs are:

	A	B	C
Capital investments	$300	$700	$1,200
Annual operating and maintenance expense	200	40	150
Terminal values	100	−300	200
Net cost to remove existing line	–	–	100
Service life, years	10	10	10

At vestcharge rates of 9 and 12 percent determine which alternative is to be preferred solely on economy of monetary costs.

7-13. Using a 7 percent vestcharge rate calculate (1) the capitalized cost and (2) the equivalent uniform annual cost for 60 years for the following three bridges:

	A	B	C
Initial cost	$60,000	$80,000	$120,000
Renewal cost, end-of-service life	30,000	45,000	60,000
Annual maintenance expense	700	500	400
Periodic repairs every 5, 10, 15... years	3,000	2,000	1,000
Terminal (scrap) value at end-of-service life	4,000	8,000	10,000
Service life, years	20	30	60

7-14. A highway paving contractor operates on a minimum attractive rate of return of 25 percent before taxes when considering new equipment devices and construction methods. What reduction in his annual costs of construction is required to make it profitable for him to purchase a new item of equipment for $40,000 which he could use for 8 years with a terminal value of $4,000?

7-15. In real life, annual cash flows for maintenance and operation or for motor vehicle operating costs, vary plus and minus year to year. The assumption of (a) uniform flow, (b) gradient increase (or decrease), or (c) exponential increase is only for the purpose of simplifying the arithmetical calculations. Compare as indicated the following annual flow sequence at 8 percent vestcharge rate on the basis of both present worth and equivalent uniform amount: (1) actual flows, (2) uniform flow of 116 each year end, (3) gradient increase of 3 each year end, and (4) an exponential growth of 3 percent per year at each year end. In each case the first effective flow is at end of year 1 (increase starts at time zero when the rate of flow is $100).

What gradient increase and what exponential growth rate are equivalent to the actual cash-flow sequence?

End of Year	Amount of Flow	End of Year	Amount of Flow
1	100	6	120
2	110	7	110
3	90	8	100
4	120	9	140
5	140	10	130

7-16. Identify that alternative of the following three mutually exclusive alternatives

A, B, and C, that has the more favorable economy at a minimum attractive rate of return of 10 percent. Consider that the service cycles and costs will stop at the end of year 40.

	A	B	C
Initial capital cost	$800	$1,200	$1,500
Equivalent uniform annual operation and maintenance expense	190	175	160
Terminal value of physical property, age 40	None	None	None
Service life of original investment, years	12	15	20

7-17. Recommend which of the three following alternative routes is most economical from the standpoint of engineering economy, using the rate-of-return method of analysis (1) when the minimum attractive rate of return (MARR) is 8 percent, and (2) when the MARR is 6 percent. Use an analysis period of 20 years and assume a zero terminal value.

Basic Data	Present Highway	North Route	Middle Route	South Route
Total construction costs	(Sunk)	$5,837,000	$6,245,000	$6,873,000
Equivalent uniform annual maintenance costs	$60,820	$142,350*	$130,380*	$131,120*
Equivalent uniform annual vehicle-miles of travel	20,865,220	18,171,554	18,143,430	17,916,548
Total user cost per vehicle-mile	$0.1811	$0.1700	$0.1690	$0.1680

* Includes required maintenance for sections of existing highway to be kept in service.

7-18. Compute the present amount (present worth) and the equivalent uniform annual ADT at 7 percent rate of vestcharge and 20 years for a proposed highway on the basis of the following forecasts: (1) uniform at 1.000 ADT at time zero to end of 20 years, (2) 1,000 ADT at time zero and increasing 100 each year, (3) 1,000 ADT at zero time and increasing compoundly at 5 percent per year, and (4) using the average traffic between zero time at 1,000 ADT and 3,000 ADT the 20th year, as a uniform annual traffic for the 20-year period.

7-19. For the cash flows listed below for three mutually exclusive alternatives and for a 25-year analysis period and vestcharge rate of 7 percent, compute the following solutions: (1) benefit/cost ratio with the maintenance and operating costs (a) in the numerator, and (b) in the denominator; (2) the rate of return; (3) the present worth of cash flows; and (4) the equivalent uniform annual transportation costs. Comment upon your results.

	A	B	C
Initial construction cost	Sunk	$1,426,000	$1,686,000
Annual highway maintenance and operation expense	$4,000	$24,200	$14,200
Uniform annual road-user costs	$1,004,000	$836,000	$820,000

7-20. The director of highways is considering constructing a 4-lane section of interstate highway in stages. His plan is to construct 2 lanes now and the remaining 2 lanes at some future date. You are asked to compare the following alternative schemes of construction scheduling to determine which is most economical. Use the present-worth method

Methods of Economic Analysis

of analysis, an analysis period of 20 years, a vestcharge rate of 6 percent, and a zero terminal value for all three schemes.

Scheme 1–construction of 4 lanes now
Total construction costs .... $ 5,888,000
Scheme 2–2 lanes now and remaining 2 lanes during the 6th year
Initial construction costs .... $ 4,566,000
6th year construction costs (end of year) 1,639,300
Scheme 3–2 lanes now, remaining 2 lanes during 12th year
Initial construction costs .... $ 4,566,000
12th year construction costs (end of year) 1,639,300

Annual maintenance costs are $ 33,000 for 4 lanes of highway and $ 23,000 for 2 lanes of highway.

Initial (zero date) road-user costs are $ 300,000 per year for 4-lane construction and $ 350,000 for 2-lane construction. Assume the road-user costs will increase at a uniform rate of $ 30,000 a year regardless of which scheme is being analyzed. First step increase is at end of first year; i.e., increase begins at zero time.

7-21. A company has a special machine to use on a contract job lasting 3 years. They desire to overload the machine and forgo normal maintenance and repairs until the contract is completed, then completely overhaul the machine. Using the following end-of-year cash-flow schedule, compute (1) the rate of return, (2) the net present value at 10 percent vestcharge.

Year	Gross Income	Cash Outlay
1	$ 1,500	$ 1,000
2	1,700	1,100
3	1,600	1,200
4	1,000	2,200
5	1,000	800
6	1,000	800
7	1,000	800

Discuss your results.

7-22. Solve for the net present value of the two following alternatives using several vestcharge rates from 0 percent upward. Plot the curves of the resulting net present values against the rates of vestcharge. What are the rates of return of the two alternatives? What is the rate of return on the difference between the alternatives? If your minimum attractive rate of return is 8 percent, which alternative would you select? Comment on your results.

Year End	Cash Flow Alternative A	Alternative B
0	−1,000	600
1	600	500
2	600	−2,000
3	600	400
4	600	200
5	−2,100	−1,000
6	700	1,386

7-23. What are the (1) advantages, (2) disadvantages, and (3) applications of continous compounding? Compare its use (4) in private industry economy studies, (5) public works economy studies, and (6) banking and loan operations.

REFERENCES

7-1. A. A. Alchian. "The Rate of Interest, Fisher's Rate of Return Over Cost, and Keynes' Internal Rate of Return," *American Economic Review*, Vol. XLV (December 1955), pp. 938-943. Recommended reading.

7-2. American Association of State Highway Officials. *Road User Benefit Analyses for Highway Improvements.* Part I, "Passenger Cars in Rural Areas," Washington, D.C., 1960.

7-3. American Society for Engineering Education, Engineering Economy Division. Final report, Committee on Standardization of Notation, *Engineering Economist*, Vol. 12, No. 4 (Summer 1967) pp. 253-263.

7-4. Norman N. Barish. *Economic Analyses.* McGraw-Hill, New York, 1962.

7-5. William J. Baumol. *Economic Theory and Operations Analysis.* Prentice-Hall, Englewood Cliffs, N.J., 1961.

7-6. Harold Bierman Jr. and Seymour Smidt. "Capital Budgeting and the Problem of Reinvesting Cash Proceeds," *Journal of Business* (October 1957), pp. 276-279.

7-7. Harold Bierman Jr. and Seymour Smidt. *The Capital Budgeting Decision-Economic Analysis and Financing of Investment Decisions.* Macmillan, New York, 1966.

7-8. John A. Canada. "Rate of Return: A Comparison Between the Discounted Cash Flow Model and a Model Which Assumes an Explicit Reinvestment Rate for the Uniform Income Flow Case," *Engineering Economist*, Vol. 9, No. 3 (Spring 1964), pp. 1-15.

7-9. David A. Curry. *Use of Marginal Cost of Time in Highway Economy Studies.* Highway Research Board, National Academy of Sciences, Washington, D.C., Highway Research Record No. 77, 1965, pp. 48-120.

7-10. Joel Dean. *Capital Budgeting.* Columbia U. P., New York, 1951.

7-11. Joel Dean. "Measuring the Productivity of Capital," *Harvard Business Review* (January-February 1954).

7-12. Otto Eckstein. *Water Resource Development–The Economics of Project Evaluation.* Harvard U. P., Cambridge, Mass., 1958.

7-13. M. S. Feldstein and J. S. Fleming. "The Problems of Time Stream Evaluation; Present Value vs. Internal Rate of Return Rates," Oxford University Institute of Economics and Statistics, Bulletin 26 (February 1964).

7-14. Gerald A. Fleischer. "On the Use of Rates of Return as Elements in Decision Matrices," *Engineering Economist*, Vol. 11, No. 1 (Fall 1965), pp. 17-27.

7-15. Eugene L. Grant and W. Grant Ireson. *Principles of Engineering Economy*, 4th ed., Ronald, New York, 1960.

7-16. Eugene L. Grant. "Concepts and Applications of Engineering Economy," in *Economic Analysis in Highway Programming, Location, and Design.* Highway Research Board, National Academy of Sciences, Special Report 56, Washington, D.C., 1960, pp. 8-18.

7-17. Eugene L. Grant and Clarkson H. Oglesby. *Economy Studies for Highways.* Highway Research Board, National Academy of Sciences, Washington, D.C., Bulletin 306, 1961, pp. 1-38.

7-18. Eugene L. Grant. "Reinvestment of Cash Flows Controversy–A Review of

Methods of Economic Analysis

Financial Analysis in Capital Budgeting by Pearson Hunt," *Engineering Economist,* Vol. 11, No. 3 (Spring 1966), pp. 23-29.

7-19. Dan G. Haney. *Use of Two Concepts of the Value of Time.* Highway Research Board, National Academy of Sciences, Washington, D.C., Highway Research Record No. 12, 1963, pp. 1-18.

7-20. P. D. Henderson. "Notes on Public Investment Criteria in the United Kingdom," *Oxford University Institute of Economics and Statistics Bulletin,* February 1965.

7-21. Highway Research Board. National Academy of Sciences, Washington, D.C., Highway Research Record No. 180, *Transportation System Analysis and Evaluation of Alternate Plans,* 1967.

    1. Bruce B. Wilson. "Policy and Procedure Review: State Highway Commission Liaison With "701" Planning in Wisconsin," pp. 1-10.

    2. Marvin L. Manheim. "Principles of Transport Systems Analysis," pp. 11-20.

    3. Morris Hill. "A Method for the Evaluation of Transportation Plans," pp. 21-34.

    4. Milton Pikarsky. "Comprehensive Planning for the Chicago Crosstown Expressway," pp. 35-51.

    5. Kozmas Balkus. "Transportation Implications of Alternative Sketch Plans," pp. 52-70.

    6. William Jessiman, Daniel Brand, Alfred Tumminia, and C. Roger Brusee. "Rational Decision-Making Technique for Transportation Planning," pp. 71-80.

    7. Stephen H. Putman. "Modeling and Evaluating the Indirect Impacts of Alternative Northeast Corridor Transportation Systems," pp. 81-93.

    8. Howard Duke Niebur. "Preliminary Engineering Economy Analysis of Puget Sound Regional Transportation Systems," pp. 94-119.

    9. W. W. Shaner. "Economic Evaluation of Investments in Agricultural Penetration Roads in Developing Countries," pp. 120-132.

7-22. Jack Hirshleifer. "On the Theory of Optimal Investment Decision," *Journal of Political Economy,* Vol. LXVI (August 1958), pp. 329-352.

7-23. Jack Hirshleifer, James C. DeHaven, and Jerome W. Milliman. *Water Supply-Economics, Technology, and Policy.* U. of Chicago Press, Chicago, 1960.

7-23A. Jack Hirshleifer. "Investment Decision Under Uncertainty: Applications of the State-Preference Approach," *Quarterly Journal of Economics,* Vol. LXXX, No. 2 (May 1966), pp. 252-277.

7-24. Pearson Hunt. *Financial Analyses in Capital Budgeting.* Graduate School of Business Administration, Harvard University. Rev. ed., 1965. (See review of this report by Eugene L. Grant, *Engineering Economist,* Vol. 11, No. 3 (Spring 1966), pp. 23-29.

7-25. Paul H. Jeynes. "The Significance of Reinvestment Rate," *Engineering Economist,* Vol. II, No. 1 (Fall 1965), pp. 1-9.

7-26. *The Journal of Business.* (Published by the School of Business of the University of Chicago). "Capital Budgeting," Vol. XXVIII, No. 4 (October 1955).

    1. James H. Lorie and Leonard J. Savage. "Three Problems in Rationing Capital," p. 229.

    2. Ezra Solomon. "Measuring a Company's Cost of Capital," p. 240.

    3. Myron J. Gordon. "The Payoff Period and the Rate of Profit," p. 253.

    4. Joel Dean and Winfield Smith. "Has Mapi a Place in a Comprehensive System of Capital Controls?" p. 261.

5. Gordon Shillinglaw. "Residual Values in Investment Analysis," p. 275.
6. Horace G. Hill, Jr. "Capital Expenditure Management," p. 285.
7. Frank E. Norton. "Administrative Organization in Capital Budgeting," p. 291.
8. Raymond Villers. "The Origin of the Break-Even Chart," p. 296.

7-27. F. H. Knight. "The Quantity of Capital and the Rate of Interest," *Journal of Political Economy*, Nos. 4-5 (1936).

7-28. John V. Krutilla and Otto Eckstein. *Multiple-Purpose River Development–Studies in Applied Economic Analysis.* Johns Hopkins Press, Baltimore, 1958.

7-29. Tillo E. Kuhn. *Public Enterprise Economics and Transport Problems.* U. of California Press, Berkeley, 1962.

7-30. Kelvin J. Lancaster. "A New Approach to Consumer Theory," *Journal of Political Economy*, LXXIV (April 1966), pp. 132-157.

7-31. I. M. D. Little. *A Critique of Welfare Economics.* Oxford U. P., 1958.

7-32. Arthur R. Maass. "Benefit/Cost Analysis: Its Relevance to Public Investment Decisions," *Quarterly Journal of Economics*, Vol. LXXX, No. 2 (May 1966).

7-33. Julius Margolis "The Economic Evaluation of Federal Water Resource Development: A Review Article," *American Economic Review*, Vol. XLIX, No. 1 (March 1959), pp. 96-111.

7-34. Roland N. McKean. *Efficiency in Government Through Systems Analysis.* Wiley, New York, 1958.

7-35. A. J. Merrett and A. Sykes. *The Finance and Analysis of Capital Projects.* Longmans, Green, 1963.

7-36. A. J. Merrett and A. Sykes. *Capital Budgeting and Company Finance.* Longmans, Green, 1966.

7-37. Herbert Mohring and Mitchell Harwitz. *Highway Benefits-An Analytical Framework.* Northwestern U. P., Evanston, Ill., 1962.

7-38. Clarkson H. Oglesby and Eugene L. Grant. *Economic Analysis–The Fundamental Approach to Decisions in Highway Planning and Design.* Highway Research Board, National Academy of Sciences, Washington, D.C., Proc. 37, 1958, pp. 45-57.

7-39. A. R. Prest and R. Turvey. "Cost-Benefit Analysis: A Survey," *Economic Journal*, 75, London (December 1965).

7-40. Edward F. Renshaw. *Toward Responsible Government–An Economic Appraisal of Federal Investment in Water Resource Programs.* Idyia Press, Chicago, 1957.

7-41. Richard W. Renshaw. *Treatment of Deferred Costs in Economic Analysis.* Highway Research Board, National Academy of Sciences, Washington, D.C., Proc. 39, 1960, pp. 9-15.

7-42. Paul A. Samuelson. *Economics–An Introductory Analysis.* 7th ed.; McGraw-Hill, New York, 1967.

7-43. Gerald W. Smith. "Benefit/Cost Ratios: A Word of Caution." Highway Research Board, National Academy of Sciences, Washington, D.C., Highway Research Record No. 12, 1963, pp. 77-90.

7-44. Ezra Solomon (ed.). *The Management of Corporate Capital.* Free Press, New York, 1959. *Note:* This book contains a collection of papers by Ezra Solomon, Joel Dean, Horace Hill, Jr., M. J. Gordon, J. H. Lorie and L. J. Savage, A. A. Alchian, Romney Robinson, Ed Renshaw, David Durand, M. J. Gordon and Eli Shapiro, F. Modigliani and M. H. Miller, H. V. Roberts, Jack Hirshleifer, A. Charnes, W. W. Cooper and M. H. Miller,

G. Shillinglaw, F. E. Norton, Joel Dean and Winfield Smith, and George Terborgh. This book is perhaps the best single reference on the subject of capital investment and its related phases.

7-45. George A. Taylor. *Managerial and Engineering Economy–Economic Decision Making.* Van Nostrand, Princeton, 1964.

7-46. R. Turvey. "Present Value versus Internal Rate of Return–An Essay in the Theory of Third Best," *Economic Journal,* 73, London (March 1963).

7-47. H. Martin Weingartner. *Mathematical Programming and the Analysis of Capital Budgeting Problems.* Prentice-Hall, Englewood Cliffs, N.J., 1963.

7-48. David W. Winch. *The Economics of Highway Planning.* U. of Toronto Press, Toronto, 1963.

7-49. Robley Winfrey. *Concepts and Applications of Engineering Economy in the Highway Field. Economic Analyses in Highway Programing, Location, and Design.* Highway Research Board, National Academy of Sciences, Washington, D.C., Special Report 56, 1960, pp. 19-34.

7-50. Martin Wohl and Brian V. Martin. *Evaluation of Mutually Exclusive Design Projects.* Highway Research Board, National Academy of Sciences, Washington, D.C., Special Report No. 92, 1967.

7-51. Martin Wohl and Brian V. Martin. *Traffic Systems Analysis for Engineers and Planners.* McGraw-Hill, New York, 1967.

7-52. Martin Wohl and B. V. Martin. "Evaluating Road Projects," *Journal of Transport Economics and Policy,* London School of Economics and Political Science, Vol. I, No. 1 (January 1967), pp. 28-45.

7-53. Richard M. Zettel. *Highway Benefit and Cost Analysis as an Aid to Investment Decision.* Institute of Transportation and Traffic Engineering, Reprint No. 49, University of California, Berkeley (1956?).

7-54. Richard M. Zettel. "A Review of Theory of Governmental Policies as Related to the Promotion of Transportation," Notes from the Short Course, *Airport Property Management.* Institute of Transportation and Traffic Engineering, University of California, Berkeley, December 1960.

Chapter 8

Depreciation Concepts

The most important concept of depreciation deals with the process of allocating the investment cost of fixed assets to production expense operations by accounting periods. This concept applies to cost accounting for estimating profit and loss. The word *depreciation* is also used to mean that a used property is worth less than it would be were it new. A third concept of depreciation, though one not used by the well-informed person, refers to a state of physical wear and tear of physical property. This chapter discusses these meanings of depreciation, methods of estimating depreciation for cost and studies, and indicates the relation of depreciation concepts to highway economics. However, engineering economy studies do not involve accounting depreciation or value depreciation.

MEANING OF THE WORD DEPRECIATION

Because of the looseness with which the word *depreciation* has been used in the literature, confusion has resulted among engineers, lawyers, businessmen, and accountants. In the cost accounting usage, the word is commonly defined by accountants to mean an allocation of a prepaid investment cost in the process of cost accounting. Common usage of the word depreciation is often in the sense of decrease in value. The common public and professional men frequently use the word to mean a state of physical wear and tear. An examination of these three meanings will help to lay a foundation for understanding of the discussions on depreciation methods.

In the sense of decrease in value or as a prepaid cost allocated to current operational expense, depreciation is an important factor in the following business operations and, at times, in personal considerations:

1. Calculating net operating profits, especially when profits are distributed to stock holders
2. Preparing income tax returns
3. Estimating bid prices as does a contractor
4. Calculating production cost as a basis of price setting as is done in regulation of public utilities
5. Deciding property valuation for tax purposes, settlement of estates, including, determining the fair value rate base in utility regulations
6. Analyzing investment securities for their investment merit.

Depreciation Concepts

DEPRECIATION IN THE SENSE OF COST ALLOCATION

Depreciation is an important element in cost accounting. Management of a business is desirous of knowing the comparative profits for different accounting periods. Obviously, in order to render operating statements for two or more time periods on a comparable basis, cash disbursements alone are not sufficient.

Consideration must be given in operating statements to two kinds of properties and their dollar cost. Inventories are taken of the working capital supplies of raw materials, maintenance materials, work in process, and finished goods; the gain or loss in the dollar inventory of these properties is taken into the operating statement and affects the computed profit accordingly. The second class of property taken into consideration in computing a profit and loss statement is the fixed capital physical properties–those long-lived properties that contribute to production over more than one accounting period. These types of properties–buildings, machines, tools, and equipment–are handled on a depreciation basis in the accounting process. That is, the original cost of depreciable property (less any terminal, or salvage, value) is allocated to the cost of production on some chosen basis which charges to operations the cost of these properties, increment by increment, over their period of usefulness. Rather than charging production with the initial cost new of the depreciable property within the accounting period during which it was purchased, the accountant spreads this initial cost over several accounting periods.

These allocated costs are known as *depreciation expense*. In this accounting sense, depreciation becomes an allocated portion of a prepaid cost. As such, and during the accounting process, there is no reference to value of the property; no attempt is made to decrease book cost of the property by its change in real value. The sole objective is to allocate this prepaid cost of long-lasting assets to the expense of production by accounting periods in a manner that will be reasonable, equitable, and expedient, and without knowingly decreasing or increasing the production expense for the one period as compared to prior or following accounting periods.

Desirability of depreciation accounting as an element of expense came about with the development of the industrial corporation or stock company. With many individual owners and with an ever-changing set of owners, an accounting scheme was essential which would fairly balance between accounting periods the expenditures for those production properties which were used for several accounting periods. In the early days of the development of the railroads, management realized the unfairness of charging the purchase price of locomotives and cars to the operating expense for the period during which they were purchased; similarly for track and track equipment. Consequently, the railroads were one of the early users of depreciation accounting in the United States. The Baltimore and Ohio Railroad Company refers to it in their annual report for 1833. Historically, the concept of depreciation can be traced to about 27 B.C. in the writings of Marcus Vitruvius Pollio [8-6].

In no sense whatsoever is cost accounting for depreciation related to replacement of the property. The accounting process is allocating a prepaid expense to current operations, and not one for accumulating money with which to purchase a replacement unit. Whether or not the unit is to be replaced upon its retirement is wholly irrelevant to the accounting process, or to determining depreciation, either in the sense of a decrease in value or as a charge to production expense.

DEPRECIATION IN THE SENSE OF VALUE

Depreciation has long been defined and used in the sense of a decrease in value. Practically all textbooks on accounting, economics, and business, as well as the dictionary, define the word to mean a loss or decrease in value. Further, these sources indicate that the lessening of value is a result of increasing age and obsolescence and decreasing usefulness of the property. True, man-made physical properties generally do decrease in value with age and use as long as the general price level remains substantially stable. Value of a property depends upon the economic laws of supply and demand; therefore, buildings, machines, structures, tools, and equipment are subject to appreciation in value as well as depreciation in value. The scarcity of many industrial goods caused certain used properties to sell for more dollars from 1941 to 1950 than they cost new some years previous. Although the value concept of the word depreciation is a correct one, it is seldom used correctly in this sense in the literature. The meaning of the word depreciation that is most commonly encountered is in the sense of amortization of a prepaid cost, which usage applies to cost accounting and to most economic studies.

DEPRECIATION IN THE SENSE OF IMPAIRED USEFULNESS

The third common usage of the word depreciation refers to the impaired service usefulness of physical property, or the state of wear and tear. The intent is to give some indication of how the remaining service usefulness of a property compares to a similar or identical property that is new. In this sense there is no direct reference to the value of the property, although it is inferred that any property highly depreciated (in the physical sense) is probably worth considerably less than it would be were it new. The use of impaired usefulness as a meaning of depreciation is not recommended. It is not a satisfactory technical description of the state of wear and tear or lack of maintenance. Continued use of the word depreciation in the sense of physical condition only adds to the existing confusion brought about by the value and cost concepts already well established.

AN APPRAISAL CONCEPT OF DEPRECIATION

Grant and Norton [8-3] present a fourth concept of depreciation which is based partly on value and partly on cost. They define depreciation in this concept as

Depreciation Concepts

"difference in value between an existing old asset and a hypothetical new asset taken as a standard of comparison."

In the appraisal of property, the replacement cost of identical, similar, or substitute property is frequently used as a measure of the value of existing property. Thus a value of an existing asset may be established by market price, economic analysis of its value continued in use in place or elsewhere, or by other means. This value of the existing property subtracted from the cost of the replacement property would be the measure of the appraisal depreciation of the existing property. In this meaning of depreciation, there is implied the economic value difference between two similar assets valued under similar circumstances, but one is old, the other new. Such comparative appraisal estimates of depreciation occur primarily in economic analyses of the desirability of replacing existing machines and equipment with newer models.

DEPRECIATION COST ACCOUNTING METHODS

Over the centuries that cost accounting for depreciation has been practiced, many methods have been used for allocating the original cost, less terminal value, to operating periods, or financial reporting periods. Newer methods and modifications of older methods have been devised from time to time as the concept of depreciation accounting has been more widely understood. For the purpose of this book, only four prime methods will be described–(1) sinking fund or present worth method, (2) straight-line method, (3) declining-balance method, and (4) sum-of-the-years-digits method. The method of sinking fund or present worth is important because of its historical significance with respect to the time value of money; the remaining three methods are in industrial use today and are accepted by the U.S. Treasury, Bureau of Internal Revenue, for income tax reporting. By description, some of the other methods no longer generally used are: accounting for the cost of depreciation only when the property is retired, arbitrary depreciation charges as determined from time to time by management, market value or appraisal methods in which the depreciation charge is determined by the reduction in value between the beginning and ending of the accounting period, and depreciation calculated as a percentage of sales income.

Figure 8-1 gives typical curves of the remaining undepreciated cost which result from use of the four main methods which are described in the next sections.

DEPRECIATION ACCOUNTING FOR SINGLE PROPERTY UNITS AND PROPERTY GROUPS

The depreciation methods now being used–and many that are no longer used–are based in concept upon a single unit of property, that is, a property asset whose cost new, whose specific service life (not average life), and whose retirement each can be identified in the property and financial records. The

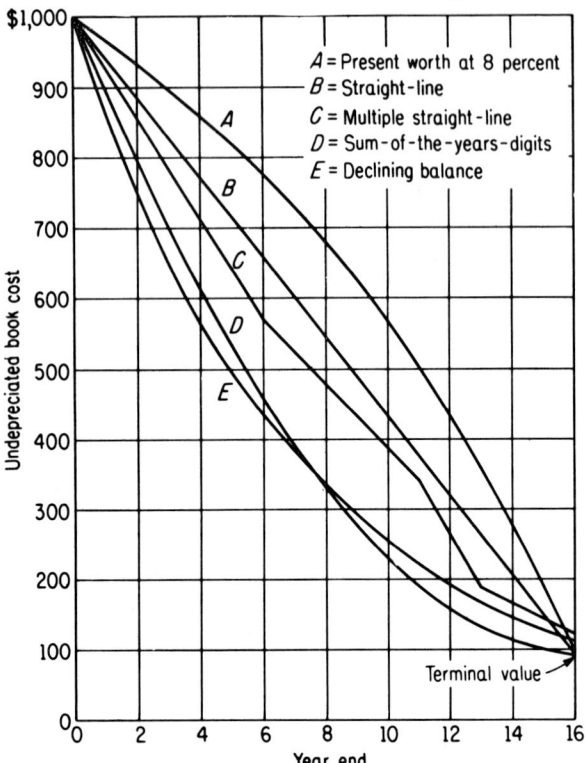

FIG. 8-1. Undepreciated book cost, or remaining cost, of a single property asset not subject to retirement by separate components, when the depreciation cost is allocated to expense by different methods.

depreciation cost accounting concepts are far more readily understood and the procedure more easily explained when applied to one property unit than when applied to groups of property units. When group depreciation accounting is involved–and it is in a high percentage of all depreciation cost accounting–application of all methods of depreciation becomes more complex. Within property groups, the retirement of the units occurs over a range of ages, so that each unit may have a different service life. The group as a whole has an average service life. This dispersion of retirement ages calls for some adjustments in the depreciation procedure developed for application to a single unit of property.

There are two types of property groups. The vintage group, or original group, is composed of units all installed in service the same year, thus the survivors at any time have the same age. The second type of property is composed of units of property of many ages, the limits of which range from the unit just installed to the age of the oldest unit in service. It is this second group to which most depreciation accounting is applied, because the property investment accounts most generally do not disclose the ages of the property units in service or the ages of the units when retired.

Depreciation Concepts

The descriptions of the four methods of depreciation cost accounting to follow are based entirely upon unit depreciation. Group depreciation cost accounting concepts and procedures are not essential to the objectives of this book. The literature may be referenced for further information on property accounting and group property depreciation. See especially Ref. 8-3, 8-5, and 8-8.

METHODS OF ALLOCATING DEPRECIATION EXPENSE

Of the several methods for allocating depreciation expense to accounting periods that have been devised over the years, four basic methods are selected for discussion. They are the sinking fund method together with its present-worth variant, the straight-line method, the declinig-balance method, and the sum-of-the-years-digits method.

The mathematics of the methods will be developed using the following notations and symbols:

B = depreciation base including terminal value. This base usually corresponds to the original cost in place of the depreciable property.

B_d = depreciable base, which is the total cost to be allocated and is equal to $B - T$

B_x = unallocated portion of the depreciation base yet unallocated to age x

T = estimated salvage, or terminal, value at age n, or at retirement

$B - B_x$ = total depreciation to age x, or the accrued depreciation

D = annual depreciation allocation

D_x = accumulated depreciation allocations to age x

x = age of the property in years

n = probable service life of the property; the period over which the depreciation is to be allocated; the number of year-end payments to the sinking fund

f = depreciation rate per year

Sinking Fund Depreciation Method

As indicated in its name, the sinking fund method of depreciation accounting is based upon the compound interest theory. Perhaps in its initial development, the objective was to accumulate a fund[1] which at compound interest would be equal to the depreciable base at the age of retirement of the property. By the sinking fund theory, the equal annual year-end deposits in a sinking fund would accumulate with compound interest thereon to the total depreciation allocation to any given date. The allocation for a specific year is, therefore, the annual

---

[1] In practice the actual monetary fund is not established as such; it is only a concept. The depreciation return, as with all other depreciation accounting methods is merged with other cash and used without identification.

deposit, or annuity, plus the compound interest increment to the fund for the year. The sinking fund method is not developed as an allocation method, but rather as a method of amortization.

From compound interest Eq. 6-4 the annual year-end deposit in a fund to accumulate to F in n years at i rate of interest is:

$$A = F \frac{i}{(1+i)^n - 1} \qquad [6\text{-}4]$$

In symbols used for the discussion of depreciation methods:

$$A = (B - T) \frac{i}{(1+i)^n - 1} \qquad (8\text{-}1)$$

In Eq. 8-1, A is the annual deposit in the fund and not the annual depreciation. In the sinking fund method, although the year-end deposit is a constant amount, the annual depreciation is an increasing amount because the compound interest earnings in the theoretical fund are an annually increasing amount.

The accumulation in the hypothetical fund to any date x is given by the standard compound interest Eq. 6-3:

$$F = A \frac{(1+i)^n - 1}{i} \qquad [6\text{-}3]$$

In terms of the depreciation symbols,

$$D_x = A \frac{(1+i)^x - 1}{i} \qquad (8\text{-}2)$$

By substituting the expression for A from Eq. 8-1 in Eq. 8-2 an expression may be developed for D_x in terms of x and n:

$$D_x = (B - T) \left(\frac{i}{(1+i)^n - 1} \right) \left(\frac{(1+i)^x - 1}{i} \right)$$

$$D_x = (B - T) \frac{(1+i)^x - 1}{(1+i)^n - 1} = B_d \frac{(1+i)^x - 1}{(1+i)^n - 1} \qquad (8\text{-}3)$$

The unallocated base at age x would be equal to the accumulation D_x in the hypothetical fund subtracted from the base B in the form of $B_d + T$:

$$B_x = (B_d + T) - B_d \frac{(1+i)^x - 1}{(1+i)^n - 1}$$

$$B_x = B_d \frac{(1+i)^n - (1+i)^x}{(1+i)^n - 1} + T \qquad (8\text{-}4)$$

From Eq. 8-4 it may be observed that the unallocated base will decrease more rapidly as x increases, thus causing the annual allocation to increase as x increases.

Depreciation Concepts 183

This feature of annually increasing depreciation allocations when the sinking fund method is applied to a single unit of property is one of the objectional features to the sinking fund method for cost accounting purposes. Much the better business practice is to have decreasing annual depreciation allocations with age for a single unit of property rather than increasing allocations. Decreasing annual depreciation charges allow recovery of the depreciable base at a more rapid rate when the property is more apt to be at its peak of usefulness and when its business value is more certain.

Another important objection to the sinking fund method for cost accounting purposes is that it is difficult to apply. Since the total depreciation is the annual deposit to the hypothetical fund plus the compound interest earning on the fund for the year, two calculations are required to arrive at the total depreciation allocation for any given year. Equation 8-4 may be used for this purpose by solving it for two consecutive values of x and taking the difference.

The sinking fund method utilizes an interest rate in determining the depreciation expense. Since the true accounting objective of depreciation is to recover a prepaid cost, there is no logical reason why an interest rate should be used in developing an allocation scheme to spread the prepaid cost to future operations. Therefore the sinking fund method has questionable value when applied to the procedure of cost accounting. But perhaps the most fundamental objection is its concept, which provides that the owner of the property must reinvest the annual depreciation deposit in order to recoup the full depreciable amount.

The sinking fund method is also difficult to apply to group accounting. In group accounting there are several items of property to account for. Each unit will have a different probable life and the units of a group will be retired over a period of years. Therefore the calculation of the sinking fund deposit is an involved procedure based upon the probable retirement distribution of the group of units.

The sinking fund method is no longer used in industry. Its greatest application was at one time in regulated public utilities, but now the regulatory commissions no longer require its use. Although the sinking fund method is unsuited to depreciation cost accounting, the method is significant in showing the development over the years of depreciation cost accounting starting from the concept of the time value of money to the present three most often used methods—straight-line, declining-balance, and sum-of-the-years-digits—which do not involve a time discount factor.

Present Worth Variant of the Sinking Fund Method

The sinking fund method requires that the total depreciable base of the property be accumulated by equal annual year-end deposits to a hypothetical sinking fund plus the compound interest accumulation thereon. The present worth method is similar in procedure, identical in result, but different in concept. The concept of the present worth method is that the decrease in value of the property in any given year, and therefore its depreciation for the year, is equal to the decrease for that year in the present value of its probable future returns. The yearly returns are composed of two factors, return *of* capital (depreciation) and the return *on* capital (net return, or net profit). Since all values are fundamentally the present worth of future income, services, and satisfactions, this present value concept applied to depreciation is logical in the sense of value.

If the future returns *of* and *on* an investment in a depreciable property be taken to be an equivalent uniform annual amount, then their present worth would be as calculated by Eq. 6-4, the present worth of a uniform series. In addition to the uniform annual return of A there would be the return of the terminal value at the end of n years. In terms of the depreciation symbols and letting the rate of return be r instead of i, the following expression may be written:

$$B = B_d + T = A\left(\frac{(1+r)^n - 1}{r(1+r)^n}\right) + \frac{T}{(1+r)^n} \tag{8-5}$$

Solving for A,

$$A = B_d\left(\frac{r(1+r)^n}{(1+r)^n - 1}\right) + rT \tag{8-6}$$

From Eq. 8-5 the present value at any age x, that is the depreciated value, would be

$$B_x = A\left(\frac{(1+r)^{n-x} - 1}{r(1+r)^{n-x}}\right) + \frac{T}{(1+r)^{n-x}} \tag{8-7}$$

Substituting in Eq. 8-7 the expression for A (Eq. 8-6):

$$B_x = \left(B_d \frac{r(1+r)^n}{(1+r)^n - 1} + rT\right)\left(\frac{(1+r)^{n-x} - 1}{r(1+r)^{n-x}}\right) + \frac{T}{(1+r)^{n-x}}$$

$$= B_d\left(\frac{(1+r)^n - (1+r)^x}{(1+r)^n - 1}\right) + T \tag{8-8}$$

This expression for B_x is exactly the same as Eq. 8-4 derived for the sinking fund method except for the substitution of r for i. Thus the two methods are equal.

The present worth method is not used in depreciation accounting for the same reason that the sinking fund method is not used. Nevertheless, it is sound in concept. It has its applications in valuation of property, in economy studies, and in cost studies. It will be recognized that the term $\frac{r(1+r)^n}{(1+r)^n - 1}$ in Eq. 8-6 is the capital recovery factor as given in Eq. 6-6.

Depreciation Concepts

Straight-Line Depreciation Method

Just prior to issue of the U.S. Treasury Internal Revenue Code of 1954 practically all depreciation accounting in industry and business was done by the straight-line method, a method which allocates equal depreciation cost to equal time periods of service. The straight-line method is a method of constant rate applied to a constant base. The name comes from the fact that as long as the estimated service life and terminal value remain unchanged, the method produces, when applied to a single unit of property, a straight line of the decreasing unallocated cost from zero age to the age of retirement. Further, the annual allocations will plot a horizontal straight line for the full service period.

As applied to a single unit of property, the straight-line method incorporates the basic plan of distributing the depreciable portion of the base in equal allotments over the service life. Therefore, the annual depreciation allocation may be expressed as follows:

$$D = \frac{B - T}{n} = \frac{B_d}{n} \tag{8-9}$$

The total allocation to age x would be:

$$D_x = Dx = B_d \left(\frac{x}{n}\right) \tag{8-10}$$

The unallocated base at age x may be developed as follows:

$$B_x = B - D_x$$

$$B_x = B - B_d \left(\frac{x}{n}\right)$$

$$B_x = (B_d + T) - B_d \left(\frac{x}{n}\right) \tag{8-11}$$

$$B_x = B_d \left(1 - \frac{x}{n}\right) + T$$

$$B_x = B_d \left(\frac{n - x}{n}\right) + T \tag{8-12}$$

The advantages of the straight-line method for cost accounting are that it is a simple, easily applied method that distributes the depreciable base according to a positive system that requires the exercise of judgment only in estimating the service life and the terminal value. Allocating the depreciation expense uniformly over the service life of the property is not preferred to allocating the expense in decreasing allotments, as is done by the declining balance and sum-of-the-years-digits methods. But the straight-line method has had a long record of acceptance in industry and business.

Depreciation Concepts

The annual depreciation sum in the straight-line method is usually determined by

$$D = \frac{B - T}{n} \quad (8\text{-}13)$$

The annual rate of depreciation in percent of the depreciation base B would be $\frac{100 - 7}{n}$, where the terminal value is expressed in percent of the depreciation base B.

Variants of the Straight-Line Method. Because the straight-line method does not result in higher annual depreciation expense in the earlier years than in the later years, it has been proposed that a shorter service life (higher depreciation rate) be used, say for the first one-half of estimated normal service life and a longer equivalent service life then used for the remaining life. Any method of adapting the straight-line method to more than one service life results in a bent line, that is, two or more straight-line segments joined at an angle (Fig. 8-1). This method is called the multiple straight-line method.

The change in depreciation rate could result from a preplanned change or from a reestimate of service life or terminal value. In either case the multiple straight line would result.

Another variant of the straight-line method is found in the use of some production unit or use basis, other than years, by which to set the depreciation rate and to allocate annual depreciation expense. Such units as vehicle-mile, ton-mile, gallon, ton, hour, unit pieces manufactured, or other measure of service may be used. The production unit has its advantage in application to short-lived property when the production unit of service is more controlling of service life than is the year. The vehicle-mile is often used for motor vehicle and the machine-hour is used for construction equipment.

This production base for the straight-line method will usually allocate depreciation expense more heavily in early years of service than in later years, because generally, new equipment is used more heavily when it is new than as it gets older. Many types of equipment are used at only a fraction of their capacity in the last year or two of their service life.

Mathematical Relationships of the Straight Line, Sinking Fund, and Present Worth Methods. There is a mathematical similarity between the straight line and sinking fund methods that is not apparent from just examining their concepts. Previously it was shown that the sinking fund and the present worth methods are equivalent when the interest rate i equals the rate of return r.

By previous derivations, the unallocated portion of the depreciation base at any age x was shown to be:

Straight-line method:

$$B_x = B_d \frac{(n - x)}{n} + T \quad (8\text{-}12)$$

Sinking fund method:

Depreciation Concepts

$$B_x = B_d \left(\frac{(1 + i)^n - (1 + i)^x}{(1 + i)^n - 1} \right) + T \qquad [8\text{-}4]$$

When the interest rate in the sinking fund method is 0 percent, Eq. 8-4 will reduce to Eq. 8-12 for the straight-line method. A direct substitution of $i = 0$ in Eq. 8-4 results in the indeterminate form 0/0. Evaluation is accomplished by differentiating the numerator and denominator separately with respect to i, and letting i tend toward zero.

$$\lim_{i \to 0} \frac{(1 + i)^n - (1 + i)^x}{(1 + i)^n - 1} = \lim_{i \to 0} \frac{n(1 + i)^{n-1} - x(1 + i)^{x-1}}{n(1 + i)^{n-1}} = \frac{n - x}{n} \qquad (8\text{-}13)$$

Thus, the straight-line method may be regarded as the special case of the sinking fund or present-worth methods when $i = 0$ or $r = 0$ percent.

Declining-Balance Depreciation Method

The declining-balance method of allocating depreciation expense uses a fixed percentage for each allocation period applied to the remaining, or unallocated, cost balance at the beginning of the period. For instance, when the depreciation rate is 12 percent per period, the successive depreciation allocations on a $100 base would be $12.00, $10.56, $9.29, $8.18, $7.20, etc., and the successive declining balances would be $100.00, $88.00, $77.44, $68.15, $59.97, $52.77, etc. The declinning-balance method is based upon no particular theory nor principle. The method is simply a scheme that allocates the cost base to depreciation expense in a series of reducing amounts.

The depreciation rate is set by decision with consideration given to the probable length of useful service of the property. The Federal Internal Revenue Code permits a maximum rate equal to 2.0 times the corresponding rate under the straight-line method.

Since the rate of depreciation in the declining balance method is constant, the unallocated portion of the base is given by

$$B_x = B(1 - f)^x \qquad (8\text{-}14)$$

Since the factor $(1 - f)^x$ is always less than unity and greater than zero, the unallocated base B_x will not reach zero. With a property unit of zero terminal value, the declining balance method will not permit of allocations which will accumulate to B, the full original cost base. Should a terminal value, or chosen end point, be considered, the depreciation rate f can be calculated for any chosen probable life n:

$$T = B(1 - f)^n \qquad (8\text{-}15)$$

$$f = 1 - \sqrt[n]{\frac{T}{B}} \qquad (8\text{-}16)$$

The expression for f is in terms of the terminal value and the base B, which is not a logical development. In practice it is preferred to establish f more or less arbitrarily, usually 1.5 to 3 times the reciprocal of the service life n.

In cost accounting practice the declining-balance method may be used to advantage when it is desired to depreciate the property in decreasing annual allocations to expense. In this respect the depreciation increments will resemble those established by the usual secondhand value curve. As with all depreciation methods, adjustment in the depreciation rate is required whenever the service life and/or terminal value ultimately appear to differ significantly from that on which the currently used depreciation rate was established.

The declining-balance method is easily applied in the accounting process to a single unit of property. The method loses its distinguishing characteristic of declining annual allocations of depreciation cost when applied to a property group to which replacements are added from time to time to replace retirements.

In the highway field, the declining-balance method of depreciation could be used in allocating depreciation expense of roadway construction and maintenance equipment when individual pieces of equipment form the accounting property unit. The method has no application at all in analyses for annual costs of highways nor for economy studies in comparing alternative proposals.

Sum-of-the-Years-Digits Method

Unlike the declining balance method, the sum-of-the-years-digits method does use a definite method of calculating the depreciation rate, but like the declining balance method, the sum-of-the-years-digits method results in decreasing annual depreciation charges to expense each following year. The decreasing charges are obtained by the arithmetical concept used. This concept is that the total depreciation to be written off is equivalent to the sum of the yearly digits of years over the service life, that is, for a service life of 10 years, the sum of the digits from 1 to 10 would be $1 + 2 + 3 + 4 + 5 + 6 + 7 + 8 + 9 + 10 = 55$. The first year's depreciation charge is 10/55 of the depreciable cost, the second year is 9/55, and so on to the tenth and last year, which would have a rate of 1/55. The depreciation rate is applied to the depreciable base, $B - T$.

Again, as in the declining-balance method, there is no conceptual relationship of the yearly depreciation charge as calculated by the sum-of-the-years-digits method to the consumption of service units of the property. It is a scheme, however, which allocates the largest annual charge to the first year of service and the smallest to the last year. From the concept of conservative business judgment, such practice in depreciation accounting is desirable. Also, as with the declining-balance method, the sum-of-the-years-digits method produces results which are not far from what would be obtained by the secondhand market value curve.

The-sum-of-the-years-digits method requires using a different rate for each succeeding year, which rates are always applied to the cost new less terminal value. The method was used only infrequently until after the U.S. Internal Revenue Code of 1954 authorized its use for income tax purposes.

Depreciation Concepts

REESTIMATING SERVICE LIFE AND TERMINAL VALUE

It is important always to keep in mind that the length of service life and the terminal value on which the depreciation expense accounting is computed are estimated factors. The estimates are usually based upon experience, the company concerned, or the experience of others, plus additional judgment to account for future factors that may differ from past experience. The objective is to estimate as closely as possible the most probable service life and terminal value. Terminal value is often estimated as a negative amount in those instances when cost of removal and disposal of the property is expected to cost more than can be realized by sale or from future use of the property. The removal of streetcar tracks and repaving of the street produces a negative terminal value, as would also the removal of a reinforced concrete structure.

When it is evident that a depreciable property will most probably last longer or less long in service than the service life corresponding to the current depreciation rate, a new estimate is in order. The resulting new depreciation rate would then be set so that the remaining depreciable expense would be allocated over the expected remaining service. A similar adjustment would be made in the depreciation rate when the estimated terminal value is materially changed.

COST ACCOUNTING AND DEPRECIATION

Some understanding of the operation of a set of financial accounts and the process of cost accounting will serve to indicate how the cost of depreciable property is recovered and turned back into cash. As previously stated, the objective of depreciation cost accounting is to charge a prepaid cost (investment cost of the property) to current operation expenses and not for building up a fund for replacement of the property. To gain this understanding, the subject will be approached with a brief statement about the accounting process and then presenting an illustration of a set of accounts for the ABC Trucking Company.

THE TWO CLASSES OF DISBURSEMENTS

In the cost accounting process for profit or loss there are two distinct classes of disbursements of money: (1) disbursements for investment in securities or fixed-capital goods (physical property to be used in service longer than the accounting period, usually one year), and (2) disbursements for labor, services, and supplies which are consumed during the accounting period. Disbursements for investments and fixed-capital goods do not reduce the assets but just transform their description. These assets are expected to earn a return on their investments. As these physical property assets are consumed in rendering service they are charged to operating expense through the system of depreciation expense accounting. By this cost accounting process the original disbursement to pay for the fixed property is ultimately, year by year, charged to expense. These two classes of disbursements are distinguished by both the accounting process and the calendar time by which they are charged to production, or operating, expense.

THE ACCOUNTING PROCESS

The accounting concepts and objectives may be stated as embodying a set of well-known and accepted procedures of classifying, recording, summarizing, and presenting the money transactions of an organization. The purpose of this total accounting process is to (1) determine monetary profits and losses resulting from operations, and (2) to show the financial condition of the organization at specific calendar dates. These two purposes are served by preparing a Statement of Income and Expense, or Operating Statement, for a specific time period, and by preparing a Statement of Assets and Liabilities, or Balance Sheet, for a specific calendar date, usually for the terminal date of the operating statement.

The operating statement develops certain entries for the balance sheet, including the profit or loss, which indicates whether there has been an increase or a decrease in net worth. The balance sheet is a presentation of the fundamental relationship that Assets = Liabilities + Ownership. Assets may be thought of as being what is owned and the liabilities as being the sources of what is owned.

Concept of Debit and Credit

The long-established double-entry system of accounting is based upon two entries to the set of accounts for each transaction or adjustment. These two entries are known as "debits" and "credits." Debits are entered on the left-hand side of the T account and the credits are entered on the right-hand side. For each debit entry there is a corresponding credit entry; thus, if all entries are correctly made, the sum of the debits must equal the sum of the credits.

The words *debit* and *credit* in double-entry accounting no longer have a specific meaning in terms of debt as being owed and credit as having paid what was owed, as they once did in simple single-entry cash bookkeeping. Now "debit" is simply the left side and "credit" the right side of the T account. It will help, however, in keeping these entries straight in mind if the debit entry to the cash account is thought of as being the "debt" owed by the cash account to the owner or organization, and the credit entry to the cash account is thought of as giving the cash account "credit" for properly discharging its debt, or properly accounting for the decrease in the cash balance on hand.

In terms of the balance sheet, the left-hand side is for assets and the entries come from the left-hand side of the corresponding T accounts. The right-hand side of the balance sheet is for liabilities and the entries come from the right-hand side of the corresponding T accounts.

ABC TRUCKING COMPANY

The following pages give the accounting for the fictitious ABC Trucking Company to illustrate, principally, the depreciation cost accounting entries and

Depreciation Concepts

the process of converting back to cash the investment in the trucks. The set of T accounts used is a briefed-down set containing the total of entries for the year, rather than daily entries. Typical accounts not presented are:

Accounts receivable	Accounts payable
Notes receivable	Notes payable
Investments (bonds held)	Accrued interest payable
Merchandise inventory	Taxes payable
	Reserve for bad debts

Advertising
Overhead expense
Sales expense
Other income

The first table is a listing of the individual cash flows and book entries over the four years. Following this table are the individual T accounts into which the A to N transactions have been entered, once as a debit and once as a credit.

TABLE 8-1
LIST OF ACCOUNTING TRANSACTIONS OF THE ABC TRUCKING COMPANY
FOR ITS FIRST FOUR YEARS

Transaction	First Year		Second Year		Third Year		Fourth Year	
	Cash	Noncash	Cash	Noncash	Cash	Noncash	Cash	Noncash
A. Founded company and sold to self capital stock	$60,000							
B. Purchased trucks 101 and 102 at $25,000 each and truck 103 at $5,000	$55,000							
C. Three-year premium paid on insurance	$1,200				$1,500			
D. General cash operating expense	$12,000		$15,000		$17,100		$19,200	
E. Insurance expense		$400		$400		$400		$500
F. Depreciation expense on trucks; 5-year life, 20 percent trade-in value		$8,800		$8,800		$8,600		$6,000
G. Cash sales income	$25,600		$31,400		$34,900		$38,200	
H. Income tax paid			$1,600		$1,800		$2,100	
J. October 2 traded truck 103 for allowance of $2,200 on truck 104 costing $6,700*					$4,500			
K. Depreciation on truck 104 ...						$268		$1,072
L. Dividends paid							$3,200	
M. July 5 traded truck 102 for allowance of $11,500 on truck costing $21,000†							$9,500	
N. Depreciation on truck 105 ...								$1,680

* Truck 103 to October had a book depreciation reserve of 2 3/4 years at $800 per year, or a total of $2,200. Actual depreciation expense was $5,000 minus $2,200, or $2,800. Therefore, earned surplus must be reduced by $600 because of failure to charge to expense in past years the full depreciation on truck 103.

† Truck 102 to July 5 had a book depreciation reserve of 3 1/2 years at $4,000 per year, or a total of $14,000. Actual depreciation expense was $25,000 minus $11,500 or $13,500. Therefore, earned surplus must be increased by $500 because of overcharging depreciation on truck 102 in past years.

ABC Trucking Company
T-Account Ledger, First Four Years

1. Cash

	Dr.		Cr.
First Year			
A	$60,000	B	$55,000
		C	1,200
G	25,600	D	12,000
Total	$85,600		$68,200
Forward	–		$17,400
Second Year			
Opening	$17,400		–
		D	$15,000
G	31,400	H	1,600
Total	$48,800		$16,600
Forward	–		$32,200
Third Year			
Opening	$32,200		–
G	34,900	C	$ 1,500
		D	17,100
		H	1,800
		J	4,500
Total	$67,100		$24,900
Forward	–		$42,200
Fourth Year			
Opening	$42,200		–
G	$38,200	D	$19,200
		H	2,100
		L	3,200
		M	9,500
Total	$80,400		$34,000
Forward	–		46,400
Fifth Year			
Opening	$46,400		–

2. Prepaid Insurance

	Dr.		Cr.
First Year			
C	$ 1,200	E	$ 400
Total	$ 1,200		$ 400
Forward	–		$ 800
Second Year			
Opening	$ 800		–
		E	$ 400
Total	$ 800		$ 400
Forward	–		400
Third Year			
Opening	$ 400		–
C	1,500	E	$ 400
Total	$ 1,900		$ 400
Forward	–		$ 1,500
Fourth Year			
Opening	$ 1,500		–
		E	$ 500
Total	$ 1,500		$ 500
Forward	–		$ 1,000
Fifth Year			
Opening	$ 1,000		–

3. Equipment (Trucks)

	Dr.		Cr.
First Year			
B	$55,000		
Total	$55,000		
Forward	–		$55,000
Second Year			
Opening	$55,000		–
Total	$55,000		
Forward	–		$55,000
Third Year			
Opening	$55,000		–
F	6,700	F	$ 5,000
Total	$61,700		$ 5,000
Forward	–		$56,700
Fourth Year			
Opening	$56,700		–
M	21,000	M	$25,000

Depreciation Concepts

	Dr.	Cr.
Total	$77,700	$25,000
Forward	–	$52,700
Fifth Year Opening	$52,700	–

4. Capital Stock

First Year		A	$60,000
Total			$60,000
Forward	$60,000		–
Second Year Opening	–		$60,000

NOTE: No change through the fourth year

5. Depreciation Reserve

First Year		F	$ 8,800
Total			$ 8,800
Forward	$ 8,800		–
Second Year Opening	–		$ 8,800
		F	8,800
Total			$17,600
Forward	$17,600		–
Third Year Opening	–		$17,600
J	$ 2,200	F	8,600
		K	268
Total	$ 2,200		$26,468
Forward	$24,268		–
Fourth Year Opening	–		$24,268
M	$14,000	F	6,000
		K	1,072
		N	1,680
Total	$14,000		$33,020
Forward	19,020		–

	Dr.	Cr.
Fifth Year Opening	–	$19,020

6. Earned Surplus (Undistributed Net Earnings)

First Year			
		C	$ 4,400
Total			$ 4,400
Forward	$ 4,400		–
Second Year Opening	–		$ 4,400
H	$ 1,600	C	7,200
Total	$ 1,600		$11,600
Forward	$10,000		–
Third Year Opening	–		$10,000
H	$ 1,800	C	8,532
J	600		
Total	$ 2,400		$18,532
Forward	16,132		–
Fourth Year Opening	–		$16,132
H	$ 2,100	C	9,748
L	3,200	M	500
Total	$ 5,300		$26,380
Forward	$21,070		–
Fifth Year Opening	–		$21,080

7. Operating Expense

First Year			
D	$12,000		
E	400		
F	8,800		
To P & L	–	a	$21,200
Total	$21,200		$21,200
Second Year			
D	$15,000		
E	400		

Depreciation Concepts

	Dr.	Cr.
F	8,800	
To P & L	–	a $24,200
Total	$24,200	$24,200

Third Year

	Dr.	Cr.
D	$17,100	
E	400	
F	8,600	
K	268	
To P & L	–	a $26,368
Total	$26,368	$26,368

Fourth Year

	Dr.	Cr.
D	$19,200	
E	500	
F	6,000	
K	1,072	
N	1,680	
To P & L	–	a $28,452
Total	$28,452	$28,452

8. Sales Income

First Year

	Dr.	Cr.
		G $25,600
To P & L b	$25,600	–
Total	$25,6000	$25,600

Second Year

	Dr.	Cr.
		G $31,400
To P & L b	$31,400	–
Total	$31,400	$31,400

Third Year

	Dr.	Cr.
		G $34,900
To P & L b	$34,900	–
Total	$34,900	$34,900

Fourth Year

	Dr.	Cr.
To P & L b	$38,200	G $38,200
Total	$38,200	$38,200

9. Summary of Expense and Income
(Profit and Loss, or Operating Statement)

First Year

	Dr.	Cr.
a	$21,200	b $25,600
To E S c	4,400	–
Total	$25,600	$25,600

Second Year

	Dr.	Cr.
a	$24,200	b $31,400
To E S c	7,200	–
Total	$31,400	$31,400

Third Year

	Dr.	Cr.
a	$26,368	b $34,900
To E S c	8,532	–
Total	$34,900	$34,900

Fourth Year

	Dr.	Cr.
a	$28,452	b $38,200
To E S c	9,748	–
Total	$38,200	$38,200

Note: The foregoing T account could be written in the familiar form:

Sales income			$38,200
Operating expense			
General expense		$19,200	
Depreciation		8,752	
Insurance		500	$28,452
Net operating income to earned surplus account			$ 9,748

Balance Sheets

From the foregoing set of T accounts, the closing balances for each year from

Depreciation Concepts

the balance sheet accounts have been brought forward to the balance sheet. A balance sheet–assets and liabilities–for the end of each of the four years follows:

ABC TRUCKING COMPANY
BALANCE SHEETS FOR FIRST FOUR YEAR ENDS

Balance Sheet at End of First Year

Assets			Liabilities	
Cash		$17,400	Capital stock	$60,000
Prepaid insurance		800	Earned surplus	4,400
Property (trucks)	$55,000		Total Liabilities	$64,400
Less depreciation	8,800			
		$46,200		
Total Assets		$64,400		

Balance Sheet at End of Second Year

Assets			Liabilities	
Cash		$32,200	Capital stock	$60,000
Prepaid insurance		400	Earned surplus	10,000
Property (trucks)	$55,000		Total Liabilities	$70,000
Less depreciation	17,600			
		$37,400		
Total Assets		$70,000		

Balance Sheet at End of Third Year

Assets			Liabilities	
Cash		$42,200	Capital stock	$60,000
Prepaid insurance		1,500	Earned surplus	16,132
Property (trucks)	$56,700		Total Liabilities	$76,132
Less depreciation	24,268			
		$32,432		
Total Assets		$76,132		

Balance Sheet at End of Fourth Year

Assets			Liabilities	
Cash		$46,400	Capital stock	$60,000
Prepaid insurance		1,000	Earned surplus	21,080
Property (trucks)	$52,700		Total Liabilities	$81,080
Less depreciation	19,020			
		$33,680		
Total Assets		$81,080		

Change in Cash Balance

After investing $55,000 in three trucks at the beginning of the first year of operation, there remained $5,000 in cash from the initial capital of $60,000. At the close of the fourth year of operations, the cash balance had increased to $46,400, or a net increase of $41,400. The source of this cash increase is accounted for as follows:

Earned surplus (undistributed net earnings, or retained profits)	$21,080
Depreciation return on trucks	19,020
Contributed capital of $60,000 less cost of trucks ($52,700) and prepaid insurance ($1,000)	6,300
Total cash	$46,400

The net earnings retained in the company result in additional capital. Such increased capital comes from cash sales income above the cash disbursements required for operations. This increase in cash plus the return of original cost through depreciation return is available for any purposes management may elect, such as working capital, expansion of fixed assets, replacement of fixed assets, investment in stocks and bonds, expansion of research and development, etc.

DEPRECIATION, TAXES, AND GOVERNMENT ACCOUNTING

Property and income taxes paid by a private enterprise are operating expenses. The government, on the other hand, receiving the tax payments treats this tax income as "sales income." This tax income is then disbursed for both fixed capital purposes and current operating expense. Usually, however, governments do not practice double-entry accounting, fixed-asset accounting, and depreciation accounting, except for some utility enterprises and special operations for purposes of establishing unit costs as a check on prices from private enterprise. Governments do not practice asset and depreciation accounting because they do not render balance sheets and profit and loss statements. Their accounting is plain cash accounting.

But for purposes of economy studies for government public works, the citizens should consider that the tax income of their government which is disbursed for fixed-capital property, such as highways, is a disbursement for investment and not current expense. Thus, the public through its governmental agencies, has large investments in public properties—highways, parks, water resource development, waterways, and public buildings. Individual citizens and individual businesses and corporations charge of their disbursements for taxes to

Depreciation Concepts

current expense rather than to investment. They charge taxes to their expense accounts because they do not keep investment records on their shares of ownership in public investments of fixed capital, which may be created by investment of tax incomes.

The fiscal management of public agencies by public officials, nevertheless, would be more meaningful to the public were it conducted on the basis of and according to the principles of investment, current expense, and cost accounting, as practiced in private enterprise.

PROBLEMS

8-1. Why do we not use just one depreciation cost allocation method in cost accounting and income tax cost accounting? Why were the different methods–straight-line, declining balance, sum-of-the-years-digits, and present worth–developed?

8-2. Read the material on depreciation in one textbook on business management and one on economics. Summarize and comment upon the definitions and concepts on depreciation that you read.

8-3. What are the short-run and long-run advantages and disadvantages of the declining-balance depreciation method as compared to the straight-line method when applied by a highway contractor?

8-4. (1) In what specific ways may an organization's cost and fiscal accounting system render useful aid to engineering economy analyses? (2) What specific data from cost and fiscal accounting systems are generally of no application to engineering economy analyses?

8-5. Why do not highway departments keep depreciation cost accounts on their highways? If a highway department does maintain depreciation cost accounts on its construction and maintenance equipment, what is its reason for doing so?

8-6. For a 5-year and a 10-year service life, work out declining-balance depreciation rates which closely approximate the annual depreciation expense as calculated by the sum-of-the-years-digits method. Use $1,000 cost new and zero terminal value. Comment on your results.

8-7. A motor truck used in highway freight cost new $8,000. On the basis of its use for 6 years and a trade-in allowance of $900 calculate each year's depreciation expense by the following methods: (1) straight line, (2) declining balance (double straight-line rate), (3) sum-of-the-years-digits, and (4) 6 percent sinking fund.

8-8. Using the declining-balance depreciation method and a service life of 12 years, plot three curves of the undepreciated balance corresponding to 1.0, 1.5, and 2.0 times the straight-line depreciation rate.

8-9. An accountant has been using the declining-balance method for depreciating a machine costing new $60,000. His rate has been 30 percent each year end. At the end of the 5th year, management reports that the machine will be retired after two years more of service at an estimated terminal value of $3,000. Calculate the book allocation depreciation expense for each of the 7 years. Show your depreciation rates for the 6th and 7th years and explain how they were determined.

8-10. From a current daily newspaper compile the advertised price of three makes of popular automobiles by year models. Plot these curves–price versus year model (age)–and then compare the resulting market value depreciation with what would be calculated for cost accounting by the straight-line, sum-of-the-years-digits, and double straight-line rate declining-balance methods. Assume a useful life of 10 years and $50 terminal value.

8-11. A small firm has a truck for which the accounting history on a year-end basis is as follows:

Year End	Market Value	Book Annual Depreciation Allocation	Annual Operating and Maintenance Expense Exclusive of Depreciation
0	$4,000	–	–
1	2,800	$500	$1,500
2	1,880	500	1,800
3	1,200	500	1,800
4	750	500	1,950
5	510	500	2,200
6	350	400	2,400
7	240	400	2,550
8	160	300	2,800
9	100	300	3,000

Mileage per year is about 20,000 and uniform. A new truck would probably closely repeat the above cost history. After taxes, the owner expects to earn 10 percent on his investment. When should a new truck be purchased, or have been purchased?

8-12. For an automobile costing $3,000 new with a $540 trade-in value at the end of 6 years, calculate the following, using a 7 percent vestcharge rate: Annual depreciation cost plus vestcharge by (1) the straight-line method, (2) the double straight-line rate declining-balance method, (3) the sum-of-the-years-digits method, and (4) the sinking fund method.

8-13. Using the following market values (used-car price) as applying to the automobile in Prob. 8-12, calculate (1) the annual capital cost (market depreciation plus vestcharge), (2) the equivalent uniform annual capital cost from (1), and (3) the annual capital cost using the capital recovery factor. Discuss your results and compare with Prob. 8-12. Year-end market values are: first, $2,275; second, $1,710; third, $1,280; fourth, $960; fifth, $720; and sixth, $540.

8-14. In discussing the possibility of buying a replacement machine, an author of a 1960 textbook wrote as follows: "According to the accounting records, the value of the machine (the presently used ABC machine) was $140. Actually, the machine had decreased in value at a faster rate than had been assumed when the depreciation account was established. If a new ABC machine were bought, there would be a sunk cost of $140 − $100 = $40. (The $100 is the allowance on the old machine offered by the salesman on the purchase of a new ABC machine). If a new XYZ machine were purchased the sunk cost would be $140 − 0 = $140, since no trade-in allowance was offered.

Comment upon (1) the first two sentences, and (2) upon the last two.

8-15. A cash disbursement for a new bridge is an investment of capital; a cash disbursement for stockpiling of sand and salt for winter maintenance is an investment in working capital; and a disbursement for labor for maintenance of the highway plant is for

Depreciation Concepts

current expense. Discuss these statements; are they acceptable as sound accounting practice? Why the distinction? What is a capital investment as compared to an expense?

8-16. The following financial data are taken from the books of a motel owner: Investment in depreciable plant, $110,000; investment in land, $6,000.

ANNUAL OPERATIONS

Year End	Gross Sales	Operating Expense	Depreciation Expense
1	$46,000	$29,000	$ 9,000
2	52,000	31,000	9,000
3	53,000	33,000	9,000
4	50,000	30,000	10,000
5	45,000	27,000	10,000

Early in the 6th year the owner sold his holdings for use as a commercial feed plant for a net of $20,000 because of relocation of the main highway. What overall rate of return did the owner realize on the venture?

8-17. What rate of return do you earn on the reduced premium paid on a risk insurance policy when you pay 3 years' annual premiums now at a price of only 2.5 times the one-year premium?

REFERENCES

8-1. Gannett Fleming Corddry and Carpenter, Inc. "Columbia Gas System Depreciation Study, Methods Used in the Estimation of Average Service Life and Accrued Depreciation." Harrisburg, Pa., 1963.

8-2. Eugene L. Grant and L. F. Bell. *Basic Accounting and Cost Accounting.* 2d ed., McGraw-Hill, New York, 1964.

8-3. Eugene L. Grant and Paul T. Norton. *Depreciation,* Rev. ed., Ronald, New York, 1955.

8-4. A. H. Kuhn. *Some Practical Aspects of Depreciation Accounting.* Edison Electric Institute, New York, Bulletin 20, 1952.

8-5. Anson Marston, Robley Winfrey, and Jean C. Hempstead. *Engineering Valuation and Depreciation.* Iowa State U. P., Ames, 1953.

8-6. Marcus Vitruvius Pollio. *The Architecture of Marcus Vitruvius Pollio* in ten books translated from the Latin by Joseph Gwilt, Book II, Chap. VIII, p. 47 (from manuscripts dated 1552, 1649, and later). Lockwood and Co., London, 1874.

8-7. U.S. Treasury Department, Internal Revenue Service, Washington, D.C. *Depreciation Guidelines and Rules, Revenue Procedure 62-21,* Publication No. 456. Rev. August 1964.

8-8. Robley Winfrey. *Depreciation of Group Properties.* Iowa State University. Engineering Experiment Station Bulletin 155, 1942.

Chapter **9**

Service Life of Physical Property

The analysis for economy of proposed facilities, depreciation cost accounting, and calculation of the economic cost of owning and operating a facility, machine, or process, requires that some future period of time be chosen over which to consider the cash flows involved and the services received. This time period may be, and most often is, different in length in the analysis for economy, in depreciation cost accounting, and in estimating the economic cost of owning and operating a property. The normal expected length of service life in years or in production units is often the best base from which to begin the considerations leading to selection of the period of time n to use in the calculations of economy, depreciation, or annual cost. A discussion of the methods used to estimate or to calculate service life of properties will aid in understanding service life as well as in the use of estimates of service life.

SOME DEFINITIONS AND CONCEPTS

Physical properties–buildings, pavements, bridges, motor equipment, machine tools, water mains, telephone poles, to name a few types–have three measures of life.

Service life of a physical property is that period of time (or of service measured in some unit of production) extending from date of installation into service to the date of retirement from service. Service life measures actual total usage, whether profitable or not.

Physical life is that period of time the property exists, not necessarily in a usable condition or used. Often property is removed from active service and kept intact some years before it is junked or destroyed, or it may be just left in its service position but never used. Many abandoned bridges are standing where they were built, unused because of relocating the approach road.

Economic life of a physical property is that period of time (or of service measured in some unit of production) extending from the date of installation into service to that date when the property is no longer economically profitable to use. The property may be kept in service after its economic life has ended because of

Service Life of Physical Property

lack of money to provide an alternative service. In this concept, economic life is determined by an analysis for economy which indicates that an alternative property or service would result in obtaining equal service at less economic cost.

Service life, physical life, or economic life in no way relate to depreciation accounting or market value. But the depreciation accounting rate most often is dependent upon the service life of the property.

Mobile property (motor vehicles) and movable property (machine tools) may have their service life broken into several ownerships or locations of installation. In such case each owner or location has a separate accounting history, including a cost of acquisition, a service period, and a terminal value.

When a number of units of like property is considered as a group it is usual to speak of *average service life* to indicate that the general class of property is intended rather than one specific unit of that property.

The life of property is a *probable service life* until it is retired from service. At any given age of property in service it has an *expectancy of life,* which is that period of time from the present to the future date of forecasted retirement.

In depreciation accounting, economic studies, and cost analyses, the average service life is the life of greatest interest because many units of property are generally considered, or when a single unit is considered, such as a bridge, it is in a general sense so far as service life is concerned.

Property units which are taken out of service for any reason whatsoever are called *retirements.* The retirements may be installed again elsewhere, used for some other purpose, junked, abandoned in place, or used in place in the same or different function after some alteration. Retirements may be or may not be replaced. If replaced, the replacement property may be in kind or wholly different.

FACTORS WHICH LEAD TO RETIREMENT OF PROPERTY

Some understanding of the factors surrounding the use of physical property and the reasons why properties are retired from service will aid in understanding the service life, estimated and factual.

SERVICE LIFE AND RETIREMENTS

Contrary to what is often implied in the literature, physical properties do not have a definite service life which always can be expected. In fact, neither does most animal or biological life. There are no "one-horse shays" in our highway systems and industries. The ending of service life of physical properties (except by disaster) is by man-made decision. Management decisions vary, management policies vary, conditions of service vary, weather varies, maintenance quality varies, service loads vary, inherent manufactured quality varies, and so on.

These factors and many other factors combine their forces to the end that management decides to retire the property from service.

The retirement of property ends service life, except when it may be retained for use (in place or other place) and reentered (or retained) in the property records at its net terminal value. But why is property retired? Or more accurately, why does management retire property? There are many reasons and conditions why property is retired. Sometimes there is just one specific factor; at other times there is a complex of factors.

When the many and varied forces of service, the environmental conditions, changing technology, inherent service qualities, and policies of management are considered it is easily and correctly concluded that the service life of man-built properties of like character may be expected to experience a wide range of service life. By observing service life by noting ages at retirements, the dispersion of service lives may be determined and better understanding obtained of the life characteristics of physical properties and how to use estimates of service life in economic studies and depreciation accounting.

REASONS FOR RETIRING HIGHWAY PROPERTY

Physical properties are said to be retired from service when for one reason or other they are removed from productive service or altered and used in a second service life. In industrial and business organizations, the physical retirement is paralleled by the removal of the corresponding dollar asset from the accounting books of the company. The physical property may or may not be removed from its place at the time of the book entries. In some instances the retired property may be abandoned in place.

For property other than automotive, shop, and office equipment, highway departments generally do not carry property-asset accounts. In other words, the physical highways and their appurtenances are not carried as assets in the accounts of highway departments and therefore are not retired from the financial books when they are rebuilt, abandoned, or transferred to another highway authority. Nevertheless, the roadway property is considered to be retired from service whenever its use is suspended or altered substantially and reused with realization of a terminal value.

Generally, highway properties are retired from useful service because of (1) physical wear and tear or deterioration brought on because of traffic use and the normal action of the weathering agents; (2) accidents and catastrophe consisting of collision of vehicles with the highway structure, storms, floods, winds, and other acts of God; (3) inadequacy in either loading capacity or traffic volume capacity, such as a bridge of inadequate structural design strength, and a two-lane pavement when a four-lane pavement is needed; (4) general obsolescence or unfitness for the current type of usage such as a rural highway designed for a maximum speed of 35 miles an hour, or because technological changes have made available

Service Life of Physical Property

improved products or methods; and (5) demand of other authorities or to make room for related improvements such as the retirement of a section of highway because of construction of a water reservoir or retirement of highway rendered useless because of construction of an expressway or bridge approach.

The likelihood of the occurrence of any of these five forces singly or in any combination is a basis of forecasting the retirement of the highway property affected. With the exception of retirement by accident and disaster, the highway property is retired only upon decision of management. Thus the useful service life of such properties comes to an end by dogmatic decision, not by reason of natural biological decay. Management makes the decision to retire the property in the belief that the property can no longer render a useful and satisfactory service to transportation, or that a more desirable service can be rendered by another facility or the same general facility when reconstructed in accordance with current design standards.

Because property is retired by a management decision and because of the unknown magnitude and direction of the five forces of retirement listed, the useful service life of highway properties, including maintenance and construction equipment and buildings, varies over a considerable range. This variation is found not only between components of the highway such as earthwork, paving, and bridges, but within a given component and for all construction and maintenance equipment. A steel bridge may be so designed, located, maintained, and used that its useful service life may reach 125 years. A similar bridge in another location, subjected to a different usage and management may be removed from service at an age of 20 years.

There are no specific or universal guide posts upon which the decisions to retire property from service are based. The action results from judgment in each individual case in much the same way that individuals decide to trade in their automobiles, refrigerators, or to discard a dress or suit of clothes. For instance, when a bridge is damaged by flood, the engineer in charge may retire the bridge and replace it with a new structure. Should the damaged bridge be adequate and well suited to its function at the time of the flood, the engineer may decide to repair the bridge rather than to build a new one, if such action is considered feasible. Circumstances of adequacy, obsolescence, availability of material and funds, and many other management factors govern the decision to retire properties from service.

Not only is the time of retirement of property governed by decision of those in charge, but the method of retirement is likewise the result of a management decision. Highway properties may be abandoned in place, completely removed as junk material, substantially rebuilt in place, or removed and reused elsewhere. During the reconstruction of a highway route, a short section on an unfavorable location would likely be abandoned in place; another section in good location but low in grade line, would be salvaged practically 100 percent on earthwork and drainage, but with zero salvage on the roadway surface and subgrade. A third section of satisfactory line and grade might be reconstructed by capping the existing pavement surface with new material. Thus in ordinary reconstruction of highways it is necessary to examine them in detail to determine the reasons for retirement, the method of retirement, and the terminal value.

The reasons for retirement have little economic significance, other than knowing the reason for retirement might make it possible to design against that factor in the future. The method of retirement has an important bearing upon the terminal value realized and the cost of the reconstruction. Knowing the probable methods of retirement of proposed facilities is a distinct aid in making the economic analysis of proposed replacement construction.

The probable length of service life and the probable terminal value are important factors in computing the economy of different materials and designs and for computing the annual economic cost of ownership and operation of highways and their appurtenances.

MAINTENANCE, BETTERMENTS, AND RETIREMENTS

Although maintenance expense and depreciation expense are distinctly separate items of cost, in many instances there is not a clear-cut distinction of what constitutes a retirement of property, a betterment of property, and a maintenance operation [9-1]. This distinction is important in the accounting classification. A maintenance operation is charged to the current expense of operation; retirements and betterments are handled through the assets accounts. The distinction is made by statement of policy and accounting classification. For instance, management decides that the replacement of more than a length of 200 feet of rigid pavement for the full width of the traveled way is to constitute a retirement of the old pavement and the construction of a new pavement. The extension of a culvert by adding full lengths of standard sections would likely be classified as a betterment to the existing structure, although the replacement of a broken section or up to half of the length of the culvert would most likely be accounted for as a maintenance operation. The classification of maintenance, betterment, and retirement operations is not a relatively important matter in highway accounting, except when costs are to be compared between areas or between highway departments. Obviously, before a comparison of expense would disclose the true relationships of maintenance efficiency, the cost accounting classifications would have to be on the same basis.

METHODS OF DETERMINING SERVICE LIVES

The physical properties used in both public works and private works cover the wide range of machinery, instruments, mechanical equipment, construction equipment, maintenance equipment, buildings, pipelines, and all forms of mobile

Service Life of Physical Property   205

units. For depreciation cost accounting purposes, planning purposes, for pricing, and for economic analyses, the probable service life of property by classes is a necessary factor. Service life of proposed works and properties is usually determined by referring to the experience with similar properties. Of particular interest, therefore, are the procedures by which past experience may be analyzed to arrive at estimates of service lives.

GENERAL CONCEPTS RELATED TO SERVICE LIFE

Specific methods of analyses for calculating average service lives and survivor curves have been developed since about 1920. The variations in the methods are associated primarily with the differences in the availability of information.

In all analyses the property is described by some type of unit. The unit may be dollars of original cost; or a physical unit such as an automobile, a telephone pole, a square yard of pavement, or a rod of snow fence. The useful life of the units of property–dollars or physical items–is most often expressed in years, but may be expressed in service units such as car-miles, lamp-hours, hours of use, or 1,000 items of products (bolts) produced. The unit of analysis is usually the same as used in the property accounting classification since such records are the source of the information required for analysis.

THE TURNOVER METHOD

The turnover method of determining average service life [9-13] is one that compares over a period of years accumulated installations or units in service with accumulated retirements (see Fig. 9-1). The average service life (turnover period) is given by that period of time it takes to accumulate future retirements which total the accumulated number of units in service at the beginning date. The average service life is the horizontal distance between the two curves starting any chosen time on either curve.

When the installations of units or retirements are not known from the beginning, the average service life may be determined by plotting the accumulated retirements curve backward from the present, or other date, to where it intersects the curve of units in service.

The turnover method does not result in a survivor curve because ages of the retirements are not used. The average service life determined will not be a true indication of the service life of the property unless the property has been continued in use at least one or two maximum life cycles, unless the annual placements have about the same potential average probable life as the retirements, and unless the property is maintained at about a constant number of units in service over the years covered by the analysis.

For comparatively new properties, growing properties, and properties in which the potential lives of the units are changing noticeably, the turnover method

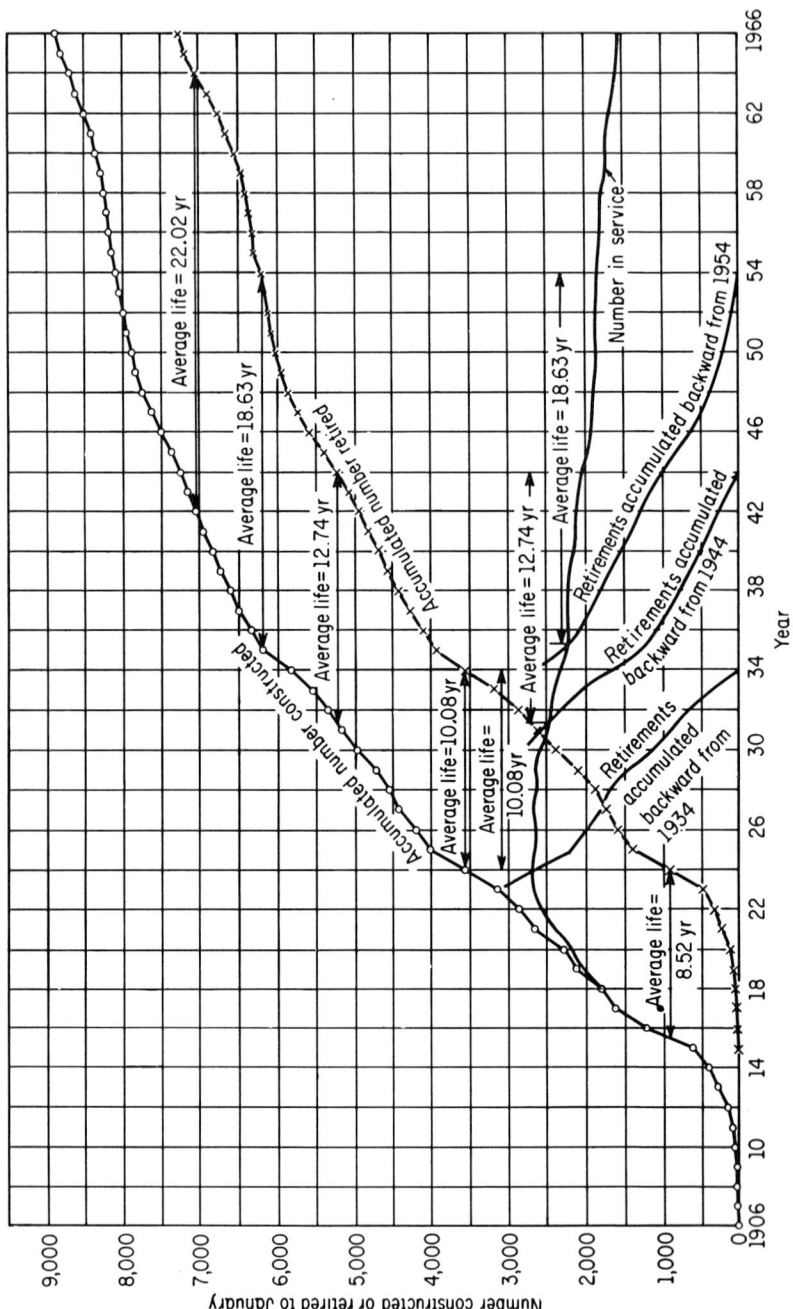

FIG. 9-1. Accumulated placement and retirement curves plotted for determining average life by the turnover method.

Service Life of Physical Property

is not to be recommended, or at least should not be used without adequate correction for these conditions (see Ref. 9-9 and 9-10). For old and stable properties continued in a normal condition year to year, the turnover method will indicate usable average service lives.

ACTUARIAL METHODS OF ANALYZING RETIREMENTS

The dispersion of the ages at retirement of physical properties leads naturally to some sort of statistical approach to the analysis of retirement histories. The life insurance business is based upon such analyses of human birth and death. The actuarial techniques developed for life insurance may be easily adopted in principle to the calculation of survivor curves, retirement frequency curves, average service lives and life expectancies of physical properties.

Explanation of the General Actuarial Procedure

The actuarial methods evolve around the yearly additions and placements (births), the property units in service at specific dates (census of population), and retirements (deaths). There are two main objectives to the actuarial analysis: To determine the average service life of the property and to determine the dispersion of the ages at retirement. For a more detailed information than given herein see Ref. 9-4 and 9-13.

When property accounts are maintained of the existing physical properties in service and when the age of the property retired year by year can be determined, it is possible to calculate survivor curves and the average service lives of the property. The statistical analysis procedure is the same in principle as that used by actuaries in computing the mortality tables and curves for humans which are used in the life insurance business. This process is the one by which the more reliable average service lives are computed for the many types of depreciable properties.

The regulated utility industry has been the leader in developing the procedures because of the necessity to establish fair depreciation rates and a fair rate base for setting the schedule of charges to customers. Beginning about 1934 the survivor curve method was applied to highway pavements. Since then some 35 states have made such studies of their state highway systems and the Bureau of Public Roads has consolidated the state data into national analyses [9-5, 9-6, and 9-8].

The survivor curve (see Fig. 9-2) is a curve which shows the number of units of property (miles, number of vehicles, etc., or original cost in dollars or percentage of units) that survive in service at given ages. The area under this curve is a direct measure of the average service life of the property units. The probable life of the surviving units at any age can also be calculated from the remaining area by dividing the remaining area by the amount surviving at that age.

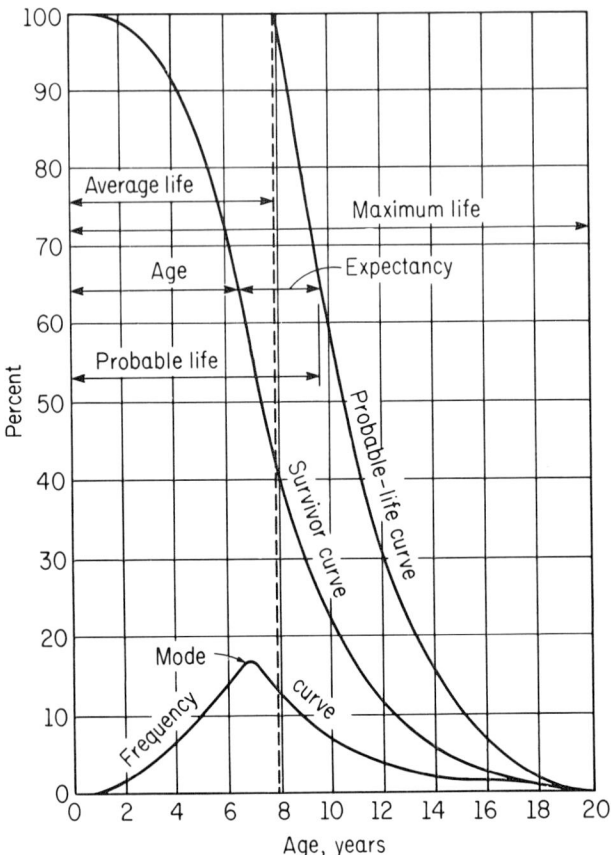

FIG. 9-2. A typical survivor curve and its derived curves.

For general usage of the survivor curves the property surviving is generally expressed in percentage of the base amount at zero age.

Individual-Unit Method

Frequently, available data show only the number of units retired during a given year or series of years together with the age of each unit at its retirement. If these several retirements are arranged in order of their ages and then summed from the oldest to the youngest, a survivor curve (Fig. 9-3) can be plotted from the successive sums. The resulting curve as usually plotted shows the percentage of the units continued in service to any given age. It should be noted that neither the method nor the original data take into account other units remaining in service during or at the end of the observation period; every unit considered has been retired. Thus the average life indicated by an individual-unit survivor curve

Service Life of Physical Property   **209**

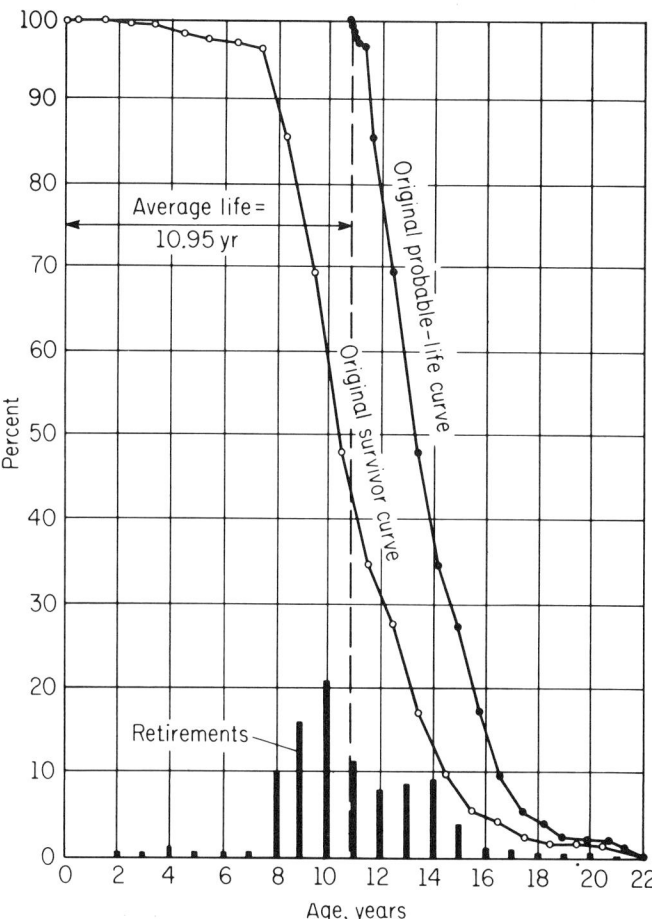

FIG. 9-3. Survivor curve calculated by the individual-unit method.

is exactly the average age at retirement of the units given consideration. Consequently, no curve need be constructed to secure this average age or average life since it can be calculated by dividing the total service in unit-years by the total number of units retired.

Since all units considered are retired from service, an individual-unit survivor curve will always range from 100 percent surviving at zero age to zero percent surviving at maximum life.

For a property long continued in service and maintained at the same number of service units by regular renewals, the replacements having practically the same potential average life as the retirements, the survivor curve, the frequency curve, and the average life when determined by the individual-unit method will approximate those determined by other methods applied to the same property.

210   **Service Life of Physical Property**

This situation is not commonly encountered because properties are frequently expanded with units that have neither the same potential service life nor service and maintenance conditions as had the retirements.

Original-Group Method

When the number of units placed in service at a given date (or during a given year) is known, together with the number of these units remaining in service at successive later observation dates, a survivor curve can be constructed covering the experience of this original group (vintage group) over the years for which the data are compiled (see Fig. 9-4). The curve will extend from 100 to

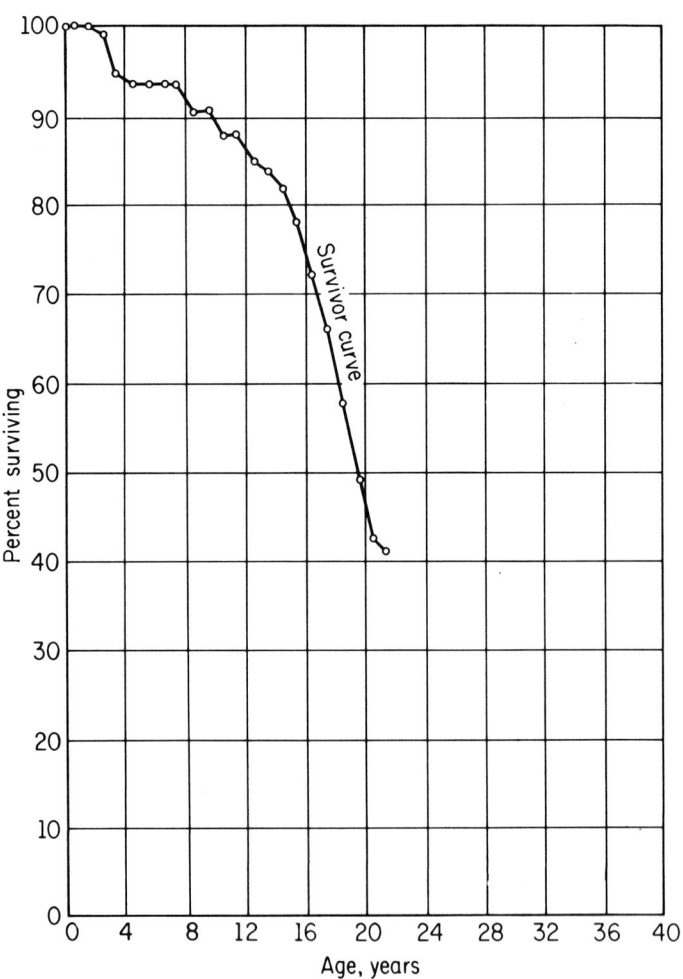

FIG. 9-4. Survivor curve calculated by the original-group method.

Service Life of Physical Property

0 percent surviving only in case all units in the original group have been retired. Should the number of units remaining in service for the first few years following installation of the group not be available, the curve will have a gap between zero age (100 percent surviving) and the age for which the first observation is known. The indicated average life applies solely to this particular original group and disregards all other groups of units in service or retired.

To determine the probable average life for an original group of units requires that the retirements be known over a period of service sufficient to extend the curve downward far enough to produce reliable results. If the curve reaches zero percent surviving, the area under the curve will give the final average life; if the curve does not reach zero, it will need to be extended by judgment (in some cases in both directions) in order to determine the probable average life. If all the units are retired, the results will be the same as secured by the individual-unit method applied to the same retirements.

The original-group method uses the data obtained by observing a single original-group over a period of years and takes into consideration the units yet in service as well as those retired from service. The individual-unit method can be applied to retirements only, but they may extend over a period of any length, and they may come from any number of original groups; however, the original-group method deals solely with one original group.

Composite Original-Group Method

When the number of units in an original group is so small as to be unreliable or when the retirements do not produce a satisfactory survivor curve, more than one vintage group may be combined and the survivor curve established by the composite original-group method. This method is especially desirable when the several original groups were installed during a series of consecutive years.

The calculation follows that of the original-group method. The combining of the several original groups into a composite group results in the final analysis, one large original group. The units surviving in each group must be combined for equal ages, and the percent surviving calculated on the basis of the combined total of the original number in each group included in the survivors at any particular age.

Multiple Original-Group Method

If the original-group method be expanded to include several original groups placed in service over a long succession of years and the period of observation condensed to one definite date at which time the number of units in each original group remaining in service is known, a survivor curve can be plotted, each original group furnishing one point on the curve. This method is known as the multiple original-group method. Instead of a single original group and a series of observation dates as in the original-group method, a series of original groups and a single observation date are utilized.

The resulting survivor curve will range from 100 percent surviving to zero percent surviving only in case all of the youngest group are yet in service and all of the oldest group have been retired. It frequently happens that no installations of units were made by the company for certain years, and thus, for the ages corresponding to these years, no point on the curve will be possible. If not all of the youngest group remain in service, the curve will need to be extended from the age of this group back to 100 percent surviving at zero age. This will usually be the case if no installations were made during the several years just preceding the observation date. Generally, the points will not follow a smooth path because the percentage of one group surviving is frequently larger than that of the next younger group (Fig. 9-5).

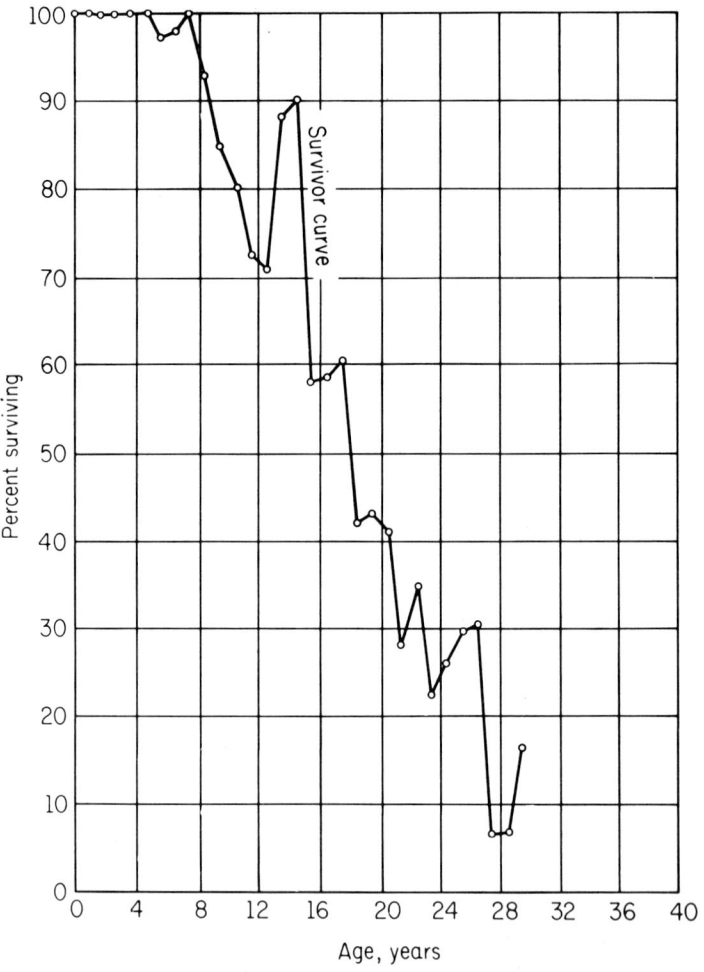

FIG. 9-5. Survivor curve calculated by the multiple original-group method.

Annual-Rate or Retirement-Rate Method

The annual-rate or retirement-rate method takes its name from the fact that the rate of retirement is calculated for each like-age group of units in service during the observation period which includes one or more years. Calculation of values for a survivor curve by this method requires two sets of data: first, the number of units retired during the period of observation and their ages at retirement (the same information as required in the individual-unit method); second, the number of units in service at the beginning of the observation period and their ages. From these two tabulations the rate of retirement is calculated for each age corresponding to the ages of the units in service. Since the rate is usually calculated for a year's retirement it is generally the annual

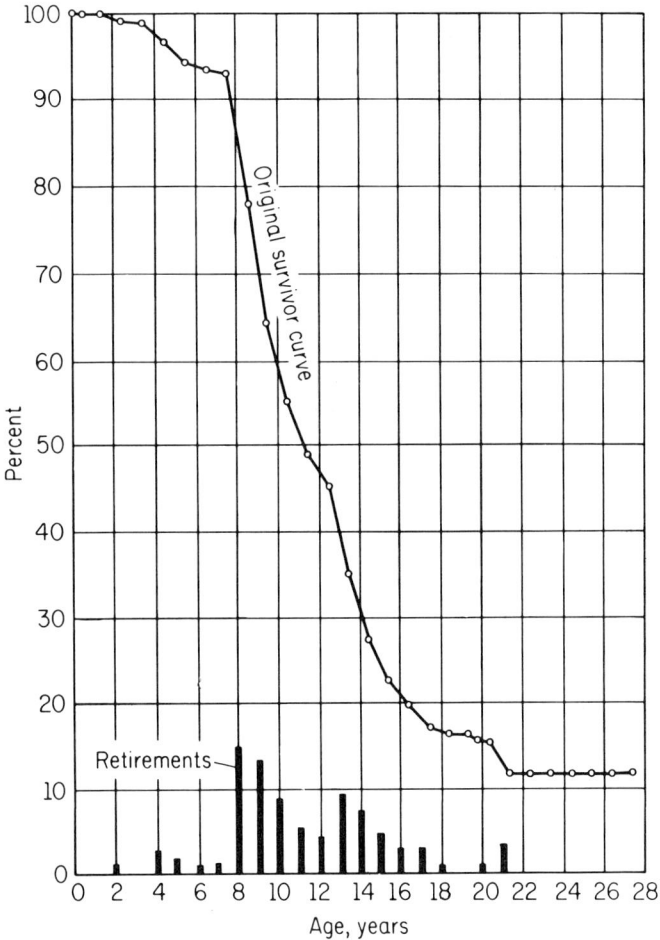

FIG. 9-6. Survivor curve calculated by the retirement-rate, or annual-rate, method.

rate, or, in other words, the percentage of the units of a given age in service at the beginning of a certain year which were retired during the following year.

If, at the beginning of the observation period, the data include the units installed the prior year and each successive preceding year, annual rates for each age-interval from 0-1 to the age of the oldest unit in service will result. If these annual rates are then applied successively to the percentage surviving at the beginning of each age-interval, starting with 100 percent at zero age, an annual-rate survivor curve will result (Fig. 9-6). These annual rates normally increase as the age increases, and, if all of the oldest units in service are removed, the last annual rate will be 100 percent. The survivor curve in this case will range from 100 percent surviving at zero age to zero percent surviving at maximum age. However, since in most cases the data are quite irregular, the annual rates will show only a general tendency to increase and will fluctuate up and down. Frequently, the survivor curve will be a stub curve not reaching zero because not all of the oldest units observed were removed during the observation period.

The probable average life (obtained by the usual method of determining the area under the completed survivor curve) for survivor curves constructed by the annual-rate method is a reflection of the average rate of retirement for the observation period chosen. It takes into consideration not only the current retirements ("deaths"), but also the units remaining in service (the "living"), and utilizes them in accordance with both their number and age.

The annual-rate method is not applicable to property groups for which placements were not made for continuous years beginning with zero age. The first annual rate must be applied to 100 percent, and if this rate is for an age-interval greater than 0-1 there is no way, other than by assumption, of determining the starting point of the curve. This objection is not a serious one for a case of this kind would exist only when the type of unit had been discontinued some years previous to the observation period. In such cases this gap at the beginning of the curve can be assumed by reference to other curves for similar property.

In many applications of the annual-rate method the retirements for any single year may not reflect normal conditions of the property, but a series, or band, of years may be chosen in order to get average conditions. The units retired for each respective age and the units exposed to retirement for each age over the band of years are totaled before the annual rate is calculated. For small property groups and for years of uneven retirements and replacements a band of years is preferable.

Simulated Plant-Balance Method

When the property records disclose the annual additions, or installations, of a class of property and the total property existing at a specific date in each year over a period of years, the average service life and the retirement dispersion may

Service Life of Physical Property

be approximated by the simulated plant-balance method [9-2]. The method consists of calculating the probable surviving property for the period of years by applying survivor curves to the yearly additions. A specific survivor curve and average life are chosen for each vintage group (original group). This choice may be made from the Iowa type curves described in the next section. Successive trials are made until the calculated total survivors for each year approximate the book records of yearly plant balances. When these totals agree within the acceptable tolerance, the survivor curves and their respective service lives used in this trial are accepted as being a good representation of the retirement dispersion and average service life of that property account.

TYPE SURVIVOR CURVES

Survivor curves of physical properties have been shown to have certain mathematical characteristics in common which permit their classification into types, despite wide variations in average service lives and retirement distributions. These type curves have practical application in depreciation, economy, and financial studies.

The Iowa 18 Type Survivor Curves

Winfrey, in 1935 [9-13], developed 18 type survivor curves (see Figs. 9-7, 9-8, and 9-9). These 18 type survivor curves through widespread usage and reference became known as the Iowa Type Curves. Since 1935 hundreds of survivor curves from a wide variety of properties have been calculated from retirement data by Winfrey and many other analysts. These subsequent studies confirmed the original 18 Iowa curves and brought to light a few additional types [9-9, 9-11, and 9-12].

The basic scheme in classifying survivor curves is to plot them with the ordinate in percent of the total number of units (at zero age 100 percent would survive), and to express the age–the abscissa–in percent of the average service life of the survivor curve. The next step is to read the survivor curve at regular age intervals, such as 0, 10, 20, 30, ... percent of average life in order that the frequency curve may be plotted. The frequency, or retirement distribution, curve will be observed to exhibit two pronounced characteristics; one, the magnitude of the modal frequency, and, two, the location (age) of the mode with respect to the 100 percent age–the average service life. These two principal characteristics are illustrated in the three families of curves in Figs. 9-7, 9-8, and 9-9, where it will be noted that the symbol designation of the 18 type curves is based upon the modal location to the left, coordinate with (symmetrical), or to the right of the age (100 percent) equal to the average life and the height of the mode, designated as a subscript number.

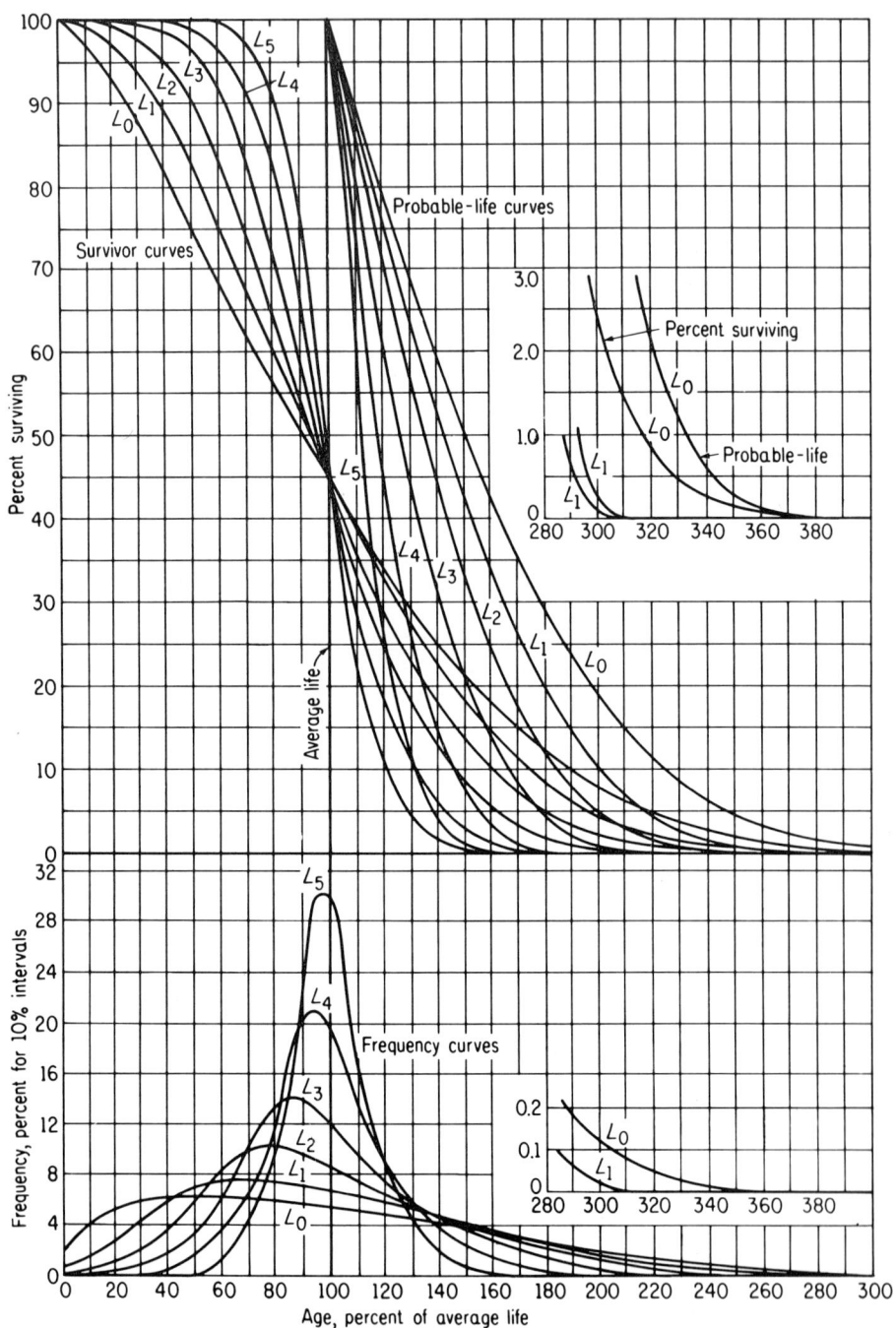

FIG. 9-7. Survivor, probable life, and frequency curves for the left-modal type Iowa curves [9-13].

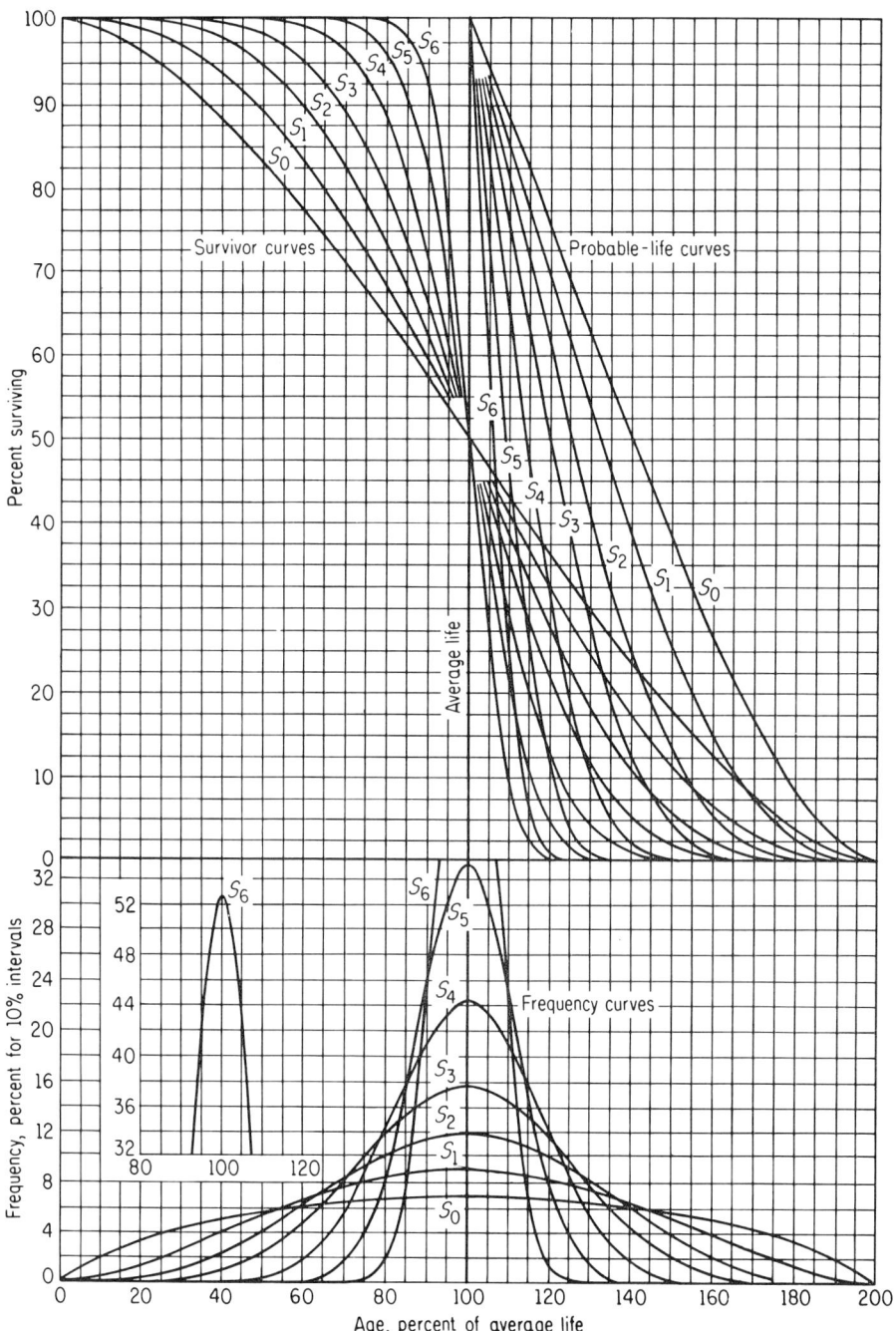

FIG. 9-8. Survivor, probable life, and frequency curves for the symmetrical type Iowa curves [9-13].

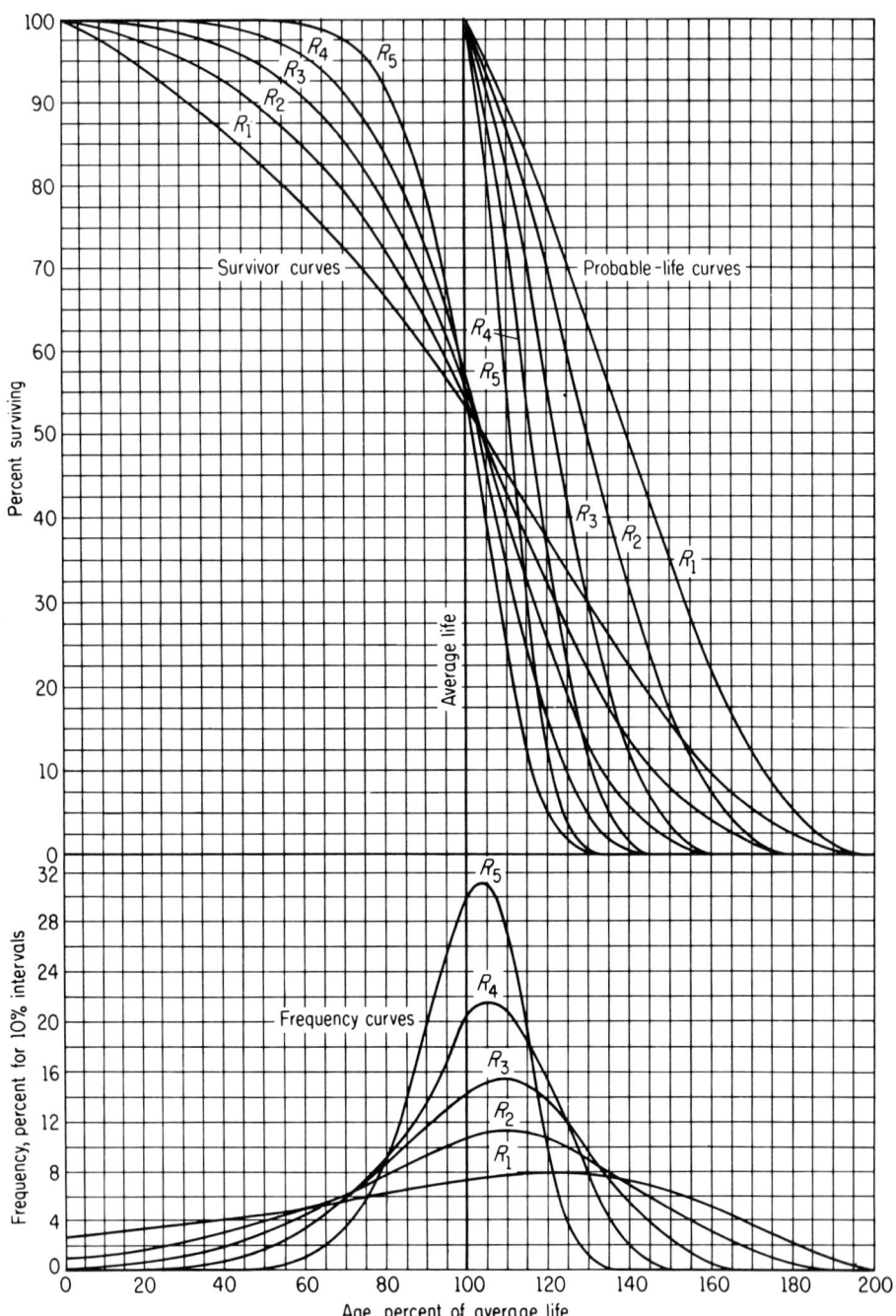

FIG. 9-9. Survivor, probable life, and frequency curves for the right-modal type Iowa curves [9-13].

Service Life of Physical Property

USES OF TYPE CURVES

Over the years many uses of the type curves have been developed in industry, public utility regulations, public works planning, statistical analysis problems, and in the military. One military use was to use the curves in the calculation of attrition of supplies, and, therefore, for estimating the necessary replacement rates.

In the simulated plant-balance method of calculating average service life, the type curves are ideal for use in the successive trials involved. The type curves simplify the process because the surviving percentages of all ages are available from prior solutions of the equations of the curves.

The original-group and retirement-rate methods of constructing survivor curves frequently result in stub curves–survivor curves that do not drop to zero percent surviving. Since the average service life is a direct function of the area under a complete survivor curve it is necessary to extend the curve to zero percent surviving to obtain the average service life. This extension may be accomplished by (1) personal judgment, (2) mathematical curve fitting, or, (3) by matching the stub curve with the type curves until the best fit is found [9-13].

Unless the stub curve is relatively long, say down to 50 to 30 percent surviving and smooth enough to fix its path without question, extension by personal judgment is associated with high probable error. Mathematical fitting and extension is sound in theory, but not so reliable in practice [9-3, 9-4, 9-9, 9-11, 9-13]. The difficulties lie in selecting the mathematical expression to fit the observed survivor curve, the statistical procedure used to find the parameters, and to judge when the best fit is found.

In the matching process, an assortment of types of mathematical expressions (18 in the Iowa type curves) are available. Although personal judgment is exercised in determining which type curve best fits the original data stub curve, the method is without major hazards and uncertainties. The matching method has the advantage of visual evidence of the whole curve so that the extended portion can readily be compared with the plotted stub portion.

This matching process is simple, fast, and as reliable as any other method yet developed [9-3]. The matching process is accomplished by preparing ahead of time plots of each type curve for a range of specific average lives, such as 5, 10, 15, 20,... years, with the age expressed in years. A common scale is used for the type curves and the stub curve.

The adequacy of the depreciation reserve account balance may be checked as to approximate agreement with the age distribution and service life of the property from prior calculations of the reserve ratio expected from the depreciation method used and the age of the account [8-8].

These type survivor curves are also used as a tool in forecasting retirements and for estimating future highway needs. When the records of the existing highway system will disclose the age of the many sections of pavements in service by

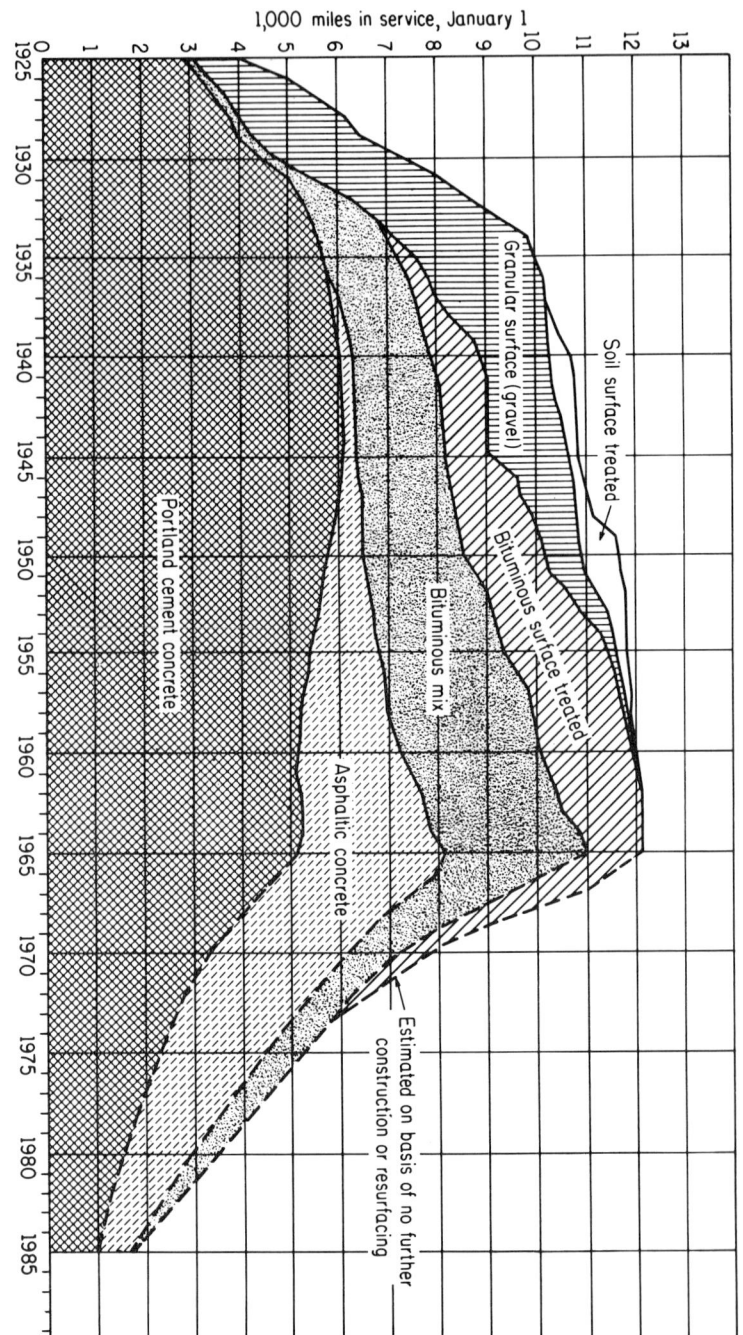

Fig. 9-10. Miles of highway by roadway surface type in service January 1, 1925 to 1965, and estimate of 1985 without further construction or reconstruction.

Service Life of Physical Property

their year of construction, the appropriate survivor curve may be applied to each of these surviving vintage groups to calculate the probable surviving amounts at future years. Based upon maintaining the same mileage in service of the type of pavement being studied, the replacement needs equal the mileage to be retired. But when traffic growth is considered, an increase in mileage of the higher types will be needed. In time, the mileage of the lower types of roadway surfaces will decrease as traffic needs lead to higher types of pavements. Figure 9-10 illustrates these changes in total miles on a state system 1925 to 1965, and forecasted to 1985.

Illustrative Survivor Curves

By way of illustrating the survivor curves that may result from analyses of retirements of highway property, Figs. 9-11 to 9-16 are presented. The average service lives indicated in these six figures have no practical use, at least they should not have, because they apply only to the particular properties they represent and for the specific calendar time or observation period indicated.

The survivor curves in Figs. 9-11 to 9-16 were calculated by the annual, or retirement, rate method. The smooth curves indicated are specific type curves as discussed. Note that the plotted original data at the lower end often tends to depart from the path of the type curve. This departure is characteristic of property and of the way management policy influences retirements. Often, the older vintages of a class of property have only a small percentage of their units

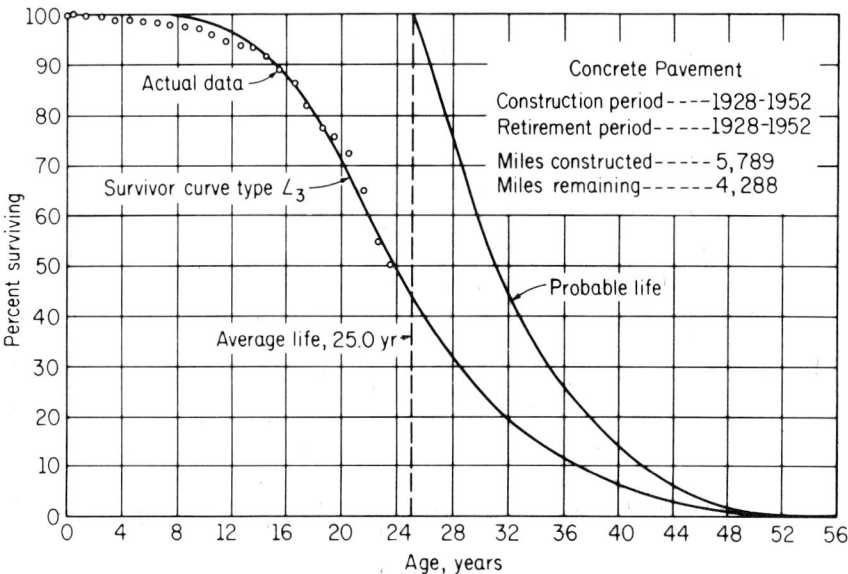

FIG. 9-11. An illustrative survivor curve and its probable life curve for concrete pavement. The type curve matching the original data is L-3 at 25 years average life.

surviving and these units are so located that they become neglected, or their use decreases to a point that first-quality service is no longer demanded.

It is to be remembered that the service lives of properties as calculated by any method using retirement data are historical lives. Whether or not existing property, or proposed property of a similar nature or of exactly identical character, will produce the same average service is not certain. Technology changes, demand for service changes, environment changes, and management policies change. Nevertheless, the only guide to the service life of new physical property is the past experience with similar property. So it is in the life and risk insurance business. The human mortality tables represent human life experience. But the standard tables for insurance purposes are updated periodically.

In the earlier period of motor vehicle highways, pavements indicated a steady increase in probable service lives, but later heavy traffic increases and new demands for bettering the geometrics of design slowed down the increasing service life materially.

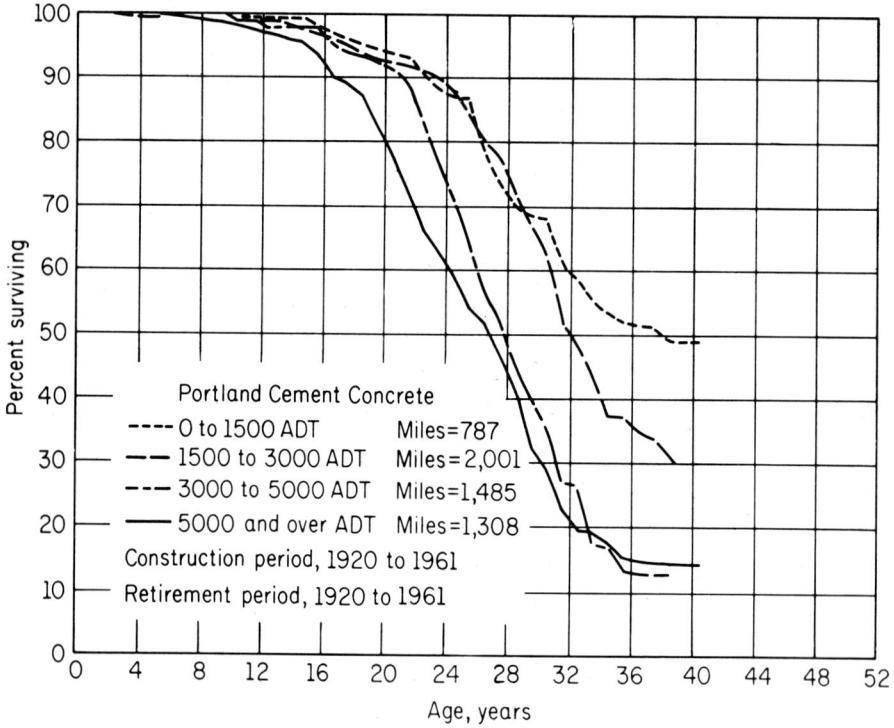

FIG. 9-12. Survivor curves for portland cement concrete pavement for four different daily traffic volumes.

Service Life of Physical Property

SOME GENERAL SERVICE LIVES OF HIGHWAY COMPONENTS

For purposes of preparing rough estimates of the annual economic cost of owning and operating a highway system, some general average service lives may prove useful. In presenting these service lives, a caution is also given. As has been stated in preceding pages, the service life of a particular property varies widely because of the wide variation in its quality, condition of service, and management policy. Therefore, any offering of specific service lives as follows in Table 9-1, is no substitute for determining particular service lives for the application at hand.

For rights-of-way (land) the service life would be measured by the period of time the land is used for highway purposes, since land is not retired in the sense that man-made properties are. As highways are abandoned, relocated, and realigned, some rights-of-way do revert to other uses. But such action has affected only a minute percentage of the nation's past and present (3.7 million miles) highways. In Colorado there are places where about five stages of historical rights-of-way may be seen. They are: (1) the pony express trail, (2)

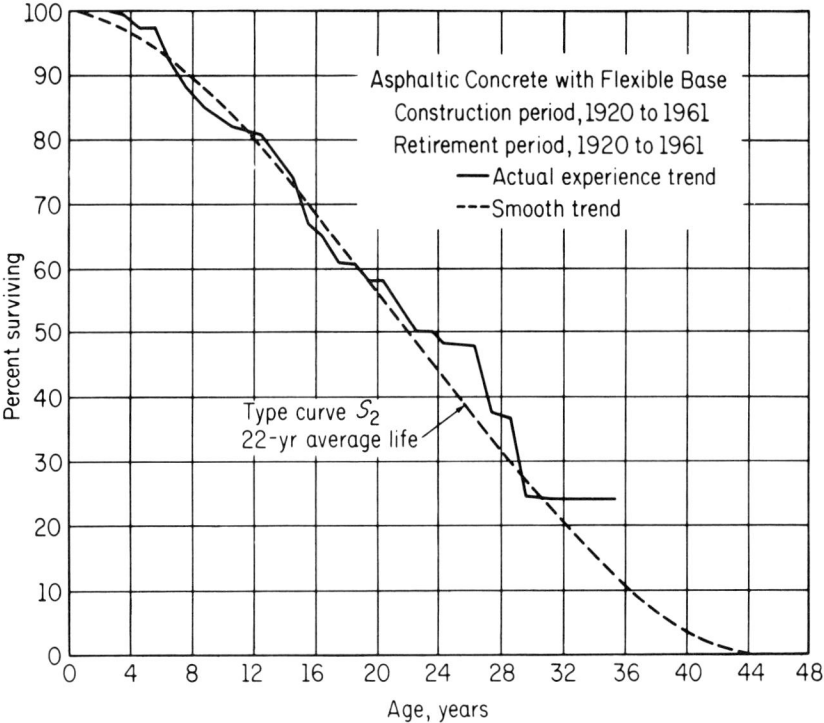

FIG. 9-13. Survivor curve for asphaltic concrete pavement on flexible base.

TABLE 9-1

GENERAL SERVICE LIVES FOR HIGHWAY COMPONENTS FOR USE
IN ESTIMATING ANNUAL HIGHWAY COSTS

Component	Years
Right-of-way land	75 to 100
Right-of-way damages (suggested write-off period)	10 to 30
Right-of-way buildings to be moved or destroyed (suggested write-off period)	10 to 30
Earthwork	60 to 100
Culverts and small drainage facilities	25 to 50
Retaining walls and general concrete work	40 to 75
Riprap and other bank protection	20 to 50
Bridge and other major structures	50 to 75
Granular roadway surfaces	3 to 10
Low-type bituminous surfaces	5 to 12
Intermediate-type bituminous surfaces	12 to 20
Rigid and flexible high-type pavements	18 to 30
Signs and traffic control devices	5 to 20

the pioneer wagon trail, (3) the first motor vehicle highway, (4) the second, or reconstructed motor vehicle highway, and (5) the current four-lane divided high-

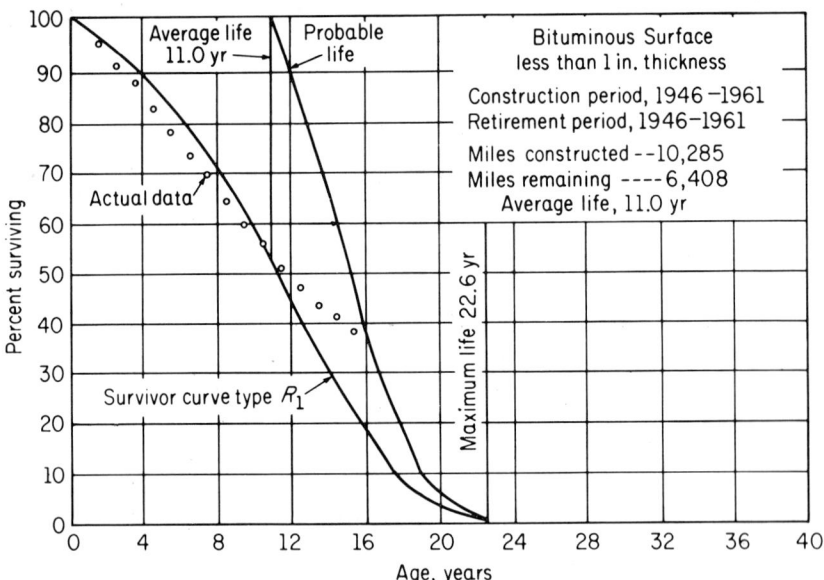

FIG. 9-14. A typical survivor curve for a bituminous surface less than one inch in thickness. No doubt the vintages represented in this population were seal-coated regularly under maintenance.

Service Life of Physical Property

speed highway. The traveled way of these early trails and roads can still be seen, though long since abandoned as public highways.

There is one aspect of rights-of-way deserving special attention. In relocating old routes and in constructing new routes, it is commonly experienced that the cost of rights-of-way includes sizable sums for buildings or other developments and for damages. Although these items of cost are normal costs, they are not costs of the land. The buildings are removed from the highway. It is better procedure to assign the acquisition costs of buildings and damages a comparatively short life, say the same as assigned to paving, so as to (at least theoretically) write this cost off early. There is no property at hand to support the investment in these items. This practice is usually followed in the public utility industry.

For economy studies and for estimating the annual economic cost of owning and operating a highway system, all analysis periods and service lives above 50 years produce about the same final result when a medium or high (6 percent and above) vestcharge rate is used. For example, the capital recovery factor at 6 percent is 0.06344 for 50 years and 0.06018 for 100 years.

The service lives in Table 9-1 are expressed in ranges to indicate that there are no fixed or exact service lives of the physical components of highways. Each analyst is expected to exercise some judgment and to select those service lives that are the most logical for the property considered and the purpose of the analysis contemplated.

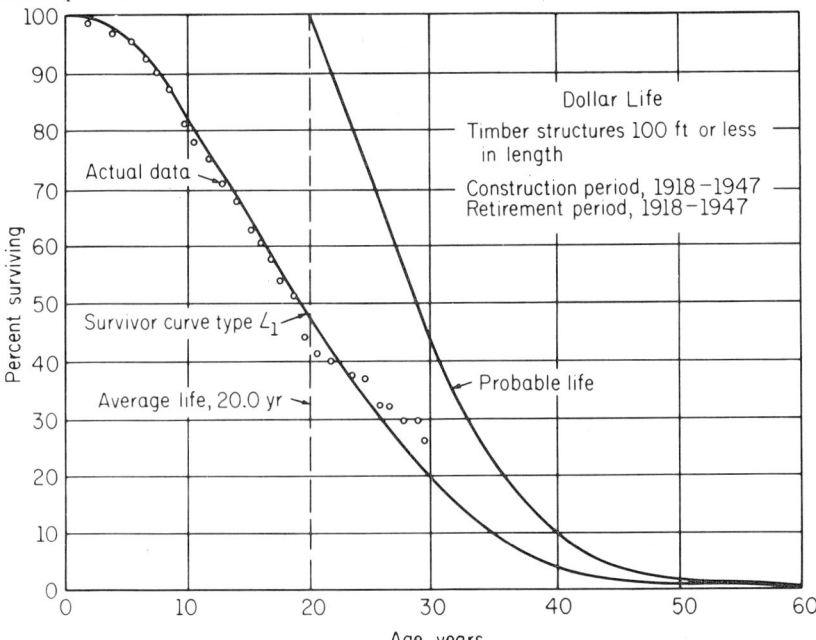

FIG. 9-15. A survivor curve for timber structures. Note that the curve was calculated from dollars rather than from the number of structures.

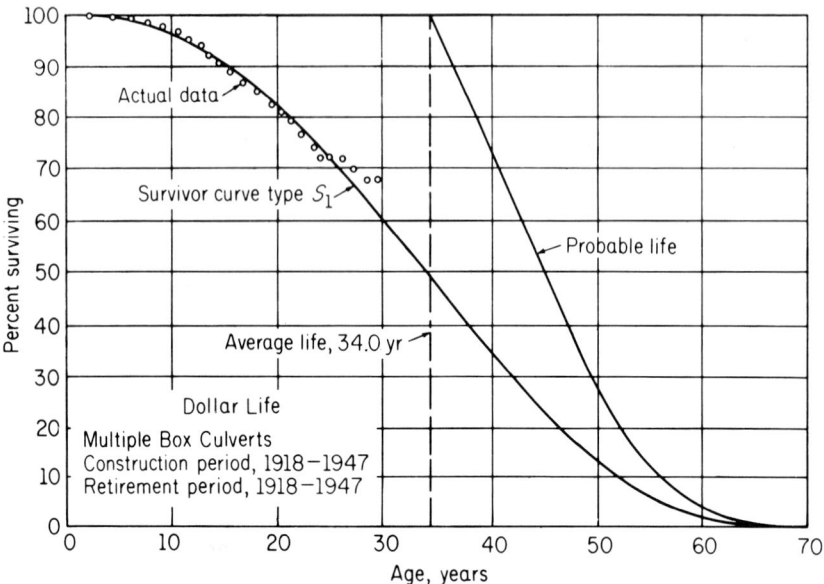

FIG. 9-16. A survivor curve for multiple box culverts calculated on the dollar unit.

PROBLEMS

9-1. List the uses, or applications, of the service lives of highway components and survivor curves that could normally be appropriate in various research, planning, designing, and management functions of a highway department.

9-2. List all the reasons, causes, or situations which result in a management decision to retire a given section of pavement. Consider that retirement results from (a) work that changes the surface type classification, (b) resurfacing to the same surface type, (c) complete reconstruction, or (d) abandonment in place as a public highway.

9-3. List all the reasons which would cause management to retire from its used and useful property (1) a motor truck, (2) a concrete-mixing plant, or (3) a highway bridge.

9-4. Consider highway pavements in a state primary system and a typical survivor curve. Discuss and compare the probable many reasons for retirement of those route sections that (1) are among the retirements between 100 and 67 percent surviving on the survivor curve, (2) 67 down to 34 percent surviving, and (3) the last 34 percent.

9-5. By judgment, extend the survivor curve in Fig. 9-12 for the 1,500 to 3,000 ADT to zero percent surviving, then calculate its service life from the area beneath the curve.

REFERENCES

9-1. American Association of State Highway Officials. *Manual of Uniform Highway Accounting Procedures.* Washington, D.C., 1958.

9-2. A. E. Bauhn. "Life Analysis of Utility Plant for Depreciation Accounting Purposes by the Simulated Plant-Record Method," Edison Electric Institute, *Methods of Estimating Utility Plant Life.* New York, 1952.

9-3. Harold A. Cowles, Jr. *Prediction of Mortality Characteristics of Industrial Property Groups.* Unpublished Ph.D. dissertation. Iowa State University, Ames, 1957.

9-4. Edison Electric Institute. *Methods of Estimating Utility Plant Life.* New York, 1952.

9-5. Fred B. Farrell. *The Capital Investment in Highways.* Highway Research Board, National Academy of Sciences, Washington, D.C., *Proc.,* Vol. 32, 1953, pp. 1-11.

9-6. Fred B. Farrell. *The Investment Analysis Approach to Estimating Highway Needs.* Highway Research Board, National Academy of Sciences, Washington, D.C., *Proc.,* Vol. 35, 1956, pp. 9-13.

9-7. Gannett Fleming Corddry and Carpenter, Inc. "Columbia Gas System Depreciation Study, Methods Used in the Estimation of Average Service Life and Accrued Depreciation," 1963.

9-8. Gordon Gronberg and Nellie B. Blosser. *Lives of Highway Surfaces–Half Century Trends.* Highway Research Board, National Academy of Sciences, Washington, D.C., *Proc.,* Vol. 35, 1956, pp. 89-101.

9-9. J. Jeming. "An Asymptotic Method of Determining Annual and Accrued Depreciation," in M. R. Scharff, F. S. Leersburger, and J. Jeming, *Depreciation of Public Utility Property.* Burstein and Chappe, New York, 1941, pp. 97-128.

9-10. P. H. Jeynes. *Determination and Forecast of Average Service Life.* Edison Electric Institute, New York, Bulletin 20, 1952.

9-11. B. T. Kimball. "A System of Life Tables for Physical Property Based on the Truncated Normal Distribution," *Econometrica,* Vol. 15 (947), pp. 342-360.

9-12. National Association of Railroad and Utility Commissioners. *Report of Special Committee on Depreciation.* State Law Reporting Co., New York, 1938.

9-13. Robley Winfrey. *Statistical Analyses of Industrial Property Retirements.* Bulletin 125, Iowa Engineering Experiment Station, Iowa State University, Ames, 1935, republished 1967.

9-14. Robley Winfrey and Phebe D. Howell. *Highway Pavements-Their Service Lines.* Highway Research Board, National Academy of Sciences, Washington, D.C., Highway Research Record No. 252, 1968, pp. 1-23.

Chapter **10**

Highway Transportation Costs

The word *cost* is a word of many meanings. It is one that demands precise definition and consistent usage, yet is used loosely and inconsistently. There is confusion between the words *cost, expense, expenditure,* and *disbursement.* There are many kinds of costs—annual cost, initial cost, prime cost, capital cost, economic cost, cash cost, overhead cost, operating cost, production cost, and other costs. For highway economy studies it is essential that the word cost be understood and used consistently with that understanding. Generally speaking, there are two broad categories of highway cost of concern to those persons interested in highway finance and economics. Cash, or disbursement cost, is the item of concern to the budget and legislative people. The annual economic cost and equivalent uniform annual cost are of greatest interest to the economist and analyst for economy. Economic cost over a specific period of time is calculated by applying the principles of compound interest as embodied in well-accepted formulas of economic cost.

ANNUAL HIGHWAY AND TRANSPORTATION COSTS

The total cost of highway transportation is made up of the highway cost and the cost of operating motor vehicles over the highways. The highway cost is that cost borne by the people through their highway department, and the motor vehicle cost is that cost which is borne directly by the owners of the motor vehicles. In order to get the total cost of highway transportation it is therefore necessary to examine two sources of basic costs. This chapter is mainly concerned with the highway costs, but the motor vehicle, or road-user costs, are included in the cost formulas when applicable.

The highway costs can be expressed in two general classifications such as cash costs and economic costs and each of these two classes can be expressed in two subclasses. Thus the breakdown may be outlined as follows:

    A. Annual highway cash costs
        1. Annual cash disbursements
        2. Annual cash or tax cost to the public

B. Annual highway economic cost
3. Annual economic cost of owning and operating highways
4. Equivalent uniform annual highway cost for economy studies.

These four concepts of highway cost are explained in the next sections.

ANNUAL HIGHWAY CASH COSTS

Highway managements, being governmental agencies, follow the custom in governmental accounting of keeping cash records only. They do not render statements of profit and loss, of operating income and expense, or statements of assets and liabilities, nor do they keep a complete record of invested capital. Depreciation cost accounting is almost totally foreign to governmental operations.

Annual Cash Disbursements

To the highway administrator, legislator, and taxpaying citizen, highway cost is nearly always taken to mean the total cash disbursement by the agency for all purposes for the time period in question. In this classification there is often no attempt to separate expenditures for capital improvement, for maintenance of highways, for debt service, or for administration. The cost is the total cash disbursement for the governmental agency being discussed. Such highway cost may be termed annual cash cost because the reported cost includes only the cash disbursement for the period. Further, no adjustment is made for increases or decreases in inventories, work in process, accounts receivable, and accounts payable. In the reported cash cost of highways there is no attempt to determine operation cost, or production cost, in the sense that applies to cost accounting in preparing a profit and loss statement.

Cash cost of a highway system has its administrative use in budgeting, setting tax rates, requesting appropriations, and comparing disbursements over the years or as between governmental agencies. Cash costs may vary considerably from year to year because of expenditures for capital improvements. Cash costs are not, therefore, a desirable basis for comparing the relative levels of fiscal efficiencies at which different units may operate.

Cash costs are used in studies of the income needed to finance a given operation for the year or other time period. For instance, a toll facility would be administered primarily on a cash basis. The necessary income would need to be sufficient to pay the administrative expense, the maintenance and operating expenses, necessary additions and betterments, and the debt service costs of bond interest and the retirement of the bonds coming due.

Annual Cash or Tax Cost to the Public

Because a highway administrative unit may not spend an amount exactly equal to its tax income during a given year, a second form of annual highway cash cost may be mentioned. This second cash cost is the amount collected in taxes from the public for the year. A highway department may spend more or less than its tax and other income for a given year. The plus difference would be made up of balances from previous years, or receipts from bond issues or other borrowings. Further, in considering a given highway administration, such as a state highway department, the tax contribution from highway users and from property owners to the highway funds may be supplemented by federal aid funds or grants. Under these conditions, the annual highway disbursements might be greatly in excess of the tax and other revenue paid by the highway users and citizens within the political area concerned.

The annual tax cost to the people is an important item for legislators, administrative boards, and others concerned with setting rates for taxes and fees. In setting a budget for operation of a highway department, both the forecasted cash cost and the tax imposts are important and may be widely different. However, in the long run and over the country as a whole, the annual highway cash cost and the annual tax income have a tendency to average out to the same total.

ANNUAL HIGHWAY ECONOMIC COST

One can logically conclude that a specific highway facility should have but one annual economic cost with due recognition given to selection of the factors used in the calculation. Yet there are two different annual economic costs, each serving a different objective: (1) the annual economic cost of owning and operating the facility, and (2) the annual economic cost as calculated for the economic analysis of alternative proposals. The differences in these two annual economic costs lie primarily in the magnitude of the factors within their respective calculations.

Explanation of Annual Economic Cost

The estimated annual economic cost of highways, or of any other establishment, takes into consideration past disbursements which are allocable to the period of time in question regardless of the accounting period that the corresponding cash disbursements were made. Such economic costs ordinarily include charges for depreciation of the physical properties involved, adjustments for inventories, and all of the cash expense for materials and services consumed in rendering the services or producing the products for that accounting period. In comparison with current disbursements of cash, economic costs may be said to be the total of short-run and long-run costs assignable to a given time period regardless of the date of disbursement.

Economic costs are used in all instances when a true comparison of the operating costs of machines and facilities and for periods of time are wanted. Profit and loss statements of businesses and enterprises are computed on the economic cost basis. Analyses of the economy of alternative proposals must be

Highway Transportation Costs

estimated on the economic basis if reliable conclusions are to be had from the analyses. The total cash expenditure necessary to construct a tunnel as compared to an open cut in no sense would disclose the economic wisdom of the choice between the two engineering methods of gaining a trafficway to the other side of the ridge.

Operating costs, maintenance costs, capital costs, and a charge for use of the capital are required. The effect on the running cost of motor vehicles is another factor to consider.

General Comparison of the Two Forms of Annual Economic Cost

The annual economic cost of owning and operating a facility may be wanted as a basis of pricing, to compare the cost of two or more types of machines or construction materials, or to determine which of two or more modes of transportation is least costly in the long run. In government, annual economic cost of owning and operating a facility may be used as a basis of setting policy or as a basis of legislation. This annual economic cost of owning and operating includes all factors of cost, each factor estimated to include its whole costs to the best of one's ability in estimating those factors that must be estimated. Often this economic cost is calculated for an existing machine or property as contrasted to estimating the annual economic cost of a proposed works.

The equivalent uniform annual cost as computed in economic analyses is used for the special purpose of comparing alternatives of accomplishing a given purpose or function. The purpose of this calculation is to aid management in reaching a decision as between the available alternatives. As long as this calculated equivalent uniform annual cost accomplishes this objective in a reliable manner, such cost need not include every cost factor or be calculated on exactly the same magnitude of factors as may be used in calculating the annual cost of owning and operating the facility. As long as the proper basis for decision making can be supplied, an economic cost obtained by a procedure less rigid and extensive than required for a full and detailed calculation of economic costs has merit in economic analysis.

The accompanying tabulation will indicate those factors and their handling that lead to the differences in the annual economic cost of owning and operating a facility and the equivalent uniform annual cost calculated for an economy study.

DISCUSSION OF THE DIFFERENCES

In calculating the economic cost of owning and operating a facility, the full probable service life should be used for each of the physical components. Also, the most probable terminal value should be used. For the economic analysis, on the other hand, a comparatively short analysis period is more appropriate and the terminal value is often taken as zero (see page 237).

COMPARISON OF THE FACTORS INVOLVED IN CALCULATING ANNUAL ECONOMIC COST
FOR THE TWO DIFFERENT OBJECTIVES

Cost Factor	Annual Economic Cost of Owning and Operating	Equivalent Uniform Annual Cost for Economy Studies
1. Analysis period	Service life, each component separately	Analysis period not to exceed period of reliable forecasts; often all components considered together
2. Terminal value	Best estimate at end of service life	Often zero
3. Vestcharge rate	When applied to an existing facility the risks and uncertainties are less	Often higher rate because of risks and uncertainties of proposed work
4. Sunk costs	Historical investments are used, or their current construction costs may be used	Not considered, because only future cash flows are pertinent
5. Common factors of like magnitude	All cost factors included because total cost is the objective	May be omitted for the alternatives involved
6. Motor vehicle operating costs	Inclusive of all running and ownership costs	Cost factors not affected by highway design may be omitted
7. Objective	Estimate of total annual cost inclusive of every factor	An estimate of comparative annual cost between alternatives to be used as an aid to decision

There is no requirement that the time period in the analysis for economy must agree with the probable service life that may be used in cost accounting. In fact, private industry regularly uses a short analysis period such as 1 to 5 years for their economy studies. In turn, when the property is built or purchased, the company will use a service life for depreciation cost accounting of 2 to 5 times the length of the analysis period used in the economic analysis leading to the choice of the alternative. However, in economic analysis for project formulation with respect to choice of materials, such as portland cement concrete versus bituminous concrete, great care is to be exercised to see that the cost of each alternative is calculated on a time basis consistent with its probable service period and terminal value.

Although the vestcharge rate is most commonly the same in the two estimates of annual economic cost, they need not be. Once a property is in operation many

Highway Transportation Costs

of the risks and uncertainties are reduced. Therefore the vestcharge rate need not be kept high to compensate for high risk and uncertainties. Further, for an operating property, some of the cash flows are a matter of book record, so are known factors rather than estimates. Operating experience is a basis for projection into the future on an existing property as compared to no direct experience on a proposed property.

One of the principles of analysis for economy is that factors of equal magnitude for all alternatives may be omitted from the analysis for economy. In the calculation of the economic cost of owning and operating a specific property all factors of cost must be included. For instance, in economic analysis the highway administration cost is often omitted because it may be the same for all alternatives. Certain cost elements of motor vehicle operation are unaffected by highway design factors, and, therefore, such cost factors are generally omitted from road-user cost in the economic analysis.

ANNUAL ECONOMIC COST COMPONENTS AND EQUATIONS

The components within the annual economic costs have their origin in the daily or yearly cash disbursements for both investment purposes (highway construction) and operations and maintenance (yearly expenses). In the end these disbursements or anticipated cash flows, with such estimated factors as may be required, are used in traditional formulas for calculating the annual economic costs for either the cost of owning and operating a facility or for the economic analysis.

ANNUAL ECONOMIC COST COMPONENTS

The annual economic cost of highways is composed of the following groups of component costs:

1. Administrative expense of the governmental agency responsible
2. Highway operating expense of traffic services and policing
3. Highway maintenance expense (costs of maintaining the physical highway structure and its appurtenances)
4. Capital costs pertaining to the physical structure and its appurtenances
   (a). Depreciation expense
   (b). Vestcharge on capital invested

Each of the items 1, 2, and 3 is composed of cost items similar to the following:

A. Wages and salaries and payroll overburdens for insurance, retirement, hospitalization, etc.

B. Materials and supplies and services purchased during the year and normally consumed during an accounting period of one year.
C. Costs of operating buildings, equipment, machinery, and other property, not a part of the physical highway
D. Capital costs of depreciation and vestcharge on buildings, equipment, and machinery

The capital costs, item 4, are those costs associated with the capital investment in long-lived property and correspond to the cost of capital recovery with vestcharge as discussed in Chapter 5.

ANNUAL ECONOMIC COST EQUATIONS

The following equations may be used in computing the annual economic transportation cost which is the highway cost plus the road-user cost:

$$C_T = -H - U \tag{10-1}$$

where C_T = total annual economic cost of highway transportation
H = total annual economic cost of the highway facility
U = total annual economic cost of the road user

The Subfactors Involved

The factors H and U will be expanded to include the appropriate subfactors. Annual economic highway cost is composed of two basic cost factors–capital cost and annual expense. Capital cost arises from the investment in long-lived property and annual expense arises from daily operating functions which do not add to the investment in property. Thus

$$H = -I_a - A - J - M \tag{10-2}$$

where I_a = annual uniform capital cost of depreciation (return of capital) and vestcharge (return on capital)
A = annual administrative expense allocable to the highway or facility
J = annual operating expense for traffic services, highway operations, and police services allocable to the highway or facility
M = annual maintenance expense of the physical components of the highway or facility

The factors A, J, and M are determined from the cost accounting records for existing highways or facilities. For proposed facilities these factors are estimated on the basis of experience and current prices.

Annual Capital Cost

The factor I_a is dependent upon three subfactors, each of which is a judgment factor. These three factors are the time period of analysis, the estimated terminal value, and the rate of vestcharge.

Highway Transportation Costs

The term I_a is calculated by either of the two following equations:

$$I_a = (-I + T)(CR\text{-}i\text{-}n) - iT \qquad (10\text{-}3)$$
$$I_a = -I(CR\text{-}i\text{-}n) + T(SF\text{-}i\text{-}n) \qquad (10\text{-}4)$$

Where I = initial investment or construction outlay or the equivalent present worth of the initial investment plus subsequent investments
T = estimated terminal value at the end of n years
i = rate per year of vestcharge
n = analysis period (years) or service life

CR = capital recovery factor, $\dfrac{i(1+i)^n}{(1+i)^n - 1}$

SF = sinking fund factor, $\dfrac{i}{(1+i)^n - 1}$

In Eq. 10-3 it is important to note that the terminal value is subtracted from the investment cost before multiplying by the capital recovery factor. Terminal value contributes only a vestcharge cost iT to the total capital cost. Equation 10-3 is of long standing and use, but it does not tie in directly with the cash-flow diagram. Equation 10-4 is preferred to Eq. 10-3 because it can be written directly from the cash-flow diagram.

Equations 10-3 and 10-4 are written on the basis that there is only one investment outlay, the initial one. There are many existing facilities which have had investment outlays at different time periods, and many proposals for new facilities will include investment outlays at different time dates. When there is more than the initial investment outlay to consider the formula is modified accordingly to give proper time discounts to each separate capital investment. For example,

$$I_a = -I_0(CR\text{-}i\text{-}n) + T(SF\text{-}i\text{-}n) - I_x(PW\text{-}i\text{-}x)(CR\text{-}i\text{-}n) \qquad (10\text{-}5)$$

where I_0 = initial investment at zero time
T = terminal value from I_0 and I_x combined
I_x = additional investment at age x
PW = single-sum present worth factor

Additional investments may be added to Eq. 10-5 in a similar manner.

Each of the factors A, J, and M in Eq. 10-2 is an annual recurring expense. When these expense items are approximately equal year after year these sums may be used in the equation directly. When the sums increase somewhat on a straight line a gradient compound factor, Eq. 6-9A, is used. For an exponential increase, the exponential Eq. 6-12 is used.

The road-user cost factor U in Eq. 10-1 is similar in character to A, J, and M as it represents annual expense, either uniform or increasing. Although A, J, M, and U are referred to as uniform or as increasing factors, they could decrease as well over time.

Basic Equation Fully Expanded

Equation 10-1 for equivalent uniform annual transportation cost will now be expanded to specifically identify each factor using only an initial investment:

$$C_T = -I(CR\text{-}i\text{-}n) + T(SF\text{-}i\text{-}n) - K - G_K(GUS\text{-}i\text{-}n) - U(EUS\text{-}i\text{-}n) \quad (10\text{-}6)$$

where K = $A + J + M$ (combined for convenience)
GUS = equivalent uniform gradient factor (see Eq. 6-9A)
EUS = equivalent uniform exponential increase factor (see Eq. 6-12)
G_K = gradient increase per year.

Just to illustrate the use of the factors GUS and EUS, one is applied to K and the other to U. The road-user costs are often first calculated from the average daily traffic volume and then expanded to include the whole year. The road-user costs, U, may include all or any selection of the three general groups of road-user costs as may be desirable for the particular analysis. The cost groups are motor vehicle running or operating costs, accident costs, and travel time costs. Motor vehicle running cost (See Chapter 14) is defined as only that portion of the total vehicle ownership and operating cost affected by highway design and traffic conditions while operating cost includes all items of cost related to ownership and use of the vehicle.

FACTORS OF ANNUAL CAPITAL COST

The annual highway capital cost is based on four separate factors I, T, n, and i. Their determination–quantitative value–is a key factor in the process of estimating annual economic cost so they are discussed each separately.

HIGHWAY CAPITAL INVESTMENT

The annual uniform capital cost of owning and operating any highway facility or other property is a direct factor of the dollars of investment in that property. The factors I_0 and I_x for use in calculating I_a, the annual uniform capital cost, are taken as the cost of constructing the facility. This cost should include all costs to acquire the property and to set it into use, including appropriate engineering costs, construction costs, and overhead costs. For existing properties this investment cost includes the cost of any additions and betterments that may have been constructed subsequent to the building of the main property. The investment cost usable here for existing facilities is historical cost, estimated historical cost if records do not disclose the actual cost, or current reproduction cost, depending upon the objective. For economic analysis of proposed alternatives, any investment cost in an existing property would be a sunk cost, so, historical cost would not be pertinent. See page 249. The investment cost is not value of the property as may be determined by estimating market value unless that

Highway Transportation Costs

is what is wanted. For proposed facilities the factor I is taken as the estimated cost to design and construct the complete facility and to open it for use.

When making a computation of the economic cost of a highway or of any other depreciable property made up of components which might not all be retired at the same time or which would have different terminal values, the desirable approach is to determine the investment cost in each component separately. For instance, the highway investment cost may be separated into items similar to the following:

- Preliminary engineering
- Right-of-way
  - Land
  - Improvements
  - Damages
- Construction engineering
- Roadway construction
  - Clearing the rights-of-way
  - Earthwork and grading
  - Drainage and related structures
  - Roadway base and surface
  - Roadside development
  - Miscellaneous structures
- Major structures
- Traffic service facilities

When the annual economic cost is to be computed on the basis of service lives it is necessary to estimate the investment cost (construction cost) by each of the above items, or at least for each component having a service life or terminal value differing materially from the others. These investment components correspond closely with those usually used by the engineer in estimating the construction cost of any proposed project.

TERMINAL VALUE

Salvage is the word commonly used to refer to the residue of property and materials on hand when depreciable properties are retired from service. The word has the connotation of "something rescued or saved." Salvage value is the term formerly most often used to express the monetary value of salvaged property. The term "terminal value" is preferred, however, because it does not infer that a property has been rescued. Another word now often used is "residual value."

Terminal value in the accounting sense is applied to the residual property, which may be the whole original property intact as for an automobile or a bridge, or to only that portion of the original property which remains as for a building after being partially destroyed by fire, or a deteriorated gravel road after years of service. In the sense of accounting, any residual property is usually given a dollar value which is known as terminal value or residual value. Terminal value is a monetary quantity representing the actual or forecasted worth of the property remaining at the current date or at the forecasted date of probable retirement from service or at the time of reconstruction.

Basic Concepts of Terminal Value

The cost concept of depreciation requires that only that net amount of the purchase price or construction cost of the depreciable property that is not recovered through other sources be charged to production expense through the depreciation expense charge. Usually the only other source of recovery of the original investment cost is through the cash sale or usable value of the property at the time of its retirement from service or reconstruction. Consequently, in calculations of the annual economic cost of highways or the analysis for economy of a proposed highway facility, it is necessary to base the capital recovery component of the capital cost on only that portion of the investment cost of the property that is not expected to be recovered through a terminal value. Terminal value is always an estimated future value in cost computations, except when making such computations on property that has been retired.

In most uses of terminal value, it is net terminal value that is wanted. Often there are costs involved in removal of the retired property, transportation of it to a useful site, sales expense, or other costs. These costs are subtracted from the value of the property in place to get the net terminal value. It follows that net terminal value can be a negative value, as would be true in the removal of a reinforced concrete bridge.

Terminal values in computations of annual economic cost and economic analyses are frequently estimated as a percentage of the original investment cost of the property. However, in many applications this value of the residual property might be first estimated directly on a dollar basis and then converted to a percentage of the cost base. An important element to keep in mind is that terminal value is always valued in terms of the dollars current at the time of retirement. For this reason, high percentage terminal values are often obtained for properties which were obtained new in a period of low prices and then retired in periods of high prices. The high-price plateau starting after World War II made it worthwhile to retain and use again highway material and construction equipment that under the pre-World War II price did not have sufficient value to warrant its further use.

The amount of net terminal value varies so widely for each particular highway and special facility that general values are hardly warranted. The analyst should estimate the terminal value for each item in each analysis according to the particular circumstances. The amount of terminal property and its net value depend upon the following characteristics and factors:

Highway Transportation Costs

1. Market price at time of disposal
2. Cost to remove and to dispose of
3. Kind or type of material
4. Geographical location in use relative to place of usage as retired property
5. Possible method of disposal or future use
    A. Sold for cash–scrap metal or a usable vehicle
    B. Removed for use elsewhere in same function–a metal culvert or steel bridge
    C. Removed for use elsewhere in a different function–broken up pavement used for bank protection.
    D. Retained in place for use in same or different function–earthwork used in reconstructed roadway; pavement structure used as foundation or base for a new surfacing.
    E. Abandoned in place and used or not used in a different function–an embankment used as a levy for water storage or a roadway pavement used as a storage yard or parking lot.

Terminal Value of Specific Properties

Terminal value considerations may be analyzed along the following lines. In all cases the price or value of the terminal property is that thought to prevail at the future date of retirement.

Properties which may be sold on the market, including trading-in on new purchases, have a terminal value of their sale price less any expense to prepare them for or to deliver them to the purchaser. Property in this category of market sales would include machines, instruments, mobile equipment, scrap metals, structural shapes, and surplus earth excavations or construction materials. It is obvious that in this category of cash sales there would be no negative net terminal value.

In contrast to sale, certain properties of a highway department have a net positive terminal value in highway department usage, but only a negative terminal value on the market. The usage may be in the same function but a different location or at a new location in a different function. Examples include the use of machinery, instruments, and structural shapes for repair parts, replacements in kind, or in substitute functions. A short length of old pavement may be used in place as a base for storage and mixing of maintenance materials; broken-up pavement may be used for bank protection or in a fill. Culverts may be moved to other locations. Terminal value in such instances would be related to current market value for new or used similar property adjusted for any lessening of value because of wear and tear or shortening of probable service life.

Material used in place for the same purpose would have a 100 percent terminal value less any loss in value of useful service because of wear and deterioration. Culverts retained in place and useful on a widening and resurfacing project would be worth their full cost less the retirement of headwalls, if any.

Earthwork, likewise, could be considered to have a terminal value equal to original cost, less a retirement for any former cuts which were filled and fills which were cut in the reconstruction work.

When road construction is continued to be used in place but in an entirely different function, its terminal value is dependent upon the value of the new function. An important example of a change in function is the use of old portland cement concrete pavement as a base for a new bituminous concrete surface. A reconstruction of this nature is a compromise between continued use of the old pavement and constructing an entirely new pavement. The net terminal value under such circumstances should be estimated on the basis of what the old concrete is worth as a base for the new surface. An estimate of the value of the old concrete pavement as a base can be arrived at by first determining what a completely new pavement with base and wearing surface would cost. The difference between this cost and the cost of the resurfacing work would be a measure of the maximum terminal value of the old pavement. This difference might be adjusted downward by an estimate of three factors: first, by a service life factor provided that the reconstructed pavement might have a potential useful life less than an entirely new pavement would have; second, by a factor to compensate for any difference in cost of maintenance work to keep the pavement in a satisfactory condition; and third, by a judgment factor to compensate for any overall less desirable characteristics of the reconstructed pavement as compared to an entirely new construction, such as in width, alignment, grade, drainage, or riding quality.

Another example of using retired property for a purpose different than its original function is an abandoned roadway fill used as a levy. In this case there may be no service to the highway function to be realized so it would be proper to consider the terminal value zero for highway purposes; but the value of the fill may be appreciable for recreation or flood protection purposes.

Right-of-way land would normally be retired at 100 percent value as land, but when the original total cost included heavy payments for damages, moving of structures, and other payments for items adding no inherent value to the land or to the highway, the reasonable procedure would be to assign these investment items zero terminal value. When a right-of-way is to be abandoned and the land reverted to farm land, the net terminal value would not be 100 percent because of the low state of fertility of the land and the cost to grade the land to a suitable contour for farming. On the other hand, the market value of the land may have so increased since its purchase for right-of-way that the costs of restoration are less than increase in value, so that the net terminal value is 100 percent or more.

Terminal Value in Economy Analyses

The allowance in the analysis for economy of a net terminal value at the end of the period of analysis is beset with uncertainties. Whether man-made highway construction consisting of land molded to highway geometrics would have a positive terminal value at some far-distant year is an uncertainty. Practically none of the highway construction, including right-of-way land, would be beneficial to any alternative use without considerable cost expended to make it suitable. Likewise, most drainage structures (bridges, culverts, and underdrainage) and miscellaneous facilities have no value except as a part of a highway. The greatest opportunity for net terminal value of highways is found in the use of prior construction in reconstruction or modernization of a highway facility.

Terminal value can be considered in connection with selecting the analysis period. A short number of years for the analysis period combined with high terminal value might produce the same equivalent uniform annual cost as would a long period of years and low or zero terminal value. For instance, compare equivalent uniform annual economic cost at 40 percent terminal value at the end of 15 years with 0 percent at the end of 20 years:

$$(-1{,}000 + 400)(CR\text{-}8\%\text{-}15 = 0.11683) - 400\,(0.08) = -\$102.10$$
$$(-1{,}000 + 0)(CR\text{-}8\%\text{-}15 = 0.10185) = -\$101.85$$

For economy studies for both economic evaluation and for project formulation, in contrast with studies of the economic cost of owning and operating a highway, it often serves the purpose of the analysis to use a zero terminal value because the terminal value is not a factor likely to alter the choice of alternative based on economy. An exception would be when one alternative, as compared to the others, would have a distinctly different equivalent uniform annual cost depending upon whether terminal value was zero or a logical value above or below zero. For example, a structural steel bridge being compared to a reinforced concrete bridge would call for consideration of terminal values, because the structural steel could have a net positive terminal value as contrasted to a net negative terminal value for the concrete bridge.

Other than for street systems in developing new land for residential or industrial uses and about 80 percent of the interstate highway system, practically all of highway construction is reconstruction, betterments, additions, and improvements to existing facilities. In such cases, sunk costs (see page 249) are involved. These sunk costs often pertain to usable portions of the existing highway, and therefore the reconstruction or replacement cost is often less than it would be without the existing work in place. But on occasion the cost is higher when existing work must be removed.

In economy studies these sunk costs should be ignored and the analysis made using the cash-flow basis. In contrast, in replacement economy studies involving salable machines, tools, vehicles, and the like, the cash terminal value and period of future usefulness of the existing property are considered. This consideration includes the decreasing market value of the existing item and possible increasing maintenance and operating expenses.

PROBABLE SERVICE LIFE IN ANNUAL ECONOMIC COST CALCULATIONS

The factor n for the period service life in the cost equation for annual economic cost of owning and operating a facility,[1] should be estimated on the basis of the experience within the area and for the type of road, road system, highway component, or specific facility being considered. The estimate for n should not exceed the number of years that the facility is most likely to continue to render a satisfactory service without major reconstruction. The reason for ending the useful service life is relatively unimportant, except as knowing the probable reason for retirement is an aid to forecasting terminal value, and, for some properties, in estimating the years of service. See also Chapter 9 on Service Lives.

The usual ranges from minimum to maximum service life for specific classes of properties make it desirable to use a chosen specific average probable service life for each proposal under study for annual economic cost when dealing with a property such as pavement from which sections may be retired over the years. When a particular existing structure, a unit facility, or item of equipment is being considered, the particular factors of location, use, design, and capacity can be used to arrive at a specific probable service life. This probable service life may be estimated greater than or less than the general probable average service life of that class of property. These service lives for annual economic cost should be the best estimate that can be made of the probable useful service in the location considered.

ANALYSIS PERIOD TO USE IN ECONOMY STUDIES

In computations for the annual economic cost of owning and operating a facility the best estimate of useful service life would be used, but in the analysis for the economy of proposals an analysis period much shorter than service life is most often used. The main exceptions are in project formulation when considering construction materials or other types of alternatives in which the choice based on economy may depend upon the analysis period and terminal value. The general rule to follow is that the analysis period should not exceed the future period of predictability.

The analysis for economy requires consistency in the treatment of benefits on the one hand and costs on the other. If the cash flows for the highway are considered for a period of 40 years, so must the motor vehicle running costs be considered for the same 40 years. The longer into the future the analysis is projected, (1) the greater is the difficulty of making reliable estimates of traffic and all cash flows and, therefore, (2) the greater is the uncertainty of the estimates.

[1] Not to be confused with selection of an analysis period used in calculation of the equivalent uniform annual cost used in economy studies.

Highway Transportation Costs

Traffic is a dynamic factor; motor vehicle running costs and travel speeds are consequences of the changing character of the changing daily traffic pattern, composition, and speed. Generally, traffic volumes increase with the years until a saturation is reached. With this increase in traffic volume comes a decrease in traffic speed and an increase in vehicular speed changes per mile. At some average speed less than design speed the vehicle running costs will begin to increase. At some point on this downward trend of traffic speed the original benefits of the new facility may be absorbed by increasing costs to traffic. This pattern of motor vehicle costs makes long-range forecasts of road-user benefits highly uncertain and justify the use of short analysis periods in economic analyses rather than service lives of the components of the highway.

In justifying the expenditure for a grade reduction or other specific highway facility on the basis of economy in transportation, the analyst should consider two important elements. First, what is the likelihood that the reductions in the running cost of the motor vehicle will continue to be realized? Certainly, motor vehicle transportation is likely to continue for another 10 to 20 years, but will it continue for another 75 to 100 years? And should motor vehicle transportation as now known, continue for another 20 years, will it be of the type that the present calculated economy caused by the grade reduction will continue? Second, is there justification in constructing a highway facility on the basis of reductions in the running cost of motor vehicle operation when such reduction must continue for 50, 65, or 100 years in order to pay for the facility? It makes logic to use a short analysis period such as 20 years, or so, because the forecasts are more reliable and there is less uncertainty of all factors. Such practice will serve the purpose of aiding in the final management decision equally well if not better than using the best estimates of service lives.

Often an analysis is to be made for single features of a complete highway such as a grade separation, a new pavement, a traffic signal installation, or a guardrail. In such applications the period of analysis to use must be realistic considering the nature of the property proposed, its physical characteristics, use, and probabilities of continued useful service. For instance, a railroad-highway grade separation could become not useful if the railroad was abandoned. Ten years or less would be a suitable period of use for economy studies of traffic signals and other traffic control devices. Studies of the economy of guardrail might be made over a period of 4 to 7 years.

The objective in the analysis for economy is to get a reliable estimate of the relative economy of the several alternatives over a time period during which all the factors involved may be reliably forecasted. The objective is to furnish managment with an aid in reaching a decision of what to do now. Unless the period of analysis would have material effect upon this decision, a relatively short analysis period is preferred.

The construction of public highway facilities in which an annual cash monetary profit need not be earned in order to satisfy the investors as is true in private industry, can be evaluated on a longer analysis period than the 1 to 5 years used in industry in their economic analyses. On the other hand, a good business judgment in managing the people's highways would call for the use of analysis periods of not more than 20 to 30 years as the basis of making a selection of total highway projects on an economy warrant basis. Further, a construction program should include, so far as possible, those projects that promise the greatest return–that is, those projects that will pay for themselves through economy to motor vehicle users in the shortest number of years. Certain projects, however, are desirable on a basis other than road-user economy in transportation.

RATE OF VESTCHARGE

In Chapter 5 a vestcharge rate of 6 to 15 percent is suggested. The rate should be selected for each proposed facility in accordance with the prevailing conditions. Projects of high risk and great uncertainty call for high rates of vestcharge. The risks and uncertainties include such factors as (1) proved soundness of design, (2) reliability of the estimate of construction and maintenance costs, (3) reliability of the forecast of traffic, and (4) climate and geographic uncertainties.

Engineering designs of long-established soundness carry a small risk so far as the design is concerned; a new type of design should carry a higher vestcharge rate. The suspension bridge at Tacoma Narrows, Washington, failed in 1940 under aerodynamic forces. This bridge departed from certain structural design practices to the point that its floor system was not rigid enough to withstand the forces of the heavy prevailing winds.

Standard designs usually will be accompanied with reliable estimates of construction cost, except when the terrain features are new, the location difficult to reach, or there exists local conditions hazardous to construction. The highway maintenance cost on the four-lane divided interstate system has been an uncertainty because of limited experience to date in maintaining this kind of highway. In fact, early estimates by the states for maintaining the rural mileage varied from $500 to $8,300 per mile.

The future traffic volume and composition is a great uncertainty calling for reasonably high vestcharge rates. If the traffic does not increase as rapidly as forecasted the aggregate road-user benefits are less. On the opposite side, if the traffic increases much faster than estimated, capacity of the facility is soon reached and congestion sets in. The benefits from many highway facilities now come about by relief of congestion through greater traffic capacity. Travel time benefits per vehicle begin to decrease as traffic volume increases. Changing land-use patterns and the shifting of population are more uncertain in some locations than others, and thus the traffic forecast is accordingly more uncertain.

When natural hazards from storms, slides, hurricanes, earthquakes, and floods prevail, the realization of the benefits as forecasted is less certain than for locations not beset with such dangers.

The selection of the vestcharge rate is a judgment decision; it cannot be calculated. The rate should be deliberately chosen for each proposed facility giving weight to all pertinent factors, including the current rates on borrowed money being paid by the local citizens.

ANNUAL RECURRING EXPENSES

The annual economic cost for highways as given in Eq. 10-2 is composed of two main types of cost: the capital cost attributed to the initial investment (and any later investments) and the annual recurring items of administration, traffic services, and highway maintenance. These latter three items are next discussed. They would be estimated in the same manner at the same magnitude for use in calculations for either annual economic cost of owning and operating a facility or for the equivalent uniform annual cost in the analysis for economy.

ADMINISTRATIVE EXPENSE

In highway department accounting practice, the annual administrative and management expenses of highway facilities are not generally allocated to individual facilities or to roadway mileage, neither are they generally allocated to the maintenance and construction functions. The administrative cost is a real cost of highway transportation and the item should be included in the overall cost of owning and operating a highway system. For analyses of the engineering economy, administrative costs can be safely omitted because usually they would be about the same for all mutually exclusive alternatives considered.

HIGHWAY OPERATING EXPENSE

Highway operating expense, as distinguished from highway maintenance expense, refers primarily to traffic services, such as lighting, snow and ice control, and highway patrol.

Again these items are a real expense of owning and operating the highway system and should be included in any statement of the total annual economic cost. They are discussed separately from highway maintenance only because they are services to traffic rather than for preservation and maintenance of the physical highway plant.

When there are distinct differences in the annual expense of these operative items, as between alternative proposals, the analysis for economy should include these items, otherwise they may be omitted.

HIGHWAY MAINTENANCE EXPENSE

The American Association of State Highway Officials Manual of Uniform Highway Accounting Procedures [10-1] defines highway maintenance as follows:

Highway maintenance is the act of preserving and keeping of the rights-of-way and each type of roadway, roadside, structure and facility as nearly as possible in its original condition as constructed or as subsequently improved, and the operation of highway facilities and services to provide satisfactory and safe highway transportation. Maintenance does not include construction or betterments.

This definition includes maintaining the physical components of the highway as well as the traffic and roadside services, including snow and ice control. The definition does not include the highway police patrol, often considered to be a service to traffic, or at least an operating function of highway administration. These maintenance and operating expenses are on a daily, monthly, seasonal, or yearly basis; they do not add to the investment value of the highway plant. Therefore they are considered in accounting to be annual expense.

The maintaining and repairing of the highway is usually performed by highway departmental employees using department-owned equipment. However, some of the larger and infrequent occurring maintenance operations may be done by contract awarded under competitive bidding. Contract work may include pavement patching, seal coating, bridge painting, chemical weed and brush control, and tree trimming.

The American Association of State Highway Officials [10-1] recommended classification of maintenance cost accounts is:[2]

400. Physical or general maintenance
    410. Routine roadway surface operations
    420. Special roadway surface operations
    440. Shoulders and side approaches
    460. Roadside and drainage
    480. Structures
500. Traffic services
    510. Snow, ice, and sand
    530. Traffic control and service facilities
    560. River crossings (ferries, drawbridges)
    570. Other services
600. Unusual or disaster maintenance
    601. Floods and washouts
    602. Tornadoes and cyclones
    603. Sandstorms
    604. Blizzards
    605. Major slides
    606. Earthquakes
    607. Settlements
    608. Accidents
    609. Military operations
    610. Other costs

[2] Copyright 1958 by AASHO. Used by permission.

Highway Transportation Costs

Despite the many years of record keeping of highway operations, there is a lack of records of maintenance expense that serves the purpose of highway economy studies. Many factors affect the total expense of maintaining highways on the one hand, and, on the other, it is difficult to record maintenance expense directly against the several features of design and of traffic which affect the maintenance expense. The difference in maintenance expense associated with pavement width, shoulder width, vertical grades, horizontal alignment, cut and embankment slopes, and types of construction materials is hard to measure under the normal processes of maintaining highways. Again, the effect of traffic volume and traffic composition on the expense of maintaining highways is not known closely enough. Weather, soil conditions, and time are factors having pronounced effects on highway maintenance expense.

The state highway departments are improving their cost reporting and cost record systems year by year. Eventually good comparative expenses of maintenance operations may become available.

For estimating the annual economic cost of owning and operating a highway system, perhaps the maintenance expense as recorded is adequate, because, in a system-wide calculation, the cost does not need to be known for specific sections of highway or by highway design or construction details. Usually a system total expense for the year is sufficient for this purpose.

For economy studies, however, the desired specific highway maintenance expenses are not available. Resource must be had to such general records as may be available plus estimates based upon judgment. Fortunately, the available maintenance reports supply usable guides for this purpose.

THE PRICE BASE AND INFLATION

The analysis for the economy of highway improvements is a study of the future. The costs and net benefits used in the analysis result from estimates of what the future consequences will be "with and without" the improvement.

The analyst is immediately confronted with a selection of prices to apply to highway construction in the future and to motor vehicle operation in the future. The pricing of construction costs is done on the same basis and in the same manner as is practiced in estimating the construction cost of work let out to bidding. Current prices are used. The same prices apply to estimates of highway costs for economy studies. Current prices are also used in determining the cost of motor vehicle operation. These prices are in most instances projected into the future without change.

The current price level may be changed for estimated future cash flows when there is substantial evidence that the real price will change up or down. By real price is meant price with a constant-value dollar. Over time, the market prices of commodities, including highway construction, have usually continued to rise, despite cost-lowering improvements in the technological processes. These price increases have been caused mainly by economic inflation–decrease in the purchasing power of the dollar. It is desirable, therefore, to settle upon how to treat economic inflation in highway engineering economy analyses.

Lee and Grant [10-7] give an excellent discussion of the factors of price and inflation as related to economy studies, with the conclusion that inflation should not be recognized when forecasting future prices, costs, and benefits, and that current prices should be used except in those situations "when there is overwhelming evidence that certain inputs or outputs, such as for land, are expected to experience significant price changes relative to the general price level." These two conclusions agree with those reached by other writers and by government agencies dealing with public works.

Inflation is perhaps more difficult to forecast than price, and, if attempted, the forecasting of inflation merely introduces another uncertainty into the analysis. Should inflation be used in the analysis, its use would contribute to more inflation should construction be undertaken as the result of such analysis. Forecasting higher costs and benefits based on economic inflation would tend to justify higher investments today and long-lived projects rather than short-lived projects. Considering the national economy, the effects of inflation as a whole are balanced out by the gains of the debtors and the losses of the creditors. Within the economy of highway transportation, the inclusion of an inflationary factor in the future highway costs of construction and maintenance would call for similar consideration of inflation in the road-user costs for motor vehicle operation and in the value of travel time. Thus both costs and benefits would be inflated, so their relative magnitude may be the same with or without the factor of inflation.

The purpose of the analysis for economy of proposed highway facilities is to assist management in reaching a decision on what to do now. The analyses for economic evaluation and for project formulation provide guides to this decision process. Certainly, when considering mutually exclusive alternatives or project priorities, inserting a factor for inflation in the analysis would not cause the results to be more reliable or lead to a different selection of alternatives. In economic evaluation, however, the inflation factor, if included, could lead to economic justification of a proposal when omitting the inflation would not lead to such justification. The question to answer, then, is how reliable is the forecast of inflation for that particular proposal. This added uncertainty, along with the others present, adds no merit to the economic analysis as a whole. As concluded by Lee and Grant, it is preferred practice and concept to omit any inflation factor.

The conclusions stated here do not infer that when considering construction in stages some years apart, different real-price levels should not be used. When it can be reliably forecast that the real price will change in the future from what it is at the time of the analysis, the forecasted price should be used. Example:

Highway Transportation Costs

The market value of land adjacent to highways usually materially increases following any major improvement in the highway. Therefore, in considering whether to acquire right-of-way now for additional traffic lanes to be constructed in the future, consideration should be given to real changes in the price of the land which would likely take place following the completion of the first stage of construction.

SUNK COST

The analysis for the economy of a proposed venture, project, machine, or structure is an analysis of the future. Therefore the past is irrelevant except as it may be used in predicting future sales, loads, events, or other factors helpful in making estimates upon which the analysis for the economy of the proposed improvement may depend. In accordance with these two principles of analysis any investment in existing property or other past cost in property involved in the proposal is not considered. Its investment cost is said to be "sunk." It is gone and it cannot be recalled.

This concept of sunk cost does not mean that the property purchased by the now sunk cost and which is still in existence is totally worthless. The property may have a terminal value in resale or in other use which is realizable. Further, it is often that existing property, that is, a past investment, makes it possible to construct the new proposal at much less cost than would be required should the existing work not be there. Therefore the future outward cash flow is less than it would be without this sunk cost.

To illustrate sunk cost, consider a highway which is being studied for reconstruction. Usually most of the earthwork, cuts and fills, will be usable except where the horizontal alignment is materially changed. Many of the culverts under the fills can be used in place and their length extended if the grade line is raised or the roadway widened. Because of this past construction, the reconstruction cost is less than it would be otherwise. But the economy study is conducted without reference to this sunk cost. The key to every analysis for economy is the difference in the costs and benefits between the alternatives. These differences are determined by the future cash flows without reference to the past cash flows.

Some writers define sunk cost as only that remainder of original investment which has not been recovered through past depreciation allocations. But past depreciation expense allocations are just as irrelevant as is the past initial cost, and neither one is an index to present value, or terminal value as of the present. It is unnecessary therefore to define sunk cost as the unrecovered original investment. Determining the amount of original cost recovered through depreciation allocation is an uncertainty itself, especially when the property at issue is part of a mass account. Depreciation allocations are arbitrary and the specific income earned by a specific property unit or segment within a company is unknown. When considering public properties such as highways for which depreciation expense accounting and profit and loss accounting are not practical, sunk cost defined as being the unrecovered original cost has no meaning whatsoever.

Whether sunk cost is defined as the total original cost or merely the unrecovered cost is unimportant because in economy studies it is unnecessary to determine the amount of sunk cost because it is an irrelevant factor. Only future cash flows are the factors of importance.

INTEREST DURING CONSTRUCTION

In public utility regulation, including setting the rates to charge customers, it is customary to allow as part of the investment in utility plant "interest during construction." The concept is that it often takes years from the beginning of planning to bring into operation large units of plants, pipelines, office buildings, and other types of property. During the stages of planning, designing, and constructing there is a steady flow of cash disbursements which do not at the time add to the service of the customers, and, therefore, the rate base and net return to the utility are exclusive of this work in process. To compensate the utility for the return forgone on investments not yet in operation, the utility is allowed to include in its investment, or rate base, the amount of computed return forgone on the same basis as though it were an actual disbursement for construction.

Economy studies often overlook the time sequence of disbursements for investment purposes on the proposed facility prior to the year of purchase or end of construction. In other words, economy studies often assume instant disbursement of the entire capital cost at time zero, the date of beginning of use of the property. Where the several alternatives considered would have about the same time requirements from the planning stage to completed construction, this failure to consider "vestcharge during construction" results in no significant error. However, in figuring total investment and for studies of the financing of the project, the disbursements by sequence of calendar time are an important consideration.

GENERAL CONSTRUCTION AND MAINTENANCE COSTS PER HIGHWAY MILE

Despite the fact that the cost of constructing and maintaining highways varies with every construction contract and every maintenance section, it is of some interest and possible use to have at hand some general average costs. Table 10-1 on construction cost and Table 10-2 on maintenance expense are presented for this purpose.

Because of the variations in design requirements, number of lanes, terrain, soil, value of land, and local prices, both highway construction costs and annual highway maintenance expenses vary widely when expressed in terms of centerline miles. Some urban sections of freeways have cost more than $25 million per mile to construct. Construction of some of the local land-service roads may cost as

Highway Transportation Costs

little as $10,000 per mile. The 41,000-mile interstate system is estimated to cost about $1,169,000 per mile exclusive of toll highways. This total breaks down into $24.143 billion for the rural portion of 32,751 miles ($737,150 per mile) and $21.043 billion for the urban portion of 5,889 miles ($3,573,400 per mile).

TABLE 10-1
Dollars per/Centerline Mile Cost of Complete Highway Construction or Reconstruction by Highway System and Census Division
(Based on Federal-Aid Projects for 1964)

Highway System and Census Division	Engineering	Right-of-Way	Earthwork and Drainage	Structures	Flexible Pavement	Total
A. Interstate Rural						
1. New England	86,389	116,987	256,698	272,134	176,893	909,101
2. Middle Atlantic	113,944	154,302	288,492	363,916	249,575	1,170,299
3. South Atlantic North	147,827	200,186	224,568	290,639	254,570	1,117,790
4. South Atlantic South	52,315	70,845	121,153	134,860	128,434	507,607
5. East North Central	76,671	103,827	188,089	225,736	194,143	788,466
6. West North Central	51,411	69,621	125,504	115,857	153,142	515,535
7. East South Central	59,407	80,448	141,234	154,266	204,122	639,477
8. West South Central	64,119	86,830	153,951	175,403	179,025	659,328
9. Mountain	49,294	66,754	193,443	86,603	91,994	488,088
10. Pacific	68,079	92,192	211,516	186,928	159,756	718,471
B. Interstate Urban						
1. New England	320,070	997,356	664,264	1,895,066	180,129	4,056,885
2. Middle Atlantic	590,662	1,840,532	746,539	2,247,594	254,678	5,680,005
3. South Atlantic North	433,368	1,350,396	581,122	2,007,453	251,906	4,624,245
4. South Atlantic South	179,028	557,860	313,511	501,801	130,234	1,682,434
5. East North Central	340,043	1,059,592	486,722	1,481,890	194,220	3,562,467
6. West North Central	187,135	583,123	324,770	1,001,001	156,144	2,252,173
7. East South Central	218,661	681,360	365,475	1,174,574	225,411	2,665,481
8. West South Central	199,334	621,134	398,885	650,445	174,394	2,044,192
9. Mountain	163,159	508,412	500,579	423,322	99,956	1,695,428
10. Pacific	355,210	1,106,853	547,346	1,559,994	160,893	3,730,296
C. Primary Rural						
1. New England	47,000	61,472	137,918	57,948	160,239	464,577
2. Middle Atlantic	52,643	68,854	155,001	46,234	216,478	539,210
3. South Atlantic North	43,852	57,356	120,656	30,777	215,898	468,539
4. South Atlantic South	35,005	45,784	65,093	23,332	118,537	287,751
5. East North Central	39,556	51,737	101,056	26,414	163,054	381,817
6. West North Central	19,407	25,384	35,504	16,485	75,314	172,094
7. East South Central	22,797	29,817	39,954	32,169	99,975	224,712
8. West South Central	19,386	25,356	43,551	21,432	84,615	194,340
9. Mountain	17,256	22,570	54,723	13,488	45,801	153,838
10. Pacific	33,982	44,446	59,836	43,658	131,530	313,452

TABLE 10-1 *(continued)*

Highway System and Census Division	Engineering	Right-of-Way	Earthwork and Drainage	Structures	Flexible Pavement	Total
D. Primary Urban						
1. New England	86,006	193,324	287,096	190,250	162,160	918,836
2. Middle Atlantic	93,936	211,150	322,656	179,210	217,910	1,024,862
3. South Atlantic North	76,025	170,889	251,162	130,440	223,839	852,355
4. South Atlantic South	57,078	128,301	135,500	111,801	118,064	550,744
5. East North Central	70,624	158,750	210,362	141,507	165,764	747,007
6. West North Central	26,782	60,201	60,550	36,549	79,173	263,255
7. East South Central	29,902	67,215	68,139	44,259	108,341	317,856
8. West South Central	28,789	64,712	74,274	55,410	86,462	309,647
9. Mountain	27,409	61,611	93,327	37,214	50,388	269,949
10. Pacific	72,442	162,836	102,047	271,490	148,722	757,537
E. Secondary Rural						
1. New England	17,475	11,979	46,160	20,338	73,756	169,708
2. Middle Atlantic	19,122	13,108	51,887	15,845	92,593	192,555
3. South Atlantic North	15,556	10,664	40,382	6,745	88,250	161,597
4. South Atlantic South	13,099	8,980	21,786	5,326	52,747	101,938
5. East North Central	14,690	10,070	33,822	6,115	71,316	136,013
6. West North Central	13,870	9,508	22,568	4,626	64,399	114,971
7. East South Central	15,503	10,627	25,397	12,463	80,071	144,061
8. West South Central	12,978	8,896	27,684	6,606	67,102	123,266
9. Mountain	12,347	8,464	34,785	6,568	41,937	104,101
10. Pacific	14,159	9,706	38,035	7,809	55,996	125,705
F. Secondary Urban						
1. New England	4,201	9,820	66,144	78,129	76,241	234,535
2. Middle Atlantic	4,510	10,542	74,337	71,782	98,107	259,278
3. South Atlantic North	3,486	8,148	57,865	36,961	101,529	207,989
4. South Atlantic South	2,870	6,708	31,218	33,817	56,066	130,679
5. East North Central	3,240	7,574	48,465	40,589	73,213	173,081
6. West North Central	2,475	5,785	32,339	13,639	62,228	116,466
7. East South Central	5,569	13,017	36,392	25,662	84,550	165,190
8. West South Central	2,496	5,834	39,669	18,090	68,918	135,007
9. Mountain	2,603	6,084	49,845	20,879	42,410	121,821
10. Pacific	3,803	8,890	54,502	68,593	66,073	201,861

Most of the highway construction for many years has been reconstruction of existing roads and highways, as contrasted to construction of new highways. About the only completely wholly new highways that have been built since 1920 are relocations of short sections of routes to improve alignment, about 80 percent of the 41,000-mile interstate system, and the urban streets in housing

TABLE 10-2
ESTIMATED 1964 MAINTENANCE EXPENSES FOR THE FEDERAL-AID SYSTEMS IN THE 48 CONTIGUOUS STATES AND THE DISTRICT OF COLUMBIA

Federal-Aid Highway Systems	System Mileage 1961	Cost of Maintenance Operations, Dollars per Mile						
		Roadside and Drainage	Surface and Base	Shoulders	Structures	Traffic Services	Snow, Ice, and Sand Control	Total
Interstate								
Rural	34,513	644	1,046	352	179	487	525	3,233
Urban	6,400	1,570	1,601	564	527	1,295	1,120	6,677
Total	40,913	789	1,132	386	233	613	619	3,772
Federal-Aid Primary								
Rural	143,800	446	823	271	121	291	406	2,358
Urban	16,083	794	1,228	420	386	796	860	4,484
Total	159,883	481	864	286	147	342	452	2,572
Federal-Aid Secondary								
Rural	589,679	129	398	79	49	73	122	850
Urban	14,978	324	648	199	118	298	392	1,979
Total	604,657	134	405	82	50	78	129	878
All Systems								
Rural	767,992	212	507	127	68	132	193	1,239
Urban	37,461	739	1,060	356	303	682	717	3,857
Total	805,453	236	533	137	79	158	218	1,361

Source: Table 49, *Supplementary Report of the Highway Cost Allocation Study* (March 1965), p. 178, House Document No. 124, 89th Congress, 1st Session.

Note: In using this table, it is assumed that expenses apply to 2-lane highways.

TABLE 10-3

Maintenance and Operating Expenses for 1965–Illinois Division of Highways [10-6]
(Expenses Are in Dollars Per Centerline Mile)

Class of Work	Urban Expressways	Interstate Highways	Regular State Highways
Maintenance			
Wearing surface	$ 7,818	$ 494	$ 768
Shoulders, ditches, cuts, and fills	553	664	397
Large bridges (100-ft length and over)	786*	92*	719*
Culverts and other drainage facilities	478	54	27
Small bridges (less than 100-ft length)	–	10	19
Miscellaneous structures and facilities	453	15	13
Service drives	–	105	1.36
Total maintenance	$ 12,242	$ 1,462	$ 1,303
Operation			
Cutting and clearing vegetation	$ 3,050	$ 1,071	$ 419
Snow removal and ice control	14,350	957	581
Clearing dirt and debris	10,514	655	257
Roadside planting maintenance	1,426	425	34
Upkeep of guardrail	1,672	147	31
Subway and drainage pumping	1,701	19	11
Electric lighting	–	–	0.40
Traffic operation	21,326	504	476
Rest area maintenance	–	56	–
Total operation	$ 54,041	$ 3,834	$ 1,809
Total maintenance and operation	$ 66,283	$ 5,296	$ 3,112
Weighted average miles maintained	105.13	626.68	14,330.71

* Cost per bridge.

and commercial developments on land formerly used for other purposes–often for farming.

The highway maintenance operations have been continually changing since the days of dragging earth roads with split logs. The coming of the motor vehicle brought the demand for all-weather all-year roads. Thus hard surfaces and snow removal followed. Now with high traffic volumes and high speeds the demand is for smoother roadway surfaces, wider shoulders, and treatment of icy coatings. Over the years the demand by the public for continuing increases in the quantity and quality of highway maintenance and traffic services

TABLE 10-4
Maintenance Expense for 1965 by Roadway Surface Type, Regular State Highway System–Illinois Division of Highways[10-6]
(Expenses Are in Dollars per Centerline Mile)

Class of Work	Portland Cement Concrete	Brick	Sheet Asphalt and Bituminous Concrete on Rigid Base	Other Bituminous Types		Bituminous Surface Treatment	Water-bound Macadam	Gravel	All Surface Types
				Rigid Base	Flexible Base				
Wearing surface	$ 991	$ 605	$ 599	$430	$ 190	$299	$198	$320	$ 768
Shoulders, ditches cuts and fills	423	414	385	273	201	295	—	221	397
Small bridges (less than 100-ft length)	22	3.52	17	—	5.49	8.21	—	—	19
Culverts and other drainage facilities	23	17	31	33	26	10.51	—	—	27
Miscellaneous items	25	—	1.75	—	2.86	1.21	—	—	13
Large bridges (100-ft length and over)	—	—	—	—	—	—	—	—	719*
Service drives	—	—	—	—	—	—	—	—	1.36
Total maintenance expense	$1,483	$1,039	$1,034	$736	$426	$613	$198	$541	$1,303
Weighted average miles maintained	6,705.02	33.83	6,936.10	30.10	171.04	400.09	7.13	47.40	—

* Cost per bridge for 1,584 bridges.

has resulted in increasing annual expense to keep the highways in an acceptable state for current heavy, high-speed traffic.

Maintenance expenses will continue to vary with the roadway and traffic character, with local conditions, and with local management policies and efficiencies. The general maintenance expenses in Table 10-2, however, furnish some information as to the general range of expense by road systems.

Tables 10-3 and 10-4 for Illinois state highways [10-6] give specific maintenance expenses for 1965. The breakdown by class of work is helpful in gaining a general impression of the relative cost of the separate items of maintenance. Table 10-4 gives the maintenance expense by type or roadway surface. Here the age of the surfacing, traffic volume, roadway width, climate, soil type, and other factors affect the maintenance operations, so this table is applicable only to the Illinois state highways in affording a comparison of the cost of maintaining different types of roadway surfaces.

PROBLEMS

10-1. Write two other logical formulas for equivalent uniform annual highway cost other than Eqs. 10-3 and 10-4.

10-2. Write the equation for the equivalent uniform annual highway cost including the following factors: (a) An initial and subsequent investment, (b) a gradient maintenance and operating annual expense, (c) a terminal value. Draw a cash-flow diagram to illustrate your equation.

10-3. What are the relative merits and demerits of allowing for risks and uncertainties in economic analyses by (1) using higher vestcharge rates, and (2) using shorter analysis periods? Compare with making the allowances for risk and uncertainties in the estimates of costs and benefits.

10-4. Compute the equivalent uniform annual cost over 30 years of the following cash flows. Investment of $10,000 now and $12,000 at the end of 5 years from now. Annual maintenance expense is $200 at end of first through 5th year, then $420 a year thereafter to end of 30th year. Use a vestcharge rate of 8 percent.

10-5. Compare the equivalent uniform annual expense of maintaining a highway for a period of 20 years at a vestcharge rate of 10 percent for (1) a gradient growth of $100 a year, and (2) an exponential growth of 4 percent per year. In both cases the time zero maintenance expense rate is $2,000.

10-6. Compare the annual costs of the following types of culvert pipe for a given service. Also compute the present worth for 60 years of service for each type of pipe. Use a vestcharge rate of 6 percent.

	Type A	Type B
First cost	$10,000	$23,000
Service life	20 years	60 years
Salvage value	None	$3,000
Annual maintenance	$600	$100

Highway Transportation Costs

10-7. A section of a new scenic highway in rough terrain has been heavily landscaped with a wide range of plantings. The maintenance expense of the right-of-way, or roadside, is estimated at $6,000 the first year and to decrease uniformly at $300 a year for 10 years after which it will stabilize at $3,300 a year. At a vestcharge rate of 10 percent, and by using the gradient concept, calculate the equivalent uniform annual maintenance expense of the right-of-way for this route section the first 10 years.

10-8. A highway department is considering the alternatives for draining an area of impounding water. *(A)* Construct a pumping house and pressure line at an initial cost of $70,000 and with $1,500 a year maintenance and operating expense. *(B)* Construct a gravity pipeline in a rock bore at a cost of $88,000, with no maintenance expense. On the basis of an analysis period of 30 years and zero terminal value, which alternative is preferred at a minimum attractive rate of 10 percent? (*Ans.:* Alternative *A*, because PW is $3,860 less than for Alternative *B*.)

10-9. Should a highway department roof a new masonry maintenance equipment garage with *(A)* 40-year life copper at a cost of $2,600, *(B)* 25-year life tar and gravel at $1,400, or *(C)* 18-year life wood shingles at $1,000. The building has an expected life of 40 years. The roof maintenance expense is nominal. Use a vestcharge rate of 7 percent.

10-10. The highway commission is investigating the following two alternative designs for a proposed maintenance garage. Using a vestcharge rate of 8 percent which alternative would you recommend?

	A Reinforced Concrete	*B* Frame and Galvanized Metal
First cost	$58,000	$30,000
Service life	50 years	25 years
Terminal value	None	None
Annual maintenance	$500	$900
Annual insurance premiums per $1,000 of first cost	$5	$15

10-11. Select the most economical alternative from the three following materials for a floor on a bridge across a small stream. Use a vestcharge rate of 10 percent.

	A Untreated Timber	*B* Treated Timber	*C* Bituminous Concrete
Original cost	$9,000	$15,000	$28,000
Annual uniform maintenance	$300	$200	$100
Expected service life, years	9	15	22
Terminal value	$ −500	Zero	$ 2,000

10-12. For peak loads and emergency use, a highway district headquarters needs to have available a dump truck. At a vestcharge rate of 8 percent under the following conditions, how many days per year on the average must the truck be used to equalize the cost of the two alternatives, *A* and *B*: *(A)* Purchase a truck at a cost of $11,000 to be used

8 years with a trade-in value of $1,000. Annual cash expense items not directly related to days or mileage of use will amount to $800 per year. Driver and other daily operating expense will be $35 a day of use. *(B)* Rent a truck with driver at $55 a day. At an anticipated use of 100 days a year, which alternative is preferred?

10-13. A proposed facility is estimated to have the two following investment costs and the two terminal values: Initial investment at time zero of $1,000 with a terminal value of $200 at end of 10th year. A second investment of $600 is required at the end of the 6th year which will have a terminal value of $100 at end of 10th year.

At a vestcharge rate of 8 percent find the combined equivalent uniform annual cost of these two investments, (1) by first transferring the $600 second investment to time zero, and then (2) by first finding the EUAC of the $600 second investment for the remaining 4 years before it is spread over the 10 years. Comment upon your solutions.

10-14. What amount of repaving cost is justified today under the following conditions? An existing pavement costs $600 a year to maintain. A new pavement would be maintenance cost free for the first 6 years, then rise uniformly to $600 a year in 10 more years, after which it could be maintained for $600 a year for 9 years before requiring major improvement.

10-15. A highway department is considering what to do with a section of old portland cement concrete pavement which is badly cracked, patched, and of poor riding quality, and too narrow for traffic. The main alternatives are: *(A)* Widen with portland cement concrete to a total width of 24 feet and surface the whole width with bituminous concrete, and *(B)* break up and dispose of the old conrete pavement and lay a wholly new portland cement concrete pavement. The main factors are:

		Alternative	
		A	B
1.	Net terminal value of the existing pavement	Sunk	$30,000
2.	Present cash outlay for resurfacing or reconstruction	$32,000	$170,000
3.	Service life of present resurfacing or reconstruction in item 2	14 years	24 years
4.	Cost of reconstruction at end of 14 years for A and resurfacing B at end of 24 years	$200,000	$28,000
5.	Service life of reconstruction and resurfacing item 4	24 years	20 years
6.	Resurfacing at end of 38 years	$24,000	–
7.	Service life of resurfacing in item 6	18 years	–
8.	Time period, years, beyond which the alternatives will not repeat their cycles	56 years	44 years

Using a vestcharge rate of 7 percent, make the calculations necessary to selecting one of these two mutually exclusive alternatives on the basis of economy.

10-16. Calculate the equivalent uniform annual costs for *(A)* rigid and *(B)* flexible highway pavements, of equal designs for the same traffic service, in such a way that the most economical pavement will be identified. Assume that their effects on road-user costs are equal and that 7 percent is an appropriate vestcharge rate. Assume a second resurfacing for the flexible pavement and adjust for the remaining service life of 6 years at age 44. Comparative cost data are:

Highway Transportation Costs

	A Rigid Pavement	B Flexible Pavement
Initial cost per mile	$ 73,000	$ 52,000
Cost of first resurfacing	24,000	19,800
Cost of second resurfacing	19,800	19,800
Maintenance expenses*	—	—
Analysis period, years	44	44
Age of initial pavement at first resurfacing, years	26	18
Service life of first and second resurfacing, years	18	18

* Annual maintenance expenses are as follows:

Service years or age		
1	$ 75	$ 50
2	88	62
3	101	74
4	114	86
5	127	98
6	140	110
6 through 18, gradient of	6	16
18	212	302
18 through 26, gradient of	5	–
26	252	–
After first or second resurfacing	200	200

10-17. List the types of sunk costs that may exist in the highways and highway department equipment and property which would be associated with economy analyses.

10-18. Calculate the cost of vestcharge (interest during construction) during planning and construction for the following urban expressway using the concept of year-end cash flow compounded annually at a rate of 6 percent.

Year-end	Activity or Work Accomplished	Cash Disbursement (Dollars)
0	Beginning of formal discussions	None
1	Preliminary studies, conferences, examinations, and surveys of alternative locations	60,000
2	Traffic studies, preliminary designs, and estimates	110,000
3	Economic analyses, public hearings, design changes, right-of-way	60,000
4	Right-of-way, final designs, and award of contracts	240,000
5	Payments to contractor and construction engineering.	1,670,000
6	Payments to contractor and construction engineering.	3,090,000
7	Final payments on contracts	4,500,000

10-19. Compare the equivalent uniform highway costs of the following cash flows at 7 percent vestcharge rate, (1) on the basis of the service lives and terminal values given, with (2) using an analysis period of 20 years for the entire highway with zero terminal value. Discuss your results in terms of (a) the economic cost of owning and operating a property, and (b) annual costs for economy studies.

Highway Component*	Investment Cost per Mile	Terminal Value, percent	Service Life, Years
Right-of-way			
Land	$ 42,000	100	100
Improvements	14,000	0	25
Damages	10,000	0	25
Roadway construction			
Clearing right-of-way	2,000	0	25
Earthwork	110,000	80	100
Drainage structures	40,000	20	50
Roadway base and surface	250,000	40	25
Roadside development	5,000	50	60
Miscellaneous structures	10,000	−10	60
Major structures	50,000	−10	75
Traffic service facilities	8,000	0	15
Total	$ 541,000		

Maintenance, operation, and administration, $ 4,500 per year.

* Engineering costs are included in the specific component.

10-20. By using the highway construction costs in Table 10-1 and the service lives in Table 9-1 for each component, (1) compute the equivalent uniform annual highway cost for system 3, primary rural, for your census division. Use a vestcharge rate of 8 percent. (2) Repeat the calculation using an analysis period of 30 years for the entire highway. Discuss your two results.

REFERENCES

10-1. American Association of State Highway Officials. *Manual of Uniform Highway Accounting Procedures.* Washington, D.C., November 1958.

10-2. Robert F. Baker, Robert Chieruzzi, and Richard W. Bletzacker. *Highway Costs and Their Relation to Vehicle Size.* Bulletin 168, Engineering Experiment Station, Ohio State University, Columbus, 1958.

10-3. R. H. Baldock. *The Annual Cost of Highways.* Highway Research Board, National Academy of Sciences, Washington, D.C., Highway Research Record No. 12, 1963.

10-4. Gottfried Haberler. *Inflation, Its Causes and Cures.* American Economic Association, Washington, D.C., 1960.

10-5. Jack Hirshliefer and others. *Water Supply.* University of Chicago Press, Chicago, 1960.

10-6. Illinois Division of Highways, Department of Public works and Buildings. Forty-eighth Annual Report (Calendar year 1965). Springfield, Ill, 1966.

10-7. Robert R. Lee and E. L. Grant. *Inflation and Hihghway Economy Studies.* Highway Research Board, National Academy of Sciences, Washington, D.C., Highway Research Record No. 100, 1965.

10-8. Ralph A. Moyer and Joseph E. Lampe. *A Study of Annual Costs of Flexible and Rigid Pavements for State Highways in California.* Highway Research Board, National Academy of Sciences, Washington, D.C., Highway Research Record No. 77, 1965.

10-9. Alfred R. Oxenfeld. *Economic Principles and Public Issues.* Holt, Rinehart and Winston, New York, 1961.

10-10. Stanford Research Institute. *Economics of Asphalt and Concrete for Highway Construction,* Menlo Park, Calif., 1961. A study prepared for the American Petroleum Institute, New York, N. Y.

10-11. Willard L. Thorp and Richard E. Quandt. *The New Inflation.* McGraw-Hill, New York, 1959.

10-12. James J. Tobin and A. R. Losh. *Highway Cost Keeping.* United States Department of Agriculture, Washington, D.C., Bulletin 660, Government Printing Office, 1918.

10-13. John W. Work, Jr. *An Economic Replacement Model for Highway Surface Determination.* Highway Research Board, National Academy of Sciences, Washington, D.C., Highway Research Record No. 12, 1963.

Chapter **11**

Road-User Consequences

The main benefits of highway improvements accrue to those who travel the highways. These benefits (and any adverse consequences) reach the road user primarily through the operating cost of motor vehicles, the reduction in highway accidents, and the reduction in travel time, all as market factors. There are the nonmarket consequences of the personal preference items of comfort, convenience, and lessening of travel strains. These two groups of factors are those considered when highway improvements are justified or not justified on an economic basis and these same factors are considered in project formulation. Chapters 12, 13, and 14 discuss the power characteristics of motor vehicles and the operating cost factors as affected by highway design. This chapter discusses travel time and the personal preferences, which are separate from the motor vehicle as a machine. Highway accidents are discussed in Chapter 15.

HIGHWAYS AND THE ROAD USER

As the driver and passengers of a motor vehicle travel over a highway, their thoughts probably are more on their own personal comfort and satisfactions than on the cost of running the vehicle or the costs to provide the highway. When riding over a highway or when not doing so, highway users as a group do not fully recognize that their motor vehicle fuel taxes and their license, or tag, fees provide most of the money for constructing and maintaining the highways. The road user probably does not sense his taxes for support of the highways in the same light as he does the direct fares he pays for bus, railroad, and airplane travel trip by trip. The highway taxes are taxes, despite the fact that they amount to only about 12 percent of the total highway transportation cost and are the one tax the road users pay which directly returns to them a personally desired service facility, and, often enough, a reduction in the running cost of their vehicles sufficient to pay, in time, the full annual economic cost of the highway improvement.

It is essential therefore for highway departments to view their highway investment programs as a business enterprise in which the economic costs of highway investments are more than matched by future economic benefits. Since the motor vehicle itself accounts for about 88 percent of the cost of highway transportation, the highway designer is obligated to give adequate recognition to

Road-User Consequences

achieving the economical highway design as it affects the running cost of motor vehicles. The interrelationships of motor vehicle running costs and highway design are discussed in Chapters 13 and 14.

Studies of the use of highways, railways, and airways as forms of personal transportation indicate that their passengers attach great weight to speed, comfort, and convenience. In fact, the major improvements in roadways, railroads, passenger motor vehicles, passenger train cars, and airplanes during the past 20 years have been mainly to increase speed, comfort, and convenience. These factors are the main subject of this chapter.

Although comfort and convenience have been designated as nonmarket factors, they are really market factors from the fact that travelers willingly and knowingly pay for these desirable factors. But here they are placed in the nonmarket group only because as yet they have not been priced in the market place.

THE FACTOR OF TRAVEL TIME

Time is consumed during travel. Time is an economic commodity that is valuable. Time is valuable because on time depends the quantity and quality of production. In an economic sense, however, time has no value unless with the passage of time desirable services and satisfactions are rendered or goods produced. Through time man is creative and productive. He serves his own needs and those of society. Time is a factor that determines the relative worth of current goods as compared to future goods. Time therefore is one of the controlling factors in computing the interest payments on loans and the discounts on future transactions. Likewise, time is a major factor in the economy of highway improvements. But time, unlike the common products found in the marketplace, has no standard price. Yet for most analyses of the economy of highway improvement it is desirable to put a price tag on highway travel time.

THE NATURE OF TIME

Time can be described in no other way except to refer to the rotation of the Earth and tilting of the Earth with respect to the sun. Time is measured with respect to the Earth and sun by the use of man-made timing devices. Time cannot be perceived by any of man's five senses (nor by his sixth sense). Time cannot be altered, transferred, changed in supply, controlled in any way, not used, recalled, reclaimed, saved, or stored up. Every person on Earth has exactly the same amount of time allotted to him each day, but each person is, more or less, free to do what he wants to do with his time. The common expression, "I did not have time to..." is an inaccurate statement. The true statement would be, "I did not take time to...," more truly, "I preferred to do... rather than to do..."

This fundamental nature of time is important to an understanding of the concept of value of time. However, time itself is not valuable. It cannot be purchased, sold, or bartered, because it is not exchangeable. Rather we buy and sell what is accomplished over a span of time, or gain satisfaction or dissatisfaction with what we accomplish over time. It should be easily realized that the "value of time" is a substitute expression for the "value of the products produced, or services gained during the passage of time." Each person uses exactly the same amount of time each day and every year. There is no way not to use time. Time is always and continuously being used. It is never wasted (although opportunities may be). Even waiting an hour for a late airplane is not wasting time. That hour is used for a purpose and was necessary to accomplish the objective of taking the airplane trip. True, we often use time at a low-value activity but we do use the time—never save it, never waste it. We just continually keep changing our activity during time.

Time, in terms of travel, is consumed—utilized—in getting from place A to place B. When a trip is made in less travel time than before, no time is actually "saved," even though that is the popular concept. Any difference in total travel time for the two trips was used merely in different occupations of time, either before starting the trip or after arriving at the destination.

In this concept is found the key to the value of travel time, particularly for passenger car travel. If passenger car travelers could price their use of time just before departure and just after arrival, a base could be established for placing values (there are many) upon passenger car travel time.

To want to "get there in a hurry" is a fundamental of human nature. It is the uncertainties, the expected future satisfactions, the curiosities, and the desire to accomplish things that drive people into the future and away from the past. And true as it is with all values, the value of time lies in the future; so man will always want to "get there in a hurry," often with little regard as to what the trip costs, or its relative importance.

Man continues to endeavor to reduce the amount of time required to accomplish a certain objective, such as travel, so he will have more time for some other activity. But when he is trying to reduce his time consumption on his present activity, he generally has in mind no other specific activity to take up next. This feature of time is an important obstacle in trying to value travel time.

TIME AND HIGHWAY TRAVEL

On highways, travelers often desire to use the route of least travel time, even at increased distance and increased running cost of the vehicle. Since the beginning of transportation, man has endeavored to increase the speeds of the different modes of travel (except walking). In addition, he has developed new modes and new power sources–sailing ship, stage coach, pony express, steamship, steam railroad, motor vehicle, airplane, and rocket. In the group of common passenger carriers–steamship, railroad, and airline–there has been in vogue from time to time extra-fast services as compared to normal speeds for which premium or extra fares are charged. The success of these extra-fare carriers and the fact that motor vehicle drivers will travel extra distance to reduce travel time is proof enough that, to the traveling public, travel time is valuable.

But how valuable is time? The value of time depends upon many factors, each of variable importance depending upon the person, the circumstance, the amount of time available, the reliability of having the time to use, and many other factors. To a family out for a short after-dinner drive to enjoy the cool of a midsummer evening, time has practically no value–that is they would pay nothing to have the present travel time between A and B reduced. But the same family, after a somewhat late start when driving to a relative's wedding, would pay a fair amount to shorten the travel time of their trip by 10 minutes. In the daily home-to-work and work-to-home movement, the workers may value their home-to-work travel time higher than the work-to-home trip, largely because of the fixed time of arrival at work, fear of penalty if late, and the desire to sleep as late as they can each morning.

In the analyses for economy of highway improvements, travel time is placed in the primary (direct) consequence group, as a market item. Time is identifiable, it has proved value, and as time it is easily measured. But travel time is not easily or soundly priced in the market. A compelling reason for including the value of time in the analysis for economy of highway improvements is the fact that the running cost of motor vehicles is directly affected by the speed of the vehicle. Fuel, tire, and oil consumption rates have positive variations with the vehicle speed. Passenger cars may consume 0.0496 gallons of gasoline per mile at 50 miles per hour and 0.07 gallons per mile at 70 miles per hour. Thus when passenger cars on four-lane rural divided access-controlled highways travel 65 to 75 miles an hour they may be doing so at 30 to 40 percent more cost for gasoline alone than when traveling certain slower-speed rural highways. The running cost of the vehicle per mile and the speed of the vehicle each contribute separately to the end result in the analysis for economy. Running cost is a direct factor of cash cost and speed is a direct factor in travel time.

The foregoing discussion is based mainly upon the family type of passenger car. When a car is used directly in business by salesmen, servicemen, taxicab service, and for professional calls, its travel time has an added element of value. In such cases the earning rate of the driver and passengers is a vital fact in determining the value of travel time.

This is also true of commercial vehicles. Cargo hauling is a business. Travel time in cargo hauling is a factor related to the investment in equipment, the investment in cargo, the time value of cargo, and the driver's wages and allowances. With commercial vehicles there exists a more firm basis of evaluating travel time than prevails with the passenger car. Commercial vehicles operate at direct cash outlays and direct cash incomes–all for a profit–so these cash flows may be used as a basis of calculating the value of travel time.

PASSENGER-CAR TRAVEL TIME

Passenger-car travel time is perhaps properly evaluated in the marketplace by determining what price people will or do pay for time. An understanding to this approach is aided somewhat by a discussion of the factors involved and their extent of involvement.

FACTORS RELATED TO THE VALUE OF TRAVEL TIME

The opening statement that time has value because of the economic goods, services, and satisfactions that can be produced in a given amount of time provides a basis for evaluating time. It is not easy, however, to determine what could be or would be produced by the passenger car occupants of a day's traffic over a specific highway if each vehicle thereon were to experience a travel time reduction of 5 minutes or 30 minutes per trip.

Some of the factors pertinent to the value of passenger-car travel time are:

A. Persons in automobile
   Ages, number, occupations, wage earnings, whether paid during time of travel
B. The trip
   Distance, number of stops, purpose (business, pleasure, etc.), regularity and frequency, total travel time, who pays the cost of trip
C. Environmental
   Day of week, hour of day, season of year, local land use, legal speed limit, rural or urban area, speed of travel, traffic volume and composition, type and design of highway.
D. Factors of value
   Activity just before starting trip, activity at end of trip, amount of time available consecutively, amount of total time (hours, minutes), hour of day that trip begins and ends, place that time may be utilized, productive time (work output), reliability of required travel time each trip, utilization of travel time decrease, value of "do-it-yourself" work, value of "time delayed" when delayed, value of leisure time, wages and earnings generally

Many factors affecting the value of travel time vary in value with each person and each trip and with the same person and his trips. Therefore, to establish any general values of travel time calls for sound reasoning and judgment.

Lacking the means of measuring increased production by passenger car travelers with a decrease in travel time of passenger cars, resort could be had to what people are willing to pay for faster travel speeds–shorter trip travel time. But here again there is no certainty that only time is being priced. There may be scenic values, items of comfort and convenience, lessened annoyances, or other personal reasons why a faster travel route is preferred to a slower one.

VALUE OF PASSENGER CAR TRAVEL TIME

There have been but few attempts to determine the value of passenger car travel time by sound statistical or market-based methods. Earlier studies involved ferry crossings compared to toll bridges or tunnels. More recent work has been related to toll highways compared to nontoll highways. In general, though, in the past the value of passenger car travel time was largely assumed, but based to some extent on the prevailing wage rate.

Toll Road versus Free Road Studies by Claffey

One good attempt to determine what highway travelers on toll highways were paying for travel time is reported by Claffey [11-2, pp. 14-22]. Claffey measured travel time, fuel consumption, and speed changes in driving sections of toll highways and parellel free highways between the same termini. Origin and destination studies were made at the same time to determine the number of drivers that chose each route–the toll highway and the parallel nontoll highway.

His mathematical analysis included the factors of running cost, accident cost, toll charge, running time, speed change, and percentage of drivers electing to use the toll highway. A speed change unit is a plus or minus change in speed of 1 mph. In the analysis, only speed changes of 3 mph and more were included. Claffey's data and the particular solution yielded a time value of 2.365 cents per vehicle-minute plus or minus 0.59 cent. The value of a speed change unit (1 mph change) was 0.048 cent plus or minus 0.062 cent. The value of a speed change unit is obviously not reliable. The speed change unit was adopted as a means of measuring the relative comfort of the drive, or one measure of impedance. The 2.365 cents per car-minute reduces to $1.42 a car-hour, or $0.89 per person-hour, on the basis of 1.6 persons per car, based on 1959 prices.

The study by Claffey is good, despite some mathematical and field data uncertainties, because it uses the marketplace as a means of pricing the traveler's time. What the traveler will pay to reduce travel time seems to be an accepted approach to the value of passenger car travel time, but so far no one has devised a way to price the traveler's willingness to pay. The Claffey study measured the price of time paid under the one condition of intercity toll highway use, but not the willingness to pay. Further, the highly important urban travel, including the home-work, work-home movement was not included.

The Thomas Study of Commuter Trips

In 1967 Haney [11-9] and Thomas [11-15] reported the final results of their work on the value of passenger-car travel time. Haney's report is especially recommended as a source of fundamental factors and concepts and references to the literature. Thomas developed a mathematical model for the value of passenger-car travel time based on field studies of commuter travel between home and workplace wherein there was opportunity to travel either a toll highway or a nontoll urban route. This approach is similar to that used by Claffey.

The report by Thomas [11-15] is important for two main results. First, he develops a reliable and mathematically sound procedure for evaluating passenger car travel time where there exists the alternatives of using a toll road or a free road for the same trip, and, second, he arrives at a value of time of $2.82 per person per hour for the commuter work trip.

It is strongly emphasized that the $2.82 is a highly specialized finding and not a generalized value. It is applicable only to the following conditions: (1) commuter work trips by employed persons, (2) work trips of more than 10 minutes and 5 miles in duration, (3) annual wage incomes of $6,000 to $8,000, (4) no separate value should be assigned to traffic impedances, and (5) the computed value of time reduction is not to be added to computed reduction in car running costs.

Although Thomas points out that the commuter is less sensitive to small amounts of reduction in travel time (such as 6 minutes or so) than to greater amounts no specific relation of value of time to total time reduction is given because of lack of sufficient statistical data.

This groundwork by Thomas now provides a satisfactory framework and procedure for additional field studies whereby the value of passenger car travel time can be measured for many other trip purposes, highway types, geographical areas, and times of the day.

The Lisco Study of Commuter Trips

The work of Thomas [11-15] is verified by the independent highly similar work of Lisco in 1967 [11-11]. Lisco, by comparing passenger car driving to travel by rail rapid transit from Skokie to the Loop in Chicago, evaluated commuter travel time at $2.50 to $2.70 per hour per person. Lisco used commuter interview data, field driving tests, and calculated travel costs when the car is used. His model is based on multiple probit regression analysis. The factors include (1) the travel time differential, (2) the driving cost differential, (3) two factors related to whether the family spouse works and drives to work, (4) a household income restraint, (5) sex variable, (6) age variable, and (7) choice of mode (car or rail transit).

In addition to the $2.50 to $2.70 value of commuter travel time Lisco

Road-User Consequences

computed a value of $1.50 to $2.50 a day for the added comfort and convenience (and satisfaction or personal preference) of driving by car over riding the rapid transit. The total trip distance is about 15 miles. The car driving time equivalent to the origin and destination of the rapid-transit trips ends, but excluding the travel by car or walking at both trip ends, varied from 18 to 42 minutes per trip, depending upon the clock time as related to the peak flow of traffic.

Another useful finding is that in the Chicago Loop area, the commuters are willing to pay about 12 cents a minute to reduce walking time from parking location to place of work. This 12 cents is composed of about 8 cents to avoid the discomfort and inconvenience of walking. On this basis, the commuter would be evaluating peak hour walking time at $2.40 per hour and the discomfort of walking at $4.80 per hour.

As with the Thomas results, these Lisco findings are restricted to peak-hour commuter travel. Lisco's work is for commuters having an average annual income of about $8,000.

Generalized Values of Passenger-Car Travel Time

The value of travel time for passenger cars usually has been assumed by each analyst of economy studies according to his own (or borrowed) idea. Usually, the basis has been the prevailing wage for a semiskilled worker, adjusted to suit the analyst's thinking of what was appropriate.

Haikalis and Hyman [11-6, p. 55] arrive at $1.17 per car-hour and $0.75 per passenger as the value of passenger-car travel time for travel in the Chicago area. Their approach was simple and typical of that used by many writers in the past. Starting with the federal minimum wage of $1.00 per hour, they reduced it to $0.75 on the basis that some car passengers are unemployed, then multiplied by 1.56 persons per car to produce $1.17 per car-hour.

Some reports have adjusted the basic hourly rate by the average number of persons per car, say, 1.6 persons, others have not. But in each case the rate per hour, once chosen, has been applied to all cars, for all hours of the day, and all days of the year, and all amounts of travel time. Certainly, some attention should be given to developing an overall rate for a particular highway section that would apply justly to each of the 8,760 hours of the year, giving weight to the classes of persons traveling, the purposes of the trips, the amount of travel time reduction by those persons actually benefiting. The value of time will always vary, person to person, place to place, and situation to situation.

There is no reason to use the same value for passenger-car travel time for all proposed highway projects subject to analysis. In fact, good judgment would call for varying the value of time to fit the type, location, and characteristics of the traffic. Reasonable values, at least according to the judgments of recent analysts, lie within the range of $1.00 to $4.00 per car-hour, depending upon the prevailing local factors.

RELATION OF CAR RUNNING COST TO TRAVEL TIME

Some light on the value of passenger-car travel time may be had by examining what it costs per hour to drive on level tangent highways at uniform speed. From Table A-1 the total running cost per mile may be converted to running cost per highway travel hour. Table 11-1, column 3, gives the results. In this table the fuel tax at 11 cents per gallon (See Table A-26 for consumption rates) has been added in to get the total of on-the-road cost, omitting all items of cost not directly affected by miles driven.

From Table 11-1 it may be observed that the car running cost varies from $0.52 an hour at 10 mph to $5.14 at 80 mph. Few persons are aware of these costs per hour and perhaps would do nothing to change their car use or driving if they did know the cost per hour, or for that matter their cost per mile.

Column 4, in Table 11-1, however, tells a different story. Starting at 35 mph the running cost of a car begins to increase in cents per mile as speed increases. Consequently, as uniform speed increases a running cost increment per mile is created for each one mile an hour of increase in speed. Offsetting this increased cost with speed is the benefit of reduced travel time. When the increment of running cost is related to the decrement in travel time the cost per hour of travel time reduction may be computed. This cost per hour of time reduction varies upward from $0.07 at 36 mph to $9.75 at 80 mph. At 80 mph the cost increase in cents per mile over driving 79 mph is 0.154 cent, and the travel time reduction is 0.00158 hour for the one mile.

At a cost of $9.75 an hour few persons would be willing to pay the price to travel 80 mph. But the public is not aware of this cost because they do not pay out the cash on a trip basis. Even at 65 mph the cost of reduced travel time is $2.69, and it is over $1 an hour for all speeds of 55 mph and above. In actual driving, the cost is probably in excess of what is given in Table 11-1, because of the speed changes—accelerating and decelerating—and horizontal movement of the vehicle path.

Table 11-1 gives realistic costs of time to the car driver, but it does not express his value of time—either minimum or maximum. Although, in fact he is paying the price of $2.69 per hour at 65 mph for travel time reduction as compared to 64 mph, he is not aware that he is doing so. It is not an open across-the-table market price. In economy studies, a price that is more knowingly paid for travel time should be used.

OTHER TIME FACTORS TO CONSIDER

In addition to placing a unit value on travel time, the analyst also must determine what specific time is to be valued. Perhaps the major consideration is to decide whether all travel time reductions by the same class of vehicle on the same highway facility have the same value in dollars per hour. For instance, does 1,200 seconds total time reduction by 120 cars at 10 seconds each have the same value as 1,200 seconds of travel time reduction by one car on one trip? Does a

Road-User Consequences

reduction of 1 minute each on 10 trips have the same value as a re[duction in] time of 10 minutes on 1 trip?

TABLE 11-1

PASSENGER CAR RUNNING COST INCLUDING FUEL TAX PER HOUR OF TRAVEL TIME AND PER HOUR OF REDUCED TRAVEL TIME AT NEXT LOWER INTEGRAL SPEED

Level Tangent Uniform Speed, mph	Running Cost		Incremental Cost of Reduced Travel Time over Next Lower Integral Speed, Dollars per Hour	Level Tangent Uniform Speed, mph	Running Cost		Incremental Cost of Reduced Travel Time over Next Lower Integral Speed, Dollars per Hour
	Cents per Car-Mile	Dollars per Car-Hour			Cents per Car-Mile	Dollars per Car-Hour	
1	2	3	4	1	2	3	4
10	5.189	0.52					
15	4.534	0.68		56	4.397	2.46	1.19
20	4.217	0.84		57	4.438	2.53	1.30
25	4.042	1.01		58	4.481	2.60	1.42
30	3.962	1.19		59	4.527	2.67	1.58
35	3.952	1.38		60	4.576	2.75	1.73
36	3.957	1.42	0.07	61	4.628	2.82	1.90
37	3.964	1.47	0.10	62	4.683	2.90	2.08
38	3.972	1.51	0.13	63	4.741	2.99	2.27
39	3.983	1.55	0.16	64	4.802	3.07	2.48
40	3.995	1.60	0.19	65	4.867	3.16	2.69
41	4.008	1.64	0.22	66	4.935	3.26	2.92
42	4.023	1.69	0.26	67	5.007	3.35	3.17
43	4.040	1.74	0.30	68	5.083	3.46	3.44
44	4.058	1.79	0.34	69	5.163	3.56	3.74
45	4.078	1.84	0.38	70	5.247	3.67	4.06
46	4.099	1.89	0.43	71	5.336	3.79	4.41
47	4.121	1.94	0.48	72	5.430	3.91	4.81
48	4.145	1.99	0.54	73	5.530	4.04	5.24
49	4.170	2.04	0.60	74	5.636	4.17	5.71
50	4.197	2.10	0.66	75	5.748	4.31	6.23
51	4.226	2.16	0.73	76	5.867	4.46	6.80
52	4.257	2.14	0.81	77	5.994	4.62	7.42
53	4.289	2.27	0.88	78	6.129	4.78	8.12
54	4.323	2.33	0.97	79	6.273	4.96	8.90
55	4.359	2.40	1.07	80	6.427	5.14	9.75

Road-User Consequences

These questions probably will be answered in time, but until then each analyst of the economy of transportation facilities must find his own answer. Perhaps a satisfactory practice would be to use the same value of time in dollars per hour for all time involved in the specific analysis, but to vary this unit value of time project to project on the basis of how usable the time might be. The unit value of time should be appropriate to the purpose of the travel, the income level of the travelers, and other factors of value. Like selecting the vestcharge rate appropriate to the improvement involved, the value of passenger car travel time (and commercial travel time) should be adapted to the local conditions and the specific proposal for highway improvement.

If drivers were to pay for their car running cost mile per mile as driven, no doubt they would change their relative values of time, distance, comfort, and convenience. The nature of car operation is such that the owner does not associate its total expense with its specific use, as he does with many daily purchases of food, entertainment, or trip by common carrier.

Thus studies of what car owners will pay to reduce travel time and to gain comfort, convenience, or uniform travel speed may not produce the comparable values as would be produced if all aspects and factors of the travel could be simultaneously priced. The choice of alternatives would no doubt be different under such a pricing system.

COMMERCIAL VEHICLE TRAVEL TIME

Establishing a unit value of travel time for commercial vehicles–commercial and industrial haulage–may be approached on a much firmer basis than is possible for passenger-car travel. Commercial travel is by paid drivers and the owners of the vehicles have reliable records of their operating costs. The factors of value are fewer and less complex also.

BASIS OF VALUE OF COMMERCIAL VEHICLE TRAVEL TIME

The value of travel time of commercial transport vehicles generally has been assumed to be equivalent to the wage rate of the vehicle driver. This assumption omits consideration of the vehicle and its cargo and the profit objective of the owners thereof. Since commercial motor vehicles are used by profit-making concerns as a part of their normal operations, and since the investment in the vehicles is also expected to earn its fair share of the overall return, it is logical to consider that the value of travel time of commercial vehicles includes cost elements other than the drivers' wages. This viewpoint was used by the authors in Refs. 11-1, 11-10, and 11-16.

VALUE OF COMMERCIAL TRAVEL TIME

The dollar per hour values of reductions in commercial vehicle travel time in

Road-User Consequences

Table 11-2 are based on the work of Adkins, Ward, and McFarland [11-1]. The cost elements included are those that would be affected by a change in travel time. This cost-reduction approach to the value of commercial travel time reductions was chosen after extensive analyses of the concepts of (1) cost of time in terms of highway costs necessary to produce the reduction in travel time, (2) willingness to pay based on the market price, and (3) revenue gain based upon the assumption that travel time reductions will result in increases in gross and net operating revenue. The concept that any reductions in transport costs are a fair measure of the value of travel time reductions was chosen in preference to the other three on the basis that travel time reductions will lead to resource reductions and that reductions in the consumption of resources will reflect the value of travel time. However, there is uncertainty in determining what resources may be conserved and how much of each may be saved by reductions in travel time. The resource elements included by the authors [11-1] in their value of travel time reductions are described next.

Return on Investment

Faster travel should result in the need for fewer transport vehicles to haul a given quantity of goods. Therefore the investment in equipment should be less when travel times are reduced than before. This reduction in investment and its corollary reduction in return on investment is over and above any reduction in depreciation expense that may be affected.

Depreciation

The cost of depreciation (the cost of consumption of capital goods) is affected by travel time because the depreciation rate applied to the transport vehicles is a resultant of hours used, miles operated, calendar time, and other factors in the service life of the vehicles. If travel time can be reduced, and if more uniform speeds can be achieved, it is most likely that a given vehicle will be used more hours or operated more miles during its useful life than it would be at greater trip travel times. It follows then, that the depreciation expense per unit of service will be less, since the original cost less trade-in value would be prorated over a greater number of units of service.

Property Tax

A consequence of reduction in capital investment in equipment would logically lead to a reduction in personal property tax. But such reduction in investment leads to a reduction in the tax base of the taxing community. If the budget of the taxing authority is held constant, there would be an increase in the tax rate. The full cycle of events is that the amount of tax reduction which results from removing from the tax rolls some of the transport vehicles is recouped by spreading that tax reduction over all remaining property in the tax base. For

TABLE 11-2
DOLLAR VALUE PER HOUR OF TRAVEL TIME FOR COMMERCIAL VEHICLES, 1965

Line No.	Vehicle Class	Interstate Commerce Commission Regions								
		New England	Middle Atlantic	Southern	Central	North-western	Mid-western	South-western	Rocky Mountain	Pacific
	A. Gasoline Power									
1	2A (4-tire) panel, pickup, etc.	3.73	3.77	3.94	4.00	4.19	3.97	4.06	3.65	3.89
2	2D (6-tire) van	3.95	3.99	4.16	4.23	4.43	4.20	4.28	3.86	4.11
3	3A van	4.37	4.42	4.66	4.65	5.14	4.73	4.94	4.26	4.55
4	2-S1 van	4.46	4.57	5.03	4.67	5.47	5.04	5.53	4.77	5.07
5	2-S1 tank	4.73	4.84	5.35	4.94	5.93	5.37	5.95	5.03	5.36
6	2-S2 van	5.04	5.16	5.71	5.26	6.41	5.75	6.40	5.34	5.70
7	2-S2 tank	5.31	5.44	6.04	5.53	6.87	6.10	6.83	5.61	5.98
8	3-S2 van	5.20	5.27	5.84	5.37	6.57	5.89	6.56	5.46	5.82
9	3-S2 tank	5.47	5.55	6.17	5.65	7.03	6.22	6.99	5.72	6.10

Road-User Consequences

	B. Diesel Power								
10	2D (6-tire) van	4.06	4.11	4.30	4.34	4.62	4.34	4.46	3.97	4.23
11	3A tank	5.00	5.07	5.41	5.29	6.22	5.53	5.94	4.89	5.22
12	2-S2 van	5.13	5.25	5.82	5.35	6.56	5.87	6.54	5.43	5.79
13	2-S2 tank	5.40	5.53	6.14	5.62	7.02	6.21	6.97	5.70	6.08
14	3-S2 van	5.58	5.67	6.38	5.76	7.22	6.37	7.16	5.83	6.22
15	3-S2 tank	5.85	5.94	6.71	6.04	7.68	6.71	7.59	6.10	6.50
16	3-2 van	—	—	—	—	—	—	—	5.79	6.18
17	3-2 tank	—	—	—	—	—	—	—	5.96	6.36
18	2-S1-2 van	—	—	—	—	—	—	—	6.58	7.02
19	3-S2-4	—	—	—	—	—	—	—	7.78	8.29
20	Composite, lines 2 to 19	4.86	5.16	5.45	5.39	6.11	5.62	6.56	5.16	5.75
	C. Bus								
21	Intercity	4.97	4.97	6.96	7.43	6.77	6.77	7.43	6.23	6.23
22	Local and suburban	3.81	3.81	5.33	5.69	5.17	5.17	5.68	4.77	4.77

States assigned to the Interstate Commerce Commission Motor Carrier Regions:

New England: Connecticut, Maine, Massachusetts, New Hampshire, Rhode Island, Vermont.
Middle Atlantic: Delaware, District of Columbia, Maryland, New Jersey, New York, Pennsylvania, West Virginia.
Southern: Alabama, Florida, Georgia, Kentucky, Mississippi, North Carolina, South Carolina, Tennessee, Virginia.
Central: Illinois, Indiana, Michigan (lower peninsula), Ohio.
Northwestern: Michigan (upper peninsula), Minnesota, North Dakota, South Dakota, Wisconsin.
Midwestern: Iowa, Kansas, Missouri, Nebraska.
Southwestern: Arkansas, Louisiana, Oklahoma, Texas.
Rocky Mountain: Colorado, Idaho, Montana, New Mexico, Utah, Wyoming.
Pacific: Arizona, California, Nevada, Oregon, Washington.

Source: Ref. 11-1, Tables 25 and 26, except lines 1 and 22 which are by author.

the highway transport industry, its increase on its remaining taxable property would be a minute share of the grand total spread to other properties. There would still be a sizable net tax reduction effected by the reduction in the investment in transport vehicles.

Drivers' Wages

It is logical to assume that any reduction in travel time would automatically result in a reduction in the total wage paid to the drivers. But such assumption has its uncertainties. Many for-hire vehicles are operated by drivers paid on a mileage basis, so unless the mileage is reduced there would be no reduction in driver wage cost because of a reduction in travel time. In the long run, however, the driver wage rates do take recognition of reductions in travel time and other consequences of highway improvements [11-5]. For the drivers paid solely on a time basis, not all of the time reductions could be fully utilized because of trip scheduling requirements. Here again there is good expectation in the long run that any reductions in travel time will result in less cost of drivers wages per ton-mile of payload haulage. The owners of transport companies welcome faster travel speeds and travel time reductions, so they must attach some business benefit to such reductions, including drivers' wages.

Drivers' Nonwage Compensations

Employee welfare compensations of (1) health, life insurance, and sick leave, (2) workmen's compensation, and (3) social security taxes are three groups of costs associated with transport operations directly related to vehicle driver wages. Since it is concluded that travel time reductions will reduce the cost of driver wages, it follows that the costs of these three groups of nondriver wage compensations would also be reduced by reductions in highway travel time.

HIGHWAY COST TO REDUCE PASSENGER-CAR TRAVEL TIME

Haney [11-8] and Curry [11-4] suggest that instead of finding only the value of travel time, the highway cost required to reduce travel time also be determined. This approach offers two advantages. First, by calculating the highway cost per vehicle-hour of travel time reduction for the projects considered in a construction program, a sound basis of priority ranking could be achieved without setting a value on travel time. Second, the analysis of the highway cost of travel time reduction would set the minimum value of time for use in economy studies. That project having the highest highway cost of time reduction of those selected on an incremental basis for a year's construction program would provide the measure of the minimum value of time.

Just for purposes of discussion, assume that a state highway department's current construction program includes projects for which the passenger-car

Road-User Consequences 277

travel time reduction in terms of the highway cost to produce the travel time reduction ranges upward to the high-cost project of $1.10 a passenger car-hour of travel time reduction. This $1.10 in effect becomes the minimum value of time on the market, for the reason that the public is buying travel time at that maximum price. But what the passenger-car traveling public would be willing to pay for an hour of travel time reduction could be substantially in excess of the $1.10.

The highway cost to reduce travel time is a factual solution that any highway department can make for a year's construction or for a series of years. This highway cost of time reduction so determined could then be used as the value of time in the analyses of economy. This procedure has material advantage over assuming a value of time, as has been done by most analysts. Certainly, decisions for adopting or rejecting proposals for highway improvements on the basis of the highway cost of time reduction used as the minimum value of time would be sound. This highway cost of travel time reduction so determined is the market price of time being paid then by the public through its highway department's improvement program.

DETERMINING THE HIGHWAY COST OF PASSENGER-CAR TRAVEL TIME

In terms of highway cost, the cost of passenger-car travel time is determined on the basis of the net cost of the highway improvement after subtracting from the highway cost all of the road-user benefits, including the commercial travel time values. The computation of the equivalent uniform annual highway cost proceeds in the normal sequence as for any analysis for economy. From this equivalent uniform annual highway cost is subtracted the equivalent uniform annual motor vehicle running cost net benefits, including accidents and commercial vehicle travel time cost reductions. This net remainder of the equivalent uniform highway cost is then divided by the equivalent uniform annual hours of reduction in passenger-car travel time for the same analysis period used in the highway cost and motor vehicle cost computations.

It follows that the highway cost of passenger-car travel time reduction will be negative for those projects in which the total dollar value of net benefits from motor vehicle running costs, accidents, and commercial vehicle travel time reductions exceed the equivalent uniform annual highway cost. In certain other types of improvements the running cost of motor vehicles is increased because of the higher uniform speed of travel. This increase in speed results in increased motor vehicle running costs, so in effect increases the highway cost of passenger-car travel time.

PERSONAL-PREFERENCE SATISFACTIONS

Man has long sought those methods of performing work and pleasure that produce the maximum of comfort, convenience, and satisfaction. He seeks body and mental comfort, directness, and uniformity. Applied to riding on roads, streets, and highways, the seeking of comfort and convenience means developing a ride which is free of physical and mental disturbances and annoyances. This condition can be partially produced by holding to a minimum the body motions and deliberate acts necessary to adapt the vehicle to the requirements of the highway and traffic, and by driving at uniform speed.

FACTORS OF PERSONAL PREFERENCE

Each driver and passenger has his own preference and evaluation of what is desirable in comfort and convenience, and what constitutes annoyances and impedances. Nevertheless, a listing of possible conditions, events, and consequences will help to clarify the meaning of these terms.

A. Items of physical comfort
   Minimum of steering effort–easy curves, no corner turns, no continuous series of reverse curves, no passing of other vehicles, no lane changes
   Eye comfort–no headlight glare, good lighting at intersections, well-painted pavement markings
   Minimum of accelerating and decelerating–no foot work on throttle or brake, no dynamic forces to resist–longitudinal or lateral
   No necessity to shift gears on grades or because of traffic
   Pavement smoothness–no bumps, dips, waves, vibrations, and no surface roughness
   Relaxed driving–no need of continuous tight grip on steering wheel
   No gas fumes, noise, dust, or shocks
B. Items of mental comfort
   Distant view of landscape, pleasing roadside development, no outdoor advertising
   Quietness–no tire-pavement noise, no air suction noise from passing close to fixed objects on shoulder or to traffic
   Minimum number of decisions affecting steering and speed of car
   Relaxed driving
   Good signing–no anxiousness about where to turn off or routing
   Freedom to set own speed without interfering with other vehicles
   Reliable travel time to destination
   Uniformity of geometric design, traffic regulations, and signing
C. Items of convenience
   No shifting of gears, minimum of steering effort
   Directness of travel to destination–no adverse distance
   Minimum changes in route numbers to follow
   Uniformity of travel direction–minimum changes of direction

Road-User Consequences 279

      Freedom of movement, choice of speed, and choice of route
      Close-at-hand motor vehicle and personal services
D. Factors producing strain, tension, and discomfort
      Recognizable accident-potential situations
      Requirement to watch other vehicles continuously to prevent accidents
      Uncertainty of behavior of other vehicles
      Uncertainty of safety to pass
      Continuous intense alertness in traffic
      Near-accidents, unexpected highway or traffic situations
      Changes in standards of design–speed, pavement width, number of lanes
      Narrow pavements, shoulders, and bridges; fixed objects close to roadway
E. Items of impedance to uniform speed
      Requirement to trail a slow-moving vehicle ahead
      Required stops and slowdowns
      High traffic volume per lane
      Traffic access from roadside property
      Necessity to slow down and to speed up to keep proper and safe position in traffic stream

Regardless of how classified and of any resulting overlapping, the above items associated with highway travel sum up the subjective evaluations of whether a given trip was pleasant, enjoyable, satisfying, and according to desires, or whether the trip was annoying, fatiguing, tiresome, disappointing, and one producing physical strain and mental anguish. These items are among those in mind when highway travelers refer to comfort, convenience, strain, tension, and impedance, and are what they often are willing to pay for or to avoid.

PERSONAL PREFERENCES, THE VEHICLE, THE HIGHWAY, AND ECONOMY

These personal preference desires of the driver and passengers have led to many changes in automobile design to produce a more comfortable ride. Such items as automatic transmissions, air conditioning, chassis springs, tires, tire air pressure, individual wheel suspension, power brakes, power steering, power window controls, seat cushions, and soundproofing are built into motor cars for the express purpose of increasing comfort and convenience. These items are costly. Car purchasers seem to desire them and pay the price accordingly.

Likewise with the highway. Highway designs can provide for a smooth pavement, properly superelevated horizontal curves, grades that can be ascended in high gear, high-speed passing distances, wide pavements, wide shoulders, wide medians, and other features that hold to the minimum the efforts required in steering, throttling, braking, and gearing the car. These same provisions of highway design and traffic control keep the number of decisions on the part of the driver to a minimum. Safety is also an important aspect, but in this discussion of personal preferences safety is not included, except as safe highways produce the minimum of mental strain and decision making for the driver.

Here again highways designed and constructed to produce these satisfactions require investments greater than would be needed without them. Thus the road user is paying for these highway comforts and conveniences just as he is paying for them in motor vehicle design.

But are these desirable items of comfort, convenience, and uniform speed in motor vehicle driving items of economy? Do they husband the dollar? They are not economy in the same sense as is applied to reducing the consumption of commodities and travel time. They are simply personal-preference items the driver (and passengers) often is willing to pay for and knowingly so. But they are paid for by trade-offs–spending less on other activities. These preference items, assuming that they could be reliably priced in the market, have no place in the analysis for economy–true economy, not "economics." They should be kept separate and given such weight as may be just and right by those officials who make the final decisions on the highway construction program.

But no doubt there is some minor contribution to economy from these personal preference items. This contribution could come from greater production on the job by reason of arriving at work less fatigued, more mentally alert, and in good humor. Perhaps also the effortless type of driving contributes to additional safety, but if so, such gain would be included in the accident factor of economy.

ECONOMY AND PUBLIC ADMINISTRATION

Comfort, convenience, and uniform speed as personal preferences are closely related to other similar aspects of life. They are also tied to personal pride, prestige, social customs, and "keeping up with the Joneses." For all ordinary driving is there transportation economy in owning and operating Cadillacs, Imperials, and Lincolns as opposed to Chevrolets, Fords, and Plymouths? Is there economy in power lawn mowers for a homestead maintaining only 7,000 sq ft of lawn? Is there economy in owning and wearning a mink coat or a $200 suit of men's clothing; or in a $15 restaurant dinner? Real economy (husbanding the dollar) is not to be found in any of these daily market actions, yet these actions exist and will continue to prevail. People desire to express themselves and to be satisfied in their own way. Therefore, highways and motor vehicles will continue to develop toward satisfying human personal preferences, all at a cost, but with little economy on some factors.

In administration of public funds a distinction is desirable between the price people will pay for a service or commodity and the price that represents economy. In business, profit is determined as the difference between all costs and sales income. The business is unconcerned as to economy on the part of the customer, so a businessman prices his goods to maximize his profit. But in public works, pricing to produce sales or calling expensive comfort an economy, is not good policy. The people should be given the truth, and allowed to make their choice as between spending for comfortable highways and spending for other items of their choice.

A family has limited income and limited spending power. What is spent for highways, high-priced comfort, and for convenience is not available for other purposes, but it is spent nevertheless. The reason that there is no economy produced in satisfying one's personal preference for comfort and convenience is that there is no savings of commodities, or money, or resources as a result. The consequence is simply a shifting of one's economic support from one activity to another.

A water-resource development cannot be justified through economy on the grounds that it will attract 10,000 man-days of fishermen per year who will spend $25 a day. Fishing may be good use of leisure time and highly satisfying to those who fish. But when a public agency uses fishing (and other water sports) as an item of economy in proving worth of a water resource project, such agency is wrong. The truth is simply that the water-resource project will attract those persons from golfing, baseball, and other leisure time activities. There is just a transfer of one's economic support from other activities to water sports according to individual preferences, so what the water-resource project gains some other economic activity loses.

And with highways, the public is entitled to the type of highway it wants to pay for, including comfort and convenience items, but the highway administrator should not treat such items as items of economy or call them "savings." Comforts and conveniences are simply personal preference factors that redistribute economic activity but produce no identifiable direct economy in the use of the dollar.

PROBLEMS

11-1. Discuss the relative merits in highway economy analyses of assigning dollar values to travel time for the following specific applications:
1. Instances where the total trip distance may reduce travel time 3 minutes or less per trip
2. Daytime and nighttime recreational travel
3. Travel by children and retired persons
4. Daytime travel by housewives on family functions
5. A scenic route, mostly summer use
6. A mountain route where both the summer and winter ADT are 75 percent vacation and sports travel
7. Driving itself as a recreation
8. Home-work-home trips

11-2. Keep a record of your driving (or riding by car) for one week noting for each trip your choice of main route and any changes in routing during the trip. Record the reasons, conditions, assumptions, and factors on which your choice of route or change in routing were based. Discuss your results with particular attention to travel time, comfort, convenience, environment, and motor vehicle costs.

11-3. For your personal and family driving or riding for a typical week by trip purpose, state what rate in dollars per hour for your own time you would have been willing to pay in cash for the following amount of travel time reduction per one-way trip: (a) 2 minutes, (b) 5 minutes, (c) 10 minutes, (d) 20 minutes, and (e) 30 minutes.

11-4. The road-user benefit from a highway project has been estimated to be $1,000 a year. Calculate the present worth of this annual benefit at a vestcharge rate of 3 percent for 5, 10, 20, 40, 100, and infinity years. Repeat the calculation at 8 percent vestcharge rate. Draw two curves of results and comment upon them.

11-5. A certain toll highway can be driven a distance of 22 miles at an average speed of 64 mph for a toll of 35 cents. The parallel free primary route of approximately the same distance and origin and destination, may be driven at an average speed of 50 mph. If the car running costs are 4.80 and 4.04 cents per mile, toll and free routes, respectively, what is the net cost per vehicle-hour (toll plus vehicle costs) for the reduced travel time?

REFERENCES

11-1. William G. Adkins, Allen W. Ward, and William F. McFarland. *Evaluation of Time Savings of Commercial Highway Vehicles.* National Cooperative Highway Research Report No. 33, Highway Research Board, National Academy of Sciences, Washington, D.C., 1967.

11-2. Paul J. Claffey. *Characteristics of Passenger Car Travel on Toll Roads and Comparable Free Roads.* Highway Research Board, National Academy of Sciences, Washington, D.C., Bulletin 306, *Studies in Highway Engineering Economy,* 1961, pp. 1-22.

11-3. Marion Clawson and Jack L. Knetsch. *Economics of Outdoor Recreation.* Published for *Resources for the Future* by Johns Hopkins Press, Baltimore, 1966.

11-4. David A. Curry. *Use of the Marginal Cost of Time in Highway Economy Studies.* Highway Research Board, National Academy of Sciences, Washinton, D.C., Highway Research Record No. 77, 1963.

11-5. Gerald A. Fleischer. *The Economic Utilization of Commercial Vehicle Time Saved as the Result of Highway Improvement.* Publication No. 3, *Engineering Economic Planning,* Stanford University, Stanford, August 1962.

11-6. George Haikalis and Joseph Hyman. *Economic Evaluation of Traffic Networks.* Highway Research Board, National Academy of Sciences, Washington, D.C., Bulletin 306, *Studies in Highway Engineering Economy,* 1961, pp. 39-63.

11-7. Dan G. Haney. *The Value of Travel Time for Passenger Cars: A Preliminary Study.* A report to the U.S. Bureau of Public Roads. A good bibliography included. Stanford Research Institute, Menlo Park, Calif., January 1963.

11-8. Dan G. Haney. *The Use of Two Concepts of the Value of Time: The Willingness to Pay and the Cost of Time.* Highway Research Board, National Academy of Sciences, Washington, D.C., Highway Research Record No. 12, January 1963.

11-9. Dan G. Haney. *The Value of Time for Passenger Cars: A Theoretical Analysis and Description of Preliminary Experiments.* Washington, D.C., May 1967. Vol. I of a final report to the U.S. Bureau of Public Roads on a research contract by the Stanford Research Institute, Menlo Park, Calif.

11-10. Charles R. Haning and William F. McFarland. *Value of Time Saved to Commercial Motor Vehicles Through Use of Improved Highways: A Report to the Bureau of Public Roads.* Bulletin 23, Texas Transportation Institute, College Station, Texas, September 1963.

11-11. Thomas Edward Lisco. *The Value of Commuter's Travel Time–A Study in Urban Transportation.* A dissertation for the degree of Doctor of Philosophy, Department of Economics, University of Chicago, June 1967.

11-12. C. B. McCullough and John Beakey. *The Economics of Highway Planning.* Oregon State Highway Commission, Salem, 1937.

11-13. G. P. St. Clair and Nathan Lieder. *Evaluation of the Unit Cost of Time and of the Strain and Discomfort Cost of Nonuniform Driving.* Highway Research Board, National Academy of Sciences, Washington, D.C., Special Report No. 56, *Economic Analysis in Highway Programming, Location, and Design,* September 1959.

11-14. Hoy Stevens. *Line-Haul Trucking Costs in Relation to Vehicle Gross Weights.* Highway Research Board, National Academy of Sciences, Washington, D.C., Bulletin 301, 1961.

11-15. Thomas C. Thomas. *The Value of Time for Passenger Cars: An Experimental Study of Commuters' Values.* Washington, D.C., May 1967. Vol. II of a final report to the U.S. Bureau of Public Roads on a research contract by the Stanford Research Institute, Menlo Park, Calif.

11-16. United States Department of Commerce. Final Report of the Highway Cost Allocation Study, House Document No. 54, 87th Congress, 1st Session. U.S. Government Printing Office, Washington, D.C., 1961.

11-17. United States Federal Aviation Agency. *Feasibility of Developing Dollar Values for Increments of Time Saved by Air Travelers.* Washington, D.C. This report written by the Systems Analysis and Research Corporation, Washington, D.C., under a research contract, February 1966.

Chapter **12**

Power Performance of Motor Vehicles

An understanding of the characteristics of the internal combustion engine and transmission of its power to the drive wheels of the vehicle is an aid to understanding vehicle performance on the road. Engines and vehicles are designed to produce a specific on-the-road performance. Variations in design between vehicles result in vehicles that have low and high accelerating rates, low and high grade climbing ability, flat and steep fuel consumption curves, and extremes in other characteristics. The reason for these variations stems from the requirements for roadway performance and from differences in opinion of vehicle designers and of vehicle users on what performance is desirable. Therefore a wide range of vehicle performance and vehicle running cost is experienced on the highway.

POWER SYSTEM OF THE ENGINE

Power is generated in the internal combustion engine by the explosion of a compressed fuel/air mixture, that drives the piston away from the center of explosion. The piston, through its connecting rod, turns the crankshaft, which, through a power linkage (power train) transmits power to the drive wheels. The center of explosion is in the combustion chamber at the head of the cylinder where the fuel/air mixture is compressed at the ratio of more or less than 8:1 by volume. The ratio is variable with the engine design. Except for the diesel engine the fuel used must be suitable to the compression ratio in order to prevent ignition of the mixture by the heat generated by compressing the mixture. Gasoline-fuel engines ignite the compressed mixture with an electrical spark, but diesel engines ignite by the pressure and heat of the compression stroke. The usual gasoline engine in motor vehicles is a four-cycle design–intake, compression, power, and exhaust.

The carburetor meters the fuel in the proper ratio with air. Liquid fuel, such as gasoline, is also atomized by the carburetor so that the fuel/air mixture passed into the manifold and then through the intake valve to the cylinder is a gaseous mixture. This gaseous mixture is rich in fuel for starting, idling, and small throttle openings. Under these conditions there is a tendency for the small incoming fuel charge to the cylinder to be mixed with heavy exhaust gases. In the middle

Power Performance of Motor Vehicles 285

range of engine speed and of power requirement, a maximum economy mixture is desired, but at wide-open throttle a fuel/air mixture for maximum power is desired. The carburetor automatically meters the fuel to produce these desirable fuel/air mixtures, which are about as follows:

	Fuel/Air Ratio (By Weight)
Low engine speed and low power	0.10 to 0.12
Intermediate engine speed and power	0.05 to 0.08
Maximum power mixture	0.07 to 0.09

By use of a float chamber to maintain a constant level of liquid fuel, a Venturi tube, and fuel jets, the carburetor continuously varies the fuel/air mixture in accordance with the pressure and velocity of air at the Venturi tube. The pressure and velocity are in turn controlled by the throttle opening and "pull" produced by the piston intake stroke. From this brief explanation it may be seen why the fuel consumption is controlled by (1) engine speed and (2) throttle opening, and not by road speed of the vehicle. On plus grades the fuel consumed per mile will vary according to the steepness of grade, even though the engine speed (gear ratio) and vehicle road speed are the same for the different percentage plus grades. These characteristics of the four-cycle gasoline engine explain why traveling down a minus grade with closed throttle, regardless of road speed, the fuel consumption is at a constant rate per hour, but at a decreasing rate per mile with increase in road speed.

HORSEPOWER AND TORQUE

The theoretical horsepower of the internal combustion engine is given by the following equation:

$$\text{Horsepower} = \frac{plan}{33{,}000} \qquad (12\text{-}1)$$

where p = mean effective pressure during power stroke, pounds per square inch
l = length of stroke, feet
a = area of the piston, square inches
n = number of explosions per minute

For the four-cycle engine, the most common type of internal combustion engine used in motor vehicles, n is the product of the number of revolutions of the engine per minute (rpm) and half the number of cylinders.

It is the brake horsepower, however, rather than the theoretical horsepower that is of more concern to the user of the engine. Brake horsepower is that developed by the engine and available at the flywheel (crankshaft output) for use. It is the theoretical horsepower less the internal absorption of power occuring in the engine itself.

The manufacturers of internal combustion engines and of motor vehicles run extensive brake-horsepower tests to determine the performance of their engines under varying conditions. From the results of the tests it is customary to prepare charts in which the brake horsepower is plotted against engine revolutions per minute. Brake-horsepower curves obtained in this manner are shown in Fig. 12-1. The maximum gross horsepower of 190 is developed at 4,150 rpm. The maximum net horsepower of 163 is developed at 3,800 rpm.

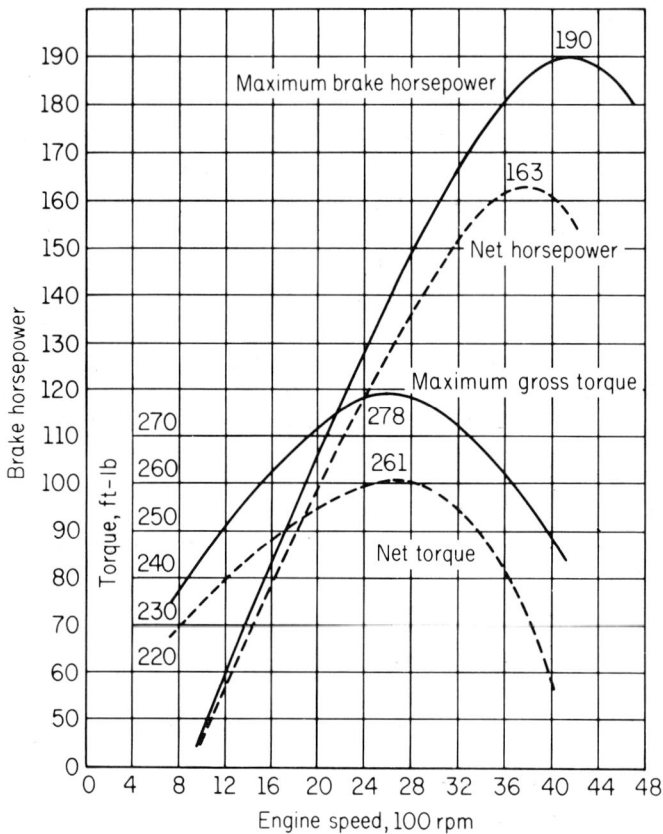

FIG. 12-1. Horsepower and torque curves for a four-cycle gasoline engine.

The gross horsepower is that developed by the "bare" engine when operated without certain accessory units. The net horsepower is that horsepower available at the drive-shaft output. The difference in gross and net horsepower is that required to drive the accessory equipment. There is no generally accepted definition of what is a "bare" engine. In general, though, the bare engine includes those items of equipment necessary to its operation such as water pump,

Power Performance of Motor Vehicles

fuel pump, oil pump, and distributor when these units are connected internally with the engine. Items that would be removed to get the gross horsepower developed would be generator, fan, air cleaner, hydraulic pumps, and muffler. These items combined may require 8 to 15 percent or an average of about 10 percent of the maximum gross power, all of which horsepower is absorbed before the power is available at the flywheel.

Torque is a measure of the pulling ability of the engine and is usually expressed in foot-pounds. It is the force in pounds that would be exerted by the engine at the rim of its flywheel with a radius of 1 ft. The relation between brake horsepower and torque is given by

$$\text{Horsepower} = \frac{T(2)3.1416\,N}{33,000} = 0.0001904\,TN \qquad (12\text{-}2)$$

where T = torque, foot-pounds
N = number of engine revolutions per minute

Torque curves are shown in Fig. 12-1. A study of these curves indicates the performance to be expected of the engine. The maximum torque of 278 ft-lb is obtained at an engine speed of 2,600 rpm.

Engines can be designed with widely different power and torque characteristics. In general, the engine for a particular vehicle is designed to develop its maximum torque and economy within the usual range of operation of the vehicle.

DRIVE-AXLE TORQUE AND TIRE-RIM PULL

Torque at the flywheel can be converted into torque at the rear axle by the following equation:

$$\text{Rear-axle torque} = T_a = kT_e G_t G_a \qquad (12\text{-}3)$$

where T_a = rear-axle torque, foot-pounds

k = efficiency of the drive mechanism, varying from 0.85 to 0.90
T_e = engine torque at the flywheel, foot-pounds
G_t = transmission gear ratio
G_a = rear-axle gear ratio

If the rear-axle torque is divided by the rolling radius of the drive tires, in feet, the result will be the tire-rim pull in pounds. Thus

$$\text{Tire-rim pull} = P = \frac{kT_e G_t G_a}{r} \qquad (12\text{-}4)$$

where P = tire-rim pull, pounds
r = rolling radius of the drive tire, feet

When a vehicle is being operated along a highway under power supplied by the engine, the tire-rim pull P is always equal to the total tire-rim resistance. The horsepower utilized at the rear wheels is

$$\text{Horsepower} = \frac{PV}{375} \tag{12-5}$$

where P = tire-rim pull (or tire-rim resistance), pounds
V = speed of the vehicle, miles per hour

The factor $V/375$ results from converting 33,000 ft-lb per min = 1 hp to a factor of V:

$$\frac{(5280 \text{ ft})V}{(33{,}000 \text{ ft-lb})(60 \text{ min})} = \frac{V}{375}$$

GEAR RATIO

The transmission gear ratio is the means by which the principle of the lever is applied by the engine to move a heavy body along the highway. The gear ratio enables the torque at the flywheel of the engine to be multiplied to such an extent that the vehicle can be started, accelerated, and moved up steep grades. In order to obtain gear reduction, and thus magnify the torque of the engine, there is a driving gear meshed with a driven gear. The ratio of the number of teeth in the driven gear to the number of teeth in the driving gear is the gear ratio. Thus if there are 38 and 7 teeth respectively in the pair of gears, the gear ratio is 5.428 to 1.

Gear reduction in motor vehicles is generally obtained in two ways: in the transmission and in the differential at the driving axle where the major reduction is effected. The total gear reduction is obtained by multiplying the transmission gear ratio by the rear-axle ratio. In direct drive there is no transmission gear reduction, and the total gear ratio is that of the differential. The gear ratio is a fundamental factor in determining the pulling ability of the vehicle, and there is a direct relation between gear ratio and the speed at which the vehicle can be operated.

USE OF POWER DEVELOPED BY THE ENGINE

It is essential that the engine develop sufficient power to move the vehicle and its load at the desired speed and to accelerate at the desired rate to the desired speed. For economical operation, as well as to keep the first cost of the power unit within reasonable limits, it is likewise desirable that the vehicle be not greatly overpowered. In other words, the motor vehicle should be equipped with an engine that will develop the power required for normal operation, but not with much excess power. This principle is illustrated in the design of motor trucks and the light foreign cars, but not with many of the high-powered domestic passenger cars.

The net power at the flywheel generated by the engine is used in the following five ways:

1. In overcoming transmission friction in the power train mechanism from the engine to the drive wheels
2. In overcoming rolling resistance; that is, the resistance of the roadway surface to the movement of the vehicle over it
3. In overcoming resistance offered by the air
4. In overcoming grade resistance
5. In accelerating the vehicle

The amounts of these various resistances and their power requirements for a particular vehicle and under specified conditions can be determined. The resulting power required to propel the vehicle can then be calculated. The first step is to consider the various resistances offered to the movement of a vehicle along the highway.

POWER TRANSMISSION RESISTANCE

To connect and to disconnect the engine from the transmission gear system, a clutch is used on manual transmissions, and a fluid coupling is used with the automatic transmission. This clutch arrangement serves two purposes: (1) it provides for a differential in revolutions per minute by allowing a slipping action while the faster-speeding engine and the lower-speeding transmission gear, whose speed is controlled by the road speed of the vehicle, are gradually coming to the same speed, and (2) it disconnects the engine from the transmission gear when shifting gears and when idling the engine. The engine must revolve at a fairly high speed as compared to the vehicle speed when starting up from zero speed in order that it can generate the horsepower required to start the vehicle in motion, then to accelerate it faster. From the clutch, the engine power is delivered to the transmission, thence through the driveshaft, with universal joints, to the rear-axle differential and on to the drive axle and wheels.

Because the engine has low power at low engine speed, a transmission is used through which to alter the engine speed relative to the road speed of the vehicle. This transmission makes it possible to develop the high engine speed required to generate the horsepower to start the vehicle from standing, and to develop the high horsepower needed at slow road speed on steep plus grades. Most passenger cars have at least three forward gears and one reverse gear. Trucks may have 4 to 10 or more forward-speed gears.

A further gear reduction is made at the differential in the drive axle. The differential permits one drive wheel to revolve faster than the other when driving a curved path, as well as to speed up the engine relative to the drive-wheel revolutions per minute.

This whole transmission train–clutch, transmission, drive shaft, differential, and axle and other bearings–absorbs power. The net horsepower at the engine crankshaft output is thus further reduced before it reaches the drive wheels. These power losses in the transmission power train are about as follows:

Clutch .... 1 percent
Transmission and differential .... 3 to 5 percent
Bearings and joints .... 1 to 2 percent
Drive-wheel slip at tire patch
(area of contact at the road surface) .... 2 to 5 percent

The total power loss is 7 to 13 percent, or approximately 10 percent as an average figure for passenger cars with manual transmission in direct drive. At lower gears, the loss will approach 15 percent. Trucks in their lowest gear will have transmission power losses of up to 25 percent.

The mechanical clutch has high efficiency because when engaged it has no moving parts, only a slight slippage. The transmission and differential have meshed gear pairs which operate in a bath of lubricant. The gear-teeth friction and the churning of the lubricant cause the power loss. The lubricant is usually of heavy viscosity which results in high resistance as compared to a low-viscosity oil. The gear resistance varies approximately as the cube of the speed. All bearings and universal joints absorb some power because of friction. The drive-wheel slip loss is dependent upon the vehicle speed, power applied, and coefficient of friction at the tire patch.

ROLLING RESISTANCE

Rolling resistance is a general term that covers the resistance of the roadway surface to the movement of the vehicle; it also includes friction in the bearings of the wheels that are not used in driving the vehicle. Rolling resistance depends to a considerable extent on the kind and size of tires, inflation pressure, tire temperature, type of roadway surface, and road speed of the vehicle. The gross weight on the tires is a direct factor in the rolling resistance.

The total rolling resistance of the vehicle is made up of the following elements:

1. Work required to compress and to deflect the roadway surface.
2. Work required to flex the tire at the tire patch.
3. Work required to overcome the friction force, or slipping force, at the tire contact patch.
4. Air friction caused by movement of the air inside of the tire, and outside air resistance of the air to the rotation of the wheel and tire.
5. Wheel-bearing friction not overcome in the power transmission.

Of these five sources, the yielding of the roadway surface and the flexing of

Power Performance of Motor Vehicles

the tire comprise the greater portion of the total rolling resistance.

For general use, it is unnecessary to determine by theoretical or experimental work the percentage of the total rolling resistance contributed by each of these elements. What is needed, however, is good average total rolling resistance values for typical vehicles and roadway surfaces. As tires and vehicles have changed over the years, rolling resistance has also changed, but not greatly. The following equations based upon Taborek's |12-17| discussion, are generally applicable for passenger cars on concrete pavement and on hard, even bituminous surfaces:

$$R_r = \left[f_0 + 3.24 f_s \left(\frac{V}{100} \right)^{2.5} \right] W \quad (12\text{-}6)$$

$$= 0.01 \left(1 + \frac{V}{100} \right) W \quad (12\text{-}7)$$

where R_r = total rolling resistance for the car in pounds
f_0 = basic coefficient of resistance–coefficient at near zero speed
f_s = coefficient to account for the speed effect
V = vehicle speed, miles per hour
W = gross or road weight of the car, pounds

Equation 12-7 is a straight-line approximation of Eq. 12-6 and is not valid above 80 mph.

Both coefficients, f_0 and f_s, vary with tire inflation. For about 24 psi, $f_0 = 0.0112$ and $f_s = 0.0068$. For a rigid smooth pavement the rolling resistance of a passenger car is then given by.

$$R_r = \left[0.0112 + 0.0220 \left(\frac{V}{100} \right)^{2.5} \right] W \quad (12\text{-}8)$$

The horsepower required for rolling resistance is:

$$R_{r\text{hp}} = R_r \frac{V}{375} \quad (12\text{-}9)$$

Table 12-1 gives the rolling resistance and horsepower for the 4,000-lb passenger car as calculated by these two equations.

The Society of Automotive Engineers Recommended Practice, TR-82 |12-15| gives the following equation for the rolling resistance of trucks:

$$R_r = (7.6 + 0.09 V) \frac{W}{1,000} \quad (12\text{-}10)$$

Table 12-1 gives the rolling resistance of a 40-kip 2-S2 tractor-semitrailer as calculated by Eq. 12-10.

TABLE 12-1
Resistances and Horsepower Requirements for a 4-Kip Passenger Car and a 40-Kip 2-S2 Tractor-Semitrailer-High-Type Pavement in Good Condition

Road Speed, mph	Resistance, Pounds		Horsepower Required					Additional Horsepower Required for a 1 percent Plus Grade
	Rolling	Air	Rolling	Air	Subtotal	Chassis	Total	
4-Kip Passenger Car								
5	44.84	0.76	0.60	0.01	0.61	0.06	0.67	0.53
10	45.08	3.04	1.20	0.08	1.28	0.13	1.41	1.07
15	45.56	6.84	1.82	0.27	2.09	0.21	2.30	1.60
20	46.36	12.17	2.47	0.65	3.12	0.31	3.43	2.13
25	47.56	19.01	3.17	1.27	4.44	0.44	4.88	2.67
30	49.12	27.38	3.93	2.19	6.12	0.61	6.73	3.20
35	51.20	37.26	4.78	3.48	8.26	0.83	9.09	3.73
40	53.72	48.67	5.73	5.19	10.92	1.09	12.01	4.27
45	56.76	61.60	6.81	7.39	14.20	1.42	15.62	4.80
50	60.36	76.05	8.05	10.14	18.19	1.82	20.01	5.33
55	64.56	92.02	9.47	13.50	22.97	2.30	25.27	5.87
60	69.36	109.51	11.10	17.52	28.62	2.86	31.48	6.40
65	74.76	128.52	12.96	22.28	35.24	3.52	38.76	6.93
70	80.88	149.06	15.10	27.82	42.92	4.29	47.21	7.47
75	87.68	171.11	17.54	34.22	51.76	5.18	56.94	8.00
80	95.20	194.69	20.31	41.53	61.84	6.18	68.02	8.53
40-Kip 2-S2 Tractor-Semitrailer								
5	322.00	4.50	4.28	0.06	4.34	8.84	13.18	5.33
10	340.00	18.00	9.08	0.48	9.56	8.39	17.95	10.67
15	358.00	40.50	14.32	1.62	15.94	8.09	24.03	16.00
20	376.00	72.00	20.04	3.84	23.88	8.00	31.88	21.33
25	394.00	112.50	26.28	7.51	33.79	8.22	42.01	26.67
30	412.00	162.00	32.96	12.96	45.92	8.63	54.55	32.00
35	430.00	220.50	40.12	20.57	60.69	9.20	69.89	37.33
40	448.00	288.00	47.80	30.77	78.52	9.90	88.42	42.67
45	466.00	364.50	55.92	43.68	99.60	10.73	110.33	48.00
50	484.00	450.00	64.52	60.00	124.52	11.69	136.21	53.33
55	502.00	544.50	73.64	79.92	153.56	12.78	166.34	58.67
60	520.00	648.00	83.20	103.68	186.88	14.00	200.88	64.00
65	538.00	760.50	93.24	131.76	225.00	15.35	240.35	69.33

Passenger car

$$R_r = \left[0.0112 + 0.0220 \left(\frac{V}{100} \right)^{2.5} \right] W$$

$$R_a = (0.002) 90 \, V^2$$

2-S2

$$R_r = (7.6 + 0.09 \, V) \frac{W}{1000}$$

$$R_a = (0.0026)(0.045) 26 \, V^2$$

The horsepower required is equal to the pounds of resistance times $V/375$.

AIR RESISTANCE

Because air has density it resists the motion of any body passing through it. The air resistance is a function of its own weight and the relative velocity of the air and the body passing through the air. For the purpose of computing the air resistance of motor vehicles the coefficients are presented in terms of standard air (60° F and 29.9 in. Hg) and at zero air velocity (still air). At this standard condition, air weighs 0.0763 lb per cu ft. Air is lighter at altitudes above sea level to the extent that at 4,000 ft its density is only 83 percent of that at sea level.

The following three factors combine to account for the total air resistance of a motor vehicle:

1. Drag resistance as related to the outside shape and size of the vehicle, protruding parts, and the shape of the rear end. The rear-end shape controls the turbulence in the wake of the vehicle.

2. Skin friction, or the resistance to the air offered by the surface of the body of the vehicle. For passenger cars, skin friction (resistance) is about 10 percent of the total air resistance.

3. Flow of air through the vehicle for purposes of ventilating and cooling.

The greatest single factor in air resistance is the projected frontal area of the vehicle. It is this area, in effect, that is driven through the air. The air is displaced continuously as the vehicle moves forward. Since the air has weight, displacing it requires work. At high speed the air in effect becomes a solid; at low speeds (up to 15 mph) its resistance is almost nil. The formula for air resistance involves factors for the weight of the air, the drag and skin friction of the vehicle, the projected frontal area, and relative speed of the vehicle and air. The equation given by Taborek |12-17| is

$$R_a = 0.0026\, c_a A\, V_r^2 \qquad (12\text{-}11)$$

where R_a = total air resistance of the vehicle, pounds
c_a = coefficient of air resistance for the particular vehicle
A = projected frontal area of the vehicle, square feet
V_r = speed of the vehicle relative to the air, miles per hour

Assuming calm air, the equation for horsepower to overcome air resistance is:

$$R_{\text{ahp}} = 0.0026\, c_a A\, V^2 \frac{V}{375} \qquad (12\text{-}12)$$

Here it is noticed that air-resistance horsepower varies as the cube of the vehicle speed. It is significant that the weight of the vehicle is not a factor in air resistance.

Taborek |12-17| gives the following range of values for the factor c_a:

        Passenger car .... 0.40 to 0.50
        Convertible .... 0.60 to 0.65

Bus .... 0.60 to 0.70
Truck .... 0.80 to 1.00
Tractor-trailer 1.30

The SAE, TR-82 |12-15| gives the following equation for the air-resistance horsepower for trucks:

$$R_{ahp} = \frac{0.002AV^3}{375} \qquad (12\text{-}13)$$

By comparison with Eq. 12-12, this SAE formula uses a c_a coefficient of $c_a = 0.002 \div 0.0026 = 0.77$, which is quite low as compared to Taborek's values of 0.80 to 1.30.

Table 12-1 gives the computed rolling, air, and internal resistances and the horsepower for the 4-kip passenger car and the 40-kip tractor-semitrailer for a level tangent.

GRADE RESISTANCE

When a vehicle is moving along a level roadway at a uniform speed, the power of the engine is used only to overcome internal resistance and tractive resistance. Tractive resistance is the sum of the rolling resistance and air resistance. When a plus grade is encountered, additional power must be supplied by the engine if the same velocity of the vehicle is to be maintained. The resistance offered to movement of a vehicle up a grade is known as *grade resistance*. It is expressed in pounds per vehicle, pounds per ton of vehicle weight, or pounds per 1,000 pounds of vehicle weight. In effect, it is the force necessary to lift the vehicle through a height equal to that attained because of the plus grade.

Grade resistance can be conveniently expressed in terms of the force necessary to lift the vehicle through a height corresponding to a horizontal travel of one unit of distance along the roadway. The unit will be taken as the foot. If the grade is 1 percent, the rise in a horizontal distance of 1 ft is 0.01 ft; on a 2-percent grade, the rise is 0.02 ft. On a grade of G percent, the rise will be 0.01 G. If the weight of the vehicle is W, then the force required to lift it through a height attained by a horizontal travel of 1 ft on any grade G will be

$$R_g = \frac{WG}{100} \qquad (12\text{-}14)$$

The horsepower required to overcome grade resistance can be determined by multiplying the force by the feet of distance the vehicle will travel in one minute of time, and dividing the result by 33,000 ft-lb.

$$R_{ghp} = \frac{WG}{100}\left(\frac{V}{375}\right) \qquad (12\text{-}15)$$

Power Performance of Motor Vehicles

To illustrate, let it be assumed that an automobile with a weight of 4,000 lb is traveling along a roadway at a speed of 40 mph and that it is desired to determine what additional horsepower will be required to maintain the same speed up a grade of 3 percent. The horsepower required will be:

$$\text{Horsepower} = \frac{4{,}000\,(3)\,(40)}{100\,(375)} = 12.8$$

Since the power required is a straight-line function, direct ratios can be used for converting one grade horsepower to any other grade, speed, or gross weight of vehicle. See Table 12-1.

POWER ACCELERATION

It is necessary that motor vehicles be powered sufficiently so that the speed can be increased at a satisfactory rate from zero to the desirable cruising speed. The internal combustion engine, because of its peculiar characteristics, cannot accelerate a vehicle from zero speed if it is connected directly to the driving wheels. Some means must be provided to accelerate the vehicle gradually from a stopped position without stalling the engine. This result is accomplished through the clutch or fluid transmission which, by slipping slightly, permits the engine to run at a nonstalling speed while the vehicle is gradually gaining speed. Thereafter, the rate of the acceleration depends upon the power developed by the engine in each particular gear. Since the horsepower of the engine increases with the number of revolutions per minute of the crankshaft up to a limiting engine speed, the introduction of reduction gears enables a greater acceleration of the vehicle at the lower speeds. That is why three or more forward speeds are used on motor vehicles.

The force necessary to accelerate a body is given by

$$R_s = \frac{W}{g} a \qquad (12\text{-}16)$$

where R_s = accelerating force, pounds
 W = weight of the body, pounds
 g = acceleration of gravity, feet per second per second
 a = acceleration of the vehicle, feet per second per second

The horsepower required to accelerate a vehicle is obtained by multiplying the force required by the distance the vehicle will travel in one second of time and dividing the result by 550 ft-lb.

The rate of acceleration of motor vehicles varies rather widely. For the passenger cars it is about 10 ft per sec per sec in first gear, 8 ft per sec per sec in second gear, and 5 ft per sec per sec in high gear (direct drive). Heavy trucks have acceleration rates of from 1 to 5 ft per sec per sec. A car equipped with an engine that gives rapid acceleration generally is not economical in fuel consumption at constant low speed.

As an illustration of the foregoing principles, let it be required to determine the power needed to accelerate a vehicle weighing 4,000 lb at the rate of 10 ft per sec per sec in first gear from a speed of 5 mph. The force required for acceleration is 4,000/32.2(5), or 621 lb. In 1 second of time the car will move a distance of 5,280/3,600(5), or 7.33 feet. The power needed is:

$$\text{Horsepower} = \frac{621 \times 7.33}{550} = 8.28$$

If the car is equipped with rear tires that have a rolling radius of 12.83 in., a rear-axle gear ratio of 3.82 to 1, and a first gear ratio of 2.78 to 1, the revolutions per minute of the engine at a vehicle speed of 5 mph will be:

$$\text{Engine speed} = \frac{(7.33)(12)(60)(3.82)(2.78)}{3.1416\,(12.83)\,(2)} = 695 \text{ rpm}$$

Here (7.33)(12)(60) is the speed of the vehicle in inches per minute; (3.82)(2.78) is the overall gear ratio; and 3.1416 (12.83) (2) is the rolling circumference of the drive tires in inches.

The net horsepower curve of Fig. 12-1 if extended downward would give about 29 hp at 695 rpm, sufficient power to overcome rolling, air, and transmission resistances at 5 mph. See Table 12-1.

TOTAL RESISTANCES AND HORSEPOWER REQUIRED

Except in the case of air resistance, the equations for the various resistances to the movement of the vehicle along the highway include a term representing the gross weight of the vehicle. Each of these resistances, thus varies directly as the weight of the vehicle. This is a fundamental principle that must be kept in mind in considering the horsepower necessary to move the vehicle under specified conditions. It is of equal importance in considering the fuel consumption and the relative economy of different grades.

Table 12-1 gives the pounds of resistance and horsepower required on level tangents on good high-type pavements for the 4-kip passenger car and the 40-kip 2-S2 combination vehicle for which the running cost tables (Appendix A) are computed. Also shown for each vehicle is the grade horsepower required for a 1 percent plus grade. Since the grade horsepower is directly proportional to steepness of grade, the 1 percent values times any other gradient will give the grade horsepower required for that grade. The grade horsepower on a minus grade is the same magnitude as on the equivalent plus grade, but is negative.

REFERENCES

12-1. Walter Bergman. "Theoretical Prediction of the Effects of Traction on Cornering Force," *Soc. Auto. Engr., Trans.* (1961), pp. 614-640.

12-2. William L. Clark and Roy B. Sawhill. *Predicting Fuel and Travel Time Consumption of Motor Transport Vehicles.* University of Washington, Seattle, Research Report No. 3A, Traffic and Operations Series, January 1962.

12-3. H. Flynne and P. Kyropoulos. "Truck Aerodynamics," *Soc. Auto. Engr.,* Preprint No. 284A, January 1961.

12-4. S.F.Hoerner. *Fluid Dynamic Drag.* Midland Park, N.J., 1958, pp. 12-1–12-9.

12-5. W. E. Lay. *Bibliography of Tractive Resistance and Allied Problems.* Highway Research Board, National Academy of Sciences, Washington, D.C., *Proc.,* Vol. 19, 1939, pp. 50-67.

12-6. C. L. McCuen. "Economic Relationship of Engine-Fuel Research." American Petroleum Institute, paper presented at midyear meeting at Tulsa, Okla. May 1951.

12-7. C. B. McCullough and John Beakey. *The Economics of Highway Planning.* Tech. Bul. No. 7, Oregon State Highway Department, Salem, 1938.

12-8. Donald L. Nordeen and Anthony D. Cortese. "Force and Moment Characteristics of Rolling Tires," *Soc. of Auto. Engr.* 713A, June 10-14, 1963.

12-9. J. R. Nothstine and F. N. Beauvais. "Laboratory Determination of Tire Forces," *Soc. of Auto. Engr.* 713B, June 10-14, 1963.

12-10. H. S. Radt, Jr. and W. F. Milliken, Jr. "Motions of Skidding Automobiles," *Soc. of Auto. Engr.,* Summer Meeting, Chicago, Ill., June 1960.

12-11. Carl C. Saal. "An Evaluation of Factors Used to Compute Truck Performance," *Soc. of Auto. Engr. Quarterly Trans.,* Vol. 3, No. 2 (April 1949), pp. 125-228.

12-12. Carl C. Saal. "Performance–Actual vs. Computed," *Soc. of Auto. Engr. Querterly Trans.,* Vol. 5, No. 1 (January 1951), pp. 19-25.

12-13. Carl C. Saal and F. William Petring. "At What Point Does High Horsepower Cease to be Practical in Transport Operations?" *Soc. of Auto. Engr.,* National West Coast Meeting, Portland, Oregon, August 1961.

12-14. R. B. Sawhill and J. C. Firey. *Predicting Fuel Consuption and Travel Time of Motor Transport Vehicles.* Highway Research Board, National Academy of Sciences, Washington, D.C., Bulletin 334, 1962, pp. 27-46.

12-15. Society of Automotive Engineers. *Truck Ability Prediction Procedure,* TR-82. New York, 1957.

12-16. A. F. Stamm and E. P. Lamb. "Predicting Road Performance of Commercial Vehicles," *Soc. of Auto. Engr. Quarterly Trans.,* Vol. 4, No. 2 (April 1950), pp. 147-160.

12-17. Jeroslav J. Taborek. "Mechanics of Vehicles," *Machine Design* (1957).

12-18. C. Fayette Taylor and Edward S. Taylor. *The Internal Combustion Engine,* Chapter 12, "Lubricating and Oils," International Textbook Co., Scranton, Pa., 1948, pp. 206-221.

12-19. Harry Tucker and Marc C. Leager. *Highway Economics.* International Textbook Company, Scranton, Pa., 1942.

Chapter **13**

Vehicle Costs and the Highway

One of the main factors in the analysis for the engineering economy of highway design is the cost of running motor vehicles thereon. The highway engineer has at his hand the data, knowledge, and experience which permits him to make reliable estimates of the construction cost and maintenance cost of the highway, including the traffic control devices. But for the running cost[1] of motor vehicles, the highway engineer needs to look beyond his own files and experience.

It is not at all an easy matter to determine motor vehicle running cost because the costs are affected by many factors and vary over wide ranges. Nevertheless, good highway design calls for making the best possible estimate of motor vehicle running costs as influenced by each feature of highway design. Operating costs and running costs are discussed in this chapter as related to the highway and vehicle owner.

VARIABLE FACTORS IN MOTOR VEHICLE COSTS

Because of the many factors bearing upon the cost of motor vehicle ownership and use, perhaps no two vehicles produce the same exact costs. Even two identical vehicles–at least identical in manufacture–vary in operating costs in the hands of different owners. The forces of time and usage soon introduce other differences in the magnitudes of the cost variables so that the two identical vehicles may ultimately exhibit widely different operating costs.

CLASSIFICATION OF FACTORS AFFECTING MOTOR VEHICLE RUNNING COST

The main factors that affect motor vehicle running cost may be grouped as follows:

---

[1] The term "running cost" is intended to include only those items of motor vehicle cost that are affected by highway design and traffic; the term "operating cost" includes all elements of cost which relate to owning and operating the vehicle. In general, running cost is the total of the cost items affected by miles driven and operating cost is the combined total of mileage items and items dependent upon time and ownership rather than upon miles driven.

Vehicle Costs and the Highway

A. The highway
  1. Distance
  2. Geometric design, transverse and longitudinal
  3. Character of roadway surface
  4. Traffic volume, composition, traffic controls, and speed changes
  5. Legal restraints
B. The vehicle
  1. Road weight and weight-horsepower ratio
  2. Engine design
  3. Transmission and rear-axle ratios
  4. Tire size and tire pressure
  5. Vehicle dimensions and vehicle dynamic characteristics
  6. Mechanical condition of engine, power transmission, and braking system
  7. Type of fuel
C. The operator
  1. Rates of acceleration and deceleration (speed changes)
  2. Number and range of speed changes
  3. Number and timing of gear changes
  4. Cruising speed
  5. Character of use, trip length, and annual mileage
  6. Care of vehicle
D. The weather and topography
  1. Air temperature, air pressure, and air humidity
  2. Wind direction and velocity
  3. Rain, snow, and ice conditions on roadway
  4. Altitude and topography

For economy studies in highway design it is required that these groups of factors be studied so that the effect of the highway on the running cost of the vehicles may be determined. There are wide ranges in variation of three principal factors–highway, vehicle, and operator. Yet, it is necessary to provide the analyst with good and reliable motor vehicle performance and cost information.

SIGNIFICANCE AND USE OF VEHICLE OPERATING COSTS

Motor vehicle operating costs as a whole are about 88 percent of the total highway transportation cost; the highway cost accounts for the remaining 12 percent. Therefore, motor vehicle operating costs are the key factor in the analysis of highway transportation costs.

Motor vehicle costs as a whole are highly variable, vehicle to vehicle, because of the many factors enumerated above, including particularly the vehicle operator and owner. Despite these many variables it is possible to develop tables of operating cost and running cost which are reliable for studies of the cost of highway transportation, and for the studies of the economic evaluation and project formulation of proposed highway facilities.

Overall motor vehicle operating costs are desired for comparison with other modes of urban travel and the intercity line-haul trucking costs are factors in common and contract carrier regulations.

MOTOR VEHICLE OPERATING COST

The overall operating cost of vehicles is difficult to determine because of the many variables involved, the length of time required to level out the periodic costs, and because of the large number of individual vehicle owners. The following discussion and reported costs are offered more to gain an understanding of vehicle operating costs than to present average operating costs.

SOURCES OF OPERATING-COST RECORDS

Obviously, the records of the cost of operating motor vehicles must come from the owners of vehicles. Unfortunately for those who seek motor vehicle operating costs, vehicle owners as a whole do not keep records of their costs. For the family or personal passenger car perhaps less than 3 percent of the owners keep complete records of cost, and of these owners perhaps only one-third of them summarize their costs over a long enough time period to make them really useful. A far greater number of owners keep partial records for short times than those who keep complete records for the full period of their ownership. Also, many owners keep the essential records for spot reference, but do not summarize them. Particularly, it is disappointing to a collector of passenger car operating-cost records not to find operating costs on the last half of a car's service life. Keepers of records of car operating costs just are not found among the owners of the older vehicles.

Another disappointing fact is that many of the few operating-cost records available do not give full description of the car and do not give much of a breakdown of the costs.

For commercial vehicles, many owners or firms keep complete motor vehicle operating-cost records. These records are often without classification of the cost items and usually are on a fleet basis where more than one vehicle is involved. Freight carrying companies subject to regulation by the Interstate Commerce Commission or a state public utilities commission, usually keep good records of the cost of operating their vehicles and report these costs in the manner prescribed by the regulatory body. For intercity highway freight (and passenger) vehicles these records are a good source of operating costs. Even so, the analyst will usually find plenty of work to do in getting from these, or other, records just what he is looking for.

It is to be kept in mind that the passenger-car records and the commercial

Vehicle Costs and the Highway

vehicle records that may be available present the cost of motor vehicle operations without respect to any features of highway design. Sometimes a rural-urban classification can be made. But for the running cost information needed in economy studies, resort must be had to controlled operation with suitable instrumentation. Even so, only fuel, oil, and tire consumption can then be measured. Maintenance and depreciation costs as affected by the highway design and traffic need to be calculated, using as a base normal operating costs.

CLASSIFICATION OF COST ITEMS

A classification of the separate items of the cost of operating a motor vehicle is an aid to a better understanding of the interactions of the highway, the vehicle, the owner, and the use of a vehicle upon its operating cost. Table 13-1 is one such grouping of cost items.

The four main groups—mileage, time, ownership, and commercial—are significant in illustrating the fact that distance driver and the time period involved are two important factors in any costs reported in cents per mile. The cash cost of the time and ownership items is directly proportional to calendar time while the cash cost of the mileage items varies directly with miles driven per year. Any report of the total cost of operating a motor vehicle is rather meaningless until three sets of explanatory information are supplied. It is essential to know the exact cost items included, the miles driven, and the time period considered. Beyond this information the character of use, the age of the vehicle, its cost new, and basis of estimating depreciation expense are factors helping to understand the significance of the reported cost in cents per mile.

The subitems in the mileage groups offer some latitude in the classification scheme, particularly for groups 1 to 11, except for 6 and 7. The exact position in the classification of many of these subitems is not important as long as they are included in one logical place or other.

The classification scheme as given in Table 13-1 is helpful in allocating running costs to features of highway design and traffic operation. For instance, the cost of brakes is wholly attributed to stops, slowdowns, and retarding the vehicle on minus grades. The body costs are related to the roughness and dirtiness of the roadway. Engine oil consumption is related to the vehicle speed, speed changes, and dirt condition of the roadway. The body costs and depreciation costs are each factors of both time and vehicle use.

OPERATING COST OF PASSENGER CARS

As already mentioned, complete and reliable records of passenger car operation are scarce and to assemble a sufficient number of them to develop a statistically sound table of average cost of operation is really difficult and expensive. People as a whole just are not record keepers, especially of the cost

TABLE 13-1
CLASSIFICATION OF MOTOR VEHICLE OPERATING-COST ITEMS

A. Mileage Items

1. *Body*
   Bumpers
   Doors
   Fenders, body dents
   Glass
   Heater
   Instruments
   Interior
   Keys, locks
   Paint
   Rattles
   Wash, polish
   Windshield wiper

2. *Brakes*
   Adjustments
   Cylinders
   Drums
   Fluid
   Inspection
   Lining
   Shoes

3. *Chassis*
   Frame
   Front end
   Gas line and tank
   Grease fittings
   Hub caps
   Muffler, tail pipe
   Shock absorbers
   Steering
   Suspension system
   Tire rims
   Wheels and axles

4. *Electrical*
   Battery and cables
   Generator and belt
   Lamps and bulbs
   Regulator
   Starter motor
   Turn signal
   Wiring

5. *Engine*
   Air cleaner
   Antifreeze
   Bug screen
   Carburetor
   Distributor
   Fuel pump
   Fan and belt
   Internal work
   Oil filter
   Radiator
   Spark plugs
   Steam clean
   Tune-ups and parts
   Water pump

6. *Engine Oil*

7. *Engine Fuel*

8. *Greasing and Lubricants*
   Chassis
   Differential
   Front wheels
   Transmission
   Universals

9. *Power Train*
   Clutch
   Differential
   Drive shaft
   Transmission
   Transmission fluid
   Universals

10. *Tires*
    Balance
    Chains
    Flats
    Rotate
    Snow tires

11. *Other*
    Radio
    Safety inspections

Vehicle Costs and the Highway

TABLE 13-1 (*continued*)

Small tools
Tow-in
B. Time Items
  12. *Accidents*
  13. *Depreciation* (also a mileage item)
  14. *Insurance*
  15. *Licenses*
  16. *Parking*
  17. *Tolls*
C. Ownership Items
  18. *Auto Club Membership*
	19. *Garage*
	20. *Personal Property Tax*
	21. *Vestcharge*
	D. Commercial Items
	  22. *Driver's Wage, Fringe Benefits, and Subsistence*
	  23. *Overheads*
	  24. *Special Road-user Taxes and Fees*

of their car, a most expensive item the operating cost of which they would rather not know. The road user is therefore simply not conscious of what it does cost him to own and to operate his car, either per mile or in dollars per year.

There is an important factor of spending psychology that affects the car driver. Not being a record keeper, and, therefore, not being conscious of his car cost per mile, he is apt to be unconcerned about his car expense except when actually disbursing his money for fuel, oil, tires, or license plates. He can make a trip to the supermarket and return, spend no money for the car during the trip, and unconsciously concludes that the trip cost nothing. This same driver can drive his car daily 10 miles one way to work in peak-hour traffic and not think about his car costs, but on the occasional trip when he must take the bus, he can complain about the high fare of 45 cents one way. The use of the automobile probably would be curtailed materially if all owners, or drivers, had to pay their total motor vehicle operating costs directly by the mile as they drive or by the trip.

Illustrative Actual Passenger Car Cost Record

The variations in car expense and use on a time basis are illustrated by Tables 13-2, 13-3, and 13-4 for a 1957 model four-door sedan in family use 1957 to 1964. These data are from the records kept by the owner of this car, and the costs and car use are not presented as being average, but as being not uncommon. Note the wide range in mileage per year and its correlation with the cost in cents per mile. Note also that the cash expense for tires was unusually heavy in 1959 and 1962, indicating the purchase of a new set of tires. Other items–brakes, electrical, engine–vary likewise year to year.

Table 13-3 shows that miles per gallon of gasoline increases with miles driven per year. The low mileage years indicate that most of the driving was

TABLE 13-2

SUMMARY OF DOLLAR COST OF OWNERSHIP AND OPERATION OF A 1957 DeSoto Firedome, 1957 to 1964
(Sedan, 4-door, 8 Cylinder, Power Steering, Automatic Transmission, Curb Weight of 4,375 Pounds
Fair Cash Price New of $ 3,200 in September, 1956. Car Based at Arlington, Virginia, and the Records Were Furnished by the Owner.)

Cost Item	Sept.–Dec. 1957	1958	1959	1960	1961	1962	1963	1964	Total	Cents per Mile
A. Mileage items, cash										
1. Body	0.00	1.75	0.00	22.90	50.92	7.70	49.45	1.30	134.02*	0.24
2. Brakes	6.50	16.70	8.50	25.15	110.10	21.10	53.55	4.50	246.10	0.45
3. Chassis	0.00	3.00	22.35	39.45	136.33	39.10	23.00	0.00	263.23	0.48
4. Electrical	1.00	6.10	18.90	16.55	1.00	16.30	2.00	0.00	61.85	0.11
5. Engine	5.34	48.04	105.16	94.08	116.71	42.85	41.15	5.75	459.08	0.83
6. Greasing	0.00	9.90	10.15	6.95	14.25	10.75	6.25	1.25	59.50	0.11
7. Power train	0.00	1.00	0.00	0.00	348.79	4.80	0.00	4.25	358.84	0.65
8. Engine oil	0.00	10.21	16.83	16.45	27.97	9.75	11.70	3.50	96.41	0.18
9. Gasoline	57.36	167.67	295.17	304.15	434.56	219.33	158.26	59.58	1,696.08	3.08
10. Tires	0.00	11.35	197.50	11.10	24.15	116.90	22.32	0.00	383.32	0.70
11. Subtotal	70.20	275.72	674.56	536.78	1,264.78	488.58	367.68	80.13	3,758.43	6.83

Vehicle Costs and the Highway

B. Time items, cash										
1. AAA membership	12.00	26.00	26.00	26.00	26.00	26.00	27.00	13.00	182.00	0.33
2. Insurance	78.90	104.00	96.01	90.77	90.72	74.00	74.00	37.00	645.40	1.17
3. Licenses	12.00	17.00	20.00	18.00	22.00	20.00	20.00	12.00	141.00	0.26
4. Parking	5.00	82.00	55.15	45.35	38.90	52.40	66.55	16.80	362.15	0.66
5. Tolls	0.00	0.00	17.30	28.50	18.60	1.50	0.00	0.00	65.90	0.12
6. Other	2.25	1.75	0.00	2.50	28.65	0.00	1.00	0.00	36.15	0.07
7. Subtotal	110.15	230.75	214.46	211.12	224.87	173.90	188.55	78.80	1,432.60	2.61
C. Total Cash	180.35	506.47	889.02	747.90	1,489.65	662.48	556.23	158.93	5,191.03	9.44
D. Depreciation	85.00	635.00	448.00	372.00	255.00	180.00	127.00	148.00†	2,250.00	4.09
E. Ownership										
1. Garage	40.00	120.00	120.00	120.00	120.00	120.00	120.00	60.00	820.00	1.49
2. Vestcharge‡, 6 percent	48.00	138.90	100.80	73.92	51.60	36.30	21.25	9.00	479.77	0.87
3. Property tax	0.00	101.63	71.74	55.44	38.70	26.32	18.49	6.60	318.92	0.58
4. Subtotal	88.00	360.53	292.54	249.36	210.30	182.62	159.74	75.60	1,618.69	2.94
F. Grand Total	353.35	1,502.00	1,629.56	1,369.26	1,954.95	1,025.10	842.97	382.53	9,059.72	16.47

* When car was sold it needed $140 worth of body work to replace rusted and bent up parts and to repaint.
† Remainder of unallocated depreciation cost when car was sold June 19, 1964.
‡ Vestcharge is computed as 6 percent of Line A-3 in Table 13-3.

TABLE 13-3

SUMMARY OF OPERATING PERFORMANCE AND UNIT COST OF A 1957 DeSoto Firedome, 1957 to 1964
(Same Vehicle as Described in Table 13-2)

Operations item	Sept.-Dec. 1957	1958	1959	1960	1961	1962	1963	1964	Total
A. Operational data									
1. Odometer (beginning)	11,545	13,698	19,199	29,565	39,021	53,714	60,435	64,731	66,527*
2. Miles driven	2,153	5,501	10,366	9,456	14,693	6,721	4,296	1,796	54,982
3. Market value, dollars (beginning)	2,400	2,315	1,680	1,232	860	605	425	298	150*
4. Gallons, gas	165.2	482.1	889.2	856.5	1,191.5	613.2	449.0	166.2	4,812.9
5. Quarts, oil	0	17	28	27	44	15	18	6	155
6. Miles per gallon, gas	13.03	11.32	11.71	11.04	12.33	10.96	9.57	10.81	11.42
7. Miles per quart, oil	—	321	372	350	334	448	239	299	355
B. Cost, cents per mile									
1. Mileage cost	3.26	5.05	6.48	5.68	8.61	7.27	8.56	4.46	6.84
2. Time items	5.11	4.23	2.06	2.23	1.53	2.59	4.39	4.39	2.61
3. Depreciation, market	3.95	11.64	4.30	3.93	1.74	2.68	2.96	8.24	4.09
4. Ownership	4.09	6.61	2.81	2.64	1.43	2.71	3.72	4.12	2.94
C. Total Cost	16.41	27.53	15.65	14.48	13.31	15.25	19.62	21.21	16.48
D. Gas cost, cents per mile	2.66	3.07	2.83	3.22	2.96	3.26	3.68	3.32	3.08
E. Gas cost, cents per gal	34.72	34.78	33.20	35.51	36.47	35.77	36.84	35.85	35.24

* Terminal mileage and market value when sold June 19, 1964, for $ 150.

Vehicle Costs and the Highway

TABLE 13-4
MONTHLY MILEAGE AND GASOLINE MILEAGE PER GALLON OF 1957 DeSoto Firedome
(Same Vehicle as Described in Table 13-2)

Month	1958		1959		1960		1961		1962		1963	
	Miles	mpg	Miles	mpg	Miles	mpg	Miles	mpg	Miles	mpg	Miles	mpg
January	575	9.2	292	9.6	583	8.3	188	6.6	380	7.96	333	7.58
February	352	10.6	562	13.3	387	8.0	333	7.1	327	7.82	315	7.57
March	390	10.4	565	11.2	281	7.2	392	7.7	625	9.57	359	8.82
April	527	12.0	550	11.0	935	12.2	405	8.4	601	9.66	510	10.55
May	550	12.3	511	12.6	707	10.5	820	11.7	816	12.74	397	11.31
June	153	10.0	2,413	14.5	1,807	14.0	386	10.3	175	10.25	295	11.44
July	329	12.6	2,324	13.3	308	11.4	326	10.9	1,697	16.37	313	12.17
August	501	14.3	477	10.4	613	11.9	361	9.8	430	11.32	327	11.81
September	415	10.9	368	10.2	429	10.2	3,997	13.8	263	10.79	410	10.82
October	435	10.8	1,016	11.1	2,804	12.2	850	11.8	403	9.79	373	10.30
November	577	12.9	610	7.9	374	8.5	2,580	15.03	321	9.05	440	9.42
December	651	10.8	724	9.7	228	7.2	4,055	13.01	683	9.43	310	8.63
Total	5,455	11.32	10,412	11.84	9,456	11.04	14,693	12.33	6,721	10.96	4,382	9.83

Note: Monthly variation in miles per gallon 1958 to June 14, 1959, partially caused by unequal tank fillings. Metered gasoline after June 14, 1959.

done around the urban area in stop-and-go traffic and slow-speed street systems. The relation of gasoline consumption to miles per month is illustrated in Table 13-4. Again, fuel consumption is shown to depend upon the character of the driving. The winter driving was productive of noticeably higher fuel consumption per mile than the summer months.

Illustrative Hypothetical Record of Car Operation

Table 13-5 presents estimates of all items of cost in owning and operating a typical standard American make passenger car of popular size, style, and weight. The information in Table 13-5 is based upon the author's collection of cost records assembled over many years. This table may be considered to be a good representation of the cars in use. Some explanation of the cost and operating items will aid in understanding the table.

Engine Oil. When in motion a new engine at a specific revolution per minute will consume less engine oil than will an engine which is well worn from use. The owner of the new car is more apt to change crankcase oil more often than is the owner of an old car. Therefore there is a tendency for the age factors related to oil consumption to balance. The other factor affecting oil consumption is the character of driving, mainly length of trip. As the annual mileage decreases with car age, the average length of trip decreases and high-speed driving decreases. Urban, short-trip driving with its high number of speed changes is hard on crankcase oil and the engine. Oil changes for short-trip driving should come at shorter intervals than for long-trip driving. These factors–car age and annual mileage–were considered in assigning the oil consumption to the years of use for Table 13-5.

Gasoline. The consumption of gasoline over the 12-year period increases per mile–decreased miles per gallon. This increase in fuel consumption is not caused by age of the car, but by the change in character of driving to shorter trips and lower speeds. Records of fuel consumption of old cars, when properly tuned, do not show an appreciable increase in fuel consumption with age.

Maintenance. As shown by Table 13-2, the expense of maintenance (adjustments, repairs, and replacements) varies year to year by a wide range. Further, the total maintenance expense is considerable for new cars. In Table 13-5 the maintenance expense is held constant a 1.00 cent per mile for the 12 years. A year-to-year variation would be more appropriate on a cash-flow basis, but too few reliable records were available on which to make a year-to-year assignment. Also, considering many vehicles, there is a strong tendency for maintenance expense to level off from year to year.

Most persons assume that the maintenance expense of a car increases materially with age. It does only in the sense of how the maintenance expense is viewed. Once a car is put into good mechanical condition it can

Vehicle Costs and the Highway

operate for another long period of mileage without a major expense. Batteries are good for two to three years, brake systems 20,000 to 40,000 miles, generators and motors for 50,000 miles or so, and the same for other items that last for more than a year or two. The state of quality of those parts of the car which are subject to periodic maintenance actions of a major character is an important factor when selling or trading a used car. Often the estimated expense of such maintenance leads to a conclusion to trade rather than to recondition, when from the single viewpoint of the cost of owning and operating the car the greater economy would be to recondition. But overall lifetime economy is found in running a car 100,000 to 300,000 miles–as is often done.

Tires. The tire expense is held constant at 0.35 cent per mile. Perhaps it should increase somewhat with the decrease in annual mileage because of more speed changes. This tendency is opposed by the change to slower speeds and the fact that owners of older cars wear off a higher percentage of the tread than do owners of newer cars; also, they may use recapped tires.

Depreciation. The yearly depreciation is based on the market value of the car, year to year, over the 12-year period. Market value is used rather than a straight line in order to more closely represent the year to year variations in overall cost of owning and operating the car. The yearly mileage by age of car is based on Ref. 13-1.

Insurance. Car insurance is a variable item on the basis that it varies with what insurance is purchased, where the car is garaged, the ages of the drivers, and the car use. The insurance costs in Table 13-5 are typical, lower than the cost in major cities and higher than in rural areas. In the sixth year the collision protection is dropped because of the low market value of the car.

Licenses. The cost of car licenses includes the license plate, or registration fee, and driver's license fee. Both of these items are variable with the state. The registration fee varies not only state to state but also with the car description and by age in some states. The range is from $2 for a compact to $75 for the larger and heavier cars. The methods of determining the fee include the following and various combinations: (1) flat fee for all passenger cars, (2) flat fee by weight blocks, (3) specific number of cents per 100 pounds of weight, (4) a percentage of retail price plus a weight factor, (5) flat fee based upon horsepower blocks, and (6) other variations. In many locations the local county or city may have a registration or wheel tax in addition to the state fee. Drivers' licenses vary state to state from $2 to $12 a year.

Vestcharge. In adherence to the general concept of the cost of money and of return forgone, a charge is made at 6 percent per year against the market value of the car at the beginning of the year. This vestcharge is a low charge considering the much higher rates the public is paying on their financing of the purchase of their cars.

Automobile Club Membership. The cost of automobile club membership is not recorded because it is not universally held. Where the membership is

TABLE 13-5
Hypothetical Cost of Owning and Operating a Typical Standard Automobile in Family Use

Cost new delivered—$3300
Curb weight—3600 pounds
Road weight—4000 pounds
Price of gasoline per gallon—35 cents
Price of oil per quart—65 cents

Cost Item	1st Year $	1st Year ¢/mi	2nd Year $	2nd Year ¢/mi	3rd Year $	3rd Year ¢/mi	4th Year $	4th Year ¢/mi	5th Year $	5th Year ¢/mi	6th Year $	6th Year ¢/mi	7th Year $	7th Year ¢/mi	8th Year $	8th Year ¢/mi	9th Year $	9th Year ¢/mi	10th Year $	10th Year ¢/mi	11th Year $	11th Year ¢/mi	12th Year $	12th Year ¢/mi	Total $	Total ¢/mi
Mileage items																										
Engine oil	26	0.18	24	0.19	22	0.19	21	0.20	21	0.21	21	0.23	21	0.24	20	0.24	20	0.26	20	0.29	18	0.29	17	0.32	251	0.22
Gasoline	336	2.35	307	2.40	279	2.45	258	2.48	245	2.50	233	2.53	222	2.55	213	2.60	200	2.63	186	2.69	168	2.75	154	2.85	2,801	2.53
Maintenance	143	1.00	128	1.00	114	1.00	104	1.00	98	1.00	92	1.00	87	1.00	82	1.00	76	1.00	69	1.00	61	1.00	54	1.00	1,108	1.00
Tires	50	0.35	45	0.35	40	0.35	36	0.35	34	0.35	32	0.35	30	0.35	29	0.35	27	0.35	24	0.35	21	0.35	19	0.35	387	0.35
Subtotal	555	3.88	504	3.94	455	3.99	419	4.03	398	4.06	378	4.11	360	4.14	344	4.19	323	4.25	299	4.33	268	4.39	244	4.52	4,547	4.10
Time items																										
Depreciation	795	5.56	622	4.86	474	4.16	351	3.37	264	2.69	198	2.15	150	1.72	115	1.40	87	1.15	68	0.99	50	0.82	37	0.68	3,211	2.90
Insurance	160	1.12	160	1.25	160	1.40	160	1.54	160	1.63	110	1.20	110	1.26	110	1.34	110	1.45	100	1.59	100	1.64	100	1.85	1,550	1.40
Licenses	30	0.21	29	0.23	28	0.25	27	0.26	26	0.27	25	0.27	24	0.27	22	0.27	20	0.26	20	0.29	20	0.33	20	0.37	291	0.26
Vestcharge, 6%	198	1.38	150	1.17	113	0.99	85	0.82	63	0.64	48	0.52	36	0.41	27	0.33	20	0.26	15	0.22	11	0.18	8	0.15	774	0.70
Subtotal	1183	8.27	961	7.51	775	6.80	623	5.99	513	5.23	381	4.14	320	3.68	274	3.34	237	3.12	213	3.09	181	2.97	165	3.05	5,826	5.26
Total mileage & time items	1738	12.15	1465	11.45	1230	10.79	1042	10.02	911	9.29	759	8.25	680	7.82	618	7.53	560	7.37	512	7.42	449	7.36	409	7.57	10,373	9.36

Other items	Yr 1		Yr 2		Yr 3		Yr 4		Yr 5		Yr 6		Yr 7		Yr 8		Yr 9		Yr 10		Yr 11		Yr 12		Total	
Auto Club	?	?	?	?	?	?	?	?	?	?	?	?	?	?	?	?	?	?	?	?	?	?	?	?	?	—
Accidents	?	?	?	?	?	?	?	?	?	?	?	?	?	?	?	?	?	?	?	?	?	?	?	?	?	—
Garage	120	0.84	120	0.94	120	1.05	120	1.15	120	1.23	120	1.30	120	1.38	120	1.47	120	1.58	120	1.74	120	1.97	120	2.22	1,440	1.30
Parking	74	0.52	73	0.57	71	0.62	70	0.67	69	0.70	69	0.75	64	0.73	60	0.73	55	0.72	50	0.72	45	0.74	40	0.74	740	0.67
Tolls	14	0.10	12	0.09	10	0.09	9	0.09	8	0.08	7	0.08	6	0.07	5	0.06	5	0.07	5	0.07	4	0.07	4	0.07	89	0.08
Property taxes, 4 percent	132	0.92	100	0.78	75	0.66	56	0.54	42	0.43	32	0.35	24	0.28	18	0.22	13	0.17	10	0.15	7	0.11	5	0.09	514	0.46
Subtotal	340	2.38	305	2.38	276	2.42	255	2.45	239	2.44	228	2.48	214	2.46	203	2.48	193	2.54	185	2.68	176	2.89	169	3.12	2,783	2.51
Grand total	2078	14.53	1770	13.83	1506	13.21	1297	12.47	1150	11.73	987	10.73	894	10.28	821	10.01	753	9.91	697	10.10	625	10.25	578	10.69	13,155	11.87

Operating data	Yr 1	Yr 2	Yr 3	Yr 4	Yr 5	Yr 6	Yr 7	Yr 8	Yr 9	Yr 10	Yr 11	Yr 12	Total
Value beg. yr., $	3300	2505	1883	1409	1058	794	596	446	331	244	176	126	
Mil. during yr.	14,300	12,800	11,400	10,400	9,800	9,200	8,700	8,200	7,600	6,900	6,100	5,400	110,800
Mil. gal., gas	14.9	14.6	14.3	14.1	14.0	13.8	13.7	13.5	13.3	13.0	12.7	12.3	13.84
Mi. qt., oil	360	345	330	315	300	285	275	260	245	230	220	210	287
Qt., oil	40	37	34	32	32	32	32	31	31	31	28	26	386
Gal., gasoline	960	877	797	737	700	666	634	609	571	532	480	440	8003

held, the fees are a proper charge against the car because they relieve other charges, particularly road services, and touring expense.

Accidents. Accident expense is not recorded because it is taken care of in the insurance expense except for any deductible items, particularly for collisions.

Garage. Not all cars are garaged–housed–at their home base by either renting or owning a garage. The home garage is generally used for storage and work purposes in addition to housing the car. The $10 a month charge is a nominal charge to include all expense items on the additional investment in the home because of the garage.

Parking. The expense of parking the car on trips away from home is a highly variable item. The nominal expense recorded in Table 13-5 is included more for the purpose of illustration than it is to represent average cost. Parking cost may vary from 5 cents to $3 a park, depending upon the place, length of time, mode (lot or garage), and size of car.

Tolls. As with parking, highway and bridge tolls vary considerably place to place, owner to owner. Some persons pay daily tolls to and from work, others pay only on out-of-town trips, while others, in the west particularly, pay only bridge tolls.

Property Taxes. In certain states and localities a personal property tax is levied on the value of the car, the same as for other personal property. In other states the registration fee is in lieu of property taxes. Some states do not tax the car at all as property. The tax shown is a typical tax based upon the value at the beginning of the year.

Variations of Operating Cost with Weight and Value of Car

Operating costs of motor vehicles increase with the road weight of the vehicle. Vehicle road weight directly affects fuel consumption, tire wear, horsepower of engine required, and other factors. In general, the vehicle weight affects the price new of the vehicle. Mileage, time, and ownership cost items are, therefore, affected by vehicle weight.

Claffey [14-8 and 14-52] and Sawhill [14-42] have reported fuel consumption with reference to vehicle weight. Stevens [13-6 and 13-7] gives truck operating cost by loaded gross vehicle weight.

But as is common with the subject, full operating costs of passenger cars by weight are scarce.

Table 13-6 reports the costs of operating the compact car, the medium-weight car (from Table 13-5), and the heavy car. This table is based on the collection of operating records of the author and bits of information from a variety of sources. A study of the literature discloses that many writers simply quote other sources. The most popular sources are those for salesmen's cars [13-5] and for the first 1 to 5 years of a car's life. Further, many of the reported operating costs do not give a sufficient description of the car, its use, and how certain items were calculated.

Vehicle Costs and the Highway

TABLE 13-6

HYPOTHETICAL COST OF OWNING AND OPERATING THE COMPACT, THE MEDIUM WEIGHT, AND THE HEAVY AMERICAN AUTOMOBILES

(Costs based upon 12 Years of Use)

Cost Item	Compact		Medium		Heavy	
	Dollars	Cents per Mile	Dollars	Cents per Mile	Dollars	Cents per Mile
Mileage items						
Engine oil	81	0.08	251	0.22	338	0.26
Gasoline	1,148	1.15	2,801	2.53	3,975	3.06
Maintenance	800	0.80	1,108	1.00	1,560	1.20
Tires	250	0.25	387	0.35	650	0.50
Subtotal	2,274	2.28	4,547	4.10	6,523	5.02
Time items						
Depreciation	2,359	2.36	3,211	2.90	6,325	4.87
Insurance	1,200	1.20	1,550	1.40	1,800	1.38
Licenses	230	0.23	291	0.26	515	0.40
Vestcharge at 6 percent	528	0.53	774	0.70	1,591	1.22
Subtotal	4,317	4.32	5,826	5.26	10,231	7.87
Total mileage and time items	6,596	6.60	10,373	9.36	16,754	12.89
Other items						
Auto club membership	?	–	?	–	?	–
Accidents	?	–	?	–	?	–
Garage	960	0.96	1,440	1.30	1,800	1.38
Parking	600	0.60	740	0.67	910	0.70
Tolls	60	0.06	89	0.08	117	0.09
Property taxes at 4 percent	352	0.35	514	0.46	1,060	0.82
Subtotal	1,972	1.97	2,783	2.51	3,887	2.99
Grand total	8,568	8.57	13,155	11.87	20,641	15.88
Informational items						
Cost new, dollars	2,400		3,300		6,500	
Terminal value, dollars	43		89		175	
Curb weight, lb	2,700		3,600		4,900	
Road weight, lb	3,000		4,000		5,500	
Total mileage	100,000		110,800		130,000	
Miles per gallon, gas	27.00		13.84		12.10	
Miles per quart, oil	800		287		250	
Gallons of gasoline	3,704		8,003		10,744	
Quarts of oil	125		386		520	

Some indication of the relative cost of owning and operating cars of different weights may be gained by reference to car rental rates. The following daily rates were in effect in the Washington, D.C., area in the fall of 1966:

Type of Car	Dollars per Day		Cents per Mile
Compact sedan	11	plus	10
Standard sedan	11	plus	11
Medium sedan	13	plus	13
Heavy sedan	16	plus	16

Air conditioning was priced at $2 per day plus 2 cents per mile additional to the above rates. These rates, of course, are for the latest model car.

COMMERCIAL VEHICLE OPERATING COSTS

Although prepared fot the purpose of a study of the desirable maximum limits of motor truck dimensions and weights Table 13-7 [13-6 and 13-7] gives line-haul trucking costs for a range of gross vehicle weights. These costs came from a nationwide analysis of the book records of common carriers, contract carriers, and private operators. The table is a weighted composite of gasoline- and diesel-powered vehicles. It does not include intracity hauling and delivery service as would be found in the local service, retail, wholesale, and construction operations, as is obvious from the minimum-loaded gross vehicle weight of 25,000 pounds.

The range in loaded gross vehicle weight up to 180,000 pounds is far beyond general practice. Most states have a gross weight limit of about 73,000 pounds, with the western states reaching up to 76,000 pounds. On the toll highways in Kansas, Indiana, Ohio, New York, and Massachusetts about 125,000 pounds gross vehicle weight is authorized by special permit to truckers using double 40-ft trailers in a combination about 100 ft in length.

The locally used small trucks as found in the service, retail, and wholesale business have relatively high operating costs because of the short-trip, stop-and-go, character of use.

HIGHWAY AND VEHICLE FACTORS COMPARED

Since about 1920 there has been more or less continuous research into the cost of operating motor vehicles and their performance on the road with respect to highway features. Even so, research has not kept pace with the constantly changing vehicles, changing use of vehicles, and traffic conditions. At no time, however, has research produced the whole of vehicle running costs necessary to estimate the effects on vehicle running cost of the various aspects of highway design. Fuel consumption has received the major attention, tire wear somewhat less attention, and the other factors but scant serious attention.

The general objective of research on motor vehicle running costs is to match in turn each of the following elements of highway design and traffic operation with each of the following factors of motor vehicle running cost:

Vehicle Costs and the Highway

TABLE 13-7

LINE-HAUL VEHICLE OPERATING COST IN CENTS PER VEHICLE-MILE
(References 13-6 and 13-7)

Loaded Gross Vehicle weight* kips	Cost Element						
	Repair, Servicing and Lubricant	Tires and Tubes	Fuel	Driver, Wage, and Subsistence	Overhead and Indirect	Depreciation and Interest	Total
25	5.00	1.43	2.67	11.50	11.22	3.12	34.94
30	5.06	1.48	2.67	11.62	11.26	3.67	35.76
35	5.17	1.55	2.68	11.74	11.30	4.20	36.64
40	5.33	1.63	2.68	11.86	11.34	4.73	37.57
45	5.53	1.73	2.69	11.98	11.39	5.24	38.56
50	5.78	1.84	2.71	12.10	11.43	5.75	39.61
55	6.07	1.97	2.72	12.23	11.48	6.25	40.72
60	6.40	2.12	2.75	12.36	11.53	6.73	41.89
65	6.78	2.28	2.77	12.49	11.59	7.21	43.12
70	7.20	2.46	2.80	12.62	11.65	7.67	44.40
75	7.67	2.66	2.83	12.75	11.71	8.12	45.74
80	8.19	2.87	2.87	12.88	11.77	8.56	47.14
85	8.74	3.10	2.91	13.02	11.83	9.00	48.60
90	9.35	3.35	2.95	13.16	11.89	9.42	50.12
95	10.00	3.61	2.99	13.30	11.96	9.83	51.69
100	10.69	3.89	3.04	13.44	12.03	10.23	53.32
105	11.42	4.18	3.10	13.59	12.10	10.62	55.01
110	12.20	4.49	3.15	13.73	12.18	11.01	56.76
115	13.03	4.81	3.21	13.88	12.26	11.38	58.57
120	13.90	5.16	3.28	14.03	12.33	11.74	60.44
125	14.81	5.52	3.35	14.18	12.42	12.08	62.36
130	15.77	5.89	3.42	14.34	12.50	12.42	64.34
135	16.78	6.28	3.49	14.49	12.59	12.75	66.38
140	17.83	6.69	3.57	14.64	12.68	13.07	68.48
145	18.92	7.12	3.65	14.81	12.76	13.38	70.64
150	20.06	7.55	3.74	14.97	12.86	13.67	72.85
155	21.24	8.01	3.83	15.13	12.95	13.96	75.12
160	22.47	8.48	3.92	15.29	13.05	14.24	77.45
165	23.74	8.97	4.02	15.46	13.15	14.50	79.84
170	25.06	9.48	4.11	15.63	13.25	14.76	82.29
175	26.42	10.00	4.22	15.80	13.36	15.00	84.80
180	27.83	10.54	4.32	15.97	13.46	15.24	87.36

* Loaded gross vehicle weight is the empty weight plus the weight of the most frequently carried payload, or approximately 80 percent of the practical maximum gross vehicle weight.

A. Highway factors
   1. Distance
   2. Minus and plus grades
   3. Horizontal curvature
   4. Speed and speed changes (stops and slowdowns)
   5. Roadway surface
B. Motor vehicle factors
   1. Fuel
   2. Tires
   3. Engine oil
   4. Maintenance
   5. Depreciation
   6. Accidents
   7. Travel time
   8. Personal preferences

The process by which appropriate unit costs of motor vehicle operation can be determined is a combination of several factors and conditions. First, suitable vehicles, representative of all vehicles of their class, must be chosen for the test work. Second, suitable measuring and recording instruments are needed. Third, highway test sections of sufficient length, proper geometrics, and adequate safety need to be selected for the test runs. Fourth, standards of vehicle and driver performance need to be selected for the purpose of getting consistency within the test program and close adherence to the performance of average traffic and drivers.

When all these conditions are met, good reliable performance and costs can be obtained for fuel consumption and tire wear only. Perhaps by extensive driving, some limited data on engine oil consumption can be obtained. Vehicle performance and running conditions such as engine and air temperatures, manifold pressure, engine revolutions per minute, tire rotations per mile, slip angles, throttle opening, and the like can be recorded during field runs. With these additional data, the vehicle can be further tested in the laboratory to gain additional information on fuel consumption, tire wear, oil consumption, brake wear, and other items.

General maintenance and repair expense is most difficult to measure with respect to the elements of highway design. By extensive breakdown of detailed maintenance operations and cost of vehicles operated under generally constant conditions, good headway can be made on this item.

Depreciation expense, another major cost item, is also hard to allocate to elements of highway design. Here again, perhaps close study of vehicles and their use could bring forth information on depreciation expense superior to what is currently available.

The Appendix A tables of motor vehicle running cost have been prepared by use of the literature, author's own records and test results, theoretical consider-

Vehicle Costs and the Highway 317

ations, and just plain judgment and estimate. These tables may be far from absolute truth, but they are somewhat relatively correct, which is the important consideration. Some of the considerations involved in the preparation of these motor vehicle running cost tables are given in the following sections.

FACTORS OF THE RUNNING COST OF MOTOR VEHICLES

Tables 13-2, 13-3, and 13-5 give the complete cost of owning and operating one specific passenger car in one ownership. Certain of these cost items, however, are unaffected by the elements of highway design. Consequently, in the following discussion attention is devoted only to the items of fuel, tires, engine oil, maintenance (including repair and replacement), and depreciation. These running cost items are discussed with reference to the highway desing elements of distance, minus and plus grades, horizontal curvature, speed changes, and roadway surface.

Fuel Consumption

The rate of fuel consumption of a motor vehicle is affected by many factors which may be classified into four groups: (1) the vehicle, (2) the highway, (3) the vehicle operator (driver), and (4) the weather and altitude.

The vehicle affects its own fuel consumption through its weight, tires, body size and design, engine design, power train, or power transmission to the driving wheels, and other lesser factors.

The weight on the wheels produces a rolling resistance directly proportional to the vehicle weight; therefore the heavier the vehicle, the more horsepower required to propel it over the roadway.

The body size controls the frontal area projected to the resisting air, and the body design controls the air friction. Air resistance in pounds varies about as the square of the vehicle speed, and in horsepower, air resistance varies as the cube of the speed. At high speed, therefore, a high percentage of the engine horsepower output is used to overcome air resistance.

Tires have their effect on fuel consumption primarily from their element of stiffness. The tires deflect at the patch of contact with the roadway surface and then expand to normal contour. This continuous deflection and expansion absorbs energy and generates heat. A tire at high pressure will have less rolling resistance than a tire at low pressure.

The engine design factors that affect the rate of fuel consumption are, among others, bore and stroke (displacement), compression ratio, carburetor design, manifold design, ignition system, cooling system, mechanical condition, and others. One of the universal characteristics of the internal combustion engine is that it builds up in horsepower developed as engine revolutions per minute increase to a maximum, then the horsepower falls off with further increase in revolutions per minute. This characteristic creates the need for a system of gear ratios in the power train whereby moderate engine speed can be obtained at low road speed in order to develop the required horsepower for starting and for steep plus grades. Also, the fuel consumption economy is low at power requirements less than about 50 percent of the potential power output at the speed. Figure 12-1 gives a typical horsepower-engine revolutions per minute curve.

The various combinations of vehicle designs and engine designs result in a variety of fuel consumption curves (expressed in miles per gallon of fuel) as shown in Fig. 13-1. Of general interest to the vehicle owner is the miles per gallon of fuel, but to the automotive engineer and highway economist gallons per mile is more significant. A third curve (Fig. 14-2) gallons per hour, is also useful, particularly in working with observed fuel consumption rates, because the curve is always concave upward against road speed.

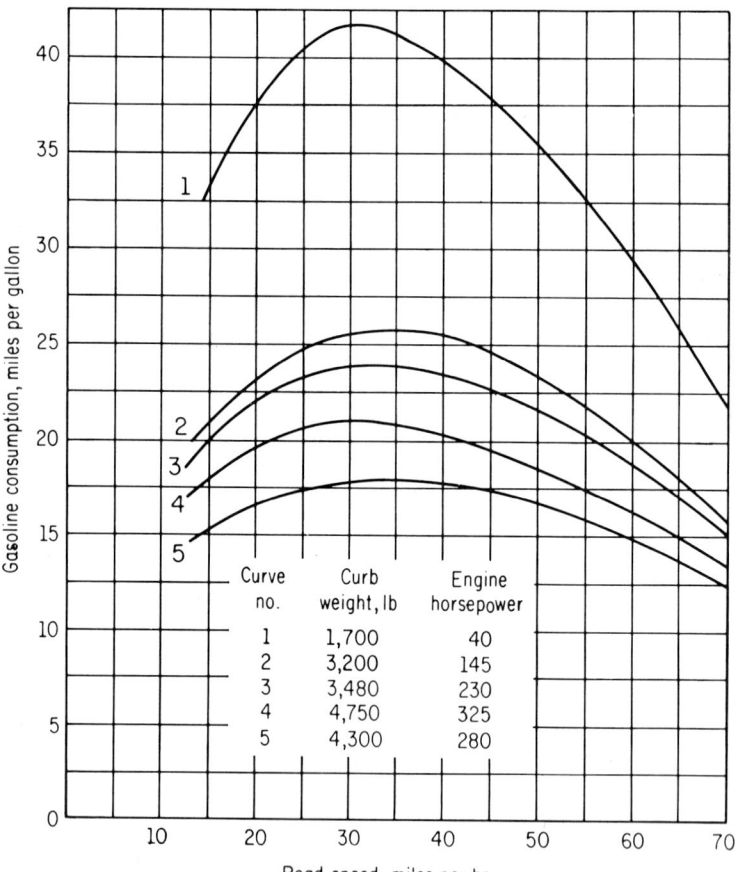

FIG. 13-1. Typical fuel consumption-speed curves of passenger cars for constant speed on level tangents.

Vehicle Costs and the Highway 319

It should be observed that fuel consumption varies with the mechanical condition and adjustment of the engine, the grade of fuel, and just how the operator controls the car. However, for the performance tables prepared for economy studies, the operator is not a factor under constant speed because the fuel consumption rate would not be changed. For speed changes, though, the operator would affect fuel consumption by his rate of deceleration and rate of acceleration.

Tire Wear

Other than the price new of motor vehicle tires, their cost in use results from three principal sources: punctures, carcass failures, and tread wear. Of these causes, only tread wear is closely associated with highway design.

Punctures are caused by foreign material in the path of the tire. Perhaps highway maintenance could keep the roadway clear of such objects, but such operation, once practiced on gravel roads by use of a magnetic device to pick up scrap iron, is no longer generally justified.

Tires, as they roll forward on the roadway surface, are in a continuous flexing action. This action causes friction within the layers of fabric. This friction generates heat and heat weakens the tire structure. Carcass failures are related to tire quality, inflation pressure, and weight on the tire. Underinflation and wheel loading above the design load cause early carcass failure. These failures to care for the tire properly and the resulting costs are not associated with highway design.

Tire tread wear results from three separate actions of the tire and roadway surface. First, as the tire rolls on the roadway surface the rolling friction causes wear of the tire tread. The second source of wear is the longitudinal and transverse forces which are resisted by frictional forces at the area (the patch) of contact of tire and roadway surface. Braking and accelerating forces and the force to hold the vehicle on a curved path cause tire wear above the nominal rolling-resistance wear. All turnings of the path of the vehicle from a straight line necessitate transverse forces to change the direction of travel from the straight line. The third main source of tire wear is the force on the drive wheels required to propel the vehicle against air resistance, against gravity when going up grade, and to retard the vehicle when deceleration is necessary. The driving force at the patch causes the drive wheels to slip as they roll. This slippage is also a source of tread wear.

These longitudinal and transverse forces result in the utilization of a higher coefficient of friction than is required solely for rolling friction and thus they account for considerable tire tread wear. The transverse forces cause the tire to actually skid on the pavement surface and grind off tread rubber. The slippage of the tire at the patch area as speed increases would be expected to result in more revolutions per mile of the drive wheels than at slower speed. But the contrary is true, at least between 35 and 60 mph, where the rear-wheel revolutions average about as follows for live sizes of passenger-car tires: 35 mph, 786.65 rpm; 45 mph, 783.39 rpm, and 60 mph 780.74 rpm. The decrease in the rate of decrease in the revolutions per mile between 45 and 60 mph as compared to 35 to 45 mph is some evidence that at some speed higher than 60 mph the revolutions per mile would begin to increase with speed.

The decrease in revolutions per mile of the drive wheels with an increase in vehicle speed, is probably accounted for by (1) higher tire temperatures with resulting higher pressure and larger rolling radius, (2) the centrifugal force on the tire which increases the rolling radius, and (3) the higher air pressure between the bottom of the car and the pavement and greater vacuum over the rear of the car both of which tend to reduce the weight on the drive wheels. These three factors act to increase the rolling radius and decrease the revolutions per mile which result opposes the tendency of the tire-pavement slippage to increase the revolutions.

At real slow speeds such as 5 mph and less, there is a creeping action of the tire tread which tends to reduce the wheel revolutions per mile from the number corresponding to the exact rolling radius of the tire.

Engine Oil Consumption

Engine oil is one of the three elements of vehicle running cost that is subject to direct quantitative measurement. But because of its low rate of consumption it is somewhat difficult to measure the rate of engine oil consumption as related to the elements of highway design and vehicle speed. By combining observations of engine oil consumption on the highway with laboratory tests, reliable rates of consumption can be established as related to engine speed and load.

The rate of consumption of engine oil is dependent upon such factors as coolant temperature, ignition timing, manifold pressure, engine revolutions per minute, engine load, fuel consumption rate, cylinder wear, piston-ring wear, crankcase ventilation, exhaust back pressure, leakage (valve stems, end bearings, and gaskets), and other factors.

Lubricating oil is consumed by burning and by leakage. Inside the engine, oil consumption results through the oxidation of the oil that passes the piston rings into the cylinder. At high vehicle speed and continuous driving, oil consumption by burning is relatively high because of the high engine temperature. At high engine speed, too, the piston rings have difficulty in following the contour of the cylinder walls, and therefore pass more oil than at low engine speed.

Oil consumption is more a factor of engine speed (piston travel) than it is of road speed. But oil consumption increases with vehicle road speed, despite the fact that, at constant gear ratio, the piston travel is the same per road mile at all road speeds. The piston travel in feet per minute does increase with road speed. At high speeds this increase in per minute speed of the engine results in "piling up" of heat in the engine block, pistons, and other engine parts, such that

Vehicle Costs and the Highway 321

the oil gets hotter. This hotter oil is of lower viscosity than cooler oil, and therefore, it passes the piston rings more readily. The cooling effect of the oil is also lost at high engine speed, partially because there is less time between engine strokes for the oil to drain from the cylinder walls back to the crankcase. The high crankcase temperature and high rate of splash induce a high rate of vaporization of the oil and more loss through crankcase ventilation.

At low vehicle road speed where the lower gears are used the engine speed is correspondingly higher. The result is that the oil consumption per road mile is higher than it is at a higher road speed in a higher gear ratio. The road speed for a passenger car favorable to low engine oil consumption is 30 to 40 mph.

Stop-and-go driving at low speed results in a cool motor which induces a high dilution rate with both fuel and water, varnish and sludge formation, and dirty oil. For slow driving and stop-and-go driving, the crankcase oil should be changed at lesser mileage intervals than is necessary at high-speed driving.

Oil consumption is measurable and can be related to the engine operation. But far more oil is "used up" by crankcase oil changes than is "burned up" in the engine. Crankcase drainage is an item of personal judgment, and not one that can be directly related to engine use. Generally, however, roadway and driving conditions that permit keeping the oil clean and at medium temperature will result in the middle range of vehicle miles per oil change. Sustained high-speed driving will result in more miles per oil change because of the additional make-up oil and the cleaner crankcase condition which results from a hot engine.

Vehicle Maintenance

Overall maintenance and repair of the vehicle result in an appreciable percentage of the total running cost of vehicle operation. Maintenance expense is defined as the monetary cost of cleaning, adjusting, repairing, replacing worn and damaged parts, lubricating (except engine oil), and antifreeze. Table 13-1 lists one classification of maintenance items and of all other costs of owning and operating a vehicle.

Maintenance expense is affected by highway design as well as by the character of vehicle use. In common with fuel, tires, and engine oil, maintenance expense of a motor vehicle is directly affected by the owner's care of the vehicle as well as by how he uses it. Unlike fuel, tires, and engine oil, there is no quantitative measure of consumption of the vehicle except as specific parts may be replaced. To further complicate the task of assigning vehicle maintenance expense to features of highway design, maintenance operations are necessitated at no regular mileage intervals. Some maintenance, such as replacement of brake linings, may not be required except at intervals of 20,000 to 40,000 miles. Yet each mile of driving contributes its share toward ultimate need for cleaning, adjusting, repairing, or replacing some part of the vehicle.

Maintenance expense may be divided into three groups: (1) that resulting directly from forward movement of the vehicle on the roadway and idling of the engine because of traffic conditions, (2) that resulting from off-highway use such as parking, garaging, engine warm-up, and engine idling waiting on loading of passengers or goods, and (3) weathering and loading and unloading of passengers and goods. For instance, the maintenance of doors, locks, seats, cargo body, and interior body finish results from causes other than driving on the highway.

Of these three groups, only the first one is affected by the geometrics of highway design and by traffic conditions. The difficulty is simply in assigning the maintenance expense to operation of the vehicle on plus and minus grades, to horizontal curvature, and to speed changes. Since the maintenance operations themselves cannot be reduced to specific quantities, resort must be had to factors of vehicle performance which bear some relation to ultimate maintenance operations. These factors are fuel consumption, oil consumption, horsepower developed, centrifugal force, brake pressure, rate of acceleration and deceleration, and time. The brake expense, for instance, is chargeable wholly to decelerating the vehicle and to retarding forward motion on minus grades. Engine expense is probably directly proportional to the factors of revolutions, speed of revolutions, and internal pressure. Thus on a roadway mile basis the engine running expense is greater per mile at speeds up to about 15 mph than it is at speeds of 20 to 40 mph because of the higher number of engine revolutions per roadway mile in the lower gears used at the low road speeds.

Because there is no means of direct measurement of the expense of maintenance as related to vehicle speed and the elements of highway design, resort must be had to allocation by judgment based upon the general factors of vehicle performance.

Vehicle Depreciation

The depreciation expense of a motor vehicle is a real cost related to time and use of the vehicle. A passenger car may cost new $3,400. Some years and several thousand miles later, the vehicle is sold or traded its last time for perhaps $25 to a used-parts dealer. There is then $3,375 of depreciation expense chargeable to the use of the vehicle. For economy studies of proposed highway improvements, at least some portion of this depreciation expense must be allocated to the use of the vehicle on particular highway designs and to particular traffic situations.

The end of the service life of the vehicle comes when the owner decides that the vehicle is no longer satisfactory for highway use. This decision is reached on the basis of three factors: (1) physical condition of the vehicle, (2) cost to restore the vehicle to an acceptable mechanical condition for highway use, and (3) styling–out-of-dateness–which makes a later model vehicle more desirable. During the years and miles of use of the vehicle, both time use and mileage use in combination, lead to the decision of the then vehicle owner not to use the vehicle any more.

Vehicle Costs and the Highway

Mileage use of the vehicle brings on a gradual wear and tear to all parts of the vehicle. In the end, the overall physical and mechanical condition becomes so inferior that personal judgment says "the vehicle is just not worth repairing, painting, and fixing; it is unsafe and unsatisfactory to drive, so I will trade it or junk it." Concurrently, with mileage use of the vehicle, time also brings a gradual physical deterioration to the car and an obsolescence. The elements of weather and time cause rusting, corrosion, and deterioration of the metal, rubber, plastic, wood, fabric, and other materials of the vehicle. Ultimately, this gradual deterioration of the vehicle, as with mileage use, reaches a state that causes the owner to conclude that the vehicle is no longer satisfactory for highway use.

Time contributes in another way to the ultimate discard of the vehicle. American passenger cars, and trucks to a lesser extent, reach at an early age a state of obsolescence that makes newer vehicles more desirable. The obsolencence factors embrace styling, performance, and capacity. Body styling is aesthetic, engine and brakes affect vehicles performance, and interior room and cargo space affect vehicle capacity and utility.

The dollar value of the vehicle on the used-vehicle market is not a factor in considering depreciation for highway economy studies because depreciation for economy studies should be based upon the entire service of the vehicle from purchase new to the "graveyard." The base for depreciation is total miles driven during its useful life. Miles are chosen rather than years because the depreciation expense must be stated in terms of miles or other unit of vehicle use in order to relate the depreciation to highway design.

The process of allocating vehicle depreciation to highway use is straightforward for overall service life; simply divide the cost new less tires, less scrap value, by the total life mileage. This answer is the overall depreciation expense per mile which is the unit of cost of concern to the owner. For economy studies, however, a unit cost of depreciation must be assigned to speeds, grades, horizontal curves, and speed changes.

Since the total vehicle depreciation charge comes about because of (1) mileage use, (2) time (weathering), (3) off-road use, and (4) passenger and goods hauling (which cause wear and tear to the interior of the vehicle) not all of the total depreciation should be charged to the features of highway design. There is not a proved just base on which to divide total depreciation between mileage, time, and nonhighway factors, so it must be done by judgment.

High annual mileage and high lifetime mileage justify that a higher percentage of the total dollar depreciation be allocated to mileage and a lesser percentage to time than would be justified in case of low annual mileage and low lifetime mileage. This judgment follows from the probability that these high mileages will cause the vehicle to be scrapped sooner in years than would be true with a low annual mileage or low lifetime mileage.

Further, it is reasonable that a high annual mileage would be obtained at high-speed driving and result in high life mileages. The allocation of some depreciation expense to time carries with it the assumption that the shorter life in years a vehicle would have, the lesser would be the percentage of the total dollar depreciation to be charged to time, weathering, and obsolescence.

Lastly, interior wear of the vehicle is proportional to neither mileage use nor time, but to the character and amount of use–that is, the amount of passengers and goods hauled and care of the vehicle.

Vehicle Speed

Road speed is an element in the running cost of a motor vehicle which is associated with distance, grades, curves, and speed changes. The consumption of fuel, oil, tires, and travel time are dependent upon the road speed of the vehicle. The basic cost information is therefore the running cost on level tangent highways for the full range of speed from zero to the maximum possible for the vehicle or an acceptable maximum, such as 80 mph, considering safety and legal limitations. Constant speed and speed changes are a factor of traffic volume and traffic composition. Roadway design, including pavement width, affects motor vehicle speed. Vehicle speed is a highly important factor in running cost, because speed affects each element of running cost.

ELEMENTS OF HIGHWAY DESIGN

The elements of highway design which affect the running cost of vehicles are distance, vertical grades, horizontal curves, speed charges, roadway surface, and all factors that influence vehicle road speed such as lane width, number of lanes, shoulder width, and traffic controls.

All features of highway design and traffic operations that affect vehicle speed and accidents also affect the running cost. If all of the highway design and traffic factors are to be evaluated in the analysis of the economy of highway improvements, the vehicle running cost must be established for distance, minus grades, plus grades, horizontal curvature, for changes in speed (including stops), and roadway surface.

Distance

The mileage elements of the cost of running a vehicle are fuel, tires, engine oil, maintenance, and depreciation. The total cost to the vehicle owner for these items is a function of the miles driven. The true cost for any one mile, however, will vary with speed, gradient, and horizontal curvature. When the speed is not uniform over the mile of distance, the change in speed causes a variation in the vehicle running cost.

Distance by itself is the major highway item in the running cost of vehicles, so the first essential is to establish the running cost of vehicles per mile of level tangent at the desirable range of speed.

Vehicle Costs and the Highway 325

Minus and Plus Grades

Vertical alignment of the highway is important in the running cost of the vehicle because of the effect of minus and plus grades on the speed of the vehicle and on the horsepower required. Minus grades have the opposite result from that of plus grades. For reasons of safety, a vehicle going down a given percentage grade should travel at a slower speed than when going up the same percentage grade. Gravity adds to the stopping distance going down and substracts from it going up. To overcome gravity going up a grade requires more horsepower than is required at the same speed on the level. Grade horsepower is given by Eq. 12-15 from which it is seen that the horsepower required for the gradient alone varies directly as the percentage grade.

This additional horsepower to overcome a plus grade resistance as compared to a level roadway, results in the consumption of more fuel, more engine oil, more tire tread, and more engine wear. A minus grade has the opposite results, except when the combination of minus grade and speed requires negative horsepower–i.e., braking. Braking, when accomplished with the wheel brakes, would cause both brake system and tire wear. When the engine is used as a brake (and it always is used when the engine and transmission are in gear except for automatic transmission) the result is to cause additional oil consumption, tire wear, and some engine wear.

On minimum grades steep enough to overcome the internal resistance in the power train and engine (when in gear), rolling resistance, and air resistance, each vehicle will coast down the grade at some definite constant speed. This speed is the "floating speed" for that vehicle for that grade. Under this condition, the engine requires no throttle and the vehicle requires no braking with the brake system to maintain the floating speed.

Horizontal Curves

Horizontal curves introduce into the operation of a vehicle a complicated set of forces which are not experienced when the vehicle is moving on a tangent. Because of continuous changing of direction when traveling a curved path, the vehicle is subjected to a centrifugal force. This force may or may not be balanced by superelevation of the roadway surface. In order to change the direction of the vehicle to the path of the curve, a coefficient of friction is developed at the contact of the front tires and roadway surface which is higher than that developed on the tangent at the same speed, and, thus, more horsepower is required.

This coefficient of friction is utilized in resisting the driving or retarding force along the line which the wheels of the car tend to travel and in opposing the centrifugal force which acts normal to the path of the vehicle. At the drive wheels, the force in the line of travel varies with the force required to drive the vehicle forward and with the braking force when decelerating. This force at the steering (front) wheels varies also with the braking action. The frictional force required to resist the centrifugal force varies with the speed, radius of curvature, and superelevation. For any given speed and radius of curve there is a superelevation that will exactly balance the centrifugal force. The superelevation introduces a gravitational force. At speeds less than and greater than the speed required to exactly balance the centrifugal and gravity forces, a lateral frictional force is introduced.

Probably more often than not, vehicles traveling on horizontal curves do not travel at the exact speed which balances the centrifugal and gravitational forces. One reason for this truth is that often design conditions, intersecting routes, and winter icing conditions compel the use of a superelevation less than the theoretical rate for the design speed. When the vehicle travels at this slower or faster speed, the front wheels develop a negative or a positive slip angle in order to create the additional friction required to hold the vehicle on the curved path. This slip angle is the difference in the true steering angle required for the particular radius of horizontal curve and the actual steering angle used. The slip angle is also defined as the angle between the horizontal diameter of the steering (front) wheel and the actual roadway path of that wheel. A similar slip angle is developed at the drive wheels, and at other wheels in combination vehicles. These wheels appear to travel in a direction which they do not; they thus skid or twist around the curve rather than roll around.

The force from the drive wheels pushes the front steering wheels ahead, but not in the true direction of the circumference of the curve. Therefore, the front wheels likewise skid along. These actions of the tires on the roadway surface and the forces developed result in higher horsepower requirements than are required at the same speed on a tangent. This additional driving horsepower requires more fuel. Likewise, the higher coefficient of friction developed between the tires and roadway surface and the skidding action, as compared to the straightforward rolling action, cause a high rate of tire tread wear. In fact, under the most unfavorable condition, the rate of tire tread wear on horizontal curves may be 500 or more times the rate of wear at the same speed on a tangent.

That additional fuel is required on horizontal curves above that at the same speed on a tangent was experienced by the author when driving a car up a 7 percent grade at a series of high speeds in fuel consumption tests. Near the top of this grade was a horizontal curve. Upon steering into the curve the car speed dropped, even though the throttle was held constant at the same point required to maintain the constant speed on the tangent portion of the grade. When making theoretical computations of the horsepower and fuel requirements on horizontal curves, it became apparent why the constant speed could not be held on horizontal curves without giving the engine additional throttle.

Another reason why additional fuel is required on horizontal curves is that, at high speed, there is a noticeable increase in air resistance over that at the

Vehicles Costs and the Highway

same speed on a tangent. As the vehicle turns into the curve, it takes somewhat of a sidewise position to the direction of travel. This new position creates additional frontal projected area which results in an increase in air resistance. Also, this new position increases the skin friction, air turbulence, and rear vacuum. Because air resistance horsepower varies as the cube of the vehicle speed, these changes in air resistance characteristics on horizontal curves result in a noticeable increase in fuel consumption at high speeds.

On the basis that horizontal curves cause an increase in horsepower it can be assumed that there will follow increases in fuel consumption and in engine oil consumption and also increased maintenance expense of the engine and of the power train. Likewise, the body and chassis maintenance expense would be increased because of the large lateral, centrifugal, and overturning forces introduced. There is no direct approach to measurement of these added oil and maintenance expenses, but they may be related to such known factors as fuel consumption, horsepower, and the magnitude of the forces themselves.

Speed Changes

The running cost of a motor vehicle is less at a specific constant, or uniform, speed than at a variable speed which averages out to the same uniform speed. The construction of the carburetor and fueling system is such that richer fuel-air mixtures pass into the manifold when the throttle is depressed. Further, the excess fuel used to accelerate above or to decelerate back to the uniform speed is more than the fuel not used during the time the throttle is not opened so wide. The whole mechanical system of the vehicle absorbs variable forces, both in direction and in magnitude during changes in vehicle speed. These forces, and the operation of the linkage system and braking system, introduce stress, strain, and wear. The overall result is that running costs are increased with speed changes over what the costs would be at uniform speed.

Fuel consumed during change in speed of the vehicle can be measured with a fuel meter. Tire tread wear can be measured by tread-depth loss or weight loss occurring during the driving of a series of speed change movements, or by special laboratory testing equipment. Engine oil consumption during speed changes is difficult to measure, but can be done by laboratory simulation tests. Mechanical maintenance due solely to speed changes, cannot be measured directly. That it does exist is well proved by maintenance cost records on urban-used vehicles in delivery services and taxicab services. These vehicles have a high number of speed changes per mile and high maintenance expense per mile as compared to similar vehicles used mainly on rural highways.

Because of the time consumed in changing vehicle speed over that consumed at uniform speed, as well as the extra wear and tear experienced, the vehicle probably would be driven somewhat fewer miles during its life than it would be under uniform driving speed. Therefore some depreciation expense is chargeable to speed changes.

Vehicle speed changes are important to highway economy studies because the running cost dollars and travel time consumed during speed changes are a direct measure of the cost of congestion under heavy traffic and of the cost of traffic control systems when vehicle stops are required.

Roadway Surface

The roadway surface has four characteristics that affect the running cost of motor vehicles: (1) flexibility of its structure, including firmness, (2) abrasiveness of the surface, (3) roughness of the surface, and (4) dustiness and looseness of the surface.

As between a rigid pavement and one that has a yielding surface, there is a difference in rolling resistance. A bituminous surfacing that becomes soft under heat from the sun is depressed by the weight of the vehicle wheel. The tire is then forced to endeavor to climb upward at the forward edge of its patch. This climbing tendency forces the depression of more soft pavement as the wheel rolls forward. This action increases the fuel consumption because of the increase in rolling resistance.

A roadway surface that is abrasive in character causes a higher rate of tire wear than does a nonabrasive surface. Pavements made with aggregates of chat, crushed gravel, or granite, and broomed concrete, result in higher tire wear than does a limestone chip aggregate, or a highly glazed surface [Moyer & Tesdall, 14-29]. At the high speeds the power wheels actually slip at the contact patch on tangents and all wheels slip or skid somewhat on horizontal curves. This slipping action grinds off rubber. The rate of grinding increases with the abrasive character of the roadway surface.

Roadway surfaces may be rough and unequal in surface contour because of cracks, settlements, or poorly made maintenance patches, and naturally so when they are unbound gravel or crushed stone. This unevenness of the roadway surface causes bouncing of the vehicle vertically, and larger movements laterally. As these vertical and horizontal movements occur, there are power losses, momentum forces to overcome, and variations in the frictional forces at the tire patch. The results are increases in the rate of fuel and tire consumption. There is also more wear and tear on the vehicle generally which ultimately results in maintenance expense and a lesser number of miles driven during the life of the vehicle.

Unbound gravel, crushed stone, and earth surfaces are dusty and loose when dry. The dust adds materially to the rate of engine and brake wear and to cleanliness of the vehicle inside and outside. These high rates of wear and dirty conditions result in higher maintenance expense than is experienced on dust-free roadways. Loose aggregate, when used as a roadway surface, induces more slip of the power and brake wheels on tangents, and of all wheels when changing direction of travel. This slipping results in power loss and an increase

Vehicle Costs and the Highway

in tire wear. For these reasons loose roadway surfaces result in materially higher vehicle running cost than applies to hard, firm pavements.

Other Design Factors

Other elements of the design of highways have little effect on the running cost of vehicles, or if they do their effect is measured in the foregoing factors. Lane width, number of lanes, medians, shoulder width, traffic controls, and sight distances each affect the running cost of vehicles, but mainly by controlling speed or by introducing speed changes. The factors of traffic volume and traffic composition also affect the vehicle speed. For these reasons, the vehicle running cost tables in Appendix A for speed and speed changes provide for these design factors.

PROBLEMS

13-1. For the overall cents per mile cost in Table 13-2, express each cost item in percentage of the cost per mile for gasoline. Make the same calculations for the total cents per mile cost in Table 13-5. Compare and discuss your results.

13-2. Analyze the data in Table 13-4 with the objective of establishing the relation of miles per gallon of gasoline consumption (1) to the month of the year, and (2) to the total miles driven per month. What conclusions do you draw?

13-3. Using the car operating cost data in Table 13-5, determine the year that the car should be traded for a new identical car, based on operating costs only.

13-4. Read the miles per gallon at each 5 mph for curve 3 of Fig. 13-1. Convert these readings to gallons of fuel consumption per hour and plot the curve of the results. By reference to Table 12-1, for the 4-kip passenger car and your gallons per hour curve compute the fuel consumption at each speed in gallons per horsepower-hour. Plot the curve of your results. Comment on all of your curves.

13-5. Annual running costs (other than driver's wages) for operation and maintenance of certain motor trucks under particular operating conditions are $2,200 at the end of the first year, $2,600 at the end of the second year, and increasing thereafter at $400 a year. The investment cost of a truck is $8,800. The estimated terminal values are: $2,300 after 4 years, $1,900 after 5 years, $1,600 after 6 years and $1,400 after 7 years. At a minimum attractive rate of return of 10 percent what would you recommend as the number of years to keep these trucks in service?

13-6. Analyze the ten mileage cost items in Table 13-2 to see if you can establish a correlation between any pair of cost items on a year to year basis. Discuss your results in terms of the degree of correlation discovered.

13-7. As you observe car drivers from day to day, what actions do you see that tend to increase the running cost of fuel, oil, tires, and vehicle maintenance? What actions do you observe that would tend to keep the costs to a minimum?

13-8. Compare the costs item by item of car operation in urban driving to rural driving. Which area of driving would be hardest on the mechanical condition and overall mileage life of the car, and why?

13-9. For identical model cars two years old and the same miles driven, how would you evaluate (used-car price) them considering that car A had been driven by an urban police department and car B by a traveling salesman?

REFERENCES

13-1. Thurley A. Bostick and Helen J. Greenhalgh. *Relationship of Passenger-Car Age and Other Factors to Miles Driven.* Highway Research Board, National Academy of Sciences, Washington, D.C., Highway Research Record No. 197, 1967, pp. 25-35.

13-2. Hermann Botzow. *An Empirical Method for Estimating Auto Commuting Costs.* Highway Research Board, National Academy of Sciences, Washington, D.C., Highway Research Record No. 197, 1967, pp. 56-69.

13-3. Chicago Area Transportation Study, Vol. III, *Automobile Operating Costs at Various Speeds,* April 1962, p. 126.

13-4. C. K. Glaze and George Von Mieghen. *Washington Motor Vehicle Cost Survey, 1952-1953.* Washington State Highway Commission, Department of Highways, Olympia, 1958.

13-5. R. E. Runzheimer, Jr. "Reimbursing Salesmen for Auto Operating Expenses," *Sales Management* (Dec. 6, 1963), pp. 38-40.

13-6. Hoy Stevens. *Line-Haul Trucking Costs in Relation to Vehicle Gross Weights.* Highway Research Board, National Academy of Sciences, Washington, D.C., Bulletin 301, 1961.

13-7. Hoy Stevens. *Line-Haul Trucking Costs Upgraded, 1964.* Highway Research Board, National Academy of Sciences, Washington, D.C., Highway Research Record 127, 1966. This publication updates the 1955-56 costs given in the preceding reference.

13-8. U.S. Department of Agriculture. *Logging Road Handbook–The Effect of Road Design on Hauling Costs.* Handbook No. 183, Forest Service, Washington, D.C., 1960.

13-9. U.S. Department of Commerce. *Automobile Driving Costs in the Northeast Corridor of the United States.* A staff report, 1967.

13-10. Robley Winfrey. *Research on Motor Vehicle Performance Related to Analyses for Transportation Economy.* Highway Research Board, National Academy of Sciences, Washigton, D.C., Highway Research Record No. 77, *Engineering Economy,* 1963.

Chapter **14**

Vehicle Running Costs for Economy Studies

Chapters 12 and 13 provide the setting from which motor vehicle running costs may be assembled for use in the analysis for the economy of proposed highway improvements, both economic evaluation and project formulation. Because on-the-road and laboratory testing of vehicles has not yet provided all of the essential costs and performances necessary to the assembly of motor vehicle running costs of highway economy analyses, resort was had to theoretical calculations and judgments where necessary of fill the gaps or to extend beyond observed values. The full set of the cost tables in Appendix A, however, provides reasonable and comparable running costs for practically every occasion.

RUNNING COST TABLES—AN EXPLANATION

Some explanation of the source of information, the assumptions made, and judgment exercised which were associated with the preparation of the motor vehicle running cost Tables A-1 to A-25 (Appendix A) and the fuel consumption Tables A-26 to A-40 will be helpful to the reader. They are tables for use in the analyses for the economy of highway improvements as contrasted to tables of the operating cost of motor vehicles, including all cost elements.

BASIC PREMISE

The running cost of motor vehicles can be determined on a reasonable basis for use in economy studies by giving proper attention to the vehicle and highway factors which cause the running cost to vary. Yet in producing these costs so that they are representative of the millions of vehicles that use the highways many possibilities for error in judgment are encountered. Nevertheless, in the interest of making analyses of the relative economy of proposed highway improvements, these risks of judgment must be taken.

The literature is short on practically all aspects of motor vehicle performance and running cost information of the type necessary to support tables of the type presented in this book. In the long, tedious process of preparing these tables, recourse was had to published sources, private files, personal experience and tests, correspondence with individuals in the automotive industry, and personal judgment. At the outset, a decision was made to complete the tables for a full range of speeds and for five specific vehicles for fuel, tires, engine oil, maintenance, and depreciation, each for minus and plus grades, horizontal curves, and speed changes regardless of the availability of test results. This decision was based on the premise that allocating cost by judgment was preferable to the decision not to allocate. For instance, considerable personal judgment was exercised in allocating maintenance expense to vehicle speed. The allocations made, even though not substantiated by test data, should produce comparative results in economy studies which are an improvement over what would be obtained under the decision to hold vehicle maintenance expense constant for all speeds and all elements of highway design.

In operating a motor vehicle there are movements and uses that cause consumption of fuel, tires, and oil, and wear and tear on the vehicle generally and are not chargeable to highway design. For instance, parking and garaging operations are not affected by highway design (except as local traffic requires a specific type of parking and idling of the engine when waiting for a passenger); speed changes incident to pickup and delivery servi services do not result from highway design; and offhighway vehicular use is not related to highway design. These factors, however, each result in added operating cost of the vehicle. These costs are not controllable through highway design, and, therefore are not included in the tables of running costs in Appendix A.

Of the 1,000, 5,000, 10,000 or 20,000 vehicles which may use a specific highway section or facility in a day, there is a wide range of speed, fuel consumption, tire tread wear, and oil consumption, vehicle-to-vehicle. These differences arise from four main sources: (1) basic differences in vehicle design, (2) differences in quality of fuel, tires, and oils, (3) differences in performance of the vehicle because of age and mechanical condition, and (4) differences in the performance of the vehicle in motion as caused by the operator. The operator's maintenance care of the vehicle also affects its running cost and performance on the road. Despite these differences, vehicle-to-vehicle, the analyst for highway transportation economy is compelled to work with good estimates of the running cost of vehicles. He does this by use of typical vehicles, since he cannot know the running cost of each vehicle in his traffic stream. Care, then, must be exercised in selecting typical vehicles, then working up the running cost tables for these typical vehicles as affected by speed, geometrics of highway design, and the requirements of operating in traffic.

TYPICAL VEHICLES

Whether the analyst of the economy of a proposed highway improvement gets his answer by using a computer program or by manual calculations, it is necessary to know in dollars the running cost of motor vehicles under the particular

Vehicle Running Costs for Economy Studies

conditions of highway design and traffic operations which are applicable. Obviously, resort must be had to average costs for typical vehicles as a means of representing the entire existing traffic or expected traffic.

These specific vehicles need to be representative of their class of vehicle on the highway. Traffic streams are usually composed of two broad classes of vehicles–those carrying people and those carrying commodities of some sort. People-carrying vehicles are passenger automobiles and buses. Since buses are few in number compared to the total traffic volume, they may be classed in the appropriate weight group of trucks or commercial vehicles. Commodity-carrying vehicles include the full range of panel, pickup, stake, platform, van, tank, and other types of commercial and industrial vehicles from two axles upward.

The whole traffic stream may be well represented by five typical vehicles, each representative of a class of vehicle, which can be identified in the traffic stream by visual means. These typical vehicles are: (1) 4-kip (4,000 lb) passenger car, (2) 5-kip commercial delivery vehicle, (3) 12-kip single-unit truck having dual rear wheels, (4) 40-kip gasoline-powered tractor-semitrailer combination having four axles (2-S2), and (5) 50-kip diesel-powered tractor-semitrailer having five axles (3-S2). These vehicles are described in detail in Table 14-1. Figure 14-1 gives a schematic silhouette of typical vehicles, including the five used as representative of their classes.

The 4-kip passenger car is representative of the whole range of passenger cars from the small light vehicle manufactured outside of the United States up through the domestic compacts to the heavy Cadillacs, Imperials, and Lincolns. Automobiles used for business and professional purposes are included in the passenger car class. The gross road weight for the typical automobile was taken as 4,000 lb. From limited actual weighings on the roadway and classification of traffic, 4,000 lb is reasonable for the typical passenger car. The weight includes about 1.7 persons, personal property, and the usual accumulation of foreign materials in, on, and under the car.

The commercial delivery vehicle is representative of the full range of light commercial vehicles used in retail delivery and service operations. Primarily, these vehicles are, panels, pickups, and various other bodies on a 2-axle, 4-tired chassis. The 5-kip gross weight recognizes the cargo carrying capacity of this type of vehicle, as well as those vehicles in the class which have an empty weight heavier than the pickup or panel vehicle. Many of these vehicles are simply commercial bodies on a passenger car type of chassis with some modifications in engine and gear ratios. For the purpose of many uses of traffic classification, these commercial delivery vehicles may be included with the passenger car group.

The 12-kip single-unit truck is typical of the large number of truck vehicles used in hauling a wide variety of commodities in a wide range of truck bodies, both intracity and intercity. The dual rear tires provide for a rear-axle load about double that which can be carried on the rear axle of the commercial delivery type of vehicle. This single-unit truck is representative of a rather wide range

TABLE 14-1

DESCRIPTION OF VEHICLES APPLICABLE TO THE RUNNING COST TABLES

Descriptive Item	4-kip Passenger Car	5-kip Commercial Delivery	12-kip Single Unit Truck, Gasoline	40-kip 2-S2, Gasoline	50-kip 3-S2, Diesel
1. Curb weight (tare weight), lb	3,500	3,600	5,745	24,000	30,000
2. Maximum rated pay load, lb	900	2,000	7,750	33,200	41,000
3. Gross road weight used in running cost tables, lb	4,000	5,000	12,000	40,000	50,000
4. Gross weight rating, lb	4,400	6,000	14,000	57,000	71,000
5. Body type	4-door sedan	Panel	Stake-12 ft	Closed van	Closed van
6. Height, ft	4.8	6.2	7.2	12.0	12.0
7. Width, ft	6.5	6.6	7.7	8.0	8.0
8. Length, ft	17.0	16.6	21.5	45.0	53.0
9. Frontal projected area, sq ft	26.0	33.0	45.0	90.0	90.0
10. Wheel base, power unit, ft	9.7	9.6	13.1	13.1	16.9
11. Wheel base, semitrailer, ft	–	–	–	19.0	23.0
12. Number of axles/wheels	2/4	2/4	2/6	4/14	5/18
13. Size of tires/ply rating	7.50 × 14/4	7 × 17.5/6	8 × 22.5/8	11 × 22.5/12	11 × 22.5/12
14. Rear axle ratio	3.92	3.73	6.17	7.17	6.167
15. Transmission gear ratios:					
First	2.78	2.94	7.06	7.25	5.19
Second	1.62	1.68	3.58	3.88	2.88
Third	1.00	1.00	1.71	2.19	1.72
Fourth	–	–	1.00	1.37	1.00
Fifth	–	–	–	1.00	–
Auxiliary-1	–	–	–	1.24	1.29
Auxiliary-2	–	–	–	1.00	1.00
Auxiliary-3	–	–	–	0.88	0.84
16. Number of cylinders	8	6	6	6	6
17. Bore and stroke, in.	3.62 × 3.66	3.88 × 3.25	3.88 × 4.12	4.87 × 3.58	4.25 × 5
18. Displacement, cu. in.	302	230	292	401	426
19. Gross horsepower and rpm	187/3800	140/4400	165/3800	210/3400	218/2100
20. Net horsepower and rpm	163/3600	120/3600	147/3600	182/3400	200/2100
21. Idling speed, rpm	450	500	600	650	600
22. Engine oil capacity, qt	5	5	6	8	18
23. Cost w/tires, power unit, $	2,510	2,500	3,600	9,000	15,000
24. Cost w/tires, semitrailer, $	–	–	–	4,500	5,000
25. Cost, set tires (no spare), $	92	128	396	1,680	2,160

of vehicle types and gross road weights, but the trucks in this class are easily identifiable and sufficient in number to warrant recognition.

The 3-axle single-unit truck is not used extensively enough to warrant developing running cost tables for it alone. This 3-axle truck is used mainly in urban areas in construction work for hauling excavation and concrete mix. It is not a practical over-the-road vehicle so is not frequently found in rural traffic. For purposes of analyses for highway economy, the 3-axle single-unit truck may be put in the 40-kip 2-S2 class.

Vehicle Running Costs for Economy Studies

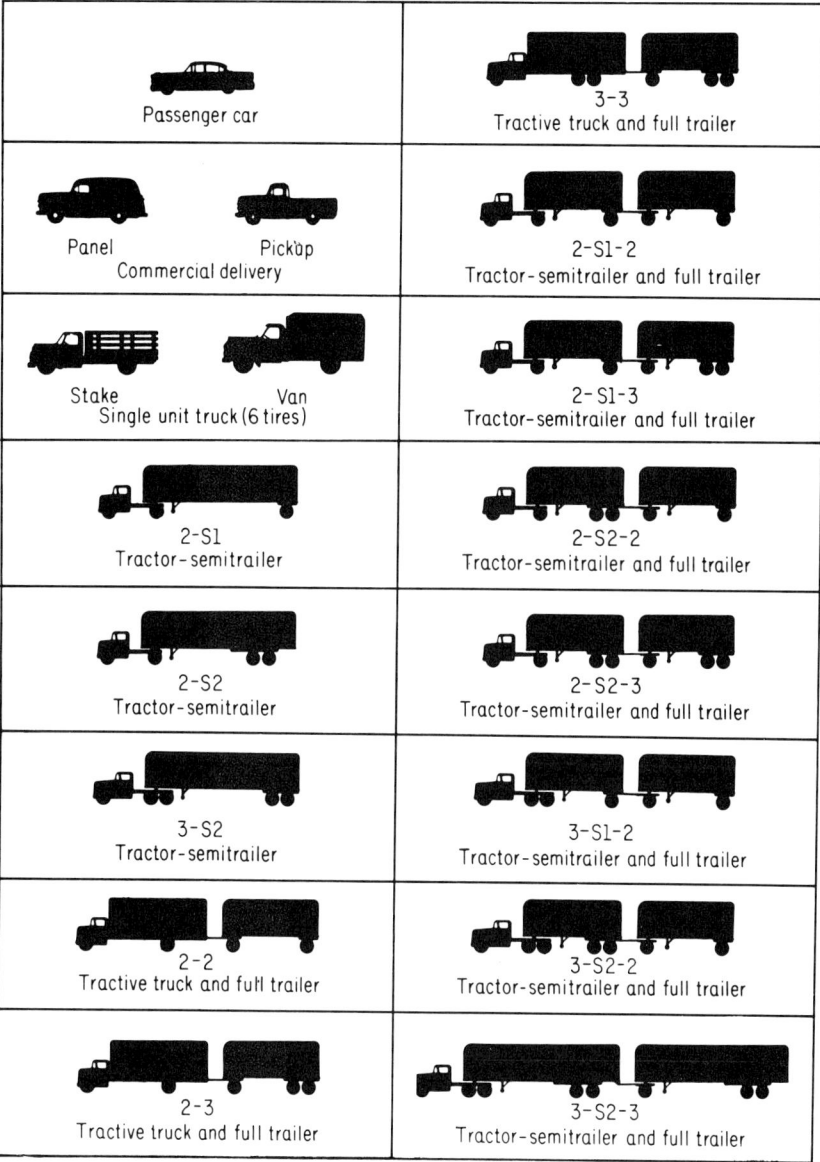

FIG. 14-1. Vehicle types and code designation based on axle arrangement.

The 40-kip 2-S2 tractor-semitrailer combination with gasoline fueled engine is chosen to represent all of the combination vehicles of 3 and 4 axles, including those having diesel power. The weight of 40 kips is close to the weighted average gross road weight of these several types of vehicles. The 40 kips is also about the average gross of the 2-S2 vehicle on the highway. The van body is the most frequently used type. Other popular bodies include the rack van (livestock), stake, platform, tank, and automobile carrier.

The 50-kip 3-S2 diesel is chosen as representative of all the combination vehicles of 5-axles and more, regardless of type of engine fuel. This 3-S2 represents a wide range of types of combinations and of gross road weights. The numbers of vehicles, however, in any one axle or gross weight classification is relatively few. Nationwide, the 3-S2 is the most popular of all combinations of 5 axles and more. The 50-kip weight is representative of the average gross road weight of the wide range of vehicles to be included in this group.

Although these five specific vehicles are chosen to represent five classes of vehicles into which the whole traffic stream may be subdivided for convenience and when traffic composition warrants, these five classes may be combined into fewer classes. When so doing, the running cost from the tables should be adjusted as may be appropriate.

Similarly, for a specific proposed highway improvement wherein the truck traffic vehicles are noticeably of lighter or heavier gross weight, the running-cost table values should be adjusted accordingly.

UNIT PRICES

The unit prices for fuel, oil, and tires on which the running-cost tables are based are given in Table 14-2.

TABLE 14-2

UNIT PRICES USED IN CALCULATING THE RUNNING-COST TABLES IN APPENDIX A

Cost Item	4-kip Passenger Car	5-kip Commercial Delivery	12-kip Single-Unit Truck	40-kip 2-S2 Gasoline	50-kip 3-S2 Diesel
Fuel, cents per gallon*	23	22	20	18	16
Engine oil, cents per quart†	60	55	40	20	20
Tires, cents per 0.001 in. of tread wear per tire	8.712	9.846	13.235	15.658	15.658

* Does not include state and federal motor fuel tax at about 11 cents per gallon, total.
† Includes local sales tax.

Gasoline, diesel fuel, engine oil, and tire prices vary widely by quality of product, geographical location, purchasing power of the purchaser, and status of the market. No attempt was made to calculate average national prices as of a given date, but instead reasonable and typical prices were adopted.

Fuel taxes, local, state, and federal, are not included in the price of motor fuel to avoid double counting as explained in the next section.

Vehicle Running Costs for Economy Studies 337

EXCLUSION OF FUEL TAX IN ECONOMY STUDIES

Past studies of the economy of highway improvements have generally included the cost of motor vehicle fuel in the cost of operating vehicles at a price including the state and federal tax on fuels. The question is properly raised whether including the fuel tax in the price is not a form of double counting, because such tax income is used mostly for constructing and maintaining highways.

The answer is that the fuel tax, to the extent that such tax income is devoted to constructing and maintaining the highway facilities being studied, should be excluded from the price of fuel used in computing the running cost of motor vehicles when the running cost is to be used in economy studies or when the total cost of highway transportation is being computed. Perhaps discussion of this answer can best be approached through the with and without concept.

EFFECT OF THE FUEL TAX ON THE SOLUTION FOR ECONOMY

With the fuel tax included, the retail price of gasoline at the service station may be assumed to be 34 cents a gallon of which about 11 cents is fuel tax, 7 cents state and 4 cents federal. Without the tax the price per gallon of gasoline for economy studies would be 23 cents. Including the tax as compared to excluding it would increase the cost of fuel, and, likewise, increase the dollar benefit from any motor vehicle fuel saving by 11 ÷ 23 or 47.8 percent. For a type of highway improvement project which increased the fuel consumption, such as an improvement greatly increasing the nominal travel speed, the dollar benefits would be understated if the fuel tax were included.

Should the fuel tax be included in the motor vehicle running costs, the ratio of benefits to costs would be overstated as compared to the ratio excluding the fuel tax. The following illustration bears out this conclusion:

1. Annual highway economic costs $100,000
2. Annual road user benefits, including fuel tax $150,000
3. Annual road user benefits, excluding fuel tax $126,000
4. Benefit/cost ratio, including fuel tax 1.50
5. Benefit/cost ratio, excluding fuel tax 1.26

Table 14-3 illlustrates the effect of the fuel tax on calculation of the rate of return.

The difference in final rate of return or benefit/cost ratio with and without the fuel tax is dependent upon the ratio of fuel costs to other motor vehicle running cost items and the relative amount of fuel consumption decrease or increase in the improvement proposals compared. But the above calculations show that there will be a difference in the calculated economy. It remains to show why the correct solution is the one without the fuel tax.

TABLE 14-3

EXAMPLE TO SHOW THE EFFECT ON THE CALCULATED RATE OF RETURN
OF INCLUDING AND EXCLUDING THE FUEL TAX IN THE RUNNING COST OF MOTOR VEHICLES

Cost or Descriptive Item	Including the Fuel Tax		Excluding the Fuel Tax	
	Existing Facility	Proposed Facility	Existing Facility	Proposed Facility
Price of fuel per gallon	$0.34	$0.34	$0.23	$0.23
Vehicle-miles per gallon of fuel	14	17	14	17
Running cost per mile, fuel only	$0.0243	$0.0200	$0.0164	$0.0135
Running cost per mile, without fuel .	$0.0450	$0.0400	$0.0450	$0.0400
Running cost per mile with fuel	$0.0693	$0.0600	$0.0614	$0.0535
Project length, miles	1.0	1.0	1.0	1.0
Vehicle-miles per year, 1000's	4,000	4,000	4,000	4,000
Annual motor vehicle running cost .	$277,200	$240,000	$245,600	$214,000
Cost to construct new facility	Sunk	$400,000	Sunk	$400,000
Rate of return when $n = 25$ years ...	–	7.9%	–	6.1%

The license, or registration fee, is also a road user tax that goes to the highway fund. Because this tax is levied against a base easily separated from the running cost of vehicles, there has been no suggestion to include it in the analysis for economy. Consistency requires that either all road user taxes be included or excluded.

Excluding the fuel tax means that the fuel tax rate could be changed at any time without altering the solution for economy, but including the fuel tax means that as the rate of tax rate changes, so will the solution for economy.

THE ROAD USER AND THE FUEL TAX

It is true that the highway user pays the fuel tax and any fuel economy achieved by a highway improvement saves the road user the fuel tax as well as the base price of the fuel. Therefore from the user's personal viewpoint it looks as though the fuel tax should be included. There are two answers to this position. First, proper authority can raise, lower, or shift the fuel tax rate at any time. For instance, if highway improvements resulted in material reduction in fuel tax income to the extent that funds for highway support became insufficient, proper authority could raise the fuel tax rate or levy additional highway taxes otherwise. The road user, then, in the long run, may not save his fuel tax. Second, the fuel taxing authority is an arm of government and the agent of the road user. Therefore, any fuel tax reduction achieved by the road user means that he is lowering his revenue an equal amount.

The state, or other collector of the motor vehicle fuel tax, and, in the end, the highway departments to which the fuel tax income is allocated, would have less revenue for highway purposes as a result of highway improvements that reduce

Vehicle Running Costs for Economy Studies   339

the fuel consumption rate per vehicle, were it not for the fact that highway use is steadily increasing. In a sales sense, the highway departments lower their sales price (fuel tax total per trip) through improved highways, but recoup at least some of the difference through increase in sales volume because of normal traffic growth and the attractiveness of the new highway. But on the basis that the highway department would need to hold the fuel tax income to the same total after the improvement as it was before, and on the basis of no increase in highway usage, it would be necessary to increase the fuel tax rate per gallon of fuel.

A mathematical analysis will illustrate how the fuel tax affects both the revenue collected and the net benefits to the road user resulting from a highway improvement which reduces or increases the fuel consumption of motor vehicles. The following symbols are used:

F_b and F_a = before and after miles per gallon rates of fuel consumption per vehicle
R_b and R_a = before and after total dollars of fuel tax revenue
T_b and T_a = before and after fuel tax rates in dollars per gallon
M = before and after vehicle-miles of travel (no change assumed for this purpose)
P = before and after price of fuel per gallon excluding the fuel tax (no change)
U_b and U_a = road user's total cost of fuel, before and after

The net fuel consumption benefit of the improvement to the road user will be equal to his before cost for fuel including tax, less his cost of fuel on the new facility, valued at the same price per gallon including the old fuel tax rate less any increase in fuel tax rate imposed when the new facility was opened, or

$$U_b - U_a = (P + T_b)\frac{M}{F_b} - (P + T_b)\frac{M}{F_a} - (T_a - T_b)\frac{M}{F_a} \quad (14\text{-}1)$$

$$U_b - U_a = \frac{PM}{F_b} + \frac{T_b M}{F_b} - \frac{PM}{F_a} - \frac{T_b M}{F_a} - \frac{T_a M}{F_a} + \frac{T_b M}{F_a}$$

$$U_b - U_a = P\left(\frac{M}{F_b} - \frac{M}{F_a}\right) + \left(\frac{T_b M}{F_b} - \frac{T_a M}{F_a}\right) \quad (14\text{-}2)$$

$$U_b - U_a = P\left(\frac{M}{F_b} - \frac{M}{F_a}\right) + (R_b - R_a) \quad (14\text{-}3)$$

Assuming that T_a was increased over T_b sufficiently to produce the same total fuel tax revenue after the improvement as before, then,

$$R_b = R_a, \text{ and, } U_b - U_a = P\left(\frac{M}{F_b} - \frac{M}{F_a}\right) \quad (14\text{-}4)$$

It will be seen that M/F_b and M/F_a are equal to the gallons of fuel consumed before and after the improvement. Therefore the net benefit to the road user is simply his price for fuel less tax, times the reduced fuel consumption. On this basis the fuel tax itself does not affect road user economy, so the tax should therefore be omitted in the analysis for economy.

In the event that the fuel tax rate was not increased from T_b to T_a, then $T_b = T_a$, and Eq. 14-1 would reduce to an apparent road user benefit of the gallons of fuel saved times the total price paid including tax. But this apparent saving is not the real net saving, because the highway fund is short by the amount of the fuel tax not paid. Since the fuel tax revenue fund belongs to the road user, he has, in effect, merely left in his pocket the fuel tax he did not pay, so his total assets are unchanged.

NONHIGHWAY USE OF THE FUEL TAX REVENUE

In some states a portion of the tax incomes on motor vehicle fuel may be used for education, general government, or other nonhighway purposes. Where such practice exists, the fuel tax not allocated to highways should be included in the price of fuel used for computing the motor vehicle running cost for economy studies in the same manner as sales taxes and property taxes are included.

This discussion applies to registration fees, operators' licenses, and third structure taxes (weight, distance, or axle taxes) on freight vehicles to the extent that they may enter the calculations for economy. Most often, however, these taxes are not included in the motor vehicle running costs used in economy studies because they are generally the same for all improvement alternatives considered, and independent of the highway design features or considerations involved.

It has been the usual practice for the states and the federal government to raise the fuel tax and other road user tax rates when authorizing an expanded highway construction program. When this increase in tax has been ordered and when it may be presumed that the legislature or the Congress authorized the enlarged program because of the potential economic benefits anticipated, it is evident that the increased fuel tax came after an evaluation of the benefits and was irrelevant to the analysis for economy.

There are types of highway facilities which are not supported by fuel tax receipts, at least not wholly supported. These proposals are most common in urban areas where the fuel tax does not support all street construction and maintenance and traffic control devices. Toll facilities are also not supported by fuel taxes. The general guide is to omit the fuel tax from the motor vehicle running cost when such tax is used to support the highway function and to include the fuel tax when it is not used to support highways.

EXPLANATION OF EACH RUNNING COST ITEM

The running cost tables in Appendix A were calculated as required or

Vehicle Running Costs for Economy Studies

composed directly from available information when possible through specially divised procedures. The detailed procedures and the working papers and curves are too voluminous to present here, but a brief explanation will supply the main features. The complete set of tables for a specific class of motor vehicle provides the whole range of running costs by which it is possible to estimate the running cost of that class of vehicle under any specific performance in traffic. Except as specifically noted, the running performance and costs are for high type pavement in good condition.

FUEL CONSUMPTION RATES

Fuel consumption of motor vehicles is the one element of running cost that has been studied far more than any other element. Fuel is a direct cash outlay item purchased frequently. Fuel has a direct relationship to miles driven, and it can be measured as used–readily in quantities and less readily in fractional gallons when the vehicle is equipped with a fuel meter. The literature therefore contains a series of reports on motor vehicle fuel consumption over the years since about 1920.

The fuel consumption rates given in Tables A-26 to A-40 for minus and plus grades, were compiled by reference to Claffey [14-8], Sawhill and Firey [14-42], personal test results, and from information obtained by correspondence with individuals in the automotive industry. Adjustments, interpolations, and extrapolations were made as necessary to meet the particular specifications of the five typical vehicles and the ranges of speeds and grades. In fact, throughout preparation of the whole series of tables such procedure was exercised.

Practically no published information was available for fuel consumption on horizontal curves. Resort was had to a few scattered bits of information and Sawhill [14-40], plus calculations of horsepower, coefficient of friction, air resistance, slip angles, and the dynamic forces generally encountered. The base fuel consumption rate used was that on level tangents and on plus grades. In general, the same rate of fuel consumption per horsepower-mile on level tangents at a given speed was used at the same speed on horizontal curves.

Fuel consumption for speed changes and on gravel surfaces is based upon the work of Claffey [14-8], Sawhill [14-42], and Moyer [14-27].

The entire series of Tables A-26 to A-40 on fuel consumption is presented so that analysts may calculate fuel costs for fuel prices differing from those used in the running cost tables, and also so that fuel tax revenues can be calculated when desired.

Figure 14-2 gives three sets of fuel consumption curves for the 4-kip passenger car and the 40-kip 2-S2. The curve for gallons of fuel consumed per hour was found to be useful in smoothing out field test data and in extrapolating test results because the gallons per hour curve does not reverse its curvature.

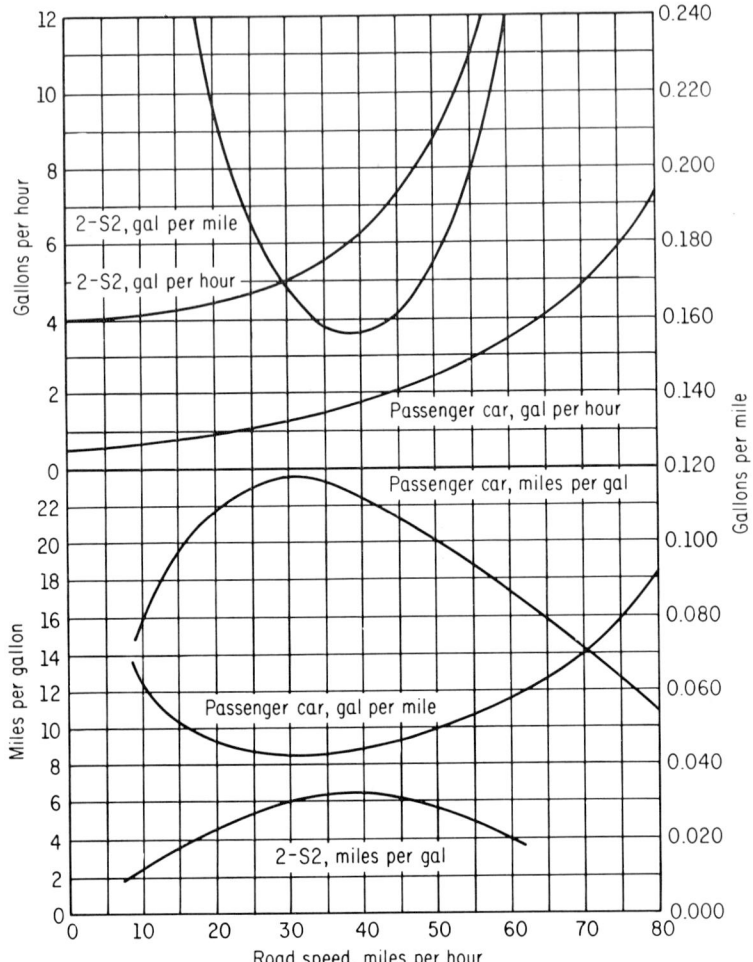

FIG. 14-2. Gasoline consumption curves in gallons per hour, miles per gallon, and gallons per mile for the 4-kip passenger car and the 40-kip 2-S2 tractor-semitrailer when operated at constant speed on level tangents.

TIRE TREAD WEAR

The literature is largely lacking in tire tread wear rates for today's tires, vehicles, and pavements. Again, scattered information and help from industry were used. Moyer and Tesdall [14-29] is the best single source, but it is for 1939 to 1942 tires. The general literature [14-20, 14-21, 14-22, 14-23, 14-26, 14-35, 14-50] offers some clues as to the magnitude of the complex set of forces involved in steering a vehicle and the interaction of the tire and roadway surface.

Vehicle Running Costs for Economy Studies

A general guide for overall control for the heavy trucks was found in reports of total mileage per tire life or tread life from Kent [14-24A].

Tire wear on horizontal curves was calculated by reference to horsepower requirements in a manner similar to that used for fuel consumption. The tire rates on horizontal curves, though seemingly rather high, are probably lower than actual at the higher speeds on the higher degrees of curve.

On speed changes, the tire wear was calculated on the basis of horsepower absorbed and developed. Weight was given to the braking action in decelerating.

Again Moyer and Tesdall [14-29] is the best reference for tire wear on gravel as well as for other surfaces.

For minus and plus grades, horizontal curves, and for speed changes, the tire wear was calculated in 0.001 in. per tire. These wear rates were then converted to dollars, using the general data in Table 14-4.

TABLE 14-4
TIRE DATA SUPPORTING THE CALCULATION OF TIRE COST PER MILE

Descriptive Item	4-kip Passenger Car	5-kip Commercial Delivery	12-kip Single-Unit Truck	40-kip 2-S2, Gasoline	50-kip 3-S2, Diesel
1. Number of tires	4	4	6	14	18
2. Type of tires	Tubeless	Tubeless	Tubeless	Tubeless	Tubeless
3. Size of tires	7.50 × 14	7 × 17.5	8 × 22.5	11 × 22.5	11 × 22.5
4. Ply rating of tires	4	6	8	12	12
5. Tread width, in.	4.50	4.75	5.60	7.40	7.40
6. Tread depth, in.	0.330	0.406	0.469	0.562	0.562
7. Rolling radius, in.	12.83	14.36	17.75	19.96	19.96
8. Revolutions per mile	786	702	568	505	505
9. Tire pressure, psi	24	40	70	75	75
10. Vehicle weight per tire, lb	1,000	1,250	2,000	2,857	2,778
11. Vehicle weight per inch of tread, lb	222	263	357	386	375
12. Percentage of new tread depth usable	80	80	75	75	75
13. Percentage of retread depth usable	–	–	70	70	70
14. Wear rate factor (item 8 times item 11)	174,665	184,735	202,855	194,980	174,999
15. Cost of one tire, $	23.90	32.00	66.00	120.00	120.00
16. Cost of one retread, $	–	–	24.00	40.00	40.00
17. Number of retreads per tire life	0	0	1	2.5	2.5
18. Cost of tire during life, $	23.00	32.00	90.00	220.00	220.00
19. Total 0.001 in. of tread depth used, per tire	264	325	680	1,405	1,405
20. Cost per 0.001 in. of tread wear, cents per tire	8.712	9.846	13.235	15.658	15.658
21. Life-end value of tire	zero	zero	zero	zero	zero

ENGINE OIL CONSUMPTION

Engine oil shows a wide variation in rate of consumption, vehicle to vehicle. Disregarding, for the moment, the changing of crankcase oil, engine oil is consumed at a variation in rate, depending upon the type of use of the vehicle and mechanical condition of the engine. There are but few reports in the literature giving oil consumption rates under known vehicle use and highway conditions. Moyer's research 1938-42 [14-27] is helpful on the factors of speed and type of roadway surface. Kent [14-24A] gives some clues for trucks. Reported oil consumption by transport companies (private data) shows wide variation in oil consumption per 1,000 miles, as well as a wide range in mileage per oil change.

To achieve oil consumption rates for level tangents, data from several sources were pieced together on a quarts-per-hour basis, giving weight to the engine speed at the known road speed. To these basic curves, converted to quarts of oil per 1,000 miles at specific speeds, was added the crankcase change oil. For the passenger car, oil change mileage was set at 1,000 miles for the 5 mph speed and increased to 8,000 miles for 80 mph. This range of mileage per oil change gives weight to the unfavorable conditions developed in the engine interior at low speed and the high rate of makeup oil at high speeds.

Oil consumption on horizontal curves was based on the oil consumption (without change of oil) on level tangents per gallon of fuel consumed. This ratio was calculated for each speed and applied directly to the fuel consumption on horizontal curves.

The oil consumption for speed changes was based on the rate of consumption per hour on level tangents. The excess time consumed on speed changes was assumed to consume the same quantity of oil as would be consumed during the same time at the initial speed.

VEHICLE MAINTENANCE

Outside of knowing that all brake expense is chargeable to decelerating and to holding speed on minus grades, there is little to base an allocation of the vehicle maintenance expense to speed or to elements of highway design. The allocation was accomplished, however, by logic and judgment.

From private information, general sources, and from Stevens [14-46] overall maintenance expenses were obtained for each typical vehicle. These overall maintenance expenses were then separated by meager evidence into (1) body, (2) braking system, (3) chassis, (4) electrical, (5) engine, and (6) power train.

By judgment, the overall maintenance expense per mile of these six items was further separated into that caused by mileage use of the vehicle and that caused by off-the-road use, loading and unloading passengers and goods, and weathering.

From reports of on-the-road vehicle speed, a speed distribution curve was developed for each vehicle. Then for each of the six subdivisions of maintenance

Vehicle Running Costs for Economy Studies

expense, a maintenance expense curve against speed was drawn so that the weighted total cost of maintenance agreed with the overall cost in cents per mile after deducting the nonmileage components.

Vehicle maintenance expenses on plus grades and on horizontal curves for all vehicles, except the 2-S2 and 3-S2 combinations, were increased over that on level tangents for each speed at the rate of 10 percent of the rate at which fuel consumption increased. For the 2-S2 and 3-S2 vehicles a factor of 20 percent was used.

Maintenance expense on minus grades was reduced from the cost on level tangents down to where negative horsepower was required by the reverse scheme applied to plus grades. The maintenance expense on minus grades when negative horsepower was required was increased as the percentage minus grade increased by the cost of the wear on the braking system.

For speed changes, the excess time consumed was used as the proration base to which was applied a cost of braking factor in cents per minute. Allowance was made for the fact that the excess time consumed in the speed change cycle includes acceleration time as well as deceleration time.

VEHICLE DEPRECIATION

The depreciation expense was assigned to speed on level tangents by the use of three curves. From general information, the average annual mileage of cars and trucks and the average years of service life can be reliably estimated. See Stevens [14-46] for line haul trucks. For vehicles in general use a total service life mileage as follows was used to establish the base for computing the overall depreciation rate in cents per mile:

Vehicle	Lifetime Mileage	Depreciable Cost, Dollars	Depreciation Cost, Cents per Mile
Passenger car	130,000	2,368	1.82
Commercial delivery	110,000	2,332	2.12
Single unit truck	150,000	3,044	2.03
2-S2 tractor-semitrailer	400,000	11,460	2.86
3-S2 tractor-semitrailer	500,000	17,380	3.48

A straight-line increase in average annual mileage was established with an increase in average road speed. A curved-line increase was established in total service life in years with a decrease in annual mileage. These two relationships were then combined to produce a lifetime mileage for each road speed, the mileage increasing with road speed increase.

A third curve was then developed which assigned increasing percentage of the total dollar depreciation to mileage and a decreasing percentage to time and weather as the life mileage increased. See Fig. 14-3 for these three curves for

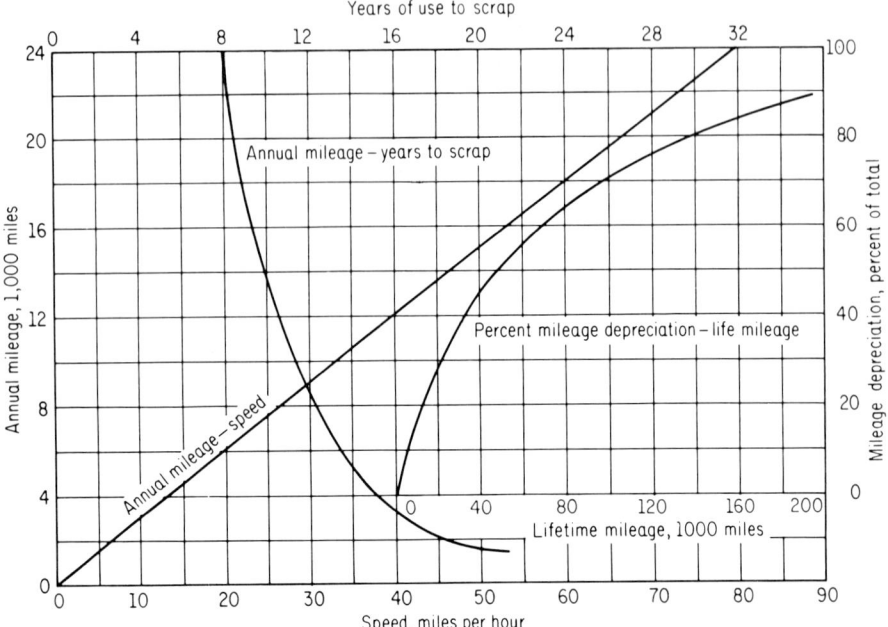

FIG. 14-3. Curves illustrative of the method of allocating depreciation expense of passenger cars to road speed. Similar curves were developed for the other typical vehicles.

the passenger car. Depreciation rates per mile were then computed for each speed. This method produces a decreasing depreciation expense per mile as road speed increases.

No additional depreciation was assigned to minus and plus grades nor to horizontal curves over that computed for level tangents at the same speed.

Because speed changes involve consumption of time over that consumed at uniform speed, an assignment of depreciation was made to speed changes. The excess time was used as the allocating base and the rate of depreciation at each speed on level tangents expressed in dollars per hour of travel time was used as the depreciation factor.

COST OF IDLING ENGINE

When legal restraints, traffic conditions, or highway design require that the vehicle be stopped with the engine running, there results a motor vehicle cost which is chargeable to the highway. The cost of decelerating and accelerating occasioned by the stop is accounted for in the cost of speed changes. The cost of the idling engine is a separate cost and dependent upon the length of time the engine is idled.

Vehicle Running Costs for Economy Studies   347

For an engine at idling speed, the cost elements are fuel, engine oil, and engine wear. Table A-41 gives these costs for all five typical vehicles. The idling fuel consumption is from a direct metering of the fuel at the idling revolutions per minute of the engine |Claffey, 14-8 and Sawhill, 14-42|. The costs of engine wear and of engine oil are based upon time and the engine speed. The basic curves for these factors were developed in connection with computing similar costs for operation on level tangents.

The value of stopped vehicle time would be computed in accordance with the value of all other travel time.

COST OF SHARP TURNS

In urban driving and on many local rural roads, intersecting streets and roads at 90-degree angles require that vehicles slow to 10 to 25 mph in order to turn the corner on a radius of 50 to 250 ft. These sharp turns involve but minor roadway distance. The vehicle running cost of making such sharp turns may, therefore, be expressed by the cost per turn. These costs for all five typical vehicles are given in Table A-42 for fuel, tires, engine oil, and vehicle maintenance. Depreciation is excluded on the basis that it is too minor and difficult to allocate on any appropriate basis.

The cost of the speed change to negotiate the sharp turn is not included in the cost of turning corners, Table A-42.

GRAVEL AND STONE ROADWAY SURFACES

Unbound, loose, and dusty roadway surfaces such as gravel, crushed stone, and natural earth, cause relatively high motor vehicle running costs as compared to firm, hard, and smooth pavements. All elements of cost—fuel, tires, oil, vehicle maintenance, and depreciation—are affected.

Rather than compute separate tables of vehicle running cost on gravel and stone surfaces, conversion factors are presented in Table A-44. The running cost on gravel and stone surfaces may be computed by multiplying the running cost on high type pavements by these conversion factors. The results are accurate enough for all practical purposes and relatively reliable. The factors should be increased for overly soft and for highly rough surfaces.

The conversion factors in Table A-44 were computed by references to the work of Moyer |14-27, 14-29|, Claffey |14-8|, Sawhill |14-42|, and to scattered sources. Again judgments were necessary, particularly for vehicle maintenance and depreciation.

LOW AND INTERMEDIATE GRADE BITUMINOUS SURFACES AND EARTH SURFACES

No running cost or conversion factors are offered for low and intermediate grade bituminous surfaces because there is no definite line of demarcation for the different qualities of bituminous surfaces. The effects of bituminous surfaces on motor vehicle running cost vary from an approximate equality with the best of gravel to an equality with the best of high type plant mix bituminous concrete and portland cement concrete. For each application of the running cost of vehicles on bituminous surfaces other than the high type in good condition, the conversion factors for gravel (Table A-44) may be reduced as judged to be applicable to the roadway under study. A factor midway between 1.000 and the gravel factor in Table A-44 would be suitable for the low type bituminous surfacings, particularly for the surface-treated gravels.

Earth roadway surfaces vary widely geographically location to location, season to season, and soil to soil in their effects on the running cost of motor vehicles. Earth surfaces are about equally bad in the dry season as they are in the rainy season, but for different reasons. The loose dusty surface is hard on fuel consumption, on the engine, brakes, all moving parts, and on the vehicle interior. When a dirt road is soaked with rain or melting snow, traction is low and tractive resistance is high, so that both tire wear and fuel consumption are high. Again, no specific conversion factors are offered for earth roads, but a reasonable solution is to double the portion of the gravel factor above 1.000 and add the 1.000 back in–that is, $E = 2G - 1$.

OTHER RUNNING COST FACTORS

The foregoing discussion relates the running costs of motor vehicles for fuel, tires, oil, maintenance, and depreciation to distance, gradients, curvature, traffic speed, and speed changes for high-type pavements in good condition; conversion factors are offered for other roadway surface types. Additional consideration is next given to rolling grades, atmospheric conditions, and superelevation of horizontal curves.

Rolling Grades

For rolling terrain there is advantage in analyzing the effect of minus and plus grades on motor vehicle running cost by the rise and fall method. This method considers the total rise and the total fall without direct attention to each specific gradient. There is only one source of information for vehicle performance on the rise and fall basis [Saal, 14-38], and this publication reports only fuel and time consumption for heavy freight vehicles.

Figures 14-4 and 14-5 reproduce two sets of curves from Saal's study. These curves show that both gross vehicle weight and gradient have a pronounced effect on fuel consumption, and that travel time is a function of the ratio of gross vehicle weight to engine net horsepower and of the rate of rise and fall. This study was conducted by running the vehicles over the Pennsylvania Turnpike and

Vehicle Running Costs for Economy Studies

the parallel primary highway route under normal traffic conditions. Other than grade (at normal speed of the test vehicles) the elements of highway design were not associated with the performance of the test vehicles. The original publication gives a series of figures similar to Figs. 14-4 and 14-5.

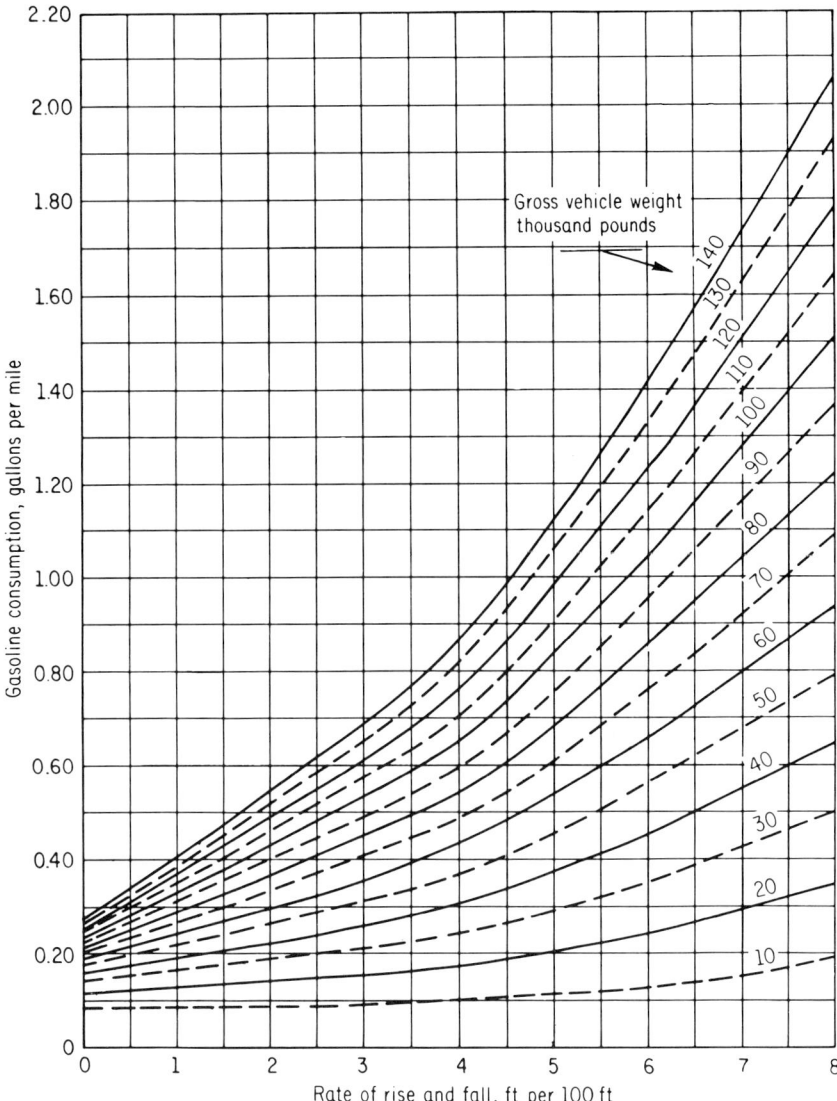

FIG. 14-4. Gasoline consumption on composite grades for a range of gross vehicle weight, based on the rise and fall when the total rise is 45 to 54 percent of the total rise and fall. From Saal [14-38, p. 39].

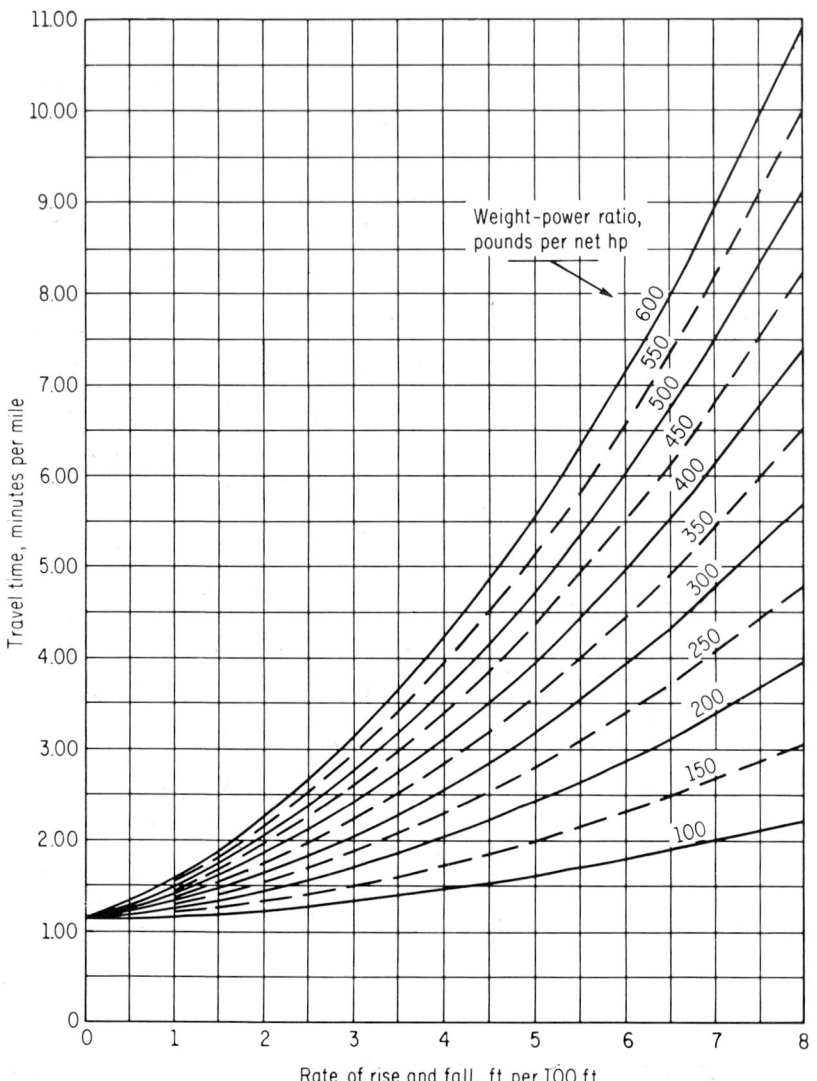

FIG. 14-5. Travel time on composite grades for a range of the ratio of gross vehicle weight to net engine horsepower, when the rate of rise and fall is 45 to 54 percent of the total rise and fall. From Saal [14-38, p. 54].

Atmospheric Conditions

The fuel consumption rates and vehicle performance for the several tables are based on low altitude, sea level to 2,000 ft or so. Altitude affects air resistance because the air weighs less as the altitude increases. This lighter air, on the other hand, decreases the power output of the engine because the engine

Vehicle Running Costs for Economy Studies

TABLE 14-5

EFFECT OF ALTITUDE ON GASOLINE
CONSUMPTION IN MILES PER GALLON FOR A 3,520-LB PASSENGER CAR

Speed, mph	3.2 Percent Plus Grades			5.6 Percent Plus Grades		
	Elevation, ft			Average Elevation, ft		
	1,000	8,843		1,000	11,630	
	Low-Altitude Jets	Low-Altitude Jets	High-Altitude Jets	Low-Altitude Jets	Low-Altitude Jets	High-Altitude Jets
	Fuel Consumption, Miles per Gallon			Fuel Consumption, Miles per Gallon		
20	25.3	21.3	22.7	24.6	15.4	20.1
30	24.2	20.6	22.7	23.6	15.0	19.7
40	21.8	18.7	20.4	21.3	13.6	18.2
50	18.7	16.1	18.0	18.2	–	–

power is proportional to the weight of the air intake into the cylinders, and increases fuel consumption because of the change in fuel-air ratio. Air temperature also affects the engine performance, the power output being approximately inversely proportional to the square root of the absolute temperature. Air humidity has only a slight effect on the engine output. [See Taborek, 12-17.]

The SAE Truck Ability Prediction Procedure [12-15] gives the following ratios of horsepower output based on altitude; sea level, 1.00; 1,000 ft, 0.96; 2,000 ft, 0.92; and similarly to a factor of 0.60 at 10,000 ft elevation.

Moyer and Jones [14-28, p. 62] give the fuel consumption rates for a test car of 3,530 lb [July 1942] as shown in Table 14-5. Even with the high altitude carburetor jets there is a marked decrease in miles per gallon of gasoline at 8,843 and 11,630 ft elevations as compared to 1,000 ft elevation.

TABLE 14-6

SUPERELEVATION OF THE HORIZONTAL CURVES REPRESENTED IN THE RUNNING-COST TABLES

Degree of Curve	Radius, ft	Superelevation, ft per ft	Degree of Curve	Radius, ft	Superelevation, ft per ft
1	5729.58	0.0197	12	477.46	0.1000
2	2864.79	0.0371	14	409.26	0.1000
3	1909.86	0.0521	16	358.10	0.1000
4	1432.40	0.0648	18	318.31	0.1000
5	1145.92	0.0752	20	286.48	0.1000
6	954.93	0.0835	25	229.18	0.1000
8	716.20	0.0946	30	190.99	0.1000
10	572.96	0.1000			

Superelevation of Horizontal Curves

All calculations of running cost on horizontal curves (Appendix A) take into consideration the superelevation and centrifugal force. Table 14-6 gives the superelevation used for the full range of the degrees of curve considered. An example of the effect of superelevation may be seen in Table A-27, for the passenger car. Fuel consumption at 20 mph, for instance, increases with an increase in degree of curve up to 10 deg, then decreases to 18 deg before again starting to increase.

PROBLEMS

14-1. For a passenger car and by using the appropriate motor vehicle Appendix A tables, compute the following: (1) average miles per hour speed, (2) average fuel consumption in miles per gallon, and (3) total running cost in cents per mile when traveling one mile at the following initial or attempted speeds and with the number of stops per mile indicated.

Initial Speed, mph	Number of Stops per Mile (Speed-Change Cycles)	Initial Speed, mph	Number of Stops per Mile (Speed-Change Cycles)
20	1 to 8	60	1 to 2
30	1 to 7	70	1
40	1 to 4	80	1
50	1 to 3		

14-2. Calculate the annual (250 trips per year) excess fuel comsumption and excess travel time on account of speed changes, as compared to a uniform speed from the following data for a home-work-home round trip of 6.3 miles in an urban area using a standard passenger car.

(*Note*: Unrestricted speed is 25 mph.)

	Travel Time, Minutes	Average Speed, mph	Fuel Consumption, Miles per Gallon
Round trip as driven	23.00	17.39	8.70
Round trip, less stopped (idle) time	16.04	21.45	9.32
Round trip, less all speed changes	15.13	25.00	12.21

Using Appendix A tables, calculate the total car running cost per year (1) without speed changes, (2) with slowdowns only, (3) with stops only, and (4) with both slowdowns and stops. From an initial speed of 25 mph the number of slowdowns is 9 of 10 mph per round trip and the number of stops is 21.

14-3. Compute the gallons of fuel consumed per year by 1,000 passenger cars per day (total for both directions) on the following section of a 2-lane bidirectional highway, before and after relocation. Before relocation the highway passed through a small urban area in addition to rural areas on either side. The relocated highway bypasses the urban area. The highway characteristics and traffic speeds before and after relocation are:

Vehicle Running Costs for Economy Studies

Mile Post	Distance, Miles	Percent Grade	Running Speed, mph	Speed Changes
		Before Relocation		
0.0-5.0	5.0	-5	50	
5.0-8.0	3.0	-1	30	3 stops
8.0-13.0	5.0	-3	30	6 slowdowns of 10 mph each
13.0-15.0	2.0	0	40	1 slowdown of 10 mph
15.0-17.0	2.0	-4	50	
		After Relocation		
0.0-4.0	4.0	-3	50	
4.0-10.0	6.0	-1	40	
10.0-15.0	5.0	-2	40	1 slowdown of 10 mph
15.0-18.0	3.0	-1	50	

14-4. Compare the daily running cost of 1,000 passenger cars in an urban area (A) on conventional urban surface streets, and (B) on a freeway. Use the following basic data:

	A Surface Streets	B Freeway
Daily mileage	20	20
Free running (attempted) speed, mph	30	45
Number of stops from running speed	22	2
Number of stops from 20 mph	28	0
Number of speed changes from running speed to 15 mph	50	15
Idling time, minutes	11	0
Number of sharp corner turns:		
50-ft radius at 15 mph	12	–
125-ft radius at 20 mph	–	2

14-5. Using the data in Prob. 14-4 and appropriate Appendix A tables, compute the travel time in hours for the 1,000 passenger cars traveling 20 miles a day. Convert the travel time to dollars at $1.00, $2.00, and $4.00 per vehicle-hour.

14-6. Using the data of Prob. 14-4 and the appropriate Appendix A tables compute the fuel tax generated by the 1,000 passenger cars traveling the 20 miles. Use rates of 4 cents and 7 cents per gallon.

14-7. A grade reduction analysis results in the following table of costs:

Constructed Grade, Percent	Construction Cost, Dollars	Annual Road User Cost Above a 0 Percent Grade, Dollars	Annual Roadway Maintenance Costs, Dollars
0	100,400	0	500
1	77,800	200	420
2	57,000	420	360
3	40,200	940	315
4	27,800	1,600	270

5	18,100	2,450	245
6	11,200	4,060	205
7	6,200	6,300	180
8	2,800	9,280	160
9	700	13,050	140
10	0	17,300	120

(1) Calculate the most economical gradient assuming that the existing grade is 10 percent. (2) Calculate the most economical gradient assuming that by prior construction the grade has been reduced to a 6 percent gradient. Use a vestcharge rate of 8 percent and an anlysis period of 30 years with zero terminal value.

14-8. The following data apply to three mutually exclusive locations for a two-lane bidirectional highway. From the sole viewpoint of transportation economy which location is preferred (1) with and (2) without travel time. Use Appendix A tables for uniform speed.

	Location A	Location B	Location C
Miles of route length	4.5	4.0	3.5
Uniform grade, percent	3	4	5
Total construction cost	$1,000,000	$850,000	$700,000
Annual maintenance cost per mile	$700	$600	$500
Resurfacing cost per mile, end 12th year	$40,000	$40,000	$40,000
Uniform traffic speed, cars and commercials, mph	60	55	50
Uniform traffic speed, 40-kip trucks, upgrade, mph	24	18	15
Uniform traffic speed, 40-kip trucks, downgrade, mph	50	50	50
ADT of passenger car	1,000	1,000	1,000
ADT of commercial deliveries	150	150	150
ADT of 40,000-lb trucks	20	20	20

Assume constant ADT for an anlysis period of 24 years. Travel time value is $1.00 for cars, $2.50 for commercial deliveries, and $3.50 for the trucks. The minimum attractive rate of return is 8 percent. Terminal value is zero.

14-9. Rework Prob. 14-8 using an annual gradient traffic growth of 10 percent of the ADT given.

14-10. The following information applies to three possible mutually exclusive locations for a new highway. For a vestcharge rate of 8 percent and an analysis period of 20 years, calculate the relative economy of the three alternatives based (1) on motor vehicle running costs, (2) on travel time value, and (3) motor vehicle and time costs combined. Assume that the services rendered by the routes are essentially equal.

	A	B	C
Length of route	12.0 miles	11.5 miles	10.8 miles
Total construction costs	$3,100,000	$3,400,000	$3,800,000
Uniform annual highway maintenance and operating costs	$12,000	$12,500	$11,000

Vehicle Running Costs for Economy Studies

	Time zero	End of 20th year
Average daily traffic, all routes		
Number of passenger cars	3,000	7,200
Number of commercial delivery vehicles	400	800
Number of 40-kip tractor semitrailers	80	140
Equivalent uniform speed, mph		
Passenger cars	60	55
Commercial delivery	55	50
40-kip tractor semitrailers	50	45
Value of travel time per vehicle-hour		
Passenger cars	$1.50	$1.50
Commercial delivery	4.00	4.00
40-kip tractor-semitrailers	5.26	5.26

Use a uniform gradient growth, zero terminal value, level tangent alignment, no speed changes, and running costs from Appendix A.

14-11. Compare the economy of one mile each of (1) gravel, and (2) bituminous pavements based on the following data:

A. Gravel surface
  (a) Average construction cost of 28-ft top width, $11,200.
  (b) Average maintenance cost of $500 per mile per year plus regraveling every four years at a cost of $2,000.
  (c) Life of 20 years, zero terminal value.
  (d) Accident rate is 450 per 100 million vehicle-miles.

B. Bituminous surface
  (a) Average construction cost of 22-ft of pavement and 6-ft of shoulders is $42,600.
  (b) Average maintenance cost of $500 per mile per year plus seal coating at a cost of $3,000 at end of years 5 and 15, and recapping at a cost of $10,000 at end of years 10 and 20.
  (c) Life of 20 years, terminal salvage value is $20,000.
  (d) Accident rate is 200 per 100 million vehicle-miles.

C. Other data
  (a) Assume a section of level tangent highway and the vehicle running costs given in Appendix A.
  (b) Speed on gravel is 35 mph; on bituminous is 50 mph.
  (c) Average cost of accident is $900 on gravel; $1,200 on bituminous.
  (d) Traffic growth is at a 3 percent exponential rate per year.
  (e) Value of time is $1.50 per vehicle-hour.
  (f) Use a vestcharge rate of 7 percent per annum.

Make your comparisons on a present worth basis for ADT's of 100, 300, and 500 passenger cars at zero date. At what initial ADT are the equivalent uniform annual transportation costs equal?

14-12. The elimination of a certain at-grade crossing of a highway and railway is under consideration. The total construction cost of this grade separation structure is estimated at $350,000. This proposal also anticipates major repairs the 10th year at a cost of $15,000 and annual maintenance costs of $300 for pumping drainage water. The ADT is 2,000 with 80 percent passenger cars, 15 percent single-unit trucks, and 5 percent 40,000-lb combination vehicles. The speed of the traffic is 50 mph for passenger cars and 40 mph for all trucks. The ADT is assumed to increase exponentially at a rate of 3 percent per annum. The railway carries 36 trains per day with no expected change. Time studies indicate that on the average each train now stops 16 vehicles of normal traffic classification, with an average delay per stopped vehicle of 2.5 min. The value of time is estimated at $1.42 per hour for passenger cars, $3.27 per hour for single-unit trucks, and $4.77 per hour for combination vehicles.

Based on past accident statistics, it is estimated that if this grade crossing is continued it will be responsible for one fatal accident every four years (firt at $n = 4$) and five nonfatal accidents per year. Accidents will not increase with ADT. Make the arbitrary assumption that the "cost" for a fatal accident is $10,000 and the cost for a nonfatal accident is $800. The grade separation will eliminate all accidents.

The railway company has maintained a watchman at this crossing for 14 hours per day at an annual cost of $7,800 while annual maintenance of the crossing has been $250.

Compute the benefit/cost ratio for this grade separation project, assuming a vest-charge rate of 8 percent and an analysis period of 20 years.

REFERENCES

14-1. American Association of State Highway Officials. *Road-User Benefit Analyses for Highway Improvements,* Part I, "Passenger Cars In Rural Areas." Washington, D.C., 1960.

14-2. John Beakey. *The Effect of Highway Design on Vehicle Speed and Fuel Consumption.* Oregon State Highway Department, Salem, Tech. Bulletin 5, 1937.

14-3. W. Beck. *An Investigation of Motor Vehicle Parasitic Power Loss by Road Test,* M.S. in M.E. thesis, University of Washington, Seattle, 1962.

14-4. Howard W. Bevis. Automobile Costs, "Chicago Area Transit Study, *Research News,* Vol. 11, No. 18 (Sept. 27, 1957), p. 3.

14-5. A. J. Bone. *Travel Time and Gasoline Consumption Studies in Boston.* Highway Research Board, National Academy of Sciences, Washington, D.C., *Proc.,* Vol. 31, 1952, pp. 440-456.

14-6. A. J. Bone. *Effect of Traffic Delays on Gasoline Consumption.* Highway Research Board, National Academy of Sciences, Washington, D.C., *Proc.,* Vol. 19, 1939, pp. 99-125.

14-7. F. R. Bryan, J. C. Neerman, and J. E. Hinsch. "Method for Determining Lubricant in Engine Exhaust," *Soc. of Auto. Engr. Trans.,* Vol. 69 (1961), pp. 204-211.

14-8. Paul J. Claffey. *Time and Fuel Consumption for Highway User Benefit Studies.* Highway Research Board, National Academy of Sciences, Washington, D.C., Bulletin 276, 1960.

14-8A. Paul J. Claffey. *Running Cost of Motor Vehicles as Affected by Highway Design–Interim Report.* National Cooperative Highway Research Program, Report 13, Highway Research Board, National Academy of Sciences, Washington, D.C., 1965.

Vehicle Running Costs for Economy Studies

14-9. William Davidson. "Tires," *Think*, Vol. 17, No. 8, pp. 25-26. "Measuring Tire Wear By Radioactive Elements," August 1951.

14-9A. David G. Davies. "The Secular Income Elasticity and Revenue Stability of Motor Fuel Taxs," *National Tax Journal*, Vol. XVIII, No. 4 (December 1965), pp. 380-387.

14-10. P. K. Dykes. "Oil Loss Past Piston Rings," *Scientific Lubrication*, Vol. 9, No. 1 (February 1957), pp. 11-14, 42.

14-11. H. A. Everett and F. C. Stewart. *Performance Tests of Lubricating Oils in Automobile Engines*. Pennsylvania State University, Engineering Experiment Station, University Park, Bulletin 44, 1935.

14-12. Factors Affecting Tire Mileage, *Commercial Car Journal*, Vol. 81, No. 2 (April 1951), pp. 120-121.

14-13. J. C. Firey and E. W. Peterson. *An Analysis of Speed Changes for Large Trasport Trucks*. Highway Research Board, National Academy of Sciences, Washington, D.C., Bulletin 334, 1962, pp. 1-26.

14-14. Joseph C. Firey and Roy B. Sawhill. *Effects of the Tire Pressure and Temperature on the Rolling Resistance of Trucks*. University of Washington, Seattle, Traffic and Operations Series, Report No. 4, June 1962.

14-15. D. Fraser, A. R. Klingel, and R. C. Tupa. "Friction and Consumption Characteristics of Motor Oils," *Industrial and Engineering Chemistry*, Vol. 45, No. 10 (October 1953), pp. 2336-2342.

14-16. Gerald Gallagher. "At What Point Do Oil Changes Pay Off for You?" *Fleet Owner*, Vol. 54, No. 12 (December 1959), pp. 74, 76, 78, 80.

14-17. L. P. Gelinas and E. B. Storey. "A Towing Device for Estimating Road Wear." *Proc.*, International Rubber Conference, Washington, D.C., Nov. 8-13, 1959, pp. 995-1003.

14-18. Carl W. Georgi. *Motor Oils and Lubrication*. Reinhold, New York, 1950, pp. 289-297.

14-19. C. K. Glaze and George Van Mieghem. *Washington Motor Vehicle Operating Cost Survey*. Highway Research Board, National Academy of Sciences, Washigton, D.C., Proc., Vol. 36, 1957, pp. 51-60, 62-63.

14-20. H. A. O. W. Geesink and C. P. Pratt. "Wear on Passenger Car Tires," *Rubber Chemistry and Technology*, Vol. 31, No. 1 (January-March 1958), pp. 166-184.

14-21. V. E. Gough. "Cornering Can Wear Tires Rapidly," *Soc. of Auto. Engr. Jour.*, Vol. 64, No. 1 (January 1956), pp. 27-29.

14-22. V. E. Gough and C. E. Shearer. "Front End Suspension and Tire Wear." *Proc.*, Inst. of Mech. Engrs., Auto. Div. (6), 1955-56, pp. 171-193.

14-23. M. P. Hershey. "Tire Degradation Is Accelerated by Modern High-Speed Driving," *Soc. Auto. Engr. Jour.*, Vol. 65, No. 8 (July 1957), pp. 85-86.

14-24. Malcolm F. Kent. *Fuel and Time Consumption Rates for Trucks in Freight Service*. Highway Research Board, National Academy of Sciences, Washington, D.C., Bulletin 276, 1960.

14-24A. Malcolm F. Kent. *AASHO Road Test Vehicle Operating Costs Related to Gros Weight*. The AASHO Road Test, Proceedings of a Conference Held May 16-18, 1962, St. Louis, Mo. Highway Research Board, National Academy of Sciences, Washington, D.C., Special Report 73, 1962, pp. 149-165.

14-25. Paul S. Lane. "Controlling Oil Consumption in Passenger Car Engines," *Soc. of Auto. Engr. Jour.*, Vol. 58, No. 11 (November 1950), pp. 18-22.

14-26. R. A. Moyer. *Skidding Characteristics of Automobile Tires on Roadway Surfaces and Their Relation to Highway Safety.* Iowa Engineering Experiment Station, Bulletin 120, Ames, 1934.

14-27. R. A. Moyer. *Motor Vehicle Operating Cost and Related Characteristics on Untreated Gravel and Portland Cement Concrete Road Surfaces.* Highway Research Board, National Academy of Sciences, Washington, D.C., Proc., Vol. 19, 1939, pp. 68-98.

14-28. Ralph A. Moyer and John Hugh Jones. *Supplementary Notes and Typical Problems for the Highway Engineering Course* (Syllabus XL). U. of California Press, Berkeley, 1958.

14-29. R. A. Moyer and Glen L. Tesdall. *Tire Wear and Cost on Selected Roadway Surfaces.* Iowa State University, Engineering Experiment Station, Ames, Bulletin 161, 1945.

14-30. New York State Thruway Authority. "Substantial Operating and Maintenance Savings and Other Benefits that Accrue to Trucking Firms Using N.Y. State Thruway," Albany, N.Y., Press Release, 58-24, 1958.

14-31. J. K. Patterson and R. C. Gregor. "Lubrication Factors Affecting Passenger Car Oil Consumption," *Soc, of Auto. Engr.,* Reprint No. 252, National Fuels and Lubricants Meeting, Cleveland, Ohio, November 1957.

14-32. J. K. Patterson, and R. C. Gregor. "Oil Consumption in Automobiles Can Be Predicted," *Soc. of Auto. Engr. Jour.,* Vol. 66, No. 4 (April 1958), pp. 63-64.

14-33. R. L. Pontius. "Determining Wear in Automotive Engines," *Lubrication Engineering,* Vol. 15, No. 3 (March 1959), pp. 110-115, 118.

14-34. William S. Powell. "Accurate Method for Determining Oil Consumption," *Automotive Industries,* Vol. 98, No. 3 (Feb. 1, 1948), pp. 35, 84, 88-90.

14-35. G. G. Richey, J. Mandel, and R. D. Stiehler. (National Bureau of Standards). "An Indoor Tester for Measuring Tire Tread Wear." *Proc.* International Rubber Conference, Washington, D.C., November 8-12, 1959, pp. 104-110.

14-36. T. A. Riehl. "Treadwear Life Down 18 Percent," *Soc. of Auto. Engr. Jour.,* Vol. 66, No. 6 (June 1958), pp. 56-57.

14-37. G. B. Roberts. "Power Wastage in Tires." *Proc.* International Rubber Conference, Washington, D.C., Nov. 8-13, 1959, pp. 57-72.

14-38. Carl Saal. *Time and Gasoline Consumption in Motor Truck Operation as Affected by the Weight and Power of Vehicles and the Rise and Fall in Highways.* Highway Research Board, National Academy of Sciences, Washington, D.C., Bulletin 9-A, 1950.

14-39. Carl C. Saal. *Operating Characteristics of a Passenger Car on Selected Routes.* Highway Research Board, National Academy of Sciences, Washington, D.C., Bulletin 107, 1955, pp. 1-35.

14-40. Roy B. Sawhill. *Fuel Consumption of Transport Vehicles on Horizontal Curves.* University of Washington, Seattle, Research Report No. 5, Traffic and Operations Series, June 1962.

14-41. Roy B. Sawhill. *Fuel Consumption of Transport Vehicles on Grades.* University of Washington, Seattle, Research Report No. 6, Traffic and Operations Series, June 1962.

14-42. Roy B. Sawhill and Joseph C. Firey. *Motor Transport Fuel Consumption Rates and Travel Time.* Highway Research Board, National Academy of Sciences, Washington, D. C., Bulletin 276, 1960.

14-43. G. M. Sprowls. "Impediments to Tire Longevity," *Soc. of Auto. Engr. Jour.,* Vol. 58, No. 11 (November 1950), pp. 40-43.

14-44. E. S. Starkman. "A Radioactive Tracer Study of Lubricating Oil Consumption" *Soc. of Auto. Engr. Trans.*, Vol. 69 (1961), pp. 86-100.

14-45. Dietrich G. Stechert and Thomas D. Bolt. "Evaluation of Treadwear," *Analytical Chemistry*, Vol. 23, No. 11 (November 1951), pp. 1641-1646.

14-46. Hoy Stevens. *Line-Haul Trucking Costs in Relation to Vehicle Gross Weight*. Highway Research Board, National Academy of Sciences, Washington, D.C., Bulletin 301, 1961.

14-47. R. D. Stiehler, M. N. Steel, G. G. Richey, J. Mandel, and R. H. Hobbs "Power Loss and Operating Temperature of Tires," *Jour. Res.*, National Bureau of Standards, 64C, No. 1 (January-March 1960), pp. 1-10. This paper also presented at the International Rubber Conference, Washington, D.C., Nov. 8-13, 1959.

14-48. R. D. Stiehler. "Factors Influencing the Road Wear of Tires," *Engineering*, Vol. 173, No. 4490 (Feb. 15, 1952), pp. 218-222.

14-49. Washington State Highway Department. *Washington Motor Vehicle Operating Cost Survey*, 1952-53.

14-50. "Why Tires Wear Faster on Curves," *Commercial Car Jour.*, Vol. 61, No. 5 (July 1946), pp. 43, 87. Based on work by Messrs. McCarthy and Shively of Goodyear Tire and Rubber Co., Akron, Ohio.

Chapter **15**

Traffic Accidents

Traffic accidents on the highway are an unfortunate and not completely unavoidable economic cost of highway transportation. The cost of accidents is included in the cost of highway travel in the same sense that motor vehicle running cost is included. Highway accidents bear a relationship to the mileage of motor vehicles, so therefore the cost of motor vehicle accidents is properly considered as a road user cost in the market class. The overall cost of accidents involves the cost of motor vehicle and medical commodities, travel time, nontravel time, professional and other personal services, and the nonpriceable pain, anguish, suffering, and sentimentalities involved. Unfortunately, for the purpose of economy studies of highway improvement, the cost of traffic accidents has not yet been reduced to a suitable basis and the accident rates have not been closely enough associated with highway design and traffic operations. The following discussion sets forth the main considerations in an effort to focus thought upon the relation of highway accidents to economy, economics, and highway design. Some accident rates are presented as guides to use in economy studies.

REPORTING AND CLASSIFYING ACCIDENTS

Certain facts on motor vehicle accidents come from official reports made within a few hours or a few days of the accidents by those persons involved and by traffic officers. The correct and complete cost of accidents usually has to be obtained by follow-up interviews and investigations. Accident facts, other than cost, are extensive in the literature, though not always in the detail desired for economy studies, but accurate and complete costs of accidents are practically nonexistent. A discussion of the reporting and classifying system is helpful in determining the cost of accidents and the traffic accident rates to use in engineering economy studies.

DEFINING ACCIDENTS

There are several definitions of the word "accident" and still wider uses of the word when restricted to highway travel. The National Safety Council |15-36, p. 97| gives the following definitions:

Accident is that occurrence in a sequence of events which usually produces unintended injury, death, or property damage.

Motor vehicle accident is any accident involving a motor vehicle in motion that results in death, injury, or property damage. However, motion of the motor vehicle is not required in a collision between a railroad train or a streetcar and a motor vehicle.

Motor vehicle traffic accident is a motor vehicle accident which occurs on a way or place, any part of which is open to the use of the public for purposes of vehicular traffic.

These three definitions narrow the meaning down to a practical basis for highway economic studies. Further classification is desirable to afford a proper base for understanding the literature, particularly certain writings on numbers of accidents and costs of accidents.

There is a difference between a motor vehicle accident and a highway traffic accident. Collisions on parking lots and at one's garage door are motor vehicle accidents, but not traffic accidents because the site of the accident is not on a public trafficway.

A traffic *accident* involves a vehicle on a trafficway and in motion (a vehicle striking a fixed object or another vehicle); a traffic *incident* involves a vehicle on a trafficway, but without the vehicle being in motion (a falling ladder striking a parked vehicle). Traffic incidents [15-25] result from such events as vandalism, riots, fires, explosions, storms, snowslides, and collisions from nonmotor vehicles such as construction equipment and falling objects.

A traffic accident *involvement* is one vehicle in an accident. An accident in which two vehicles collide consists of two involvements. For some purposes, accident cost information reported on an involvement, or single vehicle as the base unit of enumeration, is best, but collecting accident cost by involvements is an expensive expedient because the costs must come from the owner or driver of each vehicle. The number of traffic accidents is the other unit of enumeration by which accidents are reported. Accidents as such involve one or more vehicles, so a traffic accident is composed of one or more involvements. For economy analyses the per accident accounting is more useful because records and other accident information are most often available on an accident basis rather than on an involvement basis.

REPORTING ACCIDENTS

Laws and ordinances require certain motor vehicle accidents to be reported to a designated governmental agency on a prescribed form. Usually all accidents that result in personal injury and death are required to be reported, and property damage only accidents are reported when the damage is above a minimum sum, such as $50, $100, or $200. These official reports are the source of most of the summarized statistics on highway accidents, but on an accident basis, perhaps only 50 percent of all traffic accidents are reported. Although these official reports are reasonably complete in the details of the accident, they do not report the costs of the accidents other than an estimate of property damage. The medical and other costs must be collected in the weeks, months, and even years following the accident and from those persons or their survivors who suffered the costs. But such statistics on cost are often neither complete nor accurate.

Through the efforts of the National Safety Council, state departments of safety and of motor vehicles, insurance companies, and other agencies, motor vehicle accident official reporting is becoming more uniform over the country, but is still far from ideal. Accident statistics, therefore, can be summarized and analyzed on a local, county, state, or national basis. Also, they can be classified by many key factors pertaining to the vehicle driver, the vehicle, the highway, the weather, time of day, and day of the week.

In spite of the detail in which motor vehicle accidents are reported, the critical student or analyst of accidents usually finds the reports or summaries lacking. Some of the main items important to the highway designer and economic analyst most often not reported in sufficient detail are the exact location of the traffic accident with respect to the features of highway design, exact spot on the roadway, and the traffic conditions at the time of the accident. With respect to roadside objects, curbs, shoulders, ditches, sumps, crests, access points, and the like, 5 or 10 feet of distance may be important to the analyst of traffic accidents.

Because the law usually does not require minor accidents (based upon minimum property damage) to be reported, and because many lawfully reportable accidents are not reported, the total picture of motor vehicle accidents is not known. The unreported traffic accidents are more numerous than those reported, and their total cost is highly significant [15-25, p. XX and 15-4, p. 203]. See also page 400 and Table 15-9A.

More often than not, motor vehicle accident reports include all accidents involving a motor vehicle regardless of whether the accident occurred off the highway or in moving traffic. For economy studies it is only the traffic accident that is wanted because the traffic accident is affected by highway design and traffic operations. Accidents on parking lots, in parking buildings, on private land, and in repair garages are unrelated to highway design except as they affect access to the public highway. The Illinois accident cost study [15-25] separates the motor vehicle accidents into traffic and nontraffic accidents and traffic and nontraffic incidents.

CLASSIFYING ACCIDENTS

Motor vehicle accidents are classified and summarized in many ways to satisfy the many uses of the statistics. See Ref. 15-36, Accident Facts, published by the National Safety Council, Chicago. But unfortunately the many reported classifications of motor vehicle accidents lack much when applied to the analysis of highway engineering economy. The highway design features, the traffic volumes, and the exposure rates are usually not reported.

Traffic Accidents

One of the most useful classifications is the simple three-way grouping of severity into (1) fatal injury accidents, (2) nonfatal injury accidents, and (3) property damage only accidents. Property damages usually result, also, from fatal injury and nonfatal injury accidents. The fatal and nonfatal injury accidents are also reported by numbers of persons involved. When the costs of accidents are also reported by this classification and the rate, or frequency, of accident is known, the total traffic accident cost is easily calculated for that classification.

Another useful classification is (1) pedestrian accidents, (2) collision between motor vehicles, (3) other collision, and (4) noncollision.

In turn, each of these four groups is divided into the action of the vehicle or vehicles at the moment of accident, such as going straight, turning, entering intersection, one car parked, hitting fixed object, overturning, and many others.

Still another classification that bears upon economy studies is that pertaining to the action of the driver of the vehicle. In this grouping are found such items as:

Speed too fast	Improper overtaking
Drove left of center	Disregarded signal
Failed to yield right-of-way	Made improper turn
Passed stop sign	Followed too closely

These eight actions of the driver correspond to violations of traffic laws and that is the reason these actions are on the official reports. Perhaps of equal importance, but not reported, are other actions of the driver, such as reading a map, putting on sunglasses, lighting a cigarette, listening to the radio, thinking about a personal or business problem, etc.

The reported summaries of motor vehicle accidents such as found in Accident Facts, do not generally classify the accidents by highway design type, by traffic volume, by traffic characteristics, or by highway system. Rural and urban separations are given for certain types of accidents, and some overall accident totals by type of vehicle may be given. More detail is given for fatal injury accidents than for nonfatal injury accidents; property damage only accidents are published with but little detail. These general statistics are national or state totals.

To get traffic accident rates and costs as related to highway design, highway system, and traffic characteristics as needed in economy studies, recourse must be taken to detailed analyses of the original accident reports, accident by accident, personal interviews, and personal examination of accident sites. Highway departments and other organizations have reported the results of such analyses in such publications as given in the list of references at the end of this chapter.

Even such major reports as Refs. 15-25, 15-42, and 15-50 however, are short of the full information needed for economy studies. This shortage results from three primary factors: (1) the information was not available in the official reports of the accidents, (2) it was judged too costly to get the information, and (3) those making the analytical studies were not aware of the needs on accident statistics for application to economy studies. This latter factor is an important one, because perhaps no one person has ever thought sufficiently about the process to come up with a plan for motor vehicle accident studies for specific application to economy studies or to highway design. The important information not usually given is the cost of accidents, say by fatal injury, nonfatal injury, and property damage only, in relation to the traffic volume and to the design type of highway. Another example of missing information relates to intersection accidents for which traffic flow, number of lanes and their width, and the traffic control devices are important factors. Further, Wilbur Smith and Associates [15-42] report the traffic accident cost only on an involvement basis.

In general, analyses of official reports and summaries of traffic accidents and the published statistics resulting therefrom, have general use in education of drivers, as legislative background, in licensing of drivers, and in overall publicity on accidents and their cost. They also serve traffic engineering, highway design, and law enforcement, especially in those areas of concentration of accidents, either by highway location or cause of accidents.

ACCIDENTS AND ECONOMY

Highway accidents are costly in dollars and in human distress. For economy and economic studies of highway transportation, traffic accidents may be considered in respect to their consumption of commodities and consumption of time. Despite the extensive literature on highway accidents, not yet have they been reported and analyzed on an adequate basis to satisfy the objectives of the analysis for choice of alternatives in highway improvements. Some discussion from this viewpoint will help in understanding the nature of accidents and their role in attaining economy in highway transportation.

ACCIDENTS, ECONOMY, AND PUBLIC ATTITUDE

The costs of traffic accidents are unwanted, unsought costs to be avoided, but not necessarily at any expense. Economy studies of highway alternatives must necessarily be placed on a dollar market price basis. Because of the non-priceable subsequent pain, misery, and personal mental and physical sufferings from personal injury accidents, an expenditure of highway funds is warranted considerably above that level just sufficient to produce a balance between outgo for accident reduction and the accountable reduction in the cost of traffic accidents. Management acting within public policy and desires of the public, is responsible for the decision of how much money to invest above the level of dollar returns. But management first needs to know the facts of economy as a basis of making its final decision.

There always has been emphasis on highway design and traffic operations with a view toward reducing the number and severity of traffic accidents. As traffic has grown heavier and more concentrated, accidents have increased in number and in total cost. Concurrently, there has been a growing practice within highway departments to place more emphasis on and relatively more effort into accident reducing and accident prevention elements of highway design and traffic operation. These efforts parallel the general public clamor against traffic accidents.

The problem facing the highway designers and the traffic engineers is to construct a safe highway for unsafe drivers. Most drivers know how to drive safely, but they drive unsafely–too fast for conditions, take chances in turning and passing actions, drive when sleepy, drive with too much alcohol in their blood stream, drive unsafe vehicles, overtake and pass and change lanes under conditions of high risk, and generally contribute to accident frequency without getting real values in return for their unnecessary risks.

The American driving public does not want safe highways–that is, the people do not want to pay the financial cost and to suffer the restrictions necessary to produce safety in traffic. They would rather pay more for the accident costs associated with unsafe driving than to sacrifice their freedom to drive as they please. This same feeling is expressed by Hafstad [15-17]: "One quickly gains the impression that while it is generally considered that there are too many accidents, as a nation we seem unprepared to pay the price required for a further reduction of accidents." How much can be done to gain public support for safe driving and how much will be done are unanswered questions. The situation is much as expressed by Hall [15-18], when in 1962 he said: "All of these things confirm me in my belief that the one greatest reason we aren't moving faster towards safety on our streets and highways is that as a people we don't have the desire."

This attitude of the public suggests the need for research to determine the level of highway accidents we want to live with. What is a good balance socially and economically between having traffic accidents and the sacrifice to be suffered by not having traffic accidents? Perhaps if this question were answered a public policy and program would follow whereby a more positive movement toward a definite and reachable goal could be publicly agreed upon. Traffic accidents are a complex problem of society involving politics, economics, sociology, and engineering.

In the meantime, highway officials must continue their efforts to build into highways as much traffic safety as their technical knowledge and earnest endeavor can supply and as much as the public will supply money for. Economy studies furnish a good guide for carrying out these two items of policy.

USE OF TRAFFIC ACCIDENT DATA IN ECONOMY STUDIES

Considering that traffic accidents result in one element of the cost of highway transportation, the economy aspects of traffic accidents should be investigated in the same detail as are the factors of highway cost and motor vehicle running cost. For the purpose of the analysis of economy only, that is, for the moment not considering the desire to reduce human misery, accidents may be treated much in the same way as are the highway costs treated in relation to motor vehicle running costs.

Three general applications of economy analyses to improvement of traffic safety may be stated: (1) spot improvements with the single intention of reducing the accident rate or accident severity at that spot on the highway, (2) project improvement or a general improvement, reconstruction, or construction of a highway route or sizable segment thereof, and (3) transportation system planning, wherein all the highway facilities within an area (often urban) and perhaps all modes of transportation therein are considered,

For these three applications the appropriate form of accident data may vary somewhat. In general, what is needed is the cost of accidents expressed in cost per accident, per involvement, per mile of highway, or per million (or 100 million) vehicle-miles. The second set of information is the accident rate for each of the improvement alternatives being considered. The accident rate may be expressed as accidents or involvements per million (or 100 million) vehicle-miles of travel, per million vehicles (for an intersection as an example), per highway mile, or as a percentage reduction from a known base, such as for the existing facility. The literature is gradually building up a source for both cost of accidents and accident rates, but there are still wide and deep gaps of areas of wanted information.

Spot improvements to reduce accidents are common and simple in analysis. Percentage reduction in accidents and severity are good measures of the accident reducing efficiency of the improvements considered. Typical spot improvements are intersection traffic control devices, selective lighting, bridge widening, curve lengthening, lengthening sight distances, and increasing coefficient of friction of the road surface.

Route improvement projects call for accident comparison between such factors as horizontal curves, vertical grades, number of traffic lanes, medians, and degrees of access control. In these types of improvements, accident rates related to total vehicle-miles of travel for specific daily traffic volumes are most often used. Subanalyses, however, such as for curves and grades, may require more specific information or a relative factor of accident rate based upon the rate for level tangents.

For area system planning studies, overall traffic accident costs based upon annual vehicle-miles of travel are desirable. These costs may be expressed for general types of highways or streets and their general location. Typical needs are for annual accident costs on local streets serving residential and business areas, arterials, expressways, and freeways for rural, suburban, and urban locations.

COST OF TRAFFIC ACCIDENTS

The total net cost of any one single highway traffic accident is more difficult to determine than most persons believe. The first difficulty is in the planning phase of the effort to determine the costs. What cost items are to be included? What offsetting benefits are to be included? How is the time devoted to the accident by people associated therewith to be accounted for and priced? Many other similar questions must be answered. Second, there is the practical difficulty of finding the dollar costs and the clock time consumed once a decision is made on what costs and time are wanted. The official accident reports filed immediately after the accident can contain only preliminary and partial estimates of costs. It is necessary to contact appropriate records, public and private, and the people involved in the accident or associated therewith after the accident in order to get the correct total cost. For some accidents, especially serious nonfatal personal injury accidents, costs may be incurred for 1, 2, or 3 years after the accident. But in highway economy studies accident costs are an important factor and should be calculated using the best estimates available. Despite all the studies so far made on highway traffic accidents no one has yet collected the data and put them together in a form suitable for economy studies of proposed highway improvements. Two primary factors contribute to this shortcoming in accident studies. First, the analysts have not had in mind the needs of information for economy studies, and, second, the necessary information was not readily available.

CLASSIFYING ACCIDENT COST ELEMENTS

The cost elements within the total costs of traffic accidents are many and are involved. These elements may be classified by several schemes, depending upon which scheme serves the purposes to greater advantage. There are direct and indirect costs, on-site and off-site costs, immediate and future costs, cash costs and noncash costs, cost of goods and cost of personal services, costs to those involved in the accident and the costs to others, priceable costs and unpriceable costs, and tangible and intangible costs. No one classifying system will serve all purposes to equal advantage.

In the ordinary reporting system the on-site costs are limited to property damage to the vehicle and its contents, and perhaps to the highway and roadside property, ambulances, and a few other items that can be observed at the time of the accident. The off-site costs primarily are those associated with treatment of injuries, legal and court costs, and time factors. There is really no positive scheme to separate direct from indirect costs, tangible from intangible costs, and likewise for the other schemes. The important thing to know, however, is what cost elements are included in any report of traffic accident costs. Only then do the stated traffic accident costs have real meaning.

As with most transactions involving the dollar, a disbursement by one person results in an income for another; the income can be the equivalent of sales when commodities or services are purchased, and largely net income when direct labor is purchased. To the disburser, however, the full amount disbursed is a cost to him.

Special investigations of traffic accident costs have usually concentrated on the cost items of property damage, medical, legal, and wages lost for two main reasons: (1) The costs borne by the individuals directly involved in accidents are more readily obtained than the other costs, and (2) these costs usually amount to a high percentage of the total. Nevertheless, some attention should be given to the total of all economic costs, at least until some general ratio of the cost of these indirect and general items to the direct and specific cost can be established.

The following nine groupings (A to I) of traffic accident cost items are an attempt to bring to light and into sharper focus the many and varied cost elements regardless of their magnitude. The classification scheme serves the purpose of (1) emphasizing the costs for consumable goods, (2) personal services, (3) time utilization by all individuals, and (4) private costs separated from government costs. Each of the nine groupings is discussed in the sections to follow.

ELEMENTS OF COST AND BENEFITS ASSOCIATED WITH HIGHWAY TRAFFIC ACCIDENTS

A. Goods and other property consumed
 1. Parts and labor for restoration of damaged property—motor vehicles, highways, buildings, and other items
 2. Clothing and personal property damaged
 3. Cargo in freight vehicle
 4. Meals away from home
 5. Medicines, appliances, medical supplies not included in physicians' fees or hospital charges
 6. Flowers, candy, gifts, and special purchases only because of the accident
 7. Domestic animals and wildlife killed on the highway

B. Transportation and communications
 1. Ambulance and other rescue vehicles
 2. Tow truck
 3. Taxi
 4. Hire of other vehicles by all concerned, including transportation for the disabled
 5. Public transportation—from far and near—by those involved
 6. Extra travel by motor vehicles for trips pertaining to accident by all persons for all purposes, including police vehicles
 7. Commercial rentals and taxicab fares lost because of damage to rental and taxicab vehicles
 8. Lost use of damaged commercial vehicles

Traffic Accidents

    9. Motor vehicle operating expense to traffic for delays, extra distance routing, and speed changes at time of accident
   10. Extra use of mail, telephone, telegraph, and other communications
   11. Newspaper, television, magazine, and other coverage of accidents

C. Personal services rendered, including the overheads, goods, supplies and rents included in the wages and fees of those rendering the services
    1. Physicians' and dentists' fees, including total expenses
    2. Lawyers' fees
    3. Nursing fees other than received through hospital service
    4. Hospital services–entire staff–and goods and other services included therein
    5. Court costs and fines assessed to accident involved persons
    6. Witness fees and costs
    7. Additional hire at home–domestic help
    8. Additional hire at work place to replace absentees
    9. On site and other assistance from bystanders and passersby

D. Time consumed by all persons affected
    1. Lost work time of those persons–injured ones and others–involved in the accident–all purposes
    2. Lost work time of those concerned or affiliated with persons involved in accident
    3. Nonwork time of relatives, friends, associates devoted to any aspect of the accident–accident reporting, court witnesses, consultation, repair or replacement of damaged property, telephoning, etc.
    4. Probable future work time of deceased and permanently injured persons
    5. Delay time and other travel time of motor vehicles as a result of the accident

E. Unclassified, including insurance
    1. Damages awarded above the costs otherwise accounted for
    2. Present worth difference in funeral cost now and the future time of expected death
    3. Financing expense–loans and loss of returns on investment withdrawn to pay accident costs
    4. Illness and accidents to others brought on because of the case accident
    5. Motor vehicle liability insurance overhead cost–premiums paid by the insured less claims paid
    6. Workmen's compensation insurance programs–prorata share of net cost assignable to highway traffic accidents

F. Anguish, anxiety, misery, and suffering
    1. Misery, grief, distress, and suffering by all concerned
    2. Loss of sleep, fatigue, and inconvenience to all concerned
    3. Mental and physical breakdowns
    4. Pleasures and duties forgone

G. Benefits received and normal expenses avoided
   1. Future living cost of the deceased
   2. Insurance claims paid other than those associated directly with motor vehicle liability coverage. Included would be compensation, employment, medical, and health
   3. Damage and compensation court awards (previously discussed)
   4. Saving in motor vehicle use because of not traveling to work and to social and recreational functions
   5. Use of cheaper transportation or none at all while car is being repaired
   6. Food saved at home and at work and other expense because of being confined to hospital
   7. Curtailed social and recreational activities
   8. Many other small items
H. Governmental service and operations
   1. Court costs not compensated for by fines and fees
   2. Criminal and other prosecution procedures
   3. Police and highway patrol salaries and expense–overhead and operations
   4. Ambulance and rescue services (some private, some public, and some volunteer organizations)
   5. Collecting, reporting, filing, analyzing, investigating, and publishing reports and statistics related to traffic accidents
   6. Administrative costs of financial responsibility and compulsory insurance laws
   7. Future income taxes not received on probable earnings not earned because of injury or death
   8. Net of income taxes not paid because of deductions for medical expense and losses
   9. Inheritance taxes not paid because of less wealth at death–now or later–resulting from accident expenses, reduced income, or future income not earned

      Note: Items 7, 8, and 9 probably do not result in a net decrease of tax income, because (1) the governments will raise their tax rates as is necessary to raise the amount of their budgets, or (2) the incomes of others who make a financial gain from the accident cost to the others, may pay greater total sums of income tax than they would have paid had the accident not happened.
I. Activities to reduce the number of accidents and their severity
   1. Highway construction of facilities and maintenance work for specific purpose of accident prevention including spot improvements
   2. Traffic engineering surveillance and studies
   3. Traffic officer operations
   4. Education and training programs for drivers
   5. Emergency medical services

Traffic Accidents

6. Community safety programs
7. Committees, boards, councils, and group activities pertaining to safety
8. Items on the motor vehicle designed for accident prevention or reduction of accident severity–seat belts, collapsible steering columns, dual brake cylinders, signal systems, padded interiors, etc.
9. Research and development costs on traffic accidents, their prevention and treatment

Goods and Other Properties Consumed–Group A

Motor vehicle accidents on the highways, roads, and streets consume commodities in the same sense that motor vehicles consume the commodities of fuel, rubber, and mechanical parts. Only the first three groups of items are important in magnitude. Meals away from home, gifts, and animals killed are of no significant percentage of the total.

Of the three main groups, the damage to motor vehicles is by far the more costly. In fact, but incorrectly so, the other items are often not reported with normal accident reports. These nonvehicle items are real, however, and should be included. They can be obtained only by careful investigation over time, following the date of accident. The cost of repairing the damages or replacing the property consists of the usual three factors–(1) parts and supplies, (2) direct and indirect labor and services, and (3) overhead expense of suppliers. These three items are exactly of the same character as are included in the price of fuel, tires, and normal vehicle maintenance. Therefore, they may be treated exactly in the same manner in economy studies.

The off-site cost of commodities is usually never reported as such; in fact, much of the total off-site cost is never reported. The reported off-site cost is usually just that associated with medical attention to the injured. Full-scale studies of accident costs require interviews with individuals affected and other sincere efforts to assemble all costs, accident by accident. This procedure is the only way all costs can be obtained |15-25, 15-42, 15-50|.

Transportation and Communications–Group B

A traffic accident, particularly a personal injury type, brings into play a complex of travel and communication activity. In addition to the travel of emergency vehicles at the time of the accident, there is local travel by relatives, friends, and associates of the injured. Even intercity travel takes place by one or more modes. Other transportation expense occurs because of rental of vehicles to replace damaged vehicles for both personal and commercial use.

One motor vehicle expense not yet evaluated is the cost to traffic which is interfered with by the accident. This cost is composed of idling motors, speed changes, rerouting over longer distances to destinations, and often secondary minor accidents.

The last two items of use of mails, telephone, and news coverage are quite minor in magnitude.

In accident cost studies, such as those in Illinois [15-25], Utah [15-50], and Washington, D.C. [15-42], where the home interview and questionnaire procedure was used, ambulance cost and loss of use of the case vehicles were generally collected. Other items of the 12 listed were not specifically covered.

Personal Services Rendered–Group C

Medical and legal services run to high cost in fatal injury and nonfatal injury accidents. These items constitute the bulk of the cost in the personal service group.

Additional hire to replace the services at home or on the job are generally not of significant magnitude. The items in this group are the off-site variety that can be obtained only by detailed follow-up over time through interviews and questionnaires.

It is assumed that these professional services and the associated fees are scarce resources and that to reduce the cost to the road user for these services is a desirable economy. The professional services not used on traffic accidents could be devoted to other outlets. This reasoning is the same as applied to the saving in the direct running cost of motor vehicles.

Time Consumed by All Persons Affected–Group D

The hours devoted to all aspects of traffic accidents by persons not directly compensated therefor amount to a surprisingly large total. The persons so involved include the accident victims and their relatives, friends, associates, and random persons affected solely by chance. The time is devoted to telephoning, visiting the home and hospital involved, courtrooms, trips for medical service, and work related trips. In addition, there is the time taken from work by these persons and the work time lost by the injured, for which some of these persons suffer some loss of wage income. These clock times are most difficult to collect and in most cost studies such attempts have been restricted to getting only lost work time of the injured.

Wages Lost. Most studies of the cost of traffic accidents report time taken from gainful employment as a cost computed at the wage or salary rate of the injured person. This element, however, is included only for injured persons and then only if they were employed at the time of the accident, which may be assumed as an approach to getting "wages lost." This procedure is somewhat of an oddity when considered in its entirety with respect to the usual practice in economy studies of highway improvement.

First consider the person injured. His "wages lost" are included regardless of whether or not he actually suffered a decrease of income because of the accident. The fact that he may have been paid his regular wage during his period of recovery from his injuries because of sick leave or generosity of his employer is not considered. Further, he may have received compensation from employment or accident insurance to partially offset any wages actually not received.

The second consideration is that many persons injured in a traffic accident may not be employed at wages, but do perform a desirable function within society as a whole. Such persons may be donating their time to charity, to hospital service, to community functions, and the like. not counting services and chores at home. At least for many such individuals their time is as valuable to the economic system as is the time of a wage earner.

Third, there is a general theory that when an employed person does not show for work, some worthwhile economic production is lost. This theory neglects the fact that other persons, employed or not, usually step in and do whatever is necessary. These persons may sacrifice some leisure time or they may postpone other work to a later date. For short periods of a day or so, often a person's work just remains undone until he returns to work. In many work situations, increased efficiency of those at work makes up for the absent injured person.

Lastly, analysts of the economy of highway improvements have generally valued the travel time of all persons traveling at a certain price per hour. Thus if one minute of time is saved per vehicle by a change in highway design, this one minute is multiplied by the number of vehicles per year, then by the value of time per minute for the average number of persons per car, say 1.6 persons, to get the yearly value of the reduced travel time. If travel time is to be evaluated on this basis, it is just as logical to evaluate on the same basis the time of all persons devoted to accidents regardless of age, employment, economic status, and whether injured or not.

The items 1 to 4 in this group are a few of the many time-consuming actions taken by a wide group of individuals none of whom is monetarily compensated for his time devoted to the aspects of the accident. If all such time consumption could be gotten together for the countless people involved, the aggregate would be considerable. The times are not gotten together because of the size and complexities of the job; yet these times are just as valuable to the economic system and to the individuals as is the 10 seconds or 1 minute per vehicle saved at an intersection or by eliminating a stop or a slowdown. Here again this concept of time should be recognized, and users of results of studies of the cost of traffic accidents should know that the reported value of "wages lost" is not the real total consequence of time consumed because of accidents.

Time Consumed and Cost to General Traffic Because of Accident. One sizable cost of traffic accidents so far not investigated is that incurred by traffic at the site and time of accident. These costs are primarily found in five factors: (1) speed changes, (2) idling motor, (3) extra travel distance because of rerouting, (4) extra tire tread wear, tire chains, fire extinguishers, tow chains, and personal expenditures, etc., and (5) a greater travel time. Extra travel time is composed of three factors: (1) delays (stopped time), (2) slower speeds, and (3) rerouting.

On heavily traveled routes, accidents at rush hours can result in delays at the spot of a few minutes to hours and involve thousands of vehicles.

One reference in the literature [15-54] may be cited. A rear-end collision happened at 5:21 p.m., August 17, 1962, on the Northwest Freeway in San Antonio, Texas. The accident did not stop the flow of traffic, but did slow it down until the two cars involved were moved off the two-lane, one direction pavement. In all, traffic was delayed somewhat for 51 minutes, affected 1,440 vehicles for a total delay of 3,971 vehicle-minutes, or 66.2 vehicle-hours. The normal travel time for the distance involved was 47 seconds. The maximum travel time taken by any one vehicle until traffic regained its free flow was 8.75 minutes. The normal traffic flow past the site of accident was about 50 vehicles per minute. Not reported is the number of vehicles that may have turned off the freeway at a prior exit.

Unclassified and Insurance—Group E

Court actions frequently award injured persons or survivors of the fatalities dollar damages in excess of all known costs. In such cases the excess of the awards above the known costs becomes a justifiable charge against the accident. These excess awards are considered a cost despite the fact that they are a benefit above costs to the recipient, generally the injured person or survivors of the fatally injured.

Funeral costs are not included as a cost of the accident on the grounds that they are inevitable in the long run. However, there is justification of including the difference in the present worth of funeral costs at the date paid and a similar payment at the future date of normal life expectancy of the deceased.

Undoubtedly, many persons responsible for payment of the costs of traffic accidents must borrow money for this purpose or withdraw money from interest or dividend-paying investments. In such cases the interest paid on the loan or the return forgone on the withdrawn investment is a cost chargeable to the accident.

The strain, anxiety, and fatigue resulting from an accident may bring on illness to others than the accident victims or be the cause of a second accident. Although somewhat difficult to ferret out and to prove, such events are known to have happened.

A large item of traffic accident cost to the road user is liability and property damage insurance. Premiums are paid by policy holders; those having accidents have their claims paid. But claims received do not equal premiums paid. The difference is that sum required to pay for operating the insurance system. Salaries, commissions, rents, claim adjusting, court expenses, operating expenses, and communications are paid for out of premiums. All these costs are usually lumped under the heading of insurance overhead. In 1965 the general administrative and claim settlement cost of motor vehicle insurance was estimated at $ 2,850,000,000 [15-36, p. 5]. Based on 11,000,000 accidents, this overhead cost of insurance would be about $ 245 per accident and about $ 40 per registered vehicle.

One insurance cost item not directly a cost borne by traffic accident victims or vehicle owners is the net cost of workmen's compensation insurance. Accident and income compensatory insurance policies are held by many persons individually. For both of these two forms of accident and compensatory insurance some proportion of their net overhead costs is a just charge against traffic accidents.

Much more needs to be said about motor vehicle insurance, but we leave this subject to other authors to cover more appropriately. Mention can be made that motor vehicle insurance has become an important social and economic factor in society. State legislatures and congress are involved and special insurance commissioners and rate regulatory bodies are involved together with public hearings. Some states tie insurance coverage with vehicle licenses for the purpose of maintaining funds for payments when the responsible person is uninsured. Compulsory insurance is also a subject of public concern. When all these and other factors are considered, it is easily recognized that motor vehicle insurance has many ramifications involving costs, public policy, private business, and individuals.

Anguish, Anxiety, Misery, and Suffering–Group F

One of the disagreeable aspects of injury and fatal traffic accidents is the consequent human anguish, anxiety, misery, and suffering. These consequences are both mental and physical; they affect the injured, the relatives and friends of the injured, the passersby, and eyewitnesses. Individually and collectively we are willing to pay a high price to avoid these consequences, but what price has not been determined. Humans do not want to inflict pain, not even to animals.

Here again there is an element of personal preference rather than of economy. Perhaps there is some economic cost to anguish, anxiety, misery, and suffering because such factors affect production at work and safety in traffic. A deeply worried person suffers decreased attention to his work and to motor vehicle driving with consequent decreased production and increased traffic accidents. To price these two unfavorable consequences of traffic accidents is in order for economy studies, but impractical. The anguish, anxiety, misery, and suffering that result from traffic accidents become collectively another nonmarket item in the economic analysis of alternative highway improvements to which subjective judgment may be given as judged to be appropriate in each individual proposal.

These human factors are what concern many of the public and many public officials. These are the elements that justify accident reducing measures that cost more than the expected reduction in accident costs. There is no known way to put a price, cost, or value tag on these nonmarket consequences of traffic accidents. They become strictly a matter of management decision.

It is to be noted that the sacrifices of normal pleasure and duties because of accidents are placed in this group, also as a nonmarket item.

Benefits Received and Normal Cost Avoided–Group G

The tendency of most persons is to think of traffic accidents only in terms of gross cost. Practically no attention is given to monetary benefits from accidents. It is as logical to consider the monetary gains or savings from accidents as it is the costs. Gains will offset part of the costs. Court awards of damage and compensation are sometimes considered, but otherwise the gains are not usually mentioned.

The gains are not assembled because they are usually minor in dollar value, associated with scattered numbers of hard to locate persons, and are of a nature difficult to determine let alone to price.

There is no suggestion intended here that these gains should be sought out and priced; the general object of all economy studies is to get at the net costs and net benefits, so recognition that there are some gains to offset some of the costs is in order, for the same reason that, generally, with accident cost studies, there is recognition given to the fact that all of the costs are not assembled.

The handling of the living costs of the deceased is discussed in the main section ahead on death and permanent disability.

Governmental Services and Operations–Group H

All governments–city, county, state, and federal–carry on many activities directly related to traffic accidents. Although not easily assigned to specific accidents, the costs of these services and operations are a part of the total cost of traffic accidents.

In this group are two main elements of major magnitude. They are the court and legal operations and the police and traffic officer operations.

Group H includes some of the items of traffic accident costs which are borne by the public at large not including highway funds. To a certain extent, road user taxes, fees, and licenses are used in combination with general taxes in support of the cost of traffic accidents.

Courtroom costs, police investigation and reporting, and accident reporting and analysis by the several governmental agencies are examples of general cost not borne directly by those persons involved in traffic accidents. The court costs which may be assessed by the court against those persons involved in accidents do not pay the whole of the direct and indirect cost to the public agency responsible for the traffic court. From the public viewpoint, it is not likely that these costs have been fully determined. The traffic police and highway patrol devote much of their time and travel to traffic accidents, which time and travel would be otherwise utilized if accidents did not occur.

Many accidents are attended on the site by many public and private rescue and emergency personnel with special equipment. Only certain of the costs of these on the site operations may be paid for directly by the accident victims or

Traffic Accidents

vehicle owners. The remainder of the costs are borne by the public treasury or by private services.

Items 7, 8, and 9 are questionable, depending upon one's viewpoint. Considering society as a whole, tax incomes are equal to public expenditures so in effect there is no tax loss from incomes not earned, but just a redistribution of the tax responsibility.

These governmental services and operations cover a group of costs directly related to traffic accidents, but which have not been included in reports of accident costs.

Activities to Reduce the Number of Accidents and Their Severity–Group I

The cost of accidents is in effect reduced by accident prevention and accident reduction measures which are part of normal operations. Although these costs are not a part of accident costs to use in economy studies, they are of interest and a part of the total accident picture. Without accident prevention activities, such as those listed in this group, yearly accident costs would be far greater. Active and aggressive programs, efforts, and attention to accident prevention and reduction, have over the years been highly successful in reducing accident rates and holding down overall growth of the gross costs. These programs themselves are specific opportunities for economy analyses.

The U.S. Department of Health, Education, and Welfare developed in 1966 some first trial solutions of the economy of nine different traffic safety programs [15-49]. The programs, evaluated on a national basis, include the following:

1. Improve driver licensing
2. Improve driver training
3. Reduce driver drinking
4. Reduce pedestrian injury
5. Increase seat belt use
6. Improve driver environment
7. Use of improved restraint systems (in addition to lap belts)
8. Increase use of improved safety devices by motorcyclists
9. Improve emergency medical services

Benefit/cost ratios ranged 1.7 for improved driver training to 1,351 for increased seat-belt use. This report [15-49] is acknowledged by the authors to be exploratory and highly tentative. However, the report is good reading as examples of methodology and of the factors involved in predicting the result of programs to reduce traffic accident rates and their costs.

DEATH AND PERMANENT DISABILITY

How to treat death and permanent disability in accident cost reports has long offered room for different procedures. That the probable future income of these persons who die as a result of traffic accidents or whose probable future incomes are materially reduced because of work-disabling injuries represent some sort of cost chargeable to accidents is generally accepted. In principle, a fatal accident or a permanently disabling injury accident need be treated no differently than other accidents. The factor of time is present in all accidents and varies only in its duration.

VIEWPOINT TOWARD DEATH AND PERMANENT DISABILITY

The cost of fatal injury accidents and accidents that result in permanent disbility may be viewed from three aspects: (1) the cost of property damage, medical attention, and other direct expense, (2) the worth of a life to relatives and friends viewed from sentiment and love, and (3) the economic loss because a producing person is removed from his role in society as a producer. The compensation for body disfigurement may be included with this third factor.

The commodity, service, and all cash costs associated with the fatal injury accident call for treatment no different than for other accidents.

The worth of a life to those persons who have a personal care is without a sound basis for estimating. The general practice is to save a life at any cost because human life is valued highly as life and not because of one's contribution to society or to economic development. Courts and juries have placed value on life, but generally the basis of the value is not stated. The basis can be sentiment, value as support to dependents, present worth of future income, or any combination of these three factors. The value of human life as life, however, is not a cost in the sense that expense is suffered. Since there is no basis for stating a value as support to dependents, present worth of future income, or any combination of these three factors. The value of human life as life, however, is not a Likewise, economy studies of highway improvements should not include the value of life as a sentimental factor, but only as an economic factor.

Probable Future Income Forgone

When life and death are viewed from an economic focus, rather than from sentiment, there is a basis and a logic of placing a price on death.

A traffic fatality removes that person from society, and, therefore, society moves into the future without his contributions and burdens. In this respect, death is fundamentally the same as temporary disability to work; the difference being that death brings permanent absence from work and temporary disability is for one day or other comparatively short period. By logic, then, when a gainfully employed person is injured in a traffic accident and his wages lost are reported as a cost of the accident, the wages lost by a fatality should be also reported as a cost. The time period of probable future income could be fixed by human mortality and life expectancy tables.

If wages lost were to be computed on a lifetime basis, one uncertainty and one corrective factor have to be considered. A person's current wage rate,

Traffic Accidents

especially if young in age, would not remain constant throughout expected life, so some estimate should be made of his future earnings year by year. But because of the time value of money, the probable future earnings of the deceased–should he remain alive-would have to be discounted at an appropriate vestcharge rate to present worth.

This reasoning has basic logic, but two factors are yet to be considered. As stated in the discussion of time and wages lost, there is no logic to including wages lost for only those persons gainfully employed at the time of the accident; the value of time for all persons involved should be included. Likewise with death; if probable future earnings are to be reported as a cost of a fatality, then this cost should be reported for all fatalities regardless of their employment status on the day of their death.

Probable Future Living Costs Avoided

But in view of the principle that all consequences are to be considered it is in order to consider the gains from death along with the costs, so that the net cost or net gain can be determined. Man may be viewed as an economic machine. This is the viewpoint, actually, when lost wages are computed on the theory of lost production. It costs money to power, operate, and maintain a machine in profitable production. Likewise, it costs money to maintain a person in a productive work status. Food, clothing, shelter, education, and health cost money; this cost is essential to life and to keeping a person useful to society. When a person is killed, society saves the expense of sustaining that person in the future. Therefore, economically speaking, the present worth to society of a living person is the present worth of his future gross earnings (as a measure of his productivity) less the present worth of the cost of maintaining the person in society for the full period up to death.

Merits of the Concept of the Economic Man

Dublin and Lotka [15-14] develop this concept in their "The Money Value of a Man," published in 1946. There is need for an up-to-date development applicable to economy studies. A net value of man as an economic machine when developed as the present worth of his probable future net income (as a measure of his contribution to society) less the present worth of his probable future living and other expenses (as a measure of the cost incident to his contribution to society) offers a reasonable economic base for comparing highway investment alternatives in which fatalities and permanent disabilities are a factor. This scheme has the advantage that it does not require placing a value on a life simply as an individual and the scheme is in harmony with the general approach of placing a value on travel time. The net present worth as calculated by this method does result in an estimate of what would be the net present value of a wage earner to those persons who sould benefit in the future from those wages should they continue. The method avoids, however, using the court awards on account of death which awards do not result from any known systematic procedure of arriving at an answer. Court awards tend to come out of a mixture of sentiment, economics, and the ability to pay.

CALCULATING THE NET ECONOMIC WORTH OF A PERSON

Application of the concept of the economic man to traffic accidents gives rise to some practical questions and procedures worthy of discussion. There are five basic factors to establish: (1) wage, salary or fee, income for each age for the full period of working years by trades and professions, (2) age of retirement or age beyond which no further wage or salary would be earned, (3) yearly cost at each age of maintaining the individual in a working status, (4) probable life expectancy for persons of all trades and professions and at all ages, and (5) the discount rate to use in calculating the present worth of yearly incomes and expenses.

The total process of calculating the net economic worth of man based on lifetime earnings and expenses is somewhat involved. Only a brief discussion of the process is given here. Dublin and Lotka [15-14] is a good reference. The subject is discussed in Ref. 15-49 so far as life income is concerned. Wilbur Smith and Associates [15-42, pp. 205-217] applies the process to highway traffic accidents, but the procedure is somewhat at fault because of using average yearly income for the period of remaining life rather than on a yearly basis. Futher, the maintenance of man is used as a fixed sum of $2,000 a year age 11 to full life expectancy. A variation with age is more realistic.

Future Wage, Salary, and Fee Income

Current annual wage and salary incomes of individuals vary principally with five factors–age, education, type of employment or profession (trade, craft, art or skills), individual abilities, and geographical location. Income data related to these factors are available from reports of the census of the United States and other economic reports. In application of the information, it may be assumed that the future incomes of all individuals will advance or decline according to the present incomes of persons older and in the same classification with respect to education, type of employment, and other factors. Average annual incomes can be used for specific ages and types of employment or professions, but not in the sense of an average person with respect to all factors. Nationality and sex are factors bearing on income, but these factors are largely accounted for in type of employment or profession. Earned income usually begins at about age 18 and generally increases with age up to a plateau where it continues level for some years. There is often a falling off in earned incomes prior to retirement, especially for manual laborers and skilled craftsmen. No allowance is made for inflation or deflation of the value of the dollar or for probable changing technology.

Lifetime Working Expectancy

Calculation of the present worth of the future incomes is based upon only those years of productive employment or professional practice, as contrasted to life expectancy. Active wage earnings cease between age 60 and 75, depending upon many individual, employment, and geographical factors.

Yearly Expense of Maintaining Productivity

Since man is viewed as an economic force producing products and services useful to society, the usefulness being measured by the earnings of wage, salary and fees, it is appropriate to deduct from his earnings his expenses necessary to achieve the earnings. In other words, net earnings are what are wanted for each year of his working expectancy.

The expense will include such items as food, clothing, shelter, requirements for health, travel to and from work, recreation, professional and union dues, other professional expense, education, job training, taxes, and employment insurance. Each of these items is supported directly from cash earnings of the individual person, though he may not pay directly for all expense associated with his education and training when in the working class. But since this analysis is from the viewpoint of society as a whole and not as may be viewed by the legal survivors of the deceased, it is appropriate to include all the costs of sustaining a person in his working status, regardless of who pays the costs.

Income tax, property taxes, sales tax, and other taxes paid in support of the individual from his earnings are items of expense to deduct from earned income as expenses necessary to maintenance of a working status. Once collected, local and state income, property, and sales taxes and federal income and excise taxes become general funds without identification so may be treated as just taxes. A person owning no taxable property pays the equivalent of property tax in the form of rent. Where income taxes are not imposed, their equivalent income is produced by property, sales, and other forms of taxes and fees, so in the end and in theory everyone pays his share of taxes directly or indirectly. Social security tax in all of its components is a necessary expense of working, so it too, is deductible from income.

Risk and compensation insurance premiums are associated with maintaining a working status so would be another item to deduct from gross income. Life insurance is not a necessary expense to maintenance of a working status so its cost is not deducted.

For all forms of taxes and risk insurance only that paid in support of the specific traffic accident fatality would be considered. Taxes paid from his income support also his family and any other dependents, so their pro rata share would not be deducted from gross earnings to get the net probable future earnings of the deceased.

Probable Life Expectancy

From the human mortality tables as used in life insurance practice life expectancies may be obtained for persons at all ages by sex, nationality, and type of employment. The life expectancy is normally greater than working expectancy because of retirement from employment some years ahead of death. In those years of no earned income just prior to death at old age the net productive earnings would be negative by the full amount to support the retired person.

Discount Rate

The vestcharge rate, or discount rate, used to transfer the yearly net earnings for the remaining years of life expectancy to present worth should be based upon what money is worth to the individual, and most certainly would be higher than bond interest rates, rates earned on life insurance policies, but perhaps lower than the vestcharge rate used in economy studies of public works. Perhaps 5 to 8 percent would be reasonable.

ECONOMIC MAN–CONCLUDING STATEMENT

On the basis of economics, the foregoing scheme offers a logical basis of computing the cost to society of a traffic fatality. The calculation can be applied to persons of all ages for their expectancy of life and of earnings. Young persons and very old persons could have negative present worth. The young would have heavy cost in the immediate future years for living expense and education and no earnings. The earnings would be far in the future, and, therefore, would have comparatively low present value. Older persons would have normal living expense and perhaps but little future income from productive activities.

Permanent disability could be considered in the same manner as for a fatality. For total disability, the net present worth would be highly negative, indicating that, economically, death would be a lesser cost to society than disability. For total disability the costs to maintain the person would be exceedingly high for those cases needing full-time personal care. Retraining expense could be heavy for partially but permanently disabled persons.

It is to be borne in mind that the concept of the economic value of man to society is that his wage and salary income–that which he produces from his own physical and mental efforts, not including earnings on investments–is a measure of his productivity and therefore his value to society. In turn, society supplies the supports necessary to develop the earning capacity of the individual and to maintain the individual in a productive status. The analysis consists of measuring these two streams of cash flow–income earnings and outgo expenses–and finding their net present worth. In application of the calculation of the economic value of a highway traffic fatality, an average person simply does not exist, so it

Traffic Accidents

is necessary to know at least the age, sex, education, and employment status of each individual fatality.

The general present tendency is (and perhaps will continue to be) to give traffic fatalities high importance in highway design, with more thought to sentiment than to economics. But there must be some limit of cost beyond which society as a whole will not go to save a life. When viewed in the light of what vehicle drivers will not do to achieve safe driving, there is a logical conclusion that perhaps drivers value their own lives (sometimes) but not the lives of others.

UNIT ACCIDENT COSTS FOR ECONOMY STUDIES

Including traffic accidents in the analyses for the economy of proposed highway improvements requires that two factors be known: (1) the cost of traffic accidents, and (2) the comparative rate of accidents or the number of accidents associated with each alternative to be considered. The costs of accidents as available for economy studies are discussed in this section and the rates of accidents are discussed in the following section. Costs in the form ideal for economy studies are not availabe, but by judicious selection acceptable costs may be had.

BASIS OF COMPILING UNIT ACCIDENT COSTS

When determining the cost of traffic accidents it is necessary to determine the point of view from which the cost is to be measured. The viewpoint assumed by those compiling such accident cost reports as given in Refs. 15-4, 15-19-2, 15-21-3, 15-25, 15-42, and 15-50, is the combined viewpoint of the motor vehicle owner, owner of damaged property, and persons injured. The basic theory is that the cost of traffic accidents is the money value of damages and losses to persons and property plus other expenses in connection with the accident that would not have occurred without the accident. Using this combined viewpoint, traffic accident cost items may be grouped as follows as is given in the Illinois accident cost study [15-25]:

1. Property damage
    A. Damage to vehicle
    B. Damage to property in vehicle
    C. Damage to objects struck by vehicle
    D. Miscellaneous property damage
2. Treatment of injuries
    A. Ambulance service
    B. Doctor and dentist fees
    C. Hospital and treatment
    D. Miscellaneous medical needs

3. Loss of use of vehicle
4. Value of time lost
5. Legal and court expense
6. Damages awarded in excess of known cost

These six basic groups constitute what may be called the direct accident costs–that is, the money value of damages and losses to persons and property resulting directly from accidents and which would not have occurred had not the accident happened. The use of the phrase, "resulting directly from accidents," is somewhat misleading, because there are many other costs that result from accidents that are not included, as have been discussed elsewhere in this chapter.

In the Washington, D.C., area report |15-42| the classification was slightly different as may be seen in Tables 15-7, 15-8, and 15-9.

THE DETAILED COST ELEMENTS USED IN THE ILLINOIS STUDY

In the Illinois report the following is the basis that governed the collecting of the costs:

Damaged property includes the damage to motor vehicles involved, clothing and personal effects, baggage and cargo within the vehicle, roadside buildings, utility poles, landscape, and to the highway. Towing service is included.

Treatment of injuries includes all professional medical and health services, hospital care and treatment, medicines, appliances, and home treatments. These items of cost are included for fatal injury accidents up to the time of death only. Funeral and associated costs are excluded on the basis that such cost will occur anyway. But it is in order to include the difference in present worth of the cost of death based upon the date of death and the normal life expectancy of the deceased.

Loss of use of vehicle is the loss of earning or business profit during the time the vehicle is not available as a result of the accident. If other transportation was available and used there is no value lost. This value of the loss of use of the vehicle is not to be included in the value of time.

Value of work time lost is included only when the person involved in the accident was a gainfully employed member of the labor force and when such incapacitation was temporary. Permanent total or partial disability is not recognized. Loss of future income because of death is not included, not even for that time between the date of accident and date of death. All work time taken by the injured from income producing activity to have the vehicle repaired, to appear in court, to settle claims, to receive medical treatment, or for other reasons is included. The cost to a housewife who must hire help to replace her is included. The dollar value of time is computed at the rate of income for the specific person concerned. This value of time is included, regardless of whether the person actually suffered a reduction in income or was fully or partially compensated through insurance claims.

Traffic Accidents 385

Legal and court expense includes payments for legal fees, court costs, bail bond, and similar costs for any legal proceedings resulting from the accident.

Damages awarded include settlements in or out of court for only the amount in excess of the known costs. The award may include known direct cost and some compensation for past or future loss of income based upon disability. Care is taken to avoid duplicating the known costs and value of time by including only the net of awards above the known cost.

SPECIFIC UNIT COST OF ACCIDENTS

From the literature, mainly from the Illinois accident report [15-25], unit costs of traffic accidents have been compiled in a form to serve as guides in selecting accident costs for economy studies. Unit costs should be selected with careful attention to the specific alternatives being examined, to the unit costs available, and all pertinent factors. Traffic accident costs are so highly variable that broad general costs should be used only as a last resort.

TABLES OF UNIT ACCIDENT COSTS FOR ECONOMY ANALYSES

Fortunately, traffic accidents do not occur with regularity and in high frequency. But unfortunately, because of their irregularity, variability in type, place, cause, and consequence, it becomes most difficult to assemble the costs of accidents on a sound statistical basis in a form reliable in their application to proposed highway improvements. What is wanted are the accidents for the particular project, route, or system under study and in such detail that the differences in accident costs may be had for the differences in the improvement alternatives being studied. Generally, the specific accident cost data for the specific highway are not to be had, but perhaps the past number of accidents and severity are available. General average accident costs can then be applied to reach a reasonable comparison of the accident cost factors between the alternatives under study.

The traffic accident records on file within a state, city, or other jurisdiction, and most published information will give the number of accidents rather than the number of involvements (individual vehicles involved in accidents). Therefore in economy studies costs expressed in dollars per accident are preferred. The cost per accident ties in with the rate of accidents that is often given in numbers of accidents, say, per million vehicles or per milllion vehicle-miles.

Cost Tables from the Illinois and Other Studies

The Illinois 1958 accident cost study [15-25] is the source of the cost data reported in the following tables. This study is more recent than similar reports from Massachusetts, New Mexico, and Utah. The Washington, D.C., metropolitan area study of 1964-65 [15-42] is not used because all the costs are reported for numbers of involvements rather than for the number of accidents.

Table 15-1 offers general data to give the reader some orientation of the magnitude of the total accident picture and the general ratios of certain factors. The traffic accident cost in Illinois to vehicles of Illinois registry in 1958 at 1966 prices was $346,063,333, which amounts to $130 per vehicle registered and 1.21 cents per passenger car-mile and 0.45 cent per truck-mile. In urban places there was a passenger car traffic accident for each 23,000 miles of travel and a truck accident for each 22,000 miles of travel. The rural travel resulted in one

TABLE 15-1
RATIOS OF SEVERITY OF TRAFFIC ACCIDENTS
AND COST PER ACCIDENT IN ILLINOIS–ILLINOIS REGISTERED VEHICLES, 1958
(Cost Updated from 1958 to 1966 by a Factor of 1.25)

Item and Class of Vehicle	Fatal Injury	Nonfatal Injury	Property Damage Only	Total
1. Number of accidents				
Passenger car	1,169	103,306	755,871	860,346
Truck	220	9,273	112,228	121,721
Total	1,389	112,579	868,099	982,067
2. Accidents per fatal injury				
Passenger car	1.00	88.37	646.60	735.97
Truck	1.00	42.15	510.13	553.28
Total	1.00	81.05	624.98	707.03
3. Percentage of all accidents				
Passenger car	0.12	10.52	76.97	87.61
Truck	0.02	0.94	11.43	12.39
Total	0.14	11.46	88.40	100.00
4. Cost of accidents				
Passenger car	$9,869,882	$168,894,225	$144,697,912	$323,462,019
Truck	1,519,212	8,406,034	12,676,309	22,601,555
Total	11,389,095	177,300,026	157,374,212	346,063,333
5. Ratio of cost to fatal injury cost				
Passenger car	1.00	17.11	14.66	32.77
Truck	1.00	5.53	8.34	14.87
Total	1.00	15.57	13.82	30.39
6. Percentage of total costs				
Passenger car	2.85	48.80	41.82	93.47
Truck	0.44	2.43	3.66	6.53
Total	3.29	51.23	45.48	100.00
7. Cost per accident				
Passenger car	$8,442	$1,635	$191	$376
Truck	6,905	906	112	186
Total	8,200	1,575	181	352

Source: Ref. 15-25.

Traffic Accidents   **387**

passenger car traffic accident for each 68,000 miles of travel, and one truck accident for each 109,000 miles of travel.

For economy studies, Table 15-2 may be used as a guide to the cost of accidents as related to passenger cars, trucks, urban and rural highways and traffic lane arrangement. The intersection accidents, unfortunately, are shown separately rather than within each lane arrangement. Accidents are more costly per accident the more lanes of traffic there are (probably more involvements per accident) and more costly on rural highways (higher speeds) than on urban highways.

The same classification of type of highway as given in Table 15-2 is repeated in Table 15-3 where the accidents are classed by highway system. In both of these tables the "not specified" accidents are mostly in the unreported accident group.

Some idea of the relative cost of accidents by type of traffic control at intersections may be gained from Table 15-4. In using this table, two cautions should be observed: First, the number of accidents reported does not indicate the rate of accidents based upon exposure because the number of intersections and the traffic volumes are unknown, and second, in some categories the number of accidents is too few to afford reliable average costs per accident.

Table 15-5 is a helpful table in economy studies because it shows for each of the three severity groups the cost per accident for each cost element. Damage to vehicles is high in all three categories. Legal and court costs are especially high for fatal injury accidents but only $ 0.99 for property damage only accidents.

The involvements per accident in Table 15-6 show that on the average there are more vehicles involved in an accident in urban traffic than in rural traffic, and that accidents involving two or more trucks in the same accident occur in only about 6 percent of the truck accidents. In using this table it should be noted that in certain categories, especially for trucks, the number of accidents is too few to produce a reliable average number of involvements per accident. The number of accidents is given in Table 15-4.

THE ILLINOIS AND WASHINGTON, D.C., AREA REPORTS COMPARED

Tables 15-7, 15-8, and 15-9 compare the relative percentage of cost for each cost element as reported in the 1958 Illinois urban accidents and the 1964-65 Washington, D.C., area study. The reasons for some of the differences cannot be explained from information in the published reports. The Washington report included some cost elements not found in the Illinois report. Illinois combined some items reported separately in the Washington study.

Of striking significance is the fact that the Washington report shows that the net present worth of probable future earnings of the deceased persons amounted to 90.8 percent (Table 15-7) of the total cost of fatal injury accidents. Of equally striking significance is that in Illinois the legal action costs were 66.89 percent of the total cost of fatal injury accidents as compared to only 38.43 percent

TABLE 15-2

Number of and Direct Cost per Accident of Urban and Rural Traffic Accidents Classified by Type of Highway and Severity; Accidents in Illinois Involving Vehicles of Illinois Registry–1958 (Cost Updated from 1958 to 1966 by a Factor of 1.25)

Type of Highway	Passenger-Car Accidents								Truck Accidents							
	Fatal Injury		Nonfatal Injury		Property Damage Only		Total		Fatal Injury		Nonfatal Injury		Property Damage Only		Total	
	No.	$/Ac.	No.	$/Ac.	No.	$/Ac.	No.	$/Ac.	No.	$/Ac.	No.	$/Ac.	No.	$/Ac.	No.	$/Ac.
Urban Accidents																
Intersection	217	7,678	49,934	1,724	242,416	180	292,567	486	30	4,336	3,645	417	45,225	89	48,900	124
Nonintersection:																
One-way streets with																
One traffic lane	—	—	157	680	5,414	124	5,571	139	—	—	6	*	34	*	40	*
Two traffic lanes	9	7,017	163	743	17,543	77	17,715	86	—	—	17	25	1,390	23	1,407	23
Three or more traffic lanes	—	—	309	913	9,183	132	9,492	157	—	—	414	16	1,589	43	2,003	30
Undivided highways (two-way) with																
Two traffic lanes	115	6,210	22,873	996	219,927	144	242,915	227	35	5,006	1,424	699	24,128	74	25,587	116
Three traffic lanes	2	540	218	830	4,093	114	4,313	150	—	—	17	59	271	*	288	3
Four or more traffic lanes	81	7,283	7,381	1,341	76,842	234	84,304	338	11	5,248	546	426	10,379	87	10,936	109

Traffic Accidents

Divided highways with																
Four traffic lanes	24	6,788	2,182	2,454	10,234	129	12,440	549	–	–	118	1,841	1,713	116	1,831	227
Six or more traffic lanes	10	6,251	2,773	1,572	3,295	603	6,078	1,055	1	540	36	36	997	127	1,034	124
Not specified	–	–	28	2,615	38,767	73	38,795	93	1	9,600	268	86	3,370	95	3,639	97
Total	458	7,118	86,018	1,503	627,714	183	714,190	346	78	5,077	6,491	416	89,096	84	95,665	117
Rural Accidents																
Intersection	145	9,828	5,947	1,539	18,257	284	24,349	647	46	7,760	683	1,059	5,163	150	5,892	348
Nonintersection:																
Undivided highways (two-way) with																
Two traffic lanes	508	9,021	9,850	2,658	85,023	265	95,381	558	79	7,592	1,920	2,095	15,038	242	17,037	485
Three traffic lanes	9	5,039	27	76	3,963	106	3,999	117	1	972	–	–	–	–	1	972
Four or more traffic lanes ...	45	12,020	1,001	3,017	7,515	109	8,561	511	9	12,603	133	1,293	719	150	861	458
Divided highways with																
Four traffic lanes	4	3,995	381	2,077	260	575	645	1,493	6	11,160	42	2,609	300	1,340	348	1,662
Six or more traffic lanes	–	–	82	5,830	–	–	82	5,830	1	4,164	–	–	17	478	18	740
Not specified	–	–	–	–	13,139	70	13,139	70	–	–	4	361	1,895	35	1,899	44
Total	711	9,297	17,288	2,292	128,157	234	146,156	522	142	8,053	2,782	1,807	23,132	225	26,056	437

* No costs of less than $5.00 per involvement were incurred.

Source: Ref. 15-25.

TABLE 15-3

Number of and Direct Cost per Accident of Urban and Rural Accidents Classified by Type of Highway and Highway System; Accidents in Illinois Involving Vehicles of Illinois Registry–1958
(Cost Updated from 1958 to 1966 by a Factor of 1.25)

Type of Highway	Passenger-Car Accidents							Truck Accidents						
	State Highways		Local Streets and Roads		Total		State Highways		Local Streets and Roads		Total	
	Number	$/Accident	Number	$/Accident	Number	$/Accident	Number	$/Accident	Number	$/Accident	Number	$/Accident
	All Urban Highways and Streets												
Intersection	98,888	525	193,679	464	292,567	486	21,157	137	27,743	114	48,900	124
Nonintersection:												
One-way streets with												
One traffic lane	991	150	4,580	137	5,571	139	6	*	34	*	40	*
Two traffic lanes	6,102	46	11,613	108	17,715	86	419	37	988	17	1,407	23
Three or more traffic lanes	1,492	240	8,000	142	9,492	157	766	90	1,237	4	2,003	30
Undivided highways (two-way) with												
Two traffic lanes	31,745	408	211,170	200	242,915	227	5,696	290	19,891	66	25,587	116
Three traffic lanes	54	1,650	4,259	131	4,313	150	23	*	265	3	288	3
Four or more traffic lanes	48,228	323	36,076	357	84,304	338	4,916	113	6,020	106	10,936	109

Traffic Accidents

Row												
Divided highways with												
Four traffic lanes	6,443	921	5,997	150	12,440	549	1,546	248	285	112	1,831	227
Six or more traffic lanes	5,444	1,115	634	538	6,078	1,055	1,004	126	30	60	1,034	124
Not specified	2,099	111	36,786	92	38,795	93	645	64	2,994	105	3,639	97
Total	201,396	467	512,794	299	714,190	346	36,178	159	59,487	92	95,665	117
All Rural Highways and Roads												
Intersection	18,781	559	5,568	946	24,349	647	3,854	398	2,038	253	5,892	348
Nonintersection:												
Undivided highways (two-way) with												
Two traffic lanes	54,533	624	40,848	376	95,381	558	10,776	541	6,261	389	17,037	485
Three traffic lanes	1,027	94	2,972	125	3,999	117	1	972	—	—	1	972
Four or more traffic lanes	6,447	638	2,114	125	8,561	511	710	464	151	421	861	457
Divided highways with												
Four traffic lanes	645	1,493	—	—	645	1,493	348	1,662	—	—	348	1,662
Six or more traffic lanes	82	5,830	—	—	82	5,830	18	740	—	—	18	740
Not specified	7,195	34	5,944	112	13,139	70	359	34	1,540	47	1,899	44
Total	88,710	568	57,446	449	146,156	521	16,066	517	9,990	309	26,056	437

* No costs of less than $5.00 per involvement were incurred.

TABLE 15-4

NUMBER OF AND DIRECT COST PER ACCIDENT OF URBAN AND RURAL
INTERSECTION TRAFFIC ACCIDENTS IN ILLINOIS INVOLVING VEHICLES OF ILLINOIS REGISTRY–1958
(Cost Updated from 1958 to 1966 by a Factor of 1.25)

Type of Traffic Control	Cost of Passenger-Car Accidents									Cost of Truck Accidents								
	Fatal Injury		Nonfatal Injury		Property Damage Only		Total			Fatal Injury		Nonfatal Injury		Property Damage Only		Total	
	No.	$/Ac.	No.	$/Ac.	No.	$/Ac.	No.	$/Ac.		No.	$/Ac.	No.	$/Ac.	No.	$/Ac.	No.	$/Ac.
Urban Intersectional Accidents																		
1. Stop-and-go signal	79	4,093	22,015	1,539	113,355	195	135,449	415		12	5,331	1,655	445	21,926	80	23,593	109
2. Flashing beacon	—	—	163	400	520	275	683	305		—	—	20	252	17	12	37	142
3. Police officer	—	—	150	335	1,901	505	2,051	492		—	—	25	2,911	80	*	105	694
4. Stop sign, one street	67	10,331	12,234	1,484	45,423	268	57,724	538		10	4,672	872	366	8,073	81	8,955	115
5. Stop sign, both streets	5	4,430	3,991	2,669	17,891	114	21,887	581		—	—	79	105	2,003	270	2,082	311
6. Yield sign	—	—	109	1,871	260	512	369	914		—	—	11	1,846	—	—	11	1,846
7. Other control	—	—	163	1,731	3,753	195	3,916	259		—	—	35	2,347	418	24	453	204
8. No control	57	9,772	11,028	2,068	54,359	296	65,444	602		8	2,422	931	621	11,061	86	12,000	129

Traffic Accidents

9. Unknown	9	7,931	81	361	4,954	30	5,044	50	—	—	17	164	1,647	44	1,664	45
Total	217	7,714	49,934	1,725	242,416	225	292,567	486	30	4,336	3,645	521	45,225	89	48,900	124

Rural Intersectional Accidents

1. Stop-and-go signal	7	2,652	1,868	1,910	4,405	241	6,278	740	6	5,150	90	891	919	445	1,015	512
2. Flashing beacon	—	—	—	—	260	488	260	488	—	—	8	1,215	—	—	8	1,215
3. Police officer	5	26,402	—	—	—	—	5	26,402	—	—	11	661	—	—	11	661
4. Stop sign, one road	78	9,820	2,758	1,322	5,271	342	8,107	768	25	8,012	369	1,079	1,793	175	2,187	416
5. Stop sign, both roads	9	4,276	191	355	130	550	330	2,388	—	—	21	1,115	248	19	269	104
6. Yield sign	—	—	—	—	87	1,159	87	1,159	—	—	—	—	—	—	—	—
7. Other control	—	—	27	1,662	—	—	27	1,662	—	—	—	—	—	—	—	—
8. No control	46	10,219	1,103	1,100	6,125	309	7,274	491	15	8,382	184	1,110	1,564	112	1,763	288
9. Unknown	—	—	—	—	1,981	62	1,981	62	—	—	—	—	639	105	639	105
Total	145	9,827	5,947	1,539	18,257	284	24,349	648	46	7,760	683	1,059	5,163	188	5,892	348

* Less than 55 in 1958.

Source: Ref. 15-25.

TABLE 15-5
Direct Cost per Accident of Each Cost Element of the Accidents in Illinois Involving Vehicles of Illinois Registry–1958
(Costs Are Dollars and Cents per Accident, Updated from 1958 to 1966 by a Factor of 1.25)

Cost Element	Passenger-Car Traffic Accidents				Truck Traffic Accidents			
	Fatal Injury	Nonfatal Injury	Property Damage Only	Total	Fatal Injury	Nonfatal Injury	Property Damage Only	Total
Urban Traffic Accidents								
Number of accidents	458	86,108	627,714	714,190	78	6,491	86,096	95,665
Property damage								
Damage to vehicle	741.17	433.01	174.71	206.24	917.69	159.10	65.70	70.67
Damage to property in vehicle	1.67	1.06	0.33	0.42	69.04	3.82	1.85	1.98
Damage to objects struck by vehicle	32.71	5.66	2.57	2.96	5.51	29.73	2.47	4.24
Miscellaneous damage	0.38	0.82	0.26	0.33	1.99	0.64	0.10	0.13
Subtotal	775.94	440.55	177.87	209.95	994.23	193.29	70.12	77.02
Treatment of injuries								
Ambulance costs	11.29	1.08	–	0.14	2.31	0.32	–	0.02
Doctor and dentist fees	274.38	115.63	–	14.12	82.85	36.67	–	2.56
Hospital and treatment	627.01	83.03	–	10.41	139.42	31.41	–	2.24
Miscellaneous costs	17.45	2.92	–	0.36	2.88	1.08	–	0.08
Subtotal	930.12	202.66	–	25.03	227.46	69.48	–	4.90
Loss of use of vehicle	21.04	8.17	1.77	2.56	158.46	23.31	14.79	15.02
Value of time lost	584.29	186.48	1.49	24.16	195.43	60.46	1.39	5.51
Legal and court costs	1,516.34	290.10	0.99	36.82	750.54	67.73	0.39	5.56
Damage awards in excess of known costs	3,290.26	373.25	0.62	47.65	2,490.53	106.17	0.01	9.25
Total cost	7,117.99	1,501.21	182.74	346.17	4,816.65	520.44	86.70	117.26
Rural Traffic Accidents								
Number of accidents	711	17,288	128,157	146,156	142	2,782	23,132	26,056
Property damage								
Damage to vehicle	1,625.91	834.39	215.19	295.29	1,880.03	613.61	168.43	225.29
Damage to property in vehicle	13.39	6.35	4.66	4.90	298.54	27.03	2.41	6.66
Damage to objects struck by vehicle	19.75	1.18	3.88	3.64	17.82	4.68	28.86	26.22
Miscellaneous damage	1.24	0.94	0.13	0.23	14.41	1.25	0.57	0.72
Subtotal	1,660.29	842.86	223.86	304.06	2,210.80	646.57	200.27	258.89
Treatment of injuries								
Ambulance costs	26.69	7.14	–	0.97	11.89	6.99	–	0.81
Doctor and dentist fees	446.86	169.14	–	22.18	177.42	191.04	–	21.36
Hospital and treatment	803.66	267.21	–	35.52	206.67	79.28	–	9.59
Miscellaneous costs	41.23	8.53	–	1.21	9.38	2.60	–	0.33
Subtotal	1,318.44	452.02	–	59.88	405.36	279.91	–	32.09
Loss of use of vehicle	4.30	7.52	1.19	1.96	456.07	51.78	23.13	28.55
Value of time lost	742.18	319.77	0.98	42.29	451.07	617.52	1.80	69.99
Legal and court costs	1,762.17	230.78	5.78	40.94	877.42	85.87	0.07	14.02
Damage awards in excess of known costs	3,809.16	439.33	2.19	72.41	3,652.18	125.63	0.02	33.33
Total cost	9,296.54	2,292.28	234.00	521.54	8,052.90	1,807.28	225.29	436.87

Source: Ref. 15-25.

Traffic Accidents

TABLE 15-6
NUMBER OF VEHICLES INVOLVED PER ACCIDENT–ILLINOIS
ACCIDENTS INVOLVING VEHICLES OF ILLINOIS REGISTRY, 1958

Type of Highway	Passenger-Car Accidents				Truck Accidents			
	Fatal Injury	Nonfatal Injury	Property Damage Only	Total	Fatal Injury	Nonfatal Injury	Property Damage Only	Total
Urban Traffic Accidents								
Intersection	1.429	1.698	1.695	1.696	1.000	1.041	1.047	1.046
Nonintersection:								
One-way streets with								
One traffic lane	–	2.006	1.878	1.882	–	1.000	1.000	1.000
Two traffic lanes	1.000	2.006	1.518	1.523	–	1.000	1.000	1.000
Three or more traffic lanes	–	1.935	1.334	1.353	–	1.007	1.242	1.194
Undivided highways (two-way) with								
Two traffic lanes	1.226	1.443	1.492	1.487	1.029	1.032	1.096	1.093
Three traffic lanes	2.500	1.248	1.516	1.503	–	1.000	1.000	1.000
Four or more traffic lanes	1.284	1.583	1.610	1.607	1.000	1.042	1.039	1.039
Divided highways with								
Four traffic lanes	1.167	1.750	1.997	1.952	–	1.076	1.016	1.020
Six or more traffic lanes	1.400	1.181	2.308	1.792	1.000	1.000	1.008	1.008
Not specified	–	1.929	1.080	1.081	1.000	1.000	1.032	1.029
Total	1.334	1.606	1.574	1.578	1.013	1.035	1.060	1.059
Rural Traffic Accidents								
Intersection	1.552	1.893	1.664	1.719	1.109	1.073	1.024	1.030
Nonintersection								
Undivided highways (two-way) with								
Two traffic lanes	1.240	1.322	1.213	1.224	1.127	1.052	1.020	1.024
Three traffic lanes	1.556	2.037	1.500	1.504	1.000	–	–	1.000
Four or more traffic lanes	1.556	1.564	1.262	1.299	1.111	1.075	1.189	1.171
Divided highways with								
Four traffic lanes	2.500	1.142	2.000	1.496	1.000	1.238	1.047	1.069
Six or more traffic lanes	–	1.329	–	1.329	1.000	–	1.000	1.000
Not specified	–	–	1.075	1.075	–	1.000	1.000	1.000
Total	1.335	1.530	1.276	1.307	1.113	1.061	1.025	1.029

Source: Ref. 15-25.

for Washington, when the future income and funeral costs are omitted. The Washington cost for property damage to the case vehicle was 24.53 percent as compared to 11.31 for Illinois. These variations and others are further proof that the costs of traffic accidents are still in the unknown realm and all estimates and compilations contain factors of personal judgments and statistical quality shortages.

Attention is called to the inclusion of funeral costs (Table 15-7) in the Washington study. This item should not be included in economy studies because it is an item of certainty at some future date not associated with highway accidents.

TABLE 15-7
REPORTED FATAL INJURY TRAFFIC ACCIDENT COSTS FOR THE WASHINGTON, D.C., AREA, 1964-1965, COMPARED TO THE 1958 URBAN ACCIDENTS IN ILLINOIS–ALL VEHICLES

Cost Element	Illinois Direct Costs		Washington, D.C., Area, Direct Costs		
	Dollars	Percent of Total	Dollars	Percent of Total	Percent w/o Future Income and Funeral
Number of: Involvements 690 270					
Number of: Accidents .... 536 .... ?					
A. Property damage					
1. Case vehicle	328,830	11.31	243,036	1.90	24.53
2. Property in case vehicle	4,919	0.17	4,996	0.04	0.50
3. Outside vehicle–other objects	12,329	0.42	3,962	0.03	0.40
4. Miscellaneous damage	265	0.01	113	0.00	0.01
B. Deprived use of vehicle					
5. Lost use by owner	17,596	0.61	822	0.01	0.08
6. Loss of commercial vehicle rentals	Excluded	–	–	–	–
C. Worktime lost					
7. Owner/driver	9,442	0.32	18,498	0.14	1.87
8. By others	Excluded	–	264	0.00	0.03
D. Legal actions					
9. Legal and court costs	17,779	0.61	12,084	0.09	1.22
10. Damages awarded above costs	Item 23	–	700	0.01	0.07
E. Subtotal, property damage group	391,160	13.45	284,475	2.22	28.71
F. Personal Injury					
11. Ambulance	4,280	0.15	5,326	0.04	0.54
12. Other transportation	Elsewhere	–	2,056	0.02	0.21
13. Doctor and dentist	105,703	3.63	99,069	0.77	10.00
14. Private nursing	Elsewhere	–	3,459	0.03	0.35
15. Hospitalization	238,436	8.20	103,595	0.81	10.45
16. Drugs, appliances, etc.	Elsewhere	–	3,811	0.03	0.38
17. Miscellaneous items	6,572	0.22	4,285	0.03	0.43
18. Funeral expenses	Excluded	–	189,232	1.47	–
G. Subtotal, personal injury group	354,991	12.20	410,833	3.20	22.36
H. Time factors					
19. Value of time lost by injured	216,838	7.46	63,456	0.49	6.40
20. Value of time lost by others	Excluded	–	40,111	0.31	4.05
21. Special domestic services	Elsewhere	–	434	0.00	0.04
I. Subtotal, time factors	216,838	7.46	104,001	0.81	10.50
J. Legal actions					
22. Legal and court	584,642	20.10	353,772	2.76	35.70
23. Damages awarded above known costs	1,360,959	46.79	27.097	0.21	2.73
K. Subtotal, legal actions	1,945,601	66.89	380,869	2.97	38.43
24. Net present worth of future income	Excluded	–	11,639,841	90.80	–
L. Subtotal, injury and associated costs	2,517,430	86.55	12.535,544	97.78	71.29
M. Grand total	2,908,590	100.00	12,820,019	100.00	100.00

Source: Refs. 15-25 and 15-42.

Traffic Accidents

TABLE 15-8
Reported Nonfatal Injury Traffic Accident Costs for the Washington, D.C., Area, 1964-65, Compared to the 1958 Urban Accidents in Illinois–All Vehicles

Cost Element	Illinois Direct Costs		Washington, D.C., Area. Direct Costs		
	Dollars	Percent of Total	Dollars	Percent of Total	Percent w/o Future Income
Number of: Involvements	144,863		28,820		
Accidents	92,509		?		
A. Property damage					
1. Case vehicle	22,061,314	27.73	10,984,997	44.14	47.75
2. Property in case vehicle	92,745	0.12	43,484	0.17	0.19
3. Outside vehicles–other objects	544,438	0.68	143,921	0.58	0.63
4. Miscellaneous damage	20,169	0.03	35,188	0.14	0.15
B. Deprived use of vehicle					
5. Lost use by owner	564,818	0.71	266,274	1.07	1.16
6. Loss of commercial vehicle rentals	excluded	–	80,058	0.32	0.35
C. Worktime lost					
7. Owner/driver	257,620	0.032	495,104	1.99	2.15
8. By others	excluded	–	32,549	0.13	0.14
D. Legal actions					
9. Legal and court	511,419	0.64	205,804	0.83	0.90
10. Damages awarded above costs	57,102	0.07	30,676	0.12	0.13
E. Subtotal, property damage group	24,109,625	30.30	12,318,055	49.49	53.55
F. Personal injury					
11. Ambulance	76,207	0.09	148,287	0.60	0.64
12. Other transportation	elsewhere	–	78,980	0.32	0.34
13. Doctor and dentist	6,629,761	8.33	1,853,129	7.44	8.06
14. Private nursing	elsewhere	–	43,119	0.17	0.19
15. Hospitalization	5,208,901	6.55	1,390,389	5.58	6.04
16. Drugs, appliances, etc.	elsewhere	–	153,973	0.62	0.67
17. Miscellaneous items	206,642	0.26	66,034	0.27	0.29
18. Funeral expenses	–	–	–	–	–
G. Subtotal, personal injury group	12,121,511	15.23	3,733,911	15.00	16.23
H. Time factors					
19. Value of time lost by injured	9,500,608	11.94	2,131,407	8.57	9.27
20. Value of time lost by others	excluded	–	184,810	0.74	0.80
21. Special domestic services	elsewhere	–	45,420	0.18	0.20
I. Subtotal, time factors	9,500,608	11.94	2,361,637	9.49	10.27
J. Legal actions					
22. Legal and court	12,663,532	15.92	1,959,972	7.87	8.52
23. Damages awarded above known costs	21,174,396	26.62	2,630,546	10.57	11.44
K. Subtotal legal actions	33,837,928	42.53	4,590,518	18.44	19.95
24. Net present worth of future income	excluded	–	1,884,941	7.58	–
L. Subtotal, injury and associated costs	55,460,047	69.70	12,571,007	50.51	46.45
M. Grand total	79,569,672	100.00	24,889,062	100.00	100.00

Source: Refs. 15-25 and 15-42.

TABLE 15-9

REPORTED PROPERTY DAMAGE ONLY
TRAFFIC ACCIDENT COSTS FOR THE WASHINGTON, D.C., AREA, 1964-65,
COMPARED TO THE 1958 URBAN ACCIDENTS ILLINOIS–ALL VEHICLES

Cost Element	Illinois, Direct Costs		Washington, D.C., Area. Direct Costs	
	Dollars	Percent of Total	Dollars	Percent of Total
Number of: Involvements	1,227,952		67.010	
Accidents	809,855		?	
A. Property damage				
1. Case vehicle	27,441,844*	93.27	11,654,801	89.99
2. Property in case vehicle	65,054	0.22	22,324	0.17
3. Outside vehicle–other objects ...	635,283	2.16	125,759	0.97
4. Miscellaneous damage	10,750	0.04	44,572	0.34
B. Deprived use of vehicle				
5. Lost use by owner	512,919	1.74	205,373	1.59
6. Loss of commercial vehicle rentals	Excluded	–	39,708	0.31
C. Worktime lost				
7. Owner/driver	397,147	1.35	648,635	5.01
8. By others	Excluded	–	35,662	0.28
D. Legal actions				
9. Legal and court	285,402	0.97	161,252	1.24
10. Damages awarded above costs ..	73,815	0.25	13,488	0.10
E. Grand total	29,422,214	100.00	12,951,574	100.00

*Reported costs plus unreported costs total $92,259,074.
Source: Refs. 15-25 and 15-42.

A word is in order about item 24 in Table 15-7, net present worth of probable future income of the fatalities. As discussed earlier in this chapter, such calculation is one approach to determining an economic cost of traffic fatalities, but its calculation is involved, inclusive of many judgments, and the answer is affected by the procedure used. Wilbur Smith and Associates used a procedure [15-42, pp. 205-217] that is lacking in correct application of compound interest theory. They used an average rate of income for each fatality for the remaining working expectancy rather than the income for each year. For professional persons and other persons whose incomes increase with age, say up to age 60, future incomes will be discounted more heavily in a year by year summation of present worth than will be true by using the mean single sum income for each year of remaining working time.

Further, a constant subsistence yearly cost was used, $1,000 to age 10 and $2,000 from age 11 upward. It seems reasonable that the costs sustained by a

Traffic Accidents

TABLE 15-9A
Ratio of Total Cost of Traffic Accidents to Cost of Only Those Accidents Reported to Official Agencies–1958 Accidents in Illinois Involving Vehicles of Illinois Registry

Cost Item	Rural Traffic Accidents				Urban Traffic Accidents			
	Fatal Injury	Nonfatal Injury	Property Damage Only	Total	Fatal Injury	Nonfatal Injury	Property Damage Only	Total
Passenger Car Traffic Accidents								
Property damage	1.000	1.169	3.857	2.108	1.000	1.395	3.276	2.433
Treatment of injuries	1.000	1.003	–	1.003	1.000	1.187	–	1.182
Loss of use of vehicle	1.000	1.000	5.231	1.763	1.000	1.268	2.252	1.725
Value of time lost	1.000	1.009	4.807	1.025	1.000	1.361	2.019	1.377
Legal and court costs	1.000	1.000	32.547	1.157	1.000	1.480	1.776	1.467
Damage awards in excess of known costs	1.000	1.000	8.615	1.022	1.000	1.290	4.264	1.284
Total Cost	1.000	1.058	3.972	1.477	1.000	1.344	3.233	1.833
Truck Traffic Accidents								
Property damage	1.000	1.402	3.660	2.356	1.000	1.061	5.431	3.107
Treatment of injuries	1.000	2.430	–	2.212	1.000	1.000	–	1.000
Loss of use of vehicle	1.000	1.000	3.561	2.072	1.000	1.000	8.677	4.630
Value of time lost	1.000	5.096	2.146	4.336	1.000	1.000	3.457	1.192
Legal and court costs	1.000	1.000	1.000	1.000	1.000	1.011	6.086	1.065
Damage awards in excess of known costs	1.000	1.000	1.000	1.000	1.000	1.000	1.000	1.000
Total Cost	1.000	1.880	3.626	2.166	1.000	1.023	5.744	2.254
Passenger Car and Truck Traffic Accidents								
Property damage	1.000	1.191	3.828	2.137	1.000	1.381	3.344	2.457
Treatment of injuries	1.000	1.059	–	1.053	1.000	1.181	–	1.176
Loss of use of vehicle	1.000	1.000	3.833	1.975	1.000	1.210	3.723	2.384
Value of time lost	1.000	1.246	3.670	1.240	1.000	1.349	2.119	1.371
Legal and court costs	1.000	1.000	30.344	1.145	1.000	1.468	1.842	1.456
Damage awards in excess of known costs	1.000	1.000	8.516	1.021	1.000	1.283	4.221	1.275
Total Cost	1.000	1.113	3.917	1.541	1.000	1.334	3.322	1.848

Note: The state law requires the reporting of all fatal and nonfatal injury accidents and property damage only accidents of $100 or more cost. Chicago requires reporting of property damage only accidents of $50 or more cost.
Source: Ref. 15-25 plus supplementary data from Bureau of Public Roads.

worker would increase somewhat with age and with increased income, especially for all work above the labor and craft level of income.

The Wilbur Smith study used a discount rate of 4 percent per year. A higher rate, say 6 percent, would have lowered the total present worth considerably from the $11,639,841 reported.

In the Illinois accident costs the nonreported accidents are included. The nonreported accidents are primarily property damage only accidents of low cost per accident, usually less than $100. The Illinois published report [15-25] does not give the number of and cost of the nonreported traffic accidents, but the costs were obtained from original work sheets and used in developing Table 15-9A. On a total basis the ratio of total cost to reported cost is 1.541 for rural traffic accidents and 1.848 for urban accidents. The ratio for property damage only accidents is 3.233 and 3.972 for the passenger car urban and rural accidents, respectively. This ratio will vary state to state, depending upon what the law states as the minimum cost of an accident which is not required to be officially reported. A detailed examination of Table 15-9A indicates those traffic cost elements which are most often not wholly included in the official reports, such as value of time lost and legal and court costs.

In the Illinois study and other similar ones, the accident data are obtained from the vehicle owners who were selected from a statistically controlled sample drawn from the state vehicle registration files.

In the Washington metropolitan area study there were 96,100 offficially reported traffic accidents having an average cost per involvement of $527. There were 1.86 involvements not officially reported for one officially reported involvement. The involvements not reported cost an average of $95 each. The cost of the reported accidents was $50,661,000 compared to $16,570,000 for the unreported accidents. The total accident cost is therefore 1.33 times the cost of the reported accidents. On a vehicle-mile basis the costs were 0.75 cent for reported accidents and 0.24 cent for unreported accidents, which is equivalent to 14 reported involvements and 26 unreported involvements per million vehicle-miles.

ACCIDENT RATES FOR ECONOMY STUDIES

For economy studies, three statistics on traffic accidents are needed: (1) The accident rate, that is, the frequency of accidents related to vehicle-miles of driving or to number of vehicles using the facility in a given time period, (2) the cost per accident, and (3) the highway design and traffic factors involved.

For highway analyses the traffic accidents need to be converted to dollars per day, per year, or other time unit so that the accident cost can be added to the motor vehicle running cost. Traffic accident costs may be expressed in cents per vehicle-mile, dollars per accident, dollars per highway mile, or other suitable unit. The most commonly used unit for measuring traffic accidents is vehicle-miles, usually in terms of one million vehicle-miles, 10 million vehicle-miles, or 100 million vehicle-miles. For the highway improvement alternatives being considered, the accident costs may be estimated for each alternative in dollars, or as a percentage reduction from the base alternative, often the existing facility. The following pages present a variety of information to use as guide material in estimating the accident rates (and costs) when other more directly applicable information is not available.

Traffic Accidents 401

GENERAL GUIDES

The large number of factors within the highway, the vehicle, the driver, and the environment which affect the accident rate and the accident cost form such a complex of variable situations that good, solid, reliable accident rates and costs for any one particular section of highway or a specific facility are difficult to come by. When dealing with an existing facility, however, every effort should be made to get its specific accident history rather than to use general estimates. The amount of accident reduction that can be expected from an improvement to an existing facility is more dependent upon the rate of past accidents and the types of accidents than to any other single factor. Knowing the facts about the past number of accidents is not only a good guide to selecting the best means of reducing the number and severity of accidents, but it is also the best available guide to an estimate of how much reduction in accident cost may be expected by the improvement.

See Ref. 15-1, 15-12, and 15-27 for discussions and summaries of accident rates as related to highway design and traffic operations.

SOME SPECIFIC FACTORS

The main highway factors that contribute to traffic accidents are horizontal curves, vertical grades, intersections, access points, structures, roadside objects, and pavement surface condition. When these elements are unfavorable and exist in combination the accident rate goes up in some geometric ratio. Under such combination of factors the accident rate may be 6, 10, or 20 times the rate on a segment of highway which is a level tangent without structures and access.

Accidents and their location are in some form of ratio to the decision process of motor vehicle drivers and pedestrians when within a trafficway. The more complex are the factors bearing simultaneously upon a driver or pedestrian decision, the higher the accident rate. Consequently improvements to a highway facility are apt to have high accident rate reduction when one or more factors troublesome to the driver's decision can be removed to simplify the decision process.

The decision process is composed of the following steps: (1) recognizing the hazard at hand or forthcoming, (2) formulating alternative actions to avoid an accident, (3) deciding upon the most favorable alternative through a process of evaluating the merits and demerits of each, and (4) taking the action selected as being the better one. When it is realized that between step (1), recognition, and step (4), action, there may be only a fraction of a second of time or 10 to 20 ft of distance in which to avoid the hazard, it is readily realized that the decision process is a heavy tax upon most minds. Accidents are caused or are avoided by fractions of an inch in distance and hundredths of a second in time. Providing the type of highway and the type of traffic behavior that keep the decision processes simple is one good approach to accident reduction.

The speed differential between vehicles is a strong factor in producing high accident rates [15-43]. This factor is probably one reason why two-lane highways have a decreasing accident rate above ADT's of 7,000 while below 7,000 ADT the accident rate increases with increases in ADT.

Intersections are hazardous. Intersection accident rates are highly sensitive to the traffic flow on the minor legs, but not so sensitive to increase in traffic flow on the major facility.

Elimination of intersections is the factor that contributes heavily to the low accident rate on fully controlled access highways. Reduction in the number and amount of vehicle speed changes is also an important factor which contributes to the low accident rates on freeways.

The factors of speed changes, intersections, and access points probably account for the accident rates on four-lane highways without medians being as high or higher than the rates on two-lane highways. Roadside development, higher total traffic volume, and the resulting complex decision processes contribute heavily to the high accident rate on four-lane facilities without medians.

GUIDES TO ACCIDENT RATES ON SPECIFIC TYPES OF HIGHWAY ELEMENTS

In analyses for the economy of highway improvements, often, particularly in project formulation, the alternatives differ only in such elements as horizontal curves, vertical grades, intersections, and traffic controls. To make a proper comparison of such differences the accident rate resulting from such specific elements needs to be known.

Horizontal Curves and Vertical Grades

The review of the literature on accident rates on horizontal curves and vertical grades as given by the Automotive Safety Foundation [15-1, pp. 33-37] is positive in its findings that the accident rates increase with an increase in either or both of these factors of alignment. The several writers are not in agreement as to the amount of increase and their specific rates may not be attached to specific degrees of curves or percent grades. The Cornell Aeronautical Laboratory concludes that curves flatter than 4 degrees or less steep than 4 percent have no material effect on traffic accident rates [15-12]. This conclusion is based on an analysis of accident reports for Connecticut, Florida, and Ohio.

For economy analyses it seems preferable to give the weight to earlier reports. Table 15-18 is therefore presented as a guide to making some allowance for increased traffic accident costs with increase in horizontal curvature and vertical grades. Although Table 15-18 gives the appearance of authority, considerable judgment was exercised in using the evidence available. For instance, the statement "Sections with curves over 5 degrees and grades over 5 percent had 19.27 times as many accidents as the average section of highway"

Traffic Accidents 403

is not much more than a generalized conclusion despite the preciseness of the figure 19.27.

The hazardous character of horizontal curves comes from the factors of sight distance, superelevation, frequency of curves, and adherence to uniform design standards. An occasional curve is more productive of accidents than is a series of curves. Accidents on curves have been materially reduced by improved signing, increasing sight distance, reducing tendency to skid, and improving superelevation.

Vertical grades contribute to accidents primarily at their crests and sags. Even 1 and 2 percent plus grades slow down heavy trucks. All grades tend to increase vehicle speed downhill and, therefore, considerably increase stopping distance. Run away truck accidents are not uncommon on long steep grades. Restricted sight distance on grades, whether at crests or at horizontal curves, contributes greatly to accidents on grades. But as these restrictions become more frequent (more per mile of highway) the accident rate decreases.

Illumination

Many factors are involved in the relation of traffic accidents and highway illumination. Lighting does reduce night accidents on the whole, especially when applied to locations known to have a high rate of night accidents. Installing lighting on a selected basis is as effective as on a continuous basis. A uniform amount of light is preferable to high and low spots. A minimum level of illumination is desirable for effective accident control. A reflective pavement surface is a favorable factor. General rates of accident reduction effected through highway lighting are yet to be developed.

Intersections

Accident rates at intersections are dependent upon such a complex of factors that they are not detailed here. The Automotive Safety Foundation publication [15-1, pp. 47-59], Syrek [15-45A], Wenger [15-53], Solomon [15-4-4], and others offer good guides as to probable accident rates and changes in accident rates at intersections under a variety of design factors and traffic volumes.

One-way Streets

Conversion to one-way streets will reduce accidents 20 to 45 percent including a high reduction in pedestrian accidents [15-1, p. 74].

Speed

Speed of traffic per se is not a factor that contributes heavily to accident rates. On the other hand the differences between speeds of vehicles in a traffic stream do contribute heavily to the accident rate [15-39 and 15-43]. Speed, though, does contribute heavily to the severity of personal injury accidents, in number and extent of involvement, and in cost of property damage. Rural accidents are more costly per accident than urban accidents and freeway and toll highway accidents are more costly per accident than accidents on other highways where the traffic speed is much lower.

There is an omission of figures and tables in this chapter relating traffic speed to accident rate. It is thought that the factor of speed is accounted for in such factors as traffic volume, geometric design of the highway, and character of the specific facilities under examination. However, for similar types of accidents, the cost per accident should be taken at a higher amount for the alternative or facility which has the higher driving speed.

In the economic analyses of proposed highway improvements, the cost of accidents on the alternatives being considered usually can be acceptably estimated without directly considering specific speeds.

TABLES OF ACCIDENT RATES AND COST RELATED TO DESIGN AND TRAFFIC FACTORS

There are presented in this section tables and curves to be used as guide information in estimating the rate of accidents or the rate of accident cost of

TABLE 15-10
DIRECT COST OF TRAFFIC ACCIDENTS
PER 10 MILLION VEHICLE-MILES OF TRAVEL
(Based on 1958 Accidents; Cost Updated to 1966 Using an Index of 1.25)

Highway System	Illinois Registered Passenger Cars			Illinois Registered Trucks, All Types		
	Rural	Municipal	Total	Rural	Municipal	Total
Federal-aid primary and state highways ..	$72,905	$114,289	$91,955	$42,846	$52,660	$45,699
Federal-aid secondary:						
State highways	100,469	115,661	104,280	22,019	*	28,962
Local roads	54,046	*	51,426	40,711	*	38,234
Subtotal	66,928	73,632	67,969	35,660	*	35,424
Nonfederal-aid:						
State highways	63,146	252,624	196,769	36,125	95,798	69,290
Local roads	96,379	151,264	141,899	35,686	46,670	42,944
Subtotal	89,065	163,626	149,604	35,774	53,470	47,169
All roads and streets:						
State highways	73,718	144,061	108,060	41,458	61,466	47,821
Local roads	82,022	149,684	133,782	37,069	46,088	42,375
Total	76,334	147,495	120,928	40,170	52,851	45,601

* Sample was too small to provide significant data (20 or fewer sample cases).
Source: Ref. 15-4.

Traffic Accidents

proposed highway facilities or for existing facilities when historical information is unavailable. Despite the large volume of literature on the subject there is a scarcity of published accident rates and cost rates applying to specific highway designs and facilities in the detail desired for economy analyses. Often the data are reported by highway systems, as in Table 15-10, and as such has but limited application. A few reports, however, are helpful and these reports form the basis for the material to follow.

For comparisons of the effect of the degree of access control on numbers and severity of accidents, urban, suburban, and rural, Table 15-11 is useful.

TABLE 15-11
TRAFFIC ACCIDENTS AND ACCIDENT RATES
BY URBAN AND RURAL AREAS AND DEGREE OF ACCESS CONTROL

Area and Degree of Access Control	Vehicle Miles (Add 000)	Number of Traffic Accidents				Accident Rate per 100 Million Vehicle-Miles			
		Fatal Injury	Nonfatal Injury	Property Damage Only	Total	Fatal Injury	Nonfatal Injury	Property Damage Only	Total
Urban									
Full access control	2,068,597	35	1,101	2,716	3,852	1.69	53.2	131	186
Partial access control	883,328	36	871	3,478	4,385	4.08	98.6	394	496
No control	2,963,771	115	5,082	10,384	15,581	3.88	171.5	350	526
Suburban									
Full access control	198,746	7	97	169	273	3.52	48.8	85	137
Partial access control	1,045,904	46	862	2,644	3,552	4.40	82.4	253	340
No control	2,130,151	94	4,083	6,164	10,341	4.41	191.7	289	485
Rural									
Full access control	7,054,287	198	4,529	5,908	10,635	2.81	64.2	84	151
Partial access control	3,595,160	172	2,600	4,804	7,576	4.78	72.3	134	211
No control	5,562,289	394	5,986	12,070	18,450	7.08	107.6	217	332

Area and Degree of Access Control	Person Fatalities	Persons Injured	Rate per 100 Million Vehicle-Miles		Rate per 100 Accidents	
			Fatalities	Injuries	Fatalities	Injuries
Urban						
Full access control	41	1,810	1.98	88	1.06	47.0
Partial access control	41	1,430	4.64	162	0.93	32.6
No control	119	8,074	4.02	272	0.76	51.8
Suburban						
Full access control	8	180	4.03	91	2.93	65.9
Partial access control	53	1,405	5.07	134	1.49	39.6
No control	109	7,208	5.12	338	1.05	69.7
Rural						
Full access control	231	8,836	3.27	125	2.17	83.1
Partial access control	220	4,918	6.12	137	2.90	64.9
No control	484	10,875	8.70	196	2.62	58.9

Source: U.S. Bureau of Public Roads, 1958 Data.

This table is based upon a wide geographical experience on main highways and streets. The accident-reducing effect of access control is strikingly favorable.

For urban system studies or for specific urban routes, Table 15-12 for Chicago, Illinois, reports both accident rates by severity class and accident costs per 100 million vehicle-miles. Of important notice is the high cost of $ 1,020,500 per 100 million vehicle-miles of the signalized section of the Congress Street Expressway as compared to the cost of only $ 160,200 for the full standard expressway.

TABLE 15-12
ACCIDENT RATES AND ACCIDENT COSTS IN CHICAGO, ILLINOIS, 1958
(Costs Updated from 1958 to 1966 by a Factor of 1.25)

Street or Highway	Number of Accidents in 1958				Accident Rate per 100 Million Vehicle-miles	Cost of Accidents per 100 Million Vehicle-miles, ($)
	Fatal Injury	Nonfatal Injury	Property Damage Only	Total		
Twelve arterials	48	4,648	12,803	17,499*	1,424	775,200
Ten selected Park District streets	16	1,845	4,874	6,735	1,327	729,100
Lake Shore Drive						
1. Limited access (ramps)	5	449	1,571	2,025	622	312,200
2. Signalized	2	221	769	992	1,102	551,000
3. Partial access	0	75	402	477	919	391,000
Congress Expressway						
1. Full expressway	0	162	346	508	279	160,200
2. Signalized in central business district	0	28	87	115	2,018	1,020,500
Accident rate per 100 million vehicle-miles						
1. On arterial system	3.91	378	1,042	1,424	1,424	–
2. On expressways	0.00	89	190	279	279	–
Average cost per accident, 1966, dollars	7,250	1,200	281	540	–	–

* Of this total 36 percent were at intersections.

Note: In the Chicago Area Transportation Study 1.89 involvements per accident were used as a conversion factor.

Source: Ref. 15-23, pp. 351-353.

Table 15-13 is similar to Table 15-12, but covers Cook and DuPage counties of the Chicago area. These traffic accident costs and rates would not be expected to agree with the Chicago figures in Table 15-12, but the rates are of the same general order of magnitude.

The North Carolina report in Table 15-14 on accident rates as affected by access control brings out one important fact that is often overlooked. Note the high reduction in number of accidents and in the rate of accidents on the conventional arterial route after the parallel expressway was opened to traffic. Thus there is a double gain, first to the traffic that shifted to the new expressway, and second, to the traffic that remained on the older facility. This fact is

Traffic Accidents

importantly related to the principle of analysis that all consequences to whomsoever they accrue should be considered.

TABLE 15-13

TRAFFIC ACCIDENT RATES AND COSTS BY HIGHWAY TYPE
IN THE CHICAGO AREA (COOK AND DUPAGE COUNTIES), 1958
(Costs Updated from 1958 to 1966 by a Factor of 1.25)

Travel Accident Item	Local Streets	Arterial Streets	Expressways	Total
Vehicle-miles of travel, million	2,100	10,575	1,027	13,702
Number of accidents				
Fatal injury	146	332	12	490
Nonfatal injury	23,771	48,069	939	72,779
Property damage only	190,443	208,071	4,291	402,805
Total	214,360	256,472	5,242	476,074
Accident rate per 100 million vehicle-miles				
Fatal injury	6.95	3.14	1.17	3.58
Nonfatal injury	1,132	455	91	531
Property damage only	9,069	1,967	418	2,940
Total	10,208	2,425	510	3,475
Cost of accidents per 100 million vehicle-miles				
Fatal injury	$47,000	$20,700	$12,200	$24,100
Nonfatal injury	1,648,400	836,200	181,600	911,600
Property damage only	2,171,800	482,700	191,000	719,700
Total	3,867,200	1,339,600	384,800	1,655,400

Source: Ref. 15-28.

TABLE 15-14

TRAFFIC ACCIDENT RATES ON NORTH CAROLINA
ACCESS-CONTROLLED EXPRESSWAYS AND PARALLEL CONVENTIONAL HIGHWAYS

Item	Conventional Highway		New (1961) Expressway
	Before Expressway Opened	After Expressway Opened	
Miles of highway	48.6	48.6	44.3
Vehicle-miles of travel, million	104	65	72
Reported number of accidents			
Injury accidents	158	83	24
Total accidents	343	169	44
Rate per 100 million vehicle-miles			
Injury accidents	152	128	33
Total accidents	330	260	61

Source: Ref. 15-10.

In Table 15-15 is the California experience with six types of highways from the ordinary two-lane highway to the full freeway. This table has added value because it reports the single vehicle and multivehicle accidents, the latter by type of accident and whether at an intersection. Note the high rate of 0.68 accident per 100 million vehicle-miles at intersections on the two-way highways and the rate of only 0.045 on the freeways.

The accident rates for three classes of trucks on main rural highways are given in Table 15-16, by severity classes. Trucks have a traffic accident rate more favorable than passenger cars, and the truck combination has the better rate within the truck class.

Table 15-17 from the work of Roy Jorgensen and Associates [15-27] is most helpful for estimating probable accident reductions for the so-called "spot improvements" and traffic control improvements. Note the footnote that gives some guide to the reliability of the percentages. In the use of this table, judgment with respect to each specific application is a desirable factor. A reading of the full report is recommended.

Table 15-18 is an attempt to reduce to a practical form a guide to the effect of horizontal curves and vertical grades on traffic accident rates. The literature more often indicates a positive increase in accidents with an increase in the amount of each factor than it does that there is no effect. But particularly there is a noticeable increase in accident rate when curve and grade are combined at the same location. Accident rates of proved reliability will have to await more thorough observation on these two factors of design.

Figures 15-1 and 15-2 will be found most useful in studies of median opening and widening of bridge roadways. Medians are undergoing intensive study in an effort to reach more positive evidence as to their best design features. Bridge widening is well accepted in practice and with Fig. 15-2, the economy of such widening may be examined with greater certainty than before.

CORNELL AERONAUTICAL LABORATORY STUDY

As a part of the National Cooperative Highway Research Program, administered by the Highway Research Board of the National Academy of Sciences, the Cornell Aeronautical Laboratory, Buffalo, N.Y., undertook a study to determine the relationship of motor vehicle accidents to highway types and highway design elements. Some of their results are presented here as obtained from their reports [15-12 and 15-47, and 15-28A].

ANALYSIS OF THE ACCIDENT DATA

From the state records for Connecticut [1962, 1963, and 1964], Florida [9 months of 1963, 1964], and Ohio [1963, 1964] (the only states having the complete records and full details necessary to the study) the accident data for

Traffic Accidents

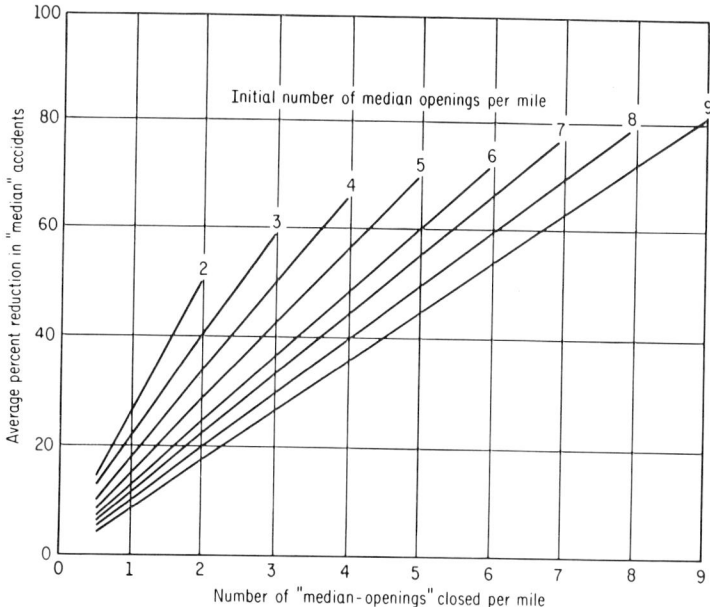

FIG. 15-1. The percentage reduction in traffic accidents to be expected from the closing of openings in medians [15-27, p. 151].

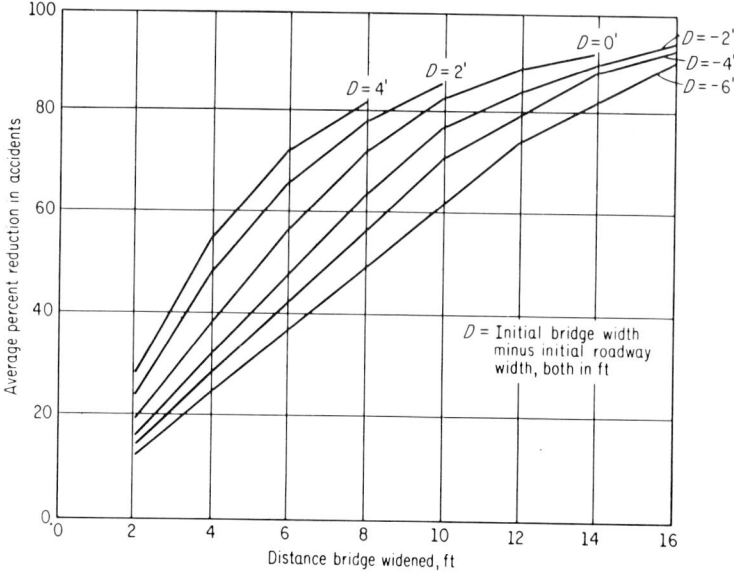

FIG. 15-2. The percentage reduction in traffic accidents to be expected as a result of the widening of the roadways on bridges [15-27, p. 151].

TABLE 15-15

ACCIDENTS AND ACCIDENT RATES ON VARIOUS CLASSES OF RURAL STATE HIGHWAYS BY KIND OF ACCIDENT–CALIFORNIA, YEAR 1959

Highway Miles and Travel	Two-Lane		Three-Lane		Four-Lane Undivided		Four-Lane* Divided		Divided† Controlled Access		Freeway	
Miles	10,450		45		167		210		794		430	
Million vehicle-miles	8,358		232		976		1,234		3,543		3,052	
Average daily traffic	2,191		14,239		15,997		16,130		12,224		19,449	
Class of Accident	Number	Rate‡	Number	Rate	Number	Rate	Number	Rate	Number	Rate	Number	Rate
Total reported accidents	19,899	2.38	597	2.57	3,995	4.09	3,591	2.91	6,011	1.69	3,066	1.00
Single-vehicle accidents	7,058	0.84	113	0.49	367	0.38	489	0.40	1,532	0.43	862	0.28
Collision between 2 or more vehicles:												
(a) Between intersections:												
1. Head-on	3,152	0.42	67	0.29	197	0.20	68	0.06	216	0.06	139	0.045
2. Nonhead-on	3,675	0.44	203	0.87	966	0.99	953	0.77	2,099	0.59	1,926	0.63
(b) At intersections	5,654	0.68	214	0.92	2,465	2.52	2,081	1.69	2,164	0.61	139§	0.045
Total excluding intersection accidents		1.70		1.65		1.57		1.22		1.08		0.95

* Four-lane divided roads have a median separating opposing traffic but roadside access is uncontrolled.
† Divided controlled-access roads are nearly all four-lane with a few miles of six-lane. Opposing traffic is separated and there is no access except at intersections. However, intersections at grade are frequent and traffic enters and exits at large angles, approximating 90 deg. All state highways except freeways require approaching traffic on cross roads to stop before entering or crossing the state highway, unless the intersection is controlled by traffic signals and the light is green.
‡ Rate is number of accidents per one million vehicle-miles.
§ Accidents at ramps.

Source: Ref. 15-34.

TABLE 15-16
Traffic Accident Rates and Costs For Two- and Four-Lane Main Rural Highways, Nonfreeway Type, 1954-58
(Costs Updated from 1954-1958 to 1966 by a Factor of 1.35)

Vehicle Class	1000 Vehicle miles	Involvements		Person Fatalities			Persons Injured			Property Damage at 1966 Dollars		
		Number	Rate*	Number	Rate*	per 100 Involvements	Number	Rate*	per 100 Involvements	Total Dollars	Rate*	per 100 Involvements
Passenger car	2,716,687	7,608	280	226	8.3	3.0	3,315	122	43.6	3,487,590	128,380	45,840
Single unit truck, four tires	259,573	801	309	17	6.5	2.1	275	106	34.3	291,262	112,210	36,360
Single unit truck, six or more tires	162,577	354	218	4	b†	b	58	36	16.4	120,285	73,990	33,980
Truck combination	505,357	908	180	5	b	b	88	17	9.7	389,980	77,170	42,950
Bus	25,710	56	218	0	b	b	19	74	33.9	18,022	70,100	32,180
Other and unknown	1,272	89	b	1	b	b	37	b	b	37,328	b	b
Total	3,671,176	9,816	267	253	6.9	2.6	3,792	103	38.6	4,344,368	118,340	44,260

* Rate is number of involvements, persons, or property damage per 100 million vehicle-miles.
† Rate not meaningful.
Source: Ref. 15-43.

TABLE 15-17

PERCENTAGE EXPECTED REDUCTIONS
IN TRAFFIC ACCIDENTS EFFECTED BY IMPROVEMENT OF FACILITY

Type of Improvement	Urban or Rural	Number of Lanes	Accident Reduction, Percent			Comments
			All Accidents	Fatal Injury Accidents	Property Damage Only Accidents	
Route sections						
Eliminate parking	U	2 plus	32	3	–	
Install/improve edge marking	R	2	14	17a	–	
Install/improve warning signs	U	2	14	14a	–	
Install/improve warning signs	U	2 plus	20a*	26a	–	
Install/improve warning signs	R	2	36	32a	–	
Install/improve warning signs	R	2 plus	18a	2	–	
Install median barriers						See Ref., p. 170.
Add painted/raised median	U	2 plus	12	–	–	
Pavement resurfacing	U	2 plus	42	46	–	See Ref., p. 182.
Pavement resurfacing	R	2	12	21	–	Depends on number of wet pavement accidents and total number per mile.
Pavement resurfacing	R	2 plus	44	59	–	
Shoulder stabilization	R	2	38	46	–	Based on accident criteria.
Widen shoulder, no dimensions	R	2	–2	7a	–	See Ref., p. 186
Widen traveled way, no dimensions	R	2	38	30	–	
Widen traveled way from 9-ft lanes	R	2	38	16	–	
Widen traveled way from 10-ft lanes	R	2	5	–65b	–37a	
Livestock fencing	R	2 or more	90	–	–	Livestock accidents only
Modernization to design standards	R	2	10	–6a	–40b	
Modernization to design standards	R	2 plus	15 b	22a	–	See Ref., p. 82.
Grades						
Add climbing lane	R	2	0	0	0	Limited data.
Centerline striping at crests	R	2	64	–	–	
Horizontal curves						
Install delineators	R	2	2b	16	–	See Ref., p. 48. Effective for night accidents.
Install delineators	R	2 plus	46a	–10b	61	
Install/improve warning signs	R	2	57	71	23a	
Install/improve warning signs	R	2 plus	52	40	–	
Reconstruct curve	R	2	88	89	96	
Install warning signs and delineators	U	2 plus	20	–27a	–	
Install warning signs and delineators	R	2	22a	41a	–	
Bridge/underpass						
Install delineators	R	2	47	–8b	–	
Install delineators	R	2 plus	53a	62	89	
Intersections						
Stop ahead sign	R	2	47	96	–	
Install yield sign	U	2	59a	80	–	Low ADT where rolling stop is safe.
Install yield sign	U	2 plus	–46	–	–	
Install minor leg stop control	U	2	48	71	–	When angle-type accidents are 50% or more
Install minor leg stop control	U	2 plus	38a	18a	22	

Traffic Accidents

TABLE 15-17 (continued)

Type of Improvement	Urban or Rural	Number of Lanes	Accident Reduction, Percent			Comments
			All Accidents	Fatal Injury Accidents	Property Damage Only Accidents	
Install minor leg stop control	R	2	65	89	–	
Install all-way stop signs	U	2	68a	67a	–	See footnote †.
Install warning signals	U	2 plus	–27 b	73a	–	
Install warning signals	R	2	56a	29a	–	See Ref., p. 52.
Install warning signals	R	2 plus	21b	–	–	
Add pedestrian signals	U	2	13	56a	–	
Add pedestrian signals	U	2 plus	3a	42a	–	
Improve signals	U	2	31	35a	–	Changes to improve
Improve signals	U	2 plus	–2	10b	–	driver's response to
Improve signals	R	2 plus	42a	45b	–	signal–vision, attention, etcetera.
Improve signals	U	2 plus	–	57	–	T-intersection.
Curtail turning movement	U	2 plus	40	39	–	
Add left turn lane w/o signal	U	2	19a	80a	–	
Add left turn lane w/o signal	U	2 plus	6	54a	–	
Add left turn lane w/o signal	R	2 plus	–6	–16	–	
Add left turn lane w/o signal	U	2	79	79	–	T-intersection.
Add left turn lane w/o signal	U	2 plus	51a	62	–	T-intersection.
Add left turn lane w/o signal	R	2	33	5	–15	Y-intersection.
Add left turn lane & signal	U	2 plus	27	1	–7b	
Add left turn lane & signal	R	2 plus	43a	58a	–	
Add left turn lane & signal	R	2	–	–	–	
Add left turn lane & signal	R	2 plus	–42a	–28b	–	
Add left turn signal w/o turn lane	U	2	–	–	–	
Add left turn signal w/o turn lane	U	2 plus	39	57	–	
Add left turn lane, signal & illumination	U	2 plus	46	76	–	
Install new traffic signals	U & R	2 plus	29	50	–	Proven hazardous intersections, 60% right angle and left turn accidents.
Deslicking	U	2 plus	20	15	–	
Rumble strips	R	2	27b	26b	24b	

*The symbols in the percentage reduction columns have the following meaning:

No symbol	Good estimate	0-30 %
a	Rough estimate	30-70 %
b	Very rough estimate	70-150 %
–	No estimate made	over 150 %

The percentage range to the right means that there is at least a 75 % certainty that the true average percentage reduction is within the percentage given in the column, plus or minus the range given above. For example, the entry 68a means that the range of 75 percent probability lies between a reduction of 98 to 38 % and a reduction of 100 % (138 %) to an increase of 2 %.

† Minor street must be 35 % or more of total intersection volume, which is less than 8,000 ADT.

Source: Ref. 15-27, pp. 146-149.

rural highways were obtained in full detail and associated with the specific roadway location by 0.1-mile segments and traffic volume. By statistical procedure the following equation was developed as one which closely fits the three-state experience:

$$\log_e \overline{A} = a + b_1 \log_e \overline{T} + b_2 \log_e^2 \overline{T} \tag{15-1}$$

Where $\overline{A}$ = mean number of accidents in the 0.3-mile length of route segments as grouped for analysis
a, b_1, b_2 = constants evaluated from the data
$\overline{T}$ = mean average daily traffic volume on the route segments as grouped for analysis

In preliminary trials the equation

$$\log_e A = a + b_1 \log_e L + b_2 \log_e T \tag{15-2}$$

was used where L is the length of route segment, but the variable length was found to have an adverse effect on the results. The final choice was to hold the segment length constant and to let the ADT be the single variable factor.

The factors of highway design that had the greatest influence on the number

TABLE 15-18
ACCIDENT RATES FOR HORIZONTAL CURVES AND VERTICAL GRADES
(Based on Refs. 15-1, pp. 33-37, 15-6, and 15-40)

Degree of Curve or Percent Grade	Ratio of Accident Rate to Rate on Level Tangents		Remarks
	Horizontal Curves	Vertical Grades	
0	1.0	1.0	For a combined horizontal curve and vertical grade use the sum of the two ratios for their respective values. Example, for a 4-deg curve and a 3 percent grade the ratio of accidents to the level tangent accidents would be 2.4 plus 2.0, or 4.4. This table is not based on highly reliable experience.
1	1.3	1.1	
2	1.6	1.5	
3	2.0	2.0	
4	2.4	2.7	
5	2.9	3.7	
6	3.6	4.9	
7	4.3	6.2	
8	5.2	7.6	
9	6.3	9.2	
10	7.6	10.8	
11	9.1	12.4	
12	10.8	14.1	
13	12.5		
14	14.4		
15	16.3		

TABLE 15-19

SAMPLE VALUES OF C-COEFFICIENTS, OR RELATIVE ACCIDENT RATES:
TOTAL ACCIDENTS ON CONVENTIONAL TWO-LANE RURAL ROADS

Geometry*	Ohio Approximate ADT			Florida Approximate ADT			Connecticut Approximate ADT		
	490	1650	5000	490	1650	5800	490	1650	5800
Baseline†	0.13	0.37	1.32	0.10	0.28	0.95	0.05	0.15	0.59
0000	1.00	1.00	1.00	1.00	1.00	1.00	1.00	1.00	1.00
0G00	1.40	1.36	1.06	–	–	–	1.52	1.06	1.11
C000	1.60	1.87	1.40	1.95	1.32	1.32	1.40	0.78	0.85
00I0	1.70	2.88	2.58	1.59	1.98	1.71	2.11	2.30	2.19
000S	1.58	1.43	1.24	1.49	1.55	1.64	3.11	0.96	1.86
CG00	1.74	2.09	1.36	–	–	–	2.12	0.91	0.93
0GI0	2.12	3.42	3.28	–	–	–	1.37	1.58	2.20
0G0S	1.77	1.72	1.22	–	–	–	1.09	1.04	2.09
C0I0	2.07	3.89	2.62	2.15	2.16	1.91	2.62	2.29	1.92
C00S	2.48	2.56	1.62	2.72	2.36	1.69	3.12	0.93	1.15
00IS	2.21	2.74	2.22	2.71	2.88	2.39	2.00	2.89	3.30
CGI0	3.05	4.41	3.32	–	–	–	1.72	2.26	2.28
CG0S	3.04	3.18	2.15	–	–	–	2.29	1.24	4.02
0GIS	2.61	3.30	2.74	–	–	–	2.67	1.98	1.95
C0IS	2.73	3.04	–	3.78	3.80	2.28	7.52	3.76	2.23
CGIS	4.17	3.65	2.76	–	–	–	6.55	2.75	2.42

* Key:
    C = Curvature present   S = Structures present
    G = Gradient present   0 = Geometric feature not present
    I = Intersections present
Example: CGI0 = Curved and graded segment with intersection(s), but no structures.
† Baseline = Number of accidents per year per 0.3-mile segment.
Source: Ref. 15-49.

of accidents within a segment were (1) highway type (number of lanes, access control, and medians), (2) horizontal curvature–yes or no, (3) vertical grades–yes or no, (4) intersections–yes or no, and (5) structures–yes or no. The elements of curves, grades, intersections and structures were identified simply as present or not present within the 0.3-mile segment. In the main results, this study gives curves showing how the accident rates per 0.3-mile segment vary with ADT for each of many highway types and design elements.

Curves on grades were found to have no marked effect on accidents below 4-deg curvature and 4 percent grades, so the segments were identified as having curves or grades only when they were 4 deg or sharper and 4 percent or steeper. This finding is questionable as based upon references in the literature [15-1 and

15-6] and as concluded in the Cornell report in the review of literature [15-12, p. 22]. Perhaps the states of Connecticut, Florida, and Ohio did not offer sufficient observation with respect to curves and grades to bring out their effect upon accident rates.

ACCIDENT RATES

The original Cornell publication [15-12] is recommended as a source of accident rates for the several basic types of highways and the four design elements and for the severity of accidents. Only a few illustrative sets of curves and one table are given here of over 200 pages of data and curves available in the original publication.

Figure 15-3 gives for Ohio the annual number of accidents by basic highway

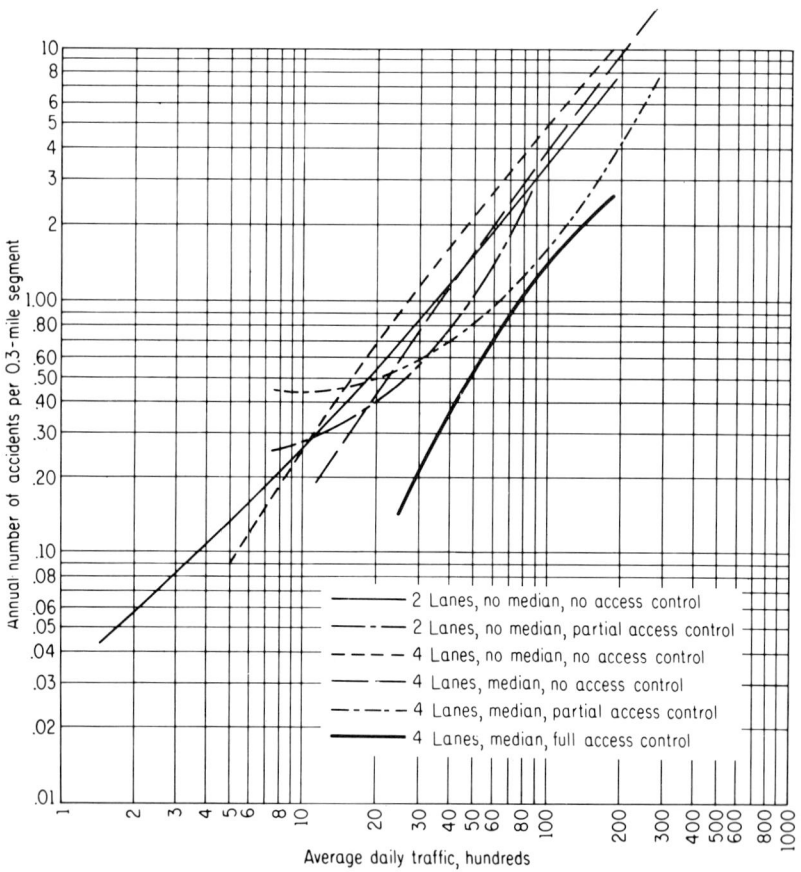

FIG. 15-3. Average number of accidents per 0.3-mile segment on rural highways in Ohio, by number of lanes, medians, and access control [15-12].

Traffic Accidents 417

type. Figure 15-4 gives the composite for Connecticut, Florida, and Ohio for four common types of highways. In both these sets of curves there are pronounced increases in the numbers of accidents with increase in the ADT. Thus, relating accident rates to vehicle-miles of travel may be accepted as a good device for

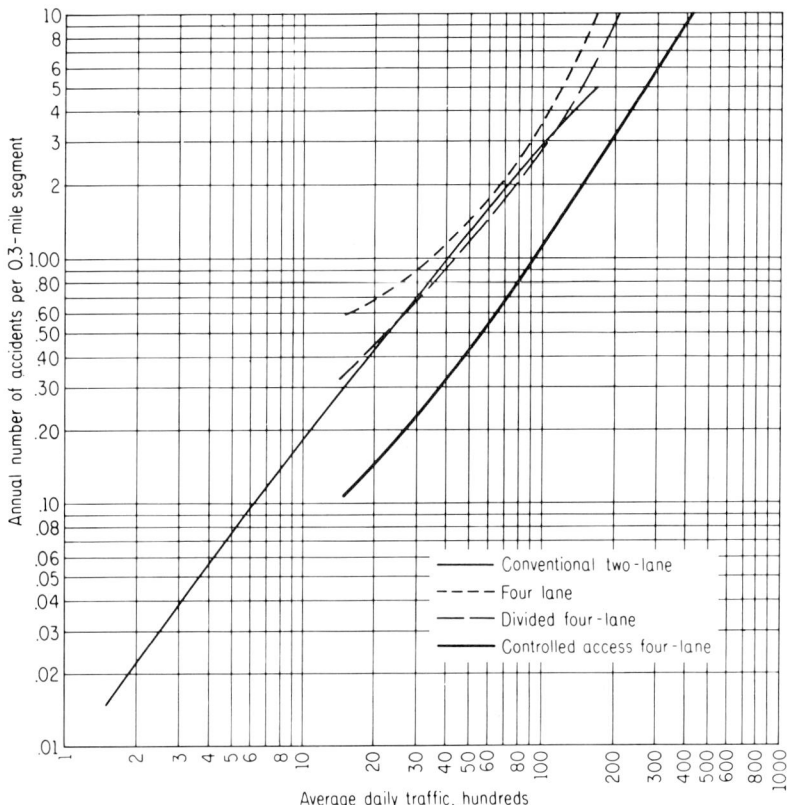

FIG. 15-4. Average number of accidents per 0.3-mile segment on rural highways in Connecticut, Florida, and Ohio [15-12].

measuring the relative safety or of the effectiveness of accident-reducing improvements.

Table 15-19 is helpful in indicating how the relative accident rates per year per 0.3-mile segment increase with combinations of the design elements of curves, grades, intersections, and structures.

Table 15-19 is supplemented by Figs. 15-5 to 15-10 for the composite of the three states to show the accident rates for two-lane rural highways for a full range of ADT and different combinations of the existence of curves, grades, intersections, and structures.

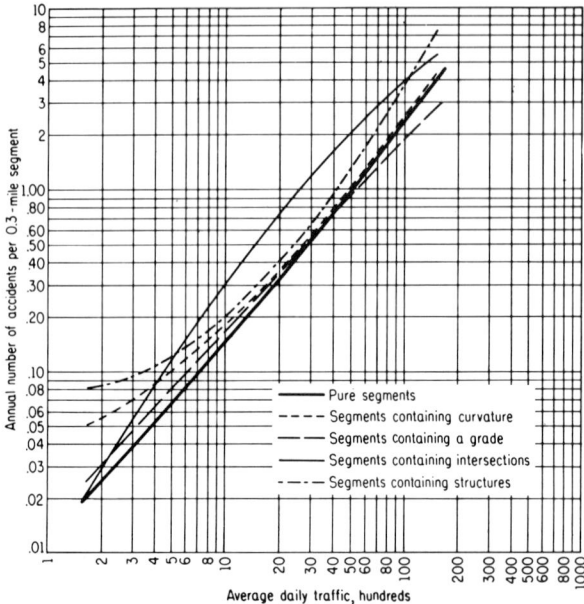

FIG. 15-5. Three-state average number of accidents per 0.3-mile segment on rural two-lane highways showing separately the effects of curves, grades, intersections, and structures [15-12].

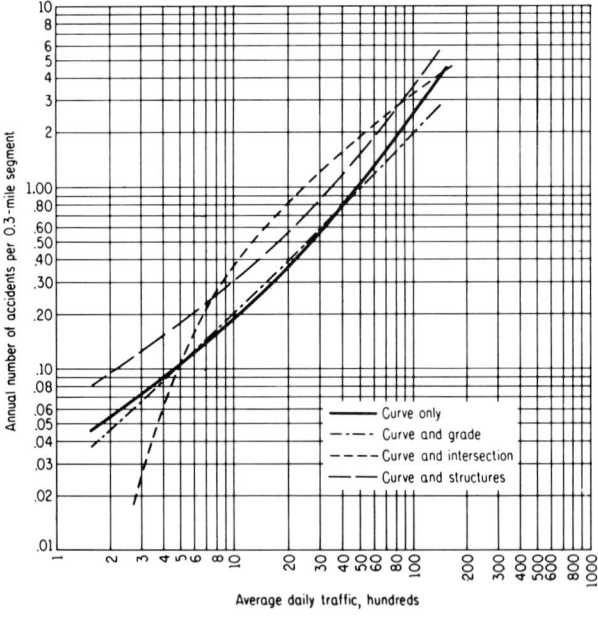

FIG. 15-6. Three-state average number of accidents per 0.3-mile segment on rural two-lane highways showing the effects of horizontal curves combined separately with grades, intersections, and structures [15-12].

Traffic Accidents

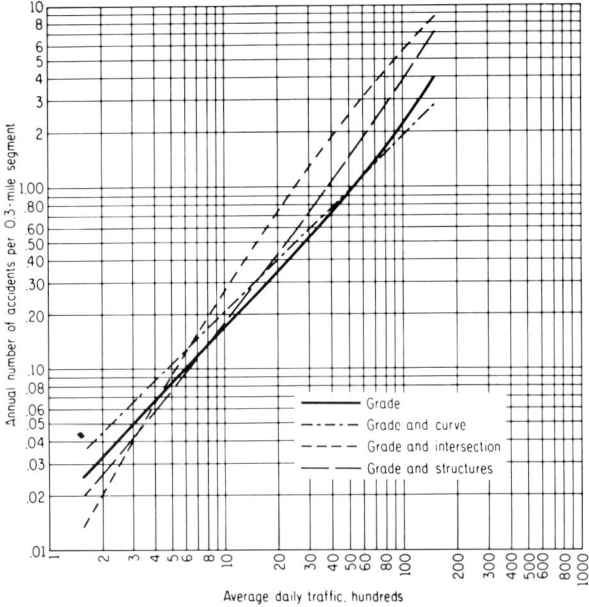

FIG. 15-7. Three-state average number of accidents per 0.3-mile segment on rural two-lane highways showing the effects of grades combined separately with curves, intersections, and structures [15-12].

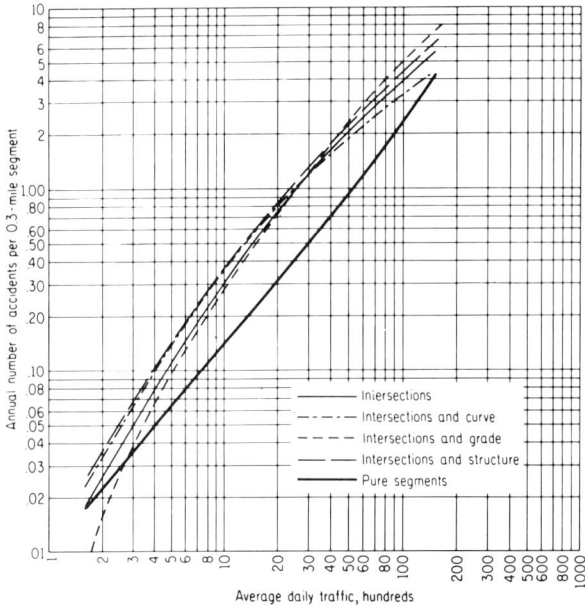

FIG. 15-8. Three-state average number of accidents per 0.3-mile segment on rural two-lane highways showing the effects of intersections combined separately with curves, grades, and structures [15-12].

420   **Traffic Accidents**

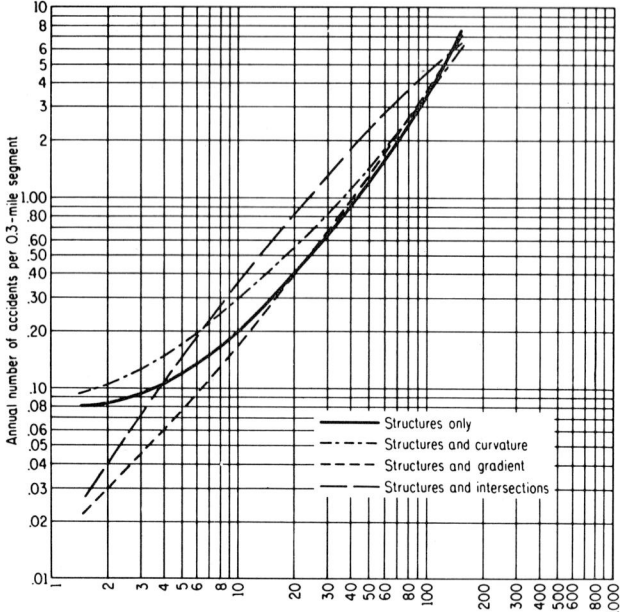

FIG. 15-9. Three-state average number of accidents per 0.3-mile segment on rural two-lane highways showing the effects of structures combined separately with curves, grades, and intersections [15-12].

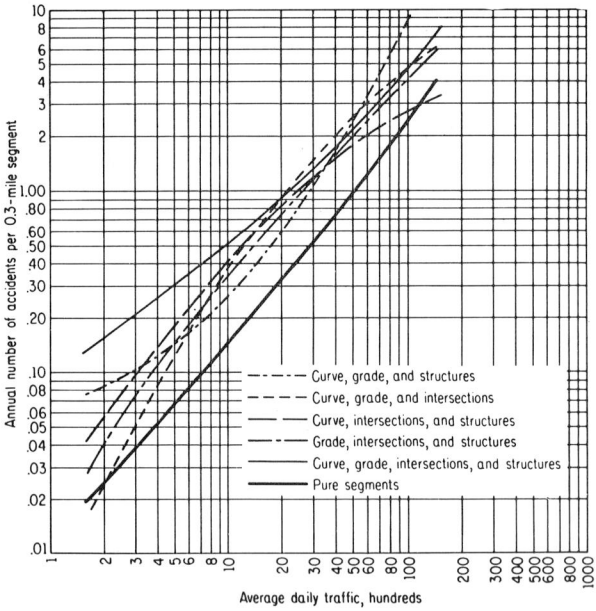

FIG. 15-10. Three-state average number of accidents per 0.3-mile segment on rural two-lane highways, showing the effects of the combined presence of curves, grades, intersections, and structures [15-12].

Traffic Accidents

PROBLEMS

15-1. From Table 15-2 for (1) the number of accidents, and (2) the cost of accidents, compute the ratios of (a) nonfatal injury, and (b) property damage only accidents to fatal accidents. Use just the total line for urban and for rural accidents. Comment on your ratios and suggest how such ratios may be used in economic analyses.

15-2. From Table 15-2 compare the cost per urban accident with the cost per rural accident. Discuss the probable reasons for the difference in costs that you find. Make a similar analysis of Table 15-3 with respect to a comparison of state highways and local streets and roads.

15-3. Discuss the relative merits of the following methods of placing a value on human life for use in economic analyses: (1) court awards in suits involving accidental death, (2) average of life insurance policies paid on accidental death, (3) cost of actual measures taken to prevent accidental death, (4) what society can afford to spend to prevent accidental death, (5) average of public opinion polls, (6) present worth of probable future earnings less present worth of cost of sustaining the individual in an earning capacity, and (7) others.

15-4. For the following conditions, compute the dollars of original construction cost per mile which would be equivalent to the accident cost reduction likely to be obtained by constructing a four-lane rural highway with full access control to replace a two-lane rural highway without access control.

Vestcharge rate 10%	Accident rate, upper section, Table 15-11
Analysis period .... 20 years	
Initial ADT 5,000	Cost per accident, nonintersection, Table 15-2 (compute combined accident cost for cars and trucks)
ADT annual growth rate 4%	
ADT composition, trucks 15%	Assume that the accident rate is unchanged by the growth in ADT

15-5. Widening an 18-ft rural two-lane highway to 24 ft probably will reduce nonfatal accidents 25 percent and fatal accidents 15 percent. The current annual accident per 100,000,000 vehicle-miles of travel on a route of 18-ft wide pavement is 200 property damage only accidents, 100 personal injury accidents, and 8.0 fatal accidents. The costs per accident are $ 250, $ 2,500, and $ 9,500 respectively. If the widening costs are $ 20,000 per mile what equivalent uniform ADT is necessary to produce a rate of return of 10 percent for the next 15 years?

15-6. To reduce the traffic accident rate at an intersection on Indiana Route US 40, the following improvements were made: (a) left turning lanes were provided, (b) lane dividers were constructed at all four approaches, (c) curb and gutter areas were reconstructed to facilitate turning of large vehicles, and (d) the traffic signals were modernized. The minor route had an ADT of 1,000 and the major route 6,000. The remodeling cost was $ 122,000. The before and after average number of traffic accidents per year and their cost are:

	Fatal	Nonfatal Injury	Property Damage Only	Total
Before	1	6	10	17
After	0	4	5	9
Before and after cost per accident	$ 8,500	$ 2,000	$ 250	–

Compute the B/C ratio on the basis of a minimum attractive rate of return of 10 percent, an analysis period of 15 years, zero terminal value, and a 3 percent per year compound increase in accident cost.

15-7. To reduce traffic accidents, the state of Colorado modified a 0.984-mile section of state Route 123, including an intersection. The main features contributing to accidents were narrow bridges, a sharp vertical downgrade combined with a horizontal curve, and a skewed T-intersection. The main changes were: (a) widened route 123 to 4 lanes near the intersection and provided a 30-ft wide median, (b) installed islands and provided separate right and left turn lanes, (c) relocated a canal, and (d) replaced two narrow bridges with wider ones. The reconstruction cost was $ 430,000. The initial ADT was 2,500 on route 123 and 800 on the minor leg, or stem. The forecasted traffic 20 years hence is 5,600 ADT on Route 123 and 2,000 on the minor leg. The accident experience data based on about two years observation, but converted to one year rates are:

	Fatal	Nonfatal Injury	Property Damage Only	Total
Before	0	4	7	11
After	0	1	2	3
Total accident cost per year, before	0	$ 16,000	$ 3,500	$ 19,500
Total accident cost per year, after	0	3,000	1,000	4,000

Using a gradient increase in traffic and in accident costs for the next 20 years, compute the rate of return to be expected from this spot improvement to reduce traffic accidents. Use zero terminal value.

15-8. In a Florida city, there was a major boulevard that had left-turn storage lanes except at one intersection. Further, the lanes for this 0.303-mile section were only 10 ft wide. This section was reconstructed by widening to four 12-ft lanes and adding left-turn lanes, following which improvements the accident rate dropped significantly. The main descriptive data are:

1. Cost of work was $ 37,600 for the 0.303 miles of street reworked.
2. Accidents, average per year:

	Fatal	Nonfatal Injury	Property Damage Only	Total
Before	0	14	33	47
After	0	6	17	23

Traffic Accidents

3. Cost per accident, cars:

	Fatal	Nonfatal Injury	Property Damage Only	Total
Before	$ 7,000	$ 1,500	$ 200	–
After	$ 7,500	2,000	250	–

4. Cost per accident, trucks:

	Fatal	Nonfatal Injury	Property Damage Only	Total
Before	$ 5,000	$ 1,000	$ 100	–
After	5,000	1,200	150	–

5. Traffic performance:

	Attempted Speed, mph	Number of Vehicle Stops per Day	Number of Slowdowns per Day from mph/to mph	Minutes Delay per Vehicle Stop	Number of Sharp Turns per Day
Before	25	7,000	5,500-25/5	0.6	3,200
After	30	6,000	4,800-30/15	0.5	3,200

Note: Assume that in the above total number of vehicles the composition is proportional to the mix in the total ADT.

6. Traffic volume and classification:

	Passenger Cars	Commercial Delivery	Single-unit Trucks	Equivalent 2-S2, 40 kip Combinations	Total
Before and after	17,800	1,700	800	700	21,000
Estimated 12 yr hence (uniform increase)	23,800	2,300	1,100	940	28,140

Calculate (1) the benefit/cost ratio for the reduction in the cost of accidents, (2) the benefit/cost ratio for the reduction in running cost of the traffic, (3) the benefit/cost ratio for the reduction in travel time, and (4) the benefit/cost ratio for the combined three types of benefits. Use a vestcharge rate of 12 percent, an analysis period of 12 years, and zero terminal value.

REFERENCES

15-1. Automotive Safety Foundation. *Traffic Control and Roadway Elements–Their Relationship to Highway Safety.* Washington, D.C., 1963. A good bibliography of over 700 references is given, plus a good discussion in the text.

15-2. D. M. Belmont. *Accidents Versus Width of Paved Shoulders on California Two-Lane Tangents, 1951 and 1952.* Highway Research Board, Washington, D.C., Bulletin 117, 1956, pp. 1-5.

15-3. D. M. Belmont. *Effect of Average Speed and Volume on Motor Vehicle Accidents on Two-Lane Tangents.* Highway Research Board, Washington, D.C., *Proc.,* Vol. 32, 1953, pp. 383-395.

15-4. Charles M. Billingsley and Dayton P. Jorgenson. "Analyses of Direct Costs and Frequencies of Illinois Motor-Vehicle Accidents, 1958," *Public Roads,* Vol. 32, No. 9 (August 1963), pp. 201-213.

15-5. C. E. Billion. *Effect of Median Dividers on Driver Behavior.* Highway Research Board, Washington, D.C., Bulletin 137, 1956, pp. 1-17.

15-6. F. Bitzel. "Accident Rates on German Expressways in Relation to Traffic Volumes and Geometric Design," *Roads and Road Construction,* (January 1957), pp. 18-20.

15-7. F. Bitzel. *Traffic Accidents on the Autobahnen.* International Course in Traffic Engineering Reports, Theme III, 1954.

15-8. R. C. Blensly and J. A. Head. *Shoulders and Accident Experience on Two-Lane Rural Highways: A Summary.* Highway Research Board, Washington, D.C., Bulletin 266, October 1959, pp. 28-33.

15-9. Mary Jean Bowman. "Human Capital: Concepts and Measures," *Festkrift* Lund, Sweden, March 1961.

15-10. James S. Burch and James R. Zook. "Expressway Accident Rates in North Carolina," *Traffic Engineering,* Vol. 32, No. 7 (April 1962), pp. 18-19.

15-11. Alfred F. Conard and others. *Automobile Accident Costs and Payments.* U. of Michigan Press, Ann Arbor, 1964.

15-12. Cornell Aeronautical Laboratory. *Motor Vehicle Accident Rates as Related to Design Elements of Rural Highways.* Highway Research Board, Washington, D.C., NCHRP Report No. 47, 1967.

15-13. Paul D. Cribbins and Garry S. Summer. *The Correlation of Accident Rates with Geometric Design Components on Various Types of Highways.* Engineering Research Department, North Carolina State University, Raleigh, 1964.

15-14. Louis I. Dublin and Alfred J. Lotka. *The Money Value of a Man.* Rev. ed., Ronald, New York, 1946.

15-15. Gerald W. Graves. "Safety and Economic Aspects of Expressway Construction in Michigan," *Traffic Engineering,* Vol. 29, No. 8 (May 1959), pp. 17-22.

15-16. Kenneth W. Haase. "Characteristics of Persons Injured in Motor Vehicle Accidents." *Traffic Quarterly,* Vol. XVII, No. 4 (October 1963), pp. 584-598.

15-17. L. R. Hafstad. "Research as Applied to Traffic Engineering," *Traffic Safety Research Review,* Vol. 1, No. 3 (December 1957), pp. 66-74.

15-18. W. E. Hall. "There's 'No Sure Cure'." *Traffic Safety,* Vol. 60, No. 4 (April 1962), pp. 31, 47.

15-19. Highway Research Board. National Academy of Sciences, National Research Council, Washington, D.C., Bulletin 208, *Traffic Studies,* 1958.

1. Bruce D. Greenshields. "Traffic Accidents and the Quality of Traffic Flow," pp. 1-15.

2. Robie Dunman. "Economic Costs of Motor Vehicle Accidents," pp. 16-28.

3. Edmund J. Cantilli. "Statistical Evaluation of Traffic Severity," pp. 29-34.

4. Richard W. Bletzacker and Thomas G. Brittenham. "An Analysis of One-Car Accidents," pp. 35-44.

5. J. A. Head. "Predicting Traffic Accidents from Roadway Elements on Urban Extensions of State Highways," pp. 45-63.

Traffic Accidents

15-20. Highway Research Board. National Academy of Sciences, National Research Council, Washington, D.C., Bulletin 240, *Highway Accident Studies*, January 1959.

    1. R. C. Blensly and J. A. Head. "Statistical Determination of Effect of Paved Shoulder Width on Traffic Accident Frequency," pp. 1-23.

    2. John Versace. "Factor Analysis of Roadway and Accident Data," pp. 24-32.

    3. A. F. Malo and H. S. Mika. "Accident Analysis of an Urban Expressway System," pp. 33-43.

    4. S. M. Breuning and A. J. Bone. "Interchange Accident Exposure," pp. 44-52.

15-21. Highway Research Board. National Academy of Sciences, National Research Council, Washington, D.C., Bulletin 263, *Economic Cost of Traffic Accidents*, January 1960.

    1. Bernard B. Twombly. "Economic Cost of Traffic Accidents in Relation to the Highway," pp. 1-22.

    2. James F. McCarthy. "Economic Cost of Traffic Accidents in Relation to the Vehicle," pp. 23-39.

    3. Robie Dunman. "Economic Cost of Traffic Accidents in Relation to the Human Element," pp. 40-49.

    4. J. Edward Johnston. "Economic Cost of Traffic Accidents in Relation to Highway Planning," pp. 50-53.

    NOTE: See also *Public Roads*, Vol. 31, No. 2 (June 1960), pp. 34-50, for these four papers.

15-22. J. A. Hiller and J. G. Wardrop. "Effect of Gradient and Curvature on Accidents on London-Birmingham Motorway," *Traffic Engineering and Control*, Vol. 7, No. 10 (February 1966), pp. 617-621.

15-23. Irving Hoch. *Accident Experience: Expressways vs. Arterials.* Chicago Area Transportation Study, Chicago, December 1949. *Note:* The main part of this paper and the appendix and some illustrations are to be found in *Traffic Quarterly*, Vol. 24, No. 3 (July 1960), pp. 340-362.

15-24. F. W. Hurd. *Accident Experience with Transversable Medians of Different Widths.* Highway Research Board, Washington, D.C., Bulletin 137, 1956, pp. 18-26.

15-25. Illinois Department of Public Works and Buildings, Division of Highways, Springfield, Ill. *Cost of Motor Vehicle Accidents to Illinois Motorists, 1958,* December 1962.

15-26. Roger T. Johnson. *Effectiveness of Median Barriers.* Highway Research Board, National Academy of Sciences, Washington, D.C., Research Record 105, 1966, p. 99.

15-27. Roy Jorgensen and Associates. *Evaluation of Criteria for Safety Improvement on the Highway.* A special report to the United States Department of Commerce, Bureau of Public Roads, Gathersburg, Md, 1966.

15-28. Dayton P. Jorgenson. "Accident Costs and Rates on Chicago Area Streets and Highways," *CATS Research News*, Vol. 4, No. 4, Chicago Area Transportation Study (March 30, 1962), pp. 2-11.

15-28A. Jaakko K. Kihlberg and K. J. Tharp. *Statistical Analysis of Accident Rates and Geometry of Highway.* Highway Research Board, National Academy of Sciences, Washington, D.C., Highway Research Record 188, 1967, p. 188. See Ref. 15-12.

15-29. J. C. Mackie. "Michigan's New Controlled Access Expressways Provide Dramatic Reduction in Death and Accident Tolls," *American Highways*, Vol. 38, No. 2 (April 1959), pp. 4, 26-28.

15-30. Maryland State Roads Commission. *Accident Experience Related to Control of Access and Design Features.* Baltimore, Md., 1961.

15-31. John W. McDonald. *Relation Between Number of Accidents and Traffic Volume at Divided-Highway Intersections.* Highway Research Board, National Academy of Sciences, Washington, D.C., Bulletin 74, 1953, pp. 7-17.

15-32. J. C. McMonagle. *Accident Analysis–Telegraph Road 1947-1948.* Highway Research Board, National Academy of Sciences, Washington, D.C., Bulletin 30, 1951, pp. 29-41.

15-33. R. M. Michaels. "Human Factors in Highway Safety," *Traffic Quarterly,* Vol. XV, No. 4 (October 1961), pp. 586-599.

15-34. Karl Moskowitz. *Accidents on Freeways in California.* World Traffic Engineering Conference, Theme IV, 1961.

15-35. Karl Moskowitz and W. E. Shaefer. *California Median Study, 1958.* Highway Research Board, National Academy of Sciences, Washington, D.C., Bulletin 266. 1960, pp. 34-62.

15-36. National Safety Council. *Accident Facts, 1966 Edition (Accidents for Calendar Year of 1965).* Chicago, 1966.

15-37. National Safety Council. *Manual on Classification of Motor Vehicle Traffic Accidents.* Chicago, 1962.

15-37A. Fletcher N. Platt. *Operations Analysis of Traffic Safety.* Ford Motor Company, Traffic Safety and Highway Improvement Department, Dearborn, Michigan, 1959. For prior publication of three parts of this publication see *International Road Safety and Traffic Review,* Vol. VI, No. 2 (Spring 1958) and No. 4 (Autumn 1958). See also *Traffic Safety Research Review.*

15-38. C. W. Prisk. "Life Saving Benefits of the Interstate System," *Public Roads,* Vol. 31, No. 11, (December 1961), pp. 219-220.

15-39. C. W. Prisk. "The Speed Factor in Highway Accidents," *Traffic Engineering,* Vol. 29, No. 11 (August 1959).

15-40. M. S. Raff. *Interstate Highway-Accident Study.* Highway Research Board, National Academy of Sciences, Washington, D.C., Bulletin 74, 1953, pp. 18-45.

15-41. D. W. Shoppert. *Predicting Traffic Accidents from Roadway Elements of Rural Two-Lane Highways with Gravel Shoulders.* Highway Research Board, National Academy of Sciences, Washington, D.C., Bulletin 158, 1957, pp. 4-26.

15-42. Wilbur Smith and Associates. *Motor Vehicle Accident Costs.* A study made for the District of Columbia Department of Highways and Traffic, Maryland State Roads Commission, and Virginia Department of Highways. New Haven, Conn., 1966.

15-43. David Solomon. *Accidents on Main Rural Highways Related to Speed, Driver, and Vehicle.* U.S. Department of Commerce, Bureau of Public Roads. Government Printing Office, Washington, D.C., 1964.

15-44. David Solomon. "Traffic Signals and Accidents in Michigan," *Public Roads,* Vol. 30, No. 10 (October 1959), pp. 234-237.

15-45. W. A. Stohner. *Relation of Highway Accidents to Shoulder Width on Two-lane Rural Highways in New York State.* Highway Research Board, National Academy of Sciences, Washington, D.C., *Proc.,* Vol. 35, 1956, pp. 500-504.

15-45A. Daniel Syrek. "Accident Rates at Intersections," *Traffic Engineering,* Vol. 25, No. 8 (May 1955), p. 312.

15-46. Thomas H. Tamburri and John C. Glennon. *Objective Criteria for Guardrail Installation.* California Division of Highways, Sacramento, July 1966.

Traffic Accidents

15-47. K. J. Tharp and Jaakko K. Kihlberg. *Motor Vehicle Accident Rates as Related to Design Elements of Rural Highways.* Highway Research Board, National Academy of Sciences. See Ref. 15-12.

15-48. Bernard B. Twombly. "The Economic Cost of Traffic Accidents in Relation to Highway Systems," *Public Roads,* Vol. 31, No. 2 (June 1960), pp. 39-43.

15-49. United States Department of Health, Education, and Welfare. *Disease Control Programs-Motor Vehicle Injury Prevention Program.* Office of the Assistant Secretary for Program Coordination, Washington, D.C., August 1966.

15-49A. United States 86th Congress, First Session. (U.S. Department of Commerce). *The Federal Role in Highway Safety.* House Document No. 93, Government Printing Office, Washington, D.C., March 1959.

15-50. Utah State Road Commission. (1) *Cost of Passenger Car Accidents to Motorists in 1955; (2) Economic Cost of Motor Vehicle Accidents: Commercial Vehicles. 1957.* Vol. 1, *Accidents and Incidents.* Vol. 2, *Direct Costs.* Vol. 3, *Involvements.* 1960.

15-51. G. M. Webb. "Median Study," *Traffic Engineering,* Vol. 30, No. 3 (December 1959), pp. 18-20.

15-52. G. M. Webb. "The Relation Between Accidents and Traffic Volumes at Signalized Intersections," Institute of Traffic Engineers, Session 3B, 1955, pp. 149-167.

15-53. Dean M. Wenger. *Accident Characteristics of Four-Way Stop Control Versus Two-Way Stop Control.* Student thesis, Yale Bureau of Highway Traffic, New Haven, Conn., May 1958.

15-54. Roy L. Wilshire and Charles J. Keese. *Effects of Traffic Accidents on Freeway Operation.* Texas Transportation Institute, Bulletin 22, College Station, Texas, April 1963.

Chapter **16**

Traffic Characteristics

In the analyses for the economy of proposed highway facilities and in the analyses of the cost of highway transportation, the behavior, or performance, of motor vehicles on the highway is an important factor. The two elements of major importance are the speeds of traffic and the volume of traffic, and these two factors are interrelated. The running cost of motor vehicles is affected greatly by motor vehicle speeds and changes in speed. The speed factor is especially important because one objective of many proposed improvements is to provide for faster travel by increasing the operating speed or overall average speed.

COMPOSITION OF TRAFFIC BY VEHICLE CLASSES

The composition of traffic, with special reference to the number of trucks and their class, is important to both geometric and structural design of highways. Also, the composition of traffic is important to economy analyses and to finance and taxation studies. The large trucks affect lane volume, running speed, speed changes, and fuel tax incomes. It is fitting to start this chapter with a table of typical traffic composition.

Most of the states each summer weigh trucks at selected roadside weighing stations. At the same time, loading information (empty or with payload), vehicle class by axle arrangement for trucks, and other information is recorded. Class of commodity hauled, body type, and length of trip, and whether common carrier or private hauler are obtained in some years.

Not all the trucks passing a weighing station are weighed so as to avoid congestion and major delays to traffic, and, further, the weighing does not operate the full 24-hour day. To provide the base for expansion of the vehicles weighed to a full 24-hour day, all traffic passing the weighing station is counted and classified for the full 24 hours. These counts and classifications are also used in financial and tax analyses, trend studies of highway use, forecast studies of highway use, and other administrative and planning work. Table 16-0 gives the results of the counts and classifications in California for 1966. Each state usually publishes a summary of each year's weighings and traffic counts.

Traffic Characteristics

TABLE 16-0
Percentage Distribution of Traffic by Vehicle Class and Highway System

Vehicle Class	Interstate Rural Highways	Interstate Urban Highways	Primary Rural Highways	Primary Urban Highways
A. Passenger vehicles				
1. Passenger cars	76.216	84.987	73.064	84.060
2. Motorcycles	0.221	0.393	0.325	0.226
3. Buses	0.740	0.171	0.360	0.649
4. Subtotal	77.177	85.551	73.749	84.935
B. Single-unit trucks				
5. Panel and pickups	10.024	10.091	12.403	8.338
6. 2-axle, 4-tire (2S)	0.385	0.400	0.425	0.535
7. 2-axle, 6-tire (2D)	2.619	1.847	2.833	2.603
8. 3-axle (3A)	0.561	0.288	1.129	0.317
9. 4-axle (4A)	0.001	0.000	0.002	0.001
10. Subtotal	13.590	12.626	16.792	11.794
C. Tractor-semitrailer combination				
11. 3-axle (2-S1)	0.703	0.235	0.632	0.576
12. 4-axle (2-S2)	0.647	0.172	0.548	0.325
13. 5-axle (3-S2)	3.061	0.562	3.045	0.954
14. 6-axle (3-S3)	0.011	0.002	0.005	0.003
15. Subtotal	4.422	0.971	4.230	1.858
D. Tractive truck and trailer combination				
16. 3-axle (2-1)	0.012	0.005	0.010	0.005
17. 4-axle (2-2)	0.042	0.008	0.048	0.016
18. 5-axle (2-3 and 3-2)	1.384	0.418	1.483	0.354
19. 6-axle and more (3-3 and ?)	0.043	0.005	0.048	0.005
20. Subtotal	1.481	0.436	1.589	0.380
E. Tractor and two-trailer combination				
21. 4-axle (2-S1-1)	0.000	0.000	0.004	0.001
22. 5-axle (2-S1-2)	3.256	0.412	3.557	1.017
23. 6-axle (3-S1-2 and 3-S2-1)	0.069	0.003	0.077	0.013
24. 7-axle and more (3-S2-2 and ?)	0.005	0.001	0.002	0.002
25. Subtotal	3.330	0.416	3.640	1.033
F. Total, all traffic	100.000	100.000	100.000	100.000
26. Total all vehicles counted	218,729	253,431	81,470	205,458
27. Number of roadside stations counted	8	3	5	3

Source: California State Truck Weight Studies for 1966.

FACTORS AFFECTING TRAFFIC SPEEDS

Motor vehicle spot speed and average overall speed for a trip are affected by many factors. To the casual driver many of these factors will not be self-evident or within his recognition at the time. A discussion of the factors that affect motor vehicle speed will aid in understanding traffic speeds and why there are wide variances in vehicular speeds. These factors will be presented under the main headings of (1) driver, (2) vehicle, (3) highway, and (4) environment.

SPEED AND THE DRIVER

The driver, or operator, of a motor vehicle on the highway has control of the vehicle speed. But there is a wide range in the speeds of vehicles at one given spot on the highway and in one given hour. Drivers just do not choose to drive at identical speeds, nor does any one driver drive the same speed on all occasions, other factors permitting. Some of the characteristics of the driver which affect his choice of speed are:

1. Age
2. Confidence
3. Driving skill
4. Mental attitude
5. Physical condition
6. Distance of trip
7. Familiarity with the area
8. Knowledge of the particular route
9. Purpose of trip
10. Travel time available

The five items in the left-hand group are somewhat interdependent. Age of the driver is often an index to attitude, physical condition, skill, and confidence. In general, elderly drivers drive more slowly than do younger drivers. Mental attitude may be considered in two ways: One's generally prevailing attitude toward driving, the legal speed limit, other drivers, and degree of selfishness; and on the other hand, one's immediate attitude is a factor affecting driving speed. A driver may be in a hurry or not, may feel elated or be depressed, may be anticipating a rewarding experience or may be driving to an event that he would rather not attend. Consciously or unconsciously, one's mental attitude does affect his driving speed.

Drivers with good driving skill and confidence are more apt to be faster drivers than those with low skill and little confidence. But on the highway will be found poor drivers with confidence. There is the driver who must always be in the lead and the driver who is a continual lane changer on multilane highways.

Familiarity with a highway network and with the particular route being traveled permits a driver to drive at higher speed than would a driver unfamiliar with the area or route. A stranger hunting for a driveway, street, turnoff road, or highway intersection will slow down sooner and more frequently than a driver who knows exactly where he is going and how to get there.

Traffic Characteristics

SPEED AND THE VEHICLE

Considering both trucks and passenger cars, the following characteristics of the vehicle affect driving speeds:

Ratio of gross vehicle road weight to net horsepower
Mechanical condition
Age of vehicle as an index of general design and use
Aerodynamic design factors

The vehicle itself has a more pronounced effect upon top limit free running speed than it does at the lower speeds controlled by traffic volume or local restrictions. Top speed of a vehicle is a factor of its aerodynamic design, gross road weight, and net horsepower of its engine. The heavier trucks of 400 or more pounds gross weight per net horsepower may not be able to travel faster than 50 to 55 mph on level tangents, in comparison with the passenger car's top speed of well over 100 mph and a weight horsepower ratio of, say, 20 pounds. The total air resistance, which varies about as the square of the speed, is a controlling factor at top speed. Internal resistance of the engine and power train are also important factors limiting top speed.

In addition to possible top speed, the weight/horsepower ratio affects the ability to accelerate. Therefore, on the highway where acceleration is required, the vehicles with the lower weight/horsepower ratio will probably be observed at higher speeds than will the vehicles with higher ratios. This comparison is observed at stops at traffic signals.

Because of the factor of riding comfort, a driver probably will drive a vehicle at such speed as produces the most physically and mentally satisfactory ride. A vehicle that has high vibrations, a rough suspension system, and a noisy operation will most likely be driven slower than will a vehicle that rides more smoothly and quietly.

The mechanical condition of a vehicle, often directly related to vehicle age, is a factor controlling top speed, acceleration, and deceleration. Because drivers in general have not the faith in old vehicles or vehicles not in good mechanical condition (engine, brakes, and body tightness), they often drive them more slowly than they would a vehicle in good mechanical condition.

SPEED AND THE HIGHWAY

Of the four groups of factors under consideration, perhaps the highway design group is the most important group of factors affecting traffic speeds. The highway design factors are important because they can be controlled by the designer and they may greatly affect the cost of highway construction and the road-user running costs. The main highway factors affecting vehicle speed on the highway are:

1. Access control
2. Geometric width
3. Gradients
4. Highway lighting
5. Horizontal curvature
6. Lane arrangement and traffic direction
7. Legal, or posted, speed limit
8. Roadway surface
9. Roadside objects
10. Scenic view with relation to design location
11. Sight distance
12. Signing
13. Traffic controls
14. Changes in number of lanes
15. Changes in horizontal and vertical alignment
16. Changes in pavement quality
17. Changes in shoulder width
18. Changes in traffic movements

The vertical alignment of a highway is a factor in vehicular speed because of its effect upon the power requirement. Trucks especially may be limited in top speed on plus grades by their engine power. Downhill movement of traffic is affected by the element of safety related to stopping distance and sight distance at horizontal curves.

The horizontal alignment controls speed through the radius of curvature and superelevation. Safe driving and comfortable riding on horizontal curves are achieved by holding the speed low enough to keep the centrifugal force at a safe and comfortable limit.

Both horizontal and vertical alignment greatly affect traffic speed when combined with sight distance. A slow-moving truck proceeding uphill or downhill will restrict other traffic, even when other vehicles do not restrict passing, when an acceptable sight distance is not available. Correct driving calls for always having ahead a clear roadway view of sufficient distance to permit the proper action to avoid a possible accident.

The roadway surface affects driving speed principally through comfort (roughness, noise, dustiness, and a lack of undulations), safety (coefficient of friction and uniformity of condition), and transverse slope, or crown.

The lane width, shoulder width, absence or presence of a curb, number of lanes, median width, and other transverse geometrics affect driving speeds. The effect is primarily through the elements of safety and psychological comfort. One-way lane arrangement and street systems permit speeds uncontrolled by opposing traffic. Where access control is provided, higher speeds result from the absence of points of vehicle conflict and turning movements.

Drivers are cautious about driving close to roadside objects and highway structures. Curbs, drainage inlets, culvert headwalls, retaining walls, bridge girders and trusses, piers, posts of all kinds, trees, and ditches near the pavement cause drivers to slow down as compared to speeds where these roadside objects do not exist.

Highway lighting is effective in permitting the driver better vision of his surroundings than can be obtained from his own headlights, especially of the roadside.

Traffic Characteristics

Highways located to afford views of buildings, unusual landscapes, mountain peaks, and other artistic features visible from the highway will result in slower speeds than would be observed on routes not so favorably located.

Both traffic control and informational signs affect traffic speed. Small, poorly worded, or poorly located signs cause traffic to slow down in order to read them. Traffic controls, including speed limit signs, greatly affect the speed and speed changes of traffic. The legal speed limit is placed in the highway group because often there is a direct relationship of the legal speed and the highway design.

There are a large number of variations in design features which cause changes in speed as well as a lowering or raising of the general running speed. Included in this class are changes in lane width, shoulder width, number of lanes, pavement quality, and many other design features as examples of factors which affect vehicular speed. Crests and sumps in vertical alignment and horizontal curves and turns usually affect traffic speed.

SPEED AND THE ENVIRONMENT

The speed at which a driver operates his motor vehicle at any given point on a highway is affected by the surroundings, or total environment. Included in the environment are such factors as: (1) maneuvering actions of traffic, (2) traffic density, (3) traffic mix by vehicle class, (4) traffic speed, (5) angle of sunshine, (6) time of day, (7) off-highway scenery, (8) on-street actions and views, and (9) weather (vision, wind, wetness).

A driver of a motor vehicle is much concerned with the world about him, or at least he should be. It is a small world to be sure, but a safe or unsafe world depending upon the attention given to it by the driver. A driver's speed is restricted by the highway design and by the traffic of which each driver is a part. Being a part of the traffic stream means that each driver is governed to some extent by the stream, and the heavier the stream of traffic the more it governs the actions of each driver. Thus has come about the well-established relationship of speed and traffic volume.

Closely associated with traffic volume and speed are vehicular movements and speed. Lane changing, turning off, merging, speedups, and slowdowns by one vehicle often affect many other vehicles. Highway design does control the opportunities for speed and speed changes, but traffic volume and density as environmental factors play an important role in determining the speed and the number and extent of speed changes.

The elements of weather–rain, snow, wind, angle of sun–affect driving speed, primarily through the factors of sight distance, visibility, and safety. Even light rain causes traffic to slow down, drivers to change routes, and people to forgo certain trips or to take previously unplanned trips. Wind has the effect of speeding up traffic as a tail wind, slowing down traffic as a head wind, and probably slowing down traffic as a cross wind. Wind affects speed both by air pressure on the vehicle and by blowing dust and light objects into the field of driving vision.

Drivers and other car occupants are attracted by what can be seen on and off the roadway. Buildings, people, animals, works of art, and just plain nature at her best and worst result in variations of vehicular speeds.

OBSERVED SPEEDS OF TRAFFIC

The running costs of motor vehicles and the travel time are directly related to the speed of the vehicles and the plus and minus changes in the speed. In an analysis for the economy of a proposed highway improvement it is necessary to use specific traffic speeds for each alternative. Specific speeds mean not average speed or average travel time, but the speed distribution and the number and range of all changes in speed for each class of vehicle. Not enough on-the-highway observations of speed are available to furnish reliable daily, let alone yearly, average speed distributions for the many types of highways and specific conditions to be encountered. Presentation and discussion of some tables and figures of traffic speeds will be helpful in providing some working data and gaining an understanding of vehicular speeds and their role in economic analyses of highway improvements.

AVAILABILITY OF SPEED INFORMATION

The speed of vehicles on the highway has long been a subject of interest to the highway designer, the traffic engineer, and the highway planner. But unfortunately these professionals have not sensed the need of traffic speed data as used in the process of economic analysis. In highway design and traffic engineering such factors as the 30th highest hour traffic volume, peak hour volume and speed, the 85th percentile speed, speed of the free-running vehicles, and overall travel time have been those factors receiving major attention.

Attention to the specific speeds of each of the vehicles in the traffic for a 24-hour day, for a week, and for a year has simply not been given. Yet the distribution of the speeds of the vehicles, not average speed and not travel time, is what is needed in economy studies. It is likewise important to know the number of changes and amount of change in speed from each initial speed. Especially in urban areas the running cost of vehicles is greatly affected by the changes in vehicular speeds. For all economic analyses, fuel tax studies, and total transportation studies, the speed distribution needs to be known if accurate estimates of costs are to be had. Fuel consumption and total running cost bear a curvilinear relationship to vehicular speed. Multiplying the average speed of traffic by the running cost at that average speed may not produce a reliable total vehicle running cost.

Traffic Characteristics

The speed trend observations taken each year by state highway departments provide good evidence of the trend over time in average speed of free-running vehicles in rural areas in daytime travel. These observations do not, however, provide the speeds of the total traffic, even for the hours observed, and certainly not for a 24-hour period, or for a week. Urban areas have been especially neglected in the observations of traffic speeds.

The relation of traffic speed and hourly lane volume is well developed, and this information affords a good source of general information. See the Highway Research Board Traffic Capacity Manual |16-12|.

The importance of the changes in speed from the attempted running speed and idling time when stopped have more recently come into recognition as important factors in economy of highway transportation. The motor vehicle costs resulting from speed changes (Appendix A Tables) are developed, but observed traffic speed changes are yet lacking for most applications of these costs of speed changes.

Often overlooked by highway designers and analysts of economy is the fact that in project formulation the differences in road user costs between alternatives may lie primarily in minor differences in running speed and speed changes. Such minor differences may, perhaps, be neglected in the overall analysis of the economic evaluation of a proposed facility, but not in the final design and choice of design alternatives. Total dollars of vehicle running cost mount up fast at high ADT times 365 days a year and spread over the years in the analysis period.

DISTRIBUTION OF VEHICLE SPEEDS

The many factors that affect the driver's choice of speed result in wide variation in the speed of vehicles highway to highway, from spot to spot and from hour to hour. It may be assumed that the speed distribution observed at a typical spot on a route segment is an acceptable index to the speed distribution for the entire segment, except for the specific effects of vertical curves, horizontal curves, traffic actions, and traffic controls. These spot speeds may be thought of as the driver's choice of attempted speed under the prevailing conditions. A few speed distributions are given to illustrate what may be often observed. For specific locations and conditions, the traffic speed should be observed sufficiently to produce the data needed for each application.

Free-Running Spot Speeds

Many of the state highway departments take seasonal spot speed observations of the traffic at specific sites on their highway systems to detect trends in speed over the years. These observations are of vehicles free to travel at the driver's choice of speed. Table 16-1 gives representative distributions of free-running spot speed on primary and interstate rural routes.

Urban expressways and freeways are highways accommodating heavy

TABLE 16-1
Spot Speed Distribution Percentages of Traffic on Primary and Interstate Rural Highways

Speed Interval, mph	Primary Rural Highways Legal Speed Limit: Day 65; Night 60				Interstate Rural Highways Legal Speed Limit: Day 70; Night 65			
	Buses	Passenger Cars	Panel and Pickups	Trucks	Buses	Passenger Cars	Panel and Pickups	Trucks
	Percent of Total Number of Vehicles				Percent of Total Number of Vehicles			
32.5-35.0			0.4					
35.0-37.5			0.9	0.4			0.3	0.5
37.5-40.0			1.5	1.0			0.8	0.9
40.0-42.5		0.3	2.3	1.7		0.3	1.3	1.5
42.5-45.0		0.7	3.6	2.6		0.5	2.0	2.1
45.0-47.5		1.3	5.1	3.8		0.7	2.9	3.1
47.5-50.0		2.0	7.0	5.9		1.1	3.9	4.3
50.0-52.5		3.0	9.2	8.6		1.6	5.6	6.2
52.5-55.0		4.6	11.7	12.3		2.3	8.0	9.2
55.0-57.5	3.7	6.8	15.4	16.0		3.1	11.8	14.6
57.5-60.0	6.2	10.9	12.9	19.7	2.4	4.3	18.1	20.8
60.0-62.5	9.5	16.9	9.5	11.6	4.8	6.7	14.7	13.6
62.5-65.0	15.6	22.6	6.4	6.2	8.9	14.2	11.0	9.5
65.0-67.5	33.0	15.7	4.8	4.0	15.3	24.3	8.2	6.5
67.5-70.0	23.3	7.5	3.5	2.8	36.1	18.0	5.7	4.2
70.0-72.5	6.4	3.6	2.6	1.9	28.1	11.8	3.4	2.2
72.5-75.0	2.3	1.9	1.7	1.2	4.4	6.0	1.7	0.8
75.0-77.5		1.2	1.0	0.3		2.5	0.6	
77.5-80.0		0.7	0.5			1.4		
80.0-82.5		0.3				0.8		
82.5-85.0						0.4		
Total	100.0	100.0	100.0	100.0	100.0	100.0	100.0	100.0
Average Speed	65.7	62.2	56.3	56.9	68.3	66.0	59.0	57.9

Source: State observations of free-running vehicles in 1966.

volumes of vehicles at relatively high speed and often at capacity volumes. There is an appreciable difference in the speed of traffic with the change in vehicular volume throughout a 24-hour day. Saturday and Sunday speed patterns are unlike those for the five weekdays.

Table 16-2 is a composite distribution of the spot speed of each vehicle for a full calendar week on a metropolitan three-lane (one direction) expressway. There is a minor mode at 30 and 31 mph, which most probably results from the evening peak hour when capacity of about 2,000 vehicles per lane per hour is reached.

Hourly Traffic Volume and Speed

The distribution of passenger car speeds at different hourly traffic volumes

Traffic Characteristics

TABLE 16-2

PERCENTAGE DISTRIBUTION BY SPEED OF THE NUMBER OF PASSENGER CARS AND TRUCKS OUTBOUND ON AN URBAN EXPRESSWAY FOR THE FULL WEEK OF 11 MARCH 1962

Spot Speed, mph	Percent of Weekly Total Vehicles		Spot Speed, mph	Percent of Weekly Total Vehicles	
	Passenger Cars	Trucks*		Passenger Cars	Trucks*
0 & 1			40 & 41	2.883	2.076
2 & 3			42 & 43	3.727	3.324
4 & 5	0.041	0.023	44 & 45	4.976	5.727
6 & 7	0.069	0.023	46 & 47	6.113	8.044
8 & 9	0.110	0.046	48 & 49	6.879	11.247
10 & 11	0.167	0.086	50 & 51	8.556	13.858
12 & 13	0.195	0.080	52 & 53	9.118	12.880
14 & 15	0.225	0.115	54 & 55	10.307	11.017
16 & 17	0.310	0.104	56 & 57	10.539	9.125
18 & 19	0.407	0.259	58 & 59	8.838	5.992
20 & 21	0.692	0.408	60 & 61	6.344	4.272
22 & 23	0.905	0.426	62 & 63	2.970	1.903
24 & 25	1.174	0.644	64 & 65	1.184	0.955
26 & 27	1.474	0.891	66 & 67	0.453	0.368
28 & 29	1.680	0.995	68 & 69	0.165	0.150
30 & 31	2.034	0.955	70 & 71	0.068	0.040
32 & 33	1.901	0.891	72 & 73	0.022	0.034
34 & 35	1.800	0.914	74 & 75	0.008	0.017
36 & 37	1.767	0.817	76 & 77	0.006	0.006
38 & 39	1.889	1.288	78 & 79	0.002	
			80 &	0.002	
			Total	100.000	100.000
Number of Vehicles	378,842	17,391	Average Speed	48.9	49.9

*Low height trucks such as panels, pickups, express, are included with the cars; tall vehicles only are included in the trucks.

on different types of highways may be approximated from the spot speed-volume relationships. One source of such information is the Highway Capacity Manual, 1965 edition [16-12, pp. 49 and 50]. Table 16-3 was prepared from three sets of curves in this manual, but adjusting the values to a different sequence of traffic volumes per hour. These tables supply good speed distributions which may be used in the absence of specific information for the highway route under study.

Distribution of Speed of a Test Car in Traffic

Seldom, if at all, has there been reported in the literature the distribution of the speed of single vehicles operating over a considerable distance under normal traffic. Usually what is reported is the speeds of all traffic passing a chosen spot

TABLE 16-3
PERCENTAGE DISTRIBUTION OF PASSENGER-CAR SPEEDS UNDER IDEAL UNINTERRUPTED FLOW CONDITIONS

Speed Interval, mph	Traffic Flow, Vehicles per Hour									
	200	400	600	800	1000	1200	1400	1600	1800	2000
Both Directions of Travel on Two-Lane Rural highways										
22.5-25.0										1.9
25.0-27.5									1.8	4.1
27.5-30.0						0.4	1.1	2.4	4.7	50.0
30.0-32.5			0.7	1.5	2.4	2.6	3.5	5.1	21.0	41.7
32.5-35.0		1.5	2.8	3.3	3.8	4.8	6.3	20.5	67.5	2.3
35.0-37.5	1.9	3.1	4.0	4.9	5.5	7.6	19.2	62.8	4.3	
37.5-40.0	2.8	4.4	5.7	7.1	8.9	18.6	54.4	7.1	0.7	
40.0-42.5	3.7	5.8	7.9	10.1	15.4	39.0	11.5	2.1		
42.5-45.0	5.0	7.9	10.2	14.5	30.6	18.3	3.4			
45.0-47.5	6.6	10.0	14.4	21.8	19.2	5.7	0.6			
47.5-50.0	7.8	13.3	18.5	18.5	8.8	2.7				
50.0-52.5	10.2	15.0	13.2	10.3	3.9	0.3				
52.5-55.0	11.8	12.0	9.6	5.0	1.5					
55.0-57.5	12.3	9.4	7.2	2.1						
57.5-60.0	9.3	7.2	3.5	0.9						
60.0-62.5	7.6	5.1	1.8							
62.5-65.0	6.0	2.7	0.5							
65.0-67.5	4.5	1.7								
67.5-70.0	3.1	0.9								
70.0-72.5	2.4									
72.5-75.0	2.2									
75.0-77.5	1.6									
77.5-80.0	1.2									
Total	100.0	100.0	100.0	100.0	100.0	100.0	100.0	100.0	100.0	100.0
One Direction of Travel on Multilane Rural Highways–Volume per Lane										
22.5-25.0										3.8
25.0-27.5								0.8	3.0	9.8
27.5-30.0							0.6	2.9	5.7	50.2
30.0-32.5					0.3	1.4	3.2	5.7	13.0	29.4
32.5-35.0			0.4	1.3	2.2	3.3	5.8	9.1	31.9	6.2
35.0-37.5	1.7	2.5	3.0	3.4	4.5	6.8	8.9	17.5	31.0	0.6
37.5-40.0	2.5	3.2	4.1	5.4	7.0	9.8	15.2	30.4	10.8	
40.0-42.5	3.6	4.3	5.4	7.5	9.6	14.6	23.2	20.0	4.2	
42.5-45.0	4.7	5.4	6.8	9.7	13.5	20.1	22.1	9.0	0.4	
45.0-47.5	5.9	7.2	9.5	12.5	18.2	18.6	12.2	4.0		

Traffic Characteristics

TABLE 16-3 (*continued*)

Speed Interval, mph	Traffic Flow, Vehicles per Hour									
	200	400	600	800	1000	1200	1400	1600	1800	2000
Both Directions of Travel on Two-Lane Rural highways										
47.5-50.0	7.6	9.1	13.0	16.0	17.6	13.0	6.4	0.6		
50.0-52.5	9.2	11.3	15.2	17.1	13.1	7.8	2.4			
52.5-55.0	10.8	13.4	13.0	12.2	7.9	3.8				
55.0-57.5	11.1	12.4	11.0	6.7	4.6	0.8				
57.5-60.0	10.2	10.6	9.3	5.1	1.5					
60.0-62.5	9.1	8.3	5.2	2.8						
62.5-65.0	7.5	5.7	2.6	0.3						
65.0-67.5	5.0	3.1	1.5							
67.5-70.0	3.6	2.1								
70.0-72.5	2.5	1.4								
72.5-75.0	2.2									
75.0-77.5	1.6									
77.5-80.0	1.2									
Total	100.0	100.0	100.0	100.0	100.0	100.0	100.0	100.0	100.0	100.0
One Direction of Travel on Freeways and Expressways-Volume per Lane										
22.5-25.0										3.6
25.0-27.5									1.0	4.9
27.5-30.0								1.2	3.0	9.5
30.0-32.5						0.4	1.6	2.5	4.7	43.0
32.5-35.0			0.2	0.5	1.2	1.9	2.6	3.8	6.7	29.6
35.0-37.5	0.5	0.9	1.2	1.8	2.3	2.9	3.9	5.1	11.5	7.2
37.5-40.0	1.2	1.6	2.0	2.4	3.1	3.8	4.9	7.9	25.6	2.2
40.0-42.5	2.1	2.3	2.7	3.0	3.8	4.7	7.3	14.5	29.0	
42.5-45.0	2.4	2.7	3.2	4.1	5.0	7.2	11.9	22.5	13.8	
45.0-47.5	2.9	3.7	4.9	6.2	9.0	13.6	18.9	25.2	3.8	
47.5-50.0	4.7	5.9	7.3	10.5	13.5	16.6	20.6	11.2	0.9	
50.0-52.5	7.6	8.9	11.2	13.6	16.1	19.5	18.0	4.7		
52.5-55.0	10.4	11.8	13.3	16.3	18.6	19.4	7.5	1.4		
55.0-57.5	12.3	13.6	15.9	18.6	18.5	7.4	2.8			
57.5-60.0	12.2	14.0	18.3	14.2	6.4	2.2				
60.0-62.5	11.4	13.9	10.2	6.1	2.2	0.4				
62.5-65.0	10.0	8.7	5.6	2.3	0.3					
65.0-67.5	7.7	5.9	2.8	0.4						
67.5-70.0	5.9	3.3	1.0							
70.0-72.5	3.7	1.7	0.2							
72.5-75.0	2.4	1.1								
75.0-77.5	1.4									
77.5-80.0	0.8									
80.0-82.5	0.4									
Total	100.0	100.0	100.0	100.0	100.0	100.0	100.0	100.0	100.0	100.0

Source: Speeds interpolated and rearranged from *Highway Capacity Manual*, pp. 49-50 [16-12]. Copyright 1965 by the Highway Research Board. Used by permission.

on the highway. Claffey |16-5| in his 1959 toll road, free road comparison, recorded the speed of his test automobile each second of time during about 14,000 miles of driving. The eleven toll highways in ten states, the parallel free highways, and selected sections of routes in travel from state to state embraced 60 route sections and a total of some 120 recorded runs over the test sections.

From a sampling of the original paper tape recordings from the traffic impedance analyzer used by Claffey, the speed of his test car was tallied for each second for each of the ten classes of highways and traffic condition identified. Claffey reported |16-5| only average speeds (and average fuel consumption) for only five general types of routes.

In Table 16-4 are given the percentages of the total number of seconds of travel time the test car was operating at each individual integral speed, zero to maximum recorded miles per hour. The 14 columns of Table 16-4 set forth three significant characteristics of car speeds, not fully or easily observed in spot speed observations. The percentage of the travel time the car was stopped is insignificant (0.30 and 0.36 percent) for rural free-moving traffic, but high (up to 23.86 percent) for urban congested traffic. Except for the rural free-moving traffic there is appreciable driving at speeds from 1 to 15 mph. The four or more lane rural divided highway with access control (class D) resulted in only 0.46 percent of the driving time from 10 to 29 mph and does not reach above 1.0 percent of the time for any specific speed until 55 mph is reached.

The modal speed of 67 mph and modal percentage of 10.20 are pronounced for the toll highway and much higher values than for other classes of routes. The two- or three-lane rural highway with free-moving traffic (class A) has a modal speed of 63 mph and a modal frequency of 6.05 percent.

The percentages of travel time in Table 16-4 were converted to percentage of travel distance at 5 mph speed groups as given in Table 16-5. It is to be noted that the time stopped, zero speed, is excluded from Table 16-5 except as used in calculating the average overall travel speed as given in the last line of the table. For the class D highway, 88.57 percent of the mileage was driven at a speed ranging from 58 to 72 mph, with 42.96 percent at 65 mph plus or minus 2 mph.

Tables 16-4 and 16-5 afford representative distribution of passenger car speeds which may be used to show how the speed distribution affects total motor vehicle running cost and travel time. In working up Tables 16-4 and 16-5 it was noticed that individual runs varied somewhat, depending upon the highway design, traffic conditions, and legal speed. The operator of the test car tried to follow normal traffic practice without any excessive or continuous exceeding of the legal speed limit.

Speed on Horizontal Curves

The speed at which a driver will travel around a horizontal curve when not restrained by a vehicle ahead is dependent primarily upon two factors–his sense

Traffic Characteristics

TABLE 16-4
PERCENTAGE DISTRIBUTION OF MILES PER HOUR SPEED FOR EACH SECOND OF TRAVEL TIME OF A TEST CAR DRIVEN IN NORMAL TRAFFIC ON TEN DIFFERENT COMBINATIONS OF TYPES OF HIGHWAYS AND TRAFFIC CONDITIONS

Speed, mph	Type of Highway and Traffic Condition											
	A	B	C	D	E	F	G	H	I	J	AB	EG	FH	IJ
0	0.30	1.17	1.40	0.36	5.57	7.47	18.84	23.86	12.14	21.37	0.56	8.06	11.60	15.90
1	0.02	0.05	0.13	0.09	0.58	0.55	1.69	1.76	1.02	1.72	0.03	0.71	0.81	1.35
2	0.02	0.06	0.09	0.06	0.49	0.47	1.45	1.70	0.91	1.56	0.03	0.62	0.78	1.13
3	0.02	0.06	0.07	0.05	0.44	0.41	1.24	1.70	0.82	1.43	0.03	0.55	0.75	0.97
4	0.02	0.07	0.06	0.04	0.41	0.38	1.04	1.83	0.73	1.29	0.04	0.50	0.74	0.87
5	0.02	0.09	0.05	0.03	0.39	0.36	0.84	1.97	0.67	1.17	0.04	0.47	0.75	0.82
6	0.02	0.10	0.06	0.03	0.39	0.36	0.61	2.10	0.61	1.23	0.05	0.48	0.78	0.82
7	0.03	0.11	0.08	0.03	0.42	0.37	0.72	2.23	0.58	1.36	0.05	0.52	0.82	0.86
8	0.03	0.12	0.09	0.03	0.45	0.40	0.89	2.35	0.56	1.49	0.06	0.58	0.88	0.94
9	0.03	0.13	0.11	0.03	0.50	0.43	1.04	2.46	0.56	1.63	0.07	0.65	0.94	1.01
10	0.04	0.14	0.13	0.02	0.55	0.48	1.20	2.56	0.58	1.76	0.07	0.73	0.99	1.09
11	0.04	0.16	0.14	0.02	0.64	0.52	1.36	2.66	0.65	1.90	0.08	0.81	1.03	1.17
12	0.04	0.18	0.15	0.02	0.73	0.57	1.54	2.74	0.71	2.04	0.09	0.91	1.08	1.25
13	0.05	0.20	0.15	0.01	0.83	0.63	1.72	2.78	0.78	2.18	0.10	1.02	1.12	1.33
14	0.05	0.22	0.15	0.01	0.93	0.69	1.91	2.80	0.85	2.32	0.11	1.15	1.16	1.42
15	0.06	0.25	0.16	0.01	1.04	0.76	2.10	2.77	0.93	2.46	0.12	1.28	1.21	1.52
16	0.07	0.28	0.16	0.02	1.18	0.83	2.30	2.74	1.01	2.61	0.13	1.44	1.25	1.61
17	0.07	0.31	0.17	0.02	1.33	0.90	2.50	2.70	1.08	2.77	0.15	1.60	1.31	1.72
18	0.08	0.34	0.18	0.02	1.49	0.98	2.72	2.64	1.19	2.91	0.16	1.78	1.36	1.84
19	0.09	0.38	0.21	0.02	1.68	1.06	2.94	2.59	1.33	3.06	0.18	1.95	1.42	1.97
20	0.10	0.42	0.23	0.02	1.88	1.14	3.16	2.52	1.48	3.22	0.20	2.13	1.48	2.12
21	0.10	0.46	0.24	0.03	2.09	1.25	3.39	2.44	1.66	3.40	0.22	2.32	1.57	2.29
22	0.11	0.52	0.27	0.03	2.33	1.37	3.60	2.36	1.89	3.60	0.24	2.51	1.62	2.52
23	0.11	0.57	0.29	0.03	2.57	1.52	3.75	2.28	2.16	3.84	0.26	2.71	1.70	2.82
24	0.12	0.63	0.30	0.03	2.83	1.68	3.70	2.19	2.50	4.20	0.29	2.94	1.80	3.22
25	0.13	0.71	0.32	0.03	3.11	1.87	3.57	2.08	2.91	5.01	0.33	3.20	1.91	3.58
26	0.15	0.80	0.34	0.03	3.42	2.08	3.42	1.97	3.48	4.44	0.37	3.46	2.04	3.84
27	0.17	0.90	0.35	0.03	3.74	2.30	3.25	1.84	4.06	3.84	0.41	3.70	2.19	4.01
28	0.19	1.01	0.37	0.03	3.98	2.53	3.06	1.71	4.59	3.00	0.46	3.87	2.35	4.08
29	0.22	1.13	0.38	0.03	4.14	2.78	2.85	1.58	4.99	2.21	0.52	3.98	2.52	4.09
30	0.25	1.27	0.39	0.04	4.18	3.03	2.63	1.44	5.31	1.60	0.59	3.90	2.69	3.99
31	0.28	1.43	0.40	0.04	4.14	3.29	2.39	1.29	5.47	1.02	0.66	3.78	2.84	3.72
32	0.33	1.62	0.41	0.05	4.02	3.50	2.15	1.13	5.15	0.71	0.74	3.63	2.99	3.37
33	0.38	1.84	0.44	0.05	3.87	3.75	1.90	0.98	4.61	0.50	0.83	3.47	3.12	2.92
34	0.44	2.10	0.48	0.05	3.70	3.96	1.65	0.84	3.95	0.35	0.93	3.29	3.22	2.33
35	0.51	2.39	0.54	0.05	3.53	4.11	1.40	0.73	2.90	0.25	1.05	3.11	3.29	1.70
36	0.59	2.73	0.63	0.05	3.34	4.20	1.17	0.63	2.10	0.19	1.18	2.91	3.32	1.33
37	0.69	3.10	0.74	0.06	3.11	4.18	0.95	0.55	1.73	0.14	1.32	2.70	3.27	1.15
38	0.79	3.53	0.88	0.06	2.89	4.06	0.78	0.47	1.49	0.10	1.49	2.48	3.17	1.00
39	0.91	3.96	1.03	0.07	2.64	3.98	0.63	0.41	1.29	0.06	1.64	2.25	3.05	0.88
40	1.03	4.25	1.19	0.07	2.39	3.60	0.51	0.35	1.13	0.04	1.78	2.01	2.89	0.78
41	1.16	4.31	1.35	0.07	2.12	3.45	0.41	0.29	1.01	0.02	1.92	1.77	2.70	0.67
42	1.30	4.21	1.52	0.07	1.85	3.19	0.32	0.26	0.91	0.00	2.05	1.54	2.49	0.59
43	1.45	4.01	1.80	0.08	1.59	2.91	0.24	0.22	0.82		2.17	1.32	2.26	0.51
44	1.63	3.81	1.90	0.10	1.34	2.61	0.18	0.18	0.73		2.29	1.12	2.01	0.45

TABLE 16-4 (*continued*)

Speed, mph	\multicolumn{13}{c}{Type of Highway and Traffic Condition}												
	A	B	C	D	E	F	G	H	I	J	AB	EG	FH	IJ
45	1.79	3.63	2.10	0.12	1.11	2.29	0.13	0.14	0.65		2.39	0.94	1.74	0.40
46	1.99	3.46	2.31	0.15	0.92	1.95	0.08	0.10	0.58		2.49	0.77	1.45	0.35
47	2.19	3.31	2.52	0.17	0.74	1.58	0.05	0.06	0.52		2.58	0.61	1.12	0.31
48	2.38	3.14	2.75	0.21	0.57	1.20	0.03	0.02	0.46		2.67	0.48	0.86	0.27
49	2.54	3.00	2.98	0.25	0.46	0.78	0.00	0.00	0.40		2.75	0.36	0.61	0.24
50	2.71	2.85	3.22	0.30	0.34	0.51			0.35		2.83	0.26	0.41	0.21
51	2.90	2.70	3.46	0.36	0.23	0.34			0.30		2.91	0.19	0.26	0.18
52	3.08	2.55	3.71	0.42	0.18	0.22			0.25		2.98	0.14	0.15	0.16
53	3.28	2.40	3.92	0.60	0.14	0.10			0.20		3.04	0.10	0.08	0.13
54	3.47	2.24	4.06	0.82	0.12	0.05			0.15		3.09	0.07	0.03	0.11
55	3.67	2.09	4.10	1.16	0.09	0.02			0.11		3.15	0.05	0.01	0.09
56	3.87	1.94	4.01	1.51	0.07	0.00			0.00		3.21	0.04	0.00	0.00
57	4.09	1.78	3.91	1.86	0.06						3.30	0.03		
58	4.32	1.62	3.80	2.13	0.05						3.41	0.02		
59	4.56	1.42	3.68	2.35	0.04						3.57	0.02		
60	4.84	1.24	3.55	2.82	0.04						3.78	0.01		
61	5.16	1.02	3.43	3.35	0.00						4.08	0.00		
62	5.64	0.82	3.30	4.10							4.56			
63	6.05	0.65	3.14	5.68							4.44			
64	5.56	0.51	2.97	7.62							3.90			
65	4.75	0.39	2.77	8.98							3.23			
66	3.12	0.31	2.53	9.81							2.41			
67	1.69	0.23	2.16	10.20							1.49			
68	1.02	0.18	1.73	9.51							0.77			
69	0.60	0.12	1.24	7.67							0.41			
70	0.30	0.07	0.85	5.63							0.23			
71	0.15	0.03	0.50	3.95							0.14			
72	0.05	0.01	0.34	2.64							0.07			
73	0.00	0.00	0.23	1.68							0.03			
74			0.15	0.90							0.00			
75			0.05	0.50										
76			0.00	0.23										
77				0.13										
78				0.07										
79				0.00										
Total	100.00	100.00	100.00	100.00	100.00	100.00	100.00	100.00	100.00	100.00	100.00	100.00	100.00	100.00
Total Seconds	24,515	10,380	15,179	41,129	9,694	14,715	2,595	4,954	1,763	1,193	34,895	12,289	19,669	2,956

* The classes of highways and traffic conditions corresponding to the letter codes are as follows:

A. Two- or three-lane rural, free-moving traffic.
B. Two-lane rural, trailing another vehicle-unable to pass.
C. Four- or more-lane rural without access control.
D. Four- or more-lane rural, divided with access control.
E. Two- or three-lane urban, free-moving traffic.
F. Four- or more-lane urban, free-moving traffic.
G. Two- or three-lane urban, congested traffic conditions.
H. Four- or more-lane urban, congested traffic conditions.
I. Two-, three-, or four-lane urban, one-way free-moving traffic conditions.
J. Two-, three-, or four-lane urban, one-way, congested traffic conditions.
AB. Combined A and B.
EG. Combined E and G.
GH. Combined G and H.
IJ. Combined I and J.

Source: Original records of Paul J. Claffey [16-5].

Traffic Characteristics

TABLE 16-5
PERCENTAGE OF MILEAGE OF TEST CAR
DRIVEN IN NORMAL TRAFFIC AT SPEEDS GROUPED BY 5-MPH INTERVALS

Mph Group	Type of Highway and Traffic Condition*													
	A	B	C	D	E	F	G	H	I	J	AB	EG	FH	IJ
1-2	†	†	0.01	†	0.06	0.05	0.26	0.38	0.12	0.33	†	0.07	0.09	0.17
3-7	0.01	0.05	0.03	0.01	0.37	0.31	1.17	3.76	0.67	2.18	0.02	0.48	0.75	1.04
8-12	0.03	0.18	0.12	0.02	1.06	0.81	3.48	9.57	1.27	6.05	0.07	1.46	1.91	2.68
13-17	0.10	0.45	0.23	0.02	2.90	1.92	8.99	15.37	2.89	12.60	0.18	3.82	3.51	5.55
18-22	0.18	1.01	0.44	0.04	6.88	3.89	17.91	18.62	6.25	21.99	0.40	8.34	5.76	10.46
23-27	0.31	2.15	0.78	0.06	14.18	7.92	24.80	19.18	15.67	36.04	0.82	15.59	9.32	21.25
28-32	0.70	4.62	1.13	0.09	22.05	15.17	21.94	15.85	31.40	16.91	1.76	22.21	15.51	27.82
33-37	1.69	10.14	1.93	0.14	22.00	23.54	13.78	9.63	21.60	3.32	3.67	20.89	21.86	15.72
38-42	3.82	19.20	4.65	0.21	16.99	24.24	5.90	5.26	9.49	0.58	6.99	15.47	21.94	7.53
43-47	7.49	19.35	9.30	0.44	9.14	16.86	1.69	2.31	6.05		10.53	8.22	14.76	4.37
48-52	12.49	16.00	15.64	1.21	3.16	4.99	0.08	0.07	3.58		13.87	2.73	4.34	2.55
53-57	18.54	13.55	21.28	5.14	0.94	0.30			1.01		17.03	0.61	0.25	0.86
58-62	26.99	8.64	20.59	13.85	0.27						22.87	0.11		
63-67	24.98	3.19	17.02	42.96							19.56			
68-72	2.67	0.67	6.24	31.76							2.19			
73-77			0.61	3.96							0.04			
Total	100.00	100.00	100.00	100.00	100.00	100.00	100.00	100.00	100.00	100.00	100.00	100.00	100.00	100.00
‡	54.80	42.79	52.42	64.48	29.48	32.51	21.91	17.66	27.80	18.83	51.32	28.11	29.40	24.61
§	54.73	42.29	51.69	64.24	27.84	30.08	17.78	13.44	24.42	14.80	51.03	25.85	25.99	20.69

* The classes of highways and traffic conditions corresponding to the letter codes are as follows:

A. Two- or three-lane rural, free-moving traffic.
B. Two-lane rural, trailing another vehicle–unable to pass.
C. Four- or more-lane rural without access control.
D. Four- or more-lane rural, divided with access control.
E. Two- or three-lane urban, free-moving traffic.
F. Four- or more-lane urban, free-moving traffic.
G. Two- or three-lane urban, congested traffic conditions.
H. Four- or more-lane urban, congested traffic conditions.
I. Two-, three-, or four-lane urban, one-way free-moving traffic conditions.
J. Two-, three-, or four-lane urban, one-way congested traffic conditions.
AB. Combined A and B.
EG. Combined E and G.
GH. Combined G and H.
IJ. Combined I and J.

† Less than 0.01 percent.
‡ Running speed–average speed while car is in motion; total distance divided by running time.
§ Overall travel speed–total distance divided by total travel time, including all traffic delays.

Source: Table 16-4 and Ref. 16-5.

of safety as judged from the sight distance ahead and his comfort as judged by the centrifugal force he feels. In addition, his speed may be influenced by roadside markings of the safe speed. Assuming that sight distance is adequate, the safe speed or comfortable speed is governed by the radius of curvature of the path of the vehicle, the superelevation of the roadway surface, and the coefficient of friction between the tires and pavement surface. For a given speed the safe minimum radius of curve may be calculated from the principles of the mechanics of forces.

On a horizontal curve the centrifugal force tending to keep the vehicle in a straight (tangent) path is opposed by the side frictional force developed at the area of contact between the tire and the roadway surface. The following equation may be written:

$$\frac{Wv^2}{gR} = fW \qquad (16\text{-}1)$$

where W = gross weight of the vehicle in pounds
  v = velocity of the vehicle in feet per second
  g = 32.2, the acceleration of gravity in feet per second per second
  R = radius of the path of the vehicle in feet
  f = coefficient of friction utilized between the tire and the roadway surface

Converting Eq. 16-1 into speed in miles per hour V, and replacing g with 32.2 results in

$$f = \frac{V^2}{15R} \qquad (16\text{-}2)$$

On a superelevated curve (a horizontal curve with pavement sloping upward towards the outside) the frictional force may be replaced by a gravitational force. Using e as the superelevation in feet per foot of horizontal distance, Eq. 16-2 becomes,

$$e = \frac{V^2}{15R} \qquad (16\text{-}3)$$

For speeds above that necessary to balance the forces for a given value of f, superelevation may be used in the place of an increased coefficient of friction. Thus e and f play exactly identical roles; as one of the factors is increased the other may be decreased an equal amount. The final equation may now be written in terms of both e and f:

$$e + f = \frac{V^2}{15R} \quad \text{or} \quad V^2 = 15R(e + f) \qquad (16\text{-}4)$$

In design of horizontal curves |16-1|, e may be varied upward with decreasing radius of curve but probably not to exceed 0.10 to 0.12. For icy conditions e must be kept low to prevent the slow speed vehicles from slipping downward and inward on the superelevated slope. The coefficient of friction is also subject to variation, but a value of 0.15 is reasonable.

Horizontal curves are driven differently by different drivers. The lowest speed and the highest speed on the curve may each come at different distances from the beginning of the curve according to the driver's choice, knowledge of the particular curve, length of curve, sight distance, approach speed, and other factors. Curves of lower radius than 35.8 ft (16 deg) are driven by many drivers at close to design speed, but curves of longer radius are more often driven at speeds less than the designed safe speed. The speed on the curve is herein

Traffic Characteristics

considered to be the minimum speed regardless of at what place on the curve the minimum speed may occur.

Table 16-6, based on Ref. 16-3, 16-14, 16-20 and 16-26, gives probable minimum speeds for passenger cars on horizontal curves for degrees of curve from 1 to 35. This table may be used as a guide to the amount of speed reduction caused by horizontal curves. The approach speed to use would be that speed at which the vehicles are traveling that highway on tangent sections.

SPEED AND TRAVEL TIME OF TRUCKS

The Factors of Truck Weight and Horsepower

In any analysis of the economy of a proposed route design involving plus grades and heavy trucks, the speed of the trucks is an important factor. Slow-moving trucks may restrict faster movement of lighter vehicles in addition to increasing their own travel time. The speed of trucks on plus grades is controlled primarily by their weight/power ratio, usually expressed in pounds of gross vehicle weight per net horsepower of the engine. For determining the most

TABLE 16-6
SPEED OF PASSENGER CARS ON HORIZONTAL CURVES

Degree of Curve	Radius, ft	Design Speed, mph $f = 0.15$ $e = 0.00$	Design Speed		Probable Minimum Road Speed, mph
			$f = 0.15$ $e =$ as below	mph	
1	5,730	113	0.000	115	62
2	2,865	80	0.046	92	59
3	1,910	65	0.066	77	56
4	1,432	57	0.079	70	54
5	1,146	51	0.086	64	51
6	955	46	0.091	59	49
7	819	43	0.095	54	47
8	716	40	0.097	52	45
9	637	38	0.099	49	44
10	573	36	0.100	46	42
12	477	33	0.100	42	39
14	409	30	0.100	39	37
16	358	28	0.100	37	35
18	318	27	0.100	35	33
20	286	25	0.100	33	32
25	229	23	0.100	29	28
30	191	21	0.100	25	24
35	164	19	0.100	23	20

Source: Refs. 16-3, 16-14, 16-20, and 16-26.

economical percentage of grade and for vertical alignment generally, some knowledge of the weight/power ratio of the trucks using the route or to use the route is essential.

TABLE 16-7

PERCENTAGE DISTRIBUTION OF GROSS VEHICLE
WEIGHT OF TRUCKS ON MAIN RURAL HIGHWAYS

Gross Weight Interval, 1,000 Pounds	2D 2-Axle Truck	3A 3-Axle Truck	2-S1 3-Axle Tractor Semitrailer	2-S2 4-Axle Tractor Semitrailer	3-S2 5-Axle Tractor	2-2 or Other 4-Axle Truck and Trailer Combination	3-2 or Other 5-Axle Truck and Trailer Combination	2-S1-2 Tractor Semitrailer and Full Trailer Combination	2-S2-2 or Other 6-Axle Tractor Semitrailer and Full Trailer Combination
Under 4	0.14	1.83							
4-10	37.70	8.99	0.35	0.03		18.52			
10-13.5	25.43	28.45	1.56	0.19	0.01	17.99			
13.5-20	23.72	4.77	20.95	9.14	0.97	26.46			
20-22	4.56	5.80	10.76	7.45	2.22	8.47	0.45		
22-24	3.96	5.63	9.12	8.43	4.52	5.82	3.62		
24-26	2.32	4.94	7.76	7.36	5.48	5.29	7.24		
26-28	1.30	2.79	6.64	5.11	5.62	3.17	3.17	3.48	1.52
28-30	0.45	3.05	6.26	3.87	4.57	4.23	4.53	3.48	0.00
30-32	0.23	4.19	5.54	3.26	3.62	2.12	2.71	3.48	1.52
32-34	0.11	3.44	5.54	2.99	3.40	1.59	3.62	2.61	6.82
34-36	0.04	4.44	6.32	3.01	2.43	1.06	4.98	4.35	1.52
36-38	0.04	4.41	5.95	3.12	2.03	0.00	1.81	1.74	2.27
38-40		9.71	5.02	3.37	1.91	0.00	1.81	1.74	2.27
40-45		5.84	6.35	8.83	4.59	0.53	2.26	6.08	4.54
45-50		1.07	1.39	10.02	5.16	2.11	1.36	3.48	4.55
50-55		0.50	0.26	9.73	7.34	0.00	0.00	5.22	9.09
55-60		0.07	0.17	6.93	10.32	2.11	0.00	7.82	7.58
60-65		0.04	0.06	3.68	12.03	0.53	2.26	8.69	12.88
65-70		0.04			1.98	10.21		10.86	16.52	20.44
70-75					1.04	7.81		19.91	18.26	20.45
75-80					0.32	3.12		25.34	5.22	3.79
80-85					0.12	1.16		1.36	2.61	0.76
85-90					0.02	0.59		0.90	2.61	
90-95						0.45		1.36	0.87	
95-100						0.30		0.45	1.74	
100-105						0.11				
105-110						0.03				
Total	100.00	100.00	100.00	100.00	100.00	100.00	100.00	100.00	100.00
Average Weight	12,672	26,580	27,170	37,284	49,488	18,704	58,104	59,043	58,811
Number of Trucks Weighed*	14,539	2,791	3,465	10,362	15,324	189	221	115	132†

* Number weighed does not indicate relative number of classs of vehicle in the traffic count because of unequal sampling.

† Includes 98 combinations weighed in 1966 in three western states, and 34 weighed in 1965.

Source: Composite of reports of 15 states of roadside weighing in 1965; Bureau of Public Roads.

Traffic Characteristics

Gross Weight of Trucks. Each year state highway departments weigh trucks at roadway stations. These weight data are reported by class of truck, axle weight, gross weight, empty weight, weight with payload, and by highway location. The reports find much use in research, forecasting, planning, roadway design, and structural design.

Table 16-7 summarizes the results of truck weighings in 1965 for selected states. The data are available in separate reports state by state.

Weight/Horsepower Ratios. The weight/horsepower ratios of the trucks in the traffic can be determined only by weighing the trucks on the highway and, at the same time, getting the net horsepower of the engine. If the net horsepower is not obtainable directly from name plate specifications, or from other available specifications, the horsepower may be obtained from the manufacturer's data books. Table 16-8 |16-33| gives weight/horsepower ratios for eight classes of trucks using typical primary highways in 1963. The variations in the ratio for a specific truck from the minimum pounds to the maximum pounds result from two factors. The net engine horsepower varies with the engine design and the gross weight of the vehicle varies with the vehicle design and load carried at the time of weighing. Empty weight of trucks of the same class varies and the payload weight will vary from zero to an overload, based on both the vehicle design and the legal gross vehicle weight or legal gross axle weight.

For most economy studies average weight/horsepower ratios for each vehicle class may be used, but from Table 16-8 the distribution of the ratio may be obtained on a percentage basis.

Speed of Trucks on Minus Grades

The downhill speed of trucks is governed by four primary factors–driver's attitude, traffic conditions, sight distance, and length of downgrade. Cautious drivers will hold down on the speed in the interest of safety, allowing for the

TABLE 16-8
Ratio of Gross Vehicle Weight in Pounds to the Engine Net Horsepower, 1963.

Truck Class	Weight/Horsepower Ratio		Weight/Horsepower Ratios by Percentiles of Vehicles Weighed						Horsepower	
	Range	Average	10th	30th	50th	70th	80th	90th	Range	Average
2	24-128	44	32	40	42	48	52	60	50-165	109
2D	42-267	97	62	74	87	100	128	160	80-198	136
3A	71-282	145	84	107	135	150	193	226	95-522	157
2-S1	84-304	149	104	122	133	169	187	219	118-230	165
2-S2	89-427	227	112	159	218	243	305	349	110-238	172
3-S2	94-701	275	140	200	272	330	361	407	128-310	184
3-2	93-511	261	–	–	–	–	–	–	128-250	184
2-S1-2	111-590	321	130	222	360	397	417	460	130-235	186

Source: Ref. 16-33, pp. 84-88.

Traffic Characteristics

TABLE 16-9
Speed (mph) of Trucks on Uniform Minus Grades
Gross Vehicle Weight/Net Horsepower Ratio: 200 Pounds

Distance, 100 ft	Uniform Minus and Plus Grade, Percent											
	−8	−7	−6	−5	−4	−3	−2	−1	0	1	2	3
0	10.0	10.0	10.0	10.0	10.0	10.0	10.0	10.0	10.0	10.0	10.0	10.0
1	19.5	18.7	18.0	17.3	16.7	15.9	15.0	14.1	13.2	12.3	11.5	10.4
2	26.0	25.0	23.9	22.8	21.7	20.5	19.2	18.0	16.5	14.9	13.2	11.1
3	31.0	29.7	28.4	27.1	25.5	24.3	22.8	21.2	19.4	17.3	15.1	12.0
4	35.4	33.8	32.4	30.9	29.2	27.7	25.9	24.0	21.9	19.5	17.1	12.9
5	39.4	37.8	36.2	34.5	32.6	30.8	27.8	26.7	24.5	21.7	18.9	14.3
6	43.2	41.5	39.7	37.9	36.0	33.7	31.5	29.2	26.9	23.8	20.7	15.8
7	46.7	44.6	42.8	40.8	38.6	36.1	33.9	31.3	28.7	25.6	22.3	17.1
8	49.6	47.3	45.5	43.2	41.1	38.4	35.9	33.2	30.4	27.2	23.7	18.9
9	51.9	49.9	47.8	45.3	43.1	40.4	37.7	34.9	32.0	28.6	25.1	19.7
10	54.0	52.0	49.6	47.2	44.8	42.1	39.3	36.4	33.4	30.0	26.2	20.8
11		54.1	52.0	50.2	47.5	44.0	41.1	38.1	35.0	31.4	27.5	22.0
12			54.2	52.2	49.5	45.7	42.7	39.5	36.3	32.6	28.6	23.1
13				54.1	51.3	47.3	44.1	40.9	37.5	33.8	29.8	24.1
14					53.0	48.9	45.6	42.1	38.7	34.8	30.8	25.0
15						50.4	47.0	43.3	39.8	35.8	31.7	25.9
16						51.8	48.4	44.5	40.9	36.8	32.6	26.8
17							49.7	45.6	41.9	37.7	33.5	27.6
18							51.0	46.7	42.8	38.6	34.3	28.3
19							52.3	47.7	43.8	39.4	35.0	29.0
20								48.8	44.8	40.2	35.7	29.6
21								49.8	45.7	41.0	36.4	30.4
22								50.7	46.5	41.7	37.1	31.0
23								51.7	47.4	42.4	37.7	31.6
24									48.2	43.1	38.2	32.1
25									49.0	43.9	38.9	32.4
26									49.8	44.4	39.3	33.1
27									50.6	45.1	39.8	33.6
28									51.4	45.7	40.3	34.0
29									52.8	46.3	40.8	34.4
30										46.8	41.2	34.8
31										47.4	41.7	35.2
32										48.0	42.1	35.6
33										48.5	42.5	36.0
34										49.1	42.9	36.3
35										49.6	43.3	36.6
36										50.2	43.7	37.0
37										50.6	44.1	37.3
38										51.0	44.4	37.6
39											44.8	37.9
40											45.2	38.2
41											45.5	38.5
42											45.8	38.8
43											46.2	39.0
44											46.5	39.3
45											46.8	39.6
46											47.2	39.8
47											47.5	40.0
48											47.8	40.3
49											48.1	40.6
50											48.4	40.8
51											48.7	41.1
52											49.1	41.3
53											49.4	41.6
54											49.7	41.8
55											50.0	42.0

Source: Firey and Peterson [16-8, pp. 9-10].

Traffic Characteristics

effect of gravity in lengthening the stopping distance as compared to a level grade and preventing the vehicle from reaching a speed that may burn out the brakes under an emergency stop. Traffic, of course, may prevent as high a

TABLE 6-10
ELAPSED TIME (SECONDS) OF TRUCKS ON UNIFORM MINUS GRADES
Gross Vehicle Weight/Horsepower Ratio: 200 Pounds
Entry Speed: 10 mph

Distance, 100 ft	Uniform Minus and Plus Grade, Percent											
	−8	−7	−6	−5	−4	−3	−2	−1	0	1	2	3
0	0.00	0.00	0.00	0.00	0.00	0.00	0.00	0.00	0.00	0.00	0.00	0.00
1	4.15	4.28	4.42	4.57	4.74	4.92	5.12	5.34	5.58	5.84	6.13	6.44
2	7.15	7.40	7.67	7.97	8.29	8.67	9.11	9.59	10.17	10.85	11.65	12.78
3	9.54	9.89	10.28	10.70	11.18	11.71	12.36	13.07	13.97	15.08	16.47	18.68
4	11.59	12.04	12.52	13.05	13.67	14.33	15.16	16.09	17.27	18.79	20.72	23.26
5	13.41	13.94	14.51	15.14	15.88	16.66	17.70	18.78	20.21	22.10	24.51	28.27
6	15.06	15.70	16.31	17.02	17.87	18.77	20.00	21.22	22.86	25.10	27.95	32.80
7	16.58	17.28	17.96	18.75	19.70	20.72	22.09	23.47	25.31	27.86	31.12	36.94
8	18.00	18.76	19.50	20.37	21.41	22.55	24.04	25.58	27.62	30.44	34.09	40.73
9	19.34	20.16	20.96	21.91	23.03	24.28	25.89	27.58	29.81	32.88	36.88	44.26
10	20.63	21.50	22.36	23.38	24.58	25.93	27.66	29.49	31.90	35.21	39.54	47.63
11		22.79	23.70	24.78	26.06	27.51	29.36	31.32	33.89	37.43	42.08	50.82
12			24.98	26.12	27.47	29.03	30.99	33.08	35.80	39.56	44.51	53.84
13				27.42	28.82	30.49	32.56	34.78	37.65	41.61	46.84	56.73
14					30.13	31.90	34.08	36.42	39.44	43.60	49.09	59.51
15						33.27	35.55	38.02	41.18	45.53	51.27	62.19
16						34.60	36.98	39.57	42.87	47.41	53.39	64.78
17							38.37	41.08	44.52	49.24	55.45	67.29
18							39.72	42.56	46.12	51.03	57.46	69.73
19							41.04	44.00	47.69	52.78	59.43	72.11
20								45.41	49.23	54.49	61.35	74.44
21								46.79	50.74	56.17	63.24	76.71
22								48.15	52.22	57.82	65.10	78.93
23								49.48	53.67	59.44	66.92	81.11
24									55.10	61.03	68.72	83.25
25									56.50	62.60	70.49	85.36
26									57.88	64.15	72.23	87.44
27									59.24	65.67	73.95	89.48
28									60.58	67.17	75.65	91.50
29									61.90	68.65	77.33	93.49
30										70...	78.99	95.46
31										71.56	80.63	97.41
32										72.99	82.25	99.44
33										74.40	83.86	101.24
34										75.80	85.46	103.13
35										77.18	87.04	105.00
36										78.55	88.61	106.85
37										79.90	90.16	108.69
38										81.24	91.70	110.51
39										82.57	93.23	112.32
40											94.75	114.11
41											96.25	115.89
42											97.74	117.65
43											99.22	119.40
44											100.69	121.14
45											102.15	122.87
46											103.60	124.59
47											105.04	126.30
48											106.47	128.00
49											107.89	129.69
50											109.30	131.37
51											110.70	133.04
52											112.09	134.69
53											113.47	136.33
54											114.85	137.97

Source: Calculated from Table 16-9.

speed as the driver may desire. Sight distance, generally on horizontal curves, is a prime factor in holding down speed on downhill travel. Two dangers may exist: overturning on the horizontal curve, and hitting another vehicle not

TABLE 16-11
SPEED (MPH) OF TRUCKS ON UNIFORM MINUS GRADES
Gross Vehicle Weight/Net Horsepower Ratio: 300 Pounds
Entry Speed: 10 mph

Distance, 100 ft	Uniform Minus and Plus Grade, Percent										
	−8	−7	−6	−5	−4	−3	−2	−1	0	1	2	3
0	10.0	10.0	10.0	10.0	10.0	10.0	10.0	10.0	10.0	10.0	10.0	
1	19.9	19.2	18.5	17.8	17.1	16.3	15.5	14.6	13.7	12.6	11.6	
2	25.8	24.5	23.4	22.4	21.3	20.1	19.0	17.6	16.3	14.8	13.1	
3	30.3	29.0	27.6	26.3	25.0	23.4	21.9	20.3	18.6	16.8	14.5	
4	34.3	32.9	31.2	29.7	28.1	26.3	24.6	22.6	20.6	18.6	15.8	
5	37.8	36.2	34.4	32.7	30.9	28.8	26.9	24.7	22.4	20.1	17.0	
6	41.1	39.4	37.3	35.4	33.4	31.2	29.0	26.6	24.1	21.5	18.3	
7	44.4	42.2	40.0	37.9	35.6	33.3	30.9	28.3	25.6	22.8	19.5	
8	47.4	45.0	42.6	40.2	37.7	35.3	32.7	29.9	27.0	24.1	20.7	
9	50.2	47.6	45.0	42.4	39.7	37.2	34.3	31.4	28.2	25.2	21.8	
10	53.3	50.2	47.3	44.5	41.6	38.9	35.9	32.8	29.4	26.3	22.8	
11		52.8	49.6	46.4	43.5	40.6	37.3	34.1	30.6	27.3	23.7	
12			51.8	48.3	45.2	42.2	38.7	35.3	31.6	28.2	24.6	
13				50.1	46.8	43.8	40.0	36.5	32.6	29.1	25.4	
14				51.8	48.7	45.3	41.3	37.6	33.5	30.0	26.1	
15					50.4	46.8	42.5	38.8	34.5	30.8	26.7	
16					52.1	48.4	43.8	39.8	35.4	31.5	27.3	
17						49.9	45.0	40.9	36.2	32.3	27.8	
18						51.3	46.2	41.9	36.9	32.9	28.4	
19							47.3	42.9	37.7	33.6	28.8	
20							48.4	43.8	38.4	34.2	29.3	
21							49.5	44.8	39.2	34.8	29.7	
22							50.6	45.8	39.9	35.3	30.1	
23							51.7	46.8	40.5	35.8	30.5	
24								47.7	41.2	36.3	30.9	
25								48.7	41.8	36.8	31.2	
26								49.6	42.5	37.3	31.6	
27								50.6	43.1	37.8	31.9	
28								51.5	43.7	38.2	32.2	
29									44.3	38.6	32.6	
30									44.9	39.0	32.9	
31									44.5	39.4	33.2	
32									46.1	39.8	33.5	
33									46.6	40.2	33.8	
34									47.2	40.5	34.0	
35									47.8	40.9	34.3	
36									48.4	41.2	34.6	
37									48.9	41.6	34.8	
38									49.5	41.9	35.1	
39									50.0	42.2	35.4	
40									50.6	42.5	35.7	
41									51.2	42.8	35.9	
42									51.7	43.2	36.2	
43									52.2	43.5	36.4	
44										43.8	36.7	
45										44.1	36.9	
46										44.4	37.2	
47										44.7	37.4	
48										45.0	37.7	
49										45.3	37.9	
50										45.6	38.2	
51										45.9	38.4	
52										46.2	38.7	
53										46.5	38.9	
54										46.7	39.0	
55										47.0	39.0	

Source: Firey and Peterson [16-8, pp. 9-10].

Traffic Characteristics

TABLE 16-12
Elapsed Time (Seconds) of Trucks on Uniform Minus Grades
Gross Vehicle Weight/Horsepower Ratio: 300 Pounds
Entry Speed: 10 mph

Distance, 100 ft	Uniform Minus and Plus Grade, Percent											
	−8	−7	−6	−5	−4	−3	−2	−1	0	1	2	3
0	0.00	0.00	0.00	0.00	0.00	0.00	0.00	0.00	0.00	0.00	0.00	
1	4.00	4.09	4.10	4.32	4.46	4.62	4.81	5.03	5.30	5.64	6.05	
2	6.98	7.21	7.35	7.71	8.01	8.37	8.76	9.26	9.85	10.62	11.57	
3	9.41	9.76	10.02	10.51	10.96	11.50	12.09	12.86	13.76	14.94	16.51	
4	11.52	11.96	12.34	12.95	13.53	14.24	15.02	16.04	17.24	18.79	21.01	
5	13.41	13.93	14.42	15.14	15.84	16.71	17.67	18.92	20.41	22.31	25.17	
6	15.14	15.73	16.32	17.14	17.96	18.98	20.11	21.58	23.34	25.59	29.03	
7	16.73	17.40	18.08	19.00	19.94	21.09	22.39	24.06	26.08	28.67	32.64	
8	18.22	18.96	19.73	20.75	21.80	23.08	24.53	26.40	28.67	31.58	36.03	
9	19.62	20.43	21.29	22.40	23.56	24.96	26.57	28.61	31.14	34.35	39.24	
10	20.94	21.82	22.77	23.97	25.24	26.75	28.51	30.74	33.51	37.00	42.30	
11	22.19	23.14	24.18	25.47	26.84	28.47	30.37	32.78	35.78	39.55	45.23	
12		24.40	25.52	26.91	28.38	30.12	32.16	34.74	37.97	42.02	48.05	
13			26.80	28.30	29.86	31.71	33.89	36.64	40.09	44.40	50.78	
14				29.64	31.29	33.24	35.57	38.48	42.15	46.71	53.43	
15				30.94	32.67	34.72	37.20	40.26	44.16	48.95	56.02	
16					34.00	36.15	38.78	41.99	46.11	51.14	58.55	
17					35.29	37.54	40.32	43.68	48.01	53.28	61.02	
18						38.89	41.82	45.33	49.88	55.37	63.44	
19						40.19	43.28	46.94	51.71	57.42	65.82	
20							44.70	48.51	53.50	59.43	68.17	
21							46.09	50.05	55.26	61.41	70.48	
22							47.45	51.56	56.98	63.36	72.76	
23							48.78	53.03	58.68	65.28	75.01	
24								54.47	60.35	67.17	77.23	
25								55.88	61.99	69.04	79.43	
26								57.27	63.61	70.88	81.60	
27								58.63	65.20	72.70	83.75	
28								59.97	66.77	74.50	85.88	
29									68.32	76.28	87.98	
30									69.85	78.04	90.06	
31									71.36	79.78	92.12	
32									72.85	81.50	94.16	
33									74.32	83.20	96.18	
34									75.77	84.89	98.19	
35									77.21	86.57	100.19	
36									78.63	88.23	102.17	
37									80.03	89.88	104.13	
38									81.42	91.51	106.08	
39									82.79	93.13	108.01	
40									84.14	94.74	109.93	
41									85.48	96.34	111.83	
42									86.81	97.92	113.72	
43									88.12	99.49	115.60	
44										101.05	117.46	
45										102.60	119.31	
46										104.14	121.15	
47										105.67	122.98	
48										107.19	124.80	
49										108.70	126.60	
50										110.20	128.39	
51										111.69	130.17	
52										113.17	131.94	
53										114.64	133.70	
54										116.10	135.45	

Source: Calculated from Table 16-11

visible within a safe stopping or slowing down distance. Observed downhill speeds of heavy trucks on long tangents in Arizona [16-31] were about as high as the speeds on the level tangents. On the contrary, the downhill speeds observed in Pennsylvania [16-21] on Route U.S. 30 in 1950 were much less than the speeds on level tangents. The difference in downhill speeds is attributed to the greater number of horizontal curves in Pennsylvania than in Arizona.

Using the Firey and Peterson formulas [16-8] downhill speeds were calculated for entry speeds of 10 mph and published in Ref. 16-8. The calculated speeds and

TABLE 16-13
SPEED (MPH) OF TRUCKS ON UNIFORM MINUS GRADES
Gross Vehicle Weight/Net Horsepower Ratio: 400 pounds
Entry Speed: 10 mph

Distance, 100 ft	Uniform Minus and Plus Grade, Percent										
	−8	−7	−6	−5	−4	−3	−2	−1	0	1	2	3
0	10.0	10.0	10.0	10.0	10.0	10.0	10.0	10.0	10.0	10.0	10.0	
1	19.2	18.5	17.7	16.9	16.3	15.5	14.6	13.7	12.9	12.2	11.2	
2	25.4	24.2	23.0	21.7	20.4	19.2	17.9	16.6	15.4	14.0	12.6	
3	29.9	28.5	27.1	25.3	23.8	22.2	20.6	19.1	17.6	15.7	13.9	
4	33.9	32.2	30.5	28.5	26.9	24.9	23.0	21.2	19.3	17.2	15.0	
5	37.5	35.6	33.6	31.5	29.6	27.4	25.2	23.2	21.0	18.5	16.2	
6	40.9	38.8	36.5	34.3	32.1	29.8	27.4	24.9	22.5	19.8	17.2	
7	44.1	41.7	39.3	37.0	34.4	31.9	29.4	26.6	24.0	21.1	18.1	
8	46.8	44.3	41.7	39.2	35.5	33.9	31.1	28.3	25.4	22.3	18.9	
9	49.1	46.6	43.9	41.3	38.4	35.6	32.7	29.9	26.7	23.4	19.6	
10	51.2	48.6	46.0	43.2	40.2	37.2	34.2	31.2	27.7	24.3	20.2	
11	53.0	50.5	47.9	45.1	42.0	38.8	35.6	32.3	28.8	25.1	20.8	
12		52.1	49.8	46.9	43.6	40.4	36.9	33.4	29.7	25.8	21.4	
13			51.4	48.7	45.2	41.7	38.0	34.3	30.5	26.5	22.0	
14				50.4	46.8	42.9	39.2	35.2	31.3	27.1	22.5	
15				51.8	48.4	44.1	40.2	36.0	32.0	27.7	23.0	
16					49.8	45.4	41.2	36.9	32.7	28.3	23.4	
17					51.1	46.7	42.2	37.8	33.4	28.8	23.9	
18						47.9	43.2	38.6	34.0	29.3	24.2	
19						49.0	44.1	39.4	34.6	29.8	24.6	
20						50.2	45.0	40.1	35.2	30.2	25.0	
21						51.3	45.9	40.9	35.8	30.6	25.3	
22							46.8	41.6	36.3	31.0	25.6	
23							47.6	42.4	36.9	31.4	25.8	
24							48.5	43.0	37.4	31.8	26.1	
25							49.3	43.7	37.9	32.1	26.4	
26							50.1	44.4	38.4	32.4	26.6	
27							50.9	45.1	38.8	32.8	26.8	
28							51.7	45.7	39.3	33.1	27.0	
29								46.4	39.8	33.4	27.2	
30								47.0	40.2	33.8	27.4	
31								47.7	40.6	34.0	27.6	
32								48.3	41.0	34.3	27.8	
33								49.0	41.5	34.6	27.9	
34								49.6	41.9	34.9	28.1	
35								50.2	42.3	35.2	28.2	
36								50.9	42.7	35.5	28.4	
37								51.5	43.1	35.8	28.5	
38								52.1	43.5	36.0	28.7	
39									43.9	36.2	28.8	
40									44.2	36.5	28.9	
41									44.6	36.8	29.0	
42									45.0	37.0	29.1	
43									45.4	37.3	29.2	
44									45.8	37.5	29.4	
45									46.2	37.8	29.5	
46									46.5	38.0	29.6	
47									46.9	38.2	29.7	
48									47.3	38.5	29.8	
49									47.6	38.7	29.8	
50									48.0	38.9	29.8	
51									48.4	39.2		
52									48.7	39.4		
53									49.0	39.6		
54										39.8		
55										40.1		

Source: Firey and Peterson [16-8, pp. 9-10].

Traffic Characteristics

distances given in Tables 16-9, 16-11, and 16-13 were read and smoothed from the original publication. Maximum speeds, of course, cannot be given, but they may be assumed to be about 50 mph or slower when legal or other restrictions control.

TABLE 16-14
ELAPSED TIME (SECONDS) OF TRUCKS ON UNIFORM MINUS GRADES
Gross Vehicles Weight/Horsepower Ratio: 400 Pounds
Entry Speed: 10 mph

Distance, 100 ft	Uniform Minus and Plus Grade, Percent											
	−8	−7	−6	−5	−4	−3	−2	−1	0	1	2	3
0	0.00	0.00	0.00	0.00	0.00	0.00	0.00	0.00	0.00	0.00	0.00	
1	4.06	4.22	4.39	4.57	4.76	4.97	5.20	5.45	5.72	6.01	6.33	
2	7.12	7.41	7.74	8.10	8.48	8.90	9.33	9.95	10.54	11.21	12.06	
3	9.59	10.00	10.46	11.00	11.57	12.19	12.83	13.77	14.67	15.80	17.21	
4	11.73	12.25	12.83	13.53	14.26	15.09	15.96	17.15	18.37	19.94	21.93	
5	13.64	14.32	14.96	15.80	16.67	17.70	18.79	20.22	21.75	23.76	26.30	
6	15.38	16.15	16.91	17.87	18.88	20.08	21.38	23.06	24.88	27.32	30.38	
7	16.98	17.84	18.71	19.78	20.93	22.29	23.78	25.71	27.81	30.65	34.24	
8	18.48	19.43	20.39	21.57	22.88	24.36	26.03	28.19	30.57	33.79	37.93	
9	19.90	20.93	21.98	23.26	24.73	26.32	28.17	30.53	33.19	36.77	41.47	
10	21.26	22.36	23.50	24.87	26.46	28.19	30.21	32.80	35.70	39.63	44.90	
11	22.57	23.74	24.95	26.41	28.12	29.98	32.16	34.95	38.11	42.39	48.23	
12		25.07	26.35	27.89	29.71	31.70	34.04	37.03	40.44	45.07	51.46	
13			27.70	29.32	31.25	33.36	35.86	39.04	42.71	47.68	54.60	
14				30.70	32.73	34.97	37.63	41.00	44.92	50.23	57.66	
15				32.03	34.16	36.54	39.35	42.92	47.07	52.72	60.66	
16					35.55	38.06	41.02	44.79	49.18	55.16	63.60	
17					36.90	39.54	42.66	46.62	51.24	57.55	66.48	
18						40.98	44.26	48.40	53.26	59.90	69.31	
19						42.39	45.82	50.15	55.26	62.21	72.10	
20						43.76	47.35	51.87	57.21	64.48	74.85	
21						45.10	48.85	53.55	59.13	66.72	77.56	
22							50.32	55.20	61.02	68.93	80.24	
23							51.76	56.82	62.88	71.12	82.89	
24							53.18	58.42	64.72	73.28	85.52	
25							54.57	59.99	66.53	75.41	88.12	
26							55.94	61.54	68.32	77.52	90.69	
27							57.29	63.06	70.09	79.61	93.24	
28							58.62	64.56	71.84	81.68	95.77	
29								66.04	73.56	83.73	98.29	
30								67.50	75.26	85.76	100.79	
31								68.94	76.95	87.77	103.27	
32								70.36	78.62	89.77	105.73	
33								71.76	80.27	91.75	108.18	
34								73.14	81.91	93.71	110.62	
35								74.51	83.53	95.66	113.04	
36								75.86	85.13	97.59	115.45	
37								77.19	86.72	99.50	117.85	
38								78.51	88.29	101.40	120.23	
39									89.85	103.29	122.60	
40									91.40	105.17	124.96	
41									92.94	107.03	127.32	
42									94.46	108.88	129.67	
43									95.97	110.72	132.01	
44									97.47	112.54	134.34	
45									98.95	114.35	136.66	
46									100.42	116.15	138.97	
47									101.88	117.94	141.27	
48									103.33	119.72	143.56	
49									104.77	121.49	145.85	
50									106.19	123.25	148.14	
51									107.60	125.00	150.43	
52									109.00	126.74	152.72	
53									110.40	128.47	155.01	
54										130.19	157.30	

Source: Calculated from Table 16-13.

Speed of Trucks on Plus Grades

The speed of trucks on plus grades is governed primarily by the weight/power ratio, the entry speed, and the length of the grade. Drivers generally attempt to go uphill as fast as the truck will go. Uphill speeds have been observed |16-7, 16-9-2, 16-21, 16-29, 16-32|, primarily in connection with studies of traffic volume, effects of trucks on slower moving vehicles, and uphill truck lanes. Firey and Peterson |16-8| developed the mathematics of truck movement on

TABLE 16-15
SPEED (MPH) AND ELAPSED TIME (SECONDS) OF TRUCKS ON UNIFORM PLUS GRADES
Gross Vehicle Weight/Net Horsepower Ratio: 200 Pounds
Entry Speed: 50 mph

Distance, 100 ft	Speed, mph							Elapsed Time, Seconds						
	Plus Grade, Percent							Plus Grade, Percent						
	0, 1, & 2	3	4	5	6	7	8	0, 1, & 2	3	4	5	6	7	8
0	50	50.00	50.00	50.00	50.00	50.00	50.00	0.00	0.00	0.00	0.00	0.00	0.00	0.00
1	50	49.85	49.47	49.08	48.77	48.38	48.17	1.36	1.37	1.37	1.38	1.38	1.39	1.39
2	50	49.70	48.93	48.16	47.52	46.74	46.19	2.73	2.74	2.76	2.78	2.80	2.82	2.83
3	50	49.55	48.40	47.24	46.27	45.08	44.11	4.09	4.12	4.16	4.21	4.25	4.31	4.34
4	50	49.40	47.87	46.32	45.00	43.45	41.95	5.45	5.50	5.58	5.67	5.74	5.85	5.92
5	50	49.25	47.33	45.40	43.73	41.76	39.72	6.82	6.88	7.01	7.16	7.28	7.45	7.59
6	50	49.10	46.80	44.48	42.44	40.04	37.47	8.18	8.27	8.46	8.68	8.86	9.12	9.36
7	50	48.94	46.27	43.56	41.16	38.32	35.12	9.55	9.66	9.93	10.23	10.49	10.86	11.24
8	50	48.79	45.73	42.64	39.86	36.58	32.77	10.91	11.06	11.41	11.81	12.17	12.68	13.25
9	50	48.64	45.20	41.72	38.56	34.82	30.39	12.27	12.46	12.91	13.43	13.91	14.59	15.41
10	50	48.49	44.67	40.80	37.26	33.10	27.98	13.64	13.86	14.43	15.08	15.71	16.60	17.75
11	50	48.34	44.13	39.88	35.96	31.31	25.56	15.00	15.27	15.97	16.77	17.57	18.72	20.30
12	50	48.19	43.60	38.96	34.65	29.52	23.12	16.36	16.68	17.52	18.50	19.50	20.96	23.10
13	50	48.04	43.07	38.04	33.34	27.70	21.02	17.73	18.10	19.09	20.27	21.51	23.34	26.19
14	50	47.89	42.53	37.12	32.02	25.88	20.90	19.09	19.52	20.68	22.08	23.60	25.89	29.44
15	50	47.74	42.00	36.20	30.71	24.04	20.90	20.45	20.95	22.29	23.94	25.77	28.63	32.70
16	50	47.59	41.47	35.28	29.39	23.30	20.90	21.82	22.38	23.92	25.85	27.98	31.52	35.96
17	50	47.44	40.93	34.36	28.07	23.30	20.90	23.18	23.81	25.57	27.81	30.35	34.45	39.23
18	50	47.29	40.40	33.44	26.75	23.30		24.55	25.25	27.25	29.82	32.84	37.37	42.49
19	50	47.14	39.87	32.52	26.40	23.30		25.91	26.69	28.95	31.89	35.41	40.30	
20	50	46.99	39.33	31.60	26.40			27.27	28.14	30.67	34.02	37.99	43.23	
21	50	46.83	38.80	30.68	26.40			28.64	29.59	32.42	36.21	40.58		
22	50	46.68	38.27	30.50	26.40			30.00	31.05	34.19	38.44	43.16		
23	50	46.53	37.73	30.50				31.36	32.51	35.98	40.68	45.74		
24	50	46.38	37.20	30.50				32.73	33.98	37.80	42.91			
25	50	46.23	36.67	30.50				34.09	35.45	39.65	45.15			
26	50	46.08	36.50					35.45	36.93	41.51	47.38			
27	50	45.93	36.50					36.82	38.41	43.38				
28	50	45.78	36.50					38.18	39.90	45.25				
29	50	45.63	36.50					39.55	41.39	47.11				
30	50	45.48						40.91	42.89	48.98				
31	50	45.33						42.27	44.39	50.85				
32	50	45.18						43.64	45.90					
33	50	45.03						45.00	47.41					
34	50	44.88						46.36	48.93					
35	50	44.80						47.73	50.45					
36	50	44.80						49.09	51.97					
37	50	44.80						50.45	53.49					
38	50	44.80						51.81	55.02					

Source: Firey and Peterson |16-8, pp. 8-9|.

Traffic Characteristics

TABLE 16-16
SPEED (MPH) AND ELAPSED TIME (SECONDS) OF TRUCKS ON UNIFORM PLUS GRADES
Gross Vehicle Weight/Net Horsepower Ratio: 300 Pounds
Entry Speed: 50 mph

Distance, 100 ft	Speed, mph								Elapsed Time, Seconds							
	Plus Grade, Percent								Plus Grade, Percent							
	0 & 1	2	3	4	5	6	7	8	1	2	3	4	5	6	7	8
0	50	50.0	50.0	50.0	50.0	50.0	50.0	50.0	0.00	0.00	0.00	0.00	0.00	0.00	0.00	0.00
1	50	49.8	49.4	49.0	48.7	48.4	48.1	47.8	1.36	1.37	1.37	1.38	1.38	1.39	1.39	1.39
2	50	49.6	48.8	48.0	47.4	46.8	46.1	45.3	2.73	2.74	2.76	2.79	2.80	2.82	2.84	2.85
3	50	49.4	48.2	47.1	46.1	45.1	44.0	42.6	4.09	4.12	4.17	4.22	4.26	4.30	4.35	4.40
4	50	49.1	47.6	46.1	44.8	43.3	41.8	39.8	5.45	5.50	5.59	5.68	5.76	5.84	5.94	6.05
5	50	48.9	47.0	45.2	43.4	41.5	39.4	36.9	6.82	6.89	7.03	7.17	7.31	7.45	7.62	7.83
6	50	48.7	46.4	44.2	42.0	39.6	37.0	33.9	8.18	8.29	8.49	8.70	8.91	9.13	9.40	9.76
7	50	48.5	45.8	43.2	40.6	37.7	34.5	30.8	9.55	9.69	9.97	10.26	10.56	10.89	11.31	11.87
8	50	48.2	45.2	42.3	39.2	35.8	31.9	27.6	10.91	11.10	11.47	11.85	12.27	12.75	13.36	14.20
9	50	48.0	44.6	41.3	37.7	33.9	29.3	24.3	12.27	12.52	12.99	13.48	14.04	14.71	15.59	16.83
10	50	47.8	44.0	40.4	36.3	31.9	26.6	20.9	13.64	13.94	14.53	15.15	15.88	16.78	18.03	19.85
11	50	47.5	43.4	39.4	34.8	29.9	23.9	17.5	15.00	15.37	16.09	16.86	17.80	18.99	20.73	23.40
12	50	47.3	42.9	38.4	33.4	27.9	21.2	14.2	16.36	16.81	17.67	18.61	19.80	21.35	23.75	27.70
13	50	47.1	42.3	37.5	31.9	25.9	18.5	13.8	17.73	18.25	19.27	20.41	21.89	23.88	27.18	32.57
14	50	46.9	41.7	36.5	30.4	23.9	15.8	13.8	19.09	19.70	20.89	22.25	24.08	26.63	31.14	37.51
15	50	46.6	41.1	35.5	29.0	21.8	15.3	13.8	20.45	21.16	22.54	24.14	26.38	29.61	35.52	42.45
16	50	46.4	40.5	34.6	27.5	19.7	15.3	13.8	21.82	22.63	24.21	26.09	28.79	32.90	39.98	47.39
17	50	46.2	39.9	33.6	26.0	17.7	15.3		23.18	24.10	25.91	28.09	31.34	36.55	44.43	52.33
18	50	45.9	39.3	32.7	24.5	17.4	15.3		24.55	25.58	27.64	30.15	34.04	40.44	48.89	
19	50	45.7	38.7	31.7	22.9	17.4			25.91	27.07	29.39	32.27	36.90	44.35		
20	50	45.5	38.1	30.7	21.4	17.4			27.27	28.57	31.17	34.46	40.00	48.27		
21	50	45.3	37.5	29.8	20.4	17.4			28.64	30.07	32.97	36.71	43.26	52.19		
22	50	45.1	36.9	28.8	20.2				30.00	31.58	34.80	39.04	46.62	56.11		
23	50	44.8	36.3	27.8	20.2				31.36	33.10	36.66	41.45	50.00			
24	50	44.6	35.7	26.9	20.2				32.73	34.63	38.55	43.94	53.37			
25	50	44.4	35.1	25.9	20.2				34.09	36.16	40.48	46.52	56.75			
26	50	44.1	34.5	25.0					35.45	37.70	42.44	49.20				
27	50	43.9	33.9	24.4					36.82	39.25	44.43	51.96				
28	50	43.7	33.3	24.4					38.18	40.81	46.46	54.75				
29	50	43.5	32.7	24.4					39.55	42.37	48.53	57.55				
30	50	43.2	32.2	24.4					40.91	43.94	50.63	60.34				
31	50	43.0	31.6						42.27	45.52	52.77	63.14				
32	50	42.8	31.0						43.64	47.11	54.95					
33	50	42.5	30.4						45.00	48.71	57.17					
34	50	42.3	30.3						46.36	50.32	59.43					
35	50	41.3	29.8						47.73	51.94	61.72					
36	50	41.8	29.8						49.09	53.57	64.01					
37	50	41.6	29.8						50.45	55.21	66.29					
38	50	41.3	29.8						51.81	56.86	68.58					
39	50	41.1							53.18	58.52						
40	50	40.9							54.55	60.19						
41	50	40.7							55.91	61.87						
42	50	40.4							57.27	63.56						
43	50	40.2							58.64	65.26						
44	50	40.0							60.00	66.97						
45	50	39.8							61.36	68.69						
46	50	39.5							62.73	70.42						
47	50	39.3							64.09	72.16						
48	50	39.1							65.45	73.91						
49	50	39.0							66.82	75.66						
50	50	39.0							68.18	77.41						
51	50	39.0							69.55	79.15						
52	50	39.0							70.91	80.89						

Source: Firey and Peterson [16-8, pp. 8-9].

grades and vertical curves. Their theoretical equations produce results in close conformity with road observations, so are presented here for general application. Tables 16-15, 16-16, and 16-17 give the speed of trucks on plus grades for weight/power ratios of 200, 300, and 400 pounds per net horsepower. These distance-speed tables result from solutions by Firey and Peterson of their formulas given in Ref. 16-8.

Truck Travel Time on Grades

From the speed-distance tables of trucks on minus and plus grades the travel times were calculated by using a straight line average speed for each 100-foot distance interval. These travel times in seconds are given in Tables 16-10, 16-12, 16-14, 16-15, 16-16, and 16-17.

SPEED CHANGES OF VEHICLES

Both motor vehicle running cost and travel time are affected by the speed at which a vehicle travels, but of equal importance is the number of the changes in speed, plus and minus, and the number of miles per hour changed. See Appendix A for the running cost of speed changes. In normal driving the driver is not particularly aware of his changes in speed other than his stops, but the speed changes are surprisingly high in number and magnitude when recorded and summarized. The literature contains but isolated references to speed changes; to supply the analyst for highway economy with the desirable information on the speed changes of traffic, extensive observations and recordings of vehicle speeds in continuous movement are required. A few results of such recordings are given in this section.

DECELERATION AND ACCELERATION–PASSENGER CAR

Although no specific evidence was found in the literature to indicate the effect of rates of deceleration and acceleration on the running cost of vehicles, the logical conclusion is that fast rates of speed changes result in higher costs than do slow rates. On the contrary, fast deceleration and acceleration will reduce the time consumption for stops and slow down cycles as compared to slow rates of deceleration and acceleration. All speed changes–stops, slowdowns and speed-ups–as explained in Chapters 13 and 14 incur running costs in excess of travel at the constant speed from which the change in speed began and to which the speed returned. Some knowledge of deceleration and acceleration rates, therefore, is important to an understanding of the cost and time consumption of speed change cycles.

Time-Speed Relationships in Deceleration and Acceleration

Again, the Claffey original data |16-5| is the source of deceleration and

Traffic Characteristics

TABLE 16-17
Speed (mph) and Elapsed Time (Seconds) of Truck on Uniform Plus Grades
Gross Vehicle Weight/Net Horsepower Ratio: 400 Pounds
Entry Speed: 50 mph

Distance, 100 ft	Speed, mph								Elapsed Time, Seconds							
	Plus Grade, Percent								Plus Grade, Percent							
	0 & 1	2	3	4	5	6	7	8	0 & 1	2	3	4	5	6	7	8
0	50	50.00	50.00	50.00	50.00	50.00	50.00	50.00	0.00	0.00	0.00	0.00	0.00	0.00	0.00	0.00
1	50	49.62	49.33	49.01	48.66	48.28	47.90	47.51	1.36	1.37	1.37	1.38	1.38	1.39	1.39	1.40
2	50	49.12	48.62	47.93	47.26	46.52	45.75	44.96	2.73	2.75	2.76	2.78	2.80	2.83	2.85	2.87
3	50	48.68	47.88	46.83	45.80	44.70	43.54	42.32	4.09	4.14	4.17	4.21	4.27	4.32	4.38	4.43
4	50	48.24	47.12	45.70	44.28	42.80	41.24	39.55	5.45	5.55	5.61	5.68	5.78	5.88	5.99	6.10
5	50	47.80	46.34	44.54	42.70	40.81	38.83	36.62	6.82	6.97	7.07	7.19	7.35	7.53	7.69	7.89
6	50	47.37	45.54	43.35	41.07	38.71	36.29	33.48	8.18	8.40	8.55	8.74	8.98	9.24	9.51	9.84
7	50	46.93	44.73	42.14	39.40	36.51	33.59	30.07	9.55	9.85	10.06	10.33	10.67	11.05	11.46	11.99
8	50	46.49	43.91	40.87	37.69	34.21	30.70	26.29	10.91	11.31	11.60	11.97	12.44	12.98	13.58	14.41
9	50	46.05	43.08	39.62	35.94	31.82	27.58	22.01	12.27	12.78	13.17	13.66	14.29	15.05	15.92	17.23
10	50	45.61	42.24	38.36	34.15	29.34	24.19	17.06	13.64	14.27	14.77	15.41	16.23	17.28	18.55	20.72
11	50	45.17	41.40	37.07	32.33	26.78	20.48	11.26	15.00	15.77	16.40	17.22	18.28	19.71	21.60	25.54
12	50	44.73	40.55	35.78	30.49	24.15	16.43	10.50	16.36	17.29	18.06	19.09	20.45	22.39	25.29	31.81
13	50	44.29	39.70	34.46	28.62	21.47	12.00	10.50	17.73	18.82	19.76	21.03	22.76	25.38	30.09	38.30
14	50	43.85	38.85	33.14	26.73	18.75	11.80	10.50	19.09	20.37	21.50	23.05	25.22	28.77	35.82	44.80
15	50	43.41	38.00	31.80	24.82	16.00	11.80	10.50	20.45	21.93	23.27	25.15	27.87	32.69	41.60	51.29
16	50	42.97	37.15	30.45	22.89	13.30	11.80		21.82	23.51	25.08	27.34	30.73	37.34	47.38	57.78
17	50	42.53	36.29	29.10	20.95	13.30	11.80		23.18	25.10	26.94	29.63	33.84	42.47	53.15	
18	50	42.10	35.43	27.73	19.00	13.30			24.55	26.71	28.84	32.03	37.25	47.59	58.93	
19	50	41.66	34.57	26.45	17.04	13.30			25.91	28.34	30.79	34.55	41.03	52.72		
20	50	41.22	33.71	24.96	15.50				27.27	29.99	32.79	37.21	45.22	57.85		
21	50	40.78	32.85	23.56	15.50				28.64	31.65	34.84	40.02	49.62			
22	50	40.34	31.99	22.16	15.50				30.00	33.33	36.94	43.00	54.02			
23	50	39.90	31.13	20.74	15.50				31.36	35.03	39.10	46.18	58.42			
24	50	39.46	30.27	19.31					32.73	36.75	41.32	49.58	62.82			
25	50	39.02	29.41	18.40					34.09	38.49	43.60	53.20				
26	50	38.58	28.55	18.40					35.45	40.25	45.95	56.91				
27	50	38.14	27.68	18.40					36.82	42.03	48.38	60.61				
28	50	37.70	26.81	18.40					38.18	43.83	50.88	64.32				
29	50	37.27	25.94						39.55	45.65	53.47	68.02				
30	50	36.83	25.07						40.91	47.49	56.14					
31	50	36.39	24.20						42.27	49.35	58.91					
32	50	35.95	23.33						43.64	51.24	61.78					
33	50	35.51	22.70						45.00	53.15	64.74					
34	50	35.07	22.70						46.36	55.08	67.74					
35	50	34.63	22.70						47.73	57.04	70.75					
36	50	34.19	22.70						49.09	59.02	73.75					
37	50	33.75							50.45	61.03	76.75					
38	50	33.31							51.82	63.06						
39	50	32.87							53.18	65.12						
40	50	32.43							54.55	67.21						
41	50	32.00							55.91	69.33						
42	50	31.56							57.27	71.48						
43	50	31.12							58.64	73.66						
44	50	30.68							60.00	75.87						
45	50	30.24							61.36	78.11						
46	50	29.80							62.73	80.38						
47	50	29.80							64.09	82.67						
48	50	29.80							65.45	84.96						
49	50	29.80							66.82	87.24						
50	50	29.80							68.18	89.53						
51									69.55							
52									70.91							

Source: Firey and Peterson [16-8, pp. 8-9].

acceleration rates for a passenger car in typical use on highways. About 200 decelerations to zero speed and about an equal number of accelerations were tabulated and plotted to arrive at the average results in Tables 16-18 to 16-23. The speed-elapsed time tables are given for three rates each of deceleration and acceleration–fast, medium, and slow. Because of the wide variation in rates, especially at speeds above 35 mph, the three levels of speed change rates were chosen. The initial speeds for deceleration and the terminal speed for acceleration range in 5 mph intervals from 15 to 70 mph.

Normal driving practice is such that at each end of a speed change there is often not a distinct beginning and ending speed or specific second of time. Many of the original field tape records exhibited both plus and minus speed changes

TABLE 16-18

FAST RATES OF DECELERATIONS FROM A STOP BY A PASSENGER CAR IN NORMAL TRAFFIC

(Vehicle: Four Door 1959 Model Standard Station Wagon Weighing 4,900 Pounds)

Elapsed Time, Seconds	Speed Decelerated from, mph											
	15	20	25	30	35	40	45	50	55	60	65	70
0	15.0	20.0	25.0	30.0	35.0	40.0	45.0	50.0	55.0	60.0	65.0	70.0
1	14.2	19.5	24.7	29.6	34.7	39.7	44.6	49.6	54.6	59.6	64.5	69.5
2	12.8	18.5	23.9	28.5	34.3	39.0	43.8	49.0	53.9	58.9	63.8	68.9
3	11.0	16.9	22.5	27.0	33.4	37.9	42.4	47.9	53.0	58.1	62.9	68.1
4	8.8	14.7	19.6	25.0	31.7	36.1	40.7	46.5	52.1	57.0	61.8	67.2
5	6.3	11.1	16.8	21.7	29.6	34.0	38.4	44.8	50.4	55.7	60.5	66.2
6	3.3	8.0	13.0	18.4	26.8	31.5	35.8	42.7	48.7	54.1	59.0	65.0
7	0.0	4.4	9.3	14.3	23.3	28.3	32.9	40.2	46.5	52.2	57.3	63.6
8		0.0	5.2	10.2	19.5	24.5	29.3	37.3	44.0	49.9	55.4	62.1
9			0.0	5.6	15.3	20.4	25.4	33.9	41.2	47.5	53.2	60.4
10				0.0	10.8	16.0	21.0	30.0	38.0	44.8	50.8	58.5
11					5.8	11.3	16.5	26.0	34.6	41.8	48.2	56.4
12					0.0	6.0	11.6	21.6	30.6	38.6	45.4	54.1
13						0.0	6.0	16.9	26.4	35.1	42.3	51.5
14							0.0	11.8	21.9	31.0	39.0	48.7
15								6.1	17.1	26.8	35.3	45.8
16								0.0	11.9	22.2	31.3	42.6
17									6.2	17.2	27.0	39.1
18									0.0	12.0	22.4	35.4
19										6.2	17.4	31.4
20										0.0	12.0	27.2
21											6.2	22.5
22											0.0	17.4
23												12.0
24												6.2
25												0.0

Source: Original record tapes from Claffey [16-5].

Traffic Characteristics

at both terminals of the deceleration and acceleration run. This characteristic necessitated the use of judgment in selecting a group of 4 to 12 runs to produce the final averages in Tables 16-18 to 16-23.

Another characteristic of the deceleration and acceleration runs, especially at speeds above 35 mph, was the wide range in the rates of speed change on any given run. When coming to an ultimate stop from 60 mph, for instance, the speed could be constant at 42 mph for 3 to 10 seconds, or even increase from 31 to 34 mph over 10 seconds or so. These types of departures from a uniformly changing rate of speed change or a constant rate of speed change, no doubt resulted from interference of vehicles ahead of the test car, misjudging the time-space relationships by the driver, and misjudging the timing of the green-red cycle of traffic signals. In the end, runs were selected for averaging which did not include extreme variations in rates of speed change. Even then the final data in Tables 16-18 to 16-23 are the result of smoothing the average so that the resulting families of curves exhibited logical similarities.

Figure 16-1 presents a number of deceleration and acceleration runs, typical of many that were extracted from the original paper tapes.

Distances Required for Deceleration to Stops and Acceleration to Operating Speed

The speed-time associations in Tables 16-18 to 16-23 were converted to speed-distance relationships as given in Table 16-24. As would be expected from the rates of deceleration and acceleration, the distances increase rapidly with increase in initial speed, and the slow rates require materially greater distances than do the fast rates of deceleration and acceleration. At 70 mph it requires 8,182 ft to stop from and regain 70 mph at the slow speed change rate, but only 4,137 ft at the fast rate. Deceleration distances are less than acceleration distances, which differences may or may not be partially attributed to the selection and grouping of the individual cases of stops from the original paper tape recordings.

The lower section of Table 16-24 is the excess travel time in hours per 1,000 stop cycles as calculated from the upper section and the deceleration and acceleration rates in Tables 16-18 to 16-23. These hours of excess travel time for the fast rates are less than those in Table A-9, Appendix, the same as for the medium rates of speed change, and greater for the slow rates of speed change. In later work by Claffey [16-6] he used 5 ft per sec per sec for both deceleration and acceleration for initial speeds to 30 mph and 2.5 ft per sec per sec above 30 mph, which deceleration and acceleration are in between the fast and slow rates in Tables 16-18 to 16-23.

SPEED CHANGES OF PASSENGER CARS

In urban-area driving, the running cost of motor vehicles is high as compared

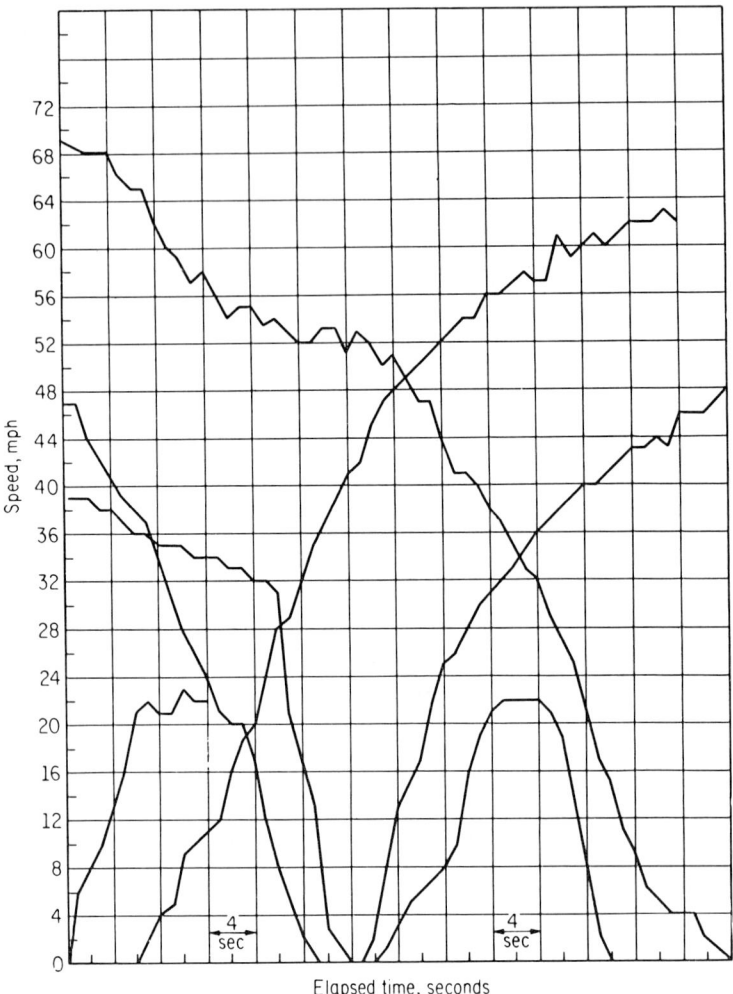

FIG. 16-1. A sample of deceleration and acceleration curves, involving stopping, taken from the original field records of the passenger car driving reported by Claffey [16-5].

to rural driving. Although the lower urban speeds should produce lower running costs, they seldom do. The one contributing factor to higher urban running costs is change in vehicle speed. As discussed in Chapter 13, changes in vehicle speed put into play a number of costly forces. The deceleration and acceleration of the vehicle consume fuel and tread rubber in excess of the consumption rates when driving at uniform speed. Brake wear and general strain and stress of all parts of the car come into play. Horizontal changes in direction are severe, unbanked sharp 90-deg turns are frequent. Further discussion of speed changes will

Traffic Characteristics

TABLE 16-19
MEDIUM RATES OF DECELERATIONS FROM A STOP BY A PASSENGER CAR IN NORMAL TRAFFIC
(Vehicle: Four Door 1959 Model Standard Station Wagon Weighing 4,900 Pounds)

Elapsed Time, Seconds	Speed Decelerated from, mph											
	15	20	25	30	35	40	45	50	55	60	65	70
0	15.0	20.0	25.0	30.0	35.0	40.0	45.0	50.0	55.0	60.0	65.0	70.0
1	14.8	19.7	24.8	29.7	34.7	39.8	44.8	49.8	54.6	59.5	64.7	69.6
2	13.9	19.0	24.3	29.2	34.3	39.4	44.5	49.3	54.1	59.1	64.3	69.1
3	12.4	17.6	23.8	28.4	33.7	38.7	44.1	48.9	53.5	58.4	63.8	68.4
4	10.3	16.1	22.7	27.4	33.0	37.9	43.4	48.0	52.9	57.8	63.3	67.8
5	8.2	14.2	21.2	26.1	32.0	36.8	42.7	47.0	52.1	57.0	62.6	67.1
6	5.8	12.0	19.5	24.7	30.9	35.5	41.4	45.9	51.2	56.1	61.8	66.3
7	3.0	9.4	17.4	22.8	29.5	34.0	40.2	44.6	50.3	55.2	60.9	65.4
8	0.0	6.4	15.1	20.8	27.9	32.6	38.8	43.2	49.3	54.2	59.9	64.5
9		3.4	12.6	18.4	26.0	31.0	37.2	41.7	48.2	53.0	58.9	63.6
10		0.0	9.9	15.9	24.0	29.2	35.4	40.1	47.0	51.7	57.8	62.5
11			6.8	13.2	21.7	27.1	33.8	38.4	45.7	50.5	56.6	61.3
12			3.6	10.2	19.2	24.8	32.0	36.4	44.2	49.2	55.4	60.1
13			0.0	7.0	16.5	22.4	30.1	34.8	42.7	47.8	54.1	58.8
14				3.7	13.6	19.8	27.9	32.8	41.0	46.5	52.7	57.6
15				0.0	10.5	16.9	25.5	30.8	39.2	45.0	51.4	56.2
16					7.2	13.9	23.0	28.5	37.3	43.4	50.0	54.8
17					3.7	10.8	20.2	26.0	35.5	41.7	48.5	53.3
18					0.0	7.4	17.2	23.4	33.5	39.9	47.1	52.2
19						3.8	14.2	20.5	31.4	38.0	45.5	50.6
20						0.0	10.9	17.6	29.0	36.1	43.9	48.9
21							7.4	14.4	26.5	34.0	42.2	47.5
22							3.8	11.1	23.8	31.8	40.4	45.8
23							0.0	7.5	20.9	29.4	38.6	44.2
24								3.8	17.8	26.9	36.6	42.6
25								0.0	14.6	24.2	34.5	40.8
26									11.2	21.1	32.2	39.2
27									7.6	18.0	29.8	37.0
28									3.9	14.8	27.3	34.9
29									0.0	11.3	24.5	32.6
30										7.7	21.6	30.2
31										3.9	18.4	27.6
32										0.0	15.0	24.7
33											11.5	21.6
34											7.8	18.4
35											3.9	15.1
36											0.0	11.6
37												7.7
38												4.0
39												0.0

Source: Original record tapes from Claffey [16-5].

TABLE 16-20

SLOW RATES OF DECELERATIONS FROM A STOP BY A PASSENGER CAR IN NORMAL TRAFFIC

Vehicle: Four Door 1959 Models Standard Station Wagon Weighing 4,900 Pounds.

Elapsed Time, Seconds	Speed Decelerated from, mph											
	15	20	25	30	35	40	45	50	55	60	65	70
0	15.0	20.0	25.0	30.0	35.0	40.0	45.0	50.0	55.0	60.0	65.0	70.0
1	14.7	19.7	24.7	29.6	34.7	39.6	44.6	49.7	54.4	59.5	64.5	69.5
2	14.2	19.3	24.2	29.1	34.4	39.2	44.2	49.3	53.8	58.9	63.9	69.1
3	13.2	18.5	23.5	28.4	34.0	38.8	43.8	48.9	53.3	58.3	63.4	68.6
4	11.8	17.6	22.7	27.8	33.4	38.3	43.2	48.4	52.7	57.8	62.8	68.1
5	10.0	16.5	21.6	26.9	32.6	37.7	42.7	47.9	52.2	57.2	62.3	67.7
6	7.9	15.2	20.4	25.9	31.9	37.0	42.1	47.4	51.6	56.6	61.7	67.1
7	5.5	13.8	19.1	24.7	30.8	36.4	41.6	46.8	51.0	56.0	61.1	66.7
8	2.9	12.1	17.6	23.4	29.8	35.6	40.9	46.2	50.3	55.4	60.5	66.1
9	0.0	10.2	16.1	21.9	28.6	34.7	40.3	45.5	49.7	54.8	59.9	65.6
10		8.0	14.3	20.2	27.4	33.8	39.5	44.8	49.1	54.1	59.3	65.0
11		5.6	12.5	18.4	26.2	32.7	38.7	44.0	48.4	53.5	58.7	64.5
12		2.9	10.4	16.7	24.8	31.5	37.8	43.2	47.7	52.8	58.1	64.0
13		0.0	8.2	14.7	23.0	30.1	36.8	42.3	47.0	52.2	57.5	63.4
14			5.7	12.7	21.1	28.7	35.7	41.3	46.2	51.5	56.9	62.8
15			3.0	10.6	19.0	27.3	34.3	40.3	45.4	50.8	56.2	62.3
16			0.0	8.3	17.2	25.8	32.8	39.2	44.6	50.1	55.5	61.6
17				5.8	15.1	23.8	31.2	38.0	43.6	49.3	54.9	61.2
18				3.1	13.0	21.8	29.7	36.6	42.6	48.5	54.2	60.6
19				0.0	10.7	19.5	28.2	37.2	41.6	47.7	53.5	60.0
20					8.4	17.7	26.6	33.7	40.5	46.8	52.8	59.4
21					5.8	15.4	24.4	32.1	39.2	45.8	52.1	58.8
22					3.1	13.2	22.3	30.4	37.8	44.7	51.3	58.2
23					0.0	10.9	20.0	28.6	36.2	43.6	50.4	57.3
24						8.5	18.0	26.7	34.5	42.5	49.6	56.6
25						5.9	15.7	24.7	32.6	41.3	48.7	56.0
26						3.2	13.4	22.6	31.1	39.9	47.8	55.3
27						0.0	11.0	20.5	29.4	38.5	46.8	54.5
28							8.6	18.2	27.5	36.9	45.7	53.9
29							5.9	15.9	25.2	35.2	44.6	53.1
30							3.2	13.7	23.1	33.3	43.2	52.3
31							0.0	11.0	20.7	31.6	41.9	51.4
32								8.5	18.5	29.7	40.5	50.5
33								6.0	16.1	27.7	39.0	49.6
34								3.2	13.8	25.5	37.4	48.6
35								0.0	11.2	23.3	35.7	47.6
36									8.8	21.0	33.9	46.5
37									6.0	18.7	32.0	45.3
38									3.2	16.3	30.0	43.8
39									0.0	14.0	27.9	42.3
40										11.4	25.7	40.7
41										8.8	23.5	39.4
42										6.0	21.2	37.8
43										3.2	18.9	36.1
44										0.0	16.5	34.3
45											14.0	32.3
46											11.5	30.2
47											8.9	28.0
48											6.1	25.8
49											3.2	23.8
50											0.0	21.4
51												19.1
52												16.6
53												14.4
54												11.6
55												9.0
56												6.2
57												3.2
58												0.0

Source: Original record tapes from Claffey [16-5].

Traffic Characteristics

TABLE 16-21

Fast Rates of Accelerations from a Stop by a Passenger Car in Normal Traffic

Vehicle: Four Door 1959 Model Standard Station Wagon Weighing 4,900 Pounds

Elapsed Time, Seconds	Speed Accelerated to, mph											
	15	20	25	30	35	40	45	50	55	60	65	70
0	0.0	0.0	0.0	0.0	0.0	0.0	0.0	0.0	0.0	0.0	0.0	0.0
1	4.4	4.6	4.8	5.0	5.2	5.3	5.5	5.6	5.7	5.8	5.9	6.0
2	7.9	8.3	8.7	9.4	10.0	10.3	10.5	10.7	10.7	10.8	10.8	11.0
3	10.4	11.4	12.6	13.2	14.0	14.4	14.7	15.0	15.1	15.2	15.2	15.2
4	12.3	14.0	15.6	16.6	17.4	18.0	18.4	18.8	18.9	19.1	19.1	19.2
5	13.7	16.1	18.2	19.6	20.5	21.2	21.7	22.1	22.3	22.6	22.7	22.8
6	14.6	17.7	20.6	22.2	23.4	24.2	24.8	25.2	25.5	25.8	26.0	26.0
7	15.0	18.9	22.8	24.5	25.9	26.7	27.4	28.0	28.5	28.8	29.1	29.3
8		19.7	24.1	26.4	27.9	28.9	29.7	30.4	30.9	31.4	31.9	32.0
9		20.0	24.7	27.9	29.8	31.2	32.1	32.9	33.4	34.2	34.6	35.0
10			25.0	29.0	31.6	33.2	34.4	35.3	35.8	36.6	37.1	37.6
11				29.7	33.0	35.0	36.6	37.4	38.0	38.8	39.4	39.8
12				30.0	34.0	36.5	38.4	39.6	40.3	40.8	41.6	42.3
13					34.7	37.8	40.2	41.5	42.3	42.9	43.6	44.2
14					35.0	38.7	41.6	43.3	44.2	44.9	45.5	46.2
15						39.7	42.9	44.9	45.9	46.8	47.3	47.9
16						40.0	43.8	46.3	47.5	48.6	49.0	49.4
17							44.6	47.5	48.7	50.2	50.6	51.0
18							45.0	48.4	50.2	51.6	52.1	52.6
19								49.2	51.4	52.9	53.6	54.0
20								49.7	52.5	54.4	55.0	55.6
21								50.0	53.5	55.6	56.3	56.8
22									54.2	56.7	57.6	58.2
23									54.7	57.7	58.8	59.5
24									55.0	58.5	60.0	60.8
25										59.2	61.1	62.0
26										59.7	62.1	63.2
27										60.0	63.0	64.4
28											63.7	65.4
29											64.3	66.4
30											64.7	67.4
31											65.0	68.2
32												68.8
33												69.3
34												69.8
35												70.0

Source: Original record tapes from Claffey [16-5].

TABLE 16-22

MEDIUM RATES OF ACCELERATIONS FROM A STOP BY A PASSENGER CAR IN NORMAL TRAFFIC
Vehicle: Four Door 1959 Model Standard Station Wagon Weighing 4,900 Pounds

Elapsed Time, Seconds	Speed Accelerated to, mph										
	15	20	25	30	35	40	45	50	55	60	65	70
0	0.0	0.0	0.0	0.0	0.0	0.0	0.0	0.0	0.0	0.0	0.0	0.0
1	3.6	3.6	3.7	3.9	4.1	4.2	4.5	4.8	5.0	5.4	5.8	6.2
2	5.8	6.1	6.4	6.8	7.2	7.7	8.1	8.6	9.2	9.7	10.4	11.2
3	8.2	8.5	8.8	9.2	9.7	10.3	11.0	11.7	12.6	13.6	14.8	16.3
4	10.3	10.8	11.2	11.8	12.4	13.2	14.0	14.9	16.0	17.2	18.5	20.2
5	12.4	12.9	13.5	14.2	14.9	15.8	16.8	17.8	19.0	20.4	22.0	23.9
6	13.8	14.6	15.3	16.1	17.0	18.0	19.1	20.3	21.6	23.2	25.2	27.5
7	14.4	16.3	17.1	18.1	19.1	20.3	21.4	22.8	24.3	26.0	28.0	30.5
8	14.7	17.6	19.5	20.4	21.4	22.6	23.8	25.2	26.7	28.5	30.6	33.2
9	15.0	18.3	21.2	22.2	23.3	24.5	25.8	27.2	28.8	30.8	33.1	36.1
10		19.4	22.6	23.8	25.2	26.5	27.9	29.7	31.2	33.2	35.4	37.9
11		20.0	23.8	25.6	26.9	28.2	29.6	31.2	32.9	34.9	37.4	40.2
12			24.4	27.0	28.4	29.8	31.4	33.1	34.9	36.9	39.3	42.0
13			24.8	28.2	29.8	31.4	33.0	34.8	36.7	38.7	41.1	43.7
14			25.0	29.1	31.6	33.2	34.8	36.6	38.4	40.5	42.8	45.3
15				29.7	32.9	34.6	36.4	38.2	40.2	42.3	44.6	47.1
16				30.0	33.8	35.7	37.8	39.7	41.7	43.9	46.3	49.2
17					34.4	36.8	38.8	40.9	43.1	45.4	48.0	50.9
18					34.8	37.8	39.9	42.0	44.3	46.7	49.3	52.3
19					35.0	38.7	41.0	43.2	45.4	47.9	50.8	53.9
20						39.4	42.1	44.4	46.4	49.4	52.3	55.6
21						39.8	43.2	45.6	48.1	50.8	53.7	57.1
22						40.0	44.0	47.1	49.5	52.1	55.1	58.5
23							44.6	48.0	50.5	53.2	56.2	59.9
24							45.0	48.9	51.5	54.2	57.2	60.8
25								49.5	52.4	55.0	58.1	61.7
26								49.8	53.1	56.0	59.2	62.8
27								50.0	53.8	56.9	60.2	63.9
28									54.4	57.8	61.1	64.7
29									54.8	58.5	61.9	65.6
30									55.0	59.1	62.6	66.4
31										59.6	63.2	67.2
32										59.9	63.8	67.8
33										60.0	64.2	68.4
34											64.6	68.8
35											64.8	69.2
36											65.0	69.5
37												69.7
38												69.9
39												70.0

Source: Original record tapes from Claffey [16-5].

Traffic Characteristics

TABLE 16-23
Slow Rates of Accelerations from a Stop by a Passenger Car in Normal Traffic
Vehicle: Four Door 1969 Model Standard Station Wagon Weighing 4,900 Pounds

Elapsed Time, Seconds	Speed Accelerated to, mph											
	15	20	25	30	35	40	45	50	55	60	65	70
0	0.0	0.0	0.0	0.0	0.0	0.0	0.0	0.0	0.0	0.0	0.0	0.0
1	3.0	3.4	3.7	3.9	4.1	4.3	4.4	4.4	4.4	4.4	4.5	4.5
2	4.9	5.5	6.0	6.4	6.8	7.1	7.4	7.6	7.8	7.9	8.0	8.1
3	6.8	7.7	8.4	9.0	9.6	10.0	10.4	10.9	11.0	11.2	11.4	11.6
4	8.2	9.2	10.1	10.8	11.6	12.2	12.8	13.2	13.6	14.1	14.4	14.8
5	9.7	10.9	11.9	12.9	13.7	14.4	15.1	15.7	16.2	16.8	17.2	17.7
6	11.3	12.6	13.8	14.9	15.8	16.8	17.4	18.0	18.7	19.2	19.7	20.3
7	12.3	14.1	16.0	16.9	17.9	19.0	19.8	20.3	21.2	21.6	22.2	22.8
8	13.2	15.3	17.5	18.6	19.8	21.2	22.0	22.6	23.4	23.9	24.3	24.9
9	14.0	16.5	19.0	20.2	21.6	23.2	24.0	24.8	25.6	25.8	26.3	27.0
10	14.4	17.4	20.2	21.6	23.3	24.9	25.8	26.6	27.2	27.6	28.1	28.9
11	14.8	18.3	21.3	23.0	24.9	26.6	27.5	28.3	28.6	29.2	29.8	30.3
12	15.0	18.9	22.3	24.1	26.4	28.0	29.0	29.9	30.2	30.9	31.3	31.8
13		19.6	23.1	25.2	27.7	29.5	30.4	31.2	31.6	32.1	32.9	33.1
14		19.8	23.8	26.2	28.9	30.8	31.6	32.6	33.0	33.4	34.0	34.5
15		20.0	24.3	27.0	30.0	31.9	32.8	33.7	34.2	34.8	35.2	35.8
16			24.7	27.8	30.9	33.0	34.0	34.8	35.3	35.8	36.5	37.0
17			24.9	28.2	31.8	33.9	35.1	35.9	36.6	37.0	37.6	38.1
18			25.0	28.9	32.6	34.8	36.1	36.9	37.6	38.0	38.9	39.2
19				29.4	33.3	35.6	37.2	37.9	38.4	38.9	39.7	40.3
20				29.7	33.9	36.3	38.2	38.8	39.4	39.9	40.6	41.3
21				29.8	34.4	36.9	39.1	39.2	40.1	40.9	41.7	42.4
22				30.0	34.7	37.5	40.0	40.9	41.4	42.0	42.6	43.3
23					34.8	38.0	41.0	41.6	42.2	42.8	43.6	44.3
24					34.9	38.4	41.6	42.4	43.1	43.7	44.5	45.3
25					35.0	38.8	42.3	43.2	44.0	44.7	45.3	46.3
26						39.2	42.9	44.0	44.9	45.5	46.3	47.2
27						39.5	43.4	44.7	45.7	46.4	47.1	48.1
28						39.8	44.0	45.4	46.6	47.2	47.9	49.0
29						40.0	44.3	46.1	47.4	48.1	48.7	49.8
30							44.6	46.7	48.2	48.9	49.5	50.7
31							44.9	47.3	49.0	49.8	50.4	51.5
32							45.0	47.8	49.8	50.5	51.2	52.4
33								48.2	50.5	51.3	52.0	53.3
34								48.6	51.2	52.1	52.8	54.1
35								48.8	51.9	52.9	53.6	55.0
36								50.0	52.5	53.7	54.5	55.8
37									53.1	54.5	55.3	56.7
38									53.7	55.3	56.0	57.5
39									54.2	56.0	56.8	58.2
40									54.6	56.8	57.5	59.1
41									55.0	57.4	58.4	59.9
42										58.2	59.1	60.6
43										58.8	59.9	61.4
44										59.4	60.6	62.1
45										59.8	61.3	62.8
46										60.0	62.0	63.6
47											62.7	64.2
48											63.2	64.9
49											63.8	65.5
50											64.3	66.1
51											64.7	66.7
52											65.0	67.3
53												67.8
54												68.2
55												68.6
56												69.0
57												69.3
58												69.6
59												69.8
60												70.0

Source: Original record tapes from Claffey [16-5].

TABLE 16-24

DISTANCE REQUIRED TO DECELERATE TO A STOP AND TO ACCELERATE BACK TO INITIAL SPEED AND EXCESS TRAVEL TIME FOR THE SPEED CHANGE CYCLE FOR A PASSENGER CAR

Rate of Deceleration or Acceleration	Speed Decelerated from or Returned to, mph											
	15	20	25	30	35	40	45	50	55	60	65	70
	Distance, Feet											
Deceleration												
Fast	91	148	212	282	410	501	598	772	961	1167	1386	1718
Medium	109	185	312	426	607	763	996	1184	1532	1835	2256	2606
Slow	127	246	374	530	764	1035	1347	1696	2063	2551	3163	4013
Acceleration												
Fast	101	174	238	346	473	614	773	1018	1283	1579	1982	2419
Medium	130	200	327	439	619	831	1011	1290	1601	1958	2363	2826
Slow	174	288	439	656	878	1173	1436	1763	2205	2679	3285	4169
Excess Travel Time over Continuing at Initial Speed, Hours per 1,000 Speed Change Cycles												
Deceleration Plus Acceleration												
Fast	1.74	1.95	2.15	2.43	2.72	3.06	3.40	3.78	4.22	4.67	5.19	5.75
Medium	2.01	2.46	2.94	3.43	3.92	4.40	4.89	5.35	5.88	6.36	6.82	7.25
Slow	2.31	3.00	3.56	4.18	4.73	5.38	6.06	6.34	7.80	8.77	9.82	10.92

Source: Calculated from Tables 16-18 to 16-23.

provide a better understanding of this factor in traffic performance and its role in economy studies.

Importance of Speed Changes

The plus and minus changes in speed from a uniform speed, usually the legal speed or attempted speed the driver wishes to maintain, affect two important factors related to highway transportation economy which are of first order concern to vehicle drivers. Changes in speed–stops, and slowdowns and regaining the initial speed–affect the running cost and the travel time. The total effect of speed changes on running cost and travel time depends upon the initial speed, speed changed from, speed changed to, and the number of speed changes per trip or given travel distance.

Table 16-25 gives the average speed, fuel consumption in miles per gallon, and total running cost for a passenger car for specific numbers of stop cycles over specific distances. In urban driving on nonaccess controlled highways and streets, stops can easily reduce average speed 10 to 40 percent, decrease miles per gallon of fuel by 50 percent, and double running cost. Slowdowns and speedups have similar effects on driving economy, but not as severe as for stops.

DEFINITION AND IDENTIFICATION OF SPEED CHANGES

A speed change of a vehicle moving on the highway literally is any increase or decrease in its speed during any interval of time or distance. This definition includes the starting and stopping of a vehicle and speed slowdowns and speedups when stops are not involved. For practical purposes, though, it is necessary

Traffic Characteristics

TABLE 16-25

SPEED, FUEL CONSUMPTION, AND RUNNING COST OF A
PASSENGER CAR AS AFFECTED BY THE NUMBER OF STOP CYCLES

Initial Speed, mph	Distance, Feet*	Number of Stops per Travel Distance Shown in Feet									
		0	1	2	3	4	5	6	7	8	10
		A. Average Speed, mph										
20	5,280	20.00	19.06	18.21	17.43	16.71	16.05	15.44	14.88	14.35	13.40
30	5,280	30.00	27.20	24.88	22.92	21.25	19.81	18.55			
40	7,770	40.00	35.88	32.28	29.44	27.06	25.03				
50	9,896	50.00	43.75	38.90	35.01	31.83					
60	11,379	60.00	50.47	44.31	39.18						
70	10,864	70.00	56.15	46.88							
		B. Average Fuel Consumption, Miles per Gallon										
20	5,280	21.79	19.88	18.28	16.92	15.75	14.73	13.83	13.04	12.33	11.12
30	5,280	23.36	19.70	17.04	15.00	13.40	12.11	11.05			
40	7,770	22.37	18.96	16.45	14.53	13.01	11.78				
50	9,896	20.16	17.21	14.88	13.15	11.79					
60	11,379	17.36	14.63	12.64	11.13						
70	10,864	14.29	11.57	9.72							
		C. Total Running Cost, Cents per Mile										
20	5,280	3.71	4.20	4.70	5.19	5.69	6.18	6.68	7.17	7.67	8.66
30	5,280	3.49	4.43	5.36	6.30	7.24	8.17	9.11			
40	7,770	3.50	4.57	5.64	6.72	7.79	8.86				
50	9,896	3.65	5.00	6.34	7.68	9.02					
60	11,379	3.94	5.76	7.57	9.38						
70	10,864	4.77	7.39	10.30							

* Distance is based on the medium deceleration and acceleration in Table 16-24.
Source: Tables 16-24, A-8, A-9, and A-28.

to restrict the definition to apply only to a change in speed above some chosen minimum of speed in miles per hour. Generally, too, the definition may be restricted to below some maximum travel time or distance. In economic analyses of alternative highway improvements or route choices, the analyst needs to know the probable speed changes which are the result of features of highway design and traffic operations. Therefore the speed changes of interest to him are those which measurably affect vehicle running cost and travel time, and which are caused by highway design and traffic controls.

In selecting guidelines by which to identify speed changes, the controlling factor should be whether the speed changes affect vehicle running cost. Speed can be changed by a small difference in vertical gradient without any change in throttle setting and without applying braking effort. Speed changes should be related to the throttle and the brake because they are the two main devices by which speed is changed. These two devices, therefore, contribute to the running cost of speed changes.

Speed changes may be caused by action of the driver or by other forces. The driver may change speed for a number of personal reasons: to stop, change routes, read roadside signs, view landscape, change the noise level, comply with legal speed limits, comply with legal signs and signals, or by unknowingly changing the foot pressure on the throttle pedal. The driver also changes speed to avoid accidents, follow the traffic lane, change directions, match car speed and curve superelevation, and overtake and pass a slower vehicle. The car speed may change without action by the driver because of vertical alignment of the roadway, resistance of horizontal curves, wind direction and pressure, and changes in the rolling resistance of the roadway surface. It is readily seen then that a vehicle is driven at absolutely uniform speed only when four conditions prevail simultaneously—and they seldom do. First, when the driver has no personal reason to change speed to achieve his purposes of the trip; second, when the vertical and horizontal alignment of the highway is constant; third, when the traffic conditions permit constant speed driving; and fourth, when weather and other environmental conditions remain constant. Even when a driver deliberately attempts to drive at a fixed constant speed, as in test-car driving on a level tangent free of other vehicles, he will find it necessary to make frequent, though slight, adjustments in throttle position.

There are three ways with respect to speed in which to drive a vehicle. A driver may drive at constant throttle position allowing the vehicle to gain and to lose speed as affected by highway and environmental forces; the driver may attempt to drive at uniform constant speed by making practically continuous changes in throttle position and by braking; or the driver may endeavor to combine the constant throttle and constant speed driving to maintain a speed between chosen minimum and maximum speeds. Driving at speeds between chosen minimum and maximum speeds is the way that traffic performs.

Recording of vehicle speed each second of time is good evidence of the fact that drivers do not drive at absolutely uniform speed, not even for, say, 10 consecutive seconds. In the work by Claffey [16-5] on his toll-road, free-road comparison, he recorded his car speed each second. The detection and recording instrument was capable of an accuracy of $\pm$ 1 mph. The instrument recorded to the nearest integral speed. Figure 16-2 gives some typical speed profiles from Claffey's records, as does Fig. 16-1. From these speed profiles two observations are significant. Driving is not at constant speed, and speed changes are difficult to define and to enumerate. Arbitrary definitions and conditions must be used in selecting and summarizing speed changes to use in economy studies.

Speed-Change Frequencies

Three objectives may be considered in an effort to delineate speed changes when the speed profile is available for each second of travel time. First, it is desired to record only those speed changes that resulted from positive action by

Traffic Characteristics

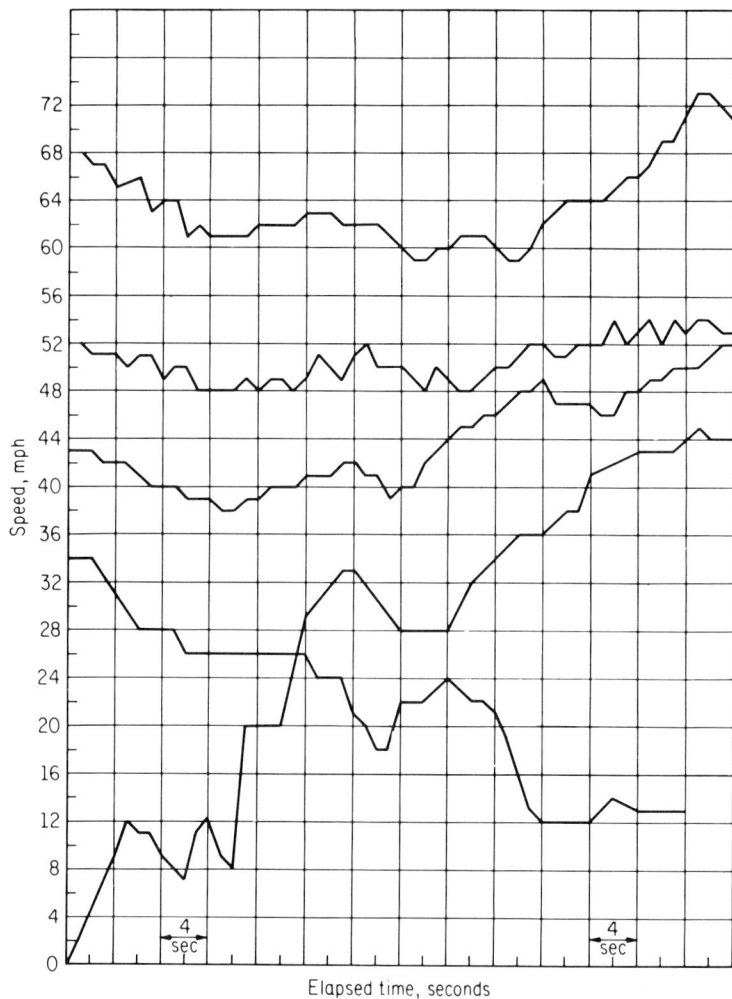

FIG. 16-2. A sample of speed profiles in miles per hour and time in seconds taken from the original field records of the passenger car driving reported by Claffey [16-5]. The profiles illustrate the difficulty of identifying a speed change when attempting to summarize the speed changes for any of the distance runs.

the driver to change speed by throttle or brake manipulation. Second, only those speed changes are wanted that can be readily identified by an observer in the car–or the driver–when specifically trying to note the speed changes. Third, it is desirable to recognize that some of the 1- and 2-mph changes given on the speed profile may be recording inaccuracies. Based on these objectives and a preliminary analysis of some of Claffey's records, the following guides and criteria were developed for use in summarizing speed changes for different classes of highways and traffic conditions from speed profiles such as those recorded by Claffey. These criteria and guides do not provide second-by-second for every possible situation found on the speed profile, but are intended to provide a consistent method of handling the more frequent and more significant speed change recordings. The guidelines are:

1. A 1-mph fluctuation (a plus-minus or minus-plus sequence) in speed is not considered to be a speed change. When consecutive one second speed readings do not vary by more than 1 mph, then a constant speed is said to exist. Note that a 1-mph increase from a base speed followed immediately by a 2-mph decrease and a 1-mph increase produces three speed changes that are discarded under this criterion.

2. Speed changes of two or more miles per hour are accepted when preceded and followed by constant speeds. No attempt is made to identify speed changes, either plus or minus, of 1 mph.

3. When a constant speed (by definition) prevails for four or more consecutive 1-sec readings, it is identified as a zero mile per hour speed change, that is, a time period of no change in speed.

4. Constant speed periods of less than four consecutive 1-sec speed readings between speed changes of opposite sign are identified as follows: When there are two consecutive speed readings, the first speed is assigned to the preceding speed change period and the point of speed change is designated at the second reading; when there are three consecutive speed readings, the middle one is chosen as the end of the preceding speed change period and the beginning of the following period. This scheme results in eliminating the constant speed period of less than four consecutive readings (3-sec interval) by assigning portions of the readings to preceding or following periods.

5. Constant speeds of less than four speed readings coming between the beginning and end of a speed change of 2 mph or more are ignored.

PROBLEMS

16-1. Using the vehicle classification distribution for primary rural highways in Table 16-0 and the appropriate Appendix A running cost tables, compute the vehicle running cost for 1,000 ADT for 1 mile of highway for 1 day. Assume level tangent operation and the speed distribution given in Table 16-1 for primary rural highways. Compute, also, the running cost based on the average speed at the bottom of Table 16-1, and comment upon your two sets of running costs with special reference to economy studies.

16-2. Use the information in Prob. 16-1 and apply it to the gallons of fuel consumption tables in Appendix A. State your answers in terms of gallons of fuel and compute the fuel tax generated at 4 cents a gallon (federal) and at 7 cents a gallon (state). Calculate the fuel consumption at the average speeds and comment upon your results with special reference to economy studies.

16-3. Use Tables 16-0 and 16-1 for primary rural highways to calculate the travel time for one mile of level tangent for the distribution of speed and for the average speeds. Comment upon your results with special reference to economy studies.

16-4. For about a week's driving of regular family trips and to class, keep a record of odometer readings and speed reading. Note the stops and idle times and the speed changes of 10 mph and greater. Work up your data to show the following: (1) Nominal or attempted cruising speed, (2) average overall speed, (3) excess travel time because of stops and slowdowns, and (4) estimated fuel consumption in miles per gallon as calculated from Appendix A tables.

REFERENCES

16-1. American Association of State Highway Officials. *A Policy on Geometric Design of Rural Highways–1965.* Washington, D.C., 1966.

16-2. American Association of State Highway Officials. *A Policy on Arterial Highways in Urban Areas.* Washington, D.C., 1957.

16-3. Georgy Bezkorovainy. *Effects of Advisory Speed Limits at Horizontal Curves of Two-lane Rural Highways.* Traffic Engineering Series Report No. 17R, Department of Civil Engineering, University of Illinois, Urbana, December 1964.

16-4. E. W. Campbell, L. E. Keefer, and R. W. Adams. *A Method for Predicting Speeds Through Signalized Street Sections.* Highway Research Board, Washington, D.C., Bulletin 230, *Trip Generation and Urban Freeway Planning,* 1959.

16-5. Paul J. Claffey. *Characteristics of Passenger Car Travel on Toll Roads and Comparable Free Roads.* Highway Research Board, Bulletin 306, *Studies in Highway Engineering Economy,* 1961, pp. 1-22.

16-6. Paul J. Claffey. *Running Cost of Motor Vehicles as Affected by Highway Design.* Highway Research Board, Washington, D.C., National Cooperative Highway Research Program Report No. 13, 1965.

16-7. Robert E. Dunn. *A Study of Motor Vehicle Performance on Ascending Grades.* (Primary State Highway No. 3, Swank Creek Vicinity, Kittitas County, Washington). Washington State Highway Commission, Department of Highways, Olympia, January 1954.

16-8. Joseph C. Firey and Edward W. Peterson. *An Analysis of Speed Changes for Large Transport Trucks.* Highway Research Board, Bulletin 334, *Vehicle Characteristics,* 1962, pp. 1-26.

16-9. Highway Research Board. *Vehicle Climbing Lanes.* Bulletin 104, 1955.
  1. T. S. Huff and F. H. Scrivner. "Simplified Climbing-Lane Design Theory and Road-Test Results." p. 1.
  2. Robert E. Dunn. "Motor-Vehicle Performance on Ascending Grades," p. 12.
  3. William E. Willey. "Truck Congestion on Uphill Grades," p. 21.

16-10. Highway Research Board. *Motor Vehicle Speeds.* Bibliography 27, Annotated, 1960. A listing of 609 annotated references, subdivided into seven main subjects and 27 minor divisions.

16-11. Highway Research Board. *Accident Analysis and Speed Characteristics,* Bulletin 341, 1962.
  1. Burton M. Rudy. "Operational Route Analysis", p. 1.

2. T. Ogawa, E. S. Fisher, and J. C. Oppenlander. "Driver Behavior Study–Influence of Speed Limits on Spot Speed Characteristics in a Series of Contiguous Rural and Urban Areas," p. 18.

3. N. J. Rowan and Charles J. Keese. "Study of Factors Influencing Traffic Speeds," p. 30.

4. J. C. Oppenlander. "A Theory on Vehicular Speed Regulation," p. 77.

5. Frank A. Haight and Walter W. Mosher, Jr. "A Practical Method for Improving the Accuracy of Vehicular Speed Distribution Measurements," p. 92.

16-12. Highway Research Board. National Academy of Sciences, Washington D.C., *Highway Capacity Manual*, National Research Council, 1965.

16-13. John W. Horn and others. *The Effect of Commercial Roadside Development on Traffic Operations*. Engineering Research Department, North Carolina State College, Raleigh, June 30, 1960.

16-13A. J. W. Korte. "Speed in Road Traffic: An Illustration of Methods Used in Road Traffic Research," *International Road Safety and Traffic Review*, Vol. 5, No. 2 (Spring 1957), pp. 11-18.

16-14. Chih-Cheng Ku. "Driver Characteristics Correlated with Speeds Observed on Horizontal Curves of Two-Lane Rural Highways," *Traffic Engineering*. University of Illinois, Urbana, May 1965.

16-15. B. A. Lefeve. *Speed Characteristics on Vertical Curves*. Highway Research Board, National Academy of Sciences, Washington, D.C., *Proc.*, Vol. 32, 1953, pp. 395-413.

16-16. B. A. Lefeve. *Speed Habits Observed on a Rural Highway*. Highway Research Board, National Academy of Sciences, Washington, D.C., *Proc.*, 33, 1954, pp. 409-248.

16-17. A. D. May, Jr. *A Friction Concept of Traffic Flow*. Highway Research Board, Washington, D.C., *Proc.*, Vol. 38, 1959, pp. 493-510.

16-18. J. C. Oppenlander. *Variables Influencing Spot-Speed Characteristics*. Highway Research Board, National Academy of Sciences, Washington, D.C., Special Report 89, 1966. (A good, brief summary of the relationship of vehicle speed to the driver, vehicle, roadway, traffic, and environment. Especially good for its 160 references.)

16-19. D. F. Petty. *Traffic Speed Report No. 69*. Joint Highway Research Project, Purdue University, Lafayette, Ind., 1960.

16-20. Earl J. Reeder. "Critical Speeds on Highway Curves," *Public Safety* (December 1934). *Also Highway Research Abstracts*, No. 19 (April 1935), pp. 5-6.

16-21. Carl Saal. *Time and Gasoline Consumption in Motor Truck Operation as Affected by the Weight and Power of Vehicles and the Rise and Fall in Highways*. Highway Research Board, Washington, D.C., Bulletin 9-A, 1950.

16-22. Roy B. Sawhill and Keith C. Crandall. *Some Measurable Qualities of Traffic Service Influenced by Freeways*. Highway Research Board, Highway Research Record 49, *Traffic and Operations–General*, 1964, pp. 30-63.

16-23. L. L. Schulman. *Traffic Speed Report 78, Truck Speed-Weight Study*. Joint Highway Research Project, Purdue University, Lafayette, Ind., January 1964.

16-24. Society of Automotive Engineers. *Truck Ability Prediction Procedure TR 82*. New York, August 1957.

16-25. W. R. Stohner. *Speeds of Passenger Cars on Wet and Dry Pavements*. Highway Research Board, Washington, D.C., Bulletin 139, *Road Roughness and Slipperiness*, 1956.

16-26. Asriel Taragin. "Driver Performance on Horizontal Curves," *Public Roads*, Vol. 28, No. 2 (June 1954), pp. 27-39.

Traffic Characteristics

16-27. Kenneth J. Tharp. *A Quantitative Evaluation of the Geometric Aspects of Highways.* Joint Highway Research Project, Purdue University, Lafayette, Ind., December 1962.

16-27A. T. B. Treadway and J. C. Oppenlander. *Statistical Modeling of Travel Speeds and Delays on a High-Volume Highway.* Highway Research Board, National Academy of Sciences, Washington, D.C., Highway Research Record No. 199, *Mathematical and Statistical Aspects of Traffic,* 1967.

16-28. Frederick Wagner, Jr. and Adolph D. May, Jr. *Volume and Speed Characteristics at Seven Study Locations (Detroit and Lansing, Michigan).* Highway Research Board, National Academy of Sciences, Washington, D.C., Bulletin 281, *Traffic Volume and Speed Studies,* 1961, pp. 48-67.

16-29. G. M. Webb. *Truck Speeds on Grades.* Traffic Bulletin 2, California State Highway Division, Traffic Department, September 26, 1961.

16-30. William E. Willey. *Survey of Uphill Speeds on Trucks on Mountain Grades.* Highway Research Board, National Academy of Sciences, Washington, D.C., *Proc.,* Vol. 29, 1949, pp. 304-310.

16-31. William E. Willey. *Survey of Downhill Truck Speeds of Trucks on Mountain Grades.* Highway Research Board, National Academy of Sciences, Washington, D.C., *Proc.,* Vol. 30, 1950, pp. 322-329.

16-32. William E. Willey. *Truck Congestion on Uphill Grades.* Highway Research Board, Bulletin 104, *Vehicle Climbing Lanes,* 1955.

16-33. John M. Wright and Samuel C. Tignor. "Relations of Gross Weights and Horsepowers of Commercial Vehicles," *Public Roads,* Vol. 34, No. 4 (October 1966), pp. 84-88.

Chapter **17**

The Traffic Volume Estimate

The analysis for the engineering economy of a proposed highway improvement is based entirely upon the future. The proposed highway facility would be constructed in the immediate future and the traffic would use the new facility from the date of construction five to thirty years in the future. The estimate, or forecast, of traffic which would use the proposed facility, once constructed, is one of the key factors in the analysis. The road user consequences are directly proportional to the traffic volume. The design of the facility itself is dependent upon this same forecast, and, therefore, the highway cost is somewhat dependent upon the estimate of future traffic volume and composition. It is not the purpose of this chapter to discuss the procedure of forecasting traffic but instead to relate the forecast to the analysis for economy.

COMPONENTS OF TOTAL TRAFFIC

A better understanding of the role of the traffic estimate in the analysis for economy may be had through a discussion of the components of traffic based upon the sources of the traffic. Whether the analysis for economy is for project evaluation or for project formulation, traffic is a consideration. Most likely the traffic engineer, planner, or other person making the traffic forecast, will also consider the traffic by certain base components, such as existing, diverted, transferred, generated, developed, normal growth, and community.

EXISTING TRAFFIC

Existing traffic is that highway traffic currently (as of the base date of the study) using the route or facility to be reconstructed or improved. Existing traffic does not include any traffic in the travel corridor or area which may be attracted to the new facility when it is completed. The before and immediately after number of trips by existing traffic is probably unchanged, but the vehicle-miles will be altered by any shortening or lengthening of distance. For a new route location, when the old routes within the area are continued in service, the

The Traffic Volume Estimate 475

existing traffic on the new facility will be zero. The term "existing traffic" is also applied to any traffic presently to be found on a facility or in any area referred to.

As a result of adding an improvement to an existing highway it is seldom that the existing traffic would decrease in volume after the improvement is completed; the usual result is an increase in overall ADT. Intersection traffic controls, change to one-way traffic flow, and the closing of access may, however, result in a decrease in existing traffic. On an area basis, existing traffic often decreases on a given existing route because of a new highway route or new facility elsewhere, such as when a new urban bypass or a grade separation structure becomes available.

DIVERTED TRAFFIC

Diverted traffic is present highway travel that is attracted to the new facility from other highway routes or facilities. This traffic is attracted to the new facility because the drivers of vehicles see advantages thereon over their former routings. Diverted traffic does not result in any increase in the number of highway vehicular trips within the area, but it does result in a redistribution of the trips between alternative routings and changes in origins and destinations. The total vehicle-miles in the affected area may increase or decrease. When a new highway is constructed on a new location and most all of the old parallel existing right-of-way is continued in use, the traffic on the new facility will be composed largely of diverted traffic.

In economy studies, diverted traffic is important because when it moves to the new facility from other highway routes, the travel speed and running cost of the traffic remaining on the old routes may be affected. Under the principle of considering all of the consequences to whomsoever they may accrue, all traffic affected by a highway improvement needs to be included in the analysis. Because of personal preference, some traffic that may be diverted to a new facility may travel adverse distance compared to its former routings. Favorable travel time is often a factor in creating adverse travel distance.

TRANSFERRED TRAFFIC

Transferred traffic is present travel attracted to the new highway facility from modes of travel other than highway, such as railway, airway, and ferry.

The analysts for economy of proposed facilities have often omitted the adverse consequences to the modes of travel that will lose travel to the new highway facility. This economic consequence, however, should not be ignored. Consideration of transferred traffic lost to other modes is a general economic factor that is considered outside the analysis for engineering economy as a part of the general group of nonuser consequences and public policy.

The transferred traffic would be included in the highway traffic using the new facility. It may or may not gain an economic benefit, depending upon the relative cost of travel by the former mode and on the new highway facility. Its transfer may be for reasons of personal preference.

One positive evidence of the magnitude of transferred traffic is in the gain of highway travel and the decrease of railway travel (and freight haul) in the past 30 years.

GENERATED TRAFFIC

Generated traffic, or induced traffic, comes into being on a new highway facility only because the facility is available. Generated traffic does not currently exist in any mode of travel, and probably would not exist in the future should not the highway facility or other attraction be constructed.

New highway facilities, especially new intercity routes, generate traffic when they are decidedly attractive in travel time and distance, in scenic values, new experiences, convenience, and pleasure. New highways stimulate people to drive 20 miles for an evening dinner, 100 miles to the seashore, 300 miles for a day in the big city, or 500 miles for a weekend in the old home town. These trips would not be taken except for the attractiveness of the new highway, at least the trips are of greater distance and more frequent than was true on the old facilities. The generated travel is difficult to estimate because it has not been studied sufficiently enough to relate it to known factors, such as land use, population density, and community characteristics.

DEVELOPED TRAFFIC

Developed traffic on the new facility is that traffic developed after the facility is completed because of any changing use of the land served by the highway facility. Consideration of the land use factor is highly important when estimating the future number of vehicle trips to the land because of the close correlation of land use and travel to and from the land. Actually, developed traffic is probably a combination of normal growth, diverted, and transferred traffic.

Developed traffic is an important component of the future traffic on highway routes on new rights-of-way and for routes that tap new land areas, new resources, or new activities, recreation for instance.

NORMAL GROWTH TRAFFIC

Normal growth traffic is that traffic currently nonexisting that comes into being in the future because of population increase and because of a generally growing increase in vehicular use. Normal growth traffic is difficult to separate from other increases in travel (transferred, diverted, developed, and generated) once the new facility has been in use a year or so. It may be considered to be that traffic that would come to the existing facility over the years, assuming that

The Traffic Volume Estimate

the new facility would not be constructed and that the existing facility had the capacity to accommodate the increased usage.

Practically all highways, roads, and streets have gained in motor vehicle traffic since the beginning of this mode of travel. This increase in traffic has resulted from an increase in population, an increase in numbers of vehicles above the rate of population increase, and an increase in the use of the vehicles in miles per year per vehicle. The diverted, transferred, and generated traffic build up rapidly on a new facility during the first year and at a slower rate thereafter. But the traffic on most facilities will continue to increase year after year up to capacity of the facility. This continuing increase in traffic is the normal growth traffic. It comprises most of the design volume traffic other than that traffic which is existing somewhere and which is immediately assigned to the new facility.

COMMUNITY TRAFFIC

Community traffic is that total current traffic on roads and streets within the geographical area to be affected by the proposed facility. Some of this community traffic will and some will not use the facility when it is completed. This community traffic not using the new facility may be affected favorably or unfavorably through area changes in travel time, running speed, and distance.

Community traffic in economy studies is considered on a before and after basis. An urban freeway or an urban bypass will cause many changes in the local area traffic pattern. Some congested streets will gain relief; streets leading to traffic interchanges may become congested; traffic may be forced to travel extra distance because of the access control on the new facility and lack of closely spaced grade separations for all crosswise traffic.

To trace out all of the consequences to community traffic resulting from a specific new highway facility would be a formidable task. About all that is practical is to estimate the consequences to the main arteries of travel which would be affected.

DISCUSSION OF THE SEVEN TRAFFIC SOURCES

Of the seven classifications of traffic, the existing traffic is the only one which is not affected in volume over the future years. Diverted and transferred traffic would show up on the new facility mainly during the first year of its use, but these two classes could continue to contribute some additional traffic to the new facility during future years. The amount of developed traffic year to year would depend upon the nature of and the rate of change in land use. Generated travel could come into being soon after the new facility was completed and thereafter at a lesser rate.

Normal growth traffic is so closely associated with developed and generated traffic that there would be difficulty in estimating it separately after the first year or so, but then it will constitute most of the year by year increase.

For highway design purposes, the total traffic only is needed, but in forecasting this total traffic some knowledge of the source of the traffic is helpful. For highway engineering economy studies the source of traffic is important only when true travel cost reductions and true net benefits need to be identified. Some attention to the source of traffic may also be desirable in economic evaluation of a project as contrasted to project formulation which often is not concerned with the source of traffic.

THE TRAFFIC FORECAST

The factors of volume, composition, and source of the traffic forecasted to use a proposed new facility are important to the analysis for economy. Volume and composition bear directly upon the road user costs on the new facility and the source of traffic determines the base cost from which the consequences are measured. The methods of forecasting or assigning traffic and the associated factors–population, incomes, land use, time, distance, and the like–are not discussed. Some aspects of the forecasts as they relate to the analysis for economy are discussed.

FORECAST OF BASE TRAFFIC

Under the "with or without" concept it is required that the road user costs be forecast for all traffic sources that contribute to the use of the new facility when making an analysis for its economic evaluation. The alternative without the proposed facility–that is, the "do nothing" alternative–requires a traffic forecast just as does the proposed facility, and it must be projected over the same future period of time. To assume that the existing traffic and sources of other traffic will hold constant in the future is as illogical as it is to assume that the first-year traffic on the new facility will continue for the whole of the analysis period. Yet few analists have forecasted the traffic for existing conditions over the analysis period, but they confined their forecast to only the proposed facility. For project formulation, however, the existing condition need not be forecast, except where any traffic changes would affect the economy of details of design.

FUTURE TRAFFIC VOLUME VERSUS RUNNING COST

The calculations of the savings and benefits to traffic should be based on the projected future conditions. As the traffic volume and composition change with future years (an increase in ADT is the usual result) the operating speed, travel time, and running cost will change. The unit cost per vehicle or per vehicle-mile will vary with future years because of changing traffic volume and consequent

The Traffic Volume Estimate

changes in speed, speed changes, and accidents. For some facilities, the traffic increase may culminate within a few years in a total traffic volume so great (as compared to the old facility) that there is no longer any realized time or running cost net benefits. The traffic which comes to the new facility in the more distant future years certainly cannot be assumed to realize the same unit benefit that is attached to the existing traffic or diverted traffic that immediately begins to use the new facility when the traffic volume is much less than capacity.

The future net benefits will depend upon the rate of increase in traffic volume with respect to the capacity of the facility to accommodate the total traffic. Many situations of the past can be cited to show that because of the rapid increase in diverted, tranferred, developed, and generated traffic, the full capacity of a new facility was soon reached so that the net benefit realized per vehicle decreased to near zero in a short period of years. Relief on other area traveled ways may have continued longer, however.

The highway trips that are diverted from other highway facilities to the proposed new facility may render to the remaining traffic a favorable consequence by reducing the volume on the formerly used route, thus permitting higher travel speed and lower running cost per vehicle-mile. The analysis of economy, therefore, should include these consequences, usually beneficial in the net, but for the forecasted volume trend rather than for just the first-year traffic.

GRADIENT AND EXPONENTIAL TRAFFIC GROWTHS

Traffic volume over past years has tended to increase geometrically (exponentially) more so than linearly (straight line). Traffic forecasters have favored the straight-line forecast. For economy studies, either type of forecast may be used. With both gradient and exponential compound interest factors available (Appendixes C and D), either forecast is equally easy to use. Whichever growth pattern is the most likely to be experienced is the one to use. Often the traffic forecaster will forecast the traffic for some future year, such as 1985, without any reference to the traffic volume for the years between the present year and 1985. The analyst for economy, then, must decide upon the year by year traffic for his period of analysis.

BASE YEAR OF THE TRAFFIC FORECAST

Traffic volume is most often expressed as the ADT, meaning the average daily traffic volume for the time period specified, usually a specific year. For economy analyses three factors are required: (1) the ADT for the beginning of the first year of use of the proposed facility, (2) the traffic volume for each year thereafter for the full analysis period, and (3) the traffic composition as between the different classes of vehicles.

These three factors are needed for the existing facility, the community area, and for the proposed facility. For certain existing facilities operating at capacity, the present traffic and the future traffic could be assumed to be the same volume and same composition.

Generally, the traffic forecast is not so precise as to state that it is the average of the past 12 months, the past 6 months and the future 6 months, or the future 12 months. The important factor to be specific about is the exact point in time any increase in traffic is to begin. Gradient and exponential compound interest factors are calculated with reference to a specific relationship between the date the increase begins and the factor n (analysis period) in the tables. For instance, in *Principles of Engineering Economy* by Grant and Ireson [2-6], the gradient factor tables are based upon the increase starting at the end of the 0 to 1 year, or at $n = 1$, which is one year after the facility is opened to traffic. On this basis and using the year-end concept, the first gradient factors given are for $n = 2$. These tables assume that any increase in the quantity (traffic, maintenance cost, or other item) between zero time and $n = 1$ is included in the base quantity.

The tables given herein for gradient (Appendix C) and exponential (Appendix D) factors are based on the increase beginning at $n = 0$ time date. Thus, the first year traffic (the base or initial traffic) would be the traffic at zero time, which is realistic with reference to when the traffic increase actually begins. The important factor, however, is to be certain of whether the initial, or base, traffic volume was estimated for $n = 0$ or $n = 1$. Usually by the time the facility is opened to traffic the base traffic estimate is more appropriately related to time $n = 0$.

SUMMARY OF TRAFFIC VOLUME ESTIMATE

In the analysis for economy it is the objective to weigh the net consequences to whomsoever they may accrue against the costs which make the consequences possible. Considering only the running cost of the motor vehicles, accident costs, and the travel time, and the fact that all vehicles which use a proposed new facility will receive a net benefit therefrom, all such vehicles should therefore be included in the analysis. This procedure would be using the sales concept.

In project formulation the pricing of the benefits, or savings, is not the important factor it is in project evaluation. Formulation of design details is a relative comparison process in which the objective is to find the most economical design for that traffic to use the facility. The benefits could be either "high" or "low" without much danger of affecting the choice of design details between design alternatives.

When computing the monetary benefits (including savings) for future traffic, due care is to be taken to use the appropriate base condition and future condition to correctly reflect the most probable consequences. In this connection the unit cost of vehicle operation, accidents, and travel time needs to be adjusted to the probable future traffic volume and traffic conditions on a year to year basis for each alternative.

When all future traffic is included in the analysis, care is to be exercised against calling the net benefits "savings."

PROBLEMS

17-1. From your highway department (state, county, or city) get the average daily traffic on a specific route or highway system for a number of past years, 5 to 25, and compute the average rate of growth. Determine particularly if it is a uniform gradient, exponential, or other form of increase.

17-2. For a given highway facility the present average daily traffic volume is 1,200. This ADT is forecasted to increase at a gradient rate of 50 ADT per year. At a vestcharge rate of 8 percent what is the equivalent uniform ADT over a 20-year period? What will be the ADT the 20th year?

17-3. Work Prob. 17-2 for traffic increase of (1) 2 percent, and (2) of 4 percent annually. What will be the ADT in the 20th year?

17-4. For certain engineering purposes the trucks and buses in a traffic stream may be expressed in equivalent numbers of passenger cars, such as 4, 5, or 6 cars per truck. Such practice may be justified in traffic volume estimates for design but the same ratios may not be good practice in economy studies. Calculate equivalent car ratios for running cost and truck travel time. Use the vehicle classification distribution in Table 16-0. Use the speed distribution in Table 16-1 for primary rural highways. For the 2-S2 and equivalent combinations of about 40,000 lb gross vehicle weight, move the column of percentages in Table 16-1 up one line, and for the 3-S2 and heavier combinations move the percentages up two lines. Use the percentages as given for the 12-kip single-unit trucks and their equivalents. For the value of travel time use $1.50 for passenger cars, and the Central Region values in Table 11-2 for commercial vehicles. Use the appropriate Appendix A tables for the running costs on level tangents. Assume 1,000 ADT for one day and one mile. Compute your passenger car equivalents for (1) running cost, (2) travel time, (3) for value of travel time, and (4) for running cost and travel time value combined.

REFERENCES

17-1. Charles F. Barnes, Jr. *Integrating Land Use and Traffic Forecasting.* Highway Research Board, Washington, D.C., Bulletin 297, 1961, pp. 1-13.

17-2. W. B. Calland. "Traffic Forecasting for Freeway Planning," *Journal American Institute of Planners,* No. 2 (1959), p. 25

17-3. E. Wilson Campbell. *A Mechanical Method for Assigning Traffic to Expressways.* Highway Research Board, Washington, D.C., Bulletin 130, 1959.

17-4. Thomas J. Fratar. "Vehicular Trip Distribution by Successive Approximations," *Traffic Quarterly* (January 1954), pp. 53-65.

17-5. K. P. Furness. *The Use of Gravity Model in the Estimation of Traffic.* Traffic Section, Department of Highways, Ontario, Canada, October 1961.

17-6. W. L. Grecco and S. M. Breuning. *Application of Systems Engineering Methods to Traffic Forecasting.* Highway Research Board, Bulletin 347, 1962, pp. 10-23.

17-7. John R. Hamburg and Robert H. Sharkey. *Land Use Forecast.* Chicago Area Transportation Study, Paper No. 3.2.6.10, August 1, 1961.

17-8. John R. Hamburg. *Land Use Projections for Predicting Future Traffic.* Highway Research Board, Bulletin 224, 1959, p. 72.

17-9. Walter G. Hansen and Alan M. Vorhees. *Evaluation of Gravity Model Trip Distribution Procedures.* Highway Research Board, Washington, D.C., Bulletin 347, 1962, pp. 67-76.

17-10. Highway Research Board. Highway Research Record No. 38, Travel Forecasting, Washington, D.C., 1963.

    1. Walter Kudlick and others. "Intermediate and Final Quality Checks–Developing a Traffic Model," p. 1.

    2. Martin Wohl. "Demand, Cost, Price, and Capacity Relationships Applied to Travel Forecasting," p. 40.

    3. Roger L. Creighton. "Perting a Transportation Study," p. 55.

    4. Donald M. Hill and Hans G. Von Cube. "Development of a Model for Forecasting Travel Mode Choice in Urban Areas," p. 78.

    5. Thomas B. Deen and others. "Application of a Modal Split Model to Travel Estimates for the Washington Area," p. 97.

    6. W. L. Grecco and S. M. Breuning. "A System Engineering Model for Trip Generation and Distribution," p. 124.

    7. *What Is Needed for a Fact Finding Study of Intercity Transportation in the United States: A Panel Discussion,* p. 146.

        A. Background, by Harold W. Hansen.

        B. Comments on a Proposed National Study, by Harmer Davis.

        C. Why a Fact-Finding Study Should Be Undertaken, by E. H. Holmes.

        D. Fundamentals and Techniques Essential to a Fact-Finding Study, by J. Douglas Carroll, Jr.

17-11. D. M. Hill and Norman Dodd. *Travel Mode Split in Assignment Programs.* Highway Research Board, Washington, D.C., Bulletin 347, 1962, pp. 290-301.

17-12. H. Hoyt. "The Effect of the Automobile on Patterns of Urban Growth," *Traffic Quarterly,* Saugatuck, Conn. (April 1963), pp. 293-301.

17-13. J. F. Kain. *A Multiple Equation Model of Household Locational and Tripmaking Behavior.* Highway Research Board, National Academy of Sciences, Washington, D.C., Highway Research Record No. 88, 1962, p. 67.

17-14. J. F. Kain. *Communting and the Residential Decisions of Chicago and Detroit Central Business District Workers.* Highway Research Board, National Academy of Sciences, Washington, D.C., Highway Research Record No. 88, 1963, p. 43.

17-15. B. V. Martin, F. W. Memmott, and A. J. Bone. *Principles and Techniques of Predicting Future Demand for Urban Area Transportation.* School of Civil Engineering, Massachusetts Institute of Technology, Cambridge, January 1963.

17-16. Robert B. Mitchell and Chester Rapkin. *Urban Traffic, a Function of Land Use.* Colombia U. P., New York, 1954.

17-17. Willa Mylroie. *Evaluation of Intercity Travel Desire.* Highway Research Board, National Academy of Sciences, Washington, D.C., Bulletin 119, 1956, pp. 69-92.

17-17A. Emory C. Parrish, Edwyn D. Peterson, and Ray Threlkeld. *Georgia's*

Program for Automated Acquisition and Analysis of Traffic-Count Data. Highway Research Board, National Academy of Sciences, Washington, D.C. Highway Research Record No. 199, *Mathematical and Statistical Aspects of Traffic,* 1967, pp. 42-61.

17-17B. Robert W. Paterson. *Forecasting Techniques–Determining the Potential Demand for Highways.* Research Center, School of Business and Public Administration, University of Missouri, Columbia, 1966.

17-18. Pittsburgh Area Transportation Study. *Forecasts and Plans.* Vol. 2, February 1963, p. 177.

17-19. Ramiro Ramírez Carril. *Traffic Forecast Based on Anticipated Land Use and Current Travel Habits.* Highway Research Board, Washington, D.C., Proc., Vol. 31, pp. 386-410. 1952.

17-20. Robert E. Schmidt and M. Earl Campbell. *Highway Traffic Estimation.* The Eno Foundation for Highway Traffic Control, Saugatuck, Conn., 1956.

17-21. G. B. Sharpe, W. G. Hansen, and L. B. Hamner. *Factors Affecting the Trip Generation of Residential Land-Use Areas.* Highway Research Board, Washington, D.C., Bulletin 203, 1958.

17-22. Frank L. Sweetser. *Projections of Greater Boston's Population to 1970 and 1980.* Boston University, Economic Base Report 9, July 1962, p. 35.

17-23. A. M. Vorhees. *Forecasting Peak Hours of Travel.* Highway Research Board, Washington, D.C., Bulletin 203, 1958.

17-24. Alan M. Vorhees and Robert Morris. *Estimating and Forecasting Travel for Baltimore by Use of a Mathematical Model.* Highway Research Board, Bulletin 224, 1959, pp. 105-114.

17-25. Melvin M. Webber. "Transportation Planning Models," Traffic Quarterly, Saugatuck, Conn., July 1961, pp. 373-390.

17-26. R. H. Wiant. *A Simplified Method for Forecasting Urban Traffic.* Highway Research Board, Washington, D.C., Bulletin 297, 1961, pp. 128-145.

17-27. Richard M. Zettel and Richard R. Carll. *Summary Review of Major Metropolitan Area Transportation Studies in the United States.* Institute of Transportation and Traffic Engineering, University of California, 1962.

Chapter **18**

Consequences of Right-of-Way Takings

If land is to be profitably used–to be a benefit to society–it must be accessible. Access to land is usually over highways, roads, and streets. Land used for these acccess highway is publicly owned and free of real estate tax. With minor exceptions, all land not held in title by an agency of some level of government is subject to the levy and payment of annual real estate taxes, whereas the land devoted to public highways is untaxed. Whether taking taxed land for tax free right-of-way land becomes a justifiable charge against highways (taxes forgone) needs clarification and solution. Further, is there an economic charge against highways when private productive land is taken for highway use?

REAL ESTATE TAXES

When land and the real property on the land is taken for highway right-of-way under the applicable statutes and compensation paid to the owner at the time of taking, the amount of real estate subject to property tax is reduced. Correspondingly, the dollars of assessed valuation against which the tax is levied is reduced at that moment. What are the cost consequences of this action? The logical quick decision by most people is that the public treasury will lose some tax income and that this tax income loss should be treated as a highway cost. But this obvious logical deduction is not the correct deduction.

TAX CONSEQUENCE OF TAKING LAND FOR RIGHT-OF-WAY

An analysis of the usual consequences that follow the removal of real estate from the tax rolls, will furnish the basis of decision as to whether taxes forgone on highway right-of-way become a highway cost.

The best answer to the question of how to handle real estate taxes can be reached by making the analysis on the with and without basis, considering all consequences to whomsoever they may accrue. The with and without basis is in reality the "before and after" taking of the property for right-of-way; the viewpoint is that of the community within the taxing district, considered as a whole; and the "to whomsoever they may accrue" is the total of all taxpaying persons

Consequences of Right-of-Way Takings

within the taxing district, considered as one group. The consequences are the cash flows of tax moneys into the public treasury.

The consequences will lie within one of three possibilities:

1. The highway money paid for the right-of-way was reinvested in newly constructed taxable property within the taxing district, and the tax rolls were restored to their former level or higher.

2. Overall events and attractiveness of the land and business opportunities within the area were such that the taxable property remaining after the right-of-way taking increased in total value enough to restore the tax base to its former level or higher.

3. The money paid for the right-of-way was not reinvested in taxable property, real or personal, within the taxing jurisdiction from which the right-of-way was taken, and all other taxable property remained at its former taxable value.

Obviously, should conditions 1 or 2 result, the property tax base (assessed valuation) on real estate would not be less than it was prior to the taking, so there could be no loss of property tax income. Should the taxable value of real estate increase because of the highway improvement (after the actual taking), there would be a tax income gain to the public treasury–unless the millage rate was reduced accordingly. Consideration, then, needs to be given only to probability 3.

Counties, cities, state governments, and their appropriate subagencies levy taxes on real estate. The levy is a tax rate (usually in mills) applied to the assessed value (value for tax purposes according to the local law or custom) of the property. This tax rate is set by the responsible officials or governing body on the basis of two factors: (1) how much tax income is necessary to meet the adopted budget for the tax year, and (2) what is the total assessed value of the property to be taxed. The quotient, budget divided by assessed value, gives the tax rate for that year. The property assessment for tax purposes is usually determined as of one specific date each year. Any in between changes in property values do not affect the millage rate nor the total tax income.

Under this process, the removal of property from the tax rolls or the addition thereto, does not by itself affect the tax dollars to be collected from real estate. The tax rate is controlling. But the tax rate is not determined until the budget is adopted and compared to the assessed value, or tax base. Therefore the taking of privately owned real estate for public highway purposes by this action alone does not alter the real estate tax income of the taxing authority.

What happens is simply that the total tax burden is redistributed. The specific taxpayers giving up their property to the highway pay a significantly less amount (or none at all) and each of the large number of other owners of taxable property pays a slightly increased tax. It is the net total tax income that is the net measure of the tax consequences and not what happens to one set or separate sets of citizens.

The Institution of Local Government

To understand the financial and tax structure of a local unit of government such as a city, it is necessary to view the governmental agency of the city as the people's self-appointed (or elected) organization created to handle the overall community business and functions. The city government is the people themselves. The government is not a separate, detached business that controls the people and their activity. Government is the operating agency within the cooperative society, or mutual society, to which the people have delegated responsibility for management of their community affairs. A city government is not unlike a cooperative marketing group, or a mutual insurance company, each operated by a board of directors for and in the best interests of their members.

Thus when the city council takes action that affects tax incomes or cash disbursements, such actions must be considered in their net of total results as they affect the total city, not as they affect an individual or selected group of individuals (except of course as to whether an action which is adverse to a few people is beneficial to all others). In the thinking and actions of the general public there is a strong tendency to view their government (on all levels) as a distinctly separate corporation over which they have no control and of which they are not a part. These persons fail to keep in mind that their government is actually themselves, and that they are not innocent bystanders who benefit–and sometimes suffer–by actions of their government.

In taking taxable land for any public function such as parks, school buildings, libraries, city halls, or streets, the decision to do so is based upon the finding that such action is in the best interest of the public community as a whole. Each fiscal year brings adjustments, trade offs, shifts, and realignments in the tax rates and structure and in the budget. The only sound measure of the financial results of the actions taken is whether the total cost of operating the city's business is increased or decreased, and whether incomes in total are increased or decreased as a result of these decisions. And this measurement must be applied to the total functions and total taxes and not by segments of activities or by separate groups of taxpayers. Thus any public action which removes land from the tax rolls does not within itself mean a reduction in tax income to the taxing district. Although in the interests of tax equity who pays how much tax is a factor, but for analysis of the total effects of a governmental action, the real measure of consequences is its effect on total income or total expense.

It is customary at all levels of property taxation to review periodically property assessments for real estate tax purposes to adjust the assessments to their proper relation to market values and to assure that as between different properties the assessments (and therefore the tax burdens) are equitable. In this tax equalization, or reassessment, procedure some individual property assessments are increased and some decreased. As a result of the reassessment, certain individuals pay more and some pay less tax, but whether the taxing district as a whole receives more or less tax revenue depends upon the overall final total tax base as compared to the former base and the new tax rate. Total tax income depends first upon the budget to be raised, and secondly upon the total property assessment from which the taxes will be produced.

Should there be a change in public policy to assess for tax purposes all property now exempt from real estate taxes, the result would be an increase in taxes for the few organizations added to the tax rolls and a decrease in taxes for many other landowners provided there was no change in the budget sum to be raised. In theory, should a new private development costing a million dollars, say an apartment building, be established within the taxing district there would be an increase in property tax income. But here again, whether or not this building alone results in an increase in total tax income for the district depends upon whether the budget for the taxing district is increased or decreased, and on the tax rate.

An examination of vacating right-of-way as public property may add some light upon the question of whether the taking of private land for right-of-way results in a tax loss. When a highway, street, or alley is vacated by the public, the land usually reverts to the adjacent property without payment therefor by the landowners. In other cases the land may be sold to the adjacent land owners or other persons. In either case the land is picked up on the tax roll and assessed at the same rate as other comparable land. If such action took place on a grand scale, say equal to 25 percent of all taxable property, the people would seemingly make a great gain in tax income. But would the tax income increase by 25 percent? In all probability not. The responsible legast legislative body would reduce the millage by 25 percent unless the budget were to be increased. Here again the transfer of land from the public to private ownership does not of itself mean a tax gain to the public individually or collectively. The answer cannot be had unless the total amount of taxes to be raised is taken into consideration.

If the theory that taxes forgone on property taken for highway purposes (or any public purpose) become an annual cost chargeable to the highway (or other public function) is accepted, it must logically follow that all existing right-of-way and other public land is currently resulting in real estate tax losses to the respective taxing jurisdictions. If this statement be true, then to restore this land to the tax rolls would result in great increases in real estate tax incomes. But such increases in tax income would not materialize unless the budget for the taxing district was increased a like amount.

The Case of Maximum Legal Tax Rates

There are a few agencies of local governments which are limited by law in their millage tax rate. When these local governments are taxing at the maximum legal rate, any reduction in the total assessed valuation of property would result in some loss of tax income.

Where the maximum millage rate is set by higher legislative authority and that maximum rate is in effect at the time of taking assessed land for right-of-way, the taking of right-of-way land as a single action will not affect the tax rate.

However, because the tax base is reduced the tax income will be less, and therefore the budget must be lowered a corresponding amount. In this case there is not a net cost to the people but an actual reduction in expenditures. The result is that the local government operations are forced to exercise greater economic efficiency to maintain their same services as before or they must reduce their total functions somewhat to match their reduced tax income. An alternative, of course, is to raise additional income through a source other than real estate taxes.

Probable Results to Property Taxes of Highway Improvements

What happens to property taxes as a general result of highway improvements and the taking of land for right-of-way? Does the assessed valuation of taxable property go down, remain the same, or go up? The more common experience is that the value of real estate affected by highway improvement increases, starting even before the highway construction is completed. Adjacent land is particularly prone to increase in price and to change to a higher value use–farm land to residential land, for instance. Further, more often than not, owners of land and buildings taken for right-of-way reinvest their right-of-way money in homes or establishments of higher value than that which they sold to the highway authority for right-of-way. The usual consequence, then, is that the taxing authority is better off afterward than before the taking, based upon assessed valuation of taxable property. In effect, this consequence is to be expected because if the community were to suffer material real loss through highway construction, the people just would not construct highways.

There is a last consideration for those persons who insist on treating right-of-way takings as a tax loss, and therefore as a highway cost. Honest analysis would call for treating any tax gain resulting from the new highway as a tax benefit, and thus as a beneficial consequence, credited to the highway. When this procedure is followed, the net result usually will be not a highway cost but a community tax benefit. To fail to consider the tax gains resulting from highway improvement, while considering the tax losses is an unacceptable, inconsistent procedure which violates the principle of economic analysis–consider all consequences to whomsoever they may accrue.

COMMUNITY CONSEQUENCES OF A NONTAX CHARACTER

In urban areas, major highway improvements–expressways and freeways–are often a device used to upgrade sectional areas and corridors in combination with urban renewal projects. Slums are cleared and substandard buildings are removed. Land use is then changed to higher order and new buildings are constructed. The end result is that not only is the tax base increased, but the city environment, health, and attitudes are greatly improved. Even widening, repaving, and general street improvement often are followed by new store fronts, remodeled residential buildings, and a general "sprucing up" of the neighborhood.

Consequences of Right-of-Way Takings

Socially, economically, and taxably the community is better off, exclusive of the benefits to transportation. In the analysis of the economy of or of the general economics of these types of urban street and highway improvements, consideration of the values is in order, the property tax factor included. Right-of-way takings for highways become a complex and important social and economic factor which is deserving of a thorough analysis and penetrating study. The offhand and freely given opinions and answers are not to be accepted except when backed up by thorough investigation of all elements by competent technical persons.

ECONOMIC COST OF LAND TAKEN FOR RIGHT-OF-WAY

The tax considerations applicable to real estate are entirely separate from the considerations of the economic and social consequences of transferring real estate from private ownership and use to public ownership and use. Certain appropriate considerations are not related to ownership at all, but to the consequences resulting from a change in the use of land as a good used in the production of other goods.

Land has value only because it is an element in the economic system that is productive of desirable products and services. These products and services, agricultural crops and areas on which to construct buildings, for instance, are necessary to sustain man on earth. These products and services are supplied by the property owners for profit, or for return on their investment.

TOTAL COST OF HIGHWAY RIGHT-OF-WAY

The costs of right-of-way as acquired by public auhtority may be classified as follows:
1. Agency overhead
2. Legal and court procedures
3. Price paid for land
4. Price paid for improvements on the land
5. Price paid for crops or work in process
6. Damages paid–severances or other just compensations

The final net cost of highway right-of-way would be the cash disbursement cost less any incomes from rent (prior to time construction began) and sale of buildings, crops, topsoil, or salvaged materials.

A cost to the owner or tenant, not compensated for in the above cost to the public, is his net relocating and reestablishing expense, if any. A resident on a given property taken for highway use must move his domicile and personal effects to another location. This he must do at his own expense, except for some allowances from public funds, recently adopted as public policy. When only a portion of a farmstead, commercial establishment, or residential land is taken, the owner often must readjust his holdings, operations, and schedule to something appropriate to less land–and often irregularly shaped land. Where these costs to the former land owner are real and can be identified, there would be justice in including them as a highway cost, making certain that these costs were not already included in damage payments.

Likewise, some owners become better off as the result of the taking of their property for highway purposes; if so, their gains should be used to offset the highway costs.

But the more important consideration is whether the taking of real estate for highway right-of-way results in an economic cost to society beyond the direct price paid for the taking. The answer is found in the analysis of the consequences pertaining to production of goods and services necessary to the maintenance of the social and economic system.

Farm Land

First, consider land used for agricultural production–crops or livestock. When 10 acres of land producing 750 bushels of corn become 10 acres of highway right-of-way, the result is that the 750 bushels of corn are not produced, at least not immediately. This is cost in the same line of reasoning that 10 gallons of gasoline not consumed because of a highway improvement is a net gain.

Assuming that the 750 bushels of corn are necessary to sustain the people, some means must be found to produce these 750 bushels on other land. Two possibilities present themselves. First, the productivity of other land may be increased at some cost by improved management (tillage, fertilizer, improved seed, irrigation, or processing). Second, other land heretofore not suitable for corn may be converted to corn by improving it, again at some expense, and at a sacrifice of the former use of the land. There are other possibilities of shifting land use among agricultural crops.

But in the end there will be incurred some monetary cost over and above the prior cost of producing the 750 bushels of corn. This increment of cost is justifiably included as an economic cost of the highway improvement.

Nonfarm Land

The considerations for land used for residential, commercial, or industrial purposes are similar to those for farm land. Assuming that the product of the land–housing of some sort–is essential to the people, some expense may be required to produce the same products or services that prevailed before the use of the land was changed to that as a highway right-of-way.

The concept here is one of preserving production up to its level prior to the taking. In a growing population and dynamic economy, such restoration of production is essential. Where this restoration of production incurs an invest-

Consequences of Right-of-Way Takings

ment or annual expense greater than that which prevailed before the taking, and when such expense was not compensated for in damages to the owner, such increment of expense is a justifiable charge against the highway.

Charging the cost of taking the right-of-way against the highway along with construction costs measures the cost of changing the land use and is an appropriate charge. The price paid for the land was undoubtedly based upon the present worth of its annual production. Therefore its price reflects the present worth of the future production from its use at the time of taking. No further charge is justified because the land is now producing highway transportation rather than salable and consumable crops, products, or services.

The price paid for buildings, damages, and items other than land does not result in tangible usable highway property. But the cost of these items is just and real. In a strict accounting sense, and in public utility management, these costs would be written off as operating expense (amortized) in a comparatively short period, the same as financing costs often are.

PROBLEMS

18-1. Get from your local town, city, or county office for the past 5 years the total budget supported by property assessments, the total assessed property valuation, and the millage tax rate. Write an analysis of these three sets of annual figures with particular reference to changes year to year and the reasons therefor.

18-2. By the proper contact with the appropriate official, inquire about the exact procedure by which the property tax rate is set—when and how—with particular relation to the budget and assessed valuation. Inquire about the procedure for setting the tax rate in case a highway removed $1,000,000, or other significant sum, from the assessed valuation subject to property tax.

18-3. Review some of the available literature on the subject of land values as affected by highway improvements. Comment upon your readings in the light of the subject matter of this chapter.

18-4. There are current trends toward fully compensating from public funds home owners and business owners for all costs incident to their forced move from land taken for highway rights-of-way. The compensation would include all costs to reestablish themselves in equal or better housing. Discuss this subject as a broad public policy, including its administration, breadth of coverage, boundaries of coverage, equity, and other factors.

REFERENCES

The literature on the economic consequences of highways is extensive. An excellent overall reference for land use and other effects together with an extensive list of references is 18-1 in the following listing.

18-1. Bureau of Public Roads, U.S. Department of Transportation. *Highways and Economic and Social Changes.* U.S. Government Printing Office, Washington, D.C., 1964.

18-2. W. G. Adkins. *Effects of the Dallas Central Expressway on Land Values and Land Use.* Texas Transportation Institute, Bulletin 6, 1957.

18-3. W. G. Adkins and A. W. Tieken. *Economic Impact of Expressways in San Antonio, Texas.* Texas Transportation Institute, Bulletin 11, 1958.

18-4. D. D. Carroll and others. *The Economic Impact of Highway Development Upon Land Use and Value.* University of Minnesota, September 1958.

18-5. W. L. Garrison and M. E. Marks. *Influence of Highway Improvements on Urban Land.* University of Washington, Highway Economic Studies, Seattle, 1958.

18-6. James H. Lemly. *Expressway Influence on Land Use and Value, Atlanta, 1941-56.* Georgia State College of Business Administration, Research Paper No. 10, November 1958.

18-7. Massachusetts Institute of Technology. *Economic Impact Study of Massachussetts Route 128.* Department of Civil and Sanitary Engineering (A.J. Bone, Project Supervisor). December 31, 1958.

18-8. Herbert Mohring. *The Nature and Measurement of Highway Benefits: An Analytical Framework.* Transportation Center, Northwestern University, Evanston, Ill., 1960.

18-9. L. V. Norris and Herbert W. Elder. *A 15-Year Study of Land Values and Land Use Along the Gulf Freeway in the City of Houston, Texas.* Houston, 1956.

18-10. University of Denver. *A Before and After Study of Effects of a Limited Access Highway upon the Business Activity of Bypassed Communities and upon Land Value and Land Use.* October 1958.

18-11. University of Kentucky, Bureau of Business Research. *The Effect of the Louisville Watterson Expressway on Land Use and Land Values.* 1960.

18-12. Paul F. Wendt. *The Influence of Transportation Changes on Urban Land Use and Values.* Highway Research Board, Bulletin 268, 1960.

Chapter **19**

Nonuser Consequences

Beginning about 1950 there emerged a second wave of intense interest in the social and economic consequences of highway improvement. The first wave started about 1913 and continued into the 1920's. The current interest stems from three main issues: (1) the desire to achieve the highest possible state of economic efficiency, (2) the trend to impose the cost of highways upon the beneficiaries thereof, and (3) concern about highway location with respect to land use, population location, social disruptions, and all forms of business activity. Much of this renewed interest comes from the urban and suburban transportation needs and from the fact that much highway construction today is reconstruction and relocation to provide greater vehicular capacity and an improved quality of service.

Since highways are everyone's business and everyone is an "expert" in this business, the social, economic, business, and community viewpoints are outwardly spoken by this large group of experts–professional and laymen alike. To treat the subject adequately would fill two books at least–but within a few pages the high points can be mentioned.

Economy and economics need to be separated; transfers and offsets have to be identified; and from whose viewpoint is vitally important.

CLASSIFICATION OF CONSEQUENCES OF HIGHWAY IMPROVEMENT

The economic and social consequences of highway improvements may be classified by more than one scheme, depending upon the objective. Possible schemes are:

Road user and nonuser of roads
Market and nonmarket
Primary and secondary (or direct and indirect)
Tangible and intangible
Beneficial and adverse
Reducible and irreducible (to dollar values)
Economy and economic
Social and community

For the purpose of this discussion, the objectives will be centered upon identifying the consequences and grouping them for two main purposes: (1) the economy and market items which can be included in the analysis for economy—both project evaluation and project formulation, and (2) the economic and nonmarket items which are considered in the final decision process, separately from the analysis for economy, but in conjunction therewith.

Generally speaking, the road-user group of consequences includes the market, primary, tangible, and dollar reducible items which are economy based. The nonuser group includes the nonmarket, secondary, intangible, and irreducible items which are social and economic in nature. The beneficial and adverse items are found in both the road-user and nonuser groups.

Although in the United States the road-user and nonuser groups are close to being the same so far as people are concerned, they need to be separated in studies of the economy, the economics, and the financing of highways. Man has many dual roles to perform in society. He is both a taxpayer and a beneficiary of taxes, he is a consumer and a producer, and he drives himself over the highway as a road user, and he is a member of the community which receives benefits from the highway.

In highway cost allocation studies for the purpose of determining tax responsibility, the community at large is one group, the land owners another, and the road users are a third group. Yet the people in the three groups are practically the same individuals. Justice and equity, however, warrant separate consideration of each group.

NONUSER FACTORS

The following discussion will be directed primarily to the aspects of land use, land value, business location, and business volume as affected by highway location and highway design. The general community aspects are considered only incidentally. The general nonuser consequences involve the complex factors of general economics. It is most difficult to isolate the beneficial and adverse consequences and the transfers to get to the net change because of the far-reaching character of the economic forces and their tendency to overlap and to offset. Here also the important factor of from whose viewpoint has to be considered. A community local to the highway improvement gains business volume and employment, but another community may have lost that business and employment. The movement of the textile industry from New England to the South is an example. Land close by the highway improvement increases in market price, but farther away land may decrease in market price.

The nonuser consequences may be classified into the following economic categories:

1. Spending for highway construction

2. Land use and land values
3. Highway travel oriented business
4. Highway-induced expansion or relocation of industry and business

SPENDING FOR HIGHWAY CONSTRUCTION

The billions of dollars spent yearly for highway construction have a great impact upon the nation. This public action sustains much direct employment and much industry through consumption of road-building materials and equipment. In the main, the financial resources to support road construction come from direct taxation on the motor vehicle and its use. The population, therefore, is getting highways as a direct result of paying for the highways.

The Gross National Product

Improved and extended highways are the desire of the public. It is the public's choice to devote this amount of its spendable income to this purchase rather than to other purchases. Economically speaking, the nation as a whole and the gross national product are affected, but affected much less than most people realize. Presumably, the public would spend what is now the highway tax dollar on other goods, services, and transportation if its choice were for those purposes rather than for highways. This choice results in an economic transfer from certain other alternatives for spending to highway transportation rather than a 100 percent new economic creation. But to describe our economic society as it would be now without the motor vehicle requires a most imaginative mind.

In times of economic depression, highway construction may be chosen for special support with general tax revenue or through borrowing. Instead of for highways, depression money could be designated for school buildings, hospitals, recreational facilities, or water resource projects, and about the same economic consequence would result. Highway transportation however, as an economic force is powerful, far-reaching, dynamic, and universal in its contact with the population.

Highway transportation does no doubt make some contribution to the economy of the country and to the gross national product. So far its specific contribution has not been identified, nor quantitatively measured. Highway transportation supports transportation by the railways, airways, waterways, and pipelines. Much of highway transportation, however, is a shift from other modes to highways. Local hauling of goods and people is such that the highway is the most economical method; much of the railway and other freight, as well as passenger service, begins and ends its travel on the highway. Rail, air, water, and pipeline haulage depend for profitability upon high concentration of travel or shipping between origin and destination; the motor vehicle is relied upon for the less dense and sparse haulage.

Cost-Reducing Factors

In the hauling of household goods, certain perishable produce, and in such industries as logging, motor vehicle service is less costly than by any other mode. In these areas, highway transportation makes a net beneficial contribution to the national economy. But as will be discussed later, not all consequences of highway transportation produce a net gain. The main and most important contribution of highway transportation to the economy of the nation comes from the lowering of transportation cost, reducing the cost of doing business, and in making it possible to reach new land and win new natural resources. Improved highways to replace existing roads provide great road-user benefits through commodity savings, accident reduction, and travel time reduction; these are the real net benefits from good highways and the contributions to the economic welfare of the nation.

Transportation brings materials together to make products and then to transport these products to points of consumption or installation. It frees men to live above mere subsistence. Transportation unites men, materials, and machinery to produce more goods and produce them more efficiently. It provides the means of trade between peoples and areas. But the production of more goods is not a gain unless transportation is available to move the goods to places of consumption.

The interstate highway system will free much capital now tied up in inventories–at the factory, in transit, in regional warehouses, at wholesalers, and at retailers–by providing faster, more reliable day-to-day deliveries. The capital released by reduction of inventories becomes available for other investments, other operations, and research and development.

LAND USE AND LAND VALUES

Construct a new highway on new right-of-way or greatly improve the roadway along an existing right-of-way and the nearby land use pattern begins to change immediately. With this change in land use, land prices move upward. These consequences are almost certain of prediction; they have been identified and measured ever since the construction of motor vehicle highways began about 1914.

This aspect of land as related to highway transportation is proof that good transportation is in demand, and that, for both goods and people, there is either net economy to good highways or a high degree of personal preference for which people are willing to pay. In the earlier days farm-land values increased somewhat as a result of the all-weather roads which provided access to the farm land and access to trading centers and railroad service every day of the year. This same result continues today with respect to farm land. But, also, there are now other aspects prevailing and, in addition, a strong urban development.

Migration of People and Business

A good highway is one of the factors that draw people from urban centers to the suburban area and to the country as a place to live. Along new highways, farm land is converted to residential land, and new communities are established. With this change in land use comes land price increases of 100 to 1,000 percent. The home-to-work commuter is trading urban living and short distance home-to-work travel for suburban and country living and longer distance home-to-work travel.

To support the new population in the suburban areas, retail shopping and service establishments follow, often in large complex trade centers. These movements of people and business outward from the core city often cause a decrease in population in the main city and a reduction of retail trade therein.

A further development that results in increases in land values along new or highly improved highways is the movement of light industry to locations along these highways. The highly satisfactory transportation for both employees and the business callers attracts industry, especially those concerns ready for a new plant or a new office building. These business developments are more concentrated and more complex along the highways in suburban areas than in rural areas, but they are evident everywhere. Even in dense urban areas new buildings soon appear in locations convenient to urban expressways.

This brief, general description of the consequences to land use and land values resulting from improved highway transportation represents the usual experience. The development is more marked and more rapid in areas of expanding economic activity and in attractive land areas. The more recent experience indicates that the same general pattern is forming at the main interchanges on the interstate system where the access is fully controlled. This development in land use and land value is a dynamic economic force, made active by the advantages of the highway. So far no way has then discovered to stop it. Not that the development should be stopped, but perhaps it should be controlled. The ultimate price paid, of course, is that the new highway and the new interchange are soon inadequate in traffic capacity. New highway construction is then called for, far sooner than expected. Some discussion of the underlying factors embodied in this attraction of population and business will help to emphasize its implications with respect to economy and economics.

Role of Motor Vehicle Operating Costs

Higher-speed highways and highways that reduce overall running cost of motor vehicles are attractive to people seeking new home sites and attractive to industry seeking new plant locations. These land-use changes are accompanied by a third economic consequence, namely, changes in the retail businesses that cater to the highway traveler, both the local commuter and the through long-distance traveler (or truck driver). Following the residential, business, and industrial developments, come the service businesses and goods merchants that are needed to support the local population and local employees.

Because of the profit nature of their business, because of the travel time reduction, and because of a reduction in motor vehicle running cost, these new landowners can afford and do pay higher prices for the land than the current market price of the land before construction of the new highway. Thus this higher price reflects the greater productivity of the land in its new uses. This is not a new discovery—merely another manifestation of what has been happening for years. The highest-priced land in an urban area has been that land at the city business center. The land price has reduced with distance outward to the city fringe adjacent to farm land. As the city has expanded outward, pressure has forced the farmer farther out in favor of the city dweller who paid higher prices for the farm land than could be justified from farming.

Economic Gains and Losses

What is the net economic gain or loss that results from these changes in land use and increases in land values, considering land alone, without its associated use or business activity?

The increase in land value represents capitalization of future profits that are expected from the business to be conducted thereon, or from the advantages of living thereon. In town, these profits and advantages result because of improved highway transportation. Highways did not create these values, but the highways were the cause of the specific local change in land use and consequent increase in land value. What then happens?

Role of Land-Use Changes

If all highways in the country were to be improved at the same time to equal high standards, would the adjacent land increase in market price? The answer is no. All land would then have equal potential so far as transportation is concerned, and there would be no concentration of new residences and new businesses simply because of an improved new highway. An individual new highway attracts population and business because of its advantages over similar locations, but locations not blessed with so favorable highway transportation.

The land-value changes result from more favorable transportation as compared to locations not so favored with equal quality of transportation. To include both the highway user net benefits and the land-value increases in a study of the economic gains from highway improvement would be double counting [Zettel, 19-25-7 and 19-46]. The gain in land value is only the reflection of the gain to the highway users because of the improved highway. If the new highway offered no advantage to the highway user, there would be no increase in the value of adjacent land as a direct result of the highway.

To also include gains in profits to industry and business would be triple counting, for such business gains are in turn reflected in the increased land values.

Windfalls and Tax Gains

There are two economic consequences not yet mentioned. The first is the fact that individual land owners may reap sizable windfalls from selling their land at these high prices brought on by the highway improvement. To such landowners their gain is unearned. Of course, through income tax provisions the public treasury will capture a share of the gain. There have been discussions about how to recapture additional portions of these gains through other schemes.

The second consequence is public gain in the form of future greater property tax income from the higher value land (and buildings thereon). But such gain is not all net, in fact, there is no assurance that there will be a net gain. The additional population and buildings will increase the community expenses for protective services, education, streets, libraries, and public services of all types. The final net change cannot be known, however, until the budget and the tax rate are examined. As discussed in Chapter 18, there may result only a shifting in the tax burden between taxpayers.

Decreases in Land Values

So far this discussion has related mainly to increases in land values. But what about decreases in land values? Decreases do occur, but often they are not discovered because they may be widely scattered and most difficult to trace to the new highway. But to the extent that they do occur they certainly offset the gains.

In certain cities in recent years, central business district real estate values have declined. This decline has often been attributed to a decline in retail trade and in downtown business generally. These business declines in the core city have been attributed to the shift of business to the suburban areas along new highways. In some instances it has been possible to trace these declines in central land values to outward movement of specific establishments [19-27].

In other instances the central business area or the total core city area has not enjoyed the normal increases associated with population increase because the increased population and associated new businesses settled in the outward areas. In these instances there may not have been a decrease in land value, but there was no gain either. Yet, in general, the community could expect gains.

An easily recognized decrease in land value and in business occurs where a major highway supporting heavy highway-oriented business is largely replaced by a new parallel highway a short distance away. These types of losses are both temporary and permanent. They are temporary when the local community has the vigor for growth and ultimately can support itself without business from the through traveler. The loss is permanent when an isolated business, such as a motel, is left without sufficient through traffic to support it.

All in all, land-value changes are not a part of economy studies for economic evaluation or project formulation of highway improvements because (1) land value is a reflection of the road-user benefits, and (2) if all consequences to land could be located and priced, the net gain, if any, could be small in proportion to the positive and material road user gains. Not including the changes in land values in the analysis for economy, however, does not preclude giving land value some weight in the final decision on highway location when the viewpoint is properly and strictly a local viewpoint.

CONSEQUENCES RELATED TO BUSINESS ACTIVITY

Apart from the changes in land use and in land value, highway improvements and traffic operations affect business volume, particularly those businesses which are adjacent to the highway or street. Highway-oriented businesses–motor vehicle services, food, and lodging–are highly sensitive to their location with respect to the highway and to traffic characteristics; other businesses, such as those catering to home and family needs, are sensitive to access, parking, traffic direction, and traffic congestion. The consequences to retail business and trade volume brought on by highway improvement and traffic operations are important factors to consider in the final decision on highway location and other aspects of design. These factors are especially responsive to whose viewpoint is applied.

Highway-Oriented Business

On the basis of demand for business services, the road users may be divided into two groups: (1) the local resident who is interested in motor vehicle services (fuel, oil, tires, and maintenance services), and (2) the through traveler who is interested in the vehicle services and, in addition, food, lodging, and personal wants. The local resident also is a road user when he uses his vehicle to get to and from all places where he purchases commodities and services, but at the moment this business will be neglected in the following discussion.

The highway-oriented business volume at any one particular location is affected by traffic volume, traffic mix, traffic origin and destination, traffic speed, highway geometrics, and other factors. Those business locations that are more or less rural, at least those that are in somewhat open space free of congested highway use and compacted business development, depend upon the through traveler more so than those businesses which can pick up some trade from local residents. Local residents are not wont to take their motor vehicle trade to congested arterial streets if other locations are available. Likewise for the through traveler; he would rather stop for service at some place free of traffic congestion.

When a new highway becomes available on a new location, the through traffic will shift almost 100 percent, and the local traffic will shift to some minor amount. The general result is that the motor vehicle service business in the former traffic-congested area may decline temporarily until local customers discover that it is now a convenient place to do their business. The rural or more isolated establishments catering to the through traveler will suffer a heavy loss in business volume, with little or no opportunity to make up the loss from local trade. The trade from the local vehicle is given to those establishments that are conveniently and safely reached without traffic annoyances and disturbances.

Central business districts have experienced more often than not increased business when their main streets have been cleared of through passenger cars and trucks. Even the elimination of curb parking and adoption of one-way traffic have improved rather than hindered trade volume in the main business centers.

Local General Trade

The net of the consequences to retail trade which results from highway improvements is probably zero gain and zero loss even for a fairly small community. Highway travel does create some net addition to local and national economy, and highway routes and highway excellence do strongly affect the specific location where the highway traveler spends money for vehicular services, food, and lodging. In local areas the exact place of expenditure by local residents may vary from one street to another as a result of street improvement and traffic operations, but certainly the nearby urban areas would not be appreciably affected. The custom of the through traveler may be moved to an adjacent town because of highway location, but here again the consequences are not far-reaching geographically. The net result is about an equal plus and minus between local merchants within a given trade area or neighborhood.

Contribution by Highways

Good highways, as contrasted to poor ones, do create some additional travel, and therefore increased purchases of motor vehicle services and personal needs of travelers. In this respect there is a net gain in trade volume in the commodities purchased by the traveler. The net overall result, however, is again an economic transfer. The good highway as such does not contribute to the road user's net income. Any additional travel expenditures which are the result of the improved highway must come from the transfer of support from the traveler's normal economic activity to support of highway travel. To finance the added travel by highway, the traveler may take any one of the following actions:

Travel less by other modes of travel
Eat less expensive meals; spend less on motels
Curtail his expenditures for recreation
Postpone buying new clothing, household needs, or appliances
Undertake more do-it-yourself projects
Give less to the community chest or to his church
Save less money for investment
Curtail his cash outgo in many other ways

These economic consequences may be far-reaching and extend backward to the source of raw material and primary labor. The net plus and minus result is still zero, because his total cash spent is the same; the traveler has simply transferred some of his daily economic support of personal and family activities to more support for highway travel. Perhaps extensive research could run out the full length of the chain of economic actions and reactions to isolate these offsets, transfers, gains, and losses. The research task is formidable and of doubtful end value. As long as the economic system is understood and offsets and transfers are recognized, the highway management can make good decisions as to highway route locations and highway design once he elects the proper viewpoint.

CONSEQUENCES TO GENERAL INDUSTRY

Most of the foregoing discussion has related to retail trade–that of a family nature and that which is oriented to highway travel. Highway improvements are now a forceful factor in determining the location of manufacturing, distributing, warehousing, and servicing types of industry, both light and heavy. This role of the highways is equal to that of the railways some years back. The essential factors are cost of land, transportation inward of essential materials and supplies; transportation outward of manufactured products; transportation facilities and travel time for employees; and available labor supply. Locations on new land close to urban centers, or even 10 miles out in the country, satisfy these factors when the modern high-speed highways is present. Therefore the construction of new highways, such as the outer belts around large urban centers, attracts a variety of establishments that cater to national, regional, or state-wide markets. But where do these companies come from?

Business Expansion

As old companies expand their business they seek enlarged facilities; when new plants are in order, they often find it more profitable to move outward to new locations and build horizontally rather than to remain in the industrial heart of the city and build vertically. When urban rehabilitation projects and urban expressway projects come along, displaced companies are inclined to move outward to the fringe areas.

Advancements in technology and new products have brought on the need for new plants to manufacture new products. Included here are the electronic industry, the plastic products industry, instrument manufacturing, and research institutes. Route 128 around Boston [19-31] is noted for this development; around other cities the same development is in evidence.

The transportation advantages of new highways are the direct factors causing the managements of these industries to locate on new highways. These new highways do not create the businesses nor cause any net gain to the gross national product (at least no more than would an equal expenditure elsewhere). No doubt

Nonuser Consequences 503

every industry that has located on a new highway soon after its construction would have come into existence anyway had that highway or other highway not been constructed.

These industries, new to their locations, certainly result in business gain and employment gain in their respective areas. To these communities such economic consequences are a gain or a loss, according to the local viewpoint. Such business is a benefit when it is desired and a detriment when it is not desired. In relation to local property taxes, the taxing district may or may not be ahead, depending upon what it will cost to support the industry, its employees, and other added population. Protective services, schools, parks, governmental services, sanitary provisions, and street transportation all cost more as population and real estate holdings increase. Before any community can justly conclude that it is better off with new industry, high-rise apartments, and added population, a thorough analysis is in order of its economic and financial position with and without these added taxable assets and population increase.

For the decision maker with respect to highway location and design, there is little he can do to control these industrial developments. Proper local authority can practice land-use zoning to attract or discourage industry, according to local policy. If the local viewpoint is to encourage new industry, perhaps one route location could be preferred to another, road-user economy being about equal. As stated with respect to local retail business, the highway authority needs to be aware of the real economic truth about the gains and losses attributed to new industry, so he can make an enlightened and proper decision.

Contributions from Highways

Perhaps much of the present chapter may leave the impression that the author believes that highways do not contribute to the economic growth of the country and to the well-being of society. Such impression is not intended–at least not wholly so–but there is present an attempt to raise some doubts about the beliefs of many persons who see only the economic changes immediately in front of them. Highways are everyone's highways and should be located, constructed, and operated for universal good, not just for the benefit of the few. It is not objectional for one community to compete for economic gain against another community, but highway authorities should understand that what may appear to be an economic advantage to one community may be offset by an economic disadvantage to another community.

All countries which have grown socially and economically have done so with powerful assistance from growing transportation facilities and their use. Economic health is achieved and made to grow through transportation which brings the essential products and forces together at the right place and time. As highways are extended and improved in quality, the country's economic growth will continue. But with this growth will come great change and hardships for some people and some economic activities. Already the small town is disappearing (the author's town of birth in Iowa has long since disappeared) and the urban areas are deep in transportation muddles.

PROBLEM

19-1. Studies of the economic consequences of highway improvements indicate that gains in the value of adjacent lands usually follow the highway improvements. Proposals have been advanced for the public to recapture some portion of this "windfall" gain to the local landowners. Discuss the pros and cons of such a proposal. Suggest how such gains could be partially recaptured by the public–state, county, or city.

REFERENCES

19-1. William G. Adkins. *Economic Effects of the Camp Creek Improvement. A Special Report to the Bureau of Public Roads.* Texas Transportation Institute, College Station. Texas, September 1958.

19-2. William G. Adkins and Alton W. Tieken. *Economic Impacts of Expressways in San Antonio.* Texas Transportation Institute, Bulletin 11, College Station, Texas, August 1958.

19-3. J. K. Allen and Richard McElyea. *Impact of Improved Highways on the Economy of the United States–A Special Report to the Bureau of Public Roads.* Stanford Research Institute, Menlo Park, Calif., December 1958.

19-4. A. J. Bone. *Economic Impact Studies of Highway Development and Their Implications for Nearby Urban Centers.* Presented at 45th Annual Meeting, American Association of State Highway Officials, Boston, October 15, 1959.

19-5. O. H. Brownlee and Walter W. Heller. "Highway Development and Financing," *American Economic Review,* Vol. 46, No. 2 (May 1956), pp. 232-250.

19-6. Christopher Brunner. *The Economic Justification of Roads.* Institution of Highway Engineers, London, November 3, 1961.

19-6A. Bureau of Public Roads, U.S. Department of Transportation. *Highways and Economic and Social Changes.* U.S. Government Printing Office, Washington, D.C., 1964. (This reference is an excellent one for summaries of the many types of economic consequences as related to types of highway improvements. Also, extended lists of references are given at the end of each chapter.)

19-7. California Department of Public Roads. *Report on Westside Freeway Route 238: Economic Effects of Six Proposed Routes in the Tracy Area.* Division of Highways, Right-of-Way Department, Land Economic Studies Section, Sacramento, Calif., July 1958.

19-8. Donald D. Carroll and others. *The Economic Impact of Highway Development upon Land Use and Value: Development of Methodology and Analysis of Selected Highway Segments in Minnesota.* U. of Minnesota Press, Minneapolis, September 1958.

19-8A. Marion Clawson and Jack L. Knetsch. *Economics of Outdoor Recreation.* Published for *Resources for the Future* by Johns Hopkins Press, Baltimore, 1966.

19-9. W. L. Garrison and Marion E. Marts. *Geographical Impact of Highway Improvement.* Highway Economic Studies, Up. of Washington Press, Seattle, 1958.

19-10. W. L. Garrison. *Influences of Highway Improvements on Urban Land: A*

Nonuser Consequences 505

Graphic Summary. Highway Economic Studies, U. of Washington Press, Seattle, 1958.

19-11. Joseph W. Harrison. *Bibliography: The Economic Effects of Limited Access Highways and Bypasses.* 2d ed., Virginia Council of Highway Investigation and Research, Charlottesville, August 1957.

Highway Research Board. National Academy of Sciences, National Research Council, Washington, D.C., —

19-12. Bulletin 189, *Land Acquisition and Economic Impact Studies,* 1958.

1. David R. Levin. "Report of Committee on Land Acquisition and Control of Highway Access and Adjacent Areas," pp. 1-56.

2. A. J. Bone and Martin Wohl. "Industrial Development Survey on Massachusetts Route 128," pp. 57-89.

3. Walter C. McKain, Jr. "Economic and Social Impact of the Connecticut Turnpike," pp. 90-95.

4. Joseph W. Harrison. "Methods Used to Study Effects of the Lexington, Virginia, Bypass on Business Volumes and Composition." pp. 96-110.

5. Robert S. Curtiss. "Tenant Relocation for Public Improvement," pp. 111-125.

19-13. Bulletin 222, *Highway Investment and Financing,* 1959.

1. Robert F. Baker. "Fundamental Problems in Relating Vehicle Size to Highway Costs," pp. 1-26.

2. O. H. Brownlee. "Pricing and Financing Highway Services," pp. 27-31.

3. M. Z. Kafoglis. "Price Theory and Tax Equity in Highway Finance," pp. 32.48.

4. Tillo E. Kuhn. "Use of Economic Criteria for Highway Investment Planning," pp. 49-74.

19-14. Bulletin 227, *Highways and Economic Development,* 1959.

1. J.' H. Lemly. "Changes in Land Use and Value along Atlanta's Expressways," pp. 1-20.

2. A. J. Bone and Martin Wohl. "Massachusetts Route 128 Impact Study," pp. 21-49.

3. William G. Adkins. "Land Value Impacts of Expressways in Dallas, Houston, and San Antonio, Texas," pp. 50-65.

4. William L. Garrison. "Approaches to Three Highway Impact Problems, pp. 66-78.

5. Stuart Parry Walsh. "Some Effects of Limited Access Highways on Adjacent Land Use," pp. 78-82.

6. Philip M. Raup. "The Land Use Map Versus the land Value Map–A Dichotomy," pp. 83-88.

19-15. Bulletin 268, *Some Evaluations of Highway Improvement Impacts.*

1. Edgar M. Horwood. "Freeway Impact on Municipal Land Planning Effort," pp. 1-12.

2. Donald J. Bowersox. "Influence of Highways on Selection of Six Industrial Locations," pp. 13-28.

3. Louis A. Vargha. "Highway Bypasses, Natural Barriers, and Community Growth in Michigan," pp. 29-36.

4. George E. Bardwell and Paul R. Merry. "Measuring the Economic Impact of a Limited-Access Highway on Communities, Land Use, and Land Value," pp. 37-73.

5. Eugene C. Holshouser. "An Investigation of Some Economic Effects of Two Kentucky Bypasses: The Methodology," pp. 74-79.

6. William W. Nash and Jerrold R. Voss. "Analyzing the Socio-Economic Impacts of Urban Highways," pp. 80-94.

7. Paul F. Wendt. "Influence of Transportation Changes on Urban Land Uses and Values," pp. 95-104.

8. A. S. Lang and Martin Wohl. "Evaluation of Highway Impact," pp. 105-119.

19-16. Bulletin, *Impact and Implications of Highway Improvements*, 1962.

1. Robert H. Stroup and Louis A. Vargha. "Reflections on Concepts for Impact Research," pp. 1-12.

2. John R. Borchert and Donald D. Carroll. "Time-Series Maps for the Projection of Land-use Patterns," pp. 13-26.

3. George W. Bliele and Leon N. Moses. "Transportation and the Spatial Distribution of Economic Activity, pp. 27-30.

4. Marion Clawson. "Implications of Recreational Needs for Highway Improvements," pp. 31-38.

19-17. Bulletin 327, *Indirect Effects of Highway Improvement*, 1962.

1. Floyd I. Thiel. "Social Effects of Modern Highway Transportation," pp. 1-20.

2. James W. Longley and Beatrice T. Goley. "A Statistical Evaluation of the Influence of Highways on Rural Land Values in the United States," pp. 21-55.

3. J. C. Frey, H. K. Dansereau, R. D. Pashek, and A. Twark. "Land-Use Planning and the Interchange Community," pp. 56-66.

4. Robert H. Stroup, Louis A. Vargha, and Robert K. Main. "Predicting the Economic Impact of Alternate Interstate Route Locations," pp. 67-72.

5. Francis R. Cella. "Highway Location and Economic Development," pp. 73-76.

19-18. Bulletin 343, *Land Acquisition*, 1962.

1. David R. Levin. "Report of Commitee on Land Acquisition and Control of Highway Access and Adjacent Areas, pp. 1-44.

2. Sidney Goldstein, William H. Stanhagen, Joseph T. Sweeney, and Carrie L. Fair. "Economic Evidence in Right-of-Way Litigation," pp. 45-91.

3. Rudolf Hess. "Relocation of People and Homes From Freeway Rights-of-Way–Community Effects," pp. 92-100.

4. Bruce C. Laing, Edgar M. Horwood, and Charles H. Graves. "Freeway Development and Quality of Local Planning," pp. 101-120.

19-19. Highway Research Record No. 2, *Community Values as Affected by Transportation*, 1963.

1. Charles J. Zwick. "The Demand for Transportation Services in a Growing Economy," pp. 3-5.

2. Robert C. Colwell. "Interactions Between Transportation and Urban Economic Growth," pp. 6-11.

3. Marvin G. Cline. "Urban Freeways and Social Structure–Some Problems and Proposals," pp. 12-20.

4. Donald Appleyard, Kevin Lynch, and John Myer. "The View from the Road," pp. 21-30.

5. Harmer E. Davis. Summary Remarks–Session 1, pp. 31-36.
6. Herbert S. Levinson and F. Houston Wynn. "Effects of Density on Urban Transportation Requirements," pp. 38-64.
7. Henry S. Shryock, Jr. "Population Distribution and Population Movements in the United States," pp. 65-78.
8. E. H. Holmes. Summary Remarks–Session II, pp. 94-100.

19-20. Highway Research Record No. 16, *Consequences of Highway Improvement*, 1963.
1. Robert H. Stroup and Louis A. Vargha. "Economic Impact of Secondary Road Improvements," pp. 1-13.
2. William C. Pendleton. "Relation of Highway Accessibility to Urban Real Estate Values," pp. 14-23.
3. David A. Grossman and Melvin R. Levin. "Area Development and Highway Transportation," pp. 24-31.
4. J. Tait Davis. "Parkways, Values, and Development in the Washington Metropolitan Region," pp. 32-43.
5. H. Kirk Dansereau, John C. Frey, and Robert D. Pashek. "Highway Development: Community Attitudes and Organization," pp. 44-59.

19-21. Highway Research Record No. 75, *Indirect and Sociological Effects of Highway Location and Improvement*, 1964.
1. P. D. Cribbins, W. T. Hill, and H. O. Seagraves. "Economic Impact of Selected Sections of Interstate Routes on Land Value and Use," pp. 1-31.
2. Robert C. Burton and Frederick D. Knapp. "Socio-Economic Change in Vicinity of Capital Beltway in Virginia," pp. 32-47.
3. Edward V. Kiley. "Highways as a Factor In Industrial Location," pp. 48-52.
4. Warren A. Pillsbury. "Economics of Highway Location: A Critique of Collateral Effect Analysis," pp. 53-61.
5. John J. Coyle, H. Kirk Dansereau, John C. Frey, and Robert D. Pashek. "Interchange Protection and Community Structure," pp. 62-74.
6. Floyd I. Thiel. "Seminar on Sociological Effects of Highway Transportation–Introductory Remarks," pp. 75-76.
7. H. Kirk Dansereau. "Five Years of Highway Research: A Sociological Perspective," pp. 76-81.
8. David R. Levin. "Informal Notes on Sociological Effects of Highways," pp. 82-83.
9. F. Houston Wynn. "Who Makes the Trips? Notes on an Exploratory Investigation of One-Worker Households in Chattanooga," pp. 84-91.
10. Barbara H. Kemp. "Social Impact of a Highway on an Urban Community," pp. 92-102.

19-22. Highway Research Record No. 96, *Indirect Effects of Highway Location and Improvement*, 1965.
1. Edgar M. Horwood. "Community Consequences of Highway Improvement," pp. 1-2.
2. Robert D. Pashek and Edgar M. Horwood. "Discussions," pp. 2-7.
3. Mark C. Flaherty. "Commercial Highway Service Districts and the Interstate: Their Proper Relationship in an Urban Setting," pp. 8-18.
4. Walter C. McKain. "Community Response to Highway Improvement," pp. 19-23.

    5. Floyd I. Thiel. "Highway Interchange Area Development," pp. 24-45.

    6. Roger H. Ashley and William F. Berard. "Interchange Development Along 180 Miles of I-94," pp. 46-58.

19-23. Highway Research Record No. 149, *Forecasting Models and Economic Impact of Highways*, 1966.

    1. C. A. Steele and others. "Will Model Building and the Computer Solve Our Economic Forecasting Problems?"

    2. David E. Boyce and Seymour E. Goldstone. "A Regional Economic Simulation Model For Urban Transportation Planning."

    3. H. B. Gamble, D. L. Raphael, and O. H. Sauerlender. "Direct and Indirect Economic Impacts of Highway Interchange Development."

    4. Robert J. Trier and Vern S. Kubitz. "The Impact of Modern Highways on American Indian Country."

    5. Roy B. Sawhill and Joseph W. Ebner. "Freeways and Residential Neighborhoods."

19-24. Special Report 28, *Economic Impact of Highway Improvement: Conference Proceedings*, March 18-19, 1957.

19-25. Special Report 56, *Economic Analysis in Highway Programming, Location and Design*, 1959.

    1. G. P. St. Clair. "Economic Analysis: A Study in Uncertainties," pp. 3-7.

    2. David S. Johnson. "Economic Analysis of Alternate Route Locations," pp. 43-46.

    3. Karl Moskowitz. "Special Problems in the Analysis of Urban Expressway Projects," pp. 47-60.

    4. R. C. Blensly. "Applications of Economic Analysis to Highway Systems and Programs," pp. 61-68.

    5. Robert G. Hennes. "Highways as an Instrument of Economic and Social Change," pp. 131-135.

    6. David R. Levin. "Identifying and Measuring Non-User Benefits," pp. 136-147.

    7. Richard M. Zettel. "The Incidence of Highway Benefits," pp. 148-164.

19-26. Special Report 75, *Benefits to Utilities from Highway Locations*, 1962.

    1. James H. Lemly. "Non-Vehicular Benefits from Utility Use of Streets and Highways," pp. 1-32.

    2. Claron E. Nelson. "Economic Implications of Utility Use of Highway Locations in Utah," pp. 33-51.

    3. R. C. Blensly. "Benefits to Utilities from Rural Highway Locations in Oregon," pp. 52-59.

19-27. Edgar M. Horwood and Ronald R. Boyce. *Studies of the Central Business District and Urban Freeway Development*. U. of Washington Press, Seattle, 1959.

19-28. James H. Lemly. *Expressway Influence on Land Use and Value: Atlanta, 1941-1956*. Georgia State College of Business Administration, Bureau of Business and Economic Research, Atlanta, November 1958.

19-29. Los Angeles, Street and Parkway Division. *The Economy of Freeways*. Los Angeles, June 1953.

19-30. Los Angeles, Street and Parkway Division. *A Study of Freeway System Benefits*. Los Angeles, September 1954.

19-31. Massachusetts Institute of Technology, Department of Civil and Sanitary Engineering, Transportation Engineering Division. *Economic Impact Study of Massachusetts Route 128.* Cambridge, December 13, 1958.

19-32. Herbert D. Mohring and Mitchell Harwitz. *Highway Benefits: An Analytical Framework.* Northwestern U. P., Evanston, Ill., 1962.

19-33. Herbert D. Mohring and Harold F. Williamson, Jr. *Economies of Scale and the Reorganization Benefits of Highway Improvements: A Supplement to "The Nature and Measurement of Highway Benefits."* Transportation Center, Northwestern U. P., Evanston, Ill., August 15, 1960.

19-34. Oregon University, Bureau of Business Research. *Economic Effects of Through Highways By-Passing Certain Oregon Communities.* Eugene, July 1956.

19-35. William C. Pendleton. *Some Economic Impacts of Highway Improvement on Land Value and Land Use.* Presented at the Land Economics Institute, Urbana, Ill., July 1958.

19-36. Warren A. Pillsbury. *The Economic and Social Effects of Highway Improvement: An Annotated Bibliography.* Virginia Council of Highway Investigation and Research, Charlottsville. May 1961.

19-37. Robert L. Shindler. *Economic Impact of Highway Location in Urban Areas.* Presented at the Northwest Regional Highway Conference sponsored by the Joint Committee on Highways of the American Municipal Association and the American Association of State Highway Officials, March 1958.

19-38. Texas Highway Department, Highway Planning Survey. "A Fifteen-Year Study of Land Values and Uses Along the Gulf Freeway in the city of Houston, Texas," 1956.

19-39. United States Department of Agriculture. *The Economic Impact of Highway Improvement.* Agricultural Research Service, Washington, D.C., March 1959.

19-40. United States Inter-Agency Committee on Water Resources, Subcommittee on Evaluation Standards. *Proposed Practices for Economic Analysis of River Basin Projects.* Rev. ed., Government Printing Office, Washington, D.C., May 1958.

19-41. United States 87th Congress, First Session. *Final Report of the Highway Cost Allocation Study.* House Document No. 54, Government Printing Office, Washington, D.C., Jan. 16, 1961. For Part VI "Studies of the Economic and Social Effects of Highway Improvement," see House Document No. 72, January 23, 1961.

19-42. Paul F. Wendt. "The Theory of Urban Land Values," *Land Economics,* Vol. 33, No. 3 (August 1957), pp. 228-240.

19-43. Richard M. Zettel. "Effect of Limited-Access Highways on Property and Business Values," *Proc.,* Twenty-Fourth Annual Meeting of Institute of Traffic Engineers, New Haven, Conn., October 1953.

19-44. Richard M. Zettel. *Highway Benefit and Cost Analysis as an Aid to Investment Decision.* Third International Study Week in Traffic Engineering, Stresa, Italy. World Touring and Automobile Organization, London, 1956. See University of California, Institute of Traffic and Traffic Engineering, Richmond, Calif. Reprint No. 49.

19-45. Richard M. Zettel. "Highway Transportation Economics," *Public Roads,* Vol. 17, No. 3 (August 1952), pp. 37-49.

19-46. Richard M. Zettel. *Ten Notes on Transportation and Economic Development.* Presented at the Phoenix Convention of the American Society of Civil Engineers, Phoenix, Arizona, April 13, 1961.

Chapter **20**

The Urban Bypass

Urban places attract the most highway travel because more people are to be found per acre in urban places than elsewhere. In the beginning of highway development the objective was to provide an all-weather road for the country people to get to town and back to their farms, but there was no consideration in the early days of getting motor vehicles through the town to the opposite side.

With the development of intercity travel, it soon became apparent that travel through the urban areas would have to have consideration along with the travel to and from the urban areas. This consideration led to the urban bypass, a provision in highway location whereby the traveler may get to the opposite side of the urban area without going through it, or at least not through the central business district. Bypasses, although highly desired by the through travelers, were not welcomed by local business interests on the basis that the community would suffer a reduction in retail trade. Some discussion of the pros and cons of bypasses and their consequences as observed from experience will shed light upon this type of local highway.

TYPES OF BYPASSES

Interpreted liberally there are four types of urban bypass highways, each so designed and located that traffic may traverse, go around, or penetrate across an urban area at much higher speed than can be achieved on the customary direct route through the central business district, and without opportunity to stop at roadside commercial establishments. These four types of bypasses are:

1. The *simple bypass* begins at the outer edge of an urban area and swings right or left away from the existing arterial route that leads to the heart of town. The bypass then continues its curved path along the fringed edge of the urban area, or nearly so, until it reaches a point on the old route at the other side of the urban area about diametrically opposite the point where it began its swing from the straight route to the city center. A long tangent on the access controlled interstate route which may pass the town by a distance of 1, 2, 3, or so miles would also be defined as a simple bypass.

2. A highway similar to the *simple bypass,* but *on the order of a full freeway or expressway,* and extending some miles from both sides of the urban area.

The Urban Bypass

3. The *circumferential, beltway,* or *outerbelt* that makes a full circle around the urban area, connecting at interchange points with the main highways leading into the city center.

4. The *expressway* or *freeway* which leads from one outer side to the other outer side of the city center and passes near to the central business district. There is question about this type of expressway or freeway being a bypass, but it is included here for the reason that it has certain aspects in common with the simple bypass of a small urban area.

The more common meaning of "bypass" is a route around a small city or town as opposed to the circumferential around a metropolitan area. Although an urban bypass may not have full access control, it will succeed for a time in accomplishing its full purposes. Eventually, however, business establishments will so populate the roadside that traffic is slowed down to nearly the same point of congestion as existed in the central business district before construction of the bypass.

To preserve the full advantages of the urban bypass, full access control should be built into the bypass initially. The early constructed part of Route 128, the circumferential around Boston, Massachusetts, was not built with controlled access, but later, after roadside business had developed heavily, the authorities bought the access rights and designed the remainder of the route with full access control. This ribbon development of business is identically the same as experienced along the main primary routes leading into urban areas.

REASONS FOR THE URBAN BYPASS

The main purpose of an urban bypass highway is to divert through traffic from the urban center and from the commercialized routes leading to the urban center to a less populated lower traffic volume area. Why is such diversion of traffic desirable? Through traffic, that traffic which has both origin and destination outside the local urban area being bypassed, finds itself mixed in with local traffic, pedestrians, and a high amount of parking and turning movement that seriously interferes with the through movement. Similarly, the local traffic finds itself mixed in with the through traffic, trucks particularly, and passenger cars whose drivers do not know where they are, nor how to go where they want to go. The result is traffic congestion, traffic delays, and confusion for everyone.

As the urban area increases in population, the percentage of approaching traffic that has a destination somewhere within the urban area also increases. The traffic to urban areas of less than 3,500 population divides about 40 percent destined within the area and 60 percent through traffic [20-3].

The natural thought of highway officials and of the road users is why not send the through traffic around the business area or even around the entire urban area on a bypass route. This suggestion is logical and practical from the viewpoint of traffic. But often this proposal meets with strong opposition from the business interests, especially from those merchants who cater to the highway traveler. A second objection usually comes from the owners of the land proposed for the bypass and often from the owners of nearby land.

The congestion of traffic and the traffic accident rate on the primary or arterial route through the urban area is evidence that greater traffic capacity needs to be provided. Streets can be widened to additional lanes, pairs of one-way routes can be established, additional traffic controls can be installed, and other schemes utilized. And such actions are common and do provide relief, but in a few years new routes are required.

Overall greater capacity for traffic at less cost for construction is generally found in the bypass on new right-of-way (usually vacant land or farm land) than can be had by widening and reconstructing existing streets. Further, the completely new right-of-way makes it easy to control access to the bypass. Access control is a must for the urban bypass to prevent the roadside from being filled with business establishments and attracting traffic that would soon bring back congestion for which the bypass was built to eliminate.

These reasons for the urban bypass are sound, logical, practical, and usually economical, considering both the road user and the local community.

CONSEQUENCES OF THE URBAN BYPASS

Although each urban bypass highway must be considered individually, their general consequences may be mentioned with confidence. Benefits may be divided into two groups—road user consequences and community consequences.

ROAD USER CONSEQUENCES

Since it is traffic congestion that usually leads to the construction of an urban bypass, it is to be expected that the road user would be the main benefactor thereof. Such consequences are the normal result.

Through traffic may be speeded up from an average speed through the urban area of 10 to 20 mph to practically a free speed of 40 to 60 mph, depending upon the legal speed limit. This higher and free speed results in two specific benefits. First, the vehicle running cost per mile is usually decreased materially by eliminating the costly slowdowns and stops. Should the bypass nominal speed be greater than 40 to 45 mph there may be some increase in running cost above this most economical speed, but any such increase is more than balanced by the extra cost of running at the 20 to 30 mph top speed which is usually the legal speed limit on arterials and other streets in urban areas.

Certain bypass routes may add some overall distance as compared to the existing urban route. If so, any added distance would add to the running cost

The Urban Bypass

distance, but it is not likely that such added cost would exceed the gain resulting from eliminating the speed changes.

Second, higher average speed on the bypass than on the urban route results in an appreciable reduction in travel time. By increasing the average speed from 15 mph to 45 mph, the time reduction per mile of travel would be 2.67 minutes, a high percentage. Through traffic, therefore, usually makes significant gains in both vehicle running cost and travel time by the use of the urban bypass highway.

Local traffic often receives the same two benefits. When the through traffic, particularly the combination vehicles, moves over to the bypass, the urban route is freed of much of its congestion, so the remaining vehicles move faster and are delayed less. These local road user gains are real. They last for some years unless the community is a rapidly growing one which again brings on traffic congestion.

A third road-user benefit is found in the fact that the bypass serves as a traffic distributor for local internal traffic and for external traffic with one trip end within the urban area. This distributor action brings relief to the main urban route and to the street system generally.

Accident costs are probably reduced by the urban bypass. The urban property damage only accident would be reduced on a vehicle-mile basis, but the injury accident severity may be increased because of the higher speed on the bypass. Additional search is necessary on the accident cost factor.

COMMUNITY CONSEQUENCES

The effects of bypasses on traffic are easily observed and measured. But the economic and social consequences of bypasses on urban communities are difficult to observe and still more difficult to measure. Further, there has been, and still is, outspoken opposition to bypasses arising mainly from the business interests within the urban area. The general experience has been quite positive in proving that, on the whole, bypasses are economically and socially desirable. The few adversities that result are minor in comparison with the benefits, and usually these adversities do not last long.

Bypasses have been studied in detail for some years on a before and after basis. The literature is rich with fully documented cases. The general approach of the investigators has been to use a variety of economic and social indicators to isolate and measure the economic activities as affected by use of the urban bypass. The most common indicators are: dollar volume of retail trade, number of retail establishments, retail trade employment, land values, land-use changes, parking conditions, traffic flow, and trip origins, destinations, and purpose. The well-established technique of using control areas is regularly employed. Control areas have been selected from comparable streets and towns within the same county and state.

General Community Consequences

The urban community as a whole usually experiences a net beneficial consequence from the urban bypass. It gains a quieter city, less air contamination, less vibration of buildings from heavy trucks, and does not have to suffer the inconvenience and disruption from widening and reconstruction of a local street to gain the added traffic capacity. And, as just discussed, the motor vehicle users will be benefited by less congestion and better access to the outside.

The effect of the bypass on the community is greatest when the population is small and when a high percentage of the total area trade comes from through traffic. Decreased trade is likely to be experienced by the motor vehicle service stations, restaurants, taverns, and those roadside outlets that cater especially to the highway traffic. General trade in the central business district is likely to increase after opening of the bypass. Even highway-oriented establishments experience no material decrease in business when located so that there is attraction of the local resident and local traffic. Most all studies of retail trade as affected by bypass routes have shown that the volume of business coming from the through traveler was much less than the local merchants and officials thought it to be.

Many of the bypassed business centers experience improved business after being bypassed because local residents find the business area free of congestion and with improved parking. Further, through traffic which really wants to do business will leave the limited-access controlled bypass and enter the business area knowing that there is only a small chance of encountering congested traffic and tight parking.

Consequences to Retail Trade

But the critical issue is local retail trade. Will it be less or more after construction of the bypass?

This question cannot be answered yes or no except for each specific urban community. The local factors–population, type of economic community, traffic mix and volume, percentage of through traffic, location of retail trade establishments with respect to the through traffic and the proposed location of the bypass–are controlling. As a general result, there is a 2-to-1 probability that retail trade in the bypassed area will improve or decrease less than surrounding areas increase after the bypass is opened. This is the experience in the United States based upon a study [20-18] of 76 bypassed urban areas of a few hundred to 200,000 population. Mostly, however, the population was in the range of 1,000 to 30,000.

It is observed that highway-oriented businesses along a major urban route or approach route to an urban area receive a major percentage of their business from local residents and local employees.

The popular conception is that an automotive service station, a restaurant,

The Urban Bypass

a specialty shop, a bar, and other retail establishments on the edge of town get their major percentage of business from the through travelers. It is common experience, however, that these establishments draw high percentages of their business from local people within a range of 5 miles or so. The local customers patronize the highway located business because of the convenience of its location. Quality restaurants at the edge of towns draw heavily from the urban area.

Figures 20-1 and 20-2 illustrate typical effects of the bypass on local retail trade volume.

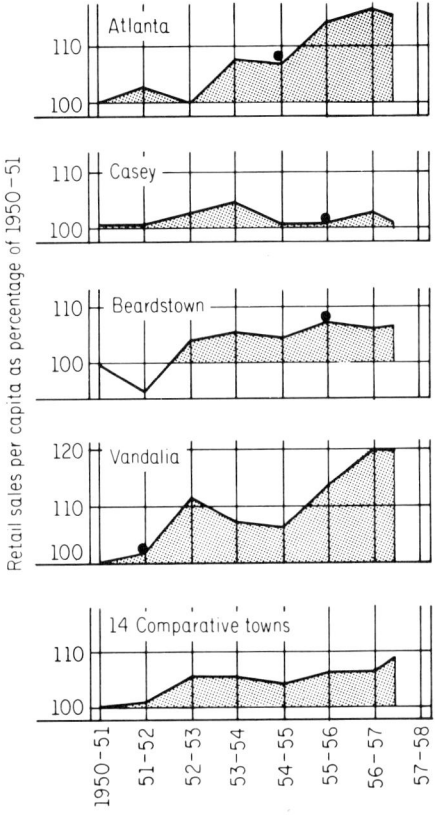

• Year bypass was completed

FIG. 20-1. Experience with bypass routes in four Illinois communities. Retail sales of four Illinois cities where a major highway bypassed the community are compared with sales in 14 cities similar in size and basic sales per capita where the highway does not bypass. State sales tax records show retail business to be sustained or improved after through traffic has been shifted into new relief routes (adapted from Ref. 20-2).

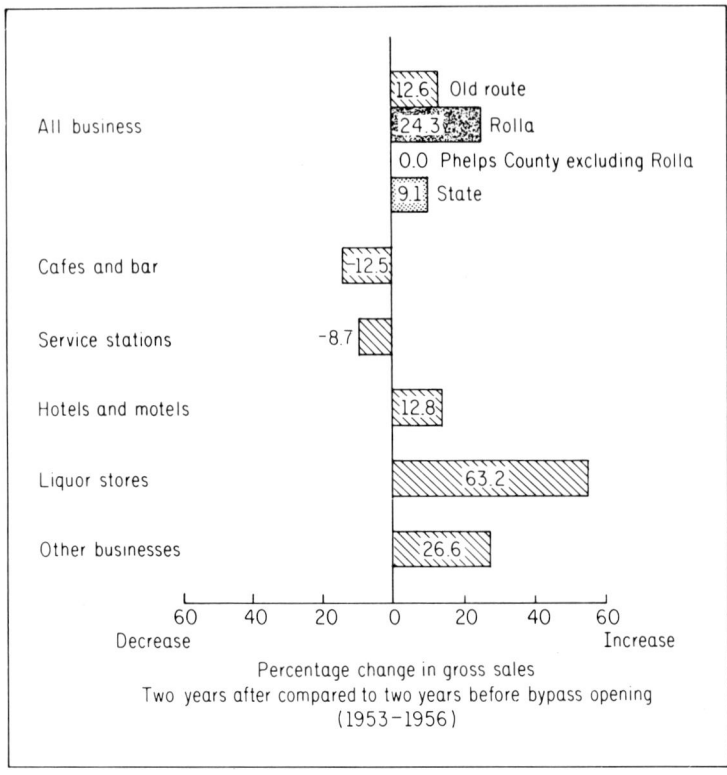

FIG. 20-2. Rolla, Missouri, bypass effects on business [20-7].

TYPICAL CASE HISTORIES AND RESULTS

Newspapers, records of public hearings, and the technical literature abound in reports, discussions, and histories of urban bypass highways. Each proposal brings out about the same story, including storms of protest. Even state legislatures have passed laws endeavoring to curtail the general construction of bypass routes without the approval of the bypassed community. In the end, but after some years in many cases, the bypass is constructed. Within a year or so after the ribbon cutting, all is forgiven, the people are pleased–at least 95 percent of them are–and the community gets back to normal.

The following first two stories are true published reports, but modified in names of persons and places and wordage. The third article taken without alteration from the *Topeka Capital Journal* (Kansas) is a good summary of results and public opinion after some experience with the urban bypass.

The Urban Bypass

REALLY COSTLY ROADS

Now at long last you can drive through Central City on Interstate 35 without having to stop at traffic signals.

It's just too bad that–thanks to politics and town boosterism–we couldn't have done that 5 to 10 years and $4 million to $5 million earlier.

Here is a classic example of how skimping on standards has cheated the public in terms of convenience and safety and has cost it dearly in highway dollars.

Now we've scotched that old highway department joke that "in 1972 when the interstate highway is completed all over the country you can drive anywhere without stopping, except at two lights in Central City."

This bending to the wishes of a community resisting a bypass goes back to the days of the Hartner administration, as does the similar situation on IH-40 West where only now is the state constructing bypasses at five towns.

Because of political pressures, the highway department delayed imposing the limited access bypass standards established by the Bureau of Public Roads for the interstate program and let the highway continue through towns, heedless of the motorists' interests.

On another stretch of the interstate, between Cellmore and Westland, it was the legislature which jimmied the works by passing a measure withholding state funds unless the route went within a mile of each of the four towns on the route.

Fortunately, by that time the U.S. Bureau of Public Roads uttered a firm "No." But planning for that route was delayed two years until the legislature repealed the unwise restrictions.

Central City still remains the classic case because the right kind of highway wasn't built in the first place. As the community spread westward, the only way to handle the problem was to build the present elevated "Chinese Wall" which splits the town.

Also, it cost the taxpayers $1.2 million more for right-of-way in Central City because of having to acquire properties that had built up in the meantime.

The bulk of the $4.5 million spent to reconstruct the 5.5 miles stretch from the Fields intersection south through Central City to the first Hillman exit went into the Central City section.

Here's why engineering and not politics should have the say-so in road planning and building. Let's hope this costly lesson will have sunk in so that such mistakes as these never happen again.

I-40 DECISION ON THE CARLOS BYPASS STANDS

Officials of the Bureau of Public Roads see little hope for reversing its decision that Interstate-40 will bypass Carlos by about five miles to the north.

A public action panel was told Friday by Harold Clarkson that the Bureau of Public Roads hoped the panel members could find a way of "easing the financial blow" to Carlos.

"Kill" Village

Clarkson, division engineer for the Bureau of Public Roads, said the bureau could see no substantial reason why I-40 should not bypass the town. Residents of Carlos take the position that loss of tourist industry from US-66 (I-40) will "kill" the village.

Clarkson said building the southern route, closer to Carlos, would cost about $2 million more than the north alignment. This figure did not include cost of acquiring right-of-way for the northern route, whereas the southern line presumably would follow existing right-of-way except for a short distance adjacent to Carlos. The town would be bypassed by either route, but the southern line would be nearer to the village.

Loss of Employment

Will Hoppe, director of the planning division of the state highway department, restated the highway department's stand endorsing the route nearer Carlos.

He predicted that "a large portion of the town's business will be hurt and there will be a resulting loss of employment." He said the northern route was in an area plagued by high winds and winter storms.

Believes Position

Hoppe said it appears that work on the bypass cannot be started before 1971 or 1972, but added that many decisions on Carlos's problems need to be made far in advance of commencement of work on the bypass.

Clarkson, not swayed by arguments favoring the southern route for I-40, said, "I believe sincerely in my position. It doesn't take any genius to see what a more popular position would have been."

Other panel members said they felt Carlos residents should try to develop other economic interests. Possibilities mentioned were a Christmas tree planting system, quarrying of flagstone from deposits near Carlos, and establishment of vegetable farms and a plant for processing and freezing vegetables for market.

Major Drawback

The lack of sufficient water in the Carlos area was noted as a major drawback in establishing new industries. Where cotton once was a major crop, the area's income now is derived from tourism, broomcorn, grain sorghums, and wheat. Almost all farming is of the dry-land variety.

The panel decided not to take a strong position on either the north or south routes proposed for I-40, but felt the highway department's support of the southern route should be honored and that agencies should assist Carlos residents in finding a way for economic survival, with or without the tourist business generated by I-40.

WHAT HAPPENS WHEN THEY OPEN THE BYPASS[1]

City fathers and businessmen used to rise up in arms and fight if their towns were to be bypassed by a major highway. Some still do.

Having a stream of tourists and trucks taken out of the heart of town was considered a major business blow.

Today in Kansas towns bypassed or in the process of being bypassed by new highways reaction is mixed. It ranges from elation to dejection.

Abilene and Salina leaders are delighted to have I-70 bypass them to the north. They look forward to a big increase in cross-country traffic and "people still have to stop to eat and sleep and buy gas."

[1] From the August 9, 1964, *Topeka Capital Journal*. Used by permission.

The Urban Bypass

Holton is adjusting to having new US-75 moved from the center of town to the west edge. "It's hurt some businesses and it's helped some others," said Mayor Vic Symons.

A supermarket, a bowling alley, a drive-in, an ice cream parlor and other businesses have sprung up along the new highway.

Smaller towns may be hurt more when they are bypassed.

Hoyt has suffered, say some of the people in the town of 318 population, 13 miles north of Topeka. US-75 used to curve into the business district and turn again leaving town.

Traffic went through at 20 miles per hour and some drivers stopped at the service stations or the cafe.

Since the first of the year cars whiz by on new US-75 a half mile west.

"The highway move just about killed me—I barely made a living before and now it's much worse," said Mrs. Mary Schweigen, operator of the Hoyt cafe.

She'd like to move her business to the new highway if someone would build a building and lease it to her, she said.

Edward M. Dickinson, operator of a Hoyt service station, was asked how business was. "What business?" he asked as he sat on an old school bench at the side of the station.

"The highway's moving has knocked the filling station pretty bad."

"We pick a few off the new highway now and then when they're running low on gas or having car trouble." The state has built a connecting spur to the town as it does in other cases where a town is bypassed.

R. L. Spencer, mayor of Hoyt and owner of the R. L. Spencer Grain Company, said Hoyt has not lost any of its businesses because of the highway move but it has "crippled what we have."

A few "strays" come through who get on old US-75 by mistake, he said.

In 1959 the people of Hoyt petitioned the state highway commission to bend the new route a half mile east to touch Hoyt.

The commission refused, replying that it would cost $21,790 extra for construction and $6,840 for right-of-way.

But even in Hoyt people see some good things about being bypassed by a major highway. "Business is bad but Hoyt is a nicer town now," Dickinson said.

"Now that all that traffic is out of here it's a much nicer place to live. It used to be you'd have to wait 10 minutes to cross the street. It was worse than Kansas Avenue in Topeka. Traffic used to pile up behind trucks and by the time they got here there'd be a long line of cars."

Mayor Spencer also said the town is quieter and may be more attractive to workers in Topeka who like to have peace and quiet and a half acre of space in a small town.

Hoyt has a number of Goodyear plant workers and others who work in Topeka. Four new homes are being built.

The mayor says he hopes more will be built, perhaps in a development between the town and the new highway. He sees Hoyt's future as that of a pleasant suburban area.

The people of Abilene feel having I-70, which bypasses the town, is a blessing, said H. W. Callahan, city manager.

"Of course, we're in a little different position from that of a lot of towns in that we have a fine tourist attraction—the Eisenhower Center," he said.

But he said he believes any town is fortunate to have a splendid, modern highway within a mile of the city limits.

It is up to the town then, he said, to develop attractions that will bring people on into the town.

Abilene has two new motels, one with 62 units and one with 54, which are filled to an average of 89 percent capacity, Callahan said.

Restaurants along old US-40 and the old 40 bypass are not suffering, he said, because they have developed their own local business clientele.

Del Hadel, president of the Abilene Chamber of Commerce, stressed the safety factor in taking heavy traffic out of downtown areas.

He said citizens interested in the safety of Abilene's children worked to get the first small bypass built 10 or 12 years ago.

Taking traffic congestion out of the heart of a city makes for better shopping conditions for local people and people in the trade area, he said.

Salina will be bypassed by two major highways under construction, I-70 on the north and I-35W on the west. I-70 is completed to US-81, which bisects Salina, and I-35W will be completed this fall.

"The majority of us think we'll be definitely benefitted," said Jim Preston, secretary of the Chamber of Commerce.

"We expect I-70 to be a top cross-country traffic carrier with lots of people who have to eat, sleep, and buy gasoline–and they'll do these things at the bigger towns," he said.

Two new motels have been built and more are in the planning stage.

US-81 and US-40, which go directly through town, will be relieved of much of their traffic, but Preston does not seem worried.

I-70 puts Salina within two hours of Topeka and three hours of Kansas City and opens the Salina area to quicker access, which is beneficial, he said. He sees speedier moving of freight by trucks as another advantage to businessmen.

Bob Slease, director of public information for the state highway commission, said the commission has made a survey of towns between Ottawa and Kansas City and along completed sections of I-70 asking for reactions to bypassing.

Replies ranged from "a great thing" to "not so sure it's good," he said.

Most reaction was favorable, Slease said, although many persons pointed out that what may be good in the long run could hurt individual businesses in the short run.

Many towns with bypasses have developed thriving motels and restaurants along their connecting roads, he said.

SUMMARY OF FINDINGS OF THE ECONOMIC CONSEQUENCES OF THE URBAN BYPASS

The following summary is from a report by I. J. Sans [20-18] of the Oklahoma Department of Highways:

> 1. The net impact upon total business activity and specifically that portion derived from travel along the highways within the various states of the Union has been one of increased sales and growth as the result of the development of freeways.
>
> 2. Of the 76 bypassed areas for which information about retail trade activity is available, 50 experienced either a greater increase or a smaller decrease than occurred in a comparable area which was not bypassed.
>
> 3. Even where the total economic impact has been good, most instances

The Urban Bypass

studied showed that a few individual businesses will suffer from highway relocations..

4. Immediately after relocation of the highway, some decline in the economic activity of the area through which the main highway formerly passed is usually experienced.

5. If the overall economy within the area where the highway bypass was accomplished is expanding, economic growth soon provides for increased business activities which overcome any adverse impact from the highway relocation.

6. The beneficial effects on business activity which are often associated with bypass routes may be due primarily to the fact that these routes result in less congestion, and, therefore, in better traffic movement and better parking conditions for local shoppers.

7. There are indications that where an isolated community derives a major portion of its income from highway traffic, the bypassed community will suffer from some decrease in business volume.

8. Sales to through travelers make up a much smaller portion of a community's total business activity than is commonly supposed.

9. The geographical location of the highway, as well as the type of signs used, will influence the economic impact that the relocation will have upon a community.

10. Wherever freeways have been constructed or where existing routes have been expanded and improved or developed into freeways, the volume of traffic in the affected area has significantly increased.

11. Studies of traffic noise created by the increased traffic has been measured. Results of these tests indicate noticeable increases in sound and noise, in some instances to the point where these sound levels have been objectionable to individuals located along the freeways.

12. Freeways may act as barriers causing neighborhood units to become divided which may have an unforeseen social and economic impact upon the affected area.

13. The interchange between the local roads and the freeway system, unless properly controlled, may become traffic congestion points as well as traffic hazards. Many states have attempted to meet this problem. California has, for the most part, bypassed the communities where engineering studies have indicated that such a bypass would be advantageous to the majority concerned.

14. Traffic counts in a number of bypassed areas, and such indirect indicators of traffic congestions as down town accident rates, show that bypasses are accomplishing their intended purpose of permitting through travelers to bypass downtown areas and thus relieve downtown streets of a portion of its traffic load.

15. Retail business in most of the bypassed cities made relatively greater gains than in the state as a whole. If business gained, it increased more than the state average; if it declined, the decline was less. In each case, the degree of gain was inverse to the size of the city.

In general, it can be concluded that the economic effect of a highway bypass on small communities involves an initial decline in total sales which is followed by an increase which is higher than the state average. Certain highway-oriented businesses are most adversely affected but many of these recover through adjustments toward local trade for stability. The claim that a bypass will "kill" business in a small town is refuted by the findings of numerous research studies of the economic impact of bypass facilities which have been constructed.

THE BASIC FACTORS TO CONSIDER

With the exception of barriers to urban development, most of the basic factors to consider in connection with the urban bypass have been mentioned. The final decision is made after weighing the probable consequences, beneficial and adverse.

The road-user economy is an important factor. A rigid analysis of the economy is in order, taking into consideration all traffic–through and local area traffic–which may be affected. To determine the probable effects on all traffic is difficult, but under the principle of considering all consequences to whomsoever they may accrue, local traffic must be included. In fact, because of high volume of local traffic (80 to 90 percent) as compared to through traffic (20 to 10 percent) benefits may be greater to local traffic than to through traffic. A high rate of return for the road user in the analysis for economy certainly would be a strong reason for approval of the bypass.

The general community consequences–other than retail trade volume–are usually wholly acceptable to the local population. These consequences, though, are often the nonpriceable, nonmarket items.

Retail trade volume is a factor that compels selecting either the viewpoint of the community as a whole or the viewpoint of those tradesmen who cater to through highway travel. On a long-time basis and for a growing community in a traffic growing area, the decision is one of timing the construction of the bypass. The experience has shown that ultimately the central business district must be relieved of through highway travel. For the growing community, the sooner the bypass is constructed, the less costly it will be, the sooner its benefits may be realized, and the sooner the growth of the community can move ahead orderly.

Rivers, railroads, and highways are barriers to and attractors of community development, depending upon the type of land use involved. When planning an urban bypass for highway travel, it is important that the bypass does not result in a barrier to growth of the community in a natural direction. No community wants to be divided by a highway wall.

The bypass should be located such that the attracted new uses of adjacent land will be in accordance with a community plan, and so that the access points to the bypass accommodate the normal development of business that is certain to come in. The bypass location may be far out to allow growth from the fringe of the present urban area out to the bypass or for larger urban areas, the bypass may be located more inward with provision for land use development beyond the bypass.

PROBLEM

20-1. From your personal knowledge of a specific urban bypass, or from reading about one, write out your analysis of the favorable and unfavorable consequences. What was the attitude of certain classes of the local population?

REFERENCES

20-1. J. K. Allen and Richard McElyea. *Impact of Improved Highways on the Economy of the United States.* Stanford Research Institute, Menlo Park, Calif., 1958.

20-2. George W. Barton and Associates. *Highways and Their Meaning to Illinois Citizens.* Prepared for the Illinois Division of Highways, Springfield, 1958.

20-3. Therel R. Black and Jerrilyn Black. *A Concept of Bypass for the Modern Highway System.* Highway Research Board, Washington, D.C., Circular No. 406, November 1959.

20-4. A. K. Branham and A. D. May, Jr. *Economic Evaluation of Two Indiana Bypasses.* Highway Research Board, Bulletin 67, *Some Economic Effects of Highway Improvement,* 1953, pp. 1-14.

20-5. Chamber of Commerce of the United States. *How Bypasses Affect Business.* Washington, D.C., 1956.

20-6. William L. Garrison and Marion E. Marts. *Geographic Impact of Highway Improvement Changes in Transportation, Land Use and Business Patterns Concurrent with the Reorientation of U.S. Highway 99 in the Vicinity of Marysville, Washington.* Highway Economic Studies, University of Washington, Seattle, July 1958.

20-7. William L. Garrison and Marion E. Marts. *Influence of Highway Improvements on Urban Land, A Graphic Summary.* Highway Economic Studies, University of Washington, Seattle, May 1958.

20-8. Carl Goldschmidt. *Effects on Businesses of a Bypass Highway.* Michigan State University, Traffic Safety Center, East Lansing, 1959.

20-9. Andreas Grotewald and Lois Grotewald. "Commercial Development of Highways in Urbanized Regions: A Case Study," *Land Economics,* Vol. 34, No. 3 (August 1958), pp. 236-244.

20-10. Joseph W. Harrison. *A Study of the Economic Effects of the U.S. Route 11 Bypass at Lexington, Virginia, on Business Volumes and Composition.* University of Virginia, Virginia Council of Highway Investigation and Research, Charlottsville. October 1958.

20-11. Raymond F. Law. "San Clementa Benefits from Freeway Project," *California Highways and Public Works,* Vol. 42 (July-August 1963), pp. 27-31.

20-12. James H. Lemly. "Highway Bypasses and Their Economic Effect upon Nearby Communities," *Atlanta Economic Review,* Vol. VI, No. 7 (July 1956), pp. 1-34.

20-13. H. L. Michael. *Bypasses–Their Use, Effect, and Control.* Purdue University, Engineering Extension Department, Purdue Road School, *Proc.,* 1953, pp. 164-185.

20-14. Missouri State Highway Commission. *Economic Study, Route U.S. 66 Bypass, Rolla, Missouri.* Jefferson City, January 1958.

20-15. North Dakota State Highway Department. *The Economic Effect on Towns Bypassed.* Bismarck, March 1959.

20-16. Pennsylvania State University. *Blairsville: A Bypass Study.* University Park, 1962.

20-17. E. Clark Rowley. *Economic and Social Effects of Highway Improvements–A Summary.* Michigan State University, Highway Traffic Safety Center, East Lansing, 1961.

20-18. I. J. Sans. *Report on Economic Impact of Freeways Bypassing Small Communities.* Oklahoma State Highway Commission, Planning Division, May 6, 1963.

20-19. South Dakota Department of Highways. *Economic Study, Route S.D. 50 Bypass.* Tyndall, 1958.

20-20. United States 87th Congress, First Session. *Final Report of the Highway Cost Allocation Study, Part VI.* House Document No. 72, Jan. 23, 1961, pp. 37-46.

20-21. University of Denver. *A "Before and After" Study of Effects of a Limited Access Highway upon the Business Activity of Bypassed Communities and upon Land Value and Land Use.* Bureau of Business and Social Research, College of Business Administration, Denver, October 1958.

20-22. University of Oregon, Bureau of Business Research. *Economic Effects of Through Highways Bypassing Certain Oregon Communities.* Eugene, 1956.

20-23. Louis Vargha. *Highway Bypasses, Natural Barriers, and Community Growth in Michigan.* Michigan State University, Highway Traffic Safety Center, East Lansing, August 1959.

20-24. Hulse Wagner. *The Economic Effects of Bypass Highways on Selected Kansas Communities.* University of Kansas, Center for Research in Business, Lawrence, 1958.

Chapter **21**

Economic Analyses in Developing Countries

> *When the Indian trail gets widened, graded and bridged, to a good road–there is a benefactor, there is a missionary, a pacificator, a wealth-bringer, a maker of markets, a vent for industry.*[1]
>
> Ralph Waldo Emerson

Of worldwide current interest are the hundreds of specific public works projects, proposed, under construction, and completed, which will have highly significant economic consequences to the independent countries concerned and to areas within the larger and older countries affected. These public works improvements include the whole range–highways, railroads, pipelines, airways, power plants, power transmission lines, harbor improvements, and water resource developments. These proposals are prime opportunities for economic analyses and, when the projects are financed by international loans or grants, the analysis is usually required. However, for many of the transportation facilities proposed it is usual to find that many of the projects are located where the existing motor vehicle traffic is practically zero and other traffic is of low volume. The economic analysis, therefore, must be placed on a different basis than is done for the majority of highway improvements in the United States. The base of economic evaluation shifts from the road user consequences to the broad economic and social developments to the country as a whole or to the area affected. This chapter is devoted to a discussion of the economic analyses as a tool by which to measure the relative worth of transportation projects as instruments to aid in developing land areas to higher productive uses in areas of low economic activity and usually low population density.

THE COMMON SITUATION

With special reference to highway transportation, public works projects to improve the economic and social status of a country or a local section thereof are no being restricted to the newly established independent countries, but such types

---

[1] *The Complete Works of Ralph Waldo Emerson*, Vol. VII, *Society and Solitude*, Chapter 11, *Civilization* (1861), p. 22. Houghton Mifflin Co., Boston, 1883.

of proposals are practically worldwide. The greater number, however, are to be found in Africa, Asia, and South America. The United States has established one large such project–the Appalachian Development extending through 12 states from New York south and southwest along the Appalachian Mountains to Georgia. The primary objective is to bring to this large land area increased incomes and higher intensity of land use through increased agriculture, manufacturing, industry, mining, recreation, and tourism. Highway transportation is the main investment in the total program by which it is hoped that these objectives will be achieved in the Appalachian region. The programs in Africa, Asia and South America have similar objectives, and transportation is one of the keys to their success.

The main difference between Appalachia and most of the land areas affected in these other countries is in the state of existing development of road networks and in motor vehicle use of these networks. The Appalachia is blessed with a good system of highways that crosses mountain ranges and which flows through the valleys southwest and northeast. But there are many inland areas without sufficient road facilities. The population is reasonably well distributed through the area and the whole area is surrounded with well-developed economic activity and population.

In Africa, Asia, and South America, much of the land being considered for development is practically without roads–perhaps trails and single-lane cart roads exist. The population in the area may exist through self-subsistence agriculture. There is but little movement of people or goods into or out of the area. Thus in these areas any proposal for new roads must be justified wholly on the basis of the overall general economic and social changes to follow. The reductions in road-user costs as such are often near zero for the reason that road users do not exist at the beginning, but where primitive roads accommodate the oxcart or mule cart, the new road will materially reduce the cost of transport.

Considering that the areas proposed for development are without local finances to pay the cost of the initial investment, and often the country itself also is without such resources, outside financing is sought. Financing is asked for from well-developed countries as international loans or as outright grants-in-aid, or from international loan agencies. In all of these cases, whether loans or grants, the supplier of the money usually wants an economic analysis of the proposal as a guide to the economic soundness of the proposed project as a whole. In the end those supplying the financing will want an analysis of the economy of proposed mutually exclusive alternatives so that their relative project design merits can be judged.

BASIC TYPES OF HIGHWAY PROPOSALS

Because each public works project, proposed as an aid to the economic development of a country or region within the country, is specific to its own loca-

Economic Analyses in Developing Countries

tion, each proposal must be examined in light of its characteristics, its objectives, and its probable influence on the future economic and social changes. Further, the project proposed cannot be isolated and considered separately. The whole of the economic forces involved is to be considered.

A brief mention of some of the general types of highway improvements to be considered and their objectives will be helpful in developing the background for a more detailed discussion of the factors involved.

TYPES OF IMPROVEMENTS

The highways, or roads, proposed for construction in a land area to be developed to a greater economic intensity, fall into two broad classes–penetration roads and road systems. In between these two extremes, strip development and area development, there are many degrees of combinations. Some proposals are to be considered simultaneously with the initial project and some proposals are natural second stage improvements to follow 3 to 15 years later.

The Penetration Road

The penetration road may be of the type that extends into land areas relatively little used and from which no, or but few, products of the land come out. Often these penetration roads will start at a seaport or town on a river. The road makes possible an access to the land as well as an access to market for the products of the land. The road may be to tap a virgin timber, a mineral deposit, or just good farming country. A two-way commerce then results. Not only is the economic activity at each end materially affected, but the whole area along the length of the route is likewise, at least wherever some profitable use of the land can be established. Penetration roads are often considered where there exists a natural barrier to land movement, such as the Andes range in Peru and Chile.

A variation of the penetration road is that type of highway that is constructed between two existing population groups, sizable enough to have established some commerce and local trade. A road connection or an improved road is desired to provide the advantages of commerce between the two areas. This type of penetration road may follow a water course–a river, lake, or seacoast–and replace much of the transport by watercraft.

A variation of the penetration road is the lateral, or feeder, road. Roads of this type come into existence after the main road has gained considerable use. The laterals are used as land service roads in their initial stages. This type of development is common in Central America along the Inter-American highway and the Rama Road in Nicaragua. These feeder roads are the beginning of a road system, often not initially so planned, but a road system may be expected to follow wherever the land is a logical prospect for agricultural use.

Road Systems

Initial planning for roads for purpose of land development may be on an area basis which calls for a road network as contrasted to a single penetration route. A combination of the single route and a road net work at the outlying end is common. The outlying network of roads may serve a scattered settlement of people, largely subsistence farmers in a potentially profitable location for volume shipment of produce to the outside.

The road system plan is especially appropriate for the small country which has a high portion of its land suitable for agricultural purposes with its population reasonably well distributed. Here the interest would be for country development or at least for regional development. The Appalachia program is an example of a network plan in an already developed country in which the Appalachia region is developed only lightly compared to its potential.

OBJECTIVES OF THE ROAD IMPROVEMENTS

Basically, there are just two reasons for the construction of penetration roads or road networks in land areas yet relatively undeveloped to a reasonable degree of their potential. These reasons are basically the same in function and may be considered to be (1) to improve the economic and social status of the area, or (2) to improve the economic and social status of the country of which the specific area is a part. Specifically, however, the immediate objective may vary with each proposal because of the existing conditions. The technical aspects of the proposal and the economic and social analysis of its merits depend upon these specific objectives.

A road may be proposed for the purpose of bringing virgin land into agricultural production, to tap the minerals of the earth, to harvest the native timber, to provide for recreational possibilities, to provide for dispersal of people from a crowded area, to give new life to a decaying area and its people, to give the existing road or trails greater capacity, to extend commerce between two specific areas, to encourage the establishment of industry, to provide for improved communication of the people, and all possible combinations of these and other objectives and purposes.

The information necessary to the economic analysis and the method of analysis need to fit the main objectives for which the road improvement is proposed. Note that the basic objectives are some form of economic or social change–economic evaluation–and not project formulation. Project formulation will enter the analysis when planning and promotion reach the stage of requiring detailed designs and final estimates of construction cost. In project formulation motor vehicle running cost may play an important role, even though it may be a minor factor in economic evaluation.

FACTORS PERTINENT TO THE ECONOMIC ANALYSIS

When considering the construction of new roads to an undeveloped area, frequently there is no existing traffic to consider, or so low a volume of traffic exists that benefits to it because of the new road are far short of providing economic support for the new road. Economic evaluation has to be based upon the products of improved land use. The new road, however, affords the means of getting people in to settle and to work the land, and it is the means of getting the products of the land to a consumer's market or to an existing transportation head. The situation is similar to the construction of logging roads into the western forests for the express purpose of harvesting the trees. The considerations are those appropriate to whether the minerals, forests, or agricultural resources of the area can be won and marketed with profitable results. The cost of the road needs to be supported by these business activities rather than by benefits to the road user.

Land development roads generally are planned first as low cost roads. They are improved to accommodate higher traffic volumes and heavier vehicles as traffic growth warrants. When these traffic conditions come about, the economy warrants a change to road-user benefits, as opposed to the initial warrant of the economics of business ventures.

THE WIDE VIEWPOINT

The main objectives of road improvements in developing countries center on economic and social improvements of the area affected in contrast to centering on road user improvements, the usual objective in areas having a highly developed highway system for the motor vehicle. The situation to be considered parallels that prevailing in the United Stated up to about 1875 and later for specific areas of some states. The situation is, also, parallel to the development of the railroad system in the United States from 1830 to about 1914. In both the wagon road extension from the Atlantic coast to the west up to 1830 and the railroad construction starting in 1830 the objectives were to open new lands to new uses and to offer new areas to population migration. Further development of roads in Africa, Asia, and South America will serve similar purposes. Current proposals for highways or railroads are to be examined in the same light as was used in the late nineteenth century in the United States.

FACTORS OF THE LOCAL SITUATION

The factors to be considered in the economic analysis will vary somewhat with the local situation and with the objectives sought. It is essential to compile a good picture of the existing transportation systems–water, road, railroad, and air because any development of additional roads will affect existing transportation either beneficially or adversely. A road system to penetrate laterally from a rail line or fanlike from a railhead would benefit the existing railroad. But a highway parallel to a railroad may have consequences adverse to the railroad. Any existing railroads probably will be government owned so there arises directly a point of government policy with respect to transportation development. Is it to be highways or railroads or airways in the future?

What are the probable future major uses of the proposed roads? Land access is important to development of agriculture, forestry, and minerals, but trunk roads are essential in collection of and distribution of goods, especially in connection with exports and imports.

When exports and imports are a factor to consider, so also are foreign trade balances and export and import tariffs.

In some countries there exists rigid controls on the use of motor trucks in order to protect the government-owned railroads. In such situations, a new highway may not stimulate the desired economic change unless the government policy toward trucking is liberalized.

The whole economic possibilities must be pictured. For instance, what is to be done with respect to electric power, water resources, irrigation, and encouragement of manufacturing and processing industries. Improved roads are just a transportation factor; they succeed only when other forces and opportunities make desirable the transport of people and goods.

Existing and probable future travel times are important factors. Time is important not from the viewpoint of its value to the traveler (which is often near zero), but from its effect upon the demand for vehicles and goods and because of the effect of transportation time on trade and industry. Often, the result is that a new higway reduces transport time 10 to 50 percent from existing transport time by rail or by a devious road.

THE COMPLEX OF FACTORS

The foregoing discussion of some of the factors involved in the economic analysis of proposed road improvements in developing countries points up the complex character of the analysis and that the proposed road itself is often only the catalyst through which other economic and social forces are brought together to create substantial developments. It is, therfore, highly important that the analyst of the proposals for road improvements understands these factors and their reactions to the extent that his report will lay before the decision makers an accurate and complete discussion of every possible consequence of each alternative. In this application, the analysis is not one of engineering economy but one of social economics.

PROBABLE CONSEQUENCES

The economic and social consequences to follow a specific road improvement in a developing country are many and complex. In general these consequences can be predicted, at least on a qualitative basis. Each situation is a case of its own because of the status of development at the beginning, potential resource development, attitude of the population, and character of the road improvement. The economic growth which may follow a road improvement almost always cannot be credited to the new road but to a combination of factors, conditions,

Economic Analyses in Developing Countries

and actions. The road may have been one immediate factor which triggered a sequence of actions, but the development came into being largely because of the other factors. Nevertheless, a general discussion of the probable and possible consequences will be helpful in gaining an understanding of what to expect overall, and therefore, a guide to the conduct of the economic analysis.

ATTITUDES OF THE PEOPLE

A new highway into undeveloped land areas by itself does not create economic change. It only enhances the opportunity for greater and different economic activity. The change itself must be achieved by the people. Their response to the opportunity presented them is the key to all the consequences. [See Wilson, Ref. 21-29, pp. 192-210.]

One of the reasons why the results from road improvement vary considerably location to location is that the local attitudes are not the same. An alert, aggressive, energetic, and trusting people will move out for pecuniary gain and greater satisfaction. A lazy, complacent, skeptic people will hold back and not act upon the new opportunities. Missionaries, Peace Corps workers, and foreign aid technicians in many less advanced regions have experienced great differences in acceptance rates of methods of improving health, farming methods, and living conditions.

The economic success of penetration roads and road systems is dependent upon their immediate use. In turn, this immediate use is dependent upon changes in land use in the direction of more produce to haul out and more manufacturing or processed goods to haul in. This economic activity results in more travel by the people. But these events do not come about unless the attitude of the people is favorable to such changes. The new areas need the presence of pioneers, entrepreneurs, and people with enterprise.

RESOURCE DEVELOPMENT AND LAND USE

Land use is usually the key to economic growth in developing areas. Improved roads encourage land owners to increase acreage of crops–cotton, rubber, grain, coffee, and the like–and to shift from subsistence farming to surplus farming for shipment outward. Shipment outward then results in shipment inward of manufactured goods–implements of agriculture and fertilizer, household goods, and clothing. Thus a two-way economic exchange is developed that generates still more transportation. There follows also the movement of people and shift in population. Where the earth minerals and forests are the object of development, the consequences of road construction are similar plus the establishment of work centers or industrial plants.

The degree to which agriculture increases and industry develops is a result of local conditions including the state of activity at the beginning. When there exists a demand for a certain product, such as cotton, and the price is rising, a road into a cotton-producing area will greatly accelerate the production of cotton. In other instances agricultural production may be greatly enhanced by importation of fertilizer and farming tools made easier to get and cheaper in cost by the new highway.

In areas where there is a latent pressure for population expansion, but a failure to expand because of physical barriers of mountains, water, or jungles, the response to settlement of new land may be quick in coming and heavy in action. No specific formula is available by which these consequences can be calculated. For each case being examined the analyst must have the proper sense of touch, observation, and insight to detect the factors and to recognize how they will react in combination.

ROAD TRAFFIC AND TOTAL TRAFFIC

Universally, the consequence of greatest certainty and one generally experienced is that traffic along the corridor of the new road will increase, or, if the road is a wholly new trafficway, a sizable volume of traffic will develop. Furthermore, the total traffic in the area most likely will have a net increase. The term "net increase" is used because some trails or roads and railroads may experience a decrease in traffic because of transfer to the new road. But surprisingly, railroad freight traffic may increase, even when the roads are parallel. The primitive forms of traffic—oxcart, mule, and load-carrying humans—generally decrease. The bus and truck soon take over. The road generates new travel of the people. The bus not only takes on this additional load, but most often will take over a sizable percentage of the railroad's passenger traffic.

As might be expected, the increased traffic in people and goods and overall improvement in mobility, result in a lowering of passenger fares and freight rates. Competition between the railroad and the highway for business is one factor leading to lower costs for travel. Another factor is the increase in travel of both people and goods. The changes often also carry back to import and export rate reductions. Rate reductions for local transportation and for export have been reduced up to 50 percent.

One reason for the shifting of transportation to the highway from primitive oxcarts and mule is that the cost reductions are high. In Nicaragua trucking costs from the field to the cotton gin were estimated at 10 cents a ton-kilometer as compared to 35 cents by oxcart. In Guatemala in 1957 the railroad charged $2 for shipping 100 pounds of general merchandise, but after the new road in 1963 the charge was reduced to $0.40 [21-29, p. 182].

One important attraction to the new road is the faster service, particularly for short hauls. Transport may be reduced from two to four days to one-half day. Even the railroads improve their service in order to hold shippers who may want to shift to the truck.

New roads or substantial improvement of oxcart trails can be certain to result in a net increase in travel of people and in shipment of goods, generally at a lowering of transport costs and with an improvement of service. These consequences are strong economic forces bearing upon economic growth of the area. But as cautioned before, each location has its own set of forces and conditions and cannot be counted upon to respond in the same direction or in the same magnitude as another area.

METHOD OF ANALYSIS

The principles of analysis and of reaching the final management decision relative to road improvements in developing countries or regions are the same as set forth in Chapters 2 and 23. Criteria for acceptance will differ from an analysis for a set of alternatives wherein the road user consequences are the major factors. The factors—market and nonmarket—and their handling will vary somewhat analysis to analysis and objective to objective. Because the most common types of road proposals in developing countries are directed toward overall and general economic and social change, it follows that the cost involved must in some way be compared to the values of the economic and social changes.

SELECTION OF FACTORS

For economic analyses of proposed road improvements in developing areas, the factors involved are somewhat the same ones as are used for the same type of analyses in developed areas. The essential differences are in the sources of information, the relative emphasis placed upon the factors, and the method of pricing.

Pricing the Facility and Consequences

Pricing the construction of the road facility proposed and the economic consequences to follow is a more difficult task to accomplish in a developing country than in a country developed to a high and stable level of economic activity. Generally, the new road plus other planned improvements bring about drastic changes in the supply and demand forces that the real prices of commodities, labor, and services change materially. Because the total economic activity in the affected area must be forecast for the full analysis period, prices must likewise be forecast. Often, too, the road facility may be constructed in stages or in segments extending over a series of years during which construction costs may vary as a result of the road improvement accomplished. The price changes referred to are real price changes effected through increased productivity, lower transportation costs, and changes in the industrial and commercial complex and not price changes resulting from deflationary or inflationary forces.

Initial construction prices at the outset of starting the road improvement may be estimated on the basis of the current market. For construction thereafter, however, prices should include the effects of the completed portion of the planned total road (and other) improvements contemplated during the analysis period. For instance, the new road may bring about exports of products of agriculture, forestry, and mining to the extent that improved harbor facilities, increased export-import trade, and population shifts reduce freight transportation tariffs, encourage the use of construction machinery, cause local industry to be established, and bring in competition. These are the results sought and when they are attained they usually modify materially the supply and demand forces with resulting major adjustments in the price structure.

In the first stage of economic evaluation, precise estimating of construction prices is not essential. When project design is undertaken and project formulation is the objective, then more accurate estimating of costs is desirable.

Shaner [21-25, pp. 34-53] and others recommend the use of "shadow" prices for the economic analyses of proposed new works in developing countries. Shadow prices are defined as those market prices at which the supply of the commodity being priced will be just sufficient to satisfy the demand. This is the price situation that generally prevails in well developed countries that have markets and competitive situations that operate to maintain the balance between price, supply, and demand. In a country, however, that may be shifting from a subsistence economy to an export-import economy, from the use of only hand labor and primitive tools to power machinery, and from manual and oxcart transportation to automotive transportation, the initial market price structure is an unsafe basis for judging the economic wisdom of major changes in transportation facilities and modes.

Justification of adjusting the local and current market prices to shadow prices for use in the economic analysis lies in the desire to base the analysis on stable economic forces and to be realistic. A parallel situation is found in the United States during the expansion to the west and the construction of both wagon roads and railroads. The new transportation facilities brought industry to the west. During the ensuing years of economic adjustment, market prices usually decreased and supplies increased.

Where railroad, bus, and truck transport exist, their charges may be obtained as a basis of initial prices of such mode of transport. For private hauling and for hauling by primitive methods transport costs are more difficult to determine. Such costs can be ferreted out by investigating the local practices, wage rates, and trends. For motor vehicle transport with both trucks and passenger cars, reference to the book by Jan DeWeille [21-16] will be helpful. This publication is directed specifically to motor vehicle operating cost in developing countries.

Analysis Period

As for any economic analysis, a time period must be chosen over which to

Economic Analyses in Developing Countries 535

measure the consequences and to account for the cash flows. Even though the effects of penetration roads and road systems into land areas not highly developed may be identified and measured for many future generations, the need to conserve the country's financial resources suggests that the best practice is to choose a relatively short period of years for the analysis period. Perhaps 5 to 20 years would cover the normal range. The exact length of time must depend upon the character of the improvement, the production of the land, the probable rate of economic change, and the period over which the improvements are scheduled to be completed.

A highly successful venture should pay for itself in 5 to 10 years, and any improvement that would not pay off in at least 20 years should not be undertaken except in the most unusual situation. Usually, the developing countries have such heavy demands on their financial resources that they must invest in fast repay ventures. Another consideration is that the economic change can be intensive and come quickly, thus bringing a new set of demands and criteria. A short analysis period is in harmony with a low cost road. Where mineral or forest resources are to be tapped, the analysis period must be kept well below the probable exhaustion life of such resources. When the objective is to settle new land or to bring additional land into agricultural production, the analysis period should be no longer than the time required to accomplish this objective.

In keeping with the principle that the economic analysis is for the purpose of aiding management in making a decision today, there is no place for a computation over a long period of time covering probable physical lives of long-lived property or to the ultimate end of consequences. The hazards of forecasting the future are great in any case, so looking 5 to 20 years hence is risky enough.

In those proposals that may be country-wide and are planned to be built over a 5- to 10-year period, the analysis, period must provide sufficient time in which to develop the benefits anticipated.

Travel Time

Reduction of travel time through improved roads in developing countries is one of the important consequences. Quicker delivery of goods and faster bus travel for passengers act to increase both truck and bus usage. In turn, general economic and social activity is accelerated. Even the added competition with both rail and air travel causes these services to be improved.

Including the value of travel time as a benefit factor is justifiable despite the fact that much of the population may "have time on their hands." The fact that faster bus and truck travel induces much additional travel and lowers the rates, is evidence that there is a real value attached to travel time. In placing a monetary value on passenger travel time, consideration should be accorded to the general wage rate of those who travel. Travel time value could be set at about the wage rate. Care is in order, however, not to let travel time reduction be the main factor in economic evaluation because such time reduction may not be put to productive use of the type that would increase monetary income when the labor supply is in excess of demand.

One distinct advantage of reduced travel time comes to the small farm operators who in their farming season can get to market and back in half a day or less as compared to the normal full day on the old road. Smaller inventories are required with faster and more frequent truck service, and emergency needs can be had in short time.

Vestcharge Rate

Many public works projects in developing countries are financed by international loans or grants, and often at more than minimum interest rates. But here again, the vestcharge rate for the economic analysis is not chosen on the basis of the interest rate to be paid on a loan. Of chief consideration is what is money worth to the people? What are the risks involved? What are the overall demands for money? What interest rates are the money lenders charging the public? What rates are charged by the banks on agricultural loans? The answers to these and other questions will furnish a base from which the vestcharge rate for the economic analysis may be decided upon.

The vestcharge rate to use in the economic analysis should be set by competent authority. Competent authority means an individual or board well informed in the financial affairs of the country, banking practices, and government policy. Knowledge of what interest rates apply to official internal and external borrowings, to industrial and agricultural loans, and what rates are charged by local money lenders is essential information. The rate of vestcharge used in the economic analysis for road projects should be consistent (but not necessarily the same) with the rates used in economic evaluation of other public works projects such as power plants and water resource projects.

By way of illustrating the range of vestcharge rates that may be considered, a few specific rates will be mentioned. The long-term lending rate of the National Bank for Development of Honduras was (about 1957) 9 percent for agricultural loans secured by farm mortgages |21-28, p. 147|. The rate of the Mortgage Bank in El Salvador was 7 percent |21-28|. At this period of time, money lenders in Honduras were charging 24 to 36 percent. The private rate of return in manufacturing was about 19 percent |21-23, p. 22| in Honduras in 1962.

Schmedtje |21-15, pp. 22-28| estimates that in the early 1960's the riskless cost of capital in Pakistan was 7 to 8 percent. Chakravarty |21-31, pp. 60-61| arrives at a rate of interest for India of 8 to 12 percent without uncertainty, but inclusive of productivity and thrift.

GENERAL PROCEDURE OF ANALYSIS

The principles of analysis and the steps of procedure are the same for proposed highways in developing countries as in fully developed motor-vehicle based economics. The differences lie in the factors involved, their magnitude, and relative weights assigned in the final decision.

In many applications there will be practically no motor vehicle travel existing to receive benefits; in other cases there will be bus and truck travel of significant value to consider, but few passenger cars. To the extent that any existing roads are used, such use is evaluated with and without the proposed improvements in the normal way.

When the proposals for road development are mainly to penetrate an undeveloped area to open its natural resources to unselfish exploitation, the analysis will be planned to compare the resulting values of such development with the road costs. This type of analysis becomes one more nearly approaching a business venture than it does a public highway improvement. Early mining developments in the United States and more recent ones in South America and Canada are examples. Real estate developments are also examples. The added income brought into the area in exchange for the products sent out becomes a direct measure of the benefits resulting from the road improvement. Social and economic benefits other than sales of products resulting from land development are real and many, but are beyond reliable monetary pricing.

A practical difficulty regularly encountered is the lack of economic records, past or present. Progressing by "dead reckoning" and "economic horse sense" is in order. Nevertheless the analysis can be kept on solid ground and within the real world. Whatever calendar time is required to complete a first class study should be taken. Much is to be lost and nothing gained by a "quickie" analysis, poorly conceived, poorly executed, and hastily completed.

THE FINAL DECISION

An analysis of the probable economic consequences to be expected from road improvement in a developing country often may be made for the guidance of two separate agencies–the country in which the improvements would be constructed and the country or financial agency which would supply the financing, either as a grant in aid or as a loan. But for either agency the report on the study should contain about the same information developed in the same manner. The decision is made much as is the decision of management in private industry or the officials in a developed country considering any proposal for a public works investment. The economic analysis is just one tool available to use in reaching a decision.

FACTORS BEARING ON THE DECISION

Of first consideration is the important question: will the proposed investments in road work accomplish the objective desired? Answer to this question cannot be reached by the decision maker until he first fixes solidly and explicitly in mind what the objective is. The objective usually is not to build a new road, but to stimulate through a road improvement the economic and social development of the area, region, or country.

A second important consideration of the decision maker is to make certain that all alternatives of reaching the objective were considered. Is there a better way, not discussed in the report? For the country itself, consideration must be given to programs other than road development which could be used to spur the desired economic and social development. Water, air, railroad, and pipeline transportation may have greater potential, or may be used in conjunction with roads. Perhaps an electric power development or a land irrigation system would provide equal or better chances of providing the stimulant to economic development.

The report itself is subject to detailed evaluation by the decision maker. Is it optimistic, pessimistic, or neutral; conservative, moderate, or risk-taking in its approach, procedures, and discusión? Did the analyst use the best information available and the most appropriate information? Are the economic and social changes forecast to take place realistic in terms of the attitudes of the people and their objectives? Other questions of similar purpose should be asked and answered by the decision maker as a means of placing the report in proper perspective.

REACHING THE DECISION

The decision maker must not be carried away by his sentiments for people or by what has been accomplished elsewhere. The proposal at hand is a specific one and is so considered. There is no basis for using generalities. For this reason he must first reach a sound appraisal of the report on the economic analysis. Much can be written on the final decision process. In the end it is an exercise in judgment, by persons qualified to judge. Wilson's discussion [21-29, pp. 174-218] is excellent reading for both the analyst and the decision maker. Wilson states on page 218[2].

> The present study suggests that there are, in fact, few magical properties in transport investments that warrant the excessive attention frequently paid to them. Transportation is merely another industry or industries; transport investment is like any other and should be judged on grounds applicable to other forms of economic activity and capital formulation. The role of transport investment in economic growth is not unique. Transport investment is no more an initiator of growth than any other form of investment or deliberate policy. Under some conditions it may turn out to be strategic but the same can be said about any specific investment or policy. The essential message of the present volume is that policy makers and analysts take a more agnostic view of transportation operations and investments.

[2] Copyright 1966 by the Brookings Institution. Used by permission.

Economic Analyses in Developing Countries

PROBLEMS

21-1. Examine some of the literature in the United States, say for 1780 to 1900, for the purpose of learning about the early development of trails, wagon roads, turnpikes, and railroads. What evidence is there indicating the reasons for the improvement of transportation facilities? Trace out the consequences of this transportation development with respect to social and economic changes. Do you find factors and objectives in this early period in the United States that might be parallel to current situations in Africa, Asia, and South America? Contrast with road building in the United States today.

21-2. Select a country that may have begun to improve its road network shortly after World War II–Turkey, Philippines, Brazil–are possibilities. Trace the development of the use of motor vehicles–vehicle-mileage, registration, imports, etc.–then to date. Comment on the results. Look, also, for any changes in the composition of the character of traffic as between oxcarts, buses, trucks, and motor cars.

21-3. Read some of the references on the social and economic changes of the developing countries in Asia, Africa, and South America. Write out a list of the factors of change which could be consequences of improvement of highways and classify them as between (1) social, (2) economic, (3) political, and (4) transportation. Indicate by system whether the change is a benefit or a detriment (1) to local areas or local business, and (2) to the country as a whole.

REFERENCES

21-1. Hans A. Adler. "Economic Evaluation of Transport Projects," in Gary Fromm (ed.), *Transport Investment and Economic Development.* Brookings Institution, Washington, D.C., 1965.

21-2. A. N. Agarwala and S. P. Singh (eds). *The Economics of Underdevelopment.* Oxford U. P., New York, 1963.

21-3. P. T. Bauer and Basil Yamey. *The Economics of Underdeveloped Countries.* U. of Chicago Press, Chicago, 1957.

21-4. R. S. P. Bonney. *The Relationship Between Roadbuilding and Economic and Social Development in Sabah.* Road Research Laboratory, Harmondsworth, England, 1964.

21-5. Gerald Karel Boon. *Economic Choice of Human and Physical Factors in Production.* North-Holland Publishing Co., Amsterdam, 1964.

21-6. Robert T. Brown. "The Railroad Decision in Chile," in Gary Fromm (ed.), *Transport Investment and Economic Development.* Brookings Institution, Washington, D.C., 1965.

21-7. Christopher T. Brunner. *The Economic Justification of Roads in Developing Countries.* Fourth World Meeting of the International Road Federation in Madrid, Spain, 1962.

27-7A. R. W. Burton and R. B. Eksteen. *Motor Vehicle Running Costs for South African Conditions.* National Institute for Road Research, Council for Scientific and Industrial Research, Pretoria, South Africa, April 1965.

21-7B. S. Chakravarty. "The Use of Shadow Prices in Program Evaluation," in P. N. Rosenstein-Rodan (ed.), *Capital Formation and Economic Development.* M. I. T. Press, Cambridge, 1964.

21-8. Paul H. Cootner. "Social Overhead Capital and Economic Growth," in W. W.

Rostow (ed.), *The Economics of Take-Off into Sustained Growth*. St. Martin's Press, New York, 1963.

21-8A. Jan DeWeille. See Ref. 21-16.

21-9. Robert W. Fogel. *Railroads and American Economic Growth: Essays in Economic History*. Johns Hopkins Press, Baltimore, 1964.

21-10. Gary Fromm (ed.). *Transport Investment and Economic Development*. Brookings Institution, Washington, D.C., 1965.

21-11. Edwin T. Haefele. *Road Construction as a Means of Developing Areas Served*. Document No. 57, Ninth Pan American Highway Congress, Organization of American States, Washington, D.C., May 6-18, 1963.

21-12. E. K. Hawkins. *Road and Road Transport in an Underdeveloped Country: A Case Study in Uganda*. Her Majesty's Stationery Office, London, 1962.

21-12A. Highway Research Board. National Academy of Sciences, Washington, D.C., Highway Research Record No. 115, *Transportation and Economic Development*, Jan. 11-15, 1965.

    1. Wilfred Owen. "Immobility: Barrier to Development," pp. 1-9.

    2. George W. Wilson. "Case Studies of Effect of Roads on Development," pp. 10-18.

    3. Nuhad J. Kanaan. "Structure and Requirements of the Transport Network of Syria," pp. 19-28.

    4. Robert T. Brown and Clell G. Harral. "Estimating Highway Benefits in Underdeveloped Countries," pp. 29-43.

    5. Richard M. Soberman. "Economic Analysis of Highway Design in Developing Countries," pp. 44-63.

    6. Holland Hunter. "The Passenger Car in the USSR," pp. 64-70.

21-13. Albert O. Hirschman. *The Strategy of Economic Development*. Yale U. P., New Haven, Conn., 1962.

21-14. International Bank for Reconstruction and Development. *The Economic Development of Uganda*. Johns Hopkins Press, Baltimore, 1962.

21-15. International Bank for Reconstruction and Development. *Estimating the Economic Cost of Capital* (with special reference to developing countries), by Jochen K. Schmedtje. Report No. EC-138, Washington, D.C., October 1965.

21-16. International Bank for Reconstruction and Development. *Quantification of Road User Savings* (prepared by Jan de Weille, Economics Department) Washington, D.C., December 1965.

21-17. International Bank for Reconstruction and Development. *The Economic Development of Venezuela*. Johns Hopkins Press, Baltimore, 1961.

21-18. Tillo E. Kuhn. *Economic Analyses for Highway Improvements in Developing Countries*. Ninth Pan American Highway Congress, Organization of American States, Washington, D.C., May 6-18, 1963.

21-19. Tillo E. Kuhn and Michael Nelson. *Economic Development and Transport Investment Planning: A Case Study in Honduras*. Ninth Pan American Highway Congress, Organization of American States, Washington, D.C., May 6-18, 1963.

21-20. Armando M. Lago. *Cost Functions for Intercity Highway Transportation Systems in Underdeveloped Countries: A Model on Optimum Technology*. Presented September 1966 at the meeting of the Institute of Management Sciences, as a version of the author's doctoral dissertation in economics at Harvard University.

21-21. Wilfred Owen. *Strategy for Mobility.* Brookings Institution, Washington, D.C., 1964.

21-22. Pan American Union. *General Problems of Transportation in Latin America.* Washington, D.C., 1963.

21-23. Republica de Honduras, Consejo Nacional de Economía, Banco Central de Honduras. *Investigación a la Industria Manufacturera.* Tegucigalpa, D.F., Julio 1964.

21-24. W. W. Rostow. *The Stages of Economic Growth.* Cambridge U. P., Cambridge, England, 1960.

21-25. Willis W. Shaner. *Economic Evaluation of Investments in Agricultural Penetration Roads in Developing Countries: A Case Study of the Tingo Maria-Tocache Project in Peru.* Engineering Economic Planning Report No. 22, Stanford University, Stanford, California, 1966. See also Highway Research Board, National Academy of Sciences, Research Record No, 180, 1967, pp. 120-132.

21-26. R. M. Soberman. *Road Transport and Economic Development in the Venezuelan Guayama: A Study in the Choice of Transportation Technology.* Unpublished Ph. D. dissertation, Massachusetts Institute of Technology, Cambridge, 1963.

21-27. Henry M. Steiner. *Criteria for Planning Rural Roads in a Developing Country: the Case of Mexico.* Engineering Economic Planning, Report No. 17, Stanford University, Stanford, California, 1965.

21-28. U. Tun Wai. *Interest Rates Outside the Organized Money Markets of Underdeveloped Countries.* International Monetary Fund, *Staff Papers,* Vol. VI, No. 1 (November 1957).

21-29. George W. Wilson and others. *The Impact of Highway Investment on Development.* Brookings Institution, Washington, D.C., 1966. This book gives the following case studies: (1) "The Cochabamba–Santa Cruz Highway in Bolivia" by Barbara R. Bergmann, (2) "The Atlantic Highway in Guatemala" by Martin S. Klein, (3) "The Littoral Highway in El Salvador," and (4) the following by George W. Wilson: (a) "The Friendship Highway in Thailand," (b) "The East-West Highway in Thailand," (c) "The Ramnad Road in Southeastern India," (d)" North Borneo (now Sabah)," (e) "The Pacific Littoral Highway in Nicaragua," (f) "The Western Montana in Central Peru," (g) "West Nile Area in Uganda", and (h) "The Tejerias-Valencia Highway in Venezuela." Wilson gives an excellent evaluation of all the case studies and discusses in detail the process of economic analysis and its significance.

21-30. Arthur Wubnig. *Measuring the Benefits of an Agricultural Feeder-Road Project.* Economic Development Institute of the International Bank for Reconstruction and Development, Washington, D.C., 1963.

Chapter **22**

Formal Report on the Economic Analysis*

All of the work on the economic analysis of proposed highway improvements is accomplished for just two important objectives. First, this analysis produces results useful to those responsible for the engineering design, and, second, the analysis furnishes the decision maker an important aid in deciding whether to build the improvement (economic evaluation) and, if so, to what design (project formulation). This second objective cannot be reached until the decision maker has the results of the analysis together with the full story of how the results were reached. Because the economic analysis is based upon probable future actions, developments, and consequences, rather than standard procedures, factors, and accepted engineering relationships, it is essential that the decision maker have the full story supporting the economic analysis. He must be in a position to pass his own judgment upon the selection, quantification, pricing, and use of each factor affecting the results and conclusions. Therefore it is appropriate to write a formal report on the economic analysis of each highway improvement considered.

THE FORMAL REPORT

A formal report on the economic analysis of a proposed highway improvement is written either by the staff or by a consultant to the highway department. It is then read by all concerned for the purpose of gaining a thorough understanding of the proposed project. The economic analysis is unlike many of the straightforward procedures in engineering design, and, therefore, is treated with appropriate differences.

NECESSITY OF THE REPORT

Within a highway department or a consulting firm, detailed reports on the designs accomplished are more often not required than are required. On the other hand, a detailed report on the economic analysis is a requirement in each case. The reason for requiring a report on the economic analysis and not for the

---
* This chapter is based on Ref. 22-19.

Formal Report on the Economic Analysis 543

general design is that the results are of a different nature and serve different purposes. A final engineering design is the end point for review and approval, prior to advertising for bids. The economic analysis is accomplished by using much judgment, selected procedures, factors, and valuation of factors by the analyst, and it is not the end point. The economic analysis is one of the aids used in making two important decisions prior to final design–should the improvement be constructed, and, if so, what is the best design. Management then must know just how the information supporting the economic analysis was collected and used. The analysis for economy resulting in the equivalent uniform annual cost, present worth of future cash flows, net present value, benefit/cost ratio, or probable rate of return cannot be understood or applied with confidence unless the users–the decision maker and others at interest–have full knowledge of how the results were obtained. It is in order, therefore, that the analysts responsible for the results present in writing the complete story about their work.

USES OF THE REPORT

Although the formal report on the economic analysis serves the management decision process as its primary function, it also serves other purposes. As discussed in Chapter 23, management is in need of the information and discussions in the report on the economic analysis. The decision maker must have the details supporting the findings about the road-user and nonroad-user consequences before he can make any intelligent progress toward his decision.

In addition, the formal report is a good reference for the staff who were involved in the project to date and those who will be responsible for the final designs. As time moves ahead, the collection of reports on economic analyses builds up a valuable reference library. It becomes a source of facts and procedures useful to studies of future project proposals, especially those nearby or related in specific objectives or aspects of design. Also, the report leaves a written record as contrasted to a memory record, possessed in part by each of those responsible for performing the analysis. This written report is left behind and available to all when any employees who worked on the analysis separate their employment.

Should the proposed improvement project become involved in controversy and public discussion, this written record becomes a valuable source of information for all persons concerned, especially for the highway department officials who must answer the many questions from a wide range of inquiry.

GENERAL CHARACTERISTICS OF THE REPORT

The formal report on the economic analysis for a proposed highway improvement follows the same general format, arrangement, functional design, and character that all good technical and business reports should. The preparation of this class of written report follows well-established customs, but the author has a wide range of freedom in composition, physical design, and choice of detail to include.

In general, the author's objective is to prepare a report that will answer the questions the decision maker and other persons will ask, and to supply the information for the permanent record which will permit anyone to follow the same paths he followed and to see the same things.

The reader, however, may not evaluate the facts, opinions, situations, and requirements in the exact way as did the author. The requirement is that the reader must have the opportunity to arrive at the same place. The relationship of the author to his reader is much the same as that of the author of detective novels to his readers. The successful writer of detective stories gives the reader every bit of information possessed by his detective. The reader plays the game also, and is cheated if the detective later discloses facts denied the reader.

The author's task is to tell the story of what he did, why he did it his way, and what material he used. He needs also to interpret his information and his results in order that the reader can better understand the significance of the results.

The length of the report is governed by the complexity of the improvement project reported upon and the requirement that the report be fully informative. A proposal that included only two alternatives and straightforward engineering can be a short, simple report. A report on the location and design of an urban freeway involving six alternative locations and a total of 24 significant variations in design would call for a formal report of hundreds of pages. However, length of the report itself is of no great importance; the important factor is the requirement that the report include all information necessary to gaining a full understanding by the reader.

There are many occasions when a sizable report is submitted on the overall aspects of a major highway improvement, such as on an urban freeway or a rural interstate location of complex problems with many alternatives. The reports, because of their total size, may contain only a briefed down summary of the economic analysis, and appropriately so. But backing up this summary, there should be a staff report or in-house report available in which full detailed information may be found.

BASIC QUALITIES OF THE REPORT

There are three basic qualities by which the formal report is judged. They are (1) technical quality, (2) reading quality, and (3) physical appearance. In other words, a good technical report must possess accurate and sound technical factors, graphic and intelligible use of English grammar and composition, and attractive physical makeup.

Any technical report is expected to adhere to well-accepted procedures of practice, concepts, and principles. Unless it does, the reader rejects the findings and the report as a whole. Novel situations and new problems call for innovations and imagination, but these can be had without violating what is regarded as the essentials to good engineering, good economics, or good statistics.

A technical report may possess sound technical aspects, but unless it can be read easily and with positive understanding its sound technical character is lost to the reader. Good writing in this concept does not mean top-quality classical writing, but just plain straightforward writing in sentences of varied lengths and common words, so arranged that the reader can get the exact meaning intended by the author. Obscure or doubtful phrases or constructions result in failure to communicate. The author must so write that there is full meeting of the minds on what is written. But this meeting of the minds does not mean agreement in principle or in practice. One cannot disagree or agree with another person until there is agreement on what the issue is and what position each takes toward the issue.

In advertising, merchandising, and social contacts the outward appearances are accepted as an index to the inward qualities. A technical report done up in an attractive cover, adorned with well-chosen preliminary parts, and made readable with a pleasing and an enticing page format will create the feeling that the report is a good report technically and in reading quality. Top-quality physical appearance of a written technical report is easy to achieve. Its cost is not much in time or money. It adds much to the acceptance and understanding of the report.

Good physical appearance, though, is not intended to cover up poor engineering or bad writing. It is used only to reinforce these two elements. When thorough reports fail to be appreciated and accepted, the cause can often be traced directly to the authors who did not provide the proper environment for their competent engineering and skillful writing. A rose in a rusty tin can is still a rose, but it may not be appreciated because of the rusty tin can.

PREPARING THE REPORT

Technical reports involve the collection of information, analysis of the information, mathematical and statistical calculations, research and investigations, some literature search, and other specialties. The material produced by these efforts must then be organized into a logical sequence. The result is a story in words, tables, and illustrations. The formal report on the economic analysis of a proposed highway improvement is a sizable undertaking involving much more than just writing. For this reason this section is called "Preparing the Report" rather than "Writing the Report." This author has seen many reports on economic analysis which were woefully inadequate. The reader was not given enough information to permit him to understand the report let alone evaluate it.

PRELIMINARY STEPS

Careful and adequate *planning of the report* prior to writing is an important early step in report preparation. Deciding where to start, what to start with, where to go, and how to get there are essential decisions to be made at the outset. Once these decisions are accomplished the report comes into being rather easily.

Preparing an outline of the report is an excellent beginning. A formal outline is not a requirement, but some listing of the subjects to cover and in what sequence and subordination is a necessity. A final outline can be prepared as the drafts are revised. A general outline is given at the end of this chapter to illustrate a possible coverage of subject matter and the sequence of the main parts.

The *writing process* will move more swiftly and orderly when all tables, illustrations, and source materials are carefully and completely prepared as steps preliminary to writing. Obviously, a report of this character is a story of the information gathered, its organization, its analysis, and results of the analysis. These materials become the objects referred to in writing so they should be available at the beginning of writing.

Before tables and figures are put into final form for typing or duplicating and before any writing is done, decisions should be made as to the mechanics of style (capitalization, use of words or numerals, abbreviations, and certain aspects of punctuation), the page layout including binding and other margins, and the schedule of headings (whether capitals, capitals and lower case, underlined, center, side, or paragraph positions). Early decision on these mechanics will save much effort in the typing and revising processes.

THE WRITING PROCESS

A good formula to follow in preparing technical reports is: Thorough Preparation + Rapid Writing + Painstaking Revision = A Good Report.

It is well-known by teachers of communications that the quality of writing and speaking performance increases with the increase of familiarity and understanding of the subject or event on the part of the author or speaker. Thus when writing or speaking about a personal experience, the writer or speaker is more at ease, more confident, and forms his thoughts quickly and logically. It follows, then that thorough preparation is the first requirement of the author of a technical report.

An author becomes thoroughly prepared by actively doing the data gathering, the estimating, the analyses, and conceptual thinking leading to the results and objectives. Thus the report may be improved by having the main sections written by the persons directly responsible for the work on them. Particularly, the general economics and social aspects should be reported by those persons having specific capabilities in these areas. Thorough preparation includes putting

Formal Report on the Economic Analysis 547

in order all source materials prior to the start of writing as mentioned above, as well as acquiring a complete knowledge and understanding of the technical subject.

Thorough preparation leads naturally to rapid writing. Rapid writing is desirable because one's mind works rapidly and logically when fed with the raw materials of knowledge about the data, their processing, and final relationships. Rapid writing is recommended because it is more apt to lead to logical thinking. In rapid writing, the author is not much concerned with the choice of the best words, the best sentence structure, correct spelling, or even best sequence of units of his subject. The author is more concerned with recording his thoughts as they flow outward. The objective of rapid writing is to record the framework of thoughts about the subject and not to produce a finished work of art. The touching-up comes in the revision process.

Professional writers depend upon revising their original drafts as the means of gaining their best production. Final wording is seldom achieved in the technical writing field on the first draft, and most certainly so under the procedure of rapid writing. Revising a draft consists of the following steps in order:

1. Rearranging all sections as desirable to produce a sound structure and logical organization.
2. Supplying missing sections and eliminating unnecessary ones
3. Recasting sentences and selecting the best words to gain clearness of expression and exactness of meaning
4. Checking all references, facts, and figures for correctness
5. Improving mechanics of style
6. Final checking of the overall excellence of the revised report.

The number of drafts that are developed depends upon the author's ability to say what he wants to say the first time and his facilities for typing. It is usual to put a technical report through 3 to 5 drafts before the final typing. When the writing has been heavily revised, it is good practice to have the pages retyped to clean them up. Clean, neat typing seems to send out a different message than comes from an interlined, crossed-out, messed-up page of revision.

The six steps listed above are usually accomplished one at a time. The author can then concentrate on one objective at a time. These several readings afford greater opportunity of catching all desirable revisions and additions.

SPECIFIC CONTENT OF THE REPORT

Whether the formal report on the economic analysis of a proposal for a highway improvement accomplishes its objectives depends upon what information and discussion are included in it, as well as upon how it is presented. Although each report must be custom designed and prepared, there are some elements of content and form that are common to many reports. There is a minimum content that a report must have if it is to afford the decision maker and the permanent record with the basic information. A full presentation of information and discussion is to be preferred to offering just the bare minimum.

A GENERAL OUTLINE

Unlike the reporting on the results of testing construction materials when the report can say "all tests performed by ASTM Standard Procedure —," the economic analysis, because it follows no prescribed procedure and uses no standard factors, must be reported in detail. The answers obtained are the result of subjective judgments, forecasts, and assumptions. The reader must be given full information on procedures, factors, sources of information, and logic used. Without this completeness of reporting, the reader cannot properly interpret or evaluate the report.

The following outline is suggestive of what is appropriate for the formal report, but there is no fixed arrangement to follow or set of subjects that is universally required.

GENERALIZED OUTLINE FOR A FORMAL REPORT ON THE ECONOMIC ANALYSIS OF A HIGHWAY IMPROVEMENT

I. Preliminary and Supplementary Parts
 1. Cover and outside title
 2. Flyleaf (blank)
 3. Inside title page
 4. Letter of transmittal
 5. Preface (when appropriate)
 6. Acknowledgments (separate part)
 7. Table of contents
 8. List of tables
 9. List of illustrations
 10. Abstract or summary

II. The Main Part, or Body
 1. Introduction
    A. General description of the improvement proposal and its scope
    B. Need for the improvement and its objectives
    C. Special problems and difficulties
    D. Relation of proposed project to overall transportation plan
    E. Basic methods, procedures, and information used
    F. Scope of the report and its organization
 2. The alternatives considered
    A. General description of all possible alternatives
    B. Discussion of main merits and demerits of each alternative

Formal Report on the Economic Analysis 549

   C. Alternatives selected for detailed analysis; reasons for rejection of those dropped
   D. Technical description and complete data for each alternative selected for detailed analysis (including such items as distances, grades, horizontal curves, traffic controls, and bridges)
   E. Design criteria and standards
   F. Traffic volume and composition forecast; gradient growth
3. Highway costs–each alternative separately
   A. Estimated cost of the complete improvement
      (a) Rights-of-way by item
      (b) Construction by highway element
   B. Cost to maintain and to operate the facility for the period of years chosen for analysis
4. Analysis factors used in the economy analysis
   A. The factors chosen and supporting discussion
      (a) Analysis period and range for sensitivity test
      (b) Vestcharge rate and range for sensitivity test
      (c) Value of travel time and range for sensitivity test
      (d) Terminal values of highway elements
   B. Motor vehicle running cost
      (a) Sources of information
      (b) Basic unit costs used
   C. Traffic accidents
      (a) Frequency of accidents by basic elements
      (b) Unit cost of traffic accidents by severity and type
   D. Analysis method chosen and why
5. Computations of all costs and benefits for each alternative
   A. Reduction of highway costs to present worth or equivalent uniform annual costs
   B. Reduction of road-user costs to present worth or equivalent uniform annual costs.
   C. Calculation of index of economy (annual costs, present worth, net present value, benefit/cost ratio, or rate of return) for each alternative using the incremental method
   D. Sensitivity analyses for selected factors
6. The nonmarket economic and social factors for each alternative
   A. Probable consequences to land use and land values
   B. Probable consequences to business and trade
   C. Offsets and transfers
   D. Community consequences
      (a) Displaced persons and establishments
      (b) Environmental consequences
      (c) Community values–protective services, aesthetics, renewal program

(d) Relationships to long-range community plan
(e) Special factors–preservation of landmarks, historical features, parks, river fronts, scenic values

III. Summary and Discussion
A bringing together of the findings and significant factors to emphasize the relative differences between the several alternatives and to point out the reasons for these differences and their probable consequences. This final section need not recommend a preferred alternative unless such recommendation is requested by higher authority.

PROBLEMS

22-1. From the library or other source, procure two or three reports on the economic feasibility analysis of some public works proposals (highway, urban transit system, toll facility, water resource project). Critically examine these reports for their adequacy, considering the suggestions in this chapter. Identify the readership to which the report is directed. Does the report reach this readership with the best suited information in the proper degree of detail–or too much or too little detail?

22-2. Examine the structural organization of one of the same reports selected for Prob. 22-1, or another report. What changes would you suggest. Consider changing the sequence of the major and minor parts, adding new material or eliminating material. Give your reason for your suggested changes.

22-3. Select any report on economic analysis of a highway facility or other public work. Write your analysis of it, setting forth your suggestions for its improvement on the following three major items:
A. Engineering, or technical, soundness
B. English composition and ease of understanding
C. Physical appearance–inside and outside–and page format

22-4. You are the chief engineer of a state highway department. Your planning section has submitted to you the economic and economy report on the location of a section of an interstate route in which four alternative locations are examined. Please state what facts, answers, and information, etc. you would want in this report to help you make the final choice of location. Public hearings indicated that each alternative had some support from certain sectors of the affected public. You need not give your reasons for wanting the information you list.

22-5. Review a book, monograph, or technical paper on engineering reports or scientific writing. Write an analysis of your findings, impressions, and evaluations from the viewpoint of the executive body or person who must read such reports in the process of decision making.

REFERENCES

22-1. Chester Reed Anderson and others. *Business Reports: Investigation and Presentation.* 3d ed., McGraw-Hill, New York, 1957.

Formal Report on the Economic Analysis

22-2. Margaret D. Blickle and Kenneth W. Houp. *Reports for Science and Industry.* Holt, New York, 1958.

22-3. L. Brown. *Effective Business Report Writing.* 2d ed., Prentice-Hall, Englewood Cliffs, N.J., 1963.

22-4. Paul Douglass. *Communication Through Reports.* Prentice-Hall, Englewood Cliffs, N.J., 1957.

22-5. Harold F. Graves and Lyne S. S. Hoffman. *Report Writing.* 4th ed., Prentice-Hall, Englewood Cliffs, N.J., 1965.

22-6. Robert D. Hay and Raymond D. Lesikar. *Business Report Writing.* Irwin, Homewood, Ill., 1957.

22-7. H. H. Holscher. *How to Organize and Write a Technical Report.* 2d ed., Littlefield, Adams, Patterson, N.J., 1965.

22-8. Daniel Marder. *The Craft of Technical Writing.* Macmillan, New York, 1960.

22-9. William S. Morgan. *Writing and Revising.* Macmillan, New York, 1957.

22-10. Porter G. Perrin, and others. *Writer's Guide and Index to English.* 4th ed., Scott, Foresman, Chicago, 1965.

22-11. Joseph Racker. *Technical Writing Techniques for Engineers.* Prentice-Hall, Englewood Cliffs, N.J., 1960.

22-12. Fred H. Rhodes. *Technical Report Writing.* 2d ed., McGraw-Hill, New York, 1961.

22-13. R. L. Shurter and others. *Business Research and Report Writing.* McGraw-Hill, New York, 1965.

22-14. James W. Souther. *Technical Report Writing.* Wiley, New York, 1957.

22-15. R. P. Turner. *Technical Report Writing.* Holt, New York, 1965.

22-16. University of Chicago Press. *A Manual of Style.* 11th ed., U. of Chicago Press, Chicago, 1949.

22-17. Willis H. Waldo. *Better Report Writing.* Reinhold, New York, 1965.

22-18. Margaret Walters. *Basic Guide to Clear and Correct Writing.* Scott, Foresman, Chicago, 1958.

22-19. Robley Winfrey. *Technical and Business Report Preparation.* 3d ed., Iowa State U. P., Ames, 1962.

Chapter **23**

The Management Decision

The analysis for economy compares each alternative proposal for highway construction projects with each other alternative. Management then has a solid market-based tool by which to compare the group of alternatives on the basis of husbanding the dollar, both for economic evaluation and for project formulation.

When at the same time the general social, economic, and community consequences are isolated, evaluated, and described, and with all the information at hand, management is ready to start the decision process—(1) to authorize construction or not, and (2) to designate which engineering design to proceed with if construction is authorized. There is no set procedure, no formula to use, and no ironclad guidelines to follow in the decision process. But some discussion of the decision-making process will help both in making a decision and in understanding a decision once it is made. Both decisions are the result of subjective judgment—not a process of applying a formula. The decision process, however, is aided by having recourse to engineering, economic, and social facts, and knowledge of local opinions, policies, and overall plans for developing the community.

MANAGEMENT DEFINED

The title of this chapter is "The Management Decision." Who is management? In the concept used herein management is that person or legal body of persons, who by law is authorized to make the decision to commit the public funds to the project and to approve the location and design. The management may be the chief engineer of a highway department, a director of public works, a city council, a county board of commissioners, a commission for state highways, a highway or bridge toll authority, or any combination of individuals and legal bodies. For each improvement considered, management authority will be fixed by law, resolution, or by internal direction. Within the total process, however, separate decisions may be made on separate phases or steps. This chapter, however, is concerned primarily with the final decision to construct or not to construct, and, if to construct the improvement, the approval of the plans and specifications.

FACTORS TO BE CONSIDERED BY MANAGEMENT

The factors to be considered may be roughly collected into four groups: (1) policy and program, (2) engineering and technical, (3) community and social, and (4) general economics. Weight, by judgment and not by formula, is accorded these groups and the factors within the groups in the decision-making process. The first decision (but not unrelated to the second decision) is, Shall the project be constructed or not? If the answer is yes, the second decision is to select the preferred engineering design.

Policy factors include the agency's stated position toward helping economically depressed areas, catering to local needs and wants rather than to a wide geographical area, intergovernmental relations, developing a certain type of industry, area redevelopment, aesthetics, broad transportation policy, and fiscal policy. *Program* factors include both short- and long-range improvement plans as to route development, system development, construction scheduling, system modernization, and improvement of highway safety.

When considering highway improvement proposals, management will give appropriate weight to policy and program factors. This weight may result in selecting an alternative not having the greatest engineering economy, but one that overall is in the best interest of the public.

Engineering and *technical* factors are more apt to affect the decision on project formulation than the decision to build or not to build. These factors include the use of locally produced products rather than imported products (this item is also a policy factor), elevated or depressed construction (a policy aesthetic factor), availability of contractors and construction equipment, force account construction rather than by contract to keep own labor busy, and construction hazards.

The *community* and *social* factors include the long list of the usual governmental services, preservation of landmarks, preservation of open space and recreation facilities, keeping up with competing communities, and the controlling of land use. The weight of these factors on the final decision to build or not and on the engineering design will vary widely, depending upon the viewpoint and community policy. A community develops its own environment as it cares to. It may choose to develop a quiet, artistic, bedroom type of town with strong attention to recreation, fine arts, and an active social life, or it may want lots of industry and people, regardless. Some counties, towns, and cities have well-controlled community plans for orderly and systematic development; others let the natural economic and social forces take over.

The *general economic* factors are primarily commercial factors related to private business. Yet these factors are strong ones in the minds of the people, civic leaders, and public officials. Mainly, they relate to the community attitude toward attracting and holding business and industry and to protecting locally established business. When considering proposals for new highway facilities which have commercial implications, highway management officials are squarely up against the from whose viewpoint question. These economic consequences are largely of the type that transfer economic gain geographically and therefore create offsetting losses elsewhere.

Thus, considering only road user consequences, the project may not be justified, but if sufficient weight is given to the local gains in general business and attracted industry, the project may be justified by the local viewpoint. From the viewpoint of local business, the engineering design factors may be the cause of local transfers in business volume. Here again the viewpoint is the key factor.

Overall, the factors of general business and land value cannot justly be divorced from the road-user consequences. To give heavy weight to both the road-user net benefits and the local gains in general business is double counting. The long-range viewpoint is best, no doubt, and the long-range viewpoint would favor the highway and road users, because the highway is more lasting than privately operated business. Further, to the extent that the highway will save in the consumption of commodities and reduce travel time, such benefits are of real net–not transfers–and not of the secondary nonmarket variety.

FIRST DECISION–BUILD OR NOT

Often the basic decision to order the project constructed or not, or to order the new equipment purchased or not, is an easy decision; at other times it is a difficult decision, especially for a marginal proposal.

When the machine is damaged beyond repair, when the bridge is washed out, when the highway project will close a gap in an important route, the economic justification is evident and the decision to build is the one easily reached. The 41,000-mile interstate system was ordered to be built by Congress. No longer is there necessity of analyzing individual segments of the routes on this highway system for their economic feasibility.[1] Comparing proposed new locations and designs with the existing highway is purely an academic exercise; the real analysis is for project formulation–comparing all proposals for individual projects on the new highway each with each other.

An individual or independent project, not a part of a major system or route then under construction, requires close scrutiny to determine its desirability.

---

[1] This statement does not violate the principle that the null alternative should be analyzed for its economic merit, nor that the proposed alternative should not be compared to the existing situation for its economic merit. The whole of the interstate highway system is one large project being constructed through the means of thousands of small contract segments. A designer of one of these contract segments is working on only a small link of the whole system. When designing his small link, the individual designer is hardly in a position to challenge the wisdom of completing the interstate system. In the normal course of his design he will (or at least should) consider all alternatives of accomplishing the function that his particular segment of the system is to accomplish. The situation can be compared to the construction of a regional interconnected electric power transmission network. Should the designer of one short link in this system network challenge either the need for constructing this one link or the need for the entire network system? Take out one link of a chain and you have two chains–not one whole system.

The Management Decision 555

Such a project could be a second river crossing, a second route between two towns, a widening and resurfacing proposal, a grade reduction of an existing roadway section, lengthening a horizontal curve and increasing passing sight distance, or the purchase of a newly developed machine for roadway construction and maintenance.

When the analysis for economy is highly favorable as indicated by a high prospective rate of return or benefit/cost ratio, and the adverse social and economic consequences are of no great importance, a decision to go ahead would be in order. When the analysis for economy indicates a marginal return (a rate of return only slightly above the minimum attractive rate of return, or a benefit/cost ratio just above 1.0) recourse must be taken to the factors related to the agency-wide program as a whole, to the community, and to the general business consequences.

Certainly it would be unwise, other factors being equal, to order construction of a marginal improvement when other unrelated projects appear to have higher priority, either because of economy or relation to the whole program. These marginal proposals so far as economy is considered, may also be justified because of their beneficial consequences to the community. They may be related to urban redevelopment, recreational needs, extension of utility services, or other public works programs. The engineering features could be somewhat uncertain and cause for not going ahead, at least until the engineering is satisfactorily resolved.

Although financing is not a factor in engineering and economic justification of proposals for highway improvements or new equipment, the availability of the funds for the purpose is a factor in management's decision to go ahead or not. This factor is especially pertinent in large-sized projects that may cost an appreciable percentage of the available funds. Perhaps more overall benefits and satisfactions would accrue by acceptance of a number of less costly projects, especially when they may have higher rates of economy.

The decision to build or not to build and when to build is wholly a management responsibility. Once given the pertinent information, including the economy and economic factors, management's sense of judgment usually leads to a good decision. The word "good" is used purposely; often there may be more than one good decision and wise decision and no positive way to arrive at the "best" decision.

SECOND DECISION–PROJECT FORMULATION

The final decision between alternatives, assuming that the first decision is to do something rather than to continue the existing situation, is one that touches but lightly on policy and program factors, but heavily on engineering design details, aesthetic, social, community, and business factors. As previously discussed, highway location and traffic operations can have marked consequences upon local land values, retail business volume, and the location of industry.

Locally, these consequences affect select individuals favorably or unfavorably. The net, however, balances out to zero more often than not. With but little possible error, it can be stated that if all economic consequences were gotten together from near and far, the net business consequences would balance out between beneficial and adverse. The faraway consequences are most difficult to observe or to predict, so usually only the local factors are considered. In the weighting of any factor the viewpoint is most important.

There are three foci from which to view the project formulation aspects of the alternative proposals: (1) from the road user's interest, (2) from the community's interest, and (3) from the businessman's interest. These three factors represent three opposed viewpoints. Many of the most difficult decisions emanate from these three viewpoints. But the decision maker must find an acceptable answer. This second decision is often the most difficult of the two decisions. Generally there is little question of the desire for a given highway improvement; the unsettled question pertains to where, how, and what to construct.

The following sections in this chapter will endeavor to reduce the whole of the management decision process to its fundamental steps, factors, and values.

CHARACTERISTICS OF THE DECISION PROCESS

The management decision process to order construction of a highway improvement project of a particular design is a common type of process experienced daily in private and public life, and in all forms of business. It differs from other economic, social, and financial decisions only in its complexity and environment.

SIMILARITY TO PRIVATE AND BUSINESS DECISIONS

In private life, a family makes economic and social decisions daily on the basis of costs and benefits—market and nonmarket. Less frequently major decisions are made as between major alternatives. Insulation of a home against heat and cold is a decision based upon market factors and nonmarket factors. The cost of the insulation and the reductions in winter fuel and summer air conditioning costs can be combined into an economy analysis leading to a rate of return solution or other index of economy. The reduction in costs of sickness can be estimated in a manner similar to that used for traffic accident reductions in highway proposals. As yet the added comforts, conveniences, and satisfactions from home insulation have not been market priced but they are considered in the decision process.

Economic analyses are common in business and industry. Top management is concerned with capital costs, operating costs, sales, and net profits. Manage-

The Management Decision

ment considers certain nonmarket factors, also, such as public goodwill, employee goodwill, competitive reactions, and community reactions.

In both private family and industry some person or persons makes the decision on the basis of judgment (though sometimes the toss of a coin may influence the decision). Highway improvements follow the same path and in the end there is a decision of the human mind. One striking difference exists, however. The decision in private affairs and in business is made by those who will enjoy or suffer the consequences; for highways, the decision maker makes the decision in behalf of the public community, which often has already spoken its opinions, pro and con. True, boards of directors make decisions in behalf of stockholders as well as for themselves, but capital investment programs of corporations are not submitted to the stockholders for approval in advance.

In all cases the responsibility of the decision maker is that of the legally constituted body or person. This responsibility cannot be shifted to others or hidden under the table. Decision makers must have the courage to make the decision, the confidence that they will reach a good and acceptable decision, and that they can stand by their decision once made.

CRITERIA AND GUIDELINES TO DECISION

In reaching a decision to accept or reject a proposal for a highway improvement, the decision maker is guided by certain criteria, past actions, plans, and policies. He must first have a clear understanding of the objectives of the improvement, its scope and limits. He needs to distinguish between factors or a combination of factors, such as local travel and through travel, traffic capacity and accident reduction, aesthetics and engineering, urban renewal and service to highway users, private values and public values, immediate and long run goals, immediate adversities and long run benefits, today's population and future population, and the viewpoint which is most responsive to the people as compared to the many viewpoints expressed by public and private individuals. A good principle for guidance of the decision maker is expressed in Ref. 23-14, p. 2, as follows:

> Well-being of all the people shall be the overriding determinant in considering the best use of water and related land resources. Hardship and basic needs of particular groups within the general public shall be of concern, but care shall be taken to avoid resource use and development for the benefit of a few or the disadvantage of many.

In other words, what is wanted is that decision which provides the most good for the most people combined with the least harm to the fewest people. A hard decision to arrive at, but this is the paramount criterion to be followed. Often, there will be found no "best" alternative and decision, but a "good" decision can always be reached.

In support of the criterion that the decision maker is making a decision in behalf of the public, the following quotation from Laski [23-9] is offered:

> Things done by government must not only appear right to the expert; their consequences must appear right to the plain average man. And there is no way of discovering his judgment save by deliberately seeking it. This, after all, is the really final test of government; for at least over any considerable period, we cannot maintain a social policy which runs counter to the wishes of the multitude."

POLICIES VERSUS TACTICS

Good decision making requires knowing when a business, a technical, a management, or a public principle must be adhered to as compared to basing the decision tactically or pragmatically upon the good and bad elements of each particular improvement proposal. In effect, the decision maker needs to be aware of what is sound and accepted public policy, what his own management policies are, and what sound engineering requires [Drucker, 23-1]. Many decisions–and all of the tough ones such as the North Ridge case in the next section–are compromises. But principles and fundamental engineering should not be compromised. The decision maker must hunt, then, for those factors among the alternatives that can be compromised from the ideal without resulting in a lasting wrong. A temporary wrong can be tolerated for its short duration, but a permanent wrong must be lived with. In the highway systems as they now exist there are many permanent wrongs–those decisions which were bad compromises.

Pragmatic decisions based upon specific local project factors can be accepted without violating principles and without setting precedents. Violate a principle in one decision and the policy is changed and the principle wiped out. This line of reasoning gets back to the objective of the proposal. Unless the decision will permit of accomplishment of the well-defined minimum objective then it is the wrong decision. That the objectives of the improvement must be met when the decision is implemented is a principle of decision making.

OBJECTIVES SOUGHT BY THE IMPROVEMENT MUST BE REACHED

The decision maker must ever keep in mind the objectives to be reached by the proposed improvement. This goal is paramount. In the long struggle to satisfy everyone and every purpose, compromises, additions, and deletions often sidetrack the original objectives to the extent that no problem is really solved. Manheim [23-10, p. 12] comments upon this point as follows:

> Once the goals have been defined, the problem of the decision maker becomes clearer; his task is to find which strategy results in values of the measures of effectiveness equivalent to his goals. We can call this process "optimization of the model." The decision maker uses the model to estimate how the real world in which he will effectuate his selected action will react to that action (or any other).

The Management Decision

OTHER PRACTICAL CONSIDERATIONS

A principle to follow in selecting from alternative designs is that when the annual costs and services are approximately equal, the alternative chosen should be that one which has the lowest capital investment cost. The combination of low investment cost and high annual maintenance and operating expense is preferred to a combination of high investment cost and low annual maintenance expense. The reasoning supporting this principle is that (1) the difference in investment costs becomes immediately available for other construction, and (2) the general support of economy is enhanced by the annual expenditures for maintenance and operation more so than it is by the one-time expenditure for capital investment.

A second principle to follow is that annual economic cost and service quality being approximately equal, the decision should be given to the project, design, or material of construction having the shortest probable service life. A short service life is preferred to a long life because greater opportunity is afforded to rebuild to current advancing standards. Advancement of technology and highway use has rendered much long-life construction inadequate or obsolete long before its physical condition would warrant remodeling or replacement. Therefore, short service life lessens the risk of obsolescence and offers opportunity to start anew more often than does long-life construction.

On certain types of projects the disruption to traffic and the extra cost to traffic during construction can be deciding factors. Detour road costs and the running cost of vehicles on the detour or through construction should be included in the analysis for economy. Most frequently they are not, however. The discomfort and annoyance values are high for detours and travel delays caused by construction. Some extra weight should be given to that alternative which has the more favorable situation for traffic during construction. Building a new highway on new right-of-way is more favorable to traffic than reconstructing an old highway on its existing right-of-way, and probably may result in lower contract price.

Favorable weight would be given to the alternative that displaces the fewest people and business establishments. Usually the private costs and personal disturbances of displaced persons do not enter the economic calculations specifically. Yet these factors are real, though of nonmarket character, and worthy of consideration from a public viewpoint.

Certain alternatives within a group of mutually exclusive ones, may possess certain engineering features of difficult construction or of high uncertainty. When these risks have not been specifically provided for in the cost estimates and other factors, such alternatives should be graded downward in the decision process when the other factors are about the equal of corresponding factors in the less risky alternatives.

DECISION MAKER'S FINAL PROBLEM

The foregoing discussion points out guides, criteria, policies, and principles involved in the decision process. They are sound and the foundation on which good decisions are reached. It is important that every decision maker who has the responsibility of resolving the many conflicting issues in public works proposals be well-informed and have full understanding of the general factors. But to apply them is difficult because of many factors, such as how much benefit is required to overbalance hardship to others and who is entitled to the benefits; the wishes of the multitude are often difficult to discover because of its silence on most matters; and is the public well enough informed to permit of an enlightened and well-intended opinion on the consequences of such complex social, economic, and technical matters as improvements to a highway system.

When the decision maker thoroughly understands the goals of the improvement project, the attitudes of the people, the engineering factors, and the non-engineering factors, he is in a position to begin to reach his decision. There is no set procedure to follow, but some discussion of the process and techniques will help in understanding the total process. Decision making is always a problem of the human mind. It cannot be otherwise. But the mind can be materially helped by knowing more about the total process.

RESUME OF A TYPICAL SITUATION FOR DECISION

Although each proposed improvement must be decided on its individual merits, public expressions, environment, and potentials, it is helpful to review a case to furnish some reality to the responsibilities, setting, and complexities in which a decision maker often finds himself. Such a case has been provided by Lash [23-8] under the title "The Case of a Town That Didn't Want a Freeway." This story follows in a much abreviated version.

MANAGEMENT DECISION CASE NO. 1

THE CASE OF A TOWN THAT DIDN'T WANT A FREEWAY

North-South Highway Through North Ridge

Many of the problems and conflicts involved in reaching a decision on the location of an urban freeway were illustrated by the case of a specific American city which, for present purposes, we shall call North Ridge. The actual name of the city and the names of participants have been fictionalized. Though the conflicts are reported as they happened they are typical of community conflicts in cities throughout the country where freeway locations are debated. It is this universality that makes these conflicts of interest to highway administrators.

The controversy over the location of the north-south highway through North Ridge raged for 15 years, 1945-1960. The conflict was a stubborn contest of wills between the state highway department, which had the responsibility for locating and constructing the highway, and the local officials who saw the highway as a threat to the preservation of their community in the form its citizens desired to keep it.

The Management Decision

Ten alternative lines for the route were studied during the 15-year period before a final decision was made: eight developed by the state highway department and two proposed by representatives from North Ridge. The final location adopted was not the line most preferred by the highway department, nor was it a line desired by the town itself. It was a location that the town fought unrelentingly to the bitter end.

The case of North Ridge raises a number of questions of interest to the student of highway administration. Recognizing the mutual interests of both the state highway agency and the government of North Ridge in the location of the freeway, is there some procedure that would have made possible a better and an earlier resolution of the conflicting points of view? Was the highway department too narrow in the factors it was willing to consider in selecting the highway location? Were the city officials of North Ridge too parochial in their own point of view, willing to sacrifice arbitrarily consideration of regional transportation needs for their own local interests? Did the state highway department adequately consult with North Ridge officials about the highway location? And most importantly, is it possible to make the highway public hearing a more effective forum for a constructive public discussion of broad questions of values and goals related to the freeway, instead of a protest rally as was the case in North Ridge?

Although the case itself will not provide answers to these questions, they should extend the reader's awareness of the nature of these crucial issues.

A Route Through North Ridge

The first discussions on the part of the state highway commissioner with the mayors of Ridge City and North Ridge about the possibility of a north-south freeway through both cities were held in 1944. This was the year before the state published its detailed report on the most urgent traffic bottlenecks in the Ridge City metropolitan area. When the report was released it showed the state's recommended location for the north-south highway, Line A, as running due north from the business center of Ridge City to the business center of North Ridge, about four miles away [Fig. 23-1]. The freeway would parallel Elm Street, the main thoroughfare between the two centers. At the southerly limits of the North Ridge business center, the route would turn east and then northeast in its eventual course toward the northern end of the state. The primary function of the proposed north-south highway as conceived in that early report was to link the centers of the two cities with a high-type traffic facility to relieve the overburdened north-south streets, particularly Elm Street. This would be only its first function, however. The route would continue northward beyond North Ridge to permit eventual connections with two important state highways running to the northeast and northwest corners of the state. In this single north-south highway, the highway department aimed to combine service for long-distance traffic traveling across the state with service for one of the most heavily traveled commuter corridors between Ridge City and North Ridge. In the succeeding 15 years of controversy over the location of the route, the state engineers never lost sight of these original goals.

When the 1945 report was released to the public, the initial reaction was not noticeably unfavorable insofar as the North Ridge section was concerned. The reason for this may have been the lack of funds in the highway budget for the project which made it seem far in the future. The portion through Ridge City, however, ran into opposition from the beginning. Large companies that were to be displaced or otherwise adversely affected by the route immediately raised objections. These objections, as well as opposition from

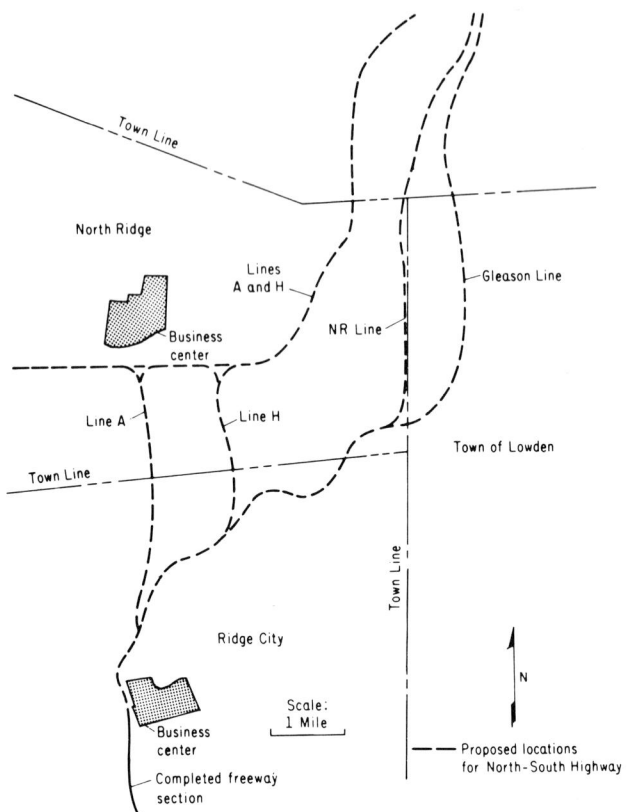

FIG. 23-1. General locations of the proposed alternative highways for a new north-south freeway connecting North Ridge and Ridge City.

downtown businessmen over the routing near the central business district of Ridge City, occupied the attention of the highway department planners from 1945 to 1948.

Public Opposition Mounts

In the fall of 1948 an informal public hearing of North Ridge homeowners and others directly in the path of the route protested its location to the attending highway department engineers. Members of the town council asked if the present streets could not be widened to accommodate the growing traffic. They were concerned over the route's dislocating 175 families and over the possibility of creating a psychological barrier across the town.

An independent engineering study of the north-south traffic problem was completed in 1945 by a consulting firm. The consultants were engaged by a committee representing five large Ridge City companies. The committee probably hoped the study would differ with the highway department's conclusions and would thus give them a basis for attacking the department's plans. As it turned out, however, the consultant's findings in most part agreed with the 1945 report.

The study did not alter the views of North Ridge town manager Andrews or the town council. In the ensuing years their convictions about the undesirability of the freeway through the developed portion of North Ridge were only to become more deep-seated.

The Management Decision

State highway commissioner Donald A. Clark continued to meet periodically with town manager Andrews in an effort to persuade him to relax his opposition to the line, but with no success.

Commissioner Clark had full legal authority to make the choice himself and to disregard the views of town officials. He wished to avoid adverse publicity as well as the political consequences that would follow from a unilateral decision.

As the issue dragged on into 1954, neither side altered its views. In an effort to give added weight to its position the highway department in the early part of 1954 released its second printed report on the north-south highway. This report showed all eight lines studied by the department and presented complete traffic and cost information for all lines. The report explained why the department preferred its proposed line (Line A) over the others.

Strangely, this extensive report failed to receive much coverage in the press. Although statistics from the report were occasionally referred to in future articles on the north-south highway, there was no large-scale coverage of the report immediately on its release. One reason for this may have been that it was a technical engineering report requiring considerable work to make it readable to the general public.

New Traffic Study

North Ridge officials still doubted the report's conclusion that it was necessary to take the route through the built-up area of the town to the edge of the business district. In Ridge City, local officials were also dubious about portions of the proposed route within their borders. In July 1954 the two cities joined together to hire a traffic engineering firm to make an entirely new study. The main question to be answered was whether it was necessary to take the line close to the center of North Ridge or whether a route bypassing North Ridge altogether might not be as good.

There must have been considerable dismay in the North Ridge city hall five months later when the consulting firm reported its final conclusions. These coincided closely with those of the highway department. However, the new report did not deter the town officials or the affected homeowners from their opposition. Indeed, by the end of 1954 their opposition increased as rumors circulated that the approaching session of the state legislature might appropriate funds for the north-south highway. Early in the following year, protest groups led the North Ridge town manager to call a public meeting on the question to which the governor and the state highway commissioner were invited. Both agreed to attend.

One speaker after another rose from the audience to object to the line. Most protested because the route would take their homes or disturb their neighborhoods. In a hand vote, only ten persons favored the state's route. All others were opposed.

To what degree this vote reflected the views of the town's total population of 52,000 is unknown, but from all indications the governor and local officials all interpreted this consensus as the prevailing view of the townspeople. The governor was visibly impressed by the amount and intensity of feeling of the opposition. He then and there sounded the death knell for the straight-line connection between the centers of Ridge City and North Ridge: "While a straight-line highway is probably the best designed, I am deeply concerned with the economic and social factors in the highway's construction." The governor instructed the highway department to look for another line.

A New Highway Commissioner Takes Office

In March, 1955, the same month as the hearing, state highway commissioner Clark left office for private reasons. The governor appointed as the new commissioner a professional engineer from the northern part of the state, Richard D. Farrell. Commissioner Farrell had been in private civil engineering practice throughout his career and had no prior experience in highway planning.

In compliance with the governor's instructions, Farrell sent North Ridge officials a compromise line (Line H) located about three-quarters of a mile east and parallel to Line A. The new line joined the original line southeast of the central business district. The state engineers considered this new proposal inferior to the original Line A primarily because of the indirection of travel between the centers of Ridge City and North Ridge, and also because the line would require an additional connector to service the North Ridge business center. It would be inferior in terms of traffic service and more costly. Also, the changed section still cut through high-type and compact residential areas.

It did not take long for the same kind of opposition to mount against Line H as had expressed itself against Line A. Whereas homeowners on Line A gave a sigh of relief and withdrew from the controversy, a new set of homeowners mobilized themselves into protest groups, doubtlessly encouraged by the success of the earlier group in defeating the highway department's proposal.

In May 1956 at a meeting of the town council more than 300 persons attended to protest Line H. Many objected to having the route in the town at all. A few days later the town council went on record against Line H and moved to ask the highway department to explore further and to consider instead the widening of present north-south arteries in the town.

Federal-Aid for North-South Highway

Meanwhile in Washington, D.C., an event was taking place that was to have a profound impact on the north-south highway. In June 1956 Congress enacted a bill providing financing to construct a 41,000-mile National System of Interstate and Defense Highways with the federal government paying 90 percent of the cost and the states 10 percent. Being a link in a cross-state route that had several years earlier been made part of the interstate highway system, the north-south highway was now brought much closer to reality. With the problem of financing solved, the state highway department now became especially anxious to reach an early decision on the route through North Ridge.

The U.S. Bureau of Public Roads instructed its field division office in each state to ask their respective state highway departments to have the interstate route locations fixed by September 15, 1956, so that a reliable estimate of cost could be prepared. If it were not possible to fix the location of a particular section of a route by that date, a tentative location could be adopted for estimating purposes.

Councilman Douglas C. Freeman, representing the area most affected by Line H, called another public hearing in North Ridge for September 11, 1956, to allow public discussion of the highway. Chief Planning Engineer James W. Killian of the highway department attended to answer questions about Line H. The mood of the 300 people in attendance was hostile. They made it plain that they feared the route would damage too many homes and would cut the town in half; they went on record as wanting no part of Line H.

The Management Decision

On September 15, the highway department submitted Line H as a tentative location through North Ridge for the official interstate route.

Introduction of Gleason Line

Richard G. Gleason, representative from North Ridge to the state legislature and a stalwart of the "out" political party, now began to take an interest in the dispute. He saw the controversy as an opportunity to become champion for a popular cause against the "in" party administration. He also believed that here was an encroachment by state bureaucracy on the rights of a town to direct its own destiny.

In December 1956 Representative Gleason announced his own proposal for a routing of the north-south highway. He sent a map to the highway department suggesting an easterly route that would virtually bypass North Ridge altogether [Gleason Line, Fig. 23-1]. He contended his route would traverse less densely developed land and thus would avoid disrupting settled residential neighborhoods in North Ridge to the same degree as did Line H.

The state's reaction to the Gleason line was cool, to say the least. As they saw it, this line would require abandoning the primary purpose of the highway, namely as a route to carry the heavy traffic between Ridge City and North Ridge to relieve the congestion on the north-south streets. The Gleason route, by being so far east, would also rule out any possibility of connecting to the cross-state route to the north, another of the original objectives. Even so, the highway commissioner asked a local engineering firm, William Lewis Associates, to make a comparative study of Line H and the Gleason line.

As the year 1957 began, the north-south highway became the top-priority issue before the North Ridge town council. In the previous year the town had hired a new town manager, Harold C. Canney, and instructed him to do everything possible to defend the town against Line H. Early in 1957 Canney became persuaded that the town's position in the debate was too negative. He believed that the town should use its own engineering and planning staffs as well as outside experts, if necessary, to make a well-planned and coordinated argument in opposition to the proposed route and to come up with a positive recommendation of its own. Also, because the federal law requires the highway department to consider the economic effects of the route, North Ridge would find evidence to show that the economic effects of Line H would be harmful to the town.

In August 1957 the state's consultant published its report comparing the state's Line H with the Gleason Line. The overall cost for the Gleason line would be less, $28,100,000 as compared to $29,900,000 for line H. The number of developed properties to be taken would be very similar, 174 on the Gleason line as opposed to 166 on Line H. However, the fatal deficiency of the Gleason line, the report contended, was that it did not serve the main north-south corridor of traffic and thus neglected the problem that the route had originally set out to solve.

North Ridge Hires a Consultant

Soon after release of the Lewis report, North Ridge Councilman Freeman expressed the sense of frustration of the town council when he said, "For every valid objection raised by the town, the commissioner and his palace guard have answers made up in advance. Until such time as we get someone on a par with the state highway department experts engineer-wise, we're not going to get satisfaction." The council thereupon voted to find the best consultant available to study the problem and to "defend the town of North Ridge in its battle with Farrell." Three months went by Town Manager Canney reported to the council that he was having trouble getting to take the assignment. Soon there after Commissioner Farrell announced that the public hearing required by the federal law would be held January 9, 1958, to receive the views of the officials and the public.

Town Manager Canney, after several meetings with Commissioner Farrell, finally persuaded him to postpone the meeting for three months, until March 4. Canney then immediately hired a consultant with whom he had been negotiating, Clarence H. Newcomb.

Three weeks before the hearing date, Canney met with Commissioner Farrell to present a new line developed by the town's consultant [Line NR, Fig. 23-1]. Canney asked that the hearing be postponed again so that more studies could be made on the new line. The state took the matter under advisement. Commissioner Farrell announced a week later that the new line "has less merit than the Gleason line," and that "a preliminary appraisal shows it does not warrant postponement of the hearing or change in the line."

Expressing disappointment at what they felt was arbitrary rejection of their line, the North Ridge officials now became determined to carry their fight to the U.S. Bureau of Public Roads if necessary. Approached on the question of an eventual appeal to the federal highway agency, one of the state's U.S. Senators forecast the outcome of the final appeal when he said that "the federal government has little choice but to accept the recommendation of the state highway commissioner. The solution must be arrived at on the state level. The federal government, as far as can be ascertained, can act only upon the final recommendation and certification of the state highway department."

Final Public Hearing Held

The public hearing on the night of March 4 was described as exciting, stormy, and turbulent. More than 1,400 persons filled the high school auditorium. Most of the people present were from North Ridge and opposed Line H. But about 500 persons were from Lowden, the town east of North Ridge through which the Gleason line would pass. The latter were there to oppose the Gleason line and to support the state's Line H.

Commissioner Farrell opened the hearing by describing the history of negotiations over the north-south highway. Using giant maps, engineers from his staff described in detail the main features of the three most prominent alternative lines. But the state's presentation was frequently interrupted by outbursts form the crowd; the audience now obviously had no interest in technical information. Gleason criticized Commissioner Farrell and his department for "bureaucratic thinking" and disregard for the rights of the town. He defended his own line, claiming it superior to all others.

Town Manager Canney presented the town's case. With the aid of two traffic engineers from the consultant firm hired by the town, he attacked the state's line as unnecessarily cutting through the built-up part of the town. "The Line NR does less damage and is more consistent with interstate highway needs," Canney argued.

Following Canney, a representative of the North Ridge Chamber of Commerce told the hearing that the 450 businessmen in the Chamber favored the Gleason line. An attorney from Lowden then took the platform to appear in favor of Line H, saying that he had submitted a petition containing 1,659 names of Lowden and North Ridge residents favoring the state's line. Representatives of a variety of groups spoke next, followed by individual citizens. The 1958 hearing proved to be the climax of the long dispute.

The Management Decision

Aftermath of the Hearing

After the stormy public hearing Commissioner Farrell was undoubtedly wary of making an immediate decision. Passions were high, and if anything, the hearing simply entrenched all interested groups further into their original positions. Farrell may also have been advised by the Governor to let things calm down before taking any other action; elections were only eight months away.

The elections came and went. The Governor was reelected. By the end of 1958, Commissioner Farrell replied to reporter's inquiries that he was still studying the information brought out at the hearing and was not yet ready to make a decision. Two months later, in February 1959, Representative Gleason publicly criticized the highway commissioner for dragging out the decision.

A month after Representative Gleason made his statement, the North Ridge Chamber of Commerce and a group of manufacturers in the town publicly called on the highway commissioner to make his decision. Still there was no response from the highway department. Finally in May 1959 the state legislature passed a resolution, introduced by Representative Gleason threee months earlier, calling on the highway commissioner to announce his decision on the route by August 1 of that year.

In reference to Farrell's turning down another proposed highway in North Ridge over which a second storm was brewing (this was an east-west route in very early stages of planning), the governor made a statement to the press criticizing Farrell for "poor public relations in not allowing the town to work with him in the selection of an east-west route." Coming at the time it did, this was interpreted by some as an indirect criticism of the highway department's handling of the north-south highway routing.

On June 17, 1959, two weeks after the governor's public criticism of him, Farrell announced his resignation. The new commissioner was Jeffrey E. Banks, a professional engineer and former Deputy State Highway Commissioner under Clark.

Upon assuming office July 1, Banks took on as his first order of business the resolution of the north-south highway controversy. He was briefed on all the issues by his engineers, he met several times with Representative Gleason and the officials and staff of North Ridge, the flew the three lines by helicopter on two different occasions, and finally he discussed the entire question with Clyde Barner, the division engineer of the U.S. Bureau of Public Roads. At a press conference in his office on July 24, 1959, after three weeks of study, he announced that he was persuaded that "the alignment which will be most advantageous to both North Ridge and the state is Line H." His press release went on to say that to disregard the local street considerations, as urged by North Ridge, would be to disregard 93 percent of the problem. "The capacity of the existing street system cannot be expanded sufficiently to handle the growing traffic loads."

On August 12, 1959, the state highway department submitted to the U.S. Bureau of Public Roads a formal request for approval of the north-south highway project for federal-aid financing. With the request was the required certification that a public hearing had been held and that the highway department had considered the economic effects before making its final decision.

Being intimately familiar with the long debate over the highway and being personally convinced that Line H, though inferior to the original Line A, was now the best line available, Public Roads Division Engineer Barner took little time to approve the state's request.

When they learned of Barner's action, the North Ridge Town Council decided in a 5 to 3 vote to carry their appeal to the U.S. Bureau of Public Roads in Washington, D.C.

North Ridge Officials Go to Washington

On November 2, 1959, six key officials from North Ridge met in Washington with the Administrator of the U.S. Bureau of Public Roads and members of his staff. The meeting lasted three hours. At its conclusion the Public Roads Washington staff agreed to review the entire record.

Two months later on January 15, 1960, the court of last resort for North Ridge rendered its verdict. "The Bureau has examined all facts of the problem. As a result of the studies we do not find any justification for withholding approval of the location selected by the state," read the Bureau's letter.

The state highway department took immediate steps to aquire right-of-way. Construction was set to begin early in 1962.

AUTHOR'S CONCLUDING STATEMENT

For many highway location proposals, especially within urban areas, the foregoing case of North Ridge illustrates the many factors that surround the decision maker. He must find his way out, knowing that any decision will not please all groups and areas. He has only his judgment, aided by all facts, opinions of others, analyses, goals, and criteria. He can use no mathematical calculations–handmade or computer–to tell him the decision he should reach.

THE TOTAL PROCESS LEADING TO THE DECISION

The entire improvement process–from recognition of its need to adoption of the final engineering plans and ordering of construction–involves an orderly process of decision making at many levels together with productive actions by a number of individuals and organizations. The total process varies according to the character of the improvement, its location, who has jurisdiction, and the authority possessed by the organization responsible. Deciding to design and build an urban freeway calls into play forces, people, and organizations far more extensively than does the simple process of remodeling a street intersection to reduce traffic accidents, yet both improvements involve many of the same factors, procedures, analyses, and decisions.

LOGICAL STEPS IN THE PROCESS

In discussing the final decision process of management it is helpful to keep in mind the whole of the process of how highway improvements come into being. How does the process move from its beginning to its end? From the beginning and for each action taken in the series of steps, the prime and final authority rests in some responsible individual, group, or organization. At the end, however, the

The Management Decision

final authority rests in that individual, commission, council, or authority which by law has the authority to say "yes" or "no" and to commit the public funds to that project. The law referred to is that city, county, state, or federal law which is applicable. The law is the source of authority because public funds are involved for public works, such as highways.

The total process of bringing an improvement to a highway system into being involves steps similar to the following sequence:

1. Conception the need or desirability of the improvement proposed and positively defining the objectives.
2. Proposition of the project to the proper authority for inclusion in the construction program.
3. Preliminary studying and analyses of the merits and feasibility of the proposal.
4. Rejection or approval of the proposed project; if approved, the project is programmed for a given year's construction budget.
5. Identification of all possible alternatives for reaching the objectives.
6. Fact gathering–preliminary surveys, analyses, decisions, cost estimates–for each alternative.
7. Preliminary analysis of the alternatives to eliminate those not worthy of further study and design.
8. Further design, detailing, and costing of the feasible alternatives, and gathering of the economic, social, and community factors and consequences involved.
9. Development of the final analyses for the economy of each alternative.
10. Writing the final report, separate staff reports, and taking other actions preparatory to submission to management for its decision.
11. Should a public hearing be in order, it may come ahead of the final analyses or just afterwards according to organizational practice.
12. Management's study and further analysis in its process of reaching a final decision.

These steps leading up to the final decision of management disclose the background actions pertaining to the improvement project and offer some perspective to management. Further discussion in this chapter, however, is devoted to step 12, the management decision process.

AIDS TO THE DECISION

As previously stated, there is no way that the decision maker can automatically select his decision from all those possible or calculate the answer. There are aids to the decision available. Judicious use of these aids and understanding of their relative values will often lead the decision maker to a clear-cut making up of his mind.

Specific Aids Available

Some of the aids to the decision are largely informational in character and some are procedural. The informational aids may be:

1. Main objectives of the improvement
2. Staff report of the economic analysis
   (a) Analysis for economy with all supporting material
   (b) Analyses of the nonmarket consequences with supporting material
3. Staff report and analysis of the engineering requirements
   (a) Standards of design
   (b) Structural and geometric limitations
   (c) The overall plan for transportation and development in the area
4. Summaries of public hearings and other official expressions of preferences from individuals and organizations
5. Listing on a qualitative and quantitative basis (when possible) the (a) beneficial factors, and (b) the adverse factors for each alternative
6. General factors as aesthetics, historical landmarks, open spaces, local traditions, long-range community plans.

The procedural aids may be:

1. Listing of preferred viewpoints–project as a whole and subfactors in particular
2. Separation of the favorable and unfavorable consequences or expressed preferences into those which are (a) publicly oriented, (b) privately oriented, (c) minority group oriented, and (d) local-area oriented.
3. Separation of consequences into (a) long-term existence and (b) short-term effects–lasting versus temporary effects
4. Comparison of identical factors by paired alternatives
5. Matching of beneficial factors against adverse factors–a trade-off analysis
6. Further calculations of unit costs or unit benefits.

The above two sets of aids are suggestive of the types of information, factors, opinions, and procedures the decision maker may have available. Some of them are more fully explained in the sections to follow.

Use of Specific Aids

The *informational aids* are guidelines and factors the decision maker uses to bring each alternative into a common focus, factor by factor–costs, benefits, and adversities–as they affect the road user, the community, and business and social groups within the community. These informational aids are used as evidence of the consequences–qualitative and quantitative measures to which the decision maker must give relative weights in accordance with his judgment of what is best for the community.

The Management Decision 571

He will remember that the analysis for economy and economic evaluation of the proposal is based upon certain judgments from the beginning. This analysis, however, when the sensitivity of the factors is known, is a good solid foundation for making the first decision as to whether the proposal is economically sound. When there are social reasons or broad economic reasons for building the proposal or not building the proposal, the decision maker still is faced with making a decision which involves weighting the analysis for economy against the other factors. In a project formulation analysis for economy, the decision maker, again, must weigh the relative merits of the road-user and nonroad-user, market and nonmarket factors. In other words, as often heretofore repeated, the analysis for economy does not of itself produce either decision–to build or not to build, and if to build, what alternative is to be selected. The analysis for economy merely gives to the decision maker an aid to exercising his judgment.

All other informational material serves as a guide. Obviously the expressions of public and private opinions cannot always be followed because they are diametrically opposing on many factors as is illustrated by the North Ridge case.

The *procedural aids* available to the decision maker afford him assistance in making up his mind. As he sifts through the available facts, opinions, reasonings, and viewpoints, the real important factors and their relative values emerge upward. The viewpoint to use and the consequences on which this viewpoint are to be focused become sharper as the aspects of the alternatives are compared, weighed, and valued.

Business volume and local trade is always a factor in urban areas and specifically in any proposal which will reroute traffic as does the bypass route. When it is remembered that a highway location change does not reduce or increase the total business volume, but merely affects its location, this factor becomes one of a private character and hardship rather than of an overall community hardship or gain. All business owners have to assume certain risks. There are risks of invention, technology, and of customer preferences and shifting of population, of land-use zoning, of competition, and of community change. Highway location, traffic routing, and traffic controls are common risks that all property owners must assume. Preservation of the business of a motel, a roadside restaurant, or of an automotive service station is not a factor that should override the community interests and the interest of the road user.

Another factor of great significance is that highways are long-time permanent facilities that will outlast both the current land use and the local population. Land use, land owners, and building tenants have rapid changeovers. In one to three years after opening a new highway, social and economic adjustments usually wipe out all hardships brought on by highway improvements, and the local people and businessmen find themselves no worse off, and usually better situated. Certainly the community as a whole is better off.

One of the often overlooked facts of highway planning is that many improvements to highway systems now desirable came about by failure years ago to make the proper decision. Too much weight was given to the wishes of local landowners and local land use.

Additional Analyses

It cannot be assumed by the decision maker that his staff has presented him with all desirable analyses and calculations useful as aids in the decision process. He may conceive other helpful ideas. For instance, in a freeway location controversy in Montgomery County, Maryland, adjacent to the District of Columbia, one alternative route was proposed mainly to reduce the number of families to be dislocated. This alternative route as compared to the original proposed location reduced the displaced families in Maryland from 570 to 175, which is a remarkable reduction. The project cost, however, was increased by $22 million and for a freeway of undesirable restriction in width. The $22 million of increased capital cost to reduce the family displacement by 395 families would result in a cost per family of $55,700.

Having made this simple calculation, the decision maker has a better insight into the merits of the two route locations. He can ask himself the question, "Is it worth to the community $55,700 per family not to move 395 families and to construct thereby an inferior freeway of such location and design that it would not solve the transportation needs for the corridor?" The answer to the decision maker's question is one of sociology involving a consideration of temporary local hardships compared to long-time community benefits.

SCALAR DEVICES AND INDEXES

The approval to construct a highway improvement and the acceptance of the details of design are the two decisions of management. Often these two decisions involve two sets of noncombinable factors–that is, noncombinable on an equivalent dollar basis. Analysts and a few decision makers have expressed a desire for some device, scheme, or procedure by which the dollar highway costs, the net road-user consequences, and the nonroad-user consequences could be combined in such a manner that the final numerical index would, in effect, be the decision as between the alternatives considered. Also, there have been desires expressed for a scheme to combine the many varied nonuser consequences into a single index expressing their overall relative value. At least the hope of certain persons is for indexes of these types. Such procedure of combining the many factors into one or two overall rating indexes is of highly doubtful merit because the decision on human and social values should not be reduced to arbitrary mathematical calculations based upon some prior fixed scale of subjective values.

The Management Decision 573

MAIN CONSIDERATIONS INVOLVED IN MERGING MARKET AND NONMARKET FACTORS

It is logical to wish for a procedure for merging the consequences related to motor vehicle operation, the personal preferences, and nonroad-user consequences into a single index measuring the relative economic and social desirabilities of the several alternatives. But even though such a system could be devised, its application has many points of doubtful merit. The only reason why such a system is proposed is to avoid the responsibility of making the decision on the basis of subjective judgment. In conformity with the principle that the analysis for economy of the market factors leading to a benefit/cost ratio or rate of return provides only a guide to the decision maker and not the decision, any procedure to merge the nonuser factors into a common index or to merge the market items and the nonmarket items into a single index would still end up with just an aid to the decision. But such an index has important objectionable features.

Any Composite Index Involves Prejudgment

Any evaluation model devised to rank highway improvement alternatives on the basis of the relative worths of the nonroad user and nonmarket consequences must start with a fairly large number of factors whose worth will vary project to project and quantity to quantity, assuming that the factor can be quantified. The model would involve such basic factors as given in the factor chart of Fig. 3-1. For each alternative, an index or a dollar value would be required for such factors, among others, as:

1. Number of business establishments displaced
2. Number of resident families displaced
3. Probable decrease and increase in local retail trade
4. Probable decrease and increase in local land value
5. Probable changes in population and employment
6. Probable changes in total real estate tax assessments
7. Aesthetic value of elements of highway design, such as elevated, depressed, and tunnel highways
8. Community value of park land versus residential land versus industrial land versus cemeteries
9. Community value of slum clearance
10. Traffic noise, air pollution, parking

The index value or dollar value assignable to these nonmarket items will be the result of subjective judgment and may vary in size with each alternative, let alone with each project location throughout a highway system. For instance, the moving of business establishments and residences from an older section of the area would not have comparable values to moving like establishments in a relatively recently built-up area. Displacing families from a slum area is not equivalent in community value to displacing families from Country Club Heights or a newly settled suburban residential development. And this statement does not in any way reflect upon the people therein involved; it is meant to refer to physical and environmental values.

In application of the model used to rank the relative desirability of different alternatives, it would be required to (1) identify each factor existing in each alternative, (2) quantify the factor when possible by some appropriate unit, and (3) assign appropriate values to each factor in each alternative. It is step three that is particularly significant.

When index values or dollar values are assigned a subjective appraisal is being made. It is at this point that the ranking of the alternatives is determined. The scheme does not avoid placing of judgmental values on these hard-to-weigh factors. A table of relative values for each of the host of factors involved could be prepared as a guide, but such a table would be a subjective measurement, and should be applied with further judgment. An important consideration in such a process is that there is little or no basic value or price from which to start. Unlike the market factors where the start is with a quantified and market priced base, these general economic, social, and community factors often have no starting point. Even with selecting the vestcharge rate the start is from market prices. But what market factor can be selected as a beginning point to assign an index or dollar value to the displacing of one family, to the riddance of traffic noise (near a hospital), to the preservation of a historical spot, a scenic view, and so on?

Composite Index Ranking Would Displace the Decision Maker

There is serious question of whether a composite index of some sort disclosing the relative preference of each alternative considered on the basis of the non-market factors could be used with any certainty or usefulness by the decision maker. He would be in a position of making a decision without exercising a judgment. The final index would be the decision because it would be a composite of the subjective judgments on all factors involved, properly weighted by the fixed weighting scheme. Not to accept the results of this weighting scheme would be admitting that the scheme is not acceptable. To accept the results as found is passing the responsibility back to the analyst who applied the model and came up with the ranking. This composite index does not allow the decision maker freedom to exercise his judgments without his starting over with the list of consequences and their general qualitative descriptions.

Use of this composite index by the decision maker would be placing him in about the same position as he would be if someone gave him a composite index on the relative desirability of his taking a vacation trip to the mountains, to the seashore, to the lakes, or touring the historical East. At the outset he would want to know how the index was arrived at, the factors considered, their relative value factors, and who assigned the basic index values. And in this process he begins to assert his differences in judgment.

The Management Decision 575

If the final index is one combining the results of the road-user analysis for economy with the general index for nonmarket consequences, the discussion is the same. Someone at some step did assign a relative worth to the market and nonmarket consequences so exercised prematurely the judgment that should be the responsibility of the decision maker.

POSSIBLE MERIT OF RANKING SCHEMES

In spite of the foregoing objections to the use of an overall index relative desirability for a group of alternatives, there can be some merit to a system which would offer help to the decision maker. Two possibilities prevail. Through research and application of ingenuity it should be possible to arrive at a process of establishing market values or relative indexes for certain factors. Perhaps noise, air pollution, scenic views, and family displacement can be market priced. By study of public opinion on a statistically sound basis, it may be possible to develop a reliable index of relative preferences among types of alternatives associated with highway location and design. Falk [23-4] has made a substantial contribution along this line in his Successive Test Attitude Measuring Scale. To be sound and usable any model by which to develop a composite-index or a single-index rating scheme will need to be based upon factors, or opinions, which can be taken collectively from the public as contrasted by the subjective judgment of the analyst or public official. For instance, Falk starts with the public preference between pairs of factors such as travel time versus appearance, construction cost versus vehicle operating cost, and influence on neighboring property versus social factors. A rating system starting with valid parameters is far different from a rating system starting with subjective values.

Computer models can be of aid to the decision maker in producing relative values based upon given conditions and weights and in predicting consequences. But all modes–computer or hand–should not be devised to produce the decision. The decision must be that of the decision maker, not that of a computer or subordinate analyst. The decision maker does not (or should not) wish to shift his responsibility to others or to a computer. Human decisions are human decisions. No amount of mathematical science or social science can make a decision involving personal judgments based upon personal preferences which in turn are based upon experience, opinions, likes, and dislikes. Computer models could be used to make the decision, however, provided the model and its inputs were designed by the decision maker especially for each problem coming up for decision, and provided that the decision maker knew beforehand on what factors, what weights, and what values his decision was to be made. These subdecisions are often made subconsciously as he studies the proposals, facts, aids, and material at hand. There is no method he can devise for building a reliable computer decision-making model until after the decision is made. It is then too late for him to gain any advantage from the decision model.

MANAGEMENT DECISION CASE NO. 2

A second case study will be helpful in developing a deeper penetration into the factors, complexities, and importance of the management decision on proposals for highway improvements. This case is presented in much greatly briefer form than in comparison to the reports and documents available. The main source has been a report to the Washington State Highway Commission, Olympia, Washington, of October 1966. This 302-page report plus a volume of aerial photo maps give almost all of the engineering and economic information than is pertinent, though the procedural information is lacking at some points.

LOCATION STUDY, INTERSTATE ROUTE 82

General Description

Interstate highway route I-82 connects to I-90 on the north and I-80N on the south to form part of the system link connecting Seattle, Washington, and Salt Lake City, Utah. The location study at hand involves the route segments southward from Granger, Washington, to the connection with I-80N, 4.5 miles west of Pendleton, Oregon. The main area and interrelations of routes and the ten alternative locations are given in Fig. 23-2. Table 23-1 gives the important numerical data on the ten main locations considered.

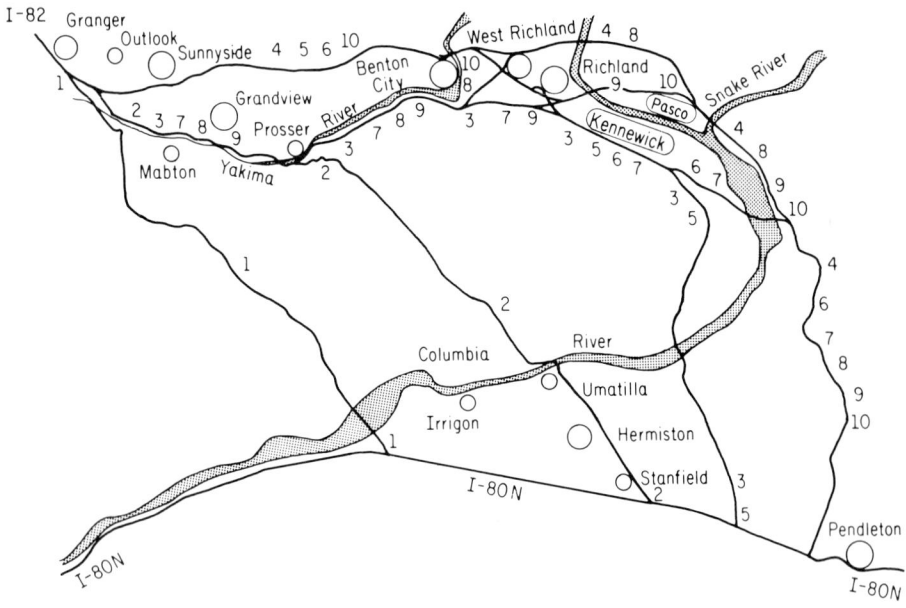

FIG. 23-2. The urban areas served and the ten alternative route locations for interstate route I-82 between Granger, Washington, and I-80N west of Pendleton, Oregon.

The Management Decision

The general geographical area is agricultural in land use, with the usual sprinkling of urban communities. Much farm land is irrigated. Post-World War II population growth was greatly accelerated by atomic energy research and development plants in the area.

Urbanwise, the area is dominated by the Tricities of Richland (25,900), Kennewick (15,200), and Pasco (14,800), having a combined population in 1965 of about 55,000. Other communities are suburban to the Tricities and scattered throughout the rural areas. Within the area directly affected by the location of the interstate route I-82 are the following numbers of towns and cities within the population range indicated:

Population Range	Number of Towns
200-500	11
700-1,000	7
1,000-1,500	5
1,500-1,700	4
3,000-4,000	4
4,000-5,000	3
7,000-8,000	1
14,000-15,000	3
26,000	2
Total	40

Within the area there are three major rivers, Yakima, Columbia, and Snake. Irrigation canal crossings total 5 to 12 for each alternative location of I-82. There are required 2 to 7 railroad crossings per alternative.

Main Factors Affecting the Interstate Location

In the final analysis there is just one main question involved in the general location of interstate I-82: To what extent should the route serve directly the Tricity area? Of the ten locations for which data are given in Table 23-1, there are just three main general locations considered. First, Route 1 to the west and south, bypassing the whole of the urban complex between Granger and the Tricities; second, Route 2, serving the urban area just east of Granger and bypassing to the west and south the whole of the eastern complex of the Tricities; and third, all other alternative routes keeping to the north and serving to some degree the whole of the urban areas in Washington.

Route I-82 had its conception as a route between Seattle and Salt Lake City as proposed by the Department of Defense to decrease the distance between the "arsenals of the Midwest and the defense establishments in Puget Sound." Later, certain interests proposed that the route be located to serve the Tricity area. Thus, new possibilities were opened up which led to an extensive study of the whole local area and its needs for highway facilities. The report to the Washington State Highway Commission deals with the whole area involved, its highway needs, interstate I-82, social and economic developments, climate, geology, and related factors.

Explanation of Procedures

Table 23-1 can be better understood when the following information on some of the procedures and concepts is kept in mind.

TABLE 23-1
Statistical Information from the Report on the Economic Analysis of the Location of Interstate Route 82, Granger, Washington, to Pendleton, Oregon

Line Number and Unit*	1	2	3	4	5	6	7	8	9	10
Travel distances										
1 Miles	83.74	81.92	99.61	104.39	98.01	101.26	102.86	109.45	105.70	103.64
2 Miles	37.97	14.02	6.94	0.00	6.94	0.00	0.00	0.00	0.00	0.00
Interstate 82										
3 Miles	45.77	67.90	92.67	104.39	91.07	101.26	102.86	109.45	105.70	103.64
4 Miles	11.4	6.2	7.7	7.0	8.3	7.8	7.3	6.4	6.2	6.5
5 Degrees	4	5	4	3	3	3	4	4	4	3
6 Percent	4.18	5.00	4.90	5.30	4.94	5.30	5.30	5.30	5.30	5.30
7 Miles	7.57	5.34	12.40	20.04	14.27	20.51	18.64	17.93	19.64	21.52
8 Feet	1,561	2,129	2,944	4,350	3,504	4,264	3,704	3,840	3,809	4,326
9 Feet	1,959	1,975	2,223	3,430	2,920	3,354	2,657	2,775	2,759	3,411
10 Number Buildings	4	20	15	16	13	15	16	18	22	21
11 Number Buildings	0	1	3	2	0	0	3	5	7	4
12 ADT	1,004	1,825	2,401	3,143	2,262	2,245	2,383	3,240	3,618	3,331
13 1000 vehicle-miles	45,942	123,909	222,463	328,152	206,001	227,643	245,127	354,666	382,358	345,220
14 $1,000	589	2,296	2,356	2,966	2,612	2,892	2,636	2,742	3,017	3,256
15 $1,000	9,377	10,934	18,733	25,317	20,943	24,207	21,997	23,729	23,450	25,661
16 $1,000	12,479	9,800	21,177	17,604	22,224	19,272	18,225	17,781	19,361	19,109
17 $1,000	7,508	9,808	13,720	16,779	14,407	16,336	15,649	16,764	17,448	18,036
18 $1,000	29,953	32,838	55,986	62,666	60,186	62,707	58,507	61,016	63,276	66,062
19 $1,000	145	4,449	8,125	15,252	7,044	7,112	8,193	16,516	16,920	15,839
Related routes										
20 Miles	85.3	55.9	53.2	28.3	52.9	49.6	51.8	22.7	22.9	24.1
21 $1,000	3,228	2,085	2,275	1,251	2,191	2,180	2,264	1,335	1,335	1,251
22 $1,000	15,931	12,589	11,252	3,285	11,047	10,079	10,284	3,490	3,418	3,213
23 $1,000	16,673	15,519	15,337	1,491	16,325	15,456	14,468	503	503	1,491
24 $1,000	14,630	11,047	9,938	4,042	9,834	9,269	9,373	4,146	3,808	3,704
25 $1,000	50,462	41,240	38,802	10,069	39,397	36,984	36,389	9,474	9,064	9,659
I-82 plus related routes										
26 Miles	131.1	123.8	145.9	132.7	144.0	150.9	154.7	132.2	128.6	127.7
27 ADT	2,750	5,000	6,577	8,612	6,197	6,152	6,529	8,878	9,911	9,126
28 1,000 vehicle-miles	123,715	274,363	379,157	282,816	343,390	356,039	395,000	414,647	425,529	398,532
29 Dollars per mile	0.65	0.27	0.25	0.19	0.29	0.28	0.24	0.17	0.17	0.19
30 $1,000	1,048	984	986	1,000	974	971	982	1,021	982	989
31 $1,000	4,358	4,017	5,114	3,941	5,366	5,382	5,129	3,825	3,925	4,103
32 $1,000	72,795	73,005	73,426	73,221	72,219	72,022	73,232	74,822	73,313	72,524
33 $1,000	6,506	6,296	5,875	6,080	7,082	7,279	6,069	4,479	5,988	6,777

The Management Decision

34 Ratio	1.48	1.58	1.16	1.55	1.33	1.36	1.19	1.17	1.54	1.66
35 Percent	8.0	8.5	6.2	8.4	7.1	7.4	6.4	6.1	8.3	9.0
36 Ratio	0.03	1.12	1.60	3.89	1.32	1.33	1.61	4.31	4.35	3.89
37 Percent	0.2	6.0	8.6	21.0	7.1	7.1	8.6	23.4	23.4	20.9
Without related route construction										
38 $1,000	76,982	77,239	77,685	74,539	76,408	76,199	77,479	76,169	74,611	73,829
39 $1,000	2,319	2,062	1,616	4,762	2,893	3,102	1,822	3,132	4,690	5,472

* Line for line description of the entries in the table:

Travel distances:

1. Travel distance, Granger, Washington, to Pendleton, Oregon. Southern end of I-82 is on I-80N about 4.5 miles west of Pendleton.
2. Travel distance on I-80N. Certain I-82 alternatives junction with I-80N and become concurrent with I-80N there to Pendleton.

Interstate 82:

3. Miles of construction on I-82.
4. Average distance between interchanges.
5. Maximum degree of horizontal curve.
6. Maximum grade.
7. Miles of grade greater than 3 percent.
8. Total feet of rise.
9. Total feet of fall.
10. Number of residential buildings to be moved.
11. Number of commercial buildings to be moved.
12. Average ADT in 1990 per mile of I-82.
13. Vehicle-miles of travel in 1990 on I-82.
14. Capital cost of right-of-way.
15. Construction cost of grading and drainage.
16. Construction cost of structures.
17. Construction cost of pavement and appurtenances.
18. Total capital cost.
19. Annual added economic benefit. Estimated annual wage income in 1978 from employment attributed to construction of I-82; does not include employment gains from construction of related routes.

Related routes:

20. Related construction (local routes improved). To provide traffic capacity, sections of certain existing routes and streets are improved to four-lanes separately from I-82 construction.
21. Capital cost of right-of-way.
22. Construction cost of grading and drainage.
23. Construction cost of structures.
24. Construction cost of pavement and appurtenances.
25. Total capital cost.

I-82 plus related routes:

26. Total miles constructed—I-82 plus related routes.
27. Average ADT in 1990 per mile of construction, I-82 plus related routes.
28. Vehicle-miles of travel in 1990 on all routes affected by construction.
29. Total construction cost per vehicle-mile of travel in 1990.
30. Annual maintenance cost.
31. Uniform annual capital cost. Calculated at 5 percent per annum vestcharge and 100 years for right-of-way, 80 years for grading and drainage, 65 years for structures, and 35 years for paving and appurtenances; related routes included.
32. Annual road-user cost, 1990 ADT, all affected traffic.
33. Annual road-user savings, 1990 ADT, all affected traffic.
34. Road-user benefit/cost ratio.
35. Road-user rate of return.
36. Added benefit/cost ratio. Calculated by using the added economic benefit, line 19, as total benefit.
37. Added rate of return. Calculated by using the added economic benefit, line 19, as total benefit.

Without related route construction:

38. Annual road-user cost, 1990 ADT, without related route construction.
39. Annual road-user savings, 1990 ADT, without related route construction.

Note: The annual maintenance cost in 1990 of the existing system of routes affected by I-82 is $1,016,000. The annual road user cost in 1990 for existing routes affected by interstate construction is $79,301,000.

General Factors. The highway designs and economic analyses are based upon the traffic forecast as of 1990. The traffic growth factor 1965 to 1990 varies from 1.62 to 9.60, averaging about 3.00. All construction cost estimates are based upon 4-lane design. Some of the non-I-82 construction is of only partial control of access. At the roadside interview stations in 1965 traffic counts varied from 340 to 14,800 ADT.

The study considers alternative highway and street networks rather than individual alternative projects and routes, but in the end the 10 alternative I-82 routes and the related route improvements are for the specific locations designated.

Social and Economic Considerations. Social and economic considerations are reviewed in the analyses. Emphasis is placed on the utilization of the various I-82 alternatives to improve the social, cultural, and educational interchange between the various communities and between the study corridor and areas beyond it. Such elements are noted as increased job opportunities, better shopping, more desirable school attendance patterns, and the influence of the highway as a catalyst in developing such patterns.

Road-User Costs. Calculated road-user benefits are based on 1990 traffic volume, speeds, and classes of highways. It is considered that the life of the facility is more than twice the 25 years, 1965 to 1990, so that the 1990 year may be considered as the midyear of a 50-year period. The 1990 traffic volume may be considered to be less than half of the 64-year period based on the weighted average life of the highway components.

The road-user costs are based on the AASHO Red Book [14-1] on road user benefit analysis. Vehicle-miles estimated are multiplied by the motor vehicle operating cost in cents per mile. However, the report does not state specifically whether travel time value and comfort and convenience values are included. Judging from the source of the motor vehicle cost data (AASHO Red Book) and the magnitude of the road user benefits, it is most probable that time value and comfort and convenience are included.

The report states that the road user costs do not include motor vehicle travel that would prevail within the highway, road, and street network and be unaffected by alternative I-82 routings and the related improvements. Local travel would be affected differently by each route location alternative. There is considerable uncertainty of just what travel is included and how that travel is priced with specific reference to speed, speed changes, and distance and volume. For instance, there is no statement relative to adjustments for the adverse distance on Routes 2 to 10 over Route 1 that would be required by all through trips.

The annual road user costs for 1990 for the existing system are given as $79,301,000. Presumably this total cost is calculated on the basis of no improvements to the 1965 network during the 25-year period. No details are given on any of the road user calculations of costs or of the $1,016,000 highway maintenance cost in 1990 for the existing system.

Added Annual Economic Benefit. The added annual economic benefit is included as a measure of the economic and social gain to the area resulting from the interstate route. This factor is intended to be a measure of the relative value of each of the ten alternative routes in terms of the gross (not net) contributions to the gross national product, over and above the direct road user net benefits. It should be specifically noted that this added economic benefit is attributed solely to the traffic induced by I-82 over and above that realized as a result of the related route improvements.

The added economic benefit is based upon the added employment in the area which is attributed to I-82. The final dollar amount for each alternative comes from multiplying the number of new employments in 1978 that can be accredited to I-82 by the average

The Management Decision

annual income of $6,200 per worker. The new employments were determined by estimating the population growth in a circular area of a 2-mile radius around each I-82 interchange. This increase in population is credited with one new employed person for each 3-person increase in population in rural areas and 3.8 persons in urban areas. Of the total increase in employment around each of the 46 interchanges the portion credited to I-82 varies from 20 to 80 percent. The higher percentages apply to the more dense populations, such as the Tricity area.

Factors and Issues Important to the Final Decision

The change in objectives of I-82 from just a link in the Seattle to Salt Lake City route to include direct service to the Tricity area in Washington gives rise to the main question to be answered by selection of the route location. Some of the principal factors involved in the main question of to what extent should I-82 serve the Tricity area may be stated as follows:

1. To what extent should through travel between Granger and Pendleton be penalized by a greater travel distance (up to 27.53 miles) to favor interstate route service to the Tricity area?

2. From the viewpoint of the interstate system as a whole, is the nation better off by allocating up to 63.68 miles to I-82 above the minimum requirement of 45.77 miles? Because only 41,000 miles are authorized to be constructed with federal-aid highway funds at 90 percent federal and 10 percent state, the length of I-82 does affect other routes not yet definitely located. This same factor is common to every route location study on the interstate system with respect to the length of each proposed route or segment.

3. To what extent is it the responsibility of the local states to finance the desired highway facilities in the Tricity area with 100 percent their funds, or where eligible, with 50 percent federal-aid funds?

4. What weight should be given to the forecasted increase in employment resulting from the interstate route I-82? How much of the forecasted employment increase, say on Route 9 with I-82, would materialize in the future somewhere in the country if I-82 were located along Route 1? The report does not provide an answer to this question. The decision is one that bears heavily on the factor of "from whose viewpoint," and the forecast of economic change with and without I-82, location by location.

5. Considering the local area as a whole and the interstate system as a whole, which is the better long-range plan: (a) Construct now I-82 through the Tricities, along, say, the Route 9 location, and sacrifice the ideal for through traffic, or (b) locate I-82 along Route 1, and then improve the area as a whole with added highway facilities as the needs arise. It is to be remembered that there is no provision for more than 41,000 miles on the interstate system and no money is provided for construction on the system after 1973.

6. Since the whole northwest United States, including the area around I-82, is an economically growing area, there is an important question to answer as to how the area would develop 1965 to 1990 if I-82 were not built at all. Certainly, the growth and economic well-being of the Northwest is not dependent upon I-82. The decision maker will need to give consideration to the probable economic and social consequences with and without I-82 and to its probable location. Economic transfers within the larger Northwest area are important factors, as they are also on other areas.

PROBLEMS

23-1. Look up a current or recent transportation proposal in your community or state on which there is some divided public opinion as to location, design, or need. Collect what information you can that will afford you background information on which you can make a reasonable adequate analysis of the pros and cons. Write your report on the decision-making process, pointing out the significant factors on which the decision should be based. If a decision has been made, comment upon why it was made as it was, or on how you think it should have been made.

23-2. From the highway department or other source, select a transcript of the record of a public hearing on a highway project. From your reading of this transcript–and any available news reports–write out your impressions and comments. What did you find that is of real significance to the final decision process? What was the motivation behind the positions taken by the main speakers?

23-3. Examine critically *Management Decision Case No. 2*, at the end of Chapter 23. What do you find here, positive or negative, that should influence the conscientious decision maker, first as a state official, and second, as a federal government official? What omissions of information do you observe?

23-4. After carefully considering *Management Decision Case No. 2*, what deficiencies do you find in the technical data and methods of analysis?

23-5. On the basis of the information in *Management Decision Case No. 2* and your general knowledge, what alternative route would you select? Give your reasons for your decision.

23-6. In the final decision of top management relative to the choice of route location among four mutually exclusive alternative projects on the interstate system, what relative weight would you give to (1) commodity savings, (2) travel time reduction and value, (3) personal preference of the road user, (4) urban renewal including displacement of residences, and (5) other socio-economic factors? Explain your position by appropriate discussion.

23-7. In the highly controversial highway improvement projects which attract wide public discussion, why do you think that the decision-making person, or public agency, delays so long in announcing the decision?

REFERENCES

23-1. Peter F. Drucker. "The Effective Decision," *Public Administration News Management Forum*, Vol. 17, No. 2 (May 1967), Section II. American Society for Public Administration, Washington, D.C.

23-2. A. L. Edwards and F. P. Kilpatrick. "Scale Analysis and the Measurement of Social Attitudes," *Psychometrika*, Vol. 13 (1948) pp. 99-114.

23-3. Ward Edwards. "The Theory of Decision Making," *Psychological Bulletin*, Vol. 51, No. 4 (1954), pp. 380-417. Includes 209 references.

23-4. Edward L. Falk. *Value Measurement and Its Application to the Decision-Making Process*. Washington State Department of Highways, Spokane Metropolitan Area Transportation Study, Olympia, March 31, 1967. See also Edward L. Falk. *Measurement of Community Values: The Spokane Experiment*. Highway Research Record No. 229 (1968), pp. 53-64.

23-5. Neal A. Irwin. *Criteria for Evaluating Alternative Transportation Systems*.

The Management Decision

Highway Research Board, Washington, D.C. Highway Research Record No. 148, *Transportation System Evaluation*, 1966, pp. 9-19.

23-6. Milton Z. Kafoglis. "Highway Policy and External Economics," *National Tax Journal* (March 1963), pp. 68-80.

23-7. Tillo E. Kuhn. *Public Enterprise Economics and Transport Problems.* U. of California Press, Berkeley, 1962.

23-8. Michael Lash. *Community Conflict and Highway Planning (The Case of a Town That Didn't Want a Freeway).* Highway Research Board, Highway Research Record No. 69, *Highway Management*, 1965, pp. 1-17.

23-9. Harold J. Laski. "The Limitations of the Expert," *Harpers*, Vol. 162 (December 1930), pp. 101-110.

23-10. Marvin L. Manheim. *Model-Building and Decision-Making.* Massachusetts Institute of Technology, School of Civil Engineering, Publication No. 165, Research Report R62-10, Cambridge, May 1962.

23-11. Marvin L. Manheim. *Transportation, Problem-Solving, and the Effective Use of Computers.* Highway Research Board, Washington, D.C., Highway Research Record No. 148, *Transportation System Evaluation*, 1966, pp. 49-58.

23-12. George Perazich and Leonard L. Fischman. *Methodology for Evaluating Costs and Benefits of Alternative Urban Transportation Systems.* Highway Research Board, Washington, D.C., Highway Research Record No. 148, *Transportation System Evaluation*, 1966, pp. 59-71.

23-13. George A. Taylor. *Managerial and Engineering Economy; Economic Decision-Making.* Van Nostrand, Princeton, N.J., 1964.

23-14. U.S. Government. *Policies, Standards, and Procedures in the Formulation, Evaluation, and Review of Plans for Use and Development of Water and Related Land Resources.* 87th Congress, 2d Session, Senate Document No. 97, U.S. Government Printing Office, Washington, D.C., 1962.

23-15. U.S. Government, Inter-Agency Committee on Water Resources. *Proposed Practices for Economic Analysis of River Basin Projects*, by the Subcommittee on Evaluation Standards (The Green Book). U.S. Government Printing Office, Washington, D.C., May 1958.

23-16. Richard Willis. "Estimating the Scalability of a Series of Items–an Application of Information Theory," *Psychological Bulletin*, Vol. 51, No. 5 (1954).

Chapter **24**

Highway Needs Studies

As a part of the process of long-range planning for highways, there has been developed a procedure of systematic study of the needs for highway improvement by highway system and by administrative jurisdiction. These studies deal with the physical needs and their cost for a specific future time period. The financial aspects are dealt with in a separate but coordinated financial study (Chapter 25). When required, and it most often is, legislative action follows to implement the program, provide the finances, and to amend the statutes to provide for desirable authority and administration. The last step is the programming of the construction on a 3-, or 5- to 10-year basis with annual program awards each year (Chapter 26).

SCOPE AND OBJECTIVES OF NEEDS STUDIES

Highway needs studies have been developed year by year until now they are well-established as a management tool. Their concept, content, and the techniques used are well understood. Although each needs study is an individual operation, the conduct of the studies now follows well-established procedures.

A highway needs study is a systematic analysis of a highway, road, or street-system physical plant to estimate the need for improvements over a given future period. The need is based upon three primary factors: (1) geometric and structural design standards, (2) existing physical condition of the highway with respect to these standards, and (3) forecast of highway use (traffic) for the period of analysis.

The objectives of this study of highway needs are (1) to formulate a broad plan for orderly development of the coordinate systems as a whole, (2) to provide a basis of adequate and systematic financing of the highway system, and (3) to provide a basis for coordinate improvement of all related systems. The needs study does not result in a construction program in the sense that specific projects are listed for construction in specific years; but the results are facts, estimates of costs, and long-range plans on which annual construction programs should be based. The execution of a needs study does not require an economic analysis of the system as a whole, or on a project by project basis. The objective is to estimate the dollar requirement to accomplish an adequate system as a basis for long-range

Highway Needs Studies

planning. The economic analysis follows when the actual construction is being specifically proposed.

CONTENT OF A HIGHWAY NEEDS STUDY

Needs studies render their greatest value when they are comprehensive of all interconnected and interdependent highway systems and all governmental agencies responsible therefor. In a state, the study should include the interstate, primary, secondary, and local highways, urban and rural–that is, all highways, roads, and streets within the state. A comprehensive state-wide study is essential when road user tax payments are shared by the state with lower levels of governments. Further, when legislative action is required, the whole highway plant within the state should be considered rather than just one selected system thereof.

An adequate appraisal of the state's highway systems and identification of desirable improvements thereto would include a full review of all factors affecting the use of and the adequacy of each highway system. These factors are illustrated from the following list by McCormack [24-18]:

1. An engineering appraisal of highway and traffic conditions and identification of needed improvements to all rural and urban highways, roads, and streets.
2. A study of all finances related to highway affairs.
3. A study of the laws related to highways.
4. A study of administrative practices and regulations of all highway agencies.

These four broadly stated subject areas encompass the whole scope of highways and highway administration within a state–and they are meant to cover the whole subject. A highway needs study conducted to penetrate into these four areas should lead to answers to the following questions which are based upon the work of Steele in Ref. 24-22:

1. What is the role of highways in the social and economic development of the state?
2. What are the present and probable future social and economic factors and their trends that cause a demand for improving the highway transportation facilities and services, and how can these factors be isolated and measured?
3. What types of highways and what mileage of each are desirable and where should they be located to supply the future need for highway service?
4. What are the long-range costs to construct, maintain, and administer the highway systems that the study indicates are desirable?
5. How best can the state and its subordinate units finance adequate highway transportation for the state?
6. What is the best administrative provision for the highway systems in the state, considering their functions, legal responsibilities, and effective use?
7. What administrative highway systems should prevail, and how should these systems be determined?

A state-wide highway needs study will require about 15 to 18 months of effort to develop the supporting material and to answer these questions. The study will, of course, point to the future, though considerable work must be done on the past and present conditions and activities. The time period chosen for study is any reasonable future time such as 10, 15, 20, or 25 years. This period must be long enough that the resources and capabilities available will permit completing the construction program in the period of time, and yet not so long that the forecast of traffic and physical needs becomes conjectural.

THE FINAL REPORT

The report on the needs study is intended primarily for members of the state legislature, the officials of the state, counties, and cities, all highway officials and professional persons, and the citizens who may be interested. The report, then, serves three classes of readers: (1) the layman who is interested in the main findings only, (2) the legislator–state, county, and city–who wants the findings plus a limited explanation supporting the findings, and (3) the professional and technical persons who want the whole store of facts and their analysis. Of recent years it has been common practice to prepare two separate reports, one a highly abbreviated, summary version for those who want only the findings, and an extended version giving the details on the whole study plus basic tables and documentary material that support the findings. Heavy use of graphic presentations is generally practiced as a means of reaching the diverse levels of readers, most of whom do not care for the precise values.

ORGANIZING FOR A NEEDS STUDY

Each agency conducting a highway needs study will devise its own organization and procedures, but the organizations will not differ materially agency to agency. The following discussion presents the general form of organization most often used.

THE STUDY COMMISSION

Many states have set up through legislative action a special agency or commission to conduct needs studies; at other times a state highway department has been directly responsible for the study. Often the study is authorized on the initiative of the state legislature as a means of getting basic information on which to base legislation, particularly as pertains to highway taxes and financing. The most effective type of agency to conduct a needs study, judged by past success in implementing the findings and recommendations, is one closely connected with the state legislature. Legislative sponsorship has been in one of two basic forms

[24-18, p. 3]: an interim committee of both houses of the legislature, or a study committee created by the legislature and composed of members of the legislature and private citizens who have an interest in highway transportation. When the needs study is conducted by the state highway department it has been general practice to get tacit approval from the leaders in the legislature and approval of the governor.

The successful conversion into law of the recommendations from the needs study is not dependent upon prior sanction of the study by the legislature, but experience has proved that the chances of success are much greater when the legislature assigns its own members to an active and responsible role in the study. The study commission is not expected to do the active work of the study, but to control the policies, scope, and direction of the agency (consultant or other organization) in active charge of the operation. The commission would accept or reject the finding and recommendations, however, in order to assure that it can support them in terms of legislation.

ADVISORY COMMITTEES

A comprehensive needs study requires the full cooperation, assistance, and understanding of all governmental units responsible for the highways, roads, and streets included in the study. Such cooperation may be achieved through advisory committees and joint staff groups. Through these units, agreements and understandings can be reached in the planning stages and the operating phases will move more smoothly. One advantage of the interagency groups, including official representation from the state legislature, is that agreement on the final report comes about with less difficulty, and in the end the recommendations are more likely to be implemented.

A citizens' advisory committee may be composed of leaders from industry, business, the professions, and community activities. Such appointments may be made by the study commission or by the governor. The advisory committee may have representatives from special interest groups such as automobile clubs, trucking organizations, chambers of commerce, and civic organizations. Often, though, to assure more general acceptance of report findings these groups should be represented on the study commission.

Often technical advisory committees are formed to offer guidance and technical assistance to the working staff conducting the needs study. These advisory committees furnish support in the areas of engineering, economics, statistics, sociology, law, and government organization and operation. The advice of these technical groups and individuals has a two-point advantage. Through their efforts the operating procedures and analyses can be improved, and in the end the advisory committee's assistance gives assurance to others of the technical soundness of the results.

ORGANIZATION OF THE WORKING STAFF

Under the broad direction of the sponsoring agency or commission, the needs study is conducted by a working staff of engineers, statisticians, clerks, and others numbering 50 to 100 at peak employment in the information gathering stage. This staff is in charge of an engineer-director or consultant who has both technical and operational responsibilities. The engineer-director or consultant also is the chief contact between the active working group, the study commission, and cooperating agencies.

Study Groups

The engineer-director or consulting firm in direct charge of the operations, may choose to divide his organization along subject lines rather than functional activities. For instance, four main working groups may be established: (1) engineering appraisal, (2) fiscal analyses, (3) law review, and (4) administrative review. Obviously these groups would be coordinated by the engineer-director or consultant but much of their work may be done independently. Certain phases of the work within any one of these four groups may be contracted to a consultant or other agency, which plan frees the needs study from hiring a large number of specialists for short-time employment.

Working Staff

The number of staff and their specialties hired or assigned to the needs study groups is dependent upon many factors. Often the state highway department will supply many services direct for both field and office activities. Too, the highway department may assume complete responsibility for all "housekeeping chores," personnel hiring, and accounting. Field work is most often done with state and local government units. Other than these general administrative and overhead activities and field work, the needs staff must furnish the personnel for the active management phases of the study, and particularly for the analyses of the information collected, and envolvement of the highway needs–the main objective of the study.

Cooperating Agencies

A highway-needs study on a state-wide basis requires a wide range of cooperation from outside the study staff. In the first place, the state highway department is the key organization. It has not only records, plans, and facilities, but both its headquarters and field staffs will be required for certain work or services. The highway department planning survey unit is the key source of information, and whether or not this unit is assigned to the needs study, it will be rendering daily service to the needs study staff.

Highway Needs Studies

The offices of the county and city highway departments of necessity will be required to cooperate and coordinate in the efforts to collect the information on their road and street systems, including their long-range plans. Other units of local governments will also be involved on financial aspects and administration. The degree of participation and amount of work furnished depend upon the particular state and staffing policies of the local governments. Here again friendly and helpful cooperation and working relationships will lead to more ready acceptance of the results because these local units of government will have had some voice in the study from its beginning.

CONDUCTING A NEEDS STUDY

To estimate the needs of a given highway system for the next 10 to 25 years is a complex undertaking, but a task not difficult to accomplish when laid out on a systematic and preplanned basis. The whole process is an engineering one coupled with a knowledge of highway administration, the relation of highway use to general economics and population, and some understanding of the dynamic character of highway transportation. The entire procedure involves many factors and detailed procedures which are discussed next.

PREPARATORY STEPS

Following the establishment of the advisory and working organizations, preparing the technical material is in order. The technical material includes many items such as the following:

1. The basis of highway system classification and the classifying of all highways, roads, and streets in accordance with the adopted system is an early step.

2. The standards of design and the minimum acceptable quality for tolerable conditions have to be adopted for each road system.

3. Since the road systems must be inspected, inventoried, and rated section by section, the criteria for establishing route-control sections must be prepared, should the state or other authority not have them already established. A control section is a limited length of a highway route with full and specifically described fixed termini. State, county, and municipal boundary lines, bridges, road intersections, marked changes in topography, and the like are the usual points of beginning and ending of control sections. For use of the needs study only, however, a less formal system of designating roadway sections may be adopted.

4. Early decision is required as to what major systems and routes will be field-inventoried mile by mile and what routes and streets may be sampled on an area basis. The needs study is a macroscopic analysis rather than a microscopic examination; the answers are reliable for the systems as a whole, but not for any particular route section, except for the high-order routes which are analyzed individually, section by section.

5. If the sufficiency or other rating scheme is to be used, its criteria and rating and weighting systems have to be established. Likewise, if the PSI (present serviceability index) is to be used for pavements, its cutoff level must be established. Other devices may be used for estimating the terminal service life of existing pavements.

6. The basis for forecasting traffic has to be established with respect to all road systems and separate routes or areas thereof.

7. All of the above material and much more should be incorporated in a series of manuals and instructional guides for the use of all workers on the study.

8. A review of the factual histories of the several road systems, their past financing, rates of improvement, and rates of traffic increase will provide valuable background for all forecasting, relative and particular needs, area by area and route by route. Population and land use histories, motor vehicle registration, and highway use data are also important to the overall study.

THE MAIN PHASES

As a whole the needs study is an extensive operation, but not particularly complex. When separated into its distinct phases its logic becomes apparent. The following sections present certain details of the several aspects of how the physical needs are determined.

Traffic Forecast

Ideally, the traffic (use of the highways by trucks and passenger cars) should be forecast to the end of the study period for each short section of highway analyzed. Practically, it is satisfactory to use general forecasts by geographical areas and highway systems except for those highway control sections that present a sound reason for individual forecast of traffic, particularly those of high traffic volume and rapidly growing traffic.

The traffic forecast is one of the key factors, because the traffic volume and vehicle classification determine the year that traffic capacity will be reached and when improvements would be required, including wholly new facilities. In addition to the time element, the forecasted traffic volume is the base factor controlling the geometric and structural designs. Further, the road-user revenue forecast is based upon the same forecast of highway use.

Highway System Classification

The cost to construct highways is directly related to the standards of engineering design. Therefore, a classification of all highways, roads, and streets to be included in the needs study is a necessary step (see Ref. 24-13).

The classification of all highways, roads, and streets serves as a basis for (1) setting highway design standards, (2) assigning administrative responsibility,

Highway Needs Studies

(3) legislating action, (4) sharing road user tax incomes, (5) differentiating vehicle regulations and highway use, and (6) policing and traffic control.

Highway systems may be classified with respect to the governmental agencies responsible for them and by the general character of their use. The financing and administration of highway systems are responsibilities of the agency having jurisdiction. The design standards are related to character and volume of traffic, to the topography, and to the land use.

Classification of highways, roads, and streets is the orderly grouping of them by their similarity of service to the traveling public. A given road system renders the same general type of travel service throughout the system. As a system, it makes possible efficient and well-directed legislation, administration, and financial management, which, in the end, all contribute to improved service to the public.

Classification of Rural Highways. Rural highways and roads may be classified into four basic systems: (1) a freeway system, such as the interstate system, (2) a primary system of state-wide importance, (3) a secondary system of the principal roads, feeding the primary system and distributing traffic, and (4) a tertiary system of roads, servicing the local land area and the local population.

The interstate, or freeway, system is the highest classification of both rural and urban highways from the standpoint of design standards and volume of traffic. Although it will carry some medium (trip length) local traffic, essentially the freeway system travelers may be assumed to have but minor interest in local areas. Certainly of all systems the interstate, or freeway, system will serve the fewest number of those trips having interest in the local land.

The rural primary system is a level below the rural interstate system and will carry more local traffic by reason of its total mileage and intended service to the larger local communities. Although local traffic will use the rural primary system, this system is characterized by its intercity travel, which predominates. The traffic on the primary system has but little interest in the land adjacent to the system or in the small towns and villages encountered along the way. The trip length for primary system travel is, therefore, greater than on the secondary system and less than on the freeway system.

The rural secondary system performs two main functions: (1) it is a collector-distributor of traffic for the primary system, and (2) an integrator system for a local area, such as a county by which the towns and villages are connected and by which traffic from the tertiary system is collected and distributed.

The rural tertiary system is the system that provides access to a high percentage of the rural land, and the origin of much traffic that feeds into the secondary and primary systems. The tertiary system is the road system that binds the rural population together, socially and economically, and provides the bridge for social and economic links to places on the secondary and primary systems.

These four rural systems may be described from the viewpoint of the interest of the people who travel them. The degree of interest in the land along the system and activity thereon increases from a low for the freeway system to a maximum for the tertiary system. The roadside areas along the primary system attract only that small percentage of the total travelers who might have a personal interest therein because of relatives, friends, or business; the center of interest of all but these few local travelers on the primary system is 20 to several hundred miles away.

The secondary system travelers have major interest in the local area, say up to 25 miles distance.

Most of the travelers on the tertiary system have their center of interest in the immediate vicinity and there are but few other travelers on this system. The center of their interest is probably not more than 5 miles away. This characteristic of the travel on the tertiary system is the reverse of the travel on the primary system; the secondary system travel is in between.

All four rural systems carry traffic to the local land, and as such serve a land access function. This land access function may be less than 1 percent of the traffic on heavily used interstate and primary routes, and almost 100 percent on the tertiary routes.

Classification of Urban Streets. The urban highways and streets may be grouped also into four systems: (1) urban freeways and expressways, (2) through arterial routes, (3) collector routes, and (4) local streets.

These urban systems parallel the rural systems in order of importance and character of traffic. The freeways and expressways are extensions of the rural interstate routes into and through the urban area; the arterials accommodate the traffic from the rural routes, into and beyond the urban areas, which do not follow the freeway system. The collector system provides communication between the internal sectors of the urban area, as well as a collector-distributor service between the local streets and the through freeways.

The local streets provide access to the property and collect and distribute traffic to and from the other systems, mainly the collector routes.

The urban systems parallel the rural systems also in the degree of interest in roadside activity of the travelers.

In both the rural and urban systems, there is represented the degree of responsibility that could be assumed by governmental levels. The rural freeways and primaries, urban expressways, and urban arterial systems indicate a state and national responsibility; the rural tertiary and urban local street systems indicate strictly a local responsibility. The rural secondary system and collector routes serve in between responsibilities.

Design Standards

The geometric design and the structural design standards for highways are important factors in a needs study. These standards directly affect the cost of all highway construction, and the decision of what condition of design and physical impairment must exist before a highway should be reconstructed or given greater capacity.

Highway Needs Studies

The design standards used in needs studies are generally those adopted by the highway authority for each highway system for current designs. The four main factors controlling the standards are: (1) the level of service (travel speed and comfort) designed for the traffic using that system, (2) the character of local land use as affecting construction costs and character of traffic, (3) the composition of the traffic with special reference to the frequency of heavy trucks, and (4) the type of terrain (level, rolling, or mountainous). The level of quality of design standards is highly important because the costing of the needs is based upon these standards. High standards result in more miles currently deficient and higher costs per mile for construction.

Minimum tolerable conditions are design standards modified to less desirable levels for use as a guide in determining whether an existing highway or structure is acceptable for future use as it stands. The needs study divides all existing highway sections into three groups: (1) those meeting the design standards, (2) those that do not meet the design standards but are tolerable at the present, and (3) those that are so deficient in design or structural condition that immediate reconstruction or traffic relief is desirable. The (1) and (2) groups are usually combined.

The roadway sections meeting the standards and the tolerable sections are listed for improvement during the appropriate future 5-year period (assuming that all needs are to be grouped by 5-year periods) should the traffic forecast indicate that the sections will not be tolerable at the end of the study period. This approach brings into the total needs and estimated cost all roadway sections currently deficient and those forecasted to become deficient within the study period. Should all deficiencies be overcome by construction during the period, and should the future events be as forecasted, the entire roadway system would be adequate at the final year of the study period.

Service Lives and Adequacy

Complete reconstruction or any improvement of a highway route section or facility result from a management decision made principally upon three factors: (1) the traffic carried with respect to the capacity of that section or facility, (2) the quality of service as measured by design geometrics, and (3) the structural condition of the highway. The traffic capacity is also a function of the geometrics of design, and structural adequacy is also partly controlling. In the sense that highway capacity is used here, both traffic safety and travel speed are included.

Should a highway section be carrying a high traffic volume relative to capacity and be far short of meeting design standards, it would be listed as currently deficient; if it is carrying a low traffic volume, it probably would be listed as tolerable. Similarly with the factor of structural condition.

Future time and future traffic, however, cause the structural adequacy to decrease, ultimately to the condition which is sufficiently bad to warrant reconstruction. In the needs study, service lives of pavements and structures are factors helpful in making the prediction of when a highway section probably will become structurally not acceptable for traffic.

States have made road life studies (see Chapter 9) over a period of years as a means of determining service lives based upon past highway use. The results of these studies are highly important to the needs studies for both the calculated service lives and for the age distribution of existing highways.

Present Serviceability Index

A device for predicting the expectancy of service life of existing pavements was developed during the AASHO road test in Illinois 1958 to 1960. This device is called the "present serviceability index" (PSI) [24-4 and 24-5]. The present serviceability index is a numerical rating of the physical condition of a pavement based on a scale of 5.0 to 0.0. A perfect pavement is rated 5.0 and a totally unsatisfactory pavement would be rated 0.0. Pavements will not be adequate for traffic when their PSI is reduced to 1.5 to 2.2 depending upon the traffic volume, traffic mix, and quality of service desired. This lower or end point is fixed by judgement.

The PSI is given by

$$PSI = 5.41 - 1.80 \log(1 + S) - 0.09 \sqrt{C + P}$$

Where S = slope variance of the pavement surface
C = length of substantial cracking, whether sealed or not, feet per 1,000 sq ft of pavement area
P = patching factor, square feet of patch per 1,000 sq ft of pavement area

The slope variance S is a measure of the undulations of the pavement as measured by a profilometer. The formula for S is:

$$S = (8.46)\left(\frac{Y^2}{N}\right) - \left(\frac{Y^2}{N} - 3\right)$$

where Y = slope of the pavement in 6-in. increments
N = number of slope measurements in a given distance

By recording the PSI for pavements when first opened to traffic and periodically thereafter, a curve can be developed which will show the decrease in PSI with the years or with traffic use. This curve may be projected to the desirable minimum value of the PSI to obtain the probable year when the pavements would be physically unsatisfactory to traffic. New pavements have a PSI of 4.0 to almost 5.0.

Sufficiency Ratings

Another device, "sufficiency rating," is used as an index of the satisfactoriness for traffic of a given highway section. A rating of 100 means that the given highway section has no deficiencies in design or structure; the rating scale proceeds downward toward zero as the deficiencies increase. An arbitrary cutoff point is chosen at about 60 to 70 as the dividing line between highways that are tolerable and those that are not tolerable.

The factors rated and their weight vary somewhat according to the judgment of the designer of the particular sufficiency rating system. The following scheme is typical:

		Maximum Rating
A.	Structural Adequacy	
	1. Surface type	7
	2. Shoulder type	3
	3. Base & surface condition	10
	4. Subgrade stability	8
	5. Drainage	7
	6. Maintenance cost	5
	Total structural adequacy	40
B.	Safety	
	1. Surface width	10
	2. Shoulder width	5
	3. Stopping sight distance	10
	4. Consistency	5
	Total safety	30
C.	Service	
	1. Alignment	8
	2. Gradient	5
	3. Passing opportunity	8
	4. Surface width	5
	5. Rideability	4
	Total service	30

D. An adjustment factor related to traffic volume and capacity

E. An adjustment factor for structures based upon their deficiencies

The instructions for determining the sufficiency rating of any section of highway usually give point reductions for specific deficiencies. Guides of this type help to obtain consistency in ratings between raters and between different highway sections.

The PSI and sufficiency rating systems furnish helpful guides to the needs studies in determining the overall needs and route sections needing some improvement. However, with respect to the sufficiency rating, often the nature of the deficiency is lost when only the tolerable cutoff level is considered. Two other drawbacks in this method are (1) a single serious deficiency can be passed up as tolerable if other items are rated high, and (2) the rating gives no evidence as to possible future deficiencies.

Pricing Construction

When the highway improvement work anticipated for the duration of the study period has been identified, pricing it in dollars of construction and rights-of-way cost is in order. Construction prices as of a date near the beginning of the study are used. There need be no allowance for future changes in construction costs through technology or inflation.

The estimates of construction cost may be reached through several approaches. For the more costly and difficult projects, unit prices based on classes of work may be preferred as a means of getting reliable estimates. Work that is more nearly standard and consistent with current construction practices and prices may be reduced to total costs by pricing each class of work on a per mile basis for each highway class. Finally, for much of the tertiary systems, for which sample information only was compiled, the total cost may be estimated per mile for all work combined. However, care must be exercised in all pricing to allow for variations in terrain, soil types, and the local labor market.

In costing the physical needs all special types of facilities, such as urban expressways, large bridges, tunnels, river protection works, and mountainous construction, should be estimated as individual projects based upon reasonably complete preliminary designs. Fortunately, for the needs study staff their goal is the ultimate total cost broken down by systems and geographical units rather than by projects and routes. But to arrive at a reliable total cost for the system it is best to start by estimating the construction on a project or route section basis for the special and more costly types of work.

Priority Grouping

As each of the highway systems is studied, section by section, the needed construction is identified by both class of work and by year of need. Inventory and evaluation of the highways by their physical status will group the existing highways into two classes as measured by their relative needs: (1) sections not adequate for current traffic, and (2) sections tolerable or more than just tolerable for current traffic.

The tolerable sections and the currently adequate sections are further classified by the 5-year period when they probably will become inadequate because of continued physical deterioration and increase in traffic. From these determinations a future construction program can be laid out on a mass basis. Generally this total program need not show the detail year by year, but, say, by immediate needs, needs for the next 1 to 5 years, and then the needs by 5-year periods thereafter. Usually special attention is given to overcoming the high current deficiencies in the next few years.

ESTIMATION AND SUMMARY OF THE TOTAL DOLLAR NEEDS

The needs study as applied to highway systems is directed to ascertaining the

Highway Needs Studies

cost to construct and reconstruct the mileage on each road system to standards adequate for expected traffic during a 15- or 20-year period of the study. The total financial needs are greater than this sum of construction needs by other costs which are to be financed from the funds for highway purposes.

Total Dollar Needs–All Purposes

Although the construction costs are the principal costs incurred in highway system management, other costs amount to sizable sums, and seem to be increasing as higher levels of services are rendered, particularly so with highway maintenance and traffic services. The main items for which support moneys are needed are:
1. Highway construction, including right-of-way and engineering
2. Highway maintenance
3. Traffic operations
4. Administration of highway departments
5. Debt service (principal and interest)
6. Highway patrol and policing when supported from highway funds
7. Cost of collection and administration of highway user funds

The sum of the cost of these items is the total financial support that must be provided. It is the base sum of concern to the legislative body responsible for setting tax bases and tax rates. The sums for these items for the study period are estimated primarily on the basis of past costs, corrected upward, of course, for increasing highway facilities, increasing traffic, and increasing levels of service.

Money Needs for Catch-Up Periods

As one step in the forecasting of highway needs, it is the usual practice to investigate the effect of the time period over which to overcome all current deficiencies. A short catch-up period may result in higher average cost per year, but would give the public an adequate highway system sooner. Long periods involve more reconstruction and replacement because of the continual wearing out of facilities under the ever increasing traffic volume. As the catch-up period is lengthened, or backlog work deferred, stop-gap work increases. At the same time, second-generation replacements decrease. In total dollar cost over the same total length of time there may be but little difference between a short catch-up period and a long catch-up period. The short catch-up period necessitates higher annual sums during the catch-up period but lesser amounts in the following years when the need is only to keep the system up to acceptable conditions. The commonly used study period is 15 to 20 years.

PRESENTATION OF FINAL DOLLAR NEEDS

Although the construction dollar needs will be worked up by roadway units and construction projects, this detail is not necessary in the official final report. Totals of various types serve the purposes better.

It is desirable, however, to show all money needs by each highway system, by political jurisdiction and highway agency, and by bands of years. Further, the amounts of needs for the different functions–right-of-way, construction, maintenance, debt service, and administration–should be given. Cost by highway routes need not be given.

It is to be kept in mind that the needs study is not a programming study; it does not result in a construction program. Its basic goal is a macroscopic picture of total need over a period of years from which a financial program (see Chapter 25) can be arrived at and then followed by a construction program (see Chapter 26).

CONTINUING NEEDS STUDIES

The foregoing description of highway needs studies is presented as though it is the first one to be conducted in a given state. Much can be said in favor of making needs studies on a continuing basis, so that each year a good picture may be had of the highway needs in relation to current financing.

By adjusting the records accumulated during a formal and extensive needs study on a year by year basis, a needs study can be kept reasonably well on a current basis. Like all other types of continuous inventories, however, sooner or later a detailed extensive check of the inventory becomes desirable. With the electronic computer it can be comparatively easy to update a needs study yearly by adjusting for construction work completed, deterioration of the highways with time and use, traffic growth, land use changes, and construction costs. Updating will also adjust the needs for any increase in design standards or tolerable conditions.

REFERENCES

See also references at the end of Chapters 25 and 26.

24-1. Arkansas State Highway Commission. Little Rock, Arkansas, December 1966.

1. "Arkansas Highways, Roads, and Streets–The Systems Plan," by The Automotive Safety Foundation, Washington, D.C.

2. "Arkansas Highways, Roads, and Streets–The Needs Study," by The Arkansas State Highway Commission.

3. "Arkansas Highways, Roads, and Streets–The Fiscal Program," by Roy Jorgensen and Associates, Gaithersburg, Md.

4. "Arkansas Highways, Roads, and Streets–Summary Report," by the Arkansas State Highway Commission.

24-2. John E. Baerwald. *Improvement Priority Ratings for Local Rural Roads in Indiana.* Highway Research Board, Proc. Vol. 35 (1956), pp. 38-62.

24-3. B. G. Bullard. *Developing and Analyzing Functionally Classified Highway Networks Utilizing Traffic Simulation–Phase 1.* Highway Planning Technical Report No. 3, U.S. Department of Transportation, Bureau of Public Roads, Washington, D.C., February 1966.

24-4. W. N. Carey, Jr. and P. E. Irick. *The Pavement Serviceability Performance Concept.* Highway Research Board, National Academy of Sciences, Washington, D.C., Bulletin 250, *Pavement Performance Concepts,* 1960, pp. 40-58.

24-5. Highway Research Board. National Academy of Sciences, Washington, D.C., Special Report 73, *The AASHO Road Test,* Proceedings of a Conference, St. Louis, Mo., May 16-18, 1962.

    1. W. N. Carey, Jr. and H. C. Huckings, and R. D. Leathers. "Slope Variance as a Measure of Roughness and the Chloe Profilometer," pp. 126-137.

    2. Bertram D. Tallamy. "New York State Thruway Use of Road Test Findings," pp. 291-298.

    3. W. E. Chastain, Sr. "Application of Road Test Formulas in Structural Design of Pavements," pp. 299-313.

24-6. Highway Research Board. National Academy of Sciences, Washington, D.C., Bulletin 53, *Highway Sufficiency Ratings,* January 1952.

    1. O. L. Kipp. "Sufficiency Ratings as an Administrative Tool," pp. 1-3.

    2. William E. Willey. "Arizona's Experience with Sufficiency Ratings," pp. 3-7.

    3. John A. Swanson. "General Comments on Sufficiency Rating Procedures," pp. 7-11.

    4. P. R. Staffeld. "Possible Areas of Improvement in Rating Procedures," pp. 11-14.

    5. Curtis J. Hooper. "Considerations in Rating Urban Streets," pp. 14-17.

    6. James O. Granum. "Graphical Presentation Procedures," pp. 17-30.

    7. M. Earl Campbell. "Elemental versus Composite Ratings," pp. 30-32.

    8. Roy E. Jorgensen. "Use of Sufficiency Ratings in Long-Range Planning," pp. 32-36.

    9. C. E. Fritts. "Relation of Sufficiency Ratings, Tolerable Standards, and Priorities," pp. 36-41.

24-7. Highway Research Board. National Academy of Sciences, Washington, D.C., Bulletin 158, *Highway Needs Studies,* 1957.

    1. David M. Baldwin, J. Stannard Baker, J. Al. Head, and C. F. McCormack. "Traffic Accident Records in Appraising Highway Needs," pp. 1-4.

    2. David W. Schoppert. "Predicting Traffic Accidents from Roadway Elements of Rural Two-Lane Highways with Gravel Shoulders," pp. 4-27.

    3. P. E. Wade, R. B. Truemner, and R. I. Wolfe. "Two New Classification Techniques," pp. 27-35.

    4. John D. Cruise. "Administrative Application of a Method of Road and Street Classification," pp. 35-43.

    5. Robert S. Scott. "Factors Influencing Rural Road Mileage," pp. 43-57.

    6. *Estimating Maintenance Needs.* Part 1, John J. Laing. "Rural State Highways," pp. 57-58; Part 2, Terry J. Owens. "City Streets," pp. 59-60; Part 3, Howard Bussard. "County and Local Roads," pp. 60-63.

7. Philip M. Donnell and Lawrence S. Tuttle. "Priorities Determination and Programing in Tennessee," pp. 63-78.
8. Fred B. Farrel. "Effect of Traffic Growth Projections upon Estimates of Highway Needs and Revenue," pp. 78-81.
9. Clint Burnes. "Methods of Estimating Improvement Costs on County FAS Systems in Minnesota," pp. 81-90.
10. Clint Burnes. "Analysis of Sampling County Road Needs in Minnesota," pp. 90-98.
11. Robert D. Jordan. "Evaluating Contract Costs in Highway Needs Studies," pp. 98-104.
12. Harold W. Hansen. "A Review of Travel Forecasts," pp. 104-109.
13. James A. Foster. "Charts for Highway Needs Studies," pp. 109-116.
14. Bertram H. Lindman. "Economic Forecasting for Statewide Highway Studies," pp. 116-126.
15. James O. Granum. "Highway Program Evaluations," pp. 126-132.
16. Forrest Cooper. "Perpetual Highway Needs Study," pp. 132-133.

24-8. Highway Research Board. National Academy of Sciences, Washington, D.C., Bulletin 194, *Highway Needs Studies*, 1958.
1. Homer A. Humphrey and James A. Foster. "What Should Highway Needs Study Reports Contain?", pp. 1-11.
2. Bertram H. Lindman. "Highway Taxation Cost/Benefit Analysis," pp. 11-14.
3. D. O. Covault. "Indiana's Highway Needs Study," pp. 14-28.
4. John B. Benson, Jr. "Analysis of County Road Management Functions," pp. 28-33.
5. Theodore F. Morf and Frank V. Houska. "Traffic Growth Patterns on Rural Highways," pp. 33-42.
6. R. N. Grunow. "Vehicle Delay at Signalized Intersections as a Factor in Determining Urban Priorities," pp. 42-49.
7. Philip M. Donnell. "Tennessee's Programing Study: First Year's Experience and Techniques for Updating," pp. 49-60.

24-9. Highway Research Board. National Academy of Sciences, Washington, D.C., Highway Research Record No. 87, *Programing and Needs, 1963 and 1964*, 1965.
1. Evan H. Gardner and James B. Chiles. "Sufficiency Rating by Investment Opportunity," pp. 1-28.
2. Marshall F. Reed and R. E. Futrell. "Multiple Project Scheduling of Preconstruction Engineering Activities," pp. 29-56.
3. Edward M. Hall and C. Dwight Hixon. "The Use of a Priority Formula in Urban Sreet Progamming," pp. 57-77.
4. James O. Granum and Ronald M. Gordon. "Flexible Analysis of Highway Needs in Manitoba," pp. 78-97.
5. R. O. Kipp and W. T. Lussky. "Appraisal of Needs and Cost Estimating Procedures: Minnesota Trunk Higway Needs Study," pp. 98-123.
6. D. G. Malcolm and D. R. Earich. "Development of an Integrated Highway Management System," pp. 124-143.

24-10. E. H. Holmes. *"Looking 25 Years Ahead in Highway Development in The United States.* (The Sixth Rees Jeffrys Triennial Lecture April 1965.) Town Planning Institute, London, May 1965.

24-11. Indiana State Highway Commission. *Guidelines for Progress-Indiana Highways, Roads, and Streets.* State Office Building, Indianapolis. Reports prepared by Cole & Williams Engineering–Joint Venture, Indianapolis, January 1967. Report is in three parts: (1) "Report on Needs and Classification," (2) "Report on Finance," and (3) "Summary Report."

24-12. Iowa State Highway Commission. *Iowa Needs and Finances 1967-1987 for Highways, Roads, and Streets.* Ames, Report prepared by Roy Jorgensen and Associates, Gaithersburg, Md., January 1967.

24-13. Joint State-County Commitee on Highways. *A Guide for Functional Highway Classification.* National Association of Counties, Washington, D.C., May 1966.

24-14. Roy E. Jorgensen. *Planning and Measuring Highway Progress.* Highway Research Board, Washington, D.C., *Proc.* Vol. 40 (1961), pp. 35-44. An application of the sufficiency rating to estimating the need for future financing.

24-15. Robert E. Livingston. *The Sufficiency Survey Method of Determining Priorities for Construction Programs.* Highway Conference, University of Colorado, 1949.

24-16. Robert E. Livingston. *The Use of Factual Data in Long-Range Planning.* Western Association of State Highway Officials, 1958, pp. 60-65.

24-17. R. S. McClough. *Statistical Requirements and Methods of Long-Range Planning.* Proc., Canadian Good Roads Association, 1955, pp. 133-144.

24-18. C. F. McCormack. *Creating, Organizing, and Reporting Highway Needs Studies.* Highway Planning Report No. 1, U.S. Bureau of Public Roads, Department of Transportation, U.S. Government Printing Office, Washington, D.C., September 1963.

24-19. J. Carl McMonagle. *A Comprehensive Method of Scientific Programing.* Highway Research Board, *Proc.,* 1956, pp. 33-37.

24-20. Nebraska State Department of Roads. *Nebraska Highway Classification and Needs–An Engineering Appraisal.* Lincoln, Report prepared by Roy Jorgensen and Associates, Gaithersburg, Md., February 1967.

24-21. Harold L. Plummer. "Advance Planning: The Big Payoff," *American Road Builder* (1959), pp. 10-11.

24-22. C. A. Steele. "Long-Range Physical and Financial Planning for Highways," *Business and Government Review* (March-April), pp. 22-32 and (July-August), pp. 11-33. University of Missouri, School of Business and Public Administration, 1961.

Chapter **25**

Highway Finance and Taxation*

The functional classification of highways, roads, and streets identifies these facilities by the level of service that each will furnish. The outgrowth of functional classification is the assignment of highway facilities to the proper jurisdictional level. Based on design standards and "tolerable conditions" criteria, the highway needs study will develop an estimate of total needs or funds required for certain time periods to achieve a stated level of adequacy of highway service. Such needs are based on the functionally classified systems and administrative jurisdictions. The next step is the development of alternative fiscal plans that will provide the necessary funds to meet physical needs. Alternative plans are a prerequisite in order for highway, road, and street administrators and state legislators to determine the size, scope, and length of the future highway construction program. The alternative fiscal programs must include an evaluation of the present highway tax structure, both user and nonuser, in terms of future tax revenues, the allocation of highway costs between users and nonusers and estimates of revenue yields from proposed new and/or increased tax rates.

THE FINANCIAL PROGRAM

Planning for an increased highway construction program, such as often follows a state-wide highway needs study, involves three main phases: (1) a financial plan for the period of the construction, (2) a thorough study of the whole tax structure by which finances would be raised, and (3) a detailed analysis and plan of the specific tax burden to be imposed on the highway users, nonusers, and specific classes within the user group. A brief discussion of highway financing and forecasting is given in this section.

HIGHWAY DEPARTMENT FINANCES

Highway departments–city, county, state–receive their highway funds from

---

* The author gratefully acknowledges the material assistance of Mr. T. R. Todd in the design and composition of this chapter.

Highway Finance and Taxation

real estate taxes, general funds, vehicle registration fees, motor vehicle fuel taxes, and aids from higher level governments. Often, when motor vehicle fees and motor vehicle fuel taxes are involved, the highway department is automatically budgeted with these tax incomes. Other sources of funds come to the highway department through direct appropriations from general funds by the proper legislative body. Most all states share with their cities and counties in some manner the state road user incomes. Borrowing is a source of current funds, and such debts are serviced through future road user tax incomes or through general funds, depending upon whether it is a city, county, or state responsibility, and the particular laws in force.

There are widespread differences, state to state, in how highway, road, and street funds are raised and how distributed. These differences, also, relate to how the responsibility for the road and street systems is assigned. Some state highway departments have major responsibilities in cities, others but little or no responsibility. In the states of Delaware, North Carolina, Virginia (with some exceptions) and West Virginia, the state has full responsibility for all rural roads. The North Carolina state highway department has responsibility for almost all urban streets. Alabama has about eight counties under state jurisdiction.

With the exception of those instances of heavy borrowing or of adoption of new highway tax laws, all highway departments can make reliable estimates of their probable incomes for one to three years in advance. These estimates are possible because such incomes are based upon continuing laws and tax rates.

The disbursements by highway departments are for the following main functions: capital outlays (highway construction), highway maintenance and operation, highway patrol (when a highway department function), administration (including research and planning), and debt service. The major portion of total disbursements is for capital outlays, but first priority on available funds is for debt service, administration, and highway maintenance and operation.

In preparing the fiscal needs to support the highway program, past expenditures for capital outlays are not an important consideration. Administration, planning, and research costs are usually estimated on a pro rata share of total projected capital outlay and maintenance needs. It is quite important, however, that a thorough analysis of the future financial needs for maintenance, operation, traffic services, and highway patrols be made. It is these costs that are increasing at a rapid rate. The public demand for high levels of service plus the increased emphasis on beautification and traffic safety, are the major causes of such increases. Although fiscal planning does not include an evaluation of these items, nevertheless, the inherent escalation of such costs makes it necessary to consider them in long-range fiscal planning.

ROLE OF HIGHWAY FINANCE STUDIES

In the conduct of a highway needs study the total program cost for all highways, roads, and streets for a definite period of time is determined without regard to how the program(s) might be financed. This is proper because if financial considerations were to play a role in determining the need for roads, the end result would not be the presentation of the costs of an adequate system of highways, roads, and streets, but something considerably less because of the bias introduced by fiscal or budgetary restraints.

The financing program is a separate function and a separate decision. Not until late in the whole planning study process are needs and revenues compared. At that time, the results of the needs study dictate the future financial program. Essentially the conduct of the financial study consists of: (1) forecasts of highway user tax revenues based on current tax rates, (2) forecasts of other revenues applicable to highway, road, and street purposes, (3) the division of program costs between users and nonusers, (4) the allocation of the user's share among the various classes of vehicles, and (5) recommendations for closing the gap between revenues and needs. Also, alternative programs are suggested if needs are obviously too great to be met by reasonable proposals for tax increase.

FORECAST OF FUTURE REVENUES

The projection of future revenues, especially those derived from the family of highway user taxes, is closely tied to motor vehicle registrations and highway use. Historically, there has been definite trends in the density of motor vehicle ownership, travel per registered vehicle, miles per gallon of fuel consumed, and total vehicle-miles of travel. As travel and fuel consumption are synonymous, the historic and projected trends are almost parallel. Density of ownership is gradually increasing and the problem of vehicle saturation disturbs many researchers [25-1]. Annual travel per registered vehicle has been approximately static for some time although it was anticipated that multicar families would cause a decrease in the average annual mileage per vehicle. This idea was dispelled to a certain degree in a recent article which showed that the second family car was driven as much or more than the vehicle of a one-car family [25-2]. Also, as density of ownership approaches saturation, there is a possibility that travel per vehicle will increase.

Population projections and indicators of future economic growth (or decline) are valuable tools in forecasting. The projected number of persons in the driving-age group and projected personal income are other factors to be considered in the revenue forecasting procedure.

Other revenue sources which must be projected are those derived from property or special assessment taxes, local highway user imposts, and general fund contributions.

If the study of highway needs is comprehensive, the financial study must also be comprehensive. This specifically implies that all governmental units having highway responsibility must be included in the financial analysis. Constitution-

Highway Finance and Taxation 605

al and statutory limitations on the power to impose taxes and issue bonds must be researched. Registration-fee schedules must be analyzed and evaluated from a standpoint of equity, revenue producing ability, and administrative complexity. Allocation of state-collected highway-user tax revenues to the several governmental units–state, county, township, municipality–must be evaluated from the standpoint of needs and equity. The basis for distributing the local share of such revenues among the recipient units poses many problems insofar as an equitable solution is concerned. The distribution formulas now in effect have evolved over the years through legislative compromise and are, in many instances, far from being equitable.

TAX SUPPORT OF HIGHWAYS

It has been universal in the United States from the Colonial days that the highways, roads, and streets are a public function to be administered by the appropriate governmental unit. With this concept of a public function is carried the concept of taxes for the support of the highways. But the concept of taxes has changed over the years from "working out" your poll tax by doing free work on the local roads to property taxes raised through millage rates, and finally to road user taxes supplemented by some general fund support. The general fund support of highways, roads, and streets has come to be known as the nonuser share of highway funds. It is now accepted at all levels, including the federal, that the main support of highways is from road-user taxes in one form or other. Construction of highways and streets in land areas newly developed for residential, commercial, and industrial purposes, however, is now generally the sole responsibility of the land developer.

EARLY CONCEPTS OF HIGHWAY TAXES

Previous to the advent of the motor vehicle, almost all of the funds available for highway, road, and street purposes were derived from specific property tax levies or from the general funds of the administering governmental units. In 1904, expenditures for rural roads in the United States totaled $79.8 million, of which $53.8 million was derived from property or poll taxes, $19.8 million from "labor" taxes, $3.5 million from bond issues, and the remainder ($2.6 million) from state aid. The $19.8 million of "labor" taxes represents the estimated cash value of such taxes if the same had been paid in money. Six states–Connecticut, Massachusetts, New Jersey, New York, Pennsylvania, and Vermont–provided state aid [25-22].

These expenditures were for a rural road system of 2.2 million miles of which 154,000 miles were improved with gravel, stone, or other material. No statistics are generally available for expenditures on municipal streets, but it is rather obvious that the larger cities at that time had a well-defined and improved street system. On the other hand, rural roads served mainly as access to property from a nearby community. The statistics for the District of Columbia included in the report [25-22] showed that 69 percent of its roads were improved. Massachusetts was next with 46 percent. According to the report, Oklahoma had no improved public highways and New Mexico, 0.01 percent.

In summation, it can be seen that property and poll taxes, and labor, provided almost all of the funds for rural roads. Similar provisions probably held for municipal streets.

The motor vehicle brought to an end this century-old and traditional concept of road financing. In 1906 it was estimated that there were about 108,000 motor vehicles in the United States which paid fees totaling $62,500, little of which was applied to road work. New York first collected registration fees in 1901; a total of $954 being realized.

By 1915, slightly over 2.4 million motor vehicles were registered and registration and other fees totaled $18.2 million of which $16.1 million was made available for road work [25-23].

In 1919, Oregon enacted the first motor fuel tax law. This tax was such a successful revenue producer that by 1929 all states had such taxes ranging from 2 to 6 cents per gallon.

With this new source of revenue and the increased reliability and use of the motor vehicle, there was a definite break with the traditional level of service concept of roads. The predominant use of roads for land access gave way to intercommunity and interregional use requiring longer trips at higher speeds and, as might be expected, a higher type of design and increased highway construction cost. Also, the whole concept of financing changed from one of property tax support to the use of highway user tax revenues as the major source of highway funds.

THE HIGHWAY USER TAX CONCEPT

The original motor vehicle tax (registration fee) was considered more of a regulatory tax than a revenue producer. However, legislators and administrators soon became aware of the revenue producing possibilities of this tax as the number of vehicles increased rapidly. From a token fee in 1906 the average gross revenue per vehicle rose to $7.46 by 1915. As noted above, such revenues were used largely for highway purposes and substantially supplemented the traditional revenue sources.

Dedicated Road-User Taxes

The most significant change in the highway user tax picture occurred in 1919 with the adoption by the Oregon state legislature of the motor fuel tax. Revenues from the combined taxes increased rapidly and with the predominance of such revenues in the state's fiscal picture, there developed the principle of dedicated,

Highway Finance and Taxation

or "earmarked," funds. In some 26 states, revenues generated by highway user taxes are now dedicated to highway use. The adoption of this principle represented a rather radical departure from the traditional general fund concept in which all tax revenues were placed in a common fund and then appropriated for specific governmental functions.

The dedication of any specific tax revenues to specific functions has many proponents and opponents. There are several distinct advantages to dedication of road-user tax funds from the highway official's viewpoint and as many disadvantages from the position of budget and finance officers.

The highway user taxes can be considered as a standby, or readiness to serve fee (registration fees), and a tax that varies with the use of the highways (motor fuel tax). To these two considerations must be added the so-called "third-structure" taxes paid by heavier vehicles which can be related to additional costs incurred for providing adequate facilities for such vehicles.

With the passage of the 1956 federal aid highway act and the establishment of the highway trust fund, the federal government adopted the concept of dedicating highway user tax revenues to highway purposes.

Sources of Highway Funds

The major portion of the tax receipts allocated to highway, road, and street purposes is presently derived from state and federal highway user tax revenues. Table 25-1 shows the source of funds available for all highways, roads, and streets in 1964.

Receipts totaled $14,056,925,000 during the year, of which 69.10 percent was derived from highway user taxes. If toll revenues were added this would increase the user contribution to 73.51 percent. Of the total $9,712,681,000 highway user tax revenues 40.52 percent was derived from state imposts, 28.02 percent from federal excise taxes, and 0.56 percent from local highway user imposts. General fund appropriations and property taxes accounted for 18.68 percent of the total and miscellaneous receipts 3.99 percent. Proceeds from the sale of bonds accounted for 7.81 percent of the total.

Distribution of current revenues shows that the highway user contributed 74.95 percent of the total; 79.74 percent if toll revenues are included. The conclusion which can be drawn is that the highway user is paying a substantial portion of the highway bill in direct user taxes. It must also be noted that the highway user, in many instances, pays a personal property tax levied on his vehicle.

There is also a substantial amount of direct highway user tax revenues which are used for other than highway purposes. In 1965 approximately $655,000,000 of state highway user tax revenues were allocated to state or local general governmental purposes. This was offset to a certain extent by general fund appropriations totaling $103,000,000. Also, of the $655,000,000 almost

TABLE 25-1

TOTAL RECEIPTS FOR HIGHWAYS,
ROADS, AND STREETS–ALL GOVERNMENTAL UNITS-1964

Item	Amount Thousand Dollars	Percent Total Receipts	Percent Revenues Only
Highway user tax revenues*			
Federal	3,938,476	28.02	30.39
State	5,696,029	40.52	43.96
Local	78,176	0.56	0.60
Subtotal	9,712,681	69.10	74.95
Road, bridge, and ferry tolls	620,336	4.41	4.79
General fund appropriations	1,019,094	7.25	7.86
Property taxes	1,045,856	7.44	8.07
Miscellaneous	560,778	3.99	4.33
Subtotal	3,246,064	23.09	25.05
Bond proceeds†	1,098,180	7.81	–
Total	14,056,925	100.00	100.00

*Excludes costs of collection expenses and nonhighway purposes.
† Excludes short-term notes and refunding bond issues.
Source: Highway Statistics (1965), U.S. Department of Transportation, Table F-1 [25-27].

$210,000,000 was "in lieu" taxes on motor vehicles and not considered as highway user taxes in the states levying such taxes as they have replaced personal property taxes on motor vehicles.

The cost of construction and maintenance of the primary systems of highways can be closely associated to the public utility concept in that the users of such facilities pay the full cost of the facility. There are many hundreds of thousands of miles of highways, roads, and streets that have little traffic on them; but the costs associated with such facilities are subsidized rather heavily by highway user tax revenues. If the general theory of the public utility concept is accepted, i.e., the user should pay the full cost of operation, it is quickly evident that substantial mileages of low volume roads could not be economically justified solely by road use. On the other hand, the public utility concept is applicable to the main arterials although there is an element of nonuser benefits accruing even on these facilities.

THE NONUSER ROLE

The concept of highway user charges for support of highways raises many questions concerning the nonuser's role in the support of and benefits derived from an adequate highway transportation system. Historically, access to land has been the main function of roads. Although at present this access function is overshadowed by the emphasis on freeways and arterials, it still remains the prime function of millions of miles of roads. It is reasonable to assume that 75 percent of the 3.7 million miles of highways, roads, and streets are still fulfilling the access function.

There is no question that land gains a benefit from access provided by highways, even on limited access facilities. If the position is taken that nonusers should pay for benefits received, the major questions which to this day remain unresolved are: (1) payments for what highways, (2) how much should be paid, and (3) on what basis should the payment be calculated. These questions are not new nor have they been resolved to the mutual satisfaction of many researchers of the problem. It is sufficient at this time to say that there is definitely a nonuser benefit accruing from the availability of a highway transportation system, and therefore a cost responsibility is incurred. Exact measurements of the extensiveness of these benefits and costs are elusive. The several approaches to the problem will be discussed later in the next major section.

HIGHWAY COST ALLOCATION TO ROAD USERS AND NONUSERS

Determining an equitable distribution of highway costs is neither a new problem nor one that is peculiar to the area of highway finance. A number of bases are recognized by tax economists as suitable for governmental financing, including the cost incurred, benefit, ability to pay, and sacrifice doctrines of taxation. In the field of highway finance, the cost incurred doctrine is most generally applied, since it is more appropriate in those areas of public finance wherein significant and distinguishable costs can be determined for identifiable segments of the tax structure. The levying of taxes on specific classes of individuals and properties is justified by the special governmental expenditures required by these classes, specifically the highway user and the nonuser.

USER AND NONUSER COST RESPONSIBILITY

The benefits derived and therefore the costs incurred by the nonuser are difficult to measure in monetary terms and are always subject to debate. Because of this unresolved problem, the usual approach to the allocation of cost responsibility is to determine the user share of program costs and allocate residual costs to the nonuser. The means of determining the user share are not wholly above argument but can be justified more thoroughly than that of the nonuser. A tremendous amount of data has been assembled over the years concerning the intensiveness of highway use by the several classes of vehicles and such data fulfill a critical need in cost allocation analyses.

The level of service concept is highly important in arriving at a reasonable allocation of costs between users and nonusers. The local rural roads and residential streets in urban areas carry relatively light volumes of traffic and these roads and streets would be there today regardless of the motor vehicle. In fact, most of the urban places had well-defined street systems before the turn of the century. The costs of such facilities if charged to the highway user would result in an exhorbitant cost per vehicle-mile. On the other hand the freeways, expressways, and other arterial facilities are designed to carry heavy volumes of traffic at relatively high speed and with relatively long trip lengths. The nonuser, obviously, should be charged with only a small portion, if any, of the costs of such facilities because of their high costs per mile in relation to the abutting property. Therefore it is necessary to find an acceptable method of allocating highway program costs between these two extremes.

METHODS OF ALLOCATING HIGHWAY COST RESPONSIBILITY

There have been, over the years, many theories developed or sponsored for allocating highway costs between users and nonusers. Some of these theories, although sound, have not been acceptable because the basic assumptions strained the credulity of highway engineers and legislators. Six methods will be discussed here: (1) the historical, (2) the public utility concept, (3) the standard cost, (4) the predominant use, (5) the relative use, and (6) the earnings-credit.

It is to be kept in mind that the objective of the financial analysis is to determine the relative highway cost (tax revenue) responsibility of road users and nonusers for the period of years used in the needs study and for the specific dollar cost of these needs. All past construction investments, past tax revenue, and traffic are irrelevant. This process is not the same at all as estimating the annual cost of owning and operating a highway system; it is estimating the fair share of future disbursements for highways to be borne by road users and by nonusers.

The Historical Method

The historical method is rarely used; perhaps the classical application of it was contained in a 1956 report [25-29]. The writers of the report traced the historic nonuser contribution for roads and streets and then extrapolated the average annual increase of this nonuser contribution to arrive at future nonuser contributions. Residual costs were considered to be the responsibility of the road users. The report acknowledged the shortcomings of the method but defended it as both practical and feasible, claiming that it would require little or no change in the existing local finance pattern. Needless to say, in recent years this approach has not been considered acceptable.

Public Utility Concept

A group of railroad economists were the principal advocates of the public utility concept [25-4]. It was their contention that highway costs should be paid

Highway Finance and Taxation

entirely by the highway user and such costs should be distributed among users on an equitable basis. They also contended that highway facilities are justified solely by transportation requirements and that there was no more reason for charging to the land a part of the highway cost than there was for making a similar charge in the case of other public utilities. This concept still has its proponents and is reasonably applicable to high-volume facilities. There is no doubt that highway tax earnings, insofar as the user share is concerned, should have some relation to highway costs. On the other hand, if highway users were expected to support all roads and streets by user tax payments a substantial portion of the highway plant would have to be abandoned.

The Standard-Cost Method

The standard-cost method is based on the cost of a standard or representative highway system(s). This standard cost(s) is expressed as the cost per ton-mile of vehicular use and is applied to the total ton-miles of travel for all roads and streets to determine the user share. The residual amount is then assumed to be the nonuser share.

The chief objections to this method are: (1) the arbitrary decision in selecting the system(s), which is considered standard or representative, and (2) the use of a rural system(s) of highways as the standard system.

Assuming that the needs study was based on a functionally classified system of highways, roads, and streets, the rural systems and the cost per ton-mile of travel might be as follows:

System	Cost per Ton-Mile of Travel, cents
A. Freeways	4.03
B. Major highways I	3.50
C. Major highways II	2.40

It is readily apparent that if system A is considered to be the standard system, the highway user will be allocated a substantial share of program costs. The same situation would hold if systems A and B were combined. If it is assumed that system B or Systems B and C were judged representative, then the nonuser would be assigned a substantial share of the costs of the freeway system.

The outstanding example of the application of the standard cost method was in a study for the Ohio program commission in 1951 (25-19).

By way of illustrating this method, assume the following:

1. Average annual costs of standard system(s) (20-year program period) .... $ 100,000,000
2. Average annual ton-miles of travel on standard system 50,000,000,000
3. Cost per ton-mile of travel .... $ 0.002
4. Average annual costs - all roads and streets .... $ 500,000,000
5. Average annual ton-miles of travel, all roads and streets   200,000,000,000

6. User share = 200,000,000,000 ($ 0.002) .... $ 400,000,000
7. Nonuser share .... $ 100,000,000

The user share of $ 400,000,000 includes an estimate of future federal aid highway funds. In earlier cost allocation studies, federal aid was deducted from program costs previous to the allocation between users and nonusers. This procedure was followed because federal aid monies were appropriated from the general fund of the United States. With the establishment of the highway trust fund in 1956, certain highway user tax revenues were dedicated to the fund and the general fund approach was no longer valid. By including the federal share of program costs, total costs are allocated. Once the user share is determined, the portion of program costs to be met by highway users of the particular state is determined by deducting federal aid.

The Predominant-Use Method

In an attempt to allocate the costs of highway improvements between users and nonusers, and lacking the necessary data to investigate the problem in detail, some researchers have relied upon the predominant use method of allocating program costs. In essence, costs are assigned in accordance with the predominant use or benefit for each level of facility.

Therefore the costs of freeways, expressways, and other high-type facilities are considered to be the sole responsibility of the highway user. The costs of the lowest class of facilities, rural access roads and residential streets, are assigned to the nonuser. The assignment of costs for the intermediate systems of roads and streets is made on the basis of judgments as to the level of service provided by these systems.

Obviously one of the major attributes of this method is its simplicity. Just as obvious is the disadvantage of the arbitrary judgments which must be made in allocating costs of intermediate systems. Some investigators have simply divided the costs of these systems evenly between users and nonusers [25-12]. Others have made a series of judgments, i.e., assigning a higher share of the costs of county and city collectors to the user and a high proportion of the costs of access roads and streets to the nonuser [25-30].

This method probably assigns to the nonuser the maximum share of program costs which could be calculated by any reasonable scheme. As noted earlier, the chief objection to the method is the judgment or series of judgments involved in determining the costs allocated to the intermediate systems. An important, and probably the most important criticism, is that it assumes that no user benefits are derived from the lowest system of roads and streets, and that nonusers derive zero benefits from the primary systems. This is not a tenable assumption.

The Relative-Use Method

The relative-use method assigns highway cost responsibility for each road and

Highway Finance and Taxation

street system in proportion to the extent each system renders three types of service: (1) land access traffic, (2) community, or neighborhood, traffic, and (3) through traffic.

One of the principal difficulties of conducting a relative use study is that the type of information needed is often lacking. The basic data source is the motor vehicle use studies [25-3]. The interview forms for such studies provide individual trip information such as origin and destination, length of trip, trip purpose, and routes traveled. This is substantially more information than can be obtained from the usual origin-destination interviews.

Trip-length data are analyzed and assigned to systems and then expanded to the total travel on each system divided into the three traffic components–access, community, and through. The resulting percentages show the allocation to the user and nonuser by the three traffic components. The through-traffic component is considered as the user share while the neighborhood and access components are assigned to the nonuser.

Perhaps the most uncertain aspect of the relative use method is the division of the traffic into the three groups–land access, community, and through. Any section of highway may carry all three types of traffic, and the classification of that traffic into each of the three groups requires setting the criteria on a judgment basis. At the extreme ends of land access and through traffic there is agreement in classification, but when land access traffic ceases and becomes community traffic, and when community traffic ceases and becomes through traffic must be based on judgment. This gray area could be substantial: shifting its upper limit up or down could materially alter the split between user and nonuser responsibilities. Trip length and trip purpose offer the best base for separating the traffic into the three groups, especially for urban traffic.

A substantial amount of writing and research has been done on the relative use method. It probably represents the best approach to the problem of user-nonuser allocation. Its principal drawback, however, is the lack of adequate data, i.e., a motor vehicle use study.

The most recent and comprehensive study of the relative-use method is contained in: Final Report of the Highway Cost Allocation Study, published in 1961 as House Document No. 54, 87th Congress, 1st Session [25-25].

The Earnings-Credit Method

The earnings-credit method in some respects is similar to both the predominant use and relative use methods. The two main hypotheses of the analysis closely resemble the predominant use approach and, in the compromise solution, give recognition to the fact that all highway, road, and street systems have access, community, and through traffic components.

In the first hypothesis, the highway user is assessed the costs of constructing, maintaining, and operating arterial highways which provide service to major through traffic movements. As related to the travel on the routes and systems serving this through movement, these costs are assumed to represent the value of highway-user benefits received on all classes of highways. Thus motor vehicle users are assessed the same charge per vehicle-mile of travel on all highway systems.

The second hypothesis assumes that general public benefits can be measured as the cost per mile for those facilities which primarily provide land access services. The general public is assessed the same charge per mile for all highway systems.

These two approaches, known as the top-drawer and bottom-drawer solutions, are not sufficient in themselves to provide a firm basis for cost allocations. The first solution discriminates against the motor vehicle user, since no share of the primary roads is assigned to the general public, although it is recognized that these facilities provide benefits to the nonuser. The second solution discriminates against the nonuser, for all the costs of local access facilities are assigned to the nonuser without regard for highway user benefits derived from these highways.

The compromise solution, achieved by averaging the results of these two approaches, reconciles some of the differences between the two hypotheses and provides a more satisfactory answer to the question of cost allocation. In effect, the compromise solution recognizes that each class of highway benefits both the motor vehicle user and nonuser.

The earnings-credit method requires separation of highways into groups of similar characteristics. Some of the criteria used for separating highways into groups include administrative systems, surface types, traffic volume groups, and service characteristics. In some studies it is possible to separate facilities into groups with similar service characteristics, based on the functional classification system. The primary group of roads and streets includes those facilities which provide the greatest benefits to motor vehicle users. The local group includes roads and streets which primarily provide access services to abutting properties. The intermediate group is composed of facilities which have a closer balance between highway user and general public benefits than do the primary and local categories.

Illustrative Earnings-Credit Solution. In order to illustrate the steps in an earnings-credit analysis, Tables 25-2 and 25-3 have been extracted from one of the more recent comprehensive highway needs and fiscal studies [25-31].

The primary group of rural roads included interstate, supplementary freeways, and major and area service highways. The intermediate group was composed of collector highways. Land access roads were placed in the third group. The municipal extensions of rural highways, classified as other than land access, plus all primary thoroughfares comprised the primary group of municipal streets. The intermediate group included streets classified as secondary thoroughfares, business access, and industrial access. The local group was composed of residential access streets.

Highway Finance and Taxation

TABLE 25-2
DATA FOR COMPUTING HIGHWAY COST RESPONSIBILITY
(Earnings-Credit Method, 1966-1985)

Highway System	Average Annual Cost,* Thousand Dollars	Miles	Annual Cost per Mile, Dollars	Average Annual Vehicle-Miles, Millions	Cost per Vehicle-Mile, Mills
Rural Roads					
Primary	380,088	10,063	37,770.84	19,892.6	19.1070
Intermediate ..	143,316	18,833	7,609.83	6,150.3	23.3023
Local	103,678	75,001	1,382.35	1,837.0	56.4387
Subtotal	627,082	103,897	6,035.61	27,879.9	22.4923
Municipal Streets					
Primary	439,624	5,121	85,847.30	24,313.8	18.0813
Intermediate ..	136,717	3,205	42,657.41	5,997.2	22.7968
Local	79,848	17,550	4,549.74	3,582.1	22.2908
Subtotal	656,189	25,876	25,358.98	33,893.1	19.3605
Total, all systems	1,283,271	129,773	9,888.58	61,773.0	20.7740

* Includes total cost for interstate system. Excludes debt service costs.

For the earnings-credit calculations, annual costs (Table 25-2) of the 20-year program were used for all highway systems. These costs totaled $627,082,000 for rural roads and $656,189,000 for municipal streets, including the total cost of construction on the interstate system. For rural roads these costs amounted to an annual requirement of $6,036 per mile or 22.5 mills per vehicle-mile. The cost for municipal streets amounted to $25,359 per year per mile of road and 19.4 mills per vehicle-mile.

The top-drawer solution (Table 25-3) for rural roads was based on a cost of 19.1 mills per vehicle-mile, representing the cost of primary highways related to the travel on these facilities. This resulted in a total user share of $532.701 million for rural roads, with the residual costs amounting to $94.381 million assigned to the general public. For municipal streets, an assignment of 18.1 mills per vehicle-mile was assessed against all motor vehicle travel and resulted in a highway user share of $612.830 million and a general public share of $43.359 million.

In the bottom-drawer solution (Table 25-3), the general public assessment for rural roads was based on a cost of $1,382 per mile and amounted to $143.622 million. The residual cost of $483.460 million was assigned to the highway user and amounted to 17.34 mills per vehicle-mile. For municipal streets, the cost per mile assessed to the general public was $4,550 resulting in a total assessment of $117.729 million. The highway-user share of $538.460 million amounted to 15.887 mills per vehicle-mile.

The compromise of the top-drawer and bottom-drawer solutions (Table 25-3) for rural roads resulted in a cost of 18.2239 mills per vehicle-mile for rural travel by motor vehicles. This resulted in a cost of $508.080 million assigned to the

TABLE 25-3
Earnings-Credit Analysis of Highway Cost Responsibility

Highway Systems	Motor Vehicle Share		General Public Share, Thousands
	Amount, Thousands	per Vehicle-Mile, Mills	
Top-Drawer Solution			
Rural roads			
Primary	380,088	19.1070	None
Intermediate	117,514	19.1070	25,802
Local	35,099	19.1070	68,579
Subtotal	532,701	19.1070	94,381
Municipal streets			
Primary	439,624	18.0813	None
Intermediate	108,437	18.0813	28,280
Local	64,769	18.0813	15,079
Subtotal	612,830	18.0813	43,359
Total, all systems	1,145,531	18.5442	137,740
Bottom-Drawer Solution			
Rural roads			
Primary	366,178	18.4077	13,910
Intermediate	117,282	19.0693	26,034
Local	None	–	103,678
Subtotal	483,460	17.3408	143,622
Municipal streets			
Primary	416,325	17.1230	23,299
Intermediate	122,135	20.3653	14,582
Local	None	–	79,848
Subtotal	538,460	15.8870	117,729
Total, all systems	1,021,920	16.5431	261,351
Compromise Solution			
Rural roads			
Primary	362,521	18.2239	17,567
Intermediate	112,082	18.2239	31,234
Local	33,477	18.2239	70,201
Subtotal	508,080	18.2239	119,002
Municipal streets			
Primary	412,950	16.9842	26,674
Intermediate	101,858	16.9842	34,859
Local	60,839	16.9842	19,009
Subtotal	575,647	16.9842	80,542
Total, all systems	1,083,727	17.5437	199,544

Highway Finance and Taxation

TABLE 25-4

SUMMARY OF HIGHWAY USER AND GENERAL PUBLIC COST RESPONSIBILITIES

Administrative Systems	Average Annual Cost, Thousand Dollars	Highway-User Share		General Public Share	
		Amount Thousand Dollars	Percent	Amount Thousand Dollars	Percent
Existing					
State	737,360	716,593	97.2	20,767	2.8
County	168,973	119,361	70.6	49,612	29.4
Township	107,495	49,568	46.1	57,927	53.9
Municipal	269,443	198,205	73.6	71,238	26.4
Recommended					
State	610,857	613,918	100.5	(3,061)	(0.5)
County	199,745	167,086	83.6	32,659	16.4
Township	103,679	33,477	32.2	70,202	67.8
Municipal	368,990	269,246	73.0	99,744	27.0
Total	1,283,271	1,083,727	84.5	199,544	15.5

highway user and of $119.002 million assigned to the general public. For municipal streets, the cost assignment was 16.9842 mills per vehicle-mile or $575.647 million. The general public share totaled $80.542 million. The total assessment to the highway user for all highway systems amounted to $1,083.727 million, or 84.5 percent of the state wide annual cost. The general public share amounted to $199.544 million, or 15.5 percent.

The results of the earnings-credit solution applied to the administrative highway systems of the state are shown in Table 25-4. For the existing state highway system, the user responsibility totals $716.593 million or 97.2 percent of the total annual cost of $737.360 million, which includes all construction costs for the interstate system. For county roads, the total annual cost amounts to $168.973 million, of which $119.361 million, or 70.6 percent, is assigned to the highway user. User responsibility for township roads amounts to $49.568 million, representing 46.1 percent of the total cost of $107.495 million. Of the $269.443 million annual cost for municipal streets, $198.205 million, or 73.6 percent, is assigned to the highway user. The residual cost on each of these systems is assigned to the general public and amounts to 2.8 percent of the state highway system cost, 29.4 percent of the cost of county roads, 53.9 percent of township roads, and 26.4 percent for municipal streets.

On the basis of the recommended division of administrative responsibility for highways (Table 25-4), $613.918 million is assigned to the highway user for state highways compared to a total cost of $610.857 million. This means that user earnings, based on travel on the system, will exceed the cost of the system slightly (0.5 percent). The highway user share of the recommended county road system amounts to $167.086 million, or 83.6 percent of the total annual cost of $199.745 million. The user share of the township road system totals $33.477

million, representing 32.2 percent of the total costs of $103.679 million. Annual costs of the municipal street system total $368.990 million, of which $269.246 million or 73.0 percent is assigned to the highway user. The general public share of the recommended highway systems amounts to a credit of 0.5 percent for state highways, and a responsibility of 16.4 percent for county roads, 67.8 percent for township roads, and 27.0 percent for municipal streets.

Admittedly, the earnings-credit method is not a perfect solution, but it does have merit over the standard cost and relative use methods. Less judgment is required and it approaches the solution from both ends. In its applications, the results have been reasonable. It gives credit to the land access type of road for its generation of road use revenue and to the higher-type facilities for the general public benefits derived.

Other Methods

There have been several other methods developed in attempts to properly allocate highway costs between users and nonusers. No attempt is made to analyze or illustrate these methods, but the chief hypothesis involved and references will be noted.

Added-Expenditure Method. The added-expenditure approach holds that motor vehicles should be responsible for all road and street expenditures in excess of the level of expenditures prevailing during the period immediately prior to the advent of motor vehicles in significant numbers [25-24, 25-26, 25-28].

Differential-Benefits Method. The method of differential benefits allocates costs to each of the beneficiary groups on the basis of the calculated benefits resulting from highway improvements. The best known application of this method was conducted in Oregon [25-14] and an excellent analysis of it is contained in a paper by G. P. St. Clair [25-20].

Basic-Access Method. The basic-access method proposes that the nonuser share should be the cost of an access road or street constructed to standards which would allow reasonably good access. All other costs are charged to the road user [25-7].

Restricted-Capacity Method. In the restricted-capacity method it is suggested that, inasmuch as most highways have unlimited access, reduced capacity results from unlimited access. This reduced capacity results in a reduction of the potential highway user tax revenue. The loss of revenue attributed to lower capacity is considered a nonuser charge. All residual costs are to be borne by the highway user [25-16].

ALLOCATION OF HIGHWAY USER SHARE OF HIGHWAY COST TO CLASSES OF VEHICLES

Once the nonuser share of program costs has been determined, the residual

Highway Finance and Taxation

or user share must be allocated among the several classes and types of vehicles. Although there are various units of measurement which might be applied in allocating the highway costs, none is really satisfactory.

The vehicle-mile, as a measurement of responsibility, allocates the same responsibility per mile traveled for a Volkswagen as for a 6-axle tractor-semitrailer-full-trailer combination, and therefore allocates substantial costs to the lighter vehicles. The ton-mile unit of measurement assigns probably the maximum allocation to the heavier vehicles. As an example, ten 4,000-pound passenger cars traveling 1 mile would generate 20 ton-miles of travel, so would one 40,000-pound truck traveling the same 1-mile distance. Yet, if it was assumed that the passenger cars operated at a fuel consumption rate of 15 miles per gallon and the truck at 6, the passenger cars would consume four times the amount of fuel. This unit of measurement would seem to penalize efficiency in transportation.

Although the problem of tax equity among highway users has been the subject of countless and endless research and writing, there has not as yet been produced a method which is acceptable to all interested parties. Probably the incremental solution, discussed later in this chapter, comes closest to an acceptable solution.

FACTORS INVOLVED IN MOTOR VEHICLE TAXING SCHEMES

Setting a tax base and a tax rate for motor vehicles is a complex undertaking for three main reasons: (1) motor vehicles as property are unlike; there is not a standard product, (2) the highway use, vehicle to vehicle, varies over a wide range, and (3) there is no one criterion or factor by which tax equity can be measured.

Motor vehicles vary in width, length, height, gross weight, axle weight, number of axles, road performance ability, and road speed. These factors affect highway construction cost through elements of highway design, and also certain elements of highway maintenance and traffic operations. The vehicles also vary in the type of engine fuel used, and the rate of fuel consumption with respect to highway use. The variation in fuel consumption rates is highly important because fuel is a tax base which accounts for a high percentage of all road-user taxes.

When considering highway construction cost and highway maintenance cost in relation to road-user tax bases and rates, the considerations are further compounded by the fact that there is no best scheme to relate the cost of right-of-way, earthwork, drainage, structures, shoulders, pavement, and traffic services to particular classes of vehicles. Some of these highway cost factors are joint costs but not directly proportional to vehicular factors; other factors may not vary with the type of vehicle or amount of use by vehicles.

Annual mileage is a factor because fuel tax payments are dependent upon miles driven, but the license or registration fee is not related to mileage. When the registration fee is large, as frequently it is for heavy trucks, for the express purpose of producing revenue rather than for regulatory purposes, a low-mileage vehicle as compared to a high-mileage vehicle may suffer great inequity. When the registration fee is a nominal sum–$5 to $15 flat rate–per vehicle, annual mileage is a less important factor in determining equitable user taxes.

THE TAX BASES

There are three commonly used levels in the highway user tax structures: (1) registration fee, (2) motor vehicle fuel tax, and (3) third structure, or special imposts on heavy trucks.

The registration fee for passenger cars is a flat fee or a computed tax on the basis of weight, horsepower, value (cost), age, or a combination thereof. The registration fee for commercial vehicles may be based on empty weight, declared gross road weight, or other factors, and is often graduated based upon increments of weight.

The motor fuel tax is a constant rate per gallon of fuel for all vehicles. There is some variation in the tax rate per gallon, however, with type of fuel. Diesel fuel may be taxed at a higher rate per gallon than gasoline.

The third structure taxes, so-called, are special imposts on the heavy gross weight vehicles, usually of high annual mileage. This tax may also be designed for the purpose of gaining a tax income from out-of-state vehicles. Third structure taxes are based on ton-miles, axle-miles, or other weight-distance factors.

METHODS OF ALLOCATING ROAD USER TAX SHARES TO VEHICLE CLASSES

Allocating road user tax responsibility among the several classes of vehicles is a process of finding a desirable base by which to measure responsibility, then pricing this base. Because of the complex nature of highway cost, vehicles, and vehicle use, no completely satisfactory scheme has been devised. The general factors considered in past years include value of highway service, benefits received, and highway cost incurred.

Such measurements as vehicle-mile, axle-mile, gross ton-mile, incremental cost (incremental with respect to vehicle size and weight), and benefits received have been used. Similar to the allocation of costs between users and nonusers, a number of theories and methods have been devised for determining an equitable allocation. The methods or theories discussed are: (1) incremental method, (2) gross ton-mile method, (3) cost function method, (4) standard cost method, (5) differential benefit method, (6) operating cost method, and (7) space-time method.

Incremental Method

The cost-incurred doctrine of highway taxation requires determining the special expenditures by governmental programs for specific vehicles. In the

Highway Finance and Taxation

1930's it became obvious to many highway administrators that highway costs were partially related to specific types of vehicles and the service such vehicles demanded. Early investigators emphasized the fact that roads and streets were necessarily constructed to accommodate the demands of heavier, wider, and longer vehicles [25-26]. A 1955 report explained the basic idea of the incremental method as follows:

> The basic cost of constructing, improving, and maintaining a given highway shall be determined from a highway design for private passenger vehicles and other vehicles commensurate with their width. All vehicles using such highways should pay their proportionate share of that total as a tax base. The total additional cost of construction, improvement, and maintenance to make a road suitable for a type of vehicle requiring such additional costs should be shared by each vehicle of that type and each vehicle of greater size. Thus, each vehicle should share in the base cost plus all increments of cost up to and including the cost required by it [25-37].

A vehicle falling in the second increment of highway cost should therefore pay only for the miles it operates at a gross weight which requires the second increment of pavement thickness and second increment of any additional highway cost on account of vehicle dimensions. The incremental method has the further attribute of attempting to determine tax responsibility for each type and weight of vehicle so that it bears the same, or as close as possible, relationship to the cost occasioned by that vehicle [25-12].

Early researchers, in attempting to conduct an incremental analysis, found themselves faced with the need to prepare a multitude of long, laborious, and complicated tabulations in order to make the analysis. There were also many gaps in the data bank which had to be filled with judgments, many of which could not be fully justified. For these reasons the incremental approach was not widely applied.

As noted earlier, the incremental approach to the problem of user responsibility for highway costs has been advanced for a number of years, but only recently has reliable application of the method been achieved. The criteria, data, and computer techniques required for application of this method were aided materially by the AASHO Road Test, which was conducted by the Highway Research Board of the National Academy of Sciences for the American Association of State Highway Officials. The site for the road test was Ottawa, Illinois.

The incremental method requires determining the cost responsibility within four major highway cost groups: pavement base and surface, grading and drainage, structure, and a category which contains all other items. Each of these categories must be allocated individually, since different vehicular elements affect each one.

Early researchers were forced to make definitions of both the basic vehicle and basic road. These two definitions varied substantially among researchers. Some felt that a 3,000-pound vehicle was about proper for the basic vehicle, others favored 6,000 pounds. Since publication of the AASHO road test results [25-36] it has been found unnecessary to define the basic road. It is assumed that a certain design criterion will be applied to the construction of a road in order to accommodate the volume of mixed traffic assigned to it. Therefore it is possible to perform the incremental solution by starting with the top increment of highway cost and working down through the various weight categories until only the cost assignment to the basic vehicle remains. This incremental variation of construction expenditures is explained as follows:

> Highways are designed to accommodate the volume and composition of traffic to which it is expected they will be subjected, with due consideration for environmental conditions. The process of determining incremental cost factors for each element of highway cost on each group of roads is simply one of starting with the design and cost of the highways that will actually be built for mixed traffic and, holding all factors except vehicle size and weight constant, ascertaining by means of design theory and engineering judgment the series of designs and costs that would accompany successive reductions in the weight of size of vehicles. In order to hold the total volume of traffic constant, the heaviest group of vehicles is considered to be replaced by a like number of the next to the heaviest vehicles. This process of transferring vehicles to the next lower weight group is continued until finally the total number of vehicles in the traffic stream is considered to be in the lightest-weight group. Since axle load and number of axles are the determinants of pavement design, the weight groups used in computing pavement increments are axle weight and the number of axles rather than the number of vehicles is held constant. Considerations other than the numbers and weights of the vehicles (or axles) in each group do not enter into the determination of the increments required and there is no reason to look upon the first increment as a road that might be built. The first increment is simply the portion of highway cost that remains when the added cost of the higher design required by the heavier vehicles because of their physical or operating characteristics, but not their numbers, is removed [25-25-7, p. 87].

In the conduct of the incremental analysis, it has been found more practical and reasonable to base the solution on a functionally classified highway network. The design and tolerable standards applied to functionally classified highway systems reflect the character of service anticipated, whether local, areawide, or regional. Therefore the highway costs associated with functional sytems are much more applicable than those found for highway systems classified by administrative or financial responsibilities.

Although there is a multiplicity of detail that must be carefully handled in the conduct of an incremental analysis, the fact that it is an engineering approach and that the judgments which must be made are based on evaluations and experiences of highway engineers has much to commend it. The judgment and experience of a highway engineer may be much more reliable than arbitrary concepts adopted for the sake of simplicity.

Highway Finance and Taxation

It must not be assumed, however, that the findings of the AASHO road test provided all the answers to the question of pavement performances. At the present time, many satellite tests are being conducted in several of the states. When data from these tests are analyzed, pavement performances under differing climatic and soil conditions will be added to the store of knowledge in this field.

Gross Ton-Mile Method

The gross ton-mile method of allocating user costs among the several classes of users is based on the theory that each weight class of vehicle should be assessed tax responsibility on the basis of total ton-miles of travel for each weight class. Total travel is then computed as an average per vehicle in each weight class. This relative use is then assumed to be proportional to benefits received, and ton-miles are assumed to measure relative use. Stated simply, ton-miles are assumed to be a measure of benefits received.

The ton-mile method does have the virtue of simplicity. Once vehicles are arrayed in weight classes, the average vehicle weight in each class is multiplied by the miles traveled, and tax responsibility distributed in proportion to this product.

Two approaches to the ton-mile solution are available. One is to compute the gross ton-miles for each weight group of vehicles, and then to allocate the entire user tax burden in proportion to the gross ton-miles of each weight group without considering the individual highway systems. The other is to calculate the gross ton-miles traveled by the vehicles in each weight group on each of the several road and street systems. Presumably, the greater the subdivision of systems the greater the accuracy of applying the method.

Advocates of the ton-mile method believe that a measure of the relative amount of service from highways is superior to the incremental approach. The Colorado Highway Planning Commission states the case for the ton-mile solution as follows:

> The theory of gross ton-mile taxation assumes that the movement of one ton-mile over a public highway constitutes a basic unit of transportation service and highway use. Given this assumption, it follows that the benefits received by each class of highway users are proportional to the use made of the highways, measured in ton-miles.
>
> If the sum total of highway benefits is considered to be measured by ton-miles of highway use, the principle of equal payments for equal benefits received requires that each highway user pay approximately the same user tax for each ton-mile of the use made of the highway system. In effect, ton-miles of highway use, or highway benefits are for sale to highway users. A vehicle traveling over a highway is utilizing ton-miles of service and receiving ton-mile benefits. Each ton-mile should logically command a price for the user in the same way that each article for sale in a department store has a price. But since it is assumed that the

cost to the government of producing facilities for one ton-mile of service to a highway user is approximately the same as the cost to any other user on the same section of road, then each ton-mile purchase should bear the same purchase price. When the user tax structure is altered to accomplish this end, all highway users... will be paying equally for equal use of the highways, and correspondingly, for equal benefits received. The relative share of highway user tax responsibility for each user would be his proportion of ton-miles of travel relative to the ton-miles of travel over the highway system [25-34].

As noted earlier, two of the main factors in favor of conducting a ton-mile solution are simplicity and ease of calculation. Another factor is that the method tends to allocate a large share of cost responsibility to heavier vehicle classes. Many authorities believe that the theory sets the upper limit to the level of taxes which should be assessed against heavier vehicles [25-37].

There are some rather fundamental weaknesses in the ton-mile theory. Probably the chief one is that ton-miles reflect relative use and, therefore, reflect relative benefit. As demonstrated earlier in this chapter, ten 4,000-pound passenger cars traveling 1 mile generate identical ton-miles of travel as one 40,000-pound truck traveling the same 1-mile distance. This indicates clearly that equal ton-miles do not necessarily result in equal benefits.

Investigators in North Dakota pointed out another weakness of the ton-mile approach:

> The other weakness of the ton-mile approach is that it assumes the same tax charge should be made for haulage over any kind of highway whether surfaced or unsurfaced. This results in vehicles which never use certain types of highways having, nevertheless, to pay for these in proportion to ton-miles traveled on the entire system [25-12, p. 20].
>
> Along the same lines, the product of weight times distance would be a valid unit of measurement of value received if all vehicles traveling on the highways were transporting goods for specific remuneration. A passenger car and a heavy commercial truck, however, travel the highways for purposes which are entirely different. It has been pointed out that a light panel delivery truck loaded with bread could easily receive the same money value as a heavy truck loaded with dirt for a trip of the same length [25-35, p. 37].

It does not appear that the ton-mile is a satisfactory measure of benefits. However, the use of the ton-mile is appropriate in some instances, but to allocate the total user share among classes of vehicles based on ton-miles of travel rests on an invalid assumption. Many highway cost factors are not related to gross weights. Pavement thickness varies with axle loads, right-of-way costs are in no way associated with gross weights, and traffic control devices and signs do not relate to vehicle weights.

Cost-Function Method

The cost-function method was developed by researchers in the motor carrier industry. The method is based on the factor of costs occasioned. At the time the method was developed, it was felt that the incremental approach was best, but because of the required elaborate calculations for the latter, the cost-function method was developed.

The cost-function method classifies all highway program costs into three categories: (1) costs that are size and weight related; (2) costs associated with highway use; and (3) costs that are neither size and weight nor use related.

Costs that are associated with traffic volumes are assigned on the basis of vehicle-miles of travel. Costs that are independent of vehicle size and weight and of traffic volume are assigned on a per vehicle basis. Costs related to vehicle size and weight are distributed on the basis of gross ton-miles of travel.

Probably the major difficulty in conducting a cost-function solution is the apportionment of costs to the proper category. The apportionment of costs to the three categories has a definite effect on the results of a cost-function solution. The solution is influenced greatly by the fineness of the cost breakdown and the decisions as to the category into which each item of cost is placed.

In the Highway Cost Allocation Study [25-25] it was determined that because the study was concerned only with highway trust funds, no costs defrayed from such funds were directly affected by the number of vehicles in existence and no expenditures were placed in that category. In this cost-function solution the cost items were placed in the following categories:

Travel Function Costs	Weight Function Costs
Right-of-way	Grading and drainage (intermediate and high type roads)
Grading and drainage (low type roads only)	Surface and base (intermediate and high type roads)
Utility adjustments	Shoulders (high type roads)
Surface and base (low type roads only)	Structures
Shoulders (low and intermediate type roads)	
Roadside development	
Traffic and pedestrian services	
Bureau of Public Roads administrative expenses	
Highway planning apportionments	

The results of this study [25-25] allocated over 77 percent of the highway costs to the weight function basis.

Other investigators have employed the cost-function analysis as an alternative to the incremental analysis [25-12, 25-38, 25-41].

Once the categorization of costs for each road system has been completed, the costs are allocated to each class of vehicles by either vehicle-miles traveled or ton-miles traveled depending on whether the costs are travel or weight related. The costs for each class are summarized and costs per vehicle and vehicle-mile of travel computed.

The cost-function method, although probably more acceptable than the ton-mile method, has a basic theoretical weakness. This weakness is the fact that the lightest-weight vehicles will be paying a share of the total costs of the heaviest pavement and highest type of bridge design. The heavy vehicles that occasion such costs are able to shift part of the highway cost burden to the lighter-weight vehicles. It is argued that the use of ton-miles to distribute the weight-function costs tends to compensate for the fact that the entire cost which they occasion is not assigned to them by the cost function method. It is doubtful that the use of ton-miles actually makes such a compensation.

As noted previously, the chief difficulty in the conduct of a cost-function solution is the categorization of costs. One investigator had this to say about the problem:

> The predicament could be partially resolved by classifying expenditures in more detail, and this is what has been done in the cost function of past expenditures made by representatives of the trucking industry. Any attempt to make an extremely detailed classification of the anticipated expenditures of a future program, however, takes on elements of unreality. For that matter, there are sound theoretical objections to the use of an extremely fine breakdown of expenditures. If each item of expenditure is completely stripped of its nonweigh components, the amount left approaches the costs that are attributable to heavy vehicles alone and for which light vehicles have no responsibility. To the extent that the costs allocated on the basis of ton-miles are restricted to costs occasioned by heavy vehicles, the cost function method assigns to light vehicles costs for which they are in no way responsible. The alternative would be to distribute weight-function costs among heavy vehicles only [25-25-5, p. 190].

In many cases, the tendency has been to overstate items in the weight-use cost category. For example, motor carrier industries have often done this. This was done in a study by the Virginia Highway-Users Association:

> It should be emphasized that the placement of an item in the weight-use category does not mean that the total cost of the item is chargeable to larger and heavier vehicles, but simply that motor vehicle weights have *some* effect on the cost of the item. It also should be pointed out that in some instances there was doubt as to whether a particular item should be placed in the weight-use category or one of the other categories. Such doubts were resolved by placing the items in the weight-use category, and to that extent the latter category has been overstated. The weight-use category has also been overstated to the extent that lack of more detailed breakdown of some costs made it necessary to place the entire item in the weight-use category [25-39, p. 47].

In the cost-function method, it is usually found that trucks are assigned somewhat lesser costs than under the gross ton-mile solution and that the incremental solution assigns higher cost to passenger cars. In other words, the cost-function assignments are somewhere between the other two solutions.

Highway Finance and Taxation   627

However, depending upon the distribution of costs, the cost-function method might be varied so as to produce results similar to the ton-mile solution.

Standard-Cost Method

The standard-cost method, discussed earlier as a method of allocating highway costs between users and nonusers, has also been used to allocate highway costs among the several classes of highway users. One such application was by Simpson in Ohio [25-19]. In this study the ton-mileage of each class and weight group was estimated for the midyear of the improvement program and the standard rate was applied to this estimated ton-miles of travel. The Ohio investigator had this to say about the method:

> ... that in any use of ton-miles for computing costs we are substantially favoring the commercial vehicles and particularly the heavier-weight types of commercial vehicles. The reason is that any average ton-mile cost for a whole highway system is based, in part, on the cost of the more expensive elements of construction required to carry heavy vehicles and heavy loads. The passenger cars and light trucks have no need and get no additional use out of these expensive features of highway construction. If these cars could be charged on the basis only of the elements of highway construction which they require and use, their cost per ton-mile would be substantially lower than any average cost... the effect, therefore, of applying an average cost to all vehicles alike is to make the passenger cars and light trucks pay a part of the cost of the heavier vehicles [25-19, p. 89].

The standard-cost method does not seem to have close adherence to either the cost incurred or the benefits received theories of taxation. Employing a standard rate in no way assures, for example, that heavy trucks will be assigned a burden consonant with the highway costs they occasion. There does not seem to be a pronounced theoretical link between the principles of allocating cost among classes of users and applying the standard-cost method. At the very least it can be said that it is difficult to defend the results that would be obtained by applying the standard-cost method to allocations of highway cost to road user classes.

Differential-Benefits Method

The concept underlying the differential-benefits method is that total user tax payment should be apportioned to the several classes of vehicles in proportion to benefits received from highway use by each class. Therefore the method uses a direct application of the benefits received theory of highway taxation. It is quite well accepted that this method is theoretically sound. However, the collection and measurement of required data is exceedingly complex. Accordingly only few applications of the differential-benefits method have been made to date.

Probably the most detailed and exhaustive application of the method was that included in Final Report of the Highway Cost Allocation Study [25-25-5]. A revision of the method, applying the findings of more recent traffic accident cost data, is included in the Supplemental Report of the Highway Cost Allocation Study [25-25-7].

The method as noted previously, is exceedingly difficult and complex and requires a substantial amount of data not readily available. The report discusses some of the more pressing problems as follows:

> The method of differential benefits for the purpose of allocating highway costs among users, while a valid application of the value-of-service principle, has suffered historically from two weaknesses: (a) a lack of experience on the use of benefits for highway cost allocation studies, and (b) incomplete data on the magnitude of certain of the benefits and on the relationship between these benefits and the associated benefit-producing improvements. Recent studies of user benefits, however, have made considerable progress in improving analysis techniques and adding to the store of basic data on benefits. Among these are the analyses of user benefits which have been developed in connection with highway planning and numerous studies of vehicle cost and user driving characteristics. While these studies did not provide examples of benefit computations for cost-allocation purposes, they do provide analysis procedures and basic data needed for a reasonably accurate determination of user benefits. It appears evident that the art of making highway-cost calculations has progressed to the point where user benefits can be satisfactorily employed to allocate highway costs [25-25-5, p. 201].

In the supplemental Report of the Highway Cost Allocation Study [25-25-7] direct user benefits were classified into four categories: (1) reduced vehicle running costs, (2) reduced accident costs, (3) reduced travel time, and (4) reduced driving strains and annoyances.

Each of these types of user benefits may arise from the following elements of highway improvements: (1) reduction of pavement surface roughness, (2) increase in lane widths, (3) increase in the number of lanes, (4) reduction in travel distance, (5) reduction in rise and fall, (6) elimination of intersections at grade, (7) elimination of residential and commercial access points; and (8) reduction in horizontal curvature. One of the most difficult problems confronting researchers in conducting a differential benefits study is quantifying the benefits stemming from different types of highway improvements, each of which may reduce (1) motor vehicle running costs, (2) traffic accident costs, (3) travel time, and (4) driver strain.

Suffice to say that the conduct of a differential benefits study is difficult and complex; but in spite of the many problems involved, it can be strongly supported by well-proven theories of highway finance. For the interested researcher, the Supplemental Report of the Highway Cost Allocation Study is a most excellent reference [25-25-7].

Evaluation of Allocation Study Results

Although two other approaches to the problem of cost allocation are yet to

Highway Finance and Taxation

be discussed, it is appropriate at this time to compare the results of the four methods (incremental, cost function, ton-mile, differential benefits) included in the highway cost-allocation studies [25-25-6 and 7] as all four methods were applied to similar costs, travel, vehicle population, and other factors. The results of the four approaches are shown in Table 25-5.

It will be noted that the differential-benefits method assigns the highest cost responsibility to automobiles for all private vehicles, but is substantially lower than the other methods for the commercial vehicles. As might be anticipated, the gross ton-mile solution assigns the least-cost responsibility to the automobile and the highest to the two classes of heavy vehicles. It shows the widest variation ($20 to $1,185) of all methods. The cost-function study shows a somewhat modified schedule of cost allocations as compared to the gross ton-mile solution. Results of the incremental solution show that it has assigned next to the highest cost responsibility to automobiles and next to the lowest cost to trucks.

Cost responsibilities assigned to for-hire vehicles follow the same general pattern as for privately owned vehicles. It will be observed that the cost assignment to the heaviest vehicle, by the differential-benefits method, seems unusually low. This result probably stems from the fact that in 1964 but few states legalized the use of full-trailer combinations, and data from a few scattered states had legalized the use of full-trailer combinations, and data from a few scattered states were not adequate to compensate for this bias.

Operating-Cost Method

The motor vehicle operating-cost method attempts to allocate costs to the various classes of vehicles on the basis of the value of highway service received.

TABLE 25-5
COMPARISON OF INCREMENTAL, GROSS TON-MILE,
COST FUNCTION, AND DIFFERENTIAL-BENEFITS METHODS
(Average Cost Responsibility per Vehicle, Dollars per Year, 1964)

Class of Vehicle	Private				For Hire			
	Incremental	Gross Ton-Mile	Cost Function	Differential Benefits	Incremental	Gross Ton-Mile	Cost Function	Differential Benefits
Automobiles	$31	$20	$23	$36	–	–	–	–
Buses, transit	–	–	–	–	$249	$326	$252	$124
Single unit trucks, 2-axle, 6-tire	48	51	46	34	80	87	77	53
Combination, tractor and semitrailer, 4-axle	741	1,185	988	552	1,035	1,734	1,439	810
Combination, truck and full trailer, 5-axle	655	1,156	944	103	985	1,827	1,459	137

Source: Final Report of the Highway Cost Allocation Study, 1961 [25-25-5]. Supplemental Report of the Highway Cost Allocation Study, 1965 [25-25-7].

In the discussion of the gross ton-mile method, it was noted that the ton-mile was the measure of value received but fell somewhat short as an appropriate measure. Nevertheless, a measure of value of service is an appropriate and theoretically sound basis for allocating highway costs among different classes of highway users. The operating-cost method was devised to accomplish this allocation on the basis of value received.

Motor vehicle operating costs rise steadily with the size and weight of the vehicle. For this reason, such costs may appropriately measure the value of the highway service provided. Operating costs bear a close relationship to the actual amount of money involved in operating motor vehicles upon highways, roads, and streets. Because it is a monetary measure, operating costs seem to be a reasonable measure of value of service.

The use of motor vehicle operating costs as a measure of value is not unlike the idea of value associated with a commodity. The relative values of goods and services are measured by the prices at which they are bought and sold. The buyer will not pay a price for a commodity greater than its value to him. The seller, on the other hand, will not sell his goods at a price that is less than the value of the goods to him at the time of sale. The major determinant of exchange value is the total cost of producing the goods and making them available to the purchaser. The value of such goods and services, in the most fundamental sense, is determined by the cost of producing them. The same is true for highway services. The price paid by a motor vehicle operator for using highway services is the cost of motor vehicle ownership and operation, including taxes. The operating-costs method measures the value of service by the relative amounts of motor vehicle operating costs, less taxes, and apportions taxes in accordance with these costs. A vehicle in any given weight and size group would pay a tax per mile of operation proportional to its own operating costs per mile.

The execution of the operating-costs method is rather simple. Given the costs of operating a basic vehicle as, for example, 10 cents per mile, and given that a certain size and weight of truck has an operating cost of, say 50 cents per mile, then the required tax payment of the truck per mile of highway use would be five times that of the basic vehicle. Further, if the annual mileage of the truck is four times that of the basic vehicle, then the total required annual tax payment of the truck would be 20 times that paid by the basic vehicle.

The difficulty is posed in deriving motor vehicle operating costs by weight groups. A number of methods for deriving such operating costs are possible, and the data requirements for the solution depend upon the manner in which these operating costs per mile are obtained. One of the more elaborate attempts at deriving operating cost data is that performed by the state of Washington [25-40].

This method does not require individual calculation of tax responsibility for each road and street system. In distributing tax responsibility based on this method, it is done such that average annual payments are proportional to average annual operating costs determined for all systems taken together. Thus an element of simplicity of calculation is inherent in the solution.

In the original operating-cost method proposal by G. P. St. Clair [25-20, p. 178] and in all subsequent discussions of the method, three categories of operating costs have been distinguished: (1) running costs, (2) vehicular costs, and (3) gross total operating costs. Running costs may be defined as those costs that vary directly with mileage travel. They include the cost of fuel, lubrication, tires, tubes, maintenance, and part of depreciation. Vehicular costs are running costs plus part of depreciation, fixed costs of insurance, interest, and garage rent. Gross total operating costs refer to all costs associated with the use and operation of a vehicle. They are the sum of vehicular costs, drivers' wages, terminal, and overhead costs.

It is usually agreed that running costs are the most appropriate concept for the purpose of cost allocation. The use of gross total operating costs distorts the relationship among vehicles of different sizes, particularly between light and heavy trucks:

> This effect is due very largely to the nature of the operations in which light trucks are generally used in business and for-hire operations. For short trips and light loads, the ratio of fixed costs and terminal costs–as for example, loading and unloading–to running cost is very high. For this reason, inclusion of terminal and overhead charges in assessing tax responsibility in proportion to operating costs would tend to discriminate against light trucks in favor of the heavier trucks and combinations [25-24, p. 102].

Studies show that of the three cost bases, running cost gives the lowest cost-allocation value for light trucks and the highest values for heavier trucks, while the reverse is true if gross total operating cost is used. Is seems logical to suppose that as a measure of the value of service, the cost elements chosen should be only those which are directly related to travel on the highway. This again suggests that running costs are the most appropriate of the three concepts.

The chief shortcomings of the motor vehicle operating-cost theory, like all concepts which use a value of service approach, is that it does not reckon with the highway costs occasioned by vehicles of different types and sizes. Only a highway cost incremental solution can do this. Consequently, some investigators have concluded that tax responsibility calculated on the basis of motor vehicle operating cost per mile will not impose a high enough burden on the heavier vehicles in relation to the costs incurred for their benefits.

It has been suggested that the operating cost and the incremental solutions can be used together as a test of equity and adequacy of each other. The results of an incremental solution can serve as a check on the accuracy of operating costs as a measure of cost responsibility. On the other hand, the results of the operating-cost analysis can act as a restraint on the possible tendency of the incremental method to give results which lead to excessive taxes on heavier vehicles.

As suggested in an earlier section, it is sometimes concluded that the operating-cost theory completely clashes with the differential-benefits method. Differential benefits is not the same concept as value of service, but this does not mean that the two solutions are totally incompatible. An application of the differential-benefits method might be likely to give results similar to those derived from an application of the operating-cost method. In applying a differential-benefits method, it is common practice to calculate the benefits for different sizes and weights of vehicle by reference to the reductions in their travel costs. Since the reductions in costs vary with the relative magnitudes of the costs themselves, it is conceivable that results of the two solutions could be quite similar.

The operating-costs theory proposes that the value of highway service in its most fundamental sense is determined by the operating costs of the vehicles using the highway. Secondly, the theory rests on the principle that taxes apportioned according to operating costs will be approximately proportional to ability-to-pay, an accepted principle of public finance.

It has been suggested that the operating-cost method might be used to suggest a lower limit of taxes on heavy vehicles, and the findings of the ton-mile solution might be used as an upper limit. It is then possible that some kind of compromise might be affected between the two. Furthermore, an incremental solution used in conjunction with these two methods might provide an adequate basis for devising an acceptable compromise.

Space-Time Method

The space-time method suggests that the use of a highway by a vehicle can be measured by the amount of highway space occupied by that vehicle and the time it takes each vehicle to cover that space. It is argued that the amount of highway surface or space needed largely determines the cost of the highway plant. It is also argued that the difference in the space required and the time spent on the highways for the largest vehicle and for the basic vehicle does not vary in direct proportion to the size of the vehicle.

Proponents of the space-time theory do not consider that the weight transported over a highway has a significant relation to the cost of that highway or to the use of that highway. Highway width, and to some extent grade standards, are determined by the volume and speed of traffic and size of vehicles that travel a given road. Standards of strength, of pavement thickness and base structure, and bridges, are determined by the needs of national defense and vehicle weight. Drainage, pavement structure, and earthwork are designed to withstand the resistances to the actions of weather. Present standards of design and construction would be changed only in part if highways were restricted to light vehicles. Therefore as a pragmatic approach to measuring highway use it is suggested that the measure used should be one that reflects the relative space required for the operation of each vehicle and the time that each would spend on the highway.

It is true that larger vehicles may derive greater monetary benefits from space and time occupied on the highways than do smaller ones, but such benefits are not easily determined. The theory deals with only one or two of the many factors of increased highway cost that must be considered in a cost incurred solution. The effects of the increased weight of heavy vehicles should not be treated lightly, as would be done in the space-time method [25-33].

All things considered, the space-time method deals with only two of the many factors which are brought out in, for example, an incremental solution. In general, highway finance authorities are not willing to dismiss all other weight-function costs simply on the ground that military needs and the action of the weather sets the standards for certain engineering features of highways.

DISCUSSION ON ALLOCATION OF HIGHWAY COST

There is widespread acceptance by legislative bodies that the highways should be financially supported (not necessarily in total) by the users of the highways and that among the road users there should be differential tax rates based upon vehicle cost, horsepower, size, weight, function, amount of highway use, or some combination of these or other factors. There are proponents who believe that the nonhighway user sectors of society (basically land owners) should also contribute toward the financial support of highways. The real problem, therefore, is not whether to tax road users and nonusers for highway support, but how much to tax each class of user and nonuser. Decision on this question mainly resolves itself into a decision on the choice of tax base and methods of computing tax rates.

Allocation of highway costs for tax purposes is quite similar to the setting of customer prices (rate schedules) for regulated public utilities. In similarity to highways, the gas utility, for instance, has residential, commercial, and industrial users, and therefore, customers who consume small to large amounts of gas. And parallel to highway use there are the daily off-peak and on-peak customers, as well as seasonal peak loads.

In utility rate schedules often provision is made for minimum (monthly) charges, reduced rates per unit of service with increasing usage, off-peak rate schedules (night rates on long-distance telephone, off-peak electric water heaters, and off-peak gas space heating or industrial fuel); in railroad tariffs differential rates are based on value of the product shipped. As with highways, there is no positive or universally accepted way to arrive at public utility rate schedules that will assure equity and justice among the different classes of customers and among users of different amounts of service. The answers lie in establishing rate schedules that are sound conceptually, that can be applied with consistency, and that seem reasonable in result as judged by both the public and by those persons or bodies having legal authority to set the rates.

In highway cost-allocation studies there are advantages in making solutions by two or more methods. The decision maker, then, has a range of values with upper and lower limits that enable him to see the relative tax rates that could be adopted. Two or more solutions in highway cost allocation serve the same purpose as do two or more solutions for highway engineering economy when made on low, medium, and high values for sensitive factors, such as vestcharge rate and traffic volume. In the end, the tax rates imposed by law upon motor vehicle owners and highway users must be the result of judgment by those qualified and legally obligated to exercise such judgment.

Not discussed in this chapter is the similar problem in allocation of state or federally collected road-user taxes to lower levels of government and to highway systems. Such an allocation involves such factors as highway needs, relative unit costs of construction, character of traffic, ability to tax and practice in raising local taxes, and extent that higher laws provide for or restrict local road-user taxes.

REFERENCES

25-1. Walter H. Bottiny and Beatrice T. Goley. *A Classification of Urbanized Areas For Transportation Analysis.* Highway Research Board, National Academy of Sciences, Washington, D.C., Highway Research Record No. 194, *Information Systems for Land Use and Transportation Planning,* 1967.

25-2. Thurley A. Bostick and Helen J. Greenhalgh. *Relationship of Passenger Car Age and Other Factors to Miles Driven.* Highway Research Board, National Academy of Sciences, Washington, D.C., Highway Research Record No. 197, *Passenger Transportation,* 1967.

25-3. Thurley Bostick, Roy T. Messer, and Clarence A. Steele. "Motor Vehicle Use Studies in Six States," *Public Roads Journal of Highway Research,* Vol. 28, No. 5 (December 1954), pp. 99-126, Washington, D.C.

25-4. C. B. Breed, Clifford Older, and W. S. Downs. *Highway Costs–A Study of Highway Costs and Motor Vehicle Payments in the United States.* Association of American Railroads, Washington, D.C., January 1939.

25-5. Don Burch. *A. Working Bibliography on Highway Finance.* Virginia Council of Highway Investigation and Research, University of Virginia, Charlottsville, June 1964. The 155 references include the subjects of (1) highway taxation, (2) highway needs studies, (3) highway programming, (4) highway accounting, (5) allocation of highway costs, and (6) general references.

25-6. Economic Research Agency (H. Russell Briggs). *Special Assessments in Theory and Practice,* prepared for the U.S. Bureau of Public Roads, Washington, D.C., by Economic Research Agency, Madison, Wis., June 1960.

25-7. Griffenhagen and Associates. *A. Highway Improvement Program for Illinois.* A report for the Illinois Division of Highways, Chicago, November 1948.

25-8. Highway Research Board. *Allocating Motor Vehicle Tax Responsibility by the Incremental Method.* A symposium of five papers and discussion. National Academy of Sciences, Washington, D.C., Bulletin 121, 1956.

25-9. Highway Research Board. *Allocating Highway Cost Responsibility–Reports on Studies in Five States.* National Academy of Sciences, Washington, D.C., Bulletin 175, 1958.

Highway Finance and Taxation

25-10. Highway Research Board. National Academy of Sciences, Washington, D.C., Highway Research Record No. 20, *Highway Financing,* 1963.
    1. William R. McCallum. "Highway Bond Financing–A Current Analysis," p. 1.
    2. H. R. Briggs. "Special Assessment Financing of Major Highway Improvements," p. 63.
    3. David McKinney. "Selection of a County Distribution Formula for State Road Assistance in Mississipi," p. 76.
    4. Everett C. Carter and Joseph R. Stowers. "Model for Funds Allocation for Urban Highway Systems Capacity Improvements," p. 84.
    5. Herbert D. Mohring. "The Place of Subsidies in an Optimum Transportation System," p. 103.
    6. Richard A. Tybout. "The Problem of Nonuser Revenues," p. 114.
    7. James M. Smith and Reed H. Winslow. "AASHO Road Test Findings Applied to State Highway Cost Allocation Studies," p. 123.
    8. C. A. Steele and T. R. Todd. "Applications of Earnings-Credit and Relative-Use Methods of Highway Cost Allocation," pp. 135-161.
    9. Sidney Goldstein. "Nonuser Benefits from Highways," pp. 162-181.
    10. Milton Kafoglis. "General Discussion," p. 182.

25-11. Highway Research Board. National Academy of Sciences, National Research Council, Washington, D.C., Highway Research Record No. 138, *Highway Finance and Benefits,* 1966.
    1. G. P. St. Clair, T. R. Todd, and Thurley A. Bostick. "The Measurement of Vehicular Benefits," pp. 1-18.
    2. Robinson Newcomb. "New Approach to Benefit-Cost Analysis," pp. 18-22.
    3. Kenneth E. Cook and Patrick A. Rush. "Consumer Awareness of Motor Fuel Tax Rates and Prices," pp. 22-32.
    4. Edward A. Gladstone and Thomas W. Cooper. "State Highway Patrols–Their Functions and Financing," pp. 32-64.

25-12. William E. Koenker and Arlyn J. Larson. *Equitable Highway Cost Allocation in North Dakota.* Prepared for the North Dakota Legislative Research Committee by the Bureau of Business and Economic Research, University of North Dakota, Grand Forks, September 1956.

25-13. William R. McCallum. *Highway Bond Financing-An Analysis 1950-1962.* Bureau of Public Roads, Department of Transportation, U.S. Government Printing Office, Washington, D.C., September 1963.

24-14. C. B. McCullough, John Beakey, and Paul Von Scoy. *An Analysis of the Highway Tax Structure in Oregon.* Oregon State Highway Commission, Salem, May 1938.

25-15. Nebraska State Department of Roads. *1966-1985 Financial Planning for Nebraska's Highways, Roads, and Streets.* Lincoln, April 1967. Report prepared by Wilbur Smith and Associates, Columbia, South Carolina.

25-16. Public Administration Service. *Financing a Proposed Highway Program in Minnesota.* A report to the Minnesota Highway Study Commission. Chicago Ill., October 1954.

25-17. John Rapp. *Cost Allocation Revisited.* Highway Research Board, National Academy of Sciences, Washington, D.C., *Highway Research News,* No. 17 (February 1965), pp. 11-23.

25-18. Linda Ritter. *Revenue Sources for Financing Virginia's Highway Program. Phase 1b: An Annotated Bibliography.* Virginia Highway Research Council, University of Virginia, Charlottesville, August 1966. Gives 290 references on highway financing.

25-19. Herbert D. Simpson. *Highway Finance.* Ohio Program Commission, Department of Highways, Columbus, September 1951.

25-20. G. P. St. Clair. *Suggested Approaches to the Problems of Highway Taxation.* U.S. Federal Works Agency, Public Roads Administration, Washington, D.C., September 1947.

25-21. Thomas R. Todd. "Financing Street and Highway Improvements in Urban Areas," *Traffic Quarterly,* October 1964, pp. 502-522.

25-22. United States Department of Agriculture, Office of Public Roads. *Public Road Mileage, Revenues, and Expenditures in the United States in 1904.* U.S. Bulletin No. 32. Government Printing Office, Washington, D.C., August 1907.

25-23. United States Department of Agriculture, Office of Public Roads and Rural Engineering. *Automobile Registrations, Licenses, and Revenues in the United States, 1916.* (Circular No. 73.) U.S. Government Printing Office, Washington, D.C., June 1917.

25-24. United States Department of Commerce, Bureau of Public Roads. *A Factual Discussion of Motor Truck Operations, Regulation, and Taxation.* Washington, D.C., June 1950.

25-25. United States Department of Commerce, Bureau of Public Roads. *Highway Cost Allocation Study.*
  1. First Progress Report, 85th Congress, 1st Sess., House Doc. No. 106, 1957.
  2. Second Progress Report, 85th Congress, 2d Sess., House Doc. No. 344, 1958.
  3. Third Progress Report, 86th Congress, 1st Sess., House Doc. No. 91, 1959.
  4. Fourth Progress Report, 86th Congress, 2d Sess., House Doc. No. 355, 1960.
  5. Final Report, Parts I-V, 87th Congress, 1st Sess., House Doc. No. 54, 1961.
  6. Final Report, Part VI *(Economic and Social Effects of Highway Improvement),* 87th Congress, 1st Sess., House Doc. No. 72, 1961.
  7. Supplementary Report *(Incremental Method of Cost Allocation),* 89th Congress, 1st Sess., House Doc. No. 124, 1965.

25-26. United States Congress, Public Aids to Domestic Transportation, Board of Investigation and Research. House Doc. No. 159,79th Congress 1st Sess., Washington, D.C., 1944.

25-27. United States Department of Transportation, Bureau of Public Roads. *Highway Statistics, 1965.* (Table F-1), Washington, D.C., April 1967.

25-28. United States Federal Coordinator of Transportation. *Public Aids to Transportation,* Vol. IV, *Public Aids to Motor Vehicle Transportation.* Washington, D.C., 1940.

25-29. University of Kentucky, College of Commerce. *Financing Kentucky's Roads and Streets.* Bureau of Business Research, University of Kentucky, Lexington, 1956.

25-30. University of Missouri, School of Business and Public Administration. *Financing Missouri's Road Needs.* Prepared by the Bureau of Business and Economic Research, Columbia, for the Missouri State Highway Department, Jefferson City, December 1960.

25-31. Wilbur Smith and Associates. *Illinois Highway Needs and Fiscal Study.* New Haven, Conn., 1967.

25-32. Robley Winfrey and Thomas R. Todd. "Financing Urban Trasportation,"

Highway Finance and Taxation

American Society of Civil Engineers, *Journal of the Highway Division*, Vol. 91, No. HW2 (December 1965), pp. 89-100.

25-33. Lewis C. Bell. *Third-Structure Taxes: Applicability for Kentucky.* Univrsity of Kentucky, Bureau of Business Research, Lexington, 1956.

25-34. Colorado Highway Planning Committee. *Highway Needs and Highway Financing.* Denver, 1950.

25-35. William L. Hall. *Financing Modern Highways for Montana.* Montana Fact-Finding Committee on Highways, Streets and Bridges, Helena, 1956.

25-36. Highway Research Board. National Academy of Sciences, Washington, D.C. *The AASHO Road Test.* Special report 61A to G. See especially Reports SR61E (No. 5) on *Pavement Research* and SR61D (No. 4) on *Bridge Research*, 1961, 1962.

25-37. D. F. Pancost. *The Ohio Incremental Study: An Experiment in Vehicle-Tax Allocation.* Highway Research Board, National Academy of Sciences, Washington, D.C., *Proc.*, 32 (1953), 68-80.

25-38. William D. Ross. *Financing Highway Improvements in Louisiana.* Louisiana State University, Baton Rouge, 1955.

25-39. Virginia Highway User Association. *Testing the Equity of Virginia's Motor Vehicle Tax Structure.* Richmond, 1953.

25-40. Washington State Highway Commission. *Washington Motor Vehicle Operating Cost Survey, 1952-1953.* Olympia, Washington (no date but probably 1958).

25-41. Wilbur Smith and Associates. Highway financial studies in Arizona, Georgia, Illinois, Kansas, New Jersey, Oklahoma, and South Dakota. New Haven, Connecticut. Various years.

Chapter **26**

Construction Programming and Scheduling

The combination of a compilation of the highway needs during the next 20 to 25 years and the devising of a plan for financing those needs is a sound approach to a good construction program. But the needs study and financial plan do not in themselves result in a year-by-year construction program. The management of a highway department must work continuously at developing and maintaining a 3- to 5- to 10-year program of construction projects. This construction program is followed with continuous effort in scheduling the construction by projects and keeping the program on schedule. The operations of programming and scheduling are management functions that are now highly developed and widely practiced. The skillful execution of these activities results in high production efficiency, financial economy, and lower construction prices.

DEFINITIONS

The following definitions taken from the glossary in Ref. 26-10 are offered as aids to the understanding of the discussions about programming and scheduling. Some of the terms defined may be used by others with somewhat different meanings, but the definitions will serve to sharpen the intentions of this chapter. Many additional terms are defined in Ref. 26-10.

1. *Advance Programming.* The programming of highway capital improvements for a period of several years in advance of actual construction.

2. *Capital Budget.* A budget which provides for the financial support of highway construction of a permanent nature as opposed to a budget for operating expenses such as maintenance.

3. *Capital Improvement.* A physical improvement to the highway system of that type of property that will be long-lasting in use, as opposed to the consumable supplies and maintenance items; land, bridges, and buildings are capital improvements, whether constructed through an original contract or through addition and betterment work.

4. *Construction Priorities.* The ranking of construction projects according to an accepted criterion of urgency.

Construction Programming and Scheduling

5. *Construction Program.* The construction program of a highway department is a detailed listing of all of the capital improvements to a highway system that are proposed to be done in a given period of time. Construction programs may be adopted for a given year, or 3-year, 5-year, 10-year, or other period of time.

6. *Construction Program Schedule.* A time schedule showing desired starting and completion dates of all major construction operations, that is, clearing, grading, structures, and paving.

7. *Lead Time.* As applied to the programming of highway improvements, the time required for activities which must precede the letting of a construction contract; specifically, location studies, preparation of surveys, plans, and estimates, acquisition of rights-of-way, negotiations with utilities, municipal agreements, and so forth.

8. *Long-Range Planning.* The continuous process of making present (risk taking) decisions systematically and with the best possible knowledge of their futurity for some 10 to 30 years, organizing systematically the efforts needed to carry out these decisions, and measuring the results of these decisions through organized, systematic feedback.

9. *Short-Range and Long-Range Program.* A timetable for meeting specifically determined present and future highway improvement needs over a definite period of years, such a period extending as far as 20 years or more into the future. In the forefront of the long-range program may be a short-range program in which projects are quite firmly scheduled for perhaps 2 to 5 years; beyond this point the long-range program will usually schedule needed improvements on a less detailed basis, with more flexible timing and a recognition of the adjustments and reanalyses that the passage of time will inevitably compel.

10. *Planning.* The formulation of a scheme of action for the future, utilizing logic, rationalization, philosophy, mathematics, observation, measurements, controlled testing–in effect, all of the tools of scientific disciplines that may be applicable.

11. *Priority Planning.* The measuring, rating, and ranking of projects by adopted procedures and standards both for physical adequacy and economic significance, balancing such ratings with each other and with administrative considerations.

12. *Priority Rating.* A numerical index applied to a proposed improvement project, denoting the urgency of its need for improvement, relative to other proposed projects, such rating being based on some form of methodical and impartial engineering and/or economic analysis.

13. *Programming.* A systematic process of setting forth a collection of things to do with due consideration given to priority and all other factors which determine the desirability of carrying out the act. Highway programming is setting forth highway improvement projects in a priority system over a specific time period.

14. *Scheduling–Control and Adjustment.* The mechanics of detailing a highway construction program for a period of years with specific time dates for specific activities to be commenced and to be completed. The schedule will provide for control and for early adjustment of programs to meet changing conditions, scheduling of precontract engineering and other operations, as well as construction expenditure rates.

15. *Short-Range Program.* A timetable, usually extending 2 to 5 years into the future, for the scheduling of specific highway improvement projects. Ideally, such a timetable may detail separately the scheduling of engineering, right-of-way acquisition, and separate contract phases of construction.

16. *Tolerable Conditions.* Geometric or structural conditions which, while not quite up to required standards for new construction, are sufficiently close to new construction standards so as not to justify reconstruction.

17. *Strategic Decision.* Management decisions going far beyond the simple answer-finding process to the matter of asking the right questions in the first place in order that the decision is "strategic" rather than "tactical."

18. *Sufficiency Rating.* A numerical index, empirically arrived at, which attempts to measure the structural and/or geometric adequacy ("hence sufficiency") of a highway section or a structure, relative to some predetermined standard of complete adequacy.

19. *System Classification.* The grouping of highways into distinct classes or systems according to the type of service, or function, they perform (functional classification); also, and secondarily, according to the agency having jurisdiction (jurisdictional classification), or the source of dedicated funds from which they are to be financed (fiscal classification). Properly, jurisdictional and fiscal classification should follow and be based on functional classification, which provides an essential tool for the orderly planning and management of highway programs and highway department operations.

20. *Tactical Decision.* The decision directed toward the accomplishment of an immediate objective of a fairly simple and evident nature.

ADVANTAGES OF A CONSTRUCTION PROGRAM

To develop a construction program year by year for 3 to 5 to 10 years,[1] is nothing more than to exercise common sense in management. Such a program offers many advantages to good management. An organization succeeds best when operating under a well-thought-out plan. A construction program (a plan) is a listing of objectives, of goals, to reach in an orderly fashion. When the objectives are stated, all units of the highway department can coordinate

---

[1] The upper limit of this range is a factor of the complexity of the projects to be undertaken. County roads and much of the rural state system can satisfactorily be programmed on a 3- to 5-year basis, but urban work, freeways, and interagency joint projects often take a total of 10 years from the early planning stage to completion of construction.

Construction Programming and Scheduling 641

their efforts toward reaching common goals. The public at large and highway contractors are also benefited by knowing the future construction program so they too can make their own future plans.

Although a highway construction program is a listing of improvement projects to be constructed on a year to year basis, the program is much more. The program is a general work order to all concerned the moment it is approved by responsible authority. Work orders for specific projects are issued just as soon as the construction program is translated into a construction schedule, giving target dates for starting and finishing each project. From a general work order, individual units in the highway department can evolve their own plans to support the total objectives and to assure coordination of interdependent units.

Financial control is tied to the construction program and to the construction schedule. The 3- to- 10- year program is based upon the anticipated funds to be available for capital improvements. Keeping the program within conservative limits of funds forecast to be available insures both safety and efficiency. Safety against obligating more funds than will be available is assured by controlling the program to the limit of funds to be available. Efficiency is achieved by systematic and coordinated planning up to the limit of funds available. Spasmodic actions, adding and subtracting projects, and overnight rush decisions are causes of high inefficiencies in the use of work time and capital funds. A realistic 3-to-10-year construction program will avoid the lumpiness of construction work and financial payments and keep the whole activity on an even keel.

Lead time–the time available for necessary activities between the date of decision to the date the scheduled activity must begin–is a factor of high importance. A long lead time is provided by early programming combined with early scheduling of construction work. A long lead time affords time to give all actions thorough consideration, to reach agreement with cooperating agencies, and to prepare throughly and in detail good highway designs and construction cost estimates. Long lead time avoids the last minute rush to get the plans and specifications finished ahead of letting dates.

The 3-to-10-year construction program, continually updated by project priority selections, combined with the detailed project schedule, provide the base for keeping the whole capital improvement effort on a level basis. All units of the department may then work on their own programmed basis toward specific common goals. Employment levels become more uniform, the necessary skills can be made available, and each construction project can be coordinated with others as desirable for engineering consistency and production efficiency.

Construction programming is just another management tool–a member of that growing family of management tools that include economic analysis, operations research, systems analysis, critical path method (CPM), program evaluation and review technique (PERT), and others–devised to give greater certainty that management decisions are sound, and, hopefully, the best. Therefore construction programming through the application of proven techniques, results in greater acceptance by the public and politicians of the decisions of highway department management.

THE FUNCTIONS OF PROGRAMMING AND SCHEDULING

As a management device, most corporate organizations and public works departments provide for organized formal methods of developing their plans for future capital investments, or construction activities. Expansion and rebuilding of physical plants require large sums of capital moneys and long calendar times to plan and to execute. An understanding of the functions of programming and scheduling of construction of plant (highways) is therefore helpful to an understanding of highway department management and highway economics.

FACTORS AND CONSIDERATIONS

Unlike the repetitive activities of operating and maintaining a highway plant, constructing new plant and reconstructing existing plant are special activities, each project of which requires detailed consideration with respect to its desirability, planning, designing, economic evaluation, and financing. To achieve the best results from construction activities and the most efficient use of resources requires advance planning, long lead times, and effective management control. The complexities of highway departments, service to the public, the large yearly dollar volume of construction, and the great number of individual projects require that a construction program be planned with great care. No longer can a good program be assembled in the off-season winter months and made ready for the following summer construction season. Good construction programs come from close attention to the subject the year around and with foresights at least 3 to 10 years ahead.

The process of formulating a highway construction program for the next 3 to 10 years on a year-to-year basis is one that brings into focus a diversity of related factors. The requirement is to relate these factors in an orderly manner so that they may be weighed one against another. High degrees of subjective judgment and conceptual ability are required to produce a program that is the overall best one in the interest of the public. The following listing mainly from Ref. 26-10, page 4, gives many of the factors involved, classified into six groups:

Group A: Long-Range Outlook
1. Long-range physical needs
2. Long-range financial needs
3. Long-range highway system plans

Construction Programming and Scheduling 643

Group B: Financial
  1. Money available–next fiscal year
  2. Money available–short-range ahead
  3. Money available–long-range ahead
  4. Feedback and probable adjustments
  5. Rate that payments to contractors will be required

Group C: Priority
  1. Rate of return, benefit/cost ratio, sufficiency rating, present serviceability index
  2. Traffic services (amount and type of service to be provided)
  3. Traffic generation and growth
  4. Physical and structural conditions of the highway
  5. Accident record and safety
  6. Comparative needs between systems, routes, areas, projects
  7. Emergency (disaster) needs
  8. Social and human values

Group D: Program Balance
  1. Distribution of work by highway systems
  2. Distribution of construction dollars by geographic areas
  3. Distribution of work by type–earthwork, pavement, structures
  4. Distribution of work by contract dollar size–small versus large contracts
  5. Adequacy of contractor supply versus competition
  6. Adequacy of labor and material supply

Group E: Project Selection–Technical and Operative
  1. Continuity of route improvement
  2. Protection of existing investment–surfacing jobs, additions, and betterments
  3. Temporary versus permanent improvements
  4. Maintenance expense
  5. Construction season length versus time to complete project
  6. Sequence of work and requirements to complete stages of work
  7. Size of project and construction time required
  8. Small versus large project versus stage construction
  9. Availability of planning and design information
  10. Study time to reach decision as to desirability and basic design
  11. Local planning and commitments
  12. Lead time for negotiations–other agencies, utilities, land development
  13. Preparation of plans versus letting dates
  14. Right-of-way acquisition versus letting dates
  15. Industrial and other land use developments
  16. Coordination with other public agencies
  17. Approvals of and agreements with other agencies

18. Adequacy of department-wide staff

Group F: Management and Policy
1. Desires of top-level public officials
2. Desires of local-level public officials
3. Demands of pressure groups
4. Demands of private citizens
5. Legislative policy and legislative outlook
6. Public defense needs
7. Requirements for public health and disaster prevention
8. Requirements of law

These 48 factors are of varying degrees of importance with respect to each other and with respect to different improvement projects. Highway department managements have long sought some scheme of getting together a list of improvement projects for a given construction year that could be proved to be the best selection on some factual basis. Facts, especially when they can be expressed in numerical form, offer a good basis to support the decision to include or exclude a proposed improvement project in a given year's construction program.

Highways are everyone's business, and, therefore, everyone has his preference for highway construction projects and their relative priority. A construction program based on facts is more apt to gain public and political acceptance than will a program based upon intuitive judgment. For this reason there has been developed various rating schemes, such as the sufficiency rating, present serviceability index, congestion index, benefit/cost ratio, rate of return, and others, for ranking priority of projects. Not all proposed improvement projects lend themselves to a numerical rating scheme, but to the extent that they do, such devices have proved helpful to highway managements. But even facts are subject to subjective evaluation; and rating schemes usually involve subjective judgment. A construction program gotten together on the basis of considering the above 48 factors has a much better chance of being accepted by those concerned than one put together by offhand unsupported judgment.

REQUIREMENTS FOR EFFECTIVE PROGRAMMING

In any highway department the effectiveness of the functions of programming and scheduling of construction is dependent upon eight primary factors [26-11-2, p. 23]: (1) legislative provisions, (2) executive support and action, (3) continuing factual surveys, (4) budget decisions and financial policies, (5) systematic priority analyses, (6) systematic coordination of all functions and organizations, (7) scheduling and control, and (8) administrative organization.

Legislative Provisions

The legislative role in programming of highway construction is mainly in providing (1) those basic laws which allow the highway department to operate with adequate authority and flexibility, (2) positive designation of highway systems and

Construction Programming and Scheduling

for jurisdiction thereof, and (3) adequate financing and positive advance allocation of state revenue to road systems. The efficient functioning of any highway department is dependent upon (1) adequate provisions of law, (2) sound personnel management, and (3) sound structural and functional organization. The legislative bodies can do much to further these three factors by well-conceived legislation and cooperative attitudes.

Executive Support

Notwithstanding the provisions of law, effective programming of construction cannot be had without leadership and direction at the top executive level. Top management must recognize that programming is an overall complex and technical function, deserving of a special staff designation. The function must be a continuing function throughout the year performed with the full cooperation and coordination of all departmental divisions. The top executive, of course, should actively exercise his responsibility as the final decision maker.

Planning Surveys

Most state highway departments operate a continuing planning survey–fact gathering and analyses–from which a store of up-to-date facts are made ready for use in planning, programming, and highway design. Although the states do not follow a uniform practice in their planning surveys, they each compile and up-date the statistics on their highway system–mileage by pavement type and geometrics, traffic counts, traffic classification, financial outlays and incomes, traffic accident records, vehicle-miles of travel by areas and systems, and many other items. These facts are used in both planning studies and highway designing. These facts, and the analyses of them, take the place of estimates, offhand opinions, and baseless decisions. These factual surveys collect and preserve in an orderly manner the engineering, physical, economic, social, and financial information so necessary to planning and programming of highway improvements.

Budgetary Provisions

Sound financial programming is dependent upon systematic and consistent budgetary practices. Although the state law will state the rate of road-user tax and the allocation of tax income to road systems, there is reponsibility within a highway department for further allocation of funds between geographical areas, road systems when so provided for, types of road work, and overall departmental functions. Programming of construction must rest upon decisions pertaining to budgets and forecasts of budgets. Thus early and consistent budgetary decisions are essential to the success of construction programming.

Construction Priorities

Because the need for improved highways generally greatly exceeds the yearly budgetary funds available to satisfy the needs, it becomes desirable in the programming of construction work to provide for early construction of the most needed projects and to delay into future budgets the less critically needed work. Setting relative priorities on proposed projects becomes a necessity. Setting priorities is a process of weighing facts about the physical condition of existing route segments and the traffic with the subjective judgments of relative importance of the facts as indicators of urgency. Rating schemes–economy measurements, ratios, and indexes–are used in one way or another to establish priorities. Further discussion is given on page 650. No single fact or function can be relied upon as the best index of priority for construction. From the 48 factors previously discussed many factors may be found that deserve attention in setting the priority on any proposal for improvement.

Coordination

Systematic coordination means that all pertinent factors are to be coordinated, that all units and functions of the highway department are to be considered, and that where other public bodies or functions are concerned their interests are also to be coordinated in setting the program of highway construction and priority of projects.

Scheduling of Construction

Since it often takes up to 10 years with 2 to 5 years most common, to move a project from its planning stage through design, right-of-way procurements, and construction, a program of construction needs to be not less than 3 years in advance and often up to 10 years. In spite of good scheduling and control, a program once firmed up must remain flexible enough to meet unforeseen changes at any stage up to the start of construction. These necessary changes come from many causes, such as legislative action, delays and speedups to construction of related projects, shortage of personnel, weather conditions, failure to procure rights-of-way on schedule, unforeseeen conditions uncovered in preliminary engineering, changes and delays resulting from public hearings, and actions or lack of actions on the part of the officials of other public agencies. The planning and programming chiefs need to have a continuous feedback from the field, design, construction, and right-of-way offices so that the program can readily be changed to meet the continuously changing events that affect programming and scheduling.

Administrative Organization

The foregoing seven primary factors that affect construction programming could be well in hand, but without the proper administrative organization and action the whole procedure would be ineffective. A highway management has as one of its principal responsibilities the programming and control of highway

Construction Programming and Scheduling 647

construction. Therefore, the right organizational spot and environment must be found for these functions. Since the decision of what highway projects will be constructed is a top-management decision, the programming activity should be closely associated with top management and not be in line with the main functions of administration, design, construction, and maintenance. See Fig. 26-1 for a typical functional chart.

ORGANIZATION AND PROCEDURES FOR PROGRAMMING

Because programming and scheduling of highway construction are management functions, it follows that there is no one best way or no one best organizational structure through which to carry on these functions. There are some generally accepted procedures and organizations, however, that may be mentioned. The details of operation are omitted, but may be found in the published literature referenced at the end of the chapter.

FUNCTIONAL ORGANIZATION OF HIGHWAY DEPARTMENTS

Whether for a state, county, or city, a highway department in charge of a highway system performs about the same functions and has about the same responsibilities. Variations in structural and functional organizations are attributed primarily to (1) laws and ordinances, (2) preference of the top, or controlling, level of management, and (3) size of the geographical area covered and mileage of the system controlled. The basic functions are (1) administration (personnel, procurement, accounting), (2) planning and research, (3) highway design, (4) right-of-way procurement, (5) construction, and (6) highway maintenance and operations. Figure 26-1 is representative functional chart for a state highway department.

Years ago it was common to find that the responsibility for the programming function was in the design or construction division; now programming is more generally performed by a staff unit responsible directly to the top level of management. Planning and programming are continuous year-round functions. For effective performance they need to be separated from the operating lines to a position of overall independent view. Programming is a subphase of overall planning so is usually found within the planning division.

SOURCE OF PROPOSED CONSTRUCTION PROJECTS

Each highway department has its own process of getting a highway construction program together and getting it approved by the proper authority. In essence, most departments follow the same general procedure.

Initial recommendations may come from the field districts. The field engineers have firsthand knowledge of the need for improvements within their areas and

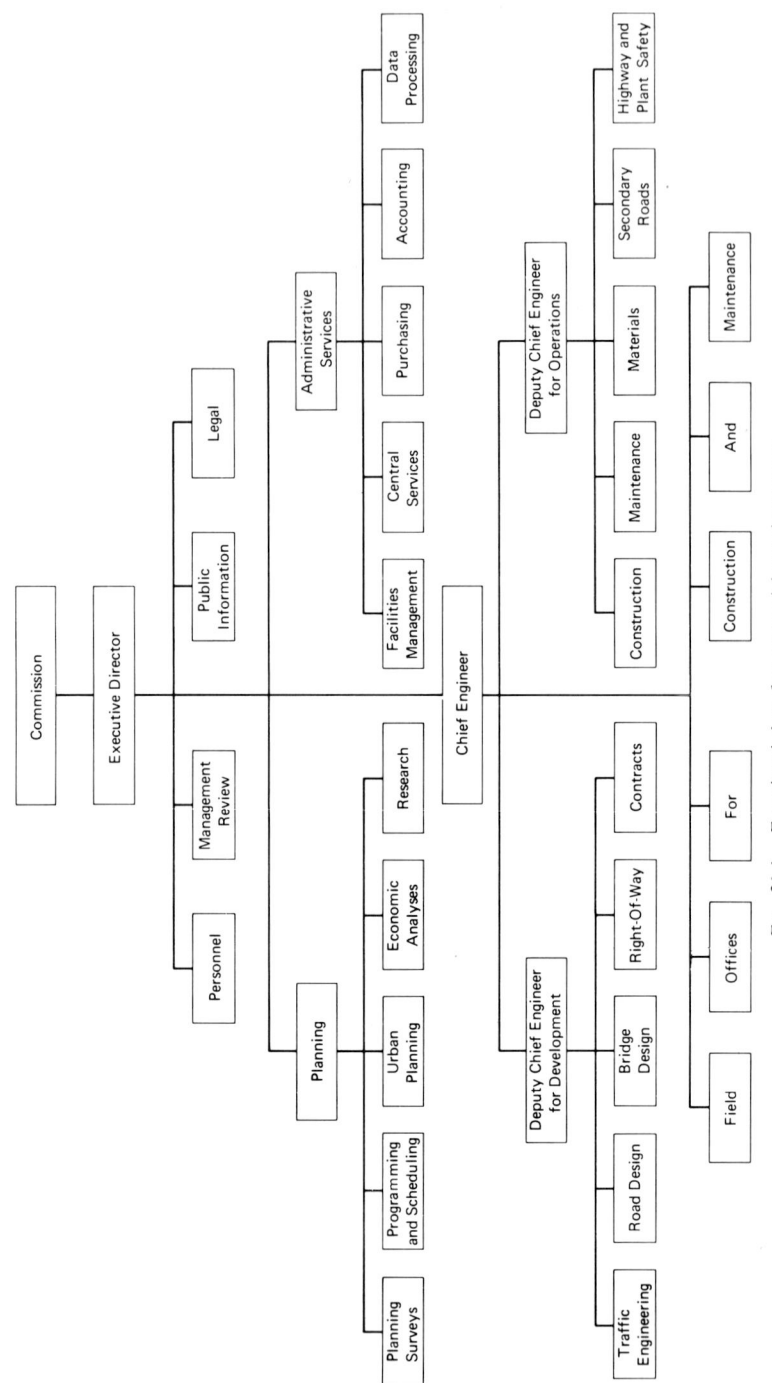

FIG. 26-1. Functional chart for a state highway department.

Construction Programming and Scheduling

of what is planned contiguous to their boundaries. Although these recommendations may not result from detailed studies of the entire mileage in each district, such recommendations are deserving of great weight. The field, or district, engineers have intimate contact with their highways, the traffic, and the local citizens. These contacts are the base of firsthand knowledge of the needs for improving the highway facilities under their jurisdiction and the knowledge leads to trusted professional judgments.

Should there have been a needs study in recent years and should it have been kept up-to-date, it is a second source of construction projects to consider for including in the current 3–to–10-year program. A third source of suggested projects for the construction program is from the planning division. Planning is a continuous process. As the professional persons responsible for overall system planning study their collection of the facts of highway use and needs they bring to light the need for new routes, major improvements of existing routes, and special systems, such as a rural freeway system, or an urban expressway system.

Lastly, all officials of the highway department may propose construction projects as a result of their personal travels and their contacts with the public and with public and private agencies. Also political persons are prone to offer their suggestions. All in all the assembly of proposals for construction projects is an easy task to accomplish. The difficult task is to reduce the whole to an orderly scheme for each year of the total program period, getting the best priorities in logical sequence and within the available funds.

Once the first 3–to–10-year program of construction has been completed, its continuous revision is an easier but still complex task, but no less important. The dynamic character of highway travel is pronounced; the demands for highway travel ever increase; higher levels of service are desired; and highways are everyone's business. To keep up with these dynamic changes a highway department is obligated to continuously review and revise its transportation plans and its advance construction program.

FORMULATING THE CONSTRUCTION PROGRAM

The real test of the effectiveness of a construction program is found in how well it balances and coordinates the many factors and resources involved and whether the more important projects get built first. But what factors determine relative importance is a judgment decision. The final step of putting a construction program into a year to year priority basis is a complex process of great importance.

THE TIME PERIOD

An up-to-date needs study covering the next 20 years (often by 5-year periods) is a good starting place from which to compile a highway construction program.

But an advance construction program by projects becomes too problematical when extended 20 years into the future. The length of time to cover in an advance construction program must be far enough into the future to afford the necessary lead times for all preliminary activities and yet kept near enough to the present that it is both realistic and current in meeting the most important needs. A program ahead for just the next year is worthless when viewed from efficiency in management and utilization of resources. Through experience, a time period of 3 to 10 years has proved to be good. The state highway departments are using 3 to 15 years for their advance and long-range planning [26-12-1, p. 5], but the firmness of the program decreases as the time period increases above 5 years. However, firm commitment of new funds may be held to a 1- or 2-year advanced basis.

CONSTRUCTION PRIORITY RATINGS

Having assembled from all proper sources the suggestions for construction projects for the stated programming period, the difficult task is at hand of selecting projects to construct during each year of the program period. This task embodies assigning relative priorities to each proposed project, not only by years but in many cases priority rank within the year. Priority is established by (1) use of rating schemes to the extent that the project factors can be measured and rated, (2) by comparing the factors not included in the rating schemes, and (3) judgment as arrived at by considering all relevant factors.

Rating Schemes

Assuming that there exists some form of highway facility, the urgency of constructing improvements to that highway facility is closely related to its adequacy to serve the existing traffic. Adequacy may be measured in terms of vehicular capacity, maximum speed, average travel speed, comfort of travel, and safety of travel. These factors in turn are measured by the structural condition and geometric design of the highway and its facilities. Thus these factors are the measurable ones that give some indication of relative importance or urgency of needed improvements.

In Chapter 24 the sufficiency index and the present serviceability index are described in connection with the needs studies. These ratings may also be used as tools in establishing construction priorities for programming purposes. Other tools are the prospective economy to be achieved by the improvement as measured by the benefit/cost ratio or rate of return solution in the economic analysis. Gardner has suggested a rating scheme based on traffic delay [26-11-1]. He calls his final rating a "modified benefit/cost ratio." Haney [26-8] and Curry [26-6], while not seeking for a programming rating scheme, developed the cost of time analysis which product can be used as an index to construction priority. Gardner's modified benefit/cost ratio is the cost of motor vehicle operation during the delayed time divided by the highway cost. Haney and Curry, as described

Construction Programming and Scheduling 651

in Chapter 11, computed the highway cost per hour of travel time reduction by dividing the equivalent uniform annual highway cost for a given time period less the motor vehicle running cost reduction by the hours of reduced travel time for the same time period.

Each of these schemes of rating the measurable factors of highway sufficiency or deficiency have their merits and demerits and their advocates and opponents. Construction programs seek a scheme for priority of construction which will minimize subjective judgment, adequately measure and rate the overall quality of the highway section, be simple and not costly to apply, be reliable, and be easily understood. Because of these hard-to-obtain specifications the hunt continues.

Rating the Nonmeasurable Projects

The currently used rating schemes are not universal in their application for two reasons. First, individually they are not designed to include all types of proposed highway improvements, and second they do not consider all factors upon which priority depends. The sufficiency rating scheme is not applicable to less than a section of a highway complete with all of its elements. As presently designed the points total 100 for such combined main features as structural adequacy, traffic safety, and traffic service. A spot type of improvement as for a bridge, horizontal curve, sight distance, pavement resurfacing, flood prevention, and the like cannot be rated by the same factors and point scales as used for a whole highway segment. The present serviceability index (PSI) applies only to roadway pavements and does not measure safety and geometrics. Both the sufficiency index and the PSI schemes omit construction cost and economy. Further, they rate an existing highway, not the merits of a new facility.

The analysis for economy and the congestion delay methods depend upon trafffic character and volume. These methods are therefore not applicable to proposed improvements for which the road-user benefits or travel time cannot be estimated. In theory the economy could be estimated for pavement resurfacing, pavement widening, bridge widening, bridge replacement, highway lighting, and other special types of construction, but from a practical viewpoint, such estimates of economy are both difficult to make and unreliable in result. As more specific information becomes available on how all features of highway design and traffic operation affect motor vehicle running costs these difficulties in estimating the economy of certain types of highways will be overcome. But every rating scheme should be considered as only a management tool to use in setting priorities and programs because many other factors, social and economic, are also involved.

Rating the Nonmeasurable Factors

The rating schemes discussed apply to factors of the highway and its traffic which can be measured or scaled. Physical dimensions, traffic volumes, travel times, construction costs, and road-user costs are among these measurable factors.

In the list of 48 factors at the beginning of this chapter there are many nonmeasurable factors. Priority of highway construction depends upon many factors and situations which can be observed and judged only by the mental process. Route continuity, progress on related construction, lead time required to reach the stage of advertising for bids, relative cost in relation to allocated funds, the balancing of the construction work load by geographical areas and type of work, and private and public works which affect the proposed highway project and vice versa, are typical of the host of factors and events that are considered in setting priorities on the individual projects in a highway construction program. And these same factors give rise to the need to change the priorities as progress is made or not made.

FINAL SELECTIONS AND APROVAL OF THE PROGRAM

The fact that highway departments have large backlogs of critically deficient projects, the construction of which would greatly benefit the public, places top importance on the construction program for the next 3 to 10 years and brings sharply into focus the necessity to set priorities on each proposed project. The department officials responsible for the selection of those projects to put into each year's construction program do their work sincerely and on an unbiased basis. But the completed program presented to top authority for approval is the culmination of much subjective judgment and weighing of alternatives. Many criteria are used as guides and every effort put forth to get a program to accomplish the most good for the most people. The final program is a meshing of such criteria as available funds, calendar time required to accomplish stages of work, changing land uses and traffic volumes, desires of the people and organized groups, the social and economic factors, and many others. In the end the program as recommended will not satisfy everyone, but will satisfy many. And in the pool of proposed projects not included in the program will be found many highly desirable projects, there to await their turn, hopefully at some interim revision or perhaps not until the next yearly heavy revision.

The construction program is approved by the top executive of the highway department or the board of commissioners as provided by the law or policy of the state. In some states the governor is required to approve the highway construction program. These top approving officials may or may not give attention to each individual project, but they will look for certain projects about which they have personal knowledge and concern because of expressed public interest. They may ask for the reasons why a certain project was not included or why a certain project was not placed in an earlier or later year. These inquiries are propei, logical, and necessary because these officials are responsible for the program and need to be conversant with the facts, reasoning, and judgment supporting the program they approve.

In a few states the construction program as approved is published |26-12-1,

Construction Programming and Scheduling 653

p. 5]. Whether published or not the intent of the highway department with respect to its future construction plans is available to the public.

CONSTRUCTION SCHEDULING AND CONTROL

Adoption of a 3-to-10-year program of highway construction is one step forward toward realistic, efficient, and effective managment of the road building phase of the responsibilities of a highway department. This well-chosen program, however, will not of itself result in an acceptable road building operation. It must be followed by specific scheduling of the work, project by project, controlling the schedule, and adjusting the schedule to meet changing conditions. This scheduling is closely tied to four factors: (1) the rate that construction money becomes available, (2) the manpower available for preconstruction activities, (3) coordination of preconstruction activities, and (4) the rate that construction can be put into place.

SCHEDULING OF CONSTRUCTION PROJECTS

The scheduling of each phase of the work on each project from the date the project is definitely programmed for construction to the date of the contractor's final estimate resembles somewhat the practice in industry in manufacturing a new product or a new model. In the automotive industry it takes 3 to 10 years to produce a new model of an automobile. Design, procurement, budget, plant organization, subcontractors, and branch and headquarters staffs are involved. In a highway department the main phases of work to be scheduled and manned are: (1) preliminary surveys including soil and construction materials surveys, (2) traffic counts and forecasts, (3) drawing final engineering plans, (4) public hearings, (5) preparing specifications and instructions to bidders, (6) advertising and letting of contracts, and (7) scheduling of construction progress. The dates for completion of these and other steps are set on the basis of experience and urgency of the project. Some of the factors inherent in this process and which collectively control the time schedule are:

1. Capabilities of the highway department staff to complete the plans and acquire right-of-way[2]
2. Rate that money will be available to pay contractors' estimates
3. Balancing of work between earthwork, structures, and paving
4. Available contractors
5. Seasonal factors, including traffic
6. Coordination of projects to permit immediate opening to traffic once the construction is completed

---

[2] When incapability of the highway department staff to perform the necessary operations is the controlling factor in executing a program, this situation should be the subject of immediate management attention. It is a factor which can be remedied.

7. Related public works and community development

This detailed schedule once prepared becomes a control for all operations leading up to the completed project. The surveys and plans division can schedule its manpower requirements accordingly, and so can the construction division. The financial division can budget its cash flows accordingly to make certain that the funds for payment to contractors will be on hand. In fact, the cash flow forecast is one of the essential factors in setting the construction schedule. If the schedule does not utilize all available money, additional projects may be added.

Contractors and public utilities may lay their plans according to the construction schedule, and thus gain advantage from advance planning and scheduling. The possible gains reach all the way to land owners who may desire to plan for different operations after the highway improvement is completed.

MONITORING AND ADJUSTING THE SCHEDULE

To develop a construction schedule is one task; to keep the many phases of related activities on schedule is another task. By skillful management, a reasonable schedule based upon experience, can be maintained with a minimum give and take. A good schedule will provide for some slack and be generous in the estimated time to complete each phase rather than restrictive. Various devices, such as CPM (critical path method), PERT (program, evaluation, and review technique), and systems analysis by use of the electronic computer may be used for controlling and adjusting the schedule, as well as in developing the schedule in the first place.

As with the construction program, the construction schedule is not a once a year task, but a continuous task, day by day. Accomplishments will not always be according to plan. Changes in the schedule are therefore in order, on a daily basis, or at least as soon as evidence is available to prove that the existing schedule no longer can be adhered to.

The schedule is monitored through several communication schemes and by having each subordinate office set up its own schedule for its subphases. The headquarters office for controlling the schedule or subschedule can be kept posted on progress through staff conferences, a progress board or chart posted weekly, telephone inquiries, periodic form reports, and other devices.

Whenever it is certain that a project is definitely substantially ahead or behind schedule, adjustments in the schedule are in order. Often departure from schedule will affect other projects, either because of their interdependency or because their progress depends upon the same manpower or facilities. The complexity associated with establishing a construction program continues through the scheduling, adjusting the schedule, and controlling the schedule.

Despite these complexities and uncertainties the whole highway construction process has made remarkable improvement from the start of formal planning surveys in 1934 on down through formal needs studies,

Construction Programming and Scheduling 655

programming, scheduling, and controlling. These management tools and techniques are resulting in improved efficiency and effectiveness to the financial and social benefit of the public.

REFERENCES

See also the references at the ends of Chapters 24 and 25.

26-1. Automotive Safety Foundation. *Multiproject Scheduling for Highway Programs.* Proceedings of a two-day workshop sponsored by the Automotive Safety Foundation, Washington, D.C., December 11-12, 1963.

26-2. James E. Burke. *A Technique of Resource Allocation for Use with the Critical Path Method.* Department of Civil Engineering, Massachusetts Institute of Technology, Cambridge, June 1965.

26-3. Clinton H. Burnes. "Three R's of Highway Programming," *Better Roads,* Vol. 30, No. 5 (May 1960), pp. 19-20, 36.

26-4. Clinton H. Burnes. *Factors Basic to Sound Programming.* U.S. Bureau of Public Roads, Washington, D.C. Presented at annual meeting of Factual Survey Committee of the American Association of State Highway Officials, Detroit, November 25 to December 2, 1960.

26-5. Robert E. Coughlin. "The Capital Programming Problem," *American Institute of Planners Journal,* Vol. 26, No. 5 (February 1960), pp. 39-48.

26-6. David A. Curry. *Use of the Marginal Cost of Time in Highway Economy Studies.* Highway Research Board, Washington, D.C., Highway Research Record No. 77, 1963.

26-7. Joel Dean. *Capital Budgeting; Top-Management Policy on Plant, Equipment and Product Development.* Columbia U. P., New York, 1951.

26-8. Dan G. Haney. *The Use of Two Concepts of the Value of Time; the Willingness to Pay and the Cost of Time.* Highway Research Board, Washington, D.C., Research Record No. 12, 1963.

26-9. Highway Research Board. National Academy of Sciences, Washington, D.C., Special Report 56, *Economic Analysis in Highway Programming, Location, and Design,* September 17-18, 1959.

  1. G. P. St. Clair. "Economic Analysis: A Study in Uncertainties," pp. 3-8.
  2. Eugene L. Grant. "Concepts and Applications of Engineering Economy," pp. 8-14.
  3. Robley Winfrey. "Concepts and Applications of Engineering Economy in the Highway Field," pp. 19-29.
  4. D. W. Loutzenheiser, W. P. Walker, and F. H. Green. "Resume of AASHO Report on Road User Benefit Analyses," pp. 36-39.
  5. David S. Johnson. "Economic Analysis of Alternate Route Locations," pp. 43-44.
  6. Karl Moskowitz. "Special Problems in the Analysis of Urban Expressway Projects," pp. 47-57.
  7. R. C. Blensly. "Applications of Economic Analysis to Highway Systems and Programs," pp. 61-63.
  8. A. S. Lang. "Research in Economic Analysis at M.I.T.," pp. 67-72.

9. Claude A. Rothrock. "Cost Elements in Economic Analysis of Highway Programming, Location and Design," pp. 75-77.
10. Eugene L. Grant. "Interest and the Rate of Return on Investments," pp. 82-86.
11. J. P. Buckley. "Highway Costs As a Factor in Engineering Needs Estimates," pp. 91-95.
12. Nathan Cherniack. "Effects of Travel Impedance Costs," pp. 99-107.
13. Paul J. Claffey. "Motor Vehicle Operating and Accident Costs and Benefits Arising from Their Reduction Through Road Improvement," pp. 109-112.
14. G. P. St. Clair and Nathan Lieder. "Evaluation of Unit Cost of Time and Strain-and-Discomfort Cost of Nonuniform Driving," pp. 116-127.
15. Robert G. Hennes. "Highways As An Instrument of Economic and Social Change," pp. 131-136.
16. David R. Levin. "Identifying and Measuring Nonuser Benefits," pp. 136-148.
17. Richard M. Zettel. "The Incidence of Highway Benefits," pp. 148-153.
18. Bibliography, p. 180: (1) "Benefit/Cost Ratio Analysis"; (2) "Value of Time, Comfort, and Convenience"; and (3) "Economics of Highway Planning"

26-10. Highway Research Board. National Academy of Sciences, Washington, D.C., Special Report 62, *Formulating Highway Construction Programs*, September 19-20, 1960.
1. James W. Martin. "Problems in Formulating Highway Construction Programs," pp. 6-15.
2. Robley Winfrey. "Concepts of and Approaches to Capital Budgeting," pp. 20-26.
3. Donald R. Lang. "Scheduling Capital Improvements," pp. 30-39.
4. J. A. Legarra. "Role of the Legislature, Executive Branch, and Other Agencies in Highway Construction Programming," pp. 41-46.
5. Philip M. Donnell. "Basic Information Needed for Sound Capital Investment Planning," pp. 50-54.
6. Eugene C. Holshouser. "Accounting and Budgeting Requirements for Advance Construction Programs," pp. 57-60.
7. Eugene C. Holshouser. "The Case for Capital Budgeting in the State Highway Departments," pp. 63-72.
8. M. Earl Campbell. "Physical and Economic Rating Methods for Priority Considerations," pp. 75-93.
9. Arthur C. England. "Balancing of Physical and Economic Ratings with other Considerations to Establish Project Priorities," pp. 95-99.
10. John A. Swanson. "Coordinating the Highway Construction Schedule With All Agencies Concerned," pp. 109-113.
11. William B. Bidell. "The Role of Time and Money as Related to Construction Schedules," pp. 119-138.
12. M. J. Walker. "Control and Adjustments of Construction Schedule," pp. 145-154.
13. David R. Levin. "Highway Programming Law," pp. 160-169.
14. W. F. Babcock. "Administrative Requirements for Highway Construction Programming," pp. 174-179.

Construction Programming and Scheduling

   15. Donald M. Brown. "Public Relations Aspects of Highway Construction Programming," pp. 183-186.
   16. Clinton H. Burnes. "Formulating Highway Construction Programs," pp. 188-197.
   17. Selected Bibliography of 55 Items, Annotated, pp. 204-208.

26-11. Highway Research Board. National Academy of Sciences, Washington, D.C., Bulletin 249, *Highway Needs and Programming Priorities,* January 11-15, 1960.
   1. Evan H. Gardner. "The Congestion Approach to Rational Programming," pp. 1-23.
   2. James O. Granum and Clinton H. Burnes. "Advance Programming Methods for State Highway Systems," pp. 23-52.
   3. Charles M. Hummel. "A Criterion Designed to Aid Highway Expenditure Programing," pp. 52-61.
   4. Donald O. Covault and Harold L. Michael. "Estimation of County Primary Road System Needs by Sample Survey Methods," pp. 61-75.

26-12. Highway Research Board. National Academy of Sciences, Washington, D.C., Highway Research Record No. 32, *Construction, Programming, and Scheduling,* 1963.
   1. "A Review of Scheduling Procedures for State Highway Construction Programs," pp. 1-37.
   2. Robert N. Grunow. "PERT and Its Application to Highway Management," pp. 38-54.

26-13. Highway Research Board. National Academy of Sciences, Washington, D.C., Highway Research Board Special Report 70, *Highway Programming–An Analysis of State Law,* 1962.

26-14. Management Technology, Inc. *Highway Management System* (For the State Road Commission of West Virginia) Management Technology Inc., Los Angeles and Washington, D.C., January 1964. A 12-chapter book of procedures and controls for all phases of highway department management, including programming and scheduling of construction.

26-15. E. S. Preston and Associates. (A) *"Manual For Applying the Critical Path Method to Highway Department Engineering and Administration."* 1963; (B) *"Application of the Critical Path Method of Management Control to Statewide Highway Programming."* Volumes I and II, 1964. Work performed under contract for the U.S. Bureau of Public Roads, Washington, D.C., by E. S. Preston and Associates, Ltd., Columbus, Ohio.

26-16. E. S. Preston and Associates. *Application of the Critical Path Method of Management Control to Highway Programming for the District of Columbia Department of Highways and Traffic.* E. S. Preston and Associates, Ltd., Columbus, Ohio, September 1965.

26-17. James W. Spencer. *Planning and Programming Local Road Improvements: An Approach Based on Economic Consequences.* Stanford University, Program in Engineering Economic Planning, EEP No. 23, Stanford, Calif., May 1967.

26-18. Jack Sternbach. *An Integrated Highway Development and Management System for New York State Through Planning, Programming, Scheduling, and Monitoring.* New York State Department of Public Works, Albany, N. Y., September 1966.

Chapter **27**

Illustrative Problems and Solutions

The working of a wide selection of the problems given at the ends of the several chapters will afford good opportunity for developing the understanding and skill to find the solutions for the economy of most of the types of proposals to be encountered in highway engineering practice. Certain problems of greater complexity, or those containing special features, are presented in Chapter 27, together with explanations and the solutions.

ILLUSTRATIVE PROBLEM 27-1

ECONOMY OF SIZE OF CULVERT

THE PROBLEM

This problem on the economy of culvert size illustrates a factor common to design of drainage ditches, culverts, river bridges, and other facilities that are designed to carry or pass storm waters. It is the usual practice to design these types of drainage facilities to pass a specific size flood flow without damage, or with only minor damage, to the facility or adjacent property. Often the flood flow designed for is that maximum to be expected in a 10-, 20-, 50-, or 100-year period, which expectancy is determined by analysis of records of past rainfalls and stream flows.

In the economic analysis for size of facility or capacity for flood flow, it is necessary, however, to consider the probability of experiencing greater than design flows. This factor has been often overlooked by analysts for economy. This illustrative problem is restricted to the analysis for the economy of the design capacity; it assumes that the hydrology and hydraulic factors will be properly handled. A circular culvert is used for illustrative purposes, but the factors and procedures involved are generally applicable to other types of facilities for passing flood flows. In this particular case the alternative of raising the highway grade line to store water on the upstream side is not considered. Such provision will increase the highway construction cost, including the cost for a longer culvert, but the temporary storage of the flood water will permit using a culvert of smaller cross section than is necessary without the temporary storage. The elevation of the grade line of the highway is another alternative associated with both culvert design and bridge design.

The procedure herein followed for the design of culverts is followed in the customary manner, except seven different sized culverts are chosen to cover the range from too small

Illustrative Problems and Solutions

to too large for the usual design maximum rate of discharge corresponding to a flood frequency such as 20-year, 40-year, or 50-year recurrence periods. In practice, perhaps only 2 to 4 discharge rates need be designed for. The construction cost and annual maintenance costs are estimated. The next step is to estimate the damages (in dollars) and extra expenses which may be caused by the design discharge and a series of discharges (recurrence periods) above the design discharge.

The flood damages include such items as follow, each estimated for one storm at each recurrence period or discharge rate:

1. Damage to private property, above and below the culvert.
2. Damage to the highway grade, pavement, shoulders, culverts, and other roadway items.
3. Expenses for flagmen, barricades, flares, and signing, including the marking of traffic detours.
4. Road-user motor vehicle running cost for detour mileage, slowdowns, stops, washing, accidents, etc.
5. Road-user travel time increases for detours, slowdowns, stops, standing delays, and reduced travel speed.

The amount of damages and costs resulting from the individual storms will depend upon the area and depth of impounded water; the depth, width, velocity, and duration of the flood flow across the roadway; and the traffic volume and composition at the time of the storm—which factor cannot be estimated specifically because of the variable character of traffic hour-by-hour and day-to-day.

ASSUMED FACTORS

Vestcharge rate of 8 percent; 40-years analysis period; and zero terminal value.

Circular culvert with head wall and end wall, 70-ft in length under a two-lane rural paved highway.

An ADT of 2,500 and a gradient increase of 60 ADT per year. Attempted speed of passenger car of 50 mph.

Table 27-1A gives the design data and the estimated costs for a range of discharges, at and above each of the seven design discharges.

SOLUTION

The solution is to compute the equivalent uniform annual cost (or present worth of costs) of the total culvert and storm damages combined for a series of sizes of culvert designs. The sizes should be chosen with the expectation that the economical size will be found between the smallest and largest diameter chosen. The most economical size is that having the lowest equivalent uniform annual cost or the lowest present worth of costs.

Table 27-1B gives the results of the calculations. In making the calculations it is suggested that curves be plotted of the estimates of damages for each item of damage and of the total damage costs in Table 27-1A. Also, the costs of damages in Table 27-1A should be plotted against the probabilities P as an aid to calculation of the equivalent uniform annual damages in Column 5 of Table 27-1B.

COMMENT

The 51-in.-diameter culvert has the lowest equivalent uniform annual cost ($310) and, therefore, would be chosen, assuming there were not other factors to the contrary.

TABLE 27-1A

ESTIMATED TOTAL DAMAGE COST, HIGHWAY COSTS, AND ROAD-USER COSTS
PER OCCURRENCE OF DISCHARGES EQUAL TO AND ABOVE THE DESIGN DISCHARGE

Recurrence Interval, Years	Probability of Recurrence, P	Maximum Discharge, cfs	Design Diameter, Inches	Total Cost ($) of Damages and Extra Expense* for Each Recurrence of Design Discharge and Greater for Recurrence Period of:							
				2.5	5	10	20	40	80	160	Max.†
2.5	0.40	105	45	50	206	352	484	600	698	850	7,000
5	0.20	128	48		66	222	368	500	616	794	6,920
10	0.10	149	51			80	236	382	514	728	6,730
20	0.05	168	54				94	250	396	644	6,410
40	0.025	187	57					108	264	542	6,000
80	0.0125	206	60						120	422	5,430
160	0.00625	225	66							140	4,250

* The totals include estimates for damage to private property, highway property, and expenses of the highway department and road users.

† The maximum cost is for a probability of recurrence of zero, or the maximum costs that could result from a maximum sized storm and discharge.

TABLE 27-1B

CALCULATION OF EQUIVALENT UNIFORM ANNUAL COSTS

Design Diameter, Inches	Total Construction Cost, Dollars	Annual Capital Costs at 8%-40-yr, Dollars	Equivalent Uniform Annual Culvert Maintenance Cost, Dollars	Equivalent Uniform Annual Damage Costs,* Dollars	Total EUAC, Dollars
1	2	3	4	5	6
45	2,430	204	20	125	349
48	2,680	225	18	76	319
51	2,930	246	16	48	310
54	3,195	268	14	34	316
57	3,470	291	12	25	328
60	3,765	316	10	20	346
66	4,440	372	10	14	396

* The equivalent uniform annual damage costs are calculated from the damage costs and probabilities given in Table 27-1A. The calculation is for the area under the curve for each culvert diameter when the damage cost as ordinate is plotted against the probability of occurrence as abscissa. The calculation for the 54-in. design for a recurrence interval of 20 years is:

$\frac{1}{2}(6410 + 644)(0.00625 - 0)$ = 22.04
$\frac{1}{2}(644 + 396)(0.0125 - 0.00625)$ = 3.25
$\frac{1}{2}(396 + 250)(0.025 - 0.0125)$ = 4.04
$\frac{1}{2}(250 + 94)(0.050 - 0.025)$ = 4.30
Total   33.63 rounded to $ 34

REFERENCES

J. B. Franzini. "Flood Control–Average Annual Benefits," *Consulting Engineer*, Vol. 16, No. 5 (May 1961), pp. 107-109.

Illustrative Problems and Solutions

James Morgali and C. H. Oglesby. *Procedures for Determining the Most Economical Design for Bridges and Roadways Crossing Flood Plains.* Highway Research Board, National Academy of Sciences, Washington, D.C., Bulletin 320, *Studies in Highway Engineering Economy*, 1962, pp. 40-65.

Harold D. Pritchett. *Application of the Principles of Engineering Economy to the Selection of Highway Culverts.* Stanford University, Institute in Engineering-Economic Systems, Report EEP-13 (August 1964).

ILLUSTRATIVE PROBLEM 27-2

ECONOMY OF INTERSECTION TRAFFIC CONTROL METHODS

THE PROBLEM

The traffic engineer has been requested to submit proposals for some type of traffic control device at a particular urban four-way intersection. The traffic data for the three types of control devices under study are listed below. Assume that the traffic volume will not be influenced by the particular control device installed, and that the attempted operating (approach) speed is 25 mph. Use an analysis period of 10 years, a vestcharge rate of 10 percent, and zero terminal value.

The average daily traffic through the intersection at zero date is 25,000; 22,000 passenger cars, 2,500 commercial delivery trucks, and 500 single-unit trucks. The major leg has 65 percent of the ADT. Traffic growth will be at a uniform annual rate of 1,250 vehicles (5 percent of zero date costs and of the ADT). The traffic signal stops 52 percent of the traffic.

Value of travel time: passenger cars, $1.50 per vehicle-hour; commercial delivery, $2.50; and single-unit trucks, $3.50.

	Two-Way Stop	Four-Way Stop	Two-Phase Traffic Signal
1. Average idle time, per stop	0.16 min.	0.10 min.	0.21 min.
2. Traffic accident frequency (accidents per year) at zero date (will increase with ADT gradient)			
(a) Personal injury ($860 cost per accident)	0.6	0.0	0.8
(b) Property damage ($200 cost per accident)	1.6	1.1	1.9
3. Control installation capital costs	$100	$200	$6,000
4. Annual maintenance and operating costs	$25	$40	$175

THE SOLUTION

	Two-Way Stop	Four-Way Stop	Signal
1. Number of vehicles stopped per day, zero date			
A. Passenger cars	7,700	22,000	11,440
B. Commercial delivery	875	2,500	1,300
C. Single-unit trucks	175	500	260

2. Excess vehicle running cost, per year, zero date
   A. Stopping costs

(a) Passenger cars ($6.96/1000 stops) (ADT stopped) (365)	$19,561	$55,889	$29,062
(b) Commercial delivery ($8.00/1000 stops) (ADT stopped) (365)	2,555	7,300	3,796
(c) Single-unit trucks ($17.65/1000 stops) (ADT stopped) (365)	1,127	3,221	1,675

   B. Idling of motor

(a) Passenger cars (114.86/1000) (0.16/60; 0.10/60; 0.21/60)	861	1,537	1,679
(b) Commercial delivery ($132.54/1000) (0.16/60; 0.10/60; 0.21/60)	113	202	220
(c) Single-unit trucks ($200.03/1000) (0.16/60; 0.10/60; 0.21/60)	34	61	66

   C. Subtotal, per year, zero date vehicle running costs — $24,251 / $68,210 / $36,498

3. Gradient growth vehicle running costs (Item 2-C) (5%) (GUS-10%-10 = 4.725461) — 5,730 / 16,116 / 8,624
4. Total EUAC of running costs, Item 2-C plus Item 3 — $29,981 / $84,326 / $45,122
5. Value of excess travel time per year, zero date
   A. Stopping time

(a) Passenger cars ($1.50) (1.91 hr/1000) (ADT stopped) (365)	$8,052	$23,006	$11,963
(b) Commercial delivery ($2.50) (2.36 hr/1000) (ADT stopped) (365)	1,884	5,384	2,800
(c) Single-unit truck ($3.50) (3.67 hr/1000) (ADT stopped) (365)	820	2,344	1,219

   B. Idling time

(a) Passenger cars ($1.50) (0.16/60; 0.10/60; 0.21/60)	$11,242	$20,075	$21,922
(b) Commercial delivery ($2.50) (0.16/60; 0.10/60; 0.21/60)	2,129	3,802	4,152
(c) Single-unit trucks ($3.50) (0.16/60; 0.10/60; 0.21/60)	596	1,065	1,163

   C. Subtotal, per year, zero date, travel time costs — $24,723 / $55,676 / $43,219

6. Gradient growth vehicle time costs (Item 5-C) (5%) (GUS-10%-10 = 4.725461) — $5,841 / $13,155 / $10,211
7. Total EUAC of travel time, Item 5-C plus Item 6 — $30,564 / $68,831 / $53,430
8. Traffic accident costs per year

A. Nonfatal injury, zero date–$860 (0.6; 0.0; 0.8)	$516	$0	$688
B. Property damage only, zero date–$200 (1.6; 1.1; 1.9)	320	220	380
C. Total first-year traffic accident cost	836	220	1,068
D. Gradient accident costs (836;220; 1,068) (4.725461) (5%)	196	52	252

9. Equivalent uniform accident costs — $1,032 / $272 / $1,320
10. Total road user EUAC–Items 4, 7, and 9 — $61,577 / $153,429 / $99,872
11. EUAC of traffic control installation

A. Capital cost ($100; $200; $6,000) (CR-10%-10=0.162745)	16	32	976
B. Equivalent uniform maintenance of installation	25	40	175

12. Total EUAC of traffic control installation — 41 / 72 / 1,151
13. Grand total EUAC–Items 10 plus 12 — $61,618 / $153,501 / $101,023

COMMENTS

This problem illustrates the effect of stopping traffic on both running costs and travel time increase. The Four-way stop is highly costly because it stops 100 percent of the traffic, often when there is little probability of traffic interference. The increased travel time because of the stops accounts for about half of the total road user costs. The two-way control is the least costly in both running costs and time costs.

REFERENCE

Stanford University, Institute in Engineering Economic Systems. *Application of the Principles of Engineering Economy to Highway Improvements.* Report EEP-8 (March 1964), pp. 12-43.

Illustrative Problems and Solutions

ILLUSTRATIVE PROBLEM 27-3

Highway managements have long practiced the construction of highways in two or more stages spread over a period of years. Such stage construction has been elected because of one or more advantages: (1) Preference to spread the available current construction budget over a larger number of projects on which some improvement could be made, (2) the traffic volume demand could be satisfied for a few years hence short of the full design capacity, and (3) improvements to be gained in the construction sequence and construction quality of the particular total project. The disadvantages of stage construction are (1) increased total cost of construction, and (2) lesser inmediate gains to the road user.

The analysis for the economy of stage construction will be illustrated by a case study[1] on constructing immediately only two lanes of a proposed four-lane divided highway with full access control. The two-lane stage has full access control.

STAGE CONSTRUCTION, TWO LANES TO FOUR LANES
THE PROBLEM

The project, a rural section of an interstate route, is 6.814 miles in length. The vertical grades do not exceed 2.5 percent and average about 1.5 percent. The grades do not appreciably alter the vehicle running cost between two-lane and four-lane operation. The horizontal curves are flat enough that there is no appreciable difference in motor vehicle running costs between the two-lane and four-lane alternatives. Changes in speed for passing and other traffic actions are included for two-lane stage construction, but speed changes on the four-lane operation are not significant. The travel time difference because of speed change is included in the stated average speed. All construction is considered to be completed on December 31—the zero date. A vest-charge rate of 8 percent is used.

THE SOLUTION

The economy of constructing immediately only two lanes of a proposed four-lane divided highway is found by projecting into the future the present worth of the highway cost and the road user cost until the year of equality is found. This projection is accomplished by assuming construction of the second two lanes during each of a series of future years. In this example, seven two-year intervals are used, but often it is necessary to solve for only two or three years. To be fully economical the year of equality of stage construction should come before the year that the ADT becomes in excess of the two-lane capacity.

[1] Robley Winfrey. *Cost Comparison of Four-Lane vs. Stage Construction on Interstate Highways.* Highway Research Board, National Academy of Sciences, Washington, D.C., Bulletin 306, *Studies in Highway Engineering Economy*, 1961, pp. 64-80. This reference gives full detailed data, calculations, and discussions for three separate projects proposed for stage construction. The problem presented herein is a modification of Project B in Bulletin 306.

TABLE 27-3A

ESTIMATED CONSTRUCTION AND MAINTENANCE COSTS

Cost Item	Two-Lane Stage Construction (6.814 Miles)		Four-Lane Construction (6.814 Miles)
	1969	Future	1969
Preliminary engineering	$35,000	$50,000	$70,000
Rights-of-way	130,000	None	130,000
Utility adjustments	49,300	None	49,300
Grade, drain, minor structures	352,000	604,000	714,000
Pavement and shoulders	400,000	493,000	696,000
Highway grade separation without ramps	None	25,000	20,600
Interchange, complete	None	280,000	251,000
Bridges	35,000	83,000	113,000
Guardrail, fencing, lighting	78,000	113,000	151,000
Roadside improvement	10,600	15,000	20,000
Other items	3,000	5,000	6,000
Construction engineering, contingencies	92,800	161,800	202,100
Total construction cost	$1,185,700	$1,829,800	$2,423,000
Annual maintenance expense	+7,400	$17,900	$17,900

TABLE 27-3B

FORECASTED AVERAGE DAILY TRAFFIC AS OF JANUARY 1, EACH YEAR

Vehicle Class	Zero Year	2nd Year	4th Year	6th Year	8th Year	10th Year	12th Year
Passenger cars	2,560	2,860	3,160	3,460	3,760	4,060	4,360
Commercial delivery	400	440	480	520	560	600	640
Single-unit trucks	160	190	220	250	280	310	340
40-kip 2-S2	80	90	100	110	120	130	140
50-kip 3-S2	200	230	260	290	320	350	380
Total	3,400	3,810	4,220	4,630	5,040	5,450	5,860

RESULTS

Table 27-3F gives the road user cost and the present worths of the highway costs for completing the project in the 2nd, 4th, ... 12th year following initial construction. Table 27-3G gives the cumulative totals of the present worths of the costs.

From Table 27-3G it is seen that the road-user costs on two-lane stage construction are higher than on the four-lane initial construction, and sufficiently so that they overbalance the vestcharge savings of the lower capital initial two-lane construction costs. For the two-lane stage construction to show economy over initial four-lane construction, the present worths in column 9 for two-lanes would have to become less at some future year

Illustrative Problems and Solutions

than the present worths for the four-lanes. Because for corresponding future years up to 12 years, the two-lane present worths exceed the four-lane present worths (column 9) the four-lane initial construction is preferred. The calculation is not carried beyond the 12th

TABLE 27-3C

AVERAGE SPEED, MPH, YEARLY AVERAGE, INCLUDING EFFECTS OF GRADES

Vehicle Class	Zero Year	2nd Year	4th Year	6th Year	8th Year	10th Year	12th Year
A. Two-lane stage							
Passenger cars	55	55	55	$52\frac{1}{2}$	$52\frac{1}{2}$	50	50
Commercial delivery	$52\frac{1}{2}$	$52\frac{1}{2}$	$52\frac{1}{2}$	50	50	50	$47\frac{1}{2}$
Single-unit trucks	50	50	$47\frac{1}{2}$	$47\frac{1}{2}$	45	45	45
40-kip 2-S2	45	45	$42\frac{1}{2}$	$42\frac{1}{2}$	40	40	$37\frac{1}{2}$
50-kip 3-S2	45	45	$42\frac{1}{2}$	$42\frac{1}{2}$	40	$37\frac{1}{2}$	$37\frac{1}{2}$
B. Four-lane construction							
Passenger cars	60	60	60	60	60	60	60
Commercial delivery	$57\frac{1}{2}$	$57\frac{1}{2}$	$57\frac{1}{2}$	$57\frac{1}{2}$	$57\frac{1}{2}$	$57\frac{1}{2}$	$57\frac{1}{2}$
Single-unit trucks	55	55	55	55	$52\frac{1}{2}$	$52\frac{1}{2}$	$52\frac{1}{2}$
40-kip 2-S2	50	50	50	$47\frac{1}{2}$	$47\frac{1}{2}$	45	45
50-kip 3-S2	50	50	50	$47\frac{1}{2}$	$47\frac{1}{2}$	45	45

TABLE 27-3D

EQUIVALENT NUMBER (100) OF YEARLY 15-MPH SPEED CHANGES FROM INITIAL SPEEDS

Vehicle Class	Initial Speed mph	Zero Year	2nd Year	4th Year	6th Year	8th Year	10th Year	12th Year
A. Two-lane stage construction								
Passenger cars	45	6,200	6,640	7,080	7,520	7,960	8,400	8,840
	50	15,300	16,300	17,300	18,300	19,300	20,300	21,300
	55	8,600	8,970	9,340	9,710	10,080	10,450	10,820
Commercial delivery	45	1,100	1,180	1,260	1,340	1,420	1,500	1,580
	50	2,500	2,640	2,780	2,920	3,060	3,200	3,340
	55	1,400	1,460	1,520	1,580	1,640	1,700	1,760
Single-unit trucks	40	325	345	365	385	405	425	445
	45	400	420	440	460	480	500	520
	50	210	220	230	240	250	260	270
40-kip 2-S2	40	200	205	210	215	220	225	230
	45	120	125	130	135	140	145	150
50-kip 3-S2	40	500	515	530	545	560	575	590
	45	300	310	320	330	340	350	360
B. Four-lane construction. Speed changes are not significant.								

TABLE 27-3E
Traffic Accident Rates and Cost of Accidents

Severity of Accident	Two-Lane Stage Construction			Four-Lane Construction		
	Accidents per 100 Million Vehicle-Miles	Cost per Accident	Cost per 100 Million Vehicle-Miles	Accidents per 100 Million Vehicle-Miles	Cost per Accident	Cost per 100 Million Vehicle-Miles
Property damage only	130	$275	$35,750	65	$500	$32,500
Nonfatal injury ..	72	2,700	194,400	45	3,000	135,000
Fatal injury ..	5	7,000	35,000	3	9,000	27,000
Total ...	207	–	$265,150	113	–	$194,500

TABLE 27-3F
Solution Calculations

Cost Item	Years Following Initial Construction						
	Zero Year	2nd Year	4th Year	6th Year	8th Year	10th Year	12th Year
1. Yearly Running Cost of Traffic							
A. Two-lane stage construction							
(a) Passenger cars	$240,100	$268,236	$296,373	$319,002	$346,661	$368,868	$396,124
(b) Commercial delivery ..	43,256	47,581	51,907	54,952	59,179	63,406	66,440
(c) Single-unit trucks	30,080	35,720	40,271	45,763	50,029	55,389	60,749
(d) 40-kip 2-S2	21,972	24,719	26.664	29,331	31,203	33,804	35,666
(e) 50-kip 3-S2	52,627	60,521	66,353	74,009	79,468	84,864	92,138
(f) Total, all traffic	$388,035	$436,777	$481,568	$523,057	$566,540	$606,331	$651,117
B. Four-lane construction							
(a) Passenger cars	$250,987	$280,400	$309,812	$339,225	$368,637	$398,050	$427,463
(b) Commercial delivery ..	45,992	50,591	55,190	59,789	64,388	68,987	73,587
(c) Single-unit trucks	31,931	37,917	43,904	49,891	54,172	59,976	65,781
(d) 40-kip 2-S2	23,618	26,570	29,522	31,251	34,092	35,705	38,451
(e) 50-kip 3-S2	56,522	65,000	73,479	78,942	87,109	92,098	99,992
(f) Total, all traffic	$409,050	$460,478	$511,907	$559,098	$608,398	$654,816	$705,274
2. Yearly Running Cost of Speed Changes							
A. Two-lane stage construction							
(a) Passenger cars							
45 mph, $10.25	$6,355	$6,806	$7,257	$7,708	$8,159	$8,610	$9,061
50 mph, $12.31	18,834	20,065	21,296	22,527	23,758	24,989	26,220
55 mph, $14.78	12,711	13,258	13,805	14,351	14,898	15,445	15,992
(b) Commercial delivery							
45 mph, $11.48	1,263	1,355	1,446	1,538	1,630	1,722	1,814
50 mph, $13.76	3,440	3,633	3,825	4,018	4,211	4,403	4,596
55 mph, $16.32	2,285	2,383	2,481	2,579	2,676	2,774	2,872

Illustrative Problems and Solutions

TABLE 27-3F (continued)

(c) Single-unit trucks							
40 mph, $21.26	691	733	756	819	861	904	946
45 mph, $25.29	1,012	1,062	1,113	1,163	1,214	1,265	1,315
50 mph, $30.33	637	667	698	728	758	789	819
(d) 40-kip, 2-S2							
40 mph, $80.05	1,601	1,641	1,681	1,721	1,761	1,801	1,841
45 mph, $97.58	1,171	1,220	1,269	1,317	1,366	1,415	1,464
(e) 50-kip, 3-S2							
40 mph, $100.04	5,002	5,152	5,302	5,452	5,602	5,752	5,902
45 mph, $123.16	3,695	3,818	3,941	4,064	4,187	4,311	4,434
(f) Total excess cost of speed changes	$58,697	$61,793	$64,870	$67,985	$71,081	$74,180	$77,276

B. Four-lane construction
(Speed change not significant)

3. Cost per Year of Traffic Accidents

A. Two-lane stage construction							
(a) Million vehicle-miles per year	8.456	9.476	10.496	11.515	12.535	13.555	10.486
(b) Total cost of accidents	$22,421	$25,126	$27,830	$30,532	$33,237	$35,941	$27,830
B. Four-lane construction							
(a) Million vehicle-miles per year		Same as for 2-lane					
(b) Total cost of accidents	$16,447	$18,431	$20,415	$22,397	$24,381	$26,364	$28,346

4. Yearly Value of Travel Time

A. Two-lane stage construction							
(a) Passenger cars $2.00/vehicle-hr	$231,527	$258,660	$285,792	$327,825	$356,249	$403,907	$433,752
(b) Commercial delivery $3.50/vehicle-hr	66,323	72,955	79,587	90,531	97,495	104,459	117,287
(c) Single-unit trucks $4.00/vehicle-hr	31,835	37,804	46,077	52,360	61,901	68,534	75,166
(d) 40-kip 2-S2 $5.00/vehicle-hr	22,108	24,871	29,260	32,186	37,307	40,416	46,426
(e) 50-kip 3-S2 $5.50/vehicle-hr	60,796	69,915	83,684	93,340	109,433	127,672	138,615
(f) Total, all traffic	$412,589	$464,205	$524,400	$596,242	$662,385	$744,988	$811,246
B. Four-lane construction							
(a) Passenger cars	$212,233	$237,104	$261,975	$286,846	$311,718	$336,589	$361,460
(b) Commercial delivery	60,556	66,611	72,667	78,722	84,778	90,834	96,889
(c) Single-unit trucks	28,941	34,367	39,794	45,220	53,058	58,743	64,428
(d) 40-kip 2-S2	19,897	22,384	24,871	28,798	31,416	35,925	38,688
(e) 50-kip 3-S2	54,716	62,924	71,131	83,515	92,154	106,393	115,512
(f) Total, all traffic	$376,343	$423,390	$470,438	$523,101	$573,124	$628,484	$676,977

5. Present Worth of Highway Construction and Maintenance

A. Two-lane stage construction							
(a) Present worth of first stage costs	$1,185,700	–	–	–	–	–	–
(b) Present worth of second stage costs	1,829,800	1,568,759	1,344,958	1,153,085	988,584	847,551	726,639
(c) Present worth of first stage maintenance	(7,400)	13,196	24,510	34,209	42,525	49,655	55,767
(d) Total present worth of stage costs	3,015,500	2,767,655	2,555,168	2,372,994	2,216,809	2,082,906	1,968,106

TABLE 27-3F *(continued)*

B. Four-lane construction							
(a) Present worth of initial construction ...	2,423,000	–	–	–	–	–	–
(b) Present worth of maintenance	(17,900)	31,920	59,287	82,750	102,865	120,110	134,896
(c) Total present worth of 4-lane construction	$2,423,000	$2,454,920	$2,482,287	$2,505,750	$2,525,865	$2,543,110	$2,557,896

year because the ADT of 5,860 the 12th year is near to that volume warranting four lanes, especially with the high percentage of trucks.

Not considered in this solution are alternatives of constructing less work on the initial two-lane stage thus increasing the costs when the second two lanes are constructed.

TABLE 27-3

SUMMARY AND COMPARISONS OF PRESENT WORTHS OF COSTS
(All Costs Are Discounted to Present Worth at 8 Percent Vestcharge)

Year from Initial Construction	Running Cost of Vehicles	Cost of Speed Changes	Cost of Traffic Accidents	Cost of Travel Time	Total Road User Costs	Cumulated Total User Costs	Highway Costs	Road User Plus Highway Costs
1	2	3	4	5	6	7	8	9
A. Two-lane stage construction								
0	$388,035	$58,697	$22,421	$412,589	$881,742		$3,015,500	$3,015,500
1					864,354	$864,354		
2	374,466	52,978	21,541	397,981	846,966	1,711,320	2,767,655	4,478,975
3					827,260	2,538,580		
4	353,967	47,681	20,456	385,450	807,554	3,346,134	2,555,168	5,901,302
5					787,492	4,133,626		
6	329,615	42,842	19,240	375,734	767,431	4,901,057	2,372,994	7,274,051
7					743,870	5,644,927		
8	306,084	38,403	17,957	357,866	720,310	6,365,237	2,216,809	8,582,046
9					698,620	7,063,857		
10	280,848	34,360	16,648	345,073	676,929	7,740,786	2,082,906	9,823,692
11					651,844	8,392,630		
12	258,568	30,687	15,346	322,157	626,758	9,019,388	1,968,106	10,987,494
B. Four-lane construction								
0	409,050	None	16,447	376,343	801,840		2,423,000	2,423,000
1					787,708	787,708		
2	394,786		15,802	362,989	773,577	1,561,285	2,454,920	4,016,205
3					755,318	2,316,603		
4	376,267		15,006	345,786	737,059	3,053,662	2,482,287	5,535,949
5					716,572	3,770,234		
6	352,327		14,114	329,643	696,084	4,466,318	2,505,750	6,972,068
7					673,798	5,140,116		
8	328,699		13,172	309,641	651,512	5,791,628	2,525,865	8,317,493
9					629,070	6,420,698		
10	303,306		12,212	291,109	606,627	7,027,325	2,543,110	9,570,435
11					583,398	7,610,723		
12	280,074		11,257	268,837	560,168	8,170,891	2,557,896	10,728,787

Note: The present worths in Column 9 for the two-lane stage construction do not get less than for the four-lane construction. Therefore, up to 12 years after initial construction the four-lane final construction is preferred to stage construction.

Illustrative Problems and Solutions 669

ILLUSTRATIVE PROBLEM 27-4

PROJECT FORMULATION FOR HIGH-LEVEL BRIDGE VS. BASCULE BRIDGE

THE PROBLEM

The original plans were to construct a high-level fixed span bridge to meet navigational clearances over the Diamond River. A bridge at the location is required to serve a new aerospace manufacturing plant. Since the river traffic to the plant from ocean traveling barges and ships is relatively light a bascule (movable span) bridge was suggested as an alternative to the high-level bridge. Compare the economy of the two types of bridges using a vestcharge rate of 7 percent, an analysis period of 25 years, zero terminal value, and the data and assumptions given in the adjoining table.

	High-Level Bridge	Low-Level Bascule
Highway costs		
Construction	$4,039,500	$2,946,000
Annual uniform maintenance	20,000	25,000
Annual operation (144 openings $10 each)		1,440

The high-level bridge will have + 3 percent and − 3 percent approaches of 0.347 mile each. The bascule low-level bridge will have a 0 percent grade and approaches.

The bascule bridge openings will average 144 per year with no appreciable increase in the number in the next 25 years. Each bridge opening will stop traffic for an average idle time of 20 minutes.

The traffic and its performance data are:

Cost or Informational Item	Passenger Cars	Commercial Delivery	Single Unit Trucks	40-kip 2-S2 Combination	50-kip 3-S2 Combination	Total
1. Zero date ADT	11,100	1,500	1,200	750	450	15,000
2. Uniform annual gradient ADT increase (3 1/3%)	370	50	40	25	15	500
3. Zero date, average number of vehicles stopped by each bridge opening	125	12	45	28	15	225
4. Uniform annual gradient increase in number of vehicles stopped	4.17	0.40	1.50	0.93	0.50	7.50
5. Approach speed, mph	50	50	50	40	40	–
6. Speed on 3% plus grade	45	45	40	30	30	–
7. Speed on 3% minus grade	50	50	45	35	35	–
8. Value of travel time, dollars/vehicle-hour	$1.00	$3.00	$4.50	$6.00	$6.50	–

Illustrative Problems and Solutions

SOLUTION

1. Vehicle running cost for 0.694 mile

   A. High-level bridge .... (ADT/1000) (miles) (¢/mi. + 3%, − 3%) = Daily cost

	(ADT/1000)	(miles)	(¢/mi. + 3%, − 3%)	= Daily cost
(a) Passenger cars	11.10	0.347	(42.17 + 30.77)	= $ 280.94
(b) Commercial deliveries	1.50	"	(49.20 + 35.34)	= 44.01
(c) Single-unit trucks	1.20	"	(93.86 + 57.45)	= 63.01
(d) 40-kip 2-S2	0.75	"	(160.45 + 69.80)	= 59.92
(e) 50-kip 3-S2	0.45	"	(146.79 + 74.40)	= 34.54
(f) Total	15.00	–	–	$ 482.42

   (g) Total annual running cost per year at zero date .. $ 176,083
   (h) Annual gradient growth at 3 1/3% .... $ 5,869

   B. Bascule bridge

(a) Passenger cars	11.10	0.694	36.53	$ 281.41
(b) Commercial deliveries	1.50	"	42.49	44.23
(c) Single-unit trucks	1.20	"	75.59	62.95
(d) 40-kip 2-S2	0.75	"	104.55	54.42
(e) 50-kip 3-S2	0.45	"	99.85	31.18
(f) Total	15.00	–	–	$ 474.19

   (g) Total running cost per year at zero date .... $ 173,079
   (h) Annual gradient growth at 3 1/3% 5,769

2. Running cost of stopping traffic at opened bascule bridge

   A. Vehicle stopping cost $/1000 stops

(a) Passenger cars	0.125	144 stops	25.15	$452.70
(b) Commercial deliveries	.012	"	29.67	51.27
(c) Single-unit trucks	.045	"	63.11	408.95
(d) 40-kip 2-S2	.028	"	141.08	568.83
(e) 50-kip 3-S2	.015	"	178.00	384.48
(f) Total	0.225	–	–	$ 1,866.23

   (g) Total stopping cost per year at zero date .... $ 1,866.23
   (h) Annual gradient growth at 3 1/3% 62.21

   B. Idling cost at 20 min per stop $/1000 hours

(a) Passenger cars	0.125	48 hours	114.86	$689.16
(b) Commercial deliveries	.012	"	132.54	76.34
(c) Single-unit trucks	.045	"	200.03	432.06
(d) 40-kip 2-S2	.028	"	249.45	335.26
(e) 50-kip 3-S2	.015	"	196.28	141.32

   (f) Total idling cost per year at zero date .... $ 1,674.14
   (g) Annual gradient growth at 3 1/3% 55.80

Illustrative Problems and Solutions

3. Excess cost of travel time due to stopping at opening of bascule bridge
   A. Excess stopping time cost

		hr/1000/stops	$/hr	Annual cost of time
(a) Passenger cars	0.125 144 stops	3.75	1.00	$67.50
(b) Commercial deliveries	.012 "	4.87	3.00	25.25
(c) Single-unit trucks	.045 "	7.33	4.50	213.74
(d) 40-kip 2-S2	.028 "	7.76	6.00	187.73
(e) 50-kip 3-S2	.015 "	11.09	6.50	155.70

   (f) Total annual cost of stopping time excess, zero date $649.92
   (g) Annual gradient growth of 3 1/3% 21.66

   B. Cost of idling motor delay time

(a) Passenger car	48 hours		1.00	$6,000
(b) Commercial deliveries	"		3.00	1,728
(c) Single-unit trucks	"		4.50	9,720
(d) 40-kip 2-S2	"		6.00	8,064
(e) 50-kip 3-S2	"		6.50	4,680

   (f) Total annual cost of idling time, zero date $30,192
   (g) Annual gradient growth at 3 1/3% 1,006

4. Summary of annual transportation costs

	High-Level Bridge		Bascule	
	Zero date	Gradient	Zero date	Gradient
A. Vehicle running cost				
(a) Distance and grade	$176,083	$5,869	$173,079	$5,769
(b) Stopping			1,866	62
(c) Idling			1,674	56
B. Travel time cost				
(a) Stopping			650	22
(b) Idling			30,192	1,006
C. Total road-user costs	$176,083	$5,869	$207,461	$6,915

5. Present worth of costs
   A. Bridge
   (a) Construction $4,039,500 $2,946,000
   (b) Operation and maintenance
   $K(SPW\text{-}7\%\text{-}25 = 11.653583)$ $233,100 291,300
   (c) Total bridge present worth $4,272,600 $3,237,300
   B. Road-user present worths
   (a) Uniform $U(SPW\text{-}7\%\text{-}25 = 11.653583)$ $2,052,000 $2,417,700
   (b) Gradient $G(GPW\text{-}7\%\text{-}25 = 112.330065)$ 659,300 776,800
   (c) Total road-user present worth $2,711,300 $3,194,500

6. Grand total present worth of costs $6,983,900 $6,431,800

The bascule has the smaller present worth of costs by a difference of $552,100, so it is the choice on the basis of engineering economy.

7. Rate of return solution

By considering the two alternatives as a pair, the differences in cash flows may be used to form the rate of return equation to be solved for the rate of return for the increase in capital cost of the high-level bridge over the bascule bridge. The solution using the EUAC procedure is:

$$0 = -(4,039,500 - 2,946,000)\,(CR\text{-}i\text{-}25) - (20,000 - 25,000) - (176,083 - 207,461)$$
$$- (5,869 - 6,915)\,(GUS\text{-}i\text{-}25)$$
$$0 = -1,093,500\,(CR\text{-}i\text{-}25) + 5,000 + 31,378 + 1,046\,(GUS\text{-}i\text{-}25)$$

Try 1% vestcharge rate:
$$0 = -1,093,500\,(0.04541) + 36,378 + 1,046\,(12.483116)$$
$$0 = -49,656 + 36,378 + 13,057 = -221$$

COMMENT

This solution indicates that the differential investment between the two alternatives will earn a rate of return of about 1 percent. Therefore, the added cost to construct the high-level bridge is not warranted.

It should be noted that this problem does not lend itself to a solution for the economic evaluation of a bridge crossing. The analyst must assume that top management has made the decision for the bridge crossing in preference to other modes, routes, or plant locations with proper regard to all factors involved. Incidentally, this problem is a real-life one, with but little modification.

ILLUSTRATIVE PROBLEM 27-5

ANALYSIS FOR ECONOMICAL MAXIMUM VERTICAL GRADE

A section of an interstate route was analyzed for the relative economy of maximum grades of 7, 6, and 5 percent. This route section crosses the Rocky Mountain Continental Divide in Montana. The following problem statement and solution come from the actual case developed during the design of this route. Minor modifications have been incorporated, particularly in condensing the vertical profile and horizontal curve data into groups. In the original solution, each section of vertical grade and each horizontal curve were analyzed individually for motor vehicle speed and running cost.

THE PROBLEM

The essential information, design data, and assumptions on the three design alternatives with the three maximum grades are:
1. Four-lane divided with full access control.
2. Same traffic volume and mix on each alternative.
3. Speed changes are not separately considered because they would result mainly from change in vertical grade and horizontal curvature. It is assumed that the running costs at the designated speed on grades and curves include the effects of the speed changes.
4. Accident costs are omitted because of lack of reliable basis for estimating their cost.
5. A vestcharge rate of 6 percent, an analysis period of 30 years, and zero terminal are appropriate.

Illustrative Problems and Solutions

		Alternatives		
		$A, 7\%$	$B, 6\%$	$C, 5\%$
6.	Construction costs	$13,900,000	$16,653,000	$16,962,000
7.	Uniform annual maintenance	143,000	145,000	147,000
8.	Daily traffic volume			
	A. Passenger cars and commercial deliveries	1,400	1,400	1,400
	B. Single-unit trucks	80	80	80
	C. Equivalent 50-kip, combinations	170	170	170
	D. Total ADT	1,650	1,650	1,650
9.	Route length, miles	25.538	25.597	25.695
10.	Lowest elevation feet	4,339	4,339	4,339
	Highest elevation, feet	6,414	6,412	6,406
11.	Total central angle, degrees	730	823	1,027

12. Value of travel time, dollars per vehicle-hour: $1.50 for passenger cars, $4.50 for single-unit trucks, and $6.00 for combinations.
13. The travel speeds, length of grade groups and of horizontal curve groups are given in Tables 27-5A and 27-5B.

TABLE 27-5A
MILES OF EACH VERTICAL GRADE GROUP
AND APPROXIMATE AVERAGE SPEED OF TRAFFIC THEREON

Grade* Group, Percent	Miles of Grade Group			Vehicle Speed† – Grade / + Grade, mph		
	Alternative A 7%	Alternative B 6%	Alternative C 5%	Passenger Cars and Commercial Delivery	Single-Unit Truck	50-kip Combination
0	5.265	2.575	2.689	65	55	50
1	6.078	7.462	7.519	65/65	55/52½	50/47½
2	3.940	5.201	4.118	60/60	52½/47½	45/40
3	3.171	3.902	2.501	55/55	45/40	35/27½
4	1.023	0.360	0.360	45/50	30/30	22½/22½
5	0.644	0.644	8.508	40/42½	22½/25	15/17½
6	1.913	5.453	–	35/37½	20/22½	12½/15
7	3.504	–	–	30/32½	10/20	10/12½
Total	25.538	25.597	25.695	–	–	–

*The percent grade is the midpoint of a range of 1 percent.

† The effect of horizontal curvature is included. In the original solution each specific section of grade was figured separately so that the effects of approach speed and horizontal curvature could be included.

SOLUTION

The equivalent uniform annual cost method of solution is used, though the benefit/cost ratio and rate of return methods are appropriate. The procedure is to calculate the motor vehicle running cost and travel time of each of the three classes of vehicles on each grade group and horizontal curve group for each of the three alternatives. To save space, the calculations herein group the grades and curves, but in the original solution the vehicle

TABLE 27-5B

MILES OF EACH DEGREE OF HORIZONTAL CURVE GROUP AND APPROXIMATE AVERAGE OF TRAFFIC THEREON

Horizontal Curve* Group, Degrees	Number and Miles of Curvature Group						Speed on Curve,† mph		
	Alternative A		Alternative B		Alternative C		Passenger Cars and Commercial Delivery	Single-unit Trucks	50-kip Combinations
	Number	Miles	Number	Miles	Number	Miles			
0	–	18.512	–	18.035	–	17.946	65	55	50
1	10	2.822	11	3.092	9	2.426	65	55	50
2	11	2.387	10	2.385	10	2.453	60	50	45
3	1	0.511	2	0.660	3	0.815	55	45	40
4	3	0.681	2	0.492	2	0.473	50	40	35
5	2	0.625	2	0.663	2	0.408	40	35	27½
6	0	–	2	0.270	2	0.397	35	27½	20
7	0	–	0	–	3	0.777	30	20	15
Total	27	25.538	29	25.597	31	25.695	–	–	–

* The degree of curve is the midpoint of a range of 1 degree.
† The effect of the plus or minus grade is included in the speed. In the original solution each horizontal curve was figured separately, so that the effects of the approach speed and the gradient speed could be considered.

speed, running cost, and travel hours on each grade and each curve were calculated separately. The maximum speed on combined grade and curves could be controlled by either design factor. Safety on downhill speed on horizontal curves is a controlling factor.

Tables 27-5C, 27-5D, and 27-5E contain the main results of calculations. The calculations are reduced to equivalent uniform annual cost and summarized in Table 27-5F.

Traffic growth after completion of the route section would normally enter this type of calculation, but has been omitted here because the solution without traffic growth indicates that the forecasted growth will still be short of being sufficient to increase the economy sufficiently to warrant the 6 or 5 percent maximum grade over the 7 percent.

TABLE 27-5C

CALCULATION OF THE DAILY VEHICLE RUNNING COST ON EACH GRADE GROUP

Grade Group, Percent	Running Cost, $/1,000 Vehicle-miles		Alternative A, 7%		Alternative B, 6%		Alternative C, 5%	
	Plus Grade	Minus Grade	Miles	Running Cost, Dollars	Miles	Running Cost, Dollars	Miles	Running Cost, Dollars
A. Passenger cars and commercial deliveries–1,400 ADT								
0	41.71	41.71	5.265	307.444	2.575	150.365	2.689	157.021
1	39.31	43.75	6.078	353.387	7.462	433.856	7.519	437.170
2	35.00	43.57	3.940	216.696	5.201	286.050	4.118	226.486
3	31.66	44.19	3.171	168.364	3.902	207.177	2.501	132.791
4	28.67	45.53	1.023	53.135	0.360	18.698	0.360	18.698
5	27.04	46.96	0.644	33.359	0.644	33.359	8.508	440.714
6	27.41	49.34	1.913	102.776	5.453	292.962	–	–
7	28.60	52.39	3.504	198.652	–	–	–	–
Total	–	–	25.538	1433.813	25.597	1422.467	25.695	1412.880

Illustrative Problems and Solutions

TABLE 27-5C (continued)

B. Single unit trucks–80 ADT									
0	80.24	80.24	5.265	33.797	2.575	16.529	2.689	17.261
1	73.08	88.66	6.078	39.322	7.462	48.276	7.519	48.645
2	66.91	92.84	3.940	25.177	5.201	33.234	4.118	26.314
3	57.45	93.86	3.171	19.192	3.902	23.616	2.501	15.137
4	50.53	92.03	1.023	5.834	0.360	2.053	0.360	2.053
5	49.97	96.09	0.644	3.763	0.644	3.763	8.508	49.707
6	51.08	102.24	1.913	11.732	5.453	33.442	–	–
7	64.97	109.65	3.504	24.475	–	–	–	–
Total	–	–	25.538	163.292	25.597	160.913	25.695	159.117
C. Equivalent 50-kip combinations–170 ADT									
0	113.63	113.63	5.265	101.705	2.575	49.742	2.689	51.944
1	100.04	132.97	6.078	120.380	7.462	147.791	7.519	148.920
2	85.20	142.28	3.940	76.183	5.201	100.565	4.118	79.625
3	74.40	143.38	3.171	58.699	3.902	72.231	2.501	46.297
4	75.03	156.21	1.023	20.107	0.360	7.076	0.360	7.076
5	88.38	167.54	0.644	14.009	0.644	14.009	8.508	185.076
6	98.57	179.36	1.913	45.193	5.453	128.822	–	–
7	111.78	190.97	3.504	90.171	–	–	–	–
Total	–	–	25.538	526.447	25.597	520.236	25.695	518.938
Total cost per day all vehicles				$ 2123.552		$ 2103.616		$ 2090.935

* The percent grade is the midpoint of a range of 1 percent.

TABLE 27-5D
Calculation of the Daily Vehicle Running Cost on Horizontal Curves Excess Above Running Cost on Tangent

Horizontal Curve Group, Degrees	Excess Running Cost Above Tangent/$1,000 Vehicle-mile	Alternative A, 7%		Alternative B, 6%		Alternative C, 5%	
		Miles	Daily Running Cost, Dollars	Miles	Daily Running Cost, Dollars	Miles	Daily Running Cost, Dollars
A. Passenger cars and commercial deliveries–1,400 ADT							
1	6.16	2.822	24.337	3.092	26.665	2.426	20.922
2	9.71	2.387	33.308	2.385	33.280	2.453	34.229
3	11.49	0.511	8.220	0.660	10.617	0.815	13.110
4	11.89	0.681	11.336	0.492	8.190	0.473	7.874
5	9.00	0.625	7.875	0.663	8.354	0.408	5.141
6	8.28	–	–	0.270	3.130	0.397	4.602
7	7.70	–	–	–	–	0.777	8.376
Total		7.026	85.076	7.562	90.236	7.749	94.254
B. Single unit trucks–80 ADT							
1	9.61	2.822	2.170	3.092	2.377	2.426	1.865
2	14.85	2.387	2.836	2.385	2.833	2.453	2.914
3	16.60	0.511	0.679	0.660	0.876	0.815	1.082
4	15.57	0.681	0.848	0.492	0.613	0.473	0.589
5	14.00	0.625	0.700	0.663	0.743	0.408	0.457
6	14.68	–	–	0.270	0.317	0.397	0.466
7	13.97	–	–	–	–	0.777	0.868
Total	–	7.026	7.233	7.562	7.759	7.749	8.241

TABLE 27-5D *(continued)*

C. Equivalent 50-kip combinations–170 ADT

1	20.59	2.822	9.878	3.092	10.823	2.426		8.492
2	29.41	2.387	11.934	2.385	11.924	2.453		12.264
3	29.10	0.511	2.528	0.660	3.265	0.815		4.032
4	30.57	0.681	3.539	0.492	2.557	0.473		2.458
5	33.43	0.625	3.552	0.663	3.768	0.408		2.319
6	32.05	–	–	0.270	1.471	0.397		2.163
7	30.36	–	–	–	–	0.777		4.010
Total	–	7.026	31.431	7.562	33.808	7.749		35.738
Total all vehicles			123.740		131.803			138.233

TABLE 27-5E

CALCULATION OF DAILY VEHICLE-HOURS OF TRAVEL

Grade* Group, Percent	Alternative A, 7%		Alternative B, 6%		Alternative C, 5%	
	Daily Vehicle-Hours		Daily Vehicle-Hours		Daily Vehicle-Hours	
	Minus Grade	Plus Grade	Minus Grade	Plus Grade	Minus Grade	Plus Grade
A. Passenger cars and commercial deliveries–1,400 ADT						
0	56.700	56.700	27.731	27.731	28.957	28.957
1	65.455	65.455	80.360	80.360	80.974	80.974
2	45.967	45.967	60.678	60.678	48.043	48.043
3	40.358	40.358	49.662	49.662	31.831	31.831
4	15.913	14.322	5.600	5.040	5.600	5.040
5	11.270	10.607	11.270	10.607	148.890	140.132
6	38.260	35.709	109.060	101.789	–	–
7	81.760	75.471	–	–	–	–
Total	355.683	344.589	344.361	335.867	344.295	334.977
B. Single unit trucks–80 ADT						
0	3.829	3.829	1.873	1.873	1.956	1.956
1	4.420	4.631	5.427	5.685	5.468	5.729
2	3.002	3.318	3.963	4.380	3.138	3.468
3	2.819	3.171	3.468	3.902	2.223	2.501
4	1.364	1.364	0.480	0.480	0.480	0.480
5	1.145	1.030	1.145	1.030	15.125	13.613
6	3.826	3.401	10.906	9.694	–	–
7	14.016	7.008	–	–	–	–
Total	34.421	27.752	27.262	27.044	28.390	27.747

Illustrative Problems and Solutions

TABLE 27-5E *(continued)*

C. Equivalent 50-kip combinations–170 ADT

0	8.950	8.950	4.378	4.378	4.571	4.571
1	10.333	10.876	12.685	13.353	12.782	13.455
2	7.442	8.372	9.824	11.052	7.778	8.751
3	7.701	9.801	9.476	12.061	6.074	7.730
4	3.865	3.865	1.360	1.360	1.360	1.360
5	3.649	3.128	3.649	3.128	48.212	41.325
6	13.008	10.840	37.080	30.900	–	–
7	29.784	23.827	–	–	–	–
Total	84.732	79.659	78.452	76.232	80.777	77.192
Total minus and plus grades....	926.836		889.218		893.378	

* The percent grade is the midpoint of a range of 1 percent.

Note: The speeds of travel in Table 27-5A include the effect of the curvature so excess travel time on curves is included herein.

TABLE 27-5F
SUMMARY OF CALCULATIONS

Calculation Item	Alternative A, 7%	Alternative B, 6%	Alternative C, 5%
1. Passenger cars and commercial deliveries			
Total travel hours, daily	700.272	680.228	679.272
Travel time value daily @ $1.50 per hour ..	$1,050.408	$1,020.342	$1,018.908
2. Single-unit trucks			
Total travel hours, daily	62.173	54.306	56.137
Travel time value daily @ $4.50 per hour	$279.778	$244.377	$252.616
3. 50-kip combinations			
Total travel hours, daily	164.391	154.684	157.969
Travel time value daily @ $6.00 per hour	$986.346	$928.104	$947.814
4. Total daily value of travel time	$2,316.532	$2,192.823	$2,219.338
5. Total yearly value of travel time	$845,534	$800,380	$810,058
6. Daily running cost, all vehicles	$2,123.552	$2,103.616	$2,090.935
7. Yearly runnig cost, all vehicles	775,096	767,820	763,191
8. Daily excess running cost on curves	123.740	131.803	138.233
9. Yearly excess running cost on curves	45,165	48,108	50,455
10. Total yearly road user running cost and time value	$1,665,795	$1,616,308	$1,623,704
11. Annual capital cost at 6% and 30 years (CR-6%-30 = 0.072649)	$1,009,821	$1,209,824	$1,232,272
12. Annual highway maintenance	143,000	145,000	147,000
13. Total annual transportation cost (10 + 11 + 12)	$2,818,616	$2,971,132	$3,002,976

DISCUSSION OF RESULTS

In total transportation cost, the 7 percent maximum grade is the most economical grade. The 6 percent grade is preferred to the 5 percent. In fact, the road user costs are higher on the 5 percent grade than on the 6 percent grade.

Although the 6 percent grade reduces the road user costs $ 49,487 from the 7 percent grade, this gain is insufficient to overcome the equivalent uniform annual highway capital cost of $ 200,003, plus $ 2,000 maintenance. It would require an equivalent uniform ADT of 4.08 times the 1,650 ADT used in the solution, or 6,732 ADT, to break even. The traffic forecast is not likely to reach this level in 20 years, let alone reach 6,732 on an equivalent uniform basis.

This case study on grade reduction illustrates several significant factors. A study of the details of design for alternatives A, B, and C discloses that (1) there is no material change in the summit elevation of 6,414 ft, (2) the reduction in maximum grade is achieved primarily by shifting mileage of lesser grades to higher grades, and (3) the reduction in maximum grade results in increasing (a) the number of horizontal curves, (b) the total degrees of curve, and (c) introducing curves of higher degrees (shorter radius).

The beneficial result of the reduction in maximum grade is therefore partially absorbed by increasing mileage of lower percentage grades and by greater horizontal curvature. These factors reduce speeds generally and increase running costs on curves. Economy of grade reduction may be achieved when it is possible to reduce the total rise and fall and when lower percentage grades can be had without introducing greater total curvature and without appreciable increase in distance.

Appendix **A**

Tables of the Running Cost of Motor Vehicles

This appendix presents a series of tables for five typical vehicles giving the dollars of running cost and gallons of fuel consumption for operation at uniform speed on minus grades, plus grades, and horizontal curves and for changes in speed.

In any analysis for the economy of a proposed highway improvement, these tables will provide the necessary information for calculation of the motor vehicle costs. It is necessary, however, to supply the traffic data for all alternatives to be analyzed. Needed are such items as traffic volume, traffic classification, vehicular speeds, and speed changes, and of course, the details of highway design.

In the running cost tables starting with Table A-1, the cost of the fuel component does not include the state or federal road-user tax. When it is desirable to include all or part of the fuel tax, it can be computed by using the fuel consumption tables, starting with Table A-26-M.

TABLE A-1

Dollars Running Cost at Uniform Speed on Level Tangents by Cost Item

Vehicle: 4-kip passenger car
Unit: Dollars per 1,000 vehicle-miles
Roadway surface: High type pavement in good condition

Speed, mph	Running Cost by Item					Total Cost
	Fuel	Tires	Engine Oil	Maintenance	Depreciation	
5	23.55	0.18	4.22	5.38	26.03	59.36
7½	17.50	0.28	3.20	5.43	23.45	49.86
10	14.56	0.38	2.64	5.49	21.86	44.93
12½	12.83	0.49	2.27	5.57	20.66	41.82
15	11.75	0.60	2.03	5.67	19.68	39.73
17½	11.04	0.71	1.86	5.79	18.81	38.21
20	10.56	0.82	1.75	5.93	18.03	37.09
22½	10.21	0.94	1.67	6.09	17.32	36.23
25	10.01	1.06	1.64	6.25	16.67	35.63
27½	9.89	1.19	1.61	6.42	16.08	35.19
30	9.84	1.32	1.60	6.60	15.55	34.91
32½	9.89	1.46	1.59	6.78	15.07	34.79
35	9.96	1.60	1.59	6.97	14.64	34.76
37½	10.10	1.75	1.58	7.16	14.25	34.84
40	10.28	1.90	1.58	7.36	13.91	35.03
42½	10.49	2.06	1.56	7.56	13.60	35.27
45	10.76	2.23	1.55	7.77	13.32	35.63
47½	11.06	2.41	1.52	7.98	13.16	36.13
50	11.41	2.61	1.49	8.19	12.83	36.53
52½	11.80	2.81	1.43	8.41	12.62	37.07
55	12.24	3.03	1.37	8.64	12.43	37.71
57½	12.72	3.27	1.38	8.88	12.25	38.50
60	13.25	3.53	1.43	9.13	12.08	39.42
62½	13.85	3.81	1.50	9.40	11.93	40.49
65	14.51	4.12	1.61	9.69	11.78	41.71
67½	15.25	4.46	1.76	10.01	11.64	43.12
70	16.10	4.85	1.93	10.37	11.51	44.76
72½	17.04	5.30	2.13	10.78	11.38	46.63
75	18.10	5.83	2.36	11.25	11.25	48.79
77½	19.34	6.45	2.64	11.79	11.12	51.34
80	20.79	7.19	2.96	12.41	11.00	54.35

Tables of the Running Cost of Motor Vehicles

TABLE A-2
DOLLARS RUNNING COST AT UNIFORM SPEED ON LEVEL TANGENTS BY COST ITEM

Vehicle: 5-kip commercial delivery
Unit: Dollars per 1,000 vehicle-miles
Roadway surface: High type pavement in good condition

Speed, mph	Running Cost by Item					Total Cost
	Fuel	Tires	Engine Oil	Maintenance	Depreciation	
5	23.72	0.22	3.53	6.56	30.32	64.35
7½	17.62	0.34	2.67	6.62	27.32	54.57
10	14.65	0.46	2.21	6.70	25.47	49.49
12½	12.94	0.59	1.91	6.80	24.07	46.31
15	11.86	0.72	1.72	6.92	22.93	44.15
17½	11.15	0.85	1.58	7.06	21.91	42.55
20	10.69	0.98	1.49	7.23	21.00	41.39
22½	10.41	1.13	1.43	7.43	20.18	40.58
25	10.25	1.27	1.39	7.62	19.42	39.95
27½	10.19	1.43	1.37	7.83	18.73	39.55
30	10.21	1.58	1.36	8.05	18.12	39.32
32½	10.32	1.75	1.34	8.27	17.56	39.24
35	10.49	1.92	1.32	8.50	17.06	39.29
37½	10.74	2.10	1.30	8.74	16.60	39.48
40	11.07	2.28	1.29	8.98	16.21	39.83
42½	11.48	2.47	1.27	9.22	15.84	40.28
45	11.97	2.68	1.23	9.48	15.52	40.88
47½	12.56	2.89	1.22	9.74	15.33	41.74
50	13.27	3.13	1.15	9.99	14.95	42.49
52½	14.08	3.37	1.07	10.26	14.70	43.48
55	15.07	3.64	0.99	10.54	14.48	44.72
57½	16.24	3.92	0.97	10.83	14.27	46.23
60	17.69	4.24	0.97	11.14	14.07	48.11
62½	19.43	4.57	1.00	11.47	13.90	50.37
65	21.56	4.94	1.06	11.82	13.72	53.10
67½	23.91	5.35	1.13	12.21	13.56	56.16
70	26.60	5.82	1.22	12.65	13.41	59.70

TABLE A-3

DOLLARS RUNNING COST AT UNIFORM SPEED ON LEVEL TANGENTS BY COST ITEM

Vehicle: 12-kip single-unit truck
Unit: Dollars per 1,000 vehicle-miles
Roadway surface: High type pavement in good condition

Speed, mph	Running Cost by Item					Total Cost
	Fuel	Tires	Engine Oil	Maintenance	Depreciation	
5	38.12	0.48	3.43	19.32	37.73	99.08
7½	29.62	0.74	2.67	18.99	33.33	85.35
10	25.46	1.00	2.17	18.92	30.13	77.68
12½	23.06	1.29	1.96	19.02	27.53	72.86
15	21.50	1.58	1.79	19.27	25.48	69.62
17½	20.46	1.87	1.67	19.64	23.73	67.37
20	19.76	2.16	1.60	20.13	22.25	65.90
22½	19.28	2.48	1.54	20.71	21.08	65.09
25	18.94	2.80	1.48	21.39	20.12	64.73
27½	18.73	3.14	1.44	22.14	19.31	64.76
30	18.64	3.48	1.40	22.98	18.61	65.11
32½	18.64	3.85	1.37	23.88	18.01	65.75
35	18.72	4.22	1.33	24.85	17.48	66.60
37½	18.86	4.62	1.29	25.88	17.01	67.66
40	19.08	5.02	1.24	26.94	16.59	68.87
42½	19.38	5.44	1.18	28.06	16.21	70.27
45	19.76	5.89	1.12	29.21	15.86	71.84
47½	20.24	6.36	1.06	30.39	15.55	73.60
50	20.80	6.89	1.05	31.59	15.26	75.59
52½	21.48	7.42	1.07	32.82	15.00	77.79
55	22.30	8.00	1.11	34.06	14.77	80.24
57½	23.24	8.63	1.13	35.30	14.56	82.86
60	24.36	9.32	1.25	36.54	14.37	85.84
62½	25.66	10.06	1.34	37.79	14.20	89.05
65	27.16	10.88	1.45	39.04	14.04	92.57

Tables of the Running Cost of Motor Vehicles

TABLE A-4
Dollars Running Cost at Uniform Speed on Level Tangents by Cost Item

Vehicle: 40-kip 2-S2, gasoline
Unit: Dollars per 1,000 vehicle-miles
Roadway surface: High type pavement in good condition

Speed, mph	Running Cost by Item					Total Cost
	Fuel	Tires	Engine Oil	Maintenance	Depreciation	
5	145.80	0.97	2.54	30.91	50.79	231.01
7½	97.87	1.48	1.92	30.38	45.01	176.66
10	74.29	2.00	1.63	30.28	40.44	148.64
12½	60.26	2.54	1.43	30.44	36.81	131.48
15	51.03	3.11	1.30	30.83	33.90	120.17
17½	44.55	3.70	1.21	31.43	31.54	112.43
20	39.82	4.34	1.15	32.21	29.58	107.10
22½	36.29	4.97	1.11	33.14	27.96	103.47
25	33.61	5.68	1.07	34.22	26.61	101.19
27½	31.61	6.39	1.04	35.43	25.46	99.93
30	30.11	7.16	1.01	36.76	24.47	99.51
32½	29.07	8.00	0.98	38.21	23.61	99.87
35	28.40	8.86	0.96	39.76	22.86	100.84
37½	28.08	9.81	0.93	41.40	22.21	102.43
40	28.10	10.82	0.89	43.11	21.63	104.55
42½	28.44	11.91	0.85	44.89	21.12	107.21
45	29.12	13.09	0.82	46.73	20.67	110.43
47½	30.19	14.36	0.78	48.62	20.28	114.23
50	31.70	15.74	0.78	50.55	19.93	118.70
52½	33.80	17.25	0.80	52.51	19.62	123.98
55	36.59	18.91	0.84	54.49	19.35	130.18
57½	40.14	20.73	0.89	56.48	19.11	137.35
60	44.55	22.68	0.95	58.47	18.90	145.55

TABLE A-5

DOLLARS RUNNING COST AT UNIFORM SPEED ON LEVEL TANGENTS BY COST ITEM

Vehicle: 50-kip 3-S2, diesel
Unit: Dollars per 1,000 vehicle-miles

Roadway surface: High type pavement in good condition

Speed, mph	Running Cost by Item					Total Cost
	Fuel	Tires	Engine Oil	Maintenance	Depreciation	
5	64.70	1.30	5.15	28.51	67.06	166.72
7½	44.16	2.00	4.01	28.20	57.98	136.35
10	34.05	2.68	3.34	28.22	51.91	120.20
12½	28.11	3.40	2.94	28.43	47.23	110.11
15	24.30	4.17	2.67	28.82	43.45	103.41
17½	21.68	4.94	2.49	29.38	40.35	98.84
20	19.86	5.77	2.37	30.05	37.76	95.81
22½	18.54	6.61	2.27	30.90	35.57	93.89
25	17.62	7.56	2.18	31.88	33.68	92.92
27½	17.02	8.49	2.10	32.97	32.06	92.64
30	16.64	9.50	2.03	34.18	30.67	93.02
32½	16.45	10.60	1.96	35.51	29.48	94.00
35	16.46	11.75	1.90	36.96	28.44	95.10
37½	16.67	12.99	1.82	38.48	27.53	97.49
40	17.06	14.33	1.71	40.04	26.71	99.85
42½	17.63	15.76	1.60	41.65	25.97	102.61
45	18.38	17.32	1.49	43.31	25.30	105.80
47½	19.34	19.01	1.37	45.03	24.70	109.45
50	20.53	20.83	1.32	46.80	24.15	113.63
52½	21.95	22.84	1.33	48.60	23.65	118.37
55	23.66	25.04	1.37	50.41	23.18	123.66
57½	25.68	27.46	1.44	52.24	22.73	129.55
60	26.43	30.06	1.54	54.09	22.48	134.60

Tables of the Running Cost of Motor Vehicles

TABLE A-6-M

DOLLARS RUNNING COST* AT UNIFORM SPEED ON MINUS GRADES

Vehicle: 4-kip passenger car
Unit: Dollars per 1,000 vehicle-miles
Roadway surface: High type pavement in good condition

Speed, mph	Minus Grade, Percent									Speed, mph
	−8	−7	−6	−5	−4	−3	−2	−1	Level	
5	57.36	56.06	55.35	54.82	54.50	54.23	54.14	57.77	59.36	5
7½	47.29	46.16	45.43	45.04	44.75	44.61	44.53	48.23	49.86	7½
10	41.94	40.92	40.24	39.88	39.62	39.50	39.42	43.25	44.93	10
12½	38.48	37.54	36.89	36.59	36.33	36.21	36.83	40.11	41.82	12½
15	36.04	35.17	34.58	34.29	34.05	33.92	35.14	37.98	39.73	15
17½	34.16	33.36	32.82	32.54	32.32	32.19	34.00	36.34	38.21	17½
20	32.69	31.95	31.44	31.16	30.96	30.82	32.95	35.24	37.09	20
22½	31.50	30.83	30.34	30.10	29.90	30.31	32.41	34.39	36.23	22½
25	30.56	29.94	29.47	29.24	29.04	29.98	31.90	33.76	35.63	25
27½	29.80	29.20	28.78	28.52	28.34	29.67	31.50	33.30	35.19	27½
30	29.15	28.60	28.20	27.94	28.04	29.46	31.24	33.00	34.91	30
32½	28.69	28.17	27.77	27.51	27.99	29.38	31.12	32.85	34.79	32½
35	28.34	27.82	27.41	27.22	28.03	29.40	31.09	32.79	34.76	35
37½	28.09	27.58	27.19	26.91	28.13	29.47	31.15	32.86	34.84	37½
40	27.93	27.44	27.02	27.04	28.30	29.67	31.31	33.00	35.03	40
42½	27.81	27.27	26.84	27.18	28.46	29.78	31.53	33.27	35.27	42½
45	27.81	27.25	26.80	27.44	28.67	30.05	31.82	33.59	35.63	45
47½	27.96	27.28	26.89	27.83	29.10	30.49	32.29	34.08	36.13	47½
50	28.07	27.33	27.02	28.08	29.33	30.77	32.58	34.43	36.53	50
52½	28.27	27.47	27.35	28.36	29.59	31.16	33.02	34.93	37.07	52½
55	28.52	27.60	27.80	28.72	30.00	31.66	33.56	35.54	37.71	55
57½	28.88	27.92	28.38	29.26	30.56	32.30	34.25	36.30	38.50	57½
60	29.31	28.33	29.07	29.87	31.21	33.03	35.00	37.17	39.42	60
62½		29.26	29.67	30.52	32.04	33.90	35.96	38.18	40.49	62½
65			30.44	31.34	32.95	34.90	37.00	39.31	41.71	65
67½				32.34	34.00	36.00	38.21	40.60	43.12	67½
70					35.24	37.30	39.58	42.10	44.76	70
72½						38.74	41.16	43.81	46.63	72½
75						40.20	43.04	45.83	48.79	75
77½							45.25	48.18	51.34	77½
80							47.74	50.93	54.35	80

* Cost includes fuel, tires, engine, oil, maintenance, and depreciation.

TABLE A-6-P
Dollars Running Cost* at Uniform Speed on Plus Grades

Vehicle: 4-kip passenger car
Unit: Dollars per 1,000 vehicle-miles
Roadway surface: High type pavement in good condition

Speed, mph	Plus Grade, Percent									Speed, mph
	Level	+1	+2	+3	+4	+5	+6	+7	+8	
5	59.36	61.27	63.29	65.46	67.68	70.33	72.53	75.23	78.49	5
7½	49.86	51.81	53.90	56.09	58.35	60.57	63.24	66.02	69.28	7½
10	44.93	46.89	49.01	51.24	53.53	55.59	58.46	61.31	64.57	10
12½	41.82	43.81	46.00	48.23	50.52	52.77	55.55	58.44	61.70	12½
15	39.73	41.74	43.94	46.20	48.50	51.00	53.58	56.49	59.78	15
17½	38.21	40.23	42.47	44.72	47.05	49.47	52.16	55.13	58.41	17½
20	37.09	39.12	41.37	43.66	45.99	48.49	51.12	54.14	57.41	20
22½	36.23	38.31	40.55	42.88	45.21	47.71	50.37	53.42	56.69	22½
25	35.63	37.72	39.96	42.27	44.62	47.12	49.83	52.93	56.20	25
27½	35.19	37.28	39.53	41.86	44.22	46.74	49.48	52.60	55.90	27½
30	34.91	37.02	39.27	41.62	43.99	46.52	49.27	52.46	55.79	30
32½	34.79	36.88	39.12	41.45	43.87	46.43	49.21	52.39	55.78	32½
35	34.76	36.84	39.07	41.43	43.83	46.39	49.23	52.47	55.89	35
37½	34.84	36.89	39.11	41.46	43.91	46.51	49.34	52.62	56.12	37½
40	35.03	37.08	39.28	41.62	44.09	46.73	49.60	52.87	56.43	40
42½	35.27	37.28	39.54	41.86	44.33	46.96	49.89	53.22	56.81	42½
45	35.63	37.63	39.88	42.17	44.64	47.31	50.28	53.65	57.32	45
47½	36.13	38.15	40.39	42.67	45.16	47.80	50.82	54.25	57.98	47½
50	36.53	38.53	40.74	43.03	45.53	48.22	51.26	54.73	58.53	50
52½	37.07	39.07	41.34	43.55	46.07	48.79	51.86	55.38	59.26	52½
55	37.71	39.72	41.89	44.19	46.70	49.47	52.57	56.14	60.11	55
57½	38.50	40.48	42.65	44.97	47.49	50.29	53.46	57.05	61.12	57½
60	39.42	41.43	43.57	45.87	48.42	51.29	54.47	58.14	62.33	60
62½	40.49	42.49	44.63	46.92	49.49	52.40	55.72	59.41		62½
65	41.71	43.75	45.87	48.19	50.77	53.74	57.13			65
67½	43.12	45.17	47.30	49.66	52.24	55.33				67½
70	44.76	46.79	48.94	51.32	53.97					70
72½	46.63	48.68	50.85	53.28						72½
75	48.79	50.86	53.07	55.59						75
77½	51.34	53.36	55.62							77½
80	54.35	56.36	58.69							80

*Cost includes fuel, tires, engine oil, maintenance, and depreciation.

TABLE A-7

Dollars Running Cost* at Uniform Speed on Horizontal Curves–Excess Above Cost on Tangents

Vehicle: 4-kip passenger car
Roadway surface: High type pavement in good condition
Unit: Dollars per 1,000 vehicle-miles

Speed, mph	Degree of Horizontal Curve													
	1	2	3	4	5	6	8	10	12	14	16	18	20	25	30
5	0.44	0.84	1.15	1.40	1.60	1.76	2.12	2.46	2.82	3.20	3.58	3.98	4.40	5.50	6.82
10	0.71	1.36	1.89	2.31	2.69	3.06	3.71	4.40	5.08	5.78	6.50	7.24	7.99	9.94	12.30
15	0.88	1.69	2.33	2.87	3.47	3.83	4.95	6.08	7.30	8.50	9.80	11.00	12.90	16.40	21.04
20	0.90	1.70	2.46	3.20	3.91	4.60	6.00	7.60	9.20	10.97	12.90	15.00	17.40	24.70	35.10
25	0.92	1.78	2.61	3.45	4.33	5.25	7.24	9.54	12.12	15.00	18.27	21.88	25.80	38.10	54.20
30	0.96	1.89	2.97	3.89	5.04	6.32	9.08	12.43	16.88	22.00	27.70	33.80	40.30	57.90	78.10
35	1.06	2.20	3.64	4.81	6.87	8.28	11.80	17.13	23.90	31.45	39.60	48.30	57.40	80.70	
40	1.39	2.84	4.75	6.47	9.00	11.34	17.70	25.40	34.06	43.50	53.60	64.10	75.40		
45	1.88	3.88	6.38	8.70	12.18	15.60	25.50	35.20	46.50	58.90	71.80				
50	2.55	5.35	8.59	11.89	16.50	21.00	33.42	46.00	59.72	74.10					
55	3.48	7.23	11.49	16.02	22.04	27.95	42.40	57.40	74.30						
60	4.68	9.71	15.25	21.26	28.89	36.31	53.40	71.80							
65	6.16	12.73	19.96	27.76	37.14	46.39	66.50	89.00							
70	7.85	16.35	25.78	35.76	46.99	58.44	82.49								
75	10.20	21.20	32.99	45.54	58.82	72.79									
80	13.20	27.45	42.49	58.30	74.84										

* Cost includes fuel, tires, engine oil, maintenance, and depreciation.

TABLE A-8

Dollars Excess Cost* of Speed-Change Cycles†–
Excess Cost Above Continuing at Initial Speed

Vehicle: 4-kip passenger car
Unit: Dollars per 1,000 cycles
Roadway surface: High type pavement in good condition

Initial Speed, mph	Speed Reduced to and Returned from, mph														
	Stop	5	10	15	20	25	30	35	40	45	50	55	60	65	70	75
5	0.86															
10	1.84	0.94														
15	3.26	2.15	1.17													
20	4.95	3.81	2.66	1.44												
25	6.96	5.80	4.62	3.28	1.79											
30	9.36	8.19	6.98	5.61	4.02	2.19										
35	12.24	11.05	9.83	8.43	6.84	4.94	2.70									
40	15.76	14.55	13.31	11.87	10.26	8.35	6.12	3.33								
45	19.99	18.78	17.50	16.05	14.40	12.46	10.25	7.39	4.05							
50	25.15	23.89	22.60	21.09	19.40	17.40	15.19	12.31	8.94	4.89						
55	31.43	30.11	28.76	27.23	25.45	23.36	21.06	18.18	14.78	10.69	5.83					
60	39.09	37.66	36.18	34.54	32.69	30.55	28.06	25.13	21.68	17.59	12.70	6.94				
65	48.48	46.87	45.27	43.47	41.46	39.19	36.56	33.55	29.99	25.80	20.88	15.09	8.26			
70	59.91	58.06	56.17	54.13	51.90	49.40	46.60	43.41	39.70	35.39	30.36	24.53	17.73	9.73		
75	73.81	71.65	69.21	66.46	63.38	60.08	56.51	52.64	48.50	44.11	39.23	33.81	27.61	20.37	11.63	
80	90.30	87.28	84.21	80.69	77.12	73.18	69.30	65.02	60.53	55.72	50.56	44.92	38.74	31.81	23.73	13.72

*Cost includes fuel, tires, engine oil, maintenance, and depreciation.
† A speed-change cycle is reducing speed from and returning to an initial speed.

TABLE A-9

Excess Hours Consumed per Speed-Change Cycle*–
Excess Hours Above Continuing at Initial Speed

Vehicle: 4-kip passenger car
Unit: Hours per 1,000 cycles
Roadway surface: High type pavement in good condition

Initial Speed, mph	Speed Reduced to and Returned from, mph														
	Stop	5	10	15	20	25	30	35	40	45	50	55	60	65	70	75
5	1.02															
10	1.51	0.62														
15	2.00	1.12	0.46													
20	2.49	1.62	0.93	0.35												
25	2.98	2.11	1.40	0.80	0.28											
30	3.46	2.60	1.87	1.24	0.70	0.23										
35	3.94	3.09	2.34	1.69	1.11	0.60	0.19									
40	4.42	3.58	2.81	2.13	1.52	0.97	0.51	0.16								
45	4.90	4.06	3.28	2.57	1.93	1.34	0.83	0.42	0.13							
50	5.37	4.54	3.75	3.01	2.34	1.71	1.15	0.68	0.35	0.11						
55	5.84	5.02	4.21	3.45	2.74	2.08	1.47	0.94	0.57	0.28	0.09					
60	6.31	5.50	4.67	3.88	3.14	2.44	1.78	1.20	0.78	0.45	0.21	0.07				
65	6.78	5.97	5.13	4.31	3.54	2.80	2.09	1.45	0.99	0.62	0.36	0.18	0.05			
70	7.25	6.44	5.58	4.74	3.94	3.16	2.40	1.70	1.19	0.78	0.50	0.30	0.16	0.04		
75	7.71	6.91	6.03	5.17	4.33	3.51	2.70	1.95	1.39	0.94	0.63	0.40	0.23	0.10	0.03	
80	8.17	7.38	6.48	5.59	4.72	3.86	3.00	2.19	1.58	1.09	0.75	0.49	0.29	0.15	0.06	0.02

* A speed-change cycle is reducing speed from and returning to an initial speed.

Tables of the Running Cost of Motor Vehicles

TABLE A-10-M
Dollars Running Cost* at Uniform Speed on Minus Grades

Vehicle: 5-kip commercial delivery
Unit: Dollars per 1,000 vehicle-miles
Roadway surface: High type pavement in good condition

Speed, mph	\-8	\-7	\-6	\-5	\-4	\-3	\-2	\-1	Level	Speed, mph
5	62.52	60.99	60.20	59.63	59.33	59.42	60.86	62.49	64.35	5
7½	52.10	50.79	49.99	49.59	49.31	49.51	51.06	52.70	54.57	7½
10	46.52	45.37	44.63	44.27	44.02	44.41	45.92	47.58	49.49	10
12½	42.90	41.85	41.17	40.88	40.62	41.16	42.74	44.36	46.31	12½
15	40.33	39.37	38.76	38.50	38.27	38.98	40.57	42.21	44.15	15
17½	38.33	37.47	36.93	36.68	36.47	37.36	39.07	40.53	42.55	17½
20	36.76	35.98	35.48	35.24	35.07	36.14	37.89	39.45	41.39	20
22½	35.48	34.81	34.32	34.14	33.96	35.30	37.09	38.58	40.58	22½
25	34.46	33.85	33.40	33.22	33.34	34.71	36.45	37.97	39.95	25
27½	33.63	33.05	32.68	32.46	32.93	34.20	36.00	37.55	39.55	27½
30	32.90	32.40	32.04	31.83	32.61	33.85	35.68	37.33	39.32	30
32½	32.38	31.91	31.57	31.36	32.46	33.67	35.52	37.22	39.24	32½
35	32.01	31.52	31.19	31.12	32.39	33.63	35.49	37.11	39.29	35
37½	31.68	31.24	30.92	31.11	32.40	33.63	35.56	37.39	39.48	37½
40	31.50	31.08	30.74	31.23	32.54	33.79	35.79	37.67	39.83	40
42½	31.31	30.89	30.54	31.26	32.59	33.96	36.08	38.03	40.28	42½
45	31.23	30.79	30.54	31.44	32.79	34.30	36.52	38.55	40.88	45
47½	31.35	30.80	30.72	31.80	33.21	34.90	37.23	39.29	41.74	47½
50	31.37	30.82	30.80	31.98	33.45	35.34	37.76	39.95	42.49	50
52½		30.86	30.91	32.15	33.76	36.01	38.49	40.83	43.48	52½
55				32.49	34.28	36.82	39.37	41.93	44.72	55
57½				32.98	34.97	37.82	40.47	43.28	46.23	57½
60						38.97	41.77	44.88	48.11	60
62½							43.47	46.90	50.37	62½
65							45.57	49.37	53.10	65
67½								52.58	56.16	67½
70								56.51	59.70	70

* Cost includes fuel, tires, engine oil, maintenance, and depreciation

TABLE A-10-P
Dollars Running Cost* at Uniform Speed on Plus Grades

Vehicle: 5-kip commercial delivery
Unit: Dollars per 1,000 vehicle-miles
Roadway surface: High type pavement in good condition

Speed, mph	Level	+1	+2	+3	+4	+5	+6	+7	+8	Speed, mph
5	64.35	66.19	68.12	70.25	72.67	75.53	78.85	82.79	87.37	5
7½	54.57	56.31	58.39	60.54	62.99	65.84	69.27	73.24	77.86	7½
10	49.49	51.35	53.42	55.70	58.26	61.14	64.49	68.43	73.10	10
12½	46.31	48.20	50.28	52.59	55.17	58.13	61.52	65.59	70.23	12½
15	44.15	46.09	48.21	50.57	53.18	56.16	59.56	63.60	68.40	15
17½	42.55	44.55	46.70	49.07	51.74	54.73	58.22	62.26	67.15	17½
20	41.39	43.43	45.60	48.01	50.72	53.74	57.26	61.39	66.30	20
22½	40.58	42.60	44.82	47.29	50.02	53.10	56.64	60.79	65.79	22½
25	39.95	42.03	44.27	46.77	49.52	52.69	56.28	60.48	65.54	25
27½	39.55	41.63	43.95	46.52	49.31	52.49	56.14	60.40	65.53	27½
30	39.32	41.45	43.82	46.41	49.27	52.48	56.87	60.54	65.75	30
32½	39.24	41.43	43.82	46.47	49.39	52.67	56.44	60.84	66.18	32½
35	39.29	41.54	43.99	46.69	49.67	53.03	58.67	61.37	66.83	35
37½	39.48	41.78	44.30	47.06	50.09	53.51	57.43	62.06	67.67	37½
40	39.83	42.22	44.79	47.62	50.72	54.23	58.24	62.99	68.86	40
42½	40.28	42.74	45.41	48.31	51.51	55.07	59.20	64.12	70.23	42½
45	40.88	43.44	46.21	49.20	52.48	56.19	60.42	65.56	72.07	45
47½	41.74	44.40	47.29	50.33	53.75	57.55	62.01	67.42	74.47	47½
50	42.49	45.26	48.22	51.45	54.99	58.97	63.66	69.50	77.33	50
52½	43.48	46.37	49.49	52.82	56.50	60.68	65.63	72.09		52½
55	44.72	47.74	50.98	54.48	58.34	62.74				55
57½	46.23	49.41	52.83	56.53	60.56	65.20				57½
60	48.11	51.46	55.09	59.01						60
62½	50.37	53.97	57.87							62½
65	53.10	56.94	61.58							65
67½	56.16	60.35								67½
70	59.70	64.28								70

* Cost includes fuel, tires, engine oil, maintenance, and depreciation.

Tables of the Running Cost of Motor Vehicles

TABLE A-11
Dollars Excess Running Cost* at Uniform Speed on Horizontal Curves – Excess Above Cost on Tangents
Vehicle: 5-kip commercial delivery
Roadway surface: High type pavement in good condition
Unit: Dollars per 1,000 vehicle-miles

Speed, mph	Degree of Horizontal Curve													
	1	2	3	4	5	6	8	10	12	14	16	18	20	25	30
5	0.56	1.00	1.33	1.60	1.80	1.98	2.38	2.81	3.30	3.75	4.18	4.62	5.08	6.34	7.92
10	0.91	1.69	2.21	2.63	2.98	3.31	4.12	5.05	6.05	7.01	8.00	9.00	9.95	12.25	14.48
15	1.08	2.00	2.70	3.25	3.84	4.60	6.00	7.40	8.70	10.00	11.40	12.80	14.40	19.00	25.00
20	1.14	2.04	2.54	3.74	4.74	5.64	7.60	9.30	11.16	13.05	15.00	17.21	19.86	27.70	39.40
25	1.17	1.97	2.97	4.22	5.27	6.29	8.68	11.32	14.16	17.18	20.39	23.91	27.83	39.80	58.50
30	1.28	2.38	3.78	5.08	6.31	7.68	10.87	14.61	18.88	23.66	28.92	34.63	40.78	61.80	94.20
35	1.52	3.02	4.82	6.70	8.53	10.56	15.27	20.78	27.12	34.21	42.16	51.07	61.82	99.20	
40	1.89	3.82	6.04	8.55	11.36	14.47	21.59	29.90	39.40	50.10	64.20	83.80			
45	2.45	5.14	8.23	11.73	15.86	20.51	30.11	41.40	54.54	70.40	93.40				
50	3.20	6.27	10.10	14.66	19.93	25.85	39.57	55.53	76.50	91.00					
55	4.19	8.90	14.00	19.90	26.57	33.97	51.08	74.30							
60	5.50	11.86	18.68	26.12	34.39	43.80	68.40								
65	7.29	15.30	24.00	33.63	44.51	57.21	93.90								
70	9.79	20.38	31.68	43.99	57.71	73.69									

* Cost includes fuel, tires, engine oil, maintenance, and depreciation.

TABLE A-12

Dollar Excess Cost* of Speed Change Cycles†–
Excess Cost Above Continuing at Initial Speed

Vehicle: 5-kip commercial delivery
Unit: Dollars per 1,000 cycles
Roadway surface: High type pavement in good condition

Initial Speed, mph	Speed Reduced to and Returned from, mph												
	Stop	5	10	15	20	25	30	35	40	45	50	55	60	65
5	1.00													
10	2.17	1.11												
15	3.70	2.58	1.39											
20	5.64	4.42	3.17	1.71										
25	8.00	6.71	5.38	3.86	2.09									
30	10.86	9.49	8.10	6.51	4.68	2.52								
35	14.33	12.89	11.43	9.77	7.87	5.68	3.07							
40	18.55	17.02	15.51	13.78	11.80	9.52	6.85	3.74						
45	23.62	22.01	20.40	18.60	16.55	14.20	11.48	8.29	4.50					
50	29.67	27.98	26.29	24.39	22.27	19.83	17.04	13.76	9.90	5.37				
55	36.80	35.01	33.23	31.23	29.02	26.50	23.58	20.25	16.32	11.69	6.30			
60	45.19	43.29	41.41	39.32	36.99	34.34	31.33	27.89	23.86	19.18	13.69	7.38		
65	55.00	53.02	51.05	48.82	46.38	43.64	40.50	36.94	32.78	27.98	22.41	15.94	8.58	
70	66.40	64.30	62.19	59.83	57.27	54.39	51.08	47.41	43.11	38.23	32.50	25.96	18.41	11.02

* Cost includes fuel, tires, engine oil, maintenance, and depreciation.
† A speed-change cycle is reducing speed from and returning to an initial speed.

TABLE A-13

Excess Hours Consumed per Speed-Change Cycle *–
Excess Hours Above Continuing at Initial Speed

Vehicle: 5-kip commercial delivery
Unit: Hours per 1,000 cycles
Roadway surface: High type pavement in good condition

Initial Speed, mph	Speed Reduced to and Returned from, mph												
	Stop	5	10	15	20	25	30	35	40	45	50	55	60	65
5	0.73													
10	1.12	0.58												
15	1.52	0.98	0.47											
20	1.93	1.37	0.86	0.41										
25	2.36	1.77	1.24	0.76	0.35									
30	2.81	2.20	1.63	1.13	0.68	0.30								
35	3.28	2.64	2.05	1.52	1.04	0.62	0.27							
40	3.78	3.11	2.50	1.93	1.42	0.96	0.57	0.25						
45	4.30	3.62	2.96	2.36	1.81	1.32	0.89	0.53	0.24					
50	4.87	4.16	3.48	2.84	2.26	1.73	1.26	0.84	0.49	0.22				
55	5.48	4.75	4.04	3.37	2.74	2.16	1.64	1.18	0.79	0.47	0.22			
60	6.15	5.40	4.67	3.96	3.28	2.65	2.07	1.56	1.11	0.73	0.43	0.21		
65	6.90	6.15	5.40	4.67	3.97	3.30	2.67	2.10	1.57	1.12	0.72	0.41	0.21	
70	7.75	6.99	6.23	5.45	4.73	4.01	3.25	2.66	2.04	1.57	1.06	0.70	0.38	0.20

* A speed-change cycle is reducing speed from and returning to an initial speed.

Tables of the Running Cost of Motor Vehicles

TABLE A-14-M

Dollars Running Cost* at Uniform Speed on Minus Grades

Vehicle: 12-kip single-unit truck
Unit: Dollars per 1,000 vehicle-miles
Roadway surface: High type pavement in good condition

Speed, mph	Minus Grade, Percent									Speed, mph
	−8	−7	−6	−5	−4	−3	−2	−1	Level	
5	87.65	87.38	87.10	86.83	86.72	89.29	92.23	95.68	99.08	5
7½	73.37	72.79	72.71	72.40	72.69	75.23	78.27	81.42	85.35	7½
10	65.31	64.97	64.58	64.20	64.78	67.19	70.27	73.51	77.68	10
12½	60.31	59.85	59.36	58.90	59.81	61.97	65.06	68.37	72.86	12½
15	56.81	56.27	55.68	55.14	56.27	58.39	61.35	64.89	69.62	15
17½	54.32	53.71	53.00	52.51	53.89	55.88	58.81	62.38	67.37	17½
20	52.55	51.84	51.08	50.89	52.21	53.95	56.94	60.74	65.90	20
22½	51.46	50.64	49.77	49.97	51.21	52.88	55.98	59.72	65.09	22½
25	50.69	49.77	48.88	49.46	50.63	51.33	55.03	59.22	64.73	25
27½		49.40	48.77	49.30	50.57	52.21	54.74	59.08	64.76	27½
30		49.15	48.76	49.27	50.53	52.21	54.68	59.30	65.11	30
32½		49.33	49.14	49.58	50.89	52.62	55.08	59.85	65.75	32½
35		50.53	50.04	50.30	51.54	53.26	55.76	60.59	66.60	35
37½			51.23	51.44	52.50	54.10	56.67	61.57	67.66	37½
40			53.30	52.82	53.62	55.10	57.81	62.71	68.87	40
42½				54.20	54.72	56.09	59.21	64.06	70.27	42½
45					56.11	57.45	60.80	65.48	71.84	45
47½						59.08	62.60	67.18	73.60	47½
50						60.94	64.64	68.99	75.59	50
52½						63.15	66.91	70.95	77.79	52½
55						65.57	69.43	73.08	80.24	55
57½							72.12	75.34	82.86	57½
60							75.06	77.77	85.84	60
62½							78.31	80.43	89.05	62½
65								83.31	92.57	65

* Cost includes fuel, tires, engine oil, maintenance, and depreciation.

TABLE A-14-P
DOLLARS RUNNING COST* AT UNIFORM SPEED ON PLUS GRADES

Vehicle: 12-kip single-unit truck
Unit: Dollars per 1,000 vehicle-miles
Roadway surface: High type pavement in good condition

Speed, mph	Plus Grade, Percent									Speed, mph
	Level	+1	+2	+3	+4	+5	+6	+7	+8	
5	99.08	102.19	105.48	109.09	113.09	117.59	122.73	128.65	135.66	5
7½	85.35	88.60	91.97	95.79	99.94	104.63	110.01	116.21	123.72	7½
10	77.68	81.04	84.59	88.56	92.94	97.86	103.43	109.97	118.06	10
12½	72.86	76.33	80.03	84.27	88.86	94.08	100.00	106.95	115.67	12½
15	69.62	73.25	77.14	81.66	86.51	92.04	98.41	106.01	115.71	15
17½	67.37	71.22	75.32	80.12	85.34	91.26	98.31	106.86	117.87	17½
20	65.90	69.96	74.30	79.43	85.14	91.64	99.57	109.65	122.13	20
22½	65.09	69.40	73.98	79.53	85.80	93.21	102.24	114.65	128.44	22½
25	64.73	69.33	74.21	80.32	87.18	96.09	106.11	121.91		25
27½	64.76	69.68	74.93	81.49	89.30	100.26	110.70			27½
30	65.11	70.42	76.02	83.17	92.03	104.56	115.81			30
32½	65.75	71.46	77.54	85.28	95.48	109.68	121.79			32½
35	66.60	72.77	79.37	87.81	99.74	115.66				35
37½	67.66	74.28	81.50	90.66	104.93	122.13				37½
40	68.87	76.09	83.95	93.86	110.86					40
42½	70.27	78.10	86.68	97.44	117.84					42½
45	71.84	80.35	89.65	101.32						45
47½	73.60	82.95	92.84	105.53						47½
50	75.59	85.74	96.27	110.14						50
52½	77.79	88.66	100.04							52½
55	80.24	91.67	104.10							55
57½	82.86	94.83								57½
60	85.84	98.17								60
62½	89.05	101.73								62½
65	92.57	105.52								65

* Cost includes fuel, tires, engine oil, maintenance, and depreciation.

TABLE A-15
Dollars Excess Running Cost* at Uniform Speed on Horizontal Curves—Excess Above Cost on Tangents
Vehicle: 12-kip single-unit truck
Roadway surface: High type pavement in good condition
Unit: Dollars per 1,000 vehicle-miles

Speed, mph	Degree of Horizontal Curve														
	1	2	3	4	5	6	8	10	12	14	16	18	20	25	30
5	1.09	2.04	2.83	3.45	3.87	4.19	4.89	5.99	7.04	8.06	9.09	10.07	11.05	13.34	15.35
10	1.96	3.59	4.89	5.88	6.47	6.99	9.10	11.15	13.24	15.29	17.32	19.38	20.89	26.25	31.04
15	2.42	4.40	5.84	6.87	8.41	9.90	12.95	15.96	19.07	22.23	25.42	28.67	31.95	40.43	55.47
20	2.34	4.19	6.21	8.21	10.21	12.05	15.88	19.75	23.79	27.95	32.54	37.13	41.93	59.91	82.08
25	2.40	4.25	7.05	9.33	11.68	13.84	18.43	23.24	28.45	34.75	45.40	56.54	68.06	100.20	148.50
30	2.50	5.25	7.82	10.35	13.03	15.51	20.92	28.72	42.73	57.46	72.84	88.90	105.86		
35	2.74	5.52	8.26	11.04	14.00	17.60	30.46	46.16	64.98	84.69	105.43	127.38			
40	2.88	5.88	10.24	15.57	22.05	29.31	46.82	67.67	92.13	118.07	145.53				
45	4.53	10.06	16.60	24.21	33.05	42.88	65.76	92.60	123.70						
50	6.84	14.85	23.90	34.14	45.75	58.43	87.78	121.78							
55	9.61	20.39	32.49	41.72	60.71	77.05									
60	12.75	26.83	42.33												
65	16.45	34.28	54.03												

* Cost includes fuel, tires, engine oil, maintenance, and depreciation.

TABLE A-16

Dollars Excess Cost* Of Speed-Change Cycles*†
Excess Cost Above Continuing At Initial Speed

Vehicle: 12-kip single-unit truck
Unit: Dollars per 1,000 cycles
Roadway surface: High type pavement in good condition

Initial Speed, mph †	Speed Reduced to and Returned from, mph										
	Stop	5	10	15	20	25	30	35	40	45	50	55	60
5	1.92												
10	4.73	2.32											
15	8.30	5.54	2.95										
20	12.57	9.70	6.90	3.67									
25	17.65	14.71	11.78	8.46	4.55								
30	23.76	20.76	17.71	14.29	10.30	5.57							
35	31.07	27.98	24.88	21.36	17.32	12.51	6.83						
40	39.92	36.75	33.58	29.96	25.82	26.26	15.17	8.32					
45	50.49	47.25	43.96	40.28	36.06	31.12	25.29	18.37	10.04				
50	63.11	59.82	56.44	52.65	48.30	43.27	37.34	30.33	21.97	11.98			
55	78.00	74.66	71.14	67.27	62.82	57.67	51.61	44.52	36.07	26.00	14.04		
60	95.45	92.01	88.46	84.47	79.92	74.66	68.51	61.28	52.71	42.56	30.54	16.49	
65	115.79	112.23	108.71	104.62	99.96	94.58	88.31	80.98	72.27	61.99	49.87	35.71	19.18

* Cost includes fuel, tires, engine oil, maintenance, and depreciation.
† A speed change cycle is reducing speed from and returning to an initial speed.

TABLE A-17

Excess Hours Consumed For Speed-Change Cycle*
Excess Hours Above Continuing At Initial Speed

Vehicle: 12-kip single-unit truck
Unit: Hours per 1,000 cycles
Roadway surface: High type pavement in good condition

Initial Speed, mph	Speed Reduced to and Returned from, mph										
	Stop	5	10	15	20	25	30	35	40	45	50	55	60
5	0.73												
10	1.47	0.69											
15	2.20	1.35	0.62										
20	2.93	2.02	1.23	0.53									
25	3.67	2.70	1.86	1.12	0.45								
30	4.40	3.40	2.50	1.72	1.01	0.39							
35	5.13	4.11	3.16	2.33	1.59	0.91	0.36						
40	5.87	4.83	3.84	2.97	2.18	1.48	0.83	0.31					
45	6.60	5.57	4.54	3.64	2.81	2.07	1.37	0.76	0.28				
50	7.33	6.31	5.26	4.33	3.47	2.68	1.93	1.27	0.72	0.28			
55	8.07	7.06	6.02	5.07	4.17	3.35	2.56	1.84	1.23	0.71	0.32		
60	8.80	7.82	6.81	5.85	4.94	4.08	3.27	2.50	1.83	1.25	0.72	0.39	
65	9.53	8.58	7.64	6.71	5.80	4.92	4.09	3.30	2.57	1.90	1.30	0.77	0.33

* A speed-change cycle is reducing speed from and returning to an initial speed.

Tables of the Running Cost of Motor Vehicles

TABLE A-18-M

DOLLARS RUNNING COST* AT UNIFORM SPEED ON MINUS GRADES

Vehicle: 40-kip 2-S2, gasoline
Unit: Dollars per 1,000 vehicle-miles
Roadway surface: High type pavement in good condition

Speed, mph	Minus Grade, Percent									Speed, mph
	−8	−7	−6	−5	−4	−3	−2	−1	Level	
5	218.35	212.50	206.65	201.96	194.63	193.37	203.19	212.25	231.01	5
7½	160.37	157.37	152.36	147.45	140.55	139.07	148.67	157.79	176.66	7½
10	132.45	128.83	123.99	118.95	112.70	110.94	120.13	129.37	148.64	10
12½	112.38	111.47	106.52	101.46	95.83	93.80	102.54	111.91	131.48	12½
15	102.35	100.33	95.03	89.79	85.01	82.80	90.87	100.38	120.17	15
17½		92.96	87.14	81.99	77.84	76.73	82.84	92.55	112.43	17½
20			81.70	76.70	73.19	71.01	77.31	87.17	107.10	20
22½			77.97	73.72	70.38	68.55	73.90	83.54	103.47	22½
25			75.64	71.00	68.87	68.00	71.20	81.30	101.19	25
27½				69.89	68.59	67.93	69.94	80.02	99.93	27½
30				69.78	68.97	68.24	69.61	79.90	99.51	30
32½					69.68	68.89	70.10	80.60	99.87	32½
35						69.80	71.31	81.98	100.84	35
37½						70.97	73.17	83.92	102.43	37½
40						72.36	75.55	86.38	104.55	40
42½						74.03	78.48	89.25	107.21	42½
45						75.97	81.99	92.47	110.43	45
47½							86.01	96.23	114.23	47½
50							90.48	100.24	118.70	50
52½							95.30	104.57	123.98	52½
55							100.47	109.22	130.18	55
57½								114.18	137.35	57½
60								119.38	145.55	60

* Cost includes fuel, tires, engine oil, maintenance, and depreciation.

TABLE A-18-P

Dollars Running Cost* at Uniform Speed on Plus Grades

Vehicle: 40-kip 2-S2, gasoline
Unit: Gallons per 1,000 vehicle-miles
Roadway surface: High type pavement in good condition

Speed, mph	Plus Grade, Percent									Speed, mph
	Level	+1	+2	+3	+4	+5	+6	+7	+8	
5	231.01	236.70	243.35	251.86	262.26	275.01	292.53	318.11	357.71	5
7½	176.66	183.28	190.90	199.97	210.83	224.53	242.12	267.64	308.99	7½
10	148.64	155.92	164.41	174.24	185.97	200.79	219.37	245.80	289.90	10
12½	131.48	139.55	148.83	159.79	172.84	189.05	209.61	238.76	285.44	12½
15	120.17	129.22	139.29	151.65	166.46	184.53	208.39	242.57		15
17½	112.43	122.53	133.52	147.49	164.58	185.44	214.08	256.24		17½
20	107.10	118.37	130.38	146.21	166.24	191.29	226.44			20
22½	103.47	116.01	129.23	147.16	170.80	202.54	245.49			22½
25	101.19	115.20	129.79	150.02	178.14	220.81				25
27½	99.93	115.53	131.71	154.45	187.84	245.27				27½
30	99.51	116.87	134.92	160.45	199.87					30
32½	99.87	119.16	139.38	167.79	214.91					32½
35	100.84	122.25	145.06	176.70						35
37½	102.43	126.18	152.02	186.36						37½
40	104.55	130.98	160.32	198.21						40
42½	107.21	136.72	169.77							42½
45	110.43	143.65	180.56							45
47½	114.23	152.40	192.73							47½
50	118.70	161.98	206.39							50
52½	123.98	172.36								52½
55	130.18	183.49								55
57½	137.35	195.50								57½
60	145.55									60

* Cost includes fuel, tires, engine oil, maintenance, and depreciation.

TABLE A-19

DOLLARS EXCESS RUNNING COST* AT UNIFORM SPEED ON HORIZONTAL CURVES-EXCESS ABOVE COST ON TANGENTS
Vehicle: 40-kip 2-S2, gasoline
Roadway surface: High type pavement in good condition
Unit: Dollars per 1,000 vehicle-miles

Speed, mph	Degree of Horizontal Curve													
	1	2	3	4	5	6	8	10	12	14	16	18	20	25	30
5	2.32	4.58	6.63	7.48	9.87	11.02	13.39	16.00	18.19	20.32	22.46	24.55	26.69	31.56	36.10
10	3.95	7.29	10.42	12.74	14.35	13.73	20.47	24.80	29.15	32.96	36.98	41.02	45.25	54.64	64.80
15	4.91	8.95	12.07	14.33	17.61	20.75	27.09	33.20	39.31	45.49	51.70	58.05	64.46	81.08	99.00
20	4.73	8.50	12.69	16.79	20.95	24.74	32.53	40.31	48.77	56.59	65.78	74.96	84.56	121.08	166.02
25	4.88	8.71	14.44	19.17	23.97	28.39	37.79	47.55	58.21	70.88	92.83	115.65	139.42	207.09	281.62
30	5.15	10.85	16.19	21.39	26.95	32.15	43.40	59.59	88.60	119.34	151.45	185.15	220.77		
35	5.78	11.67	17.47	23.34	29.58	37.32	64.55	97.80	137.85	180.01	224.66	272.09			
40	6.23	12.84	22.30	33.87	48.03	63.81	102.12	147.91	202.00	259.90	321.74				
45	10.12	22.56	37.28	54.32	74.30	96.38	148.27	209.75	281.61						
50	15.82	34.31	55.41	79.24	106.36	136.15	205.70	287.60							
55	23.01	48.85	78.09	100.77	146.78	186.70									
60	31.59	66.53	105.47												

* Cost includes fuel, tires, engine oil, maintenance, and depreciation.

Tables of the Running Cost of Motor Vehicles

TABLE A-20
Dollars Excess Cost* of Speed-Change Cycles† –
Excess Cost Above Continuing at Initial Speed

Vehicle: 40-kip 2-S2, gasoline
Unit: Dollars per 1,000 cycles
Roadway surface: High type pavement in good condition

Initial Speed, mph	Speed Reduced to and Returned from, mph										
	Stop	5	10	15	20	25	30	35	40	45	50	55
5	7.36										
10	15.80	8.48									
15	26.88	19.42	11.08								
20	41.06	33.22	28.43	14.07							
25	58.85	50.70	42.19	31.47	17.57						
30	80.92	72.59	63.69	52.77	38.94	21.63					
35	107.94	99.54	90.50	79.27	65.22	47.84	26.42				
40	141.08	132.56	123.41	111.84	97.60	80.05	58.51	32.20			
45	181.23	172.52	163.14	151.47	137.04	119.29	97.58	71.14	38.82		
50	229.62	220.76	211.17	199.27	184.58	166.65	144.75	118.16	85.71	46.52	
55	287.13	278.09	268.26	256.12	241.24	223.04	200.87	173.93	141.18	101.78	54.80
60	355.49	346.16	336.07	323.65	308.37	289.76	267.05	239.57	206.35	166.66	119.60	64.45

* Cost includes fuel, tires, engine oil, maintenance, and depreciation.
† A speed-change cycle is reducing speed from and returning to an initial speed.

TABLE A-21
Excess Hours Consumed per Speed Change Cycle*
Excess Hours Above Continuing at Initial Speed

Vehicle: 40-kip 2-S2, gasoline
Unit: Hours per 1,000 cycles
Roadway surface: High type pavement in good condition

Initial Speed, mph	Speed Reduced to and Returned from, mph										
	Stop	5	10	15	20	25	30	35	40	45	50	55
5	0.67										
10	1.47	0.61									
15	2.30	1.34	0.55								
20	3.19	2.08	1.19	0.50							
25	4.16	2.97	1.95	1.12	0.47						
30	5.22	3.98	2.85	1.88	1.07	0.44					
35	6.41	5.14	3.94	2.84	1.88	1.07	0.44				
40	7.76	6.48	5.24	4.05	2.96	1.98	1.16	0.49			
45	9.35	8.07	6.80	5.57	4.37	3.25	2.21	1.31	0.57		
50	11.34	10.02	8.70	7.40	6.12	4.88	3.72	2.64	1.66	0.78	
55	13.94	12.52	11.12	9.74	8.38	7.05	5.76	4.50	3.30	2.14	1.04
60	17.47	15.87	14.29	12.73	11.18	9.67	8.18	6.73	5.30	3.91	2.56	1.26

* A speed-change cycle is reducing speed from and returning to an initial speed.

Tables of the Running Cost of Motor Vehicles

TABLE A-22-M

DOLLARS RUNNING COST* AT UNIFORM SPEED ON MINUS GRADES

Vehicle: 50-kip 3-S2, diesel
Unit: Dollars per 1,000 vehicle-miles
Roadway surface: High type pavement in good condition

Speed, mph	Minus Grade, Percent									Speed, mph
	−8	−7	−6	−5	−4	−3	−2	−1	Level	
5	161.84	158.90	156.00	153.69	150.13	149.42	153.82	157.86	166.72	5
7½	129.98	128.21	125.53	122.96	119.46	118.56	122.96	127.08	136.35	7½
10	113.96	111.78	109.07	106.30	102.95	101.83	106.17	110.50	120.20	10
12½	102.37	101.46	98.57	95.67	92.51	91.23	95.41	99.95	110.11	12½
15	96.27	94.76	91.51	88.38	85.54	84.07	88.05	92.76	103.41	15
17½		89.71	86.10	82.84	80.28	79.32	82.35	87.79	98.84	17½
20			82.64	79.37	77.04	75.47	78.65	84.36	95.81	20
22½			80.35	77.59	75.03	73.59	76.44	82.10	93.89	22½
25			79.03	75.70	73.98	72.73	74.57	80.79	92.92	25
27½				75.06	73.78	72.43	73.76	80.18	92.64	27½
30				75.11	73.64	72.73	73.58	80.38	93.02	30
32½					74.42	73.41	74.27	81.27	94.00	32½
35						74.40	75.48	82.75	95.10	35
37½						75.70	77.25	84.69	97.49	37½
40						77.29	79.47	87.07	99.85	40
42½						79.37	82.11	89.79	102.61	42½
45						81.83	85.20	92.77	105.80	45
47½							88.69	96.25	109.45	47½
50							92.76	100.04	113.63	50
52½							97.19	104.11	118.37	52½
55							101.92	108.32	123.66	55
57½								112.98	129.55	57½
60									134.60	60

* Cost includes fuel, tires, engine oil, maintenance, and depreciation.

TABLE A-22-P

DOLLARS RUNNING COST*AT UNIFORM SPEED ON PLUS GRADES

Vehicle: 50-kip 3-S2, diesel
Unit: Dollars per 1,000 vehicle-miles
Roadway surface: High type pavement in good condition

Speed, mph	Plus Grade, Percent									Speed, mph
	Level	+1	+2	+3	+4	+5	+6	+7	+8	
5	166.72	171.45	176.40	181.21	186.01	190.64	195.25	199.85	205.37	5
7½	136.35	144.82	153.22	161.46	169.51	177.45	185.29	193.02	201.65	7½
10	120.20	130.60	140.90	151.05	161.04	170.93	180.66	190.69	201.65	10
12½	110.11	121.74	133.34	144.83	156.24	167.63	178.65	190.97	203.83	12½
15	103.41	115.97	128.51	141.17	153.86	166.66	179.36	193.66	208.54	15
17½	98.84	112.17	125.54	139.24	153.22	167.54	182.22	198.84	214.81	17½
20	95.81	109.81	123.95	138.70	154.07	170.17	187.30	207.02	223.23	20
22½	93.89	108.54	123.45	139.28	156.21	174.65	194.76	218.71		22½
25	92.92	108.27	123.94	140.91	159.75	181.46	204.95			25
27½	92.64	108.64	125.17	143.38	164.46	190.43				27½
30	93.02	109.75	127.14	146.79	170.57					30
32½	94.00	111.48	129.89	151.20	178.03					32½
35	95.10	113.76	133.34	156.46	187.58					35
37½	97.49	116.55	137.43	162.56						37½
40	99.85	119.87	142.28	169.96						40
42½	102.61	123.57	147.63							42½
45	105.80	127.88	153.72							45
47½	109.45	132.97	160.68							47½
50	113.63	138.77								50
52½	118.37	145.20								52½
55	123.66	152.23								55
57½	129.55	160.20								57½
60	134.60									60

* Cost includes fuel, tires, engine oil, maintenance and depreciation.

TABLE A-23

Dollars Excess Running Cost* at Uniform Speed on Horizontal Curves—Excess Above Cost on Tangents

Vehicle: 50-kip 3-S2, diesel
Roadway surface: High type pavement in good condition
Unit: Dollars per 1,000 vehicle-miles

Speed, mph	Degree of Horizontal Curve													
	1	2	3	4	5	6	8	10	12	14	16	18	20	25	30
5	2.96	5.66	7.89	9.77	11.09	12.09	14.24	17.40	20.77	23.06	25.83	28.54	31.19	37.51	43.13
10	5.21	9.59	13.14	15.80	17.47	18.93	24.66	30.14	35.72	41.11	46.51	51.89	57.56	70.16	85.30
15	6.50	11.66	15.57	18.22	22.36	26.31	34.40	42.45	50.73	59.10	67.50	76.11	84.77	107.17	135.00
20	6.22	11.12	16.54	21.78	27.11	32.05	42.23	52.57	63.80	74.43	86.67	98.85	111.56	159.51	218.45
25	6.47	11.45	19.05	25.12	31.43	37.24	49.63	62.59	76.70	93.42	122.33	152.29	183.38	269.73	366.72
30	6.78	14.26	21.28	28.11	35.42	42.23	56.99	78.22	116.30	156.43	198.62	241.72	287.62		
35	7.59	15.31	22.90	30.57	38.69	48.81	84.47	127.85	179.74	233.97	290.93	351.01			
40	8.15	16.75	29.10	44.22	62.67	83.23	132.89	191.78	260.63	333.29	410.07				
45	13.19	29.41	48.45	70.69	96.59	125.00	191.30	268.84	357.94						
50	20.59	44.59	71.88	102.55	137.22	174.94	261.74	361.67							
55	29.92	63.31	100.63	129.09	187.52	237.06									
60	40.96	85.63	135.04												

* Cost includes fuel, tires, engine oil, maintenance, and depreciation.

TABLE A-24

**Dollars Excess Cost* of Speed-Change Cycles † –
Excess Cost Above Continuing at Initial Speed**

Vehicle: 50-kip 3-S2, diesel
Unit: Dollars per 1,000 cycles
Roadway surface: High type pavement in good condition

Initial Speed, mph	Speed Reduced to and Returned from, mph										
	Stop	5	10	15	20	25	30	35	40	45	50	55
5	9.08											
10	21.00	10.03										
15	34.50	23.91	13.14									
20	52.51	41.64	35.27	16.83								
25	74.75	63.71	52.30	38.35	21.26							
30	102.21	90.96	79.37	65.07	47.92	26.61						
35	136.05	124.55	112.67	98.29	80.85	59.46	33.03					
40	178.00	166.14	154.19	139.28	121.56	100.04	73.60	40.99				
45	229.46	217.31	204.75	189.71	171.64	149.80	123.16	90.57	50.24			
50	292.31	279.81	266.88	251.45	232.93	210.64	183.58	150.68	110.26	60.91		
55	368.52	355.53	342.04	326.05	306.91	283.92	256.20	222.67	191.69	132.28	72.45	
60	462.48	448.35	433.75	416.54	396.26	372.26	343.57	309.11	267.42	217.78	165.57	85.65

* Cost includes fuel, tires, engine oil, maintenance, and depreciation.
† A speed-change cycle is reducing speed from and returning to an initial speed.

TABLE A-25

**Excess Hours Consumed per Speed-Change Cycle* –
Excess Hours Above Continuing at Initial Speed**

Vehicle: 50-kip 3-S2, diesel
Unit: Hours per 1,000 cycles
Roadway surface: High type pavement in good condition

Initial Speed, mph	Speed Reduced to and Returned from, mph										
	Stop	5	10	15	20	25	30	35	40	45	50	55
5	1.10											
10	2.27	0.95										
15	3.48	1.96	0.81									
20	4.76	3.05	1.71	0.69								
25	6.10	4.25	2.72	1.49	0.60							
30	7.56	5.59	3.90	2.45	1.36	0.54						
35	9.19	7.12	5.29	3.66	2.35	1.31	0.52					
40	11.09	8.94	6.99	5.20	3.66	2.40	1.36	0.58				
45	13.39	11.20	9.12	7.19	5.45	3.95	2.65	1.58	0.71			
50	16.37	14.13	11.95	9.88	7.95	6.19	4.60	3.18	1.95	0.89		
55	20.72	18.33	15.98	13.71	11.53	9.45	7.48	5.66	3.98	2.48	1.15	
60	27.94	24.99	22.10	19.28	16.55	13.93	11.44	9.10	6.92	4.92	3.10	1.46

* A speed-change cycle is reducing speed from and returning to an initial speed.

Tables of the Running Cost of Motor Vehicles

TABLE A-26-M
Gasoline Consumption at Uniform Speed on Tangent Minus Grades

Vehicle: 4-kip passenger car
Unit: Gallons per 1,000 vehicle-miles
Roadway surface: High type pavement in good condition

Speed, mph	\-8	\-7	\-6	\-5	\-4	\-3	\-2	\-1	Level	Speed, mph
5	80.0	80.0	80.0	80.0	80.0	80.0	80.0	95.6	102.4	5
7½	53.3	53.3	53.3	53.3	53.3	53.3	53.3	69.2	76.1	7½
10	40.0	40.0	40.0	40.0	40.0	40.0	40.0	56.3	63.3	10
12½	32.0	32.0	32.0	32.0	32.0	32.0	32.0	48.8	55.8	12½
15	26.7	26.7	26.7	26.7	26.7	26.7	32.3	44.0	51.1	15
17½	22.9	22.9	22.9	22.9	22.9	22.9	30.7	40.8	48.0	17½
20	20.0	20.0	20.0	20.0	20.0	20.0	29.1	38.6	45.9	20
22½	17.8	17.8	17.8	17.8	17.8	20.0	28.9	37.2	44.4	22½
25	16.0	16.0	16.0	16.0	16.0	20.2	28.6	36.2	43.5	25
27½	14.5	14.5	14.5	14.5	14.5	20.5	28.4	35.7	43.0	27½
30	13.3	13.3	13.3	13.3	14.5	20.9	28.5	35.4	42.8	30
32½	12.3	12.3	12.3	12.3	15.1	21.4	28.8	35.5	43.0	32½
35	11.4	11.4	11.4	11.4	15.8	22.0	29.2	35.8	43.3	35
37½	10.7	10.7	10.7	10.7	16.5	22.7	29.8	36.3	43.9	37½
40	10.0	10.0	10.0	11.2	17.3	23.6	30.5	37.0	44.7	40
42½	9.4	9.4	9.4	12.0	18.2	24.4	31.4	38.0	45.6	42½
45	8.9	8.9	8.9	12.9	19.1	25.4	32.4	39.1	46.8	45
47½	8.4	8.4	8.4	13.8	20.1	26.5	33.6	40.4	48.1	47½
50	8.0	8.0	8.7	14.8	21.2	27.7	34.8	41.8	49.6	50
52½	7.6	7.6	9.8	15.9	22.4	29.0	36.2	43.4	51.3	52½
55	7.3	7.3	11.0	17.0	23.6	30.4	37.8	45.2	53.2	55
57½	7.0	7.0	12.2	18.3	25.0	32.0	39.5	47.2	55.3	57½
60	6.7	8.2	13.6	19.7	26.4	33.7	41.3	49.4	57.6	60
62½		9.5	15.0	21.2	28.0	35.5	43.4	51.7	60.2	62½
65			16.5	22.8	29.8	37.5	45.6	54.3	63.1	65
67½				24.5	31.7	39.7	48.1	57.1	66.3	67½
70					33.9	42.1	50.9	60.3	70.0	70
72½						44.9	54.1	63.9	74.1	72½
75						48.3	57.6	67.9	78.7	75
77½							61.7	72.6	84.1	77½
80							66.4	78.1	90.4	80

TABLE A-26-P
Gasoline Consumption at Uniform Speed on Tangent Plus Grades

Vehicle: 4-kip passenger car
Unit: Gallons per 1,000 vehicle-miles
Roadway surface: High type pavement in good condition

Speed, mph	Level	+1	+2	+3	+4	+5	+6	+7	+8	Speed, mph
5	102.4	110.4	118.8	127.8	137.0	148.0	157.0	168.2	181.8	5
7½	76.1	84.1	92.7	101.7	110.9	120.0	130.9	142.3	155.7	7½
10	63.3	71.3	79.9	89.0	98.2	106.5	118.2	129.7	143.0	10
12½	55.8	63.8	72.6	81.6	90.8	99.6	110.9	122.5	135.7	12½
15	51.1	59.1	67.9	76.9	86.1	96.0	106.3	117.9	131.1	15
17½	48.0	56.0	64.9	73.8	83.0	92.6	103.2	115.0	128.1	17½
20	45.9	53.9	62.8	71.8	80.9	90.8	101.1	113.0	126.0	20
22½	44.4	52.5	61.3	70.4	79.5	893	99.7	111.7	124.6	22½
25	43.5	51.6	60.4	69.4	78.5	88.2	98.8	110.9	123.8	25
27½	43.0	51.1	59.8	68.8	77.9	87.6	98.2	110.4	123.3	27½
30	42.8	50.9	59.5	68.5	77.6	87.3	97.9	110.3	123.2	30
32½	43.0	51.0	59.5	68.4	77.6	87.4	98.0	110.3	123.4	32½
35	43.3	51.2	59.7	68.6	77.7	87.4	98.2	110.6	123.8	35
37½	43.9	51.7	60.1	68.9	78.1	87.9	98.6	111.1	124.5	37½
40	44.7	52.4	60.7	69.4	78.6	88.5	99.3	111.8	125.3	40
42½	45.6	53.1	61.5	70.1	79.3	89.1	100.1	112.7	126.3	42½
45	46.8	54.2	62.5	71.0	80.1	89.9	101.0	113.7	127.6	45
47½	48.1	55.5	63.7	72.1	81.2	90.9	102.1	114.9	129.0	47½
50	49.6	56.9	65.0	73.3	82.4	92.2	103.4	116.3	130.6	50
52½	51.3	58.5	66.5	74.7	83.8	93.7	104.9	118.0	132.5	52½
55	53.2	60.4	68.2	76.4	85.4	95.4	106.7	119.9	134.7	55
57½	55.3	62.4	70.1	78.3	87.3	97.3	108.8	122.0	137.1	57½
60	57.6	64.8	72.3	80.4	89.4	99.6	111.1	124.5	140.0	60
62½	60.2	67.3	74.8	82.8	91.8	102.1	114.0	127.4		62½
65	63.1	70.2	77.6	85.6	94.6	105.1	117.2			65
67½	66.3	73.4	80.7	88.8	97.7	108.5				67½
70	70.0	76.9	84.2	92.3	101.4					70
72½	74.1	81.0	88.2	96.4						72½
75	78.7	85.5	92.8	101.3						75
77½	84.1	90.7	98.1							77½
80	90.4	96.8	104.3							80

TABLE A-27
GALLONS EXCESS GASOLINE CONSUMED AT UNIFORM SPEED ON HORIZONTAL CURVES–EXCESS ABOVE CONSUMPTION ON TANGENTS

Vehicle: 4-kip passenger car
Roadway surface: High type pavement in good condition
Unit: Gallons per 1,000 vehicle-miles

Speed, mph	Degree of Horizontal Curve														
	1	2	3	4	5	6	8	10	12	14	16	18	20	25	30
5	0.03	0.20	0.36	0.61	0.87	1.12	1.51	1.77	1.80	1.80	1.86	1.98	2.07	2.34	2.83
10	0.04	0.18	0.35	0.60	0.79	0.97	1.28	1.48	1.58	1.54	1.44	1.41	1.41	1.41	1.41
15	0.03	0.16	0.29	0.46	0.61	0.73	0.93	1.06	0.97	0.95	0.80	0.69	0.69	0.59	0.46
20	0.03	0.12	0.21	0.34	0.46	0.55	0.63	0.64	0.60	0.51	0.43	0.39	0.39	0.58	1.26
25	0.04	0.10	0.16	0.23	0.29	0.33	0.39	0.39	0.39	0.51	0.74	1.11	1.65	3.51	5.70
30	0.04	0.10	0.14	0.19	0.24	0.27	0.35	0.51	0.80	1.26	1.94	2.87	4.01	8.35	16.07
35	0.05	0.10	0.15	0.22	0.31	0.40	0.67	1.13	2.00	3.20	4.80	6.89	9.49	19.10	
40	0.08	0.17	0.27	0.39	0.55	0.76	1.44	2.54	4.32	6.72	9.85	13.81	18.72		
45	0.13	0.28	0.47	0.72	1.06	1.49	2.81	4.91	8.08	12.22	17.62				
50	0.20	0.45	0.80	1.25	1.84	2.61	4.93	8.33	13.38	20.02					
55	0.32	0.77	1.37	2.16	3.16	4.48	7.99	13.90	21.89						
60	0.49	1.18	2.11	3.35	4.97	7.04	12.86	21.36							
65	0.72	1.74	3.16	5.02	7.46	10.57	19.13								
70	1.03	2.52	4.60	7.23	10.90	15.40									
75	1.44	3.57	6.52	10.37	15.50										
80	1.97	4.92	9.51	15.73											

TABLE A-28

EXCESS GALLONS OF GASOLINE CONSUMED DURING SPEED-CHANGE CYCLES*–
EXCESS CONSUMPTION ABOVE CONTINUING AT INITIAL SPEED

Vehicle: 4-kip passenger car
Unit: Gallons per 1,000 vehicle-cycles
Roadway surface: High type pavement in good condition

Initial Speed, mph	Speed Reduced to and Returned from, mph														
	Stop	5	10	15	20	25	30	35	40	45	50	55	60	65	70	75
5	0.26															
10	1.00	0.51														
15	2.69	1.50	0.78													
20	4.40	3.20	2.02	1.06												
25	6.16	5.00	3.79	2.56	1.35											
30	7.95	6.87	5.68	4.48	3.13	1.66										
35	9.83	8.88	7.70	6.53	5.28	3.75	2.00									
40	11.83	11.00	9.91	8.75	7.59	6.20	4.80	2.38								
45	14.02	13.30	12.35	11.22	10.13	8.78	7.60	5.21	2.81							
50	16.51	15.84	15.10	14.02	12.99	11.62	10.60	8.44	6.11	3.34						
55	19.50	18.93	18.25	17.43	16.32	14.93	13.92	12.01	9.87	7.21	4.03					
60	23.17	22.59	21.88	21.00	20.02	18.81	17.40	15.70	13.75	11.44	8.55	4.92				
65	27.83	27.19	26.46	25.62	24.67	23.53	22.14	20.55	18.65	16.44	13.79	10.50	6.26			
70	33.80	32.98	32.04	31.08	30.00	28.75	27.36	25.80	23.94	21.81	19.31	16.38	12.80	7.95		
75	41.58	40.35	38.96	37.48	36.00	34.45	32.77	31.00	29.03	26.84	24.42	21.61	18.31	14.25	8.95	
80	51.65	49.83	47.96	46.10	44.15	42.12	40.00	37.88	35.68	33.36	30.84	28.04	24.96	21.33	16.90	11.09

* A speed-change cycle is reducing speed from and returning to an initial speed.

Tables of the Running Cost of Motor Vehicles

TABLE A-29-M

GASOLINE CONSUMPTION AT UNIFORM SPEED ON TANGENT MINUS GRADES

Vehicle: 5-kip commercial delivery
Unit: Gallons per 1,000 vehicle-miles
Roadway surface: High type pavement in good condition

Speed, mph	Minus Grade, Percent									Speed, mph
	−8	−7	−6	−5	−4	−3	−2	−1	Level	
5	84.0	84.0	84.0	84.0	84.0	85.4	92.4	99.6	107.8	5
7½	56.0	56.0	56.0	56.0	56.0	57.5	64.8	71.9	80.1	7½
10	42.0	42.0	42.0	42.0	42.0	44.2	51.4	58.4	66.6	10
12½	33.6	33.6	33.6	33.6	33.6	36.5	43.8	50.6	58.8	12½
15	28.0	28.0	28.0	28.0	28.0	31.7	39.1	45.8	53.9	15
17½	24.0	24.0	24.0	24.0	24.0	28.5	36.1	42.6	50.7	17½
20	21.0	21.0	21.0	21.0	21.0	26.4	34.1	40.6	48.6	20
22½	18.7	18.7	18.7	18.7	18.7	25.1	32.9	39.2	47.3	22½
25	16.8	16.8	16.8	16.8	18.1	24.3	32.2	38.5	46.6	25
27½	15.3	15.3	15.3	15.3	18.1	23.9	31.9	38.3	46.3	27½
30	14.0	14.0	14.0	14.0	18.2	23.8	31.9	38.4	46.4	30
32½	12.9	12.9	12.9	12.9	18.4	24.0	32.1	38.8	46.9	32½
35	12.0	12.0	12.0	12.6	18.7	24.4	32.6	38.9	47.7	35
37½	11.2	11.2	11.2	13.0	19.2	24.9	33.3	40.5	48.8	37½
40	10.5	10.5	10.5	13.6	19.8	25.7	34.3	41.8	50.3	40
42½	9.9	9.9	9.9	14.0	20.4	26.8	35.6	43.3	52.2	42½
45	9.3	9.3	9.7	14.6	21.2	28.2	37.2	45.2	54.4	45
47½	8.8	8.8	9.6	15.3	22.2	29.9	39.4	47.5	57.1	47½
50	8.4	8.4	9.8	16.0	23.4	31.9	41.7	50.3	60.3	50
52½		8.0	10.1	16.8	24.8	34.3	44.3	53.6	64.0	52½
55				17.8	26.5	37.1	47.3	57.5	68.5	55
57½				19.0	28.5	40.3	50.9	62.1	73.8	57½
60						43.9	55.2	67.6	80.4	60
62½							60.6	74.4	88.3	62½
65							67.7	83.1	98.0	65
67½								94.3	108.7	67½
70								108.7	120.9	70

TABLE A-29-P

Gasoline Consumption At Uniform Speed On Tangent Plus Grades

Vehicle: 5-kip commercial delivery
Unit: Gallons per 1,000 vehicle-miles
Roadway surface: High type pavement in good condition

Speed, mph	Level	+1	+2	+3	+4	+5	+6	+7	+8	Speed, mph
5	107.8	115.8	124.2	133.4	144.0	156.4	171.0	188.4	208.6	5
7½	80.1	88.0	96.4	105.6	116.1	128.5	143.3	160.7	180.9	7½
10	66.6	74.5	83.2	92.9	103.8	116.2	130.6	147.5	167.8	10
12½	58.8	66.7	75.4	85.1	96.0	108.5	123.0	140.5	160.5	12½
15	53.9	61.9	70.7	80.5	91.5	104.0	118.5	135.7	156.3	15
17½	50.7	58.8	67.7	77.5	88.6	101.2	115.8	133.0	153.9	17½
20	48.6	56.8	65.7	75.6	86.8	99.4	114.2	131.6	152.6	20
22½	47.3	55.5	64.5	74.6	85.8	98.6	113.4	130.9	152.2	22½
25	46.6	54.8	63.9	74.1	85.4	98.5	113.5	131.1	152.6	25
27½	46.3	54.7	64.0	74.3	85.8	98.9	114.1	131.9	153.6	27½
30	46.4	54.9	64.4	74.8	86.5	99.7	115.1	133.2	155.2	30
32½	46.9	55.6	65.2	75.8	87.7	101.1	116.7	135.0	157.5	32½
35	47.7	56.6	66.4	77.2	89.3	103.0	118.8	137.5	160.4	35
37½	48.8	58.0	68.1	79.1	91.3	105.3	121.4	140.5	164.1	37½
40	50.3	59.8	70.1	81.4	93.9	108.2	124.6	144.3	168.9	40
42½	52.2	62.0	72.6	84.2	97.1	111.6	128.5	148.8	174.5	42½
45	54.4	64.6	75.6	87.6	100.8	115.8	133.2	154.4	181.8	45
47½	57.1	67.7	79.1	91.5	105.2	120.7	138.9	161.3	191.1	47½
50	60.3	71.3	83.2	96.1	110.3	126.5	145.7	170.0	203.2	50
52½	64.0	75.5	87.9	101.3	116.1	133.1	153.4	180.4		52½
55	68.5	80.5	93.5	107.5	123.0	140.9				55
57½	73.8	86.5	100.2	115.0	131.3	150.2				57½
60	80.4	93.9	108.4	124.2						60
62½	88.3	102.8	118.5							62½
65	98.0	113.4	132.5							65
67½	108.7	125.6								67½
70	120.9	139.5								70

Tables of the Running Cost of Motor Vehicles

TABLE A-30
Gallons Excess Gasoline Consumed at Uniform Speed on Horizontal Curves–Excess Above Consumption on Tangents

Vehicle: 5-kip commercial delivery truck
Unit: Gallons per 1,000 vehicle-miles
Roadway surface: High type pavement in good condition

Speed, mph	Degree of Horizontal Curve													
	1	2	3	4	5	6	8	10	12	14	16	18	20	25	30
5	0.04	0.17	0.31	0.53	0.74	0.97	1.31	1.52	1.56	1.56	1.62	1.72	1.80	2.03	2.45
10	0.03	0.16	0.31	0.53	0.70	0.86	1.13	1.31	1.40	1.37	1.28	1.25	1.25	1.25	1.25
15	0.03	0.14	0.26	0.42	0.55	0.66	0.84	0.94	0.88	0.80	0.73	0.62	0.63	0.53	0.42
20	0.03	0.11	0.20	0.32	0.43	0.51	0.59	0.60	0.57	0.48	0.40	0.36	0.36	0.55	1.18
25	0.03	0.10	0.16	0.22	0.28	0.32	0.38	0.38	0.38	0.50	0.73	1.09	1.61	3.43	5.57
30	0.04	0.10	0.15	0.20	0.25	0.28	0.36	0.53	0.83	1.34	1.99	2.95	4.12	8.59	16.52
35	0.05	0.10	0.17	0.25	0.34	0.44	0.74	1.25	2.21	3.54	5.31	7.62	10.50	21.13	
40	0.10	0.20	0.32	0.48	0.67	0.92	1.75	3.09	5.25	8.19	11.98	16.80	22.76		
45	0.18	0.37	0.64	0.96	1.42	2.00	3.79	6.59	11.76	17.78	25.64				
50	0.30	0.77	1.22	1.91	2.80	3.99	7.51	13.74	22.06	33.00					
55	0.53	1.26	2.23	3.54	5.18	7.33	14.12	24.56	38.70						
60	0.89	2.18	3.88	6.16	9.15	12.96	25.43	42.23							
65	1.58	3.83	6.96	11.07	18.01	25.48	46.13								
70	2.57	6.28	11.42	17.96	32.38	45.76									

TABLE A-31
Excess Gallons of Gasoline Consumed During Speed-Change Cycles*– Excess Consumption Above Continuing at Initial Speed

Vehicle: 5-kip commercial delivery truck
Unit: Gallons per 1,000 vehicle-cycles
Roadway surface: High type pavement in good condition

Initial Speed, mph	Speed Reduced to and Returned from, mph												
	Stop	5	10	15	20	25	30	35	40	45	50	55	60	65
5	0.28													
10	1.18	0.54												
15	2.32	1.59	0.79											
20	3.69	2.86	1.96	1.00										
25	5.25	4.31	3.32	2.29	1.18									
30	6.98	5.92	4.84	3.72	2.56	1.32								
35	8.86	7.68	6.51	5.29	4.02	2.75	1.43							
40	10.86	9.56	8.29	6.98	5.62	4.29	2.88	1.49						
45	12.96	11.53	10.16	8.75	7.31	5.89	4.44	2.94	1.52					
50	15.12	13.55	12.08	10.57	9.03	7.52	6.02	4.46	2.93	1.50				
55	17.32	15.60	14.02	12.40	10.74	9.12	7.46	5.92	4.34	2.84	1.42			
60	19.54	17.66	15.95	14.20	12.41	10.67	8.92	7.27	5.65	4.12	2.64	1.29		
65	21.77	19.72	17.87	15.93	14.01	12.12	10.27	8.49	6.80	5.23	3.72	2.25	1.08	
70	24.01	21.84	19.70	17.61	15.57	13.60	11.69	9.86	8.10	6.43	4.86	3.40	2.08	0.93

* A speed-change cycle is reducing speed from and returning to an initial speed.

Tables of the Running Cost of Motor Vehicles

TABLE A-32-M
GASOLINE CONSUMPTION AT UNIFORM SPEED ON TANGENT MINUS GRADES

Vehicle: 12-kip single-unit truck
Unit: Gallons per 1,000 vehicle-miles
Roadway surface: High type pavement in good condition

Speed, mph	\-8	\-7	\-6	\-5	\-4	\-3	\-2	\-1	Level	Speed, mph
5	130.0	130.0	130.0	130.0	131.0	145.2	159.6	176.0	190.6	5
7½	86.7	86.7	86.7	86.7	90.1	103.6	117.5	132.4	148.1	7½
10	65.0	65.0	65.0	65.0	70.0	83.1	96.7	111.1	127.3	10
12½	52.0	52.0	52.0	52.0	59.0	71.1	84.5	98.8	115.3	12½
15	43.3	43.3	43.3	43.3	51.5	63.5	76.6	91.0	107.5	15
17½	37.1	37.1	37.1	37.7	47.1	58.3	71.2	85.8	102.3	17½
20	32.5	32.5	32.5	35.0	44.0	54.6	67.4	82.3	98.8	20
22½	28.9	28.9	28.9	33.4	42.0	52.1	64.7	79.9	96.4	22½
25	26.0	26.0	26.0	32.4	40.6	50.4	62.8	78.3	94.7	25
27½	23.6	23.6	25.5	31.9	39.9	49.4	61.6	77.3	93.7	27½
30	21.7	21.7	25.1	31.5	39.3	49.0	61.0	76.9	93.2	30
32½	20.0	20.0	25.0	31.2	38.9	49.0	60.9	76.9	93.2	32½
35	18.6	22.6	26.3	31.9	39.6	49.6	61.2	77.2	93.6	35
37½	17.3	25.0	28.5	33.8	41.1	50.7	62.0	77.8	94.3	37½
40	16.2	27.8	31.2	36.1	43.2	52.2	63.2	78.8	95.4	40
42½	15.3	30.9	34.5	38.9	45.6	54.2	65.0	80.2	96.9	42½
45		34.6	38.3	42.3	48.7	56.8	67.2	81.9	98.8	45
47½			42.4	46.4	52.3	59.9	69.9	83.9	101.2	47½
50				51.1	56.5	63.6	73.2	86.3	104.0	50
52½					61.4	68.3	77.2	89.1	107.4	52½
55						73.3	82.0	92.4	111.5	55
57½							87.6	96.2	116.2	57½
60							94.1	100.7	121.8	60
62½							101.6	106.0	128.3	62½
65								112.2	135.8	65

TABLE A-32-P

Gasoline Consumption at Uniform Speed on Tangent Plus Grades

Vehicle: 12-kip single-unit truck
Unit: Gallons per 1,000 vehicle-miles
Roadway surface: High type pavement in good condition

Speed, mph	Level	+1	+2	+3	+4	+5	+6	+7	+8	Speed, mph
5	190.6	205.2	220.4	237.0	255.4	275.8	298.8	324.8	354.4	5
7½	148.1	162.7	177.9	194.8	213.2	233.7	256.8	282.7	312.3	7½
10	127.3	141.8	157.1	174.3	192.8	213.4	236.5	262.3	291.9	10
12½	115.3	129.7	145.1	162.6	181.1	201.8	225.0	250.7	280.3	12½
15	107.5	122.0	137.5	155.2	173.9	194.7	217.9	243.6	273.1	15
17½	102.3	116.9	132.5	150.4	169.2	190.1	213.4	239.1	268.5	17½
20	98.8	113.4	129.2	147.2	166.2	187.1	210.6	236.2	265.6	20
22½	96.4	111.1	126.9	145.2	164.3	185.3	208.9	234.6	263.8	22½
25	94.7	109.5	125.4	144.5	163.2	184.3	208.1	233.8	262.9	25
27½	93.7	108.6	124.7	143.6	162.8	183.9	207.9	233.6	262.7	27½
30	93.2	108.3	124.4	143.6	162.9	184.1	208.3	234.0	263.1	30
32½	93.2	108.4	124.7	144.1	163.5	184.9	209.2	235.0	264.2	32½
35	93.6	108.9	125.4	145.0	164.5	186.0	210.5	236.4	265.7	35
37½	94.3	109.8	126.5	146.3	166.0	187.6	212.3	238.3	267.7	37½
40	95.4	111.2	128.0	148.0	167.8	189.6	214.5	240.7	270.3	40
42½	96.9	112.9	129.9	150.1	170.1	192.1	217.2	243.6	273.5	42½
45	98.8	115.0	132.3	152.7	172.9	195.1	220.4	247.2		45
47½	101.2	117.6	135.1	155.7	176.1	198.6	224.2			47½
50	104.0	120.7	138.4	159.3	179.9	202.6				50
52½	107.4	124.4	142.4	163.4	184.4					52½
55	111.5	128.7	147.0	168.2						55
57½	116.2	133.8	152.4							57½
60	121.8	139.8	158.8							60
62½	128.3	146.8								62½
65	135.8	155.2								65

Tables of the Running Cost of Motor Vehicles

TABLE A-33
GALLONS EXCESS GASOLINE CONSUMED AT UNIFORM SPEED ON HORIZONTAL CURVES—EXCESS ABOVE CONSUMPTION ON TANGENTS

Vehicle: 12-kip single-unit truck
Unit: Gallons per 1,000 vehicle-miles
Roadway surface: High type pavement in good condition

Speed, mph	\multicolumn{14}{c}{Degree of Horizontal Curve}
	1	2	3	4	5	6	8	10	12	14	16	18	20	25	30
5	0.090	0.320	0.640	1.000	1.372	1.720	2.301	2.651	2.741	2.831	2.951	3.071	3.191	3.572	4.081
10	0.075	0.288	0.570	0.889	1.206	1.499	1.960	2.203	2.220	2.189	2.184	2.179	2.184	2.200	2.199
15	0.069	0.241	0.470	0.775	0.980	1.194	1.489	1.532	1.480	1.369	1.266	1.164	1.065	0.859	0.702
20	0.052	0.194	0.367	0.550	0.719	0.853	0.997	0.958	0.956	0.728	0.643	0.618	0.636	1.198	1.941
25	0.043	0.150	0.270	0.392	0.509	0.566	0.622	0.605	0.603	0.871	1.134	1.561	2.176	6.106	9.380
30	0.044	0.132	0.220	0.306	0.386	0.444	0.568	0.785	1.276	2.074	3.234	4.801	6.844		
35	0.072	0.161	0.265	0.390	0.532	0.648	1.112	1.827	3.494	5.794	8.601	12.386	17.168		
40	0.127	0.282	0.461	0.684	0.988	1.371	2.602	4.699	8.084	12.701	18.695	26.299			
45	0.234	0.522	0.893	1.362	2.016	2.885	5.504	9.716	16.134	24.522					
50	0.409	0.934	1.636	2.573	3.830	5.488	10.505	17.884	28.947						
55	0.670	1.584	3.007	4.471	6.735	9.599	17.304	30.350							
60	1.050	2.534	4.583	7.338	10.992	15.662	28.891								
65	1.573	3.873	7.076	11.365	17.016	24.224									
70	2.288	5.705	10.492	16.886	25.262										

TABLE A-34

Excess Gallons of Gasoline Consumed During Speed-Change Cycles* —
Excess Consumption Above Continuing at Initial Speed

Vehicle: 12-kip single-unit truck
Unit: Gallons per 1,000 vehicle-cycles
Roadway surface: High type pavement in good condition

Initial Speed, mph	Speed Reduced to and Returned from, mph											
	Stop	5	10	15	20	25	30	35	40	45	50	55	60
5	—												
10	3.33	1.41											
15	7.56	4.46	2.06										
20	11.79	8.65	5.54	2.70									
25	16.02	12.85	9.72	6.58	3.33								
30	20.25	17.05	13.89	10.76	7.50	3.95							
35	24.48	21.25	18.05	14.91	11.70	8.27	4.47						
40	28.71	25.44	22.20	19.03	15.87	12.56	8.87	4.86					
45	32.94	29.62	26.35	23.12	19.89	16.55	13.00	9.25	5.08				
50	37.17	33.82	30.50	27.20	23.89	20.45	16.97	13.30	9.45	5.16			
55	41.40	38.03	34.65	31.26	27.85	24.39	20.86	17.19	13.44	9.50	5.11		
60	45.63	42.22	38.80	35.36	31.89	28.37	24.78	21.10	17.33	13.45	9.45	4.99	
65	49.86	46.42	42.95	39.44	35.89	32.30	28.66	24.96	21.19	17.32	13.33	9.17	4.78

* A speed-change cycle is reducing speed from and returning to an initial speed.

Tables of the Running Cost of Motor Vehicles

TABLE A-35-M
Gasoline Consumption at Uniform Speed on Tangent Minus Grades

Vehicle: 40-kip 2-S2, gasoline
Unit: Gallons per 1,000 vehicle-miles
Roadway surface: High type pavement in good condition

Speed, mph	Minus Grade, Percent									Speed, mph
	−8	−7	−6	−5	−4	−3	−2	−1	Level	
5	736.4	706.4	676.4	652.4	614.2	608.4	661.6	710.8	810.0	5
7½	449.9	435.9	411.2	386.9	352.1	345.7	396.9	446.4	543.7	7½
10	318.6	302.1	279.4	255.2	224.7	217.2	265.8	314.8	412.7	10
12½	226.0	224.0	201.3	177.8	151.3	142.7	188.5	237.4	334.8	12½
15	181.2	174.1	150.4	126.9	105.3	96.2	138.1	187.1	283.5	15
17½		140.7	115.3	92.9	75.0	71.7	103.5	152.6	247.5	17½
20			90.0	69.0	54.8	46.8	79.1	128.0	221.2	20
22½			71.5	52.0	41.9	35.9	61.7	110.2	201.6	22½
25			57.8	40.0	33.2	32.0	49.4	97.2	186.7	25
27½				31.7	29.1	29.1	41.0	87.7	175.6	27½
30				26.7	26.7	26.7	35.8	82.3	167.3	30
32½					24.6	24.6	33.1	79.4	161.5	32½
35						22.9	32.7	78.6	157.8	35
37½						21.3	34.2	79.4	156.0	37½
40						20.0	37.3	81.7	156.1	40
42½						19.1	41.9	85.1	158.0	42½
45						18.6	47.8	89.6	161.8	45
47½							54.8	95.1	167.7	47½
50							62.9	101.5	176.1	50
52½							71.9	108.7	187.8	52½
55							81.7	116.9	203.3	55
57½							92.2	125.6	223.0	57½
60								135.0	247.5	60
62½								144.9		62½

TABLE A-35-P

Gasoline Consumption at Uniform Speed on Tangent Plus Grades

Vehicle: 40-kip 2-S2, gasoline
Unit: Gallons per 1,000 vehicle-miles
Roadway surface: High type pavement in good condition

Speed, mph	\multicolumn{9}{c}{Plus Grade, Percent}	Speed, mph
	Level	+1	+2	+3	+4	+5	+6	+7	+8	
5	810.0	839.6	874.4	919.0	973.6	1040.8	1133.2	1268.6	1478.4	5
7½	543.7	577.3	616.1	662.4	718.1	788.5	879.2	1011.3	1225.9	7½
10	412.7	448.8	491.0	540.1	598.9	673.4	767.2	901.0	1125.4	10
12½	334.8	374.0	419.2	472.9	537.0	617.0	718.8	863.6	1096.7	12½
15	283.5	326.5	374.7	434.1	505.5	593.0	709.0	875.9		15
17½	247.5	294.7	346.2	412.2	493.3	592.5	729.4	931.9		17½
20	221.2	273.0	328.3	401.8	495.1	612.4	777.7			20
22½	201.6	258.3	318.1	399.9	508.4	654.7	853.5			22½
25	186.7	249.0	313.9	404.7	531.7	725.8				25
27½	175.6	243.9	314.8	415.3	563.9	821.5				27½
30	167.3	242.2	320.1	431.2	604.2					30
32½	161.5	243.6	329.8	451.7	655.5					32½
35	157.8	247.9	344.0	478.1						35
37½	156.0	255.1	363.1	507.2						37½
40	156.1	265.4	387.0	544.5						40
42½	158.0	279.5	415.7							42½
45	161.8	298.2	450.0							45
47½	167.7	324.6	490.4							47½
50	176.1	354.8	537.4							50
52½	187.8	389.0								52½
55	203.3	427.4								55
57½	223.0	470.2								57½
60	247.5									60

Tables of the Running Cost of Motor Vehicles

TABLE A-36
GALLONS EXCESS GASOLINE CONSUMED AT UNIFORM SPEED ON HORIZONTAL CURVES—EXCESS ABOVE CONSUMPTION ON TANGENTS

Vehicle: 40-kip 2-S2, gasoline
Roadway surface: High type pavement in good condition
Unit: Gallons per 1,000 vehicle-miles

Speed, mph	Degree of Horizontal Curve													
	1	2	3	4	5	6	8	10	12	14	16	18	20	25	30
5	1.02	3.61	7.16	11.21	15.30	19.00	25.25	28.97	29.72	30.47	31.36	32.37	33.45	36.74	41.14
10	0.71	1.74	5.38	8.35	11.50	13.98	18.20	20.47	21.73	20.36	20.29	20.29	20.30	20.33	20.50
15	0.58	1.55	3.75	6.11	7.66	9.39	11.80	12.72	12.03	11.26	10.54	9.92	9.28	7.99	7.16
20	0.37	1.23	2.40	3.67	4.84	5.78	6.89	6.98	6.92	5.78	5.50	5.58	5.98	10.30	15.88
25	0.26	0.86	1.30	2.39	3.12	3.53	4.09	4.31	4.69	5.77	7.59	10.32	14.01	31.65	54.93
30	0.26	0.76	1.32	1.86	2.41	2.89	3.91	5.51	8.47	12.96	19.17	27.38	37.39	76.69	143.45
35	0.44	0.99	1.63	2.47	3.32	4.14	6.93	11.68	19.60	30.59	45.39	64.17			
40	0.75	1.66	2.75	4.08	5.86	8.05	14.84	25.94	43.61	68.23	100.31				
45	1.30	2.92	5.03	7.69	11.32	16.03	30.52	53.81	89.93						
50	2.22	5.20	9.23	14.69	22.12	32.06	62.86	110.92							
55	3.77	9.17	17.56	27.18	41.85	61.13									
60	6.23	16.13	29.69												

TABLE A-37
Excess Gallons of Gasoline Consumed During Speed-Change Cycles*
Excess Consumption Above Continuing at Initial Speed

Vehicle: 40-kip 2-S2 gasoline
Unit: Gallons per 1,000 vehicle-cycles
Roadway surface: High type pavement in good condition

Initial Speed, mph	Speed Reduced to and Returned from, mph										
	Stop	5	10	15	20	25	30	35	40	45	50	55
5	4.3											
10	11.0	8.3										
15	21.2	18.5	12.1									
20	34.8	31.0	25.6	15.7								
25	51.4	46.8	41.0	32.8	19.0							
30	70.8	65.8	58.9	51.0	38.2	21.8						
35	92.5	87.5	80.7	71.7	59.2	43.5	24.1					
40	116.0	110.9	103.2	94.1	81.7	66.3	47.6	25.6				
45	140.9	135.3	127.8	118.0	105.6	90.4	72.3	51.2	26.5			
50	166.7	161.2	153.5	143.4	130.7	115.6	97.8	77.2	53.7	27.2		
55	193.1	187.7	180.0	169.9	157.3	142.1	124.2	103.5	80.2	54.7	27.7	
60	219.8	214.8	207.4	197.8	185.0	169.4	150.6	129.0	105.3	80.4	54.1	28.0

* A speed-change cycle is reducing speed from and returning to an initial speed.

TABLE A-38-M
Diesel Fuel Consumption at Uniform Speed on Tangent Minus Grades

Vehicle: 50-kip 3-S2, diesel
Unit: Gallons per 1,000 vehicle-miles
Roadway surface: High type pavement in good condition

Speed, mph	Minus Grade, Percent									Speed, mph
	−8	−7	−6	−5	−4	−3	−2	−1	Level	
5	367.6	352.6	337.6	325.8	306.6	303.8	330.4	354.8	404.4	5
7½	228.4	221.2	208.7	196.4	178.8	175.5	201.5	226.5	276.0	7½
10	164.3	155.8	144.1	131.6	115.9	112.0	137.1	162.3	212.8	10
12½	118.6	117.5	105.6	93.3	79.4	75.0	98.9	124.6	175.7	12½
15	97.1	93.3	80.5	68.0	56.4	51.5	74.0	100.2	151.9	15
17½		77.0	63.1	50.6	41.1	39.1	56.7	83.5	135.5	17½
20		65.4	50.4	38.6	30.8	26.3	44.4	71.8	124.1	20
22½			41.1	29.9	24.1	20.6	35.5	63.4	115.9	22½
25			34.1	23.6	19.6	16.9	29.1	57.3	110.1	25
27½				19.2	17.6	14.5	24.8	53.2	106.4	27½
30				15.9	13.3	13.3	21.7	51.1	104.0	30
32½					12.3	12.3	21.1	50.6	102.8	32½
35					11.4	11.4	21.3	51.3	102.9	35
37½					11.4	10.7	22.9	53.0	104.2	37½
40						10.5	25.5	55.8	106.6	40
42½						12.1	29.2	59.4	110.2	42½
45						14.5	33.9	63.6	114.9	45
47½							39.5	68.5	120.9	47½
50							45.8	73.9	128.3	50
52½							52.6	79.5	137.2	52½
55							59.9	84.9	147.9	55
57½								91.0	160.5	57½
60									165.2	60

Tables of the Running Cost of Motor Vehicles

TABLE A-38-P
Diesel Fuel Consumption at Uniform Speed on Tangent Plus Grades

Vehicle: 50-kip 3-S2, diesel
Unit: Gallons per 1,000 vehicle-miles
Roadway surface: High type pavement in good condition

Speed, mph	Level	+1	+2	+3	+4	+5	+6	+7	+8	Speed, mph
5	404.4	431.6	460.0	487.0	513.4	538.2	561.5	582.8	606.0	5
7½	276.0	325.2	373.6	420.4	465.3	508.3	549.5	586.8	624.0	7½
10	212.8	272.7	331.3	388.4	443.4	496.3	546.5	594.6	640.3	10
12½	175.7	241.8	307.0	370.6	432.3	492.1	547.0	604.9	655.8	12½
15	151.9	222.0	291.5	360.1	427.1	492.6	553.5	617.4	677.2	15
17½	135.5	208.6	281.4	353.9	425.7	496.6	563.8	632.9	699.3	17½
20	124.1	199.3	274.8	350.8	427.0	503.5	577.7	651.4	727.0	20
22½	115.9	192.9	270.7	350.0	430.7	513.2	595.6	673.3		22½
25	110.1	188.6	268.5	350.9	436.7	525.9	617.2			25
27½	106.4	185.9	268.0	353.6	445.4	541.8				27½
30	104.0	184.7	268.9	358.2	457.1	561.5				30
32½	102.8	184.6	271.1	365.1	472.3					32½
35	102.9	185.7	274.7	374.2	492.0					35
37½	104.2	187.8	279.6	386.3						37½
40	106.6	191.0	286.4	402.2						40
42½	110.2	195.3	294.6							42½
45	114.9	201.1	305.3							45
47½	120.9	208.9	319.5							47½
50	128.3	219.2								50
52½	137.2	232.5								52½
55	147.9	249.5								55
57½	160.5	271.0								57½
60	165.2									60

TABLE A-39

Gallons Excess Diesel Fuel Consumed at Uniform Speed on Horizontal Curves—Excess Above Consumption on Tangents

Vehicle: 50-kip 3-S2, diesel
Unit: Gallons per 1,000 vehicle-miles
Roadway surface: High type pavement in good condition

Speed, mph	Degree of Horizontal Curve													
	1	2	3	4	5	6	8	10	12	14	16	18	20	25	30
5	0.63	2.19	4.39	6.90	9.45	11.79	15.78	18.24	18.87	19.49	20.20	21.02	21.89	24.53	28.05
10	0.43	1.63	3.21	5.00	6.77	8.40	10.95	12.34	12.74	12.40	12.30	12.32	12.34	12.47	12.70
15	0.35	1.17	2.25	3.66	4.60	5.64	7.08	7.62	7.19	6.72	6.28	5.89	5.50	4.68	4.15
20	0.22	0.76	1.48	2.25	2.97	3.55	4.22	4.26	4.19	3.48	3.28	3.30	3.53	6.17	9.62
25	0.16	0.54	1.01	1.50	1.97	2.22	2.57	2.69	2.90	3.59	4.77	6.57	9.06	18.48	37.51
30	0.17	0.49	0.85	1.21	1.55	1.86	2.52	3.56	5.57	8.66	15.99	18.81	26.36	54.90	103.85
35	0.28	0.65	1.06	1.60	2.06	2.68	4.52	7.67	13.02	20.52	30.52	43.45			
40	0.50	1.11	1.81	2.68	3.83	5.26	9.62	16.78	28.17	43.45	63.86				
45	0.89	1.98	3.36	5.09	7.40	10.43	19.40	33.61	55.27	83.86					
50	1.54	3.51	6.07	9.43	13.90	19.71	37.11	63.02							
55	2.59	6.07	11.24	16.83	25.17	35.73									
60	4.26	10.11	18.04	28.65											

Tables of the Running Cost of Motor Vehicles

TABLE A-40

EXCESS GALLONS OF DIESEL FUEL CONSUMED DURING SPEED-CHANGE CYCLES*–
EXCESS CONSUMPTION ABOVE CONTINUING AT INITIAL SPEED

Vehicle: 50-kip 3-S2, diesel
Unit: Gallons per 1,000 vehicle-cycles
Roadway surface: High type pavement in good condition

Initial speed, mph	Speed Reduced to and Returned from, mph										
	Stop	5	10	15	20	25	30	35	40	45	50	55
5	–											
10	5.7	2.4										
15	16.0	10.7	5.5									
20	26.5	21.5	15.6	8.6								
25	37.3	32.9	27.4	20.4	11.7							
30	48.5	44.4	39.3	32.9	25.0	14.7						
35	60.2	55.8	50.9	45.1	38.1	29.3	17.5					
40	72.5	67.6	62.4	56.7	50.2	42.6	33.2	20.3				
45	85.5	80.0	74.3	68.4	62.1	55.0	46.7	36.6	22.8			
50	99.3	93.2	87.0	80.6	73.9	66.8	59.1	50.3	39.4	25.1		
55	113.9	107.1	100.2	93.1	85.8	78.2	70.3	61.8	52.4	41.5	27.1	
60	129.4	121.8	114.1	106.0	97.7	89.7	81.6	73.3	64.3	54.4	42.9	27.5

* A speed-change cycle is reducing speed from and returning to an initial speed.

TABLE A-41

COST OF IDLING ENGINE WITH STATIONARY VEHICLE
Dollars cost per 1,000 vehicle-hours

Cost Item	4-kip Passenger Car	5-kip Commercial Delivery	12-kip Single-Unit Truck	40-kip 2-S2	50-kip 3-S2 Diesel
Fuel	84.64	92.40	130.00	144.00	64.00
Engine oil	3.48	1.92	3.20	2.60	3.46
Engine maintenance	7.54	9.42	11.03	17.65	16.02
Depreciation	19.20	28.80	55.80	85.20	112.80
Total	114.86	132.54	200.03	249.45	196.28
Performance Data					
Fuel consumption, gallons per 1,000 hr	368	420	650	800	400
Engine oil consumption, quarts per 1,000 hr	5.80	3.50	8.00	13.00	17.30
Engine rpm	450	500	600	650	600

TABLE A-42

VEHICLE RUNNING COST* AND FUEL CONSUMPTION RESULTING FROM TURNING 90-DEGREE CORNERS

Unit: Running cost in dollars and fuel consumption in gallons for 1,000 vehicle-turns.

Roadway surface: High type pavement in good condition.

Speed, mph	4-kip Passenger Car		5-kip Commercial Delivery		12-kip Single Unit Truck		40-kip 2-S2 Gasoline		50-kip 3-S2 Diesel	
	Dollars	Gallons	Dollars	Gallons	Dollars	Gallons	Dollars	Gallons	Dollars	Gallons
25-ft radius										
5	0.267	0.081	0.306	0.069	0.378	0.116	0.677	0.825	0.878	0.562
10	.718	.061	.937	.039	1.800	.068	2.179	.384	3.481	.238
15	1.435	.056	8.75	.052	9.64	.78	8.628	.205	15.62	.358
20	7.177	.142	–	.133	–	.218	–	1.145	–	.694
25	–	.608	–	.594	–	1.001	–	3.423	–	2.338
50-ft radius										
5	0.344	0.126	0.400	0.104	0.553	0.181	1.158	1.331	1.369	0.907
10	.763	.068	.906	.060	1.866	.106	2.916	.621	4.462	.385
15	1.577	.062	4.98	.057	7.08	.095	8.152	.299	14.28	.506
20	4.195	.154	36.40	.144	19.88	.238	43.44	1.618	75.27	.980
25	23.60	.653	–	638	–	1.074	–	4.845	–	3.309
75-ft radius										
5	0.366	0.150	0.429	0.129	0.652	0.216	1.401	1.662	1.691	1.133
10	.754	.079	.881	.070	1.832	.122	3.168	.772	4.657	.479
15	1.468	.058	3.213	.053	5.556	.094	7.296	.330	12.14	546
20	3.191	.138	11.60	.129	13.02	.212	25.54	1.739	41.95	1.054
25	12.40	.577	33.80	.564	75.20	.950	185.19	5.184	–	3.540
100-ft radius										
5	0.372	0.162	0.446	0.139	0.729	0.233	1.606	1.887	1.932	1.287
10	.735	.083	.861	.073	1.788	.129	3.332	.874	4.730	.541
15	1.392	.051	2.329	.047	4.525	.080	6.604	.338	10.50	.528
20	2.621	.121	5.682	.113	9.16	.187	17.285	1.691	27.96	1.025
25	8.181	.502	4.458	4.91	26.77	.826	61.58	5.823	81.51	3.549
125-ft radius										
5	0.376	0.167	0.454	0.144	0.777	0.241	1.778	2.045	2.115	1.395
10	.716	.082	.848	.073	1.760	.128	3.452	.955	4.773	.591
15	1.316	.045	1.897	.041	3.893	.060	6.099	.335	9.260	.477
20	2.298	.104	3.756	.097	7.18	.161	13.57	1.504	20.94	.959
25	5.355	.428	7.363	.418	17.24	704	33.84	4.922	45.74	3.362
150-ft radius										
5	0.379	0.166	0.457	0.143	0.820	0.240	1.897	2.167	2.254	1.477
10	.708	.081	.839	.072	1.755	.126	3.548	1.041	4.803	.645
15	1.263	.037	1.642	.033	3.423	.044	5.801	.341	8.434	.402
20	2.053	.090	2.896	.084	5.89	.139	11.34	1.245	16.82	.816
25	4.097	.379	4.937	.370	12.21	.623	23.29	4.314	31.55	2.946
175-ft radius										
5	0.381	0.166	0.456	0.142	0.855	0.238	2.004	2.261	2.370	1.541
10	.702	.079	.835	.075	1.760	.122	3.624	1.126	4.829	.698
15	1.218	.032	1.473	.030	3.191	.037	5.675	.370	7.966	.298
20	1.890	.081	2.428	.075	5.08	.119	9.94	1.025	4.16	.621
25	3.394	.327	3.801	.336	9.47	.566	18.12	3.560	24.83	2.487

Tables of the Running Cost of Motor Vehicles

TABLE A-42 (*continued*)

200-ft radius

5	0.382	0.161	0.452	0.138	0.881	0.234	2.088	2.380	2.467	1.606
10	.698	.084	.836	.080	1.777	.131	3.683	1.219	4.849	.753
15	1.183	.030	1.368	.027	3.046	.046	5.593	.437	7.675	.264
20	1.771	.067	2.118	.060	4.51	.101	9.04	.851	12.49	.508
25	2.939	.278	3.165	.283	7.91	.508	15.05	2.841	20.77	1.915

225-ft radius

5	0.383	0.162	0.444	0.141	0.902	0.241	2.162	2.507	2.540	1.667
10	.696	.095	.837	.086	1.794	.147	3.734	1.371	4.863	.836
15	1.158	.037	1.305	.033	2.932	.059	5.542	.525	7.430	.315
20	1.680	.043	1.928	.042	4.10	.086	8.50	.716	11.38	.435
25	2.637	.232	2.791	.222	6.96	.437	13.32	2.252	18.34	1.320

250-ft radius

5	0.383	0.165	0.435	0.144	0.928	0.251	2.231	2.644	2.593	1.740
10	.695	.106	.839	.093	1.800	.163	3.778	1.521	4.878	.925
15	1.139	.047	1.264	.042	2.826	.072	5.511	.625	7.300	.372
20	1.606	.037	1.800	.040	3.87	.072	7.96	.625	10.49	.387
25	2.439	.192	2.596	.179	6.33	.348	12.35	1.752	16.96	1.035

*Running cost includes fuel, tires, engine oil, and vehicle maintenance. Running cost and fuel consumption are for travel of one-fourth of a full circle and are exclusive of any speed changes required in making the 90-deg. turn.

TABLE A-43

GALLONS OF FUEL CONSUMED BY VEHICLES
ON LEVEL TANGENT ROADWAY SURFACES OF GRAVEL OR STONE
Unit: Gallons per 1,000 vehicle-miles

Speed, mph	4-kip Passenger Car	5-kip Commercial Delivery	12-kip Single-Unit Truck	40-kip 2-S2 Gasoline	50-kip 3-S2 Diesel
5	111.4	117.1	210.2	910.4	471.1
7½	84.9	88.7	166.8	626.9	332.0
10	71.8	74.7	145.9	486.2	262.2
12½	64.3	66.7	134.2	402.1	220.9
15	59.6	61.7	126.7	346.4	194.4
17½	56.5	58.4	121.9	306.9	176.4
20	54.4	56.2	118.9	278.0	164.3
22½	53.0	54.9	117.0	256.6	155.9
25	52.2	54.2	115.8	240.3	150.3
27½	51.8	54.0	115.3	228.6	147.4
30	51.7	54.2	115.5	220.0	146.3
32½	52.0	54.8	116.1	214.3	146.5
35	52.5	55.8	117.3	211.5	148.6
37½	53.3	57.2	118.8	210.9	152.4
40	54.0	59.0	120.8	212.9	158.0
42½	55.5	61.2	123.3	217.2	165.3
45	57.0	63.8	126.2	224.2	174.7
47½	58.6	66.9	129.7	234.4	186.1
50	60.4	70.7	133.9	248.1	199.9
52½	62.5	75.2	138.9	266.7	216.4
55	64.9	80.5	144.9	290.9	236.0
57½	67.6	86.8			
60	70.7	94.4			

Tables of the Running Cost of Motor Vehicles

TABLE A-44

FACTORS*TO CONVERT MOTOR VEHICLE RUNNING COST
ON HIGH TYPE PAVEMENTS TO COST ON GRAVEL AND STONE ROADWAY SURFACES

Speed, mph	4-kip Passenger Car	5-kip Commercial Delivery	12-kip Single-Unit Truck	40-kip 2-S2 Gasoline	50-kip 3-S2 Diesel
5	1.079	1.074	1.090	1.114	1.129
7½	1.106	1.100	1.122	1.148	1.172
10	1.132	1.125	1.152	1.181	1.210
12½	1.157	1.149	1.180	1.212	1.245
15	1.181	1.172	1.207	1.241	1.278
17½	1.205	1.193	1.232	1.267	1.307
20	1.228	1.215	1.256	1.291	1.334
22½	1.250	1.237	1.279	1.314	1.359
25	1.272	1.259	1.300	1.336	1.382
27½	1.294	1.281	1.321	1.357	1.404
30	1.315	1.302	1.341	1.377	1.424
32½	1.337	1.323	1.361	1.397	1.444
35	1.358	1.344	1.381	1.417	1.464
37½	1.380	1.365	1.401	1.437	1.483
40	1.402	1.387	1.421	1.457	1.501
42½	1.424	1.409	1.441	1.476	1.518
45	1.447	1.431	1.462	1.495	1.535
47½	1.471	1.455	1.484	1.515	1.553
50	1.498	1.479	1.507	1.536	1.572
52½	1.526	1.504	1.531	1.557	1.592
55	1.557	1.531	1.557	1.579	1.614
57½	1.592	1.561			
60	1.631	1.592			

*Factors include fuel, tires, engine oil, maintenance, and depreciation and may be used for level tangents, grades, horizontal curves, and speed changes. These factors may be used for bituminous surfaces of less than high type in good condition by an appropriate reduction to somewhere less than given, but more than 1.000. Also, they may be used for earth surfaces by adjustment, say $E = 2G - 1$.

Appendix **B**

Standard Compound Interest Factors

Tables B-1 through B-45 give the six compound interest factors for rates of interest from 1/4 percent to 100 percent. The factors are for the end-of-period step convention and correspond to Eqs. 6-1 to 6-6.

Standard Compound Interest Factors

TABLE B-1
1/4 Percent Compound Interest Factors for One Dollar

n	Single Amount		Uniform Series (Periodic Equal Amounts)				n
	Compound Amount	Present Worth	Compound Amount	Sinking Fund	Present Worth	Capital Recovery	
	CA	PW	SCA	SF	SPW	CR	
1	1.0025 0000	0.997506	1.000000	1.000000	0.997506	1.002500	1
2	1.0050 0625	0.995019	2.002500	0.499376	1.992525	0.501876	2
3	1.0075 1877	0.992537	3.007506	0.332501	2.985062	0.335001	3
4	1.0100 3756	0.990062	4.015025	0.249064	3.975124	0.251564	4
5	1.0125 6266	0.987593	5.025063	0.199002	4.962718	0.201502	5
6	1.0150 9406	0.985130	6.037625	0.165628	5.947848	0.168128	6
7	1.0176 3180	0.982674	7.052719	0.141789	6.930522	0.144289	7
8	1.0201 7588	0.980223	8.070351	0.123910	7.910745	0.126410	8
9	1.0227 2632	0.977779	9.090527	0.110005	8.888524	0.112505	9
10	1.0252 8313	0.975340	10.113253	0.098880	9.863864	0.101380	10
11	1.0278 4634	0.972908	11.138536	0.089778	10.836772	0.092278	11
12	1.0304 1596	0.970482	12.166383	0.082194	11.807254	0.084694	12
13	1.0329 9200	0.968062	13.196799	0.075776	12.775316	0.078276	13
14	1.0355 7448	0.965648	14.229791	0.070275	13.740963	0.072775	14
15	1.0381 6341	0.963239	15.265365	0.065508	14.704203	0.068008	15
16	1.0407 5882	0.960837	16.303529	0.061336	15.665040	0.063836	16
17	1.0433 6072	0.958441	17.344287	0.057656	16.623481	0.060156	17
18	1.0459 6912	0.956051	18.387648	0.054384	17.579533	0.056884	18
19	1.0485 8404	0.953667	19.433617	0.051457	18.533200	0.053957	19
20	1.0512 0550	0.951289	20.482201	0.048823	19.484488	0.051323	20
21	1.0538 3352	0.948916	21.533407	0.046439	20.433405	0.048939	21
22	1.0564 6810	0.946550	22.587240	0.044273	21.379955	0.046773	22
23	1.0591 0927	0.944190	23.643708	0.042295	22.324145	0.044795	23
24	1.0617 5704	0.941835	24.702818	0.040481	23.265980	0.042981	24
25	1.0644 1144	0.939486	25.764575	0.038813	24.205466	0.041313	25
26	1.0670 7247	0.937143	26.828986	0.037273	25.142609	0.039773	26
27	1.0697 4015	0.934806	27.896059	0.035847	26.077416	0.038347	27
28	1.0724 1450	0.932475	28.965799	0.034523	27.009891	0.037023	28
29	1.0750 9553	0.930150	30.038213	0.033291	27.940041	0.035791	29
30	1.0777 8327	0.927830	31.113309	0.032141	28.867871	0.034641	30
31	1.0804 7773	0.925517	32.191092	0.031064	29.793388	0.033564	31
32	1.0831 7892	0.923209	33.271570	0.030056	30.716596	0.032556	32
33	1.0858 8687	0.920906	34.354749	0.029108	31.637503	0.031608	33
34	1.0886 0159	0.918610	35.440636	0.028216	32.556112	0.030716	34
35	1.0913 2309	0.916319	36.529237	0.027375	33.472431	0.029875	35
36	1.0940 5140	0.914034	37.620560	0.026581	34.386465	0.029081	36
37	1.0967 8653	0.911754	38.714612	0.025830	35.298220	0.028330	37
38	1.0995 2850	0.909481	39.811398	0.025118	36.207700	0.027618	38
39	1.1022 7732	0.907213	40.910927	0.024443	37.114913	0.026943	39
40	1.1050 3301	0.904950	42.013204	0.023802	38.019863	0.026302	40
45	1.1189 1516	0.893723	47.566064	0.021023	42.510876	0.023523	45
50	1.1329 7171	0.882635	53.188683	0.018801	46.946170	0.021301	50
55	1.1472 0484	0.871684	58.881937	0.016983	51.326437	0.019483	55
60	1.1616 1678	0.860869	64.646713	0.015469	55.652358	0.017969	60
65	1.1762 0977	0.850188	70.483910	0.014188	59.924608	0.016688	65
70	1.1909 8609	0.839640	76.394437	0.013090	64.143853	0.015590	70
75	1.2059 4804	0.829223	82.379217	0.012139	68.310751	0.014639	75
80	1.2210 9795	0.818935	88.439181	0.011307	72.425952	0.013807	80
90	1.2519 7114	0.798740	100.788454	0.009922	80.503816	0.012422	90
100	1.2836 2489	0.779044	113.449955	0.008814	88.382483	0.011314	100
48	1.1273 2802	0.887053	50.931208	0.019634	45.178695	0.022134	48
72	1.1969 4847	0.835458	78.779387	0.012694	65.816858	0.015194	72
84	1.2333 5480	0.810797	93.341920	0.010713	75.681321	0.013213	84
96	1.2708 6847	0.786863	108.347387	0.009230	85.254603	0.011730	96
108	1.3095 2315	0.763637	123.809259	0.008077	94.545300	0.010577	108
120	1.3493 5355	0.741096	139.741419	0.007156	103.561753	0.009656	120
144	1.4326 8563	0.697990	173.074254	0.005778	120.804069	0.008278	144
180	1.5674 3172	0.637986	226.972690	0.004406	144.805471	0.006906	180
240	1.8207 5500	0.549223	328.301998	0.003046	180.310914	0.005546	240
300	2.1150 1956	0.472809	446.007823	0.002242	210.876453	0.004742	300

Standard Compound Interest Factors

TABLE B-2
1/3 Percent Compound Interest Factors for One Dollar

n	Single Amount		Uniform Series (Periodic Equal Amounts)				n
	Compound Amount	Present Worth	Compound Amount	Sinking Fund	Present Worth	Capital Recovery	
	CA	PW	SCA	SF	SPW	CR	
1	1.0033 3333	0.996678	1.000000	1.000000	0.996678	1.003333	1
2	1.0066 7778	0.993367	2.003333	0.499168	1.990044	0.502501	2
3	1.0100 3337	0.990066	3.010011	0.332225	2.980111	0.335558	3
4	1.0134 0015	0.986777	4.020044	0.248753	3.966888	0.252087	4
5	1.0167 7815	0.983499	5.033445	0.198671	4.950386	0.202004	5
6	1.0201 6741	0.980231	6.050223	0.165283	5.930618	0.168617	6
7	1.0235 6797	0.976975	7.070390	0.141435	6.907592	0.144768	7
8	1.0269 7986	0.973729	8.093958	0.123549	7.881321	0.126882	8
9	1.0304 0313	0.970494	9.120938	0.109638	8.851815	0.112971	9
10	1.0338 3780	0.967270	10.151341	0.098509	9.819085	0.101842	10
11	1.0372 8393	0.964056	11.185179	0.089404	10.783141	0.092737	11
12	1.0407 4154	0.960853	12.222463	0.081817	11.743994	0.085150	12
13	1.0442 1068	0.957661	13.263204	0.075397	12.701656	0.078730	13
14	1.0476 9138	0.954480	14.307415	0.069894	13.656135	0.073227	14
15	1.0511 8369	0.951309	15.355106	0.065125	14.607444	0.068458	15
16	1.0546 8763	0.948148	16.406290	0.060952	15.555592	0.064286	16
17	1.0582 0326	0.944998	17.460978	0.057271	16.500590	0.060604	17
18	1.0617 3060	0.941859	18.519181	0.053998	17.442448	0.057331	18
19	1.0652 6971	0.938729	19.580912	0.051070	18.381178	0.054403	19
20	1.0688 2060	0.935611	20.646181	0.048435	19.316788	0.051768	20
21	1.0723 8334	0.932502	21.715002	0.046051	20.249291	0.049384	21
22	1.0759 5795	0.929404	22.787385	0.043884	21.178695	0.047217	22
23	1.0795 4448	0.926317	23.863343	0.041905	22.105012	0.045239	23
24	1.0831 4296	0.923239	24.942888	0.040092	23.028251	0.043425	24
25	1.0867 5344	0.920172	26.026031	0.038423	23.948423	0.041756	25
26	1.0903 7595	0.917115	27.112784	0.036883	24.865538	0.040216	26
27	1.0940 1053	0.914068	28.203160	0.035457	25.779606	0.038790	27
28	1.0976 5724	0.911031	29.297171	0.034133	26.690637	0.037466	28
29	1.1013 1609	0.908005	30.394828	0.032900	27.598641	0.036234	29
30	1.1049 8715	0.904988	31.496144	0.031750	28.503629	0.035083	30
31	1.1086 7044	0.901981	32.601131	0.030674	29.405611	0.034007	31
32	1.1123 6601	0.898985	33.709802	0.029665	30.304595	0.032998	32
33	1.1160 7389	0.895998	34.822168	0.028717	31.200593	0.032051	33
34	1.1197 9414	0.893021	35.938241	0.027826	32.093615	0.031159	34
35	1.1235 2679	0.890054	37.058036	0.026985	32.983669	0.030318	35
36	1.1272 7187	0.887097	38.181562	0.026191	33.870766	0.029524	36
37	1.1310 2945	0.884150	39.308834	0.025440	34.754917	0.028773	37
38	1.1347 9955	0.881213	40.439864	0.024728	35.636130	0.028061	38
39	1.1385 8221	0.878285	41.574663	0.024053	36.514415	0.027386	39
40	1.1423 7748	0.875367	42.713245	0.023412	37.389782	0.026745	40
45	1.1615 4446	0.860923	48.463339	0.020634	41.723189	0.023967	45
50	1.1810 3303	0.846716	54.309909	0.018413	45.985089	0.021746	50
55	1.2008 4858	0.832744	60.254573	0.016596	50.176662	0.019930	55
60	1.2209 9659	0.819003	66.298978	0.015083	54.299069	0.018417	60
65	1.2414 8266	0.805488	72.444797	0.013804	58.353451	0.017137	65
70	1.2623 1244	0.792197	78.693731	0.012707	62.340930	0.016041	70
75	1.2834 9170	0.779125	85.047511	0.011758	66.262611	0.015091	75
80	1.3050 2632	0.766268	91.507895	0.010928	70.119578	0.014261	80
90	1.3491 8554	0.741188	104.755663	0.009546	77.643630	0.012879	90
100	1.3948 3902	0.716929	118.451705	0.008442	84.921417	0.011776	100
48	1.1731 9867	0.852371	51.959601	0.019246	44.288834	0.022579	48
72	1.2707 4188	0.786942	81.222564	0.012312	63.917437	0.015645	72
84	1.3225 1386	0.756136	96.754159	0.010335	73.159278	0.013669	84
96	1.3763 9512	0.726536	112.918536	0.008856	82.039332	0.012189	96
108	1.4324 7158	0.698094	129.741474	0.007708	90.571761	0.011041	108
120	1.4908 3268	0.670766	147.249805	0.006791	98.770175	0.010125	120
144	1.6147 8492	0.619278	184.435477	0.005422	114.216744	0.008755	144
180	1.8203 0163	0.549506	246.090488	0.004064	135.192149	0.007397	180
240	2.2225 8209	0.449927	366.774626	0.002726	165.021858	0.006060	240
300	2.7137 6516	0.368492	514.129547	0.001945	189.452483	0.005278	300

Standard Compound Interest Factors

TABLE B-3
5/12 Percent Compound Interest Factors for One Dollar

n	Single Amount		Uniform Series (Periodic Equal Amounts)				n
	Compound Amount	Present Worth	Compound Amount	Sinking Fund	Present Worth	Capital Recovery	
	CA	PW	SCA	SF	SPW	CR	
1	1.0041 6667	0.995851	1.000000	1.000000	0.995851	1.004167	1
2	1.0083 5069	0.991718	2.004167	0.498960	1.987569	0.503127	2
3	1.0125 5216	0.987603	3.012517	0.331948	2.975173	0.336115	3
4	1.0167 7112	0.983506	4.025070	0.248443	3.958678	0.252610	4
5	1.0210 0767	0.979425	5.041841	0.198340	4.938103	0.202507	5
6	1.0252 6187	0.975361	6.062848	0.164939	5.913463	0.169106	6
7	1.0295 3379	0.971313	7.088110	0.141081	6.884777	0.145248	7
8	1.0338 2352	0.967283	8.117644	0.123188	7.852060	0.127355	8
9	1.0381 3111	0.963269	9.151467	0.109272	8.815329	0.113439	9
10	1.0424 5666	0.959272	10.189599	0.098139	9.774602	0.102306	10
11	1.0468 0023	0.955292	11.232055	0.089031	10.729894	0.093198	11
12	1.0511 6190	0.951328	12.278855	0.081441	11.681222	0.085607	12
13	1.0555 4174	0.947381	13.330017	0.075019	12.628603	0.079185	13
14	1.0599 3983	0.943450	14.385559	0.069514	13.572053	0.073681	14
15	1.0643 5625	0.939535	15.445499	0.064744	14.511588	0.068910	15
16	1.0687 9106	0.935637	16.509855	0.060570	15.447224	0.064737	16
17	1.0732 4436	0.931754	17.578646	0.056887	16.378978	0.061054	17
18	1.0777 1621	0.927888	18.651891	0.053614	17.306867	0.057781	18
19	1.0822 0670	0.924038	19.729607	0.050685	18.230904	0.054852	19
20	1.0867 1589	0.920204	20.811814	0.048050	19.151108	0.052216	20
21	1.0912 4387	0.916385	21.898529	0.045665	20.067494	0.049832	21
22	1.0957 9072	0.912583	22.989773	0.043498	20.980077	0.047664	22
23	1.1003 5652	0.908796	24.085564	0.041519	21.888873	0.045685	23
24	1.1049 4134	0.905025	25.185921	0.039705	22.793898	0.043871	24
25	1.1095 4526	0.901270	26.290862	0.038036	23.695169	0.042203	25
26	1.1141 6836	0.897530	27.400407	0.036496	24.592699	0.040662	26
27	1.1188 1073	0.893806	28.514575	0.035070	25.486505	0.039236	27
28	1.1234 7244	0.890097	29.633386	0.033746	26.376603	0.037912	28
29	1.1281 5358	0.886404	30.756859	0.032513	27.263007	0.036680	29
30	1.1328 5422	0.882726	31.885012	0.031363	28.145733	0.035529	30
31	1.1375 7444	0.879063	33.017866	0.030287	29.024796	0.034453	31
32	1.1423 1434	0.875416	34.155441	0.029278	29.900212	0.033445	32
33	1.1470 7398	0.871783	35.297755	0.028330	30.771995	0.032497	33
34	1.1518 5346	0.868166	36.444829	0.027439	31.640161	0.031605	34
35	1.1566 5284	0.864564	37.596683	0.026598	32.504725	0.030765	35
36	1.1614 7223	0.860976	38.753336	0.025804	33.365701	0.029971	36
37	1.1663 1170	0.857404	39.914808	0.025053	34.223105	0.029220	37
38	1.1711 7133	0.853846	41.081119	0.024342	35.076951	0.028509	38
39	1.1760 5121	0.850303	42.252291	0.023667	35.927254	0.027834	39
40	1.1809 5142	0.846775	43.428342	0.023026	36.774029	0.027193	40
45	1.2057 6046	0.829352	49.382511	0.020250	40.955490	0.024417	45
50	1.2310 9068	0.812288	55.461763	0.018030	45.050916	0.022197	50
55	1.2569 5302	0.795575	61.668726	0.016216	49.062077	0.020382	55
60	1.2833 5868	0.779205	68.006083	0.014705	52.990706	0.018871	60
65	1.3103 1905	0.763173	74.476573	0.013427	56.838502	0.017594	65
70	1.3378 4580	0.747470	81.082993	0.012333	60.607129	0.016500	70
75	1.3659 5082	0.732091	87.828198	0.011386	64.298214	0.015553	75
80	1.3946 4627	0.717028	94.715104	0.010558	67.913353	0.014725	80
90	1.4538 5829	0.687825	108.925990	0.009181	74.922013	0.013347	90
100	1.5155 8426	0.659812	123.740222	0.008081	81.645228	0.012248	100
48	1.2208 9536	0.819071	53.014885	0.018863	43.422956	0.023029	48
72	1.3490 1774	0.741280	83.764259	0.011938	62.092777	0.016105	72
84	1.4180 3605	0.705201	100.328663	0.009967	70.751835	0.014134	84
96	1.4905 8547	0.670877	117.740512	0.008493	78.989441	0.012660	96
108	1.5668 4665	0.638225	136.043196	0.007351	86.826108	0.011517	108
120	1.6470 0950	0.607161	155.282279	0.006440	94.281350	0.010607	120
144	1.8198 4887	0.549496	196.763730	0.005082	108.120917	0.009249	144
180	2.1137 0393	0.473103	267.288944	0.003741	126.455243	0.007908	180
240	2.7126 4029	0.368645	411.033669	0.002433	151.525313	0.006600	240
300	3.4812 9045	0.287250	595.509708	0.001679	171.060047	0.005846	300

TABLE B-4
1/2 Percent Compound Interest Factors for One Dollar

n	Single Amount		Uniform Series (Periodic Equal Amounts)				n
	Compound Amount	Present Worth	Compound Amount	Sinking Fund	Present Worth	Capital Recovery	
	CA	PW	SCA	SF	SPW	CR	
1	1.0050 0000	0.995025	1.000000	1.000000	0.995025	1.005000	1
2	1.0100 2500	0.990075	2.005000	0.498753	1.985099	1.503753	2
3	1.0150 7513	0.985149	3.015025	0.331672	2.970248	0.336672	3
4	1.0201 5050	0.980248	4.030100	0.248133	3.950496	0.253133	4
5	1.0252 5125	0.975371	5.050251	0.198010	4.925866	0.203010	5
6	1.0303 7751	0.970518	6.075502	0.164595	5.896384	0.169595	6
7	1.0355 2940	0.965690	7.105879	0.140729	6.862074	0.145729	7
8	1.0407 0704	0.960885	8.141409	0.122829	7.822959	0.127829	8
9	1.0459 1058	0.956105	9.182116	0.108907	8.779064	0.113907	9
10	1.0511 4013	0.951348	10.228026	0.097771	9.730412	0.102771	10
11	1.0563 9583	0.946615	11.279167	0.088659	10.677027	0.093659	11
12	1.0616 7781	0.941905	12.335562	0.081066	11.618932	0.086066	12
13	1.0669 8620	0.937219	13.397240	0.074642	12.556151	0.079642	13
14	1.0723 2113	0.932556	14.464226	0.069136	13.488708	0.074136	14
15	1.0776 8274	0.927917	15.536548	0.064364	14.416625	0.069364	15
16	1.0830 7115	0.923300	16.614230	0.060189	15.339925	0.065189	16
17	1.0884 8651	0.918707	17.697301	0.056506	16.258632	0.061506	17
18	1.0939 2894	0.914136	18.785788	0.053232	17.172768	0.058232	18
19	1.0993 9858	0.909588	19.879717	0.050303	18.082356	0.055303	19
20	1.1048 9558	0.905063	20.979115	0.047666	18.987419	0.052666	20
21	1.1104 2006	0.900560	22.084011	0.045282	19.887979	0.050282	21
22	1.1159 7216	0.896080	23.194431	0.043114	20.784059	0.048114	22
23	1.1215 5202	0.891622	24.310403	0.041135	21.675681	0.046135	23
24	1.1271 5978	0.887186	25.431955	0.039321	22.562866	0.044321	24
25	1.1327 9558	0.882772	26.559115	0.037652	23.445638	0.042652	25
26	1.1384 5955	0.878380	27.691911	0.036112	24.324018	0.041112	26
27	1.1441 5185	0.874010	28.830370	0.034686	25.198028	0.039686	27
28	1.1498 7261	0.869662	29.974522	0.033362	26.067689	0.038362	28
29	1.1556 2197	0.865335	31.124395	0.032129	26.933024	0.037129	29
30	1.1614 0008	0.861030	32.280017	0.030979	27.794054	0.035979	30
31	1.1672 0708	0.856746	33.441417	0.029903	28.650800	0.034903	31
32	1.1730 4312	0.852484	34.608624	0.028895	29.503284	0.033895	32
33	1.1789 0833	0.848242	35.781667	0.027947	30.351526	0.032947	33
34	1.1848 0288	0.844022	36.960575	0.027056	31.195548	0.032056	34
35	1.1907 2689	0.839823	38.145378	0.026215	32.035371	0.031215	35
36	1.1966 8052	0.835645	39.336105	0.025422	32.871016	0.030422	36
37	1.2026 6393	0.831487	40.532785	0.024671	33.702504	0.029671	37
38	1.2086 7725	0.827351	41.735449	0.023960	34.529854	0.028960	38
39	1.2147 2063	0.823235	42.944127	0.023286	35.353089	0.028286	39
40	1.2207 9424	0.819139	44.158847	0.022646	36.172228	0.027646	40
45	1.2516 2082	0.798964	50.324164	0.019871	40.207196	0.024871	45
50	1.2832 2581	0.779286	56.645163	0.017654	44.142786	0.022654	50
55	1.3156 2887	0.760093	63.125775	0.015841	47.981445	0.020841	55
60	1.3488 5015	0.741372	69.770031	0.014333	51.725561	0.019333	60
65	1.3829 1031	0.723113	76.582062	0.013058	55.377461	0.018058	65
70	1.4178 3053	0.705303	83.566105	0.011967	58.939418	0.016967	70
75	1.4536 3252	0.687932	90.726505	0.011022	62.413645	0.016022	75
80	1.4903 3857	0.670988	98.067714	0.010197	65.802305	0.015197	80
90	1.5665 5468	0.638344	113.310936	0.008825	72.331300	0.013825	90
100	1.6466 6849	0.607287	129.333698	0.007732	78.542645	0.012732	100
48	1.2704 8916	0.787098	54.097832	0.018485	42.580318	0.023485	48
72	1.4320 4428	0.698302	86.408856	0.011573	60.339514	0.016573	72
84	1.5203 6964	0.657735	104.073927	0.009609	68.453042	0.014609	84
96	1.6141 4271	0.619524	122.828542	0.008141	76.095218	0.013141	96
108	1.7136 9950	0.583533	142.739900	0.007006	83.293424	0.012006	108
120	1.8193 9673	0.549633	163.879347	0.006102	90.073453	0.011102	120
144	2.0507 5082	0.487626	210.150163	0.004759	102.474743	0.009759	144
180	2.4540 9356	0.407482	290.818712	0.003439	118.503515	0.008439	180
240	3.3102 0448	0.302096	462.040895	0.002164	139.580772	0.007164	240
300	4.4649 6981	0.223966	692.993962	0.001443	155.206864	0.006443	300

Standard Compound Interest Factors

TABLE B-5
7/12 Percent Compound Interest Factors for One Dollar

n	Single Amount		Uniform Series (Periodic Equal Amounts)				n
	Compound Amount	Present Worth	Compound Amount	Sinking Fund	Present Worth	Capital Recovery	
	CA	PW	SCA	SF	SPW	CR	
1	1.0058 3333	0.994200	1.000000	1.000000	0.994200	1.005833	1
2	1.0117 0069	0.988435	2.005833	0.498546	1.982635	0.504379	2
3	1.0176 0228	0.982702	3.017534	0.331396	2.965337	0.337230	3
4	1.0235 3830	0.977003	4.035136	0.247823	3.942340	0.253656	4
5	1.0295 0894	0.971337	5.058675	0.197680	4.913677	0.203514	5
6	1.0355 1440	0.965704	6.088184	0.164253	5.879381	0.170086	6
7	1.0415 5490	0.960103	7.123698	0.140377	6.839484	0.146210	7
8	1.0476 3064	0.954535	8.165253	0.122470	7.794019	0.128304	8
9	1.0537 4182	0.948999	9.212883	0.108544	8.743018	0.114377	9
10	1.0598 8865	0.943495	10.266625	0.097403	9.686513	0.103236	10
11	1.0660 7133	0.938024	11.326514	0.088288	10.624537	0.094122	11
12	1.0722 9008	0.932583	12.392585	0.080693	11.557120	0.086527	12
13	1.0785 4511	0.927175	13.464875	0.074267	12.484295	0.080101	13
14	1.0848 3662	0.921798	14.543420	0.068760	13.406093	0.074593	14
15	1.0911 6483	0.916452	15.628257	0.063987	14.322545	0.069820	15
16	1.0975 2996	0.911137	16.719422	0.059811	15.233682	0.065644	16
17	1.1039 3222	0.905853	17.816952	0.056126	16.139534	0.061960	17
18	1.1103 7182	0.900599	18.920884	0.052852	17.040133	0.058685	18
19	1.1168 4899	0.895376	20.031256	0.049922	17.935510	0.055755	19
20	1.1233 6395	0.890183	21.148105	0.047286	18.825693	0.053119	20
21	1.1299 1690	0.885021	22.271469	0.044900	19.710714	0.050734	21
22	1.1365 0808	0.879888	23.401386	0.042733	20.590602	0.048566	22
23	1.1431 3771	0.874785	24.537894	0.040753	21.465387	0.046587	23
24	1.1498 0602	0.869712	25.681032	0.038939	22.335099	0.044773	24
25	1.1565 1322	0.864668	26.830838	0.037271	23.199767	0.043104	25
26	1.1632 5955	0.859653	27.987351	0.035730	24.059421	0.041564	26
27	1.1700 4523	0.854668	29.150610	0.034305	24.914089	0.040138	27
28	1.1768 7049	0.849711	30.320656	0.032981	25.763800	0.038814	28
29	1.1837 3557	0.844783	31.497526	0.031749	26.608583	0.037582	29
30	1.1906 4069	0.839884	32.681262	0.030599	27.448467	0.036432	30
31	1.1975 8610	0.835013	33.871902	0.029523	28.283480	0.035356	31
32	1.2045 7202	0.830170	35.069488	0.028515	29.113650	0.034348	32
33	1.2115 9869	0.825356	36.274060	0.027568	29.939006	0.033401	33
34	1.2186 6634	0.820569	37.485659	0.026677	30.759575	0.032510	34
35	1.2257 7523	0.815810	38.704325	0.025837	31.575385	0.031670	35
36	1.2329 2559	0.811079	39.930101	0.025044	32.386464	0.030877	36
37	1.2401 1765	0.806375	41.163026	0.024294	33.192840	0.030127	37
38	1.2473 5167	0.801699	42.403144	0.023583	33.994538	0.029416	38
39	1.2546 2789	0.797049	43.650496	0.022909	34.791587	0.028743	39
40	1.2619 4655	0.792427	44.905124	0.022269	35.584014	0.028103	40
45	1.2991 8525	0.769713	51.288900	0.019497	39.477742	0.025331	45
50	1.3375 2283	0.747651	57.861056	0.017283	43.259864	0.023116	50
55	1.3769 9170	0.726221	64.627149	0.015473	46.933579	0.021307	55
60	1.4176 2526	0.705405	71.592902	0.013968	50.501994	0.019801	60
65	1.4594 5787	0.685186	78.764207	0.012696	53.968126	0.018529	65
70	1.5025 2492	0.665546	86.147129	0.011608	57.334909	0.017441	70
75	1.5468 6283	0.646470	93.747914	0.010667	60.605189	0.016500	75
80	1.5925 0910	0.627940	101.572989	0.009845	63.781732	0.015678	80
90	1.6878 8232	0.592458	117.922684	0.008480	69.864280	0.014313	90
100	1.7889 6731	0.558982	135.251539	0.007394	75.603136	0.013227	100
48	1.3220 5388	0.756399	55.209236	0.018113	41.760201	0.023946	48
72	1.5201 0550	0.657849	89.160944	0.011216	58.654444	0.017049	72
84	1.6299 9405	0.613499	107.998981	0.009259	66.257285	0.015093	84
96	1.7478 2646	0.572139	128.198821	0.007800	73.347569	0.013634	96
108	1.8741 7697	0.533568	149.858909	0.006673	79.959850	0.012506	108
120	2.0096 6138	0.497596	173.084807	0.005778	86.126314	0.011611	120
144	2.3107 2074	0.432765	173.084807	0.004450	97.240216	0.010284	144
180	2.8489 4673	0.351007	316.962297	0.003155	111.255958	0.008988	180
240	4.0387 3885	0.247602	520.926660	0.001920	128.982506	0.007753	240
300	5.7254 1821	0.174660	818.071693	0.001234	141.486903	0.007068	300

TABLE B-6
2/3 Percent Compound Interest Factors for One Dollar

n	Single Amount		Uniform Series (Periodic Equal Amounts)				n
	Compound Amount	Present Worth	Compound Amount	Sinking Fund	Present Worth	Capital Recovery	
	CA	PW	SCA	SF	SPW	CR	
1	1.0066 6667	0.993377	1.000000	1.000000	0.993377	1.006667	1
2	1.0133 7778	0.986799	2.006667	0.498339	1.980176	0.505006	2
3	1.0201 3363	0.980264	3.020044	0.331121	2.960440	0.337788	3
4	1.0269 3452	0.973772	4.040178	0.247514	3.934212	0.254181	4
5	1.0337 8075	0.967323	5.067113	0.197351	4.901535	0.204018	5
6	1.0406 7262	0.960917	6.100893	0.163910	5.862452	0.170577	6
7	1.0476 1044	0.954553	7.141566	0.140025	6.817005	0.146692	7
8	1.0545 9451	0.948232	8.189176	0.122112	7.765237	0.128779	8
9	1.0616 2514	0.941952	9.243771	0.108181	8.707189	0.114848	9
10	1.0687 0264	0.935714	10.305396	0.097037	9.642903	0.103703	10
11	1.0758 2732	0.929517	11.374099	0.087919	10.572420	0.094586	11
12	1.0829 9951	0.923361	12.449926	0.080322	11.495782	0.086988	12
13	1.0902 1950	0.917246	13.532926	0.073894	12.413028	0.080561	13
14	1.0974 8763	0.911172	14.623145	0.068385	13.324200	0.075051	14
15	1.1048 0422	0.905138	15.720633	0.063611	14.229338	0.070277	15
16	1.1121 6958	0.899143	16.825437	0.059434	15.128481	0.066100	16
17	1.1195 8404	0.893189	17.937606	0.055749	16.021670	0.062415	17
18	1.1270 4794	0.887274	19.057191	0.052474	16.908944	0.059140	18
19	1.1345 6159	0.881398	20.184238	0.049544	17.790342	0.056210	19
20	1.1421 2533	0.875561	21.318800	0.046907	18.665902	0.053574	20
21	1.1497 3950	0.869762	22.460925	0.044522	19.535665	0.051188	21
22	1.1574 0443	0.864002	23.610665	0.042354	20.399667	0.049020	22
23	1.1651 2046	0.858280	24.768069	0.040375	21.257947	0.047041	23
24	1.1728 8793	0.852596	25.933190	0.038561	22.110544	0.045227	24
25	1.1807 0718	0.846950	27.106078	0.036892	22.957494	0.043559	25
26	1.1885 7857	0.841341	28.286785	0.035352	23.798835	0.042019	26
27	1.1965 0242	0.835769	29.475363	0.033927	24.634604	0.040593	27
28	1.2044 7911	0.830234	30.671866	0.032603	25.464838	0.039270	28
29	1.2125 0897	0.824736	31.876345	0.031371	26.289575	0.038038	29
30	1.2205 9236	0.819274	33.088854	0.030222	27.108849	0.036888	30
31	1.2287 2964	0.813849	34.309446	0.029146	27.922698	0.035813	31
32	1.2369 2117	0.808459	35.538176	0.028139	28.731157	0.034805	32
33	1.2451 6731	0.803105	36.775097	0.027192	29.534262	0.033859	33
34	1.2534 6843	0.797786	38.020264	0.026302	30.332048	0.032968	34
35	1.2618 2489	0.792503	39.273733	0.025462	31.124551	0.032129	35
36	1.2702 3705	0.787255	40.535558	0.024670	31.911806	0.031336	36
37	1.2787 0530	0.782041	41.805795	0.023920	32.693847	0.030587	37
38	1.2872 3000	0.776862	43.084500	0.023210	33.470708	0.029877	38
39	1.2958 1153	0.771717	44.371730	0.022537	34.242426	0.029204	39
40	1.3044 5028	0.766606	45.667542	0.021897	35.009032	0.028564	40
45	1.3485 1559	0.741556	52.277338	0.019129	38.766580	0.025795	45
50	1.3940 6946	0.717324	59.110418	0.016917	42.401344	0.023584	50
55	1.4411 6217	0.693884	66.174325	0.015112	45.917334	0.021778	55
60	1.4898 4571	0.671210	73.476856	0.013610	49.318433	0.020276	60
65	1.5401 7381	0.649277	81.026072	0.012342	52.608395	0.019008	65
70	1.5922 0204	0.628061	88.830306	0.011257	55.790851	0.017924	70
75	1.6459 8782	0.607538	96.898173	0.010320	58.869314	0.016987	75
80	1.7015 9053	0.587685	105.238579	0.009502	61.847182	0.016169	80
90	1.8184 9429	0.549905	122.774143	0.008145	67.514176	0.014812	90
100	1.9434 2965	0.514554	141.514447	0.007066	72.816861	0.013733	100
48	1.3756 6610	0.726921	56.349915	0.017746	40.961913	0.024413	48
72	1.6135 0217	0.619770	92.025325	0.010867	57.034522	0.017533	72
84	1.7474 2205	0.572272	112.113308	0.008920	64.159261	0.015586	84
96	1.8924 5722	0.528414	133.868583	0.007470	70.737970	0.014137	96
108	2.0495 3024	0.487917	157.429535	0.006352	76.812497	0.013019	108
120	2.2196 4023	0.450523	182.946035	0.005466	82.421481	0.012133	120
144	2.6033 8924	0.384115	240.508387	0.004158	92.382800	0.010825	144
180	3.3069 2148	0.302396	346.038202	0.002890	104.640592	0.009557	180
240	4.9268 0277	0.202971	589.020416	0.001698	119.554292	0.008364	240
300	7.3401 7596	0.136237	951.026395	0.001051	129.564523	0.007718	300

Standard Compound Interest Factors

TABLE B-7
3/4 Percent Compound Interest Factors for One Dollar

n	Single Amount		Uniform Series (Periodic Equal Amounts)				n
	Compound Amount	Present Worth	Compound Amount	Sinking Fund	Present Worth	Capital Recovery	
	CA	PW	SCA	SF	SPW	CR	
1	1.0075 0000	0.992556	1.000000	1.000000	0.992556	1.007500	1
2	1.0150 5625	0.985167	2.007500	0.498132	1.977723	0.505632	2
3	1.0226 6917	0.977833	3.022556	0.330846	2.955556	0.338346	3
4	1.0303 3919	0.970554	4.045225	0.247205	3.926110	0.254705	4
5	1.0380 6673	0.963329	5.075565	0.197022	4.889440	0.204522	5
6	1.0458 5224	0.956158	6.113631	0.163569	5.845598	0.171069	6
7	1.0536 9613	0.949040	7.159484	0.139675	6.794638	0.147175	7
8	1.0615 9885	0.941975	8.213180	0.121756	7.736613	0.129256	8
9	1.0695 6084	0.934963	9.274779	0.107819	8.671576	0.115319	9
10	1.0775 8255	0.928003	10.344339	0.096671	9.599580	0.104171	10
11	1.0856 6441	0.921095	11.421922	0.087551	10.520675	0.095051	11
12	1.0938 0690	0.914238	12.507586	0.079951	11.434913	0.087451	12
13	1.1020 1045	0.907432	13.601395	0.073522	12.342345	0.081022	13
14	1.1102 7553	0.900677	14.703404	0.068011	13.243022	0.075511	14
15	1.1186 0259	0.893973	15.813679	0.063236	14.136995	0.070736	15
16	1.1269 9211	0.887318	16.932282	0.059059	15.024313	0.066559	16
17	1.1354 4455	0.880712	18.052274	0.055373	15.905025	0.062873	17
18	1.1439 6039	0.874156	19.194718	0.052098	16.779181	0.059598	18
19	1.1525 4009	0.867649	20.338679	0.049167	17.646830	0.056667	19
20	1.1611 8414	0.861190	21.491219	0.046531	18.508020	0.054031	20
21	1.1698 9302	0.854779	22.652403	0.044145	19.362799	0.051645	21
22	1.1786 6722	0.848416	23.822296	0.041977	20.211215	0.049477	22
23	1.1875 0723	0.842100	25.000963	0.039998	21.053315	0.047498	23
24	1.1964 1353	0.835831	26.188471	0.038185	21.889146	0.045685	24
25	1.2053 8663	0.829609	27.384884	0.036516	22.718755	0.044016	25
26	1.2144 2703	0.823434	28.590271	0.034977	23.542189	0.042477	26
27	1.2235 3523	0.817304	29.804698	0.033552	24.359493	0.041052	27
28	1.2327 1175	0.811220	31.028233	0.032229	25.170713	0.039729	28
29	1.2419 5709	0.805181	32.260945	0.030997	25.975893	0.038497	29
30	1.2512 7176	0.799187	33.502902	0.029848	26.775080	0.037348	30
31	1.2606 5630	0.793238	34.754174	0.028774	27.568318	0.036274	31
32	1.2701 1122	0.787333	36.014830	0.027766	28.355650	0.035266	32
33	1.2796 3706	0.781472	37.284941	0.026820	29.137122	0.034320	33
34	1.2892 3434	0.775654	38.564578	0.025931	29.912776	0.033431	34
35	1.2989 0359	0.769880	39.853813	0.025092	30.682656	0.032592	35
36	1.3086 4537	0.764149	41.152716	0.024300	31.446805	0.031800	36
37	1.3184 6021	0.758461	42.461361	0.023551	32.205266	0.031051	37
38	1.3283 4866	0.752814	43.779822	0.022842	32.958080	0.030342	38
39	1.3383 1128	0.747210	45.108170	0.022169	33.705290	0.029669	39
40	1.3483 4861	0.741648	46.446482	0.021530	34.446938	0.029030	40
45	1.3996 7584	0.714451	53.290112	0.018765	38.073181	0.026265	45
50	1.4529 5693	0.688252	60.394257	0.016558	41.566447	0.024058	50
55	1.5082 6626	0.663013	67.768834	0.014756	44.931612	0.022256	55
60	1.5656 8103	0.638700	75.424137	0.013258	48.173374	0.020758	60
65	1.6252 8139	0.615278	83.370852	0.011995	51.296257	0.019495	65
70	1.6871 5055	0.592715	91.620073	0.010915	54.304622	0.018415	70
75	1.7513 7486	0.570980	100.183314	0.009982	57.202668	0.017482	75
80	1.8180 4398	0.550042	109.072531	0.009168	59.994440	0.016668	80
90	1.9590 9246	0.510440	127.878995	0.007820	65.274609	0.015320	90
100	2.1110 8384	0.473690	148.144512	0.006750	70.174623	0.014250	100
48	1.4314 0533	0.698614	57.520711	0.017385	40.184782	0.024885	48
72	1.7125 5271	0.583924	95.007028	0.010526	55.476849	0.018026	72
84	1.8732 0196	0.533845	116.426928	0.008589	62.153965	0.016089	84
96	2.0489 2123	0.488062	139.856164	0.007150	68.258439	0.014650	96
108	2.2411 2417	0.446205	165.483223	0.006043	73.839382	0.013543	108
120	2.4513 5708	0.407937	193.514277	0.005168	78.941693	0.012668	120
144	2.9328 3677	0.340967	257.711570	0.003880	87.871092	0.011380	144
180	3.8380 4327	0.260549	378.405769	0.002643	98.593409	0.010143	180
240	6.0091 5152	0.166413	667.886870	0.001497	111.144954	0.008997	240
300	9.4084 1453	0.106288	1121.121937	0.000892	119.161622	0.008392	300

TABLE B-8
5/6 Percent Compound Interest Factors for One Dollar

n	Single Amount		Uniform Series (Periodic Equal Amounts)				n
	Compound Amount	Present Worth	Compound Amount	Sinking Fund	Present Worth	Capital Recovery	
	CA	PW	SCA	SF	SPW	CR	
1	1.0083 3333	0.991736	1.000000	1.000000	0.991736	1.008333	1
2	1.0167 3611	0.983539	2.008333	0.497925	1.975275	0.506259	2
3	1.0252 0891	0.975411	3.025069	0.330571	2.950686	0.338904	3
4	1.0337 5232	0.967350	4.050278	0.246897	3.918036	0.255230	4
5	1.0423 6692	0.959355	5.084031	0.196694	4.877391	0.205028	5
6	1.0510 5331	0.951426	6.126398	0.163228	5.828817	0.171561	6
7	1.0598 1209	0.943563	7.177451	0.139325	6.772381	0.147659	7
8	1.0686 4386	0.935765	8.237263	0.121400	7.708146	0.129733	8
9	1.0775 4922	0.928032	9.305907	0.107459	8.636178	0.115792	9
10	1.0865 2880	0.920362	10.383456	0.096307	9.556540	0.104640	10
11	1.0955 8321	0.912755	11.469985	0.087184	10.469296	0.095517	11
12	1.1047 1307	0.905212	12.565568	0.079583	11.374508	0.087916	12
13	1.1139 1901	0.897731	13.670281	0.073151	12.272240	0.081485	13
14	1.1232 0167	0.890312	14.784200	0.067640	13.162552	0.075973	14
15	1.1325 6168	0.882954	15.907402	0.062864	14.045506	0.071197	15
16	1.1419 9970	0.875657	17.039964	0.058686	14.921163	0.067019	16
17	1.1515 1636	0.868420	18.181963	0.055000	15.789583	0.063333	17
18	1.1611 1233	0.861243	19.333480	0.051724	16.650826	0.060057	18
19	1.1707 8827	0.854125	20.494592	0.048793	17.504952	0.057127	19
20	1.1805 4483	0.847066	21.665380	0.046157	18.352018	0.054490	20
21	1.1903 8271	0.840066	22.845925	0.043771	19.192084	0.052105	21
22	1.2003 0256	0.833123	24.036308	0.041604	20.025207	0.049937	22
23	1.2103 0509	0.826238	25.236610	0.039625	20.851445	0.047958	23
24	1.2203 9096	0.819410	26.446915	0.037812	21.670855	0.046145	24
25	1.2305 6089	0.812638	27.667306	0.036144	22.483492	0.044477	25
26	1.2408 1556	0.805922	28.897867	0.034605	23.289414	0.042938	26
27	1.2511 5569	0.799261	30.138683	0.033180	24.088675	0.041513	27
28	1.2615 8199	0.792656	31.389838	0.031857	24.881331	0.040191	28
29	1.2720 9517	0.786105	32.651420	0.030627	25.667435	0.038960	29
30	1.2826 9596	0.779608	33.923516	0.029478	26.447043	0.037811	30
31	1.2933 8510	0.773165	35.206212	0.028404	27.220208	0.036737	31
32	1.3041 6331	0.766775	36.499597	0.027398	27.986983	0.035731	32
33	1.3150 3133	0.760438	37.803760	0.026452	28.747422	0.034786	33
34	1.3259 8993	0.754154	39.118791	0.025563	29.501575	0.033896	34
35	1.3370 3984	0.747921	40.444781	0.024725	30.249496	0.033058	35
36	1.3481 8184	0.741740	41.781821	0.023934	30.991236	0.032267	36
37	1.3594 1669	0.735610	43.130003	0.023186	31.726845	0.031519	37
38	1.3707 4516	0.729530	44.489420	0.022477	32.456375	0.030811	38
39	1.3821 6804	0.723501	45.860165	0.021805	33.179876	0.030139	39
40	1.3936 8611	0.717522	47.242333	0.021167	33.897398	0.029501	40
45	1.4527 3230	0.688358	54.327876	0.018407	37.397032	0.026740	45
50	1.5142 8009	0.660379	61.713611	0.016204	40.754423	0.024537	50
55	1.5784 3548	0.633539	69.412258	0.014407	43.975353	0.022740	55
60	1.6453 0893	0.607789	77.437072	0.012914	47.065369	0.021247	60
65	1.7150 1561	0.583085	85.801873	0.011655	50.029789	0.019988	65
70	1.7876 7554	0.559386	94.521065	0.010580	52.873725	0.018913	70
75	1.8634 1385	0.536649	103.609663	0.009652	55.602067	0.017985	75
80	1.9423 6096	0.514837	113.083316	0.008843	58.219516	0.017176	80
90	2.1104 3113	0.473837	133.251736	0.007505	63.139580	0.015838	90
100	2.2930 4420	0.436101	155.165304	0.006445	67.667821	0.014778	100
48	1.4893 5410	0.671432	58.722492	0.017029	39.428160	0.025363	48
72	1.8175 9428	0.550178	98.111314	0.010193	53.978665	0.018526	72
84	2.0079 2015	0.498028	120.950418	0.008268	60.236667	0.016601	84
96	2.2181 7563	0.450821	146.181076	0.006841	65.901488	0.015174	96
108	2.4504 4761	0.408089	174.053713	0.005745	71.029355	0.014079	108
120	2.7070 4149	0.369407	204.844979	0.004882	75.671163	0.013215	120
144	3.3036 4897	0.302696	276.437876	0.003617	83.676528	0.011951	144
180	4.4539 1955	0.224521	414.470346	0.002413	93.057139	0.010746	180
240	7.3280 7363	0.136462	759.368836	0.001317	103.624619	0.009650	240
300	12.0569 4502	0.082940	1326.833403	0.000754	110.047230	0.009087	300

Standard Compound Interest Factors

TABLE B-9
1 Percent Compound Interest Factors for One Dollar

	Single Amount		Uniform Series (Periodic Equal Amounts)				
	Compound Amount	Present Worth	Compound Amount	Sinking Fund	Present Worth	Capital Recovery	
n	Given P to find F $(1+i)^n$	Given F to find P $\dfrac{1}{(1+i)^n}$	Given A to find F $\dfrac{(1+i)^n-1}{i}$	Given F to find A $\dfrac{i}{(1+i)^n-1}$	Given A to find P $\dfrac{(1+i)^n-1}{i(1+i)^n}$	Given P to find A $\dfrac{i(1+i)^n}{(1+i)^n-1}$	n
	CA	PW	SCA	SF	SPW	CR	
1	1.0100 0000	0.990099	1.000000	1.000000	0.990099	1.010000	1
2	1.0201 0000	0.980296	2.010000	0.497512	1.970395	0.507512	2
3	1.0303 0100	0.970590	3.030100	0.330022	2.940985	0.340022	3
4	1.0406 0401	0.960980	4.060401	0.246281	3.901966	0.256281	4
5	1.0510 1005	0.951466	5.101005	0.196040	4.853431	0.206040	5
6	1.0615 2015	0.942045	6.152015	0.162548	5.795476	0.172548	6
7	1.0721 3535	0.932718	7.213535	0.138628	6.728195	0.148628	7
8	1.0828 5671	0.923483	8.285671	0.120690	7.651678	0.130690	8
9	1.0936 8527	0.914340	9.368527	0.106740	8.566018	0.116740	9
10	1.1046 2213	0.905287	10.462213	0.095582	9.471305	0.105582	10
11	1.1156 6835	0.896324	11.566835	0.086454	10.367628	0.096454	11
12	1.1268 2503	0.887449	12.682503	0.078849	11.255077	0.088849	12
13	1.1380 9328	0.878663	13.809328	0.072415	12.133740	0.082415	13
14	1.1494 7421	0.869963	14.947421	0.066901	13.003703	0.076901	14
15	1.1609 6896	0.861349	16.096896	0.062124	13.865053	0.072124	15
16	1.1725 7864	0.852821	17.257864	0.057945	14.717874	0.067945	16
17	1.1843 0443	0.844377	18.430443	0.054258	15.562251	0.064258	17
18	1.1961 4748	0.836017	19.614748	0.050982	16.398269	0.060982	18
19	1.2081 0895	0.827740	20.810895	0.048052	17.226008	0.058052	19
20	1.2201 9004	0.819544	22.019004	0.045415	18.045553	0.055415	20
21	1.2323 9194	0.811430	23.239194	0.043031	18.856983	0.053031	21
22	1.2447 1586	0.803396	24.471586	0.040864	19.660379	0.050864	22
23	1.2571 6302	0.795442	25.716302	0.038886	20.455821	0.048886	23
24	1.2697 3465	0.787566	26.973465	0.037073	21.243387	0.047073	24
25	1.2824 3200	0.779768	28.243200	0.035407	22.023156	0.045407	25
26	1.2952 5631	0.772048	29.525631	0.033869	22.795204	0.043869	26
27	1.3082 0888	0.764404	30.820888	0.032446	23.559608	0.042446	27
28	1.3212 9097	0.756836	32.129097	0.031124	24.316443	0.041124	28
29	1.3345 0388	0.749342	33.450388	0.029895	25.065785	0.039895	29
30	1.3478 4892	0.741923	34.784892	0.028748	25.807708	0.038748	30
31	1.3613 2740	0.734577	36.132740	0.027676	26.542285	0.037676	31
32	1.3749 4068	0.727304	37.494068	0.026671	27.269589	0.036671	32
33	1.3886 9009	0.720103	38.869009	0.025727	27.989693	0.035727	33
34	1.4025 7699	0.712973	40.257699	0.024840	28.702666	0.034840	34
35	1.4166 0276	0.705914	41.660276	0.024004	29.408580	0.034004	35
36	1.4307 6878	0.698925	43.076878	0.023214	30.107505	0.033214	36
37	1.4450 7647	0.692005	44.507647	0.022468	30.799510	0.032468	37
38	1.4595 2724	0.685153	45.952724	0.021761	31.484663	0.031761	38
39	1.4741 2251	0.678370	47.412251	0.021092	32.163033	0.031092	39
40	1.4888 6373	0.671653	48.886373	0.020456	32.834686	0.030456	40
45	1.5648 1075	0.639055	56.481075	0.017705	36.094508	0.027705	45
50	1.6446 3182	0.608039	64.463182	0.015513	39.196118	0.025513	50
55	1.7285 2457	0.578528	72.852457	0.013726	42.147192	0.023726	55
60	1.8166 9670	0.550450	81.669670	0.012244	44.955038	0.022244	60
65	1.9093 6649	0.523734	90.936649	0.010997	47.626608	0.020997	65
70	2.0067 6337	0.498315	100.676337	0.009933	50.168514	0.019933	70
75	2.1091 2847	0.474129	110.912847	0.009016	52.587051	0.019016	75
80	2.2167 1522	0.451118	121.671522	0.008219	54.888206	0.018219	80
90	2.4486 3267	0.408391	144.863267	0.006903	59.160881	0.016903	90
100	2.7048 1383	0.369711	170.481383	0.005866	63.028879	0.015866	100

Standard Compound Interest Factors

TABLE B-10
$1\frac{1}{4}$ Percent Compound Interest Factors for One Dollar

	Single Amount		Uniform Series (Periodic Equal Amounts)				
	Compound Amount	Present Worth	Compound Amount	Sinking Fund	Present Worth	Capital Recovery	
n	Given P to find F $(1+i)^n$	Given F to find P $\dfrac{1}{(1+i)^n}$	Given A to find F $\dfrac{(1+i)^n-1}{i}$	Given F to find A $\dfrac{i}{(1+i)^n-1}$	Given A to find P $\dfrac{(1+i)^n-1}{i(1+i)^n}$	Given P to find A $\dfrac{i(1+i)^n}{(1+i)^n-1}$	n
	CA	PW	SCA	SF	SPW	CR	
1	1.0125 0000	0.987654	1.000000	1.000000	0.987654	1.012500	1
2	1.0251 5625	0.975461	2.012500	0.496894	1.963115	0.509394	2
3	1.0379 7070	0.963418	3.037656	0.329201	2.926534	0.341701	3
4	1.0509 4534	0.951524	4.075627	0.245361	3.878058	0.257861	4
5	1.0640 8215	0.939777	5.126572	0.195062	4.817835	0.207562	5
6	1.0773 8318	0.928175	6.190654	0.161534	5.746010	0.174034	6
7	1.0908 5047	0.916716	7.268038	0.137589	6.662726	0.150089	7
8	1.1044 8610	0.905398	8.358888	0.119633	7.568124	0.132133	8
9	1.1182 9218	0.894221	9.463374	0.105670	8.462345	0.118171	9
10	1.1322 7083	0.883181	10.581666	0.094503	9.345526	0.107003	10
11	1.1464 2422	0.872277	11.713937	0.085368	10.217803	0.097868	11
12	1.1607 5452	0.861509	12.860361	0.077758	11.079312	0.090258	12
13	1.1752 6395	0.850873	14.021116	0.071321	11.930185	0.083821	13
14	1.1899 5475	0.840368	15.196380	0.065805	12.770553	0.078305	14
15	1.2048 2918	0.829993	16.386335	0.061026	13.600546	0.073526	15
16	1.2198 8955	0.819746	17.591164	0.056847	14.420292	0.069347	16
17	1.2351 3817	0.809626	18.811053	0.053160	15.229918	0.065660	17
18	1.2505 7739	0.799631	20.046192	0.049885	16.029549	0.062385	18
19	1.2662 0961	0.789759	21.296769	0.046955	16.819308	0.059455	19
20	1.2820 3723	0.780009	22.562979	0.044320	17.599316	0.056820	20
21	1.2980 6270	0.770379	23.845016	0.041937	18.369695	0.054437	21
22	1.3142 8848	0.760868	25.143078	0.039772	19.130563	0.052272	22
23	1.3307 1709	0.751475	26.457367	0.037797	19.882037	0.050297	23
24	1.3473 5105	0.742197	27.788084	0.035987	20.624235	0.048487	24
25	1.3641 9294	0.733034	29.135435	0.034322	21.357269	0.046822	25
26	1.3812 4535	0.723984	30.499628	0.032787	22.081253	0.045287	26
27	1.3985 1092	0.715046	31.880873	0.031367	22.796299	0.043867	27
28	1.4159 9230	0.706219	33.279384	0.030049	23.502518	0.042549	28
29	1.4336 9221	0.697500	34.695377	0.028822	24.200018	0.041322	29
30	1.4516 1336	0.688889	36.129069	0.027679	24.888906	0.040179	30
31	1.4697 5853	0.680384	37.580682	0.026609	25.569290	0.039109	31
32	1.4881 3051	0.671984	39.050441	0.025608	26.241274	0.038108	32
33	1.5067 3214	0.663688	40.538571	0.024668	26.904962	0.037168	33
34	1.5255 6629	0.655494	42.045303	0.023784	27.560456	0.036284	34
35	1.5446 3587	0.647402	43.570870	0.022951	28.207858	0.035451	35
36	1.5639 4382	0.639409	45.115505	0.022165	28.847267	0.034665	36
37	1.5834 9312	0.631515	46.679449	0.021423	29.478783	0.033923	37
38	1.6032 8678	0.623719	48.262942	0.020720	30.102501	0.033220	38
39	1.6233 2787	0.616019	49.866229	0.020054	30.718520	0.032554	39
40	1.6436 1946	0.608413	51.489557	0.019421	31.326933	0.031921	40
45	1.7489 4614	0.571773	59.915691	0.016690	34.258168	0.029190	45
50	1.8610 2237	0.537339	68.881790	0.014518	37.012876	0.027018	50
55	1.9802 8070	0.504979	78.422456	0.012751	39.601687	0.025251	55
60	2.1071 8135	0.474568	88.574508	0.011290	42.034592	0.023790	60
65	2.2422 1407	0.445988	99.377125	0.010063	44.320980	0.022563	65
70	2.3858 9997	0.419129	110.871998	9.019410-3	46.469676	0.021519	70
75	2.5387 9358	0.393888	123.103486	8.123247-3	48.488970	0.020623	75
80	2.7014 8494	0.370167	136.118795	7.346524-3	50.386657	0.019847	80
90	3.0588 1260	0.326934	164.705008	6.071461-3	53.846060	0.018571	90
100	3.4634 0427	0.288733	197.072342	5.074279-3	56.901339	0.017574	100

Standard Compound Interest Factors

TABLE B-11
1½ Percent Compound Interest Factors for One Dollar

	Single Amount		Uniform Series (Periodic Equal Amounts)				
	Compound Amount	Present Worth	Compound Amount	Sinking Fund	Present Worth	Capital Recovery	
n	Given P to find F $(1+i)^n$	Given F to find P $\dfrac{1}{(1+i)^n}$	Given A to find F $\dfrac{(1+i)^n-1}{i}$	Given F to find A $\dfrac{i}{(1+i)^n-1}$	Given A to find P $\dfrac{(1+i)^n-1}{i(1+i)^n}$	Given P to find A $\dfrac{i(1+i)^n}{(1+i)^n-1}$	n
	CA	PW	SCA	SF	SPW	CR	
1	1.0150 0000	0.985222	1.000000	1.000000	0.985222	1.015000	1
2	1.0302 2500	0.970662	2.015000	0.496278	1.955883	0.511278	2
3	1.0456 7838	0.956317	3.045225	0.328383	2.912200	0.343383	3
4	1.0613 6355	0.942184	4.090903	0.244445	3.854385	0.259445	4
5	1.0772 8400	0.928260	5.152267	0.194089	4.782645	0.209089	5
6	1.0934 4326	0.914542	6.229551	0.160525	5.697187	0.175525	6
7	1.1098 4491	0.901027	7.322994	0.136556	6.598214	0.151556	7
8	1.1264 9259	0.887711	8.432839	0.118584	7.485925	0.133584	8
9	1.1433 8998	0.874592	9.559332	0.104610	8.360517	0.119610	9
10	1.1605 4083	0.861667	10.702722	0.093434	9.222185	0.108434	10
11	1.1779 4894	0.848933	11.863262	0.084294	10.071118	0.099294	11
12	1.1956 1817	0.836387	13.041211	0.076680	10.907505	0.091680	12
13	1.2135 5244	0.824027	14.236830	0.070240	11.731532	0.085240	13
14	1.2317 5573	0.811849	15.450382	0.064723	12.543382	0.079723	14
15	1.2502 3207	0.799852	16.682138	0.059944	13.343233	0.074944	15
16	1.2689 8555	0.788031	17.932370	0.055765	14.131264	0.070765	16
17	1.2880 2033	0.776385	19.201355	0.052080	14.907649	0.067080	17
18	1.3073 4064	0.764912	20.489376	0.048806	15.672561	0.063806	18
19	1.3269 5075	0.753607	21.796716	0.045878	16.426168	0.060878	19
20	1.3468 5501	0.742470	23.123667	0.043246	17.168639	0.058246	20
21	1.3670 5783	0.731498	24.470522	0.040865	17.900137	0.055865	21
22	1.3875 6370	0.720688	25.837580	0.038703	18.620824	0.053703	22
23	1.4083 7715	0.710037	27.225144	0.036731	19.330861	0.051731	23
24	1.4295 0281	0.699544	28.633521	0.034924	20.030405	0.049924	24
25	1.4509 4535	0.689206	30.063024	0.033263	20.719611	0.048263	25
26	1.4727 0953	0.679021	31.513969	0.031732	21.398632	0.046732	26
27	1.4948 0018	0.668986	32.986678	0.030315	22.067617	0.045315	27
28	1.5172 2218	0.659099	34.481479	0.029001	22.726717	0.044001	28
29	1.5399 8051	0.649359	35.998701	0.027779	23.376076	0.042779	29
30	1.5630 8022	0.639762	37.538681	0.026639	24.015838	0.041639	30
31	1.5865 2642	0.630308	39.101762	0.025574	24.646146	0.040574	31
32	1.6103 2432	0.620993	40.688288	0.024577	25.267139	0.039577	32
33	1.6344 7918	0.611816	42.298612	0.023641	25.878954	0.038641	33
34	1.6589 9637	0.602774	43.933092	0.022762	26.481728	0.037762	34
35	1.6838 8132	0.593866	45.592088	0.021934	27.075595	0.036934	35
36	1.7091 3954	0.585090	47.275969	0.021152	27.660684	0.036152	36
37	1.7347 7663	0.576443	48.985109	0.020414	28.237127	0.035414	37
38	1.7607 9828	0.567924	50.719885	0.019716	28.805052	0.034716	38
39	1.7872 1025	0.559531	52.480684	0.019055	29.364583	0.034055	39
40	1.8140 1841	0.551262	54.267894	0.018427	29.915845	0.033427	40
45	1.9542 1301	0.511715	63.614201	0.015720	32.552337	0.030720	45
50	2.1052 4242	0.475005	73.682828	0.013572	34.999688	0.028572	50
55	2.2679 4398	0.440928	84.529599	0.011830	37.271467	0.026830	55
60	2.4432 1978	0.409269	96.214652	0.010393	39.380269	0.025393	60
65	2.6320 4158	0.379933	108.802772	0.009191	41.337786	0.024191	65
70	2.8354 5629	0.352677	122.363753	0.008172	43.154872	0.023172	70
75	3.0545 9171	0.327376	136.972781	0.007301	44.841600	0.022301	75
80	3.2906 6279	0.303890	152.710852	0.006548	46.407323	0.021548	80
90	3.8189 4851	0.261852	187.929900	0.005321	49.209855	0.020321	90
100	4.4320 4565	0.225629	228.803043	0.004371	51.624704	0.019371	100

TABLE B-12
1 ¾ Percent Compound Interest Factors for One Dollar

	Single Amount		Uniform Series (Periodic Equal Amounts)				
	Compound Amount	Present Worth	Compound Amount	Sinking Fund	Present Worth	Capital Recovery	
n	Given P to find F $(1+i)^n$	Given F to find P $\dfrac{1}{(1+i)^n}$	Given A to find F $\dfrac{(1+i)^n - 1}{i}$	Given F to find A $\dfrac{i}{(1+i)^n - 1}$	Given A to find P $\dfrac{(1+i)^n - 1}{i(1+i)^n}$	Given P to find A $\dfrac{i(1+i)^n}{(1+i)^n - 1}$	n
	CA	PW	SCA	SF	SPW	CR	
1	1.0175 0000	0.982801	1.000000	1.000000	0.982801	1.017500	1
2	1.0353 0625	0.965898	2.017500	0.495663	1.948699	0.513163	2
3	1.0534 2411	0.949285	3.052806	0.327567	2.897984	0.345067	3
4	1.0718 5903	0.932959	4.106230	0.243532	3.830943	0.261032	4
5	1.0906 1656	0.916913	5.178089	0.193121	4.747855	0.210621	5
6	1.1097 0235	0.901143	6.268706	0.159523	5.648998	0.177023	6
7	1.1291 2215	0.885644	7.378408	0.135531	6.534641	0.153031	7
8	1.1488 8178	0.870412	8.507530	0.117543	7.405053	0.135043	8
9	1.1689 8721	0.855441	9.656412	0.103558	8.260494	0.121058	9
10	1.1894 4449	0.840729	10.825399	0.092375	9.101223	0.109875	10
11	1.2102 5977	0.826269	12.014844	0.083230	9.927492	0.100730	11
12	1.2314 3931	0.812058	13.225104	0.075614	10.739550	0.093114	12
13	1.2529 8950	0.798091	14.456543	0.069173	11.537641	0.086673	13
14	1.2749 1682	0.784365	15.709533	0.063656	12.322006	0.081156	14
15	1.2972 2786	0.770875	16.984449	0.058877	13.092880	0.076377	15
16	1.3199 2935	0.757616	18.281677	0.054700	13.850497	0.072200	16
17	1.3430 2811	0.744586	19.601607	0.051016	14.595083	0.068516	17
18	1.3665 3111	0.731780	20.944635	0.047745	15.326863	0.065245	18
19	1.3904 4540	0.719194	22.311166	0.044821	16.046057	0.062321	19
20	1.4147 7820	0.706825	23.701611	0.042191	16.752881	0.059691	20
21	1.4395 3681	0.694668	25.116389	0.039815	17.447549	0.057315	21
22	1.4647 2871	0.682720	26.555926	0.037656	18.130269	0.055156	22
23	1.4903 6146	0.670978	28.020655	0.035688	18.801248	0.053188	23
24	1.5164 4279	0.659438	29.511016	0.033886	19.460686	0.051386	24
25	1.5429 8054	0.648096	31.027459	0.032230	20.108782	0.049730	25
26	1.5699 8269	0.636950	32.570440	0.030703	20.745732	0.048203	26
27	1.5974 5739	0.625995	34.140422	0.029291	21.371726	0.046791	27
28	1.6254 1290	0.615228	35.737880	0.027982	21.986955	0.045482	28
29	1.6538 5762	0.604647	37.363293	0.026764	22.591602	0.044264	29
30	1.6828 0013	0.594248	39.017150	0.025630	23.185849	0.043130	30
31	1.7122 4913	0.584027	40.699950	0.024570	23.769877	0.042070	31
32	1.7422 1349	0.573982	42.412200	0.023578	24.343859	0.041078	32
33	1.7727 0223	0.564111	44.154413	0.022648	24.907970	0.040148	33
34	1.8037 2452	0.554408	45.927115	0.021774	25.462378	0.039274	34
35	1.8352 8970	0.544873	47.730840	0.020951	26.007251	0.038451	35
36	1.8674 0727	0.535502	49.566129	0.020175	26.542753	0.037675	36
37	1.9000 8689	0.526292	51.433537	0.019443	27.069045	0.036943	37
38	1.9333 3841	0.517240	53.333624	0.018750	27.586285	0.036250	38
39	1.9671 7184	0.508344	55.266962	0.018094	28.094629	0.035594	39
40	2.0015 9734	0.499601	57.234134	0.017472	28.594230	0.034972	40
45	2.1829 7522	0.458090	67.598584	0.014793	30.966263	0.032293	45
50	2.3807 8893	0.420029	78.902225	0.012674	33.141209	0.030174	50
55	2.5965 2785	0.385130	91.230163	0.010961	35.135445	0.028461	55
60	2.8318 1628	0.353130	104.675216	9.553360-3	36.963986	0.027053	60
65	3.0884 2574	0.323790	119.338614	8.379517-3	38.640597	0.025880	65
70	3.3682 8827	0.296887	135.330758	7.389303-3	40.177903	0.024889	70
75	3.6735 1098	0.272219	152.772056	6.545700-3	41.587478	0.024046	75
80	4.0063 9192	0.249601	171.793824	5.820931-3	42.879935	0.023321	80
90	4.7653 8080	0.209847	215.164617	4.647604-3	45.151610	0.022148	90
100	5.6681 5594	0.176424	266.751768	3.748804-3	47.061473	0.021249	100

Standard Compound Interest Factors

TABLE B-13
2 Percent Compound Interest Factors for One Dollar

	Single Amount		Uniform Series (Periodic Equal Amounts)				
	Compound Amount	Present Worth	Compound Amount	Sinking Fund	Present Worth	Capital Recovery	
n	Given P to find F $(1+i)^n$	Given F to find P $\dfrac{1}{(1+i)^n}$	Given A to find F $\dfrac{(1+i)^n-1}{i}$	Given F to find A $\dfrac{i}{(1+i)^n-1}$	Given A to find P $\dfrac{(1+i)^n-1}{i(1+i)^n}$	Given P to find A $\dfrac{i(1+i)^n}{(1+i)^n-1}$	n
	CA	PW	SCA	SF	SPW	CR	
1	1.0200 0000	0.980392	1.000000	1.000000	0.980392	1.020000	1
2	1.0404 0000	0.961169	2.020000	0.495050	1.941561	0.515050	2
3	1.0612 0800	0.942322	3.060400	0.326755	2.883883	0.346755	3
4	1.0824 3216	0.923845	4.121608	0.242624	3.807729	0.262624	4
5	1.1040 8080	0.905731	5.204040	0.192158	4.713460	0.212158	5
6	1.1261 6242	0.887971	6.308121	0.158526	5.601431	0.178526	6
7	1.1486 8567	0.870560	7.434283	0.134512	6.471991	0.154512	7
8	1.1716 5938	0.853490	8.582969	0.116510	7.325481	0.136510	8
9	1.1950 9257	0.836755	9.754628	0.102515	8.162237	0.122515	9
10	1.2189 9442	0.820348	10.949721	0.091327	8.982585	0.111327	10
11	1.2433 7431	0.804263	12.168715	0.082178	9.786848	0.102178	11
12	1.2682 4179	0.788493	13.412090	0.074560	10.575341	0.094560	12
13	1.2936 0663	0.773033	14.680332	0.068118	11.348374	0.088118	13
14	1.3194 7876	0.757875	15.973938	0.062602	12.106249	0.082602	14
15	1.3458 6834	0.743015	17.293417	0.057825	12.849264	0.077825	15
16	1.3727 8571	0.728446	18.639285	0.053650	13.577709	0.073650	16
17	1.4002 4142	0.714163	20.012071	0.049970	14.291872	0.069970	17
18	1.4282 4625	0.700159	21.412312	0.046702	14.992031	0.066702	18
19	1.4568 1117	0.686431	22.840559	0.043782	15.678462	0.063782	19
20	1.4859 4740	0.672971	24.297370	0.041157	16.351433	0.061157	20
21	1.5156 6634	0.659776	25.783317	0.038785	17.011209	0.058785	21
22	1.5459 7967	0.646839	27.298984	0.036631	17.658048	0.056631	22
23	1.5768 9926	0.634156	28.844963	0.034668	18.292204	0.054668	23
24	1.6084 3725	0.621721	30.421862	0.032871	18.913926	0.052871	24
25	1.6406 0599	0.609531	32.030300	0.031220	19.523456	0.051220	25
26	1.6734 1811	0.597579	33.670906	0.029699	20.121036	0.049699	26
27	1.7068 8648	0.585862	35.344324	0.028293	20.706898	0.048293	27
28	1.7410 2421	0.574375	37.051210	0.026990	21.281272	0.046990	28
29	1.7758 4469	0.563112	38.792235	0.025778	21.844385	0.045778	29
30	1.8113 6158	0.552071	40.568079	0.024650	22.396456	0.044650	30
31	1.8475 8882	0.541246	42.379441	0.023596	22.937702	0.043596	31
32	1.8845 4059	0.530633	44.227030	0.022611	23.468335	0.042611	32
33	1.9222 3140	0.520229	46.111570	0.021687	23.988564	0.041687	33
34	1.9606 7603	0.510028	48.033802	0.020819	24.498592	0.040819	34
35	1.9998 8955	0.500028	49.994478	0.020002	24.998619	0.040002	35
36	2.0398 8734	0.490223	51.994367	0.019233	25.488842	0.039233	36
37	2.0806 8509	0.480611	54.034255	0.018507	25.969453	0.038507	37
38	2.1222 9879	0.471187	56.114940	0.017821	26.440641	0.037821	38
39	2.1647 4477	0.461948	58.237238	0.017171	26.902589	0.037171	39
40	2.2080 3966	0.452890	60.401983	0.016556	27.355479	0.036556	40
45	2.4378 5421	0.410197	71.892710	0.013910	29.490160	0.033910	45
50	2.6915 8803	0.371528	84.579401	0.011823	31.423606	0.031823	50
55	2.9717 3067	0.336504	98.586534	0.010143	33.174788	0.030143	55
60	3.2810 3079	0.304782	114.051539	0.008768	34.760887	0.028768	60
65	3.6225 2311	0.276051	131.126155	0.007626	36.197466	0.027626	65
70	3.9995 5822	0.250028	149.977911	0.006668	37.498619	0.026668	70
75	4.4158 3546	0.226438	170.791773	0.005855	38.677114	0.025855	75
80	4.8754 3916	0.205110	193.771958	0.005161	39.744514	0.025161	80
90	5.9431 3313	0.168261	247.156656	0.004046	41.586929	0.024046	90
100	7.2446 4612	0.138033	312.232306	0.003203	43.098352	0.023203	100

TABLE B-14
2½ Percent Compound Interest Factors for One Dollar

	Single Amount		Uniform Series (Periodic Equal Amounts)				
	Compound Amount	Present Worth	Compound Amount	Sinking Fund	Present Worth	Capital Recovery	
n	Given P to find F $(1+i)^n$	Given F to find P $\dfrac{1}{(1+i)^n}$	Given A to find F $\dfrac{(1+i)^n-1}{i}$	Given F to find A $\dfrac{i}{(1+i)^n-1}$	Given A to find P $\dfrac{(1+i)^n-1}{i(1+i)^n}$	Given P to find A $\dfrac{i(1+i)^n}{(1+i)^n-1}$	n
	CA	PW	SCA	SF	SPW	CR	
1	1.0250 0000	0.975610	1.000000	1.000000	0.975610	1.025000	1
2	1.0506 2500	0.951814	2.025000	0.493827	1.927424	0.518827	2
3	1.0768 9063	0.928599	3.075625	0.325137	2.856024	0.350137	3
4	1.1038 1289	0.905951	4.152516	0.240818	3.761974	0.265818	4
5	1.1314 0821	0.883854	5.256329	0.190247	4.645828	0.215247	5
6	1.1596 9342	0.862297	6.387737	0.156550	5.508125	0.181550	6
7	1.1886 8575	0.841265	7.547430	0.132495	6.349391	0.157495	7
8	1.2184 0290	0.820747	8.736116	0.114467	7.170137	0.139467	8
9	1.2488 6297	0.800728	9.954519	0.100457	7.970866	0.125457	9
10	1.2800 8454	0.781198	11.203382	0.089259	8.752064	0.114259	10
11	1.3120 8666	0.762145	12.483466	0.080106	9.514209	0.105106	11
12	1.3448 8882	0.743556	13.795553	0.072487	10.257765	0.097487	12
13	1.3785 1104	0.725420	15.140442	0.066048	10.983185	0.091048	13
14	1.4129 7382	0.707727	16.518953	0.060537	11.690912	0.085537	14
15	1.4482 9817	0.690466	17.931927	0.055766	12.381378	0.080766	15
16	1.4845 0562	0.673625	19.380225	0.051599	13.055003	0.076599	16
17	1.5216 1826	0.657195	20.864730	0.047928	13.712198	0.072928	17
18	1.5596 5872	0.641166	22.386349	0.044670	14.353364	0.069670	18
19	1.5986 5019	0.625528	23.946007	0.041761	14.978891	0.066761	19
20	1.6386 1644	0.610271	25.544658	0.039147	15.589162	0.064147	20
21	1.6795 8185	0.595386	27.183274	0.036787	16.184549	0.061787	21
22	1.7215 7140	0.580865	28.862856	0.034647	16.765413	0.059647	22
23	1.7646 1068	0.566697	30.584427	0.032696	17.332110	0.057696	23
24	1.8087 2595	0.552875	32.349038	0.030913	17.884986	0.055913	24
25	1.8539 4410	0.539391	34.157764	0.029276	18.424376	0.054276	25
26	1.9002 9270	0.526235	36.011708	0.027769	18.950611	0.052769	26
27	1.9478 0002	0.513400	37.912001	0.026377	19.464011	0.051377	27
28	1.9964 9502	0.500878	39.859801	0.025088	19.964889	0.050088	28
29	2.0464 0739	0.488661	41.856296	0.023891	20.453550	0.048891	29
30	2.0975 6758	0.476743	43.902703	0.022778	20.930293	0.047778	30
31	2.1500 0677	0.465115	46.000271	0.021739	21.395407	0.046739	31
32	2.2037 5694	0.453771	48.150278	0.020768	21.849178	0.045768	32
33	2.2588 5086	0.442703	50.354034	0.019859	22.291881	0.044859	33
34	2.3153 2213	0.431905	52.612885	0.019007	22.723786	0.044007	34
35	2.3732 0519	0.421371	54.928207	0.018206	23.145157	0.043206	35
36	2.4325 3532	0.411094	57.301413	0.017452	23.556251	0.042452	36
37	2.4933 4870	0.401067	59.733948	0.016741	23.957318	0.041741	37
38	2.5556 8242	0.391285	62.227297	0.016070	24.348603	0.041070	38
39	2.6195 7448	0.381741	64.782979	0.015436	24.730344	0.040436	39
40	2.6850 6384	0.372431	67.402554	0.014836	25.102775	0.039836	40
45	3.0379 0328	0.329174	81.516131	0.012268	26.833024	0.037268	45
50	3.4371 0872	0.290942	97.484349	0.010258	28.362312	0.035258	50
55	3.8887 7303	0.257151	115.550921	0.008654	29.713979	0.033654	55
60	4.3997 8975	0.227284	135.991590	0.007353	30.908656	0.032353	60
65	4.9779 5826	0.200886	159.118330	0.006285	31.964577	0.031285	65
70	5.6321 0286	0.177554	185.284114	0.005397	32.897857	0.030397	70
75	6.3722 0743	0.156931	214.888297	0.004654	33.722740	0.029654	75
80	7.2095 6782	0.138705	248.382713	0.004026	34.451817	0.029026	80
90	9.2288 5633	0.108356	329.154253	0.003038	35.665768	0.028038	90
100	11.8137 1635	0.084647	432.548654	0.002312	36.614105	0.027312	100

Standard Compound Interest Factors

TABLE B-15
3 Percent Compound Interest Factors for One Dollar

	Single Amount		Uniform Series (Periodic Equal Amounts)				
	Compound Amount	Present Worth	Compound Amount	Sinking Fund	Present Worth	Capital Recovery	
n	Given P to find F $(1+i)^n$	Given F to find P $\dfrac{1}{(1+i)^n}$	Given A to find F $\dfrac{(1+i)^n-1}{i}$	Given F to find A $\dfrac{i}{(1+i)^n-1}$	Given A to find P $\dfrac{(1+i)^n-1}{i(1+i)^n}$	Given P to find A $\dfrac{i(1+i)^n}{(1+i)^n-1}$	n
	CA	PW	SCA	SF	SPW	CR	
1	1.0300 0000	0.970874	1.000000	1.000000	0.970874	1.030000	1
2	1.0609 0000	0.942596	2.030000	0.492611	1.913470	0.522611	2
3	1.0927 2700	0.915142	3.090900	0.323530	2.828611	0.353530	3
4	1.1255 0881	0.888487	4.183627	0.239027	3.717098	0.269027	4
5	1.1592 7407	0.862609	5.309136	0.188355	4.579707	0.218355	5
6	1.1940 5230	0.837484	6.468410	0.154598	5.417191	0.184598	6
7	1.2298 7387	0.813092	7.662462	0.130506	6.230283	0.160506	7
8	1.2667 7008	0.789409	8.892336	0.112456	7.019692	0.142456	8
9	1.3047 7318	0.766417	10.159106	0.098434	7.786109	0.128434	9
10	1.3439 1638	0.744094	11.463879	0.087231	8.530203	0.117231	10
11	1.3842 3387	0.722421	12.807796	0.078077	9.252624	0.108077	11
12	1.4257 6089	0.701380	14.192030	0.070462	9.954004	0.100462	12
13	1.4685 3371	0.680951	15.617790	0.064030	10.634955	0.094030	13
14	1.5125 8972	0.661118	17.086324	0.058526	11.296073	0.088526	14
15	1.5579 6742	0.641862	18.598914	0.053767	11.937935	0.083767	15
16	1.6047 0644	0.623167	20.156881	0.049611	12.561102	0.079611	16
17	1.6528 4763	0.605016	21.761588	0.045953	13.166118	0.075953	17
18	1.7024 3306	0.587395	23.414435	0.042709	13.753513	0.072709	18
19	1.7535 0605	0.570286	25.116868	0.039814	14.323799	0.069814	19
20	1.8061 1123	0.553676	26.870374	0.037216	14.877475	0.067216	20
21	1.8602 9457	0.537549	28.676486	0.034872	15.415024	0.064872	21
22	1.9161 0341	0.521893	30.536780	0.032747	15.936917	0.062747	22
23	1.9735 8651	0.506692	32.452884	0.030814	16.443608	0.060814	23
24	2.0327 9411	0.491934	34.426470	0.029047	16.935542	0.059047	24
25	2.0937 7793	0.477606	36.459264	0.027428	17.413148	0.057428	25
26	2.1565 9127	0.463695	38.553042	0.025938	17.876842	0.055938	26
27	2.2212 8901	0.450189	40.709634	0.024564	18.327031	0.054564	27
28	2.2879 2768	0.437077	42.930923	0.023293	18.764108	0.053293	28
29	2.3565 6551	0.424346	45.218850	0.022115	19.188455	0.052115	29
30	2.4272 6247	0.411987	47.575416	0.021019	19.600441	0.051019	30
31	2.5000 8035	0.399987	50.002678	0.019999	20.000428	0.049999	31
32	2.5750 8276	0.388337	52.502759	0.019047	20.388766	0.049047	32
33	2.6523 3524	0.377026	55.077841	0.018156	20.765792	0.048156	33
34	2.7319 0530	0.366045	57.730177	0.017322	21.131837	0.047322	34
35	2.8138 6245	0.355383	60.462082	0.016539	21.487220	0.046539	35
36	2.8982 7833	0.345032	63.275944	0.015804	21.832252	0.045804	36
37	2.9852 2668	0.334983	66.174223	0.015112	22.167235	0.045112	37
38	3.0747 8348	0.325226	69.159449	0.014459	22.492462	0.044459	38
39	3.1670 2698	0.315754	72.234233	0.013844	22.808215	0.043844	39
40	3.2620 3779	0.306557	75.401260	0.013262	23.114772	0.043262	40
45	3.7815 9584	0.264439	92.719861	0.010785	24.518713	0.040785	45
50	4.3839 0602	0.228107	112.796867	0.008865	25.729764	0.038865	50
55	5.0821 4859	0.196767	136.071620	0.007349	26.774428	0.037349	55
60	5.8916 0310	0.169733	163.053437	0.006133	27.675564	0.036133	60
65	6.8299 8273	0.146413	194.332758	0.005146	28.452892	0.035146	65
70	7.9178 2191	0.126297	230.594064	0.004337	29.123421	0.034337	70
75	9.1789 2567	0.108945	272.630856	0.003668	29.701826	0.033668	75
80	10.6408 9056	0.093977	321.363019	0.003112	30.200763	0.033112	80
90	14.3004 6711	0.069928	443.348904	0.002256	31.002407	0.032256	90
100	19.2186 3198	0.052033	607.287733	0.001647	31.598905	0.031647	100

TABLE B-16
3½ Percent Compound Interest Factors for One Dollar

	Single Amount		Uniform Series (Periodic Equal Amounts)				
n	Compound Amount	Present Worth	Compound Amount	Sinking Fund	Present Worth	Capital Recovery	n
	Given P to find F $(1 + i)^n$	Given F to find P $\dfrac{1}{(1 + i)^n}$	Given A to find F $\dfrac{(1 + i)^n - 1}{i}$	Given F to find A $\dfrac{i}{(1 + i)^n - 1}$	Given A to find P $\dfrac{(1 + i)^n - 1}{i(1 + i)^n}$	Given P to find A $\dfrac{i(1 + i)^n}{(1 + i)^n - 1}$	
	CA	PW	SCA	SF	SPW	CR	
1	1.0350 0000	0.966184	1.000000	1.000000	0.966184	1.035000	1
2	1.0712 2500	0.933511	2.035000	0.491400	1.899694	0.526400	2
3	1.1087 1788	0.901943	3.106225	0.321934	2.801637	0.356934	3
4	1.1475 2300	0.871442	4.214943	0.237251	3.673079	0.272251	4
5	1.1876 8631	0.841973	5.362466	0.186481	4.515052	0.221481	5
6	1.2292 5533	0.813501	6.550152	0.152668	5.328553	0.187668	6
7	1.2722 7926	0.785991	7.779408	0.128544	6.114544	0.163544	7
8	1.3168 0904	0.759412	9.051687	0.110477	6.873956	0.145477	8
9	1.3628 9735	0.733731	10.368496	0.096446	7.607687	0.131446	9
10	1.4105 9876	0.708919	11.731393	0.085241	8.316605	0.120241	10
11	1.4599 6972	0.684946	13.141992	0.076092	9.001551	0.111092	11
12	1.5110 6866	0.661783	14.601962	0.068484	9.663334	0.103484	12
13	1.5639 5606	0.639404	16.113030	0.062062	10.302738	0.097062	13
14	1.6186 9452	0.617782	17.676986	0.056571	10.920520	0.091571	14
15	1.6753 4883	0.596891	19.295681	0.051825	11.517411	0.086825	15
16	1.7339 8604	0.576706	20.971030	0.047685	12.094117	0.082685	16
17	1.7946 7555	0.557204	22.705016	0.044043	12.651321	0.079043	17
18	1.8574 8920	0.538361	24.499691	0.040817	13.189682	0.075817	18
19	1.9225 0132	0.520156	26.357180	0.037940	13.709837	0.072940	19
20	1.9897 8886	0.502566	28.279682	0.035361	14.212403	0.070361	20
21	2.0594 3147	0.485571	30.269471	0.033037	14.697974	0.068037	21
22	2.1315 1158	0.469151	32.328902	0.030932	15.167125	0.065932	22
23	2.2061 1448	0.453286	34.460414	0.029019	15.620410	0.064019	23
24	2.2833 2849	0.437957	36.666528	0.027273	16.058368	0.062273	24
25	2.3632 4498	0.423147	38.949857	0.025674	16.481515	0.060674	25
26	2.4459 5856	0.408838	41.313102	0.024205	16.890352	0.059205	26
27	2.5315 6711	0.395012	43.759060	0.022852	17.285365	0.057852	27
28	2.6201 7196	0.381654	46.290627	0.021603	17.667019	0.056603	28
29	2.7118 7798	0.368748	48.910799	0.020445	18.035767	0.055445	29
30	2.8067 9370	0.356278	51.622677	0.019371	18.392045	0.054371	30
31	2.9050 3148	0.344230	54.429471	0.018372	18.736276	0.053372	31
32	3.0067 0759	0.332590	57.334502	0.017442	19.068865	0.052442	32
33	3.1119 4235	0.321343	60.341210	0.016572	19.390208	0.051572	33
34	3.2208 6033	0.310476	63.453152	0.015760	19.700684	0.050760	34
35	3.3335 9045	0.299977	66.674013	0.014998	20.000661	0.049998	35
36	3.4502 6611	0.289833	70.007603	0.014284	20.290494	0.049284	36
37	3.5710 2543	0.280032	73.457869	0.013613	20.570525	0.048613	37
38	3.6960 1132	0.270562	77.028895	0.012982	20.841087	0.047982	38
39	3.8253 7171	0.261413	80.724906	0.012388	21.102500	0.047388	39
40	3.9592 5972	0.252572	84.550278	0.011827	21.355072	0.046827	40
45	4.7023 5855	0.212659	105.781673	0.009453	22.495450	0.044453	45
50	5.5849 2686	0.179053	130.997910	0.007634	23.455618	0.042634	50
55	6.6331 4114	0.150758	160.946890	0.006213	24.264053	0.041213	55
60	7.8780 9090	0.126924	196.516883	0.005089	24.944734	0.040089	60
65	9.3567 0068	0.106875	238.762876	0.004188	25.517849	0.039188	65
70	11.1128 2526	0.089986	288.937865	0.003461	26.000397	0.038461	70
75	13.1985 5038	0.075766	348.530011	0.002869	26.406689	0.037869	75
80	15.6757 3754	0.063793	419.306787	0.002385	26.748776	0.037385	80
90	22.1121 7595	0.045224	603.205027	0.001658	27.279316	0.036658	90
100	31.1914 0798	0.032060	862.611657	0.001159	27.655425	0.036159	100

Standard Compound Interest Factors

TABLE B-17
4 Percent Compound Interest Factors for One Dollar

	Single Amount		Uniform Series (Periodic Equal Amounts)				
	Compound Amount	Present Worth	Compound Amount	Sinking Fund	Present Worth	Capital Recovery	
n	Given P to find F $(1+i)^n$	Given F to find P $\dfrac{1}{(1+i)^n}$	Given A to find F $\dfrac{(1+i)^n-1}{i}$	Given F to find A $\dfrac{i}{(1+i)^n-1}$	Given A to find P $\dfrac{(1+i)^n-1}{i(1+i)^n}$	Given P to find A $\dfrac{i(1+i)^n}{(1+i)^n-1}$	n
	CA	PW	SCA	SF	SPW	CR	
1	1.0400 0000	0.961538	1.000000	1.000000	0.961538	1.040000	1
2	1.0816 0000	0.924556	2.040000	0.490196	1.886095	0.530196	2
3	1.1248 6400	0.888996	3.121600	0.320349	2.775091	0.360349	3
4	1.1698 5856	0.854804	4.246464	0.235490	3.629895	0.275490	4
5	1.2166 5290	0.821927	5.416323	0.184627	4.451822	0.224627	5
6	1.2653 1902	0.790315	6.632975	0.150762	5.242137	0.190762	6
7	1.3159 3178	0.759918	7.898294	0.126610	6.002055	0.166610	7
8	1.3685 6905	0.730690	9.214226	0.108528	6.732745	0.148528	8
9	1.4233 1181	0.702587	10.582795	0.094493	7.435332	0.134493	9
10	1.4802 4428	0.675564	12.006107	0.083291	8.110896	0.123291	10
11	1.5394 5406	0.649581	13.486351	0.074149	8.760477	0.114149	11
12	1.6010 3222	0.624597	15.025805	0.066552	9.385074	0.106552	12
13	1.6650 7351	0.600574	16.626838	0.060144	9.985648	0.100144	13
14	1.7316 7645	0.577475	18.291911	0.054669	10.563123	0.094669	14
15	1.8009 4351	0.555265	20.023588	0.049941	11.118387	0.089941	15
16	1.8729 8125	0.533908	21.824531	0.045820	11.652296	0.085820	16
17	1.9479 0050	0.513373	23.697512	0.042199	12.165669	0.082199	17
18	2.0258 1652	0.493628	25.645413	0.038993	12.659297	0.078993	18
19	2.1068 4918	0.474642	27.671229	0.036139	13.133939	0.076139	19
20	2.1911 2314	0.456387	29.778079	0.033582	13.590326	0.073582	20
21	2.2787 6807	0.438834	31.969202	0.031280	14.029160	0.071280	21
22	2.3699 1879	0.421955	34.247970	0.029199	14.451115	0.069199	22
23	2.4647 1554	0.405726	36.617889	0.027309	14.856842	0.067309	23
24	2.5633 0416	0.390121	39.082604	0.025587	15.246963	0.065587	24
25	2.6658 3633	0.375117	41.645908	0.024012	15.622080	0.064012	25
26	2.7724 6978	0.360689	44.311745	0.022567	15.982769	0.062567	26
27	2.8833 6858	0.346817	47.084214	0.021239	16.329586	0.061239	27
28	2.9987 0332	0.333477	49.967533	0.020013	16.663063	0.060013	28
29	3.1186 5145	0.320651	52.966286	0.018880	16.983715	0.058880	29
30	3.2433 9751	0.308319	56.084938	0.017830	17.292033	0.057830	30
31	3.3731 3341	0.296460	59.328335	0.016855	17.588494	0.056855	31
32	3.5080 5875	0.285058	62.701469	0.015949	17.873551	0.055949	32
33	3.6483 8110	0.274094	66.209527	0.015104	18.147646	0.055104	33
34	3.7943 1634	0.263552	69.857909	0.014315	18.411198	0.054315	34
35	3.9460 8899	0.253415	73.652225	0.013577	18.664613	0.053577	35
36	4.1039 3255	0.243669	77.598314	0.012887	18.908282	0.052887	36
37	4.2680 8986	0.234297	81.702246	0.012240	19.142579	0.052240	37
38	4.4388 1345	0.225285	85.970336	0.011632	19.367864	0.051632	38
39	4.6163 6599	0.216621	90.409150	0.011061	19.584485	0.051061	39
40	4.8010 2063	0.208289	95.025516	0.010523	19.792774	0.050523	40
45	5.8411 7568	0.171198	121.029392	0.008262	20.720040	0.048262	45
50	7.1066 8335	0.140713	152.667084	0.006550	21.482185	0.046550	50
55	8.6463 6692	0.115656	191.159173	0.005231	22.108612	0.045231	55
60	10.5196 2741	0.095060	237.990685	0.004202	22.623490	0.044202	60
65	12.7987 3522	0.078133	294.968380	0.003390	23.046682	0.043390	65
70	15.5716 1835	0.064219	364.290459	0.002745	23.394515	0.042745	70
75	18.9452 5466	0.052784	448.631367	0.002229	23.680408	0.042229	75
80	23.0497 9907	0.043384	551.244977	0.001814	23.915392	0.041814	80
90	34.1193 3334	0.029309	827.983334	0.001208	24.267278	0.041208	90
100	50.5049 4818	0.019800	1,237.623705	0.000808	24.504999	0.040808	100

TABLE B-18
4½ Percent Compound Interest Factors for One Dollar

	Single Amount		Uniform Series (Periodic Equal Amounts)				
n	Compound Amount	Present Worth	Compound Amount	Sinking Fund	Present Worth	Capital Recovery	n
	Given P to find F $(1+i)^n$	Given F to find P $\dfrac{1}{(1+i)^n}$	Given A to find F $\dfrac{(1+i)^n - 1}{i}$	Given F to find A $\dfrac{i}{(1+i)^n - 1}$	Given A to find P $\dfrac{(1+i)^n - 1}{i(1+i)^n}$	Given P to find A $\dfrac{i(1+i)^n}{(1+i)^n - 1}$	
	CA	PW	SCA	SF	SPW	CR	
1	1.0450 0000	0.956938	1.000000	1.000000	0.956938	1.045000	1
2	1.0920 2500	0.915730	2.045000	0.488998	1.872668	0.533998	2
3	1.1411 6613	0.876297	3.137025	0.318773	2.748964	0.363773	3
4	1.1925 1860	0.838561	4.278191	0.233744	3.587526	0.278744	4
5	1.2461 8194	0.802451	5.470710	0.182792	4.389977	0.227792	5
6	1.3022 6012	0.767896	6.716892	0.148878	5.157872	0.193878	6
7	1.3608 6183	0.734828	8.019152	0.124701	5.892701	0.169701	7
8	1.4221 0061	0.703185	9.380014	0.106610	6.595886	0.151610	8
9	1.4860 9514	0.672904	10.802114	0.092574	7.268790	0.137574	9
10	1.5529 6942	0.643928	12.288209	0.081379	7.912718	0.126379	10
11	1.6228 5305	0.616199	13.841179	0.072248	8.528917	0.117248	11
12	1.6958 8143	0.589664	15.464032	0.064666	9.118581	0.109666	12
13	1.7721 9610	0.564272	17.159913	0.058275	9.682852	0.103275	13
14	1.8519 4492	0.539973	18.932109	0.052820	10.222825	0.097820	14
15	1.9352 8244	0.516720	20.784054	0.048114	10.739546	0.093114	15
16	2.0223 7015	0.494469	22.719337	0.044015	11.234015	0.089015	16
17	2.1133 7681	0.473176	24.741707	0.040418	11.707191	0.085418	17
18	2.2084 7877	0.452800	26.855084	0.037237	12.159992	0.082237	18
19	2.3078 6031	0.433302	29.063562	0.034407	12.593294	0.079407	19
20	2.4117 1402	0.414643	31.371423	0.031876	13.007936	0.076876	20
21	2.5202 4116	0.396787	33.783137	0.029601	13.404724	0.074601	21
22	2.6336 5201	0.379701	36.303378	0.027546	13.784425	0.072546	22
23	2.7521 6635	0.363350	38.937030	0.025682	14.147775	0.070682	23
24	2.8760 1383	0.347703	41.689196	0.023987	14.495478	0.068987	24
25	3.0054 3446	0.332731	44.565210	0.022439	14.828209	0.067439	25
26	3.1406 7901	0.318402	47.570645	0.021021	15.146611	0.066021	26
27	3.2820 0956	0.304691	50.711324	0.019719	15.451303	0.064719	27
28	3.4296 9999	0.291571	53.993333	0.018521	15.742874	0.063521	28
29	3.5840 3649	0.279015	57.423033	0.017415	16.021889	0.062415	29
30	3.7453 1813	0.267000	61.007070	0.016392	16.288889	0.061392	30
31	3.9138 5745	0.255502	64.752388	0.015443	16.544391	0.060443	31
32	4.0899 8104	0.244500	68.666245	0.014563	16.788891	0.059563	32
33	4.2740 3018	0.233971	72.756226	0.013745	17.022862	0.058745	33
34	4.4663 6154	0.223896	77.030256	0.012982	17.246758	0.057982	34
35	4.6673 4781	0.214254	81.496618	0.012270	17.461012	0.057270	35
36	4.8773 7846	0.205028	86.163966	0.011606	17.666041	0.056606	36
37	5.0968 6049	0.196199	91.041344	0.010984	17.862240	0.055984	37
38	5.3262 1921	0.187750	96.138205	0.010402	18.049990	0.055402	38
39	5.5658 9908	0.179665	101.464424	0.009856	18.229656	0.054856	39
40	5.8163 6454	0.171929	107.030323	0.009343	18.401584	0.054343	40
45	7.2482 4843	0.137964	138.849965	0.007202	19.156347	0.052202	45
50	9.0326 3627	0.110710	178.503028	0.005602	19.762008	0.050602	50
55	11.2563 0817	0.088839	227.917959	0.004388	20.248021	0.049388	55
60	14.0274 0793	0.071289	289.497954	0.003454	20.638022	0.048454	60
65	17.4807 0239	0.057206	366.237831	0.002730	20.950979	0.047730	65
70	21.7841 3558	0.045905	461.869680	0.002165	21.202112	0.047165	70
75	27.1469 9629	0.036836	581.044362	0.001721	21.403634	0.046721	75
80	33.8300 9643	0.029559	729.557699	0.001371	21.565345	0.046371	80
90	52.5371 0530	0.019034	1,145.269007	0.000873	21.799241	0.045873	90
100	81.5885 1803	0.012257	1,790.855956	0.000558	21.949853	0.045558	100

Standard Compound Interest Factors

TABLE B-19
5 Percent Compound Interest Factors for One Dollar

	Single Amount		Uniform Series (Periodic Equal Amounts)				
	Compound Amount	Present Worth	Compound Amount	Sinking Fund	Present Worth	Capital Recovery	
n	Given P to find F $(1+i)^n$	Given F to find P $\dfrac{1}{(1+i)^n}$	Given A to find F $\dfrac{(1+i)^n-1}{i}$	Given F to find A $\dfrac{i}{(1+i)^n-1}$	Given A to find P $\dfrac{(1+i)^n-1}{i(1+i)^n}$	Given P to find A $\dfrac{i(1+i)^n}{(1+i)^n-1}$	n
	CA	PW	SCA	SF	SPW	CR	
1	1.0500 0000	0.952381	1.000000	1.000000	0.952381	1.050000	1
2	1.1025 0000	0.907029	2.050000	0.487805	1.859410	0.537805	2
3	1.1576 2500	0.863838	3.152500	0.317209	2.723248	0.367209	3
4	1.2155 0625	0.822702	4.310125	0.232012	3.545951	0.282012	4
5	1.2762 8156	0.783526	5.525631	0.180975	4.329477	0.230975	5
6	1.3400 9564	0.746215	6.801913	0.147017	5.075692	0.197017	6
7	1.4071 0042	0.710681	8.142008	0.122820	5.786373	0.172820	7
8	1.4774 5544	0.676839	9.549109	0.104722	6.463213	0.154722	8
9	1.5513 2822	0.644609	11.026564	0.090690	7.107822	0.140690	9
10	1.6288 9463	0.613913	12.577893	0.079505	7.721735	0.129505	10
11	1.7103 3936	0.584679	14.206787	0.070389	8.306414	0.120389	11
12	1.7958 5633	0.556837	15.917127	0.062825	8.863252	0.112825	12
13	1.8856 4914	0.530321	17.712983	0.056456	9.393573	0.106456	13
14	1.9799 3160	0.505068	19.598632	0.051024	9.898641	0.101024	14
15	2.0789 2818	0.481017	21.578564	0.046342	10.379658	0.096342	15
16	2.1828 7459	0.458112	23.657492	0.042270	10.837770	0.092270	16
17	2.2920 1832	0.436297	25.840366	0.038699	11.274066	0.088699	17
18	2.4066 1923	0.415521	28.132385	0.035546	11.689587	0.085546	18
19	2.5269 5020	0.395734	30.539004	0.032745	12.085321	0.082745	19
20	2.6532 9771	0.376889	33.065954	0.030243	12.462210	0.080243	20
21	2.7859 6259	0.358942	35.719252	0.027996	12.821153	0.077996	21
22	2.9252 6072	0.341850	38.505214	0.025971	13.163003	0.075971	22
23	3.0715 2376	0.325571	41.430475	0.024137	13.488574	0.074137	23
24	3.2250 9994	0.310068	44.501999	0.022471	13.798642	0.072471	24
25	3.3863 5494	0.295303	47.727099	0.020952	14.093945	0.070952	25
26	3.5556 7269	0.281241	51.113454	0.019564	14.375185	0.069564	26
27	3.7334 5632	0.267848	54.669126	0.018292	14.643034	0.068292	27
28	3.9201 2914	0.255094	58.402583	0.017123	14.898127	0.067123	28
29	4.1161 3560	0.242946	62.322712	0.016046	15.141074	0.066046	29
30	4.3219 4238	0.231377	66.438848	0.015051	15.372451	0.065051	30
31	4.5380 3949	0.220359	70.760790	0.014132	15.592811	0.064132	31
32	4.7649 4147	0.209866	75.298829	0.013280	15.802677	0.063280	32
33	5.0031 8854	0.199873	80.063771	0.012490	16.002549	0.062490	33
34	5.2533 4797	0.190355	85.066959	0.011755	16.192904	0.061755	34
35	5.5160 1537	0.181290	90.320307	0.011072	16.374194	0.061072	35
36	5.7918 1614	0.172657	95.836323	0.010434	16.546852	0.060434	36
37	6.0814 0694	0.164436	101.628139	0.009840	16.711287	0.059840	37
38	6.3854 7729	0.156605	107.709546	0.009284	16.867893	0.059284	38
39	6.7047 5115	0.149148	114.095023	0.008765	17.017041	0.058765	39
40	7.0399 8871	0.142046	120.799774	0.008278	17.159086	0.058278	40
45	8.9850 0779	0.111297	159.700156	0.006262	17.774070	0.056262	45
50	11.4673 9979	0.087204	209.347996	0.004777	18.255925	0.054777	50
55	14.6356 3092	0.068326	272.712618	0.003667	18.633472	0.053667	55
60	18.6791 8589	0.053536	353.583718	0.002828	18.929290	0.052828	60
65	23.8399 0056	0.041946	456.798011	0.002189	19.161070	0.052189	65
70	30.4264 2554	0.032866	588.528511	0.001699	19.342677	0.051699	70
75	38.8326 8592	0.025752	756.653718	0.001322	19.484970	0.051322	75
80	49.5614 4107	0.020177	971.228821	0.001030	19.596460	0.051030	80
90	80.7303 6505	0.012387	1,594.607301	0.000627	19.752262	0.050627	90
100	131.5012 5785	0.007604	2,610.025157	0.000383	19.847910	0.050383	100

Standard Compound Interest Factors

TABLE B-20
5½ Percent Compound Interest Factors for One Dollar

	Single Amount		Uniform Series (Periodic Equal Amounts)				
	Compound Amount	Present Worth	Compound Amount	Sinking Fund	Present Worth	Capital Recovery	
n	Given P to find F $(1+i)^n$	Given F to find P $\dfrac{1}{(1+i)^n}$	Given A to find F $\dfrac{(1+i)^n - 1}{i}$	Given F to find A $\dfrac{i}{(1+i)^n - 1}$	Given A to find P $\dfrac{(1+i)^n - 1}{i(1+i)^n}$	Given P to find A $\dfrac{i(1+i)^n}{(1+i)^n - 1}$	n
	CA	PW	SCA	SF	SPW	CR	
1	1.0550 0000	0.947867	1.000000	1.000000	0.947867	1.055000	1
2	1.1130 2500	0.898452	2.055000	0.486618	1.846320	0.541618	2
3	1.1742 4138	0.851614	3.168025	0.315654	2.697933	0.370654	3
4	1.2388 2465	0.807217	4.342266	0.230294	3.505150	0.285294	4
5	1.3069 6001	0.765134	5.581091	0.179176	4.270284	0.234176	5
6	1.3788 4281	0.725246	6.888051	0.145179	4.995530	0.200179	6
7	1.4546 7916	0.687437	8.266894	0.120964	5.682967	0.175964	7
8	1.5346 8651	0.651599	9.721573	0.102864	6.334566	0.157864	8
9	1.6190 9427	0.617629	11.256260	0.088839	6.952195	0.143839	9
10	1.7081 4446	0.585431	12.875354	0.077668	7.537626	0.132668	10
11	1.8020 9240	0.554911	14.583498	0.068571	8.092536	0.123571	11
12	1.9012 0749	0.525982	16.385591	0.061029	8.618518	0.116029	12
13	2.0057 7390	0.498561	18.286798	0.054684	9.117079	0.109684	13
14	2.1160 9146	0.472569	20.292572	0.049279	9.589648	0.104279	14
15	2.2324 7649	0.447933	22.408663	0.044626	10.037581	0.099626	15
16	2.3552 6270	0.424581	24.641140	0.040583	10.462162	0.095583	16
17	2.4848 0215	0.402447	26.996403	0.037042	10.864609	0.092042	17
18	2.6214 6627	0.381466	29.481205	0.033920	11.246074	0.088920	18
19	2.7656 4691	0.361579	32.102671	0.031150	11.607654	0.086150	19
20	2.9177 5749	0.342729	34.868318	0.028679	11.950382	0.083679	20
21	3.0782 3415	0.324862	37.786076	0.026465	12.275244	0.081465	21
22	3.2475 3703	0.307926	40.864310	0.024471	12.583170	0.079471	22
23	3.4261 5157	0.291873	44.111847	0.022670	12.875042	0.077670	23
24	3.6145 8990	0.276657	47.537998	0.021036	13.151699	0.076036	24
25	3.8133 9235	0.262234	51.152588	0.019549	13.413933	0.074549	25
26	4.0231 2893	0.248563	54.965981	0.018193	13.662495	0.073193	26
27	4.2444 0102	0.235605	58.989109	0.016952	13.898100	0.071952	27
28	4.4778 4307	0.223322	63.233510	0.015814	14.121422	0.070814	28
29	4.7241 2444	0.211679	67.711354	0.014769	14.333101	0.069769	29
30	4.9839 5129	0.200644	72.435478	0.013805	14.533745	0.068805	30
31	5.2580 6861	0.190184	77.419429	0.012917	14.723929	0.067917	31
32	5.5472 6238	0.180269	82.677498	0.012095	14.904198	0.067095	32
33	5.8523 6181	0.170871	88.224760	0.011335	15.075069	0.066335	33
34	6.1742 4171	0.161963	94.077122	0.010630	15.237033	0.065630	34
35	6.5138 2501	0.153520	100.251364	0.009975	15.390552	0.064975	35
36	6.8720 8538	0.145516	106.765189	0.009366	15.536068	0.064366	36
37	7.2500 5008	0.137930	113.637274	0.008800	15.673999	0.063800	37
38	7.6488 0283	0.130739	120.887324	0.008272	15.804738	0.063272	38
39	8.0694 8699	0.123924	128.536127	0.007780	15.928662	0.062780	39
40	8.5133 0877	0.117463	136.605614	0.007320	16.046125	0.062320	40
45	11.1265 5409	0.089875	184.119165	0.005431	16.547726	0.060431	45
50	14.5419 6120	0.068767	246.217476	0.004061	16.931518	0.059061	50
55	19.0057 6171	0.052616	327.377486	0.003055	17.225170	0.058055	55
60	24.8397 7045	0.040258	433.450372	0.002307	17.449854	0.057307	60
65	32.4645 8654	0.030803	572.083392	0.001748	17.621767	0.056748	65
70	42.4299 1623	0.023568	753.271204	0.001328	17.753304	0.056328	70
75	55.4542 0359	0.018033	990.076429	0.001010	17.853947	0.056010	75
80	72.4764 2628	0.013798	1,299.571387	0.000769	17.930953	0.055769	80
90	123.8002 0591	0.008078	2,232.731017	0.000448	18.034954	0.055448	90
100	211.4686 3567	0.004729	3,826.702467	0.000261	18.095839	0.055261	100

Standard Compound Interest Factors

TABLE B-21
6 Percent Compound Interest Factors for One Dollar

	Single Amount		Uniform Series (Periodic Equal Amounts)				
	Compound Amount	Present Worth	Compound Amount	Sinking Fund	Present Worth	Capital Recovery	
n	Given P to find F $(1+i)^n$	Given F to find P $\dfrac{1}{(1+i)^n}$	Given A to find F $\dfrac{(1+i)^n - 1}{i}$	Given F to find A $\dfrac{i}{(1+i)^n - 1}$	Given A to find P $\dfrac{(1+i)^n - 1}{i(1+i)^n}$	Given P to find A $\dfrac{i(1+i)^n}{(1+i)^n - 1}$	n
	CA	PW	SCA	SF	SPW	CR	
1	1.0600 0000	0.943396	1.000000	1.000000	0.943396	1.060000	1
2	1.1236 0000	0.889996	2.060000	0.485437	1.833393	0.545437	2
3	1.1910 1600	0.839619	3.183600	0.314110	2.673012	0.374110	3
4	1.2624 7696	0.792094	4.374616	0.228591	3.465106	0.288591	4
5	1.3382 2558	0.747258	5.637093	0.177396	4.212364	0.237396	5
6	1.4185 1911	0.704961	6.975319	0.143363	4.917324	0.203363	6
7	1.5036 3026	0.665057	8.393838	0.119135	5.582381	0.179135	7
8	1.5938 4807	0.627412	9.897468	0.101036	6.209794	0.161036	8
9	1.6894 7896	0.591898	11.491316	0.087022	6.801692	0.147022	9
10	1.7908 4770	0.558395	13.180795	0.075868	7.360087	0.135868	10
11	1.8982 9856	0.526788	14.971643	0.066793	7.886875	0.126793	11
12	2.0121 9647	0.496969	16.869941	0.059277	8.383844	0.119277	12
13	2.1329 2826	0.468839	18.882138	0.052960	8.852683	0.112960	13
14	2.2609 0396	0.442301	21.015066	0.047585	9.294984	0.107585	14
15	2.3965 5819	0.417265	23.275970	0.042963	9.712249	0.102963	15
16	2.5403 5168	0.393646	25.672528	0.038952	10.105895	0.098952	16
17	2.6927 7279	0.371364	28.212880	0.035445	10.477260	0.095445	17
18	2.8543 3915	0.350344	30.905653	0.032357	10.827603	0.092357	18
19	3.0255 9950	0.330513	33.759902	0.029621	11.158116	0.089621	19
20	3.2071 3547	0.311805	36.785591	0.027185	11.469921	0.087185	20
21	3.3995 6360	0.294155	39.992727	0.025005	11.764077	0.085005	21
22	3.6035 3742	0.277505	43.392290	0.023046	12.041582	0.083046	22
23	3.8197 4966	0.261797	46.995828	0.021278	12.303379	0.081278	23
24	4.0489 3464	0.246979	50.815577	0.019679	12.550358	0.079679	24
25	4.2918 7072	0.232999	54.864512	0.018227	12.783356	0.078227	25
26	4.5493 8296	0.219810	59.156383	0.016904	13.003166	0.076904	26
27	4.8223 4594	0.207368	63.705766	0.015697	13.210534	0.075697	27
28	5.1116 8670	0.195610	68.528112	0.014593	13.406164	0.074593	28
29	5.4183 8790	0.184557	73.639798	0.013580	13.590721	0.073580	29
30	5.7434 9117	0.174110	79.058186	0.012649	13.764831	0.072649	30
31	6.0881 0064	0.164255	84.801677	0.011792	13.929086	0.071792	31
32	6.4533 8668	0.154957	90.889778	0.011002	14.084043	0.071002	32
33	6.8405 8988	0.146186	97.343165	0.010273	14.230230	0.070273	33
34	7.2510 2528	0.137912	104.183755	0.009598	14.368141	0.069598	34
35	7.6860 8679	0.130105	111.434780	0.008974	14.498246	0.068974	35
36	8.1472 5200	0.122741	119.120867	0.008395	14.620987	0.068395	36
37	8.6360 8712	0.115793	127.268119	0.007857	14.736780	0.067857	37
38	9.1542 5235	0.109239	135.904206	0.007358	14.846019	0.067358	38
39	9.7035 0749	0.103056	145.058458	0.006894	14.949075	0.066894	39
40	10.2857 1794	0.097222	154.761966	0.006462	15.046297	0.066462	40
45	13.7646 1083	0.072650	212.743514	0.004700	15.455832	0.064700	45
50	18.4201 5427	0.054288	290.335905	0.003444	15.761861	0.063444	50
55	24.6503 2159	0.040567	394.172027	0.002537	15.990543	0.062537	55
60	32.9876 9085	0.030314	533.128181	0.001876	16.161428	0.061876	60
65	44.1449 7165	0.022653	719.082861	0.001391	16.289123	0.061391	65
70	59.0759 3018	0.016927	967.932170	0.001033	16.384544	0.061033	70
75	79.0569 2079	0.012649	1,300.948680	0.000769	16.455848	0.060769	75
80	105.7959 9348	0.009452	1,746.599891	0.000573	16.509131	0.060573	80
90	189.4645 1123	0.005278	3,141.075187	0.000318	16.578699	0.060318	90
100	339.3020 8351	0.002947	5,638.368059	0.000177	16.617546	0.060177	100

TABLE B-22
6½ Percent Compound Interest Factors for One Dollar

	Single Amount		Uniform Series (Periodic Equal Amounts)				
	Compound Amount	Present Worth	Compound Amount	Sinking Fund	Present Worth	Capital Recovery	
n	Given P to find F $(1+i)^n$	Given F to find P $\dfrac{1}{(1+i)^n}$	Given A to find F $\dfrac{(1+i)^n - 1}{i}$	Given F to find A $\dfrac{i}{(1+i)^n - 1}$	Given A to find P $\dfrac{(1+i)^n - 1}{i(1+i)^n}$	Given P to find A $\dfrac{i(1+i)^n}{(1+i)^n - 1}$	n
	CA	PW	SCA	SF	SPW	CR	
1	1.0650 0000	0.938967	1.000000	1.000000	0.938967	1.065000	1
2	1.1342 2500	0.881659	2.065000	0.484262	1.820626	0.549262	2
3	1.2079 4963	0.827849	3.199225	0.312576	2.648476	0.377576	3
4	1.2864 6635	0.777323	4.407175	0.226903	3.425799	0.291903	4
5	1.3700 8666	0.729881	5.693641	0.175635	4.155679	0.240635	5
6	1.4591 4230	0.685334	7.063728	0.141568	4.841014	0.206568	6
7	1.5539 8655	0.643506	8.522870	0.117331	5.484520	0.182331	7
8	1.6549 9567	0.604231	10.076856	0.099237	6.088751	0.164237	8
9	1.7625 7039	0.567353	11.731852	0.085238	6.656104	0.150238	9
10	1.8771 3747	0.532726	13.494423	0.074105	7.188830	0.139105	10
11	1.9991 5140	0.500212	15.371560	0.065055	7.689042	0.130055	11
12	2.1290 9624	0.469683	17.370711	0.057568	8.158725	0.122568	12
13	2.2674 8750	0.441017	19.499808	0.051283	8.599742	0.116283	13
14	2.4148 7418	0.414100	21.767295	0.045940	9.013842	0.110940	14
15	2.5718 4101	0.388827	24.182169	0.041353	9.402669	0.106353	15
16	2.7390 1067	0.365095	26.754010	0.037378	9.767764	0.102378	16
17	2.9170 4637	0.342813	29.493021	0.033906	10.110577	0.098906	17
18	3.1066 5438	0.321890	32.410067	0.030855	10.432466	0.095855	18
19	3.3085 8691	0.302244	35.516722	0.028156	10.734710	0.093156	19
20	3.5236 4506	0.283797	38.825309	0.025756	11.018507	0.090756	20
21	3.7526 8199	0.266476	42.348954	0.023613	11.284983	0.088613	21
22	3.9966 0632	0.250212	46.101636	0.021691	11.535196	0.086691	22
23	4.2563 8573	0.234941	50.098242	0.019961	11.770137	0.084961	23
24	4.5330 5081	0.220602	54.354628	0.018398	11.990739	0.083398	24
25	4.8276 9911	0.207138	58.887679	0.016981	12.197877	0.081981	25
26	5.1414 9955	0.194496	63.715378	0.015695	12.392373	0.080695	26
27	5.4756 9702	0.182625	68.856877	0.014523	12.574998	0.079523	27
28	5.8316 1733	0.171479	74.332574	0.013453	12.746477	0.078453	28
29	6.2106 7245	0.161013	80.164192	0.012474	12.907490	0.077474	29
30	6.6143 6616	0.151186	86.374864	0.011577	13.058676	0.076577	30
31	7.0442 9996	0.141959	92.989230	0.010754	13.200635	0.075754	31
32	7.5021 7946	0.133295	100.033530	0.009997	13.333929	0.074997	32
33	7.9898 2113	0.125159	107.535710	0.009299	13.459088	0.074299	33
34	8.5091 5950	0.117520	115.525531	0.008656	13.576609	0.073656	34
35	9.0622 5487	0.110348	124.034690	0.008062	13.686957	0.073062	35
36	9.6513 0143	0.103613	133.096945	0.007513	13.790570	0.072513	36
37	10.2786 3603	0.097289	142.748247	0.007005	13.887859	0.072005	37
38	10.9467 4737	0.091351	153.026883	0.006535	13.979210	0.071535	38
39	11.6582 8595	0.085776	163.973630	0.006099	14.064986	0.071099	39
40	12.4160 7453	0.080541	175.631916	0.005694	14.145527	0.070694	40
45	17.0110 9813	0.058785	246.324587	0.004060	14.480228	0.069060	45
50	23.3066 7868	0.042906	343.179672	0.002914	14.724521	0.067914	50
55	31.9321 6963	0.031316	475.879533	0.002101	14.902825	0.067101	55
60	43.7498 3974	0.022857	657.689842	0.001520	15.032966	0.066520	60
65	59.9410 7195	0.016683	906.785722	0.001103	15.127953	0.066103	65
70	82.1244 6327	0.012177	1,248.068666	0.000801	15.197282	0.065801	70
75	112.5176 3187	0.008887	1,715.655875	0.000583	15.247885	0.065583	75
80	154.1589 0683	0.006487	2,356.290874	0.000424	15.284818	0.065424	80
90	289.3774 5961	0.003456	4,436.576302	0.000225	15.331451	0.065225	90
100	543.2012 7103	0.001841	8,341.558016	0.000120	15.356293	0.065120	100

Standard Compound Interest Factors

TABLE 3-23
7 Percent Compound Interest Factors for One Dollar

	Single Amount		Uniform Series (Periodic Equal Amounts)				
	Compound Amount	Present Worth	Compound Amount	Sinking Fund	Present Worth	Capital Recovery	
n	Given P to find F $(1 + i)^n$	Given F to find P $\dfrac{1}{(1 + i)^n}$	Given A to find F $\dfrac{(1 + i)^n - 1}{i}$	Given F to find A $\dfrac{i}{(1 + i)^n - 1}$	Given A to find P $\dfrac{(1 + i)^n - 1}{i(1 + i)^n}$	Given P to find A $\dfrac{i(1 + i)^n}{(1 + i)^n - 1}$	n
	CA	PW	SCA	SF	SPW	CR	
1	1.0700 0000	0.934579	1.000000	1.000000	0.934579	1.070000	1
2	1.1449 0000	0.873439	2.070000	0.483092	1.808018	0.553092	2
3	1.2250 4300	0.816298	3.214900	0.311052	2.624316	0.381052	3
4	1.3107 9601	0.762895	4.439943	0.225228	3.387211	0.295228	4
5	1.4025 5173	0.712986	5.750739	0.173891	4.100197	0.243891	5
6	1.5007 3035	0.666342	7.153291	0.139796	4.766540	0.209796	6
7	1.6057 8148	0.622750	8.654021	0.115553	5.389289	0.185553	7
8	1.7181 8618	0.582009	10.259803	0.097468	5.971299	0.167468	8
9	1.8384 5921	0.543934	11.977989	0.083486	6.515232	0.153486	9
10	1.9671 5136	0.508349	13.816448	0.072378	7.023582	0.142378	10
11	2.1048 5195	0.475093	15.783599	0.063357	7.498674	0.133357	11
12	2.2521 9159	0.444012	17.888451	0.055902	7.942686	0.125902	12
13	2.4098 4500	0.414964	20.140643	0.049651	8.357651	0.119651	13
14	2.5785 3415	0.387817	22.550488	0.044345	8.745468	0.114345	14
15	2.7590 3154	0.362446	25.129022	0.039795	9.107914	0.109795	15
16	2.9521 6375	0.338735	27.888054	0.035858	9.446649	0.105858	16
17	3.1588 1521	0.316574	30.840217	0.032425	9.763223	0.102425	17
18	3.3799 3228	0.295864	33.999033	0.029413	10.059087	0.099413	18
19	3.6165 2754	0.276508	37.378965	0.026753	10.335595	0.096753	19
20	3.8696 8446	0.258419	40.995492	0.024393	10.594014	0.094393	20
21	4.1405 6237	0.241513	44.865177	0.022289	10.835527	0.092289	21
22	4.4304 0174	0.225713	49.005739	0.020406	11.061240	0.090406	22
23	4.7405 2986	0.210947	53.436141	0.018714	11.272187	0.088714	23
24	5.0723 6695	0.197147	58.176671	0.017189	11.469334	0.087189	24
25	5.4274 3264	0.184249	63.249038	0.015811	11.653583	0.085811	25
26	5.8073 5292	0.172195	68.676470	0.014561	11.825779	0.084561	26
27	6.2138 6763	0.160930	74.483823	0.013426	11.986709	0.083426	27
28	6.6488 3836	0.150402	80.697691	0.012392	12.137111	0.082392	28
29	7.1142 5705	0.140563	87.346529	0.011449	12.277674	0.081449	29
30	7.6122 5504	0.131367	94.460786	0.010586	12.409041	0.080586	30
31	8.1451 1290	0.122773	102.073041	0.009797	12.531814	0.079797	31
32	8.7152 7080	0.114741	110.218154	0.009073	12.646555	0.079073	32
33	9.3253 3975	0.107235	118.933425	0.008408	12.753790	0.078408	33
34	9.9781 1354	0.100219	128.258765	0.007797	12.854009	0.077797	34
35	10.6765 8148	0.093663	138.236878	0.007234	12.947672	0.077234	35
36	11.4239 4219	0.087535	148.913460	0.006715	13.035208	0.076715	36
37	12.2236 1814	0.081809	160.337402	0.006237	13.117017	0.076237	37
38	13.0792 7141	0.076457	172.561020	0.005795	13.193473	0.075795	38
39	13.9948 2041	0.071455	185.640292	0.005387	13.264928	0.075387	39
40	14.9744 5784	0.066780	199.635112	0.005009	13.331709	0.075009	40
45	21.0024 5176	0.047613	285.749311	0.003500	13.605522	0.073500	45
50	29.4570 2506	0.033945	406.528929	0.002460	13.800746	0.072460	50
55	41.3150 0148	0.024204	575.928593	0.001736	13.939939	0.071736	55
60	57.9464 2683	0.017257	813.520383	0.001229	14.039181	0.071229	60
65	81.2728 6124	0.012304	1,146.755161	0.000872	14.109940	0.070872	65
70	113.9893 9220	0.008773	1,614.134174	0.000620	14.160389	0.070620	70
75	159.8760 1931	0.006255	2,269.657419	0.000441	14.196359	0.070441	75
80	224.2343 8758	0.004460	3,189.062680	0.000314	14.222005	0.070314	80
90	441.1029 7988	0.002267	6,287.185427	0.000159	14.253328	0.070159	90
100	867.7163 2557	0.001152	12,381.661794	0.000081	14.269251	0.070081	100

TABLE B-24
7½ Percent Compound Interest Factors for One Dollar

	Single Amount		Uniform Series (Periodic Equal Amounts)				
	Compound Amount	Present Worth	Compound Amount	Sinking Fund	Present Worth	Capital Recovery	
n	Given P to find F $(1+i)^n$	Given F to find P $\dfrac{1}{(1+i)^n}$	Given A to find F $\dfrac{(1+i)^n-1}{i}$	Given F to find A $\dfrac{i}{(1+i)^n-1}$	Given A to find P $\dfrac{(1+i)^n-1}{i(1+i)^n}$	Given P to find A $\dfrac{i(1+i)^n}{(1+i)^n-1}$	n
	CA	PW	SCA	SF	SPW	CR	
1	1.0750 0000	0.930233	1.000000	1.000000	0.930233	1.075000	1
2	1.1556 2500	0.865333	2.075000	0.481928	1.795565	0.556928	2
3	1.2422 9688	0.804961	3.230625	0.309538	2.600526	0.384538	3
4	1.3354 6914	0.748801	4.472922	0.223568	3.349326	0.298568	4
5	1.4356 2933	0.696559	5.808391	0.172165	4.045885	0.247165	5
6	1.5433 0153	0.647962	7.244020	0.138045	4.693846	0.213045	6
7	1.6590 4914	0.602755	8.787322	0.113800	5.296601	0.188800	7
8	1.7834 7783	0.560702	10.446371	0.095727	5.857304	0.170727	8
9	1.9172 3866	0.521583	12.229849	0.081767	6.378887	0.156767	9
10	2.0610 3156	0.485194	14.147087	0.070686	6.864081	0.145686	10
11	2.2156 0893	0.451343	16.208119	0.061697	7.315424	0.136697	11
12	2.3817 7960	0.419854	18.423728	0.054278	7.735278	0.129278	12
13	2.5604 1307	0.390562	20.805508	0.048064	8.125840	0.123064	13
14	2.7524 4405	0.363313	23.365921	0.042797	8.489154	0.117797	14
15	2.9588 7735	0.337966	26.118365	0.038287	8.827120	0.113287	15
16	3.1807 9315	0.314387	29.077242	0.034391	9.141507	0.109391	16
17	3.4193 5264	0.292453	32.258035	0.031000	9.433960	0.106000	17
18	3.6758 0409	0.272049	35.677388	0.028029	9.706009	0.103029	18
19	3.9514 8940	0.253069	39.353192	0.025411	9.959078	0.100411	19
20	4.2478 5110	0.235413	43.304681	0.023092	10.194491	0.098092	20
21	4.5664 3993	0.218989	47.552532	0.021029	10.413480	0.096029	21
22	4.9089 2293	0.203711	52.118972	0.019187	10.617191	0.094187	22
23	5.2770 9215	0.189498	57.027895	0.017535	10.806689	0.092535	23
24	5.6728 7406	0.176277	62.304987	0.016050	10.982967	0.091050	24
25	6.0983 3961	0.163979	67.977862	0.014711	11.146946	0.089711	25
26	6.5557 1508	0.152539	74.076201	0.013500	11.299485	0.088500	26
27	7.0473 9371	0.141896	80.631916	0.012402	11.441381	0.087402	27
28	7.5759 4824	0.131997	87.679310	0.011405	11.573378	0.086405	28
29	8.1441 4436	0.122788	95.255258	0.010498	11.696165	0.085498	29
30	8.7549 5519	0.114221	103.399403	0.009671	11.810386	0.084671	30
31	9.4115 7683	0.106252	112.154358	0.008916	11.916638	0.083916	31
32	10.1174 4509	0.098839	121.565935	0.008226	12.015478	0.083226	32
33	10.8762 5347	0.091943	131.683380	0.007594	12.107421	0.082594	33
34	11.6919 7248	0.085529	142.559633	0.007015	12.192950	0.082015	34
35	12.5688 7042	0.079562	154.251606	0.006483	12.272511	0.081483	35
36	13.5115 3570	0.074011	166.820476	0.005994	12.346522	0.080994	36
37	14.5249 0088	0.068847	180.332012	0.005545	12.415370	0.080545	37
38	15.6142 6844	0.064044	194.856913	0.005132	12.479414	0.080132	38
39	16.7853 3858	0.059576	210.471181	0.004751	12.538989	0.079751	39
40	18.0442 3897	0.055419	227.256520	0.004400	12.594409	0.079400	40
45	25.9048 3863	0.038603	332.064515	0.003011	12.818629	0.078011	45
50	37.1897 4603	0.026889	482.529947	0.002072	12.974812	0.077072	50
55	53.3906 9004	0.018730	698.542534	0.001432	13.083602	0.076432	55
60	76.6492 4036	0.013046	1,008.656538	0.000991	13.159381	0.075991	60
65	110.0398 9729	0.009088	1,453.865297	0.000688	13.212165	0.075688	65
70	157.9765 0360	0.006330	2,093.020048	0.000478	13.248933	0.075478	70
75	226.7957 0141	0.004409	3,010.609352	0.000332	13.274543	0.075332	75
80	325.5945 6000	0.003071	4,327.927467	0.000231	13.292383	0.075231	80
90	671.0606 6463	0.001490	8,934.142195	0.000112	13.313464	0.075112	90
100	1,383.0772 0993	0.000723	18,427.696132	0.000054	13.323693	0.075054	100

Standard Compound Interest Factors

TABLE B-25
8 Percent Compound Interest Factors for One Dollar

	Single Amount		Uniform Series (Periodic Equal Amounts)				
	Compound Amount	Present Worth	Compound Amount	Sinking Fund	Present Worth	Capital Recovery	
n	Given P to find F $(1+i)^n$	Given F to find P $\dfrac{1}{(1+i)^n}$	Given A to find F $\dfrac{(1+i)^n - 1}{i}$	Given F to find A $\dfrac{i}{(1+i)^n - 1}$	Given A to find P $\dfrac{(1+i)^n - 1}{i(1+i)^n}$	Given P to find A $\dfrac{i(1+i)^n}{(1+i)^n - 1}$	n
	CA	PW	SCA	SF	SPW	CR	
1	1.0800 0000	0.925926	1.000000	1.000000	0.925926	1.080000	1
2	1.1664 0000	0.857339	2.080000	0.480769	1.783265	0.560769	2
3	1.2597 1200	0.793832	3.246400	0.308034	2.577097	0.388034	3
4	1.3604 8896	0.735030	4.506112	0.221921	3.312127	0.301921	4
5	1.4693 2808	0.680583	5.866601	0.170456	3.992710	0.250456	5
6	1.5868 7432	0.630170	7.335929	0.136315	4.622880	0.216315	6
7	1.7138 2427	0.583490	8.922803	0.112072	5.206370	0.192072	7
8	1.8509 3021	0.540269	10.636628	0.094015	5.746639	0.174015	8
9	1.9990 0463	0.500249	12.487558	0.080080	6.246888	0.160080	9
10	2.1589 2500	0.463193	14.486562	0.069029	6.710081	0.149029	10
11	2.3316 3900	0.428883	16.645487	0.060076	7.138964	0.140076	11
12	2.5181 7012	0.397114	18.977126	0.052695	7.536078	0.132695	12
13	2.7196 2373	0.367698	21.495297	0.046522	7.903776	0.126522	13
14	2.9371 9362	0.340461	24.214920	0.041297	8.244237	0.121297	14
15	3.1721 6911	0.315242	27.152114	0.036830	8.559479	0.116830	15
16	3.4259 4264	0.291890	30.324283	0.032977	8.851369	0.112977	16
17	3.7000 1805	0.270269	33.750226	0.029629	9.121638	0.109629	17
18	3.9960 1950	0.250249	37.450244	0.026702	9.371887	0.106702	18
19	4.3157 0106	0.231712	41.446263	0.024128	9.603599	0.104128	19
20	4.6609 5714	0.214548	45.761964	0.021852	9.818147	0.101852	20
21	5.0338 3372	0.198656	50.422921	0.019832	10.016803	0.099832	21
22	5.4365 4041	0.183941	55.456755	0.018032	10.200744	0.098032	22
23	5.8714 6365	0.170315	60.893296	0.016422	10.371059	0.096422	23
24	6.3411 8074	0.157669	66.764759	0.014978	10.528758	0.094978	24
25	6.8484 7520	0.146018	73.105940	0.013679	10.674776	0.093679	25
26	7.3963 5321	0.135202	79.954415	0.012507	10.809978	0.092507	26
27	7.9880 6147	0.125187	87.350768	0.011448	10.935165	0.091448	27
28	8.6271 0639	0.115914	95.338830	0.010489	11.051078	0.090489	28
29	9.3172 7490	0.107328	103.965936	0.009619	11.158406	0.089619	29
30	10.0626 5689	0.099377	113.283211	0.008827	11.257783	0.088827	30
31	10.8676 6944	0.092016	123.345868	0.008107	11.349799	0.088107	31
32	11.7370 8300	0.085200	134.213637	0.007451	11.434999	0.087451	32
33	12.6760 4964	0.078889	145.950620	0.006852	11.513888	0.086852	33
34	13.6901 3361	0.073045	158.626670	0.006304	11.586934	0.086304	34
35	14.7853 4429	0.067635	172.316804	0.005803	11.654568	0.085803	35
36	15.9681 7184	0.062625	187.102148	0.005345	11.717193	0.085345	36
37	17.2456 2558	0.057986	203.070320	0.004924	11.775179	0.084924	37
38	18.6252 7563	0.053690	220.315945	0.004539	11.828869	0.084539	38
39	20.1152 9768	0.049713	238.941221	0.004185	11.878582	0.084185	39
40	21.7245 2150	0.046031	259.056519	0.003860	11.924613	0.083860	40
45	31.9204 4939	0.031328	386.505617	0.002587	12.108402	0.082587	45
50	46.9016 1251	0.021321	573.770156	0.001743	12.233485	0.081743	50
55	68.9138 5611	0.014511	848.923201	0.001178	12.318614	0.081178	55
60	101.2570 6367	0.009876	1,253.213296	0.000798	12.376552	0.080798	60
65	148.7798 4662	0.006721	1,847.248083	0.000541	12.415983	0.080541	65
70	218.6064 0590	0.004574	2,720.080074	0.000368	12.442820	0.080368	70
75	321.2045 2996	0.003113	4,002.556624	0.000250	12.461084	0.080250	75
80	471.9548 3426	0.002119	5,886.935428	0.000170	12.473514	0.080170	80
90	1,018.9150 8928	0.000981	12,723.938616	0.000079	12.487732	0.080079	90
100	2,199.7612 5634	0.000455	27,484.515704	0.000036	12.494318	0.080036	100

TABLE B-26

8½ Percent Compound Interest Factors for One Dollar

	Single Amount		Uniform Series (Periodic Equal Amounts)				
	Compound Amount	Present Worth	Compound Amount	Sinking Fund	Present Worth	Capital Recovery	
n	Given P to find F $(1+i)^n$	Given F to find P $\dfrac{1}{(1+i)^n}$	Given A to find F $\dfrac{(1+i)^n - 1}{i}$	Given F to find A $\dfrac{i}{(1+i)^n - 1}$	Given A to find P $\dfrac{(1+i)^n - 1}{i(1+i)^n}$	Given P to find A $\dfrac{i(1+i)^n}{(1+i)^n - 1}$	n
	CA	PW	SCA	SF	SPW	CR	
1	1.0850 0000	0.921659	1.000000	1.000000	0.921659	1.085000	1
2	1.1772 2500	0.849455	2.085000	0.479616	1.771114	0.564616	2
3	1.2772 8913	0.782908	3.262225	0.306539	2.554022	0.391539	3
4	1.3858 5870	0.721574	4.539514	0.220288	3.275597	0.305288	4
5	1.5036 5669	0.665045	5.925373	0.168766	3.940642	0.253766	5
6	1.6314 6751	0.612945	7.429030	0.134607	4.553587	0.219607	6
7	1.7701 4225	0.564926	9.060497	0.110369	5.118514	0.195369	7
8	1.9206 0434	0.520669	10.830639	0.092331	5.639183	0.177331	8
9	2.0838 5571	0.479880	12.751244	0.078424	6.119063	0.163424	9
10	2.2609 8344	0.442285	14.835099	0.067408	6.561348	0.152408	10
11	2.4531 6703	0.407636	17.096083	0.058493	6.968984	0.143493	11
12	2.6616 8623	0.375702	19.549250	0.051153	7.344686	0.136153	12
13	2.8879 2956	0.346269	22.210936	0.045023	7.690955	0.130023	13
14	3.1334 0357	0.319142	25.098866	0.039842	8.010097	0.124842	14
15	3.3997 4288	0.294140	28.232269	0.035420	8.304237	0.120420	15
16	3.6887 2102	0.271097	31.632012	0.031614	8.575333	0.116614	16
17	4.0022 6231	0.249859	35.320733	0.028312	8.825192	0.113312	17
18	4.3424 5461	0.230285	39.322995	0.025430	9.055476	0.110430	18
19	4.7115 6325	0.212244	43.665450	0.022901	9.267720	0.107901	19
20	5.1120 4612	0.195616	48.377013	0.020671	9.463337	0.105671	20
21	5.5465 7005	0.180292	53.489059	0.018695	9.643628	0.103695	21
22	6.0180 2850	0.166167	59.035629	0.016939	9.809796	0.101939	22
23	6.5295 6092	0.153150	65.053658	0.015372	9.962945	0.100372	23
24	7.0845 7360	0.141152	71.583219	0.013970	10.104097	0.098970	24
25	7.6867 6236	0.130094	78.667792	0.012712	10.234191	0.097712	25
26	8.3401 3716	0.119902	86.354555	0.011580	10.354093	0.096580	26
27	9.0490 4881	0.110509	94.694692	0.010560	10.464602	0.095560	27
28	9.8182 1796	0.101851	103.743741	0.009639	10.566453	0.094639	28
29	10.6527 6649	0.093872	113.561959	0.008806	10.660326	0.093806	29
30	11.5582 5164	0.086518	124.214725	0.008051	10.746844	0.093051	30
31	12.5407 0303	0.079740	135.772977	0.007365	10.826584	0.092365	31
32	13.6066 6279	0.073493	148.313680	0.006742	10.900078	0.091742	32
33	14.7632 2913	0.067736	161.920343	0.006176	10.967813	0.091176	33
34	16.0181 0360	0.062429	176.683572	0.005660	11.030243	0.090660	34
35	17.3796 4241	0.057539	192.701675	0.005189	11.087781	0.090189	35
36	18.8569 1201	0.053031	210.081318	0.004760	11.140812	0.089760	36
37	20.4597 4953	0.048876	228.938230	0.004368	11.189689	0.089368	37
38	22.1988 2824	0.045047	249.397979	0.004010	11.234736	0.089010	38
39	24.0857 2865	0.041518	271.596808	0.003682	11.276255	0.088682	39
40	26.1330 1558	0.038266	295.682536	0.003382	11.314520	0.088382	40
45	39.2950 8371	0.025448	450.530397	0.002220	11.465312	0.087220	45
50	59.0863 1551	0.016924	683.368418	0.001463	11.565595	0.086463	50
55	88.8455 3362	0.011255	1,033.476866	0.000968	11.632288	0.085968	55
60	133.5931 8102	0.007485	1,559.919777	0.000641	11.676642	0.085641	60
65	200.8782 8041	0.004978	2,351.509181	0.000425	11.706140	0.085425	65
70	302.0519 7024	0.003311	3,541.787885	0.000282	11.725757	0.085282	70
75	454.1824 6584	0.002202	5,331.558422	0.000188	11.738803	0.085188	75
80	682.9345 0332	0.001464	8,022.758863	0.000125	11.747479	0.085125	80
90	1,544.1036 0392	0.000648	18,154.160046	0.000055	11.757087	0.085055	90
100	3,491.1926 8107	0.000286	41,061.090366	0.000024	11.761336	0.085024	100

Standard Compound Interest Factors

TABLE B-27
9 Percent Compound Interest Factors for One Dollar

	Single Amount		Uniform Series (Periodic Equal Amounts)				
	Compound Amount	Present Worth	Compound Amount	Sinking Fund	Present Worth	Capital Recovery	
n	Given P to find F $(1+i)^n$	Given F to find P $\dfrac{1}{(1+i)^n}$	Given A to find F $\dfrac{(1+i)^n - 1}{i}$	Given F to find A $\dfrac{i}{(1+i)^n - 1}$	Given A to find P $\dfrac{(1+i)^n - 1}{i(1+i)^n}$	Given P to find A $\dfrac{i(1+i)^n}{(1+i)^n - 1}$	n
	CA	PW	SCA	SF	SPW	CR	
1	1.0900 0000	0.917431	1.000000	1.000000	0.917431	1.090000	1
2	1.1881 0000	0.841680	2.090000	0.478469	1.759111	0.568469	2
3	1.2950 2900	0.772183	3.278100	0.305055	2.531295	0.395055	3
4	1.4115 8161	0.708425	4.573129	0.218669	3.239720	0.308669	4
5	1.5386 2395	0.649931	5.984711	0.167092	3.889651	0.257092	5
6	1.6771 0011	0.596267	7.523335	0.132920	4.485919	0.222920	6
7	1.8280 3912	0.547034	9.200435	0.108691	5.032953	0.198691	7
8	1.9925 6264	0.501866	11.028474	0.090674	5.534819	0.180674	8
9	2.1718 9328	0.460428	13.021036	0.076799	5.995247	0.166799	9
10	2.3673 6367	0.422411	15.192930	0.065820	6.417658	0.155820	10
11	2.5804 2641	0.387533	17.560293	0.056947	6.805191	0.146947	11
12	2.8126 6478	0.355535	20.140720	0.049651	7.160725	0.139651	12
13	3.0658 0461	0.326179	22.953385	0.043567	7.486904	0.133567	13
14	3.3417 2703	0.299246	26.019189	0.038433	7.786150	0.128433	14
15	3.6424 8246	0.274538	29.360916	0.034059	8.060688	0.124059	15
16	3.9703 0588	0.251870	33.003399	0.030300	8.312558	0.120300	16
17	4.3276 3341	0.231073	36.973705	0.027046	8.543631	0.117046	17
18	4.7171 2042	0.211994	41.301338	0.024212	8.755625	0.114212	18
19	5.1416 6125	0.194490	46.018458	0.021730	8.950115	0.111730	19
20	5.6044 1077	0.178431	51.160120	0.019546	9.128546	0.109546	20
21	6.1088 0774	0.163698	56.764530	0.017617	9.292244	0.107617	21
22	6.6586 0043	0.150182	62.873338	0.015905	9.442425	0.105905	22
23	7.2578 7447	0.137781	69.531939	0.014382	9.580207	0.104382	23
24	7.9110 8317	0.126405	76.789813	0.013023	9.706612	0.103023	24
25	8.6230 8066	0.115968	84.700896	0.011806	9.822580	0.101806	25
26	9.3991 5792	0.106393	93.323977	0.010715	9.928972	0.100715	26
27	10.2450 8213	0.097608	102.723135	0.009735	10.026580	0.099735	27
28	11.1671 3952	0.089568	112.968217	0.008852	10.116128	0.098852	28
29	12.1721 8208	0.082155	124.135356	0.008056	10.198283	0.098056	29
30	13.2676 7847	0.075371	136.307539	0.007336	10.273654	0.097336	30
31	14.4617 6953	0.069148	149.575217	0.006686	10.342802	0.096686	31
32	15.7633 2879	0.063438	164.036987	0.006096	10.406240	0.096096	32
33	17.1820 2838	0.058200	179.800315	0.005562	10.464441	0.095562	33
34	18.7284 1093	0.053395	196.982344	0.005077	10.517835	0.095077	34
35	20.4139 6792	0.048986	215.710755	0.004636	10.566821	0.094636	35
36	22.2512 2503	0.044941	236.124723	0.004235	10.611763	0.094235	36
37	24.2538 3528	0.041231	258.375948	0.003870	10.652993	0.093870	37
38	26.4366 8046	0.037826	282.629783	0.003538	10.690820	0.093538	38
39	28.8359 8170	0.034703	309.066463	0.003236	10.725523	0.093236	39
40	31.4094 2005	0.031838	337.882445	0.002960	10.757360	0.092960	40
45	48.3272 8610	0.020692	525.858734	0.001902	10.881197	0.091902	45
50	74.3575 2008	0.013449	815.083556	0.001227	10.961683	0.091227	50
55	114.4082 6162	0.008741	1,260.091796	0.000794	11.013993	0.090794	55
60	176.0312 9196	0.005681	1,944.792133	0.000514	11.047991	0.090514	60
65	270.8459 6262	0.003692	2,998.288474	0.000334	11.070087	0.090334	65
70	416.7300 8618	0.002400	4,619.223180	0.000216	11.084449	0.090216	70
75	641.1908 9332	0.001560	7,113.232148	0.000141	11.093782	0.090141	75
80	986.5516 6813	0.001014	10,950.574090	0.000091	11.099849	0.090091	80
90	2,335.5265 8223	0.000428	25,939.184247	0.000039	11.106354	0.090039	90
100	5,529.0407 9183	0.000181	61,422.675465	0.000016	11.109102	0.090016	100

TABLE B-28

10 Percent Compound Interest Factors for One Dollar

	Single Amount		Uniform Series (Periodic Equal Amounts)				
	Compound Amount	Present Worth	Compound Amount	Sinking Fund	Present Worth	Capital Recovery	
n	Given P to find F $(1+i)^n$	Given F to find P $\dfrac{1}{(1+i)^n}$	Given A to find F $\dfrac{(1+i)^n-1}{i}$	Given F to find A $\dfrac{i}{(1+i)^n-1}$	Given A to find P $\dfrac{(1+i)^n-1}{i(1+i)^n}$	Given P to find A $\dfrac{i(1+i)^n}{(1+i)^n-1}$	n
	CA	PW	SCA	SF	SPW	CR	
1	1.1000 0000	0.909091	1.000000	1.000000	0.909091	1.100000	1
2	1.2100 0000	0.826446	2.100000	0.476190	1.735537	0.576190	2
3	1.3310 0000	0.751315	3.310000	0.302115	2.486852	0.402115	3
4	1.4641 0000	0.683013	4.641000	0.215471	3.169865	0.315471	4
5	1.6105 1000	0.620921	6.105100	0.163797	3.790787	0.263797	5
6	1.7715 6100	0.564474	7.715610	0.129607	4.355261	0.229607	6
7	1.9487 1710	0.513158	9.487171	0.105405	4.868419	0.205405	7
8	2.1435 8881	0.466507	11.435888	0.087444	5.334926	0.187444	8
9	2.3579 4769	0.424098	13.579477	0.073641	5.759024	0.173641	9
10	2.5937 4246	0.385543	15.937425	0.062745	6.144567	0.162745	10
11	2.8531 1671	0.350494	18.531167	0.053963	6.495061	0.153963	11
12	3.1384 2838	0.318631	21.384284	0.046763	6.813692	0.146763	12
13	3.4522 7121	0.289664	24.522712	0.040779	7.103356	0.140779	13
14	3.7974 9834	0.263331	27.974983	0.035746	7.366687	0.135746	14
15	4.1772 4817	0.239392	31.772482	0.031474	7.606080	0.131474	15
16	4.5949 7299	0.217629	35.949730	0.027817	7.823709	0.127817	16
17	5.0544 7028	0.197845	40.544703	0.024664	8.021553	0.124664	17
18	5.5599 1731	0.179859	45.599173	0.021930	8.201412	0.121930	18
19	6.1159 0904	0.163508	51.159090	0.019547	8.364920	0.119547	19
20	6.7274 9995	0.148644	57.274999	0.017460	8.513564	0.117460	20
21	7.4002 4994	0.135131	64.002499	0.015624	8.648694	0.115624	21
22	8.1402 7494	0.122846	71.402749	0.014005	8.771540	0.114005	22
23	8.9543 0243	0.111678	79.543024	0.012572	8.883218	0.112572	23
24	9.8497 3268	0.101526	88.497327	0.011300	8.984744	0.111300	24
25	10.8347 0594	0.092296	98.347059	0.010168	9.077040	0.110168	25
26	11.9181 7654	0.083905	109.181765	0.009159	9.160945	0.109159	26
27	13.1099 9419	0.076278	121.099942	0.008258	9.237223	0.108258	27
28	14.4209 9361	0.069343	134.209936	0.007451	9.306567	0.107451	28
29	15.8630 9297	0.063039	148.630930	0.006728	9.369606	0.106728	29
30	17.4494 0227	0.057309	164.494023	0.006079	9.426914	0.106079	30
31	19.1943 4250	0.052099	181.943425	0.005496	9.479013	0.105496	31
32	21.1137 7675	0.047362	201.137767	0.004972	9.526376	0.104972	32
33	23.2251 5442	0.043057	222.251544	0.004499	9.569432	0.104499	33
34	25.5476 6986	0.039143	245.476699	0.004074	9.608575	0.104074	34
35	28.1024 3685	0.035584	271.024368	0.003690	9.644159	0.103690	35
36	30.9126 8053	0.032349	299.126805	0.003343	9.676508	0.103343	36
37	34.0039 4859	0.029408	330.039486	0.003030	9.705917	0.103030	37
38	37.4043 4344	0.026735	364.043434	0.002747	9.732651	0.102747	38
39	41.1447 7779	0.024304	401.447778	0.002491	9.756956	0.102491	39
40	45.2592 5557	0.022095	442.592556	0.002259	9.779051	0.102259	40
45	72.8904 8367	0.013719	718.904837	0.001391	9.862808	0.101391	45
50	117.3908 5288	0.008519	1,163.908529	0.000859	9.914814	0.100859	50
55	189.0591 4247	0.005289	1,880.591425	0.000532	9.947106	0.100532	55
60	304.4816 3954	0.003284	3,034.816395	0.000330	9.967157	0.100330	60
65	490.3707 2530	0.002039	4,893.707253	0.000204	9.979607	0.100204	65
70	789.7469 5680	0.001266	7,887.469568	0.000127	9.987338	0.100127	70
75	1,271.8953 7140	0.000786	12,708.953714	0.000079	9.992138	0.100079	75
80	2,048.4002 1459	0.000488	20,474.002146	0.000049	9.995118	0.100049	80
90	5,313.0226 1185	0.000188	53,120.226118	0.000019	9.998118	0.100019	90
100	13,780.6123 3982	0.000073	137,796.123398	0.000007	9.999274	0.100007	100

Standard Compound Interest Factors

TABLE B-29
12 Percent Compound Interest Factors for One Dollar

	Single Amount		Uniform Series (Periodic Equal Amounts)				
	Compound Amount	Present Worth	Compound Amount	Sinking Fund	Present Worth	Capital Recovery	
n	Given P to find F $(1+i)^n$	Given F to find P $\dfrac{1}{(1+i)^n}$	Given A to find F $\dfrac{(1+i)^n - 1}{i}$	Given F to find A $\dfrac{i}{(1+i)^n - 1}$	Given A to find P $\dfrac{(1+i)^n - 1}{i(1+i)^n}$	Given P to find A $\dfrac{i(1+i)^n}{(1+i)^n - 1}$	n
	CA	PW	SCA	SF	SPW	CR	
1	1.1200 0000	0.892857	1.00000	1.000000	0.89286	1.120000	1
2	1.2544 0000	0.797194	2.12000	0.471698	1.69005	0.591698	2
3	1.4049 2800	0.711780	3.37440	0.296349	2.40183	0.416349	3
4	1.5735 1936	0.635518	4.77933	0.209234	3.03735	0.329234	4
5	1.7623 4168	0.567427	6.35285	0.157410	3.60477	0.277410	5
6	1.9738 2269	0.506631	8.11519	0.123226	4.11140	0.243226	6
7	2.2106 8141	0.452349	10.08901	0.099118	4.56375	0.219118	7
8	2.4759 6318	0.403883	12.29969	0.081303	4.96764	0.201303	8
9	2.7730 7876	0.360610	14.77566	0.067679	5.32825	0.187679	9
10	3.1058 4821	0.321973	17.54874	0.056984	5.65023	0.176984	10
11	3.4785 4999	0.287476	20.65458	0.048415	5.93771	0.168415	11
12	3.8959 7599	0.256675	24.13313	0.041437	6.19437	0.161437	12
13	4.3634 9311	0.229174	28.02911	0.035677	6.42356	0.155677	13
14	4.8871 1229	0.204620	32.39260	0.030871	6.62818	0.150871	14
15	5.4735 6576	0.182696	37.27971	0.026824	6.81088	0.146824	15
16	6.1303 9365	0.163122	42.75328	0.023390	6.97399	0.143390	16
17	6.8660 4089	0.145644	48.88367	0.020457	7.11962	0.140457	17
18	7.6899 6580	0.130040	55.74972	0.017937	7.24969	0.137937	18
19	8.6127 6169	0.116107	63.43968	0.015763	7.36578	0.135763	19
20	9.6462 9309	0.103667	72.05244	0.013879	7.46943	0.133879	20
21	10.8038 4826	0.092560	81.69874	0.012240	7.56201	0.132240	21
22	12.1003 1006	0.082643	92.50258	0.010811	7.64462	0.130811	22
23	13.5523 4726	0.073788	104.60289	0.009560	7.71843	0.129560	23
24	15.1786 2893	0.065882	118.15524	0.008463	7.78434	0.128463	24
25	17.0000 6441	0.058823	133.33387	0.007500	7.84314	0.127500	25
26	19.0400 7214	0.052521	150.33393	0.006652	7.89565	0.126652	26
27	21.3248 8079	0.046894	169.37401	0.005904	7.94256	0.125904	27
28	23.8838 6649	0.041869	190.69889	0.005244	7.98441	0.125244	28
29	26.7499 3047	0.037383	214.58275	0.004660	8.02182	0.124660	29
30	29.9599 2212	0.033378	241.33268	0.004144	8.05516	0.124144	30
31	33.5551 1278	0.029802	271.29261	0.003686	8.08499	0.123686	31
32	37.5817 2631	0.026609	304.84772	0.003280	8.11162	0.123280	32
33	42.0915 3347	0.023758	342.42945	0.002920	8.13537	0.122920	33
34	47.1425 1748	0.021212	384.52098	0.002601	8.15654	0.122601	34
35	52.7996 1958	0.018940	431.66350	0.002317	8.17548	0.122317	35
36	59.1355 7393	0.016910	484.46312	0.002064	8.19242	0.122064	36
37	66.2318 4280	0.015098	543.59869	0.001840	8.20749	0.121840	37
38	74.1796 6394	0.013481	609.83053	0.001640	8.22098	0.121640	38
39	83.0812 2361	0.012036	684.01020	0.001462	8.23303	0.121462	39
40	93.0509 7044	0.010747	767.09142	0.001304	8.24375	0.121304	40
41	104.2170 8689	0.009595	860.14239	0.001163	8.25334	0.121163	41
42	116.7231 3732	0.008567	964.35948	0.001037	8.26194	0.121037	42
43	130.7299 1380	0.007649	1,081.08262	0.000925	8.26959	0.120925	43
44	146.4175 0346	0.006830	1,211.81253	0.000825	8.27643	0.120825	44
45	163.9876 0387	0.006098	1,358.23003	0.000736	8.28253	0.120736	45
50	289.0021 8983	0.003460	2,400.01825	0.000417	8.30448	0.120417	50
55	509.3206 0567	0.001963	4,236.00505	0.000236	8.31698	0.120236	55
60	897.5969 3349	0.001114	7,471.64111	0.000134	8.32404	0.120134	60

TABLE B-30
15 Percent Compound Interest Factors for One Dollar

	Single Amount		Uniform Series (Periodic Equal Amounts)				
	Compound Amount	Present Worth	Compound Amount	Sinking Fund	Present Worth	Capital Recovery	
n	Given P to find F $(1+i)^n$	Given F to find P $\dfrac{1}{(1+i)^n}$	Given A to find F $\dfrac{(1+i)^n - 1}{i}$	Given F to find A $\dfrac{i}{(1+i)^n - 1}$	Given A to find P $\dfrac{(1+i)^n - 1}{i(1+i)^n}$	Given P to find A $\dfrac{i(1+i)^n}{(1+i)^n - 1}$	n
	CA	PW	SCA	SF	SPW	CR	
1	1.1500 0000	0.869565	1.00000	1.000000	0.86957	1.150000	1
2	1.3225 0000	0.756144	2.15000	0.465116	1.62571	0.615116	2
3	1.5208 7500	0.657516	3.47250	0.287977	2.28322	0.437977	3
4	1.7490 0625	0.571753	4.99338	0.200265	2.85498	0.350265	4
5	2.0113 5719	0.497177	6.74238	0.148316	3.35215	0.298316	5
6	2.3130 6077	0.432328	8.75374	0.114237	3.78448	0.264237	6
7	2.6600 1988	0.375937	11.06680	0.090360	4.16043	0.240360	7
8	3.0590 2286	0.326902	13.72682	0.072850	4.48732	0.222850	8
9	3.5178 7629	0.284262	16.78584	0.059574	4.77158	0.209574	9
10	4.0455 5774	0.247185	20.30372	0.049252	5.01877	0.199252	10
11	4.6523 9140	0.214943	24.34928	0.041069	5.23371	0.191069	11
12	5.3502 5011	0.186907	29.00167	0.034481	5.42061	0.184481	12
13	6.1527 8762	0.162528	34.35192	0.029110	5.58316	0.179110	13
14	7.0757 0576	0.141329	40.50471	0.024688	5.72449	0.174688	14
15	8.1370 6163	0.122894	47.58041	0.021017	5.84737	0.171017	15
16	9.3576 2087	0.106865	55.71747	0.017948	5.95422	0.167948	16
17	10.7612 6400	0.092926	65.07509	0.015367	6.04716	0.165367	17
18	12.3754 5361	0.080805	75.83636	0.013186	6.12798	0.163186	18
19	14.2317 7165	0.070265	88.21181	0.011336	6.19824	0.161336	19
20	16.3665 3739	0.061100	102.44358	0.009761	6.25935	0.159761	20
21	18.8215 1800	0.053131	118.81012	0.008417	6.31245	0.158417	21
22	21.6447 4570	0.046201	137.63164	0.007266	6.35865	0.157266	22
23	24.8914 5756	0.040174	159.27638	0.006278	6.39885	0.156278	23
24	28.6251 7619	0.034934	184.16784	0.005430	6.43376	0.155430	24
25	32.9189 5262	0.030378	212.79302	0.004699	6.46417	0.154699	25
26	37.8567 9551	0.026415	245.71197	0.004070	6.49056	0.154070	26
27	43.5353 1484	0.022970	283.56877	0.003526	6.51355	0.153526	27
28	50.0656 1207	0.019974	327.10408	0.003057	6.53351	0.153057	28
29	57.5754 5388	0.017369	377.16969	0.002651	6.55089	0.152651	29
30	66.2117 7199	0.015103	434.74515	0.002300	6.56599	0.152300	30
31	76.1435 3775	0.013133	500.95692	0.001996	6.57912	0.151996	31
32	87.5650 6841	0.011420	577.10046	0.001733	6.59052	0.151733	32
33	100.6998 2867	0.009931	664.66552	0.001505	6.60044	0.151505	33
34	115.8048 0298	0.008635	765.36535	0.001307	6.60908	0.151307	34
35	133.1755 2342	0.007509	881.17016	0.001135	6.61660	0.151135	35
36	153.1518 5194	0.006529	1,014.34568	0.000986	6.62313	0.150986	36
37	176.1246 2973	0.005678	1,167.49753	0.000857	6.62879	0.150857	37
38	202.5433 2419	0.004937	1,343.62216	0.000744	6.63376	0.150744	38
39	232.9248 2281	0.004293	1,546.16549	0.000647	6.63803	0.150647	39
40	267.8635 4623	0.003733	1,779.09031	0.000562	6.64178	0.150562	40
41	308.0430 7817	0.003246	2,046.95385	0.000489	6.64500	0.150489	41
42	354.2495 3990	0.002823	2,354.99693	0.000425	6.64783	0.150425	42
43	407.3869 7088	0.002455	2,709.24647	0.000369	6.65031	0.150369	43
44	468.4950 1651	0.002134	3,116.63344	0.000321	6.65243	0.150321	44
45	538.7692 6899	0.001856	3,585.12846	0.000279	6.65429	0.150279	45
50	1,083.6574 4158	0.000923	7,217.71628	0.000139	6.66049	0.150139	50

Standard Compound Interest Factors

TABLE B-31
17 Percent Compound Interest Factors for One Dollar

	Single Amount		Uniform Series (Periodic Equal Amounts)				
	Compound Amount	Present Worth	Compound Amount	Sinking Fund	Present Worth	Capital Recovery	
n	Given P to find F $(1+i)^n$	Given F to find P $\dfrac{1}{(1+i)^n}$	Given A to find F $\dfrac{(1+i)^n-1}{i}$	Given F to find A $\dfrac{i}{(1+i)^n-1}$	Given A to find P $\dfrac{(1+i)^n-1}{i(1+i)^n}$	Given P to find A $\dfrac{i(1+i)^n}{(1+i)^n-1}$	n
	CA	PW	SCA	SF	SPW	CR	
1	1.1700 0000	0.854701	1.000000	1.000000	0.854701	1.170000	1
2	1.3689 0000	0.730514	2.170000	0.460829	1.585214	0.630829	2
3	1.6016 1300	0.624371	3.538900	0.282574	2.209585	0.452574	3
4	1.8738 8721	0.533650	5.140513	0.194533	2.743235	0.364533	4
5	2.1924 4804	0.456111	7.014400	0.142564	3.199346	0.312564	5
6	2.5651 6420	0.389839	9.206848	0.108615	3.589185	0.278615	6
7	3.0012 4212	0.333195	11.772012	0.084947	3.922380	0.254947	7
8	3.5114 5328	0.284782	14.773255	0.067690	4.207163	0.237690	8
9	4.1084 0033	0.243404	18.284708	0.054691	4.450566	0.224691	9
10	4.8068 2839	0.208037	22.393108	0.044657	4.658604	0.214657	10
11	5.6239 8922	0.177810	27.199937	0.036765	4.836413	0.206765	11
12	6.5800 6738	0.151974	32.823926	0.030466	4.988387	0.200466	12
13	7.6986 7884	0.129892	39.403993	0.025378	5.118280	0.195378	13
14	9.0074 5424	0.111019	47.102672	0.021230	5.229299	0.191230	14
15	10.5387 2146	0.094888	56.110126	0.017822	5.324187	0.187822	15
16	12.3303 0411	0.081101	66.648848	0.015004	5.405288	0.185004	16
17	14.4264 5581	0.069317	78.979152	0.012662	5.474605	0.182662	17
18	16.8789 5329	0.059245	93.405608	0.010706	5.533851	0.180706	18
19	19.7483 7535	0.050637	110.284561	0.009067	5.584488	0.179067	19
20	23.1055 9916	0.043280	130.032936	0.007690	5.627767	0.177690	20
21	27.0335 5102	0.036991	153.138535	0.006530	5.664758	0.176530	21
22	31.6292 5470	0.031616	180.172086	0.005550	5.696375	0.175550	22
23	37.0062 2799	0.027022	211.801341	0.004721	5.723397	0.174721	23
24	43.2972 8675	0.023096	248.807569	0.004019	5.746493	0.174019	24
25	50.6578 2550	0.019740	292.104856	0.003423	5.766234	0.173423	25
26	59.2696 5584	0.016872	342.762681	0.002917	5.783106	0.172917	26
27	69.3454 9733	0.014421	402.032337	0.002487	5.797526	0.172487	27
28	81.1342 3187	0.012325	471.377835	0.002121	5.809851	0.172121	28
29	94.9270 5129	0.010534	552.512066	0.001810	5.820386	0.171810	29
30	111.0646 5001	0.009004	647.439118	0.001545	5.829390	0.171545	30
31	129.9456 4051	0.007696	758.503768	0.001318	5.837085	0.171318	31
32	152.0363 9940	0.006577	888.449408	0.001126	5.843663	0.171126	32
33	177.8825 8730	0.005622	1040.485808	0.000961	5.849284	0.170961	33
34	208.1226 2714	0.004805	1218.368395	0.000821	5.854089	0.170821	34
35	243.5034 7375	0.004107	1426.491022	0.000701	5.858196	0.170701	35
36	284.8990 6429	0.003510	1669.994496	0.000599	5.861706	0.170599	36
37	333.3319 0522	0.003000	1954.893560	0.000512	5.864706	0.170512	37
38	389.9983 2910	0.002564	2288.225465	0.000437	5.867270	0.170437	38
39	456.2980 4505	0.002192	2678.223794	0.000373	5.869461	0.170373	39
40	533.8687 1271	0.001873	3134.521839	0.000319	5.871335	0.170319	40

TABLE B-32
20 Percent Compound Interest Factors for One Dollar

	Single Amount		Uniform Series (Periodic Equal Amounts)				
	Compound Amount	Present Worth	Compound Amount	Sinking Fund	Present Worth	Capital Recovery	
n	Given P to find F $(1+i)^n$	Given F to find P $\dfrac{1}{(1+i)^n}$	Given A to find F $\dfrac{(1+i)^n-1}{i}$	Given F to find A $\dfrac{i}{(1+i)^n-1}$	Given A to find P $\dfrac{(1+i)^n-1}{i(1+i)^n}$	Given P to find A $\dfrac{i(1+i)^n}{(1+i)^n-1}$	n
	CA	PW	SCA	SF	SPW	CR	
1	1.2000 0000	0.833333	1.00000	1.000000	0.83333	1.200000	1
2	1.4400 0000	0.694444	2.20000	0.454545	1.52778	0.654545	2
3	1.7280 0000	0.578704	3.64000	0.274725	2.10648	0.474725	3
4	2.0736 0000	0.482253	5.36800	0.186289	2.58874	0.386289	4
5	2.4883 2000	0.401878	7.44160	0.134380	2.99061	0.334380	5
6	2.9859 8400	0.334898	9.92992	0.100706	3.32551	0.300706	6
7	3.5831 8080	0.279082	12.91590	0.077424	3.60459	0.277424	7
8	4.2998 1696	0.232568	16.49908	0.060609	3.83717	0.260609	8
9	5.1597 8035	0.193807	20.79890	0.048079	4.03097	0.248079	9
10	6.1917 3642	0.161506	25.95868	0.038523	4.19247	0.238523	10
11	7.4300 8371	0.134588	32.15042	0.031104	4.32706	0.231104	11
12	8.9161 0045	0.112157	39.58050	0.025265	4.43922	0.225265	12
13	10.6993 2054	0.093464	48.49660	0.020620	4.53268	0.220620	13
14	12.8391 8465	0.077887	59.19592	0.016893	4.61057	0.216893	14
15	15.4070 2157	0.064905	72.03511	0.013882	4.67548	0.213882	15
16	18.4884 2589	0.054088	87.44213	0.011436	4.72956	0.211436	16
17	22.1861 1107	0.045073	105.93056	0.009440	4.77464	0.209440	17
18	26.6233 3328	0.037561	128.11667	0.007805	4.81220	0.207805	18
19	31.9479 9994	0.031301	154.74000	0.006462	4.84351	0.206462	19
20	38.3375 9992	0.026084	186.68800	0.005357	4.86957	0.205357	20
21	46.0051 1991	0.021737	225.02560	0.004444	4.89131	0.204444	21
22	55.2061 4389	0.018114	271.03072	0.003690	4.90942	0.203690	22
23	66.2473 7267	0.015095	326.23686	0.003065	4.92453	0.203065	23
24	79.4968 4720	0.012579	392.48424	0.002548	4.93710	0.202548	24
25	95.3962 1664	0.010483	471.98108	0.002119	4.94758	0.202119	25
26	114.4754 5997	0.008735	567.37730	0.001762	4.95633	0.201762	26
27	137.3705 5197	0.007280	681.85276	0.001467	4.96359	0.201467	27
28	164.8446 6236	0.006066	819.22331	0.001221	4.96966	0.201221	28
29	197.8135 9483	0.005055	984.06797	0.001016	4.97473	0.201016	29
30	237.3763 1380	0.004213	1,181.88157	0.000846	4.97894	0.200846	30
31	284.8515 7656	0.003511	1,419.25788	0.000705	4.98244	0.200705	31
32	341.8218 9187	0.002926	1,704.10946	0.000587	4.98537	0.200587	32
33	410.1862 7025	0.002438	2,045.93135	0.000489	4.98780	0.200489	33
34	492.2235 2430	0.002032	2,456.11762	0.000407	4.98985	0.200407	34
35	590.6682 2915	0.001693	2,948.34115	0.000339	4.99154	0.200339	35
36	708.8018 7499	0.001411	3,539.00938	0.000283	4.99293	0.200283	36
37	850.5622 4998	0.001176	4,247.81125	0.000235	4.99413	0.200235	37
38	1,020.6746 9998	0.000980	5,098.37350	0.000196	4.99510	0.200196	38
39	1,224.8096 3997	0.000816	6,119.04820	0.000163	4.99593	0.200163	39
40	1,469.7715 6797	0.000680	7,343.85784	0.000136	4.99660	0.200136	40
41	1,763.7258 8156	0.000567	8,813.62941	0.000113	4.99718	0.200113	41

Standard Compound Interest Factors

TABLE B-33
25 Percent Compound Interest Factors for One Dollar

	Single Amount		Uniform Series (Periodic Equal Amounts)				
	Compound Amount	Present Worth	Compound Amount	Sinking Fund	Present Worth	Capital Recovery	
n	Given P to find F $(1+i)^n$	Given F to find P $\dfrac{1}{(1+i)^n}$	Given A to find F $\dfrac{(1+i)^n - 1}{i}$	Given F to find A $\dfrac{i}{(1+i)^n - 1}$	Given A to find P $\dfrac{(1+i)^n - 1}{i(1+i)^n}$	Given P to find A $\dfrac{i(1+i)^n}{(1+i)^n - 1}$	n
	CA	PW	SCA	SF	SPW	CR	
1	1.2500 0000	0.800000	1.00000	1.000000	0.80000	1.250000	1
2	1.5625 0000	0.640000	2.25000	0.444444	1.44000	0.694444	2
3	1.9531 2500	0.512000	3.81250	0.262295	1.95200	0.512295	3
4	2.4414 0625	0.409600	5.76562	0.173442	2.36160	0.423442	4
5	3.0517 5781	0.327680	8.20703	0.121847	2.68928	0.371847	5
6	3.8146 9727	0.262144	11.25879	0.088819	2.95143	0.338819	6
7	4.7683 7158	0.209715	15.07349	0.066342	3.16114	0.316342	7
8	5.9604 6448	0.167772	19.84186	0.050399	3.32891	0.300399	8
9	7.4505 8060	0.134218	25.80232	0.038756	3.46313	0.288756	9
10	9.3132 2575	0.107374	33.25290	0.030073	3.57050	0.280073	10
11	11.6415 3218	0.085899	42.56613	0.023493	3.65640	0.273493	11
12	14.5519 1523	0.068719	54.20766	0.018448	3.72512	0.268448	12
13	18.1898 9404	0.054976	68.75958	0.014543	3.78010	0.264543	13
14	22.7373 6754	0.043980	86.94947	0.011501	3.82408	0.261501	14
15	28.4217 0943	0.035184	109.68684	0.009117	3.85926	0.259117	15
16	35.5271 3679	0.028147	138.10855	0.007241	3.88741	0.257241	16
17	44.4089 2099	0.022518	173.63568	0.005759	3.90993	0.255759	17
18	55.5111 5123	0.018014	218.04460	0.004586	3.92795	0.254586	18
19	69.3889 3904	0.014412	273.55576	0.003656	3.94235	0.253656	19
20	86.7361 7380	0.011529	342.94470	0.002916	3.95388	0.252916	20
21	108.4202 1725	0.009223	429.68087	0.002327	3.96311	0.252327	21
22	135.5252 7156	0.007379	538.10109	0.001858	3.97049	0.251858	22
23	169.4065 8945	0.005903	673.62636	0.001485	3.97638	0.251485	23
24	211.7582 3681	0.004722	843.03295	0.001186	3.98111	0.251186	24
25	264.6977 9602	0.003778	1,054.79118	0.000948	3.98489	0.250948	25
26	330.8722 4502	0.003022	1,319.48898	0.000758	3.98791	0.250758	26
27	413.5903 0628	0.002418	1,650.36123	0.000606	3.99033	0.250606	27
28	516.9878 8285	0.001934	2,063.95153	0.000485	3.99226	0.250485	28
29	646.2348 5356	0.001547	2,580.93941	0.000387	3.99382	0.250387	29
30	807.7935 6695	0.001238	3,227.17427	0.000310	3.99505	0.250310	30
31	1,009.7419 5868	0.000990	4,034.96783	0.000248	3.99604	0.250248	31
32	1,262.1774 4835	0.000792	5,044.70979	0.000198	3.99683	0.250198	32
33	1,577.7218 1044	0.000634	6,306.88724	0.000159	3.99746	0.250159	33
34	1,972.1522 6305	0.000507	7,884.60905	0.000127	3.99797	0.250127	34
35	2,465.1903 2882	0.000406	9,856.76132	0.000101	3.99838	0.250101	35

TABLE B-34

30 Percent Compound Interest Factors for One Dollar

	Single Amount		Uniform Series (Periodic Equal Amounts)				
	Compound Amount	Present Worth	Compound Amount	Sinking Fund	Present Worth	Capital Recovery	
n	Given P to find F $(1+i)^n$	Given F to find P $\dfrac{1}{(1+i)^n}$	Given A to find F $\dfrac{(1+i)^n-1}{i}$	Given F to find A $\dfrac{i}{(1+i)^n-1}$	Given A to find P $\dfrac{(1+i)^n-1}{i(1+i)^n}$	Given P to find A $\dfrac{i(1+i)^n}{(1+i)^n-1}$	n
	CA	PW	SCA	SF	SPW	CR	
1	1.3000 0000	0.769231	1.000000	1.000000	0.769231	1.300000	1
2	1.6900 0000	0.591716	2.300000	0.434783	1.360947	0.734783	2
3	2.1970 0000	0.455166	3.990000	0.250627	1.816113	0.550627	3
4	2.8561 0000	0.350128	6.187000	0.161629	2.166241	0.461629	4
5	3.7129 3000	0.269329	9.043100	0.110582	2.435570	0.410582	5
6	4.8268 0900	0.207176	12.756030	0.078394	2.642746	0.378394	6
7	6.2748 5170	0.159366	17.582839	0.056874	2.802112	0.356874	7
8	8.1573 0721	0.122589	23.857691	0.041915	2.924702	0.341915	8
9	10.6044 9937	0.094300	32.014998	0.031235	3.019001	0.331235	9
10	13.7858 4918	0.072538	42.619497	0.023463	3.091539	0.323463	10
11	17.9216 0394	0.055799	56.405346	0.017729	3.147338	0.317729	11
12	23.2980 8512	0.042922	74.326950	0.013454	3.190260	0.313454	12
13	30.2875 1066	0.033017	97.625036	0.010243	3.223277	0.310243	13
14	39.3737 6386	0.025398	127.912546	0.007818	3.248675	0.307818	14
15	51.1858 9301	0.019537	167.286310	0.005978	3.268211	0.305978	15
16	66.5416 6092	0.015028	218.472203	0.004577	3.283239	0.304577	16
17	86.5041 5919	0.011560	285.013864	0.003509	3.294800	0.303509	17
18	112.4554 0695	0.008892	371.518023	0.002692	3.303692	0.302692	18
19	146.1920 2904	0.006840	483.973430	0.002066	3.310532	0.302066	19
20	190.0496 3775	0.005262	630.165459	0.001587	3.315794	0.301587	20
21	247.0645 2907	0.004048	820.215097	0.001219	3.319842	0.301219	21
22	321.1838 8780	0.003113	1067.279626	0.000937	3.322955	0.300937	22
23	417.5390 5413	0.002395	1388.463514	0.000720	3.325350	0.300720	23
24	542.8007 7037	0.001842	1806.002568	0.000554	3.327192	0.300554	24
25	705.6410 0149	0.001417	2348.803338	0.000426	3.328609	0.300426	25
26	917.3333 0193	0.001090	3054.444340	0.000327	3.329700	0.300327	26
27	1192.5332 9251	0.000839	3971.777642	0.000252	3.330538	0.300252	27
28	1550.2932 8027	0.000645	5164.310934	0.000194	3.331183	0.300194	28
29	2015.3812 6435	0.000496	6714.604214	0.000149	3.331679	0.300149	29
30	2619.9956 4365	0.000382	8729.985479	0.000115	3.332061	0.300115	30
31	3405.9943 3674	0.000294	11349.981122	0.000088	3.332355	0.300088	31
32	4427.7926 3777	0.000226	14755.975459	0.000068	3.332581	0.000068	32
33	5756.1304 2910	0.000174	19183.768097	0.000052	3.332754	0.000052	33
34	7482.9695 5783	0.000134	24939.898526	0.000040	3.332888	0.300040	34
35	9727.8604 2518	0.000103	32422.868084	0.000031	3.332991	0.300031	35
∞	∞	0.000000	∞	0.000000	3.333333	0.300000	∞

Standard Compound Interest Factors

TABLE B-35
35 Percent Compound Interest Factors for One Dollar

	Single Amount		Uniform Series (Periodic Equal Amounts)				
	Compound Amount	Present Worth	Compound Amount	Sinking Fund	Present Worth	Capital Recovery	
n	Given P to find F $(1+i)^n$	Given F to find P $\dfrac{1}{(1+i)^n}$	Given A to find F $\dfrac{(1+i)^n-1}{i}$	Given F to find A $\dfrac{i}{(1+i)^n-1}$	Given A to find P $\dfrac{(1+i)^n-1}{i(1+i)^n}$	Given P to find A $\dfrac{i(1+i)^n}{(1+i)^n-1}$	n
	CA	PW	SCA	SF	SPW	CR	
1	1.3500 0000	0.740741	1.000000	1.000000	0.740741	1.350000	1
2	1.8225 0000	0.548697	2.350000	0.425532	1.289438	0.775532	2
3	2.4603 7500	0.406442	4.172500	0.239664	1.695880	0.589664	3
4	3.3215 0625	0.301068	6.632875	0.150764	1.996948	0.500764	4
5	4.4840 3344	0.223014	9.954381	0.100458	2.219961	0.450458	5
6	6.0534 4514	0.165195	14.438415	0.069260	2.385157	0.419260	6
7	8.1721 5094	0.122367	20.491860	0.048800	2.507523	0.398800	7
8	11.0324 0377	0.090642	28.664011	0.034887	2.598165	0.384887	8
9	14.8937 4509	0.067142	39.696415	0.025191	2.665308	0.375191	9
10	20.1065 5587	0.049735	54.590160	0.018318	2.715043	0.368318	10
11	27.1438 5042	0.036841	74.696715	0.013387	2.751884	0.363387	11
12	36.6441 9807	0.027289	101.840566	0.009819	2.779173	0.359819	12
13	49.4696 6740	0.020214	138.484764	0.007221	2.799387	0.357221	13
14	66.7840 5098	0.014974	187.954431	0.005320	2.814361	0.355320	14
15	90.1584 6883	0.011092	254.738482	0.003926	2.825453	0.353926	15
16	121.7139 3292	0.008216	344.896951	0.002899	2.833669	0.352899	16
17	164.3138 0944	0.006086	466.610884	0.002143	2.839755	0.352143	17
18	221.8236 4274	0.004508	630.924694	0.001585	2.844263	0.351585	18
19	299.4619 1770	0.003339	852.748336	0.001173	2.847602	0.351173	19
20	404.2735 8890	0.002474	1152.210254	0.000868	2.850076	0.350868	20
21	545.7693 4501	0.001832	1556.483843	0.000642	2.851908	0.350642	21
22	736.7886 1577	0.001357	2102.253188	0.000476	2.853265	0.350476	22
23	994.6646 3128	0.001005	2839.041804	0.000352	2.854270	0.350352	23
24	1342.7972 5223	0.000745	3833.706435	0.000261	2.855015	0.350261	24
25	1812.7762 9052	0.000552	5176.503687	0.000193	2.855567	0.350193	25
26	2447.2479 9220	0.000409	6989.279978	0.000143	2.855975	0.350143	26
27	3303.7847 8947	0.000303	9436.527970	0.000106	2.856278	0.350106	27
28	4460.1094 6578	0.000224	12740.312759	0.000078	2.856502	0.350078	28
29	6021.1477 7880	0.000166	17200.422225	0.000058	2.856668	0.350058	29
30	8128.5495 0138	0.000123	23221.570004	0.000043	2.856791	0.350043	30
∞	∞	0.000000	∞	0.000000	2.857142	0.350000	∞

TABLE B-36
40 PERCENT COMPOUND INTEREST FACTORS FOR ONE DOLLAR

	Single Amount		Uniform Series (Periodic Equal Amounts)				
	Compound Amount	Present Worth	Compound Amount	Sinking Fund	Present Worth	Capital Recovery	
n	Given P to find F $(1+i)^n$	Given F to find P $\dfrac{1}{(1+i)^n}$	Given A to find F $\dfrac{(1+i)^n-1}{i}$	Given F to find A $\dfrac{i}{(1+i)^n-1}$	Given A to find P $\dfrac{(1+i)^n-1}{i(1+i)^n}$	Given P to find A $\dfrac{i(1+i)^n}{(1+i)^n-1}$	n
	CA	PW	SCA	SF	SPW	CR	
1	1.4000 0000	0.714286	1.000000	1.000000	0.714286	1.400000	1
2	1.9600 0000	0.510204	2.400000	0.416667	1.224490	0.816667	2
3	2.7440 0000	0.364431	4.360000	0.229358	1.588921	0.629358	3
4	3.8416 0000	0.260308	7.104000	0.140766	1.849229	0.540766	4
5	5.3782 4000	0.185934	10.945600	0.091361	2.035164	0.491361	5
6	7.5295 3600	0.132810	16.323840	0.061260	2.167974	0.461260	6
7	10.5413 5040	0.094865	23.853376	0.041923	2.262839	0.441923	7
8	14.7578 9056	0.067760	34.394726	0.029074	2.330599	0.429074	8
9	20.6610 4678	0.048400	49.152617	0.020345	2.378999	0.420345	9
10	28.9254 6550	0.034572	69.813664	0.014324	2.413571	0.414324	10
11	40.4956 5170	0.024694	98.739129	0.010128	2.438265	0.410128	11
12	56.6939 1238	0.017639	139.234781	0.007182	2.455904	0.407182	12
13	79.3714 7733	0.012599	195.928693	0.005104	2.468503	0.405104	13
14	111.1200 683	0.008999	275.300170	0.003632	2.477502	0.403632	14
15	155.5680 956	0.006428	386.420239	0.002588	2.483930	0.402588	15
16	217.7953 338	0.004591	541.988334	0.001845	2.488521	0.401845	16
17	304.9134 673	0.003280	759.783668	0.001316	2.491801	0.401316	17
18	426.8788 542	0.002343	1064.69714	0.000939	2.494144	0.400939	18
19	597.6303 959	0.001673	1491.57599	0.000670	2.495817	0.400670	19
20	836.6825 542	0.001195	2089.20639	0.000479	2.497012	0.400479	20
21	1171.3555 76	0.000854	2925.88894	0.000342	2.497866	0.400342	21
22	1639.8978 06	0.000610	4097.24452	0.000244	2.498476	0.400244	22
23	2295.8569 29	0.000436	5737.14232	0.000174	2.498911	0.400174	23
24	3214.1997 00	0.000311	8032.99925	0.000124	2.499222	0.400124	24
25	4499.8795 81	0.000222	11247.1989	0.000089	2.499444	0.400089	25
26	6299.8314 13	0.000159	15747.0785	0.000064	2.499603	0.400064	26
27	8819.7639 78	0.000113	22046.9099	0.000045	2.499717	0.400045	27
∞	∞	0.000000	∞	0.000000	2.500000	0.400000	∞

Standard Compound Interest Factors

TABLE B-37
45 Percent Compound Interest Factors for One Dollar

	Single Amount		Uniform Series (Periodic Equal Amounts)				
	Compound Amount	Present Worth	Compound Amount	Sinking Fund	Present Worth	Capital Recovery	
n	Given P to find F $(1+i)^n$	Given F to find P $\dfrac{1}{(1+i)^n}$	Given A to find F $\dfrac{(1+i)^n-1}{i}$	Given F to find A $\dfrac{i}{(1+i)^n-1}$	Given A to find P $\dfrac{(1+i)^n-1}{i(1+i)^n}$	Given P to find A $\dfrac{i(1+i)^n}{(1+i)^n-1}$	n
	CA	PW	SCA	SF	SPW	CR	
1	1.4500 0000	0.689655	1.000000	1.000000	0.689655	1.450000	1
2	2.1025 0000	0.475624	2.450000	0.408163	1.165279	0.858163	2
3	3.0486 2500	0.328017	4.552500	0.219660	1.493296	0.669660	3
4	4.4205 0625	0.226218	7.601125	0.131559	1.719515	0.581559	4
5	6.4097 3406	0.156013	12.021631	0.083183	1.875527	0.533183	5
6	9.2941 1439	0.107595	18.431365	0.054255	1.983122	0.504255	6
7	13.4764 6587	0.074203	27.725480	0.036068	2.057326	0.486068	7
8	19.5408 7551	0.051175	41.201946	0.024271	2.108500	0.474271	8
9	28.3342 6948	0.035293	60.742821	0.016463	2.143793	0.466463	9
10	41.0846 9075	0.024340	89.077091	0.011226	2.168133	0.461226	10
11	59.5728 0159	0.016786	130.161781	0.007683	2.184920	0.457683	11
12	86.3805 6230	0.011577	189.734583	0.005271	2.196496	0.455271	12
13	125.2518 153	0.007984	276.115146	0.003622	2.204480	0.453622	13
14	181.6151 322	0.005506	401.366961	0.002491	2.209986	0.452491	14
15	263.3419 418	0.003797	582.982093	0.001715	2.213784	0.451715	15
16	381.8458 155	0.002619	846.324035	0.001182	2.216403	0.451182	16
17	553.6764 325	0.001806	1228.16985	0.000814	2.218209	0.450814	17
18	802.8308 272	0.001246	1781.84628	0.000561	2.219454	0.450561	18
19	1164.1046 99	0.000859	2584.67711	0.000387	2.220313	0.450387	19
20	1687.9518 14	0.000592	3748.78181	0.000267	2.220906	0.450267	20
21	2447.5301 31	0.000409	5436.73362	0.000184	2.221314	0.450184	21
22	3548.9186 89	0.000282	7884.26375	0.000127	2.221596	0.450127	22
23	5145.9320 99	0.000194	11433.1824	0.000087	2.221790	0.450087	23
24	7461.6015 44	0.000134	16579.1145	0.000060	2.221924	0.450060	24
∞	∞	0.000000	∞	0.000000	2.222222	0.450000	∞

TABLE B-38
50 Percent Compound Interest Factors for One Dollar

n	Single Amount		Uniform Series (Periodic Equal Amounts)				n
	Compound Amount	Present Worth	Compound Amount	Sinking Fund	Present Worth	Capital Recovery	
	Given P to find F $(1+i)^n$	Given F to find P $\dfrac{1}{(1+i)^n}$	Given A to find F $\dfrac{(1+i)^n-1}{i}$	Given F to find A $\dfrac{i}{(1+i)^n-1}$	Given A to find P $\dfrac{(1+i)^n-1}{i(1+i)^n}$	Given P to find A $\dfrac{i(1+i)^n}{(1+i)^n-1}$	
	CA	PW	SCA	SF	SPW	CR	
1	1.5000 0000	0.666667	1.000000	1.000000	0.666667	1.500000	1
2	2.2500 0000	0.444444	2.500000	0.400000	1.111111	0.900000	2
3	3.3750 0000	0.296296	4.750000	0.210526	1.407407	0.710526	3
4	5.0625 0000	0.197531	8.125000	0.123077	1.604938	0.623077	4
5	7.5937 5000	0.131687	13.187500	0.075829	1.736626	0.575829	5
6	11.3906 2500	0.087791	20.781250	0.048120	1.824417	0.548120	6
7	17.0859 3750	0.058528	32.171875	0.031083	1.882945	0.531083	7
8	25.6289 0625	0.039018	49.257812	0.020301	1.921963	0.520301	8
9	38.4433 5938	0.026012	74.886719	0.013354	1.947975	0.513354	9
10	57.6650 3906	0.017342	113.330078	0.008824	1.965317	0.508824	10
11	86.4975 5859	0.011561	170.995117	0.005848	1.976878	0.505848	11
12	129.7463 3789	0.007707	257.492676	0.003884	1.984585	0.503884	12
13	194.6195 0684	0.005138	387.239014	0.002582	1.989724	0.502582	13
14	291.9292 6025	0.003425	581.858521	0.001719	1.993149	0.501719	14
15	437.8938 9038	0.002284	873.787781	0.001144	1.995433	0.501144	15
16	656.8408 3557	0.001522	1311.68167	0.000762	1.996955	0.500762	16
17	985.2612 5336	0.001015	1968.52251	0.000508	1.997970	0.500508	17
18	1477.8918 8003	0.000677	2953.78376	0.000339	1.998647	0.500339	18
19	2216.8378 2005	0.000451	4431.67564	0.000226	1.999098	0.500226	19
20	3325.2567 3007	0.000301	6648.51346	0.000150	1.999399	0.500150	20
21	4987.8850 9511	0.000200	9973.77019	0.000100	1.999599	0.500100	21
22	7481.8276 4267	0.000134	14961.6553	0.000067	1.999733	0.500067	22
∞	∞	0.000000	∞	0.000000	2.000000	0.500000	∞

Standard Compound Interest Factors

TABLE B-39
55 Percent Compound Interest Factors for One Dollar

	Single Amount		Uniform Series (Periodic Equal Amounts)				
	Compound Amount	Present Worth	Compound Amount	Sinking Fund	Present Worth	Capital Recovery	
n	Given P to find F $(1+i)^n$	Given F to find P $\dfrac{1}{(1+i)^n}$	Given A to find F $\dfrac{(1+i)^n - 1}{i}$	Given F to find A $\dfrac{i}{(1+i)^n - 1}$	Given A to find P $\dfrac{(1+i)^n - 1}{i(1+i)^n}$	Given P to find A $\dfrac{i(1+i)^n}{(1+i)^n - 1}$	n
	CA	PW	SCA	SF	SPW	CR	
1	1.5500 0000	0.645161	1.000000	1.000000	0.645161	1.550000	1
2	2.4025 0000	0.416233	2.550000	0.392157	1.061394	0.942157	2
3	3.7238 7500	0.268537	4.952500	0.201918	1.329932	0.751918	3
4	5.7720 0625	0.173250	8.676375	0.115256	1.503182	0.665256	4
5	8.9466 0969	0.111774	14.448381	0.069212	1.614956	0.619212	5
6	13.8672 4502	0.072112	23.394991	0.042744	1.687068	0.592744	6
7	21.4942 2977	0.046524	37.262236	0.026837	1.733593	0.576837	7
8	33.3160 5615	0.030016	58.756466	0.017019	1.763608	0.567019	8
9	51.6398 8703	0.019365	92.072522	0.010861	1.782973	0.560861	9
10	80.0418 2490	0.012493	143.712409	0.006958	1.795466	0.556958	10
11	124.0648 286	0.008060	223.754234	0.004469	1.803527	0.554469	11
12	192.3004 843	0.005200	347.819062	0.002875	1.808727	0.552875	12
13	298.0657 507	0.003355	540.119547	0.001851	1.812082	0.551851	13
14	462.0019 136	0.002164	838.185297	0.001193	1.814246	0.551193	14
15	716.1029 660	0.001396	1300.18721	0.000769	1.815643	0.550769	15
16	1109.9595 97	0.000901	2016.29018	0.000496	1.816544	0.550496	16
17	1720.4373 76	0.000581	3126.24977	0.000320	1.817125	0.550320	17
18	2666.6779 33	0.000375	4846.68715	0.000206	1.817500	0.550206	18
19	4133.3507 96	0.000242	7513.36508	0.000133	1.817742	0.550133	19
20	6406.6937 33	0.000156	11646.7159	0.000086	1.817898	0.550086	20
21	9930.3752 87	0.000101	18053.4096	0.000055	1.817999	0.550055	21
∞	∞	0.000000	∞	0.000000	1.818182	0.550000	∞

TABLE B-40
60 Percent Compound Interest Factors for One Dollar

	Single Amount		Uniform Series (Periodic Equal Amounts)				
	Compound Amount	Present Worth	Compound Amount	Sinking Fund	Present Worth	Capital Recovery	
n	Given P to find F $(1+i)^n$	Given F to find P $\dfrac{1}{(1+i)^n}$	Given A to find F $\dfrac{(1+i)^n - 1}{i}$	Given F to find A $\dfrac{i}{(1+i)^n - 1}$	Given A to find P $\dfrac{(1+i)^n - 1}{i(1+i)^n}$	Given P to find A $\dfrac{i(1+i)^n}{(1+i)^n - 1}$	n
	CA	PW	SCA	SF	SPW	CR	
1	1.6000 0000	0.625000	1.000000	1.000000	0.625000	1.600000	1
2	2.5600 0000	0.390625	2.600000	0.384615	1.015625	0.984615	2
3	4.0960 0000	0.244141	5.160000	0.193798	1.259766	0.793798	3
4	6.5536 0000	0.152588	9.256000	0.108038	1.412354	0.708038	4
5	10.4857 6000	0.095367	15.809600	0.063253	1.507721	0.663253	5
6	16.7772 1600	0.059605	26.295360	0.038030	1.567326	0.638030	6
7	26.8435 4560	0.037253	43.072576	0.023217	1.604578	0.623217	7
8	42.9496 7296	0.023283	69.916122	0.014303	1.627862	0.614303	8
9	68.7194 7674	0.014552	112.865795	0.008860	1.642413	0.608860	9
10	109.9511 628	0.009095	181.585271	0.005507	1.651508	0.605507	10
11	175.9218 604	0.005684	291.536434	0.003430	1.657193	0.603430	11
12	281.4749 767	0.003553	467.458295	0.002139	1.660745	0.602139	12
13	450.3599 627	0.002220	748.933271	0.001335	1.662966	0.601335	13
14	720.5759 404	0.001388	1199.29323	0.000834	1.664354	0.600834	14
15	1152.9215 05	0.000867	1919.86917	0.000521	1.665221	0.600521	15
16	1844.6744 07	0.000542	3072.79068	0.000325	1.665763	0.600325	16
17	2951.4790 52	0.000339	4917.46509	0.000203	1.666102	0.600203	17
18	4722.3664 83	0.000212	7868.94414	0.000127	1.666314	0.600127	18
19	7555.7863 73	0.000132	12591.3106	0.000079	1.666446	0.600079	19
∞	∞	0.000000	∞	0.000000	1.666667	0.600000	∞

Standard Compound Interest Factors

TABLE B-41
65 Percent Compound Interest Factors for One Dollar

	Single Amount		Uniform Series (Periodic Equal Amounts)				
	Compound Amount	Present Worth	Compound Amount	Sinking Fund	Present Worth	Capital Recovery	
n	Given P to find F $(1+i)^n$	Given F to find P $\dfrac{1}{(1+i)^n}$	Given A to find F $\dfrac{(1+i)^n - 1}{i}$	Given F to find A $\dfrac{i}{(1+i)^n - 1}$	Given A to find P $\dfrac{(1+i)^n - 1}{i(1+i)^n}$	Given P to find A $\dfrac{i(1+i)^n}{(1+i)^n - 1}$	n
	CA	PW	SCA	SF	SPW	CR	
1	1.6500 0000	0.606061	1.000000	1.000000	0.606061	1.650000	1
2	2.7225 0000	0.367309	2.650000	0.377358	0.973370	1.027358	2
3	4.4921 2500	0.222612	5.372500	0.186133	1.195982	0.836133	3
4	7.4120 0625	0.134916	9.864625	0.101372	1.330898	0.751372	4
5	12.2298 1031	0.081767	17.276631	0.057882	1.412666	0.707882	5
6	20.1791 8702	0.049556	29.506442	0.033891	1.462222	0.683891	6
7	33.2956 5858	0.030034	49.685629	0.020127	1.492255	0.670127	7
8	54.9378 3665	0.018202	82.981287	0.012051	1.510458	0.662051	8
9	90.6474 3047	0.011032	137.919124	0.007251	1.521490	0.657251	9
10	149.5682 603	0.006686	228.566554	0.004375	1.528176	0.654375	10
11	246.7876 295	0.004052	378.134815	0.002645	1.532228	0.652645	11
12	407.1995 886	0.002456	624.922444	0.001600	1.534683	0.651600	12
13	671.8793 212	0.001488	1032.12203	0.000969	1.536172	0.650969	13
14	1108.6008 80	0.000902	1704.001350	0.000587	1.537074	0.650587	14
15	1829.1914 52	0.000547	2812.60223	0.000356	1.537620	0.650356	15
16	3018.1658 96	0.000331	4641.79369	0.000215	1.537952	0.650215	16
17	4979.9737 28	0.000201	7659.95958	0.000131	1.538153	0.650131	17
18	8216.9566 51	0.000122	12639.9333	0.000079	1.538274	0.650079	18
∞	∞	0.000000	∞	0.000000	1.538462	0.650000	∞

TABLE B-42
70 Percent Compound Interest Factors for One Dollar

	Single Amount		Uniform Series (Periodic Equal Amounts)				
	Compound Amount	Present Worth	Compound Amount	Sinking Fund	Present Worth	Capital Recovery	
n	Given P to find F $(1 + i)^n$	Given F to find P $\dfrac{1}{(1 + i)^n}$	Given A to find F $\dfrac{(1 + i)^n - 1}{i}$	Given F to find A $\dfrac{i}{(1 + i)^n - 1}$	Given A to find P $\dfrac{(1 + i)^n - 1}{i(1 + i)^n}$	Given P to find A $\dfrac{i(1 + i)^n}{(1 + i)^n - 1}$	n
	CA	PW	SCA	SF	SPW	CR	
1	1.7000 0000	0.588235	1.000000	1.000000	0.588235	1.700000	1
2	2.8900 0000	0.346021	2.700000	0.370370	0.934256	1.070370	2
3	4.9130 0000	0.203542	5.590000	0.178891	1.137798	0.878891	3
4	8.3521 0000	0.119730	10.503000	0.095211	1.257528	0.795211	4
5	14.1985 7000	0.070430	18.855100	0.053036	1.327958	0.753036	5
6	24.1375 6900	0.041429	33.053670	0.030254	1.369387	0.730254	6
7	41.0338 6730	0.024370	57.191239	0.017485	1.393757	0.717485	7
8	69.7575 7441	0.014335	98.225106	0.010181	1.408092	0.710181	8
9	118.5878 765	0.008433	167.982681	0.005953	1.416525	0.705953	9
10	201.5993 900	0.004960	286.570557	0.003490	1.421485	0.703490	10
11	342.7189 631	0.002918	488.169947	0.002048	1.424403	0.702048	11
12	582.6222 372	0.001716	830.888910	0.001204	1.426119	0.701204	12
13	990.4578 033	0.001010	1413.51115	0.000707	1.427129	0.700707	13
14	1683.7782 66	0.000594	2403.96895	0.000416	1.427723	0.700416	14
15	2862.4230 52	0.000349	4087.74722	0.000245	1.428072	0.700245	15
16	4866.1191 88	0.000206	6950.17027	0.000144	1.428278	0.700144	16
17	8272.4026 19	0.000121	11816.2895	0.000085	1.428399	0.700085	17
∞	∞	0.000000	∞	0.000000	1.428571	0.700000	∞

Standard Compound Interest Factors

TABLE B-43
75 Percent Compound Interest Factors for One Dollar

n	Single Amount		Uniform Series (Periodic Equal Amounts)				n
	Compound Amount	Present Worth	Compound Amount	Sinking Fund	Present Worth	Capital Recovery	
	CA	PW	SCA	SF	SPW	CR	
1	1.7500 0000	0.571429	1.000000	1.000000	0.571429	1.750000	1
2	3.0625 0000	0.326531	2.750000	0.363636	0.897959	1.113636	2
3	5.3593 7500	0.186589	5.812500	0.172043	1.084548	0.922043	3
4	9.3789 0625	0.106622	11.171875	0.089510	1.191170	0.839510	4
5	16.4130 8594	0.060927	20.550781	0.048660	1.252097	0.798660	5
6	28.7229 0039	0.034815	36.963867	0.027053	1.286913	0.777053	6
7	50.2650 7568	0.019895	65.686768	0.015224	1.306807	0.765224	7
8	87.9638 8245	0.011368	115.951843	0.008624	1.318176	0.758624	8
9	153.9367 943	0.006496	203.915726	0.004904	1.324672	0.754904	9
10	269.3893 900	0.003712	357.852520	0.002794	1.328384	0.752794	10
11	471.4314 325	0.002121	627.241910	0.001594	1.330505	0.751594	11
12	825.0050 069	0.001212	1098.67334	0.000910	1.331717	0.750910	12
13	1443.7587 62	0.000693	1923.67835	0.000520	1.332410	0.750520	13
14	2526.5778 33	0.000396	3367.43711	0.000297	1.332806	0.750297	14
15	4421.5112 09	0.000226	5894.01494	0.000170	1.333032	0.750170	15
16	7737.6446 15	0.000129	10315.5262	0.000097	1.333161	0.750097	16
∞	∞	0.000000	∞	0.000000	1.333333	0.750000	∞

TABLE B-44
80 Percent Compound Interest Factors for One Dollar

n	Single Amount		Uniform Series (Periodic Equal Amounts)				n
	Compound Amount	Present Worth	Compound Amount	Sinking Fund	Present Worth	Capital Recovery	
	CA	PW	SCA	SF	SPW	CR	
1	1.8000 0000	0.555556	1.000000	1.000000	0.555556	1.800000	1
2	3.2400 0000	0.308642	2.800000	0.357143	0.864198	1.157143	2
3	5.8320 0000	0.171468	6.040000	0.165563	1.035665	0.965563	3
4	10.4976 0000	0.095260	11.872000	0.084232	1.130925	0.884232	4
5	18.8956 8000	0.052922	22.369600	0.044704	1.183847	0.844704	5
6	34.0122 '2400	0.029401	41.265280	0.024233	1.213249	0.824233	6
7	61.2220 0320	0.016334	75.277504	0.013284	1.229583	0.813284	7
8	110.1996 058	0.009074	136.499507	0.007326	1.238657	0.807326	8
9	198.3592 904	0.005041	246.699113	0.004054	1.243698	0.804054	9
10	357.0467 227	0.002801	445.058403	0.002247	1.246499	0.802247	10
11	642.6841 008	0.001556	802.105126	0.001247	1.248055	0.801247	11
12	1156.8313 81	0.000864	1444.78923	0.000692	1.248919	0.800692	12
13	2082.2964 87	0.000480	2601.62061	0.000384	1.249400	0.800384	13
14	3748.1336 76	0.000267	4683.91709	0.000213	1.249667	0.800213	14
15	6746.6406 16	0.000148	8432.05077	0.000119	1.249815	0.800119	15
∞	∞	0.000000	∞	0.000000	1.250000	0.800000	∞

TABLE B-45
90 and 100 Percent Compound Interest Factors for One Dollar

n	Single Amount		Uniform Series (Periodic Equal Amounts)				n
	Compound Amount	Present Worth	Compound Amount	Sinking Fund	Present Worth	Capital Recovery	
	Given P to find F $(1+i)^n$	Given F to find P $\dfrac{1}{(1+i)^n}$	Given A to find F $\dfrac{(1+i)^n-1}{i}$	Given F to find A $\dfrac{i}{(1+i)^n-1}$	Given A to find P $\dfrac{(1+i)^n-1}{i(1+i)^n}$	Given P to find A $\dfrac{i(1+i)^n}{(1+i)^n-1}$	
	CA	PW	SCA	SF	SPW	CR	

90 Percent

n	CA	PW	SCA	SF	SPW	CR	n
1	1.9000 0000	0.526316	1.000000	1.000000	0.526316	1.900000	1
2	3.6100 0000	0.277008	2.900000	0.344828	0.803324	1.244828	2
3	6.8590 0000	0.145794	6.510000	0.153610	0.949118	1.053610	3
4	13.0321 0000	0.076734	13.369000	0.074800	1.025852	0.974800	4
5	24.7609 9000	0.040386	26.401100	0.037877	1.066238	0.937877	5
6	47.0458 8100	0.021256	51.162090	0.019546	1.087494	0.919546	6
7	89.3871 7390	0.011187	98.207971	0.010182	1.098681	0.910182	7
8	169.8356 304	0.005888	187.595145	0.005331	1.104569	0.905331	8
9	322.6876 978	0.003099	357.430775	0.002798	1.107668	0.902798	9
10	613.1066 258	0.001631	680.118473	0.001470	1.109299	0.901470	10
11	1164.9025 89	0.000858	1293.22510	0.000773	1.110157	0.900773	11
12	2213.3149 19	0.000452	2458.12769	0.000407	1.110609	0.900407	12
13	4205.2983 46	0.000238	4671.44261	0.000214	1.110847	0.900214	13
14	7990.0668 58	0.000125	8876.74095	0.000113	1.110972	0.900113	14
∞	∞	0.000000	∞	0.000000	1.111111	0.900000	∞

100 Percent

n	CA	PW	SCA	SF	SPW	CR	n
1	2.0000 0000	0.500000	1.000000	1.000000	0.500000	2.000000	1
2	4.0000 0000	0.250000	3.000000	0.333333	0.750000	1.333333	2
3	8.0000 0000	0.125000	7.000000	0.142857	0.875000	1.142857	3
4	16.0000 0000	0.062500	15.000000	0.066667	0.937500	1.066667	4
5	32.0000 0000	0.031250	31.000000	0.032258	0.968750	1.032258	5
6	64.0000 0000	0.015625	63.000000	0.015873	0.984375	1.015873	6
7	128.0000 0000	0.007812	127.000000	0.007874	0.992188	1.007874	7
8	256.0000 0000	0.003906	255.000000	0.003922	0.996094	1.003922	8
9	512.0000 0000	0.001953	511.000000	0.001957	0.998047	1.001957	9
10	1024.0000 0000	0.000977	1023.000000	0.000978	0.999023	1.000978	10
11	2048.0000 0000	0.000489	2047.000000	0.000489	0.999512	1.000489	11
12	4096.0000 0000	0.000244	4095.000000	0.000244	0.999756	1.000244	12
13	8192.0000 0000	0.000122	8191.000000	0.000122	0.999878	1.000122	13
∞	∞	0.000000	∞	0.000000	1.000000	1.000000	∞

Appendix **C**

Arithmetic Gradient Factors

The following two sets of tables [Tables C-1 to C-5, the uniform gradient conversion factors (GUS) and Tables C-6 to C-10, the present worth of uniform gradients (GPW)] are calculated for the uniform gradient increase starting at time zero ($n = 0$), so that the first gradient ($1G$) is effective at $n = 1$. The factors are based upon the end-of-period step convention as applied to the standard compound interest tables in Appendix B. The two formulas from which Tables C-1 to C-10 were calculated are

$$GUS = \left[\left(\frac{(1+i)^{n+1}-1}{i}\right)-(n+1)\right]\frac{1}{(1+i)^n-1} \quad (6\text{-}9A)$$

$$GPW = \left[\left(\frac{(1+i)^{n+1}-1}{i}\right)-(n+1)\right]\frac{1}{i(1+i)^n} \quad (6\text{-}8A)$$

Arithmetic Gradient Factors

FACTORS (GUS) TO CONVERT AN ARITHMETIC GRADIENT
TO AN EQUIVALENT UNIFORM SERIES FOR INTEREST RATES FROM 1% TO 2½%

n	1%	1¼%	1½%	1¾%	2%	2½%	n
1	1.000000	1.000000	1.000000	1.000000	1.000000	1.000000	1
2	1.497512	1.496894	1.496278	1.495663	1.495050	1.493827	2
3	1.993367	1.991719	1.990075	1.988435	1.986799	1.983540	3
4	2.487562	2.484473	2.481390	2.478316	2.475249	2.469140	4
5	2.980100	2.975157	2.970226	2.965307	2.960401	2.950628	5
6	3.470980	3.463771	3.456581	3.449409	3.442256	3.428007	6
7	3.960202	3.950316	3.940457	3.930623	3.920815	3.901280	7
8	4.447766	4.434793	4.421854	4.408949	4.396080	4.370449	8
9	4.933673	4.917201	4.900773	4.884390	4.868053	4.835520	9
10	5.417923	5.397541	5.377215	5.356946	5.336736	5.296495	10
11	5.900517	5.875814	5.851181	5.826620	5.802131	5.753379	11
12	6.381454	6.352020	6.322672	6.293412	6.264242	6.206179	12
13	6.860734	6.826161	6.791690	6.757326	6.723071	6.654899	13
14	7.338360	7.298236	7.258236	7.218363	7.178621	7.099546	14
15	7.814330	7.768248	7.722311	7.676525	7.630896	7.540126	15
16	8.288645	8.236196	8.183917	8.131816	8.079899	7.976647	16
17	8.761306	8.702082	8.643056	8.584237	8.525635	8.409116	17
18	9.232314	9.165906	9.099729	9.033792	8.968108	8.837542	18
19	9.701668	9.627671	9.553938	9.480484	9.407322	9.261933	19
20	10.169370	10.087377	10.005686	9.924315	9.843282	9.682297	20
21	10.635420	10.545025	10.454974	10.365289	10.275993	10.098645	21
22	11.099819	11.000616	10.901804	10.803410	10.705459	10.510987	22
23	11.562568	11.454153	11.346180	11.238682	11.131688	10.919332	23
24	12.023667	11.905636	11.788104	11.671137	11.554683	11.323692	24
25	12.483116	12.355067	12.227577	12.100691	11.974452	11.724079	25
26	12.940918	12.802447	12.664603	12.527437	12.391000	12.120503	26
27	13.397073	13.247778	13.099184	12.951350	12.804334	12.512978	27
28	13.851580	13.691062	13.531324	13.372434	13.214460	12.901515	28
29	14.304443	14.132301	13.961025	13.790694	13.621385	13.286129	29
30	14.755660	14.571496	14.388290	14.206134	14.025117	13.666831	30
31	15.205234	15.008649	14.813123	14.618761	14.425662	14.043637	31
32	15.653166	15.443762	15.235526	15.028578	14.823028	14.416560	32
33	16.099455	15.876837	15.655504	15.435591	15.217224	14.785616	33
34	16.544104	16.307875	16.073059	15.839805	15.608256	15.150819	34
35	16.987114	16.736881	16.488196	16.241227	15.996134	15.512185	35
36	17.428485	17.163854	16.900917	16.639861	16.380865	15.869729	36
37	17.868218	17.588798	17.311227	17.035715	16.762459	16.223469	37
38	18.306316	18.011714	17.719130	17.428793	17.140924	16.573421	38
39	18.742779	18.432606	18.124629	17.819102	17.516269	16.919601	39
40	19.177608	18.851475	18.527729	18.206649	17.888504	17.262027	40
45	21.327295	20.915572	20.507385	20.103176	19.703364	18.918481	45
50	23.436345	22.929500	22.427723	21.931675	21.441976	20.483886	50
55	25.504949	24.893621	24.289356	23.693098	23.105724	21.960775	55
60	27.533314	26.808336	26.092957	25.388481	24.696103	23.351850	60
65	29.521668	28.674075	27.839250	27.018935	26.214708	24.659959	65
70	31.470255	30.491304	29.529011	28.585644	27.663230	25.888073	70
75	33.379339	32.260519	31.163064	30.089858	29.043439	27.039258	75
80	35.249199	33.982246	32.742277	31.532887	30.357178	28.116656	80
90	38.872449	37.285482	35.739868	34.240892	32.792924	30.062879	90
100	42.342569	40.405770	38.529525	36.721122	34.986282	31.752485	100

Arithmetic Gradient Factors

TABLE C-2
Factors (GUS) to Convert an Arithmetic Gradient to an Equivalent Uniform Series for Interest Rates from 3% to 5½%

n	3%	3½%	4%	4½%	5%	5½%	n
1	1.000000	1.000000	1.000000	1.000000	1.000000	1.000000	1
2	1.492611	1.491400	1.490196	1.488998	1.487805	1.486618	2
3	1.980297	1.977070	1.973860	1.970665	1.967486	1.964323	3
4	2.463061	2.457013	2.450995	2.445009	2.439053	2.433128	4
5	2.940905	2.931232	2.921611	2.912040	2.902520	2.893051	5
6	3.413833	3.399736	3.385715	3.371771	3.357904	3.344115	6
7	3.881851	3.862530	3.843318	3.824216	3.805225	3.786347	7
8	4.344963	4.319624	4.294434	4.269395	4.244510	4.219780	8
9	4.803176	4.771027	4.739077	4.707328	4.675786	4.644452	9
10	5.256498	5.216752	5.177264	5.138040	5.099085	5.060406	10
11	5.704936	5.656811	5.609014	5.561556	5.514444	5.467688	11
12	6.148499	6.091217	6.034348	5.977905	5.921902	5.866350	12
13	6.587198	6.519987	6.453288	6.387120	6.321501	6.256448	13
14	7.021042	6.943137	6.865859	6.789235	6.713289	6.638043	14
15	7.450043	7.360685	7.272087	7.184286	7.097314	7.011201	15
16	7.874214	7.772649	7.672000	7.572313	7.473629	7.375989	16
17	8.293567	8.179050	8.065628	7.953357	7.842292	7.732481	17
18	8.708116	8.579910	8.453002	8.327463	8.203360	8.080755	18
19	9.117876	8.975252	8.834156	8.694677	8.556896	8.420890	19
20	9.522862	9.365099	9.209125	9.055047	8.902965	8.752971	20
21	9.923090	9.749476	9.577945	9.408624	9.241635	9.077086	21
22	10.318577	10.128411	9.940654	9.755462	9.572976	9.393325	22
23	10.709341	10.501929	10.297292	10.095615	9.897062	9.701784	23
24	11.095401	10.870059	10.647901	10.429140	10.213968	10.002558	24
25	11.476774	11.232832	10.992523	10.756096	10.523771	10.295749	25
26	11.853482	11.590277	11.331203	11.076543	10.826553	10.581457	26
27	12.225544	11.942426	11.663985	11.390545	11.122396	10.859789	27
28	12.592982	12.289312	11.990917	11.698166	11.411383	11.130851	28
29	12.955818	12.630969	12.312048	11.999470	11.693601	11.394753	29
30	13.314074	12.967430	12.627426	12.294527	11.969139	11.651606	30
31	13.667774	13.298732	12.937102	12.583404	12.238085	11.901522	31
32	14.016941	13.624910	13.241128	12.866172	12.500532	12.144617	32
33	14.361599	13.946002	13.539558	13.142902	12.756571	12.381006	33
34	14.701775	14.262046	13.832444	13.413667	13.006297	12.610807	34
35	15.037493	14.573081	14.119843	13.678541	13.249805	12.834138	35
36	15.368780	14.879147	14.401810	13.937599	13.487191	13.051117	36
37	15.695663	15.180283	14.678402	14.190916	13.718552	13.261866	37
38	16.018169	15.476532	14.949677	14.438571	13.943987	13.466504	38
39	16.336326	15.767935	15.215693	14.680641	14.163593	13.665152	39
40	16.650163	16.054535	15.476511	14.917203	14.377471	13.857932	40
45	18.155570	17.417014	16.704737	16.020204	15.364439	14.738056	45
50	19.557509	18.666129	17.812249	16.997616	16.223265	15.489591	50
55	20.860036	19.807782	18.807041	17.859669	16.966450	16.127240	55
60	22.067416	20.848078	19.697232	18.616648	17.606179	16.665015	60
65	23.184072	21.793239	20.490935	19.278216	18.154103	17.116004	65
70	24.214541	22.649525	21.196141	19.854269	18.621186	17.492218	70
75	25.163425	23.423157	21.820622	20.353824	19.017587	17.804514	75
80	26.035345	24.120254	22.371849	20.785434	19.352602	18.062568	80
90	27.566654	25.308486	23.282554	21.475908	19.871195	18.448920	90
100	28.844447	26.259228	23.980000	21.981351	20.233724	18.706688	100

TABLE C-3
Factors (GUS) to Convert an Arithmetic Gradient to an Equivalent Uniform Series for Interest Rates from 6% to 8½%

n	6%	6½%	7%	7½%	8%	8½%	n
1	1.000000	1.000000	1.000000	1.000000	1.000000	1.000000	1
2	1.485437	1.484262	1.483092	1.481928	1.480769	1.479616	2
3	1.961176	1.958045	1.954929	1.951828	1.948743	1.945674	3
4	2.427234	2.421370	2.415536	2.409733	2.403960	2.398217	4
5	2.883633	2.874266	2.864950	2.855685	2.846472	2.837309	5
6	3.330404	3.316771	3.303217	3.289742	3.276346	3.263029	6
7	3.767581	3.748929	3.730392	3.711971	3.693665	3.675476	7
8	4.195208	4.170794	4.146541	4.122451	4.098524	4.074762	8
9	4.613331	4.582426	4.551740	4.521274	4.491033	4.461018	9
10	5.022007	4.983894	4.946071	4.908543	4.871314	4.834388	10
11	5.421295	5.375273	5.329629	5.284371	5.239503	5.195032	11
12	5.811261	5.756646	5.702516	5.648880	5.595747	5.543126	12
13	6.191977	6.128104	6.064842	6.002206	5.940207	5.878856	13
14	6.563521	6.489743	6.416727	6.344491	6.273051	6.202422	14
15	6.925976	6.841665	6.758295	6.675886	6.594460	6.514036	15
16	7.279428	7.183982	7.089681	6.996553	6.904626	6.813921	16
17	7.623972	7.516807	7.411025	7.306660	7.203746	7.102309	17
18	7.959705	7.840262	7.722474	7.606383	7.492028	7.379442	18
19	8.286728	8.154473	8.024182	7.895905	7.769688	7.645569	19
20	8.605148	8.459571	8.316307	8.175416	8.036948	7.900947	20
21	8.915075	8.755692	8.599014	8.445109	8.294034	8.145839	21
22	9.216625	9.042977	8.872471	8.705185	8.541181	8.380514	22
23	9.509914	9.321570	9.136853	8.955848	8.778626	8.605244	23
24	9.795065	9.591619	9.392336	9.197308	9.006612	8.820305	24
25	10.072201	9.853277	9.639101	9.429776	9.225382	9.025976	25
26	10.341450	10.106696	9.877332	9.653468	9.435184	9.222538	26
27	10.602942	10.352035	10.107217	9.868600	9.636268	9.410272	27
28	10.856809	10.589454	10.328943	10.075392	9.828883	9.589461	28
29	11.103187	10.819115	10.542701	10.274065	10.013281	9.760386	29
30	11.342211	11.041181	10.748684	10.464839	10.189712	9.923326	30
31	11.574020	11.255816	10.947084	10.647936	10.358427	10.078561	31
32	11.798753	11.463189	11.138096	10.823578	10.519675	10.226366	32
33	12.016552	11.663465	11.321912	10.991985	10.673702	10.367013	33
34	12.227559	11.856812	11.498727	11.153378	10.820753	10.500772	34
35	12.431916	12.043398	11.668734	11.307973	10.961072	10.627907	35
36	12.629766	12.223392	11.832126	11.455989	11.094897	10.748680	36
37	12.821253	12.396960	11.989095	11.597639	11.222464	10.863345	37
38	13.006521	12.564271	12.139829	11.733135	11.344005	10.972154	38
39	13.185715	12.725490	12.284519	11.862686	11.459749	11.075350	39
40	13.358976	12.880784	12.423349	11.986499	11.569919	11.173174	40
45	14.141295	13.574065	13.035990	12.526456	12.044652	11.589620	45
50	14.796428	14.143134	13.528679	12.951727	12.410714	11.903918	50
55	15.341117	14.606531	13.921458	13.283528	12.690151	12.138607	55
60	15.790945	14.981101	14.232092	13.540199	12.901538	12.312194	60
65	16.160118	15.281819	14.475976	13.737221	13.060157	12.439508	65
70	16.461348	15.521744	14.666187	13.887407	13.178318	12.532188	70
75	16.705829	15.712076	14.813648	14.001175	13.265775	12.599210	75
80	16.903279	15.862282	14.927347	14.086872	13.330132	12.647393	80
90	17.189123	16.072524	15.081217	14.199017	13.411584	12.706382	90
100	17.371073	16.200182	15.170336	14.260978	13.454520	12.736054	100

Arithmetic Gradient Factors

TABLE C-4
Factors (GUS) to Convert an Arithmetic Gradient to an Equivalent Uniform Series for Interest Rates from 9% to 20%

n	9%	10%	11%	12%	15%	20%	n
1	1.000000	1.000000	1.000000	1.000000	1.000000	1.000000	1
2	1.478469	1.476190	1.473934	1.471698	1.465116	1.454545	2
3	1.942619	1.936556	1.930553	1.924609	1.907127	1.879121	3
4	2.392504	2.381168	2.369951	2.358852	2.326257	2.274218	4
5	2.828197	2.810126	2.792259	2.774595	2.722815	2.640507	5
6	3.249792	3.223557	3.197642	3.172047	3.097190	2.978828	6
7	3.657404	3.621615	3.586301	3.551465	3.449850	3.290163	7
8	4.051166	4.004479	3.958469	3.913144	3.781329	3.575623	8
9	4.431231	4.372351	4.314409	4.257417	4.092226	3.836424	9
10	4.797768	4.725461	4.654416	4.584653	4.383196	4.073862	10
11	5.150964	5.064054	4.978808	4.895255	4.654941	4.289291	11
12	5.491023	5.388402	5.287932	5.189653	4.908205	4.484102	12
13	5.818164	5.698792	5.582155	5.468304	5.143760	4.659700	13
14	6.132618	5.995529	5.861865	5.731688	5.362408	4.817486	14
15	6.434631	6.278933	6.127467	5.980303	5.564961	4.958841	15
16	6.724460	6.549341	6.379382	6.214664	5.752246	5.085109	16
17	7.002375	6.807097	6.618043	6.435297	5.925089	5.197588	17
18	7.268653	7.052560	6.843894	6.642737	6.084312	5.297515	18
19	7.523580	7.286095	7.057386	6.837524	6.230729	5.386067	19
20	7.767450	7.508075	7.258975	7.020203	6.365137	5.464347	20
21	8.000563	7.718878	7.449122	7.191317	6.488316	5.533386	21
22	8.223224	7.918886	7.628289	7.351407	6.601020	5.594142	22
23	8.435742	8.108483	7.796935	7.501007	6.703979	5.647495	23
24	8.638428	8.288054	7.955518	7.640645	6.797894	5.694255	24
25	8.831597	8.457982	8.104490	7.770840	6.883433	5.735159	25
26	9.015563	8.618650	8.244300	7.892097	6.961234	5.770876	26
27	9.190639	8.770437	8.375387	8.004912	7.031900	5.802010	27
28	9.357141	8.913716	8.498181	8.109764	7.096002	5.829106	28
29	9.515378	9.048858	8.613103	8.207117	7.154077	5.852652	29
30	9.665661	9.176226	8.720564	8.297419	7.206627	5.873084	30
31	9.808293	9.296174	8.820961	8.381102	7.254123	5.890788	31
32	9.943578	9.409051	8.914681	8.458580	7.297003	5.906109	32
33	10.071812	9.515196	9.002095	8.530248	7.335673	5.919352	33
34	10.193286	9.614940	9.083565	8.596486	7.370512	5.930785	34
35	10.308285	9.708603	9.159435	8.657653	7.401867	5.940645	35
36	10.417091	9.796497	9.230038	8.714091	7.430061	5.949138	36
37	10.519976	9.878922	9.295690	8.766126	7.455389	5.956448	37
38	10.617205	9.956169	9.356697	8.814063	7.478122	5.962733	38
39	10.709039	10.028516	9.413347	8.858194	7.498509	5.968132	39
40	10.795729	10.096234	9.465918	8.898791	7.516777	5.972766	40
45	11.160285	10.374048	9.676278	9.057239	7.582988	5.987692	45
50	11.429518	10.570413	9.818526	9.159724	7.620484	5.994505	50
55	11.626138	10.707539	9.913491	9.225134	7.641421	5.997571	55
60	11.768315	10.802294	9.976199	9.266414	7.652977	5.998935	60
65	11.870233	10.867176	10.017219	9.292217	7.659294	5.999536	65
70	11.942733	10.911252	10.043835	9.308215	7.662720	5.999799	70
75	11.993959	10.940986	10.060986	9.318065	7.664564	5.999914	75
80	12.029938	10.960026	10.071970	9.324093	7.665552	5.999963	80
90	12.072559	10.983057	10.083407	9.329986	7.666357	5.999993	90
100	12.093022	10.992743	10.087973	9.332136	7.666582	5.999999	100

TABLE C-5

Factors (GUS) to Convert an Arithmetic Gradient to an Equivalent Uniform Series for Interest Rates from 25% to 50%

n	25%	30%	35%	40%	45%	50%	n
1	1.000000	1.000000	1.000000	1.000000	1.000000	1.000000	1
2	1.444444	1.434783	1.425532	1.416667	1.408163	1.400000	2
3	1.852459	1.827068	1.802876	1.779817	1.757825	1.736842	3
4	2.224932	2.178277	2.134124	2.092342	2.052805	2.015385	4
5	2.563065	2.490308	2.422025	2.357989	2.297962	2.241706	5
6	2.868332	2.765447	2.669834	2.581099	2.498818	2.422556	6
7	3.142434	3.006282	2.881146	2.766351	2.661166	2.564837	7
8	3.387248	3.215595	3.059727	2.918516	2.790743	2.675178	8
9	3.604777	3.396273	3.209369	3.042242	2.892965	2.759637	9
10	3.797098	3.551219	3.333762	3.141904	2.972750	2.823524	10
11	3.966314	3.683277	3.436394	3.221488	3.034422	2.871341	11
12	4.114516	3.795171	3.520482	3.284537	3.081675	2.906793	12
13	4.243742	3.889458	3.588934	3.334123	3.117596	2.932858	13
14	4.355948	3.968501	3.644325	3.372866	3.144709	2.951878	14
15	4.452988	4.034445	3.688903	3.402955	3.165045	2.965667	15
16	4.536596	4.089214	3.724598	3.426198	3.180210	2.975604	16
17	4.608375	4.134513	3.753049	3.444063	3.191463	2.982728	17
18	4.669792	4.171834	3.775630	3.457734	3.199774	2.987812	18
19	4.722177	4.202472	3.793483	3.468154	3.205887	2.991425	19
20	4.766726	4.227541	3.807549	3.476067	3.210367	2.993984	20
21	4.804506	4.247990	3.818594	3.482057	3.213639	2.995789	21
22	4.836462	4.264623	3.827243	3.486576	3.216021	2.997059	22
23	4.863426	4.278116	3.833996	3.489978	3.217752	2.997950	23
24	4.886125	4.289037	3.839256	3.492531	3.219005	2.998574	24
25	4.905195	4.297854	3.843344	3.494443	3.219911	2.999010	25
26	4.921182	4.304959	3.846514	3.495872	3.220565	2.999314	26
27	4.934560	4.310673	3.848968	3.496938	3.221035	2.999525	27
28	4.945735	4.315261	3.850864	3.497732	3.221373	2.999671	28
29	4.955055	4.318937	3.852326	3.498322	3.221616	2.999773	29
30	4.962816	4.321879	3.853452	3.498760	3.221790	2.999844	30
31	4.969269	4.324229	3.854318	3.499085	3.221914	2.999892	31
32	4.974627	4.326105	3.854983	3.499325	3.222003	2.999926	32
33	4.979070	4.327599	3.855493	3.499503	3.222066	2.999949	33
34	4.982751	4.328789	3.855884	3.499634	3.222111	2.999965	34
35	4.985797	4.329735	3.856183	3.499731	3.222143	2.999976	35
36	4.988314	4.330486	3.856411	3.499802	3.222166	2.999984	36
37	4.990392	4.331083	3.856586	3.499855	3.222183	2.999989	37
38	4.992106	4.331555	3.856719	3.499894	3.222194	2.999992	38
39	4.993519	4.331930	3.856821	3.499922	3.222202	2.999995	39
40	4.994682	4.332226	3.856898	3.499943	3.222208	2.999996	40
45	4.998040	4.332998	3.857081	3.499988	3.222220	2.999999	45
50	4.999286	4.333233	3.857128	3.499998	3.222222	3.000000	50
55	4.999743	4.333304	3.857139	3.499999	3.222222	3.000000	55
60	4.999908	4.333325	3.857142	3.500000	3.222222	3.000000	60
65	4.999967	4.333331	3.857143	3.500000	3.222222	3.000000	65
70	4.999988	4.333333	3.857143	3.500000	3.222222	3.000000	70
75	4.999996	4.333333	3.857143	3.500000	3.222222	3.000000	75
80	4.999999	4.333333	3.857143	3.500000	3.222222	3.000000	80
90	5.000000	4.333333	3.857143	3.500000	3.222222	3.000000	90
100	5.000000	4.333333	3.857143	3.500000	3.222222	3.000000	100

Arithmetic Gradient Factors

TABLE C-6
Factors (GPW) to Convert an Arithmetic Gradient to Present Worth for Interest Rates from 1% to 2½%

n	1%	1¼%	1½%	1¾%	2%	2½%	n
1	0.990099	0.987654	0.985222	0.982801	0.980392	0.975610	1
2	2.950691	2.938576	2.926545	2.914597	2.902730	2.879239	2
3	5.862462	5.828831	5.795496	5.762452	5.729697	5.665037	3
4	9.706383	9.634929	9.564233	9.494286	9.425078	9.288839	4
5	14.463711	14.333814	14.205535	14.078849	13.953732	13.708111	5
6	20.115983	19.902863	19.692788	19.485704	19.281561	18.881892	6
7	26.645009	26.319875	25.999975	25.685211	25.375482	24.770749	7
8	34.032875	33.563062	33.101664	32.648503	32.203405	31.336721	8
9	42.261933	41.611048	40.972995	40.347475	39.734202	38.543276	9
10	51.314803	50.442858	49.589667	48.754761	47.937685	46.355260	10
11	61.174364	60.037910	58.927932	57.843719	56.784579	54.738853	11
12	71.823755	70.376013	68.964581	67.588414	66.246497	63.661524	12
13	83.246368	81.437358	79.676933	77.963601	76.295920	73.091989	13
14	95.425850	93.202511	91.042823	88.944709	86.906170	83.000169	14
15	108.346092	105.652409	103.040595	100.507828	98.051391	93.357153	15
16	121.991232	118.768350	115.649092	112.629689	109.706524	104.135152	16
17	136.345649	132.531993	128.847641	125.287652	121.847288	115.307468	17
18	151.393961	146.925344	142.616050	138.459690	134.450156	126.848454	18
19	167.121020	161.930759	156.934592	152.124376	147.492341	138.733481	19
20	183.511909	177.530930	171.784000	166.260868	160.951767	150.938899	20
21	200.551942	193.708885	187.145457	180.848893	174.807060	163.442011	21
22	218.226659	210.447980	203.000585	195.868740	189.037518	176.221034	22
23	236.521820	227.731894	219.331438	211.301237	203.623105	189.255071	23
24	255.423407	245.544624	236.120492	227.127749	218.544420	202.524079	24
25	274.917618	263.870477	253.350638	243.330157	233.782692	216.008844	25
26	294.990865	282.694070	271.005171	259.890849	249.319753	229.690947	26
27	315.629771	302.000319	289.067786	276.792709	265.138029	243.552739	27
28	336.821167	321.774438	307.522565	294.019101	281.220516	257.577317	28
29	358.552089	342.001931	326.353972	311.553863	297.550773	271.748494	29
30	380.809777	362.668592	345.546845	329.381292	314.112900	286.050774	30
31	403.581669	383.760492	365.086387	347.486134	330.891525	300.469333	31
32	426.855400	405.263982	384.958161	365.853573	347.871790	314.989991	32
33	450.618801	427.165685	405.148078	384.469221	365.039338	329.599189	33
34	474.859895	449.452491	425.642397	403.319106	382.380296	344.283971	34
35	499.566892	472.111553	446.427710	422.389665	399.881263	359.031958	35
36	524.728190	495.130283	467.490940	441.667730	417.529296	373.831332	36
37	550.332371	518.496346	488.819335	461.140524	435.311900	388.670813	37
38	576.368199	542.197658	510.400455	480.795645	453.217014	403.539640	38
39	602.824617	566.222379	532.222174	500.621061	471.232994	418.427554	39
40	629.690742	590.558913	554.272667	520.605130	489.348611	433.324779	40
45	769.798232	716.529200	667.563322	622.520229	581.055348	507.640052	45
50	918.613747	848.686724	784.963297	726.842223	673.784195	580.970363	50
55	1074.961977	985.829388	905.299929	832.467548	766.527477	652.542011	55
60	1237.761183	1126.877443	1027.547661	938.459432	858.458421	721.774305	60
65	1406.016889	1270.863112	1150.812967	1044.027775	948.906004	788.245168	65
70	1578.815953	1416.921019	1274.320688	1148.511242	1037.332935	851.662124	70
75	1755.321018	1564.279362	1397.401648	1251.361318	1123.316605	911.837801	75
80	1934.765305	1712.251787	1519.481436	1352.128127	1206.531281	968.669884	80
90	2299.728364	2007.676321	1758.753729	1546.031422	1363.757012	1072.215671	90
100	2668.804632	2299.142424	1989.075315	1728.150094	1507.851098	1162.588842	100

TABLE C-7
FACTORS (GPW) TO CONVERT AN ARITHMETIC GRADIENT TO PRESENT WORTH FOR INTEREST RATES FROM 3% TO 5½%

n	3%	3½%	4%	4½%	5%	5½%	n
1	0.970874	0.966184	0.961538	0.956938	0.952381	0.947867	1
2	2.856066	2.833205	2.810651	2.788398	2.766440	2.744772	2
3	5.601491	5.539033	5.477640	5.417288	5.357953	5.299613	3
4	9.155439	9.024802	8.896857	8.771533	8.648763	8.528480	4
5	13.468483	13.234668	13.006492	12.783788	12.566393	12.354152	5
6	18.493388	18.115672	17.748379	17.391163	17.043686	16.705627	6
7	24.185029	23.617608	23.067804	22.534962	22.018455	21.517685	7
8	30.500303	29.692901	28.913326	28.160443	27.433170	26.730475	8
9	37.398053	36.296480	35.236606	34.216583	33.234650	32.289139	9
10	44.838992	43.385668	41.992248	40.655859	39.373783	38.143445	10
11	52.785626	50.920071	49.137638	47.434046	45.805255	44.247460	11
12	61.202185	58.861470	56.632803	54.510012	52.487304	50.559238	12
13	70.054552	67.173724	64.440266	61.845543	59.381482	57.040527	13
14	79.310202	75.822669	72.524917	69.405163	66.452433	63.656498	14
15	88.938131	84.776029	80.853885	77.155970	73.667689	70.375494	15
16	98.908802	94.003323	89.396416	85.067479	80.997474	77.168791	16
17	109.194082	103.475787	98.123761	93.111478	88.414517	84.010382	17
18	119.767184	113.166288	107.009067	101.261884	95.893889	90.876769	18
19	130.602619	123.049246	116.027273	109.494618	103.412834	97.746771	19
20	141.676134	133.100564	125.155012	117.787476	110.950624	104.601350	20
21	152.964669	143.297553	134.370518	126.120011	118.488414	111.423443	21
22	164.446304	153.618867	143.653536	134.473431	126.009111	118.197808	22
23	176.100214	164.044436	152.985242	142.830484	133.497251	124.910879	23
24	187.906624	174.555407	162.348157	151.175367	140.938881	131.550637	24
25	199.846763	185.134082	171.726077	159.493632	148.321450	138.106479	25
26	211.902826	195.763861	181.103997	167.772097	155.633709	144.569111	26
27	224.057930	206.429192	190.468045	175.998764	162.865614	150.930432	27
28	236.296080	217.115514	199.805414	184.162743	170.008236	157.183443	28
29	248.602124	227.809210	209.104305	192.254179	177.053679	163.322147	29
30	260.961727	238.497562	218.353865	200.264179	183.995002	169.341467	30
31	273.361328	249.168703	227.544133	208.184754	190.826146	175.237168	31
32	285.788113	259.811574	236.665987	216.008751	197.541863	181.005779	32
33	298.229980	270.415883	245.711095	223.729801	204.137657	186.644528	33
34	310.675506	280.972069	254.671866	231.342261	210.609720	192.151277	34
35	323.113925	291.471259	263.541407	238.841167	216.954880	197.524464	35
36	335.535092	301.905237	272.313481	246.222181	223.170547	202.763049	36
37	347.929461	312.266407	280.982465	253.481552	229.254666	207.866462	37
38	360.288055	322.547760	289.543311	260.616068	235.205670	212.834560	38
39	372.602443	332.742848	297.991515	267.623023	241.022440	217.667581	39
40	384.864717	342.845747	306.323076	274.500171	246.704268	222.366106	40
45	445.151195	391.803577	346.122819	306.888591	273.088608	243.881305	45
50	503.210098	437.825593	382.646031	335.907014	296.170708	262.262284	50
55	558.515536	480.617074	415.797587	361.622942	316.143869	277.794465	55
60	610.728173	520.049756	445.620138	384.208723	333.272452	290.802089	60
65	659.653891	556.116592	472.248061	403.897492	347.852048	301.614235	65
70	705.210294	588.896637	495.873438	420.952429	360.183575	310.544662	70
75	747.399673	618.528016	516.721243	435.645789	370.557116	317.880856	75
80	786.287287	645.187277	535.031576	448.245049	379.242509	323.879057	80
90	854.632621	690.398170	565.004201	468.158480	392.501053	332.725430	90
100	911.452950	726.210120	587.629874	482.487410	401.597134	338.513222	100

Arithmetic Gradient Factors

TABLE C-8
Factors (GPW) to Convert an Arithmetic Gradient to Present Worth for Interest Rates from 6% to 8½%

n	6%	6½%	7%	7½%	8%	8½%	n
1	0.943396	0.938967	0.934579	0.930233	0.925926	0.921659	1
2	2.723389	2.702286	2.681457	2.660898	2.640604	2.620570	2
3	5.242247	5.185833	5.130351	5.075779	5.022100	4.969294	3
4	8.410622	8.295125	8.181931	8.070982	7.962220	7.855591	4
5	12.146912	11.944530	11.746862	11.553775	11.365136	11.180818	5
6	16.376676	16.056534	15.744916	15.441544	15.146153	14.858489	6
7	21.032076	20.561078	20.104164	19.660828	19.230586	18.812973	7
8	26.051374	25.394927	24.760237	24.146446	23.552737	22.978329	8
9	31.378461	30.501106	29.655640	28.840697	28.054978	27.297246	9
10	36.962408	35.828367	34.739133	33.692637	32.686913	31.720100	10
11	42.757071	41.330701	39.965154	38.657412	37.404624	36.204100	11
12	48.720704	46.966896	45.293298	43.695661	42.169989	40.712520	12
13	54.815611	52.700113	50.687835	48.772967	46.950062	45.214015	13
14	61.007824	58.497517	56.117277	53.859356	51.716517	49.682000	14
15	67.266800	64.329915	61.553967	58.928846	56.445143	54.094098	15
16	73.565141	70.171440	66.973721	63.959038	61.115390	58.431645	16
17	79.878336	75.999253	72.355485	68.930739	65.709962	62.679242	17
18	86.184524	81.793267	77.681036	73.827627	70.214445	66.824363	18
19	92.464271	87.535900	82.934694	78.635940	74.616974	70.856995	19
20	98.700366	93.211841	88.103074	83.344203	78.907938	74.769323	20
21	104.877629	98.807838	93.174849	87.942972	83.079709	78.555447	21
22	110.982741	104.312509	98.140539	92.424607	87.126400	82.211129	22
23	117.004078	109.716154	102.992317	96.783068	91.043652	85.733571	23
24	122.931564	115.010602	107.723836	101.013727	94.828436	89.121213	24
25	128.756529	120.189052	112.330065	105.113204	98.478883	92.373558	25
26	134.471590	125.245943	116.807148	109.079209	101.994129	95.491012	26
27	140.070525	130.176822	121.152268	112.910413	105.374173	98.474751	27
28	145.548169	134.978234	125.363530	116.606320	108.619757	101.326593	28
29	150.900314	139.647616	129.439852	120.167160	111.732255	104.048890	29
30	156.123618	144.183198	133.380865	123.593791	114.713575	106.644439	30
31	161.215518	148.583919	137.186828	126.887607	117.566073	109.116389	31
32	166.174155	152.849346	140.858544	130.050461	120.292474	111.468178	32
33	170.998300	156.979601	144.397289	133.084594	122.895809	113.703462	33
34	175.687292	160.975296	147.804747	135.992572	125.379349	115.826060	34
35	180.240975	164.837469	151.082950	138.777229	127.746558	117.839910	35
36	184.659643	168.567536	154.234226	141.441619	130.001043	119.749025	36
37	188.943991	172.167235	157.261153	143.988969	132.146515	121.557453	37
38	193.095067	175.638586	160.166514	146.422640	134.186753	123.269255	38
39	197.114232	178.983846	162.953259	148.746097	136.125576	124.888472	39
40	201.003120	182.205476	165.624475	150.962871	137.966813	126.419102	40
45	218.565478	196.555560	177.361444	160.571986	145.841488	132.878613	45
50	233.219236	208.250872	186.705865	168.046211	151.826275	137.675898	50
55	245.312789	217.678578	194.064271	173.796399	156.325068	141.199776	55
60	255.204218	225.210383	199.806924	178.180636	159.676559	143.765083	60
65	263.234145	231.182641	204.255148	181.498436	162.154685	145.618616	65
70	269.711679	235.888324	207.678920	183.993316	163.975437	146.949385	70
75	274.908591	239.575923	210.299871	185.859199	165.305933	147.899638	75
80	279.058439	242.452099	212.296804	187.248093	166.273598	148.574981	80
90	284.973307	246.415118	214.957527	189.038108	167.480267	149.390034	90
100	288.664606	248.774746	216.469332	190.008899	168.105044	149.793013	100

TABLE C-9
Factors (GPW) to Convert an Arithmetic Gradient to Present Worth for Interest Rates from 9% to 20%

n	9%	10%	11%	12%	15%	20%	n
1	0.917431	0.909091	0.900901	0.892857	0.869565	0.833333	1
2	2.600791	2.561983	2.524146	2.487245	2.381853	2.222222	2
3	4.917342	4.815928	4.717720	4.622586	4.354401	3.958333	3
4	7.751042	7.547982	7.352644	7.164658	6.641414	5.887346	4
5	11.000699	10.652588	10.319900	10.001792	9.127298	7.896734	5
6	14.578303	14.039432	13.527745	13.041579	11.721263	9.906121	6
7	18.407543	17.631539	16.899354	16.208023	14.352823	11.859693	7
8	22.422473	21.363598	20.370766	19.439089	16.968037	13.720237	8
9	26.566323	25.180476	23.889089	22.684580	19.526399	15.464498	9
10	30.790431	29.035909	27.410934	25.904312	21.998246	17.079553	10
11	35.053293	32.891342	30.901050	29.066549	24.362621	18.560021	11
12	39.319709	36.714912	34.331140	32.146650	26.605507	19.905901	12
13	43.560032	40.480549	37.678826	35.125915	28.718370	21.120931	13
14	47.749482	44.167186	40.926753	37.990592	30.696972	22.211343	14
15	51.867553	47.758067	44.061818	40.731036	32.540389	23.184925	15
16	55.897469	51.240133	47.074494	43.340982	34.250225	24.050332	16
17	59.825713	54.603493	49.958248	45.816936	35.829965	24.816577	17
18	63.641601	57.840951	52.709047	48.157649	37.284457	25.492676	18
19	67.336904	60.947603	55.324922	50.363678	38.619499	26.087392	19
20	70.905522	63.920475	57.805601	52.437013	39.841504	26.609073	20
21	74.343181	66.758217	60.152188	54.380765	40.957248	27.065544	21
22	77.647179	69.460829	62.366899	56.198900	41.973661	27.464050	22
23	80.816151	72.029426	64.452828	57.896023	42.897673	27.811234	23
24	83.849869	74.466041	66.413747	59.477194	43.736096	28.113133	24
25	86.749065	76.773441	68.253949	60.947776	44.495537	28.375198	25
26	89.515271	78.954982	69.978103	62.313317	45.182336	28.602321	26
27	92.150681	81.014480	71.591136	63.579444	45.802522	28.798869	27
28	94.658038	82.956094	73.098141	64.751783	46.361788	28.968726	28
29	97.040520	84.784237	74.504290	65.835898	46.865475	29.115329	29
30	99.301654	86.503493	75.814775	66.837236	47.318566	29.241710	30
31	101.445236	88.118552	77.034746	67.761089	47.725692	29.350539	31
32	103.475264	89.634150	78.169272	68.612567	48.091135	29.444155	32
33	105.395876	91.055024	79.223309	69.396573	48.418841	29.524606	33
34	107.211300	92.385869	80.201666	70.117790	48.712439	29.593681	34
35	108.925812	93.631313	81.108993	70.780673	48.975250	29.652935	35
36	110.543700	94.795883	81.949759	71.389444	49.210311	29.703725	36
37	112.069232	95.883992	82.728247	71.948088	49.420389	29.747226	37
38	113.506629	96.899917	83.448542	72.460358	49.608003	29.784456	38
39	114.860045	97.847790	84.114533	72.929778	49.775439	29.816298	39
40	116.133548	98.731587	84.729909	73.359650	49.924769	29.843513	40
45	121.437267	102.317241	87.163043	75.016729	50.459420	29.930276	45
50	125.286749	104.803684	88.775705	76.066910	50.756345	29.969232	50
55	128.050199	106.509029	89.832866	76.725177	50.919437	29.986531	55
60	130.016241	107.668168	90.519658	77.134084	51.008212	29.994143	60
65	131.404516	108.450152	90.962499	77.386188	51.056172	29.997468	65
70	132.378605	108.974355	91.246230	77.540633	51.081917	29.998911	70
75	133.058364	109.323844	91.427029	77.634735	51.095662	29.999533	75
80	133.530493	109.555751	91.541694	77.691799	51.102966	29.999801	80
90	134.082114	109.809901	91.659691	77.746996	51.108868	29.999964	90
100	134.342604	109.919452	91.706157	77.766869	51.110500	29.999994	100

Arithmetic Gradient Factors

TABLE C-10
Factors (GPW) to Convert an Arithmetic Gradient to Present Worth for Interest Rates from 25% to 50%

n	25%	30%	35%	40%	45%	50%	n
1	0.800000	0.769231	0.740741	0.714286	0.689655	0.666667	1
2	2.080000	1.952663	1.838134	1.734694	1.640904	1.555556	2
3	3.616000	3.318161	3.057461	2.827988	2.624954	2.444444	3
4	5.254400	4.718672	4.261734	3.869221	3.529828	3.234568	4
5	6.892800	6.065318	5.376801	4.798893	4.309891	3.893004	5
6	8.465664	7.308375	6.367972	5.595755	4.955461	4.419753	6
7	9.933670	8.423939	7.224540	6.259807	5.474885	4.829447	7
8	11.275848	9.404655	7.949677	6.801890	5.884283	5.141594	8
9	12.483807	10.253351	8.553957	7.237492	6.201920	5.375705	9
10	13.557549	10.978733	9.051307	7.583208	6.445320	5.549120	10
11	14.502442	11.592517	9.456556	7.854842	6.629968	5.676291	11
12	15.327076	12.107581	9.784029	8.066505	6.768888	5.768780	12
13	16.041758	12.536801	10.046816	8.230292	6.872679	5.835577	13
14	16.657485	12.892367	10.256447	8.356282	6.949765	5.883533	14
15	17.185250	13.185417	10.422821	8.452703	7.006725	5.917788	15
16	17.635610	13.425868	10.554277	8.526166	7.048627	5.942147	16
17	18.018416	13.622390	10.657737	8.581920	7.079330	5.959402	17
18	18.342675	13.782454	10.738883	8.624086	7.101751	5.971581	18
19	18.616494	13.912420	10.802330	8.655878	7.118073	5.980152	19
20	18.847078	14.017655	10.851801	8.679782	7.129921	5.986166	20
21	19.040769	14.102653	10.890279	8.697710	7.138501	5.990377	21
22	19.203101	14.171150	10.920138	8.711126	7.144700	5.993317	22
23	19.338869	14.226235	10.943262	8.721144	7.149170	5.995367	23
24	19.452205	14.270450	10.961135	8.728611	7.152386	5.996792	24
25	19.546653	14.305879	10.974926	8.734166	7.154697	5.997782	25
26	19.625233	14.334222	10.985550	8.738293	7.156354	5.998469	26
27	19.690515	14.356862	10.993723	8.741355	7.157541	5.998944	27
28	19.744675	14.374924	11.000001	8.743622	7.158390	5.999272	28
29	19.789550	14.389313	11.004817	8.745300	7.158997	5.999499	29
30	19.826688	14.400763	11.008508	8.746539	7.159429	5.999656	30
31	19.857389	14.409865	11.011333	8.747454	7.159738	5.999764	31
32	19.882742	14.417092	11.013493	8.748129	7.159957	5.999838	32
33	19.903659	14.422825	11.015143	8.748626	7.160113	5.999889	33
34	19.920899	14.427369	11.016402	8.748992	7.160224	5.999924	34
35	19.935096	14.430967	11.017362	8.749261	7.160303	5.999948	35
36	19.946779	14.433813	11.018094	8.749458	7.160359	5.999964	36
37	19.956385	14.436064	11.018651	8.749603	7.160398	5.999976	37
38	19.964277	14.437842	11.019075	8.749710	7.160426	5.999983	38
39	19.970757	14.439246	11.019397	8.749788	7.160446	5.999989	39
40	19.976074	14.440353	11.019641	8.749845	7.160460	5.999992	40
45	19.991289	14.443218	11.020218	8.749968	7.160488	5.999999	45
50	19.996860	14.444081	11.020361	8.749993	7.160493	6.000000	50
55	19.998878	14.444337	11.020397	8.749999	7.160494	6.000000	55
60	19.999602	14.444413	11.020405	8.750000	7.160494	6.000000	60
65	19.999859	14.444435	11.020407	8.750000	7.160494	6.000000	65
70	19.999951	14.444442	11.020408	8.750000	7.160494	6.000000	70
75	19.999983	14.444444	11.020408	8.750000	7.160494	6.000000	75
80	19.999994	14.444444	11.020408	8.750000	7.160494	6.000000	80
90	19.999999	14.444444	11.020408	8.750000	7.160494	6.000000	90
100	20.000000	14.444444	11.020408	8.750000	7.160494	6.000000	100

Appendix **D**

Exponential Growth Factors

Appendix D gives the factors for converting exponential growth rates from 1 percent to 10 percent per period to equivalent uniform series (EUS) and to present worth (EPW) for interest rates from 1 percent to 50 percent. The factors are calculated on the basis that the exponential growth starts at time zero ($n = 0$), so that the first increase is effective at $n = 1$. The factors are based upon the end-of-period step convention as applied to the standard compound interest tables in Appendix B. The two formulas from which the exponential tables were calculated are

$$E_t US = \left[\frac{\left(\frac{1+t}{1+i}\right)^{n+1} - 1}{\left(\frac{1+t}{1+i}\right) - 1} - 1 \right] \left(\frac{i(1+i)^n}{(1+i)^n - 1} \right) \qquad (6\text{-}12)$$

$$E_t PW = \left[\frac{\left(\frac{1+t}{1+i}\right)^{n+1} - 1}{\left(\frac{1+t}{1+i}\right) - 1} - 1 \right] \qquad (6\text{-}11)$$

Exponential Growth Factors

TABLE D-1
Factors (EUS) to Convert an Exponential Growth to an Equivalent Uniform Series for Interest Rates from 1% to 2½%
Growth Rate of 1% per Period

n	1%	1¼%	1½%	1¾%	2%	2½%	n
1	1.010000	1.010000	1.010000	1.010000	1.010000	1.010000	1
2	1.015025	1.015019	1.015012	1.015006	1.015000	1.014988	2
3	1.020066	1.020050	1.020033	1.020016	1.020000	1.019967	3
4	1.025124	1.025093	1.025061	1.025030	1.024999	1.024936	4
5	1.030199	1.030148	1.030098	1.030047	1.029997	1.029897	5
6	1.035290	1.035216	1.035142	1.035068	1.034994	1.034847	6
7	1.040398	1.040296	1.040194	1.040092	1.039990	1.039788	7
8	1.045522	1.045387	1.045253	1.045118	1.044985	1.044718	8
9	1.050663	1.050491	1.050319	1.050148	1.049977	1.049637	9
10	1.055821	1.055607	1.055393	1.055180	1.054968	1.054545	10
11	1.060995	1.060734	1.060474	1.060214	1.059956	1.059441	11
12	1.066185	1.065873	1.065562	1.065251	1.064942	1.064326	12
13	1.071393	1.071024	1.070656	1.070290	1.069925	1.069198	13
14	1.076616	1.076186	1.075758	1.075330	1.074905	1.074058	14
15	1.081857	1.081360	1.080866	1.080373	1.079881	1.078904	15
16	1.087114	1.086546	1.085980	1.085416	1.084855	1.083738	16
17	1.092387	1.091743	1.091101	1.090461	1.089824	1.088558	17
18	1.097677	1.096951	1.096228	1.095507	1.094790	1.093364	18
19	1.102983	1.102170	1.101361	1.100554	1.099751	1.098156	19
20	1.108306	1.107401	1.106499	1.105602	1.104708	1.102933	20
21	1.113646	1.112643	1.111644	1.110650	1.109660	1.107695	21
22	1.119002	1.117896	1.116794	1.115698	1.114607	1.112443	22
23	1.124374	1.123160	1.121950	1.120747	1.119549	1.117175	23
24	1.129763	1.128434	1.127111	1.125795	1.124486	1.121891	24
25	1.135169	1.133720	1.132278	1.130844	1.129417	1.126591	25
26	1.140591	1.139016	1.137450	1.135892	1.134343	1.131274	26
27	1.146029	1.144323	1.142626	1.140939	1.139262	1.135941	27
28	1.151484	1.149641	1.147808	1.145985	1.144175	1.140591	28
29	1.156956	1.154969	1.152994	1.151031	1.149081	1.145224	29
30	1.162443	1.160308	1.158185	1.156075	1.153981	1.149839	30
31	1.167948	1.165657	1.163380	1.161118	1.158873	1.154437	31
32	1.173468	1.171016	1.168579	1.166160	1.163759	1.159016	32
33	1.179005	1.176386	1.173783	1.171200	1.168637	1.163578	33
34	1.184559	1.181765	1.178991	1.176238	1.173507	1.168120	34
35	1.190129	1.187155	1.184203	1.181273	1.178369	1.172644	35
36	1.195715	1.192555	1.189418	1.186307	1.183224	1.177149	36
37	1.201318	1.197965	1.194637	1.191338	1.188070	1.181634	37
38	1.206937	1.203384	1.199860	1.196367	1.192907	1.186100	38
39	1.212572	1.208813	1.205086	1.201392	1.197736	1.190546	39
40	1.218224	1.214252	1.210315	1.206415	1.202556	1.194972	40
45	1.246727	1.241590	1.236505	1.231478	1.226514	1.216793	45
50	1.275637	1.269157	1.262755	1.256439	1.250217	1.238080	50
55	1.304951	1.296943	1.289048	1.281277	1.273641	1.258811	55
60	1.334667	1.324939	1.315368	1.305971	1.296762	1.278963	60
65	1.364783	1.353133	1.341697	1.330500	1.319560	1.298521	65
70	1.395297	1.381514	1.368019	1.354844	1.342013	1.317470	70
75	1.426207	1.410071	1.394318	1.378984	1.364102	1.335798	75
80	1.457508	1.438794	1.420576	1.402901	1.385810	1.353498	80
90	1.521276	1.496692	1.472910	1.450000	1.428020	1.386994	90
100	1.586574	1.555117	1.524896	1.496010	1.468532	1.417951	100

TABLE D-2

Factors (EUS) to Convert an Exponential Growth
to an Equivalent Uniform Series for Interest Rates from 3% to 5½%
Growth Rate of 1% per Period

n	3%	3½%	4%	4½%	5%	5½%	n
1	1.010000	1.010000	1.010000	1.010000	1.010000	1.010000	1
2	1.014975	1.014963	1.014951	1.014939	1.014927	1.014915	2
3	1.019934	1.019901	1.019868	1.019836	1.019804	1.019772	3
4	1.024874	1.024813	1.024751	1.024690	1.024630	1.024569	4
5	1.029797	1.029698	1.029599	1.029501	1.029404	1.029307	5
6	1.034701	1.034556	1.034412	1.034268	1.034125	1.033983	6
7	1.039587	1.039387	1.039188	1.038990	1.038794	1.038598	7
8	1.044453	1.044189	1.043927	1.043667	1.043408	1.043151	8
9	1.049299	1.048963	1.048629	1.048297	1.047968	1.047640	9
10	1.054125	1.053707	1.053293	1.052881	1.052472	1.052066	10
11	1.058930	1.058422	1.057918	1.057417	1.056920	1.056427	11
12	1.063714	1.063107	1.062504	1.061906	1.061312	1.060724	12
13	1.068476	1.067760	1.067050	1.066345	1.065647	1.064955	13
14	1.073217	1.072383	1.071556	1.070736	1.069924	1.069120	14
15	1.077935	1.076974	1.076021	1.075078	1.074143	1.073218	15
16	1.082630	1.081533	1.080446	1.079369	1.078304	1.077251	16
17	1.087303	1.086059	1.084828	1.083611	1.082406	1.081216	17
18	1.091951	1.090553	1.089169	1.087801	1.086449	1.085115	18
19	1.096576	1.095013	1.093467	1.091941	1.090433	1.088946	19
20	1.101176	1.099439	1.097723	1.096029	1.094357	1.092710	20
21	1.105752	1.103832	1.101935	1.100065	1.098222	1.096407	21
22	1.110303	1.108189	1.106104	1.104050	1.102026	1.100036	22
23	1.114828	1.112512	1.110230	1.107982	1.105771	1.103598	23
24	1.119328	1.116800	1.114311	1.111862	1.109455	1.107093	24
25	1.123801	1.121053	1.118348	1.115689	1.113079	1.110521	25
26	1.128249	1.125269	1.122340	1.119464	1.116644	1.113881	26
27	1.132669	1.129450	1.126288	1.123186	1.120148	1.117176	27
28	1.137063	1.133594	1.130190	1.126855	1.123592	1.120404	28
29	1.141429	1.137702	1.134048	1.130471	1.126976	1.123566	29
30	1.145768	1.141773	1.137860	1.134034	1.130300	1.126662	30
31	1.150079	1.145807	1.141627	1.137545	1.133565	1.129693	31
32	1.154362	1.149803	1.145348	1.141002	1.136771	1.132659	32
33	1.158616	1.153762	1.149023	1.144406	1.139917	1.135561	33
34	1.162842	1.157683	1.152652	1.147758	1.143005	1.138400	34
35	1.167039	1.161567	1.156236	1.151057	1.146034	1.141175	35
36	1.171207	1.165412	1.159774	1.154303	1.149005	1.143887	36
37	1.175346	1.169219	1.163266	1.157497	1.151919	1.146538	37
38	1.179455	1.172988	1.166712	1.160638	1.154775	1.149127	38
39	1.183534	1.176718	1.170112	1.163728	1.157573	1.151655	39
40	1.187583	1.180410	1.173467	1.166766	1.160316	1.154124	40
45	1.207375	1.198288	1.189554	1.181187	1.173198	1.165593	45
50	1.226394	1.215194	1.204510	1.194359	1.184749	1.175682	50
55	1.244624	1.231130	1.218360	1.206331	1.195047	1.184498	55
60	1.262057	1.246103	1.231135	1.217163	1.204178	1.192155	60
65	1.278689	1.260130	1.242875	1.226919	1.212234	1.198767	65
70	1.294519	1.273232	1.253625	1.235670	1.219306	1.204447	70
75	1.309554	1.285435	1.263434	1.243487	1.225488	1.209301	75
80	1.323802	1.296770	1.272357	1.250445	1.230868	1.213431	80
90	1.349992	1.316977	1.287764	1.262070	1.239561	1.219882	90
100	1.373233	1.334156	1.300291	1.271101	1.246007	1.224445	100

Exponential Growth Factors

TABLE D-3
Factors (EUS) to Convert an Exponential Growth to an Equivalent Uniform Series for Interest Rates from 6% to 8½%
Growth Rate of 1% per Period

n	6%	6½%	7%	7½%	8%	8½%	n
1	1.010000	1.010000	1.010000	1.010000	1.010000	1.010000	1
2	1.014903	1.014891	1.014879	1.014867	1.014856	1.014844	2
3	1.019740	1.019708	1.019676	1.019645	1.019613	1.019582	3
4	1.024509	1.024449	1.024390	1.024331	1.024272	1.024213	4
5	1.029210	1.029114	1.029019	1.028924	1.028829	1.028736	5
6	1.033842	1.033702	1.033562	1.033424	1.033286	1.033149	6
7	1.038404	1.038211	1.038020	1.037829	1.037640	1.037451	7
8	1.042896	1.042642	1.042390	1.042139	1.041891	1.041644	8
9	1.047315	1.046993	1.046672	1.046354	1.046038	1.045725	9
10	1.051663	1.051263	1.050866	1.050473	1.050082	1.049695	10
11	1.055938	1.055453	1.054972	1.054495	1.054023	1.053554	11
12	1.060140	1.059562	1.058989	1.058421	1.057859	1.057302	12
13	1.064269	1.063589	1.062916	1.062251	1.061592	1.060940	13
14	1.068323	1.067535	1.066755	1.065983	1.065221	1.064467	14
15	1.072303	1.071399	1.070504	1.069620	1.068747	1.067885	15
16	1.076209	1.075180	1.074164	1.073161	1.072171	1.071195	16
17	1.080041	1.078880	1.077736	1.076607	1.075494	1.074397	17
18	1.083798	1.082499	1.081218	1.079957	1.078715	1.077493	18
19	1.087480	1.086035	1.084613	1.083214	1.081837	1.080485	19
20	1.091087	1.089491	1.087920	1.086377	1.084861	1.083373	20
21	1.094621	1.092865	1.091140	1.089447	1.087787	1.086159	21
22	1.098080	1.096159	1.094274	1.092426	1.090616	1.088844	22
23	1.101465	1.099373	1.097322	1.095315	1.093351	1.091432	23
24	1.104776	1.102507	1.100286	1.098115	1.095993	1.093923	24
25	1.108014	1.105562	1.103166	1.100826	1.098543	1.096319	25
26	1.111179	1.108539	1.105963	1.103451	1.101004	1.098623	26
27	1.114272	1.111439	1.108678	1.105990	1.103376	1.100836	27
28	1.117293	1.114262	1.111313	1.108446	1.105662	1.102961	28
29	1.120243	1.117010	1.113868	1.110819	1.107863	1.105000	29
30	1.123122	1.119682	1.116345	1.113111	1.109981	1.106955	30
31	1.125931	1.122281	1.118745	1.115324	1.112019	1.108828	31
32	1.128670	1.124807	1.121070	1.117460	1.113978	1.110623	32
33	1.131342	1.127261	1.123320	1.119520	1.115860	1.112340	33
34	1.133945	1.129644	1.125497	1.121505	1.117668	1.113983	34
35	1.136482	1.131957	1.127603	1.123418	1.119402	1.115553	35
36	1.138952	1.134202	1.129639	1.125261	1.121067	1.117053	36
37	1.141357	1.136380	1.131606	1.127034	1.122662	1.118486	37
38	1.143699	1.138492	1.133506	1.128741	1.124191	1.119854	38
39	1.145976	1.140539	1.135341	1.130382	1.125656	1.121158	39
40	1.148192	1.142522	1.137112	1.131959	1.127058	1.122402	40
45	1.158371	1.151529	1.145061	1.138955	1.133201	1.127784	45
50	1.167150	1.159142	1.151641	1.144626	1.138075	1.131963	50
55	1.174666	1.165523	1.157040	1.149179	1.141903	1.135174	55
60	1.181056	1.170833	1.161434	1.152803	1.144883	1.137616	60
65	1.186455	1.175220	1.164985	1.155666	1.147183	1.139459	65
70	1.190990	1.178823	1.167834	1.157910	1.148945	1.140839	70
75	1.194778	1.181765	1.170107	1.159660	1.150287	1.141865	75
80	1.197928	1.184154	1.171911	1.161016	1.151304	1.142624	80
90	1.202684	1.187640	1.174455	1.162865	1.152643	1.143592	90
100	1.205892	1.189883	1.176016	1.163947	1.153391	1.144108	100

TABLE D-4

Factors (EUS) to Convert an Exponential Growth to an Equivalent Uniform Series for Interest Rates from 9% to 20%

Growth Rate of 1% per Period

n	9%	10%	11%	12%	15%	20%	n
1	1.010000	1.010000	1.010000	1.010000	1.010000	1.010000	1
2	1.014833	1.014810	1.014787	1.014764	1.014698	1.014591	2
3	1.019551	1.019490	1.019429	1.019368	1.019191	1.018907	3
4	1.024155	1.024039	1.023925	1.023812	1.023479	1.022949	4
5	1.028642	1.028457	1.028274	1.028093	1.027563	1.026720	5
6	1.033012	1.032742	1.032475	1.032212	1.031442	1.030224	6
7	1.037265	1.036894	1.036529	1.036169	1.035119	1.033469	7
8	1.041399	1.040914	1.040436	1.039965	1.038597	1.036463	8
9	1.045414	1.044800	1.044195	1.043601	1.041878	1.039214	9
10	1.049311	1.048553	1.047809	1.047078	1.044969	1.041733	10
11	1.053090	1.052175	1.051278	1.050399	1.047872	1.044033	11
12	1.056751	1.055666	1.054604	1.053566	1.050595	1.046125	12
13	1.060295	1.059027	1.057789	1.056581	1.053142	1.048021	13
14	1.063722	1.062260	1.060836	1.059449	1.055520	1.049735	14
15	1.067034	1.065367	1.063746	1.062172	1.057736	1.051279	15
16	1.070232	1.068349	1.066523	1.064755	1.059797	1.052666	16
17	1.073318	1.071209	1.069170	1.067200	1.061711	1.053909	17
18	1.076292	1.073949	1.071689	1.069513	1.063483	1.055019	18
19	1.079156	1.076572	1.074085	1.071698	1.065123	1.056009	19
20	1.081913	1.079079	1.076361	1.073759	1.066637	1.056888	20
21	1.084564	1.081475	1.078520	1.075701	1.068032	1.057668	21
22	1.087111	1.083761	1.080567	1.077528	1.069316	1.058358	22
23	1.089557	1.085941	1.082504	1.079245	1.070495	1.058968	23
24	1.091903	1.088018	1.084337	1.080857	1.071577	1.059505	24
25	1.094152	1.089994	1.086068	1.082369	1.072568	1.059977	25
26	1.096307	1.091874	1.087702	1.083785	1.073474	1.060392	26
27	1.098370	1.093661	1.089243	1.085110	1.074302	1.060755	27
28	1.100343	1.095357	1.090695	1.086349	1.075057	1.061073	28
29	1.102230	1.096966	1.092062	1.087505	1.075745	1.061351	29
30	1.104032	1.098491	1.093347	1.088584	1.076371	1.061593	30
31	1.105752	1.099936	1.094555	1.089589	1.076939	1.061805	31
32	1.107393	1.101303	1.095688	1.090525	1.077456	1.061988	32
33	1.108957	1.102596	1.096752	1.091396	1.077924	1.062148	33
34	1.110448	1.103818	1.097749	1.092205	1.078348	1.062287	34
35	1.111867	1.104972	1.098682	1.092957	1.078732	1.062407	35
36	1.113218	1.106061	1.099556	1.093654	1.079079	1.062511	36
37	1.114502	1.107088	1.100372	1.094300	1.079393	1.062601	37
38	1.115723	1.108057	1.101136	1.094899	1.079676	1.062679	38
39	1.116883	1.108968	1.101849	1.095454	1.079931	1.062746	39
40	1.117984	1.109827	1.102514	1.095967	1.080160	1.062804	40
45	1.122690	1.113405	1.105218	1.098000	1.081006	1.062994	45
50	1.126265	1.116007	1.107098	1.099352	1.081499	1.063083	50
55	1.128951	1.117875	1.108389	1.100239	1.081781	1.063124	55
60	1.130948	1.119202	1.109265	1.100815	1.081941	1.063142	60
65	1.132421	1.120137	1.109855	1.101184	1.082031	1.063151	65
70	1.133498	1.120790	1.110248	1.101420	1.082081	1.063155	70
75	1.134281	1.121243	1.110508	1.101569	1.082109	1.063157	75
80	1.134847	1.121555	1.110680	1.101663	1.082125	1.063157	80
90	1.135545	1.121916	1.110866	1.101759	1.082137	1.063158	90
100	1.135900	1.122083	1.110944	1.101796	1.082141	1.063158	100

Exponential Growth Factors

TABLE D-5
Factors (EUS) to Convert an Exponential Growth to an Equivalent Uniform Series for Interest Rates from 25% to 50%
Growth Rate of 1% per Period

n	25%	30%	35%	40%	45%	50%	n
1	1.010000	1.010000	1.010000	1.010000	1.010000	1.010000	1
2	1.014489	1.014391	1.014298	1.014208	1.014122	1.014040	2
3	1.018636	1.018379	1.018133	1.017899	1.017676	1.017463	3
4	1.022446	1.021971	1.021521	1.021095	1.020693	1.020311	4
5	1.025927	1.025182	1.024483	1.023828	1.023214	1.022639	5
6	1.029088	1.028031	1.027049	1.026137	1.025293	1.024510	6
7	1.031944	1.030539	1.029249	1.028066	1.026982	1.025991	7
8	1.034510	1.032732	1.031119	1.029659	1.028338	1.027145	8
9	1.036803	1.034636	1.032695	1.030961	1.029414	1.028034	9
10	1.038843	1.036278	1.034013	1.032016	1.030259	1.028709	10
11	1.040648	1.037686	1.035106	1.032864	1.030915	1.029218	11
12	1.042237	1.038885	1.036007	1.033539	1.031420	1.029597	12
13	1.043632	1.039901	1.036744	1.034072	1.031807	1.029877	13
14	1.044849	1.040758	1.037343	1.034491	1.032100	1.030083	14
15	1.045907	1.041476	1.037829	1.034819	1.032321	1.030233	15
16	1.046825	1.042076	1.038219	1.035073	1.032486	1.030341	16
17	1.047616	1.042575	1.038533	1.035269	1.032610	1.030419	17
18	1.048298	1.042989	1.038782	1.035420	1.032702	1.030475	18
19	1.048882	1.043330	1.038981	1.035536	1.032770	1.030515	19
20	1.049381	1.043611	1.039138	1.035624	1.032820	1.030544	20
21	1.049807	1.043841	1.039262	1.035692	1.032856	1.030564	21
22	1.050169	1.044029	1.039360	1.035743	1.032883	1.030579	22
23	1.050477	1.044183	1.039437	1.035781	1.032903	1.030589	23
24	1.050737	1.044307	1.039497	1.035810	1.032917	1.030596	24
25	1.050957	1.044409	1.039544	1.035832	1.032928	1.030601	25
26	1.051142	1.044491	1.039580	1.035849	1.032935	1.030604	26
27	1.051297	1.044557	1.039609	1.035861	1.032941	1.030607	27
28	1.051428	1.044611	1.039631	1.035870	1.032944	1.030608	28
29	1.051538	1.044654	1.039648	1.035877	1.032947	1.030610	29
30	1.051630	1.044689	1.039661	1.035882	1.032949	1.030610	30
31	1.051706	1.044717	1.039672	1.035886	1.032951	1.030611	31
32	1.051771	1.044739	1.039680	1.035889	1.032952	1.030611	32
33	1.051824	1.044757	1.039686	1.035891	1.032953	1.030612	33
34	1.051868	1.044771	1.039690	1.035893	1.032953	1.030612	34
35	1.051905	1.044783	1.039694	1.035894	1.032954	1.030612	35
36	1.051936	1.044792	1.039697	1.035895	1.032954	1.030612	36
37	1.051962	1.044799	1.039699	1.035896	1.032954	1.030612	37
38	1.051983	1.044805	1.039701	1.035896	1.032954	1.030612	38
39	1.052000	1.044810	1.039702	1.035896	1.032954	1.030612	39
40	1.052015	1.044813	1.039703	1.035897	1.032954	1.030612	40
45	1.052057	1.044823	1.039705	1.035897	1.032955	1.030612	45
50	1.052074	1.044826	1.039706	1.035897	1.032955	1.030612	50
55	1.052080	1.044827	1.039706	1.035897	1.032955	1.030612	55
60	1.052082	1.044827	1.039706	1.035897	1.032955	1.030612	60
65	1.052083	1.044828	1.039706	1.035897	1.032955	1.030612	65
70	1.052083	1.044828	1.039706	1.035897	1.032955	1.030612	70
75	1.052083	1.044828	1.039706	1.035897	1.032955	1.030612	75
80	1.052083	1.044828	1.039706	1.035897	1.032955	1.030612	80
90	1.052083	1.044828	1.039706	1.035897	1.032955	1.030612	90
100	1.052083	1.044828	1.039706	1.035897	1.032955	1.030612	100

TABLE D-6

Factors (EUS) to Convert an Exponential Growth to an Equivalent Uniform Series for Interest Rates from 1% to 2½%

Growth Rate of 2% per Period

n	1%	1¼%	1½%	1¾%	2%	2½%	n
1	1.020000	1.020000	1.020000	1.020000	1.020000	1.020000	1
2	1.030149	1.030137	1.030124	1.030112	1.030099	1.030074	2
3	1.040399	1.040365	1.040332	1.040298	1.040264	1.040197	3
4	1.050751	1.050687	1.050623	1.050559	1.050495	1.050368	4
5	1.061206	1.061102	1.060998	1.060895	1.060792	1.060587	5
6	1.071765	1.071611	1.071459	1.071307	1.071155	1.070852	6
7	1.082428	1.082216	1.082005	1.081794	1.081584	1.081165	7
8	1.093197	1.092916	1.092636	1.092357	1.092078	1.091524	8
9	1.104074	1.103713	1.103354	1.102996	1.102639	1.101928	9
10	1.115058	1.114608	1.114159	1.113711	1.113265	1.112377	10
11	1.126151	1.125600	1.125051	1.124503	1.123957	1.122871	11
12	1.137355	1.136692	1.136031	1.135372	1.134715	1.133408	12
13	1.148670	1.147883	1.147099	1.146317	1.145539	1.143989	13
14	1.160097	1.159175	1.158256	1.157340	1.156428	1.154613	14
15	1.171638	1.170568	1.169502	1.168440	1.167382	1.165278	15
16	1.183294	1.182064	1.180838	1.179618	1.178402	1.175985	16
17	1.195065	1.193662	1.192265	1.190873	1.189487	1.186734	17
18	1.206953	1.205364	1.203782	1.202207	1.200638	1.197522	18
19	1.218959	1.217171	1.215391	1.213618	1.211854	1.208350	19
20	1.231084	1.229083	1.227091	1.225108	1.223134	1.219217	20
21	1.243330	1.241102	1.238884	1.236676	1.234480	1.230123	21
22	1.255697	1.253228	1.250770	1.248324	1.245891	1.241066	22
23	1.268168	1.265461	1.262769	1.260050	1.257366	1.252047	23
24	1.280802	1.277804	1.274821	1.271855	1.268906	1.263064	24
25	1.293541	1.290256	1.286989	1.283740	1.280511	1.274117	25
26	1.306407	1.302819	1.299251	1.295704	1.292180	1.285205	26
27	1.319401	1.315493	1.311608	1.307748	1.303913	1.296327	27
28	1.332524	1.328280	1.324062	1.319872	1.315711	1.307484	28
29	1.345777	1.341180	1.336612	1.332075	1.327572	1.318674	29
30	1.359162	1.354194	1.349259	1.344359	1.339498	1.329896	30
31	1.372679	1.367323	1.362003	1.356724	1.351487	1.341150	31
32	1.386331	1.380568	1.374846	1.369169	1.363539	1.352435	32
33	1.400118	1.393929	1.387787	1.381694	1.375656	1.363751	33
34	1.414043	1.407409	1.400827	1.394301	1.387835	1.375097	34
35	1.428105	1.421007	1.413966	1.406988	1.400077	1.386471	35
36	1.442307	1.434724	1.427206	1.419757	1.412383	1.397875	36
37	1.456650	1.448563	1.440546	1.432607	1.424751	1.409305	37
38	1.471136	1.462522	1.453988	1.445539	1.437182	1.420763	38
39	1.485765	1.476605	1.467531	1.458552	1.449675	1.432248	39
40	1.500540	1.490810	1.481177	1.471647	1.462230	1.443757	40
45	1.576642	1.563722	1.550956	1.538355	1.525933	1.501666	45
50	1.656597	1.639866	1.623370	1.607129	1.591160	1.560109	50
55	1.740600	1.719370	1.698488	1.677984	1.657886	1.618997	55
60	1.828856	1.802364	1.776377	1.750935	1.726078	1.678242	60
65	1.921579	1.888986	1.857106	1.825995	1.795706	1.737756	65
70	2.018996	1.979377	1.940745	1.903175	1.866735	1.797451	70
75	2.121342	2.073684	2.027364	1.982485	1.939131	1.857243	75
80	2.228869	2.172057	2.117036	2.063935	2.012856	1.917051	80
90	2.460521	2.381637	2.305826	2.233281	2.164142	2.036393	90
100	2.716208	2.609442	2.507705	2.411260	2.320274	2.154880	100

Exponential Growth Factors

TABLE D-7
Factors (EUS) to Convert an Exponential Growth to an Equivalent Uniform Series for Interest Rates from 3% to 5½%
Growth Rate of 2% per Period

n	3%	3½%	4%	4½%	5%	5½%	n
1	1.020000	1.020000	1.020000	1.020000	1.020000	1.020000	1
2	1.030049	1.030025	1.030000	1.029976	1.029951	1.029927	2
3	1.040130	1.040064	1.039997	1.039932	1.039866	1.039801	3
4	1.050241	1.050116	1.049990	1.049866	1.049742	1.049619	4
5	1.060382	1.060179	1.059977	1.059776	1.059576	1.059377	5
6	1.070552	1.070253	1.069955	1.069659	1.069365	1.069073	6
7	1.080749	1.080335	1.079923	1.079514	1.079107	1.078703	7
8	1.090972	1.090424	1.089879	1.089338	1.088800	1.088265	8
9	1.101221	1.100519	1.099821	1.099128	1.098439	1.097756	9
10	1.111495	1.110618	1.109747	1.108882	1.108024	1.107172	10
11	1.121791	1.120720	1.119655	1.118599	1.117551	1.116511	11
12	1.132111	1.130822	1.129544	1.128276	1.127018	1.125771	12
13	1.142451	1.140925	1.139412	1.137911	1.136423	1.134949	13
14	1.152812	1.151026	1.149256	1.147501	1.145763	1.144043	14
15	1.163192	1.161124	1.159075	1.157045	1.155037	1.153049	15
16	1.173590	1.171217	1.168867	1.166541	1.164241	1.161967	16
17	1.184005	1.181304	1.178631	1.175987	1.173374	1.170793	17
18	1.194436	1.191383	1.188364	1.185380	1.182434	1.179526	18
19	1.204883	1.201454	1.198065	1.194720	1.191418	1.188163	19
20	1.215343	1.211514	1.207733	1.204003	1.200326	1.196703	20
21	1.225815	1.221562	1.217365	1.213228	1.209154	1.205145	21
22	1.236300	1.231597	1.226960	1.222394	1.217901	1.213485	22
23	1.246795	1.241617	1.236517	1.231499	1.226566	1.221724	23
24	1.257300	1.251621	1.246033	1.240540	1.235147	1.229859	24
25	1.267813	1.261608	1.255507	1.249517	1.243643	1.237889	25
26	1.278334	1.271576	1.264939	1.258428	1.252051	1.245812	26
27	1.288861	1.281524	1.274325	1.267272	1.260371	1.253628	27
28	1.299393	1.291450	1.283665	1.276046	1.268601	1.261336	28
29	1.309930	1.301354	1.292957	1.284750	1.276740	1.268935	29
30	1.320470	1.311234	1.302200	1.293382	1.284787	1.276423	30
31	1.331011	1.321088	1.311393	1.301941	1.292740	1.283801	31
32	1.341554	1.330915	1.320534	1.310425	1.300600	1.291067	32
33	1.352097	1.340715	1.329622	1.318834	1.308364	1.298221	33
34	1.362639	1.350485	1.338655	1.327166	1.316032	1.305263	34
35	1.373179	1.360225	1.347633	1.335421	1.323604	1.312191	35
36	1.383715	1.369933	1.356554	1.343597	1.331078	1.319007	36
37	1.394247	1.379609	1.365417	1.351693	1.338454	1.325710	37
38	1.404775	1.389251	1.374221	1.359709	1.345731	1.332299	38
39	1.415296	1.398857	1.382965	1.367643	1.352909	1.338776	39
40	1.425809	1.408428	1.391648	1.375495	1.359988	1.345139	40
45	1.478234	1.455699	1.434109	1.413498	1.393882	1.375270	45
50	1.530330	1.501910	1.474908	1.449358	1.425271	1.402637	50
55	1.581980	1.546945	1.513955	1.483037	1.454178	1.427335	55
60	1.633074	1.590703	1.551189	1.514526	1.480660	1.449493	60
65	1.683509	1.633102	1.586566	1.543843	1.504803	1.469262	65
70	1.733188	1.674070	1.620067	1.571029	1.526712	1.486809	70
75	1.782026	1.713553	1.651691	1.596146	1.546510	1.502312	75
80	1.829943	1.751511	1.681453	1.619269	1.564332	1.515951	80
90	1.922748	1.822755	1.735526	1.659902	1.594603	1.538337	90
100	2.011150	1.887732	1.782671	1.693732	1.618653	1.555300	100

TABLE D-8

Factors (EUS) to Convert an Exponential Growth to an Equivalent Uniform Series for Interest Rates from 6% to 8½%

Growth Rate of 2% per Period

n	6%	6½%	7%	7½%	8%	8½%	n
1	1.020000	1.020000	1.020000	1.020000	1.020000	1.020000	1
2	1.029903	1.029879	1.029855	1.029831	1.029808	1.029784	2
3	1.039736	1.039672	1.039607	1.039544	1.039480	1.039417	3
4	1.049496	1.049374	1.049253	1.049132	1.049012	1.048893	4
5	1.059179	1.058983	1.058787	1.058592	1.058399	1.058206	5
6	1.068782	1.068493	1.068206	1.067920	1.067636	1.067354	6
7	1.078301	1.077902	1.077505	1.077111	1.076719	1.076330	7
8	1.087734	1.087206	1.086682	1.086162	1.085645	1.085132	8
9	1.097077	1.096402	1.095733	1.095069	1.094410	1.093756	9
10	1.106326	1.105487	1.104655	1.103829	1.103010	1.102198	10
11	1.115480	1.114457	1.113443	1.112438	1.111442	1.110456	11
12	1.124535	1.123310	1.122097	1.120895	1.119705	1.118527	12
13	1.133489	1.132044	1.130613	1.129196	1.127795	1.126409	13
14	1.142340	1.140654	1.138988	1.137340	1.135711	1.134102	14
15	1.151084	1.149141	1.147221	1.145324	1.143451	1.141603	15
16	1.159720	1.157500	1.155309	1.153147	1.151014	1.148911	16
17	1.168245	1.165730	1.163251	1.160806	1.158398	1.156027	17
18	1.176658	1.173830	1.171045	1.168303	1.165604	1.162949	18
19	1.184956	1.181798	1.178690	1.175634	1.172630	1.169679	19
20	1.193138	1.189632	1.186185	1.182800	1.179477	1.176217	20
21	1.201203	1.197330	1.193529	1.189800	1.186144	1.182563	21
22	1.209148	1.204893	1.200721	1.196634	1.192633	1.188719	22
23	1.216974	1.212319	1.207761	1.203302	1.198943	1.194686	23
24	1.224677	1.219606	1.214648	1.209804	1.205076	1.200466	24
25	1.232258	1.226755	1.221382	1.216141	1.211034	1.206061	25
26	1.239716	1.233766	1.227965	1.222315	1.216817	1.211473	26
27	1.247049	1.240637	1.234394	1.228324	1.222427	1.216705	27
28	1.254257	1.247368	1.240673	1.234172	1.227867	1.221759	28
29	1.261340	1.253961	1.246800	1.239858	1.233138	1.226638	29
30	1.268298	1.260414	1.252776	1.245385	1.238242	1.231345	30
31	1.275129	1.266729	1.258604	1.250755	1.243182	1.235884	31
32	1.281834	1.272905	1.264283	1.255968	1.247961	1.240257	32
33	1.288412	1.278943	1.269815	1.261028	1.252581	1.244469	33
34	1.294865	1.284844	1.275201	1.265936	1.257044	1.248523	34
35	1.301192	1.290609	1.280444	1.270694	1.261355	1.252422	35
36	1.307393	1.296239	1.285543	1.275305	1.265516	1.256170	36
37	1.313469	1.301734	1.290503	1.279771	1.269531	1.259772	37
38	1.319421	1.307096	1.295323	1.284095	1.273401	1.263230	38
39	1.325248	1.312326	1.300006	1.288279	1.277132	1.266550	39
40	1.330952	1.317426	1.304555	1.292326	1.280725	1.269734	40
45	1.357655	1.341023	1.325350	1.310606	1.296758	1.283765	45
50	1.381429	1.361606	1.343115	1.325895	1.309881	1.295002	50
55	1.402444	1.379418	1.358159	1.338562	1.320515	1.303907	55
60	1.420896	1.394720	1.370800	1.348969	1.329058	1.310900	60
65	1.437001	1.407779	1.381347	1.357456	1.335865	1.316345	65
70	1.450977	1.418857	1.390091	1.364331	1.341253	1.320556	70
75	1.463046	1.428204	1.397298	1.369867	1.345492	1.323792	75
80	1.473419	1.436053	1.403209	1.374303	1.348809	1.326266	80
90	1.489870	1.448079	1.411961	1.380648	1.353396	1.329573	90
100	1.501758	1.456365	1.417711	1.384625	1.356137	1.331460	100

Exponential Growth Factors

TABLE D-9
Factors (EUS) to Convert an Exponential Growth to an Equivalent Uniform Series for Interest Rates from 9% to 20%
Growth Rate of 2% per Period

n	9%	10%	11%	12%	15%	20%	n
1	1.020000	1.020000	1.020000	1.020000	1.020000	1.020000	1
2	1.029761	1.029714	1.029668	1.029623	1.029488	1.029273	2
3	1.039354	1.039229	1.039105	1.038983	1.038623	1.038046	3
4	1.048774	1.048538	1.048305	1.048074	1.047396	1.046315	4
5	1.058015	1.057636	1.057261	1.056890	1.055804	1.054077	5
6	1.067073	1.066517	1.065968	1.065426	1.063841	1.061337	6
7	1.075944	1.075178	1.074423	1.073679	1.071508	1.068100	7
8	1.084623	1.083615	1.082622	1.081645	1.078804	1.074376	8
9	1.093107	1.091824	1.090563	1.089323	1.085732	1.080180	9
10	1.101393	1.099804	1.098243	1.096712	1.092295	1.085528	10
11	1.109478	1.107551	1.105663	1.103813	1.098500	1.090437	11
12	1.117361	1.115066	1.112820	1.110626	1.104353	1.094930	12
13	1.125039	1.122346	1.119718	1.117154	1.109862	1.099026	13
14	1.132512	1.129393	1.126355	1.123400	1.115037	1.102750	14
15	1.139779	1.136206	1.132736	1.129368	1.119888	1.106125	15
16	1.146839	1.142787	1.138861	1.135061	1.124427	1.109175	16
17	1.153692	1.149137	1.144734	1.140486	1.128666	1.111923	17
18	1.160340	1.155258	1.150360	1.145648	1.132616	1.114392	18
19	1.166782	1.161152	1.155742	1.150553	1.136291	1.116605	19
20	1.173021	1.166823	1.160885	1.155208	1.139704	1.118583	20
21	1.179057	1.172274	1.165795	1.159620	1.142869	1.120348	21
22	1.184893	1.177507	1.170476	1.163796	1.145798	1.121918	22
23	1.190531	1.182528	1.174935	1.167746	1.148504	1.123313	23
24	1.195973	1.187341	1.179178	1.171475	1.151002	1.124549	24
25	1.201223	1.191950	1.183211	1.174994	1.153303	1.125642	25
26	1.206282	1.196360	1.187041	1.178309	1.155419	1.126608	26
27	1.211156	1.200576	1.190674	1.181430	1.157364	1.127459	27
28	1.215846	1.204602	1.194117	1.184364	1.159148	1.128208	28
29	1.220357	1.208446	1.197378	1.187121	1.160783	1.128866	29
30	1.224693	1.212111	1.200463	1.189708	1.162280	1.129443	30
31	1.228858	1.215604	1.203379	1.192133	1.163648	1.129949	31
32	1.232855	1.218930	1.206134	1.194404	1.164897	1.130392	32
33	1.236689	1.222094	1.208733	1.196529	1.166036	1.130779	33
34	1.240364	1.225103	1.211183	1.198517	1.167075	1.131117	34
35	1.243885	1.227961	1.213492	1.200373	1.168020	1.131412	35
36	1.247256	1.230676	1.215666	1.202106	1.168879	1.131668	36
37	1.250482	1.233251	1.217711	1.203722	1.169660	1.131892	37
38	1.253567	1.235693	1.219634	1.205228	1.170369	1.132086	38
39	1.256515	1.238008	1.221440	1.206631	1.171012	1.132255	39
40	1.259331	1.240199	1.223135	1.207937	1.171595	1.132401	40
45	1.271586	1.249499	1.230150	1.213202	1.173776	1.132888	45
50	1.281186	1.256470	1.235178	1.216811	1.175084	1.133123	50
55	1.288628	1.261632	1.238737	1.219252	1.175858	1.133235	55
60	1.294338	1.265417	1.241230	1.220886	1.176311	1.133287	60
65	1.298683	1.268167	1.242960	1.221969	1.176573	1.133312	65
70	1.301966	1.270151	1.244152	1.222683	1.176724	1.133324	70
75	1.304431	1.271573	1.244968	1.223149	1.176810	1.133329	75
80	1.306272	1.272587	1.245523	1.223452	1.176860	1.133331	80
90	1.308652	1.273814	1.246153	1.223775	1.176903	1.133333	90
100	1.309947	1.274422	1.246438	1.223908	1.176917	1.133333	100

TABLE D-10

Factors (EUS) to Convert an Exponential Growth to an Equivalent Uniform Series for Interest Rates from 25% to 50%

Growth Rate of 2% per Period

n	25%	30%	35%	40%	45%	50%	n
1	1.020000	1.020000	1.020000	1.020000	1.020000	1.020000	1
2	1.029067	1.028870	1.028681	1.028500	1.028327	1.028160	2
3	1.037497	1.036974	1.036476	1.036002	1.035549	1.035117	3
4	1.045291	1.044322	1.043405	1.042538	1.041717	1.040941	4
5	1.052454	1.050930	1.049500	1.048160	1.046905	1.045729	5
6	1.059001	1.056828	1.054810	1.052939	1.051205	1.049599	6
7	1.064951	1.062052	1.059391	1.056953	1.054721	1.052679	7
8	1.070329	1.066646	1.063308	1.060288	1.057559	1.055095	8
9	1.075164	1.070658	1.066628	1.063030	1.059823	1.056964	9
10	1.079489	1.074139	1.069419	1.065264	1.061610	1.058394	10
11	1.083340	1.077139	1.071747	1.067068	1.063007	1.059476	11
12	1.086751	1.079710	1.073677	1.068513	1.064088	1.060286	12
13	1.089759	1.081901	1.075265	1.069662	1.064919	1.060889	13
14	1.092401	1.083759	1.076564	1.070569	1.065554	1.061333	14
15	1.094712	1.085325	1.077622	1.071282	1.066035	1.061659	15
16	1.096725	1.086641	1.078477	1.071838	1.066397	1.061896	16
17	1.098473	1.087742	1.079167	1.072270	1.066669	1.062068	17
18	1.099986	1.088658	1.079720	1.072605	1.066872	1.062192	18
19	1.101290	1.089419	1.080162	1.072862	1.067023	1.062281	19
20	1.102412	1.090048	1.080515	1.073060	1.067134	1.062345	20
21	1.103373	1.090567	1.080794	1.073211	1.067217	1.062390	21
22	1.104196	1.090994	1.081015	1.073327	1.067278	1.062422	22
23	1.104898	1.091344	1.081190	1.073414	1.067322	1.062445	23
24	1.105495	1.091630	1.081328	1.073481	1.067355	1.062462	24
25	1.106002	1.091864	1.081436	1.073531	1.067379	1.062473	25
26	1.106432	1.092054	1.081520	1.073569	1.067396	1.062481	26
27	1.106796	1.092209	1.081587	1.073598	1.067409	1.062487	27
28	1.107103	1.092334	1.081638	1.073620	1.067418	1.062491	28
29	1.107363	1.092436	1.081679	1.073636	1.067425	1.062494	29
30	1.107581	1.092519	1.081710	1.073648	1.067429	1.062496	30
31	1.107764	1.092585	1.081735	1.073657	1.067433	1.062497	31
32	1.107918	1.092639	1.081754	1.073664	1.067435	1.062498	32
33	1.108047	1.092682	1.081768	1.073669	1.067437	1.062498	33
34	1.108155	1.092717	1.081780	1.073673	1.067439	1.062499	34
35	1.108246	1.092745	1.081789	1.073676	1.067439	1.062499	35
36	1.108321	1.092767	1.081795	1.073678	1.067440	1.062499	36
37	1.108385	1.092785	1.081801	1.073680	1.067441	1.062500	37
38	1.108437	1.092800	1.081805	1.073681	1.067441	1.062500	38
39	1.108481	1.092811	1.081808	1.073682	1.067441	1.062500	39
40	1.108518	1.092821	1.081810	1.073682	1.067441	1.062500	40
45	1.108626	1.092845	1.081816	1.073684	1.067442	1.062500	45
50	1.108669	1.092853	1.081818	1.073684	1.067442	1.062500	50
55	1.108685	1.092856	1.081818	1.073684	1.067442	1.062500	55
60	1.108692	1.092857	1.081818	1.073684	1.067442	1.062500	60
65	1.108694	1.092857	1.081818	1.073684	1.067442	1.062500	65
70	1.108695	1.092857	1.081818	1.073684	1.067442	1.062500	70
75	1.108695	1.092857	1.081818	1.073684	1.067442	1.062500	75
80	1.108696	1.092857	1.081818	1.073684	1.067442	1.062500	80
90	1.108696	1.092857	1.081818	1.073684	1.067442	1.062500	90
100	1.108696	1.092857	1.081818	1.073684	1.067442	1.062500	100

Exponential Growth Factors

TABLE D-11
Factors (EUS) to Convert an Exponential Growth to an Equivalent Uniform Series for Interest Rates from 1% to 2½%
Growth Rate of 3% per Period

n	1%	1¼%	1½%	1¾%	2%	2½%	n
1	1.030000	1.030000	1.030000	1.030000	1.030000	1.030000	1
2	1.045373	1.045354	1.045335	1.045316	1.045297	1.045259	2
3	1.061001	1.060949	1.060898	1.060846	1.060795	1.060693	3
4	1.076888	1.076790	1.076692	1.076594	1.076496	1.076302	4
5	1.093039	1.092879	1.092720	1.092561	1.092403	1.092087	5
6	1.109459	1.109222	1.108986	1.108751	1.108517	1.108050	6
7	1.126151	1.125822	1.125494	1.125167	1.124841	1.124191	7
8	1.143123	1.142684	1.142247	1.141811	1.141377	1.140511	8
9	1.160377	1.159812	1.159249	1.158687	1.158127	1.157013	9
10	1.177920	1.177211	1.176503	1.175798	1.175095	1.173695	10
11	1.195757	1.194884	1.194014	1.193146	1.192282	1.190561	11
12	1.213893	1.212837	1.211784	1.210735	1.209690	1.207610	12
13	1.232332	1.231073	1.229818	1.228568	1.227322	1.224843	13
14	1.251082	1.249598	1.248120	1.246647	1.245180	1.242263	14
15	1.270147	1.268417	1.266694	1.264977	1.263267	1.259868	15
16	1.289533	1.287534	1.285543	1.283560	1.281585	1.277662	16
17	1.309246	1.306954	1.304672	1.302400	1.300137	1.295643	17
18	1.329291	1.326683	1.324085	1.321499	1.318925	1.313815	18
19	1.349675	1.346724	1.343786	1.340862	1.337952	1.332177	19
20	1.370404	1.367085	1.363780	1.360492	1.357220	1.350730	20
21	1.391485	1.387768	1.384070	1.380391	1.376732	1.369475	21
22	1.412923	1.408782	1.404662	1.400564	1.396490	1.388414	22
23	1.434725	1.430129	1.425559	1.421014	1.416496	1.407547	23
24	1.456897	1.451817	1.446766	1.441745	1.436755	1.426875	24
25	1.479447	1.473851	1.468288	1.462759	1.457267	1.446399	25
26	1.502381	1.496236	1.490129	1.484062	1.478037	1.466121	26
27	1.525706	1.518979	1.512294	1.505656	1.499066	1.486040	27
28	1.549430	1.542085	1.534789	1.527545	1.520357	1.506157	28
29	1.573559	1.565560	1.557616	1.549733	1.541913	1.526475	29
30	1.598102	1.589410	1.580783	1.572224	1.563737	1.546992	30
31	1.623065	1.613643	1.604294	1.595021	1.585831	1.567712	31
32	1.648456	1.638263	1.628153	1.618129	1.608199	1.588633	32
33	1.674283	1.663278	1.652366	1.641552	1.630842	1.609757	33
34	1.700554	1.688694	1.676938	1.665293	1.653765	1.631086	34
35	1.727278	1.714518	1.701875	1.689356	1.676969	1.652619	35
36	1.754462	1.740757	1.727182	1.713746	1.700459	1.674358	36
37	1.782115	1.767417	1.752864	1.738467	1.724236	1.696303	37
38	1.810245	1.794505	1.778927	1.763523	1.748303	1.718455	38
39	1.838862	1.822029	1.805377	1.788918	1.772664	1.740816	39
40	1.867974	1.849996	1.832219	1.814656	1.797322	1.763385	40
45	2.021286	1.996738	1.972520	1.948655	1.925166	1.879391	45
50	2.188422	2.155718	2.123537	2.091916	2.060889	2.000734	50
55	2.370704	2.327995	2.286087	2.245038	2.204896	2.127507	55
60	2.569591	2.514722	2.461049	2.408655	2.357609	2.259798	60
65	2.786684	2.717154	2.649368	2.583441	2.519469	2.397694	65
70	3.023744	2.936658	2.852063	2.770114	2.690937	2.541276	70
75	3.282709	3.174721	3.070228	2.969432	2.872494	2.690626	75
80	3.565707	3.432963	3.305041	3.182202	3.064641	2.845820	80
90	4.213418	4.017192	3.829784	3.651581	3.482833	3.174049	90
100	4.988589	4.705418	4.437640	4.185761	3.950009	3.526569	100

TABLE D-12

Factors (EUS) to Convert an Exponential Growth to an Equivalent Uniform Series for Interest Rates from 3% to 5½%

Growth Rate of 3% per Period

n	3%	3½%	4%	4½%	5%	5½%	n
1	1.030000	1.030000	1.030000	1.030000	1.030000	1.030000	1
2	1.045222	1.045184	1.045147	1.045110	1.045073	1.045036	2
3	1.060591	1.060490	1.060389	1.060289	1.060189	1.060090	3
4	1.076108	1.075916	1.075724	1.075534	1.075344	1.075156	4
5	1.091773	1.091460	1.091150	1.090841	1.090533	1.090228	5
6	1.107585	1.107123	1.106664	1.106207	1.105753	1.105301	6
7	1.123544	1.122902	1.122263	1.121628	1.120997	1.120370	7
8	1.139651	1.138756	1.137946	1.137102	1.136263	1.135429	8
9	1.155905	1.154804	1.153710	1.152624	1.151545	1.150474	9
10	1.172305	1.170924	1.169553	1.168191	1.166839	1.165498	10
11	1.188852	1.187155	1.185471	1.183800	1.182141	1.180497	11
12	1.205545	1.203456	1.201463	1.199446	1.197447	1.195465	12
13	1.222384	1.219945	1.217525	1.215128	1.212751	1.210398	13
14	1.239369	1.236500	1.233656	1.230840	1.228050	1.225289	14
15	1.256499	1.253160	1.249853	1.246579	1.243339	1.240135	15
16	1.273774	1.269923	1.266112	1.262342	1.258614	1.254930	16
17	1.291193	1.286788	1.282432	1.278125	1.273871	1.269670	17
18	1.308757	1.303754	1.298809	1.293925	1.289105	1.284349	18
19	1.326464	1.320817	1.315241	1.309738	1.304311	1.298964	19
20	1.344314	1.337978	1.331725	1.325561	1.319487	1.313508	20
21	1.362307	1.355234	1.348259	1.341389	1.334628	1.327979	21
22	1.380443	1.372582	1.364840	1.357220	1.349729	1.342371	22
23	1.398720	1.390023	1.381464	1.373050	1.364787	1.356680	23
24	1.417138	1.407553	1.398129	1.388875	1.379798	1.370903	24
25	1.435697	1.425171	1.414833	1.404692	1.394757	1.385035	25
26	1.454396	1.442875	1.431573	1.420498	1.409662	1.399072	26
27	1.473234	1.460664	1.448345	1.436290	1.424508	1.413010	27
28	1.492211	1.478535	1.465148	1.452063	1.439292	1.426847	28
29	1.511325	1.496487	1.481978	1.467814	1.454010	1.440577	29
30	1.530578	1.514517	1.498832	1.483541	1.468659	1.454199	30
31	1.549967	1.532624	1.515709	1.499240	1.483235	1.467708	31
32	1.569492	1.550806	1.532604	1.514908	1.497735	1.481102	32
33	1.589152	1.569061	1.549516	1.530541	1.512156	1.494377	33
34	1.608947	1.587387	1.566442	1.546137	1.526495	1.507532	34
35	1.628875	1.605783	1.583378	1.561693	1.540748	1.520562	35
36	1.648937	1.624245	1.600323	1.577205	1.554914	1.533467	36
37	1.669130	1.642772	1.617274	1.592670	1.568988	1.546242	37
38	1.689455	1.661363	1.634227	1.608087	1.582968	1.558886	38
39	1.709910	1.680015	1.651182	1.623451	1.596851	1.571398	39
40	1.730495	1.698726	1.668133	1.638761	1.610636	1.583774	40
45	1.835333	1.793100	1.752770	1.714394	1.677993	1.643568	45
50	1.943275	1.888658	1.836978	1.788277	1.742553	1.699765	50
55	2.054199	1.985155	1.920465	1.860138	1.804113	1.752267	55
60	2.167978	2.082348	2.002963	1.929750	1.862531	1.801047	60
65	2.284478	2.180003	2.084229	1.996925	1.917720	1.846139	65
70	2.403564	2.277897	2.164045	2.061520	1.969646	1.887631	70
75	2.525097	2.375815	2.242224	2.123431	2.018319	1.925651	75
80	2.648940	2.473555	2.318602	2.182590	2.063786	1.960357	80
90	2.903000	2.667770	2.465440	2.292547	2.145443	2.020570	90
100	3.164667	2.859215	2.603773	2.391444	2.215517	2.069850	100

TABLE D-13
Factors (EUS) to Convert an Exponential Growth to an Equivalent Uniform Series for Interest Rates from 6% to 8½%
Growth Rate of 3% per Period

n	6%	6½%	7%	7½%	8%	8½%	n
1	1.030000	1.030000	1.030000	1.030000	1.030000	1.030000	1
2	1.045000	1.044964	1.044928	1.044892	1.044856	1.044820	2
3	1.059992	1.059893	1.059796	1.059698	1.059602	1.059505	3
4	1.074968	1.074782	1.074596	1.074411	1.074228	1.074045	4
5	1.089924	1.089621	1.089321	1.089022	1.088725	1.088429	5
6	1.104852	1.104406	1.103962	1.103521	1.103082	1.102647	6
7	1.119746	1.119127	1.118511	1.117900	1.117292	1.116689	7
8	1.134601	1.133779	1.132962	1.132151	1.131345	1.130546	8
9	1.149410	1.148354	1.147306	1.146266	1.145233	1.144209	9
10	1.164167	1.162846	1.161536	1.160237	1.158948	1.157671	10
11	1.178866	1.177248	1.175645	1.174057	1.172483	1.170923	11
12	1.193501	1.191555	1.189627	1.187719	1.185829	1.183959	12
13	1.208067	1.205759	1.203475	1.201216	1.198981	1.196771	13
14	1.222557	1.219855	1.217183	1.214542	1.211932	1.209354	14
15	1.236967	1.233837	1.230744	1.227691	1.224676	1.221702	15
16	1.251292	1.247699	1.244154	1.240657	1.237209	1.233811	16
17	1.265525	1.261436	1.257406	1.253435	1.249524	1.245674	17
18	1.279662	1.275043	1.270495	1.266020	1.261617	1.257290	18
19	1.293697	1.288514	1.283417	1.278407	1.273486	1.268654	19
20	1.307627	1.301846	1.296167	1.290593	1.285125	1.279764	20
21	1.321446	1.315033	1.308741	1.302573	1.296532	1.290617	21
22	1.335150	1.328071	1.321135	1.314345	1.307704	1.301212	22
23	1.348735	1.340956	1.333345	1.325905	1.318639	1.311548	23
24	1.362197	1.353684	1.345367	1.337251	1.329336	1.321623	24
25	1.375531	1.366252	1.357200	1.348380	1.339792	1.331439	25
26	1.388735	1.378656	1.368840	1.359290	1.350008	1.340994	26
27	1.401803	1.390893	1.380285	1.369981	1.359982	1.350290	27
28	1.414734	1.402961	1.391532	1.380450	1.369716	1.359328	28
29	1.427524	1.414857	1.402581	1.390697	1.379208	1.368110	29
30	1.440169	1.426578	1.413428	1.400722	1.388459	1.376636	30
31	1.452668	1.438122	1.424074	1.410525	1.397472	1.384911	31
32	1.465017	1.449488	1.434517	1.420104	1.406246	1.392936	32
33	1.477215	1.460674	1.444757	1.429462	1.414783	1.400714	33
34	1.489258	1.471678	1.454792	1.438598	1.423087	1.408248	34
35	1.501144	1.482499	1.464624	1.447514	1.431158	1.415542	35
36	1.512873	1.493136	1.474251	1.456210	1.438999	1.422600	36
37	1.524442	1.503588	1.483675	1.464689	1.446613	1.429426	37
38	1.535850	1.513856	1.492895	1.472952	1.454003	1.436023	38
39	1.547095	1.523938	1.501914	1.481001	1.461173	1.442396	39
40	1.558176	1.533834	1.510730	1.488838	1.468124	1.448550	40
45	1.611096	1.580538	1.551839	1.524934	1.499746	1.476195	45
50	1.659840	1.622678	1.588160	1.556151	1.526508	1.499084	50
55	1.704435	1.660416	1.619985	1.582908	1.548942	1.517847	55
60	1.744982	1.693981	1.647668	1.605664	1.567593	1.533093	60
65	1.781642	1.723653	1.671590	1.624882	1.582986	1.545389	65
70	1.814619	1.749740	1.692141	1.641015	1.595609	1.555240	70
75	1.844150	1.772564	1.709707	1.654484	1.605905	1.563089	75
80	1.870487	1.792449	1.724655	1.665679	1.614264	1.569314	80
90	1.914620	1.824633	1.748026	1.682591	1.626467	1.578098	90
100	1.949063	1.848583	1.764611	1.694037	1.634345	1.583509	100

TABLE D-14

Factors (EUS) to Convert an Exponential Growth to an Equivalent Uniform Series for Interest Rates from 9% to 20%

Growth Rate of 3% per Period

n	9%	10%	11%	12%	15%	20%	n
1	1.030000	1.030000	1.030000	1.030000	1.030000	1.030000	1
2	1.044785	1.044714	1.044645	1.044575	1.044372	1.044045	2
3	1.059410	1.059220	1.059031	1.058845	1.058297	1.057420	3
4	1.073864	1.073503	1.073146	1.072793	1.071757	1.070104	4
5	1.088135	1.087552	1.086976	1.086406	1.084737	1.082086	5
6	1.102213	1.101355	1.100508	1.099671	1.097225	1.093362	6
7	1.116089	1.114902	1.113731	1.112577	1.109211	1.103931	7
8	1.129752	1.128182	1.126635	1.125113	1.120690	1.113801	8
9	1.143193	1.141186	1.139212	1.137272	1.131656	1.122983	9
10	1.156405	1.153907	1.151454	1.149048	1.142110	1.131494	10
11	1.169379	1.166336	1.163354	1.160434	1.152054	1.139355	11
12	1.182109	1.178467	1.174907	1.171428	1.161492	1.146591	12
13	1.194587	1.190296	1.186109	1.182028	1.170430	1.153229	13
14	1.206809	1.201817	1.196957	1.192233	1.178877	1.159300	14
15	1.218769	1.213026	1.207450	1.202044	1.186845	1.164834	15
16	1.230463	1.223921	1.217587	1.211462	1.194344	1.169865	16
17	1.241867	1.234500	1.227368	1.220491	1.201390	1.174424	17
18	1.253038	1.244762	1.236795	1.229136	1.207997	1.178546	18
19	1.263913	1.254707	1.245869	1.237402	1.214181	1.182261	19
20	1.274511	1.264334	1.254595	1.245295	1.219959	1.185603	20
21	1.284831	1.273645	1.262977	1.252822	1.225349	1.188601	21
22	1.294871	1.282643	1.271019	1.259993	1.230368	1.191285	22
23	1.304632	1.291329	1.278727	1.266815	1.235034	1.193683	23
24	1.314115	1.299708	1.286107	1.273298	1.239366	1.195821	24
25	1.323320	1.307782	1.293166	1.279451	1.243381	1.197724	25
26	1.332248	1.315556	1.299912	1.285286	1.247097	1.199414	26
27	1.340903	1.323035	1.306352	1.290812	1.250533	1.200913	27
28	1.349286	1.330225	1.312494	1.296041	1.253704	1.202239	28
29	1.357401	1.337130	1.318346	1.300983	1.256628	1.203412	29
30	1.365250	1.343757	1.323918	1.305650	1.259321	1.204448	30
31	1.372837	1.350113	1.329218	1.310053	1.261798	1.205361	31
32	1.380166	1.356203	1.334255	1.314202	1.264073	1.206165	32
33	1.387241	1.362035	1.339039	1.318110	1.266162	1.206871	33
34	1.394067	1.367615	1.343579	1.321787	1.268077	1.207492	34
35	1.400649	1.372951	1.347883	1.325243	1.269831	1.208037	35
36	1.406992	1.378051	1.351961	1.328490	1.271437	1.208515	36
37	1.413100	1.382920	1.355822	1.331538	1.272905	1.208933	37
38	1.418980	1.387568	1.359476	1.334397	1.274246	1.209299	38
39	1.424636	1.392000	1.362930	1.337076	1.275470	1.209619	39
40	1.430073	1.396226	1.366194	1.339585	1.276587	1.209898	40
45	1.454195	1.414496	1.379952	1.349896	1.280840	1.210843	45
50	1.473731	1.428649	1.390142	1.357197	1.283476	1.211314	50
55	1.489387	1.439489	1.397601	1.362305	1.285088	1.211546	55
60	1.501822	1.447712	1.403005	1.365841	1.286063	1.211660	60
65	1.511620	1.453899	1.406889	1.368269	1.286649	1.211714	65
70	1.519291	1.458523	1.409661	1.369924	1.286998	1.211741	70
75	1.525261	1.461959	1.411628	1.371047	1.287205	1.211753	75
80	1.529886	1.464500	1.413018	1.371804	1.287327	1.211759	80
90	1.536198	1.467744	1.414680	1.372654	1.287441	1.211764	90
100	1.539908	1.469483	1.415493	1.373034	1.287480	1.211764	100

Exponential Growth Factors

TABLE D-15

FACTORS (EUS) TO CONVERT AN EXPONENTIAL GROWTH
TO AN EQUIVALENT UNIFORM SERIES FOR INTEREST RATES FROM 25% TO 50%
Growth Rate of 3% per Period

n	25%	30%	35%	40%	45%	50%	n
1	1.030000	1.030000	1.030000	1.030000	1.030000	1.030000	1
2	1.043733	1.043435	1.043149	1.042875	1.042612	1.042360	2
3	1.056584	1.055789	1.055031	1.054309	1.053620	1.052964	3
4	1.068539	1.067058	1.065657	1.064331	1.063078	1.061892	4
5	1.079594	1.077255	1.075061	1.073005	1.071080	1.069277	5
6	1.089760	1.086411	1.083302	1.080421	1.077752	1.075281	6
7	1.099056	1.094570	1.090455	1.086686	1.083239	1.080086	7
8	1.107508	1.101788	1.096606	1.091922	1.087693	1.083876	8
9	1.115154	1.108129	1.101850	1.096253	1.091267	1.086826	9
10	1.122033	1.113661	1.106284	1.099800	1.094103	1.089094	10
11	1.128193	1.118458	1.110006	1.102681	1.096332	1.090820	11
12	1.133682	1.122592	1.113106	1.105001	1.098068	1.092121	12
13	1.138552	1.126136	1.115673	1.106857	1.099410	1.093093	13
14	1.142854	1.129157	1.117784	1.108330	1.100439	1.093814	14
15	1.146639	1.131720	1.119512	1.109494	1.101224	1.094345	15
16	1.149955	1.133885	1.120919	1.110408	1.101819	1.094734	16
17	1.152852	1.135706	1.122059	1.111122	1.102268	1.095018	17
18	1.155372	1.137232	1.122979	1.111677	1.102605	1.095224	18
19	1.157559	1.138505	1.123718	1.112107	1.102856	1.095372	19
20	1.159450	1.139565	1.124311	1.112439	1.103044	1.095479	20
21	1.161081	1.140443	1.124783	1.112695	1.103183	1.095555	21
22	1.162484	1.141170	1.125160	1.112891	1.103286	1.095610	22
23	1.163688	1.141770	1.125459	1.113041	1.103363	1.095650	23
24	1.164719	1.142263	1.125695	1.113156	1.103419	1.095677	24
25	1.165600	1.142668	1.125882	1.113243	1.103460	1.095697	25
26	1.166351	1.143000	1.126030	1.113309	1.103490	1.095711	26
27	1.166990	1.143271	1.126146	1.113359	1.103512	1.095721	27
28	1.167533	1.143493	1.126237	1.113397	1.103528	1.095728	28
29	1.167994	1.143674	1.126309	1.113426	1.103540	1.095733	29
30	1.168384	1.143821	1.126365	1.113448	1.103549	1.095737	30
31	1.168714	1.143940	1.126408	1.113464	1.103555	1.095739	31
32	1.168993	1.144037	1.126443	1.113477	1.103560	1.095741	32
33	1.169228	1.144116	1.126469	1.113486	1.103563	1.095742	33
34	1.169426	1.144180	1.126490	1.113493	1.103565	1.095743	34
35	1.169593	1.144231	1.126506	1.113498	1.103567	1.095743	35
36	1.169733	1.144273	1.126519	1.113502	1.103568	1.095744	36
37	1.169851	1.144306	1.126529	1.113505	1.103569	1.095744	37
38	1.169950	1.144333	1.126536	1.113507	1.103570	1.095744	38
39	1.170033	1.144355	1.126542	1.113509	1.103570	1.095744	39
40	1.170103	1.144373	1.126547	1.113510	1.103571	1.095744	40
45	1.170313	1.144421	1.126558	1.113513	1.103571	1.095745	45
50	1.170398	1.144437	1.126561	1.113513	1.103571	1.095745	50
55	1.170432	1.144442	1.126562	1.113513	1.103571	1.095745	55
60	1.170446	1.144444	1.126562	1.113514	1.103571	1.095745	60
65	1.170451	1.144444	1.126562	1.113514	1.103571	1.095745	65
70	1.170453	1.144444	1.126562	1.113514	1.103571	1.095745	70
75	1.170454	1.144444	1.126562	1.113514	1.103571	1.095745	75
80	1.170454	1.144444	1.126562	1.113514	1.103571	1.095745	80
90	1.170455	1.144444	1.126562	1.113514	1.103571	1.095745	90
100	1.170455	1.144444	1.126562	1.113514	1.103571	1.095745	100

TABLE D-16

Factors (EUS) to Convert an Exponential Growth to an Equivalent Uniform Series for Interest Rates from 1% to 2½%

Growth Rate of 4% per Period

n	1%	1¼%	1½%	1¾%	2%	2½%	n
1	1.040000	1.040000	1.040000	1.040000	1.040000	1.040000	1
2	1.060697	1.060671	1.060645	1.060620	1.060594	1.060543	2
3	1.081873	1.081803	1.081734	1.081664	1.081595	1.081456	3
4	1.103542	1.103409	1.103275	1.103142	1.103010	1.102745	4
5	1.125717	1.125498	1.125281	1.125064	1.124847	1.124416	5
6	1.146409	1.148084	1.147760	1.147437	1.147115	1.146474	6
7	1.171632	1.171178	1.170725	1.170274	1.169823	1.168927	7
8	1.195400	1.194793	1.194186	1.193582	1.192979	1.191779	8
9	1.219728	1.218940	1.218155	1.217372	1.216592	1.215038	9
10	1.244629	1.243634	1.242643	1.241655	1.240670	1.238710	10
11	1.270117	1.268888	1.267662	1.266441	1.265224	1.262801	11
12	1.296210	1.294715	1.293225	1.291741	1.290261	1.287319	12
13	1.322921	1.321129	1.319344	1.317565	1.315793	1.312269	13
14	1.350267	1.348145	1.346032	1.343926	1.341828	1.337658	14
15	1.378265	1.375778	1.373301	1.370834	1.368377	1.363494	15
16	1.406931	1.404042	1.401166	1.398301	1.395449	1.389783	16
17	1.436283	1.432954	1.429639	1.426339	1.423055	1.416532	17
18	1.466339	1.462529	1.458736	1.454961	1.451205	1.443749	18
19	1.497116	1.492783	1.488471	1.484179	1.479910	1.471441	19
20	1.528635	1.523734	1.518857	1.514006	1.509181	1.499614	20
21	1.560914	1.555398	1.549912	1.544455	1.539030	1.528277	21
22	1.593972	1.587794	1.581649	1.575539	1.569466	1.557437	22
23	1.627832	1.620938	1.614084	1.607272	1.600503	1.587102	23
24	1.662513	1.654851	1.647235	1.639668	1.632151	1.617279	24
25	1.698038	1.689550	1.681117	1.672741	1.664423	1.647977	25
26	1.734427	1.725056	1.715744	1.706505	1.697331	1.679202	26
27	1.771705	1.761389	1.751145	1.740976	1.730888	1.710964	27
28	1.809895	1.798569	1.787325	1.776170	1.765106	1.743271	28
29	1.849020	1.836616	1.824308	1.812100	1.799998	1.776131	29
30	1.889106	1.875554	1.862111	1.848784	1.835578	1.809552	30
31	1.930178	1.915404	1.900755	1.886237	1.871859	1.843542	31
32	1.972261	1.956188	1.940257	1.924477	1.909854	1.878112	32
33	2.015383	1.997931	1.980640	1.963519	1.946578	1.913268	33
34	2.059572	2.040656	2.021923	2.003382	1.985046	1.949021	34
35	2.104855	2.084388	2.064127	2.044084	2.024271	1.985378	35
36	2.151263	2.129152	2.107273	2.085641	2.064269	2.022351	36
37	2.198824	2.174974	2.151385	2.128074	2.105054	2.059946	37
38	2.247569	2.221880	2.196484	2.171400	2.146643	2.098175	38
39	2.297531	2.269898	2.242594	2.215639	2.189051	2.137046	39
40	2.348741	2.319056	2.289739	2.260811	2.232294	2.176569	40
45	2.624726	2.582974	2.541848	2.501386	2.461623	2.384316	45
50	2.937353	2.880026	2.823724	2.768507	2.714431	2.609894	50
55	3.291839	3.214631	3.139047	3.065181	2.993118	2.854671	55
60	3.694169	3.591817	3.491969	3.394765	3.300324	3.120110	60
65	4.151221	4.017307	3.887166	3.760996	3.638961	3.407782	65
70	4.670892	4.497618	4.329912	4.168048	4.012238	3.719368	70
75	5.262261	5.040170	4.826153	4.620570	4.423690	4.056672	75
80	5.935765	5.653417	5.382591	5.123750	4.877211	4.421628	80
90	7.578993	7.131856	6.707267	6.305892	5.928072	5.242951	90
100	9.719955	9.027272	8.376780	7.769231	7.204686	6.201827	100

Exponential Growth Factors

TABLE D-17
Factors (EUS) to Convert an Exponential Growth to an Equivalent Uniform Series for Interest Rates from 3% to 5½%
Growth Rate of 4% per Period

n	3%	3½%	4%	4½%	5%	5½%	n
1	1.040000	1.040000	1.040000	1.040000	1.040000	1.040000	1
2	1.060493	1.060442	1.060392	1.060342	1.060293	1.060243	2
3	1.081319	1.081182	1.081046	1.080910	1.080775	1.080641	3
4	1.102482	1.102220	1.101960	1.101701	1.101444	1.101187	4
5	1.123987	1.123560	1.123136	1.122713	1.122294	1.121876	5
6	1.145836	1.145202	1.144571	1.143944	1.143321	1.142701	6
7	1.168035	1.167148	1.166267	1.165391	1.164521	1.163656	7
8	1.190586	1.189401	1.188223	1.187052	1.185889	1.184734	8
9	1.213494	1.211961	1.210437	1.208924	1.207421	1.205929	9
10	1.236763	1.234829	1.232909	1.231003	1.229112	1.227235	10
11	1.260396	1.258009	1.255639	1.253288	1.250956	1.248644	11
12	1.284398	1.281500	1.278626	1.275776	1.272950	1.270150	12
13	1.308773	1.305306	1.301868	1.298462	1.295088	1.291746	13
14	1.333524	1.329426	1.325366	1.321345	1.317364	1.313425	14
15	1.358655	1.353862	1.349117	1.344420	1.339774	1.335181	15
16	1.384171	1.378616	1.373120	1.367685	1.362313	1.357006	16
17	1.410076	1.403689	1.397375	1.391136	1.384975	1.378894	17
18	1.436374	1.429083	1.421880	1.414769	1.407754	1.400838	18
19	1.463068	1.454797	1.446634	1.438582	1.430646	1.422830	19
20	1.490163	1.480835	1.471635	1.462570	1.453645	1.444864	20
21	1.517663	1.507196	1.496882	1.486730	1.476745	1.466933	21
22	1.545572	1.533881	1.522374	1.511058	1.499941	1.489030	22
23	1.573895	1.560893	1.548108	1.535550	1.523228	1.511149	23
24	1.602634	1.588232	1.574084	1.560203	1.546599	1.533282	24
25	1.631796	1.615898	1.600299	1.585012	1.570050	1.555422	25
26	1.661383	1.643893	1.626752	1.609974	1.593575	1.577564	26
27	1.691400	1.672218	1.653441	1.635085	1.617167	1.599701	27
28	1.721851	1.700874	1.680363	1.660340	1.640823	1.621825	28
29	1.752741	1.729861	1.707518	1.685736	1.664535	1.643931	29
30	1.784074	1.759180	1.734903	1.711269	1.688299	1.666012	30
31	1.815853	1.788832	1.762516	1.736934	1.712110	1.688063	31
32	1.848084	1.818818	1.790355	1.762727	1.735961	1.710076	32
33	1.880770	1.849139	1.818418	1.788644	1.759847	1.732046	33
34	1.913917	1.879794	1.846702	1.814682	1.783763	1.753967	34
35	1.947527	1.910785	1.875206	1.840835	1.807704	1.775833	35
36	1.981607	1.942112	1.903928	1.867100	1.831663	1.797639	36
37	2.016159	1.973776	1.932864	1.893473	1.855637	1.819379	37
38	2.051189	2.005778	1.962013	1.919948	1.879620	1.841047	38
39	2.086701	2.038117	1.991372	1.946523	1.903607	1.862639	39
40	2.122699	2.070795	2.020940	1.973193	1.927592	1.884149	40
45	2.310145	2.239272	2.171811	2.107823	2.047328	1.990303	45
50	2.510432	2.416263	2.327510	2.244210	2.166319	2.093729	50
55	2.724144	2.601796	2.487718	2.381856	2.284029	2.193955	55
60	2.951882	2.795890	2.652111	2.520288	2.399981	2.290604	60
65	3.194265	2.998553	2.820363	2.659062	2.513761	2.383392	65
70	3.451931	3.209787	2.992154	2.797767	2.625021	2.472118	70
75	3.725544	3.429588	3.167175	2.936028	2.733473	2.556658	75
80	4.015790	3.657954	3.345126	3.073509	2.838889	2.636956	80
90	4.649075	4.140365	3.708698	3.344975	3.039963	2.784874	90
100	5.357883	4.657039	4.080800	3.610219	3.227408	2.916388	100

TABLE D-18

Factors (EUS) to Convert an Exponential Growth to an Equivalent Uniform Series for Interest Rates from 6% to 8½%

Growth Rate of 4% per Period

n	6%	6½%	7%	7½%	8%	8½%	n
1	1.040000	1.040000	1.040000	1.040000	1.040000	1.040000	1
2	1.060194	1.060145	1.060097	1.060048	1.060000	1.059952	2
3	1.080508	1.080375	1.080243	1.080111	1.079980	1.079850	3
4	1.100932	1.100679	1.100427	1.100176	1.099926	1.099678	4
5	1.121461	1.121048	1.120637	1.120229	1.119823	1.119419	5
6	1.142085	1.141472	1.140863	1.140258	1.139656	1.139058	6
7	1.162796	1.161942	1.161093	1.160250	1.159412	1.158580	7
8	1.183586	1.182447	1.181315	1.180191	1.179076	1.177968	8
9	1.204448	1.202978	1.201519	1.200071	1.198634	1.197209	9
10	1.225372	1.223524	1.221692	1.219875	1.218074	1.216288	10
11	1.246350	1.244077	1.241825	1.239592	1.237381	1.235191	11
12	1.267375	1.264627	1.261905	1.259210	1.256544	1.253905	12
13	1.288437	1.285163	1.281923	1.278718	1.275549	1.272417	13
14	1.309529	1.305676	1.301867	1.298103	1.294386	1.290715	14
15	1.330641	1.326156	1.321727	1.317356	1.313042	1.308787	15
16	1.351766	1.346595	1.341494	1.336464	1.331507	1.326623	16
17	1.372896	1.366983	1.361157	1.355419	1.349770	1.344213	17
18	1.394023	1.387311	1.380706	1.374209	1.367822	1.361546	18
19	1.415137	1.407571	1.400133	1.392826	1.385653	1.378615	19
20	1.436232	1.427752	1.419427	1.411261	1.403255	1.395410	20
21	1.457299	1.447847	1.438581	1.429504	1.420618	1.411925	21
22	1.478331	1.467848	1.457586	1.447548	1.437736	1.428152	22
23	1.499320	1.487746	1.476433	1.465384	1.454601	1.444086	23
24	1.520258	1.507534	1.495116	1.483006	1.471207	1.459721	24
25	1.541138	1.527204	1.513625	1.500406	1.487548	1.475053	25
26	1.561953	1.546748	1.531955	1.517578	1.503619	1.490078	26
27	1.582695	1.566160	1.550099	1.534517	1.519414	1.504791	27
28	1.603359	1.585432	1.568050	1.551216	1.534931	1.519191	28
29	1.623936	1.604558	1.585802	1.567671	1.550164	1.533276	29
30	1.644420	1.623531	1.603350	1.583878	1.565111	1.547043	30
31	1.664805	1.642346	1.620689	1.599831	1.579769	1.560493	31
32	1.685086	1.660997	1.637813	1.615528	1.594136	1.573624	32
33	1.705254	1.679478	1.654718	1.630966	1.608211	1.586437	33
34	1.725306	1.697785	1.671400	1.646141	1.621992	1.598934	34
35	1.745235	1.715911	1.687855	1.661051	1.635480	1.611114	35
36	1.765036	1.733853	1.704079	1.675695	1.648673	1.622980	36
37	1.784704	1.751606	1.720071	1.690071	1.661572	1.634534	37
38	1.804233	1.769167	1.735826	1.704177	1.674178	1.645778	38
39	1.823619	1.786530	1.751343	1.718013	1.686491	1.656715	39
40	1.842856	1.803694	1.766619	1.731580	1.698514	1.667349	40
45	1.936692	1.886408	1.839343	1.795366	1.754337	1.716104	45
50	2.026275	1.963752	1.905921	1.852521	1.803280	1.757920	50
55	2.111275	2.035577	1.966412	1.903314	1.845814	1.793453	55
60	2.191479	2.101867	2.021005	1.948130	1.882499	1.823406	60
65	2.266782	2.162717	2.069988	1.987427	1.913933	1.848484	65
70	2.337172	2.218306	2.113713	2.021700	1.940717	1.869361	70
75	2.402713	2.268876	2.152571	2.051454	1.963430	1.886656	75
80	2.463531	2.314714	2.186975	2.077184	1.982616	1.900926	80
90	2.571715	2.393453	2.244051	2.118419	2.012324	1.922282	90
100	2.663439	2.457115	2.288061	2.148746	2.033169	1.936581	100

Exponential Growth Factors

TABLE D-19

Factors (EUS) to Convert an Exponential Growth to an Equivalent Uniform Series for Interest Rates from 9% to 20%

Growth Rate of 4% per Period

n	9%	10%	11%	12%	15%	20%	n
1	1.040000	1.040000	1.040000	1.040000	1.040000	1.040000	1
2	1.059904	1.059810	1.059716	1.059623	1.059349	1.058909	2
3	1.079721	1.079463	1.079209	1.078957	1.078216	1.077029	3
4	1.099431	1.098941	1.098456	1.097977	1.096569	1.094322	4
5	1.119018	1.118222	1.117435	1.116657	1.114379	1.110761	5
6	1.138464	1.137287	1.136124	1.134976	1.131622	1.126326	6
7	1.157753	1.156117	1.154503	1.152912	1.148276	1.141006	7
8	1.176869	1.174695	1.172553	1.170446	1.164325	1.154798	8
9	1.195795	1.193003	1.190257	1.187559	1.179753	1.167708	9
10	1.214518	1.211026	1.207599	1.204238	1.194553	1.179748	10
11	1.233022	1.228749	1.224564	1.220467	1.208718	1.190936	11
12	1.251294	1.246158	1.241139	1.236237	1.222245	1.201297	12
13	1.269321	1.263241	1.257312	1.251537	1.235136	1.210860	13
14	1.287091	1.279986	1.273075	1.266359	1.247395	1.219658	14
15	1.304592	1.296382	1.288417	1.280699	1.259029	1.227728	15
16	1.321814	1.312421	1.303333	1.294552	1.270048	1.235107	16
17	1.338747	1.328095	1.317818	1.307918	1.280464	1.241836	17
18	1.355383	1.343396	1.331866	1.320795	1.290293	1.247955	18
19	1.371712	1.358319	1.345477	1.333186	1.299550	1.253506	19
20	1.387729	1.372860	1.358649	1.345093	1.308253	1.258528	20
21	1.403426	1.387014	1.371381	1.356521	1.316421	1.263062	21
22	1.418798	1.400779	1.383676	1.367476	1.324076	1.267146	22
23	1.433840	1.414155	1.395536	1.377966	1.331237	1.270817	23
24	1.448549	1.427140	1.406966	1.387998	1.337927	1.274110	24
25	1.462921	1.439735	1.417968	1.397582	1.344168	1.277059	25
26	1.476953	1.451942	1.428551	1.406727	1.349980	1.279694	26
27	1.490644	1.463762	1.438719	1.415445	1.355388	1.282046	27
28	1.503994	1.475199	1.448480	1.423748	1.360412	1.284142	28
29	1.517001	1.486256	1.457842	1.431646	1.365073	1.286006	29
30	1.529666	1.496938	1.466813	1.439154	1.369394	1.287662	30
31	1.541991	1.507250	1.475404	1.446282	1.373394	1.289131	31
32	1.553976	1.517198	1.483622	1.453046	1.377093	1.290433	32
33	1.565625	1.526787	1.491479	1.459457	1.380510	1.291586	33
34	1.576939	1.536024	1.498984	1.465529	1.383664	1.292605	34
35	1.587922	1.544917	1.506148	1.471275	1.386572	1.293505	35
36	1.598578	1.553471	1.512981	1.476710	1.389251	1.294299	36
37	1.608909	1.561697	1.519495	1.481845	1.391716	1.294999	37
38	1.618922	1.569600	1.525700	1.486693	1.393984	1.295616	38
39	1.628621	1.577189	1.531606	1.491269	1.396068	1.296158	39
40	1.638010	1.584473	1.537225	1.495584	1.397982	1.296636	40
45	1.680510	1.616609	1.561386	1.513664	1.405415	1.298279	45
50	1.716164	1.642390	1.579918	1.526922	1.410183	1.299128	50
55	1.745783	1.662857	1.593975	1.536534	1.413204	1.299561	55
60	1.770185	1.678962	1.604542	1.543437	1.415102	1.299780	60
65	1.790149	1.691543	1.612425	1.548357	1.416284	1.299891	65
70	1.806385	1.701311	1.618271	1.551843	1.417017	1.299946	70
75	1.819526	1.708858	1.622585	1.554301	1.417468	1.299973	75
80	1.830118	1.714666	1.625755	1.556027	1.417746	1.299987	80
90	1.845442	1.722526	1.629774	1.558079	1.418020	1.299997	90
100	1.855236	1.727106	1.631911	1.559075	1.418122	1.299999	100

TABLE D-20

Factors (EUS) to Convert an Exponential Growth to an Equivalent Uniform Series for Interest Rates from 25% to 50%

Growth Rate of 4% per Period

n	25%	30%	35%	40%	45%	50%	n
1	1.040000	1.040000	1.040000	1.040000	1.040000	1.040000	1
2	1.058489	1.058087	1.057702	1.057333	1.056980	1.056640	2
3	1.075899	1.074823	1.073798	1.072822	1.071891	1.071003	3
4	1.092195	1.090184	1.088281	1.086481	1.084780	1.083170	4
5	1.107360	1.104169	1.101177	1.098374	1.095749	1.093292	5
6	1.121390	1.116802	1.112545	1.108601	1.104949	1.101570	6
7	1.134296	1.128127	1.122471	1.117293	1.112559	1.108233	7
8	1.146103	1.138205	1.131056	1.124599	1.118773	1.113518	8
9	1.156847	1.147111	1.138418	1.130676	1.123786	1.117655	9
10	1.166572	1.154927	1.144680	1.135683	1.127788	1.120854	10
11	1.175332	1.161744	1.149965	1.139772	1.130951	1.123302	11
12	1.183185	1.167655	1.154394	1.143085	1.133428	1.125158	12
13	1.190194	1.172750	1.158082	1.145749	1.135354	1.126552	13
14	1.196421	1.177119	1.161134	1.147878	1.136839	1.127592	14
15	1.201932	1.180849	1.163645	1.149568	1.137978	1.128362	15
16	1.206791	1.184017	1.165702	1.150902	1.138847	1.128930	16
17	1.211060	1.186697	1.167378	1.151951	1.139506	1.129346	17
18	1.214796	1.188955	1.168739	1.152772	1.140003	1.129650	18
19	1.218057	1.190852	1.169839	1.153412	1.140377	1.129870	19
20	1.220895	1.192439	1.170725	1.153909	1.140657	1.130030	20
21	1.223357	1.193764	1.171437	1.154293	1.140867	1.130145	21
22	1.225487	1.194866	1.172007	1.154590	1.141023	1.130228	22
23	1.227327	1.195780	1.172463	1.154818	1.141138	1.130287	23
24	1.228912	1.196538	1.172826	1.154993	1.141224	1.130330	24
25	1.230274	1.197163	1.173114	1.155128	1.141288	1.130360	25
26	1.231443	1.197679	1.173343	1.155230	1.141334	1.130382	26
27	1.232444	1.198103	1.173524	1.155309	1.141369	1.130397	27
28	1.233299	1.198452	1.173667	1.155368	1.141394	1.130408	28
29	1.234030	1.198738	1.173780	1.155414	1.141413	1.130416	29
30	1.234653	1.198972	1.173869	1.155448	1.141426	1.130422	30
31	1.235183	1.199164	1.173940	1.155475	1.141436	1.130425	31
32	1.235633	1.199320	1.173995	1.155494	1.141444	1.130428	32
33	1.236016	1.199448	1.174038	1.155509	1.141449	1.130430	33
34	1.236340	1.199552	1.174072	1.155521	1.141453	1.130432	34
35	1.236615	1.199637	1.174099	1.155529	1.141456	1.130432	35
36	1.236848	1.199705	1.174119	1.155536	1.141458	1.130433	36
37	1.237045	1.199761	1.174136	1.155541	1.141459	1.130434	37
38	1.237211	1.199807	1.174149	1.155544	1.141461	1.130434	38
39	1.237351	1.199844	1.174158	1.155547	1.141461	1.130434	39
40	1.237470	1.199874	1.174166	1.155549	1.141462	1.130434	40
45	1.237834	1.199957	1.174186	1.155554	1.141463	1.130435	45
50	1.237987	1.199985	1.174191	1.155555	1.141463	1.130435	50
55	1.238051	1.199995	1.174193	1.155555	1.141463	1.130435	55
60	1.238077	1.199998	1.174193	1.155556	1.141463	1.130435	60
65	1.238088	1.199999	1.174194	1.155556	1.141463	1.130435	65
70	1.238092	1.200000	1.174194	1.155556	1.141463	1.130435	70
75	1.238094	1.200000	1.174194	1.155556	1.141463	1.130435	75
80	1.238095	1.200000	1.174194	1.155556	1.141463	1.130435	80
90	1.238095	1.200000	1.174194	1.155556	1.141463	1.130435	90
100	1.238095	1.200000	1.174194	1.155556	1.141463	1.130435	100

TABLE D-21
FACTORS (EUS) TO CONVERT AN EXPONENTIAL GROWTH TO AN EQUIVALENT UNIFORM SERIES FOR INTEREST RATES FROM 1% TO 2½%
Growth Rate of 5% per Period

n	1%	1¼%	1½%	1¾%	2%	2½%	n
1	1.050000	1.050000	1.050000	1.050000	1.050000	1.050000	1
2	1.076119	1.076087	1.076055	1.076022	1.075990	1.075926	2
3	1.103018	1.102929	1.102841	1.102753	1.102665	1.102489	3
4	1.130722	1.130551	1.130381	1.130212	1.130043	1.129706	4
5	1.159257	1.158978	1.158699	1.158421	1.158144	1.157591	5
6	1.188652	1.188234	1.187818	1.187402	1.186988	1.186162	6
7	1.218935	1.218348	1.217762	1.217178	1.216596	1.215436	7
8	1.250136	1.249346	1.248558	1.247772	1.246988	1.245428	8
9	1.282286	1.281256	1.280230	1.279208	1.278188	1.276158	9
10	1.315415	1.314109	1.312808	1.311510	1.310217	1.307643	10
11	1.349558	1.347935	1.346317	1.344705	1.343098	1.339901	11
12	1.384748	1.382764	1.380788	1.378818	1.376855	1.372952	12
13	1.421021	1.418631	1.416249	1.413877	1.411514	1.406815	13
14	1.458413	1.455567	1.452733	1.449910	1.447098	1.441509	14
15	1.496961	1.493609	1.490270	1.486944	1.483634	1.477056	15
16	1.536706	1.532791	1.528893	1.525012	1.521148	1.513476	16
17	1.577688	1.573152	1.568636	1.564141	1.559668	1.550789	17
18	1.619949	1.614729	1.609535	1.604365	1.599223	1.589019	18
19	1.663532	1.657563	1.651624	1.645716	1.639840	1.628188	19
20	1.708482	1.701694	1.694942	1.688227	1.681551	1.668318	20
21	1.754847	1.747166	1.739527	1.731933	1.724385	1.709433	21
22	1.802674	1.794021	1.785419	1.776870	1.768375	1.751557	22
23	1.852015	1.842307	1.832659	1.823073	1.813553	1.794716	23
24	1.902920	1.892069	1.881289	1.870582	1.859952	1.838933	24
25	1.955443	1.943356	1.931352	1.919434	1.907607	1.884235	25
26	2.009641	1.996220	1.982895	1.969671	1.956553	1.930649	26
27	2.065571	2.050711	2.035964	2.021334	2.006827	1.978201	27
28	2.123293	2.106885	2.090607	2.074465	2.058466	2.026920	28
29	2.182870	2.164796	2.146873	2.129109	2.111510	2.076834	29
30	2.244364	2.224503	2.204815	2.185311	2.165997	2.127972	30
31	2.307843	2.286065	2.264486	2.243117	2.221968	2.180365	31
32	2.373376	2.349543	2.325940	2.302577	2.279466	2.234042	32
33	2.441034	2.415003	2.389234	2.363740	2.338535	2.289036	33
34	2.510892	2.482510	2.454427	2.426657	2.399217	2.345378	34
35	2.583026	2.552132	2.521578	2.491382	2.461560	2.403101	35
36	2.657516	2.623941	2.590752	2.557968	2.525610	2.462239	36
37	2.734444	2.698009	2.662011	2.626473	2.591417	2.522827	37
38	2.813896	2.774413	2.735424	2.696954	2.659029	2.584901	38
39	2.895960	2.853231	2.811058	2.769471	2.728499	2.648495	39
40	2.980729	2.934543	2.888985	2.844086	2.799879	2.713649	40
45	3.448592	3.381580	3.315672	3.250926	3.187394	3.064159	45
50	3.999921	3.904930	3.811807	3.720648	3.631539	3.459758	50
55	4.650654	4.518474	4.389357	4.263454	4.140895	3.906226	55
60	5.419887	5.238710	5.062420	4.891243	4.725367	4.410086	60
65	6.330523	6.085256	5.847606	5.617899	5.396393	4.978695	65
70	7.410044	7.081452	6.764496	6.459625	6.167177	5.620350	70
75	8.691449	8.255079	7.836176	7.435338	7.052973	6.344413	75
80	10.214379	9.639218	9.089880	8.567121	8.071395	7.161443	80
90	14.185069	13.204234	12.277676	11.406398	10.590664	9.123575	90
100	19.831530	18.191593	16.660951	15.240301	13.928672	11.621529	100

TABLE D-22

Factors (EUS) to Convert an Exponential Growth to an Equivalent Uniform Series for Interest Rates from 3% to 5½%

Growth Rate of 5% per Period

n	3%	3½%	4%	4½%	5%	5½%	n
1	1.050000	1.050000	1.050000	1.050000	1.050000	1.050000	1
2	1.075862	1.075799	1.075735	1.075672	1.075610	1.075547	2
3	1.102315	1.102141	1.101969	1.101797	1.101626	1.101456	3
4	1.129371	1.129037	1.128706	1.128376	1.128047	1.127721	4
5	1.157042	1.156496	1.155952	1.155412	1.154874	1.154339	5
6	1.185342	1.184525	1.183714	1.182907	1.182105	1.181307	6
7	1.214283	1.213136	1.211997	1.210864	1.209739	1.208620	7
8	1.243878	1.242337	1.240806	1.239285	1.237775	1.236274	8
9	1.274142	1.272138	1.270149	1.268173	1.266211	1.264263	9
10	1.305087	1.302549	1.300029	1.297528	1.295046	1.292583	10
11	1.336728	1.333578	1.330453	1.327353	1.324278	1.321229	11
12	1.369079	1.365237	1.361427	1.357649	1.353905	1.350195	12
13	1.402155	1.397534	1.392955	1.388418	1.383925	1.379476	13
14	1.435970	1.430481	1.425045	1.419662	1.414336	1.409066	14
15	1.470540	1.464087	1.457700	1.451382	1.445134	1.438959	15
16	1.505879	1.498362	1.490928	1.483579	1.476319	1.469149	16
17	1.542004	1.533317	1.524733	1.516254	1.507885	1.499630	17
18	1.578931	1.568963	1.559120	1.549409	1.539832	1.530395	18
19	1.616675	1.605309	1.594096	1.583043	1.572155	1.561437	19
20	1.655254	1.642367	1.629666	1.617159	1.604852	1.592751	20
21	1.694684	1.680148	1.665835	1.651756	1.637918	1.624329	21
22	1.734982	1.718662	1.702609	1.686836	1.671351	1.656165	22
23	1.776166	1.757920	1.739993	1.722398	1.705147	1.688250	23
24	1.818254	1.797934	1.777992	1.758443	1.739302	1.720579	24
25	1.861263	1.838715	1.816612	1.794972	1.773811	1.753144	25
26	1.905213	1.880275	1.855858	1.831984	1.808672	1.785938	26
27	1.950123	1.922625	1.895735	1.869480	1.843880	1.818952	27
28	1.996011	1.965776	1.936250	1.907460	1.879431	1.852181	28
29	2.042896	2.009741	1.977406	1.945923	1.915320	1.885617	29
30	2.090800	2.054531	2.019209	1.984870	1.951543	1.919251	30
31	2.139742	2.100159	2.061665	2.024300	1.988096	1.953077	31
32	2.189744	2.146637	2.104778	2.064212	2.024973	1.987087	32
33	2.240825	2.193977	2.148555	2.104607	2.062171	2.021274	33
34	2.293008	2.242192	2.192999	2.145484	2.099685	2.055629	34
35	2.346314	2.291294	2.238117	2.186842	2.137510	2.090146	35
36	2.400766	2.341296	2.283913	2.228680	2.175640	2.124818	36
37	2.456386	2.392211	2.330392	2.270998	2.214072	2.159635	37
38	2.513199	2.444052	2.377561	2.313796	2.252801	2.194592	38
39	2.571226	2.496832	2.425423	2.357072	2.291820	2.229681	39
40	2.630492	2.550566	2.473984	2.400825	2.331126	2.264894	40
45	2.946285	2.834002	2.727456	2.626716	2.531778	2.442574	45
50	3.296930	3.143356	2.999175	2.864379	2.738837	2.622307	50
55	3.685984	3.480492	3.289786	3.113654	2.951678	2.803270	55
60	4.117350	3.847388	3.599946	3.374371	3.169691	2.984708	60
65	4.595317	4.246154	3.930328	3.646351	3.392295	3.165944	65
70	5.124597	4.679041	4.281624	3.929422	3.618941	3.346382	70
75	5.710368	5.148450	4.654557	4.223421	3.849121	3.525504	75
80	6.358320	5.656954	5.049883	4.528199	4.082370	3.702872	80
90	7.866422	6.802452	5.910938	5.169602	4.556440	4.050943	90
100	9.706828	8.140035	6.871727	5.852886	5.038314	4.388413	100

Exponential Growth Factors

TABLE D-23
Factors (EUS) to Convert an Exponential Growth to an Equivalent Uniform Series for Interest Rates from 6% to 8½%
Growth Rate of 5% per Period

n	6%	6½%	7%	7½%	8%	8½%	n
1	1.050000	1.050000	1.050000	1.050000	1.050000	1.050000	1
2	1.075485	1.075424	1.075362	1.075301	1.075240	1.075180	2
3	1.101286	1.101118	1.100950	1.100784	1.100618	1.100453	3
4	1.127396	1.127073	1.126751	1.126432	1.126114	1.125797	4
5	1.153808	1.153279	1.152753	1.152231	1.151711	1.151194	5
6	1.180514	1.179726	1.178943	1.178164	1.177391	1.176622	6
7	1.207509	1.206404	1.205307	1.204217	1.203135	1.202059	7
8	1.234783	1.233303	1.231833	1.230374	1.228925	1.227487	8
9	1.262330	1.260411	1.258507	1.256617	1.254743	1.252884	9
10	1.290140	1.287717	1.285314	1.282932	1.280570	1.278230	10
11	1.318206	1.315211	1.312243	1.309302	1.306389	1.303505	11
12	1.346520	1.342881	1.339278	1.335711	1.332182	1.328690	12
13	1.375072	1.370715	1.366405	1.362143	1.357930	1.353766	13
14	1.403855	1.398703	1.393612	1.388583	1.383617	1.378714	14
15	1.432858	1.426832	1.420884	1.415015	1.409225	1.403516	15
16	1.462072	1.455091	1.448207	1.441422	1.434737	1.428154	16
17	1.491490	1.483468	1.475568	1.467791	1.460138	1.452612	17
18	1.521100	1.511952	1.502953	1.494105	1.485411	1.476873	18
19	1.550894	1.540530	1.530348	1.520351	1.510541	1.500921	19
20	1.580863	1.569191	1.557740	1.546513	1.535513	1.524742	20
21	1.610996	1.597923	1.585116	1.572578	1.560312	1.548321	21
22	1.641284	1.626714	1.612462	1.598531	1.584925	1.571645	22
23	1.671717	1.655554	1.639767	1.624361	1.609338	1.594701	23
24	1.702286	1.684430	1.667017	1.650053	1.633539	1.617477	24
25	1.732981	1.713331	1.694201	1.675595	1.657515	1.639962	25
26	1.763792	1.742246	1.721306	1.700975	1.681256	1.662147	26
27	1.794711	1.771164	1.748320	1.726182	1.704750	1.684020	27
28	1.825726	1.800075	1.775233	1.751205	1.727986	1.705574	28
29	1.856829	1.828966	1.802034	1.776032	1.750957	1.726801	29
30	1.888010	1.857828	1.828710	1.800654	1.773652	1.747694	30
31	1.919259	1.886651	1.855253	1.825061	1.796064	1.768245	31
32	1.950569	1.915424	1.881653	1.849245	1.818184	1.788450	32
33	1.981928	1.944138	1.907899	1.873196	1.840007	1.808304	33
34	2.013328	1.972783	1.933982	1.896906	1.861525	1.827802	34
35	2.044761	2.001349	1.959895	1.920368	1.882733	1.846942	35
36	2.076217	2.029828	1.985627	1.943576	1.903626	1.865720	36
37	2.107687	2.058211	2.011172	1.966522	1.924199	1.884134	37
38	2.139164	2.086489	2.036521	1.989199	1.944448	1.902183	38
39	2.170638	2.114654	2.061668	2.011604	1.964371	1.919865	39
40	2.202102	2.142697	2.086605	2.033730	1.983963	1.937182	40
45	2.358983	2.280837	2.207935	2.140045	2.076918	2.018293	45
50	2.514464	2.414914	2.323218	2.238905	2.161488	2.090479	50
55	2.667725	2.544253	2.432015	2.330155	2.237821	2.154182	55
60	2.818071	2.668349	2.534085	2.413847	2.306252	2.209998	60
65	2.964933	2.786857	2.629359	2.490189	2.367250	2.258614	65
70	3.107860	2.899570	2.717900	2.559509	2.421362	2.300749	70
75	3.246511	3.006399	2.799883	2.622209	2.469175	2.337120	75
80	3.380641	3.107351	2.875556	2.678740	2.511284	2.368411	80
90	3.634770	3.292020	3.009226	2.775182	2.580684	2.418244	90
100	3.869752	3.454870	3.121656	2.852564	2.633814	2.454653	100

TABLE D-24

Factors (EUS) to Convert an Exponential Growth to an Equivalent Uniform Series for Interest Rates from 9% to 20%

Growth Rate of 5% per Period

n	9%	10%	11%	12%	15%	20%	n
1	1.050000	1.050000	1.050000	1.050000	1.050000	1.050000	1
2	1.075120	1.075000	1.074882	1.074764	1.074419	1.073864	2
3	1.100288	1.099962	1.099639	1.099320	1.098380	1.096875	3
4	1.125483	1.124859	1.124241	1.123630	1.121836	1.118975	4
5	1.150680	1.149661	1.148654	1.147659	1.144743	1.140114	5
6	1.175857	1.174343	1.172848	1.171372	1.167060	1.160253	6
7	1.200991	1.198877	1.196792	1.194737	1.188750	1.179365	7
8	1.226059	1.223237	1.220458	1.217723	1.209782	1.197432	8
9	1.251040	1.247398	1.243818	1.240301	1.230129	1.214447	9
10	1.275910	1.271335	1.266846	1.262444	1.249769	1.230413	10
11	1.300649	1.295025	1.289518	1.284129	1.268684	1.245340	11
12	1.325237	1.318446	1.311811	1.305334	1.286862	1.259249	12
13	1.349652	1.341575	1.333703	1.326038	1.304293	1.272165	13
14	1.373876	1.364394	1.355176	1.346224	1.320973	1.284122	14
15	1.397889	1.386883	1.376212	1.365879	1.336903	1.295155	15
16	1.421674	1.409025	1.396796	1.384988	1.352086	1.305307	16
17	1.445213	1.430803	1.416913	1.403543	1.366530	1.314622	17
18	1.468491	1.452203	1.436551	1.421535	1.380245	1.323146	18
19	1.491492	1.473211	1.455701	1.438960	1.393245	1.330925	19
20	1.514201	1.493815	1.474355	1.455814	1.405545	1.338008	20
21	1.536605	1.514004	1.492505	1.472095	1.417163	1.344443	21
22	1.558692	1.533769	1.510146	1.487804	1.428121	1.350276	22
23	1.580449	1.553101	1.527276	1.502945	1.438439	1.355552	23
24	1.601867	1.571994	1.543893	1.517520	1.448140	1.360315	24
25	1.622935	1.590443	1.559997	1.531536	1.457248	1.364608	25
26	1.643645	1.608442	1.575588	1.545000	1.465789	1.368469	26
27	1.663989	1.625990	1.590670	1.557921	1.473785	1.371938	27
28	1.683961	1.643084	1.605246	1.570308	1.481264	1.375048	28
29	1.703553	1.659723	1.619320	1.582172	1.488250	1.377834	29
30	1.722763	1.675907	1.632900	1.593525	1.494768	1.380325	30
31	1.741585	1.691638	1.645991	1.604379	1.500843	1.382550	31
32	1.760016	1.706917	1.658602	1.614746	1.506499	1.384535	32
33	1.778053	1.721749	1.670741	1.624642	1.511760	1.386303	33
34	1.795696	1.736135	1.682417	1.634079	1.516649	1.387878	34
35	1.812943	1.750082	1.693640	1.643072	1.521187	1.389278	35
36	1.829794	1.763594	1.704420	1.651635	1.525398	1.390522	36
37	1.846249	1.776676	1.714768	1.659784	1.529300	1.391626	37
38	1.862309	1.789336	1.724695	1.667533	1.532914	1.392606	38
39	1.877977	1.801581	1.734212	1.674898	1.536259	1.393474	39
40	1.893255	1.813416	1.743331	1.681892	1.539353	1.394243	40
45	1.963902	1.866749	1.783370	1.711815	1.551614	1.396943	45
50	2.025394	1.911140	1.815231	1.734579	1.559773	1.398390	50
55	2.078445	1.947735	1.840328	1.751715	1.565142	1.399157	55
60	2.123874	1.977665	1.859936	1.764507	1.568647	1.399561	60
65	2.162537	2.001989	1.875155	1.773994	1.570920	1.399772	65
70	2.195276	2.021654	1.886906	1.780993	1.572387	1.399882	70
75	2.222886	2.037486	1.895943	1.786136	1.573330	1.399939	75
80	2.246092	2.050191	1.902870	1.789903	1.573934	1.399969	80
90	2.281814	2.068480	1.912206	1.794663	1.574567	1.399992	90
100	2.306728	2.080112	1.917625	1.797188	1.574825	1.399998	100

Exponential Growth Factors

TABLE D-25
Factors (EUS) to Convert an Exponential Growth to an Equivalent Uniform Series for Interest Rates from 25% to 50%
Growth Rate of 5% per Period

n	25%	30%	35%	40%	45%	50%	n
1	1.050000	1.050000	1.050000	1.050000	1.050000	1.050000	1
2	1.073333	1.072826	1.072340	1.071875	1.071429	1.071000	2
3	1.095443	1.094079	1.092780	1.091542	1.090362	1.089237	3
4	1.116267	1.113705	1.111283	1.108992	1.106826	1.104778	4
5	1.135764	1.131683	1.127858	1.124276	1.120922	1.117783	5
6	1.153913	1.148021	1.142558	1.137497	1.132813	1.128480	6
7	1.170709	1.162756	1.155467	1.148800	1.142706	1.137141	7
8	1.186169	1.175947	1.166701	1.158355	1.150831	1.144050	8
9	1.200321	1.187672	1.176390	1.166350	1.157424	1.149488	9
10	1.213210	1.198025	1.184679	1.172975	1.162717	1.153718	10
11	1.224889	1.207107	1.191716	1.178418	1.166924	1.156974	11
12	1.235421	1.215028	1.197648	1.182852	1.170239	1.159455	12
13	1.244878	1.221898	1.202617	1.186439	1.172830	1.161330	13
14	1.253332	1.227824	1.206752	1.189321	1.174841	1.162737	14
15	1.260859	1.232912	1.210176	1.191624	1.176392	1.163786	15
16	1.267535	1.237260	1.212996	1.193453	1.177581	1.164562	16
17	1.273435	1.240961	1.215309	1.194898	1.178489	1.165135	17
18	1.278632	1.244098	1.217197	1.196037	1.179178	1.165556	18
19	1.283195	1.246749	1.218733	1.196929	1.179699	1.165863	19
20	1.287190	1.248981	1.219978	1.197626	1.180092	1.166086	20
21	1.290678	1.250855	1.220984	1.198169	1.180388	1.166249	21
22	1.293716	1.252424	1.221795	1.198590	1.180609	1.166366	22
23	1.296355	1.253734	1.222446	1.198917	1.180774	1.166451	23
24	1.298643	1.254825	1.222969	1.199169	1.180898	1.166512	24
25	1.300622	1.255733	1.223387	1.199363	1.180989	1.166556	25
26	1.302331	1.256486	1.223720	1.199513	1.181058	1.166588	26
27	1.303805	1.257109	1.223986	1.199628	1.181108	1.166611	27
28	1.305072	1.257625	1.224198	1.199716	1.181145	1.166627	28
29	1.306161	1.258051	1.224366	1.199784	1.181173	1.166638	29
30	1.307096	1.258402	1.224499	1.199835	1.181193	1.166646	30
31	1.307897	1.258691	1.224605	1.199875	1.181208	1.166652	31
32	1.308582	1.258928	1.224689	1.199905	1.181219	1.166656	32
33	1.309168	1.259124	1.224755	1.199928	1.181228	1.166659	33
34	1.309668	1.259284	1.224807	1.199945	1.181234	1.166662	34
35	1.310095	1.259415	1.224848	1.199958	1.181238	1.166663	35
36	1.310458	1.259523	1.224881	1.199968	1.181241	1.166664	36
37	1.310768	1.259611	1.224906	1.199976	1.181244	1.166665	37
38	1.311032	1.259682	1.224926	1.199982	1.181245	1.166665	38
39	1.311256	1.259741	1.224942	1.199986	1.181247	1.166666	39
40	1.311446	1.259789	1.224955	1.199990	1.181247	1.166666	40
45	1.312043	1.259925	1.224987	1.199997	1.181249	1.166667	45
50	1.312304	1.259974	1.224996	1.199999	1.181250	1.166667	50
55	1.312416	1.259991	1.224999	1.200000	1.181250	1.166667	55
60	1.312464	1.259997	1.225000	1.200000	1.181250	1.166667	60
65	1.312485	1.259999	1.225000	1.200000	1.181250	1.166667	65
70	1.312494	1.260000	1.225000	1.200000	1.181250	1.166667	70
75	1.312497	1.260000	1.225000	1.200000	1.181250	1.166667	75
80	1.312499	1.260000	1.225000	1.200000	1.181250	1.166667	80
90	1.312500	1.260000	1.225000	1.200000	1.181250	1.166667	90
100	1.312500	1.260000	1.225000	1.200000	1.181250	1.166667	100

TABLE D-26

Factors (EUS) to Convert an Exponential Growth to an Equivalent Uniform Series for Interest Rates from 1% to 2½%

Growth Rate of 6% per Period

n	1%	1¼%	1½%	1¾%	2%	2½%	n
1	1.060000	1.060000	1.060000	1.060000	1.060000	1.060000	1
2	1.091642	1.091602	1.091563	1.091524	1.091485	1.091407	2
3	1.124437	1.124330	1.124222	1.124114	1.124007	1.123794	3
4	1.158434	1.158226	1.158018	1.157810	1.157603	1.157191	4
5	1.193680	1.193337	1.192994	1.192652	1.192311	1.191632	5
6	1.230227	1.229711	1.229197	1.228683	1.228171	1.227152	6
7	1.268129	1.267399	1.266672	1.265947	1.265224	1.263784	7
8	1.307440	1.306454	1.305470	1.304489	1.303512	1.301565	8
9	1.348219	1.346928	1.345641	1.344358	1.343079	1.340534	9
10	1.390526	1.388880	1.387239	1.385603	1.383972	1.380728	10
11	1.434425	1.432368	1.430318	1.428275	1.426239	1.422189	11
12	1.479982	1.477455	1.474936	1.472427	1.469927	1.464957	12
13	1.527265	1.524204	1.521154	1.518116	1.515090	1.509075	13
14	1.576346	1.572682	1.569033	1.565399	1.561779	1.554588	14
15	1.627301	1.622960	1.618638	1.614335	1.610051	1.601542	15
16	1.680207	1.675111	1.670038	1.664988	1.659962	1.649984	16
17	1.735147	1.729210	1.723301	1.717421	1.711571	1.699962	17
18	1.792206	1.785337	1.778503	1.771704	1.764941	1.751528	18
19	1.851472	1.843574	1.835718	1.827904	1.820135	1.804734	19
20	1.913040	1.904008	1.895026	1.886095	1.877219	1.859634	20
21	1.977007	1.966728	1.956509	1.946354	1.936263	1.916284	21
22	2.043473	2.031827	2.020255	2.008757	1.997337	1.974742	22
23	2.112545	2.099405	2.086351	2.073388	2.060517	2.035067	23
24	2.184333	2.169562	2.154893	2.140330	2.125878	2.097321	24
25	2.258954	2.242405	2.225976	2.209674	2.193502	2.161569	25
26	2.336529	2.318044	2.299702	2.281509	2.263470	2.227875	26
27	2.417182	2.396596	2.376177	2.355932	2.335868	2.296309	27
28	2.501047	2.478181	2.455510	2.433042	2.410786	2.366941	28
29	2.588261	2.562924	2.537815	2.512943	2.488317	2.439844	29
30	2.678968	2.650958	2.623212	2.595741	2.568557	2.515093	30
31	2.773318	2.742418	2.711824	2.681548	2.651606	2.592767	31
32	2.871469	2.837448	2.803780	2.770481	2.737566	2.672946	32
33	2.973584	2.936197	2.899216	2.862660	2.826547	2.755712	33
34	3.079836	3.038821	2.998272	2.958211	2.918658	2.841154	34
35	3.190403	3.145481	3.101093	3.057264	3.014017	2.929359	35
36	3.305473	3.256347	3.207831	3.159955	3.112744	3.020419	36
37	3.425241	3.371595	3.318645	3.266425	3.214963	3.114430	37
38	3.549913	3.491411	3.433701	3.376821	3.320805	3.211491	38
39	3.679702	3.615986	3.553170	3.491296	3.430403	3.311703	39
40	3.814832	3.745522	3.677231	3.610009	3.543898	3.415171	40
45	4.579158	4.475272	4.373245	4.273161	4.175097	3.985297	45
50	5.516964	5.364699	5.215690	5.070081	4.927997	4.654821	50
55	6.670222	6.450957	6.237216	6.029229	5.827192	5.441569	55
60	8.091438	7.780150	7.477988	7.185302	6.902375	6.366607	60
65	9.846358	9.405541	8.987456	8.580614	8.189403	7.454835	65
70	12.017385	11.410300	10.826536	10.266802	9.731592	8.735689	70
75	14.707883	13.870951	13.070313	12.306903	11.581288	10.243966	75
80	18.047614	16.901657	15.811385	14.777903	13.801763	12.020806	80
90	27.368820	25.256004	23.269232	21.409717	19.677061	16.583609	90
100	41.857134	38.029209	34.475283	31.194630	28.182613	22.929719	100

Exponential Growth Factors

TABLE D-27
Factors (EUS) to Convert an Exponential Growth to an Equivalent Uniform Series for Interest Rates from 3% to 5½%
Growth Rate of 6% per Period

n	3%	3½%	4%	4½%	5%	5½%	n
1	1.060000	1.060000	1.060000	1.060000	1.060000	1.060000	1
2	1.091330	1.091253	1.091176	1.091100	1.091024	1.090949	2
3	1.123581	1.123370	1.123160	1.122951	1.122743	1.122535	3
4	1.156781	1.156373	1.155968	1.155564	1.155163	1.154763	4
5	1.190957	1.190285	1.189617	1.188953	1.188292	1.187635	5
6	1.226138	1.225129	1.224127	1.223131	1.222140	1.221155	6
7	1.262352	1.260929	1.259515	1.258109	1.256712	1.255324	7
8	1.299631	1.297709	1.295799	1.293902	1.292018	1.290146	8
9	1.338005	1.335494	1.332999	1.330523	1.328064	1.325623	9
10	1.377507	1.374309	1.371134	1.367983	1.364857	1.361756	10
11	1.418169	1.414180	1.410223	1.406298	1.402405	1.398547	11
12	1.460025	1.455135	1.450285	1.445479	1.440716	1.435997	12
13	1.503111	1.497200	1.491342	1.485540	1.479795	1.474109	13
14	1.547462	1.540403	1.533413	1.526495	1.519650	1.512881	14
15	1.593115	1.584773	1.576520	1.568357	1.560288	1.552316	15
16	1.640108	1.630340	1.620682	1.611140	1.601716	1.592413	16
17	1.688481	1.677133	1.665923	1.654857	1.643939	1.633172	17
18	1.738273	1.725182	1.712263	1.699522	1.686964	1.674594	18
19	1.789527	1.774521	1.759725	1.745149	1.730797	1.716678	19
20	1.842284	1.825180	1.808332	1.791751	1.775446	1.759423	20
21	1.896589	1.877192	1.858106	1.839343	1.820915	1.802829	21
22	1.952487	1.930591	1.909070	1.887939	1.867210	1.846895	22
23	2.010025	1.985103	1.961249	1.937553	1.914338	1.891619	23
24	2.069250	2.041690	2.014667	1.988199	1.962305	1.937000	24
25	2.130212	2.099462	2.069347	2.039891	2.011115	1.983036	25
26	2.192961	2.158763	2.125315	2.092644	2.060774	2.029726	26
27	2.257549	2.219632	2.182596	2.146472	2.111289	2.077067	27
28	2.324032	2.282108	2.241215	2.201390	2.162663	2.125058	28
29	2.392462	2.346231	2.301200	2.257413	2.214903	2.173695	29
30	2.462898	2.412040	2.362577	2.314555	2.268013	2.222978	30
31	2.535399	2.479578	2.425372	2.372832	2.321999	2.272902	31
32	2.610023	2.548887	2.489613	2.432258	2.376866	2.323466	32
33	2.686834	2.620011	2.555328	2.492848	2.432618	2.374666	33
34	2.765894	2.692995	2.622546	2.554618	2.489260	2.426500	34
35	2.847271	2.767883	2.691295	2.617584	2.546798	2.478964	35
36	2.931031	2.844722	2.761606	2.681760	2.605236	2.532055	36
37	3.017243	2.923562	2.833507	2.747162	2.664578	2.585770	37
38	3.105980	3.004449	2.907029	2.813807	2.724830	2.640106	38
39	3.197316	3.087436	2.982204	2.881709	2.785995	2.695058	39
40	3.291325	3.172572	3.059062	2.950886	2.848079	2.750624	40
45	3.804290	3.632382	3.469753	3.316455	3.172432	3.037533	45
50	4.396819	4.154369	3.927592	3.716377	3.520416	3.339231	50
55	5.081212	4.746463	4.437190	4.152855	3.892586	3.655250	55
60	5.871665	5.417586	5.003588	4.628219	4.289512	3.985137	60
65	6.784564	6.177772	5.632300	5.144949	4.711793	4.328463	65
70	7.838825	7.038315	6.329368	5.705693	5.160069	4.684840	70
75	9.056284	8.011921	7.101417	6.313290	5.635033	5.053929	75
80	10.462146	9.112895	7.955716	6.970285	6.137444	5.435445	80
90	13.959952	11.763395	9.943792	8.448848	7.227999	6.234890	90
100	18.623158	15.144582	12.367449	10.169304	8.439339	7.081996	100

TABLE D-28

Factors (EUS) to Convert an Exponential Growth to an Equivalent Uniform Series for Interest Rates from 6% to 8½

Growth Rate of 6% per Period

n	6%	6½%	7%	7½%	8%	8½%	n
1	1.060000	1.060000	1.060000	1.060000	1.060000	1.060000	1
2	1.090874	1.090799	1.090725	1.090651	1.090577	1.090504	2
3	1.122329	1.122124	1.121920	1.121717	1.121516	1.121315	3
4	1.154366	1.153971	1.153578	1.153187	1.152798	1.152411	4
5	1.186982	1.186332	1.185686	1.185044	1.184405	1.183770	5
6	1.220176	1.219203	1.218235	1.217274	1.216318	1.215369	6
7	1.253945	1.252575	1.251214	1.249861	1.248518	1.247184	7
8	1.288288	1.286442	1.284609	1.282790	1.280984	1.279192	8
9	1.323200	1.320796	1.318411	1.316044	1.313696	1.311368	9
10	1.358680	1.355629	1.352604	1.349606	1.346634	1.343689	10
11	1.394722	1.390933	1.387178	1.383459	1.379776	1.376129	11
12	1.431324	1.426698	1.422118	1.417585	1.413101	1.408666	12
13	1.468481	1.462915	1.457410	1.451968	1.446589	1.441274	13
14	1.506189	1.499575	1.493041	1.486588	1.480218	1.473931	14
15	1.544441	1.536668	1.528996	1.521428	1.513966	1.506610	15
16	1.583234	1.574183	1.565261	1.556470	1.547813	1.539291	16
17	1.622562	1.612110	1.601821	1.591696	1.581738	1.571948	17
18	1.662418	1.650439	1.638660	1.627086	1.615719	1.604560	18
19	1.702796	1.689157	1.675765	1.662624	1.649736	1.637104	19
20	1.743691	1.728255	1.713120	1.698290	1.683768	1.669558	20
21	1.785095	1.767720	1.750709	1.734067	1.717796	1.701901	21
22	1.827003	1.807541	1.788517	1.769936	1.751800	1.734112	22
23	1.869405	1.847706	1.826530	1.805880	1.785760	1.766173	23
24	1.912296	1.888204	1.864731	1.841881	1.819658	1.798062	24
25	1.955668	1.929022	1.903105	1.877922	1.853475	1.829763	25
26	1.999513	1.970148	1.941638	1.913986	1.887193	1.861257	26
27	2.043823	2.011570	1.980313	1.950055	1.920795	1.892526	27
28	2.088591	2.053276	2.019117	1.986115	1.954264	1.923556	28
29	2.133809	2.095254	2.058034	2.022147	1.987584	1.954331	29
30	2.179467	2.137491	2.097050	2.058137	2.020739	1.984836	30
31	2.225559	2.179976	2.136150	2.094069	2.053714	2.015057	31
32	2.272075	2.222695	2.175320	2.129929	2.086495	2.044983	32
33	2.319007	2.265638	2.214546	2.165702	2.119069	2.074600	33
34	2.366346	2.308792	2.253814	2.201373	2.151421	2.103897	34
35	2.414085	2.352145	2.293111	2.236930	2.183540	2.132866	35
36	2.462214	2.395686	2.332423	2.272359	2.215414	2.161494	36
37	2.510725	2.439402	2.371738	2.307649	2.247032	2.189775	37
38	2.559609	2.483282	2.411043	2.342785	2.278384	2.217700	38
39	2.608857	2.527315	2.450326	2.377757	2.309459	2.245263	39
40	2.658461	2.571489	2.489574	2.412555	2.340248	2.272456	40
45	2.911522	2.794099	2.684910	2.583563	2.489639	2.402706	45
50	3.172214	3.018661	2.877798	2.748813	2.630875	2.523156	50
55	3.439533	3.243997	3.067144	2.907462	2.763462	2.633709	55
60	3.712543	3.469075	3.252071	3.058936	2.887199	2.734546	60
65	3.990393	3.693015	3.431906	3.202893	3.002113	2.826054	65
70	4.272319	3.915080	3.606161	3.339182	3.108407	2.908752	70
75	4.557650	4.134670	3.774501	3.467805	3.206406	2.983241	75
80	4.845803	4.351304	3.936727	3.588885	3.296523	3.050157	80
90	5.428053	4.774300	4.242547	3.809296	3.454974	3.163831	90
100	6.017736	5.182081	4.523780	4.002677	3.587632	3.254667	100

Exponential Growth Factors

TABLE D-29
Factors (EUS) to Convert an Exponential Growth to an Equivalent Uniform Series for Interest Rates from 9% to 20%
Growth Rate of 6% per Period

n	9%	10%	11%	12%	15%	20%	n
1	1.060000	1.060000	1.060000	1.060000	1.060000	1.060000	1
2	1.090431	1.090286	1.090142	1.090000	1.089581	1.088909	2
3	1.121115	1.120718	1.120325	1.119936	1.118792	1.116960	3
4	1.152026	1.151263	1.150508	1.149760	1.147567	1.144069	4
5	1.183139	1.181887	1.180650	1.179427	1.175845	1.170159	5
6	1.214425	1.212556	1.210711	1.208889	1.203567	1.195171	6
7	1.245859	1.243237	1.240652	1.238103	1.230681	1.219053	7
8	1.277413	1.273896	1.270433	1.267026	1.257138	1.241769	8
9	1.309059	1.304500	1.300019	1.295617	1.282894	1.263295	9
10	1.340770	1.335016	1.329371	1.323838	1.307912	1.283617	10
11	1.372520	1.365412	1.358455	1.351650	1.332158	1.302736	11
12	1.404280	1.395658	1.387238	1.379021	1.355607	1.320661	12
13	1.436025	1.425723	1.415686	1.405919	1.378235	1.337410	13
14	1.467728	1.455577	1.443772	1.432313	1.400027	1.353010	14
15	1.499363	1.485194	1.471464	1.458178	1.420971	1.367497	15
16	1.530904	1.514544	1.498739	1.483490	1.441061	1.380910	16
17	1.562329	1.543604	1.525571	1.508228	1.460296	1.393294	17
18	1.593611	1.572349	1.551937	1.532374	1.478678	1.404698	18
19	1.624729	1.600756	1.577818	1.555912	1.496215	1.415173	19
20	1.655659	1.628802	1.603196	1.578829	1.512916	1.424772	20
21	1.686381	1.656469	1.628054	1.601114	1.528795	1.433548	21
22	1.716873	1.683738	1.652378	1.622762	1.543870	1.441554	22
23	1.747117	1.710592	1.676157	1.643765	1.558159	1.448844	23
24	1.777092	1.737015	1.699380	1.664121	1.571683	1.455469	24
25	1.806782	1.762993	1.722039	1.683829	1.584467	1.461478	25
26	1.836170	1.788513	1.744128	1.702890	1.596533	1.466921	26
27	1.865240	1.813566	1.765642	1.721308	1.607909	1.471842	27
28	1.893978	1.838140	1.786579	1.739086	1.618620	1.476285	28
29	1.922370	1.862228	1.806937	1.756233	1.628695	1.480291	29
30	1.950403	1.885823	1.826716	1.772755	1.638160	1.483898	30
31	1.978066	1.908920	1.845919	1.788661	1.647043	1.487142	31
32	2.005348	1.931514	1.864549	1.803963	1.655371	1.490056	32
33	2.032241	1.953602	1.882608	1.818671	1.663172	1.492671	33
34	2.058734	1.975182	1.900104	1.832799	1.670473	1.495016	34
35	2.084822	1.996253	1.917043	1.846359	1.677300	1.497116	35
36	2.110496	2.016816	1.933431	1.859365	1.683678	1.498995	36
37	2.135753	2.036872	1.949277	1.871831	1.689633	1.500675	37
38	2.160585	2.056423	1.964590	1.883773	1.695189	1.502176	38
39	2.184991	2.075472	1.979380	1.895205	1.700368	1.503516	39
40	2.208966	2.094022	1.993656	1.906143	1.705194	1.504712	40
45	2.322324	2.179476	2.057726	1.953969	1.724733	1.508999	45
50	2.424848	2.253376	2.110695	1.991769	1.738241	1.511387	50
55	2.516836	2.316736	2.154090	2.021364	1.747488	1.512704	55
60	2.598842	2.370684	2.189386	2.044365	1.753773	1.513426	60
65	2.671558	2.416365	2.217934	2.062141	1.758021	1.513820	65
70	2.735766	2.454879	2.240923	2.075820	1.760881	1.514034	70
75	2.792270	2.487240	2.259374	2.086311	1.762801	1.514150	75
80	2.841863	2.514360	2.274145	2.094338	1.764086	1.514212	80
90	2.923281	2.555981	2.295362	2.105142	1.765520	1.514264	90
100	2.985392	2.584940	2.308839	2.111413	1.766158	1.514280	100

TABLE D-30

FACTORS (EUS) TO CONVERT AN EXPONENTIAL GROWTH
TO AN EQUIVALENT UNIFORM SERIES FOR INTEREST RATES FROM 25% TO 50%
Growth Rate of 6% per Period

n	25%	30%	35%	40%	45%	50%	n
1	1.060000	1.060000	1.060000	1.060000	1.060000	1.060000	1
2	1.088267	1.087652	1.087064	1.086500	1.085959	1.085440	2
3	1.115217	1.113558	1.111977	1.110472	1.109036	1.107667	3
4	1.140758	1.137628	1.134667	1.131869	1.129223	1.126720	4
5	1.164819	1.159810	1.155117	1.150722	1.146608	1.142758	5
6	1.187353	1.180091	1.173360	1.167127	1.161361	1.156028	6
7	1.208335	1.198492	1.189477	1.181234	1.173706	1.166833	7
8	1.227764	1.215064	1.203584	1.193231	1.183903	1.175502	8
9	1.245659	1.229882	1.215824	1.203327	1.192226	1.182365	9
10	1.262054	1.243044	1.226358	1.211742	1.198946	1.187734	10
11	1.277001	1.254661	1.235353	1.218695	1.204319	1.191890	11
12	1.290564	1.264853	1.242981	1.224394	1.208577	1.195075	12
13	1.302814	1.273745	1.249408	1.229031	1.211925	1.197497	13
14	1.313833	1.281463	1.254789	1.232780	1.214538	1.199325	14
15	1.323704	1.288128	1.259272	1.235791	1.216566	1.200695	15
16	1.332514	1.293860	1.262986	1.238198	1.218130	1.201716	16
17	1.340348	1.298768	1.266050	1.240113	1.219331	1.202474	17
18	1.347291	1.302955	1.268567	1.241629	1.220248	1.203033	18
19	1.353426	1.306515	1.270628	1.242825	1.220947	1.203444	19
20	1.358832	1.309531	1.272309	1.243765	1.221477	1.203746	20
21	1.363581	1.312079	1.273675	1.244502	1.221877	1.203966	21
22	1.367744	1.314226	1.274784	1.245078	1.222179	1.204126	22
23	1.371384	1.316030	1.275680	1.245527	1.222407	1.204243	23
24	1.374560	1.317544	1.276403	1.245876	1.222577	1.204327	24
25	1.377326	1.318810	1.276986	1.246146	1.222705	1.204388	25
26	1.379730	1.319868	1.277454	1.246356	1.222800	1.204433	26
27	1.381816	1.320750	1.277830	1.246518	1.222871	1.204464	27
28	1.383623	1.321484	1.278131	1.246644	1.222924	1.204487	28
29	1.385186	1.322094	1.278371	1.246740	1.222964	1.204504	29
30	1.386537	1.322600	1.278564	1.246814	1.222993	1.204516	30
31	1.387702	1.323020	1.278717	1.246872	1.223015	1.204524	31
32	1.388706	1.323368	1.278839	1.246915	1.223031	1.204530	32
33	1.389570	1.323655	1.278937	1.246949	1.223043	1.204535	33
34	1.390314	1.323893	1.279014	1.246975	1.223052	1.204538	34
35	1.390953	1.324090	1.279076	1.246995	1.223059	1.204540	35
36	1.391501	1.324251	1.279125	1.247010	1.223063	1.204542	36
37	1.391971	1.324385	1.279163	1.247021	1.223067	1.204543	37
38	1.392374	1.324494	1.279194	1.247030	1.223070	1.204543	38
39	1.392720	1.324585	1.279218	1.247037	1.223072	1.204544	39
40	1.393015	1.324659	1.279238	1.247042	1.223073	1.204544	40
45	1.393961	1.324874	1.279288	1.247055	1.223076	1.204545	45
50	1.394390	1.324954	1.279304	1.247058	1.223077	1.204545	50
55	1.394583	1.324983	1.279308	1.247059	1.223077	1.204545	55
60	1.394668	1.324994	1.279310	1.247059	1.223077	1.204545	60
65	1.394707	1.324998	1.279310	1.247059	1.223077	1.204545	65
70	1.394724	1.324999	1.279310	1.247059	1.223077	1.204545	70
75	1.394731	1.325000	1.279310	1.247059	1.223077	1.204545	75
80	1.394734	1.325000	1.279310	1.247059	1.223077	1.204545	80
90	1.394736	1.325000	1.279310	1.247059	1.223077	1.204545	90
100	1.394737	1.325000	1.279310	1.247059	1.223077	1.204545	100

Exponential Growth Factors

TABLE D-31
Factors (EUS) to Convert an Exponential Growth to an Equivalent Uniform Series for Interest Rates from 1% to 2½%
Growth Rate of 7% per Period

n	1%	1¼%	1½%	1¾%	2%	2½%	n
1	1.070000	1.070000	1.070000	1.070000	1.070000	1.070000	1
2	1.107264	1.107217	1.107171	1.107125	1.107079	1.106988	2
3	1.146133	1.146006	1.145878	1.145751	1.145624	1.145372	3
4	1.186667	1.186439	1.186192	1.185945	1.185699	1.185209	4
5	1.229005	1.228594	1.228185	1.227776	1.227369	1.226558	5
6	1.273173	1.272553	1.271935	1.271319	1.270704	1.269479	6
7	1.319262	1.318402	1.317524	1.316649	1.315776	1.314038	7
8	1.367426	1.366229	1.365036	1.363847	1.362661	1.360300	8
9	1.417704	1.416130	1.414561	1.412997	1.411437	1.408334	9
10	1.470222	1.468203	1.466192	1.464187	1.462189	1.458213	10
11	1.525088	1.522553	1.520027	1.517510	1.515001	1.510013	11
12	1.582419	1.579288	1.576169	1.573062	1.569966	1.563811	12
13	1.642337	1.638525	1.634727	1.630944	1.627177	1.619690	13
14	1.704970	1.700382	1.695813	1.691264	1.686734	1.677736	14
15	1.770452	1.764988	1.759547	1.754131	1.748740	1.738036	15
16	1.838926	1.832475	1.826054	1.819663	1.813304	1.800683	16
17	1.910541	1.902983	1.895463	1.887981	1.880539	1.865776	17
18	1.985453	1.976661	1.967914	1.959215	1.950564	1.933413	18
19	2.063829	2.053661	2.043550	2.033496	2.023503	2.003701	19
20	2.145843	2.134148	2.122522	2.110967	2.099485	2.076749	20
21	2.231677	2.218292	2.204991	2.191775	2.178648	2.152671	21
22	2.321525	2.306273	2.291121	2.276074	2.261133	2.231587	22
23	2.415590	2.398280	2.381090	2.364025	2.347089	2.313620	23
24	2.514086	2.494511	2.475080	2.455799	2.436672	2.398901	24
25	2.617238	2.595176	2.573286	2.551573	2.530045	2.487564	25
26	2.725284	2.700495	2.675909	2.651534	2.627378	2.579750	26
27	2.838473	2.810698	2.783163	2.755878	2.728851	2.675607	27
28	2.957068	2.926029	2.895272	2.864809	2.834650	2.775287	28
29	3.081347	3.046743	3.012470	2.978542	2.944971	2.878950	29
30	3.211602	3.173110	3.135005	3.097303	3.060020	2.986764	30
31	3.348141	3.305412	3.263135	3.221328	3.180009	3.098901	31
32	3.491287	3.443947	3.397132	3.350865	3.305165	3.215544	32
33	3.641382	3.585029	3.537284	3.486174	3.435722	3.336881	33
34	3.798786	3.740986	3.683889	3.627527	3.571926	3.463109	34
35	3.963879	3.900165	3.837264	3.775210	3.714034	3.594434	35
36	4.137059	4.066932	3.997741	3.929524	3.862318	3.731072	36
37	4.318748	4.241671	4.165667	4.090782	4.017059	3.873244	37
38	4.509390	4.424784	4.341409	4.259315	4.178552	4.021186	38
39	4.709453	4.616658	4.525351	4.435470	4.347109	4.175141	39
40	4.919430	4.817861	4.717858	4.619608	4.523053	4.335362	40
45	6.137240	5.979790	5.825347	5.674066	5.526045	5.240154	45
50	7.694123	7.455172	7.221717	6.993966	6.772102	6.346617	50
55	9.690254	9.333651	8.986707	8.649764	8.323104	7.701475	55
60	12.256450	11.731316	11.222716	10.731177	10.257113	9.362491	60
65	15.563754	14.798750	14.061419	13.352537	12.672660	11.401095	65
70	19.836077	18.731469	17.672303	16.659653	15.694194	13.905647	70
75	25.366450	23.783571	22.273692	20.836501	19.478971	16.985474	75
80	32.541018	30.285592	28.147048	26.126994	24.225804	20.775874	80
90	54.000364	49.491091	45.267741	41.331033	37.678211	31.198160	90
100	90.485180	81.607695	73.402884	65.864091	58.975620	47.050111	100

TABLE D-32

Factors (EUS) to Convert an Exponential Growth to an Equivalent Uniform Series for Interest Rates from 3% to 5½%

Growth Rate of 7% per Period

n	3%	3½%	4%	4½%	5%	5½%	n
1	1.070000	1.070000	1.070000	1.070000	1.070000	1.070000	1
2	1.106897	1.106806	1.106716	1.106626	1.106537	1.106448	2
3	1.145121	1.144870	1.144622	1.144374	1.144128	1.143883	3
4	1.184721	1.184236	1.183754	1.183274	1.182797	1.182322	4
5	1.225751	1.224948	1.224150	1.223356	1.222567	1.221782	5
6	1.268262	1.267051	1.265848	1.264651	1.263462	1.262279	6
7	1.312310	1.310593	1.308887	1.307191	1.305505	1.303831	7
8	1.357954	1.355623	1.353307	1.351007	1.348722	1.346453	8
9	1.405252	1.402190	1.399151	1.396132	1.393136	1.390163	9
10	1.454267	1.450348	1.446460	1.442601	1.438773	1.434976	10
11	1.505063	1.500151	1.495279	1.490448	1.485658	1.480910	11
12	1.557707	1.551654	1.545653	1.539707	1.533816	1.527981	12
13	1.612269	1.604914	1.597629	1.590414	1.583272	1.576205	13
14	1.668821	1.659993	1.651254	1.642607	1.634055	1.625599	14
15	1.727438	1.716950	1.706578	1.696323	1.686189	1.676179	15
16	1.788197	1.775851	1.763650	1.751599	1.739701	1.727962	16
17	1.851181	1.836761	1.822524	1.808475	1.794620	1.780963	17
18	1.916472	1.899749	1.883253	1.866991	1.850972	1.835200	18
19	1.984156	1.964884	1.945891	1.927188	1.908785	1.890689	19
20	2.054329	2.032240	2.010495	1.989108	1.968088	1.947445	20
21	2.127079	2.101892	2.077124	2.052792	2.028908	2.005485	21
22	2.202507	2.173917	2.145837	2.118285	2.091276	2.064826	22
23	2.280714	2.248297	2.216696	2.185630	2.155221	2.125483	23
24	2.361804	2.325415	2.289763	2.254875	2.220772	2.187473	24
25	2.445888	2.405056	2.365105	2.326064	2.287959	2.250812	25
26	2.533078	2.487410	2.442787	2.399245	2.356814	2.315517	26
27	2.623494	2.572569	2.522880	2.474468	2.427366	2.381603	27
28	2.717257	2.660627	2.605452	2.551780	2.499649	2.449086	28
29	2.814495	2.751683	2.690578	2.631234	2.573693	2.517984	29
30	2.915341	2.845839	2.778332	2.712881	2.649530	2.588313	30
31	3.019931	2.943199	2.868791	2.796773	2.727194	2.660088	31
32	3.128408	3.043873	2.962034	2.882965	2.806719	2.733327	32
33	3.240920	3.147972	3.058143	2.971513	2.888136	2.808045	33
34	3.357621	3.255613	3.157201	3.062472	2.971482	2.884260	34
35	3.478672	3.366915	3.259294	3.155900	3.056791	2.961988	35
36	3.604237	3.482003	3.364511	3.251857	3.144098	3.041245	36
37	3.734490	3.601005	3.472942	3.350403	3.233438	3.122049	37
38	3.869609	3.724053	3.584682	3.451600	3.324949	3.204417	38
39	4.009781	3.851284	3.699827	3.555511	3.418367	3.288366	39
40	4.155199	3.982841	3.818475	3.662199	3.514031	3.373912	40
45	4.968276	4.710808	4.467941	4.239683	4.025877	3.826221	45
50	5.946140	5.571113	5.221572	4.897193	4.597345	4.321144	50
55	7.122931	6.587726	6.095346	5.644628	5.233884	4.861032	55
60	8.539940	7.788952	7.107656	6.493082	5.941443	5.448393	60
65	10.247099	9.208214	8.279680	7.455004	6.726525	6.085926	65
70	12.304803	10.884983	9.635811	8.544369	7.596265	6.776545	70
75	14.786101	12.865864	11.204144	9.776882	8.558497	7.523415	75
80	17.779351	15.205897	13.017050	11.170198	9.621834	8.329978	80
90	25.751653	21.235147	17.531452	14.521147	12.090710	10.137479	90
100	37.374892	29.646900	23.550943	18.787815	15.090853	12.233074	100

Exponential Growth Factors

TABLE D-33

FACTORS (EUS) TO CONVERT AN EXPONENTIAL GROWTH
TO AN EQUIVALENT UNIFORM SERIES FOR INTEREST RATES FROM 6% TO 8½%
Growth Rate of 7% per Period

n	6%	6½%	7%	7½%	8%	8½%	n
1	1.070000	1.070000	1.070000	1.070000	1.070000	1.070000	1
2	1.106359	1.106271	1.106184	1.106096	1.106010	1.105923	2
3	1.143639	1.143396	1.143155	1.142915	1.142676	1.142438	3
4	1.181850	1.181380	1.180912	1.180448	1.179985	1.179525	4
5	1.221001	1.220225	1.219453	1.218686	1.217923	1.217165	5
6	1.261104	1.259936	1.258775	1.257621	1.256474	1.255335	6
7	1.302167	1.300514	1.298873	1.297242	1.295622	1.294013	7
8	1.344200	1.341963	1.339742	1.337537	1.335349	1.333177	8
9	1.387212	1.384283	1.381378	1.378496	1.375638	1.372803	9
10	1.431210	1.427477	1.423775	1.420106	1.416470	1.412867	10
11	1.476205	1.471543	1.466926	1.462353	1.457825	1.453343	11
12	1.522203	1.516484	1.510824	1.505224	1.499685	1.494207	12
13	1.569213	1.562298	1.555461	1.548704	1.542027	1.535431	13
14	1.617241	1.608984	1.600829	1.592778	1.584831	1.576991	14
15	1.666296	1.656542	1.646919	1.637430	1.628077	1.618859	15
16	1.716383	1.704969	1.693722	1.682646	1.671741	1.661010	16
17	1.767510	1.754264	1.741228	1.728407	1.715802	1.703416	17
18	1.819682	1.804423	1.789427	1.774697	1.760238	1.746050	18
19	1.872906	1.855444	1.838307	1.821500	1.805026	1.788887	19
20	1.927188	1.907324	1.887859	1.868797	1.850143	1.831899	20
21	1.982532	1.960058	1.938069	1.916571	1.895567	1.875060	21
22	2.038945	2.013643	1.988927	1.964803	1.941275	1.918345	22
23	2.096430	2.068073	2.040420	2.013477	1.987245	1.961727	23
24	2.154994	2.123346	2.092536	2.062572	2.033454	2.005182	24
25	2.214640	2.179454	2.145263	2.112071	2.079880	2.048685	25
26	2.275372	2.236393	2.198587	2.161956	2.126500	2.092211	26
27	2.337195	2.294157	2.252495	2.212208	2.173292	2.135737	27
28	2.400113	2.352741	2.306974	2.262809	2.220236	2.179239	28
29	2.464129	2.412138	2.362011	2.313740	2.267309	2.222696	29
30	2.529248	2.472342	2.417592	2.364983	2.314491	2.266085	30
31	2.595471	2.533347	2.473704	2.416520	2.361762	2.309385	31
32	2.662803	2.595145	2.530333	2.468334	2.409101	2.352577	32
33	2.731247	2.657731	2.587466	2.520406	2.456488	2.395640	33
34	2.800806	2.721097	2.645089	2.572719	2.503905	2.438555	34
35	2.871482	2.785236	2.703189	2.625255	2.551333	2.481305	35
36	2.943278	2.850141	2.761751	2.677999	2.598755	2.523873	36
37	3.016198	2.915806	2.820763	2.730932	2.646151	2.566241	37
38	3.090243	2.982222	2.880212	2.784040	2.693507	2.608395	38
39	3.165417	3.049383	2.940084	2.837305	2.740804	2.650319	39
40	3.241721	3.117281	3.000366	2.890712	2.788028	2.691999	40
45	3.640294	3.467578	3.307415	3.159358	3.022532	2.896307	45
50	4.067513	3.835233	3.622992	3.429429	3.253168	3.092850	50
55	4.523712	4.219400	3.945498	3.699413	3.478616	3.280684	55
60	5.009273	4.619304	4.273754	3.968052	3.697885	3.459239	60
65	5.524654	5.034269	4.606682	4.234330	3.910275	3.628246	65
70	6.070405	5.463718	4.943367	4.497445	4.115328	3.787672	70
75	6.647183	5.907184	5.283045	4.756782	4.312781	3.937659	75
80	7.255759	6.364305	5.625086	5.011885	4.502527	4.078473	80
90	8.572006	7.318540	6.314315	5.508189	4.859022	4.334032	90
100	10.027782	8.325330	7.008076	5.984863	5.185786	4.557631	100

TABLE D-34
Factors (EUS) to Convert an Exponential Growth to an Equivalent Uniform Series for Interest Rates from 9% to 20%
Growth Rate of 7% per Period

n	9%	10%	11%	12%	15%	20%	n
1	1.070000	1.070000	1.070000	1.070000	1.070000	1.070000	1
2	1.105837	1.105667	1.105498	1.105330	1.104837	1.104045	2
3	1.142202	1.141732	1.141267	1.140807	1.139454	1.137287	3
4	1.179068	1.178160	1.177263	1.176375	1.173768	1.169609	4
5	1.216410	1.214915	1.213437	1.211977	1.207700	1.200912	5
6	1.254202	1.251959	1.249744	1.247559	1.241175	1.231106	6
7	1.292415	1.289254	1.286137	1.283065	1.274121	1.260114	7
8	1.331022	1.326761	1.322568	1.318442	1.306471	1.287878	8
9	1.369993	1.364443	1.358991	1.353636	1.338164	1.314350	9
10	1.409298	1.402260	1.395359	1.388597	1.369143	1.339497	10
11	1.448907	1.440174	1.431629	1.423274	1.399357	1.363303	11
12	1.488791	1.478147	1.467757	1.457622	1.428764	1.385760	12
13	1.528918	1.516140	1.503698	1.491595	1.457324	1.406877	13
14	1.569258	1.554117	1.539413	1.525150	1.485005	1.426670	14
15	1.609780	1.592040	1.574861	1.558248	1.511781	1.445166	15
16	1.650454	1.629874	1.610006	1.590851	1.537632	1.462400	16
17	1.691250	1.667584	1.644810	1.622927	1.562545	1.478415	17
18	1.732137	1.705136	1.679240	1.654443	1.586509	1.493257	18
19	1.773085	1.742498	1.713263	1.685371	1.609522	1.506978	19
20	1.814066	1.779638	1.746851	1.715687	1.631585	1.519634	20
21	1.855051	1.816526	1.779975	1.745368	1.652702	1.531281	21
22	1.896012	1.853133	1.812610	1.774394	1.672885	1.541978	22
23	1.936922	1.889433	1.844733	1.802750	1.692145	1.551782	23
24	1.977753	1.925400	1.876322	1.830422	1.710499	1.560752	24
25	2.018481	1.961008	1.907360	1.857400	1.727966	1.568945	25
26	2.059080	1.996237	1.937829	1.883674	1.744568	1.576415	26
27	2.099527	2.031065	1.967715	1.909240	1.760329	1.583216	27
28	2.139798	2.065472	1.997005	1.934094	1.775274	1.589400	28
29	2.179870	2.099441	2.025690	1.958235	1.789429	1.595014	29
30	2.219724	2.132954	2.053760	1.981663	1.802823	1.600105	30
31	2.259339	2.165999	2.081209	2.004382	1.815483	1.604717	31
32	2.298696	2.198560	2.108033	2.026396	1.827439	1.608889	32
33	2.337776	2.230626	2.134227	2.047711	1.838720	1.612661	33
34	2.376563	2.262187	2.159791	2.068335	1.849355	1.616067	34
35	2.415040	2.293234	2.184723	2.088277	1.859372	1.619140	35
36	2.453193	2.323758	2.209027	2.107547	1.868802	1.621911	36
37	2.491008	2.353754	2.232703	2.126157	1.877671	1.624406	37
38	2.528472	2.383216	2.255756	2.144118	1.886008	1.626653	38
39	2.565572	2.412141	2.278190	2.161443	1.893839	1.628674	39
40	2.602297	2.440524	2.300011	2.178146	1.901192	1.630492	40
45	2.779984	2.574295	2.400188	2.252845	1.931627	1.637148	45
50	2.947154	2.694633	2.486341	2.314260	1.953517	1.641006	50
55	3.103337	2.802066	2.559836	2.364331	1.969125	1.643223	55
60	3.248440	2.897404	2.622143	2.404896	1.980184	1.644490	60
65	3.382650	2.981616	2.674717	2.437603	1.987984	1.645211	65
70	3.506353	3.055735	2.718918	2.463882	1.993468	1.645621	70
75	3.620063	3.120794	2.755980	2.484940	1.997315	1.645853	75
80	3.724369	3.177782	2.786996	2.501784	2.000008	1.645984	80
90	3.907280	3.271167	2.834546	2.525972	2.003208	1.646100	90
100	4.060078	3.342329	2.867630	2.541351	2.004769	1.646137	100

Exponential Growth Factors

TABLE D-35
Factors (EUS) to Convert an Exponential Growth to an Equivalent Uniform Series for Interest Rates from 25% to 50%
Growth Rate of 7% per Period

n	25%	30%	35%	40%	45%	50%	n
1	1.070000	1.070000	1.070000	1.070000	1.070000	1.070000	1
2	1.103289	1.102565	1.101872	1.101208	1.100571	1.099960	2
3	1.135224	1.133261	1.131392	1.129611	1.127913	1.126293	3
4	1.165676	1.161956	1.158440	1.155115	1.151973	1.149001	4
5	1.194538	1.188562	1.182963	1.177721	1.172817	1.168228	5
6	1.221734	1.213034	1.204971	1.197509	1.190608	1.184228	6
7	1.247213	1.235371	1.224531	1.214625	1.205582	1.197331	7
8	1.270949	1.255608	1.241753	1.229266	1.218024	1.207905	8
9	1.292943	1.273814	1.256785	1.241660	1.228238	1.216325	9
10	1.313219	1.290082	1.269797	1.252051	1.236533	1.222950	10
11	1.331816	1.304527	1.280977	1.260688	1.243204	1.228108	11
12	1.348795	1.317277	1.290513	1.267809	1.248522	1.232085	12
13	1.364226	1.328468	1.298596	1.273638	1.252728	1.235126	13
14	1.378192	1.338241	1.305406	1.278378	1.256031	1.237435	14
15	1.390781	1.346734	1.311112	1.282210	1.258609	1.239177	15
16	1.402086	1.354083	1.315870	1.285291	1.260610	1.240483	16
17	1.412204	1.360415	1.319820	1.287757	1.262156	1.241457	17
18	1.421228	1.365851	1.323085	1.289722	1.263344	1.242181	18
19	1.429253	1.370501	1.325775	1.291282	1.264255	1.242717	19
20	1.436369	1.374467	1.327982	1.292516	1.264950	1.243112	20
21	1.442663	1.377839	1.329789	1.293489	1.265479	1.243403	21
22	1.448215	1.380699	1.331264	1.294255	1.265880	1.243616	22
23	1.453103	1.383119	1.332465	1.294856	1.266184	1.243771	23
24	1.457396	1.385162	1.333441	1.295326	1.266414	1.243885	24
25	1.461160	1.386883	1.334232	1.295693	1.266587	1.243968	25
26	1.464453	1.388330	1.334872	1.295980	1.266717	1.244028	26
27	1.467331	1.389545	1.335389	1.296203	1.266815	1.244072	27
28	1.469842	1.390564	1.335806	1.296376	1.266888	1.244104	28
29	1.472029	1.391416	1.336142	1.296511	1.266943	1.244127	29
30	1.473931	1.392129	1.336412	1.296615	1.266984	1.244143	30
31	1.475585	1.392724	1.336629	1.296696	1.267015	1.244155	31
32	1.477020	1.393220	1.336803	1.296759	1.267038	1.244164	32
33	1.478264	1.393633	1.336943	1.296807	1.267055	1.244170	33
34	1.479342	1.393977	1.337055	1.296844	1.267068	1.244175	34
35	1.480275	1.394264	1.337145	1.296873	1.267078	1.244178	35
36	1.481082	1.394502	1.337217	1.296896	1.267085	1.244180	36
37	1.481780	1.394699	1.337274	1.296913	1.267090	1.244182	37
38	1.482382	1.394863	1.337320	1.296926	1.267094	1.244183	38
39	1.482902	1.394999	1.337356	1.296936	1.267097	1.244184	39
40	1.483350	1.395112	1.337386	1.296944	1.267099	1.244184	40
45	1.484816	1.395444	1.337463	1.296963	1.267104	1.244186	45
50	1.485508	1.395572	1.337488	1.296968	1.267105	1.244186	50
55	1.485831	1.395622	1.337496	1.296969	1.267105	1.244186	55
60	1.485981	1.395641	1.337499	1.296970	1.267105	1.244186	60
65	1.486051	1.395648	1.337500	1.296970	1.267105	1.244186	65
70	1.486083	1.395651	1.337500	1.296970	1.267105	1.244186	70
75	1.486098	1.395652	1.337500	1.296970	1.267105	1.244186	75
80	1.486105	1.395652	1.337500	1.296970	1.267105	1.244186	80
90	1.486110	1.395652	1.337500	1.296970	1.267105	1.244186	90
100	1.486111	1.395652	1.337500	1.296970	1.267105	1.244186	100

TABLE D-36

Factors (EUS) to Convert an Exponential Growth to an Equivalent Uniform Series for Interest Rates from 1% to 2½%

Growth Rate of 8% per Period

n	1%	1¼%	1½%	1¾%	2%	2½%	n
1	1.080000	1.080000	1.080000	1.080000	1.080000	1.080000	1
2	1.122985	1.122932	1.122878	1.122825	1.122772	1.122667	2
3	1.168108	1.167960	1.167812	1.167665	1.167518	1.167225	3
4	1.215488	1.215159	1.214911	1.214624	1.214337	1.213767	4
5	1.265251	1.264770	1.264291	1.263813	1.263336	1.262386	5
6	1.317530	1.316801	1.316073	1.315348	1.314625	1.313185	6
7	1.372468	1.371427	1.370389	1.369354	1.368322	1.366268	7
8	1.430213	1.428791	1.427373	1.425960	1.424551	1.421746	8
9	1.490926	1.489046	1.487172	1.485303	1.483441	1.479735	9
10	1.554775	1.552352	1.549936	1.547529	1.545131	1.540359	10
11	1.621938	1.618878	1.615829	1.612791	1.609764	1.603745	11
12	1.692605	1.688806	1.685021	1.681250	1.677495	1.670029	12
13	1.766976	1.762325	1.757691	1.753077	1.748483	1.739353	13
14	1.845265	1.839637	1.834033	1.828453	1.822898	1.811866	14
15	1.927698	1.920957	1.914246	1.907568	1.900921	1.887726	15
16	2.014513	2.006510	1.998546	1.990622	1.982738	1.967096	16
17	2.105965	2.096537	2.087158	2.077829	2.068551	2.050151	17
18	2.202324	2.191293	2.180322	2.169413	2.158567	2.137073	18
19	2.303875	2.291045	2.278290	2.265611	2.253010	2.228054	19
20	2.410923	2.396081	2.381330	2.366673	2.352114	2.323295	20
21	2.523789	2.506701	2.489726	2.472865	2.456124	2.423009	21
22	2.642815	2.623228	2.603777	2.584467	2.565300	2.527417	22
23	2.768363	2.746001	2.723802	2.701773	2.679919	2.636756	23
24	2.900820	2.875379	2.850136	2.825097	2.800268	2.751270	24
25	3.040594	3.011746	2.983134	2.954768	2.926655	2.871219	25
26	3.188118	3.155505	3.123174	3.091137	3.059402	2.996877	26
27	3.343855	3.307087	3.270655	3.234572	3.198849	3.128528	27
28	3.508293	3.466946	3.425958	3.385463	3.345356	3.266476	28
29	3.681953	3.635566	3.589650	3.544223	3.499302	3.411038	29
30	3.865386	3.813458	3.762086	3.711289	3.661088	3.562546	30
31	4.059179	4.001167	3.943806	3.887122	3.831137	3.721352	31
32	4.263956	4.199267	4.135342	4.072209	4.009895	3.887825	32
33	4.480378	4.408371	4.337257	4.267067	4.197834	4.062354	33
34	4.709148	4.629128	4.550146	4.472241	4.395451	4.245347	34
35	4.951013	4.862224	4.774641	4.688309	4.603272	4.437233	35
36	5.206769	5.108391	5.011411	4.915881	4.821852	4.638466	36
37	5.477259	5.368404	5.261165	5.155603	5.051776	4.849522	37
38	5.763379	5.643084	5.524653	5.408158	5.293663	5.070900	38
39	6.066084	5.933304	5.802673	5.674269	5.548167	5.303127	39
40	6.386387	6.239992	6.096066	5.954701	5.815979	5.546759	40
45	8.292054	8.057056	7.826877	7.601691	7.381648	6.957496	45
50	10.831773	10.462118	10.101525	9.750293	9.408679	8.755096	50
55	14.228397	13.656002	13.100115	12.561234	12.039761	11.050160	55
60	18.785711	17.910324	17.064241	16.248242	15.462909	13.985605	60
65	24.918465	23.592976	22.319383	21.095835	19.926092	17.746203	65
70	33.193729	31.202902	29.298851	27.483144	25.756615	22.570963	70
75	44.387881	41.417607	38.592949	35.915835	33.386834	28.769189	75
80	59.565197	55.158096	50.992024	47.068916	43.388265	36.741313	80
90	108.258122	98.683705	89.747362	81.446974	73.773029	60.234877	90
100	198.833689	178.314041	159.422002	142.131874	126.397231	99.326014	100

Exponential Growth Factors

TABLE D-37
Factors (EUS) to Convert an Exponential Growth to an Equivalent Uniform Series for Interest Rates from 3% to 5½%
Growth Rate of 8% per Period

n	3%	3½%	4%	4½%	5%	5½%	n
1	1.080000	1.080000	1.080000	1.080000	1.080000	1.080000	1
2	1.122562	1.122457	1.122353	1.122249	1.122146	1.122044	2
3	1.166934	1.166644	1.166356	1.166069	1.165783	1.165499	3
4	1.213199	1.212634	1.212072	1.211513	1.210957	1.210404	4
5	1.261442	1.260503	1.259569	1.258640	1.257716	1.256797	5
6	1.311753	1.310329	1.308914	1.307507	1.306108	1.304718	6
7	1.364226	1.362196	1.360179	1.358175	1.356184	1.354205	7
8	1.418959	1.416150	1.413439	1.410708	1.407994	1.405300	8
9	1.476055	1.472400	1.468771	1.465169	1.461593	1.458045	9
10	1.535622	1.530920	1.526255	1.521626	1.517034	1.512480	10
11	1.597773	1.591849	1.585973	1.580148	1.574373	1.568651	11
12	1.662626	1.655287	1.648013	1.640806	1.633668	1.626599	12
13	1.730305	1.721341	1.712464	1.703675	1.694976	1.686370	13
14	1.800940	1.790123	1.779418	1.768830	1.758359	1.748010	14
15	1.874666	1.861748	1.848974	1.836350	1.823878	1.811564	15
16	1.951626	1.936336	1.921231	1.906316	1.891597	1.877079	16
17	2.031969	2.014014	1.996293	1.978813	1.961582	1.944605	17
18	2.115851	2.094913	2.074269	2.053928	2.033898	2.014188	18
19	2.203436	2.179171	2.155271	2.131749	2.108616	2.085881	19
20	2.294894	2.266929	2.239416	2.212371	2.185805	2.159732	20
21	2.390406	2.358338	2.326826	2.295888	2.265539	2.235795	21
22	2.490159	2.453554	2.417626	2.382399	2.347892	2.314121	22
23	2.594350	2.552737	2.511947	2.472007	2.432939	2.394764	23
24	2.703186	2.656059	2.609925	2.564817	2.520761	2.477780	24
25	2.816882	2.763695	2.711701	2.660938	2.611437	2.563225	25
26	2.935666	2.875830	2.817421	2.760483	2.705051	2.651154	26
27	3.059774	2.992657	2.927238	2.863568	2.801687	2.741627	27
28	3.189455	3.114376	3.041309	2.970313	2.901433	2.834702	28
29	3.324970	3.241197	3.159799	3.080843	3.004378	2.930441	29
30	3.466591	3.373338	3.282878	3.195286	3.110616	3.028904	30
31	3.614605	3.511026	3.410722	3.313774	3.220239	3.130154	31
32	3.769311	3.654501	3.543516	3.436444	3.333347	3.234257	32
33	3.931022	3.804010	3.681450	3.563438	3.450037	3.341276	33
34	4.100069	3.959812	3.824722	3.694902	3.570414	3.451280	34
35	4.276796	4.122177	3.973538	3.830987	3.694583	3.564337	35
36	4.461564	4.291387	4.128111	3.971848	3.822651	3.680516	36
37	4.654752	4.467736	4.288664	4.117648	3.954730	3.799888	37
38	4.856757	4.651532	4.455426	4.268552	4.090935	3.922526	38
39	5.067994	4.843093	4.628637	4.424732	4.231382	4.048504	39
40	5.288899	5.042754	4.808546	4.586366	4.376193	4.177898	40
45	6.555222	6.175306	5.817929	5.482995	5.170163	4.878885	45
50	8.141915	7.569590	7.037949	6.546267	6.093355	5.677651	50
55	10.132641	9.287145	8.512394	7.806086	7.165111	6.585760	55
60	12.633245	11.404086	10.294146	9.297675	8.407637	7.616159	60
65	15.777665	14.014555	12.447063	11.062529	9.846448	8.783368	65
70	19.735432	17.235006	15.048265	13.149538	11.510877	10.103697	70
75	24.721168	21.209489	18.190882	15.616321	13.434661	11.595483	75
80	31.006621	26.116156	21.987386	18.530786	15.656622	13.279353	80
90	48.944918	39.660263	32.113551	26.037257	21.180544	17.319132	90
100	77.557504	60.335002	46.888178	36.496938	28.527449	22.446020	100

TABLE D-38

Factors (EUS) to Convert an Exponential Growth to an Equivalent Uniform Series for Interest Rates from 6% to 8½%

Growth Rate of 8% per Period

n	6%	6½%	7%	7½%	8%	8½%	n
1	1.080000	1.080000	1.080000	1.080000	1.080000	1.080000	1
2	1.121942	1.121840	1.121739	1.121639	1.121538	1.121439	2
3	1.165217	1.164936	1.164656	1.164377	1.164101	1.163825	3
4	1.209854	1.209307	1.208763	1.208222	1.207683	1.207148	4
5	1.255884	1.254976	1.254073	1.253175	1.252282	1.251395	5
6	1.303336	1.301962	1.300597	1.299240	1.297892	1.296553	6
7	1.352239	1.350286	1.348347	1.346420	1.344507	1.342607	7
8	1.402625	1.399969	1.397332	1.394715	1.392118	1.389540	8
9	1.454523	1.451030	1.447564	1.444126	1.440717	1.437337	9
10	1.507965	1.503488	1.499050	1.494653	1.490295	1.485978	10
11	1.562980	1.557364	1.551801	1.546293	1.540840	1.535443	11
12	1.619601	1.612675	1.605823	1.599044	1.592340	1.585712	12
13	1.677858	1.669443	1.661124	1.652904	1.644783	1.636764	13
14	1.737784	1.727684	1.717711	1.707868	1.698156	1.688576	14
15	1.799409	1.787417	1.775591	1.763932	1.752443	1.741126	15
16	1.862766	1.848661	1.834768	1.821091	1.807630	1.794389	16
17	1.927887	1.911434	1.895250	1.879337	1.863700	1.848341	17
18	1.994805	1.975753	1.957039	1.938666	1.920638	1.902957	18
19	2.063552	2.041636	2.020140	1.999068	1.978425	1.958213	19
20	2.134161	2.109100	2.084557	2.060537	2.037044	2.014081	20
21	2.206666	2.178163	2.150294	2.123064	2.096477	2.070536	21
22	2.281100	2.248841	2.217352	2.186639	2.156706	2.127553	22
23	2.357498	2.321152	2.285735	2.251253	2.217710	2.185104	23
24	2.435893	2.395111	2.355444	2.316897	2.279471	2.243163	24
25	2.516320	2.470736	2.426481	2.383560	2.341969	2.301705	25
26	2.598813	2.548042	2.498848	2.451230	2.405185	2.360702	26
27	2.683409	2.627047	2.572544	2.519898	2.469099	2.420128	27
28	2.770143	2.707766	2.647572	2.589552	2.533689	2.479958	28
29	2.859050	2.790216	2.723931	2.660181	2.598938	2.540166	29
30	2.950168	2.874412	2.801622	2.731772	2.664823	2.600726	30
31	3.043533	2.960371	2.880644	2.804314	2.731326	2.661614	31
32	3.139183	3.048109	2.960998	2.877795	2.798426	2.722805	32
33	3.237154	3.137641	3.042683	2.952202	2.866104	2.784275	33
34	3.337486	3.228984	3.125698	3.027525	2.934340	2.846000	34
35	3.440218	3.322155	3.210043	3.103750	3.003114	2.907958	35
36	3.545388	3.417168	3.295718	3.180865	3.072408	2.970125	36
37	3.653036	3.514041	3.382722	3.258859	3.142203	3.032481	37
38	3.763204	3.612790	3.471053	3.337719	3.212480	3.095002	38
39	3.875932	3.713432	3.560712	3.417433	3.283220	3.157670	39
40	3.991261	3.815982	3.651697	3.497990	3.354406	3.220463	40
45	4.608438	4.357962	4.126497	3.913010	3.716428	3.535657	45
50	5.297308	4.950279	4.634395	4.347426	4.087143	3.851357	50
55	6.063928	5.595285	5.175427	4.800000	4.464788	4.165785	55
60	6.914934	6.295551	5.749762	5.269673	4.847877	4.477539	60
65	7.857625	7.053918	6.357725	5.755574	5.235187	4.785549	65
70	8.900051	7.873525	6.999813	6.257016	5.625735	5.089024	70
75	10.051104	8.757850	7.676700	6.773482	6.018738	5.387399	75
80	11.320611	9.710731	8.389239	7.304609	6.413589	5.680291	80
90	14.259554	11.835469	9.925540	8.410047	7.207073	6.248745	90
100	17.818063	14.298566	11.618865	9.572751	8.003638	6.793516	100

TABLE D-39
FACTORS (EUS) TO CONVERT AN EXPONENTIAL GROWTH TO AN EQUIVALENT UNIFORM SERIES FOR INTEREST RATES FROM 9% TO 20%
Growth Rate of 8% per Period

n	9%	10%	11%	12%	15%	20%	n
1	1.080000	1.080000	1.080000	1.080000	1.080000	1.080000	1
2	1.121340	1.121143	1.120948	1.120755	1.120186	1.119273	2
3	1.163551	1.163007	1.162468	1.161935	1.160366	1.157855	3
4	1.206615	1.205558	1.204513	1.203479	1.200444	1.195603	4
5	1.250512	1.248763	1.247034	1.245326	1.240324	1.232386	5
6	1.295222	1.292585	1.289982	1.287414	1.279913	1.268085	6
7	1.340720	1.336986	1.333305	1.329679	1.319121	1.302596	7
8	1.386983	1.381927	1.376952	1.372058	1.357863	1.335831	8
9	1.433985	1.427369	1.420870	1.414489	1.396059	1.367716	9
10	1.481701	1.473271	1.465007	1.456910	1.433631	1.398195	10
11	1.530102	1.519591	1.509310	1.499260	1.470512	1.427229	11
12	1.579160	1.566288	1.553727	1.541481	1.506636	1.454791	12
13	1.628846	1.613319	1.598207	1.583514	1.541946	1.480872	13
14	1.679130	1.660643	1.642699	1.625303	1.576393	1.505474	14
15	1.729981	1.708216	1.687154	1.666797	1.609931	1.528611	15
16	1.781369	1.755998	1.731522	1.707943	1.642525	1.550309	16
17	1.833261	1.803945	1.775757	1.748695	1.674142	1.570603	17
18	1.885627	1.852018	1.819813	1.789006	1.704759	1.589534	18
19	1.938433	1.900175	1.863648	1.828834	1.734357	1.607152	19
20	1.991649	1.948377	1.907217	1.868140	1.762925	1.623510	20
21	2.045242	1.996586	1.950483	1.906889	1.790456	1.638665	21
22	2.099180	2.044763	1.993407	1.945046	1.816947	1.652678	22
23	2.153433	2.092871	2.035954	1.982582	1.842403	1.665609	23
24	2.207968	2.140876	2.078090	2.019471	1.866831	1.677522	24
25	2.262755	2.188743	2.119785	2.055689	1.890242	1.688478	25
26	2.317764	2.236440	2.161008	2.091214	1.912650	1.698538	26
27	2.372963	2.283935	2.201735	2.126029	1.934075	1.707762	27
28	2.428325	2.331198	2.241939	2.160120	1.954537	1.716208	28
29	2.483821	2.378201	2.281600	2.193474	1.974058	1.723933	29
30	2.539422	2.424916	2.320696	2.226081	1.992663	1.730988	30
31	2.595101	2.471320	2.359211	2.257935	2.010379	1.737426	31
32	2.650832	2.517386	2.397129	2.289029	2.027234	1.743294	32
33	2.706590	2.563094	2.434435	2.319363	2.043255	1.748637	33
34	2.762349	2.608423	2.471118	2.348934	2.058472	1.753499	34
35	2.818086	2.653352	2.507168	2.377744	2.072916	1.757919	35
36	2.873777	2.697864	2.542576	2.405797	2.086614	1.761935	36
37	2.929401	2.741943	2.577336	2.433096	2.099598	1.765580	37
38	2.984936	2.785573	2.611444	2.459648	2.111898	1.768887	38
39	3.040362	2.828741	2.644895	2.485460	2.123542	1.771885	39
40	3.095660	2.871435	2.677688	2.510541	2.134559	1.774602	40
45	3.369609	3.077434	2.831789	2.625338	2.181220	1.784778	45
50	3.637956	3.270376	2.969776	2.723611	2.216167	1.790920	50
55	3.899236	3.449856	3.092462	2.807122	2.242143	1.794601	55
60	4.152432	3.616076	3.200966	2.877705	2.261349	1.796797	60
65	4.396889	3.769312	3.296549	2.937125	2.275496	1.798103	65
70	4.632242	3.910201	3.380504	2.987006	2.285889	1.798877	70
75	4.858344	4.039460	3.454091	3.028793	2.293512	1.799336	75
80	5.075210	4.157861	3.518492	3.063749	2.299095	1.799608	80
90	5.481824	4.365226	3.623972	3.117355	2.306170	1.799863	90
100	5.853900	4.538342	3.704384	3.154705	2.309952	1.799952	100

TABLE D-40

Factors (EUS) to Convert an Exponential Growth to an Equivalent Uniform Series for Interest Rates from 25% to 50%
Growth Rate of 8% per Period

n	25%	30%	35%	40%	45%	50%	n
1	1.080000	1.080000	1.080000	1.080000	1.080000	1.080000	1
2	1.118400	1.117565	1.116766	1.116000	1.115265	1.114560	2
3	1.155465	1.153191	1.151025	1.148961	1.146994	1.145118	3
4	1.191025	1.186696	1.182605	1.178737	1.175082	1.171625	4
5	1.224935	1.217950	1.211408	1.205286	1.199558	1.194200	5
6	1.257083	1.246872	1.237413	1.228662	1.220572	1.213096	6
7	1.287384	1.273429	1.260662	1.249001	1.238363	1.228660	7
8	1.315786	1.297635	1.281255	1.266502	1.253230	1.241293	8
9	1.342265	1.319543	1.299336	1.281405	1.265508	1.251411	9
10	1.366824	1.339237	1.315082	1.293974	1.275537	1.259419	10
11	1.389490	1.356832	1.328691	1.304483	1.283651	1.265689	11
12	1.410311	1.372456	1.340371	1.313200	1.290158	1.270553	12
13	1.429353	1.386256	1.350330	1.320378	1.295335	1.274295	13
14	1.446695	1.398381	1.358773	1.326251	1.299425	1.277153	14
15	1.462426	1.408984	1.365892	1.331028	1.302638	1.279322	15
16	1.476643	1.418216	1.371865	1.334893	1.305147	1.280958	16
17	1.489448	1.426222	1.376854	1.338006	1.307097	1.282187	17
18	1.500943	1.433139	1.381006	1.340503	1.308606	1.283106	18
19	1.511233	1.439055	1.384447	1.342498	1.309769	1.283790	19
20	1.520417	1.444208	1.387291	1.344086	1.310663	1.284298	20
21	1.528594	1.448584	1.389633	1.345347	1.311348	1.284674	21
22	1.535857	1.452321	1.391559	1.346346	1.311871	1.284952	22
23	1.542293	1.455504	1.393136	1.347134	1.312270	1.285156	23
24	1.547985	1.458209	1.394427	1.347756	1.312573	1.285306	24
25	1.553010	1.460504	1.395481	1.348245	1.312803	1.285416	25
26	1.557439	1.462447	1.396339	1.348629	1.312978	1.285497	26
27	1.561335	1.464090	1.397038	1.348930	1.313110	1.285556	27
28	1.564759	1.465477	1.397605	1.349166	1.313210	1.285599	28
29	1.567762	1.466646	1.398066	1.349350	1.313285	1.285631	29
30	1.570395	1.467631	1.398439	1.349494	1.313342	1.285654	30
31	1.572699	1.468459	1.398741	1.349607	1.313385	1.285670	31
32	1.574714	1.469155	1.398985	1.349694	1.313417	1.285682	32
33	1.576474	1.469739	1.399183	1.349763	1.313441	1.285691	33
34	1.578010	1.470229	1.399342	1.349816	1.313459	1.285697	34
35	1.579350	1.470640	1.399470	1.349857	1.313473	1.285702	35
36	1.580518	1.470984	1.399574	1.349889	1.313483	1.285705	36
37	1.581535	1.471272	1.399658	1.349914	1.313491	1.285708	37
38	1.582420	1.471513	1.399725	1.349933	1.313496	1.285710	38
39	1.583190	1.471714	1.399779	1.349948	1.313501	1.285711	39
40	1.583860	1.471882	1.399822	1.349960	1.313504	1.285712	40
45	1.586096	1.472388	1.399941	1.349989	1.313511	1.285714	45
50	1.587195	1.472592	1.399980	1.349997	1.313513	1.285714	50
55	1.587731	1.472673	1.399994	1.349999	1.313513	1.285714	55
60	1.587991	1.472706	1.399998	1.350000	1.313513	1.285714	60
65	1.588117	1.472719	1.399999	1.350000	1.313514	1.285714	65
70	1.588178	1.472724	1.400000	1.350000	1.313514	1.285714	70
75	1.588208	1.472726	1.400000	1.350000	1.313514	1.285714	75
80	1.588222	1.472727	1.400000	1.350000	1.313514	1.285714	80
90	1.588232	1.472727	1.400000	1.350000	1.313514	1.285714	90
100	1.588235	1.472727	1.400000	1.350000	1.313514	1.285714	100

TABLE D-41

Factors (EUS) to Convert an Exponential Growth to an Equivalent Uniform Series for Interest Rates from 1% to 2½%

Growth Rate of 10% per Period

n	1%	1¼%	1½%	1¾%	2%	2½%	n
1	1.100000	1.100000	1.100000	1.100000	1.100000	1.100000	1
2	1.154726	1.154658	1.154591	1.154523	1.154455	1.154321	2
3	1.212901	1.212710	1.212520	1.212331	1.212142	1.211766	3
4	1.274766	1.274391	1.274018	1.273645	1.273273	1.272532	4
5	1.340585	1.339955	1.339327	1.338701	1.338076	1.336832	5
6	1.410640	1.409674	1.408712	1.407752	1.406795	1.404888	6
7	1.485232	1.483841	1.482453	1.481069	1.479690	1.476943	7
8	1.564690	1.562768	1.560853	1.558944	1.557040	1.553252	8
9	1.649362	1.646795	1.644237	1.641687	1.639146	1.634090	9
10	1.739628	1.736285	1.732953	1.729633	1.726326	1.719747	10
11	1.835894	1.831627	1.827376	1.823141	1.818923	1.810537	11
12	1.938597	1.933242	1.927908	1.922595	1.917305	1.906792	12
13	2.048210	2.041580	2.034979	2.028407	2.021864	2.008868	13
14	2.165239	2.157129	2.149055	2.141020	2.133022	2.117145	14
15	2.290233	2.280409	2.270634	2.260908	2.251231	2.232030	15
16	2.423780	2.411985	2.400251	2.388580	2.376973	2.353956	16
17	2.566516	2.552460	2.538482	2.524584	2.510767	2.483385	17
18	2.719126	2.702486	2.685945	2.669505	2.653169	2.620814	18
19	2.882347	2.862765	2.843307	2.823975	2.804774	2.766771	19
20	3.056976	3.034052	3.011282	2.988670	2.966220	2.921823	20
21	3.243870	3.217159	3.190639	3.164315	3.138193	3.086572	21
22	3.443956	3.412963	3.382206	3.351692	3.321426	3.261666	22
23	3.658230	3.622407	3.586874	3.551638	3.516707	3.447794	23
24	3.887771	3.846508	3.805598	3.765053	3.724881	3.645696	24
25	4.133739	4.086360	4.039411	3.992904	3.946853	3.856161	25
26	4.397389	4.343144	4.289419	4.236231	4.183595	4.080033	26
27	4.680074	4.618132	4.556818	4.496151	4.436150	4.318215	27
28	4.983255	4.912694	4.842889	4.773863	4.705638	4.571673	28
29	5.308508	5.228310	5.149017	5.070658	4.993259	4.841439	29
30	5.657536	5.566572	5.476689	5.387922	5.300302	5.128619	30
31	6.032177	5.929199	5.827509	5.727146	5.628150	5.434394	31
32	6.434416	6.318044	6.203201	6.089934	5.978290	5.760028	32
33	6.866399	6.735107	6.605624	6.478009	6.352314	6.106875	33
34	7.330441	7.182542	7.036781	6.893225	6.751936	6.476381	34
35	7.829046	7.662677	7.498827	7.337575	7.178993	6.870096	35
36	8.364916	8.178021	7.994086	7.813205	7.635460	7.289677	36
37	8.940975	8.731282	8.525061	8.322420	8.123456	7.736898	37
38	9.560381	9.325383	9.094449	8.867703	8.645258	8.213657	38
39	10.226546	9.963478	9.705156	9.451724	9.203312	8.721987	39
40	10.943161	10.648972	10.360315	10.077357	9.800245	9.264063	40
45	15.434519	14.926820	14.430721	13.946546	13.474574	12.568120	45
50	21.945489	21.085036	20.248021	19.435035	18.646560	17.144512	50
55	31.427855	29.989479	28.597058	27.251603	25.953868	23.503336	55
60	45.295128	42.916173	40.625127	38.423591	36.312604	32.363769	60
65	65.651015	61.749348	58.012422	54.442541	51.040830	44.740879	65
70	95.632559	89.276404	83.223670	77.477205	72.037453	62.068738	70
75	139.926168	129.627730	119.879967	110.685326	102.041470	86.375138	75
80	205.544478	188.933170	173.308728	158.670339	145.007905	120.529575	80
90	448.057099	405.293267	365.600904	328.942388	295.247330	236.329849	90
100	987.771478	878.857185	779.185354	688.527480	606.547760	466.866277	100

TABLE D-42

Factors (EUS) to Convert an Exponential Growth to an Equivalent Uniform Series for Interest Rates from 3% to 5½%

Growth Rate of 10% per Period

n	3%	3½%	4%	4½%	5%	5½%	n
1	1.100000	1.100000	1.100000	1.100000	1.100000	1.100000	1
2	1.154187	1.154054	1.153922	1.153790	1.153659	1.153528	2
3	1.211392	1.211019	1.210648	1.210280	1.209913	1.209548	3
4	1.271796	1.271063	1.270334	1.269609	1.268887	1.268170	4
5	1.335594	1.334363	1.333139	1.331922	1.330712	1.329509	5
6	1.402993	1.401110	1.399237	1.397375	1.395525	1.393686	6
7	1.474214	1.471501	1.468806	1.466128	1.463468	1.460825	7
8	1.549489	1.545751	1.542039	1.538352	1.534691	1.531057	8
9	1.629069	1.624085	1.619137	1.614226	1.609352	1.604517	9
10	1.713218	1.706739	1.700312	1.693938	1.687616	1.681348	10
11	1.802218	1.793970	1.785792	1.777686	1.769654	1.761697	11
12	1.896371	1.886043	1.875813	1.865680	1.855648	1.845718	12
13	1.995994	1.983246	1.970627	1.958139	1.945787	1.933571	13
14	2.101430	2.085880	2.070500	2.055295	2.040268	2.025424	14
15	2.213039	2.194265	2.175713	2.157390	2.139301	2.121450	15
16	2.331209	2.308742	2.286564	2.264681	2.243102	2.221832	16
17	2.456350	2.429673	2.403365	2.377438	2.351899	2.326759	17
18	2.588899	2.557439	2.526450	2.495943	2.465932	2.436427	18
19	2.729323	2.692449	2.656168	2.620497	2.585451	2.551043	19
20	2.878118	2.835133	2.792890	2.751412	2.710717	2.670821	20
21	3.035813	2.985949	2.937009	2.889021	2.842006	2.795984	21
22	3.202971	3.145383	3.088938	3.033670	2.979605	2.926766	22
23	3.380192	3.313950	3.249116	3.185726	3.123815	3.063408	23
24	3.568115	3.492200	3.418005	3.345576	3.274952	3.206164	24
25	3.767422	3.680712	3.596094	3.513625	3.433348	3.355297	25
26	3.978839	3.880103	3.783902	3.690300	3.599348	3.511082	26
27	4.203138	4.091028	3.981974	3.876051	3.773315	3.673805	27
28	4.441145	4.314182	4.190890	4.071352	3.955630	3.843766	28
29	4.693738	4.550304	4.411261	4.276701	4.146692	4.021274	29
30	4.961853	4.800177	4.643732	4.492622	4.346916	4.206654	30
31	5.246487	5.064632	4.888988	4.719667	4.556742	4.400245	31
32	5.548705	5.344554	5.147750	4.958418	4.776626	4.602399	32
33	5.869641	5.640879	5.420783	5.209485	5.007049	4.813482	33
34	6.210503	5.954603	5.708895	5.473514	5.248513	5.033877	34
35	6.572580	6.286784	6.012940	5.751181	5.501545	5.263984	35
36	6.957246	6.638544	6.333819	6.043200	5.766697	5.504219	36
37	7.365966	7.011076	6.672490	6.350321	6.044546	5.755015	37
38	7.800301	7.405646	7.029960	6.673335	6.335697	6.016823	38
39	8.261917	7.823598	7.407298	7.013075	6.640786	6.290116	39
40	8.752590	8.266360	7.805633	7.370414	6.960476	6.575383	40
45	11.712666	10.908817	10.156519	9.455132	8.803501	8.200042	45
50	15.743553	14.443744	13.243696	12.140771	11.131303	10.210827	50
55	21.246657	19.181540	17.302688	15.602354	14.070919	12.697454	55
60	28.776634	25.542053	22.645019	20.066065	17.782646	15.770410	60
65	39.100445	34.093396	29.682797	25.824204	22.468745	19.565786	65
70	53.279283	45.604564	38.961258	33.254524	28.384439	24.251226	70
75	72.782250	61.116755	51.201842	42.845210	35.851775	30.033266	75
80	99.644111	82.040057	67.359207	55.227164	45.277129	37.166376	80
90	187.810623	148.437895	116.886282	91.864627	72.187321	56.812781	90
100	356.092279	269.741901	203.388667	152.988561	115.048870	86.677807	100

Exponential Growth Factors

TABLE D-43
FACTORS (EUS) TO CONVERT AN EXPONENTIAL GROWTH TO AN EQUIVALENT UNIFORM SERIES FOR INTEREST RATES FROM 6% TO 8½%
Growth Rate of 10% per Period

n	6%	6½%	7%	7½%	8%	8½%	n
1	1.100000	1.100000	1.100000	1.100000	1.100000	1.100000	1
2	1.153398	1.153269	1.153140	1.153012	1.152885	1.152758	2
3	1.209185	1.208823	1.208464	1.208106	1.207750	1.207396	3
4	1.267456	1.266746	1.266040	1.265338	1.264639	1.263945	4
5	1.328313	1.327123	1.325940	1.324764	1.323595	1.322433	5
6	1.391858	1.390041	1.388236	1.386442	1.384660	1.382889	6
7	1.458199	1.455592	1.453002	1.450429	1.447875	1.445339	7
8	1.527448	1.523867	1.520312	1.516784	1.513283	1.509809	8
9	1.599720	1.594962	1.590243	1.585563	1.580923	1.576323	9
10	1.675135	1.668976	1.662874	1.656827	1.650837	1.644905	10
11	1.753816	1.746011	1.738284	1.730635	1.723066	1.715577	11
12	1.835891	1.826170	1.816555	1.807047	1.797649	1.788360	12
13	1.921495	1.909560	1.897769	1.886124	1.874625	1.863274	13
14	2.010764	1.996293	1.982013	1.967925	1.954033	1.940338	14
15	2.103842	2.086482	2.069371	2.052514	2.035913	2.019570	15
16	2.200877	2.180243	2.159933	2.139952	2.120303	2.100988	16
17	2.302022	2.277698	2.253789	2.230302	2.207241	2.184607	17
18	2.407437	2.378969	2.351031	2.323628	2.296764	2.270442	18
19	2.517285	2.484185	2.451753	2.419993	2.388911	2.358509	19
20	2.631738	2.593478	2.556051	2.519463	2.483719	2.448820	20
21	2.750972	2.706982	2.664023	2.622103	2.581225	2.541389	21
22	2.875172	2.824837	2.775770	2.727980	2.681466	2.636228	22
23	3.004526	2.947185	2.891395	2.837160	2.784480	2.733349	23
24	3.139233	3.074176	3.011002	2.949712	2.890303	2.832764	24
25	3.279496	3.205962	3.134699	3.065706	2.998972	2.934482	25
26	3.425528	3.342698	3.262595	3.185211	3.110526	3.038514	26
27	3.577546	3.484547	3.394804	3.308298	3.225001	3.144872	27
28	3.735780	3.631676	3.531440	3.435041	3.342435	3.253564	28
29	3.900464	3.784256	3.672620	3.565512	3.462865	3.364600	29
30	4.071843	3.942463	3.818647	3.699786	3.586330	3.477991	30
31	4.250171	4.106481	3.969103	3.837940	3.712869	3.593746	31
32	4.435710	4.276497	4.124655	3.980051	3.842520	3.711874	32
33	4.628733	4.452705	4.285253	4.126198	3.975322	3.832386	33
34	4.829522	4.635305	4.451031	4.276461	4.111317	3.955291	34
35	5.038370	4.824503	4.622124	4.430923	4.250544	4.080599	35
36	5.255580	5.020512	4.798673	4.589667	4.393044	4.208321	36
37	5.481469	5.223551	4.980822	4.752778	4.538860	4.338468	37
38	5.716361	5.433846	5.168718	4.920345	4.688034	4.471051	38
39	5.960596	5.651630	5.362513	5.092456	4.840611	4.606081	39
40	6.214526	5.877145	5.562361	5.269203	4.996633	4.743570	40
45	7.642825	7.129655	6.658148	6.225803	5.830062	5.468361	45
50	9.374294	8.616277	7.931146	7.313222	6.756901	6.256751	50
55	11.470227	10.377156	9.406175	8.545524	7.783968	7.110946	55
60	14.004294	12.459411	11.111737	9.938592	8.918954	8.033631	60
65	17.065082	14.918397	13.080608	11.510390	10.170533	9.028009	65
70	20.759154	17.819183	15.350507	13.281239	11.548458	10.097809	70
75	25.214721	21.238284	17.964845	15.274112	13.063649	11.247296	75
80	30.586064	25.265683	20.973596	17.514954	14.728291	12.481262	80
90	44.856566	35.587349	28.412881	22.861325	18.561541	15.224466	90
100	65.557273	49.874726	38.239847	29.601722	23.174751	18.376459	100

TABLE D-44

Factors (EUS) to Convert an Exponential Growth to an Equivalent Uniform Series for Interest Rates from 9% to 20%

Growth Rate of 10% per Period

n	9%	10%	11%	12%	15%	20%	n
1	1.100000	1.100000	1.100000	1.100000	1.100000	1.100000	1
2	1.152632	1.152381	1.152133	1.151887	1.151163	1.150000	2
3	1.207044	1.206344	1.205652	1.204967	1.202952	1.199725	3
4	1.263254	1.261883	1.260527	1.259186	1.255251	1.248975	4
5	1.321278	1.318987	1.316724	1.314488	1.307941	1.297558	5
6	1.381129	1.377644	1.374205	1.370811	1.360904	1.345293	6
7	1.442821	1.437838	1.432929	1.428092	1.414019	1.392013	7
8	1.506362	1.499552	1.492852	1.486264	1.467168	1.437565	8
9	1.571763	1.562765	1.553930	1.545259	1.520235	1.481817	9
10	1.639030	1.627454	1.616112	1.605005	1.573108	1.524651	10
11	1.708168	1.693595	1.679349	1.665433	1.625676	1.565971	11
12	1.779181	1.761160	1.743588	1.726470	1.677837	1.605700	12
13	1.852072	1.830121	1.808776	1.788041	1.729492	1.643776	13
14	1.926841	1.900447	1.874858	1.850076	1.780548	1.680159	14
15	2.003487	1.972107	1.941777	1.912500	1.830919	1.714824	15
16	2.082009	2.045066	2.009478	1.975243	1.880528	1.747761	16
17	2.162403	2.119290	2.077904	2.038234	1.929301	1.778977	17
18	2.244665	2.194744	2.146997	2.101404	1.977176	1.808489	18
19	2.328789	2.271390	2.216701	2.164685	2.024094	1.836325	19
20	2.414768	2.349192	2.286959	2.228011	2.070006	1.862525	20
21	2.502595	2.428112	2.357717	2.291320	2.114869	1.887134	21
22	2.592262	2.508111	2.428918	2.354550	2.158649	1.910206	22
23	2.683761	2.589152	2.500509	2.417643	2.201314	1.931798	23
24	2.777080	2.671195	2.572438	2.480544	2.242844	1.951972	24
25	2.872210	2.754202	2.644652	2.543200	2.283221	1.970792	25
26	2.969141	2.838135	2.717102	2.605561	2.322433	1.988324	26
27	3.067861	2.922956	2.789740	2.667580	2.360475	2.004635	27
28	3.168360	3.008628	2.862518	2.729214	2.397346	2.019792	28
29	3.270625	3.095114	2.935392	2.790420	2.433048	2.033859	29
30	3.374645	3.182377	3.008319	2.851162	2.467589	2.046902	30
31	3.480410	3.270383	3.081258	2.911404	2.500978	2.058984	31
32	3.587906	3.359095	3.154169	2.971114	2.533230	2.070166	32
33	3.697123	3.448480	3.227014	3.030262	2.564362	2.080506	33
34	3.808049	3.538506	3.299758	3.088821	2.594391	2.090060	34
35	3.920674	3.629140	3.372368	3.146768	2.623339	2.098883	35
36	4.034987	3.720350	3.444811	3.204081	2.651228	2.107025	36
37	4.150976	3.812108	3.517058	3.260740	2.678083	2.114534	37
38	4.268633	3.904383	3.589080	3.316729	2.703928	2.121456	38
39	4.387948	3.997148	3.660852	3.372032	2.728790	2.127833	39
40	4.508911	4.090377	3.732348	3.426638	2.752696	2.133706	40
45	5.138170	4.562595	4.084954	3.688876	2.858847	2.156743	45
50	5.807584	5.042959	4.427891	3.932730	2.945234	2.171860	50
55	6.516666	5.529246	4.759685	4.158253	3.015143	2.181729	55
60	7.265321	6.019771	5.079491	4.366021	3.071506	2.188151	60
65	8.053836	6.513282	5.386920	4.556921	3.116835	2.192321	65
70	8.882848	7.008875	5.681900	4.732004	3.153233	2.195026	70
75	9.753301	7.505901	5.964570	4.892383	3.182430	2.196779	75
80	10.666412	8.003907	6.235208	5.039172	3.205834	2.197915	80
90	12.626633	9.001694	6.741868	5.296239	3.239609	2.199126	90
100	14.777488	10.000726	7.205174	5.511110	3.261277	2.199634	100

Exponential Growth Factors

TABLE D-45
FACTORS (EUS) TO CONVERT AN EXPONENTIAL GROWTH TO AN EQUIVALENT UNIFORM SERIES FOR INTEREST RATES FROM 25% TO 50%
Growth Rate of 10% per Period

n	25%	30%	35%	40%	45%	50%	n
1	1.100000	1.100000	1.100000	1.100000	1.100000	1.100000	1
2	1.146889	1.147826	1.146809	1.145833	1.144898	1.144000	2
3	1.196656	1.192734	1.190953	1.188303	1.185777	1.183368	3
4	1.243042	1.237433	1.232134	1.227126	1.222393	1.217920	4
5	1.287817	1.278689	1.270145	1.262152	1.254678	1.247690	5
6	1.330782	1.317327	1.304872	1.293358	1.282722	1.272899	6
7	1.371777	1.353237	1.336292	1.320833	1.306743	1.293905	7
8	1.410675	1.386364	1.364456	1.344754	1.327054	1.311155	8
9	1.447388	1.416712	1.389483	1.365367	1.344025	1.325133	9
10	1.481362	1.444329	1.411543	1.382962	1.358055	1.336327	10
11	1.514077	1.469305	1.430842	1.397851	1.369541	1.345198	11
12	1.544042	1.491762	1.447610	1.410352	1.378864	1.352162	12
13	1.571794	1.511844	1.462085	1.420774	1.386373	1.357585	13
14	1.597392	1.529713	1.474511	1.429407	1.392380	1.361778	14
15	1.620912	1.545539	1.485121	1.436518	1.397157	1.365001	15
16	1.642446	1.559497	1.494137	1.442346	1.400936	1.367463	16
17	1.662097	1.571760	1.501768	1.447100	1.403910	1.369336	17
18	1.679973	1.582494	1.508200	1.450963	1.406243	1.370755	18
19	1.696189	1.591861	1.513603	1.454090	1.408065	1.371825	19
20	1.710860	1.600011	1.518128	1.456615	1.409484	1.372631	20
21	1.724101	1.607083	1.521907	1.458646	1.410586	1.373235	21
22	1.736024	1.613204	1.525055	1.460277	1.411439	1.373688	22
23	1.746740	1.618491	1.527672	1.461583	1.412099	1.374025	23
24	1.756352	1.623049	1.529843	1.462627	1.412608	1.374277	24
25	1.764958	1.626971	1.531641	1.463460	1.413000	1.374464	25
26	1.772653	1.630340	1.533127	1.464124	1.413301	1.374604	26
27	1.779523	1.633230	1.534353	1.464653	1.413533	1.374707	27
28	1.785648	1.635707	1.535365	1.465072	1.413710	1.374783	28
29	1.791102	1.637825	1.536198	1.465406	1.413846	1.374840	29
30	1.795954	1.639637	1.536883	1.465670	1.413950	1.374882	30
31	1.800266	1.641183	1.537446	1.465879	1.414030	1.374913	31
32	1.804095	1.642503	1.537909	1.466045	1.414091	1.374936	32
33	1.807491	1.643628	1.538289	1.466176	1.414137	1.374953	33
34	1.810502	1.644586	1.538600	1.466279	1.414172	1.374965	34
35	1.813169	1.645403	1.538855	1.466361	1.414199	1.374974	35
36	1.815531	1.646097	1.539064	1.466426	1.414220	1.374981	36
37	1.817621	1.646687	1.539235	1.466477	1.414236	1.374986	37
38	1.819469	1.647189	1.539375	1.466517	1.414248	1.374990	38
39	1.821103	1.647616	1.539489	1.466549	1.414257	1.374993	39
40	1.822546	1.647978	1.539583	1.466574	1.414264	1.374994	40
45	1.827592	1.649115	1.539849	1.466639	1.414280	1.374999	45
50	1.830288	1.649614	1.539945	1.466658	1.414284	1.375000	50
55	1.831721	1.649832	1.539980	1.466664	1.414285	1.375000	55
60	1.832481	1.649927	1.539993	1.466666	1.414286	1.375000	60
65	1.832883	1.649968	1.539997	1.466666	1.414286	1.375000	65
70	1.833095	1.649986	1.539999	1.466667	1.414286	1.375000	70
75	1.833208	1.649994	1.540000	1.466667	1.414286	1.375000	75
80	1.833267	1.649997	1.540000	1.466667	1.414286	1.375000	80
90	1.833315	1.650000	1.540000	1.466667	1.414286	1.375000	90
100	1.833328	1.650000	1.540000	1.466667	1.414286	1.375000	100

TABLE D-46

Factors (EPW) to Convert an Exponential Growth to Present Worth for Interest Rates from 1% to 2½%

Growth Rate of 1% per Period

n	1%	1¼%	1½%	1¾%	2%	2½%	n
1	1.000000	0.997531	0.995074	0.992629	0.990196	0.985366	1
2	2.000000	1.992599	1.985246	1.977941	1.970684	1.956312	2
3	3.000000	2.985210	2.970540	2.955991	2.941560	2.913049	3
4	4.000000	3.975370	3.950981	3.926831	3.902917	3.855784	4
5	5.000000	4.963085	4.926592	4.890516	4.854849	4.784724	5
6	6.000000	5.948361	5.897397	5.847097	5.797449	5.700070	6
7	7.000000	6.931205	6.863420	6.796627	6.730807	6.602020	7
8	8.000000	7.911621	7.824684	7.739158	7.655015	7.490771	8
9	9.000000	8.889617	8.781212	8.674741	8.570162	8.366516	9
10	10.000000	9.865199	9.733029	9.603429	9.476337	9.229445	10
11	11.000000	10.838371	10.680157	10.525271	10.373628	10.079746	11
12	12.000000	11.809140	11.622619	11.440318	11.262122	10.917603	12
13	13.000000	12.777513	12.560439	12.348620	12.141905	11.743199	13
14	14.000000	13.743494	13.493638	13.250227	13.013062	12.556713	14
15	15.000000	14.707091	14.422241	14.145189	13.875679	13.358322	15
16	16.000000	15.668308	15.346270	15.033553	14.729839	14.148200	16
17	17.000000	16.627151	16.265746	15.915370	15.575625	14.926519	17
18	18.000000	17.583627	17.180693	16.790687	16.413119	15.693448	18
19	19.000000	18.537742	18.091133	17.659551	17.242402	16.449154	19
20	20.000000	19.489501	18.997088	18.522012	18.063555	17.193801	20
21	21.000000	20.438909	19.898580	19.378115	18.876658	17.927550	21
22	22.000000	21.385974	20.795632	20.227908	19.681789	18.650561	22
23	23.000000	22.330700	21.688264	21.071437	20.479026	19.362992	23
24	24.000000	23.273093	22.576499	21.908748	21.268447	20.064997	24
25	25.000000	24.213159	23.460359	22.739887	22.050129	20.756729	25
26	26.000000	25.150905	24.339864	23.564900	22.824147	21.438338	26
27	27.000000	26.086335	25.215037	24.383832	23.590577	22.109972	27
28	28.000000	27.019455	26.085899	25.196728	24.349493	22.771777	28
29	29.000000	27.950271	26.952471	26.003632	25.100969	23.423897	29
30	30.000000	28.878789	27.814774	26.804588	25.845077	24.066475	30
31	31.000000	29.805014	28.672830	27.599640	26.581890	24.699648	31
32	32.000000	30.728952	29.526658	28.388832	27.311479	25.323556	32
33	33.000000	31.650609	30.376280	29.172206	28.033916	25.938333	33
34	34.000000	32.569990	31.221718	29.949807	28.749269	26.544113	34
35	35.000000	33.487102	32.062990	30.721676	29.457610	27.141029	35
36	36.000000	34.401948	32.900118	31.487855	30.159006	27.729209	36
37	37.000000	35.314536	33.733122	32.248387	30.853526	28.308781	37
38	38.000000	36.224870	34.562023	33.003313	31.541236	28.879872	38
39	39.000000	37.132957	35.386841	33.752674	32.222204	29.442606	39
40	40.000000	38.038802	36.207595	34.496512	32.896496	29.997104	40
45	45.000000	42.534585	40.251118	38.134262	36.170083	32.650227	45
50	50.000000	46.975138	44.196022	41.639905	39.286316	35.114817	50
55	55.000000	51.361140	48.044714	45.018238	42.252758	37.404273	55
60	60.000000	55.693260	51.799540	48.273884	45.076609	39.531040	60
65	65.000000	59.972160	55.462790	51.411299	47.764725	41.506678	65
70	70.000000	64.198495	59.036696	54.434777	50.323628	43.341927	70
75	75.000000	68.372910	62.523438	57.348455	52.759531	45.046763	75
80	80.000000	72.496043	65.925141	60.156320	55.078345	46.630452	80
90	90.000000	80.590975	72.481679	65.469846	59.386949	49.468216	90
100	100.000000	88.488237	78.722300	70.404450	63.291299	51.917005	100

Exponential Growth Factors

TABLE D-47
Factors (EPW) to Convert an Exponential Growth to Present Worth for Interest Rates from 3% to 5½%
Growth Rate of 1% per Period

n	3%	3½%	4%	4½%	5%	5½%	n
1	0.980583	0.975845	0.971154	0.966507	0.961905	0.957346	1
2	1.942125	1.928120	1.914294	1.900643	1.887166	1.873857	2
3	2.884996	2.857392	2.830227	2.803493	2.777178	2.751276	3
4	3.809559	3.764218	3.719740	3.676103	3.633286	3.591269	4
5	4.716170	4.649141	4.583594	4.519487	4.456780	4.395433	5
6	5.605176	5.512688	5.422525	5.334624	5.248902	5.165296	6
7	6.476920	6.355377	6.237263	6.122459	6.010849	5.902321	7
8	7.331737	7.177711	7.028496	6.883908	6.743769	6.607909	8
9	8.169956	7.980181	7.796905	7.619854	7.448768	7.283401	9
10	8.991899	8.763269	8.543148	8.331151	8.126910	7.930081	10
11	9.797881	9.527441	9.267865	9.018624	8.779219	8.549177	11
12	10.588214	10.273155	9.971676	9.683072	9.406677	9.141866	12
13	11.363200	11.000857	10.655186	10.325266	10.010232	9.709275	13
14	12.123138	11.710981	11.318979	10.945951	10.590795	10.252481	14
15	12.868319	12.403952	11.963623	11.545847	11.149241	10.772517	15
16	13.599032	13.080185	12.589673	12.125651	11.686412	11.270372	16
17	14.315555	13.740084	13.197663	12.686036	12.203120	11.746991	17
18	15.018166	14.384044	13.788115	13.227652	12.700144	12.203281	18
19	15.707134	15.012448	14.361535	13.751128	13.178234	12.640107	19
20	16.382723	15.625674	14.918414	14.257071	13.638111	13.058302	20
21	17.045195	16.224088	15.459229	14.746069	14.080469	13.458659	21
22	17.694802	16.808047	15.984443	15.218688	14.505975	13.841939	22
23	18.331797	17.377901	16.494507	15.675479	14.915271	14.208870	23
24	18.956422	17.933990	16.989858	16.116970	15.308975	14.560151	24
25	19.568919	18.476648	17.470920	16.543674	15.687681	14.896448	25
26	20.169522	19.006197	17.938105	16.956087	16.051959	15.218400	26
27	20.758963	19.522956	18.391813	17.354687	16.402361	15.526620	27
28	21.335969	20.027232	18.832434	17.739937	16.739414	15.821693	28
29	21.902261	20.519328	19.260345	18.112283	17.063627	16.104180	29
30	22.457557	20.999537	19.675912	18.472159	17.375488	16.374618	30
31	23.002070	21.468148	20.079491	18.819981	17.675470	16.633520	31
32	23.536011	21.925439	20.471429	19.156154	17.964023	16.881380	32
33	24.059583	22.371684	20.852061	19.481068	18.241584	17.118667	33
34	24.572989	22.807151	21.221713	19.795099	18.508572	17.345833	34
35	25.076426	23.232099	21.580702	20.098612	18.765388	17.563309	35
36	25.570088	23.646782	21.929335	20.391960	19.012421	17.771509	36
37	26.054164	24.051450	22.267912	20.675483	19.250043	17.970829	37
38	26.528840	24.446342	22.596723	20.949510	19.478613	18.161646	38
39	26.994300	24.831696	22.916048	21.214359	19.698475	18.344325	39
40	27.450721	25.207742	23.226162	21.470337	19.909962	18.519212	40
45	29.603282	26.956033	24.647598	22.627228	20.852511	19.287914	45
50	31.554816	28.503134	25.875512	23.602932	21.628695	19.906077	50
55	33.324097	29.872199	26.936249	24.425825	22.267877	20.403181	55
60	34.928146	31.083714	27.852571	25.118960	22.794240	20.802934	60
65	36.382394	32.155811	28.644139	25.705161	23.227697	21.124401	65
70	37.700830	33.104534	29.327938	26.199811	23.584646	21.382913	70
75	38.896138	33.944080	29.918640	26.615148	23.878591	21.590799	75
80	39.979817	34.687012	30.428920	26.966280	24.120653	21.757974	80
90	41.853013	35.926231	31.250520	27.512178	24.484142	22.000518	90
100	43.392673	36.896648	31.863634	27.900474	24.730640	22.157366	100

TABLE D-48

Factors (EPW) to Convert an Exponential Growth to Present Worth for Interest Rates from 6% to 8½%
Growth Rate of 1% per Period

n	6%	6½%	7%	7½%	8%	8½%	n
1	0.952830	0.948357	0.943925	0.939535	0.935185	0.930876	1
2	1.860716	1.847737	1.834920	1.822261	1.809757	1.797405	2
3	2.725776	2.700671	2.675953	2.651612	2.627643	2.604036	3
4	3.550032	3.509557	3.469824	3.430817	3.392518	3.354909	4
5	4.335408	4.276669	4.219180	4.162907	4.107817	4.053878	5
6	5.083738	5.004165	4.926516	4.850732	4.776755	4.704532	6
7	5.796769	5.694091	5.594188	5.496966	5.402336	5.310210	7
8	6.476167	6.348386	6.224420	6.104127	5.987370	5.874020	8
9	7.123517	6.968892	6.819313	6.674575	6.534485	6.398857	9
10	7.740332	7.557353	7.380846	7.210531	7.046138	6.887415	10
11	8.328053	8.115424	7.910892	7.714080	7.524629	7.342202	11
12	8.888050	8.644675	8.411216	8.187182	7.972107	7.765553	12
13	9.421633	9.146593	8.883484	8.631678	8.390582	8.159639	13
14	9.930046	9.622590	9.329270	9.049298	8.781933	8.526484	14
15	10.414478	10.074006	9.750059	9.441666	9.147919	8.867971	15
16	10.876059	10.502109	10.147252	9.810309	9.490183	9.185853	16
17	11.315868	10.908103	10.522172	10.156662	9.810264	9.481762	17
18	11.734931	11.293131	10.876069	10.482074	10.109599	9.757216	18
19	12.134226	11.658274	11.210121	10.787809	10.389532	10.013630	19
20	12.514687	12.004561	11.525442	11.075057	10.651322	10.252319	20
21	12.877202	12.332964	11.823081	11.344938	10.896143	10.474509	21
22	13.222617	12.644407	12.104029	11.598500	11.125097	10.681340	22
23	13.551739	12.939766	12.369224	11.836730	11.339211	10.873874	23
24	13.865336	13.219872	12.619548	12.060555	11.539447	11.053100	24
25	14.164141	13.485512	12.855835	12.270847	11.726706	11.219936	25
26	14.448851	13.737434	13.078872	12.468424	11.901826	11.375240	26
27	14.720132	13.976346	13.289403	12.654054	12.065597	11.519809	27
28	14.978616	14.202920	13.488128	12.828460	12.218753	11.654384	28
29	15.224908	14.417792	13.675709	12.992321	12.361982	11.779657	29
30	15.459582	14.621568	13.852772	13.146274	12.495927	11.896271	30
31	15.683187	14.814821	14.019907	13.290917	12.621191	12.004823	31
32	15.896244	14.998093	14.177669	13.426815	12.738336	12.105872	32
33	16.099251	15.171900	14.326585	13.554496	12.847889	12.199937	33
34	16.292683	15.336732	14.467150	13.674457	12.950340	12.287499	34
35	16.476990	15.493051	14.599833	13.787164	13.046152	12.369008	35
36	16.652604	15.641297	14.725076	13.893057	13.135753	12.444883	36
37	16.819934	15.781887	14.843296	13.992546	13.219547	12.515513	37
38	16.979371	15.915217	14.954887	14.086020	13.297909	12.581261	38
39	17.131288	16.041661	15.060221	14.173842	13.371193	12.642464	39
40	17.276038	16.161575	15.159647	14.256354	13.439727	12.699437	40
45	17.903591	16.674409	15.579147	14.599843	13.721251	12.930396	45
50	18.396456	17.067809	15.893502	14.851308	13.922625	13.091830	50
55	18.783540	17.369592	16.129065	15.035404	14.066668	13.204667	55
60	19.087546	17.601093	16.305586	15.170178	14.169702	13.283537	60
65	19.326306	17.778680	16.437863	15.268845	14.243402	13.338665	65
70	19.513822	17.914909	16.536985	15.341078	14.296120	13.377197	70
75	19.661092	18.019412	16.611263	15.393959	14.333829	13.404130	75
80	19.776755	18.099578	16.666924	15.432672	14.360802	13.422956	80
90	19.938936	18.208248	16.739889	15.481763	14.393897	13.445311	90
100	20.038972	18.272197	16.780861	15.508074	14.410830	13.456233	100

Exponential Growth Factors

TABLE D-49

FACTORS (EPW) TO CONVERT AN EXPONENTIAL GROWTH
TO PRESENT WORTH FOR INTEREST RATES FROM 9% TO 20%
Growth Rate of 1% per Period

n	9%	10%	11%	12%	15%	20%	n
1	0.926606	0.918182	0.909910	0.901786	0.878261	0.841667	1
2	1.785203	1.761240	1.737846	1.715003	1.649603	1.550069	2
3	2.580785	2.535320	2.491193	2.448351	2.327043	2.146308	3
4	3.317975	3.246067	3.176671	3.109674	2.922011	2.648143	4
5	4.001059	3.898661	3.800395	3.706045	3.444549	3.070520	5
6	4.634009	4.497862	4.367927	4.243844	3.903674	3.426021	6
7	5.220504	5.048037	4.884330	4.728824	4.306529	3.725235	7
8	5.763953	5.553197	5.354210	5.166171	4.660517	3.977072	8
9	6.267516	6.017027	5.781758	5.560565	4.971410	4.189036	9
10	6.734120	6.442906	6.170789	5.916224	5.244456	4.367439	10
11	7.166479	6.833941	6.524772	6.236952	5.484261	4.517594	11
12	7.567104	7.192982	6.846865	6.526180	5.694873	4.643975	12
13	7.938326	7.522647	7.139940	6.787002	5.879845	4.750346	13
14	8.282302	7.825340	7.406612	7.022207	6.042299	4.839874	14
15	8.601032	8.103267	7.649260	7.234312	6.184975	4.915228	15
16	8.896369	8.358454	7.870047	7.425584	6.310283	4.978650	16
17	9.170030	8.592762	8.070944	7.598072	6.420335	5.032030	17
18	9.423606	8.807900	8.253742	7.753618	6.516990	5.076959	18
19	9.658571	9.005435	8.420071	7.893888	6.601878	5.114774	19
20	9.876290	9.186809	8.571416	8.020381	6.676432	5.146601	20
21	10.078030	9.353343	8.709126	8.134451	6.741910	5.173389	21
22	10.264964	9.506251	8.834430	8.237317	6.799417	5.195936	22
23	10.438178	9.646649	8.948446	8.330081	6.849922	5.214913	23
24	10.598678	9.775559	9.052189	8.413733	6.894280	5.230885	24
25	10.747399	9.893923	9.146587	8.489170	6.933237	5.244328	25
26	10.885205	10.002602	9.232480	8.557198	6.967452	5.255643	26
27	11.012896	10.102389	9.310635	8.618545	6.997501	5.265166	27
28	11.131216	10.194011	9.381749	8.673866	7.023892	5.273181	28
29	11.240851	10.278138	9.446456	8.723754	7.047070	5.279928	29
30	11.342440	10.355381	9.505334	8.768743	7.067427	5.285606	30
31	11.436573	10.426304	9.558907	8.809313	7.085306	5.290385	31
32	11.523797	10.491425	9.607654	8.845898	7.101007	5.294407	32
33	11.604619	10.551217	9.652010	8.878890	7.114798	5.297793	33
34	11.679510	10.606118	9.692369	8.908642	7.126909	5.300642	34
35	11.748903	10.656526	9.729093	8.935472	7.137547	5.303041	35
36	11.813204	10.702811	9.762508	8.959667	7.146889	5.305059	36
37	11.872785	10.745308	9.792913	8.981485	7.155094	5.306758	37
38	11.927994	10.784328	9.820578	9.001161	7.162300	5.308188	38
39	11.979150	10.820156	9.845751	9.018904	7.168628	5.309392	39
40	12.026552	10.853052	9.868657	9.034904	7.174187	5.310405	40
45	12.216209	10.981304	9.955706	9.094203	7.193332	5.313515	45
50	12.345762	11.065000	10.010000	9.129567	7.203337	5.314829	50
55	12.434257	11.119620	10.043865	9.150657	7.208565	5.315384	55
60	12.494706	11.155265	10.064987	9.163235	7.211296	5.315618	60
65	12.535996	11.178526	10.078162	9.170735	7.212724	5.315717	65
70	12.564204	11.193706	10.086379	9.175209	7.213469	5.315759	70
75	12.583471	11.203613	10.091504	9.177877	7.213859	5.315777	75
80	12.596632	11.210078	10.094701	9.179467	7.214063	5.315784	80
90	12.611764	11.217050	10.097939	9.180982	7.214225	5.315789	90
100	12.618824	11.220020	10.099198	9.181521	7.214269	5.315789	100

TABLE D-50

Factors (EPW) to Convert an Exponential Growth to Present Worth for Interest Rates from 25% to 50%
Growth Rate of 1% per Period

n	25%	30%	35%	40%	45%	50%	n
1	0.808000	0.776923	0.748148	0.721429	0.696552	0.673333	1
2	1.460664	1.380533	1.307874	1.241888	1.181736	1.126711	2
3	1.988378	1.849491	1.726632	1.617362	1.519692	1.431985	3
4	2.414610	2.213835	2.039924	1.888240	1.755096	1.637537	4
5	2.759004	2.496903	2.274314	2.083659	1.919067	1.775942	5
6	3.037276	2.716824	2.449672	2.224639	2.033281	1.869134	6
7	3.262119	2.887687	2.580866	2.326347	2.112837	1.931884	7
8	3.443792	3.020433	2.679018	2.399722	2.168252	1.974135	8
9	3.590584	3.123568	2.752450	2.452656	2.206851	2.002584	9
10	3.709192	3.203695	2.807389	2.490845	2.233738	2.021740	10
11	3.805027	3.265947	2.848491	2.518395	2.252466	2.034638	11
12	3.882462	3.314313	2.879241	2.538271	2.265511	2.043323	12
13	3.945029	3.351889	2.902247	2.552610	2.274597	2.049171	13
14	3.995584	3.381083	2.919459	2.562954	2.280926	2.053108	14
15	4.036431	3.403765	2.932336	2.570417	2.285335	2.055760	15
16	4.069437	3.421386	2.941970	2.575801	2.288406	2.057545	16
17	4.096105	3.435077	2.949177	2.579685	2.290545	2.058747	17
18	4.117653	3.445714	2.954570	2.582487	2.292035	2.059556	18
19	4.135063	3.453978	2.958604	2.584508	2.293072	2.060101	19
20	4.149131	3.460398	2.961622	2.585967	2.293795	2.060468	20
21	4.160496	3.465386	2.963880	2.587019	2.294299	2.060715	21
22	4.169682	3.469262	2.965570	2.587778	2.294649	2.060882	22
23	4.177103	3.472272	2.966834	2.588326	2.294894	2.060994	23
24	4.183100	3.474612	2.967779	2.588721	2.295064	2.061069	24
25	4.187944	3.476429	2.968487	2.589006	2.295182	2.061120	25
26	4.191859	3.477841	2.969016	2.589211	2.295265	2.061154	26
27	4.195022	3.478938	2.969412	2.589359	2.295323	2.061177	27
28	4.197578	3.479790	2.969708	2.589466	2.295363	2.061193	28
29	4.199643	3.480452	2.969930	2.589544	2.295390	2.061203	29
30	4.201311	3.480967	2.970096	2.589599	2.295410	2.061210	30
31	4.202660	3.481367	2.970220	2.589640	2.295423	2.061215	31
32	4.203749	3.481677	2.970313	2.589669	2.295433	2.061218	32
33	4.204629	3.481918	2.970382	2.589689	2.295439	2.061220	33
34	4.205340	3.482106	2.970434	2.589705	2.295444	2.061222	34
35	4.205915	3.482251	2.970473	2.589715	2.295447	2.061222	35
36	4.206379	3.482365	2.970502	2.589723	2.295449	2.061223	36
37	4.206755	3.482452	2.970524	2.589729	2.295451	2.061224	37
38	4.207056	3.482521	2.970540	2.589733	2.295452	2.061224	38
39	4.207303	3.482574	2.970552	2.589736	2.295453	2.061224	39
40	4.207500	3.482615	2.970561	2.589738	2.295453	2.061224	40
45	4.208047	3.482718	2.970582	2.589743	2.295454	2.061224	45
50	4.208235	3.482747	2.970587	2.589743	2.295455	2.061224	50
55	4.208299	3.482755	2.970588	2.589744	2.295455	2.061224	55
60	4.208322	3.482758	2.970588	2.589744	2.295455	2.061224	60
65	4.208329	3.482758	2.970588	2.589744	2.295455	2.061224	65
70	4.208332	3.482759	2.970588	2.589744	2.295455	2.061224	70
75	4.208333	3.482759	2.970588	2.589744	2.295455	2.061224	75
80	4.208333	3.482759	2.970588	2.589744	2.295455	2.061224	80
90	4.208333	3.482759	2.970588	2.589744	2.295455	2.061224	90
100	4.208333	3.482759	2.970588	2.589744	2.295455	2.061224	100

Exponential Growth Factors

TABLE D-51
Factors (EPW) to Convert an Exponential Growth to Present Worth for Interest Rates from 1% to 2½%
Growth Rate of 2% per Period

n	1%	1¼%	1½%	1¾%	2%	2½%	n
1	1.009901	1.007407	1.004926	1.002457	1.000000	0.995122	1
2	2.029801	2.022277	2.014803	2.007377	2.000000	1.985390	2
3	3.059799	3.044664	3.029654	3.014766	3.000000	2.970827	3
4	4.099995	4.074625	4.049504	4.024630	4.000000	3.951457	4
5	5.150490	5.112215	5.074379	5.036976	5.000000	4.927303	5
6	6.211386	6.157490	6.104302	6.051809	6.000000	5.898390	6
7	7.282786	7.210509	7.139298	7.069135	7.000000	6.864739	7
8	8.364794	8.271327	8.179303	8.088961	8.000000	7.826374	8
9	9.457514	9.340004	9.224612	9.111293	9.000000	8.783319	9
10	10.561054	10.416596	10.274980	10.136136	10.000000	9.735596	10
11	11.675520	11.501164	11.330521	11.163498	11.000000	10.683227	11
12	12.801020	12.593765	12.391263	12.193383	12.000000	11.626235	12
13	13.937664	13.694460	13.457230	13.225800	13.000000	12.564644	13
14	15.085562	14.803307	14.528448	14.260752	14.000000	13.498475	14
15	16.244825	15.920369	15.604942	15.298248	15.000000	14.427751	15
16	17.415565	17.045705	16.686740	16.338293	16.000000	15.352493	16
17	18.597898	18.179377	17.773867	17.380893	17.000000	16.272725	17
18	19.791936	19.321446	18.866349	18.426055	18.000000	17.188468	18
19	20.997797	20.471976	19.964213	19.473785	19.000000	18.099744	19
20	22.215597	21.631027	21.067485	20.524089	20.000000	19.006574	20
21	23.445454	22.798665	22.176192	21.576974	21.000000	19.908981	21
22	24.687489	23.974951	23.290360	22.632445	22.000000	20.806986	22
23	25.941820	25.159951	24.410017	23.690510	23.000000	21.700611	23
24	27.208571	26.353728	25.535190	24.751175	24.000000	22.589876	24
25	28.487864	27.556348	26.665905	25.814446	25.000000	23.474803	25
26	29.779823	28.767877	27.802190	26.880329	26.000000	24.355414	26
27	31.084574	29.988379	28.944073	27.948831	27.000000	25.231729	27
28	32.402243	31.217923	30.091580	29.019958	28.000000	26.103769	28
29	33.732958	32.456574	31.244741	30.093717	29.000000	26.971556	29
30	35.076849	33.704401	32.403582	31.170115	30.000000	27.835109	30
31	36.434045	34.961470	33.568132	32.249157	31.000000	28.694450	31
32	37.804679	36.227852	34.738418	33.330850	32.000000	29.549599	32
33	39.188884	37.503614	35.914469	34.415201	33.000000	30.400577	33
34	40.586794	38.788826	37.096314	35.502216	34.000000	31.247403	34
35	41.998544	40.083558	38.283981	36.591902	35.000000	32.090099	35
36	43.424272	41.387880	39.477498	37.684266	36.000000	32.928684	36
37	44.864116	42.701864	40.676894	38.779313	37.000000	33.763178	37
38	46.318217	44.025582	41.882199	39.877051	38.000000	34.593602	38
39	47.786714	45.359105	43.093442	40.977486	39.000000	35.419974	39
40	49.269751	46.702506	44.310651	42.080625	40.000000	36.242316	40
45	56.906117	53.570259	50.487231	47.637108	45.000000	40.294242	45
50	64.932182	60.696169	56.317449	53.262190	50.000000	44.248301	50
55	73.361423	68.089942	63.305128	58.956716	55.000000	48.106855	55
60	82.216298	75.761644	69.954184	64.721543	60.000000	51.872212	60
65	91.518300	83.721725	76.768632	70.557541	65.000000	55.546623	65
70	101.290007	91.981024	83.752585	76.465587	70.000000	59.132284	70
75	111.555136	100.550788	90.910259	82.446572	75.000000	62.631339	75
80	122.338602	109.442688	98.245977	88.501394	80.000000	66.045879	80
90	145.566564	128.241787	113.469365	100.836212	90.000000	72.629534	90
100	171.199544	148.480758	129.459518	113.477470	100.000000	78.898993	100

TABLE D-52

FACTORS (EPW) TO CONVERT AN EXPONENTIAL GROWTH TO PRESENT WORTH FOR INTEREST RATES FROM 3% TO 5½%

Growth Rate of 2% per Period

n	3%	3½%	4%	4½%	5%	5½%	n
1	0.990291	0.985507	0.980769	0.976077	0.971429	0.966825	1
2	1.970968	1.956732	1.942678	1.928802	1.915102	1.901575	2
3	2.942124	2.913881	2.886088	2.858735	2.831813	2.805314	3
4	3.903851	3.857158	3.811355	3.766421	3.722333	3.679071	4
5	4.856240	4.786764	4.718829	4.652392	4.587409	4.523841	5
6	5.799364	5.702898	5.608852	5.517167	5.427769	5.340586	6
7	6.733370	6.605755	6.481758	6.361254	6.244118	6.130235	7
8	7.658289	7.495526	7.337878	7.185147	7.037144	6.893687	8
9	8.574228	8.372403	8.177534	7.989330	7.807511	7.631811	9
10	9.481274	9.236571	9.001043	8.774275	8.555868	8.345447	10
11	10.379514	10.088215	9.808716	9.540440	9.282843	9.035409	11
12	11.269034	10.927516	10.600856	10.288277	9.989047	9.702481	12
13	12.149917	11.754653	11.377762	11.018222	10.675075	10.347422	13
14	13.022248	12.569803	12.139728	11.730705	11.341501	10.970967	14
15	13.886109	13.373139	12.887041	12.426143	11.988887	11.573826	15
16	14.741584	14.164833	13.619983	13.104943	12.617776	12.156685	16
17	15.588753	14.945053	14.338829	13.767504	13.228696	12.720207	17
18	16.427697	15.713965	15.043852	14.414215	13.822162	13.265035	18
19	17.258496	16.471734	15.735316	15.045453	14.398672	13.791787	19
20	18.081229	17.218520	16.413483	15.661591	14.958710	14.301064	20
21	18.895975	17.954484	17.078609	16.262988	15.502747	14.793446	21
22	19.702810	18.679781	17.730943	16.849998	16.031240	15.269493	22
23	20.501812	19.394567	18.370733	17.422965	16.544633	15.729747	23
24	21.293056	20.098993	18.998219	17.982224	17.043358	16.174731	24
25	22.076619	20.793211	19.613637	18.528104	17.527833	16.604953	25
26	22.852574	21.477367	20.217221	19.060924	17.998466	17.020903	26
27	23.620996	22.151608	20.809198	19.580998	18.455653	17.423053	27
28	24.381957	22.816078	21.389790	20.088629	18.899777	17.811862	28
29	25.135530	23.470917	21.959217	20.584117	19.331212	18.187771	29
30	25.881787	24.116266	22.517694	21.067750	19.750320	18.551210	30
31	26.620799	24.752262	23.065431	21.539814	20.157654	18.902592	31
32	27.352636	25.379041	23.602634	22.000584	20.552955	19.242316	32
33	28.077367	25.996736	24.129506	22.450331	20.937157	19.570770	33
34	28.795060	26.605479	24.646246	22.889318	21.310381	19.888328	34
35	29.505790	27.205400	25.153049	23.317803	21.672941	20.195350	35
36	30.209618	27.796626	25.650106	23.736037	22.025143	20.492187	36
37	30.906612	28.379283	26.137604	24.144266	22.367282	20.779176	37
38	31.596839	28.953497	26.615727	24.542729	22.699645	21.056644	38
39	32.280365	29.519388	27.084655	24.931659	23.022512	21.324907	39
40	32.957254	30.077078	27.544566	25.311284	23.336155	21.584270	40
45	36.244393	32.746609	29.714805	27.077450	24.774961	22.757589	45
50	39.375031	35.228221	31.684240	28.642221	26.019639	23.748773	50
55	42.356618	37.535145	33.471451	30.028563	27.096379	24.586096	55
60	45.196252	39.679675	35.093299	31.256822	28.027842	25.293441	60
65	47.900689	41.673244	36.565083	32.345023	28.833628	25.890985	65
70	50.476368	43.526479	37.900689	33.309137	29.530695	26.395773	70
75	52.929419	45.249258	39.112717	34.163314	30.133711	26.822202	75
80	55.265679	46.850765	40.212600	34.920090	30.655365	27.182437	80
90	59.609806	49.723506	42.116480	36.184599	31.497020	27.743830	90
100	63.550130	52.206037	43.684342	37.177167	32.126877	28.144460	100

TABLE D-53

Factors (EPW) to Convert an Exponential Growth to Present Worth for Interest Rates from 6% to 8½%

Growth Rate of 2% per Period

n	6%	6½%	7%	7½%	8%	8½%	n
1	0.962264	0.957746	0.953271	0.948837	0.944444	0.940092	1
2	1.888216	1.875025	1.861997	1.849129	1.836420	1.823865	2
3	2.779227	2.753545	2.728259	2.703360	2.678841	2.654694	3
4	3.636615	3.594944	3.554041	3.513886	3.474461	3.435749	4
5	4.461648	4.400792	4.341235	4.282943	4.225880	4.170013	5
6	5.255548	5.172589	5.091645	5.012653	4.935553	4.860289	6
7	6.019490	5.911776	5.806988	5.705028	5.605800	5.509211	7
8	6.754603	6.619729	6.488905	6.361981	6.238811	6.119259	8
9	7.461977	7.297768	7.138956	6.985321	6.836655	6.692759	9
10	8.142657	7.947159	7.753631	7.576770	7.401285	7.231903	10
11	8.797651	8.569110	8.349349	8.137958	7.934547	7.738747	11
12	9.427928	9.164781	8.912464	8.670435	8.438183	8.215228	12
13	10.034422	9.735283	9.449264	9.175668	8.913840	8.663164	13
14	10.618028	10.281680	9.960981	9.655053	9.363071	9.084264	14
15	11.179612	10.804989	10.448786	10.109911	9.787345	9.480138	15
16	11.720004	11.306187	10.913796	10.541497	10.188048	9.852296	16
17	12.240004	11.786207	11.357076	10.951001	10.566490	10.202158	17
18	12.740361	12.245945	11.779643	11.339555	10.923907	10.531061	18
19	13.221876	12.686257	12.182463	11.708229	11.261468	10.840260	19
20	13.685202	13.107964	12.566460	12.058040	11.580275	11.130936	20
21	14.131043	13.511853	12.932514	12.389954	11.881371	11.404198	21
22	14.560060	13.898676	13.281462	12.704887	12.165739	11.661089	22
23	14.972886	14.269155	13.614104	13.003707	12.434309	11.902591	23
24	15.370138	14.623979	13.931201	13.287238	12.687959	12.129624	24
25	15.752397	14.963811	14.233482	13.556263	12.927517	12.343057	25
26	16.120231	15.289284	14.521637	13.811524	13.153766	12.543703	26
27	16.474184	15.601004	14.796327	14.053725	13.367445	12.732329	27
28	16.814781	15.899553	15.058181	14.283535	13.569254	12.909655	28
29	17.142525	16.185488	15.307798	14.501586	13.759851	13.076358	29
30	17.457902	16.459340	15.545752	14.708482	13.939859	13.233074	30
31	17.761377	16.721622	15.772586	14.904792	14.109867	13.380401	31
32	18.053401	16.972821	15.988820	15.091059	14.270430	13.518903	32
33	18.334404	17.213406	16.194950	15.267795	14.422073	13.649107	33
34	18.604804	17.443825	16.391448	15.435489	14.565291	13.771510	34
35	18.865000	17.664509	16.578763	15.594604	14.700553	13.886581	35
36	19.115378	17.875868	16.757326	15.745578	14.828300	13.994758	36
37	19.356307	18.078296	16.927544	15.888827	14.948950	14.096655	37
38	19.588144	18.272171	17.089808	16.024748	15.062897	14.192059	38
39	19.811233	18.457853	17.244490	16.153714	15.170514	14.281936	39
40	20.025904	18.635651	17.391944	16.276082	15.272152	14.366428	40
45	20.983688	19.418315	18.032075	16.800178	15.701662	14.718762	45
50	21.773691	20.048991	18.535983	17.203237	16.024405	14.977468	50
55	22.425836	20.557219	18.932656	17.513213	16.266919	15.167427	55
60	22.963712	20.966774	19.244915	17.751602	16.449149	15.306907	60
65	23.407478	21.296813	19.490724	17.934937	16.586080	15.409322	65
70	23.773599	21.562773	19.684223	18.075931	16.688973	15.484521	70
75	24.075662	21.777097	19.836544	18.184364	16.766288	15.539738	75
80	24.324873	21.949809	19.956451	18.267756	16.824385	15.580281	80
90	24.700115	22.201147	20.125144	18.381210	16.900842	15.631909	90
100	24.955534	22.364363	20.229678	18.448312	16.944013	15.659744	100

TABLE D-54

Factors (EPW) to Convert an Exponential Growth to Present Worth for Interest Rates from 9% to 20%

Growth Rate of 2% per Period

n	9%	10%	11%	12%	15%	20%	n
1	0.935780	0.927273	0.918919	0.910714	0.886957	0.850000	1
2	1.811464	1.787107	1.763331	1.740115	1.673648	1.572500	2
3	2.630911	2.584409	2.539277	2.495462	2.371410	2.186625	3
4	3.397733	3.323724	3.252309	3.183367	2.990294	2.708631	4
5	4.115310	4.009272	3.907527	3.809852	3.539217	3.152337	5
6	4.786804	4.644961	4.509619	4.380401	4.026088	3.529486	6
7	5.415174	5.234418	5.062893	4.900008	4.457922	3.850063	7
8	6.003191	5.781006	5.571307	5.373222	4.840939	4.122554	8
9	6.553444	6.287842	6.038499	5.804184	5.180659	4.354171	9
10	7.068361	6.757817	6.467810	6.196667	5.481976	4.551045	10
11	7.550209	7.193612	6.862312	6.554108	5.749231	4.718388	11
12	8.001113	7.597713	7.224827	6.879634	5.986274	4.860630	12
13	8.423060	7.972425	7.557949	7.176095	6.196522	4.981536	13
14	8.817909	8.319885	7.864061	7.446087	6.383002	5.084305	14
15	9.187401	8.642075	8.145354	7.691972	6.548402	5.171659	15
16	9.533165	8.940833	8.403838	7.915903	6.695104	5.245911	16
17	9.856723	9.217864	8.641365	8.119840	6.825223	5.309024	17
18	10.159502	9.474746	8.859633	8.305569	6.940632	5.362670	18
19	10.442837	9.712947	9.060203	8.474714	7.042996	5.408270	19
20	10.707976	9.933823	9.244511	8.628758	7.133787	5.447029	20
21	10.956087	10.138636	9.413875	8.769047	7.214316	5.479975	21
22	11.188265	10.328553	9.569507	8.896811	7.285741	5.507979	22
23	11.405533	10.504659	9.712520	9.013167	7.349092	5.531782	23
24	11.608847	10.667956	9.843937	9.119134	7.405282	5.552015	24
25	11.799105	10.819378	9.964699	9.215640	7.455119	5.569212	25
26	11.977144	10.959786	10.075669	9.303529	7.499323	5.583831	26
27	12.143749	11.089984	10.177642	9.383571	7.538530	5.596256	27
28	12.299655	11.210712	10.271347	9.456467	7.573305	5.606818	28
29	12.445549	11.322660	10.357454	9.522854	7.604149	5.615795	29
30	12.582073	11.426467	10.436579	9.583313	7.631506	5.623426	30
31	12.709830	11.522724	10.509289	9.638375	7.655770	5.629912	31
32	12.829382	11.611980	10.576103	9.688520	7.677292	5.635425	32
33	12.941257	11.694745	10.637500	9.734188	7.696381	5.640111	33
34	13.045947	11.771491	10.693919	9.775778	7.713312	5.644095	34
35	13.143913	11.842655	10.745764	9.813655	7.728329	5.647480	35
36	13.235589	11.908644	10.793404	9.848150	7.741648	5.650358	36
37	13.321377	11.969834	10.837182	9.879565	7.753462	5.652805	37
38	13.401655	12.026573	10.877411	9.908175	7.763960	5.654884	38
39	13.476778	12.079186	10.914378	9.934231	7.773234	5.656651	39
40	13.547077	12.127972	10.948347	9.957961	7.781477	5.658154	40
45	13.836378	12.323571	11.081083	10.048366	7.810651	5.662889	45
50	14.043973	12.457663	11.168054	10.105003	7.826666	5.664991	50
55	14.192939	12.549589	11.225040	10.140486	7.835456	5.665923	55
60	14.299833	12.612609	11.262377	10.162715	7.840282	5.666337	60
65	14.376538	12.655812	11.286842	10.176642	7.842930	5.666520	65
70	14.431580	12.685429	11.302871	10.185366	7.844384	5.666602	70
75	14.471076	12.705734	11.313374	10.190832	7.845183	5.666638	75
80	14.499418	12.719653	11.320256	10.194257	7.845621	5.666654	80
90	14.534349	12.735738	11.327719	10.197745	7.845993	5.666664	90
100	14.552336	12.743297	11.330923	10.199115	7.846105	5.666666	100

Exponential Growth Factors

TABLE D-55
Factors (EPW) to Convert an Exponential Growth to Present Worth for Interest Rates from 25% to 50%
Growth Rate of 2% per Period

n	25%	30%	35%	40%	45%	50%	n
1	0.816000	0.784615	0.755556	0.728571	0.703448	0.680000	1
2	1.481856	1.400237	1.326420	1.259388	1.198288	1.142400	2
3	2.025194	1.883263	1.757739	1.646125	1.546382	1.456832	3
4	2.468559	2.262252	2.083625	1.927891	1.791248	1.670646	4
5	2.830344	2.559613	2.329850	2.133178	1.963498	1.816039	5
6	3.125561	2.792927	2.515987	2.282744	2.084668	1.914907	6
7	3.366457	2.975989	2.656448	2.391713	2.169904	1.982136	7
8	3.563029	3.119622	2.762649	2.471106	2.229864	2.027853	8
9	3.723432	3.232319	2.842891	2.528948	2.272042	2.058940	9
10	3.854320	3.320743	2.903517	2.571091	2.301712	2.080079	10
11	3.961125	3.390121	2.949324	2.601795	2.322584	2.094454	11
12	4.048278	3.444557	2.983934	2.624165	2.337266	2.104229	12
13	4.119395	3.487267	3.010083	2.640463	2.347594	2.110875	13
14	4.177426	3.520779	3.029841	2.652337	2.354859	2.115395	14
15	4.224780	3.547073	3.044769	2.660989	2.359970	2.118469	15
16	4.263420	3.567703	3.056047	2.667292	2.363565	2.120559	16
17	4.294951	3.583890	3.064569	2.671884	2.366094	2.121980	17
18	4.320680	3.596591	3.071008	2.675230	2.367873	2.122946	18
19	4.341675	3.606556	3.075873	2.677667	2.369124	2.123604	19
20	4.358807	3.614375	3.079548	2.679443	2.370005	2.124050	20
21	4.372786	3.620509	3.082325	2.680737	2.370624	2.124354	21
22	4.384194	3.625323	3.084424	2.681680	2.371060	2.124561	22
23	4.393502	3.629099	3.086009	2.682367	2.371366	2.124701	23
24	4.401098	3.632063	3.087207	2.682867	2.371582	2.124797	24
25	4.407296	3.634388	3.088112	2.683232	2.371733	2.124862	25
26	4.412353	3.636212	3.088796	2.683498	2.371840	2.124906	26
27	4.416480	3.637643	3.089312	2.683691	2.371915	2.124936	27
28	4.419846	3.638766	3.089703	2.683832	2.371968	2.124957	28
29	4.422596	3.639647	3.089997	2.683935	2.372005	2.124970	29
30	4.424838	3.640339	3.090220	2.684010	2.372031	2.124980	30
31	4.426668	3.640881	3.090389	2.684064	2.372049	2.124986	31
32	4.428161	3.641307	3.090516	2.684104	2.372062	2.124991	32
33	4.429379	3.641641	3.090612	2.684133	2.372071	2.124994	33
34	4.430374	3.641903	3.090685	2.684154	2.372078	2.124996	34
35	4.431185	3.642108	3.090739	2.684169	2.372082	2.124997	35
36	4.431847	3.642270	3.090781	2.684180	2.372086	2.124998	36
37	4.432367	3.642396	3.090812	2.684189	2.372088	2.124999	37
38	4.432828	3.642495	3.090836	2.684195	2.372089	2.124999	38
39	4.433188	3.642573	3.090854	2.684199	2.372090	2.124999	39
40	4.433481	3.642634	3.090867	2.684202	2.372091	2.125000	40
45	4.434312	3.642791	3.090899	2.684209	2.372093	2.125000	45
50	4.434612	3.642837	3.090907	2.684210	2.372093	2.125000	50
55	4.434721	3.642851	3.090908	2.684210	2.372093	2.125000	55
60	4.434760	3.642855	3.090909	2.684211	2.372093	2.125000	60
65	4.434775	3.642857	3.090909	2.684211	2.372093	2.125000	65
70	4.434780	3.642857	3.090909	2.684211	2.372093	2.125000	70
75	4.434782	3.642857	3.090909	2.684211	2.372093	2.125000	75
80	4.434782	3.642857	3.090909	2.684211	2.372093	2.125000	80
90	4.434783	3.642857	3.090909	2.684211	2.372093	2.125000	90
100	4.434783	3.642857	3.090909	2.684211	2.372093	2.125000	100

TABLE D-56

Factors (EPW) to Convert an Exponential Growth to Present Worth for Interest Rates from 1% to 2½%

Growth Rate of 3% per Period

n	1%	1¼%	1½%	1¾%	2%	2½%	n
1	1.019802	1.017284	1.014778	1.012285	1.009804	1.004878	1
2	2.059798	2.052151	2.044553	2.037006	2.029508	2.014658	2
3	3.120368	3.104904	3.089547	3.074316	3.059209	3.029364	3
4	4.201960	4.175853	4.149983	4.124369	4.099005	4.049019	4
5	5.304989	5.265312	5.226092	5.187322	5.148995	5.073648	5
6	6.429841	6.373601	6.318103	6.263333	6.209280	6.103276	6
7	7.576966	7.501046	7.426252	7.352563	7.279959	7.137926	7
8	8.746807	8.647978	8.550778	8.455174	8.361135	8.177623	8
9	9.939813	9.814733	9.691922	9.571331	9.452911	9.222392	9
10	11.156443	11.001654	10.849931	10.701200	10.555390	10.272257	10
11	12.397165	12.209090	12.025053	11.844950	11.668678	11.327244	11
12	13.662455	13.437396	13.217542	13.002750	12.792881	12.387377	12
13	14.952801	14.686931	14.427653	14.174774	13.928105	13.452681	13
14	16.268698	15.958063	15.655648	15.361196	15.074459	14.523182	14
15	17.610652	17.251165	16.901791	16.562194	16.232052	15.598905	15
16	18.979180	18.566618	18.166349	17.777945	17.400994	16.679875	16
17	20.374808	19.904806	19.449596	19.008633	18.581396	17.766119	17
18	21.798071	21.266124	20.751806	20.254439	19.773370	18.857661	18
19	23.249518	22.650970	22.073262	21.515550	20.977031	19.954527	19
20	24.729706	24.059752	23.414246	22.792154	22.192492	21.056744	20
21	26.239206	25.492884	24.775048	24.084441	23.419869	22.164338	21
22	27.778596	26.950786	26.155960	25.392603	24.659280	23.277335	22
23	29.348465	28.433886	27.557279	26.716837	25.910841	24.395761	23
24	30.949429	29.942620	28.979308	28.057339	27.174673	25.519643	24
25	32.582091	31.477430	30.422352	29.414308	28.450895	26.649007	25
26	34.247083	33.038769	31.886722	30.787949	29.739630	27.783880	26
27	35.945045	34.627093	33.372732	32.178464	31.040998	28.924289	27
28	37.676630	36.242870	34.880704	33.586062	32.355726	30.070262	28
29	39.442504	37.886574	36.410960	35.010952	33.682137	31.221824	29
30	41.243345	39.558687	37.963832	36.453347	35.022158	32.379003	30
31	43.079847	41.259702	39.539652	37.913462	36.375316	33.541828	31
32	44.952715	42.990116	41.138760	39.391514	37.741741	34.710325	32
33	46.862670	44.750439	42.761500	40.887724	39.121562	35.884521	33
34	48.810446	46.541188	44.408222	42.402316	40.514911	37.064446	34
35	50.796791	48.362887	46.079280	43.935514	41.921920	38.250126	35
36	52.822470	50.216073	47.775032	45.487547	43.342723	39.441590	36
37	54.888262	52.101289	49.495846	47.058647	44.777455	40.638866	37
38	56.994960	54.019089	51.242090	48.649048	46.226254	41.841982	38
39	59.143375	55.970036	53.014190	50.258987	47.689256	43.050968	39
40	61.334333	57.954704	54.812379	51.888704	49.166602	44.265850	40
45	72.957342	68.404596	64.210138	60.342572	56.773448	50.429738	45
50	85.777631	79.789327	74.323143	69.328635	64.760562	56.745439	50
55	99.918531	92.192527	85.205830	78.880398	73.146954	63.216691	55
60	115.516679	105.705307	96.916778	89.033472	81.952583	69.847327	60
65	132.720323	120.426930	109.519024	99.825711	91.198408	76.641270	65
70	151.696767	136.465537	123.080404	111.297354	100.906435	83.602544	70
75	172.627965	153.938944	137.673915	123.491170	111.099767	90.735269	75
80	195.715276	172.975508	153.378110	136.452623	121.802667	98.043670	80
90	249.269513	216.309972	188.463118	164.874780	144.840336	113.204907	90
100	314.425202	267.744570	229.091863	196.988092	170.238894	129.122164	100

Exponential Growth Factors

TABLE D-57
Factors (EPW) to Convert an Exponential Growth to Present Worth for Interest Rates from 3% to 5½%
Growth Rate of 3% per Period

n	3%	3½%	4%	4½%	5%	5½%	n
1	1.000000	0.995169	0.990385	0.985646	0.980952	0.976303	1
2	2.000000	1.985531	1.971246	1.957144	1.943220	1.929471	2
3	3.000000	2.971108	2.942677	2.914497	2.887159	2.860053	3
4	4.000000	3.951924	3.904766	3.858505	3.813118	3.768582	4
5	5.000000	4.928001	4.857605	4.788766	4.721439	4.655583	5
6	6.000000	5.899364	5.801282	5.705673	5.612459	5.521564	6
7	7.000000	6.866033	6.735885	6.609420	6.486508	6.367025	7
8	8.000000	7.828033	7.661501	7.500194	7.343908	7.192451	8
9	9.000000	8.785386	8.578218	8.378181	8.184976	7.998317	9
10	10.000000	9.738113	9.486120	9.243566	9.010024	8.785087	10
11	11.000000	10.686238	10.385291	10.096529	9.819357	9.553212	11
12	12.000000	11.629783	11.275818	10.937249	10.613274	10.303136	12
13	13.000000	12.568770	12.157781	11.765901	11.392069	11.035289	13
14	14.000000	13.503220	13.031264	12.582658	12.156029	11.750093	14
15	15.000000	14.433156	13.896348	13.387692	12.905438	12.447958	15
16	16.000000	15.358600	14.753114	14.181170	13.640573	13.129286	16
17	17.000000	16.279573	15.601641	14.963259	14.361705	13.794469	17
18	18.000000	17.196097	16.442010	15.734121	15.069101	14.443889	18
19	19.000000	18.108193	17.274299	16.493918	15.763023	15.077920	19
20	20.000000	19.015883	18.098584	17.242909	16.443727	15.696927	20
21	21.000000	19.919188	18.914944	17.980951	17.111466	16.301265	21
22	22.000000	20.818129	19.723454	18.708497	17.766485	16.891282	22
23	23.000000	21.712727	20.524190	19.425600	18.409028	17.467318	23
24	24.000000	22.603004	21.317227	20.132409	19.039333	18.029704	24
25	25.000000	23.488980	22.102638	20.829073	19.657631	18.578763	25
26	26.000000	24.370675	22.880497	21.515737	20.264152	19.114812	26
27	27.000000	25.248112	23.650877	22.192545	20.859121	19.638157	27
28	28.000000	26.121309	24.413849	22.859638	21.442757	20.149101	28
29	29.000000	26.990288	25.169485	23.517155	22.015276	20.647938	29
30	30.000000	27.855070	25.917856	24.165234	22.576889	21.134954	30
31	31.000000	28.715673	26.659030	24.804010	23.127806	21.610429	31
32	32.000000	29.572119	27.393078	25.433618	23.668229	22.074636	32
33	33.000000	30.424428	28.120068	26.054188	24.198358	22.527844	33
34	34.000000	31.272619	28.840067	26.665850	24.718389	22.970312	34
35	35.000000	32.116713	29.553143	27.268733	25.228515	23.402295	35
36	36.000000	32.956728	30.259363	27.862962	25.728924	23.824042	36
37	37.000000	33.792686	30.958792	28.448661	26.219802	24.235795	37
38	38.000000	34.624606	31.651496	29.025953	26.701329	24.637790	38
39	39.000000	35.452506	32.337540	29.594959	27.173685	25.030259	39
40	40.000000	36.276407	33.016986	30.155796	27.637043	25.413429	40
45	45.000000	40.336592	36.317470	32.841518	29.824762	27.197305	45
50	50.000000	44.299648	39.462299	35.339939	31.811913	28.779603	50
55	55.000000	48.167898	42.459814	37.664121	33.516986	30.183103	55
60	60.000000	51.943611	45.314010	39.826217	35.256382	31.428008	60
65	65.000000	55.628999	48.034551	41.837530	36.745571	32.532239	65
70	70.000000	59.226225	50.626789	43.708576	38.098234	33.511692	70
75	75.000000	62.737397	53.096773	45.449136	39.326887	34.380468	75
80	80.000000	66.164310	55.450270	47.069310	40.442899	35.151072	80
90	90.000000	72.774928	59.829517	49.975774	42.377360	36.440886	90
100	100.000000	79.072797	63.805443	52.491949	43.973383	37.455673	100

TABLE D-58

Factors (EPW) to Convert an Exponential Growth to Present Worth for Interest Rates from 6% to 8½%

Growth Rate of 3% per Period

n	6%	6½%	7%	7½%	8%	8½%	n
1	0.971698	0.967136	0.962617	0.958140	0.953704	0.949309	1
2	1.915895	1.902488	1.889248	1.876171	1.863254	1.850496	2
3	2.833370	2.807102	2.781239	2.755773	2.730696	2.706001	3
4	3.724878	3.681986	3.639884	3.598555	3.557979	3.518139	4
5	4.591155	4.528117	4.466430	4.406057	4.346961	4.289109	5
6	5.432915	5.346442	5.262078	5.179757	5.099417	5.020997	6
7	6.250852	6.137874	6.027981	5.921069	5.817036	5.715785	7
8	7.045639	6.903296	6.765253	6.631350	6.501433	6.375354	8
9	7.817932	7.643563	7.474963	7.311898	7.154144	7.001488	9
10	8.568368	8.355502	8.158142	7.963958	7.776638	7.595883	10
11	9.297565	9.051913	8.815782	8.588723	8.370312	8.160147	11
12	10.006125	9.721568	9.449837	9.187335	8.936501	8.695807	12
13	10.694630	10.369216	10.058226	9.760888	9.476478	9.204315	13
14	11.363650	10.995580	10.644834	10.310432	9.991456	9.687046	14
15	12.013736	11.601359	11.209513	10.836972	10.482592	10.145306	15
16	12.645422	12.187230	11.753083	11.341471	10.950991	10.580336	16
17	13.259231	12.753847	12.276332	11.824851	11.397704	10.993315	17
18	13.855668	13.301843	12.780021	12.287997	11.823736	11.385359	18
19	14.435225	13.831829	13.264880	12.731755	12.230045	11.757529	19
20	14.998379	14.344398	13.731613	13.156938	12.617543	12.110834	20
21	15.545594	14.840122	14.180899	13.564322	12.987101	12.446230	21
22	16.077323	15.319555	14.613389	13.954652	13.339550	12.764624	22
23	16.594002	15.783231	15.029710	14.328644	13.675682	13.066878	23
24	17.096059	16.231670	15.430469	14.686980	13.996252	13.353810	24
25	17.583906	16.665371	15.816246	15.030315	14.301981	13.626198	25
26	18.057947	17.084818	16.187601	15.359279	14.593556	13.884778	26
27	18.518571	17.490482	16.545074	15.674472	14.871632	14.130250	27
28	18.966158	17.882813	16.889183	15.976471	15.136834	14.363279	28
29	19.401078	18.262251	17.220429	16.265828	15.389759	14.584495	29
30	19.823689	18.629220	17.539291	16.543072	15.630974	14.794497	30
31	20.234340	18.984128	17.846234	16.808711	15.861021	14.993855	31
32	20.633368	19.327373	18.141702	17.063230	16.080418	15.183106	32
33	21.021103	19.659337	18.426124	17.307095	16.289658	15.362765	33
34	21.397864	19.980391	18.699914	17.540751	16.489211	15.533316	34
35	21.763962	20.290895	18.963468	17.764627	16.679525	15.695221	35
36	22.119699	20.591194	19.217170	17.979131	16.861029	15.848920	36
37	22.465368	20.881625	19.461388	18.184656	17.034129	15.994827	37
38	22.801254	21.162510	19.696477	18.381577	17.199216	16.133338	38
39	23.127633	21.434165	19.922777	18.570255	17.356660	16.264828	39
40	23.444776	21.696892	20.140617	18.751035	17.506814	16.389652	40
45	24.900826	22.896550	21.113584	19.547562	18.159531	16.925038	45
50	26.162169	23.893158	21.917767	20.190760	18.674513	17.337804	50
55	27.254841	24.744882	22.582499	20.710143	19.080824	17.656034	55
60	28.201398	25.465552	23.131913	21.129547	19.401307	17.901380	60
65	29.021379	26.075334	23.586030	21.468216	19.654323	18.090534	65
70	29.731709	26.591290	23.961378	21.741692	19.853878	18.236366	70
75	30.347052	27.027857	24.271621	21.962525	20.011323	18.348799	75
80	30.880110	27.397251	24.528051	22.140848	20.135544	18.435481	80
90	31.741910	27.974269	24.915190	22.401121	20.310879	18.553833	90
100	32.388636	28.387379	25.179676	22.570835	20.420024	18.624191	100

Exponential Growth Factors

TABLE D-59
Factors (EPW) to Convert an Exponential Growth to Present Worth for Interest Rates from 9% to 20%
Growth Rate of 3% per Period

n	9%	10%	11%	12%	15%	20%	n
1	0.944954	0.936364	0.927928	0.919643	0.895652	0.858333	1
2	1.837892	1.813140	1.788978	1.765386	1.697845	1.595069	2
3	2.681678	2.634122	2.587971	2.543167	2.416331	2.227435	3
4	3.479017	3.402860	3.329378	3.258449	3.059844	2.770215	4
5	4.232466	4.122678	4.017351	3.916252	3.636208	3.236101	5
6	4.944440	4.796690	4.655740	4.521196	4.152430	3.635987	6
7	5.617223	5.427809	5.248119	5.077528	4.614785	3.979222	7
8	6.252972	6.018767	5.797804	5.589155	5.028894	4.273832	8
9	6.853726	6.572118	6.307872	6.059670	5.399792	4.526706	9
10	7.421411	7.090256	6.781179	6.492375	5.731988	4.743756	10
11	7.957847	7.575422	7.220373	6.890309	6.029520	4.930057	11
12	8.464754	8.029713	7.627914	7.256266	6.296005	5.089966	12
13	8.943759	8.455095	8.006082	7.592816	6.534682	5.227221	13
14	9.396396	8.853407	8.356995	7.902322	6.748455	5.345031	14
15	9.824117	9.226372	8.682617	8.186957	6.939920	5.446152	15
16	10.228294	9.575603	8.984771	8.448719	7.111407	5.532947	16
17	10.610223	9.902610	9.265148	8.689447	7.264999	5.607446	17
18	10.971108	10.208807	9.525317	8.910831	7.402564	5.671391	18
19	11.312167	10.495520	9.766736	9.114425	7.525775	5.726277	19
20	11.634433	10.763987	9.990755	9.301659	7.636129	5.773388	20
21	11.938960	11.015369	10.198628	9.473847	7.734968	5.813825	21
22	12.226723	11.250755	10.391520	9.632198	7.823493	5.848533	22
23	12.498647	11.471161	10.570510	9.777825	7.902781	5.878324	23
24	12.755602	11.677542	10.736599	9.911750	7.973795	5.903895	24
25	12.998413	11.870789	10.890718	10.034913	8.037399	5.925843	25
26	13.227858	12.051739	11.033729	10.148179	8.094366	5.944682	26
27	13.444673	12.221174	11.166434	10.252343	8.145389	5.960852	27
28	13.649554	12.379826	11.289573	10.348137	8.191087	5.974731	28
29	13.843156	12.528383	11.403838	10.436233	8.232017	5.986644	29
30	14.026102	12.667486	11.509868	10.517250	8.268676	5.996870	30
31	14.198977	12.797737	11.608256	10.591757	8.301510	6.005647	31
32	14.362336	12.919699	11.699553	10.660276	8.330918	6.013180	32
33	14.516703	13.033900	11.784270	10.723290	8.357277	6.019646	33
34	14.662572	13.140834	11.862881	10.781240	8.380847	6.025196	34
35	14.800412	13.240962	11.935826	10.834533	8.401976	6.029960	35
36	14.930665	13.334719	12.003515	10.883544	8.420901	6.034049	36
37	15.053748	13.422510	12.066324	10.928616	8.437850	6.037559	37
38	15.170055	13.504714	12.124607	10.970067	8.453031	6.040571	38
39	15.279960	13.581687	12.178690	11.008186	8.466628	6.043157	39
40	15.383816	13.653761	12.228874	11.043243	8.478806	6.045376	40
45	15.823380	13.950901	12.430483	11.180533	8.523087	6.052559	45
50	16.154569	14.164788	12.569184	11.270842	8.548610	6.055905	50
55	16.404103	14.318747	12.664607	11.330250	8.563320	6.057464	55
60	16.592114	14.429570	12.730256	11.369326	8.571798	6.058190	60
65	16.733771	14.509342	12.775420	11.395031	8.576685	6.058528	65
70	16.840502	14.566764	12.806492	11.411940	8.579501	6.058686	70
75	16.920718	14.608097	12.827868	11.423063	8.581125	6.058759	75
80	16.981508	14.637849	12.842574	11.430380	8.582060	6.058794	80
90	17.061555	14.674681	12.859653	11.438359	8.582910	6.058817	90
100	17.106996	14.693765	12.867736	11.441811	8.583193	6.058822	100

TABLE D-60

Factors (EPW) to Convert an Exponential Growth to Present Worth for Interest Rates from 25% to 50%

Growth Rate of 3% per Period

n	25%	30%	35%	40%	45%	50%	n
1	0.824000	0.792308	0.762963	0.735714	0.710345	0.686667	1
2	1.502976	1.420059	1.345075	1.276990	1.214935	1.158178	2
3	2.062452	1.917431	1.789206	1.675214	1.573367	1.481949	3
4	2.523461	2.311503	2.128061	1.968193	1.827978	1.704271	4
5	2.903332	2.623730	2.386594	2.183742	2.008840	1.856933	5
6	3.216345	2.871109	2.583846	2.342325	2.137314	1.961761	6
7	3.474268	3.067109	2.734342	2.458996	2.228575	2.033742	7
8	3.686797	3.222402	2.849165	2.544833	2.293401	2.083170	8
9	3.861921	3.345442	2.936770	2.607984	2.339451	2.117110	9
10	4.006223	3.442927	3.003610	2.654445	2.372161	2.140415	10
11	4.125128	3.520165	3.054606	2.688628	2.395397	2.156419	11
12	4.223105	3.581362	3.093514	2.713776	2.411903	2.167407	12
13	4.303839	3.629848	3.123200	2.732278	2.423628	2.174953	13
14	4.370363	3.668264	3.145849	2.745890	2.431956	2.180134	14
15	4.425179	3.698702	3.163129	2.755905	2.437872	2.183692	15
16	4.470348	3.722817	3.176313	2.763273	2.442075	2.186135	16
17	4.507566	3.741925	3.186372	2.768694	2.445060	2.187813	17
18	4.538235	3.757063	3.194047	2.772682	2.447181	2.188965	18
19	4.563505	3.769058	3.199903	2.775616	2.448687	2.189756	19
20	4.584328	3.778561	3.204370	2.777775	2.449757	2.190299	20
21	4.601487	3.786091	3.207779	2.779363	2.450517	2.190672	21
22	4.615625	3.792057	3.210379	2.780531	2.451057	2.190928	22
23	4.627275	3.796783	3.212363	2.781391	2.451440	2.191104	23
24	4.636875	3.800528	3.213877	2.782023	2.451713	2.191225	24
25	4.644785	3.803495	3.215032	2.782488	2.451906	2.191308	25
26	4.651303	3.805846	3.215914	2.782831	2.452044	2.191365	26
27	4.656673	3.807709	3.216586	2.783083	2.452141	2.191404	27
28	4.661099	3.809185	3.217099	2.783268	2.452211	2.191431	28
29	4.664745	3.810354	3.217490	2.783404	2.452260	2.191449	29
30	4.667750	3.811281	3.217789	2.783505	2.452295	2.191462	30
31	4.670226	3.812015	3.218017	2.783578	2.452320	2.191470	31
32	4.672266	3.812596	3.218191	2.783633	2.452338	2.191476	32
33	4.673947	3.813057	3.218323	2.783673	2.452350	2.191480	33
34	4.675333	3.813422	3.218424	2.783702	2.452359	2.191483	34
35	4.676474	3.813711	3.218502	2.783724	2.452365	2.191485	35
36	4.677415	3.813941	3.218560	2.783740	2.452370	2.191486	36
37	4.678190	3.814122	3.218605	2.783751	2.452373	2.191487	37
38	4.678828	3.814266	3.218640	2.783760	2.452375	2.191488	38
39	4.679355	3.814380	3.218666	2.783766	2.452377	2.191488	39
40	4.679788	3.814470	3.218686	2.783771	2.452378	2.191489	40
45	4.681047	3.814707	3.218733	2.783781	2.452380	2.191489	45
50	4.681525	3.814781	3.218746	2.783783	2.452381	2.191489	50
55	4.681707	3.814804	3.218749	2.783784	2.452381	2.191489	55
60	4.681776	3.814812	3.218750	2.783784	2.452381	2.191489	60
65	4.681802	3.814814	3.218750	2.783784	2.452381	2.191489	65
70	4.681812	3.814814	3.218750	2.783784	2.452381	2.191489	70
75	4.681816	3.814815	3.218750	2.783784	2.452381	2.191489	75
80	4.681817	3.814815	3.218750	2.783784	2.452381	2.191489	80
90	4.681818	3.814815	3.218750	2.783784	2.452381	2.191489	90
100	4.681818	3.814815	3.218750	2.783784	2.452381	2.191489	100

Exponential Growth Factors

TABLE D-61
Factors (EPW) to Convert an Exponential Growth to Present Worth for Interest Rates from 1% to 2½%
Growth Rate of 4% per Period

n	1%	1¼%	1½%	1¾%	2%	2½%	n
1	1.029703	1.027160	1.024631	1.022113	1.019608	1.014634	1
2	2.089991	2.082219	2.074498	2.066828	2.059208	2.044117	2
3	3.181773	3.165934	3.150225	3.134645	3.119192	3.088665	3
4	4.305984	4.279083	4.252447	4.226074	4.199961	4.148499	4
5	5.463588	5.422465	5.381818	5.341639	5.301921	5.223843	5
6	6.655575	6.596902	6.539005	6.481871	6.425488	6.314923	6
7	7.882969	7.803238	7.724695	7.647318	7.571086	7.421971	7
8	9.146819	9.042338	8.939589	8.838537	8.739146	8.545219	8
9	10.448210	10.315093	10.184407	10.056096	9.930110	9.684905	9
10	11.788256	11.622417	11.459885	11.300580	11.144426	10.841270	10
11	13.168105	12.965248	12.766778	12.572583	12.382552	12.014557	11
12	14.588940	14.344551	14.105862	13.872714	13.644955	13.205014	12
13	16.051978	15.761316	15.477927	15.201595	14.932111	14.412892	13
14	17.558472	17.216562	16.883787	16.559861	16.244505	15.638446	14
15	19.109714	18.711333	18.324275	17.948162	17.582633	16.881936	15
16	20.707032	20.246702	19.800242	19.367164	18.946998	18.143623	16
17	22.351795	21.823773	21.312563	20.817543	20.338116	19.423773	17
18	24.045413	23.443678	22.862134	22.299995	21.756510	20.722658	18
19	25.789336	25.107581	24.449871	23.815228	23.202716	22.040550	19
20	27.585059	26.816675	26.076715	25.363968	24.677279	23.377729	20
21	29.434120	28.572190	27.743629	26.946955	26.180755	24.734476	21
22	31.338104	30.375385	29.451601	28.564946	27.713711	26.111079	22
23	33.298642	32.227556	31.201640	30.218717	29.276725	27.507826	23
24	35.317413	34.130033	32.994784	31.909057	30.870387	28.925014	24
25	37.396146	36.084182	34.832094	33.636776	32.495296	30.362941	25
26	39.536628	38.091407	36.714658	35.402699	34.152067	31.821911	26
27	41.740686	40.153149	38.643590	37.207673	35.841323	33.302231	27
28	44.010211	42.270889	40.620033	39.052560	37.563702	34.804215	28
29	46.347148	44.446147	42.645157	40.938243	39.319853	36.328179	29
30	48.753499	46.680487	44.720161	42.865625	41.110438	37.874445	30
31	51.231326	48.975513	46.846273	44.835626	42.936133	39.443340	31
32	53.782752	51.332872	49.024753	46.849190	44.797626	41.035193	32
33	56.409962	53.754259	51.256890	48.907281	46.695618	42.650343	33
34	59.115208	56.241412	53.544005	51.010882	48.630827	44.289128	34
35	61.900809	58.796117	55.887454	53.160999	50.603980	45.951896	35
36	64.769150	61.420209	58.288623	55.358663	52.615823	47.638997	36
37	67.722689	64.115572	60.748933	57.604923	54.667113	49.350787	37
38	70.763957	66.884144	63.269843	59.900855	56.758625	51.087628	38
39	73.895559	69.727910	65.852844	62.247557	58.891148	52.849886	39
40	77.120180	72.648915	68.499466	64.646152	61.065484	54.637933	40
45	94.738177	88.487963	82.743102	77.458577	72.593643	63.978409	45
50	115.132832	106.598052	98.829446	91.751662	85.297213	74.022634	50
55	138.741756	127.304812	116.996891	107.696517	99.296039	84.823629	55
60	166.071529	150.980558	137.514668	125.484028	114.722186	96.438414	60
65	197.708569	178.050991	160.686844	145.327141	131.721183	108.928301	65
70	234.331717	209.002850	186.856818	167.463409	150.453398	122.359225	70
75	276.726801	244.392665	216.412418	192.157839	171.095551	136.802087	75
80	325.803499	284.856759	249.791645	219.706063	193.842377	152.333130	80
90	448.379879	384.022357	330.063631	284.721195	246.530303	186.993889	90
100	612.637848	513.663875	432.448783	365.631468	310.510089	227.074360	100

TABLE D-62

Factors (EPW) to Convert an Exponential Growth to Present Worth for Interest Rates from 3% to 5½%

Growth Rate of 4% per Period

n	3%	3½%	4%	4½%	5%	5½%	n
1	1.009709	1.004831	1.000000	0.995215	0.990476	0.985782	1
2	2.029220	2.014516	2.000000	1.985669	1.971519	1.957548	2
3	3.058630	3.029079	3.000000	2.971383	2.943219	2.915498	3
4	4.098035	4.048543	4.000000	3.952381	3.905665	3.859827	4
5	5.147530	5.072932	5.000000	4.928686	4.858944	4.790730	5
6	6.207215	6.102270	6.000000	5.900319	5.803145	5.708397	6
7	7.277188	7.136581	7.000000	6.867303	6.738353	6.613017	7
8	8.357549	8.175888	8.000000	7.829660	7.664654	7.504775	8
9	9.448399	9.220216	9.000000	8.787413	8.582134	8.383854	9
10	10.549840	10.269589	10.000000	9.740584	9.490875	9.250435	10
11	11.661974	11.324031	11.000000	10.689193	10.390962	10.104694	11
12	12.784906	12.383568	12.000000	11.633264	11.282477	10.946807	12
13	13.918740	13.448222	13.000000	12.572818	12.165501	11.776947	13
14	15.063582	14.518021	14.000000	13.507876	13.040115	12.595285	14
15	16.219539	15.592987	15.000000	14.438460	13.906400	13.401987	15
16	17.386719	16.673146	16.000000	15.364592	14.764434	14.197219	16
17	18.565231	17.758524	17.000000	16.286293	15.614296	14.981145	17
18	19.755185	18.849145	18.000000	17.203583	16.456065	15.753925	18
19	20.956691	19.945034	19.000000	18.116485	17.289817	16.515717	19
20	22.169863	21.046218	20.000000	19.025018	18.115628	17.266679	20
21	23.394813	22.152721	21.000000	19.929205	18.933574	18.006963	21
22	24.631656	23.264570	22.000000	20.829065	19.743731	18.736722	22
23	25.880507	24.381790	23.000000	21.724620	20.546172	19.456105	23
24	27.141483	25.504408	24.000000	22.615890	21.340970	20.165260	24
25	28.414701	26.632448	25.000000	23.502895	22.128199	20.864332	25
26	29.700281	27.765939	26.000000	24.385656	22.907930	21.553465	26
27	30.998342	28.904904	27.000000	25.264194	23.680236	22.232799	27
28	32.309005	30.049373	28.000000	26.138528	24.445186	22.902475	28
29	33.632393	31.199370	29.000000	27.008678	25.202851	23.562630	29
30	34.968630	32.354922	30.000000	27.874665	25.953300	24.213398	30
31	36.317840	33.516057	31.000000	28.736509	26.696602	24.854913	31
32	37.680149	34.682801	32.000000	29.594229	27.432825	25.487308	32
33	39.055685	35.855182	33.000000	30.447845	28.162036	26.110711	33
34	40.444575	37.033226	34.000000	31.297377	28.884302	26.725251	34
35	41.846950	38.216962	35.000000	32.142844	29.599690	27.331053	35
36	43.262939	39.406416	36.000000	32.984266	30.308264	27.928242	36
37	44.692677	40.601616	37.000000	33.821662	31.010090	28.516940	37
38	46.136295	41.802590	38.000000	34.655051	31.705232	29.097268	38
39	47.593929	43.009365	39.000000	35.484453	32.393754	29.669344	39
40	49.065714	44.221971	40.000000	36.309886	33.075718	30.233287	40
45	56.641776	50.373437	45.000000	40.378187	36.389344	32.934986	45
50	64.592818	56.674932	50.000000	44.350087	39.548155	35.450005	50
55	72.937400	63.130115	55.000000	48.227871	42.559385	37.791244	55
60	81.695002	69.742733	60.000000	52.013768	45.429929	39.970710	60
65	90.886066	76.516628	65.000000	55.709957	48.166358	41.999581	65
70	100.532045	83.455732	70.000000	59.318561	50.774941	43.888263	70
75	110.655457	90.564074	75.000000	62.841658	53.261648	45.646443	75
80	121.279932	97.845782	80.000000	66.281273	55.632175	47.283137	80
90	144.132501	112.946313	90.000000	72.917922	60.046146	50.225069	90
100	169.303252	128.792402	100.000000	79.243779	64.057305	52.774488	100

Exponential Growth Factors

TABLE D-63
Factors (EPW) to Convert an Exponential Growth to Present Worth for Interest Rates from 6% to 8½%
Growth Rate of 4% per Period

n	6%	6½%	7%	7½%	8%	8½%	n
1	0.981132	0.976526	0.971963	0.967442	0.962963	0.958525	1
2	1.943752	1.930129	1.916674	1.903386	1.890261	1.877296	2
3	2.888210	2.861346	2.834898	2.808857	2.783214	2.757961	3
4	3.814847	3.770704	3.727378	3.684847	3.643095	3.602101	4
5	4.724001	4.658716	4.594834	4.532319	4.471128	4.411231	5
6	5.616001	5.525882	5.437970	5.352196	5.268494	5.186802	6
7	6.491171	6.372692	6.257466	6.145380	6.036328	5.930206	7
8	7.349828	7.199625	7.053986	6.912740	6.775723	6.642778	8
9	8.192284	8.007145	7.828173	7.655116	7.487733	7.325797	9
10	9.018845	8.795710	8.580654	8.373321	8.173373	7.980487	10
11	9.829810	9.565764	9.312038	9.068143	8.833618	8.608025	11
12	10.625474	10.317741	10.022915	9.740343	9.469410	9.209535	12
13	11.406125	11.052066	10.713861	10.390658	10.081654	9.786098	13
14	12.172048	11.769154	11.385435	11.019799	10.671222	10.338748	14
15	12.923516	12.469408	12.038180	11.628457	11.238955	10.868478	15
16	13.660810	13.153225	12.672624	12.217298	11.785660	11.376237	16
17	14.384191	13.820990	13.289279	12.786967	12.312117	11.862937	17
18	15.093924	14.473079	13.888645	13.338089	12.819076	12.329451	18
19	15.790265	15.109861	14.471206	13.871268	13.307258	12.776616	19
20	16.473467	15.731696	15.037434	14.387087	13.777360	13.205236	20
21	17.143779	16.338933	15.587787	14.886112	14.230050	13.616079	21
22	17.801444	16.931916	16.122708	15.368890	14.665974	14.009882	22
23	18.446699	17.510979	16.642633	15.835949	15.085753	14.387352	23
24	19.079781	18.076449	17.147979	16.287802	15.489984	14.749167	24
25	19.700917	18.628645	17.639157	16.724943	15.879244	15.095976	25
26	20.310334	19.167878	18.116564	17.147852	16.254087	15.428401	26
27	20.908252	19.694454	18.580586	17.556992	16.615047	15.747039	27
28	21.494889	20.208669	19.031597	17.952811	16.962638	16.052461	28
29	22.070457	20.710813	19.469964	18.335742	17.297355	16.345216	29
30	22.635165	21.201169	19.896040	18.706206	17.619675	16.625829	30
31	23.189219	21.680015	20.310169	19.064609	17.930057	16.894804	31
32	23.732818	22.147620	20.712688	19.411343	18.228944	17.152623	32
33	24.266161	22.604249	21.103921	19.746787	18.516761	17.399749	33
34	24.789441	23.050159	21.484185	20.071311	18.793918	17.636626	34
35	25.302848	23.485601	21.853787	20.385268	19.060810	17.863679	35
36	25.806568	23.910821	22.213027	20.689003	19.317817	18.081314	36
37	26.300784	24.326060	22.562194	20.982850	19.565305	18.289923	37
38	26.785674	24.731552	22.901572	21.267129	19.803627	18.489880	38
39	27.261416	25.127525	23.231434	21.542153	20.033123	18.681544	39
40	27.728182	25.514203	23.552048	21.808222	20.254118	18.865259	40
45	29.933184	27.315625	25.025217	23.014133	21.242217	19.675667	45
50	31.937870	28.915309	26.303126	24.036107	22.060396	20.331392	50
55	33.760436	30.335846	27.411657	24.902199	22.737876	20.861959	55
60	35.417432	31.597300	28.373260	25.636186	23.298852	21.291256	60
65	36.923895	32.717486	29.207409	26.258219	23.763359	21.638612	65
70	38.293501	33.712225	29.930997	26.785372	24.147986	21.919669	70
75	39.538684	34.595564	30.558678	27.232119	24.466471	22.147080	75
80	40.670747	35.379979	31.103165	27.610724	24.730186	22.331085	80
90	42.635685	36.695111	31.985199	28.203496	25.129365	22.600436	90
100	44.259826	37.732178	32.648913	28.629228	25.403059	22.776777	100

TABLE D-64

Factors (EPW) to Convert an Exponential Growth to Present Worth for Interest Rates from 9% to 20%

Growth Rate of 4% per Period

n	9%	10%	11%	12%	15%	20%	n
1	0.954128	0.945455	0.936937	0.928571	0.904348	0.866667	1
2	1.864490	1.839339	1.814788	1.790816	1.722193	1.617778	2
3	2.733091	2.684466	2.637279	2.591472	2.461809	2.268741	3
4	3.561848	3.483495	3.407901	3.334939	3.130680	2.832909	4
5	4.352589	4.238941	4.129925	4.025300	3.735571	3.321854	5
6	5.107058	4.953180	4.806416	4.666350	4.282603	3.745607	6
7	5.826917	5.628461	5.440246	5.261611	4.777311	4.112859	7
8	6.513756	6.266909	6.034104	5.814353	5.224699	4.431145	8
9	7.169088	6.870532	6.590512	6.327613	5.629293	4.706992	9
10	7.794359	7.441230	7.111831	6.804212	5.995186	4.946060	10
11	8.390948	7.980800	7.600274	7.246769	6.326082	5.153252	11
12	8.960171	8.490938	8.057915	7.657714	6.625326	5.332818	12
13	9.503282	8.973250	8.486695	8.039306	6.895947	5.488443	13
14	10.021480	9.429255	8.888435	8.393641	7.140682	5.623317	14
15	10.515908	9.860386	9.264840	8.722667	7.362008	5.740208	15
16	10.987655	10.268002	9.617507	9.028190	7.562164	5.841514	16
17	11.437763	10.653383	9.947935	9.311891	7.743175	5.929312	17
18	11.867223	11.017744	10.257525	9.575327	7.906871	6.005404	18
19	12.276984	11.362231	10.547591	9.819947	8.054909	6.071350	19
20	12.667948	11.687927	10.819364	10.047094	8.188788	6.128503	20
21	13.040978	11.995859	11.073999	10.258015	8.309860	6.178036	21
22	13.396896	12.286994	11.312576	10.453872	8.419352	6.220965	22
23	13.736488	12.562249	11.536107	10.635738	8.518370	6.258169	23
24	14.060502	12.822490	11.745542	10.804614	8.607917	6.290413	24
25	14.369654	13.068536	11.941769	10.961427	8.688899	6.318358	25
26	14.664624	13.301161	12.125621	11.107039	8.762135	6.342577	26
27	14.946063	13.521098	12.297879	11.242251	8.828366	6.363567	27
28	15.214592	13.729038	12.459274	11.367804	8.888261	6.381758	28
29	15.470803	13.925636	12.610491	11.484390	8.942427	6.397524	29
30	15.715262	14.111510	12.752172	11.592648	8.991413	6.411187	30
31	15.948507	14.287246	12.884918	11.693173	9.035712	6.423029	31
32	16.171052	14.453396	13.009292	11.786518	9.075775	6.433292	32
33	16.383390	14.610484	13.125823	11.873195	9.112005	6.442186	33
34	16.585986	14.759003	13.235006	11.953681	9.144770	6.449895	34
35	16.779290	14.899421	13.337303	12.028418	9.174400	6.456575	35
36	16.963726	15.032180	13.433148	12.097817	9.201197	6.462365	36
37	17.139702	15.157697	13.522950	12.162258	9.225430	6.467383	37
38	17.307605	15.276368	13.607088	12.222097	9.247346	6.471732	38
39	17.467807	15.388566	13.685920	12.277662	9.267165	6.475501	39
40	17.620660	15.494644	13.759781	12.329257	9.285088	6.478768	40
45	18.285966	15.944304	14.064821	12.536945	9.352042	6.489619	45
50	18.812050	16.283997	14.285067	12.680325	9.392542	6.494924	50
55	19.228047	16.540617	14.444090	12.779308	9.417040	6.497518	55
60	19.556992	16.734479	14.558909	12.847643	9.431859	6.498787	60
65	19.817103	16.880931	14.641811	12.894819	9.440822	6.499407	65
70	20.022783	16.991568	14.701668	12.927387	9.446245	6.499710	70
75	20.185423	17.075148	14.744886	12.949871	9.449524	6.499858	75
80	20.314029	17.138289	14.776091	12.965393	9.451508	6.499931	80
90	20.496136	17.222022	14.814889	12.983506	9.453434	6.499983	90
100	20.610003	17.269808	14.835115	12.992139	9.454139	6.499996	100

Exponential Growth Factors

TABLE D-65
Factors (EPW) to Convert an Exponential Growth to Present Worth for Interest Rates from 25% to 50%
Growth Rate of 4% per Period

n	25%	30%	35%	40%	45%	50%	n
1	0.832000	0.800000	0.770370	0.742857	0.717241	0.693333	1
2	1.524224	1.440000	1.363841	1.294694	1.231677	1.174044	2
3	2.100154	1.952000	1.821033	1.704630	1.600651	1.507337	3
4	2.579328	2.361600	2.173240	2.009154	1.865294	1.738421	4
5	2.978001	2.689280	2.444570	2.235371	2.055108	1.898638	5
6	3.309697	2.951424	2.653595	2.403419	2.191250	2.009723	6
7	3.585668	3.161139	2.814621	2.528254	2.288896	2.086741	7
8	3.815276	3.328911	2.938671	2.620989	2.358933	2.140140	8
9	4.006309	3.463129	3.034236	2.689877	2.409165	2.177164	9
10	4.165249	3.570503	3.107856	2.741052	2.445194	2.202834	10
11	4.297488	3.656403	3.164570	2.779067	2.471036	2.220631	11
12	4.407510	3.725122	3.208261	2.807307	2.489571	2.232971	12
13	4.499048	3.780098	3.241920	2.828285	2.502864	2.241527	13
14	4.575208	3.824078	3.267849	2.843869	2.512399	2.247458	14
15	4.638573	3.859263	3.287825	2.855445	2.519238	2.251571	15
16	4.691293	3.887410	3.303213	2.864045	2.524143	2.254423	16
17	4.735156	3.909928	3.315068	2.870434	2.527661	2.256400	17
18	4.771649	3.927942	3.324200	2.875179	2.530185	2.257770	18
19	4.802012	3.942354	3.331236	2.878705	2.531995	2.258721	19
20	4.827274	3.953883	3.336656	2.881323	2.533293	2.259380	20
21	4.848292	3.963107	3.340831	2.883269	2.534224	2.259837	21
22	4.865779	3.970485	3.344048	2.884714	2.534891	2.260153	22
23	4.880328	3.976388	3.346526	2.885788	2.535370	2.260373	23
24	4.892433	3.981111	3.348435	2.886585	2.535714	2.260525	24
25	4.902504	3.984888	3.349905	2.887177	2.535960	2.260631	25
26	4.910884	3.987911	3.351038	2.887618	2.536137	2.260704	26
27	4.917855	3.990329	3.351911	2.887944	2.536264	2.260755	27
28	4.923655	3.992263	3.352583	2.888187	2.536355	2.260790	28
29	4.928481	3.993810	3.353101	2.888368	2.536420	2.260814	29
30	4.932496	3.995048	3.353500	2.888502	2.536467	2.260831	30
31	4.935837	3.996039	3.353807	2.888601	2.536500	2.260843	31
32	4.938616	3.996831	3.354044	2.888675	2.536524	2.260851	32
33	4.940929	3.997465	3.354227	2.888730	2.536542	2.260857	33
34	4.942853	3.997972	3.354367	2.888771	2.536554	2.260861	34
35	4.944454	3.998377	3.354476	2.888801	2.536563	2.260863	35
36	4.945785	3.998702	3.354559	2.888824	2.536569	2.260865	36
37	4.946893	3.998962	3.354623	2.888841	2.536574	2.260867	37
38	4.947815	3.999169	3.354673	2.888853	2.536577	2.260868	38
39	4.948582	3.999335	3.354711	2.888862	2.536579	2.260868	39
40	4.949221	3.999468	3.354740	2.888869	2.536581	2.260869	40
45	4.951121	3.999826	3.354812	2.888884	2.536585	2.260869	45
50	4.951879	3.999943	3.354831	2.888888	2.536585	2.260870	50
55	4.952181	3.999981	3.354837	2.888889	2.536585	2.260870	55
60	4.952301	3.999994	3.354838	2.888889	2.536585	2.260870	60
65	4.952349	3.999998	3.354839	2.888889	2.536585	2.260870	65
70	4.952368	3.999999	3.354839	2.888889	2.536585	2.260870	70
75	4.952376	4.000000	3.354839	2.888889	2.536585	2.260870	75
80	4.952379	4.000000	3.354839	2.888889	2.536585	2.260870	80
90	4.952381	4.000000	3.354839	2.888889	2.536585	2.260870	90
100	4.952381	4.000000	3.354839	2.888889	2.536585	2.260870	100

TABLE D-66

Factors (EPW) to Convert an Exponential Growth to Present Worth for Interest Rates from 1% to 2½%

Growth Rate of 5% per Period

n	1%	1¼%	1½%	1¾%	2%	2½%	n
1	1.039604	1.037037	1.034483	1.031941	1.029412	1.024390	1
2	2.120380	2.112483	2.104637	2.096843	2.089100	2.073766	2
3	3.243960	3.227760	3.211694	3.195760	3.179956	3.148736	3
4	4.412037	4.384344	4.356925	4.329777	4.302896	4.249924	4
5	5.626376	5.583764	5.541646	5.500015	5.458864	5.377971	5
6	6.888806	6.827607	6.767220	6.707632	6.648830	6.533531	6
7	8.201234	8.117518	8.035055	7.953822	7.873796	7.717276	7
8	9.565640	9.455204	9.346609	9.239816	9.134790	8.929893	8
9	10.984081	10.842434	10.703389	10.566887	10.432872	10.172085	9
10	12.458698	12.281043	12.106954	11.936345	11.769133	11.444575	10
11	13.991715	13.772933	13.558918	13.349545	13.144696	12.748101	11
12	15.585447	15.320079	15.060949	14.807884	14.560716	14.083421	12
13	17.242296	16.924526	16.614775	16.312805	16.018384	15.451309	13
14	18.964763	18.588397	18.222181	17.865793	17.518925	16.852560	14
15	20.755447	20.313894	19.885015	19.468386	19.063599	18.287989	15
16	22.617049	22.103297	21.605188	21.122168	20.653705	19.758427	16
17	24.552378	23.958975	23.384677	22.828773	22.290579	21.264731	17
18	26.564353	25.883381	25.225528	24.589888	23.975596	22.807773	18
19	28.656010	27.879062	27.129857	26.407256	25.710172	24.388450	19
20	30.830506	29.948657	29.099852	28.282672	27.495765	26.007681	20
21	33.091120	32.094904	31.137778	30.217990	29.333876	27.666405	21
22	35.441263	34.320641	33.245977	32.215125	31.226049	29.365585	22
23	37.884482	36.628813	35.426873	34.276051	33.173874	31.106209	23
24	40.424461	39.022472	37.682972	36.402804	35.178988	32.889287	24
25	43.065034	41.504786	40.016867	38.597488	37.243076	34.715855	25
26	45.810184	44.079037	42.431242	40.862273	39.367872	36.586974	26
27	48.664052	46.748631	44.928871	43.199397	41.555162	38.503729	27
28	51.630946	49.517099	47.512625	45.611171	43.806785	40.467235	28
29	54.715339	52.388103	50.185474	48.099980	46.124631	42.478631	29
30	57.921868	55.365440	52.950491	50.668284	48.510650	44.539085	30
31	61.255428	58.453049	55.810852	53.318623	50.966846	46.649795	31
32	64.720969	61.655014	58.769847	56.053616	53.495282	48.811985	32
33	68.323801	64.975570	61.830877	58.875967	56.098085	51.026911	33
34	72.069298	68.419109	64.997459	61.788467	58.777440	53.295860	34
35	75.963131	71.990187	68.273233	64.793996	61.535600	55.620150	35
36	80.011176	75.693528	71.661965	67.895524	64.374882	58.001129	36
37	84.219539	79.534029	75.167550	71.096118	67.297673	60.440181	37
38	88.594571	83.516717	78.794017	74.398942	70.306428	62.938722	38
39	93.142871	87.647021	82.545535	77.807262	73.403676	65.498203	39
40	97.871301	91.930244	86.426416	81.324448	76.592019	68.120110	40
45	124.475227	115.846750	107.932882	100.669020	93.996751	82.220652	45
50	156.781388	144.532689	133.412057	123.306784	114.116056	98.126726	50
55	196.011990	178.939181	163.597783	149.798370	137.373320	116.069523	55
60	243.651206	220.207021	199.359463	180.799844	164.257961	136.309840	60
65	301.501321	269.704499	241.727096	217.078965	195.335734	159.141868	65
70	371.750922	329.072771	291.920944	259.534192	231.260630	184.897467	70
75	457.057689	400.280272	351.386676	309.216935	272.788644	213.950992	75
80	560.648948	485.687983	421.837009	367.357597	320.793680	246.724731	80
90	839.201198	710.995983	604.182668	515.017255	440.433196	325.399297	90
100	1249.959098	1035.126025	860.116675	717.230998	600.302808	425.511887	100

Exponential Growth Factors

TABLE D-67
Factors (EPW) to Convert an Exponential Growth to Present Worth for Interest Rates from 3% to 5½%
Growth Rate of 5% per Period

n	3%	3½%	4%	4½%	5%	5½%	n
1	1.019417	1.014493	1.009615	1.004785	1.000000	0.995261	1
2	2.058629	2.043688	2.028939	2.014377	2.000000	1.985804	2
3	3.118020	3.087800	3.058063	3.028800	3.000000	2.971654	3
4	4.197982	4.147043	4.097083	4.048076	4.000000	3.952831	4
5	5.298914	5.221638	5.146093	5.072230	5.000000	4.929358	5
6	6.421223	6.311807	6.205190	6.101284	6.000000	5.901256	6
7	7.565324	7.417775	7.274471	7.135261	7.000000	6.868549	7
8	8.731641	8.535772	8.354033	8.174186	8.000000	7.831257	8
9	9.920605	9.678029	9.443976	9.218081	9.000000	8.789403	9
10	11.132655	10.832783	10.544399	10.266972	10.000000	9.743008	10
11	12.368241	12.004273	11.655403	11.320881	11.000000	10.692093	11
12	13.627818	13.192741	12.777089	12.379832	12.000000	11.636680	12
13	14.911854	14.398432	13.909561	13.443851	13.000000	12.576791	13
14	16.220822	15.621598	15.052922	14.512960	14.000000	13.512446	14
15	17.555207	16.862491	16.207277	15.587185	15.000000	14.443666	15
16	18.915502	18.121368	17.372732	16.666549	16.000000	15.370474	16
17	20.302211	19.398489	18.549393	17.751078	17.000000	16.292888	17
18	21.715846	20.694119	19.737368	18.840796	18.000000	17.210932	18
19	23.156930	22.008527	20.936765	19.935728	19.000000	18.124624	19
20	24.625997	23.341984	22.147696	21.035899	20.000000	19.033986	20
21	26.123589	24.694766	23.370270	22.141334	21.000000	19.939038	21
22	27.650261	26.067154	24.604599	23.252058	22.000000	20.839801	22
23	29.206576	27.459431	25.850797	24.368097	23.000000	21.736295	23
24	30.793112	28.871887	27.108978	25.489475	24.000000	22.628540	24
25	32.410454	30.304813	28.379257	26.616219	25.000000	23.516556	25
26	34.059201	31.758506	29.661750	27.748354	26.000000	24.400364	26
27	35.739962	33.233267	30.956574	28.885906	27.000000	25.279983	27
28	37.453359	34.729401	32.263849	30.028901	28.000000	26.155433	28
29	39.200026	36.247218	33.583694	31.177364	29.000000	27.026735	29
30	40.980609	37.787033	34.916229	32.331323	30.000000	27.893907	30
31	42.795767	39.349164	36.261578	33.490803	31.000000	28.756969	31
32	44.646170	40.933935	37.619862	34.655831	32.000000	29.615940	32
33	46.532503	42.541673	38.991207	35.826433	33.000000	30.470841	33
34	48.455465	44.172712	40.375738	37.002636	34.000000	31.321690	34
35	50.415765	45.827389	41.773581	38.184467	35.000000	32.168507	35
36	52.414129	47.506046	43.184866	39.371952	36.000000	33.011310	36
37	54.451297	49.209033	44.609720	40.565119	37.000000	33.850119	37
38	56.528021	50.936700	46.048275	41.763996	38.000000	34.684953	38
39	58.645070	52.689405	47.500662	42.968608	39.000000	35.515830	39
40	60.803227	54.467513	48.967015	44.178984	40.000000	36.342769	40
45	72.239113	63.752146	56.513002	50.318290	45.000000	40.419048	45
50	84.829242	73.729364	64.428821	56.605881	50.000000	44.399644	50
55	98.690114	84.450831	72.732597	63.045340	55.000000	48.286803	55
60	113.949987	95.972065	81.443345	69.640334	60.000000	52.082718	60
65	130.750065	108.352722	90.581010	76.394620	65.000000	55.789531	65
70	149.245803	121.656913	100.166516	83.312045	70.000000	59.409334	70
75	169.608347	135.953528	110.221811	90.396551	75.000000	62.944168	75
80	192.026118	151.316596	120.769921	97.652171	80.000000	66.396029	80
90	243.878023	185.566235	143.442379	112.693391	90.000000	73.058575	90
100	306.725154	225.116144	168.391669	128.469981	100.000000	79.412010	100

TABLE D-68

FACTORS (EPW) TO CONVERT AN EXPONENTIAL GROWTH TO PRESENT WORTH FOR INTEREST RATES FROM 6% TO 8½%

Growth Rate of 5% per Period

n	6%	6½%	7%	7½%	8%	8½%	n
1	0.990566	0.985915	0.981308	0.976744	0.972222	0.967742	1
2	1.971787	1.957945	1.944275	1.930773	1.917438	1.904266	2
3	2.943751	2.916284	2.889241	2.862616	2.836398	2.810580	3
4	3.906546	3.861125	3.816545	3.772788	3.729832	3.687658	4
5	4.860258	4.792658	4.726516	4.661793	4.598447	4.536444	5
6	5.804973	5.711071	5.619479	5.530123	5.442935	5.357849	6
7	6.740775	6.616549	6.495750	6.378260	6.263965	6.152757	7
8	7.667749	7.509274	7.355643	7.206672	7.062188	6.922023	8
9	8.585977	8.389425	8.199462	8.015819	7.838238	7.666474	9
10	9.495544	9.257180	9.027510	8.806149	8.592732	8.386910	10
11	10.396529	10.112712	9.840080	9.578099	9.326267	9.084106	11
12	11.289015	10.956195	10.637461	10.332097	10.039426	9.758813	12
13	12.173080	11.787758	11.419939	11.068560	10.732775	10.411754	13
14	13.048806	12.607688	12.187790	11.787896	11.406865	11.043633	14
15	13.916270	13.416031	12.941290	12.490503	12.062230	11.655129	15
16	14.775551	14.212988	13.680705	13.176770	12.699390	12.246899	16
17	15.626725	14.998721	14.406299	13.847078	13.318851	12.819579	17
18	16.469869	15.773386	15.118331	14.501797	13.921106	13.373787	18
19	17.305059	16.537142	15.817054	15.141290	14.506630	13.910116	19
20	18.132370	17.290140	16.502716	15.765911	15.075891	14.429145	20
21	18.951875	18.032532	17.175563	16.376006	15.629338	14.931430	21
22	19.763650	18.764468	17.835833	16.971913	16.167412	15.417513	22
23	20.567767	19.486095	18.483761	17.553962	16.690540	15.887916	23
24	21.364297	20.197559	19.119578	18.122474	17.199136	16.343144	24
25	22.153313	20.899002	19.743512	18.677765	17.693604	16.783688	25
26	22.934886	21.590565	20.355782	19.220143	18.174337	17.210021	26
27	23.709085	22.272388	20.956609	19.749907	18.641717	17.622601	27
28	24.475980	22.944608	21.545205	20.267351	19.096114	18.021872	28
29	25.235641	23.607360	22.124781	20.772761	19.537888	18.408263	29
30	25.988135	24.260777	22.692542	21.266418	19.967391	18.782190	30
31	26.733530	24.904992	23.249690	21.748594	20.384964	19.144055	31
32	27.471893	25.540133	23.796425	22.219557	20.790937	19.494247	32
33	28.203290	26.166328	24.332941	22.679568	21.185633	19.833142	33
34	28.927787	26.783704	24.859428	23.128880	21.569366	20.161105	34
35	29.645450	27.392384	25.376074	23.567743	21.942439	20.478489	35
36	30.356342	27.992491	25.883063	23.996400	22.305149	20.785634	36
37	31.060527	28.584146	26.380576	24.415089	22.657784	21.082872	37
38	31.758069	29.167468	26.868790	24.824040	23.000623	21.370521	38
39	32.449031	29.742574	27.347878	25.223481	23.333939	21.648891	39
40	33.133474	30.309580	27.818011	25.613633	23.657996	21.918282	40
45	36.460044	33.027046	30.040103	27.432439	25.148156	23.140357	45
50	39.632633	35.558457	32.062142	29.049366	26.442533	24.177637	50
55	42.658369	37.916550	33.902140	30.486822	27.566849	25.058064	55
60	45.544049	40.113193	35.576484	31.764728	28.543448	25.805358	60
65	48.296157	42.159440	37.100090	32.900793	29.391737	26.439650	65
70	50.920874	44.065587	38.486526	33.910760	30.128574	26.978027	70
75	53.424098	45.841226	39.748144	34.808626	30.768603	27.434994	75
80	55.811452	47.495294	40.896179	35.606933	31.324541	27.822861	80
90	60.259757	50.471438	42.391486	36.947289	32.226891	28.431510	90
100	64.305785	53.054001	44.543695	38.006692	32.907707	28.870003	100

Exponential Growth Factors

TABLE D-69
Factors (EPW) to Convert an Exponential Growth to Present Worth for Interest Rates from 9% to 20%
Growth Rate of 5% per Period

n	9%	10%	11%	12%	15%	20%	n
1	0.963303	0.954545	0.945946	0.937500	0.913043	0.875000	1
2	1.891255	1.865702	1.840760	1.816406	1.746692	1.640625	2
3	2.785154	2.735443	2.687205	2.640381	2.507849	2.310547	3
4	3.646249	3.565650	3.487897	3.412857	3.202819	2.896729	4
5	4.475745	4.358121	4.245308	4.137053	3.837356	3.409637	5
6	5.274800	5.114570	4.961778	4.815988	4.416717	3.858433	6
7	6.044532	5.836635	5.639519	5.452488	4.945698	4.251129	7
8	6.786017	6.525879	6.280626	6.049208	5.428681	4.594738	8
9	7.500292	7.183793	6.887079	6.608632	5.869665	4.895395	9
10	8.188354	7.811803	7.460750	7.133093	6.272303	5.158471	10
11	8.851167	8.411266	8.003413	7.624775	6.639929	5.388662	11
12	9.489656	8.983481	8.516742	8.085726	6.975587	5.590079	12
13	10.104715	9.529687	9.002323	8.517868	7.282058	5.766319	13
14	10.697202	10.051065	9.461657	8.923002	7.561879	5.920529	14
15	11.267947	10.548744	9.896162	9.302814	7.817368	6.055463	15
16	11.817747	11.023801	10.307180	9.658888	8.050640	6.173530	16
17	12.347371	11.477264	10.695981	9.992708	8.263628	6.276839	17
18	12.857559	11.910116	11.063766	10.305663	8.458095	6.367234	18
19	13.349025	12.323292	11.411671	10.599059	8.635652	6.446330	19
20	13.822455	12.717688	11.740770	10.874118	8.797769	6.515539	20
21	14.278512	13.094157	12.052079	11.131986	8.945789	6.576096	21
22	14.717833	13.453513	12.346562	11.373737	9.080938	6.629084	22
23	15.141032	13.796536	12.625126	11.600378	9.204335	6.675449	23
24	15.548700	14.123956	12.888632	11.812854	9.317001	6.716018	24
25	15.941408	14.436513	13.137896	12.012051	9.419871	6.751515	25
26	16.319705	14.734853	13.373685	12.198798	9.513795	6.782576	26
27	16.684120	15.019633	13.596729	12.373873	9.599552	6.809754	27
28	17.035161	15.291467	13.807717	12.538006	9.677852	6.833535	28
29	17.373320	15.550946	14.007300	12.691881	9.749343	6.854343	29
30	17.699070	15.798630	14.196094	12.836138	9.814617	6.872550	30
31	18.012866	16.035056	14.374684	12.971379	9.874216	6.888481	31
32	18.315146	16.260736	14.543620	13.098168	9.928632	6.902421	32
33	18.606333	16.476157	14.703424	13.217033	9.978316	6.914618	33
34	18.886835	16.681786	14.854590	13.328468	10.023680	6.925291	34
35	19.157043	16.878068	14.997585	13.432939	10.065099	6.934630	35
36	19.417335	17.065429	15.132851	13.530880	10.102917	6.942801	36
37	19.668075	17.244273	15.260805	13.622700	10.137446	6.949951	37
38	19.909613	17.414988	15.381843	13.708781	10.168972	6.956207	38
39	20.142288	17.577943	15.496338	13.789483	10.197757	6.961681	39
40	20.366424	17.733491	15.604644	13.865140	10.224039	6.966471	40
45	21.369601	18.411388	16.064436	14.178138	10.324892	6.982803	45
50	22.201731	18.948602	16.412688	14.404811	10.388888	6.991179	50
55	22.891979	19.374327	16.676458	14.568966	10.429495	6.995476	55
60	23.464537	19.711703	16.876240	14.687947	10.455262	6.997680	60
65	23.939471	19.979062	17.027558	14.773940	10.471612	6.998810	65
70	24.333427	20.190937	17.142167	14.836288	10.481987	6.999390	70
75	24.660211	20.358842	17.228973	14.881440	10.488570	6.999687	75
80	24.931278	20.491901	17.294721	14.914140	10.492747	6.999839	80
90	25.342636	20.680909	17.382238	14.954970	10.497080	6.999958	90
100	25.625677	20.799608	17.432443	14.976383	10.498824	6.999989	100

TABLE D-70

Factors (EPW) to Convert an Exponential Growth to Present Worth for Interest Rates from 25% to 50%

Growth Rate of 5% per Period

n	25%	30%	35%	40%	45%	50%	n
1	0.840000	0.807692	0.777778	0.750000	0.724138	0.700000	1
2	1.545600	1.460059	1.382716	1.312500	1.248514	1.190000	2
3	2.136304	1.586971	1.853224	1.734375	1.628234	1.533000	3
4	2.636175	2.412553	2.219174	2.050781	1.903204	1.773100	4
5	3.054387	2.756293	2.503802	2.288086	2.102320	1.941170	5
6	3.405685	3.033929	2.725179	2.466064	2.246508	2.058819	6
7	3.700776	3.258173	2.897362	2.599548	2.350919	2.141173	7
8	3.948652	3.439254	3.031281	2.699661	2.426528	2.198821	8
9	4.156867	3.585584	3.135441	2.774746	2.481279	2.239175	9
10	4.331769	3.703741	3.216454	2.831059	2.520926	2.267422	10
11	4.478686	3.799175	3.279464	2.873295	2.549636	2.287196	11
12	4.602096	3.876257	3.328472	2.904971	2.570426	2.301037	12
13	4.705761	3.938515	3.366590	2.928728	2.585481	2.310726	13
14	4.792839	3.988801	3.396236	2.946546	2.596383	2.317508	14
15	4.865985	4.029416	3.419295	2.959910	2.604277	2.322256	15
16	4.927427	4.062221	3.437229	2.969932	2.609994	2.325579	16
17	4.979039	4.098717	3.451178	2.977449	2.614133	2.327905	17
18	5.022393	4.110117	3.462028	2.983087	2.617131	2.329534	18
19	5.058810	4.127402	3.470466	2.987315	2.619302	2.330674	19
20	5.069400	4.141363	3.477029	2.990486	2.620874	2.331472	20
21	5.115096	4.152640	3.482134	2.992865	2.622012	2.332030	21
22	5.136681	4.161747	3.486104	2.994649	2.622836	2.332421	22
23	5.154812	4.169104	3.489192	2.995986	2.623433	2.332695	23
24	5.170042	4.175045	3.491594	2.996990	2.623865	2.332886	24
25	5.182835	4.179844	3.493462	2.997742	2.624178	2.333020	25
26	5.193582	4.183720	3.494915	2.998307	2.624405	2.333114	26
27	5.202609	4.186851	3.496045	2.998730	2.624569	2.333180	27
28	5.210191	4.189380	3.496924	2.999048	2.624688	2.333226	28
29	5.216561	4.191422	3.497607	2.999286	2.624774	2.333258	29
30	5.221911	4.193072	3.498139	2.999464	2.624836	2.333281	30
31	5.226405	4.194404	3.498553	2.999598	2.624882	2.333297	31
32	5.230180	4.195480	3.498874	2.999699	2.624914	2.333308	32
33	5.233351	4.196349	3.499124	2.999774	2.624938	2.333315	33
34	5.236015	4.197051	3.499319	2.999830	2.624955	2.333321	34
35	5.238253	4.197618	3.499470	2.999873	2.624967	2.333324	35
36	5.240132	4.198076	3.499588	2.999905	2.624976	2.333327	36
37	5.241711	4.198446	3.499680	2.999928	2.624983	2.333329	37
38	5.243037	4.198745	3.499751	2.999946	2.624988	2.333330	38
39	5.244151	4.198986	3.499806	2.999960	2.624991	2.333331	39
40	5.245087	4.199181	3.499849	2.999970	2.624994	2.333332	40
45	5.247945	4.199719	3.499957	2.999993	2.624999	2.333333	45
50	5.249141	4.199903	3.499988	2.999998	2.625000	2.333333	50
55	5.249641	4.199967	3.499997	3.000000	2.625000	2.333333	55
60	5.249850	4.199989	3.499999	3.000000	2.625000	2.333333	60
65	5.249937	4.199996	3.500000	3.000000	2.625000	2.333333	65
70	5.249974	4.199999	3.500000	3.000000	2.625000	2.333333	70
75	5.249989	4.200000	3.500000	3.000000	2.625000	2.333333	75
80	5.249995	4.200000	3.500000	3.000000	2.625000	2.333333	80
90	5.249999	4.200000	3.500000	3.000000	2.625000	2.333333	90
100	5.250000	4.200000	3.500000	3.000000	2.625000	2.333333	100

Exponential Growth Factors

TABLE D-71
Factors (EPW) to Convert an Exponential Growth to Present Worth for Interest Rates from 1% to 2½%
Growth Rate of 6% per Period

n	1%	1¼%	1½%	1¾%	2%	2½%	n
1	1.049505	1.046914	1.044335	1.041769	1.039216	1.034146	1
2	2.150966	2.142942	2.134971	2.127052	2.119185	2.103605	2
3	3.306954	3.290388	3.273959	3.257666	3.241506	3.209582	3
4	4.520170	4.491666	4.463445	4.435504	4.407839	4.353324	4
5	5.793445	5.749299	5.705667	5.662540	5.619912	5.536120	5
6	7.129754	7.065933	7.002963	6.940828	6.879516	6.759305	6
7	8.532218	8.444335	8.357774	8.272509	8.188517	8.024256	7
8	10.004109	9.887403	9.772650	9.659813	9.548851	9.332402	8
9	11.548867	11.398170	11.250256	11.105063	10.962531	10.685215	9
10	13.170098	12.979812	12.793370	12.610680	12.431650	12.084223	10
11	14.871588	14.635655	14.404899	14.179185	13.958381	13.531001	11
12	16.657311	16.369180	16.087875	15.813205	15.544984	15.027182	12
13	18.531435	18.184030	17.845645	17.515674	17.193807	16.574451	13
14	20.498338	20.084022	19.680979	19.288850	18.907290	18.174555	14
15	22.562612	22.073149	21.597869	21.136296	20.687968	19.829295	15
16	24.729078	24.155593	23.599745	23.060908	22.538477	21.540540	16
17	27.002794	26.335732	25.690374	25.065909	24.461554	23.310217	17
18	29.389071	28.618149	27.873691	27.154657	26.460046	25.140322	18
19	31.893481	31.007642	30.153806	29.330650	28.536911	27.032918	19
20	34.521871	33.509235	32.535009	31.597532	30.695221	28.990140	20
21	37.280379	36.128187	35.021783	33.959100	32.938171	31.014193	21
22	40.175448	38.870003	37.618808	36.419308	35.269080	33.107361	22
23	43.213836	41.740448	40.330972	38.982276	37.691397	35.272002	23
24	46.402640	44.745555	43.163379	41.652298	40.208706	37.510558	24
25	49.749305	47.891643	46.121362	44.433843	42.824734	39.825553	25
26	53.261647	51.185325	49.210486	47.331571	45.543351	42.219596	26
27	56.947867	54.633525	52.436567	50.350335	48.368580	44.695387	27
28	60.816574	58.243493	55.805675	53.495189	51.304603	47.255718	28
29	64.876800	62.022818	59.324154	56.771401	54.355764	49.903474	29
30	69.138028	65.979444	62.998624	60.184457	57.526578	52.641641	30
31	73.610207	70.121689	66.836001	63.740073	60.821738	55.473307	31
32	78.303782	74.458262	70.843508	67.444204	64.246120	58.401664	32
33	83.229712	78.998279	75.028689	71.303053	67.804792	61.430014	33
34	88.399500	83.751285	79.399419	75.323082	71.503019	64.561770	34
35	93.825217	88.727271	83.963925	79.511024	75.346274	67.800465	35
36	99.519535	93.936659	88.730798	83.873892	79.340246	71.149749	36
37	105.495750	99.390519	93.709011	88.418993	83.490844	74.613399	37
38	111.767817	105.100198	98.907933	93.153939	87.804210	78.195320	38
39	118.350382	111.077738	104.337349	98.086659	92.286728	81.899550	39
40	125.258816	117.335706	110.007477	103.225414	96.945031	85.730266	40
45	165.282455	153.314638	142.359344	132.323819	123.124283	106.937572	45
50	216.243582	198.562955	182.547534	168.028604	154.855424	132.021497	50
55	281.131148	255.468763	232.470177	211.839655	193.315856	161.690673	55
60	363.750909	327.035426	294.485168	265.597382	239.932674	196.783268	60
65	468.948646	417.040072	371.521536	331.560038	296.435621	238.290664	65
70	602.894370	530.232950	467.217784	412.498580	364.921268	287.385463	70
75	773.444208	672.588126	586.093742	511.813068	447.930812	345.454615	75
80	990.601175	851.618794	733.764037	633.675502	548.544368	414.138600	80
90	1619.163489	1359.936314	1145.075495	966.683209	818.308528	591.467171	90
100	2638.208234	2163.912922	1779.776294	1468.065232	1214.624144	839.551148	100

TABLE D-72

Factors (EPW) to Convert an Exponential Growth to Present Worth for Interest Rates from 3% to 5½%

Growth Rate of 6% per Period

n	3%	3½%	4%	4½%	5%	5½%	n
1	1.029126	1.024155	1.019231	1.014354	1.009524	1.004739	1
2	2.088227	2.073047	2.058062	2.043268	2.028662	2.014240	2
3	3.178175	3.147275	3.116871	3.086952	3.057507	3.028526	3
4	4.299870	4.247451	4.196042	4.145616	4.096149	4.047619	4
5	5.454235	5.374201	5.295965	5.219476	5.144684	5.071541	5
6	6.642222	6.528167	6.417042	6.308751	6.203205	6.100316	6
7	7.864811	7.710007	7.559677	7.413661	7.271807	7.133967	7
8	9.123010	8.920394	8.724286	8.534432	8.350586	8.172516	8
9	10.417855	10.160017	9.911292	9.671290	9.439639	9.215988	9
10	11.750414	11.429582	11.121124	10.824466	10.539064	10.264405	10
11	13.121785	12.729814	12.354223	11.994195	11.648960	11.317791	11
12	14.533099	14.061452	13.611035	13.180715	12.769427	12.376169	12
13	15.985519	15.425255	14.892016	14.384266	13.900564	13.439563	13
14	17.480243	16.822000	16.197632	15.605092	15.042474	14.507997	14
15	19.018503	18.252484	17.528356	16.843443	16.195260	15.581495	15
16	20.601566	19.717519	18.884670	18.099569	17.359024	16.660080	16
17	22.230738	21.217943	20.267068	19.373726	18.533872	17.743777	17
18	23.907361	22.754608	21.676050	20.666171	19.719909	18.832610	18
19	25.632818	24.328391	23.112128	21.977169	20.917241	19.926604	19
20	27.408531	25.940188	24.575823	23.306985	22.125977	21.025782	20
21	29.235964	27.590917	26.067665	24.655889	23.346224	22.130169	21
22	31.116623	29.281519	27.588197	26.024155	24.578093	23.239791	22
23	33.052059	31.012956	29.137970	27.412062	25.821694	24.354671	23
24	35.043867	32.786216	30.717547	28.819890	27.077139	25.474836	24
25	37.093688	34.602308	32.327499	30.247927	28.344540	26.600309	25
26	39.203213	36.462267	33.968413	31.696462	29.624012	27.731116	26
27	41.374180	38.367153	35.640882	33.165789	30.915669	28.867283	27
28	43.608380	40.318050	37.345515	34.656207	32.219628	30.008834	28
29	45.907653	42.316071	39.082928	36.168019	33.536005	31.155795	29
30	48.273895	44.362353	40.853754	37.701531	34.864919	32.308192	30
31	50.709057	46.458062	42.658634	39.257055	36.206490	33.466051	31
32	53.215146	48.604392	44.498223	40.834908	37.560838	34.629397	32
33	55.794228	50.802566	46.373189	42.435409	38.928084	35.798257	33
34	58.448429	53.053835	48.284212	44.058884	40.308351	36.972656	34
35	61.179937	55.359483	50.231985	45.705662	41.701764	38.152621	35
36	63.991003	57.720824	52.217216	47.376078	43.108447	39.338179	36
37	66.883945	60.139201	54.240624	49.070472	44.528528	40.529355	37
38	69.861147	62.615993	56.302943	50.789187	45.962133	41.726176	38
39	72.925064	65.152611	58.404923	52.532572	47.409391	42.928670	39
40	76.078221	67.750501	60.547325	54.300982	48.870433	44.136863	40
45	93.276281	81.712080	71.893414	63.531157	56.387032	50.264262	45
50	113.129122	97.443289	84.373248	73.443077	64.268447	56.538243	50
55	136.046549	115.168430	98.100116	84.087097	72.532385	62.962312	55
60	162.501629	135.140233	113.198625	95.501790	81.197411	69.540060	60
65	193.040450	157.643462	129.805836	107.791723	90.282991	76.275163	65
70	228.293401	182.598976	148.072504	120.972751	99.809539	83.171386	70
75	268.988178	211.568303	168.164449	135.127338	109.798454	90.232583	75
80	315.964811	243.758752	190.264058	150.327389	120.272180	97.462701	80
90	432.792126	320.897366	241.308772	184.178464	142.769322	112.445958	90
100	588.471401	418.829847	303.064328	223.214725	167.503246	128.154659	100

TABLE D-73

Factors (EPW) to Convert an Exponential Growth to Present Worth for Interest Rates from 6% to 8½%

Growth Rate of 6% per Period

n	6%	6½%	7%	7½%	8%	8½%	n
1	1.000000	0.995305	0.990654	0.986047	0.981481	0.976959	1
2	2.000000	1.985938	1.972050	1.958334	1.944787	1.931406	2
3	3.000000	2.971919	2.944274	2.917055	2.890254	2.863863	3
4	4.000000	3.953272	3.907411	3.862399	3.818213	3.774833	4
5	5.000000	4.930017	4.861548	4.794551	4.728986	4.664814	5
6	6.000000	5.902176	5.806767	5.713697	5.622894	5.534289	6
7	7.000000	6.869772	6.743152	6.620017	6.500248	6.383729	7
8	8.000000	7.832824	7.670786	7.513692	7.361354	7.213597	8
9	9.000000	8.791356	8.589751	8.394896	8.206514	8.024344	9
10	10.000000	9.745387	9.500127	9.263804	9.036023	8.816409	10
11	11.000000	10.694939	10.401995	10.120588	9.850171	9.590225	11
12	12.000000	11.640033	11.295434	10.965417	10.649242	10.346210	12
13	13.000000	12.580690	12.180524	11.798458	11.433515	11.084777	13
14	14.000000	13.516931	13.057341	12.619875	12.203265	11.806326	14
15	15.000000	14.448777	13.925964	13.429830	12.958760	12.511249	15
16	16.000000	15.376247	14.786469	14.228484	13.700265	13.199930	16
17	17.000000	16.299363	15.638932	15.015993	14.428038	13.872743	17
18	18.000000	17.218146	16.483428	15.792514	15.142333	14.530053	18
19	19.000000	18.132615	17.320032	16.558200	15.843401	15.172218	19
20	20.000000	19.042790	18.148816	17.313202	16.531486	15.799586	20
21	21.000000	19.948693	18.969856	18.057669	17.206829	16.412499	21
22	22.000000	20.850342	19.783221	18.791748	17.869666	17.011289	22
23	23.000000	21.747758	20.588986	19.515584	18.520227	17.596282	23
24	24.000000	22.640961	21.387219	20.229320	19.158742	18.167797	24
25	25.000000	23.529971	22.177993	20.933097	19.785432	18.726142	25
26	26.000000	24.414806	22.961376	21.627054	20.400516	19.271623	26
27	27.000000	25.295488	23.737438	22.311328	21.004210	19.804535	27
28	28.000000	26.172035	24.506247	22.986053	21.596725	20.325168	28
29	29.000000	27.044467	25.267871	23.651364	22.178267	20.833804	29
30	30.000000	27.912803	26.022377	24.307392	22.749040	21.330721	30
31	31.000000	28.777062	26.769831	24.954265	23.309243	21.816189	31
32	32.000000	29.637263	27.510300	25.592113	23.859072	22.290470	32
33	33.000000	30.493427	28.243849	26.221060	24.398719	22.753823	33
34	34.000000	31.345570	28.970542	26.841231	24.928372	23.206500	34
35	35.000000	32.193713	29.690443	27.452749	25.448217	23.648747	35
36	36.000000	33.037874	30.403617	28.055734	25.958435	24.080803	36
37	37.000000	33.878072	31.110125	28.650305	26.459205	24.502904	37
38	38.000000	34.714325	31.810030	29.236580	26.950701	24.915280	38
39	39.000000	35.546652	32.503395	29.814674	27.433095	25.318154	39
40	40.000000	36.375071	33.190279	30.384702	27.906557	25.711745	40
45	45.000000	40.459193	36.529599	33.117729	30.145548	27.547773	45
50	50.000000	44.448340	39.715766	35.665328	32.184763	29.181803	50
55	55.000000	48.344719	42.755803	38.040078	34.042027	30.636059	55
60	60.000000	52.150490	45.656413	40.253708	35.733574	31.930318	60
65	65.000000	55.867757	48.423991	42.317169	37.274191	33.082184	65
70	70.000000	59.498581	51.064637	44.240591	38.677344	34.107323	70
75	75.000000	63.044970	53.584174	46.033532	39.955300	35.019677	75
80	80.000000	66.508889	55.988156	47.704827	41.119228	35.831654	80
90	90.000000	73.196942	60.470409	50.714930	43.144793	37.197438	90
100	100.000000	79.577553	64.550946	53.330435	44.825017	38.279228	100

TABLE D-74

Factors (EPW) to Convert an Exponential Growth to Present Worth for Interest Rates from 9% to 20%

Growth Rate of 6% per Period

n	9%	10%	11%	12%	15%	20%	n
1	0.972477	0.963636	0.954955	0.946429	0.921739	0.883333	1
2	1.918189	1.892231	1.866894	1.842156	1.771342	1.663611	2
3	2.837872	2.787059	2.737755	2.689897	2.554455	2.352856	3
4	3.732242	3.649348	3.569387	3.492224	3.276280	2.961690	4
5	4.601997	4.480281	4.363559	4.251569	3.941614	3.499493	5
6	5.447813	5.280998	5.121957	4.970235	4.554879	3.974552	6
7	6.270351	6.052598	5.846193	5.650401	5.120150	4.394188	7
8	7.070249	6.796140	6.537806	6.294130	5.641181	4.764866	8
9	7.848132	7.512644	7.198265	6.903373	6.121437	5.092298	9
10	8.604606	8.203093	7.828974	7.479978	6.564107	5.381530	10
11	9.340259	8.868435	8.431273	8.025693	6.972133	5.637018	11
12	10.055665	9.509583	9.006441	8.542174	7.348227	5.862699	12
13	10.751380	10.127416	9.555700	9.030986	7.694888	6.062051	13
14	11.427948	10.722783	10.080218	9.493612	8.014418	6.238145	14
15	12.085894	11.296500	10.581109	9.931454	8.308942	6.393695	15
16	12.725732	11.849355	11.059438	10.345840	8.580416	6.531097	16
17	13.347959	12.382105	11.516220	10.738028	8.830644	6.652469	17
18	13.953062	12.895483	11.952426	11.109205	9.061290	6.759681	18
19	14.541509	13.390193	12.368983	11.460497	9.273884	6.854385	19
20	15.113761	13.866913	12.766777	11.792971	9.469841	6.938040	20
21	15.670263	14.326298	13.146652	12.107633	9.650462	7.011935	21
22	16.211449	14.768978	13.509415	12.405438	9.816948	7.077210	22
23	16.737739	15.195561	13.855838	12.687290	9.970404	7.134868	23
24	17.249545	15.606631	14.186656	12.954042	10.111851	7.185800	24
25	17.747263	16.002754	14.502572	13.206504	10.242228	7.230790	25
26	18.231284	16.384472	14.804258	13.445441	10.362401	7.270532	26
27	18.701982	16.752309	15.092355	13.671579	10.473170	7.305636	27
28	19.159726	17.106771	15.367474	13.885601	10.575270	7.336645	28
29	19.604871	17.448343	15.630200	14.088158	10.669379	7.364037	29
30	20.037765	17.777494	15.881092	14.279864	10.756123	7.388232	30
31	20.458744	18.094676	16.120683	14.461300	10.836079	7.409605	31
32	20.868136	18.400324	16.349481	14.633016	10.909777	7.428485	32
33	21.266261	18.694858	16.567973	14.795533	10.977707	7.445161	33
34	21.653428	18.978681	16.776623	14.949344	11.040322	7.459893	34
35	22.029939	19.252184	16.975874	15.094915	11.098036	7.472905	35
36	22.396087	19.515741	17.166150	15.232687	11.151233	7.484400	36
37	22.752158	19.769714	17.347855	15.363079	11.200267	7.494553	37
38	23.098429	20.014451	17.521375	15.486485	11.245463	7.503522	38
39	23.435170	20.250290	17.687079	15.603281	11.287123	7.511444	39
40	23.762642	20.477552	17.845318	15.713819	11.325522	7.518442	40
45	25.269669	21.495750	18.535807	16.183781	11.476876	7.542932	45
50	26.580413	22.341805	19.084173	16.540645	11.577578	7.556103	50
55	27.720439	23.044819	19.519670	16.811628	11.644578	7.563187	55
60	28.711982	23.628977	19.865529	17.017397	11.689155	7.566996	60
65	29.574381	24.114373	20.140201	17.173647	11.718814	7.569045	65
70	30.324457	24.517704	20.358338	17.292294	11.738547	7.570147	70
75	30.976839	24.852845	20.531576	17.382389	11.751676	7.570739	75
80	31.544251	25.131325	20.669156	17.450802	11.760412	7.571058	80
90	32.466994	25.554999	20.865192	17.542198	11.770090	7.571321	90
100	33.165023	25.847525	20.988834	17.594897	11.774375	7.571398	100

Exponential Growth Factors

TABLE D-75
Factors (EPW) to Convert an Exponential Growth to Present Worth for Interest Rates from 25% to 50%
Growth Rate of 6% per Period

n	25%	30%	35%	40%	45%	50%	n
1	0.848000	0.815385	0.785185	0.757143	0.731034	0.706667	1
2	1.567104	1.480237	1.401701	1.330408	1.265446	1.206044	2
3	2.176904	2.022347	1.885780	1.764452	1.656119	1.558938	3
4	2.694015	2.464375	2.265872	2.093085	1.941715	1.808316	4
5	3.132525	2.824798	2.564314	2.341907	2.150495	1.984543	5
6	3.504381	3.118682	2.798647	2.530301	2.303120	2.109077	6
7	3.819715	3.358310	2.982641	2.672942	2.414695	2.197081	7
8	4.087118	3.553699	3.127111	2.780942	2.496260	2.259271	8
9	4.313876	3.713016	3.240546	2.862713	2.555886	2.303218	9
10	4.506167	3.842921	3.329614	2.924626	2.599476	2.334274	10
11	4.669230	3.948843	3.399549	2.971502	2.631341	2.356220	11
12	4.807507	4.035210	3.454461	3.006995	2.654635	2.371729	12
13	4.924766	4.105633	3.497576	3.033867	2.671664	2.382689	13
14	5.024201	4.163055	3.531430	3.054214	2.684113	2.390433	14
15	5.108523	4.209875	3.558012	3.069619	2.693214	2.395906	15
16	5.180027	4.248052	3.578884	3.081283	2.699867	2.399774	16
17	5.240663	4.279181	3.595271	3.090114	2.704730	2.402507	17
18	5.292082	4.304563	3.608139	3.096801	2.708285	2.404438	18
19	5.335686	4.325259	3.618243	3.101863	2.710885	2.405803	19
20	5.372662	4.342134	3.626176	3.105697	2.712785	2.406767	20
21	5.404017	4.355894	3.632405	3.108599	2.714174	2.407449	21
22	5.430606	4.367114	3.637295	3.110796	2.715189	2.407931	22
23	5.453154	4.376262	3.641136	3.112460	2.715931	2.408271	23
24	5.472275	4.383721	3.644151	3.113720	2.716474	2.408511	24
25	5.488489	4.389803	3.646519	3.114674	2.716871	2.408681	25
26	5.502239	4.394763	3.648378	3.115396	2.717161	2.408802	26
27	5.513898	4.398807	3.649837	3.115942	2.717373	2.408886	27
28	5.523786	4.402104	3.650983	3.116356	2.717528	2.408946	28
29	5.532170	4.404792	3.651883	3.116670	2.717641	2.408989	29
30	5.539281	4.406985	3.652590	3.116907	2.717724	2.409019	30
31	5.545310	4.408772	3.653145	3.117087	2.717784	2.409040	31
32	5.550423	4.410229	3.653580	3.117223	2.717828	2.409055	32
33	5.554759	4.411418	3.653922	3.117326	2.717861	2.409065	33
34	5.558435	4.412387	3.654191	3.117404	2.717884	2.409073	34
35	5.561553	4.413177	3.654402	3.117463	2.717902	2.409078	35
36	5.564197	4.413821	3.654567	3.117508	2.717914	2.409082	36
37	5.566439	4.414347	3.654697	3.117542	2.717924	2.409085	37
38	5.568340	4.414775	3.654799	3.117567	2.717930	2.409086	38
39	5.569953	4.415124	3.654879	3.117587	2.717935	2.409088	39
40	5.571320	4.415409	3.654942	3.117601	2.717939	2.409089	40
45	5.575603	4.416213	3.655104	3.117636	2.717947	2.409091	45
50	5.577481	4.416503	3.655152	3.117644	2.717948	2.409091	50
55	5.578304	4.416608	3.655166	3.117646	2.717949	2.409091	55
60	5.578665	4.416645	3.655171	3.117647	2.717949	2.409091	60
65	5.578824	4.416659	3.655172	3.117647	2.717949	2.409091	65
70	5.578893	4.416664	3.655172	3.117647	2.717949	2.409091	70
75	5.578924	4.416666	3.655172	3.117647	2.717949	2.409091	75
80	5.578937	4.416666	3.655172	3.117647	2.717949	2.409091	80
90	5.578945	4.416667	3.655172	3.117647	2.717949	2.409091	90
100	5.578947	4.416667	3.655172	3.117647	2.717949	2.409091	100

TABLE D-76

Factors (EPW) to Convert an Exponential Growth to Present Worth for Interest Rates from 1% to 2½%

Growth Rate of 7% per Period

n	1%	1¼%	1½%	1¾%	2%	2½%	n
1	1.059406	1.056790	1.054187	1.051597	1.049020	1.043902	1
2	2.181747	2.173595	2.165498	2.157453	2.149462	2.133635	2
3	3.370762	3.353824	3.337027	3.320369	3.303847	3.271209	3
4	4.630411	4.601079	4.572039	4.543287	4.514820	4.458725	4
5	5.964891	5.919165	5.873972	5.829304	5.785154	5.698377	5
6	7.378646	7.312105	7.246453	7.181676	7.117760	6.992452	6
7	8.876388	8.784150	8.693305	8.603827	8.515689	8.343340	7
8	10.463104	10.339793	10.218558	10.099356	9.982145	9.753535	8
9	12.144080	11.983782	11.826460	11.672050	11.520485	11.225642	9
10	13.924917	13.721132	13.521490	13.325890	13.134234	12.762377	10
11	15.811546	15.557147	15.309369	15.065064	14.827089	14.366579	11
12	17.810251	17.497429	17.192074	16.893974	16.602927	16.041210	12
13	19.927692	19.547901	19.177851	18.817250	18.465815	17.789360	13
14	22.170921	21.714819	21.271232	20.839762	20.420022	19.614259	14
15	24.547412	24.004796	23.478048	22.966629	22.470023	21.519275	15
16	27.065080	26.424821	25.804444	25.203236	24.620514	23.507926	16
17	29.732312	28.982280	28.256902	27.555246	26.876422	25.583884	17
18	32.557994	31.684978	30.842251	30.028613	29.242913	27.750981	18
19	35.551538	34.541162	33.567693	32.629598	31.725409	30.013220	19
20	38.722917	37.559549	36.440820	35.364786	34.329596	32.374776	20
21	42.082694	40.749350	39.469632	38.241101	37.061439	34.840010	21
22	45.642062	44.120301	42.662568	41.265827	39.927195	37.413474	22
23	49.412877	47.682688	46.028520	44.446619	42.933430	40.099919	23
24	53.407702	51.447384	49.576864	47.791530	46.087030	42.904305	24
25	57.639842	55.425878	53.317482	51.309029	49.395218	45.831811	25
26	62.123397	59.630310	57.260794	55.008021	52.865571	48.887842	26
27	66.873302	64.073513	61.417783	58.897870	56.506041	52.078040	27
28	71.905380	68.769046	65.800027	62.988423	60.324964	55.408296	28
29	77.236392	73.731239	70.419733	67.290037	64.331090	58.884757	29
30	82.884099	78.975235	75.289768	71.813602	68.533594	62.513844	30
31	88.867313	84.517038	80.423696	76.570569	72.942104	66.302257	31
32	95.205965	90.373562	85.835817	81.572981	77.566717	70.256990	32
33	101.921171	96.562677	91.541206	86.833504	82.418026	74.385346	33
34	109.035300	103.103274	97.555755	92.365454	87.507145	78.694946	34
35	116.572050	110.015312	103.896214	98.182836	92.845731	83.193749	35
36	124.556528	117.319885	110.580246	104.300378	98.446012	87.890060	36
37	133.015332	125.039286	117.626466	110.733567	104.320816	92.792550	37
38	141.976639	133.197072	125.054501	117.498689	110.483602	97.910272	38
39	151.470300	141.818141	132.885040	124.612872	116.948484	103.252674	39
40	161.527942	150.928801	141.139895	132.094126	123.730272	108.829621	40
45	221.520674	204.856295	189.628676	175.704631	162.963965	140.609164	45
50	301.579766	275.937368	252.757827	231.788493	212.803871	180.004719	50
55	408.417010	369.628331	334.947759	303.913308	276.117214	228.841471	55
60	550.989160	493.121072	441.953555	396.667066	356.546327	289.382018	60
65	741.248820	655.895119	581.267945	515.949999	458.718188	364.431190	65
70	995.146503	870.445302	762.645964	669.349910	588.510609	457.465997	70
75	1333.967843	1153.240882	998.788015	866.624845	753.390376	572.796722	75
80	1786.118093	1525.989762	1306.229175	1120.323817	962.842802	715.766613	80
90	3194.709142	2664.900296	2227.618951	1866.162685	1566.921104	1112.706338	90
100	5703.179442	4643.587148	3789.402139	3099.661140	2541.752025	1722.697705	100

Exponential Growth Factors

TABLE D-77

Factors (EPW) to Convert an Exponential Growth
to Present Worth for Interest Rates from 3% to 5½%
Growth Rate of 7% per Period

n	3%	3½%	4%	4½%	5%	5½%	n
1	1.038835	1.033816	1.028846	1.023923	1.019048	1.014218	1
2	2.118013	2.102593	2.087371	2.072343	2.057506	2.042856	2
3	3.239101	3.207511	3.176429	3.145844	3.115744	3.086120	3
4	4.403726	4.349794	4.296903	4.245027	4.194139	4.144216	4
5	5.613580	5.530705	5.449699	5.370506	5.293075	5.217357	5
6	6.870418	6.751550	6.635748	6.522910	6.412943	6.305755	6
7	8.176065	8.013680	7.856009	7.702884	7.554142	7.409628	7
8	9.532417	9.318491	9.111471	8.911087	8.717078	8.529196	8
9	10.941443	10.667425	10.403148	10.148194	9.902165	9.664683	9
10	12.405188	12.061976	11.732085	11.414898	11.109826	10.816313	10
11	13.925778	13.503685	13.099357	12.711905	12.340489	11.984318	11
12	15.505420	14.994148	14.506069	14.039941	13.594593	13.168929	12
13	17.146407	16.535013	15.953360	15.399748	14.872586	14.370383	13
14	18.851122	18.127984	17.442399	16.792086	16.174921	15.588919	14
15	20.622039	19.774824	18.974391	18.217734	17.502062	16.824780	15
16	22.461730	21.477354	20.550576	19.677489	18.854482	18.078213	16
17	24.372865	23.237458	22.172227	21.172165	20.232663	19.349468	17
18	26.358219	25.057082	23.840656	22.702600	21.637094	20.638796	18
19	28.420674	26.938240	25.557214	24.269648	23.068277	21.946457	19
20	30.563225	28.883011	27.323287	25.874185	24.526721	23.272710	20
21	32.788981	30.893548	29.140305	27.517108	26.012944	24.617820	21
22	35.101175	32.972074	31.009737	29.199336	27.527476	25.982054	22
23	37.503162	35.120888	32.933095	30.921808	29.070857	27.365685	23
24	39.998430	37.342367	34.911934	32.685487	30.643635	28.768989	24
25	42.590602	39.638969	36.947855	34.491360	32.246371	30.192244	25
26	45.283441	42.013233	39.042505	36.340436	33.879635	31.635736	26
27	48.080857	44.467787	41.197577	38.233748	35.544009	33.099751	27
28	50.986909	47.005345	43.414815	40.172354	37.240085	34.584582	28
29	54.005818	49.628714	45.696012	42.157339	38.968468	36.090524	29
30	57.141967	52.340796	48.043012	44.189811	40.729772	37.617877	30
31	60.399907	55.144592	50.457714	46.270907	42.524625	39.166947	31
32	63.784370	58.043201	52.942071	48.401790	44.353665	40.738041	32
33	67.300267	61.039831	55.498093	50.583651	46.217545	42.331472	33
34	70.952705	64.137796	58.127845	52.817709	48.116926	43.947650	34
35	74.746985	67.340524	60.833456	55.105214	50.052487	45.586625	35
36	78.688615	70.651556	63.617114	57.447444	52.024915	47.248994	36
37	82.783319	74.074555	66.481069	59.845709	54.034914	48.934998	37
38	87.037040	77.613308	69.427638	62.301347	56.083198	50.644975	38
39	91.455954	81.271729	72.459205	64.815734	58.170497	52.379263	39
40	96.046477	85.053865	75.578220	67.390273	60.297554	54.138210	40
45	121.815740	105.971744	92.575914	81.216850	71.556226	63.315263	45
50	152.992772	130.673895	112.170776	96.778373	83.928787	73.163530	50
55	190.712412	159.844935	134.759641	114.292542	97.525439	83.732105	55
60	236.347648	194.293335	160.799977	134.004373	112.467289	95.073671	60
65	291.559595	234.973827	190.819154	156.189628	128.887425	107.244764	65
70	358.357960	283.013864	225.425120	181.158668	146.932104	120.306058	70
75	439.174191	339.744869	265.318714	209.260801	166.762051	134.322661	75
80	536.949987	406.735132	311.307862	240.889164	188.553893	149.364446	80
90	798.363237	579.280278	425.440612	316.549976	238.818874	182.828973	90
100	1181.005668	819.897639	577.115844	412.389770	299.521896	221.367746	100

TABLE D-78

FACTORS (EPW) TO CONVERT AN EXPONENTIAL GROWTH
TO PRESENT WORTH FOR INTEREST RATES FROM 6% TO 8½%
Growth Rate of 7% per Period

n	6%	6½%	7%	7½%	8%	8½%	n
1	1.009434	1.004695	1.000000	0.995349	0.990741	0.986175	1
2	2.028391	2.014107	2.000000	1.986068	1.972308	1.958716	2
3	3.056961	3.028257	3.000000	2.972179	2.944787	2.917813	3
4	4.095234	4.047169	4.000000	3.953704	3.908261	3.863649	4
5	5.143302	5.070865	5.000000	4.930664	4.862814	4.796410	5
6	6.201258	6.095367	6.000000	5.903079	5.808529	5.716275	6
7	7.269194	7.132697	7.000000	6.870972	6.745487	6.623423	7
8	8.347205	8.170879	8.000000	7.834363	7.673769	7.518030	8
9	9.435387	9.213934	9.000000	8.793273	8.593457	8.400270	9
10	10.533834	10.261887	10.000000	9.747723	9.504628	9.270312	10
11	11.642643	11.314760	11.000000	10.697733	10.407363	10.128326	11
12	12.761914	12.372576	12.000000	11.643325	11.301739	10.974478	12
13	13.891743	13.435358	13.000000	12.584519	12.187834	11.808933	13
14	15.032231	14.503129	14.000000	13.521335	13.065725	12.631851	14
15	16.183479	15.575914	15.000000	14.453794	13.935487	13.443392	15
16	17.345587	16.653735	16.000000	15.381916	14.797195	14.243714	16
17	18.518658	17.736617	17.000000	16.305721	15.650925	15.032971	17
18	19.702797	18.824582	18.000000	17.225229	16.496750	15.811317	18
19	20.898106	19.917655	19.000000	18.140461	17.334743	16.578902	19
20	22.104692	21.015860	20.000000	19.051435	18.164976	17.335876	20
21	23.322661	22.119221	21.000000	19.958173	18.987523	18.082385	21
22	24.552120	23.227762	22.000000	20.860693	19.802453	18.818573	22
23	25.793178	24.341507	23.000000	21.759015	20.609838	19.544584	23
24	27.045943	25.460481	24.000000	22.653159	21.409747	20.260557	24
25	28.310528	26.584709	25.000000	23.543145	22.202249	20.966632	25
26	29.587042	27.714215	26.000000	24.428991	22.987414	21.662946	26
27	30.875599	28.849023	27.000000	25.310716	23.765308	22.349633	27
28	32.176312	29.989159	28.000000	26.188341	24.536000	23.026828	28
29	33.489297	31.134648	29.000000	27.061883	25.299555	23.694659	29
30	34.814667	32.285515	30.000000	27.931363	26.056041	24.353259	30
31	36.152541	33.441785	31.000000	28.796799	26.805522	25.002753	31
32	37.503037	34.603484	32.000000	29.658209	27.548063	25.643268	32
33	38.866273	35.770636	33.000000	30.515612	28.283729	26.274928	33
34	40.242370	36.943268	34.000000	31.369028	29.012584	26.897855	34
35	41.631449	38.121406	35.000000	32.218475	29.734689	27.512170	35
36	43.033633	39.305074	36.000000	33.063970	30.450109	28.117993	36
37	44.449044	40.494300	37.000000	33.905533	31.158904	28.715440	37
38	45.877809	41.689109	38.000000	34.743182	31.861137	29.304627	38
39	47.320052	42.889527	39.000000	35.576934	32.556867	29.885669	39
40	48.775902	44.095582	40.000000	36.406809	33.246155	30.458678	40
45	56.263776	50.211319	45.000000	40.498642	36.598036	33.207065	45
50	64.111579	56.471973	50.000000	44.496198	39.797584	35.770652	50
55	72.336612	62.880977	55.000000	48.401647	42.851723	38.161865	55
60	80.957008	69.441846	60.000000	52.217114	45.767060	40.392295	60
65	89.991771	76.158179	65.000000	55.944670	48.549905	42.472751	65
70	99.460820	83.033660	70.000000	59.586341	51.206278	44.413319	70
75	109.385029	90.072060	75.000000	63.144107	53.741928	46.223405	75
80	119.786277	97.277239	80.000000	66.619900	56.162340	47.911781	80
90	142.112715	112.203837	90.000000	73.333079	60.678163	50.955591	90
100	166.637126	127.846206	100.000000	79.740472	64.792853	53.603834	100

Exponential Growth Factors

TABLE D-79
Factors (EPW) to Convert an Exponential Growth to Present Worth for Interest Rates from 9% to 20%
Growth Rate of 7% per Period

n	9%	10%	11%	12%	15%	20%	n
1	0.961651	0.972727	0.963964	0.955357	0.930435	0.891667	1
2	1.945291	1.918926	1.893190	1.868064	1.796144	1.686736	2
3	2.891249	2.839319	2.788931	2.740026	2.601629	2.395673	3
4	3.819850	3.734610	3.652393	3.573060	3.351081	3.027808	4
5	4.731412	4.605484	4.484739	4.368906	4.048397	3.591463	5
6	5.626249	5.452607	5.287091	5.129223	4.697204	4.094054	6
7	6.504666	6.276627	6.060529	5.855597	5.300877	4.542198	7
8	7.366966	7.078174	6.806096	6.549543	5.862555	4.941793	8
9	8.213443	7.857860	7.524795	7.212510	6.385160	5.298099	9
10	9.044389	8.616282	8.217595	7.845880	6.871410	5.615805	10
11	9.860089	9.354020	8.885430	8.450975	7.323834	5.899093	11
12	10.660821	10.071637	9.529198	9.029056	7.744784	6.151691	12
13	11.446861	10.769683	10.149767	9.581330	8.136451	6.376925	13
14	12.218478	11.448692	10.747974	10.108950	8.500872	6.577758	14
15	12.975937	12.109182	11.324624	10.613014	8.839942	6.756834	15
16	13.719498	12.751659	11.880493	11.094576	9.155424	6.916510	16
17	14.449416	13.376614	12.416331	11.554640	9.448960	7.058888	17
18	15.165940	13.984524	12.932860	11.994165	9.722076	7.185842	18
19	15.869317	14.575856	13.430775	12.414068	9.976192	7.299043	19
20	16.559789	15.151060	13.910747	12.815226	10.212631	7.399980	20
21	17.237591	15.710576	14.373423	13.198475	10.432622	7.489982	21
22	17.902956	16.254833	14.819425	13.564614	10.637309	7.570234	22
23	18.556113	16.784247	15.249356	13.914408	10.827757	7.641792	23
24	19.197285	17.299222	15.663794	14.248586	11.004957	7.705598	24
25	19.826693	17.800152	16.063297	14.567946	11.169829	7.762491	25
26	20.444551	18.287421	16.448403	14.872853	11.323232	7.813221	26
27	21.051073	18.761400	16.819632	15.164243	11.465964	7.858456	27
28	21.646467	19.222453	17.177483	15.442625	11.598767	7.898790	28
29	22.230935	19.670931	17.522438	15.708580	11.722331	7.934754	29
30	22.804679	20.107179	17.854963	15.962661	11.837299	7.966822	30
31	23.367896	20.531528	18.175505	16.205399	11.944269	7.995417	31
32	23.920779	20.944305	18.484496	16.437301	12.043799	8.020913	32
33	24.463517	21.345624	18.782352	16.658850	12.136404	8.043648	33
34	24.996296	21.736392	19.069474	16.870509	12.222567	8.063919	34
35	25.519300	22.116309	19.346250	17.072718	12.302736	8.081995	35
36	26.032707	22.485864	19.613052	17.265900	12.377329	8.098112	36
37	26.536694	22.845341	19.870239	17.450458	12.446732	8.112483	37
38	27.031434	23.195013	20.118158	17.626777	12.511307	8.125297	38
39	27.517096	23.535149	20.357144	17.795225	12.571390	8.136723	39
40	27.993846	23.866009	20.587517	17.956152	12.627293	8.146912	40
45	30.249550	25.389777	21.620681	18.659229	12.853612	8.183503	45
50	32.305764	26.716788	22.480631	19.218770	13.011428	8.204127	50
55	34.180131	27.872448	23.196407	19.664078	13.121475	8.215752	55
60	35.886734	28.878882	23.792180	20.018474	13.198213	8.222305	60
65	37.446233	29.755359	24.288070	20.300519	13.251724	8.225998	65
70	38.865989	30.518661	24.700822	20.524983	13.289037	8.228080	70
75	40.160186	31.183401	25.044374	20.703621	13.315057	8.229253	75
80	41.339928	31.762306	25.330329	20.845790	13.333201	8.229915	80
90	43.395637	32.705514	25.766452	21.048980	13.354675	8.230498	90
100	45.103820	33.420863	26.068598	21.177675	13.365117	8.230683	100

TABLE D-80

Factors (EPW) to Convert an Exponential Growth to Present Worth for Interest Rates from 25% to 50%

Growth Rate of 7% per Period

n	25%	30%	35%	40%	45%	50%	n
1	0.856000	0.823077	0.792593	0.764286	0.737931	0.713333	1
2	1.588736	1.500533	1.420796	1.349418	1.282473	1.222178	2
3	2.215958	2.058131	1.918705	1.794863	1.684308	1.585153	3
4	2.752860	2.517077	2.313344	2.136074	1.980834	1.844076	4
5	3.212448	2.894825	2.626132	2.396856	2.199650	2.028774	5
6	3.605856	3.205740	2.874045	2.596169	2.361121	2.160526	6
7	3.942612	3.461648	3.070539	2.748500	2.480276	2.254508	7
8	4.230876	3.672279	3.226279	2.864925	2.568203	2.321549	8
9	4.477630	3.845645	3.349718	2.953907	2.633088	2.369372	9
10	4.688851	3.988339	3.447554	3.021915	2.680968	2.403485	10
11	4.869657	4.105787	3.525098	3.073892	2.716301	2.427819	11
12	5.024426	4.202455	3.586559	3.113617	2.742374	2.445178	12
13	5.156909	4.282021	3.635273	3.143979	2.761614	2.457560	13
14	5.270314	4.347509	3.673883	3.167184	2.775811	2.466393	14
15	5.367389	4.401412	3.704485	3.184919	2.786288	2.472694	15
16	5.450485	4.445777	3.728740	3.198474	2.794020	2.477188	16
17	5.521615	4.482294	3.747964	3.208834	2.799725	2.480394	17
18	5.582502	4.512349	3.763201	3.216751	2.803935	2.482681	18
19	5.634622	4.537088	3.775278	3.222803	2.807042	2.484313	19
20	5.679236	4.557449	3.784850	3.227428	2.809334	2.485476	20
21	5.717426	4.574208	3.792437	3.230963	2.811026	2.486306	21
22	5.750117	4.588002	3.798450	3.233664	2.812274	2.486899	22
23	5.778100	4.599355	3.803216	3.235729	2.813196	2.487321	23
24	5.802054	4.608700	3.806993	3.237307	2.813875	2.487622	24
25	5.822558	4.616392	3.809987	3.238513	2.814377	2.487837	25
26	5.840110	4.622722	3.812360	3.239435	2.814747	2.487991	26
27	5.855134	4.627933	3.814241	3.240140	2.815020	2.488100	27
28	5.867995	4.632222	3.815732	3.240678	2.815222	2.488178	28
29	5.879003	4.635752	3.816913	3.241090	2.815371	2.488234	29
30	5.888427	4.638657	3.817850	3.241404	2.815480	2.488273	30
31	5.896493	4.641049	3.818592	3.241645	2.815561	2.488302	31
32	5.903398	4.643017	3.819180	3.241829	2.815621	2.488322	32
33	5.909309	4.644637	3.819647	3.241969	2.815665	2.488336	33
34	5.914369	4.645971	3.820016	3.242076	2.815698	2.488347	34
35	5.918699	4.647068	3.820309	3.242158	2.815722	2.488354	35
36	5.922407	4.647971	3.820541	3.242221	2.815740	2.488359	36
37	5.925580	4.648715	3.820725	3.242269	2.815753	2.488363	37
38	5.928297	4.649327	3.820871	3.242306	2.815762	2.488365	38
39	5.930622	4.649831	3.820987	3.242333	2.815769	2.488367	39
40	5.932612	4.650245	3.821078	3.242355	2.815775	2.488369	40
45	5.939007	4.651445	3.821319	3.242406	2.815786	2.488371	45
50	5.941945	4.651899	3.821394	3.242420	2.815789	2.488372	50
55	5.943296	4.652070	3.821418	3.242423	2.815789	2.488372	55
60	5.943917	4.652135	3.821425	3.242424	2.815789	2.488372	60
65	5.944202	4.652159	3.821428	3.242424	2.815789	2.488372	65
70	5.944333	4.652168	3.821428	3.242424	2.815789	2.488372	70
75	5.944393	4.652172	3.821428	3.242424	2.815789	2.488372	75
80	5.944421	4.652173	3.821429	3.242424	2.815789	2.488372	80
90	5.944439	4.652174	3.821429	3.242424	2.815789	2.488372	90
100	5.944443	4.652174	3.821429	3.242424	2.815789	2.488372	100

Exponential Growth Factors

TABLE D-81
Factors (EPW) to Convert an Exponential Growth to Present Worth for Interest Rates from 1% to 2½%
Growth Rate of 8% per Period

n	1%	1¼%	1½%	1¾%	2%	2½%	n
1	1.069307	1.066667	1.064039	1.061425	1.058824	1.053659	1
2	2.212724	2.204444	2.196219	2.188048	2.179931	2.163855	2
3	3.435388	3.418074	3.400903	3.383874	3.366986	3.333623	3
4	4.742791	4.712612	4.682735	4.653154	4.623867	4.566159	4
5	6.140807	6.093453	6.046653	6.000399	5.954683	5.864830	5
6	7.635714	7.566350	7.497917	7.430399	7.363782	7.233187	6
7	9.234229	9.137440	9.042119	8.948237	8.855769	8.674968	7
8	10.943532	10.813269	10.685210	10.559308	10.435520	10.194113	8
9	12.771301	12.600821	12.433524	12.269339	12.108198	11.794772	9
10	14.725748	14.507542	14.293799	14.084409	13.879268	13.481321	10
11	16.815651	16.541378	16.273205	16.010970	15.754519	15.258368	11
12	19.050400	18.710803	18.379370	18.055870	17.740079	17.130768	12
13	21.440031	21.024857	20.620414	20.226378	19.842437	19.103638	13
14	23.995281	23.493181	23.004972	22.530210	22.068462	21.182370	14
15	26.727627	26.126059	25.542237	24.975554	24.425431	23.372644	15
16	29.649344	28.534463	28.241986	27.571104	26.921044	25.680444	16
17	32.773556	31.530094	31.114625	30.326086	29.563459	28.112078	17
18	36.114297	35.125434	34.171227	33.250293	32.361309	30.674189	18
19	39.686575	38.533796	37.423571	36.354119	35.323739	33.373780	19
20	43.506437	42.169383	40.884194	39.648598	38.460430	36.218226	20
21	47.591041	46.047341	44.566433	43.145441	41.781632	39.215302	21
22	51.958737	50.183831	48.484481	46.857077	45.298198	42.373196	22
23	56.629145	54.596086	52.653438	50.796701	49.021621	45.700538	23
24	61.623244	59.302492	57.089372	54.978317	52.964070	49.206421	24
25	66.963469	64.322658	61.809381	59.416788	57.138427	52.900424	25
26	72.673808	69.677502	66.831657	64.127893	61.558334	56.792642	26
27	78.779914	75.389336	72.175556	69.128378	66.238236	60.893710	27
28	85.309215	81.481958	77.861675	74.436018	71.193427	65.214836	28
29	92.291042	87.980755	83.911930	80.069680	76.440099	69.767828	29
30	99.756757	94.912805	90.349640	86.049390	81.995399	74.565126	30
31	107.739899	102.306993	97.199617	92.396404	87.877481	79.619840	31
32	116.276328	110.194125	104.488263	99.133284	94.105568	84.945782	32
33	125.404390	118.607067	112.243668	106.283977	100.700013	90.557507	33
34	135.165090	127.580872	120.495720	113.873902	107.682367	96.470349	34
35	145.602275	137.152930	129.276241	121.930039	115.075448	102.700465	35
36	156.762828	147.363125	138.619054	130.481024	122.903415	109.264881	36
37	168.696886	158.254000	148.560176	139.557254	131.191851	116.181533	37
38	181.458056	169.870933	159.137921	149.190992	139.967842	123.469322	38
39	195.103664	182.262329	170.393059	159.416483	149.260069	131.148164	39
40	209.695007	195.479817	182.368969	170.270075	159.098896	139.239041	40
45	299.297624	276.019965	254.783148	235.395963	217.685993	186.690663	45
50	424.563449	387.233089	353.550219	323.136512	295.654622	248.314753	50
55	599.686983	540.800710	488.260500	441.344557	399.416498	328.344227	55
60	844.512354	752.853151	671.994406	600.599795	537.504432	432.276259	60
65	1186.781955	1045.663818	922.592564	815.155670	721.274025	567.249869	65
70	1665.280077	1449.588740	1264.388142	1104.215072	965.837486	742.536308	70
75	2334.227748	2008.297099	1730.569612	1493.649002	1291.306381	970.175884	75
80	3269.426793	2779.232072	2366.403357	2018.312046	1724.445500	1265.805008	80
90	6404.645945	5313.728750	4415.454628	3677.462032	3067.993737	2148.323185	90
100	12532.264471	10146.307749	8230.113604	6688.935377	5447.512320	3636.733119	100

TABLE D-82

Factors (EPW) to Convert an Exponential Growth to Present Worth for Interest Rates from 3% to 5½%

Growth Rate of 8% per Period

n	3%	3½%	4%	4½%	5%	5½%	n
1	1.048544	1.043478	1.038462	1.033493	1.028571	1.023697	1
2	2.147988	2.132325	2.116864	2.101600	2.086531	2.071652	2
3	3.300802	3.268513	3.236743	3.205482	3.174717	3.144440	3
4	4.509579	4.454101	4.399695	4.346335	4.293995	4.242649	4
5	5.777035	5.691236	5.607376	5.525399	5.445252	5.366882	5
6	7.106017	6.982159	6.861505	6.743953	6.629402	6.517756	6
7	8.499513	8.329209	8.163871	8.003320	7.847385	7.695902	7
8	9.960654	9.734827	9.516328	9.304866	9.100167	8.901966	8
9	11.492725	11.201559	10.920802	10.650005	10.388743	10.136610	9
10	13.059168	12.732061	12.379294	12.040197	11.714136	11.400511	10
11	14.783593	14.329107	13.893882	13.476950	13.077397	12.694362	11
12	16.549787	15.995590	15.466724	14.961824	14.479608	14.018873	12
13	18.401719	17.734529	17.100059	16.496430	15.921883	15.374770	13
14	20.343550	19.549074	18.796216	18.082435	17.405365	16.762798	14
15	22.379644	21.442512	20.557608	19.721560	18.931233	18.183717	15
16	24.514579	23.418273	22.386747	21.415583	20.500697	19.638308	16
17	26.753150	25.479937	24.286238	23.166345	22.115002	21.127367	17
18	29.100391	27.631239	26.258785	24.975744	23.775431	22.651712	18
19	31.561575	29.876075	28.307200	26.845745	25.483300	24.212179	19
20	34.142234	32.218513	30.434400	28.778377	27.239966	25.809624	20
21	36.848167	34.662796	32.643415	30.775739	29.046822	27.444924	21
22	39.685457	37.213353	34.937393	32.839998	30.905303	29.118974	22
23	42.660479	39.874803	37.319600	34.973396	32.816883	30.832694	23
24	45.779920	42.651968	39.793431	37.178246	34.783080	32.587023	24
25	49.050790	45.549880	42.362409	39.456943	36.805453	34.382924	25
26	52.480440	48.573788	45.030194	41.811961	38.885609	36.221382	26
27	56.076578	51.729170	47.800596	44.245854	41.025198	38.103405	27
28	59.847285	55.021742	50.677532	46.761265	43.225918	40.030026	28
29	63.801037	58.457470	53.665129	49.360925	45.489516	42.002302	29
30	67.946718	62.042578	56.767634	52.047655	47.817788	44.021313	30
31	72.293646	65.783559	59.989466	54.824370	50.212582	46.088169	31
32	76.851590	69.687192	63.335215	57.694086	52.675798	48.204003	32
33	81.630794	73.760549	66.809646	60.659917	55.209392	50.369974	33
34	86.641997	78.011007	70.417710	63.725081	57.815375	52.587272	34
35	91.896463	82.446268	74.164545	66.892907	60.495814	54.857113	35
36	97.406000	87.074367	78.055489	70.166832	63.252838	57.180741	36
37	103.182991	91.903687	82.096084	73.550410	66.088633	59.559432	37
38	109.240417	96.942978	86.292088	77.047314	69.005451	61.994489	38
39	115.591894	102.201368	90.649476	80.661339	72.005607	64.487250	39
40	122.251695	107.688384	95.174455	84.396408	75.091481	67.039080	40
45	160.725594	138.916285	120.547717	105.034152	91.894845	80.734455	45
50	209.489513	177.549408	151.190511	129.367371	111.239835	96.131252	50
55	271.295665	225.343789	188.197219	158.057790	133.510890	113.440846	55
60	349.632165	284.471882	232.889501	191.885626	159.150598	132.900869	60
65	448.920177	357.621305	286.863502	231.770804	188.668491	154.778476	65
70	574.763295	448.117002	352.046855	278.797976	222.651172	179.374013	70
75	734.263823	560.072379	430.767524	334.246015	261.773971	207.025137	75
80	936.423613	658.576274	525.936946	399.622793	306.814367	238.111454	80
90	1517.410283	1081.904828	779.308468	567.592435	418.363649	312.349747	90
100	2450.732223	1668.590158	1148.994750	801.102414	566.210249	406.179578	100

Exponential Growth Factors

TABLE D-83
Factors (EPW) to Convert an Exponential Growth to Present Worth for Interest Rates from 6% to 8½%
Growth Rate of 8% per Period

n	6%	6½%	7%	7½%	8%	8½%	n
1	1.018868	1.014085	1.009346	1.004651	1.000000	0.995392	1
2	2.056960	2.042452	2.028125	2.013975	2.000000	1.986196	2
3	3.114638	3.085303	3.056425	3.027994	3.000000	2.972435	3
4	4.192273	4.142843	4.094335	4.046728	4.000000	3.954129	4
5	5.290240	5.215277	5.141946	5.070202	5.000000	4.931299	5
6	6.408924	6.302816	6.199347	6.098435	6.000000	5.903966	6
7	7.548715	7.405673	7.266631	7.131451	7.000000	6.872150	7
8	8.710012	8.524063	8.343889	8.169272	8.000000	7.835873	8
9	9.893219	9.658204	9.431215	9.211920	9.000000	8.795155	9
10	11.098752	10.808320	10.528703	10.259417	10.000000	9.750016	10
11	12.327030	11.974634	11.636448	11.311786	11.000000	10.700476	11
12	13.578484	13.157376	12.754546	12.369050	12.000000	11.646557	12
13	14.853549	14.356775	13.883093	13.431232	13.000000	12.588278	13
14	16.152673	15.573068	15.022187	14.498354	14.000000	13.525659	14
15	17.476308	16.806491	16.171927	15.570439	15.000000	14.458721	15
16	18.824918	18.057287	17.332413	16.647511	16.000000	15.387482	16
17	20.198973	19.325700	18.503744	17.729593	17.000000	16.311964	17
18	21.598954	20.611977	19.686022	18.816707	18.000000	17.232185	18
19	23.025349	21.916371	20.879349	19.908878	19.000000	18.148166	19
20	24.478657	23.239137	22.083829	21.006128	20.000000	19.059926	20
21	25.959387	24.580533	23.299566	22.108482	21.000000	19.967484	21
22	27.468054	25.940822	24.526664	23.215964	22.000000	20.870859	22
23	29.005188	27.320271	25.765231	24.328596	23.000000	21.770072	23
24	30.571323	28.719148	27.015374	25.446404	24.000000	22.665141	24
25	32.167008	30.137727	28.277200	26.569410	25.000000	23.556085	25
26	33.792801	31.576287	29.550818	27.697640	26.000000	24.442923	26
27	35.449269	33.035108	30.836340	28.831117	27.000000	25.325675	27
28	37.136991	34.514475	32.133876	29.969867	28.000000	26.204358	28
29	38.856557	36.014679	33.443538	31.113913	29.000000	27.078993	29
30	40.608567	37.536013	34.765440	32.263280	30.000000	27.949596	30
31	42.393635	39.078773	36.099697	33.417993	31.000000	28.816188	31
32	44.212383	40.643263	37.446423	34.578076	32.000000	29.678786	32
33	46.065446	42.229788	38.805735	35.743556	33.000000	30.537409	33
34	47.953474	43.838658	40.177752	36.914456	34.000000	31.392076	34
35	49.877124	45.470189	41.562590	38.090802	35.000000	32.242803	35
36	51.837070	47.124698	42.960372	39.272620	36.000000	33.089611	36
37	53.833996	48.802511	44.371216	40.459934	37.000000	33.932516	37
38	55.868599	50.503955	45.795246	41.652771	38.000000	34.771536	38
39	57.941592	52.229363	47.232585	42.851156	39.000000	35.606691	39
40	60.053697	53.979072	48.683357	44.055115	40.000000	36.437996	40
45	71.227243	63.104251	56.143147	50.159429	45.000000	40.537412	45
50	83.495432	72.890490	63.958103	56.407031	50.000000	44.543239	50
55	96.965504	83.385547	72.145134	62.801284	55.000000	48.457611	55
60	111.755204	94.640807	80.721954	69.345630	60.000000	52.282619	60
65	127.993813	106.711334	89.707121	76.043592	65.000000	56.020301	65
70	145.823276	119.656184	99.120078	82.898777	70.000000	59.672651	70
75	165.399446	133.538686	108.981192	89.914874	75.000000	63.241618	75
80	186.893442	148.426758	119.311799	97.095662	80.000000	66.729105	80
90	236.404860	181.516243	141.471971	111.966860	90.000000	73.467039	90
100	296.092480	219.572975	165.792493	127.544401	100.000000	79.900829	100

TABLE D-84

Factors (EPW) to Convert an Exponential Growth to Present Worth for Interest Rates from 9% to 20%
Growth Rate of 8% per Period

n	9%	10%	11%	12%	15%	20%	n
1	0.990826	0.981818	0.972973	0.964286	0.939130	0.900000	1
2	1.972561	1.945785	1.919649	1.894133	1.921096	1.710000	2
3	2.945290	2.892225	2.840740	2.790771	2.649377	2.439000	3
4	3.909095	3.821458	3.736936	3.655386	3.427241	3.095100	4
5	4.864057	4.733795	4.608911	4.489122	4.157757	3.685590	5
6	5.810258	5.629544	5.457319	5.293082	4.843807	4.217031	6
7	6.747779	6.509007	6.282797	6.068329	5.488097	4.695328	7
8	7.676698	7.372479	7.085964	6.815889	6.093169	5.125795	8
9	8.597096	8.220253	7.867425	7.536750	6.661411	5.513216	9
10	9.509049	9.052612	8.627765	8.231866	7.195064	5.861894	10
11	10.412636	9.869837	9.367555	8.902157	7.696234	6.175705	11
12	11.307933	10.672203	10.087350	9.548508	8.166898	6.458134	12
13	12.195016	11.459982	10.787692	10.171776	8.608913	6.712321	13
14	13.073961	12.233436	11.469106	10.772784	9.024023	6.941089	14
15	13.944842	12.992828	12.132103	11.352327	9.413865	7.146980	15
16	14.807733	13.738413	12.777181	11.911173	9.779977	7.332282	16
17	15.662708	14.470442	13.404825	12.450059	10.123805	7.499054	17
18	16.509639	15.189161	14.015506	12.969700	10.446704	7.649148	18
19	17.349198	15.894813	14.609681	13.470782	10.749948	7.784233	19
20	18.180857	16.587635	15.187798	13.953969	11.034734	7.905810	20
21	19.004866	17.267860	15.750290	14.419898	11.302185	8.015229	21
22	19.821355	17.935717	16.297579	14.869188	11.553356	8.113706	22
23	20.630333	18.591431	16.830077	15.302431	11.789239	8.202336	23
24	21.431890	19.235223	17.348183	15.720201	12.010763	8.282102	24
25	22.226093	19.867310	17.852286	16.123051	12.218804	8.353892	25
26	23.013009	20.487904	18.342765	16.511514	12.414181	8.418503	26
27	23.792706	21.097215	18.819988	16.886102	12.597666	8.476652	27
28	24.565250	21.695448	19.284312	17.247313	12.769982	8.528987	28
29	25.330707	22.282803	19.736088	17.595623	12.931809	8.576088	29
30	26.089141	22.859479	20.175653	17.931494	13.083786	8.618480	30
31	26.840616	23.425671	20.603338	18.255369	13.226512	8.656632	31
32	27.585198	23.981568	21.019464	18.567677	13.360550	8.690968	32
33	28.322948	24.527357	21.424343	18.868832	13.486430	8.721872	33
34	29.053931	25.063223	21.818280	19.159231	13.604647	8.749684	34
35	29.778206	25.589347	22.201570	19.439258	13.715669	8.774716	35
36	30.495838	26.105904	22.574500	19.709285	13.819932	8.797244	36
37	31.206885	26.613069	22.937352	19.969667	13.917849	8.817520	37
38	31.911409	27.111014	23.290396	20.220751	14.009806	8.835768	38
39	32.609469	27.599904	23.633899	20.462867	14.096166	8.852191	39
40	33.301126	28.079906	23.968118	20.696336	14.177269	8.866972	40
45	36.665381	30.352136	25.508503	21.744407	14.514475	8.921448	45
50	39.878119	32.425175	26.851680	22.618222	14.760810	8.953616	50
55	42.946162	34.316486	28.022897	23.346753	14.940761	8.972611	55
60	45.876029	36.041999	29.044168	23.954155	15.072218	8.983827	60
65	48.673941	37.616249	29.934690	24.460569	15.168250	8.990450	65
70	51.345843	39.052496	30.711203	24.882785	15.238402	8.994361	70
75	53.897408	40.362837	31.388303	25.234801	15.289650	8.996670	75
80	56.334057	41.558310	31.978717	25.528289	15.327087	8.998034	80
90	60.883076	43.644047	32.942458	25.976990	15.374414	8.999314	90
100	65.031568	45.380128	33.675229	26.288890	15.399670	8.999761	100

Exponential Growth Factors

TABLE D-85
Factors (EPW) to Convert an Exponential Growth to Present Worth for Interest Rates from 25% to 50%
Growth Rate of 8% per Period

n	25%	30%	35%	40%	45%	50%	n
1	0.864000	0.830769	0.800000	0.771429	0.744828	0.720000	1
2	1.610496	1.520947	1.440000	1.366531	1.299596	1.238400	2
3	2.255469	2.094325	1.952000	1.825609	1.712802	1.611648	3
4	2.812725	2.570670	2.361600	2.179756	2.020570	1.880387	4
5	3.294194	2.966403	2.689280	2.452954	2.249804	2.073878	5
6	3.710184	3.295165	2.951424	2.663708	2.420544	2.213192	6
7	4.069599	3.568291	3.161139	2.826289	2.547715	2.313499	7
8	4.380133	3.795156	3.328911	2.951709	2.642436	2.385719	8
9	4.648435	3.983701	3.463129	3.048461	2.712987	2.437718	9
10	4.880248	4.140306	3.570503	3.123098	2.765535	2.475157	10
11	5.080534	4.270408	3.656403	3.180676	2.804674	2.502113	11
12	5.253582	4.378493	3.725122	3.225093	2.833826	2.521521	12
13	5.403095	4.468286	3.780098	3.259357	2.855540	2.535495	13
14	5.532274	4.542884	3.824078	3.285790	2.871712	2.545557	14
15	5.643884	4.604857	3.859263	3.306181	2.883758	2.552801	15
16	5.740316	4.656343	3.887410	3.321911	2.892730	2.558017	16
17	5.823633	4.699116	3.909928	3.334046	2.899413	2.561772	17
18	5.895619	4.734650	3.927942	3.343407	2.904390	2.564476	18
19	5.957815	4.764171	3.942354	3.350628	2.908098	2.566423	19
20	6.011552	4.788656	3.953883	3.356199	2.910859	2.567824	20
21	6.057981	4.809070	3.963107	3.360496	2.912916	2.568833	21
22	6.098096	4.825997	3.970485	3.363811	2.914447	2.569560	22
23	6.132755	4.840059	3.976388	3.366369	2.915588	2.570083	23
24	6.162700	4.851741	3.981111	3.368342	2.916438	2.570460	24
25	6.188573	4.861447	3.984888	3.369864	2.917071	2.570731	25
26	6.210927	4.869509	3.987911	3.371038	2.917543	2.570926	26
27	6.230241	4.876208	3.990329	3.371943	2.917894	2.571067	27
28	6.246928	4.881773	3.992263	3.372642	2.918155	2.571168	28
29	6.261346	4.886396	3.993810	3.373181	2.918350	2.571241	29
30	6.273803	4.890236	3.995048	3.373597	2.918495	2.571294	30
31	6.284566	4.893427	3.996039	3.373917	2.918603	2.571331	31
32	6.293865	4.896078	3.996831	3.374165	2.918684	2.571359	32
33	6.301899	4.898280	3.997465	3.374356	2.918744	2.571378	33
34	6.308841	4.900110	3.997972	3.374503	2.918789	2.571392	34
35	6.314838	4.901630	3.998377	3.374617	2.918822	2.571402	35
36	6.320020	4.902892	3.998702	3.374704	2.918847	2.571410	36
37	6.324498	4.903941	3.998962	3.374772	2.918865	2.571415	37
38	6.328366	4.904813	3.999169	3.374824	2.918879	2.571419	38
39	6.331708	4.905537	3.999335	3.374864	2.918889	2.571422	39
40	6.334596	4.906138	3.999468	3.374895	2.918897	2.571424	40
45	6.344108	4.907922	3.999826	3.374971	2.918914	2.571428	45
50	6.348689	4.908629	3.999943	3.374992	2.918918	2.571428	50
55	6.350894	4.908908	3.999981	3.374998	2.918919	2.571429	55
60	6.351955	4.909018	3.999994	3.374999	2.918919	2.571429	60
65	6.352467	4.909062	3.999998	3.375000	2.918919	2.571429	65
70	6.352713	4.909080	3.999999	3.375000	2.918919	2.571429	70
75	6.352831	4.909086	4.000000	3.375000	2.918919	2.571429	75
80	6.352888	4.909089	4.000000	3.375000	2.918919	2.571429	80
90	6.352929	4.909091	4.000000	3.375000	2.918919	2.571429	90
100	6.352938	4.909091	4.000000	3.375000	2.918919	2.571429	100

TABLE D-86

Factors (EPW) to Convert an Exponential Growth
to Present Worth for Interest Rates from 1% to 2½%
Growth Rate of 10% per Period

n	1%	1¼%	1½%	1¾%	2%	2½%	n
1	1.089109	1.086420	1.083744	1.081081	1.078431	1.073171	1
2	2.275267	2.266728	2.258245	2.249817	2.241446	2.224866	2
3	3.567123	3.549037	3.531102	3.513316	3.495677	3.460832	3
4	4.974094	4.942164	4.910554	4.879261	4.848279	4.787234	4
5	6.506439	6.455684	6.405527	6.355957	6.306967	6.210690	5
6	8.175330	8.100003	8.025694	7.952386	7.880063	7.738302	6
7	9.992933	9.886423	9.781541	9.678256	9.576538	9.377690	7
8	11.972501	11.827225	11.684428	11.544060	11.406071	11.137033	8
9	14.128467	13.935750	13.746671	13.561146	13.379096	13.025109	9
10	16.476548	16.226494	15.981614	15.741780	15.506868	15.051336	10
11	19.033864	18.715204	18.403719	18.099221	17.801524	17.225824	11
12	21.819060	21.418987	21.028661	20.647807	20.276154	19.559421	12
13	24.852442	24.356430	23.873426	23.403034	22.944872	22.063769	13
14	28.156125	27.547726	26.956422	26.381659	25.822901	24.751362	14
15	31.754195	31.014814	30.297601	29.601793	28.926658	27.635608	15
16	35.672886	34.781526	33.918582	33.083020	32.273846	30.730896	16
17	39.940767	38.873757	37.942798	36.846508	35.883560	34.052669	17
18	44.588954	43.319637	42.095643	40.915143	39.776388	37.617498	18
19	49.651336	48.149729	46.704638	45.313668	43.974536	41.443169	19
20	55.164821	53.397237	51.699608	50.068931	48.501951	45.548767	20
21	61.169607	59.098233	57.112876	55.209547	53.384457	49.954774	21
22	67.709473	65.291907	62.979471	60.767078	58.649904	54.683172	22
23	74.832100	72.020837	69.337358	66.775219	64.328328	59.757551	23
24	82.589416	79.331280	76.227678	73.270507	70.452119	65.203225	24
25	91.037977	87.273489	83.695021	80.292440	77.056207	71.047364	25
26	100.239381	95.902062	91.787707	87.883719	84.178262	77.319122	26
27	110.260712	105.276315	100.558107	96.090507	91.858910	84.049789	27
28	121.175033	115.460688	110.062973	104.962710	100.141962	91.272945	28
29	133.061917	126.525191	120.363813	114.554281	109.074665	99.024624	29
30	146.008029	138.545887	131.527285	124.923547	118.707972	107.343498	30
31	160.107754	151.605408	143.625629	136.133565	129.096832	116.271071	31
32	175.463891	165.793530	156.737135	148.252502	140.300505	125.851882	32
33	192.188396	181.207785	170.946648	161.354057	152.382898	136.133727	33
34	210.403203	197.954137	186.346121	175.517899	165.412929	147.167902	34
35	230.241112	216.147705	203.035205	190.830161	179.464923	159.009455	35
36	251.846756	235.913556	221.121897	207.383958	194.619035	171.717464	36
37	275.377655	257.387567	240.723239	225.279955	210.961704	185.355328	37
38	301.005367	280.717356	261.966071	244.626978	228.586152	199.991083	38
39	328.916736	306.063301	284.987861	265.542679	247.592909	215.697748	39
40	359.315257	333.599635	309.937583	288.154248	268.090392	232.553681	40
45	557.101379	511.365524	469.753701	431.872415	397.367346	337.240662	45
50	860.177954	780.417814	708.674410	644.100575	585.942156	486.257991	50
55	1324.595829	1187.633932	1065.854307	957.497221	861.014067	698.377627	55
60	2036.244241	1803.963816	1599.828444	1420.289047	1262.258308	1000.320632	60
65	3126.735130	2736.791641	2398.105113	2103.692261	1847.548678	1430.123265	65
70	4797.743396	4148.645543	3591.506812	3112.871606	2701.305026	2041.928474	70
75	7358.304506	6285.515164	5375.609576	4603.123523	3946.669616	2912.806374	75
80	11281.967680	9519.710834	8042.794210	6803.773774	5763.268654	4152.462891	80
90	26507.452903	21823.445714	17991.167313	14852.278557	12278.429791	8428.885691	90
100	62258.128714	50008.150944	40225.213000	32403.117426	26141.208652	17093.891017	100

Exponential Growth Factors

TABLE D-87
Factors (EPW) to Convert an Exponential Growth to Present Worth for Interest Rates from 3% to 5½%
Growth Rate of 10% per Period

n	3%	3½%	4%	4½%	5%	5½%	n
1	1.067961	1.062802	1.057692	1.052632	1.047619	1.042654	1
2	2.208502	2.192350	2.176405	2.160665	2.145125	2.129781	2
3	3.426556	3.392836	3.359659	3.327016	3.294893	3.263279	3
4	4.727390	4.668714	4.611178	4.554753	4.499411	4.445125	4
5	6.116630	6.024720	5.934900	5.847109	5.761288	5.677382	5
6	7.600284	7.465886	7.334991	7.207483	7.083254	6.962199	6
7	9.184770	8.997560	8.815855	8.639456	8.468171	8.301819	7
8	10.876938	10.625427	10.382155	10.146795	9.919036	9.698579	8
9	12.684109	12.355526	12.038817	11.733469	11.438990	11.154916	9
10	14.614097	14.194279	13.791057	13.403651	13.031323	12.673373	10
11	16.675249	16.148509	15.644387	15.161738	14.699481	14.256597	11
12	18.876480	18.225468	17.604640	17.012356	16.447076	15.907352	12
13	21.227308	20.432865	19.677985	18.960375	18.277889	17.628519	13
14	23.737902	22.778890	21.870946	21.010921	20.195884	19.423100	14
15	26.419119	25.272250	24.190423	23.169390	22.205211	21.294228	15
16	29.282554	27.922198	26.643717	25.441464	24.310222	23.245167	16
17	32.340592	30.738568	29.238547	27.833120	26.515470	25.279321	17
18	35.606457	33.731812	31.983078	30.350652	28.825731	27.400239	18
19	39.094275	36.913037	34.885948	33.000687	31.246004	29.611624	19
20	42.819128	40.294049	37.956291	35.790196	33.781527	31.917333	20
21	46.797127	43.887395	41.203770	38.726522	36.437791	34.321390	21
22	51.045476	47.706410	44.638602	41.817392	39.220543	36.827990	22
23	55.582547	51.765266	48.271595	45.070939	42.135807	39.441506	23
24	60.427963	56.079027	52.114191	48.495725	45.189893	42.166499	24
25	65.602678	60.663700	56.178471	52.100763	48.389411	45.007724	25
26	71.129074	65.536300	60.477229	55.895540	51.741288	47.970139	26
27	77.031050	70.714908	65.023992	59.890043	55.252778	51.058912	27
28	83.334131	76.218743	69.833069	64.094782	58.931482	54.279435	28
29	90.065577	82.068229	74.919592	68.520823	62.785362	57.637325	29
30	97.254500	88.285074	80.299569	73.179814	66.822760	61.138443	30
31	104.931990	94.892350	85.989928	78.084014	71.052415	64.788898	31
32	113.131251	101.914574	92.008578	83.246331	75.483483	68.595060	32
33	121.867744	109.377809	98.374457	88.680348	80.125553	72.563570	33
34	131.239338	117.309748	105.107599	94.400367	84.988675	76.701352	34
35	141.226478	125.739829	112.229191	100.421438	90.083373	81.015628	35
36	151.892355	134.699335	119.761645	106.759409	95.420677	85.513925	36
37	163.283097	144.221516	127.728663	113.430957	101.012138	90.204092	37
38	175.447968	154.341708	136.155316	120.453739	106.869859	95.094314	38
39	188.439578	165.097467	145.069123	127.845935	113.006519	100.193124	39
40	202.314112	176.528709	154.495130	135.627300	119.435400	105.509418	40
45	287.179502	245.398745	210.443476	181.125788	156.474040	135.692051	45
50	405.077902	338.786941	284.503522	239.926005	203.212237	172.884798	50
55	568.867083	465.421915	382.538421	315.916785	262.190081	218.715802	55
60	796.409578	637.139725	512.309357	414.123882	336.612849	275.191357	60
65	1112.520731	869.990141	684.089977	541.042359	430.525208	344.783728	65
70	1551.675011	1185.736750	911.479733	705.066142	549.031028	430.539381	70
75	2161.765736	1613.891110	1212.480516	917.043169	698.570764	536.212345	75
80	3009.328220	2194.471080	1610.921821	1190.992832	887.271477	666.428544	80
90	5822.581393	4049.284183	2836.511853	2002.579121	1425.862863	1024.615895	90
100	11252.126206	7459.827020	4984.039081	3358.076394	2283.479631	1568.507681	100

Exponential Growth Factors

TABLE D-88

Factors (EPW) to Convert an Exponential Growth to Present Worth for Interest Rates from 6% to 8½%

Growth Rate of 10% per Period

n	6%	6½%	7%	7½%	8%	8½%	n
1	1.037736	1.032864	1.028037	1.023256	1.018519	1.013825	1
2	2.114632	2.099672	2.084898	2.070308	2.055898	2.041666	2
3	3.232165	3.201539	3.171391	3.141711	3.112489	3.083716	3
4	4.391869	4.339617	4.288346	4.238030	4.188646	4.140173	4
5	5.595336	5.515098	5.436617	5.359844	5.284732	5.211236	5
6	6.844216	6.729209	6.617083	6.507748	6.401116	6.297105	6
7	8.140225	7.983221	7.830646	7.682346	7.538174	7.397987	7
8	9.485139	9.278444	9.078234	8.884261	8.696288	8.514088	8
9	10.880804	10.616233	10.360801	10.114128	9.875849	9.645619	9
10	12.329137	11.997987	11.679329	11.372596	11.077254	10.792794	10
11	13.832123	13.425151	13.034824	12.660331	12.300907	11.955828	11
12	15.391826	14.899217	14.428324	13.978013	13.547220	13.134941	12
13	17.010385	16.421727	15.860893	15.326339	14.816613	14.330355	13
14	18.690022	17.994272	17.333629	16.706021	16.109513	15.542295	14
15	20.433042	19.618496	18.847656	18.117789	17.426356	16.770990	15
16	22.241836	21.296100	20.404132	19.562389	18.767585	18.016672	16
17	24.118887	23.028835	22.004248	21.040584	20.133651	19.279575	17
18	26.066769	24.818515	23.649227	22.553156	21.525015	20.559938	18
19	28.088157	26.667011	25.340327	24.100903	22.942145	21.858002	19
20	30.185823	28.576256	27.078840	25.684645	24.385518	23.174011	20
21	32.362646	30.548245	28.866098	27.305219	25.855620	24.508214	21
22	34.621614	32.585042	30.703465	28.963479	27.352946	25.860862	22
23	36.965826	34.688776	32.592347	30.660305	28.878001	27.232210	23
24	39.398499	36.861646	34.534188	32.396591	30.431297	28.622518	24
25	41.922970	39.105926	36.530474	34.173256	32.013358	30.032045	25
26	44.542705	41.423961	38.582730	35.991238	33.624717	31.461060	26
27	47.261298	43.818176	40.692527	37.851500	35.265915	32.909830	27
28	50.082479	46.291073	42.861476	39.755023	36.937506	34.378630	28
29	53.010119	48.845240	45.091237	41.702814	38.640053	35.867735	29
30	56.048237	51.483347	47.383515	43.695903	40.374128	37.377427	30
31	59.201001	54.208151	49.740062	45.735342	42.140315	38.907991	31
32	62.472577	57.022504	52.162680	47.822211	43.939210	40.459714	32
33	65.867934	59.929347	54.653223	49.957611	45.771418	42.032890	33
34	69.391253	62.931720	57.213594	52.142672	47.637555	43.627815	34
35	73.047526	66.032762	59.845750	54.378548	49.538251	45.244789	35
36	76.841772	69.235717	62.551706	56.666421	51.474144	46.884118	36
37	80.779198	72.543933	65.333529	59.007501	53.445888	48.546111	37
38	84.865205	75.960869	68.193348	61.403024	55.454145	50.231080	38
39	89.105402	79.490100	71.133348	63.854257	57.499592	51.939344	39
40	93.505606	83.135314	74.155779	66.362496	59.582918	53.671224	40
45	118.126224	103.239029	90.587575	79.806264	70.592737	62.696461	45
50	147.756320	126.870552	109.455736	94.887679	82.660446	72.363051	50
55	183.415156	154.648943	131.121505	111.806233	95.887694	82.716572	55
60	226.329379	187.301897	155.999687	130.785713	110.385896	93.805838	60
65	277.975220	225.684804	184.566592	152.077172	126.277172	105.683133	65
70	340.129274	270.803159	217.369151	175.962243	143.695379	118.404455	70
75	414.929611	323.838900	255.035390	202.756865	162.787232	132.029784	75
80	504.949331	386.181373	298.286454	232.815465	183.713545	146.623361	80
90	743.663533	545.605697	404.978115	304.363430	231.791556	178.995369	90
100	1089.401009	765.890916	545.653965	394.404262	289.552699	216.131709	100

TABLE D-89
Factors (EPW) to Convert an Exponential Growth to Present Worth for Interest Rates from 9% to 20%
Growth Rate of 10% per Period

n	9%	10%	11%	12%	15%	20%	n
1	1.009174	1.000000	0.990991	0.982143	0.956522	0.916667	1
2	2.027607	2.000000	1.973054	1.946747	1.871456	1.756944	2
3	3.055383	3.000000	2.946270	2.894127	2.746610	2.527199	3
4	4.092589	4.000000	3.910718	3.824589	3.583714	3.233266	4
5	5.139310	5.000000	4.866477	4.738436	4.384422	3.880494	5
6	6.195634	6.000000	5.813626	5.635964	5.150316	4.473786	6
7	7.261649	7.000000	6.752242	6.517464	5.882911	5.017637	7
8	8.337444	8.000000	7.682402	7.383224	6.583654	5.516167	8
9	9.423108	9.000000	8.604182	8.233523	7.253930	5.973153	9
10	10.518733	10.000000	9.517658	9.068639	7.895064	6.392057	10
11	11.624409	11.000000	10.422904	9.888842	8.508322	6.776052	11
12	12.740230	12.000000	11.319995	10.694398	9.094917	7.128048	12
13	13.866287	13.000000	12.209004	11.485570	9.656007	7.450711	13
14	15.002675	14.000000	13.090004	12.262613	10.192702	7.746485	14
15	16.149488	15.000000	13.963067	13.025781	10.706063	8.017611	15
16	17.306823	16.000000	14.828265	13.775320	11.197104	8.266144	16
17	18.474776	17.000000	15.685668	14.511475	11.666795	8.493965	17
18	19.653443	18.000000	16.535347	15.234485	12.116065	8.702801	18
19	20.842924	19.000000	17.37737C	15.944583	12.545801	8.894234	19
20	22.043318	20.000000	18.211809	16.642001	12.956853	9.069715	20
21	23.254725	21.000000	19.038729	17.326966	13.350034	9.230572	21
22	24.477245	22.000000	19.858200	17.999698	13.726119	9.378024	22
23	25.710981	23.000000	20.670288	18.660418	14.085853	9.513189	23
24	26.956036	24.000000	21.475061	19.309339	14.429946	9.637090	24
25	28.212514	25.000000	22.272583	19.946672	14.759079	9.750666	25
26	29.480518	26.000000	23.062920	20.572625	15.073902	9.854777	26
27	30.760156	27.000000	23.846137	21.187399	15.375036	9.950212	27
28	32.051534	28.000000	24.622298	21.791196	15.663078	10.037694	28
29	33.354759	29.000000	25.391466	22.384210	15.938597	10.117887	29
30	34.669940	30.000000	26.153705	22.966635	16.202136	10.191396	30
31	35.997187	31.000000	26.909077	23.538659	16.454217	10.258780	31
32	37.336611	32.000000	27.657644	24.100469	16.695338	10.320548	32
33	38.688323	33.000000	28.399467	24.652246	16.925976	10.377169	33
34	40.052436	34.000000	29.134607	25.194170	17.146585	10.429072	34
35	41.429064	35.000000	29.863124	25.726417	17.357603	10.476649	35
36	42.818321	36.000000	30.585078	26.249160	17.559447	10.520262	36
37	44.220324	37.000000	31.300528	26.762568	17.752514	10.560240	37
38	45.635190	38.000000	32.009532	27.266808	17.937187	10.596886	38
39	47.063035	39.000000	32.712149	27.762043	18.113831	10.630479	39
40	48.503981	40.000000	33.408436	28.248435	18.282795	10.661273	40
45	55.909445	45.000000	36.796900	30.553173	19.023604	10.780766	45
50	63.660899	50.000000	40.035457	32.659350	19.616776	10.858106	50
55	71.774510	55.000000	43.130737	34.584072	20.091733	10.908162	55
60	80.267196	60.000000	46.089080	36.342973	20.472035	10.940560	60
65	89.156669	65.000000	48.916544	37.950339	20.776546	10.961529	65
70	98.461467	70.000000	51.618919	39.419225	21.020371	10.975100	70
75	108.200995	75.000000	54.201739	40.761561	21.215603	10.983884	75
80	118.395563	80.000000	56.670293	41.988251	21.371927	10.989569	80
90	140.235853	90.000000	61.284602	44.133684	21.597321	10.995631	90
100	164.164618	100.000000	65.499661	45.925370	21.741829	10.998170	100

TABLE D-90

Factors (EPW) to Convert an Exponential Growth to Present Worth for Interest Rates from 25% to 50%
Growth Rate of 10% per Period

n	25%	30%	35%	40%	45%	50%	n
1	0.880000	0.846154	0.814815	0.785714	0.758621	0.733333	1
2	1.654400	1.562130	1.478738	1.403061	1.334126	1.271111	2
3	2.335872	2.167956	2.019712	1.888120	1.770716	1.665481	3
4	2.935567	2.680578	2.460506	2.269237	2.101923	1.954686	4
5	3.463299	3.114336	2.819672	2.568686	2.353183	2.166770	5
6	3.927703	3.481361	3.112325	2.803968	2.543794	2.322298	6
7	4.336379	3.791921	3.350784	2.988832	2.688395	2.436352	7
8	4.696013	4.054702	3.545083	3.134082	2.798093	2.519991	8
9	5.012492	4.277056	3.703401	3.248207	2.881312	2.581327	9
10	5.290993	4.465201	3.832401	3.337877	2.944444	2.626306	10
11	5.536074	4.624401	3.937512	3.408332	2.992336	2.659291	11
12	5.751745	4.759108	4.023158	3.463689	3.028669	2.683480	12
13	5.941535	4.873092	4.092943	3.507185	3.056232	2.701219	13
14	6.108551	4.969539	4.149806	3.541359	3.077141	2.714227	14
15	6.255525	5.051148	4.196138	3.568211	3.093004	2.723767	15
16	6.384862	5.120203	4.233890	3.589309	3.105037	2.730762	16
17	6.498679	5.178633	4.264651	3.605885	3.114166	2.735892	17
18	6.598837	5.228074	4.289716	3.618910	3.121092	2.739654	18
19	6.686977	5.269909	4.310139	3.629143	3.126345	2.742413	19
20	6.764540	5.305307	4.326780	3.637184	3.130331	2.744436	20
21	6.832795	5.335260	4.340339	3.643502	3.133355	2.745920	21
22	6.892859	5.360605	4.351387	3.648466	3.135648	2.747008	22
23	6.945716	5.382050	4.360390	3.652366	3.137388	2.747806	23
24	6.992230	5.400196	4.367725	3.655430	3.138708	2.748391	24
25	7.033163	5.415551	4.373702	3.657838	3.139710	2.748820	25
26	7.069183	5.428543	4.378572	3.659730	3.140470	2.749135	26
27	7.100881	5.439536	4.382540	3.661216	3.141046	2.749365	27
28	7.126775	5.448838	4.385773	3.662384	3.141483	2.749535	28
29	7.153322	5.456709	4.388408	3.663302	3.141815	2.749659	29
30	7.174924	5.463370	4.390555	3.664023	3.142066	2.749750	30
31	7.193933	5.469005	4.392304	3.664589	3.142257	2.749816	31
32	7.210661	5.473773	4.393729	3.665035	3.142402	2.749865	32
33	7.225382	5.477808	4.394890	3.665384	3.142512	2.749901	33
34	7.238336	5.481222	4.395837	3.665659	3.142595	2.749928	34
35	7.249736	5.484111	4.396608	3.665875	3.142658	2.749947	35
36	7.259767	5.486556	4.397236	3.666045	3.142706	2.749961	36
37	7.268595	5.488624	4.397748	3.666178	3.142743	2.749971	37
38	7.276364	5.490374	4.398165	3.666283	3.142770	2.749979	38
39	7.283200	5.491855	4.398505	3.666365	3.142791	2.749985	39
40	7.289216	5.493108	4.398782	3.666430	3.142807	2.749989	40
45	7.310051	5.497011	4.399562	3.666596	3.142845	2.749998	45
50	7.321047	5.498703	4.399843	3.666645	3.142854	2.749999	50
55	7.326849	5.499438	4.399944	3.666660	3.142856	2.750000	55
60	7.329911	5.499756	4.399980	3.666665	3.142857	2.750000	60
65	7.331528	5.499894	4.399993	3.666666	3.142857	2.750000	65
70	7.332380	5.499954	4.399997	3.666666	3.142857	2.750000	70
75	7.332830	5.499980	4.399999	3.666667	3.142857	2.750000	75
80	7.333068	5.499991	4.400000	3.666667	3.142857	2.750000	80
90	7.333255	5.499998	4.400000	3.666667	3.142857	2.750000	90
100	7.333313	5.500000	4.400000	3.666667	3.142857	2.750000	100

Appendix **E**

Continuous Compounding Factors

Tables E-1 and E-2 were calculated from the following formulas:

$$\text{CCA} = e^{rn} \qquad \text{Continuous single sum compound amount}$$

$$\text{CPW} = \frac{1}{e^{rn}} \qquad \text{Continuous single sum present worth}$$

$$\text{CSCA} = \frac{e^{rn} - 1}{r} \qquad \text{Continuous uniform series compound amount}$$

$$\text{CSF} = \frac{r}{e^{rn} - 1} \qquad \text{Continuous sinking fund}$$

$$\text{GSPW} = \frac{e^{rn} - 1}{re^{rn}} \qquad \text{Continuous uniform series present worth}$$

$$\text{CCR} = \frac{re^{rn}}{e^{rn} - 1} \qquad \text{Continuous capital recovery}$$

Table E-1 is calculated using nominal integral values of r from 1% through 25 percent. The effective rates corresponding to these nominal rates are given in parentheses.

Table E-2 is calculated using the nominal rates as shown which correspond to the effective integral rates given in parentheses.

TABLE E-1
Continuous Compounding Factors for Nominal Rates from 1% to 25%

n	CCA	CPW	CSCA	CSF	CSPW	CCR	n
\multicolumn{7}{c}{1% (effective rate of 1.0050 1671%)}							

n	CCA	CPW	CSCA	CSF	CSPW	CCR	n
1	1.0100 5017	0.990050	1.005017	0.995008	0.995017	1.005008	1
2	1.0202 0134	0.980199	2.020134	0.495017	1.980133	0.505017	2
3	1.0304 5453	0.970446	3.045453	0.328358	2.955447	0.338358	3
4	1.0408 1077	0.960789	4.081077	0.245033	3.921056	0.255033	4
5	1.0512 7110	0.951229	5.127110	0.195042	4.877058	0.205042	5
6	1.0618 3655	0.941765	6.183655	0.161717	5.823547	0.171717	6
7	1.0725 0818	0.932394	7.250818	0.137915	6.760618	0.147915	7
8	1.0832 8707	0.923116	8.328707	0.120067	7.688365	0.130067	8
9	1.0941 7428	0.913931	9.417428	0.106186	8.606881	0.116186	9
10	1.1051 7092	0.904837	10.517092	0.095083	9.516258	0.105083	10
11	1.1162 7807	0.895834	11.627807	0.086001	10.416586	0.096001	11
12	1.1274 9685	0.886920	12.749685	0.078433	11.307956	0.088433	12
13	1.1388 2838	0.878095	13.882838	0.072031	12.190457	0.082031	13
14	1.1502 7380	0.869358	15.027380	0.066545	13.064176	0.076545	14
15	1.1618 3424	0.860708	16.183424	0.061792	13.929202	0.071792	15
16	1.1735 1087	0.852144	17.351087	0.057633	14.785621	0.067633	16
17	1.1853 0485	0.843665	18.530485	0.053965	15.633518	0.063965	17
18	1.1972 1736	0.835270	19.721736	0.050705	16.472979	0.060705	18
19	1.2092 4960	0.826959	20.924960	0.047790	17.304087	0.057790	19
20	1.2214 0276	0.818731	22.140276	0.045167	18.126925	0.055167	20
21	1.2336 7806	0.810584	23.367806	0.042794	18.941575	0.052794	21
22	1.2460 7673	0.802519	24.607673	0.040638	19.748120	0.050638	22
23	1.2586 0001	0.794534	25.860001	0.038670	20.546640	0.048670	23
24	1.2712 4915	0.786628	27.124915	0.036866	21.337214	0.046866	24
25	1.2840 2542	0.778801	28.402542	0.035208	22.119922	0.045208	25
26	1.2969 3009	0.771052	29.693009	0.033678	22.894841	0.043678	26
27	1.3099 6445	0.763379	30.996445	0.032262	23.662051	0.042262	27
28	1.3231 2981	0.755784	32.312981	0.030947	24.421626	0.040947	28
29	1.3364 2749	0.748264	33.642749	0.029724	25.173643	0.039724	29
30	1.3498 5881	0.740818	34.985881	0.028583	25.918178	0.038583	30
31	1.3634 2511	0.733447	36.342511	0.027516	26.655304	0.037516	31
32	1.3771 2776	0.726149	37.712776	0.026516	27.385096	0.036516	32
33	1.3909 6813	0.718924	39.096813	0.025578	28.107627	0.035578	33
34	1.4049 4759	0.711770	40.494759	0.024695	28.822968	0.034695	34
35	1.4190 6755	0.704688	41.906755	0.023863	29.531191	0.033863	35
36	1.4333 2941	0.697676	43.332941	0.023077	30.232367	0.033077	36
37	1.4477 3461	0.690734	44.773461	0.022335	30.926567	0.032335	37
38	1.4622 8459	0.683861	46.228459	0.021632	31.613859	0.031632	38
39	1.4769 8079	0.677057	47.698079	0.020965	32.294313	0.030965	39
40	1.4918 2470	0.670320	49.182470	0.020332	32.967995	0.030332	40
45	1.5683 1219	0.637628	56.831219	0.017596	36.237185	0.027596	45
50	1.6487 2127	0.606531	64.872127	0.015415	39.346934	0.025415	50
55	1.7332 5302	0.576950	73.325302	0.013638	42.305019	0.023638	55
60	1.8221 1880	0.548812	82.211880	0.012164	45.118836	0.022164	60
65	1.9155 4083	0.522046	91.554083	0.010923	47.795422	0.020923	65
70	2.0137 5271	0.496585	101.375271	0.009864	50.341470	0.019864	70
75	2.1170 0002	0.472367	111.700002	0.008953	52.763345	0.018953	75
80	2.2255 4093	0.449329	122.554093	0.008160	55.067104	0.018160	80
90	2.4596 0311	0.406570	145.960311	0.006851	59.343034	0.016851	90
100	2.7182 8183	0.367879	171.828183	0.005820	63.212056	0.015820	100

Continuous Compounding Factors

TABLE E-1 (*continued*)

n	CCA	CPW	CSCA	CSF	CSPW	CCR	n
2% (effective rate of 2.0201 3400%)							
1	1.0202 0134	0.980199	1.010067	0.990033	0.990066	1.010033	1
2	1.0408 1077	0.960789	2.040539	0.490067	1.960528	0.510067	2
3	1.0618 3655	0.941765	3.091827	0.323433	2.911773	0.343433	3
4	1.0832 8707	0.923116	4.164353	0.240133	3.844183	0.260133	4
5	1.1051 7092	0.904837	5.258546	0.190167	4.758129	0.210167	5
6	1.1274 9685	0.886920	6.374843	0.156867	5.653978	0.176867	6
7	1.1502 7380	0.869358	7.513690	0.133090	6.532088	0.153090	7
8	1.1735 1087	0.852144	8.675544	0.115267	7.392811	0.135267	8
9	1.1972 1736	0.835270	9.860868	0.101411	8.236489	0.121411	9
10	1.2214 0276	0.818731	11.070138	0.090333	9.063462	0.110333	10
11	1.2460 7673	0.802519	12.303837	0.081275	9.874060	0.101275	11
12	1.2712 4915	0.786628	13.562458	0.073733	10.668607	0.093733	12
13	1.2969 3009	0.771052	14.846504	0.067356	11.447421	0.087356	13
14	1.3231 2981	0.755784	16.156491	0.061895	12.210813	0.081895	14
15	1.3498 5881	0.740818	17.492940	0.057166	12.959089	0.077166	15
16	1.3771 2776	0.726149	18.856388	0.053032	13.692548	0.073032	16
17	1.4049 4759	0.711770	20.247380	0.049389	14.411484	0.069389	17
18	1.4333 2941	0.697676	21.666471	0.046154	15.116184	0.066154	18
19	1.4622 8459	0.683861	23.114229	0.043263	15.806930	0.063263	19
20	1.4918 2470	0.670320	24.591235	0.040665	16.483998	0.060665	20
21	1.5219 6156	0.657047	26.098078	0.038317	17.147659	0.058317	21
22	1.5527 0722	0.644036	27.635361	0.036186	17.798179	0.056186	22
23	1.5840 7399	0.631284	29.203699	0.034242	18.435818	0.054242	23
24	1.6160 7440	0.618783	30.803720	0.032464	19.060830	0.052464	24
25	1.6487 2127	0.606531	32.436064	0.030830	19.673467	0.050830	25
26	1.6820 2765	0.594521	34.101382	0.029324	20.273973	0.049324	26
27	1.7160 0686	0.582748	35.800343	0.027933	20.862587	0.047933	27
28	1.7506 7250	0.571209	37.533625	0.026643	21.439547	0.046643	28
29	1.7860 3843	0.559898	39.301922	0.025444	22.005082	0.045444	29
30	1.8221 1880	0.548812	41.105940	0.024327	22.559418	0.044327	30
31	1.8589 2804	0.537944	42.946402	0.023285	23.102778	0.043285	31
32	1.8964 8088	0.527292	44.824044	0.022309	23.635379	0.042309	32
33	1.9347 9233	0.516851	46.739617	0.021395	24.157433	0.041395	33
34	1.9738 7773	0.506617	48.693887	0.020536	24.669150	0.040536	34
35	2.0137 5271	0.496585	50.687635	0.019729	25.170735	0.039729	35
36	2.0544 3321	0.486752	52.721661	0.018968	25.662387	0.038968	36
37	2.0959 3551	0.477114	54.796776	0.018249	26.144304	0.038249	37
38	2.1382 7622	0.467666	56.913811	0.017570	26.616679	0.037570	38
39	2.1814 7227	0.458406	59.073613	0.016928	27.079699	0.036928	39
40	2.2255 4093	0.449329	61.277046	0.016319	27.533552	0.036319	40
45	2.4596 0311	0.406570	72.980156	0.013702	29.671517	0.033702	45
50	2.7182 8183	0.367879	85.914091	0.011640	31.606028	0.031640	50
55	3.0041 6602	0.332871	100.208301	0.009979	33.356146	0.029979	55
60	3.3201 1692	0.301194	116.005846	0.008620	34.940289	0.028620	60
65	3.6692 9667	0.272532	133.464833	0.007493	36.373410	0.027493	65
70	4.0551 9997	0.246597	152.759998	0.006546	37.670152	0.026546	70
75	4.4816 8907	0.223130	174.084454	0.005744	38.843492	0.025744	75
80	4.9530 3242	0.201897	197.651621	0.005059	39.905174	0.025059	80
90	6.0496 4746	0.165299	252.482373	0.003961	41.735056	0.023961	90
100	7.3890 5610	0.135335	319.455305	0.003130	43.233236	0.023130	100

Continuous Compounding Factors

TABLE E-1 (*continued*)

n	CCA	CPW	CSCA	CSF	CSPW	CCR	n
\multicolumn{7}{c	}{3% (effective rate of 3.0454 5340%)}	

n	CCA	CPW	CSCA	CSF	CSPW	CCR	n
1	1.0304 5453	0.970446	1.015151	0.985075	0.985149	1.015075	1
2	1.0618 3655	0.941765	2.061218	0.485150	1.941182	0.515150	2
3	1.0941 7428	0.913931	3.139143	0.318558	2.868960	0.348558	3
4	1.1274 9685	0.886920	4.249895	0.235300	3.769319	0.265300	4
5	1.1618 3424	0.860708	5.394475	0.185375	4.643067	0.215375	5
6	1.1972 1736	0.835270	6.573912	0.152116	5.490993	0.182116	6
7	1.2336 7806	0.810584	7.789269	0.128382	6.313858	0.158382	7
8	1.2712 4915	0.786628	9.041638	0.110599	7.112405	0.140599	8
9	1.3099 6445	0.763379	10.332148	0.096785	7.887350	0.126785	9
10	1.3498 5881	0.740818	11.661960	0.085749	8.639393	0.115749	10
11	1.3909 6813	0.718924	13.032271	0.076733	9.369209	0.106733	11
12	1.4333 2941	0.697676	14.444314	0.069231	10.077456	0.099231	12
13	1.4769 8079	0.677057	15.899360	0.062896	10.764771	0.092896	13
14	1.5219 6156	0.657047	17.398719	0.057475	11.431773	0.087475	14
15	1.5683 1219	0.637628	18.943740	0.052788	12.079062	0.082788	15
16	1.6160 7440	0.618783	20.535813	0.048695	12.707220	0.078695	16
17	1.6652 9119	0.600496	22.176373	0.045093	13.316814	0.075093	17
18	1.7160 0686	0.582748	23.866895	0.041899	13.908392	0.071899	18
19	1.7682 6705	0.565525	25.608902	0.039049	14.482485	0.069049	19
20	1.8221 1880	0.548812	27.403960	0.036491	15.039612	0.066491	20
21	1.8776 1058	0.532592	29.253686	0.034184	15.580273	0.064184	21
22	1.9347 9233	0.516851	31.159744	0.032093	16.104956	0.062093	22
23	1.9937 1553	0.501576	33.123851	0.030190	16.614131	0.060190	23
24	2.0544 3321	0.486752	35.147774	0.028451	17.108258	0.058451	24
25	2.1170 0002	0.472367	37.233334	0.026858	17.587782	0.056858	25
26	2.1814 7227	0.458406	39.382409	0.025392	18.053133	0.055392	26
27	2.2479 0799	0.444858	41.596933	0.024040	18.504731	0.054040	27
28	2.3163 6698	0.431711	43.878899	0.022790	18.942983	0.052790	28
29	2.3869 1085	0.418952	46.230362	0.021631	19.368282	0.051631	29
30	2.4596 0311	0.406570	48.653437	0.020554	19.781011	0.050554	30
31	2.5345 0918	0.394554	51.150306	0.019550	20.181543	0.049550	31
32	2.6116 9647	0.382893	53.723216	0.018614	20.570237	0.048614	32
33	2.6912 3447	0.371577	56.374482	0.017739	20.947444	0.047739	33
34	2.7731 9476	0.360595	59.106492	0.016919	21.313502	0.046919	34
35	2.8576 5112	0.349938	61.921704	0.016149	21.668742	0.046149	35
36	2.9446 7955	0.339596	64.822652	0.015427	22.013482	0.045427	36
37	3.0343 5839	0.329559	67.811946	0.014747	22.348035	0.044747	37
38	3.1267 6837	0.319819	70.892279	0.014106	22.672699	0.044106	38
39	3.2219 9264	0.310367	74.066421	0.013501	22.987769	0.043501	39
40	3.3201 1692	0.301194	77.337231	0.012930	23.293526	0.042930	40
45	3.8574 2553	0.259240	95.247518	0.010499	24.691991	0.040499	45
50	4.4816 8907	0.223130	116.056302	0.008617	25.895661	0.038617	50
55	5.2069 7983	0.192050	140.232661	0.007131	26.931670	0.037131	55
60	6.0496 4746	0.165299	168.321582	0.005941	27.823370	0.035941	60
65	7.0286 8758	0.142274	200.956253	0.004976	28.590864	0.034976	65
70	8.1661 6991	0.122456	238.872330	0.004186	29.251452	0.034186	70
75	9.4877 3584	0.105399	282.924528	0.003535	29.820026	0.033535	75
80	11.0231 7638	0.090718	334.105879	0.002993	30.309402	0.032993	80
90	14.8797 3172	0.067206	462.557724	0.002161	31.093150	0.032161	90
100	20.0855 3692	0.049787	636.184564	0.001572	31.673764	0.031572	100

Continuous Compounding Factors

TABLE E-1 (*continued*)

n	CCA	CPW	CSCA	CSF	CSPW	CCR	n
4% (effective rate of 4.0810 7742%)							
1	1.0408 1077	0.960789	1.020269	0.980133	0.980264	1.020133	1
2	1.0832 8707	0.923116	2.082177	0.480267	1.922091	0.520267	2
3	1.1274 9685	0.886920	3.187421	0.313733	2.826989	0.353733	3
4	1.1735 1087	0.852144	4.337772	0.230533	3.696405	0.270533	4
5	1.2214 0276	0.818731	5.535069	0.180666	4.531731	0.220666	5
6	1.2712 4915	0.786628	6.781229	0.147466	5.334303	0.187466	6
7	1.3231 2981	0.755784	8.078245	0.123789	6.105406	0.163789	7
8	1.3771 2776	0.726149	9.428194	0.106065	6.846274	0.146065	8
9	1.4333 2941	0.697676	10.833235	0.092309	7.558092	0.132309	9
10	1.4918 2470	0.670320	12.295617	0.081330	8.241999	0.121330	10
11	1.5527 0722	0.644036	13.817680	0.072371	8.899089	0.112371	11
12	1.6160 7440	0.618783	15.401860	0.064927	9.530415	0.104927	12
13	1.6820 2765	0.594521	17.050691	0.058649	10.136986	0.098649	13
14	1.7506 7250	0.571209	18.766813	0.053286	10.719773	0.093286	14
15	1.8221 1880	0.548812	20.552970	0.048655	11.279709	0.088655	15
16	1.8964 8088	0.527292	22.412022	0.044619	11.817689	0.084619	16
17	1.9738 7773	0.506617	24.346943	0.041073	12.334575	0.081073	17
18	2.0544 3321	0.486752	26.360830	0.037935	12.831194	0.077935	18
19	2.1382 7622	0.467666	28.456906	0.035141	13.308339	0.075141	19
20	2.2255 4093	0.449329	30.638523	0.032639	13.766776	0.072639	20
21	2.3163 6698	0.431711	32.909174	0.030387	14.207237	0.070387	21
22	2.4108 9971	0.414783	35.272493	0.028351	14.630427	0.068351	22
23	2.5092 9039	0.398519	37.732260	0.026503	15.037024	0.066503	23
24	2.6116 9647	0.382893	40.292412	0.024819	15.427678	0.064819	24
25	2.7182 8183	0.367879	42.957046	0.023279	15.803014	0.063279	25
26	2.8292 1701	0.353455	45.730425	0.021867	16.163633	0.061867	26
27	2.9446 7955	0.339596	48.616989	0.020569	16.510112	0.060569	27
28	3.0648 5420	0.326280	51.621355	0.019372	16.843005	0.059372	28
29	3.1899 3328	0.313486	54.748332	0.018265	17.162845	0.058265	29
30	3.3201 1692	0.301194	58.002923	0.017241	17.470145	0.057241	30
31	3.4556 1346	0.289384	61.390337	0.016289	17.765395	0.056289	31
32	3.5966 3973	0.278037	64.915993	0.015405	18.049067	0.055405	32
33	3.7434 2138	0.267135	68.585534	0.014580	18.321617	0.054580	33
34	3.8961 9330	0.256661	72.404833	0.013811	18.583481	0.053811	34
35	4.0551 9997	0.246597	76.379999	0.013092	18.835076	0.053092	35
36	4.2206 9582	0.236928	80.517395	0.012420	19.076806	0.052420	36
37	4.3929 4568	0.227638	84.823642	0.011789	19.309058	0.051789	37
38	4.5722 2520	0.218712	89.305630	0.011198	19.532203	0.051198	38
39	4.7588 2125	0.210136	93.970531	0.010642	19.746598	0.050642	39
40	4.9530 3242	0.201897	98.825811	0.010119	19.952587	0.050119	40
45	6.0496 4746	0.165299	126.241187	0.007921	20.867528	0.047921	45
50	7.3890 5610	0.135335	159.726402	0.006261	21.616618	0.046261	50
55	9.0250 1350	0.110803	200.625337	0.004984	22.229921	0.044984	55
60	11.0231 7638	0.090718	250.579410	0.003991	22.732051	0.043991	60
65	13.4637 3804	0.074274	311.593451	0.003209	23.143161	0.043209	65
70	16.4446 4677	0.060810	386.116169	0.002590	23.479748	0.042590	70
75	20.0855 3692	0.049787	477.138423	0.002096	23.755323	0.042096	75
80	24.5325 3020	0.040762	588.313255	0.001700	23.980945	0.041700	80
90	36.5982 3444	0.027324	889.955861	0.001124	24.316907	0.041124	90
100	54.5981 5003	0.018316	1339.953751	0.000746	24.542109	0.040746	100

TABLE E-1 (continued)

n	CCA	CPW	CSCA	CSF	CSPW	CCR	n
\multicolumn{7}{c}{5% (effective rate of 5.1271 0964%)}							
1	1.0512 7110	0.951229	1.025422	0.975208	0.975412	1.025208	1
2	1.1051 7092	0.904837	2.103418	0.475417	1.903252	0.525417	2
3	1.1618 3424	0.860708	3.236685	0.308958	2.785840	0.358958	3
4	1.2214 0276	0.818731	4.428055	0.225833	3.625385	0.275833	4
5	1.2840 2542	0.778801	5.680508	0.176041	4.423984	0.226041	5
6	1.3498 5881	0.740818	6.997176	0.142915	5.183636	0.192915	6
7	1.4190 6755	0.704688	8.381351	0.119313	5.906238	0.169313	7
8	1.4918 2470	0.670320	9.836494	0.101662	6.593599	0.151662	8
9	1.5683 1219	0.637628	11.366244	0.087980	7.247437	0.137980	9
10	1.6487 2127	0.606531	12.974425	0.077075	7.869387	0.127075	10
11	1.7332 5302	0.576950	14.665060	0.068189	8.461004	0.118189	11
12	1.8221 1880	0.548812	16.442376	0.060818	9.023767	0.110818	12
13	1.9155 4083	0.522046	18.310817	0.054613	9.559084	0.104613	13
14	2.0137 5271	0.496585	20.275054	0.049322	10.068294	0.099322	14
15	2.1170 0002	0.472367	22.340000	0.044763	10.552669	0.094763	15
16	2.2255 4093	0.449329	24.510819	0.040798	11.013421	0.090798	16
17	2.3396 4685	0.427415	26.792937	0.037323	11.451701	0.087323	17
18	2.4596 0311	0.406570	29.192062	0.034256	11.868607	0.084256	18
19	2.5857 0966	0.386741	31.714193	0.031532	12.265180	0.081532	19
20	2.7182 8183	0.367879	34.365637	0.029099	12.642411	0.079099	20
21	2.8576 5112	0.349938	37.153022	0.026916	13.001245	0.076916	21
22	3.0041 6602	0.332871	40.083320	0.024948	13.342578	0.074948	22
23	3.1581 9291	0.316637	43.163858	0.023168	13.667265	0.073168	23
24	3.3201 1692	0.301194	46.402338	0.021551	13.976116	0.071551	24
25	3.4903 4296	0.286505	49.806859	0.020078	14.269904	0.070078	25
26	3.6692 9667	0.272532	53.385933	0.018732	14.549364	0.068732	26
27	3.8574 2553	0.259240	57.148511	0.017498	14.815195	0.067498	27
28	4.0551 9997	0.246597	61.103999	0.016366	15.068061	0.066366	28
29	4.2631 1452	0.234570	65.262290	0.015323	15.308594	0.065323	29
30	4.4816 8907	0.223130	69.633781	0.014361	15.537397	0.064361	30
31	4.7114 7018	0.212248	74.229404	0.013472	15.755041	0.063472	31
32	4.9530 3242	0.201897	79.060648	0.012649	15.962070	0.062649	32
33	5.2069 7983	0.192050	84.139597	0.011885	16.159002	0.061885	33
34	5.4739 4739	0.182684	89.478948	0.011176	16.346330	0.061176	34
35	5.7546 0268	0.173774	95.092054	0.010516	16.524521	0.060516	35
36	6.0496 4746	0.165299	100.992949	0.009902	16.694022	0.059902	36
37	6.3598 1952	0.157237	107.196390	0.009329	16.855257	0.059329	37
38	6.6858 9444	0.149569	113.717889	0.008794	17.008628	0.058794	38
39	7.0286 8758	0.142274	120.573752	0.008294	17.154519	0.058294	39
40	7.3890 5610	0.135335	127.781122	0.007826	17.293294	0.057826	40
45	9.4877 3584	0.105399	169.754717	0.005891	17.892016	0.055891	45
50	12.1824 9396	0.082085	223.649879	0.004471	18.358300	0.054471	50
55	15.6426 3188	0.063928	292.852638	0.003415	18.721443	0.053415	55
60	20.0855 3692	0.049787	381.710738	0.002620	19.004259	0.052620	60
65	25.7903 3992	0.038774	495.806798	0.002017	19.224516	0.052017	65
70	33.1154 5196	0.030197	642.309039	0.001557	19.396052	0.051557	70
75	42.5210 8200	0.023518	830.421640	0.001204	19.529645	0.051204	75
80	54.5981 5003	0.018316	1071.963001	0.000933	19.633687	0.050933	80
90	90.0171 3130	0.011109	1780.342626	0.000562	19.777820	0.050562	90
100	148.4131 5910	0.006738	2948.263182	0.000339	19.865241	0.050339	100

Continuous Compounding Factors

TABLE E-1 (continued)

n	CCA	CPW	CSCA	CSF	CSPW	CCR	n
\multicolumn{7}{c}{6% (effective rate of 6.1836 5465%)}							
1	1.0618 3655	0.941765	1.030609	0.970300	0.970591	1.030300	1
2	1.1274 9685	0.886920	2.124948	0.470600	1.884659	0.530600	2
3	1.1972 1736	0.835270	3.286956	0.304233	2.745496	0.364233	3
4	1.2712 4915	0.786628	4.520819	0.221199	3.556202	0.281199	4
5	1.3498 5881	0.740818	5.830980	0.171498	4.319696	0.231498	5
6	1.4333 2941	0.697676	7.222157	0.138463	5.038728	0.198463	6
7	1.5219 6156	0.657047	8.699359	0.114951	5.715886	0.174951	7
8	1.6160 7440	0.618783	10.267907	0.097391	6.353610	0.157391	8
9	1.7160 0686	0.582748	11.933448	0.083798	6.954196	0.143798	9
10	1.8221 1880	0.548812	13.701980	0.072982	7.519806	0.132982	10
11	1.9347 9233	0.516851	15.579872	0.064185	8.052478	0.124185	11
12	2.0544 3321	0.486752	17.573887	0.056903	8.554129	0.116903	12
13	2.1814 7227	0.458406	19.691204	0.050784	9.026566	0.110784	13
14	2.3163 6698	0.431711	21.939450	0.045580	9.471491	0.105580	14
15	2.4596 0311	0.406570	24.326719	0.041107	9.890506	0.101107	15
16	2.6116 9647	0.382893	26.861608	0.037228	10.285119	0.097228	16
17	2.7731 9476	0.360595	29.553246	0.033837	10.656751	0.093837	17
18	2.9446 7955	0.339596	32.411326	0.030853	11.006741	0.090853	18
19	3.1267 6837	0.319819	35.446139	0.028212	11.336350	0.088212	19
20	3.3201 1692	0.301194	38.668615	0.025861	11.646763	0.085861	20
21	3.5254 2149	0.283654	42.090358	0.023758	11.939100	0.083758	21
22	3.7434 2138	0.267135	45.723690	0.021871	12.214412	0.081871	22
23	3.9749 0163	0.251579	49.581694	0.020169	12.473691	0.080169	23
24	4.2206 9582	0.236928	53.678264	0.018630	12.717871	0.078630	24
25	4.4816 8907	0.223130	58.028151	0.017233	12.947831	0.077233	25
26	4.7588 2125	0.210136	62.647021	0.015962	13.164399	0.075962	26
27	5.0530 9032	0.197899	67.551505	0.014804	13.368355	0.074804	27
28	5.3655 5597	0.186374	72.759264	0.013744	13.560434	0.073744	28
29	5.6973 4342	0.175520	78.289057	0.012773	13.741327	0.072773	29
30	6.0496 4746	0.165299	84.160791	0.011882	13.911685	0.071882	30
31	6.4237 7677	0.155673	90.395613	0.011062	14.072123	0.071062	31
32	6.8209 5847	0.146607	97.015974	0.010308	14.223217	0.070308	32
33	7.2427 4299	0.138069	104.045716	0.009611	14.365513	0.069611	33
34	7.6906 0920	0.130029	111.510153	0.008968	14.499521	0.068968	34
35	8.1661 6991	0.122456	119.436165	0.008373	14.625726	0.068373	35
36	8.6711 3766	0.115325	127.852294	0.007822	14.744581	0.067822	36
37	9.2073 3087	0.108609	136.788848	0.007311	14.856515	0.067311	37
38	9.7766 8041	0.102284	146.278007	0.006836	14.961930	0.066836	38
39	10.3812 3656	0.096328	156.353943	0.006396	15.061206	0.066396	39
40	11.0231 7638	0.090718	167.052940	0.005986	15.154701	0.065986	40
45	14.8797 3172	0.067206	231.328862	0.004323	15.546575	0.064323	45
50	20.0855 3692	0.049787	318.092282	0.003144	15.836882	0.063144	50
55	27.1126 3892	0.036883	435.210649	0.002298	16.051947	0.062298	55
60	36.5982 3444	0.027324	593.303907	0.001685	16.211271	0.061685	60
65	49.4024 4911	0.020242	806.707485	0.001240	16.329301	0.061240	65
70	66.6863 3104	0.014996	1094.772184	0.000913	16.416740	0.060913	70
75	90.0171 3130	0.011109	1483.618855	0.000674	16.481517	0.060674	75
80	121.5104 1752	0.008230	2008.506959	0.000498	16.529504	0.060498	80
90	221.4064 1620	0.004517	3673.440270	0.000272	16.591390	0.060272	90
100	403.4287 9349	0.002479	6707.146558	0.000149	16.625354	0.060149	100

TABLE E-1 (*continued*)

n	CCA	CPW	CSCA	CSF	CSPW	CCR	n
7% (effective rate of 7.2508 1813%)							
1	1.0725 0818	0.932394	1.035831	0.965408	0.965803	1.035408	1
2	1.1502 7380	0.869358	2.146769	0.465816	1.866311	0.535816	2
3	1.2336 7806	0.810584	3.338258	0.299557	2.705939	0.369557	3
4	1.3231 2981	0.755784	4.616140	0.216631	3.488804	0.286631	4
5	1.4190 6755	0.704688	5.986679	0.167038	4.218742	0.237038	5
6	1.5219 6156	0.657047	7.456594	0.134109	4.899331	0.204109	6
7	1.6323 1622	0.612626	9.033089	0.110704	5.533909	0.180704	7
8	1.7506 7250	0.571209	10.723893	0.093250	6.125585	0.163250	8
9	1.8776 1058	0.532592	12.537294	0.079762	6.677260	0.149762	9
10	2.0137 5271	0.496585	14.482182	0.069050	7.191639	0.139050	10
11	2.1597 6625	0.463013	16.568089	0.060357	7.671242	0.130357	11
12	2.3163 6698	0.431711	18.805243	0.053177	8.118421	0.123177	12
13	2.4843 2253	0.402524	21.204608	0.047160	8.535368	0.117160	13
14	2.6644 5624	0.375311	23.777946	0.042056	8.924127	0.112056	14
15	2.8576 5112	0.349938	26.537873	0.037682	9.286604	0.107682	15
16	3.0648 5420	0.326280	29.497917	0.033901	9.624574	0.103901	16
17	3.2870 8121	0.304221	32.672589	0.030607	9.939696	0.100607	17
18	3.5254 2149	0.283654	36.077450	0.027718	10.233514	0.097718	18
19	3.7810 4339	0.264477	39.729191	0.025170	10.507468	0.095170	19
20	4.0551 9997	0.246597	43.645714	0.022912	10.762901	0.092912	20
21	4.3492 3514	0.229925	47.846216	0.020900	11.001064	0.090900	21
22	4.6645 9027	0.214381	52.351290	0.019102	11.223127	0.089102	22
23	5.0028 1123	0.199888	57.183018	0.017488	11.430177	0.087488	23
24	5.3655 5597	0.186374	62.365085	0.016035	11.623229	0.086035	24
25	5.7546 0268	0.173774	67.922895	0.014723	11.803229	0.084723	25
26	6.1718 5845	0.162026	73.883692	0.013535	11.971061	0.083535	26
27	6.6193 6868	0.151072	80.276695	0.012457	12.127546	0.082457	27
28	7.0993 2707	0.140858	87.133244	0.011477	12.273451	0.081477	28
29	7.6140 8636	0.131336	94.486948	0.010583	12.409493	0.080583	29
30	8.1661 6991	0.122456	102.373856	0.009768	12.536337	0.079768	30
31	8.7582 8404	0.114178	110.832629	0.009023	12.654605	0.079023	31
32	9.3933 3129	0.106459	119.904733	0.008340	12.764879	0.078340	32
33	10.0744 2466	0.099261	129.634638	0.007714	12.867696	0.077714	33
34	10.8049 0286	0.092551	140.070041	0.007139	12.963563	0.077139	34
35	11.5883 4672	0.086294	151.262096	0.006611	13.052949	0.076611	35
36	12.4285 9666	0.080460	163.265667	0.006125	13.136291	0.076125	36
37	13.3297 7160	0.075020	176.139594	0.005677	13.213999	0.075677	37
38	14.2962 8910	0.069948	189.946987	0.005265	13.286454	0.075265	38
39	15.3328 8702	0.065219	204.755529	0.004884	13.354010	0.074884	39
40	16.4446 4677	0.060810	220.637811	0.004532	13.416999	0.074532	40
45	23.3360 6458	0.042852	319.086637	0.003134	13.673541	0.073134	45
50	33.1154 5196	0.030197	458.792171	0.002180	13.854323	0.072180	50
55	46.9930 6323	0.021280	657.043760	0.001522	13.981718	0.071522	55
60	66.6863 3104	0.014996	938.376158	0.001066	14.071492	0.071066	60
65	94.6324 0831	0.010567	1337.605833	0.000748	14.134754	0.070748	65
70	134.2897 7968	0.007447	1904.139710	0.000525	14.179335	0.070525	70
75	190.5662 6846	0.005248	2708.089549	0.000369	14.210750	0.070369	75
80	270.4264 0743	0.003698	3848.948678	0.000260	14.232888	0.070260	80
90	544.5719 1013	0.001836	7765.313002	0.000129	14.259431	0.070129	90
100	1096.6331 5843	0.000912	15651.902263	0.000064	14.272687	0.070064	100

Continuous Compounding Factors

TABLE E-1 (*continued*)

8% (effective rate of 8.3287 0677%)

n	CCA	CPW	CSCA	CSF	CSPW	CCR	n
1	1.0832 8707	0.923116	1.041088	0.960533	0.961046	1.040533	1
2	1.1735 1087	0.852144	2.168886	0.461066	1.848203	0.541066	2
3	1.2712 4915	0.786628	3.390614	0.294932	2.667152	0.374932	3
4	1.3771 2776	0.726149	4.714097	0.212130	3.423137	0.292130	4
5	1.4918 2470	0.670320	6.147809	0.162660	4.120999	0.242660	5
6	1.6160 7440	0.618783	7.700930	0.129854	4.765208	0.209854	6
7	1.7506 7250	0.571209	9.383406	0.106571	5.359887	0.186571	7
8	1.8964 8088	0.527292	11.206011	0.089238	5.908845	0.169238	8
9	2.0544 3321	0.486752	13.180415	0.075870	6.415597	0.155870	9
10	2.2255 4093	0.449329	15.319262	0.065277	6.883388	0.145277	10
11	2.4108 9971	0.414783	17.636246	0.056701	7.315214	0.136701	11
12	2.6116 9647	0.382893	20.146206	0.049637	7.713839	0.129637	12
13	2.8292 1701	0.353455	22.865213	0.043735	8.081816	0.123735	13
14	3.0648 5420	0.326280	25.810678	0.038744	8.421503	0.118744	14
15	3.3201 1692	0.301194	29.001462	0.034481	8.735072	0.114481	15
16	3.5966 3973	0.278037	32.457997	0.030809	9.024534	0.110809	16
17	3.8961 9330	0.256661	36.202416	0.027622	9.291740	0.107622	17
18	4.2206 9582	0.236928	40.258698	0.024839	9.538403	0.104839	18
19	4.5722 2520	0.218712	44.652815	0.022395	9.766101	0.102395	19
20	4.9530 3242	0.201897	49.412905	0.020238	9.976294	0.100238	20
21	5.3655 5597	0.186374	54.569450	0.018325	10.170325	0.098325	21
22	5.8124 3739	0.172045	60.155467	0.016624	10.349439	0.096624	22
23	6.2965 3826	0.158817	66.206728	0.015104	10.514782	0.095104	23
24	6.8209 5847	0.146607	72.761981	0.013743	10.667413	0.093743	24
25	7.3890 5610	0.135335	79.363201	0.012521	10.808309	0.092521	25
26	8.0044 6891	0.124930	87.555861	0.011421	10.938372	0.091421	26
27	8.6711 3766	0.115325	95.889221	0.010429	11.058436	0.090429	27
28	9.3933 3129	0.106459	104.916641	0.009531	11.169269	0.089531	28
29	10.1756 7431	0.098274	114.695929	0.008719	11.271580	0.088719	29
30	11.0231 7638	0.090718	125.289705	0.007982	11.366026	0.087982	30
31	11.9412 6442	0.083743	136.765805	0.007312	11.453210	0.087312	31
32	12.9358 1732	0.077305	149.197716	0.006703	11.533691	0.086703	32
33	14.0132 0361	0.071361	162.665045	0.006148	11.607984	0.086148	33
34	15.1803 2224	0.065875	177.254028	0.005642	11.676566	0.085642	34
35	16.4446 4677	0.060810	193.058085	0.005180	11.739874	0.085180	35
36	17.8142 7318	0.056135	210.178415	0.004758	11.798315	0.084758	36
37	19.2979 7176	0.051819	228.724647	0.004372	11.852264	0.084372	37
38	20.9052 4324	0.047835	248.815540	0.004019	11.902064	0.084019	38
39	22.6463 7964	0.044157	270.579746	0.003696	11.948035	0.083696	39
40	24.5325 3020	0.040762	294.156627	0.003400	11.990472	0.083400	40
45	36.5982 3444	0.027324	444.977931	0.002247	12.158453	0.082247	45
50	54.5981 5003	0.018316	669.976875	0.001493	12.271055	0.081493	50
55	81.4508 6866	0.012277	1005.635858	0.000994	12.346533	0.080994	55
60	121.5104 1752	0.008230	1506.380219	0.000664	12.397128	0.080664	60
65	181.2722 4188	0.005517	2253.403023	0.000444	12.431043	0.080444	65
70	270.4264 0743	0.003698	3367.830093	0.000297	12.453777	0.080297	70
75	403.4287 9349	0.002479	5030.359919	0.000199	12.469016	0.080199	75
80	601.8450 3787	0.001662	7510.562973	0.000133	12.479231	0.080133	80
90	1339.4307 6439	0.000747	16730.384555	0.000060	12.490668	0.080060	90
100	2980.9579 8704	0.000335	37249.474838	0.000027	12.495807	0.080027	100

TABLE E-1 (continued)

9% (effective rate of 9.4174 2837%)

n	CCA	CPW	CSCA	CSF	CSPW	CCR	n
1	1.0941 7428	0.913931	1.046381	0.955675	0.956320	1.045675	1
2	1.1972 1736	0.835270	2.191304	0.456349	1.830331	0.546349	2
3	1.3099 6445	0.763379	3.444049	0.290356	2.629117	0.380356	3
4	1.4333 2941	0.697676	4.814771	0.207694	3.359152	0.297694	4
5	1.5683 1219	0.637628	6.314580	0.158364	4.026354	0.248364	5
6	1.7160 0686	0.582748	7.955632	0.125697	4.636131	0.215697	6
7	1.8776 1058	0.532592	9.751229	0.102551	5.193424	0.192551	7
8	2.0544 3321	0.486752	11.715925	0.085354	5.702753	0.175354	8
9	2.2479 0799	0.444858	13.865644	0.072121	6.168244	0.162121	9
10	2.4596 0311	0.406570	16.217812	0.061661	6.593670	0.151661	10
11	2.6912 3447	0.371577	18.791494	0.053216	6.982481	0.143216	11
12	2.9446 7955	0.339596	21.607551	0.046280	7.337827	0.136280	12
13	3.2219 9264	0.310367	24.688807	0.040504	7.662590	0.130504	13
14	3.5254 2149	0.283654	28.060239	0.035638	7.959400	0.125638	14
15	3.8574 2553	0.259240	31.749173	0.031497	8.230664	0.121497	15
16	4.2206 9582	0.236928	35.785509	0.027944	8.478580	0.117944	16
17	4.6181 7682	0.216536	40.201965	0.024874	8.705159	0.114874	17
18	5.0530 9032	0.197899	45.034337	0.022205	8.912237	0.112205	18
19	5.5289 6148	0.180866	50.321794	0.019872	9.101491	0.109872	19
20	6.0496 4746	0.165299	56.107194	0.017823	9.274457	0.107823	20
21	6.6193 6868	0.151072	62.437430	0.016016	9.432535	0.106016	21
22	7.2427 4299	0.138069	69.363811	0.014417	9.577008	0.104417	22
23	7.9248 2312	0.126186	76.942479	0.012997	9.709047	0.102997	23
24	8.6711 3766	0.115325	85.234863	0.011732	9.829721	0.101732	24
25	9.4877 3584	0.105399	94.308176	0.010604	9.940009	0.100604	25
26	10.3812 3656	0.096328	104.235962	0.009594	10.040804	0.099594	26
27	11.3588 8208	0.088037	115.098690	0.008688	10.132924	0.098688	27
28	12.4285 9666	0.080460	126.984407	0.007875	10.217115	0.097875	28
29	13.5990 5085	0.073535	139.989454	0.007143	10.294061	0.097143	29
30	14.8797 3172	0.067206	154.219241	0.006484	10.364383	0.096484	30
31	16.2810 1980	0.061421	169.789109	0.005890	10.428653	0.095890	31
32	17.8142 7318	0.056135	186.825258	0.005353	10.487392	0.095353	32
33	19.4919 1960	0.051303	205.465773	0.004867	10.541074	0.094867	33
34	21.3275 5716	0.046888	225.861746	0.004427	10.590137	0.094427	34
35	23.3360 6458	0.042852	248.178495	0.004029	10.634976	0.094029	35
36	25.5337 2175	0.039164	272.596908	0.003668	10.675957	0.093668	36
37	27.9383 4170	0.035793	299.314908	0.003341	10.713410	0.093341	37
38	30.5694 1502	0.032712	328.549056	0.003044	10.747640	0.093044	38
39	33.4482 6778	0.029897	360.536309	0.002774	10.778923	0.092774	39
40	36.5982 3444	0.027324	395.535938	0.002528	10.807514	0.092528	40
45	57.3974 5705	0.017422	626.638412	0.001596	10.917529	0.091596	45
50	90.0171 3130	0.011109	989.079237	0.001011	10.987678	0.091011	50
55	141.1749 6392	0.007083	1557.499599	0.000642	11.032407	0.090642	55
60	221.4064 1620	0.004517	2448.960180	0.000408	11.060927	0.090408	60
65	347.2343 8048	0.002880	3847.048672	0.000260	11.079112	0.090260	65
70	544.5719 1013	0.001836	6039.687890	0.000166	11.090708	0.090166	70
75	854.0587 6253	0.001171	9478.430695	0.000106	11.098101	0.090106	75
80	1339.4307 6439	0.000747	14871.452938	0.000067	11.102816	0.090067	80
90	3294.4680 7528	0.000304	36594.089725	0.000027	11.107738	0.090027	90
100	8103.0839 2757	0.000123	90023.154751	0.000011	11.109740	0.090011	100

Continuous Compounding Factors

TABLE E-1 (*continued*)

n	CCA	CPW	CSCA	CSF	CSPW	CCR	n
\multicolumn{7}{c}{10% (effective rate of 10.5170 9181%)}	
1	1.1051 7092	0.904837	1.051709	0.950833	0.951626	1.050833	1
2	1.2214 0276	0.818731	2.214028	0.451666	1.812692	0.551666	2
3	1.3498 5881	0.740818	3.498588	0.285830	2.591818	0.385830	3
4	1.4918 2470	0.670320	4.918247	0.203324	3.296800	0.303324	4
5	1.6487 2127	0.606531	6.487213	0.154149	3.934693	0.254149	5
6	1.8221 1880	0.548812	8.221188	0.121637	4.511884	0.221637	6
7	2.0137 5271	0.496585	10.137527	0.098643	5.034147	0.198643	7
8	2.2255 4093	0.449329	12.255409	0.081597	5.506710	0.181597	8
9	2.4596 0311	0.406570	14.596031	0.068512	5.934303	0.168512	9
10	2.7182 8183	0.367879	17.182818	0.058198	6.321206	0.158198	10
11	3.0041 6602	0.332871	20.041660	0.049896	6.671289	0.149896	11
12	3.3201 1692	0.301194	23.201169	0.043101	6.988058	0.143101	12
13	3.6692 9667	0.272532	26.692967	0.037463	7.274682	0.137463	13
14	4.0551 9997	0.246597	30.552000	0.032731	7.534030	0.132731	14
15	4.4816 8907	0.223130	34.816891	0.028722	7.768698	0.128722	15
16	4.9530 3242	0.201897	39.530324	0.025297	7.981035	0.125297	16
17	5.4739 4739	0.182684	44.739474	0.022352	8.173165	0.122352	17
18	6.0496 4746	0.165299	50.496475	0.019803	8.347011	0.119803	18
19	6.6858 9444	0.149569	56.858944	0.017587	8.504314	0.117587	19
20	7.3890 5610	0.135335	63.890561	0.015652	8.646647	0.115652	20
21	8.1661 6991	0.122456	71.661699	0.013954	8.775436	0.113954	21
22	9.0250 1350	0.110803	80.250135	0.012461	8.891968	0.112461	22
23	9.9741 8245	0.100259	89.741825	0.011143	8.997412	0.111143	23
24	11.0231 7638	0.090718	100.231764	0.009977	9.092820	0.109977	24
25	12.1824 9396	0.082085	111.824940	0.008943	9.179150	0.108943	25
26	13.4637 3804	0.074274	124.637380	0.008023	9.257264	0.108023	26
27	14.8797 3172	0.067206	138.797317	0.007205	9.327945	0.107205	27
28	16.4446 4677	0.060810	154.446468	0.006475	9.391899	0.106475	28
29	18.1741 4537	0.055023	171.741454	0.005823	9.449768	0.105823	29
30	20.0855 3692	0.049787	190.855369	0.005240	9.502129	0.105240	30
31	22.1979 5128	0.045049	211.979513	0.004717	9.549508	0.104717	31
32	24.5325 3020	0.040762	235.325302	0.004249	9.592378	0.104249	32
33	27.1126 3892	0.036883	261.126389	0.003830	9.631168	0.103830	33
34	29.9641 0005	0.033373	289.641000	0.003453	9.666267	0.103453	34
35	33.1154 5196	0.030197	321.154520	0.003114	9.698026	0.103114	35
36	36.5982 3444	0.027324	355.982344	0.002809	9.726763	0.102809	36
37	40.4473 0436	0.024724	394.473044	0.002535	9.752765	0.102535	37
38	44.7011 8449	0.022371	437.011845	0.002288	9.776292	0.102288	38
39	49.4024 4911	0.020242	484.024491	0.002066	9.797581	0.102066	39
40	54.5981 5003	0.018316	535.981500	0.001866	9.816844	0.101866	40
45	90.0171 3130	0.011109	890.171313	0.001123	9.888910	0.101123	45
50	148.4131 5910	0.006738	1474.131591	0.000678	9.932621	0.100678	50
55	244.6919 3226	0.004087	2436.919323	0.000410	9.959132	0.100410	55
60	403.4287 9349	0.002479	4024.287935	0.000248	9.975212	0.100248	60
65	665.1416 3304	0.001503	6641.416330	0.000151	9.984966	0.100151	65
70	1096.6331 5843	0.000912	10956.331584	0.000091	9.990881	0.100091	70
75	1808.0424 1446	0.000553	18070.424145	0.000055	9.994469	0.100055	75
80	2980.9579 8704	0.000335	29799.579870	0.000034	9.996645	0.100034	80
90	8103.0839 2758	0.000123	81020.839276	0.000012	9.998766	0.100012	90

TABLE E-1 (continued)

n	CCA	CPW	CSCA	CSF	CSPW	CCR	n
12% (effective rate of 12.7496 8516%)							
1	1.1274 9685	0.886920	1.062474	0.941200	0.942330	1.061200	1
2	1.2712 4915	0.786628	2.260410	0.442398	1.778101	0.562398	2
3	1.4333 2941	0.697676	3.611078	0.276926	2.519364	0.396926	3
4	1.6160 7440	0.618783	5.133953	0.194782	3.176805	0.314782	4
5	1.8221 1880	0.548812	6.850990	0.145964	3.759903	0.265964	5
6	2.0544 3321	0.486752	8.786943	0.113805	4.277065	0.233805	6
7	2.3163 6698	0.431711	10.969725	0.091160	4.735746	0.211160	7
8	2.6116 9647	0.382893	13.430804	0.074456	5.142559	0.194456	8
9	2.9446 7955	0.339596	16.205663	0.061707	5.503371	0.181707	9
10	3.3201 1692	0.301194	19.334308	0.051722	5.823382	0.171722	10
11	3.7434 2138	0.267135	22.861845	0.043741	6.107206	0.163741	11
12	4.2206 9582	0.236928	26.839132	0.037259	6.358935	0.157259	12
13	4.7588 2125	0.210136	31.323510	0.031925	6.582199	0.151925	13
14	5.3655 5597	0.186374	36.379633	0.027488	6.780217	0.147488	14
15	6.0496 4746	0.165299	42.080396	0.023764	6.955843	0.143764	15
16	6.8209 5847	0.146607	48.507987	0.020615	7.111609	0.140615	16
17	7.6906 0920	0.130029	55.755077	0.017936	7.249761	0.137936	17
18	8.6711 3766	0.115325	63.926147	0.015643	7.372291	0.135643	18
19	9.7766 8041	0.102284	73.139003	0.013673	7.480965	0.133673	19
20	11.0231 7638	0.090718	83.526470	0.011972	7.577350	0.131972	20
21	12.4285 9666	9.080460	95.238306	0.010500	7.662837	0.130500	21
22	14.0132 0361	0.071361	108.443363	0.009221	7.738656	0.129221	22
23	15.7998 4295	0.063292	123.332025	0.008108	7.805902	0.128108	23
24	17.8142 7318	0.056135	140.118943	0.007137	7.865544	0.127137	24
25	20.0855 3692	0.049787	159.046141	0.006287	7.918441	0.126287	25
26	22.6463 7964	0.044157	180.386497	0.005544	7.965357	0.125544	26
27	25.5337 2175	0.039164	204.447681	0.004891	8.006968	0.124891	27
28	28.7891 9088	0.034735	231.576591	0.004318	8.043873	0.124318	28
29	32.4597 2208	0.030807	262.164351	0.003814	8.076605	0.123814	29
30	36.5982 3444	0.027324	296.651954	0.003371	8.105636	0.123371	30
31	41.2643 9411	0.024234	335.536618	0.002980	8.131384	0.122980	31
32	46.5254 7444	0.021494	379.378954	0.002636	8.154220	0.122636	32
33	52.4573 2595	0.019063	428.811050	0.002332	8.174474	0.122332	33
34	59.1454 6985	0.016907	484.545582	0.002064	8.192438	0.122064	34
35	66.6863 3104	0.014996	547.386092	0.001827	8.208370	0.121827	35
36	75.1886 2829	0.013300	618.238569	0.001617	8.222501	0.121617	36
37	84.7749 4167	0.011796	698.124514	0.001432	8.235034	0.121432	37
38	95.5834 7983	0.010462	788.195665	0.001269	8.246150	0.121269	38
39	107.7700 7257	0.009279	889.750605	0.001124	8.256008	0.121124	39
40	121.5104 1752	0.008230	1004.253479	0.000996	8.264752	0.120996	40
45	221.4064 1620	0.004517	1836.720135	0.000544	8.295695	0.120544	45
50	403.4287 9349	0.002479	3353.573279	0.000298	8.312677	0.120298	50
55	735.0951 8924	0.001360	6117.459910	0.000163	8.321997	0.120163	55
60	1339.4307 6439	0.000747	11153.589703	0.000090	8.327112	0.120090	60
65	2440.6019 7762	0.000410	20330.016480	0.000049	8.329919	0.120049	65
70	4447.0667 4770	0.000225	37050.556231	0.000027	8.331459	0.120027	70
75	8103.0839 2757	0.000123	67517.366063	0.000015	8.332305	0.120015	75

Continuous Compounding Factors

TABLE E-1 (*continued*)

n	CCA	CPW	CSCA	CSF	CSPW	CCR	n
			15% (effective rate of 16.1834 2427%)				
1	1.1618 3424	0.860708	1.078895	0.926874	0.928613	1.076874	1
2	1.3498 5881	0.740818	2.332392	0.428744	1.727879	0.578744	2
3	1.5683 1219	0.637628	3.788748	0.263939	2.415812	0.413939	3
4	1.8221 1880	0.548812	5.480792	0.182455	3.007922	0.332455	4
5	2.1170 0002	0.472367	7.446667	0.134288	3.517556	0.284288	5
6	2.4596 0311	0.406570	9.730687	0.102768	3.956202	0.252768	6
7	2.8576 5112	0.349938	12.384341	0.080747	4.333748	0.230747	7
8	3.3201 1692	0.301194	15.467446	0.064652	4.658705	0.214652	8
9	3.8574 2553	0.259240	19.049504	0.052495	4.938398	0.202495	9
10	4.4816 8907	0.223130	23.211260	0.043083	5.179132	0.193083	10
11	5.2069 7983	0.192050	28.046532	0.035655	5.386334	0.185655	11
12	6.0496 4746	0.165299	33.664316	0.029705	5.564674	0.179705	12
13	7.0286 8758	0.142274	40.191251	0.024881	5.718173	0.174881	13
14	8.1661 6991	0.122456	47.774466	0.020932	5.850290	0.170932	14
15	9.4877 3584	0.105399	56.584906	0.017673	5.964005	0.167673	15
16	11.0231 7638	0.090718	66.821176	0.014965	6.061880	0.164965	16
17	12.8071 0378	0.078082	78.714025	0.012704	6.146122	0.162704	17
18	14.8797 3172	0.067206	92.531545	0.010807	6.218630	0.160807	18
19	17.2877 8184	0.057844	108.585212	0.009209	6.281038	0.159209	19
20	20.0855 3692	0.049787	127.236913	0.007859	6.334753	0.157859	20
21	23.3360 6458	0.042852	148.907097	0.006716	6.380986	0.156716	21
22	27.1126 3892	0.036883	174.084259	0.005744	6.420779	0.155744	22
23	31.5003 9231	0.031746	203.335949	0.004918	6.455029	0.154918	23
24	36.5982 3444	0.027324	237.321563	0.004214	6.484509	0.154214	24
25	42.5210 8200	0.023518	276.807213	0.003613	6.509882	0.153613	25
26	49.4024 4911	0.020242	322.682994	0.003099	6.531721	0.153099	26
27	57.3974 5705	0.017422	375.983047	0.002660	6.550518	0.152660	27
28	66.6863 3104	0.014996	437.908874	0.002284	6.566696	0.152284	28
29	77.4784 6293	0.012907	509.856420	0.001961	6.580621	0.151961	29
30	90.0171 3130	0.011109	593.447542	0.001685	6.592607	0.151685	30
31	104.5849 8558	0.009562	690.566571	0.001448	6.602923	0.151448	31
32	121.5104 1752	0.008230	803.402783	0.001245	6.611802	0.151245	32
33	141.1749 6392	0.007083	934.499759	0.001070	6.619444	0.151070	33
34	164.0219 0730	0.006097	1086.812715	0.000920	6.626022	0.150920	34
35	190.5662 6846	0.005248	1263.775123	0.000791	6.631683	0.150791	35
36	221.4064 1620	0.004517	1469.376108	0.000681	6.636556	0.150681	36
37	257.2375 5591	0.003887	1708.250373	0.000585	6.640750	0.150585	37
38	298.8674 0097	0.003346	1985.782673	0.000504	6.644360	0.150504	38
39	347.2343 8048	0.002880	2308.229203	0.000433	6.647467	0.150433	39
40	403.4287 9349	0.002479	2682.858623	0.000373	6.650142	0.150373	40
45	854.0587 6253	0.001171	5687.058417	0.000176	6.658861	0.150176	45
50	1808.0424 1446	0.000553	12046.949430	0.000083	6.662979	0.150083	50
55	3827.6258 2144	0.000261	25510.838810	0.000039	6.664925	0.150039	55
60	8103.0839 2757	0.000123	54013.892850	0.000019	6.665844	0.150019	60

TABLE E-1 (*continued*)

n	CCA	CPW	CSCA	CSF	CSPW	CCR	n
20% (effective rate of 22.1402 7582%)							
1	1.2214 0276	0.818731	1.107014	0.903331	0.906346	1.103331	1
2	1.4918 2470	0.670320	2.459123	0.406649	1.648400	0.606649	2
3	1.8221 1880	0.548812	4.110594	0.243274	2.255942	0.443274	3
4	2.2255 4093	0.449329	6.127705	0.163193	2.753355	0.363193	4
5	2.7182 8183	0.367879	8.591409	0.116395	3.160603	0.316395	5
6	3.3201 1692	0.301194	11.600585	0.086203	3.494029	0.286203	6
7	4.0551 9997	0.246597	15.276000	0.065462	3.767015	0.265462	7
8	4.9530 3242	0.201897	19.765162	0.050594	3.990517	0.250594	8
9	6.0496 4746	0.165299	25.248237	0.039607	4.173506	0.239607	9
10	7.3890 5610	0.135335	31.945280	0.031304	4.323324	0.231304	10
11	9.0250 1350	0.110803	40.125067	0.024922	4.445984	0.224922	11
12	11.0231 7638	0.090718	50.115882	0.019954	4.546410	0.219954	12
13	13.4637 3804	0.074274	62.318690	0.016047	4.628632	0.216047	13
14	16.4446 4677	0.060810	77.223234	0.012949	4.695950	0.212949	14
15	20.0855 3692	0.049787	95.427685	0.010479	4.751065	0.210479	15
16	24.5325 3020	0.040762	117.662651	0.008499	4.796189	0.208499	16
17	29.9641 0005	0.033373	144.820500	0.006905	4.833134	0.206905	17
18	36.5982 3444	0.027324	177.991172	0.005618	4.863381	0.205618	18
19	44.7011 8449	0.022371	218.505922	0.004577	4.888146	0.204577	19
20	54.5981 5003	0.018316	267.990750	0.003731	4.908422	0.203731	20
21	66.6863 3104	0.014996	328.431655	0.003045	4.925022	0.203045	21
22	81.4508 6867	0.012277	402.254343	0.002486	4.938613	0.202486	22
23	99.4843 1564	0.010052	492.421578	0.002031	4.949741	0.202031	23
24	121.5104 1752	0.008230	602.552088	0.001660	4.958851	0.201660	24
25	148.4131 5910	0.006738	737.065796	0.001357	4.966310	0.201357	25
26	181.2722 4188	0.005517	901.361209	0.001109	4.972417	0.201109	26
27	221.4064 1620	0.004517	1102.032081	0.000907	4.977417	0.200907	27
28	270.4264 0743	0.003698	1347.132037	0.000742	4.981511	0.200742	28
29	330.2995 5991	0.003028	1646.497800	0.000607	4.984862	0.200607	29
30	403.4287 9349	0.002479	2012.143967	0.000497	4.987606	0.200497	30
31	492.7490 4109	0.002029	2458.745205	0.000407	4.989853	0.200407	31
32	601.8450 3787	0.001662	3004.225189	0.000333	4.991692	0.200333	32
33	735.0951 8924	0.001360	3670.475946	0.000272	4.993198	0.200272	33
34	897.8472 9165	0.001114	4484.236458	0.000223	4.994431	0.200223	34
35	1096.6331 5843	0.000912	5478.165792	0.000183	4.995441	0.200183	35
36	1339.4307 6439	0.000747	6692.153822	0.000149	4.996267	0.200149	36
37	1635.9844 3000	0.000611	8174.922150	0.000122	4.996944	0.200122	37
38	1998.1958 9510	0.000500	9985.979476	0.000100	4.997498	0.200100	38
39	2440.6019 7762	0.000410	12198.009888	0.000082	4.997951	0.200082	39
40	2980.9579 8704	0.000335	14899.789935	0.000067	4.998323	0.200067	40
45	8103.0839 2758	0.000123	40510.419638	0.000025	4.999383	0.200025	45

Continuous Compounding Factors

TABLE E-1 (*continued*)

n	CCA	CPW	CSCA	CSF	CSPW	CCR	n
\multicolumn{7}{c}{25% (effective rate of 28.4025 4167%)}	

n	CCA	CPW	CSCA	CSF	CSPW	CCR	n
1	1.2840 2542	0.778801	1.136102	0.880203	0.884797	1.130203	1
2	1.6487 2127	0.606531	2.594885	0.385374	1.573877	0.635374	2
3	2.1170 0002	0.472367	4.468000	0.223814	2.110534	0.473814	3
4	2.7182 8183	0.367879	6.873127	0.145494	2.528482	0.395494	4
5	3.4903 4296	0.286505	9.961372	0.100388	2.853981	0.350388	5
6	4.4816 8907	0.223130	13.926756	0.071804	3.107479	0.321804	6
7	5.7546 0268	0.173774	19.018411	0.052581	3.304904	0.302581	7
8	7.3890 5610	0.135335	25.556224	0.039129	3.458659	0.289129	8
9	9.4877 3584	0.105399	33.950943	0.029454	3.578403	0.279454	9
10	12.1824 9396	0.082085	44.729976	0.022356	3.671660	0.272356	10
11	15.6426 3188	0.063928	58.570528	0.017073	3.744289	0.267073	11
12	20.0855 3692	0.049787	76.342148	0.013099	3.800852	0.263099	12
13	25.7903 3992	0.038774	99.161360	0.010085	3.844903	0.260085	13
14	33.1154 5196	0.030197	128.461808	0.007784	3.879210	0.257784	14
15	42.5210 8200	0.023518	166.084328	0.006021	3.905929	0.256021	15
16	54.5981 5003	0.018316	214.392600	0.004664	3.926737	0.254664	16
17	70.1054 1235	0.014264	276.421649	0.003618	3.942943	0.253618	17
18	90.0171 3130	0.011109	356.068525	0.002808	3.955564	0.252808	18
19	115.5842 8453	0.008652	458.337138	0.002182	3.965393	0.252182	19
20	148.4131 5910	0.006738	589.652636	0.001696	3.973048	0.251696	20
21	190.5662 6846	0.005248	758.265074	0.001319	3.979010	0.251319	21
22	244.6919 3226	0.004087	974.767729	0.001026	3.983653	0.251026	22
23	314.1906 6029	0.003183	1252.762641	0.000798	3.987269	0.250798	23
24	403.4287 9349	0.002479	1609.715174	0.000621	3.990085	0.250621	24
25	518.0128 2467	0.001930	2068.051299	0.000484	3.992278	0.250484	25
26	665.1416 3304	0.001503	2656.566532	0.000376	3.993986	0.250376	26
27	854.0587 6253	0.001171	3412.235050	0.000293	3.995316	0.250293	27
28	1096.6331 5843	0.000912	4382.532634	0.000228	3.996352	0.250228	28
29	1408.1048 4820	0.000710	5628.419393	0.000178	3.997159	0.250178	29
30	1808.0424 1446	0.000553	7228.169658	0.000138	3.997788	0.250138	30
31	2321.5724 1461	0.000431	9282.289658	0.000108	3.998277	0.250108	31
32	2980.9579 8704	0.000335	11919.831948	0.000084	3.998658	0.250084	32
33	3827.6258 2144	0.000261	15306.503286	0.000065	3.998955	0.250065	33
34	4914.7688 4030	0.000203	19655.075361	0.000051	3.999186	0.250051	34
35	6310.6881 0809	0.000158	25238.752432	0.000040	3.999366	0.250040	35
36	8103.0839 2757	0.000123	32408.335710	0.000031	3.999506	0.250031	36

TABLE E-2
Continuous Compounding Factors for Effective Rates of 1% to 25%

n	CCA	CPW	CSCA	CSF	CSPW	CCR	n
\multicolumn{7}{c}{0.9950 3308 5316 81% (effective rate of 1%)}	
1	1.0100 0000	0.990099	1.004992	0.995033	0.995041	1.004983	1
2	1.0201 0000	0.980296	2.020033	0.495041	1.980231	0.504992	2
3	1.0303 0100	0.970590	3.045225	0.328383	2.955666	0.338333	3
4	1.0406 0401	0.960980	4.080669	0.245058	3.921443	0.255008	4
5	1.0510 1005	0.951466	5.126468	0.195066	4.877658	0.205016	5
6	1.0615 2015	0.942045	6.182724	0.161741	5.824406	0.171691	6
7	1.0721 3535	0.932718	7.249543	0.137940	6.761780	0.147890	7
8	1.0828 5671	0.923483	8.327030	0.120091	7.689873	0.130041	8
9	1.0936 8527	0.914340	9.415292	0.106210	8.608777	0.116161	9
10	1.1046 2213	0.905287	10.514437	0.095107	9.518583	0.105058	10
11	1.1156 6835	0.896324	11.624573	0.086025	10.419380	0.095975	11
12	1.1268 2503	0.887449	12.745810	0.078457	11.311260	0.088407	12
13	1.1380 9328	0.878663	13.878260	0.072055	12.194308	0.082005	13
14	1.1494 7421	0.869963	15.022034	0.066569	13.068614	0.076519	14
15	1.1609 6896	0.861349	16.177247	0.061815	13.934263	0.071766	15
16	1.1725 7864	0.852821	17.344011	0.057657	14.791341	0.067607	16
17	1.1843 0443	0.844377	18.522443	0.053989	15.639933	0.063939	17
18	1.1961 4748	0.836017	19.712659	0.050729	16.480124	0.060679	18
19	1.2081 0895	0.827740	20.914777	0.047813	17.311996	0.057763	19
20	1.2201 9004	0.819544	22.128916	0.045190	18.135631	0.055140	20
21	1.2323 9194	0.811430	23.355197	0.042817	18.951112	0.052767	21
22	1.2447 1586	0.803396	24.593741	0.040661	19.758518	0.050611	22
23	1.2571 6302	0.795442	25.844670	0.038693	20.557931	0.048643	23
24	1.2697 3465	0.787566	27.108109	0.036889	21.349428	0.046840	24
25	1.2824 3200	0.779768	28.384181	0.035231	22.133089	0.045181	25
26	1.2952 5632	0.772048	29.673015	0.033701	22.908991	0.043651	26
27	1.3082 0888	0.764404	30.974737	0.032284	23.677210	0.042235	27
28	1.3212 9097	0.756836	32.289476	0.030970	24.437824	0.040920	28
29	1.3345 0388	0.749342	33.617362	0.029747	25.190906	0.039697	29
30	1.3478 4892	0.741923	34.958528	0.028605	25.936533	0.038556	30
31	1.3613 2740	0.734577	36.313105	0.027538	26.674777	0.037489	31
32	1.3749 4068	0.727304	37.681227	0.026538	27.405711	0.036489	32
33	1.3886 9009	0.720103	39.063031	0.025600	28.129409	0.035550	33
34	1.4025 7699	0.712973	40.458653	0.024717	28.845941	0.034667	34
35	1.4166 0276	0.705914	41.868232	0.023884	29.555379	0.033835	35
36	1.4307 6878	0.698925	43.291906	0.023099	30.257793	0.033049	36
37	1.4450 7647	0.692005	44.729816	0.022356	30.953252	0.032307	37
38	1.4595 2724	0.685153	46.182106	0.021653	31.641826	0.031604	38
39	1.4741 2251	0.678370	47.648919	0.020987	32.323581	0.030937	39
40	1.4888 6373	0.671653	49.130400	0.020354	32.998587	0.030304	40
45	1.5648 1075	0.639055	56.763012	0.017617	36.274682	0.027567	45
50	1.6446 3182	0.608039	64.784964	0.015436	39.391773	0.025386	50
55	1.7285 2457	0.578528	73.216116	0.013658	42.357579	0.023609	55
60	1.8166 9670	0.550450	82.077341	0.012184	45.179441	0.022134	60
65	1.9093 6649	0.523734	91.390578	0.010942	47.864346	0.020892	65
70	2.0067 6337	0.498315	101.178884	0.009883	50.418941	0.019834	70
75	2.1091 2847	0.474129	111.466491	0.008971	52.849550	0.018922	75
80	2.2167 1522	0.451118	122.278870	0.008178	55.162192	0.018128	80
90	2.4486 3267	0.408391	145.586383	0.006869	59.456195	0.016819	90
100	2.7048 1383	0.369711	171.332376	0.005837	63.343501	0.015787	100

Continuous Compounding Factors

TABLE E-2 (*continued*)

n	CCA	CPW	CSCA	CSF	CSPW	CCR	n
\multicolumn{7}{c}{1.9802 6272 9617 97% (effective rate of 2%)}	
1	1.0200 0000	0.980392	1.009967	0.990131	0.990164	1.009934	1
2	1.0404 0000	0.961169	2.040133	0.490164	1.960912	0.509967	2
3	1.0612 0800	0.942322	3.090903	0.323530	2.912627	0.343333	3
4	1.0824 3216	0.923845	4.162688	0.240229	3.845680	0.260032	4
5	1.1040 8080	0.905731	5.255909	0.190262	4.760439	0.210065	5
6	1.1261 6242	0.887971	6.370994	0.156961	5.657260	0.176764	6
7	1.1486 8567	0.870560	7.508381	0.133185	6.536497	0.152987	7
8	1.1716 5938	0.853490	8.668515	0.115360	7.398494	0.135163	8
9	1.1950 9257	0.836755	9.851853	0.101504	8.243590	0.121306	9
10	1.2189 9442	0.820348	11.058857	0.090425	9.072114	0.110228	10
11	1.2433 7431	0.804263	12.290001	0.081367	9.884394	0.101170	11
12	1.2682 4179	0.788493	13.545768	0.073824	10.680746	0.093626	12
13	1.2936 0663	0.773033	14.826650	0.067446	11.461483	0.087249	13
14	1.3194 7876	0.757875	16.133150	0.061984	12.226912	0.081787	14
15	1.3458 6834	0.743015	17.465780	0.057255	12.977332	0.077057	15
16	1.3727 8571	0.728446	18.825063	0.053121	13.713038	0.072923	16
17	1.4002 4142	0.714163	20.211531	0.049477	14.434319	0.069279	17
18	1.4282 4625	0.700159	21.625729	0.046241	15.141457	0.066044	18
19	1.4568 1117	0.686431	23.068210	0.043350	15.834729	0.063152	19
20	1.4859 4740	0.672971	24.539542	0.040751	16.514408	0.060553	20
21	1.5156 6634	0.659776	26.040299	0.038402	17.180760	0.058205	21
22	1.5459 7967	0.646839	27.571072	0.036270	17.834046	0.056073	22
23	1.5768 9926	0.634156	29.132461	0.034326	18.474522	0.054129	23
24	1.6084 3725	0.621721	30.725077	0.032547	19.102441	0.052349	24
25	1.6406 0599	0.609531	32.349546	0.030912	19.718047	0.050715	25
26	1.6734 1811	0.597579	34.006503	0.029406	20.321582	0.049209	26
27	1.7068 8648	0.585862	35.696601	0.028014	20.913283	0.047816	27
28	1.7410 2421	0.574375	37.420500	0.026723	21.493383	0.046526	28
29	1.7758 4469	0.563112	39.178877	0.025524	22.062108	0.045327	29
30	1.8113 6158	0.552071	40.972421	0.024407	22.619681	0.044209	30
31	1.8475 8882	0.541246	42.801836	0.023363	23.166321	0.043166	31
32	1.8845 4059	0.530633	44.667840	0.022387	23.702244	0.042190	32
33	1.9222 3140	0.520229	46.571164	0.021473	24.227657	0.041275	33
34	1.9606 7603	0.510028	48.512554	0.020613	24.742769	0.040416	34
35	1.9998 8955	0.500028	50.492772	0.019805	25.247780	0.039607	35
36	2.0398 8734	0.490223	52.512595	0.019043	25.742890	0.038846	36
37	2.0806 8509	0.480611	54.572814	0.018324	26.228291	0.038127	37
38	2.1222 9879	0.471187	56.674237	0.017645	26.704174	0.037447	38
39	2.1647 4477	0.461948	58.817689	0.017002	27.170727	0.036804	39
40	2.2080 3966	0.452890	61.004009	0.016392	27.628131	0.036195	40
45	2.4378 5421	0.410197	72.609265	0.013772	29.784088	0.033575	45
50	2.6915 8803	0.371528	85.422404	0.011707	31.736805	0.031509	50
55	2.9717 3067	0.336504	99.569145	0.010043	33.505440	0.029846	55
60	3.2810 3079	0.304782	115.188291	0.008681	35.107348	0.028484	60
65	3.6225 2311	0.276051	132.433089	0.007551	36.558246	0.027354	65
70	3.9995 5822	0.250028	151.472740	0.006602	37.872368	0.026404	70
75	4.4158 3546	0.226458	172.494054	0.005797	39.062609	0.025600	75
80	4.8754 3916	0.205110	195.703282	0.005110	40.140647	0.024912	80
90	5.9431 3313	0.168261	249.620066	0.004006	42.001426	0.023809	90
100	7.2446 4612	0.138033	315.344324	0.003171	43.527913	0.022974	100

Continuous Compounding Factors

TABLE E-2 (*continued*)

n	CCA	CPW	CSCA	CSF	CSPW	CCR	n
		2.9558 8022 4154 44% (effective rate of 3%)					
1	1.0300 0000	0.970874	1.014926	0.985293	0.985365	1.014852	1
2	1.0609 0000	0.942596	2.060300	0.485366	1.942030	0.514925	2
3	1.0927 2700	0.915142	3.137035	0.318772	2.870832	0.348331	3
4	1.1255 0881	0.888487	4.246072	0.235512	3.772580	0.265071	4
5	1.1592 7407	0.862609	5.388381	0.185585	4.648064	0.215143	5
6	1.1940 5230	0.837484	6.564958	0.152324	5.498049	0.181883	6
7	1.2298 7387	0.813092	7.776833	0.128587	6.323277	0.158146	7
8	1.2667 7008	0.789409	9.025064	0.110803	7.124469	0.140361	8
9	1.3047 7318	0.766417	10.310742	0.096986	7.902325	0.126545	9
10	1.3439 1638	0.744094	11.634990	0.085948	8.657526	0.115506	10
11	1.3842 3387	0.722421	12.998966	0.076929	9.390730	0.106488	11
12	1.4257 6089	0.701380	14.403861	0.069426	10.102578	0.098985	12
13	1.4685 3371	0.680951	15.850903	0.063088	10.793694	0.092647	13
14	1.5125 8972	0.661118	17.341356	0.057666	11.464680	0.087224	14
15	1.5579 6742	0.641862	18.876523	0.052976	12.116122	0.082535	15
16	1.6047 0644	0.623167	20.457745	0.048881	12.748590	0.078440	16
17	1.6528 4763	0.605016	22.086403	0.045277	13.362637	0.074836	17
18	1.7024 3306	0.587395	23.763922	0.042081	13.958799	0.071639	18
19	1.7535 0605	0.570286	25.491765	0.039228	14.537598	0.068787	19
20	1.8061 1123	0.553676	27.271444	0.036668	15.099538	0.066227	20
21	1.8602 9457	0.537549	29.104514	0.034359	15.645110	0.063918	21
22	1.9161 0341	0.521893	30.992575	0.032266	16.174793	0.061825	22
23	1.9735 8651	0.506692	32.937279	0.030361	16.689047	0.059920	23
24	2.0327 9411	0.491934	34.940323	0.028620	17.188324	0.053179	24
25	2.0937 7793	0.477606	37.003459	0.027025	17.673058	0.056583	25
26	2.1565 9127	0.463695	39.128489	0.025557	18.143674	0.055116	26
27	2.2212 8901	0.450189	41.317270	0.024203	18.600583	0.053762	27
28	2.2879 2768	0.437077	43.571714	0.022951	19.044183	0.052509	28
29	2.3565 6551	0.424346	45.893791	0.021789	19.474863	0.051348	29
30	2.4272 6247	0.411987	48.285531	0.020710	19.893000	0.050269	30
31	2.5000 8035	0.399987	50.749023	0.019705	20.298957	0.049264	31
32	2.5750 8276	0.388337	53.286420	0.018767	20.693090	0.048325	32
33	2.6523 3524	0.377026	55.899939	0.017889	21.075744	0.047448	33
34	2.7319 0530	0.366045	58.591863	0.017067	21.447253	0.046626	34
35	2.8138 6245	0.355383	61.364545	0.016296	21.807941	0.045855	35
36	2.8982 7833	0.345032	64.220408	0.015571	22.158123	0.045130	36
37	2.9852 2668	0.334983	67.161946	0.014889	22.498106	0.044448	37
38	3.0747 8348	0.325226	70.191730	0.014247	22.828186	0.043805	38
39	3.1670 2698	0.315754	73.312408	0.013640	23.148653	0.043199	39
40	3.2620 3779	0.306557	76.526707	0.013067	23.459785	0.042626	40
45	3.7815 9584	0.264439	94.103808	0.010627	24.884681	0.040185	45
50	4.3839 0602	0.228107	114.480485	0.008735	26.113809	0.038294	50
55	5.0821 4859	0.196767	138.102639	0.007241	27.174066	0.036800	55
60	5.8916 0310	0.169733	165.487189	0.006043	28.088652	0.035602	60
65	6.8299 8273	0.146413	197.233389	0.005070	28.877582	0.034629	65
70	7.9178 2191	0.126297	234.035935	0.004273	29.558121	0.033832	70
75	9.1789 2567	0.108945	276.700172	0.003614	30.145159	0.033173	75
80	10.6408 9056	0.093977	326.159716	0.003066	30.651543	0.032625	80
90	14.3004 6711	0.069928	449.966376	0.002222	31.465152	0.031781	90
100	19.2186 3198	0.052033	616.352173	0.001622	32.070554	0.031181	100

Continuous Compounding Factors

TABLE E-2 (*continued*)

n	CCA	CPW	CSCA	CSF	CSPW	CCR	n
\multicolumn{7}{c}{3.9220 7131 5328 13% (effective rate of 4%)}	
1	1.0400 0000	0.961538	1.019869	0.980518	0.980644	1.019739	1
2	1.0816 0000	0.924556	2.080533	0.480646	1.923570	0.519867	2
3	1.1248 6400	0.888996	3.183624	0.314107	2.830230	0.353328	3
4	1.1698 5856	0.854804	4.330838	0.230902	3.702019	0.270123	4
5	1.2166 5290	0.821927	5.523941	0.181030	4.540277	0.220251	5
6	1.2653 1902	0.790315	6.764768	0.147825	5.346294	0.187045	6
7	1.3159 3178	0.759918	8.055228	0.124143	6.121311	0.163364	7
8	1.3685 6905	0.730690	9.397306	0.106413	6.866520	0.145634	8
9	1.4233 1181	0.702587	10.793068	0.092652	7.583066	0.131873	9
10	1.4802 4428	0.675564	12.244660	0.081668	8.272053	0.120889	10
11	1.5394 5406	0.649581	13.754315	0.072704	8.934541	0.111925	11
12	1.6010 3222	0.624597	15.324357	0.065256	9.571548	0.104476	12
13	1.6650 7351	0.600574	16.957201	0.058972	10.184055	0.098193	13
14	1.7316 7645	0.577475	18.655358	0.053604	10.773004	0.092825	14
15	1.8009 4351	0.555265	20.421442	0.048968	11.339302	0.088189	15
16	1.8729 8125	0.533908	22.258169	0.044927	11.883818	0.084148	16
17	1.9479 0050	0.513373	24.168365	0.041376	12.407392	0.080597	17
18	2.0258 1652	0.493628	26.154968	0.038234	12.910828	0.077454	18
19	2.1068 4918	0.474642	28.221036	0.035435	13.394901	0.074655	19
20	2.1911 2314	0.456387	30.369747	0.032928	13.860356	0.072148	20
21	2.2787 6807	0.438834	32.604406	0.030671	14.307909	0.069891	21
22	2.3699 1879	0.421955	34.928452	0.028630	14.738248	0.067851	22
23	2.4647 1554	0.405726	37.345459	0.026777	15.152036	0.065998	23
24	2.5633 0416	0.390121	39.859147	0.025088	15.549909	0.064309	24
25	2.6658 3633	0.375117	42.473382	0.023544	15.932479	0.062765	25
26	2.7724 6978	0.360689	45.192187	0.022128	16.300335	0.061348	26
27	2.8833 6858	0.346817	48.019743	0.020825	16.654043	0.060045	27
28	2.9987 0332	0.333477	50.960402	0.019623	16.994146	0.058844	28
29	3.1186 5145	0.320651	54.018688	0.018512	17.321169	0.057733	29
30	3.2433 9751	0.308319	57.199304	0.017483	17.635613	0.056703	30
31	3.3731 3341	0.296460	60.507146	0.016527	17.937964	0.055748	31
32	3.5080 5875	0.285058	63.947301	0.015638	18.228686	0.054859	32
33	3.6483 8110	0.274094	67.525062	0.014809	18.508226	0.054030	33
34	3.7943 1634	0.263552	71.245934	0.014036	18.777015	0.053257	34
35	3.9460 8899	0.253415	75.115641	0.013313	19.035465	0.052534	35
36	4.1039 3255	0.243669	79.140136	0.012636	19.283976	0.051857	36
37	4.2680 8986	0.234297	83.325610	0.012001	19.522928	0.051222	37
38	4.4388 1345	0.225285	87.678504	0.011405	19.752690	0.050626	38
39	4.6163 6599	0.216621	92.205513	0.010845	19.973614	0.050066	39
40	4.8010 2063	0.208289	96.913603	0.010318	20.186042	0.049539	40
45	5.8411 7568	0.171198	123.434157	0.008101	21.131732	0.047322	45
50	7.1066 8335	0.140713	155.700467	0.006423	21.909020	0.045643	50
55	8.6463 6692	0.115656	194.957366	0.005129	22.547894	0.044350	55
60	10.5196 2741	0.095060	242.719386	0.004120	23.073002	0.043341	60
65	12.7987 3522	0.078133	300.829186	0.003324	23.504603	0.042545	65
70	15.5716 1835	0.064219	371.528643	0.002692	23.859347	0.041912	70
75	18.9452 5466	0.052784	457.545343	0.002186	24.150921	0.041406	75
80	23.0497 9907	0.043384	562.197811	0.001779	24.390573	0.040999	80
90	34.1193 3334	0.029309	844.434756	0.001184	24.749451	0.040405	90
100	50.5049 4818	0.019800	1262.214381	0.000792	24.991895	0.040013	100

TABLE E-2 (continued)

n	CCA	CPW	CSCA	CSF	CSPW	CCR	n
\multicolumn{7}{c}{4.8790 1641 6943 20% (effective rate of 5%)}	

n	CCA	CPW	CSCA	CSF	CSPW	CCR	n
1	1.0500 0000	0.952381	1.024797	0.975803	0.975997	1.024593	1
2	1.1025 0000	0.907029	2.100833	0.476002	1.905518	0.524792	2
3	1.1576 2500	0.863838	3.230672	0.309533	2.790776	0.358323	3
4	1.2155 0625	0.822702	4.417002	0.226398	3.633878	0.275188	4
5	1.2762 8156	0.783526	5.662649	0.176596	4.436833	0.225386	5
6	1.3400 9564	0.746215	6.970578	0.143460	5.201553	0.192250	6
7	1.4071 0042	0.710681	8.343904	0.119848	5.929856	0.168638	7
8	1.4774 5544	0.676839	9.785895	0.102188	6.623479	0.150978	8
9	1.5513 2822	0.644609	11.299987	0.088496	7.284072	0.137286	9
10	1.6288 9463	0.613913	12.889783	0.077581	7.913209	0.126371	10
11	1.7103 3936	0.584679	14.559069	0.068686	8.512386	0.117476	11
12	1.7958 5633	0.556837	16.311819	0.061305	9.083031	0.110095	12
13	1.8856 4914	0.530321	18.152207	0.055090	9.626503	0.103880	13
14	1.9799 3160	0.505068	20.084614	0.049789	10.144095	0.098580	14
15	2.0789 2818	0.481017	22.113641	0.045221	10.637039	0.094011	15
16	2.1828 7459	0.458112	24.244120	0.041247	11.106511	0.090037	16
17	2.2920 1832	0.436297	26.481123	0.037763	11.553626	0.086553	17
18	2.4066 1923	0.415521	28.829975	0.034686	11.979450	0.083476	18
19	2.5269 5020	0.395734	31.296271	0.031953	12.384997	0.080743	19
20	2.6532 9771	0.376889	33.885881	0.029511	12.771232	0.078301	20
21	2.7859 6259	0.358942	36.604972	0.027319	13.139075	0.076109	21
22	2.9252 6072	0.341850	39.460017	0.025342	13.489402	0.074132	22
23	3.0715 2376	0.325571	42.457815	0.023553	13.823046	0.072343	23
24	3.2250 9994	0.310068	45.605502	0.021927	14.140803	0.070717	24
25	3.3863 5494	0.295303	48.910574	0.020445	14.443428	0.069236	25
26	3.5556 7269	0.281241	52.380900	0.019091	14.731643	0.067881	26
27	3.7334 5632	0.267848	56.024741	0.017849	15.006133	0.066639	27
28	3.9201 2914	0.255094	59.850775	0.016708	15.267552	0.065498	28
29	4.1161 3560	0.242946	63.868110	0.015657	15.516522	0.064447	29
30	4.3219 4238	0.231377	68.086313	0.014687	15.753637	0.063477	30
31	4.5380 3949	0.220359	72.515425	0.013790	15.979461	0.062580	31
32	4.7649 4147	0.209866	77.165993	0.012959	16.194531	0.061749	32
33	5.0031 8854	0.199873	82.049089	0.012188	16.399360	0.060978	33
34	5.2533 4797	0.190355	87.176341	0.011471	16.594435	0.060261	34
35	5.5160 1537	0.181290	92.559954	0.010804	16.780221	0.059594	35
36	5.7918 1614	0.172657	98.212749	0.010182	16.957159	0.058972	36
37	6.0814 0694	0.164436	104.148183	0.009602	17.125672	0.058392	37
38	6.3854 7729	0.156605	110.380389	0.009060	17.286161	0.057850	38
39	6.7047 5115	0.149148	116.924205	0.008553	17.439007	0.057343	39
40	7.0399 8871	0.142046	123.795212	0.008078	17.584575	0.056868	40
45	8.9850 0779	0.111297	163.660195	0.006110	18.214808	0.054900	45
50	11.4673 9979	0.087204	214.539138	0.004661	18.708612	0.053451	50
55	14.6356 3092	0.068326	279.474996	0.003578	19.095521	0.052368	55
60	18.6791 8589	0.053536	362.351433	0.002760	19.398674	0.051550	60
65	23.8399 0056	0.041946	468.125102	0.002136	19.636202	0.050926	65
70	30.4264 2554	0.032866	603.122085	0.001658	19.822312	0.050448	70
75	38.8326 8592	0.025752	775.416246	0.001290	19.968133	0.050080	75
80	49.5614 4107	0.020177	995.312106	0.001005	20.082388	0.049795	80
90	80.7303 6505	0.012387	1634.148325	0.000612	20.242053	0.049402	90
100	131.5012 5785	0.007604	2674.745209	0.000374	20.340073	0.049164	100

Continuous Compounding Factors

TABLE E-2 (*continued*)

n	CCA	CPW	CSCA	CSF	CSPW	CCR	n
		5.8268 9081 2397 58%	(effective rate of 6%)				
1	1.0600 0000	0.943396	1.029709	0.971148	0.971423	1.029417	1
2	1.1236 0000	0.889996	2.121200	0.471431	1.887860	0.529700	2
3	1.1910 1600	0.839619	3.278181	0.305047	2.752424	0.363316	3
4	1.2624 7696	0.792094	4.504580	0.221996	3.568049	0.280265	4
5	1.3382 2558	0.747258	5.804564	0.172278	4.337508	0.230547	5
6	1.4185 1911	0.704961	7.182546	0.139226	5.063412	0.197495	6
7	1.5036 3026	0.665057	8.643207	0.115698	5.748227	0.173967	7
8	1.5938 4807	0.627412	10.191509	0.098121	6.394279	0.156390	8
9	1.6894 7896	0.591898	11.832708	0.084512	7.003762	0.142780	9
10	1.7908 4770	0.558395	13.572379	0.073679	7.578745	0.131948	10
11	1.8982 9856	0.526788	15.416430	0.064866	8.121183	0.123135	11
12	2.0121 9647	0.496969	17.371125	0.057567	8.632917	0.115836	12
13	2.1329 2826	0.468839	19.443101	0.051432	9.115584	0.109701	13
14	2.2609 0396	0.442301	21.639396	0.046212	9.571126	0.104481	14
15	2.3965 5819	0.417265	23.967468	0.041723	10.000787	0.099992	15
16	2.5403 5168	0.393646	26.435225	0.037828	10.406128	0.096097	16
17	2.6927 7279	0.371364	29.051047	0.034422	10.788525	0.092691	17
18	2.8543 3915	0.350344	31.823818	0.031423	11.149277	0.089692	18
19	3.0255 9950	0.330513	34.762956	0.028766	11.489609	0.087035	19
20	3.2071 3547	0.311805	37.878442	0.026400	11.810677	0.084669	20
21	3.3995 6360	0.294155	41.180857	0.024283	12.113572	0.082552	21
22	3.6035 3742	0.277505	44.681418	0.022381	12.399321	0.080650	22
23	3.8197 4966	0.261797	48.392011	0.020665	12.668896	0.078933	23
24	4.0489 3464	0.246979	52.325241	0.019111	12.923212	0.077380	24
25	4.2918 7072	0.232999	56.494464	0.017701	13.163133	0.075970	25
26	4.5493 8296	0.219810	60.913840	0.016417	13.389473	0.074686	26
27	4.8223 4594	0.207368	65.598379	0.015244	13.603002	0.073513	27
28	5.1116 8670	0.195630	70.563991	0.014172	13.804444	0.072440	28
29	5.4183 8790	0.184557	75.827539	0.013188	13.994483	0.071457	29
30	5.7434 9117	0.174110	81.406900	0.012284	14.173766	0.070553	30
31	6.0881 0064	0.164255	87.321023	0.011452	14.342901	0.069721	31
32	6.4533 8668	0.154957	93.589993	0.010685	14.502462	0.068954	32
33	6.8405 8988	0.146186	100.235101	0.009977	14.652991	0.068245	33
34	7.2510 2528	0.137912	107.278916	0.009321	14.795000	0.067590	34
35	7.6860 8679	0.130105	114.745359	0.008715	14.928970	0.066984	35
36	8.1472 5200	0.122741	122.659789	0.008153	15.055357	0.066422	36
37	8.6360 8712	0.115793	131.049085	0.007631	15.174590	0.065900	37
38	9.1542 5235	0.109239	139.941739	0.007146	15.287075	0.065415	38
39	9.7035 0749	0.103056	149.367952	0.006695	15.393192	0.064964	39
40	10.2857 1794	0.097222	159.359738	0.006275	15.493302	0.064544	40
45	13.7646 1083	0.072650	219.063841	0.004565	15.915004	0.062834	45
50	18.4201 5428	0.054288	298.961399	0.003345	16.230125	0.061614	50
55	24.6503 2159	0.040567	405.882354	0.002464	16.465601	0.060733	55
60	32.9876 9085	0.030314	548.966711	0.001822	16.641562	0.060091	60
65	44.1449 7165	0.022653	740.445858	0.001351	16.773051	0.059619	65
70	59.0759 3018	0.016927	996.688149	0.001003	16.871307	0.059272	70
75	79.0569 2079	0.012649	1339.598137	0.000746	16.944729	0.059015	75
80	105.7959 9348	0.009452	1798.489005	0.000556	16.999595	0.058825	80
90	189.4645 1123	0.005278	3234.392359	0.000309	17.071231	0.058578	90
100	339.3020 8351	0.002947	5805.876486	0.000172	17.111231	0.058441	100

Continuous Compounding Factors

TABLE E-2 (*continued*)

n	CCA	CPW	CSCA	CSF	CSPW	CCR	n
6.7658 6484 7381 48% (effective rate of 7%)							
1	1.0700 0000	0.934579	1.034605	0.966552	0.966921	1.034211	1
2	1.1449 0000	0.873439	2.141633	0.466933	1.870585	0.534592	2
3	1.2250 4300	0.816298	3.326153	0.300648	2.715131	0.368306	3
4	1.3107 9601	0.762895	4.593589	0.217695	3.504427	0.285353	4
5	1.4025 5173	0.712986	5.949745	0.168074	4.242086	0.235733	5
6	1.5007 3035	0.666342	7.400833	0.135120	4.931487	0.202779	6
7	1.6057 8148	0.622750	8.953497	0.111688	5.575788	0.179347	7
8	1.7181 8618	0.582009	10.614847	0.094208	6.177937	0.161866	8
9	1.8384 5921	0.543934	12.392491	0.080694	6.740694	0.148353	9
10	1.9671 5136	0.508349	14.294571	0.069957	7.266635	0.137615	10
11	2.1048 5195	0.475093	16.329796	0.061238	7.758169	0.128896	11
12	2.2521 9159	0.444012	18.507487	0.054032	8.217546	0.121691	12
13	2.4098 4500	0.414964	20.837617	0.047990	8.646870	0.115649	13
14	2.5785 3415	0.387817	23.330855	0.042862	9.048108	0.110520	14
15	2.7590 3154	0.362446	25.998621	0.038464	9.423097	0.106122	15
16	2.9521 6375	0.338735	28.853130	0.034658	9.773553	0.102317	16
17	3.1588 1521	0.316574	31.907454	0.031341	10.101083	0.098999	17
18	3.3799 3228	0.295864	35.175581	0.028429	10.407185	0.096087	18
19	3.6165 2754	0.276508	38.672477	0.025858	10.693262	0.093517	19
20	3.8696 8446	0.258419	42.414156	0.023577	10.960624	0.091236	20
21	4.1405 6237	0.241513	46.417752	0.021543	11.210495	0.089202	21
22	4.4304 0174	0.225713	50.701600	0.019723	11.444019	0.087382	22
23	4.7405 2986	0.210947	55.285318	0.018088	11.662265	0.085747	23
24	5.0723 6695	0.197147	60.189895	0.016614	11.866234	0.084273	24
25	5.4274 3264	0.184249	65.437793	0.015282	12.056860	0.082940	25
26	5.8073 5292	0.172195	71.053044	0.014074	12.235014	0.081733	26
27	6.2138 6763	0.160930	77.061362	0.012977	12.401513	0.080635	27
28	6.6488 3836	0.150402	83.490263	0.011977	12.557120	0.079636	28
29	7.1142 5705	0.140563	90.369187	0.011066	12.702547	0.078724	29
30	7.6122 5504	0.131367	97.729635	0.010232	12.838460	0.077891	30
31	8.1451 1290	0.122773	105.605315	0.009469	12.965482	0.077128	31
32	8.7152 7080	0.114741	114.032293	0.008769	13.084194	0.076428	32
33	9.3253 3975	0.107235	123.049158	0.008127	13.195139	0.075785	33
34	9.9781 1354	0.100219	132.697205	0.007536	13.298827	0.075195	34
35	10.6765 8148	0.093663	143.020615	0.006992	13.395731	0.074651	35
36	11.4239 4219	0.087535	154.066663	0.006491	13.486296	0.074149	36
37	12.2236 1814	0.081809	165.885935	0.006028	13.570936	0.073687	37
38	13.0792 7141	0.076457	178.532555	0.005601	13.650038	0.073260	38
39	13.9948 2041	0.071455	192.064440	0.005207	13.723966	0.072865	39
40	14.9744 5784	0.066780	206.543556	0.004842	13.793057	0.072500	40
45	21.0024 5176	0.047613	295.637767	0.003383	14.076345	0.071041	45
50	29.4570 2506	0.033948	420.597007	0.002378	14.278326	0.070036	50
55	41.3150 0148	0.024204	595.858896	0.001678	14.422335	0.069337	55
60	57.9464 2683	0.017257	841.672545	0.001188	14.525012	0.068847	60
65	81.2728 6124	0.012304	1186.439030	0.000843	14.598219	0.068502	65
70	113.9893 9220	0.008773	1669.991860	0.000599	14.650415	0.068257	70
75	159.8760 1931	0.006255	2348.199719	0.000426	14.687629	0.068085	75
80	224.2343 8758	0.004460	3299.421325	0.000303	14.714163	0.067962	80
90	441.1029 7988	0.002267	6504.755708	0.000154	14.746569	0.067812	90
100	867.7163 2557	0.001152	12810.133591	0.000078	14.763043	0.067737	100

Continuous Compounding Factors

TABLE E-2 (continued)

7.6961 0411 3612 83% (effective rate of 8%)

n	CCA	CPW	CSCA	CSF	CSPW	CCR	n
1	1.0800 0000	0.925926	1.039487	0.962013	0.962488	1.038974	1
2	1.1664 0000	0.857339	2.162133	0.462506	1.853680	0.539467	2
3	1.2597 1200	0.793832	3.374591	0.296332	2.678859	0.373293	3
4	1.3604 8896	0.735030	4.684045	0.213491	3.442913	0.290452	4
5	1.4693 2808	0.680583	6.098255	0.163981	4.150370	0.240942	5
6	1.5868 7432	0.630170	7.625603	0.131137	4.805423	0.208098	6
7	1.7138 2427	0.583490	9.275138	0.107815	5.411954	0.184776	7
8	1.8509 3021	0.540269	11.056636	0.090443	5.973556	0.167404	8
9	1.9990 0463	0.500249	12.980654	0.077038	6.493559	0.153999	9
10	2.1589 2500	0.463193	15.058593	0.066407	6.975042	0.143368	10
11	2.3316 3900	0.428883	17.302767	0.057794	7.420860	0.134755	11
12	2.5181 7012	0.397114	19.726476	0.050693	7.833655	0.127654	12
13	2.7196 2373	0.367698	22.344081	0.044755	8.215872	0.121716	13
14	2.9371 9362	0.340461	25.171094	0.039728	8.569777	0.116689	14
15	3.1721 6911	0.315242	28.224269	0.035431	8.897467	0.112392	15
16	3.4259 4264	0.291890	31.521697	0.031724	9.200883	0.108685	16
17	3.7000 1805	0.270269	35.082920	0.028504	9.481824	0.105465	17
18	3.9960 1950	0.250249	38.929041	0.025688	9.741955	0.102649	18
19	4.3157 0106	0.231712	43.082851	0.023211	9.982816	0.100172	19
20	4.6609 5714	0.214548	47.568966	0.021022	10.205836	0.097983	20
21	5.0338 3372	0.198656	52.413970	0.019079	10.412336	0.096040	21
22	5.4365 4041	0.183941	57.646575	0.017347	10.603540	0.094308	22
23	5.8714 6365	0.170315	63.297788	0.015798	10.780581	0.092759	23
24	6.3411 8074	0.157699	69.401098	0.014409	10.944507	0.091370	24
25	6.8484 7520	0.146018	75.992673	0.013159	11.096291	0.090120	25
26	7.3963 5321	0.135202	83.111573	0.012032	11.236831	0.088993	26
27	7.9880 6147	0.125187	90.799986	0.011013	11.366961	0.087974	27
28	8.6271 0639	0.115914	99.103472	0.010090	11.487452	0.087052	28
29	9.3172 7490	0.107328	108.071237	0.009253	11.599018	0.086214	29
30	10.0626 5689	0.099377	117.756423	0.008492	11.702319	0.085453	30
31	10.8676 6944	0.092016	128.216423	0.007799	11.797969	0.084760	31
32	11.7370 8300	0.085200	139.513224	0.007168	11.886533	0.084129	32
33	12.6760 4964	0.078889	151.713769	0.006591	11.968537	0.083552	33
34	13.6901 3361	0.073045	164.890358	0.006065	12.044467	0.083026	34
35	14.7853 4429	0.067635	179.121073	0.005583	12.114772	0.082544	35
36	15.9681 7184	0.062625	194.490246	0.005142	12.179869	0.082103	36
37	17.2456 2558	0.057986	211.088953	0.004737	12.240145	0.081698	37
38	18.6252 7563	0.053690	229.015556	0.004367	12.295955	0.081328	38
39	20.1152 9768	0.049713	248.376288	0.004026	12.347632	0.080987	39
40	21.7245 2150	0.046031	269.285878	0.003714	12.395480	0.080675	40
45	31.9204 4939	0.031328	401.767556	0.002489	12.586526	0.079450	45
50	46.9016 1251	0.021321	596.426605	0.001677	12.716548	0.078638	50
55	68.9138 5611	0.014511	882.444612	0.001133	12.805039	0.078094	55
60	101.2570 6367	0.009876	1302.698900	0.000768	12.865264	0.077729	60
65	148.7798 4662	0.006721	1920.190325	0.000521	12.906253	0.077482	65
70	218.6064 0590	0.004574	2827.487813	0.000354	12.934149	0.077315	70
75	321.2045 2996	0.003113	4160.605486	0.000240	12.953135	0.077201	75
80	471.9548 3426	0.002119	6119.392712	0.000163	12.966056	0.077124	80
90	1018.9150 8928	0.000981	13226.368488	0.000076	12.980835	0.077037	90
100	2199.7612 5634	0.000455	28569.796145	0.000035	12.987680	0.076996	100

TABLE E-2 (*continued*)

n	CCA	CPW	CSCA	CSF	CSPW	CCR	n
8.6177 6962 4105 23% (effective rate of 9%)							
1	1.0900 0000	0.917431	1.044354	0.957530	0.958123	1.043708	1
2	1.1881 0000	0.841680	2.182699	0.458148	1.837134	0.544326	2
3	1.2950 2900	0.772183	3.423496	0.292099	2.643567	0.378277	3
4	1.4115 8161	0.708425	4.775964	0.209382	3.383414	0.295559	4
5	1.5386 2395	0.649931	6.250155	0.159996	4.062172	0.246174	5
6	1.6771 0011	0.596267	7.857023	0.127275	4.684886	0.213452	6
7	1.8280 3912	0.547034	9.608508	0.104074	5.256183	0.190252	7
8	1.9925 6264	0.501866	11.517628	0.086823	5.780309	0.173001	8
9	2.1718 9328	0.460428	13.598568	0.073537	6.261159	0.159715	9
10	2.3673 6367	0.422411	15.866793	0.063025	6.702305	0.149202	10
11	2.5804 2641	0.387533	18.339158	0.054528	7.107026	0.140706	11
12	2.8126 6478	0.355535	21.034036	0.047542	7.478330	0.133720	12
13	3.0658 0461	0.326179	23.971453	0.041716	7.818976	0.127894	13
14	3.3417 2703	0.299246	27.173238	0.036801	8.131495	0.122979	14
15	3.6424 8246	0.274538	30.663183	0.032612	8.418210	0.118790	15
16	3.9703 0588	0.251870	34.467223	0.029013	8.681251	0.115191	16
17	4.3276 3341	0.231073	38.613627	0.025898	8.922573	0.112075	17
18	4.7171 2042	0.211994	43.133207	0.023184	9.143970	0.109362	18
19	5.1416 6125	0.194490	48.059549	0.020808	9.347086	0.106985	19
20	5.6044 1077	0.178431	53.429263	0.018716	9.533431	0.104894	20
21	6.1088 0774	0.163698	59.282250	0.016868	9.704390	0.103046	21
22	6.6586 0043	0.150182	65.662006	0.015230	9.861232	0.101407	22
23	7.2578 7447	0.137781	72.615941	0.013771	10.005125	0.099949	23
24	7.9110 8317	0.126405	80.195729	0.012469	10.137136	0.098647	24
25	8.6230 8066	0.115968	88.457698	0.011305	10.258248	0.097483	25
26	9.3991 5792	0.106393	97.463245	0.010260	10.369359	0.096438	26
27	10.2450 8213	0.097608	107.279291	0.009321	10.471296	0.095499	27
28	11.1671 3952	0.089548	117.978781	0.008476	10.564817	0.094654	28
29	12.1721 8208	0.082155	129.641225	0.007714	10.650615	0.093891	29
30	13.2676 7847	0.075371	142.353289	0.007025	10.729329	0.093202	30
31	14.4617 6953	0.069148	156.209438	0.006402	10.801544	0.092579	31
32	15.7633 2879	0.063438	171.312642	0.005837	10.867796	0.092015	32
33	17.1820 2838	0.058200	187.775133	0.005326	10.928578	0.091503	33
34	18.7284 1093	0.053395	205.719249	0.004861	10.984341	0.091039	34
35	20.4139 6792	0.048986	225.278335	0.004439	11.035500	0.090617	35
36	22.2512 2503	0.044941	246.597739	0.004055	11.082434	0.090233	36
37	24.2538 3528	0.041231	269.835289	0.003706	11.125494	0.089884	37
38	26.4366 8046	0.037826	295.165473	0.003388	11.164998	0.089566	38
39	28.8159 8170	0.034703	322.774719	0.003098	11.201240	0.089276	39
40	31.4094 2005	0.031838	352.868798	0.002834	11.234489	0.089012	40
45	48.3272 8610	0.020692	549.182540	0.001821	11.363819	0.087999	45
50	74.3575 2008	0.013449	851.235566	0.001175	11.447875	0.087352	50
55	114.4082 6162	0.008741	1315.981589	0.000760	11.502505	0.086938	55
60	176.0312 9196	0.005681	2031.050952	0.000492	11.538011	0.086670	60
65	270.8459 6262	0.003692	3131.273803	0.000319	11.561087	0.086497	65
70	416.7300 8618	0.002400	4824.103037	0.000207	11.576085	0.086385	70
75	641.1609 9332	0.001560	7428.730649	0.000135	11.585853	0.086312	75
80	986.5516 6813	0.001014	11436.273086	0.000087	11.592168	0.086265	80
90	2335.5265 8223	0.000428	27089.684269	0.000037	11.598962	0.086215	90
100	5529.0407 9183	0.000181	64147.001289	0.000016	11.601832	0.086193	100

Continuous Compounding Factors

TABLE E-2 (continued)

9.5310 1798 0432 49% (effective rate of 10%)

n	CCA	CPW	CSCA	CSF	CSPW	CCR	n
1	1.1000 0000	0.909091	1.049206	0.953102	0.953824	1.048412	1
2	1.2100 0000	0.826446	2.203332	0.453858	1.820936	0.549168	2
3	1.3310 0000	0.751315	3.472871	0.287946	2.609220	0.383256	3
4	1.4641 0000	0.683013	4.869364	0.205366	3.325841	0.300676	4
5	1.6105 1000	0.620921	6.405507	0.156116	3.977316	0.251426	5
6	1.7715 6100	0.564474	8.095263	0.123529	4.569565	0.218839	6
7	1.9487 1710	0.513158	9.953995	0.100462	5.107974	0.195772	7
8	2.1435 8881	0.466507	11.998601	0.083343	5.597436	0.178653	8
9	2.3579 4769	0.424098	14.247667	0.070187	6.042402	0.165497	9
10	2.5937 4246	0.385543	16.721639	0.059803	6.446916	0.155113	10
11	2.8531 1671	0.350494	19.443009	0.051432	6.814656	0.146743	11
12	3.1384 2838	0.318631	22.436516	0.044570	7.148965	0.139880	12
13	3.4522 7121	0.289664	25.729373	0.038866	7.452883	0.134176	13
14	3.7974 9834	0.263331	29.351517	0.034070	7.729172	0.129380	14
15	4.1772 4817	0.239392	33.335874	0.029998	7.980343	0.125308	15
16	4.5949 7299	0.217629	37.718668	0.026512	8.208681	0.121822	16
17	5.0544 7029	0.197845	42.539740	0.023507	8.416261	0.118818	17
18	5.5599 1731	0.179859	47.842920	0.020902	8.604970	0.116212	18
19	6.1159 0904	0.163508	53.676418	0.018630	8.776523	0.113940	19
20	6.7274 9995	0.148644	60.093266	0.016641	8.932481	0.111951	20
21	7.4002 4994	0.135131	67.151798	0.014892	9.074261	0.110202	21
22	8.1402 7494	0.122846	74.916184	0.013348	9.203152	0.108658	22
23	8.9543 0243	0.111678	83.457008	0.011982	9.320325	0.107292	23
24	9.8497 3268	0.101526	92.851915	0.010770	9.426846	0.106080	24
25	10.8347 0594	0.092296	103.186312	0.009691	9.523684	0.105001	25
26	11.9181 7654	0.083905	114.554149	0.008729	9.611718	0.104040	26
27	13.1099 9419	0.076278	127.058770	0.007870	9.691749	0.103181	27
28	14.4209 9361	0.069343	140.813853	0.007102	9.764504	0.102412	28
29	15.8630 9297	0.063039	155.944444	0.006413	9.830646	0.101723	29
30	17.4494 0227	0.057309	172.588094	0.005794	9.890774	0.101104	30
31	19.1943 4250	0.052099	190.896109	0.005238	9.945436	0.100545	31
32	21.1137 7675	0.047362	211.034926	0.004739	9.995129	0.100049	32
33	23.2251 5442	0.043057	233.187625	0.004288	10.040305	0.099599	33
34	25.5476 6986	0.039143	257.555593	0.003883	10.081373	0.099193	34
35	28.1024 3685	0.035584	284.360358	0.003517	10.118708	0.098827	35
36	30.9126 8053	0.032349	313.845600	0.003186	10.152649	0.098496	36
37	34.0039 4859	0.029408	346.279365	0.002888	10.183505	0.098198	37
38	37.4043 4344	0.026735	381.956508	0.002618	10.211555	0.097928	38
39	41.1447 7779	0.024304	421.201365	0.002374	10.237055	0.097684	39
40	45.2592 5557	0.022095	464.370707	0.002153	10.260237	0.097464	40
45	72.8904 8369	0.013719	754.279174	0.001326	10.348116	0.096636	45
50	117.3908 5288	0.008519	1221.179659	0.000819	10.402682	0.096129	50
55	189.0591 4247	0.005289	1973.127559	0.000507	10.436563	0.095817	55
60	304.4816 3954	0.003284	3184.147173	0.000314	10.457600	0.095624	60
65	490.3707 2530	0.002039	5134.506370	0.000195	10.470663	0.095505	65
70	789.7469 5680	0.001266	8275.579360	0.000121	10.478773	0.095431	70
75	1271.8953 7140	0.000786	13334.308822	0.000075	10.483810	0.095385	75
80	2048.4002 1459	0.000488	21481.443208	0.000047	10.486937	0.095357	80
90	5313.0226 1185	0.000188	55734.052992	0.000018	10.490084	0.095328	90

TABLE E-2 (*continued*)

n	CCA	CPW	CSCA	CSF	CSPW	CCR	n
\multicolumn{7}{c}{11.3328 6853 0700 32% (effective rate of 12%)}	
1	1.1200 0000	0.892857	1.058867	0.944406	0.945417	1.057734	1
2	1.2544 0000	0.797194	2.244798	0.445474	1.789539	0.558803	2
3	1.4049 2800	0.711780	3.573041	0.279874	2.543220	0.393202	3
4	1.5735 1936	0.635518	5.060672	0.197602	3.216149	0.310931	4
5	1.7623 4168	0.567427	6.726820	0.148659	3.816978	0.261987	5
6	1.9738 2269	0.506631	8.592906	0.116375	4.353433	0.229704	6
7	2.2106 8141	0.452349	10.682921	0.093607	4.832411	0.206936	7
8	2.4759 6318	0.403883	13.023739	0.076783	5.260070	0.190112	8
9	2.7730 7876	0.360610	15.645454	0.063916	5.641908	0.177245	9
10	3.1058 4821	0.321973	18.581776	0.053816	5.982834	0.167145	10
11	3.4785 4999	0.287476	21.870456	0.045724	6.287233	0.159052	11
12	3.8959 7599	0.256675	25.553777	0.039133	6.559018	0.152462	12
13	4.3634 9311	0.229174	29.679098	0.033694	6.801683	0.147022	13
14	4.8871 1229	0.204620	34.299456	0.029155	7.018348	0.142484	14
15	5.4735 6576	0.182696	39.474258	0.025333	7.211799	0.138662	15
16	6.1303 9365	0.163122	45.270036	0.022090	7.384523	0.135418	16
17	6.8660 4089	0.145644	51.761307	0.019319	7.538741	0.132648	17
18	7.6899 6580	0.130040	59.031531	0.016940	7.676436	0.130269	18
19	8.6127 6169	0.116107	67.174182	0.014887	7.799378	0.128215	19
20	9.6462 9309	0.103667	76.293950	0.013107	7.909147	0.126436	20
21	10.8038 4826	0.092560	86.508091	0.011560	8.007155	0.124888	21
22	12.1003 1006	0.082643	97.947929	0.010210	8.094663	0.123538	22
23	13.5523 4726	0.073788	110.760648	0.009028	8.172794	0.122357	23
24	15.1786 2893	0.065882	125.110680	0.007993	8.242555	0.121322	24
25	17.0000 6441	0.058823	141.182829	0.007083	8.304841	0.120412	25
26	19.0400 7214	0.052521	159.183636	0.006282	8.360453	0.119611	26
27	21.3248 8079	0.046894	179.344539	0.005576	8.410107	0.118905	27
28	23.8838 6649	0.041869	201.924750	0.004952	8.454441	0.118281	28
29	26.7499 3047	0.037383	227.214587	0.004401	8.494025	0.117730	29
30	29.9599 2212	0.033378	255.539205	0.003913	8.529368	0.117242	30
31	33.5551 1278	0.029802	287.262776	0.003481	8.560924	0.116810	31
32	37.5817 2631	0.026609	322.793176	0.003098	8.589099	0.116427	32
33	42.0915 3347	0.023758	362.587225	0.002758	8.614256	0.116087	33
34	47.1425 1748	0.021212	407.156558	0.002456	8.636716	0.115785	34
35	52.7996 1958	0.018940	457.074212	0.002188	8.656771	0.115517	35
36	59.1355 7393	0.016910	512.981985	0.001949	8.674677	0.115278	36
37	66.2318 4280	0.015098	575.598690	0.001737	8.690664	0.115066	37
38	74.1796 6394	0.013481	645.729400	0.001549	8.704938	0.114877	38
39	83.0812 2361	0.012036	724.275795	0.001381	8.717683	0.114709	39
40	93.0509 7044	0.010747	812.247757	0.001231	8.729063	0.114560	40
45	163.9876 0387	0.006098	1438.184899	0.000695	8.770083	0.114024	45
50	289.0021 8983	0.003460	2541.300016	0.000393	8.793359	0.113722	50
55	509.3206 0567	0.001963	4485.365769	0.000223	8.806566	0.113552	55
60	897.5969 3349	0.001114	7911.473878	0.000126	8.814061	0.113455	60
65	1581.8724 9060	0.000632	13949.447012	0.000072	8.818313	0.113400	65
70	2787.7998 2770	0.000359	24590.418747	0.000041	8.820726	0.113369	70
75	4913.0558 4077	0.000204	43343.446785	0.000023	8.822095	0.113352	75
80	8658.4831 0008	0.000115	76392.689782	0.000013	8.822872	0.113342	80

Continuous Compounding Factors

TABLE E-2 (continued)

13.9761 9423 7515 87% (effective rate of 15%)

n	CCA	CPW	CSCA	CSF	CSPW	CCR	n
1	1.1500 0000	0.869565	1.073254	0.931746	0.933264	1.071508	1
2	1.3225 0000	0.756144	2.307495	0.433370	1.744798	0.573132	2
3	1.5208 7500	0.657516	3.726873	0.268321	2.450479	0.408083	3
4	1.7490 0625	0.571753	5.359157	0.186596	3.064116	0.326358	4
5	2.0113 5719	0.497177	7.236285	0.138192	3.597712	0.277954	5
6	2.3130 6077	0.432328	9.394981	0.106440	4.061709	0.246202	6
7	2.6600 1988	0.375937	11.877481	0.084193	4.465185	0.223955	7
8	3.0590 2286	0.326902	14.732357	0.067878	4.816034	0.207640	8
9	3.5178 7629	0.284262	18.015464	0.055508	5.121119	0.195270	9
10	4.0455 5774	0.247185	21.791038	0.045890	5.386411	0.185652	10
11	4.6523 9140	0.214943	26.132947	0.038266	5.617100	0.178028	11
12	5.3502 5011	0.186907	31.126142	0.032127	5.817699	0.171889	12
13	6.1527 8762	0.162528	36.868317	0.027124	5.992132	0.166886	13
14	7.0757 0576	0.141329	43.471818	0.023003	6.143814	0.162765	14
15	8.1370 6163	0.122894	51.065845	0.019583	6.275711	0.159345	15
16	9.3576 2087	0.106865	59.798975	0.016723	6.390404	0.156485	16
17	10.7612 6400	0.092926	69.842075	0.014318	6.490137	0.154080	17
18	12.3754 5361	0.080805	81.391639	0.012286	6.576861	0.152048	18
19	14.2317 7165	0.070265	94.673639	0.010563	6.652274	0.150325	19
20	16.3665 3739	0.061100	109.947938	0.009095	6.717850	0.148857	20
21	18.8215 1800	0.053131	127.513382	0.007842	6.774872	0.147604	21
22	21.6447 4570	0.046201	147.713643	0.006770	6.824457	0.146532	22
23	24.8914 5756	0.040174	170.943943	0.005850	6.867575	0.145612	23
24	28.6251 7619	0.034934	197.658788	0.005059	6.905068	0.144821	24
25	32.9189 5262	0.030378	228.380860	0.004379	6.937671	0.144141	25
26	37.8567 9551	0.026415	263.711243	0.003792	6.966021	0.143554	26
27	43.5353 1484	0.022970	304.341183	0.003286	6.990674	0.143048	27
28	50.0656 1207	0.019974	351.065614	0.002848	7.012111	0.142610	28
29	57.5754 5388	0.017369	404.798709	0.002470	7.030752	0.142232	29
30	66.2117 7196	0.015103	466.591769	0.002143	7.046961	0.141905	30
31	76.1435 3775	0.013133	537.653788	0.001860	7.061056	0.141622	31
32	87.5650 6841	0.011420	619.375110	0.001615	7.073313	0.141376	32
33	100.6998 2867	0.009931	713.354630	0.001402	7.083971	0.141164	33
34	115.8048 0298	0.008635	821.431078	0.001217	7.093238	0.140979	34
35	133.1755 2342	0.007509	945.718993	0.001057	7.101297	0.140819	35
36	153.1518 5194	0.006529	1088.650096	0.000919	7.108305	0.140681	36
37	176.1246 2973	0.005678	1253.020864	0.000798	7.114399	0.140560	37
38	202.5433 2419	0.004937	1442.047247	0.000693	7.119698	0.140455	38
39	232.9248 2281	0.004293	1659.427587	0.000603	7.124305	0.140365	39
40	267.8635 4623	0.003733	1909.414979	0.000524	7.128312	0.140286	40
45	538.7692 6899	0.001856	3847.751826	0.000260	7.141743	0.140022	45
50	1083.6574 4158	0.000923	7746.439576	0.000129	7.148421	0.139891	50
55	2179.6221 8392	0.000459	15588.093203	0.000064	7.151741	0.139826	55
60	4383.9987 4566	0.000228	31360.459587	0.000032	7.153392	0.139794	60
65	8817.7873 8707	0.000113	63084.322078	0.000016	7.154212	0.139778	65

TABLE E-2 (*continued*)

n	CCA	CPW	CSCA	CSF	CSPW	CCR	n
18.2321 5567 9395 46% (effective rate of 20%)							
1	1.2000 0000	0.833333	1.096963	0.911608	0.914136	1.093929	1
2	1.4400 0000	0.694444	2.413319	0.414367	1.675916	0.596689	2
3	1.7280 0000	0.578704	3.992945	0.250442	2.310732	0.432763	3
4	2.0736 0000	0.482253	5.888497	0.169823	2.839746	0.352144	4
5	2.4883 2000	0.401878	8.163160	0.122502	3.280591	0.304823	5
6	2.9859 8400	0.334898	10.892755	0.091804	3.647962	0.274126	6
7	3.5831 8080	0.279082	14.168269	0.070580	3.954104	0.252902	7
8	4.2998 1696	0.232568	18.098885	0.055252	4.209222	0.237574	8
9	5.1597 8035	0.193807	22.815625	0.043830	4.421821	0.226151	9
10	6.1917 3642	0.161506	28.475714	0.035118	4.598987	0.217439	10
11	7.4300 8371	0.134588	35.267819	0.028354	4.746625	0.210676	11
12	8.9161 0045	0.112157	43.418346	0.023032	4.869656	0.205353	12
13	10.6993 2054	0.093464	53.198978	0.018797	4.972183	0.201119	13
14	12.8391 8465	0.077887	64.935737	0.015400	5.057622	0.197721	14
15	15.4070 2157	0.064905	79.019847	0.012655	5.128820	0.194977	15
16	18.4884 2589	0.054088	95.920780	0.010425	5.188153	0.192747	16
17	22.1861 1107	0.045073	116.201899	0.008606	5.237597	0.190927	17
18	26.6233 3328	0.037561	140.539241	0.007115	5.278800	0.189437	18
19	31.9479 9994	0.031301	169.744053	0.005891	5.313136	0.188213	19
20	38.3375 9992	0.026084	204.789826	0.004883	5.341749	0.187205	20
21	46.0051 1991	0.021737	246.844754	0.004051	5.365593	0.186373	21
22	55.2061 4389	0.018114	297.310668	0.003363	5.385463	0.185685	22
23	66.2473 7267	0.015095	357.869765	0.002794	5.402022	0.185116	23
24	79.4968 4720	0.012579	430.540681	0.002323	5.415821	0.184644	24
25	95.3962 1664	0.010483	517.745780	0.001931	5.427320	0.184253	25
26	114.4754 5997	0.008735	622.391899	0.001607	5.436902	0.183928	26
27	137.3705 5197	0.007280	747.967242	0.001337	5.444888	0.183659	27
28	164.8446 6236	0.006066	898.657653	0.001113	5.451542	0.183434	28
29	197.8135 9483	0.005055	1079.486147	0.000926	5.457088	0.183248	29
30	237.3763 1380	0.004213	1296.480339	0.000771	5.461709	0.183093	30
31	284.8515 7656	0.003511	1556.873370	0.000642	5.465560	0.182964	31
32	341.8218 9187	0.002926	1869.345007	0.000535	5.468769	0.182857	32
33	410.1862 7025	0.002438	2244.310971	0.000446	5.471443	0.182767	33
34	492.2235 2430	0.002032	2694.270129	0.000371	5.473672	0.182693	34
35	590.6682 2915	0.001693	3234.221117	0.000309	5.475529	0.182631	35
36	708.8018 7499	0.001411	3882.162304	0.000258	5.477077	0.182579	36
37	850.5622 4998	0.001176	4659.691728	0.000215	5.478366	0.182536	37
38	1020.6746 9998	0.000980	5592.727036	0.000179	5.479441	0.182500	38
39	1224.8096 3997	0.000816	6712.369407	0.000149	5.480337	0.182471	39
40	1469.7715 6797	0.000680	8055.940251	0.000124	5.481083	0.182446	40
45	3657.2619 8801	0.000273	20053.920405	0.000050	5.483315	0.182371	45
50	9100.4381 5000	0.000110	49908.734381	0.000020	5.484212	0.182342	50

Continuous Compounding Factors

TABLE E-2 (*continued*)

n	CCA	CPW	CSCA	CSF	CSPW	CCR	n
\multicolumn{7}{c}{22.3143 5513 1420 98% (effective rate of 25%)}	
1	1.2500 0000	0.800000	1.120355	0.892574	0.896284	1.115718	1
2	1.5625 0000	0.640000	2.520799	0.396700	1.613311	0.619843	2
3	1.9531 2500	0.512000	4.271354	0.234118	2.186933	0.457261	3
4	2.4414 0625	0.409600	6.459547	0.154810	2.645830	0.377953	4
5	3.0517 5781	0.327680	9.194789	0.108757	3.012948	0.331901	5
6	3.8146 9727	0.262144	12.613841	0.079278	3.306643	0.302422	6
7	4.7683 7158	0.209715	16.887656	0.059215	3.541598	0.282358	7
8	5.9604 6448	0.167772	22.229925	0.044984	3.729563	0.268128	8
9	7.4505 8060	0.134218	28.907762	0.034593	3.879934	0.257736	9
10	9.3132 2575	0.107374	37.255057	0.026842	4.000231	0.249986	10
11	11.6415 3218	0.085899	47.689176	0.020969	4.096469	0.244113	11
12	14.5519 1523	0.068719	60.731826	0.016466	4.173459	0.239609	12
13	18.1898 9404	0.054976	77.035137	0.012981	4.235051	0.236125	13
14	22.7373 6754	0.043980	97.414276	0.010265	4.284325	0.233409	14
15	28.4217 0943	0.035184	122.888200	0.008137	4.323744	0.231281	15
16	35.5271 3679	0.028147	154.730605	0.006463	4.355279	0.229606	16
17	44.4089 2099	0.022518	194.533612	0.005140	4.380508	0.228284	17
18	55.5111 5123	0.018014	244.287370	0.004094	4.400690	0.227237	18
19	69.3389 3904	0.014412	306.479567	0.003263	4.416836	0.226406	19
20	86.7361 7380	0.011529	384.219814	0.002603	4.429753	0.225746	20
21	108.4202 1725	0.009223	481.395123	0.002077	4.440086	0.225221	21
22	135.5252 7156	0.007379	602.864258	0.001659	4.448353	0.224802	22
23	169.4065 8945	0.005903	754.700678	0.001325	4.454966	0.224469	23
24	211.7582 3681	0.004722	944.496202	0.001059	4.460257	0.224202	24
25	264.6977 9602	0.003778	1181.740608	0.000846	4.464490	0.223990	25
26	330.8722 4502	0.003022	1478.296115	0.000676	4.467876	0.223820	26
27	413.5903 0628	0.002418	1848.990499	0.000541	4.470585	0.223684	27
28	516.9878 8285	0.001934	2312.358479	0.000432	4.472752	0.223576	28
29	646.2348 5356	0.001547	2891.568453	0.000346	4.474485	0.223489	29
30	807.7935 6695	0.001238	3615.580922	0.000277	4.475872	0.223420	30
31	1009.7419 5868	0.000990	4520.596507	0.000221	4.476982	0.223365	31
32	1262.1774 4835	0.000792	5651.865989	0.000177	4.477870	0.223320	32
33	1577.7218 1044	0.000634	7065.952841	0.000142	4.478580	0.223285	33
34	1972.1522 6305	0.000507	8833.561407	0.000113	4.479148	0.223257	34
35	2465.1903 2882	0.000406	11043.072113	0.000091	4.479602	0.223234	35
36	3081.4879 1102	0.000325	13804.960497	0.000072	4.479966	0.223216	36
37	3851.8598 8877	0.000260	17257.320976	0.000058	4.480257	0.223201	37
38	4814.8248 6097	0.000208	21572.771575	0.000046	4.480489	0.223190	38
39	6018.5310 7621	0.000166	26967.084824	0.000037	4.480676	0.223181	39
40	7523.1638 4526	0.000133	33709.976385	0.000030	4.480824	0.223173	40

Index

AASHO; *see* American Association of State Highway Officials
Abandonment alternative, 140
Abandonment of property, 202
ABC Trucking Company, 190-197
Abilene, Kansas, 519
Ability to pay, 609
Acceleration horsepower, 296
Acceleration in speed, 456, 460, 463, 465
Access to land, 608, 609
Accident facts, 362, 363
Accidents; *see* Traffic accidents
Accounting
  books, 202
  concepts, 190
  objectives, 190
  receivables, 229
Actuarial analysis, 207
Added expenditure cost allocation, 618
Additions, investment, 241
Adkins, William G., 273
Administration, 229, 233, 647
  and economy, 280
  public, 280
Administrative expense, 26, 245
Administrators, 6, 602
Advance planning, 642
Advance programming, 638
Adverse consequences, 27
Adverse travel distance, 26, 475
Advertising, 545
Aerospace manufacturing plant, 669
Aesthetic requirements, 22
Africa, 526, 529
Agriculture
  crops, 489, 531
  subsistence, 526
Air conditioning, 556
Air resistance, 293
Alabama, 603

Allocating depreciation expense, 181
Alternatives
  all possible, 20, 22
  choice of, 14
  do-nothing, 23, 25, 478
  incremental solution, 25
  significant differences, 20
  with and without, 478
Altitude, effect on fuel consumption, 351
American Association of State Highway Officials
  manual of uniform accounting, 245
  road test, 621, 623
  road user benefit analysis, 146, 149, 580
American driving public, 365
American Society for Engineering Education, Engineering Economy Division, 86
American Society for Testing Materials, 548
Amortization, 178, 182, 491
Analysis for economy, 10, 14
  applications, 40
  common factors omitted, 20, 26
  economic evaluation, 40
  economic justification, 40
  factor chart, 45
  market factors, 20, 23
  money based, 20, 23
  nonmarket factors, 20, 23
  not the decision, 20, 21
  objectives, 40
  project formulation, 40
  sequence of steps, 46
  study of future, 20, 46
Analysis period, 25, 29, 534
  in cost of owning and operating, 242
  in economy studies, 242
  not to exceed period of predictability, 242
  restricted to period of forecast, 20, 25
  same for inputs and outputs, 20, 25
  sensitivity, 165
  unequal, 156
  use in methods of analysis, 156

Andes Mountains, 527
Anguish, anxiety, misery, 375
Annual cost
   capital, 234, 236
   cash, 228
   economic, 228, 230-236
   tax, 228
Annual depreciation allocation, 181
Annual depreciation charges, 188
Annual expense, 143
   recurring, 245-247
Annual travel per vehicle, 604
Appalachia, 526, 528
Arnold, Dr. Thomas, 12
Arterial facilities, 610
Asia, 526, 529
Assessed value for taxation, 485
Asset accounts, 204
Assets, 190
Atmospheric conditions, effect on running cost, 350
Automotive industry, 653
Automotive Safety Foundation, 402, 403
Average daily traffic (ADT), 143, 155
Average service life, 201
Axle classification of vehicles, 335

Babylonia, 74
Balance sheet, 71, 190, 194
Baltimore and Ohio Railroad, 177
Bascule bridge economy, 669
Basic access cost allocation, 618
Beautification, 61, 603
Benefactors, 148
Beneficiaries, 148
Benefit/cost ratio method, 51, 146, 543, 644
   annual expense factor position, 148
   consistency with other methods, 150
   definition, 127
   illustrative solution, 132
Benefits, 49, 65
   accidents, 376
   difference in costs, 60
   economists' definition, 50
   engineers' definition, 51
   highway, 55
      definition, 49
      identification, 49
      measurement, 49
      understanding of, 49
   nonuser, 55
   relation to savings, 51, 481

Benefits (continued)
   road user, 55
Best alternative, 557
Betterments, 241
Bias or subjective bent, 30
Bierman, Harold J., 75
Biological decay of property, 203
Bituminous surfaces, 159, 347
Bond interest rate related to vestcharge rate, 76
Borrowed money, 70
Boston, Massachusetts, 502, 511
Boulding, Kenneth, 79
Brake
   horsepower, 285, 286, 287
   linings, 321
   wear, 460
Break-even analysis, 153
Bridges
   accident rates when widened, 409, 417
   economic value of null alternative, 138
   high level economy, 669
Brookings Institution, 538
Budget, 230
   payment of interest, 73
   preparation of, 73
   public tax, 487
Bus, 534, 535
Business, 39
   bypassed, 514
   consequences, 555
   establishments, 511
   reports, 543
   and trade, 39
   use of time, 265
Businessmen, 17
Bypass
   circumferential, 511
   effect on Kansas towns, 518
   freeway, 510
   Rolla, Missouri, 516
   simple, 510
   urban, 510-523

Cadillacs, 280
California accident rates on rural highways, 408, 410
California bypasses, 521
Capacity manual, 435, 437
Capital budgets, 153, 638
Capital improvement, 229, 638
Capital investment, 559
Capital stock account, 193

Index

Capitalization of profits, 498
Capitalized cost method, 152
Carburetor, 284
Carlos, 517
Car rental rates, 313
Cart roads, 526
Cash account, 192
Cash cost, 229
Cash flow
  cycle repetition, 156
  discount to same time date, 25
  irregular, 159
  negative and positive, 126
  reversals of sign, 159
  time dates, 73
  uniformily increasing, 90
Cash flow diagrams
  continuous compounding
    exponential series, 114
    gradient series, 108, 109
    uniform series, 103
  step compounding
    exponential series, 94
    gradient series, 90
    uniform series, 83
Catastrophe to property, 202
Catch-up period, 597
Central America, 527
Central business district, 499, 501, 511
Central City, 517
Chakravarty, S., 536
Challenger, 145
  alternative, 136
Chamber of Commerce, 587
Chicago, Illinois, 268
  accident rates, 406
Chicago area (Cook and Dupage Counties)
  accident rates and costs, 407
Chile, 527
Civic organizations, 587
Claffey, Paul J., 267, 268, 312, 341, 347, 440, 459, 468
Classical economics, 124
Coefficient of friction, 444
Collector-distributor system, 591
Colorado, early roads, 223
Colorado Highway Planning Commission, 623
Commercial travel, 29
Commercial vehicle
  costs
    line-haul by vehicle weight, 315
    operating, 314

Comercial vehicle *(continued)*
  depreciation, 273
  property tax, 273
  travel time
    drivers' wages, 276
    value, 265, 272-276
Commodity savings, 36, 496
Commodity surplus, 15
Common stocks, 125
Communication, 528
Community
  consequences, 488
    keep separate, 40
    listing of, 39
    not reducible to dollars, 39
    point of view, 40
    protective services, 40
  development, 654
  traffic, 477
    pricing benefits, 65
    road-user benefits, 65
Commuter work trip, 268
Composite grades
  fuel consumption, 349
  time consumption, 350
Composite index of consequences, 573, 574
Composite original group method, 211
Compound interest; *see also* Interest, compound
  applications, 87
  period-end step convention, 84
  premise, 100
  principles and concepts, 126
  theory, 82
Compound interest equations, 82, 87, 88
  derivation, 84
  exponential series, 93
  interrelations, 85, 87
  mnemonic symbols, 86
  notations, 82
  standard, 71
  uniformily increasing series, 88
Compound interest factor tables
  continuous compounding, 875-903
  exponential growth, 784-874
  gradient growth, 773-783
  year-end step convention, 728-772
Compromises, 558
Conceptual ability, 642
Congress, 554
Congress Street Expressway (Chicago), 406
Connecticut, 605
  accident rates, 408, 417
Consequences, 493; *see also* Highways, improvements

Consequences *(continued)*
  business, 556
  classification, 36
  commercial, 30
  community, 494
  economic, 23, 556
  far away, 556
  grouping, 494
  intangible, 494
  nonmarket, 494
  on other transportation modes, 30
  to public functions, 30
  road-user, 35, 556
  of saving, 16
  secondary, 494
  social, 23
  of wastefulness, 16
  to whomsoever they accrue, 14, 17, 21, 30, 480, 484
Conservation, 14, 15, 16
Constitutional limitations, 605
Construction
  equipment, 44
  materials, 18
  prices, 534, 596
  priorities, 638
  program, 2, 244, 584, 638, 641
    adjusting, 654
    advance planning, 650
    advantages of, 640
    approval of, 652
    communication, 654
    factors involved, 644
    formulation, 649
    monitoring, 654
    nonmeasurable factors, 651
    present serviceability index, 651
    priority ratings, 650
    publication, 652
    rating schemes, 650
    reviewing, 649
    time period, 649
  project, sources of, 647
  scheduling, 638, 639, 641, 653
  stages, 663
  work load by geographical area, 652
Consumer's surplus, 56, 57, 59, 60
Consumption of commodities, 554
Continuing needs studies, 598
Continuous compounding, 100-117
  derivation of equations, 102-117
  exponential series, 117
  gradient series, 108-117
  tables, 875-903
Contractors, 641
Controlled access, 5
Conversion factors for running cost on gravel and stone, 727

Cook, Kenneth E., 62
Cooperative marketing, 486
Corn production, 490
Cornell Aeronautical Laboratory, 402, 408, 416
Cost accounting, 150, 176, 179, 189
Cost allocation, 41, 604, 609
  botton dramer, 614, 616
  public share of, 617
Cost allocation to vehicle classes, 618-629
  of capital, 68, 124, 125
  of highways
    allocation report, 627, 628
    components, 237
    expense factors, 55
  many kinds of, 228
  methods
    cost function, 620, 624
    differential benefit, 620, 627
    incremental, 620
    operating cost, 620, 629
    space-time, 620, 632
    standard cost, 620, 627
    ton-mile, 620, 623
  of travel time, 650
Cost, price, and value, 56
Cost responsibility allocation, 609, 610
Cotton, 532
Country Club Heights, 573
Courts and juries, 378
Crankcase oil changes, 321
Credit organizations, 98
Credits (accounting), 190
Critical path method, 641, 654
Critically deficient projects, 652
Culverts
  economic value (null alternative), 138
  economical size, 658, 659
  survivor curve, 226
Current expenses, 72
Curry, David A., 276, 650

Debit and credit, 190
Debits (accounting), 190
Debt capital, 125
Debt service, 229, 603
Decaying area, 528
Deceleration in speed of passenger car, 456, 458-462
Decision case No. 1, 560
Decision case No. 2, 576
Decision maker, 21, 141, 537, 542, 543, 557, 574

Index

Decision maker *(continued)*
  qualifications, 557
Decision-making process
  criteria, 20, 29
  factors to be considered, 553
  principles, 558
Decision process, 19, 552-581
Declining balance depreciation
  method, 187
Defender, 145
Deficiencies, 593
Deflation, 533
Delaware, 603
Demand curves, 56-60
Demand for highway travel, elasticity of, 60
Depreciable base, 181
Depreciation, 71, 88, 124, 230
  commercial vehicles, 273
  concept, 176
  cost amounting, 177, 179-184, 189, 205
  economic value, 179
  expense, 204
    allocating methods, 179-188
  group property, 181
  impaired usefulness, 178
  market value, 179
  rate, 187
  reserve account, 193
  reserve balance, 219
  return, 181
  sense of value, 178
  sunk cost, 249
  unit property, 181
  vehicle, 322
Derived curves, 208
Design, highways
  economy, 8
  factors affecting traffic speed, 432
  effect on motor vehicle operating cost
    distance, 324
    grades, 325
    horizontal curves, 325
    speed changes, 327
    roadway surface, 328
  factors affecting traffic speed, 432
  geometrics, 43
  grade line, 658
  lowest construction cost, 4
  proposals for, 43
  standards for, 583, 584, 602
  versus economy, 4
Design maximum flood discharge, 659
DeSoto Firedome, cost record, 304
Detective stories, 544
Detour costs, 559
Developing countries, 525-538

DeWeille, Jan, 534
Diamond River, 669
Diesel engine, 284
Difference in alternatives, 141, 142
Differential analysis, 135
Differential benefits cost allocation
  to vehicles, 618, 620, 627
Disbursements, 189, 228
Discomfort, 5
Discount factor, 67
Discount rate, 124
  economic value of man, 382
Discounted cash flow, 127
Displacement of people, 559
Distributing goods, 502
District of Columbia, 572, 606
Dividends, 67
Double counting, 26, 498, 554
Drag resistance, 293
Drainage facilities, 658
Drive axle torque, 287
Driver of vehicle, 430
Drivers' nonwage compensations, 276
Driving methods, 468
Driving strains and annoyances, 628
Drucker, Peter F., 558
Dublin, Louis I., 379, 380

Earned surplus account, 193
Earning-credit cost allocation, 610, 613, 616
Earnings-credit cost responsibility, 615, 616
Earth, 263
Earth roads, log drags, 254
Economic activity, balancing of trade-offs and shifting, 15
Economic analysis
  an art, 124
  application
    of equations of solutions, 130
    to individual alternatives, 140
    in project formulation, 140
    types of improvements, 42
  characteristics, 142
  developing countries, 533-537
  discount factors to same time date, 20, 25
  equations for solution, 129
  formal report on, 542
  index of economic feasibility, 141
  limitations, 142

Economic analysis *(continued)*
  methods, 123, 126
    comparison, 140
    definitions, 126
    evolution, 126
    mutually exclusive alternatives, 134
    using unequal analysis periods, 156
  misconceptions, 43
  mutually exclusive alternatives, 141
  numerical solution, 141
  objective, 149
  objectives of, 542
  range of factors, 24
  study of the future, 24
Economic cost
  of money, 67
  of owning and operating a facility, 25
  of taking land and improvements, 490
Economic depression, 16, 495
Economic efficiency of proposals, 14
Economic evaluation, 28, 142, 478, 499, 528, 552
Economic feasibility, 18, 141, 554
Economic forces, 5
Economic gains and losses, 498
Economic impacts, 27
Economic inflation, 248
Economic justification, 23
Economic life, 200
Economic support
  transfers of, 17
Economic transfers, 495, 501
Economics
  definition, 8
Economists, 144
Economy, 6
  bridge type, 669
  definition, 9
  intersection traffic controls, 661
  objective of analysis for, 40
  public administration, 280
  stage construction, 663
  vertical grades, 672
Effective interest rates, 96-99, 102
Effective programming, 644
Eisenhower Center, 519
Elasticity, 57
Electronic industry, 502
El Salvador, 536
Emerson, Ralph Waldo, 525
Employment gain, 503
Employment levels, 641
Engine horsepower
  rolling resistance, 290

Engine horsepower *(continued)*
  transmission resistance, 289
  utilization of, 288
Engine oil consumption, 344
  measurement, 316
Engineering costs, 236
Engineering design, 542
Engineering economy
  concepts, 8, 13, 19
  defined, 9
  guide to design, 6
  guidelines, 20
  history, 10
  literature, 13
  premise, 14
  principles, 6, 8, 19, 34
  public works, 34
  theorics, 13
  tool, 14
Engineering judgment, 22
Engineering practice, 658
Engineers vs. economists, 124, 126
Engineer's responsibilities, 13
Entrepreneurs, 531
Environment and speed, 433
Equipment account, 192
Equity capital, 125
Equivalent uniform annual net return method, 127, 132, 144
Equivalent uniform cost analysis, 228, 659
  definition, 126
  method, 126, 131, 142
  repetition of cash flow cycle, 156
Estates, 176
Estimates, 18
Exhaust gases, 285
Existing condition alternative, 137
Existing facilities and situation, 23, 24
Expenditures, 228, 605
Expense, 228
  of transportation, 11
Exponential series compounding
  derivation of equations, 93

Fair net return, 10
Falk, Edward L., 575
Family
  displacement, 572
  income, 281
  needs, 500
  trade, 502
Farm land, 490
Farming methods, 531

Index

Fatal injury accident, worth
    of a life, 377-383
Federal Internal Revenue Code, 187, 188
Federal officials, 28
Feeder roads, 527
Financial
    condition, 190
    control, 641
    policy, 41
    program, 602
Firey, Joseph C., 341
Fixed capital physical property, 177
Fixed property investment, 72
Floods, 27
    damage, 659
    flow, 658
    water storage, 658
Florida accident rates, 408, 417
Forecast of highway use, 584
Forecasting prices
    with real price level, 248
    without inflation, 248
Forecasts, 25
Formal reports, 542-550
Four-cycle engine, 285
Franzini, J. B., 660
Freeways, 250
Fuel/air mixture, 284, 285
Fuel consumption, 303
    composite grades, 349
    gallons per hour, 318
    gallons per mile, 318
    gravel and stone, 726
    horizontal curves, 342
    measurement, 316
    miles per gallon, 318
    90-degree turns, 724
    rates, 341
    speed changes, 466
    stopping, 467
    tables, 705-723
Fuel taxes, 434
    effect on calculated rate of return, 338
    exclusion in economy studies, 337-340
    nonhighway use of, 340
    omitted in running costs, 336
    origin of tax law, 606
    and the road user, 338
Funeral costs, 374, 395
Future earnings, present worth of, 387

Gain, defined, 50
Gardner, Evan H., 650

Gasoline
    sales, 62
    tax rates, 62
Gear ratio, 288, 318
General fund, 604
Generated traffic, 476, 477, 479
Gillespie, William M., 10, 11, 12
Good alternative, 557
Good decision, 555
Goodwill, 557
Government
    accounting, 196, 229
    attitude toward vestcharge (interest), 76
    county, city, state, 485
    highway function, 14
    local institution, 485
    operations under reduced tax income, 488
    people's agent, 486
    subsidy, 16
Grade horsepower, 296
Grade reduction, 26, 243
Grade resistance, 294
Grade separation, 243
Gradient compound interest, 90-93
    factor tables, 773-783
Gradient growth, time of beginning, 480
Granger, Washington, 576
Grant, Eugene L., 78, 91, 178, 248, 480
Gravel or stone surfaces, 347
    fuel consumption, 726
    running cost conversion factors, 726
Gross national product, 495
Gross weight of vehicle, 296
Ground transportation, 4
Group accounting, 183
Guardrail, 243
Guatemala, 532

Hafstad, L.R., 365
Haikalis, George, 269
Hall, W.E., 365
Hammurabi, 74
Haney, Dan G., 268, 276, 650
Harbor improvements, 525
Highway Research Board, National Academy of Sciences, 408, 435, 621
Highways
    accidents; see Traffic accidents

Highways *(continued)*
  allocation of cost, 609, 633
  annual construction, 3
  comfort factors, 262, 278, 281
  construction costs per mile, 250, 251
  contribution to social growth, 503
  convenience, 263, 278
  cost to reduce travel time, 276-277
  department, 262, 552, 602
    agents of public, 35
    control of motor vehicle use, 56
    functional chart, 648
    functional organization, 647
    officials, 543
    planning division, 6
    responsibilities, 8
  economic contribution of, 503
  financing, 555, 584, 602
    current, 598
    developing countries, 526
    not a factor in economic analysis, 20, 27
    road improvements, 4
    separate function, 27
    studying, 603
  funds, 605-609
  improvements; *see also* Consequences
    consequences, 36
    economic analysis, 42
    effects on improving local property, 488
    effects on property taxes, 488
    objectives, 52
    social consequences, 8, 23
    types, 43
  maintenance, 229, 559, 603
    cost accounts, 246
    cost per mile, 250, 253-255
    definition, 246
    equipment, 44
    expense, 103, 204
    expense in B/C ratio method, 51, 148
    pavements, 255
    quality, 201
  managements, 229, 663
  mileage, 3
  modernization, 44
  officials, 9, 586
  officials and accidents, 365
  oriented business, 499, 500, 514
  patrol, 603
  policing, 26
  public function, 53
  as a public utility, 42
  receipts for 1964, 608
  service character, 34
  services, 53, 632
  tax support, 605
  transportation, 4, 495
    cost 228
    comfort, 281

Highways *(continued)*
  demand, 62
  developing countries, 526
  economy, 9
  effective supply, 61
  inelastic character, 62
  reduction in cost, 35
  social forces of, 5
  traveler, 501
  trust fund, 607
  unique character, 42
Hirshleifer, Jack, 77
Historical lives, 222
Historical method of cost allocation, 610
Honduras, 536
Horizontal curves
  accident rates, 414
  alignment, 5, 14
  centrifugal force, 443
  design for speed, 444
  on grades, 678
  radius, 444
  speed on, 440, 445
  superelevation, 351, 352
Horse-days, 12
Horse-drawn vehicles, 10
Horsepower, 285-292
Household goods, 496
Hoyt, Kansas, 519
Hunch decisions, 20, 22
Hunt, Pearson, 162
Husbanding the dollar, 38, 280
Husbanding public monies, 42
Hyman, Joseph, 269
Hypothetical new asset, 179

Idling engine running cost, 346, 723
Illinois accident study, 362, 384, 385, 387, 399
Illinois Division of Highways, 254, 255
Illinois traffic accident report
  cost of car and truck accidents, 388
  cost of rural and urban accidents, 388
  direct cost by cost element, 394
  fatality compared to Washington, D. C., 396
  nonfatal injury compared to Washington, D. C., 397
  number and unit cost by highway type, 390
  number of and unit cost of intersection accidents, 392
  property damage only, 398
  vehicles involved per accident, 395
Impedance, 267, 279

Index 913

Imports, 532
Inadequacy of property, 202
Inadequacy, risk of, 77
Income, real, 63
Income tax, 176, 499
Inconvenience, 5
Incremental
  benefit/cost ratio, 135
  cost allocation to vehicles, 620
  rate of return, 137
  solution for economy, 25
Independent countries, 525
Index of profitability, 137
Indian trail, 525
Individual preferences, 28
Individual unit method, 208
Induced traffic; see Traffic, generated
Industry depreciation, 182
Industry, general, 502
Infinitesimal flows, 102, 105
Inflation, 247, 533
Ingenieria Internacional, 13
In lieu taxes, 608
Insulation of houses, 556
Insurance, 375-381
Integrator system, 591
Intellectual honesty, 21
Inter-American Highway, 527
Interchanges, 497
Interdiciplinary discussions, 124
Internal rate of return, 127
International loans, 526
Intersection
  economic value of, 138
  urban four-way, 661
Intersection traffic control economy, 661
Interest; see also Compound interest; Vestcharge
  bonds, 70
  charges, 88
  compared to profit, 70
  compensation for foregoing, 69
  compound, 82
  concept, 67
  defined, 67, 68
  delayed consumption, 69
  denial of satisfactions, 69
  during construction, 250
  early recognition, 70
  gaining immediate satisfaction, 69
  imputed, 68, 69
  on investment, 10
  justification of use, 69

Interest *(continued)*
  mathematical concept, 69, 71
  mortgages, 70
  notes, 70
  simple, 82
Interest rate, 124
  department store credit, 74
  developing countries, 536
  effective, 101
  fixed by law, 74
  household finance, 74
  loan shark loans, 74
  nominal, 96, 97, 101, 102
Internal combustion engine, 284
International loans, 536
Interstate Commerce Commission, 300
Interstate highway system, 241, 254, 554
Interstate route, 40, 517
Interstate route 82, location case, 576
Inventories, 177, 229, 496
  of needs, 598
Investment programs, 262
Involvements (accident), 361, 387
Ireson, Grant, 91, 480
Irregular cash flows, 125, 145, 163
Irregularly shaped parcels, 490

James, E. W., 13
Jamestown, Virginia, 1
Jones, John Hugh, 351
Jorgensen, Roy, and Associates, 408

Kansas City, 520
Kent, Malcolm F., 343, 344
Kuhn, Tillo E., 59

Labor, 502
  annual saving, 11
  mechanics, 16
  taxes, 605
Land
  developer, 605
  development, 527
  economic cost of right-of-way taking, 489
  farm, 490, 496
  nonfarm, 490
  owners, 28, 498
  price, 496
  resources, 557
  service roads, 250

Land *(continued)*
 transportation, 1
 use, 484, 496-498
 value, 496-500
Lash, Michael, 560
Laski, Harold J., 558
Lateral roads, 527
Laws and ordinances, 361
Lawyer's search of law, 22
Lead time, 639, 641, 642, 652
Lee, Robert R., 248
Legislative authority, 54
Legislators, 230, 586
Legislature, 587
Lending organizations, 98
Level of service, 143
Liabilities, 190
Lieder, Nathan, 82
Life insurance, 207
Line-haul trucking cost, 300
Lisco, Thomas E., 268
Literature of economic analysis, 123
Loans, 98-100
Local injustices, 17
Local residences, 500
Long-lifed property, 72
Long-range outlook, 642
Long-range plans, 584, 639
Long run cost, 230
Lotka, Alfred J., 379, 380

Management, 364, 543, 553, 556, 640, 644
 control, 642
 decision, 21, 45, 55, 124, 125, 201, 243, 552, 555, 593, 647
 decision case No. 1, 560
 decision case No. 2, 576
 expense, 245
 highways, 638
 policy, 44
 programming, 644
 separate decision levels, 20, 27
 tool, 584, 655
Manheim, Marvin L., 558
Manual of Uniform Accounting, 245
Manufacturing, 502
Marginal price, 64
Marginal rate of return, 127
Marginal return, 555
Market and nonmarket merging, 573
Market factors, 10, 45

Market price, 179
Martin, Brian, 52, 162
Massachusetts, 605, 606
 accident costs, 385
Materials of construction, 44
Maximum grade economy, 672
McCormack, C. F., 585
McFarland, William F., 273
Medians, reduction in accident rates from closing openings, 409
Mental comfort, 278
Merchandising, 545
Methods of economic analysis, 125-133
Millage tax rate, 487
Minerals, 527, 528
Minimum attractive rate of return, 68, 70, 79, 124, 125
Mink coat, 280
Missionaries, 531
Mnemonic symbols, 86
Modal characteristics, 215
Modified benefit/cost ratio, 650
Monetary profit, 243
Money, time value, 71
Montgomery County, Maryland, 572
Morgali, James, 661
Mortgage interest, related to vestcharge rate, 76
Motor vehicle; *see also* Vehicle
 accidents, 360-420; *see also* Traffic accidents
 axle arrangement, 335
 cost allocation, 618
 description for running cost, 332, 334
 driver spending psychology, 303
 fuel consumption tables, 705-727
 fuel tax, 262, 607
 monthly fuel consumption, 307
 monthly mileage, 307
 operating cost, 34, 298, 631
  accidents, 312
  automobile club, 311
  classification of items, 302
  commercial vehicles, 300
  compact, medium, heavy cars, 313
  depreciation, 311
  engine oil, 308
  factors, 299
  garage, 312
  gasoline, 308
  highway design attention, 5
  insurance, 311
  licenses, 311
  maintenance, 308
  ownership, 34, 304

Motor vehicle *(continued)*
  parking, 312
  passenger cars, 301, 308
  percentage of transportation cost, 262, 299
  property taxes, 312
  records, 300, 303, 308
  tires, 311
  tolls, 312
  variation with vehicle weight, 312
  vehicle owners, 300
 owners, 300
 ownership, 5
 readiness-to-serve charge, 42
 registration growth, 2
 running cost, 36, 243, 298, 497
  depreciation, 322
  engine oil, 320
  fuel, 317
  gravel or stone, 727
  highway factors, 314
  incremental travel time, 270
  maintenance, 322
  90-degree turns, 724
  per hour, 270
  speed, 324
  tables, 679-704
  tire wear, 316
 taxes, 42
 typical vehicles, 333
Moyer, Ralph A., 328, 341, 342, 343, 344, 347, 351
Mule carts, 526
Multiple original group method, 211
Multiple rates of return
 in rate of return method, 160
 with reversal of sign of cash flow, 160
Multiple straight line depreciation method, 186
Mutually exclusive alternatives, 55, 134-136, 143, 155, 559
Mutually exclusive proposals, 156

Napierian logarithm system, 101
National Safety Council, 360, 362
Natural resources, 496
Navigable waterways, 23
Needs studies, 584-599, 602, 604, 638, 649
Negative benefits, 148
Net consequences, 20, 26
Net costs, 20, 26
Net present value method, 127, 132, 144, 160
New England, 494
New Jersey, 605

New Mexico, 606
 accident costs, 385
New York, 605, 606
Nicaragua, 527, 532
Nonmarket factors, 10, 45, 555, 559
 items, 26
 weight in decision process, 21, 30
Nonuser consequences, 493
 benefits, 609
 factors, 494
 pricing, 35
 transfers, 494
Normal growth traffic, 476, 477
North Carolina, 603
 expressway accident rates, 406, 407
North Ridge decision case, 558
Northwest United States, 581
Norton, Paul T., 178
Null alternative, 23, 554
 economic desirability, 137
 illustrative example, 139
Numerator vs denominator benefit/cost ratio method, 148

Objectives of project, 558
Objectivity, 20, 21
Obsolescence, 559
 risk of, 77
Offsetting consequences, 14
Oglesby, C. H., 661
Ohio
 accident rates, 408, 417
 cost allocation, 611
On-site traffic accident cost, 367
One-horse shay, 201
Operating
 cost, 230, 236
  ownership, 34
  method of cost allocation, 620, 629
 expense, 245
 expense account, 193
 statement, 190
Operation of highways, 603
Opportunity costs and gains, 26
Oregon, 606
Original group method, 210
Original property group, 181
Ottawa, Kansas, 520
Outer belts, 502
Overdesigning, 28
Overhead costs, 236
Ownership, 190

Oxcarts, 526, 532

Passenger car
  acceleration, 456, 460-465
  deceleration, 456, 458, 460
  persons per car, 269, 459-470
  speed changes, 459-470
  speed profiles, 469
  travel, 29
  travel time value, 269
Past events (investments) irrelevant, 20, 24
Pavements, 243
  service lives, 220-223
Payback period, 153
Peace Corps, 531
Pendleton, Oregon, 576
Penetration roads, 527
Pennsylvania, 605
Period-end step covention, 90, 93, 108
Perishable produce, 496
Permanent disability, 377
Permanent record, 544
Personal preferences, 23, 277, 475, 573
  aesthetics, 38
  comfort and convenience, 38
  economy of, 279, 280
  highway related, 279
  uniform speed, 38
  vehicle related, 279
Peru, 527
Physical comfort, 278
Physical condition of highway, 584
Physical plant rebuilding, 642
Physical property
  additions to, 207
  ages at retirement, 207
  economic life, 200
  physical life, 200
  placements, 207
  remaining service, 178
  replacements, 209
  retirements, 209
  service life, 200
  service load, 201
  survivor curves, 207-226
  wear, 176, 202
Physician's diagnosis, 22
Pioneers, 531
Piston travel, 320
Placement and retirement curves, 206
Planning, 639

Planning division, 649
Planning, programming, budgeting systems (PPBS), 124
Plus grades, 285, 318
Political acceptance, 644
Political persons, 649
Poll tax, 605
Pollio, Marcus Vitruvius, 177
Pony express trail, 223
Population
  expansion, 532
  per motor vehicle, 2, 3
  United States, 3
Portland cement concrete pavements, 159
Post World War II, 577
Power acceleration, 295
Power performance, 284
Power transmission, 525
Pragmatic decisions, 558
Predominant use cost allocation, 610, 612
Preferred stocks, 125
Preliminary analysis, 23
Preliminary designs, 18
Premium rate, 127
Prepaid cost, 176
Prepaid insurance account, 192
Preparing reports, 545
Present serviceability index, 590, 594, 595, 651
Present worth of costs, 143, 659
  definition, 126
  illustrative solution, 131
Present worth depreciation method, 183
Price base, 247
Price, cost, and value, 56
Price-demand curves, 57, 58
Price elasticity, 57
Price inflation, 247
Price-volume (demand) reversal of relationship, 62
Pricing, 533
  construction, 596
  nonmarket factors, 573
  project formulation, 55
  unequal traffic volumes, 55
Primitive roads, 526
Priority
  construction, 596
  planning, 639
  programming, 643
  rating, 639
  selection, 641

Index

Pritchett, Harold D., 661
Private business, 53
Private industry, 17, 49, 55, 124
Probability of flood flow, 658
Probable service life, 201
Probit analysis, 268
Problems and solutions, 658-678
Production cost, 229
Production unit depreciation, 186
Professional fees, 16
Professional writers, 547
Profit, defined, 50
Profit on investment. 70
Profit and loss statement and
    account, 71, 177, 190, 194, 230
Program
  costs, 604
  evaluation and review technique, 641, 654
  of highway construction, 153
  public acceptance, 644
Programming
  balance in, 643
  construction, 41, 639-655
  financial factors in, 643
Project evaluation, 17, 18, 43, 54
  public policy in, 53
Project formulation
  applied to subfeatures of design, 43
  cost methods of analysis, 140
  decision levels, 28
  defined, 17-19
  in developing countries, 528
  doing something, 23
  in equivalent uniform annual cost, 142
  management tool, 552
  not concerned with source of traffic, 54, 478
  not related to land value, 499
  in pricing benefits, 480
  second management decision, 555
Project priority, 41
Project selection in programming, 643
Projects
  public attitudes towards, 76
Property
  accounting, 181
  assessment, 485
  asset account, 202
  biological decay of, 203
  damage accidents, 399
  expectancy of life of, 201
  group, 188
  owners, 230
  records, 202

Property (continued)
  taxes, 42, 484, 606
  valuation, 176
Proposed new facilities, 24
Protective services, 503
Public hearings, 516
Public officials, 67, 143, 148
  agents of people, 35
  budget requests, 73
  understanding of interest, 72
Public policy to save, 15
Public preferences in spending, 281
Public sector, 124
Public tax policy, 487
Public treasury, 485
Public utilities
  advantages to, of construction scheduling, 654
  commissions, 300
  concept in cost allocation, 610
  concept related to highways, 608
  depreciation and production cost, 176
  highways thought of as, 42
  interest during construction, 250
  management, 491
  sinking fund depreciation concept, 183
  use of type survivor curves, 219
Public ventures, 10
Public works
  application of economic analysis, 34
  benefit/cost ratio adapted to, 146, 147
  consideration of economy, 6
  developing countries, 535
  different from private business, 53
  director as a decision maker, 552
  effect on highway construction schedule, 654
  managers, 7
  pricing to produce sales, 280
  in sense of a business investment, 148
  use of survivor curves, 219

Quality of highway service, 60, 593

Railhead, 529
Railroads
  capitalized cost analysis, 152
  compared to wagon roads, 11
  consequences from airways, highways, 15
  dependence upon highways, 495
  depreciation accounting, early days, 177
  developing countries
    decision making, 538
    government owned, 529
    highways, decrease use of, 532

Railroads *(continued)*
  economy of, by Wellington, 12
  factor in plant location, 502
  ownership, 5
  United States
    early building, 1
    farm land service, 496
    land use development, 529
  Western expansion, 534
Rama Road, 527
Ranking
  alternatives under fluctuating cash flows, 162
  independent projects, 153
  projects, 644
  schemes, 575
Rapid writing, 546
Rate of return, 124, 543
Rate of return method
  basis of project priority ranking, 644
  definition, 127
  description, 151
  illustrative solution, 133
  multiple solution with cash flow reversal, 160
  reinvestment rate, 162
  under fluctuation cash flow, 160-162
Rating and ranking nonmarket factors, 572
Rating schemens, 644, 650
  for construction priority, 650
Raw material, 502
Real estate taxes, 484, 486
Real price, 247
Real value, 177
Reconstructing new plant, 642
Reconstruction, 241
Recreation, 528
Recurrence period, 659
Redevelopment
  urban, 555
Reinvestment rate, 162
Relative use cost allocation, 610, 612
Replacement cost, 179
  of property, 178
  of highway construction, 597
Report; *see also* Formal report
  on economic analysis, 542
Residential streets, 610
Residual value, 237; *see also* Terminal value
Restaurants, 515
Restricted capacity cost allocation. 618
Retail business, 497
  local, 503

Retail shopping, 497
Retail trade, 501, 513
Retirement of property, 201-204
  demand of other authorities, 203
  dispersion, 180
  forecasting, 219
  highway property, 202
  managment decision, 203
  methods, 203
  reasons for, 202, 204
  related to depreciation accounting, 180, 181
Return of capital, 184
Return on capital, 184
Return on investment, 67, 70
Revenues, 604
Reversals of sign of cash flow, 145
Reverse cash flows, 143
Revision of report, 546
Ribbon cutting, 516
Ribbon development (land use), 511
Ridge City, 561
Right-of-way
  buildings on, 225
  damages paid, 225
  economic value (null alternative), 139
  service life, 225
  takings, 26, 484-491
Rigid pavement, 328
Risk
  decision making, 559
  in economic analysis, 74-77
  and uncertainty, 74-77
  in use of money, 73
Road user
  benefits, 52
    identification, 38
    long-range forecast, 243
    pricing, 38
    traffic components, 63
  consequences, 35, 262
  cost and pricing, 36, 54
  flood costs, 659
  negative cost reduction, 60
  nonpriceable satisfactions, 60
  personal preferences, 38, 277-280
  share of cost allocation, 617
  taxes, 52, 605-607
Roads; *see also* Highways
  center of interest of travelers, 592
  cheapest, 11
  improvement in developing countries, 530
  mileage, 2
  systems, 527, 591

Index 919

Roadway property, 202
Roadway surfaces, 328
Rolling grades, 348
Rolling radius of tire, 296
Rolling resistance, 13, 290, 328
Route
  improvement continuity, 652
  location, 23, 43, 55
  numbers
    Boston 128; 502, 511
    Interstate, 70, 520; 80N, 576; 82, 576; 90, 576
    U.S. 40, 520; 81, 520
  safety improvements, 366
Running cost of vehicles
  change with future traffic, 478
  detour roads, 559
  source of, 298
  tables, 679-703

Safety improvements for areas, 366
Sales concept, 53, 64
Sales income, 49, 52
Sales viewpoint, 54
Salina, Kansas, 520
Salt Lake City, Utah, 576
Salvage value; *see* Terminal value
San Antonio Freeway, 374
Sans, I. J., 520
Saving, 15, 51
Sawhill, Roy B., 312, 341, 347
Scalar devices and indexes, 572
Scheduling construction, 41, 638-655
Schmedtje, Jochen K., 536
Seattle, Washington, 576
Secondary road
  economic value of (null alternative), 138
Secondhand value curve, 188
Sensitivity of factors, 125, 163-167
Service life, 200-226; *see also* Analysis period
  annual economic cost, 242
  decision making, 559
  depreciation accounting, 180, 189
  dispersion, 202
  highway components, 224
  length, 204
  methods of determining, 204-222
  needs studies, 593
Shadow prices, 534
Shaner, Willis W., 534

Sharp turns, cost of, 347
Short-lived property, 186
Short-range program, 640
Short-run cost, 230
Sight distances
  on horizontal curves and grades, 451
Signal installation, 243
Simpson, Herbert D., 627
Simulated plant balance method, 214, 219
Single unit property, 185
Sinking fund and capital recovery
  factors, relation of, 89
Sinking fund depreciation method, 181-183
Skin friction, 293
Slip angle, 326
Slum clearance, 488
Small towns, 503
Smidt, Seymour, 75
Smith, Wilbur, and Associates, 364, 398, 399
Snow removal, 254
Social and economic considerations, 580
Society of Automotive Engineers, 291, 294, 351
Solomon, David, 403
Solomon, Ezra, 125, 162
South America, 526, 529
Space-time cost allocation to vehicles, 620, 632
Speed, 431-433, 479; *see also* Traffic speed
  changes, 456, 467-470, 513
  change unit, 267
  profiles, 469
Spending for highway construction, 495
Spot traffic accident improvements, 366, 412
St. Clair, G. P., 631
Stage construction, 44, 663
Standard cost method of cost allocation, 610, 611
  to vehicles, 620, 627
Stanford University, 662
State legislators, 602
State legislatures, 516, 602
Statement
  of assets and liabilities, 190
  of income and expense, 190
Statistical data for economic analysis, 43
Step compounding convention, 100

Stevens, Hoy, 344, 345
Stock company, 177
Stone roadway surfaces, 347
Stop-and-go driving, 321
Stopping distance, passenger car, 459, 466
Straight line depreciation method, 185
Straine, driving, 279
Strategic decision, 640
Structural adequacy, 593, 595
Subalternatives, 28
Subjective
 appraisals, 574
 evaluation, 644
 judgment, 642
 values, 572
Subjectivity, 20
Suboptimizations, 18
Subsistence
 economic cost of man, 398
 farming, 528
 living, 496
Suburban living, 497
Sufficiency rating, 590, 594, 595, 640, 651
Sum-of-the-years digits depreciation method, 188
Summit elevation, 678
Sunk cost, 241, 249
Superelevation, 351, 352, 444
Survivor curves, 205-226
 box culverts, 226
 by method of calculation, 209-214
 extensión of stubs, 219
 matching process, 219
 pavements, 220-224
 timber structures, 225
 type curves (Iowa), 215-219
Syrek, Daniel, 403
System classification of highways, 640

T-accounts, 190, 191
Taborek, Jeroslav J., 291, 293
Tacoma Narrows bridge, 244
Tactical decision, 640
Tax
 base, 486, 609
 concepts, 605, 609
 economists, 609
 equity, 486
 and financing, 586
 forgone, 487
 fuel, 336-340, 606
 gains, 488, 499
 income, 381, 645

Tax *(continued)*
 maximum legal rate, 487
 paid through labor, 605
 revenues, 604
 sales, 381
 structure, 602
Taxable property, 381
 removed from tax roll, 485
Taxpaying citizen, 229
Technical reports, 543
Technology, 27, 202, 502, 559
Terminal value, 237-241
 as reduction in capital cost, 142
 estimating for depreciation, 189
 in declining balance method, 188
 in present worth depreciation, 184
 position in benefit/cost ratio method, 151
 retained in property asset account, 202
 sensitivity in economic analysis, 164
 under unequal analysis periods, 159
Tertiary system, 591
Tesdall, Glen L., 328, 342, 343
Textile industry, 494
Theory of value, 124
Third structure taxes, 607
Thomas, Thomas C., 268
Through traffic, 511
Timber resource, 527, 528
Timber structure survivor curve, 225
Time
 commercial vehicle, 272-276
 consumption on grades, 350
 economic commodity, 263
 factors of value, 266
 highway travel, 264-266
 toll road travel, 267
 value of, 264-267
 willingness to pay, 267
Time-discount factor, 69
Tire
 cost per mile, data for, 343
 punctures, 319
 rim pull, 287
 wear measurement, 316
 wear in speed changes, 460
Tolerable road conditions, 593, 596, 602, 640
Toll facilities, 229
 Indiana, 314
 Kansas, 314
 Massachusetts, 314
 New York, 314
 Ohio, 314
 revenues, 607
 travel time, 267

Index   921

Ton-mile cost allocation to vehicles, 620, 623
Top drawer cost allocation, 614, 616
Topeka, Kansas, 519, 520
Topeka Capital Journal, 516
Torque, 285, 286, 287
Town That Didn't Want A Freeway, 560
Trade
 centers 496, 497
 export-import, 532, 534
 general, 501
 local, 501
Trade-in period, 164
Trade-offs in spending, 280
Trade volume, 500
Traffic
 annoyances, 501
 characteristics, 428-470
 components, 474-478
  community, 65, 477
  developed, 64, 156, 476, 477, 479
  diverted, 64, 475, 476, 477, 479
  existing, 63, 474, 477, 478
  generated, 54, 63, 156, 476, 477
  growth, 64, 156, 476, 477
  transferred, 54, 60, 63, 64, 156, 475, 477
 composition, 428, 479
 congested area, 500
 congestion, 511, 512
 control economy, 661
 controls, 44
 cost to, caused by accidents, 374
 distribution, 513
 exponential growth, 479
 flow, 103
 forecast, 25, 54, 474, 478
 gradient growth, 479
 incident (accident class), 361
 local, 513
 percentage distribution by vehicle type, 429
 relief improvements, 44
 service expense (maintenance), 245
 thirtieth highest hour, 434
 unequal volumes in alternatives, 55
Traffic accident
 benefits from, 368, 376
 classifying, 360, 362
 courts and juries, 378
 death (fatalities), 377
 driver decision process, 401
 economic man, 378
 economy studies, 364, 365
 future income forgone, 378
 future living cost, 378
 geographical area improvements, 366

Traffic accident *(continued)*
 laws and ordinances, 361
 prevention activities, 377
 property, 202
 public attitude toward, 364
 reduction, 496
 reporting, 360, 361
 route improvement, 366
 spot improvements, 366
 statistics, 362
 unreported, 362
 viewpoint toward, 378
 wages lost, 378
Traffic accident costs
 Chicago streets, 406
 classifying cost elements, 367, 368
 communication, 371
 compiling, 383
 court, 376
 damage awards, 385
 damaged property, 384
 deaths, 377
 elements of cost, 368
 funeral, 374
 general, 236, 373, 513, 628
 goods and property, 371
 government services, 376
 highway system, 404
 Illinois, 385
 injuries, 384
 insurance, 374
 investigations, 368
 involvement, 387
 legal and court, 385
 loss of vehicle use, 384
 Massachusetts, 385
 New Mexico, 385
 on-site, 367
 per accident, 386
 permanent disability, 377
 personal services, 372
 rural, 386
 safety programs, 377
 severity, 386
 strain, anguish, anxiety, 374
 time consumed, 372, 373
 transportation, 311
 truck, 386
 unreported, 387
 urban, 386
 Utah, 385
 wages lost, 372
 Washington, D. C., 385
Traffic accident fatality, 377-383
Traffic accident rate
 access control, 402, 405
 availability, 400, 401
 bridge roadway, 409, 417
 California, 410

Traffic accident rate *(continued)*
 Chicago streets, 406
 driven decisions, 401
 general, 512
 highway design, 363
 horizontal curves, 402, 414
 illumination, 403
 intersections, 402, 403, 415
 lanes of traffic, 411
 medians, 409
 one-way streets, 403
 reduction from spot improvements, 412
 rural highways, 410
 speed, 402
 vertical grades, 402, 414, 415
Traffic speed
 average speed, 434
 changes, 434, 435, 456
 factors, 430
 hourly speed and volume, 436
 modal, 440
 observed, 434
 on different types of highways, 437-441
 on horizontal curves, 440, 445
 passenger car distribution, 440
 passenger car effects of stopping, 467
 percentage of travel time at speeds, 440
 percentage of vehicle-miles at each speed, 443
 spot speed distribution, 435, 436
 test car speed distribution, 437
 trucks, 445
  minus grades, 447, 452
  plus grades, 454-457
 urban expressway, 437
Trails, 526
Transfer (and offset) consequences, 14, 26, 493, 554
Transferred traffic, 475, 479
Transmission, 288
 resistance, 289
Transportation, 502
 cost differences between alternatives, 53
 cost on grades, 678
 modes, 538
 planning, 7
 supply, 59
Travel modes, 265
Travel speed; *see* Traffic speed
Travel time; *see also* Time
 cost, 236
 in developing countries, 530
 excess in stopping, 466
 highway cost to reduce, 276
 incremental cost of time reduction, 271
 reduction, 496, 498, 554, 628
  compared to commodity saving, 37
  conversion to dollars, 36, 37

Travel time *(continued)*
 trucks, 445
  on grades, 454, 455, 457
 urban, 267
Traveler center of interest, 592
Tricities of Washington, 577
Trip length, 591
Triple counting, 498
Trucks
 ability prediction procedure, 351
 gross weight by class, 446
 roadside weighing, 428
 speed and travel time
  on minus grades, 447-453
  on plus grades, 454-457
 transport, 534, 535
 weight/horsepower ratio, 447
Turnover period, 205

Uncertainty, 20, 27, 559
 economic analysis, 73-77
Undepreciated book cost, 180
Underdesigning, 28
Unemployed persons, 16
Unequal analysis periods; *see also* Analysis periods
 adjusting for, 159
 conclusions on use of, 158
Unequal levels of service, 155
Unit prices of vehicle commodities, 337
United States, 526, 529, 534, 605, 606, 612
U.S. Bureau of Public Roads, 517, 564, 566, 567, 568
U.S. Department of Health, Education, and Welfare, 377
U.S. Flood Control Act, 146
U.S. Government, 124
U.S. Senator, 566
U.S. Treasury, Bureau of Internal Revenue, 179, 185
Unreported accidents, 362, 399
Urban renewal, 488
Used car value, 323
User benefits, 498
Utah, accident costs, 385
Utility industry, 207
Utility services, 555

Value
 cost and time, 56
 passenger car travel time, 267-272
 propriety, 178

Index

Value *(continued)*
  time, 28
    passenger car travel, 267-272
    truck travel, 272-276
Vehicle; *see also* Motor vehicle
  depreciation expense, 316
  dirver and speed, 430
  maintenance and repair, 316, 344
  methods of allocating depreciation to speed, 345, 346
  ownership density, 604
  registration fee, 606
  running cost, 331-333, 679
  service life, 322
  speed change running time, 688-704
Vehicular traffic, 18
Venturi tube, 285
Vermont, 605
Vertical alignment, 14
  accidents, 414
  economy, 5, 672
  vehicle performance, 294, 325
Vestcharge, 67-79
  concept and definition, 68-71
  during construction, 250
  penalty for not consuming goods, 72
  rate, 68, 73-78, 124, 166, 244, 245, 536, 574
  sensitivity, 166
Viewpoint
  community, 17, 40
  decision maker, 21, 28, 553-557
  private, 17
  seller and purchaser, 56
  whose, 20, 28, 30
Vintage group method, 210; *see also* Original group method
Vintage property group, 181
Virgin land, 528
Virginia, 603
Virginia Highway Users Association, 626

Wage rate of drivers, 272, 276
Wagon roads, 1, 529
Ward, Allen W., 273
Warehouses, 496, 502
Washington, D. C., 564, 568
  accident study compared to Illinois, 387, 395, 396
Washington state highway commission, 576, 630
Water-resource development, 10, 123, 124, 281, 525, 538, 557
Wealth, use of and interest, 69
Weather and traffic speed, 433
Welfare economics, 124
Wenger, Dean M., 403
West Virginia, 603
Wheel revolutions per mile, 320
Widening streets, 512
Wilson, George W., 531, 538
Windfalls, 499
Winfrey, Robley, 215, 542, 663
Wise decision, 555
With and without, 20, 23, 137, 140
Wohl, Martin, 52, 59, 162
Work horses, 17
Work in process, 229
Working capital, 177
World War II, 238
Writing reports, 545

Year end deposit, 182
Yield rate, 127

Zettel, Richard M., 498